SHOCK COMPRESSION OF CONDENSED MATTER—1995

SHOCK COMPRESSION OF CONDENSED MATTER—1995

Proceedings of the Conference of the American Physical Society
Topical Group on Shock Compression of Condensed Matter
held at Seattle, Washington, August 13–18, 1995

Edited by:

S. C. SCHMIDT
Los Alamos National Laboratory
Los Alamos, New Mexico, USA

W. C. TAO
Lawrence Livermore National Laboratory
Livermore, California, USA

American Institute of Physics
AIP Conference Proceedings 370
PART 2

Woodbury, New York

CONTENTS

PART 1

CHAPTER I: PLENARY

CHAPTER II: EQUATION OF STATE

CHAPTER III: ATOMIC AND MOLECULAR STUDIES

CHAPTER IV: PHASE TRANSITIONS

CHAPTER V: MODELING AND SIMULATION: NONREACTIVE MATERIALS

CHAPTER VI: MODELING AND SIMULATION: REACTIVE MATERIALS

CHAPTER VII: MECHANICAL PROPERTIES: MACROSCOPIC ASPECTS

CHAPTER X: EXPLOSIVE AND INITIATION STUDIES

CHAPTER XI: ELECTRICAL, MAGNETIC, AND OPTICAL STUDIES

CHAPTER XII: EXPERIMENTAL TECHNIQUES

CHAPTER XIII: INTERFEROMETRY

CHAPTER XIV: GAUGES AND APPLICATIONS

CHAPTER XV: IMPACT PHENOMENA AND HYPERVELOCITY STUDIES

CHAPTER XVI: HIGH VELOCITY LAUNCHERS AND SHAPED CHARGES

CHAPTER XVII: LASER AND PARTICLE BEAM-MATTER INTERACTION

CHAPTER IX

SHOCK-INDUCED MODIFICATION AND MATERIAL SYNTHESIS

DISCRETE MESO-ELEMENT DYNAMIC SIMULATION OF SHOCK RESPONSE OF REACTIVE POROUS SOLIDS

Z.P. Tang[*] , Y. Horie

North Carolina State University, Raleigh, NC 27695-7908

S.G. Psakhie

Russian Materials Science Center, Tomsk, Russia

A new computer program, Discrete Meso-dynamics Method (DM^2) is being developed to describe heterogeneous flow of reactive porous solids under high rates of loading. A simulated sample of Ni/Al powder mixture shows that the chemical reaction occurs for impact velocity above 1 Km/s. Fragmentation and pore collapse generate mechanical mass mixing and hot spots which appear to be the principle mechanisms for chemical reactions in porous solids.

INTRODUCTION

Recently, a new numerical simulation technique has been developed to investigate the shock response of porous materials on the meso-mechanical level, with special focus on understanding the behavior and mechanisms of powders under dynamic loading. To present, work has been based on continuum physics. For example, Flinn et al. (1) studied the shock compaction of stainless steel powder using a Eulerian hydrocode, CTH, and Benson et al. (2) used another hydrocode to investigate copper powder. Now, the discrete element dynamics method has been introduced into this field (3).

At North Carolina State University (NCSU) a new computer program is under development that will describe heterogeneous flow of reactive porous solids under high rates of loading. The program is based on an assembly of interacting elements governed by the classical equations of motion. The preliminary simulation (4) has demonstrated that the program is capable of predicting unique phenomenon of mass penetration and mixing. In

____

* On leave from University of Science and Technology of China.

this paper, the DM^2 model will be described briefly. An example calculation using a nickel-aluminum powder mixture and its chemical reaction will be presented.

THE DM^2 MODEL

Interaction Forces and Equations of Motion

For DM^2, each interacting element-pair may have linked and/or contact states, depending on the distance and the interaction history (5). The interaction includes: central potential force, friction force and shear force. Presently, the following Lennard-Jones function is adopted for the central interaction

$$\bar{p}^{ij} = \frac{-\alpha^{ij}mn}{r_o^{ij}(n-m)}\left\{\left(\frac{r^{ij}}{r_o^{ij}}\right)^{-(n+1)} - \left(\frac{r^{ij}}{r_o^{ij}}\right)^{-(m+1)}\right\}\frac{\bar{r}^{ij}}{r^{ij}}\,(1)$$

where $\bar{p}^{ij}$ is the force acting on element i by element j; α^{ij}, m, n are material constants; r_o^{ij}, r^{ij} are initial and current distances of the two elements. Two critical parameters, $r_s^{ij}(T), r_{max}^{ij}$, are introduced, where the latter is bond break distance and the former is softening distance and temperature, T, related.

Viscous friction interaction includes tangential friction $\bar{f}_{vt}^{ij}$ and central damping $\bar{f}_{vn}^{ij}$

$$\bar{f}_{vt}^{ij} = -\left[\eta_1^{ij}v_t^{ij} + \eta_2^{ij}\left(v_t^{ij}\right)^2\right]\bar{t}^{ij} \qquad (2)$$

$$\bar{f}_{vn}^{ij} = -\alpha\bar{v}_n^{ij} \qquad (3)$$

where v_n^{ij}, v_t^{ij} are relative velocities along central and tangent directions, η_1, η_2, α are friction coefficients, and $\bar{t}^{ij} = -\bar{v}_t^{ij}/v_t^{ij}$.

Dry friction is given by

$$\bar{f}_d^{ij} = -\mu^{ij}p^{ij}\bar{t}^{ij} \qquad (4)$$

where μ^{ij} is Coulomb frictional coefficient, and p^{ij} is central pressure. Dry friction only occurs in case of contact.

Shear resistance force $\bar{f}_{sh}^{ij}$ is calculated

$$\bar{f}_{sh}^{ij} = -A^{ij}G^{ij}\gamma^{ij}\bar{t}^{ij} \qquad (5)$$

where γ^{ij}, G^{ij} and A^{ij} are shear strain, effective shear modulus, and effective link area of elements i and j. When γ^{ij} is greater than yield limit γ_y, a perfect plasticity model is applied.

The governing equations of motion are

$$m^i d^2\bar{r}^i/dt^2 = \sum \bar{F}^{ij} = \sum\left(\bar{p}^{ij} + \bar{f}_{vt}^{ij} + \bar{f}_{vn}^{ij} + \bar{f}_d^{ij} + \bar{f}_{sh}^{ij}\right)$$

$$\bar{J}d^2\bar{\theta}^i/dt^2 = \sum \bar{q}^{ij} \times \bar{F}^{ij} \quad (i = 1,2,...,N) \qquad (6)$$

where $m^i, \bar{J}^i$ are mass and inertial moment of element i; $\bar{r}, \bar{\theta}$ are the location and angle; and $\bar{q}^{ij}$ is the moment arm. N is the total number of elements in the system.

Phase Transition and Chemical Reaction

Only solid-liquid phase transition is considered in the present model. Melting temperature T_m is related to pressure through use of Simon's equation.

In the case of chemical reaction, only the stoichiometric case is discussed in this paper

$$n_i A + n_j B = n_c C + Q \qquad (7)$$

where A and B are reactants, C is product, and Q is heat release. $n_i = m_i/M_A$, $n_j = m_j/M_B$, m_i and m_j are mass of i and j, M_A and M_B are mass/mole of reactants A and B. The mole number of product is n_c. For stoichiometric reaction, $n_c = m_c/M_c = \left(m_i + m_j\right)/M_c$; M_c is mass/mole of product C.

Assuming that the number of atoms transferred for each species of reactants is equal, one obtains

$$\begin{cases} m_i' = m_i + \Delta n\left(M_B - M_A\right) \\ m_j' = m_j + \Delta n\left(M_A - M_B\right) \end{cases} \qquad (8)$$

where $\Delta n = n_i n_j/\left(n_i + n_j\right)$ is the mole number of mass transfer, m_i' and m_j' are the masses of product in element i and j. After reaction, the velocity, temperature, and element size are calculated based on momentum transfer, heat release and physical properties of the product.

MODEL SIMULATION

Nickel/aluminum powder mixture is chosen for a 2-D sample simulation. The initial geometry is shown in Fig. 1, where a piston at top moves down at a constant velocity of 1.0 Km/s. The diameters of powder grains are 16, 20, 24 μm for Al and 6, 10, 14 μm for Ni.

The distribution and orientation of powder grains are both randomly arranged. The radius of elements of Al and Ni are 0.6 μm and 0.844372 μm, for stoichiometric need. Elements are arranged in a rectangle. The radius of elements for 304 stainless steel is 2 μm. The initial density of Ni/Al powder mixture, ρ_o, is 3.86 g/cm^3. This is

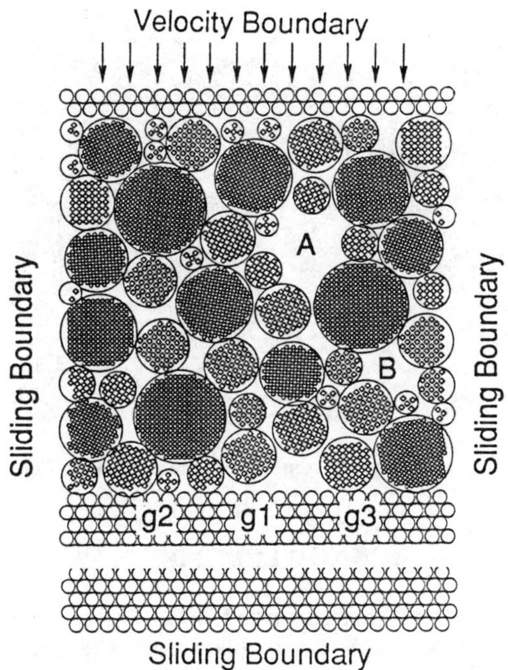

FIGURE 1. The initial configuration of Ni/Al powder mixture. ◊—Ni, o—Al, O—304 stainless steel.

67.65% of the theoretical density. The area of the powder sample is 100 μm X 100 μm. The imaginary gauge locations are g1—g3.

The reaction product is assumed to be Ni_3Al. The threshold temperature and heat release of reaction are set at 1000 K and 7.29×10^9 erg/g, respectively. Viscous coefficients are $\eta_1 = 0.2$, $\eta_2 = 0$, and $\alpha = 0.05$. Other principle material parameters used in the simulation are listed in Table 1. Time increment is 0.03 ns.

RESULTS

Fig. 2 shows powder configurations at t = 36 ns and 60 ns. Reaction product Ni_3Al (solid dots) is found in two locations. One location is along the interfaces of Ni/Al grains near the impact and bottom surfaces of the powder sample. The other reaction location, more remarkably, is in the regions that were originally occupied by large pores, labeled A and B in Fig. 1.

It is observed that fragmentation and mass penetration happen during grain collision and pore collapse. Primarily the collapse of larger (relative to powder sizes) pores generates stronger mechanical mass mixing of reactants and hot spots, which are essential ingredients of solid state chemical reaction.

A parallel simulation showed no reaction for $V_o = 0.5$ Km/s and a stronger reaction for $V_o = 1.5$ Km/s (Fig. 3). Reaction in Fig. 3 occurred along most of the grain boundaries of reactants. This result is comparable to those of Bennett et al. (6) who found that the impact velocity for chemical initiation in a similar geometry was 1.07 Km/s.

Fig. 4 shows the pressure profiles at back surface of powder sample for the impact velocity of $V_o = 1$ Km/s. The shock wave speed with chemical reaction is 2.20 Km/s, which is about 2.3% higher than the shock speed of 2.15 Km/s without reaction. The corresponding jump pressure with chemical reaction is about 23 GPa, which is 9.5% higher than the pressure of 21 GPa without reaction. This is attributed to the density change due to the reaction, since thermal expansion is not considered in this simulation.

TABLE 1. Principle Material Parameters Used in Simulation

	$\rho_o\left(g/cm^3\right)$	$\alpha_*^{ij}\left(10^{11}\ dyne/cm^2\right)^a$	m	n	$c_v\left(10^7\ erg/g.k\right)$	γ	$T_{mo}(K)$	$\Delta H_m\left(10^9\ erg/g\right)$
Ni	8.9	9.5604	1	2	0.4414	1.93	1726	2.976
Al	2.7	1.2212	2	3	0.9008	2.0	993.2	3.999
Ni_3Al	6.82	5.9648	1	2	0.485	1.94	1663	2.0^b
304 ss	7.9	8.2492	1	2	0.423	1.93	2380	2.473

a is $\alpha_*^{ij} = \alpha^{ij}$ per unit volume; b is arbitrarily chosen

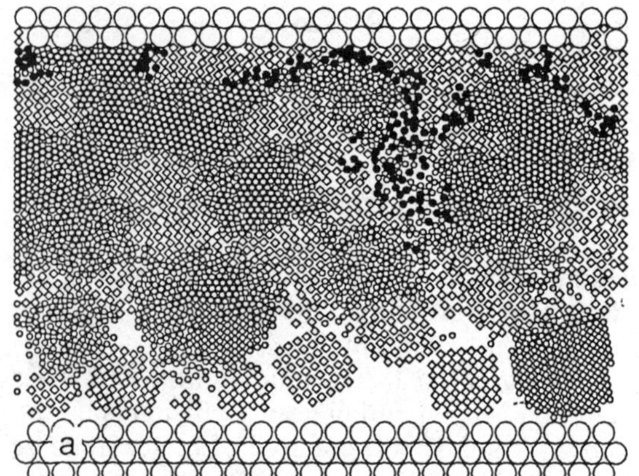
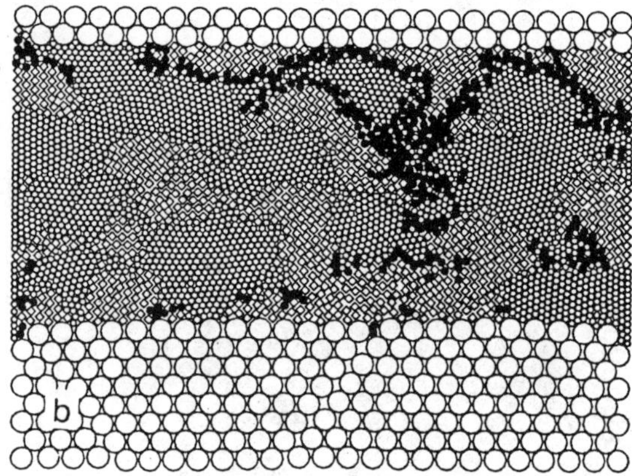

FIGURE 2 (a, b) Results for $V_o = 1\,Km/s$. (a) t=36 ns, (b) t=60 ns, ● Ni_3Al. The other symbols are the same as in Fig. 1.

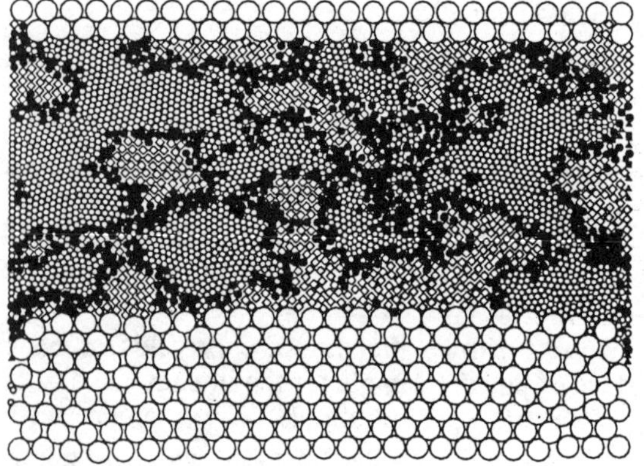

FIGURE 3. Results for $V_o = 1.5\,Km/s$, t = 40 ns. Symbols are the same as in Figs. 1, 2.

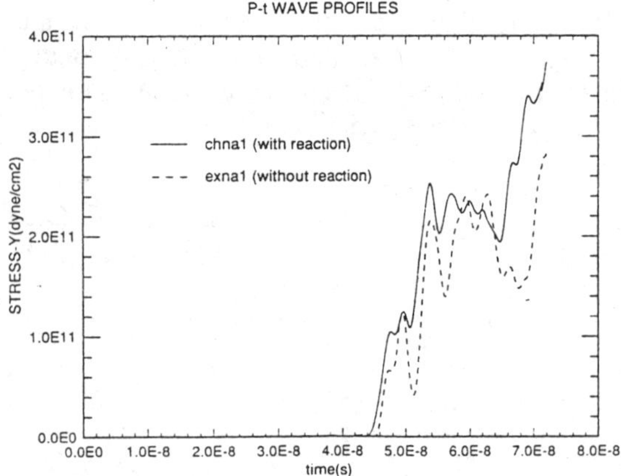

FIGURE 4. Comparison of pressure profiles with and without reaction for $V_o = 1\,Km/s$.

ACKNOWLEDGMENT

This work was supported, in part, by the Office of Naval Research (Dr. J. Goldwasser, N00014-94-1-0240-02). The authors gratefully acknowledge the computing support of the North Carolina Supercomputing Center.

REFERENCES

1. Flinn, J.E., Williamson, R.L., et al., *J. Appl. Phys.*, **64**, 1446 (1988).
2. Benson, D.J., and Nellis, W.J., "Numerical simulation of the shock compaction of copper powder," in *High-Pressure Science and Technology-1993*, pp. 1243-1246, 1994.
3. Negreskul, S.I., Psakhie, S.G. and Korostelev, S.Y., "The simulation of explosive compaction of powders by the element dynamics," in *Shock Compression of Condensed Matter-1989*,1990, pp. 233-236.
4. Horie, Y, Tang, Z.P., and Psakhie, S.G., "Computer simulation of shock compaction of powders by element dymanics," presented at the TMS Materials Week '94, Chicago, Il. Oct. 3-6, 1994.
5. Tang, Z.P., Horie, Y., and Psakhie, S.G., "Discrete meso-element modeling of shock processes in porous materials," in *High-Pressure Shock Compression of Solids, IV, Response of Highly Porous Solids to Shock Loading*, Springer-Verlag, 1996.
6. Bennett, L.S., Horie, Y., and Hwang, M.M., *J. Appl. Phys.*, **76**, 3394 (1994).

CALCULATION OF NONEQUILIBRIUM VELOCITIES IN POWDER MIXTURES UNDER SHOCK LOADING

K. Yano   Y. Horie   Z. P. Tang [†]

North Carolina State University, Raleigh, North Carolina 27695

An equation is derived for non-equilibrium particle velocities difference in the shock loading of powder mixtures within the framework of a continuum mixture theory. The equation is then applied to estimate the velocity difference for Ni/Al and Ti/Si mixtures. Results show that the energy associated with the difference is large enough to trigger mass mixing and shock-induced chemical reaction.

INTRODUCTION

Mass mixing and chemical reactions are known to occur in shock compression of powder mixtures that may last only a few microseconds. In such a short period of time diffusional mass transport mechanisms cannot explain the initiation of strong chemical reactions that are observed experimentally [1]. This paper explores one of possible origins for the fast mass mixing : the existence of non-equilibrium particle velocities in the shock front.

There are several mathematical models for heterogeneous mixtures [2,3]. However, in their applications to solids often homobaric (common pressure) and homokinetic assumptions (common particle velocities) are made in final analyses [2]. It is also pointed out [4] that the interactions between particles are not taken into account properly in [2]. Therefore in this study a continuum mixture theory [5] which is said to be more mathematically consistent is used to re-examine the question of non-equilibrium particle velocities in shock loading of mixed powders.

BALANCE EQUATIONS AND JUMP CONDITIONS

One of the metaphysical principles of the continuum mixture theory assumes that "so as to describe the motion of a constituent, we may in imagination isolate it from the rest of the mixture, provided we allow properly for the actions of the other constituents upon it" [6]. Based on this postulate the balance equations are developed for each constituent. In Eulerian coordinates they are :

Mass

$$\frac{\partial \rho_i}{\partial t} + \frac{\partial}{\partial x}[\rho_i u_i] = 0 \tag{1}$$

Momentum

$$\frac{\partial}{\partial t}[\rho_i u_i] + \frac{\partial}{\partial x}[\rho_i u_i{}^2] = -\frac{\partial p_i}{\partial x} + \sum_{\substack{j=1 \\ i \neq j}}^{N} R_{ij} \tag{2}$$

Energy

$$\frac{\partial}{\partial t}[\rho_i E_i] + \frac{\partial}{\partial x}[\rho_i u_i E_i] = -\frac{\partial}{\partial x}[p_i u_i] + \sum_{\substack{j=1 \\ i \neq j}}^{N} R_{ij} u_i \tag{3}$$

$$E_i = \epsilon_i + \frac{u_i^2}{2}$$

where "i" denotes constituent, ρ_i, u_i, p_i, R_{ij} and ϵ_i represent partial density, particle velocity, partial pressure, interaction force between i-th and j-th components, and internal energy per unit mass of i- th component, respectively. The effect

[†]On leave from University of Science and Technology of China

of heat conductivity is not considered in this formulation. The partial density and partial pressure are related to intrinsic density ρ_i^o and solid pressure p_i^o by: $\rho_i = \alpha_i \rho_i^o$ and $p_i = \alpha_i p_i^o$ where α_i is the volume fraction of i-th component.

The jump conditions across a shock front in a mixture are derived based on three assumptions. The first is that each constituent has common shock velocity U. The second is that wave profiles are steady in the coordinate attached to the shock front: $\xi = x - Ut$. The third is that shock front is infinitely thin. Then, the jump conditions are obtained as follows:

Mass

$$\rho_i u_{*i} = \rho_{i0} u_{*i0} = Const. \tag{4}$$

Momentum

$$p_i + \rho_i u_{*i}^2 = p_{i0} + \rho_{i0} u_{*i0}^2 = Const \tag{5}$$

Energy

$$h_{t*i} = h_{t*i0} = Const \tag{6}$$

where subscript "0" denotes initial state, u_{*i} and h_{t*i} are the particle velocity and total enthalpy per unit mass of i-th component in the moving coordinate. They are given by $u_{*i} = u_i - U$ and $h_{t*i} = \epsilon_i + p_i/\rho_i + u_{*i}^2/2$. Since the shock front is assumed to be a surface discontinuity, there is no time for the interaction among constituents to take place. This is the reason why the interaction force does not appear in the right hand sides of equation (5) and (6). However, in the downstream of the shock front , where interactions are taking place, (5) and (6) are no longer valid.

DERIVATION OF VELOCITY DIFFERENCE EQUATION

Another metaphysical principle of the continuum mixture theory assumes that "the motion of mixture is governed by the same equations as is a single body" [6]. This principle necessitates certain relations to hold between the over all mixture physical properties and those of each constituent. The relations are known as summation rules in the continuum mixture theory. For the system of equations (1) $\sim$ (3), the summation rules are:

$$\rho = \sum_{i=1}^{N} \rho_i \tag{7}$$

$$\rho w = \sum_{i=1}^{N} \rho_i u_i \tag{8}$$

$$p + \rho w^2 = \sum_{i=1}^{N} (p_i + \rho_i u_i^2) \tag{9}$$

$$\rho(\epsilon + \frac{w^2}{2}) = \sum_{i=1}^{N} \rho_i(\epsilon_i + \frac{u_i^2}{2}) \tag{10}$$

$$pw + \rho w(\epsilon + \frac{w^2}{2}) = \sum_{i=1}^{N} \left[p_i u_i + \rho_i u_i(\epsilon_i + \frac{u_i^2}{2}) \right] \tag{11}$$

$$Q = \sum_{\substack{j=1 \\ i \neq j}}^{N} R_{ij} u_i \tag{12}$$

where ρ, w, p, ϵ, and Q are the density, velocity, pressure, internal energy and dissipated heat for the mixture. Physically, equations (7) $\sim$ (10) and (12) should be understood as the definitions of ρ, w, p, ϵ, and Q. After some manipulations the definitions (7) $\sim$ (10) and equation (11) yield the following relation.

$$\sum_{\substack{i,j=1 \\ i>j}}^{N} x_i x_j (u_i - u_j) \left[(u_i - u_j) - \frac{h_{ti} - h_{tj}}{\sum_{i=1}^{N} x_i u_i} \right] = 0 \tag{13}$$

where x_i and h_{ti} are mass fraction of i-th component and total enthalpy of i-th component in laboratory coordinate. They are defined by $x_i = \rho_i/\rho$ and $h_{ti} = \epsilon_i + p_i/\rho_i + u_i^2/2$ respectively.

It can be shown that if equations (7) $\sim$ (11) hold in Eulerian coordinate, then the identical equations in form hold in the moving coordinate. In the moving coordinate equation (13) has the following form.

$$\sum_{\substack{i,j=1 \\ i>j}}^{N} x_i x_j (u_{*i} - u_{*j}) \left[(u_{*i} - u_{*j}) - \frac{h_{t*i} - h_{t*j}}{\sum_{i=1}^{N} x_i u_{*i}} \right] = 0 \tag{14}$$

Equation (14), when applied to the point immediately behind the shock front, reduces to

$$\sum_{\substack{i,j=1 \\ i>j}}^{N} x_i x_j (u_{*i} - u_{*j}) \left[(u_{*i} - u_{*j}) - \frac{h_{t*i0} - h_{t*j0}}{\sum_{i=1}^{N} x_i u_{*i}} \right] = 0 \tag{15}$$

where equation (6) is used. In particular, if $p_{i0} = 0$ and $u_{*i0} = -U$ (for $i = 1, 2, \cdots, N$), it follows that $h_{t*i0} = \epsilon_{i0}$. Thus equation (15) can be written as

$$\sum_{\substack{i,j=1 \\ i>j}}^{N} x_i x_j (u_{*i} - u_{*j}) \left[(u_{*i} - u_{*j}) - \frac{\epsilon_{i0} - \epsilon_{j0}}{\sum_{i=1}^{N} x_i u_{*i}} \right] = 0 \tag{16}$$

In the case of a two component mixture the above equation reduces to

$$(u_{*1} - u_{*2}) \left[(u_{*1} - u_{*2}) - \frac{\epsilon_{10} - \epsilon_{20}}{w_*} \right] = 0 \tag{17}$$

where $w_* = x_1 u_{*1} + x_2 u_{*2}$.

There are two possible solutions to equation (17). The first is the equilibrium solution in which $u_{*1} = u_{*2}$. The solution of interest is

$$\Delta u = u_{*1} - u_{*2} = \frac{\epsilon_{10} - \epsilon_{20}}{w_*} \tag{18}$$

This equation determines the particle velocity difference (slip velocity) immediately behind the shock front in a two component mixture.

APPLICATION TO SOLID AND POROUS MIXTURE

Ni/Al and Ti/Si mixtures are chosen for model calculations. The adiabatic equation of state used for solid constituents is of Murnaghan type given by (19).

$$p_i^o = \frac{K_i}{s_i} \left[\left(\frac{\rho_i^o}{\rho_{i0}^o} \right)^{s_i} - 1 \right] \tag{19}$$

where ρ_{i0}^o denotes the initial intrinsic density of i-th constituent. K_i and s_i are material constants.

For a porous mixture, a model equation of state shown in Fig. 1 is used. In this model, compression of the mixture below the consolidation pressure P_s obeys a linear model. Above P_s, the compression of each constituent follows equation (19). The values of P_s in calculation were set close to experimentally observed values. Parameters used in this study are listed in Table 1. Figures 2 and 3 summarize the results for Ni/Al and Ti/Si respectively.

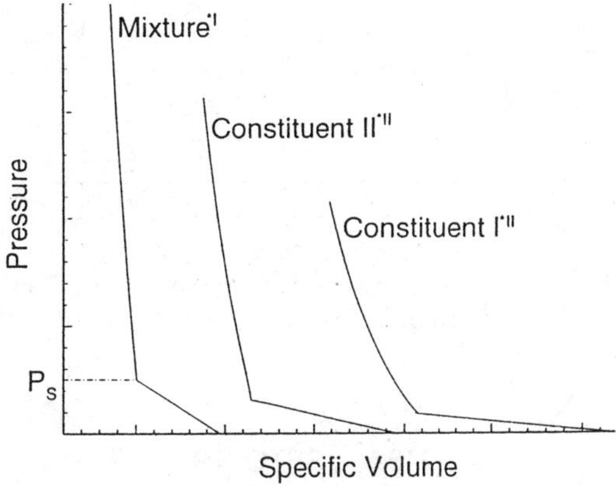

FIGURE 1. A model equation of state for porous material. *I Mixture pressure is defined by eqn. (9). **II Plotted based on partial density and partial pressure of each constituent.

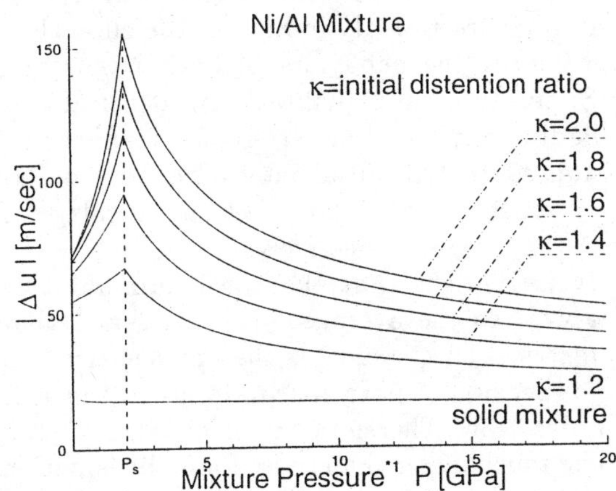

FIGURE 2. Calculated $P - |\Delta u|$ results for Ni/Al mixture. *1 Defined by eqn. (9).

TABLE 1. Material parameters used in the calculation

	Ni	Al	Ti	Si
ρ_{i0}^{o} (g/cm^3)	8.90	2.79	4.53	2.33
K_i $(Mbar)$	1.9250	0.7921	1.077	0.929
s_i	4.620	5.000	2.92	4.19
ϵ_{i0} (cal/g)	41.2	20.0	23.6	27.8

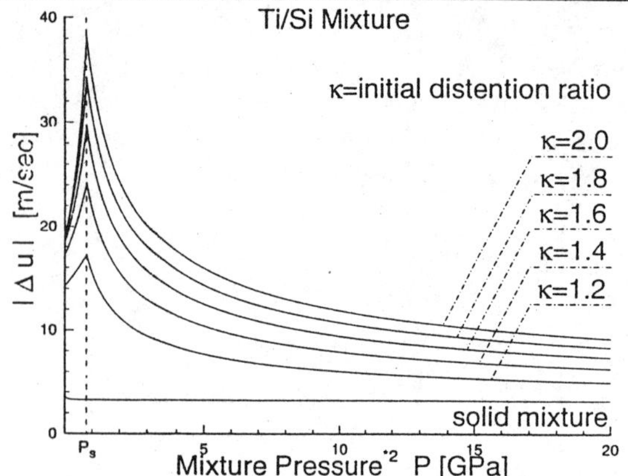

FIGURE 3. Calculated $P - |\Delta u|$ results for Ti/Si mixture. *2 Defined by eqn. (9).

DISCUSSIONS

Several observations can be made from Figs. 2 and 3. First, there is a maximum value of $|\Delta u|$ at $p = P_s$ for the porous mixtures, while $|\Delta u|$ for the solid mixtures is almost constant, but decreases very gradually. Second, the porosity of mixture strongly influences the slip velocity. Third, the behaviors of both Ni/Al and Ti/Si mixtures are qualitatively identical. A main difference between Ni/Al and Ti/Si can be attributed to the initial internal energy difference, which is about 20 cal/g and 4 cal/g for Ni/Al and Ti/Si respectively.

It is thought that the continuum mixture theory gives the averaged quantities and that geometrical effects such as the particle size and the dimension of pore space are not taken into consideration. Therefore the local value of $|\Delta u|$ can be much larger than those found by equation (18). $|\Delta u|$ on the order of hundreds of m/sec is large enough to initiate chemical reactions and

mass mixing [7].

CONCLUSION

Within the framework of a continuum mixture theory, non-equilibrium particle velocities were obtained for two component mixtures under the assumption of steady state discontinuity shock wave propagating with common shock velocity.

Slip velocities for solid Ni/Al and Ti/Si mixtures are 20 m/sec and 4m/sec respectively. For their porous mixtures the slip velocities are a function of porosity and become maximum at the pore collapse pressure.

The slip velocity is averaged quantity. For mixtures of 50 % theoretical , the values are 150 m/sec and 40 m/sec for Ni/Al and Ti/Si respectively. Hence local values can be much bigger.

ACKNOWLEDGMENTS

This work is partially supported by U.S. Army Research Office (DAAH 04-94-G-0034) and North Carolina Super Computing Center.

REFERENCES

1. Horie, Y., "Kinetic Modeling of Shock Chemistry," in Proc. of the Workshop at Georgia Institute of Technology, *Shock Synthesis of Materials*, May 24-26, 1994

2. Nikolaevskii, V. N., *Zh. Prinkl. Mekhan. i Tech. Fiz.* No. 3, pp.406 ~ 411, 1969

3. Nigmatulin, R. I., *Dynamics of Multiphase Media*, Hemisphere Publishing Corporation, 1991, Vol. 1.

4. Alekseev, Yu. F., Al'tshuler, L. V., and Krupnikova, V. P., *Zh. Prinkl. Mekhan. i Tech. Fiz.* No. 4, pp.152 ~ 155 (1971)

5. Truesdell et al., *The Classical Field Theories*, Handbuch der Physik, Springer, 1960, Vol II / 1

6. Truesdell, C., *Rational Thermodynamics*, Springer-Verlag, 1984, Second ed., p221

7. A.Yu.Dologoborodov et all., *Combustion, Explosion and Shock Waves*, **28**, pp.308 ~ 314 (1992)

ANALYTICAL MODEL FOR COMPACTION OF POWDER BY MEANS OF SHOCK WAVES

A. Ferreira and A.M.Costa Junior

Military Institute of Engineering, Praça General Tibúrcio 80, 22290-270 Rio de Janeiro, RJ - Brasil

A model for prediction of the shock required to consolidate a general porous material is presented. In this work it is considered a double tube cylindrical configuration, and a finite difference technique is applied in order to numerically simulate the formation and propagation of the detonation wave, and the convergent shock wave. The mathematical models of the literature are reviewed, modified, developed and combined. A model of porous collapse, dependent of temperature, is also developed to eliminate pressure singularities at the center of the model. The results are in good agreement with the experiments, and they are discussed in an extensive way.

INTRODUCTION

The utilization of waves, generated by the detonation of explosives or by impact, has had an increasing application to obtain new materials and pieces that are very difficult to get by using the traditional processes. Some applications, in which cylindrical configuration with concentric tubes were employed, are the obtaining of metallic pieces recovered with different materials, and the most successful experiment, the consolidation of powder materials.

The purpose of this study is to develop an analytical model that can predict the necessary pressure to consolidate any powder material. The finite difference method is used to solve the equations numerically.

Figure 1 shows the double tube experimental configuration, whose dimensions were used in the numerical simulation. The explosive is contained inside a PVC tube and also surrounds a steel tube, named flyer tube. The powder material is placed inside a stainless steel tube that is assembled inside the steel tube, and there is a small gap between the two tubes. The detonation wave accelerates the flyer tube towards the center of the system an hits the stainless steel tube. The impact creates the shock wave that will propagate the powder material and causes its campaction.

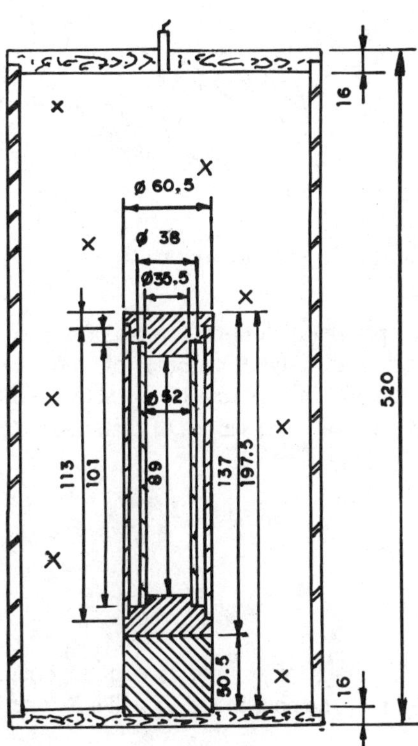

Figure 1 - View and dimensions in mm of the experimental double tube configuration.

VOID COLLAPSE MODEL

The p- $\propto$ model (1-3) was selected as a basis to simulate the void collapse of the powder material under dynamic loading, in which a new configuration was added, so that there is a solid nucleus in the inner part of the spherical shell (4), therefore one can prevent the pressure singularities that arises in the center of the previous model. Figure 2 presents a schematic view of the model, in which the concentric spheres were used.

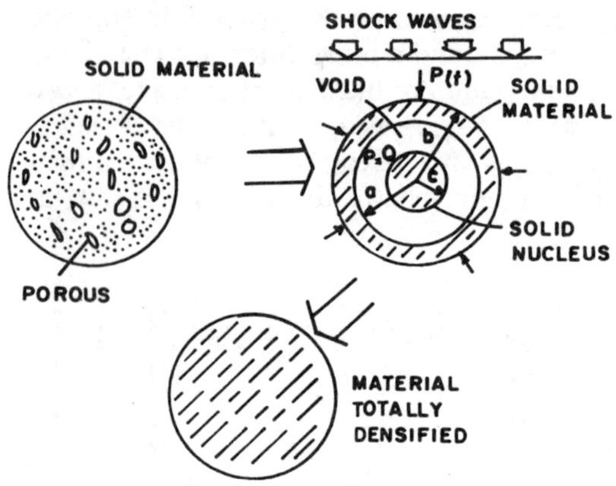

Figure 2 - Model of void sphere with solid nucleus

The powder distention ($\propto$) is defined as a relation between the total volume of the real porous material and the volume of the solid matrix of the same material. The values of distention ($\propto$) and its initial value ($\propto_0$) are:

$$\alpha = b^3/(b^3 - a^3 + c_0^3) \tag{1}$$

$$\alpha_0 = b_0^3/(b_0^3 - a_0^3 + c_0^3) \tag{2}$$

The values of $\propto$ and b can be obtained as a function of the porous radius **a**. Then, one has:

$$\alpha = 1 + (\alpha_0 - 1)(a^3 - c_0^3)/(a_0^3 - c_0^3) \tag{3}$$

The radial equilibrium equation in polar spherical coordinates is written in the following way (5):

$$\frac{\partial \sigma_{rr}}{\partial r} + \frac{2}{r}\sigma = \rho\frac{\partial \psi}{\partial r} \tag{4}$$

where σ_{rr} is the radial stress and σ is the shear stress. Solving the equation (4) and taken into consideration the boundary conditions:

$$\sigma_{rr} = -P(t), \qquad \text{at } r = b$$

$$\sigma = 0, \qquad \text{at } r = a \tag{5}$$

one has:

$$P(t) = 2\int_a^b \sigma\frac{dr}{r} - \rho\left[(a\ddot{a} + 2\dot{a}^2)\left(1 - \frac{a}{b}\right) + \right.$$

$$\left. - \frac{1}{2}\dot{a}^2\left(1 - \frac{a^4}{b^4}\right)\right] \tag{6}$$

Taking in account that the material is viscoplastic, with yield stress Y and viscosity η, both temperature dependent, the shear stress can be written as (6):

$$\sigma = Y(T) - 6\eta(T)\frac{\dot{r}}{r} \tag{7}$$

By using an linearly decreasing yield stress until the fusion temperature of the material, and taking for viscosity an exponential temperature dependence (5), with the help of the equations (6) and (7), one has for $T \leq T_m$:

$$P(t) = 2\int_a^b \left\{ Y(1 - T/T_m) + \right.$$

$$\left. - 6\eta_m \exp\left[B\left(\frac{1}{T} - \frac{1}{T_m}\right)\right]\frac{\dot{r}}{r}\right\}\frac{dr}{r} +$$

$$- \rho\left[(a\ddot{a} + 2\dot{a}^2)(1 - \frac{a}{b}) - \frac{1}{2}\dot{a}^2(1 - \frac{a^4}{b^4})\right] \tag{8}$$

If $T > T_m$ it is sufficient to do $Y=0$ in the equation (8).

In order to obtain all equations governing the dynamic compaction, it is still necessary an equation based on the energy included during the process:

$$\rho \dot{E} = -2 \frac{\sigma}{r} \dot{r} \qquad (9)$$

Taking into consideration that the calorific capacity has a constant value C_v in the range of $T_0 < T < T_m$ one has:

$$\dot{E} = C_v \dot{T} \qquad (10)$$

To simulate the void it is imperative to determine the distention of the powder material at each interval of time. From equation (3) one has:

$$\ddot{\alpha} = (6a\dot{a}^2 + 3a^2\ddot{a}) \frac{(\alpha_0 - 1)}{(a_0^3 - c_0^3)} \qquad (11)$$

By using the equations (7), (9) and (10), and the formulations of yield stress and viscosity, one obtains:

$$\rho C_v \dot{T} = -2Y\left(1 - T/T_m\right)\frac{\dot{r}}{r} +$$

$$+ 12\eta_m \exp\left[B\left(\frac{1}{T} - \frac{1}{T_m}\right)\right]\left(\frac{\dot{r}}{r}\right)^2 \qquad (12)$$

Introducing the equation (12) into equation (7), solving the integral and separating the value of a, one has:

$$\ddot{a} = \frac{1}{a}\left\{\left[C_v(T_a - T_b) - \frac{P(t)}{\rho_s} + \right.\right.$$

$$\left.\left. + \frac{1}{2}\dot{a}^2\left(1 - \frac{a^4}{b^4}\right)\right]\frac{1}{(1 - a/b)} - 2\dot{a}^2\right\} \qquad (13)$$

The values of the distention can be determined as function of pressure. It is sufficient to solve the system of equations (11) and (13), and using a finite difference scheme, where:

$$\alpha_{n+1} = 2\alpha_n - \alpha_{n-1} +$$

$$+ \Delta t^2(6a\dot{a}^2 + 3a^2\ddot{a})\frac{(\alpha_0 - 1)}{(a_0^3 - c_0^3)} \qquad (14)$$

The values of pressure $P(t)$ are given by the solution of the equations of convergent wave propagation in tubes or cylinders (7). It is just necessary to use a finite difference method. The values of $\ddot{a}$ and $\dot{a}$ are given by:

$$\ddot{a}_n = (a_{n+1} - 2a_n + a_{n-1})/\Delta t^2 \qquad (15)$$

$$\dot{a}_n = (a_n - a_{n-1})/\Delta t \qquad (16)$$

When the value of the distention is equal to one ($\alpha=1$), it means that the radii a and c_0 are equals, and the material is compacted.

RESULTS AND DISCUSSION

Figure 3 presentes a plot of the pressure generated by the shock wave, induced by the impact between the tubes, versus a radial coordinate (R), represented by the diameter of the PVC tube, in which the explosive is enclosed (8). This result is obtained by the discretization of the tube configuration and by utilization of the finite difference scheme (8).

Although the purpose of this work is not to explain the formation and propagation of the shock wave, Fig. 3 indicates the value of the pressure P(t) that are used in the equation (13). One can see in Figure 3 several intervals of time in which the pressure peak is travelling along the radius R. At initial time, $t=0\mu s$, that represents the impact between the tubes, the values of pressure is fixed as 0.0001 GPa. As the wave propagates towards the convergence line, center of cylindrical configuration (R=0), the peak pressure increases considerably, showing the convergence effect. At $t= 17,98 \mu s$, the maximum pressure is localized at R= 1,0 mm and its value is equal to 11,25 GPa.

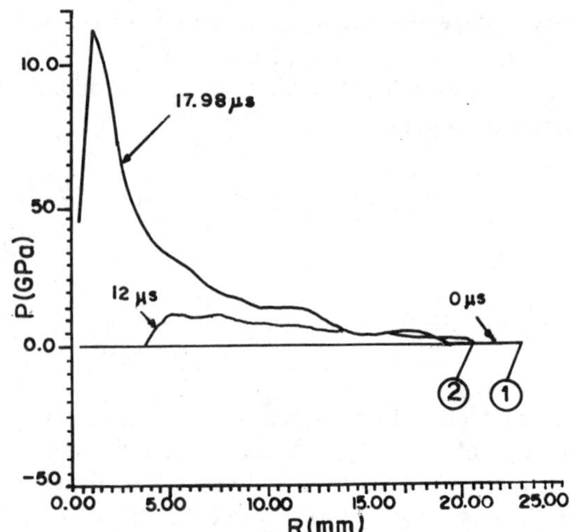

Figure 3 - Plot of pressure (P) versus radial coordinate (R).

Figure 4 shows a plot of distention ($\propto$), that was obtained from equations (13) to (16), against pressure, P = P(t), for compactions of copper powder, at central cell of the discretization. The results were obtained for initial radius $a_0=20\mu m$ and for different values of c_0 radius of the solid nucleus ($c_0 = 15\mu m$, 10 μm, and $5\mu m$). Other parameters are: initial temperature of 293 K, fusion temperature (T_m) equal to 1356 K, yield stress (Y) of 0,4 GPa, specific mass of porous material is equal to 65% of the specific mass of the solid material.

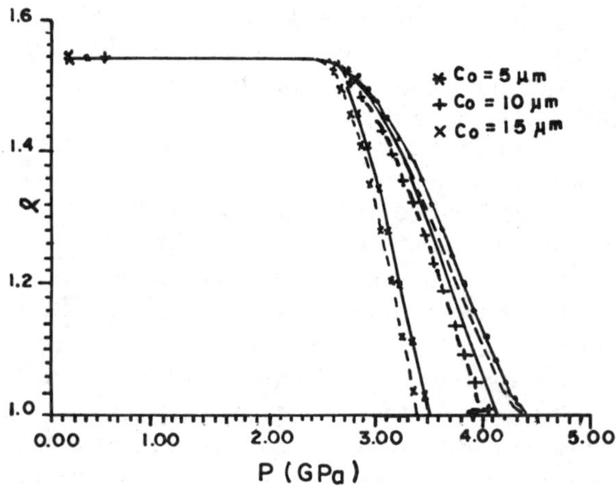

Figure 4 - Distention of powder material x applied pressure.

The full lines are for 130 mm diameter of the PVC tube, while the dashed lines are for 150 mm diameter.

One can see that the larger is the amount of explosive the less is the historic of pressures and, as a result, less pressure will be required for compation will be required. The material is rapidly compacted due to inertial effect that becomes more dominant than the dissipate viscoplastic effect. The increase of temperature reduces the mechanical strength and the viscosity of the material.

As the radius of the nucleus increases, the initial external radius (b_0) decreases, as shown in equation (2), once the volume of the solid part must be constant. Therefore, the thickness of the spherical shell of the model become thinner, and as a result, offers less resistance to the collapse. Thus, the compaction occurs more rapidly with less pressure.

CONCLUSIONS

The model of the void sphere with solid nucleus, developed to simulate the void collapse, shows to be efficient and eliminates the pressure singularities.

A rigorous analysis is in progress to confirm the validity of model with experimental results. The procedure that is now presented can serve as a guide to predict the pressure of compaction.

REFERENCES

1. Herrmann, W. J., J.Appl.Phys., vol. 40, pp. 2490-2499

2. Carroll, M. M. and Holt, A. C. J., J. Appl. Phys., 1972, vol.43 (4), pp. 759-761.

3. Carroll, M. M. and Holt, A. C. J., J.Appl. Phys., 1972, vol 43 (4), pp. 1626-1635.

4. Nesterenko, V. F. Procedings of Novolibirsk Conference on Dynamic Compaction, Novosibirsk, Jun 1988.

5. Carroll, M. M., Kim, K. T. and Nesterenko, V. F., J. Appl. Phys., 1986, 186, vol 59, pp. 1962-1967.

6. Dunnin, S. Z. and Surkov, V. V., J. Appl. Phys., 1982 vol 23, pp123-128

7. Costa Junior, A. C., Master Thesys, Military Institute of Engineering, 1992, Rio de Janeiro, RJ, Brasil, (In Portuguese).

8. Costa Junior, A. C. and Ferreira, A., Anais do XXII Congresso Ibero Latino Americano sobre Métodos Computacionais, Porto Alegre, RS, Nov 92, Brasil, (In Portuguese).

DEFORMATION MECHANISMS OF POWDER PARTICLES DURING DYNAMIC CONSOLIDATION

J. D. Walker and R. D. Young

Materials and Structures Division, Southwest Research Institute, San Antonio, TX 78228

Analytical models and numerical simulations focused on understanding the accommodation mechanisms involved in densification of spherical ceramic particles subjected to a high-pressure dynamic compressive wave are presented. A highly-resolved numerical model of a limited number of particles was used to understand the deformation mechanisms involved in dynamic compaction. Deformation is shown to occur when the steep compression wave front travels through the sample. Particle rearrangement is shown to be unlikely.

INTRODUCTION

Ceramic particles, normally considered brittle in nature, are known to plastically deform when subjected to intense dynamic loading. Such particle deformation is required for consolidation of the powder into a highly dense compact. For example, 30 μm spherical α-alumina powder has been dynamically consolidated to >90% of theoretical density [1] by explosive shock loading in a "Baby Bear" [2] fixture using baratol explosives (Fig. 1). The dynamic loading is characterized by maximum pressures between 13 and 26 GPa and durations on the order of 10's of microseconds [3]. Powder densification was found to occur primarily through plastic deformation, with only an occasional fractured particle.

Of interest is when and how particle deformation evolves during the dynamic consolidation process, and the mechanisms that are principally responsible for the deformation. The interaction associated with the consolidation of discrete particles is the subject of this investigation.

This investigation studies spherical alumina particles ranging in size from 10 to 100 μm in diameter. The range of particle velocity of interest is between 0.2 to 1.0 km/s. The initial packing density of the spheres is analogous to 60% of theoretical density.

SINGLE PARTICLE MODEL

As a first step, an analytical model is presented to examine the deformation and acceleration of a single

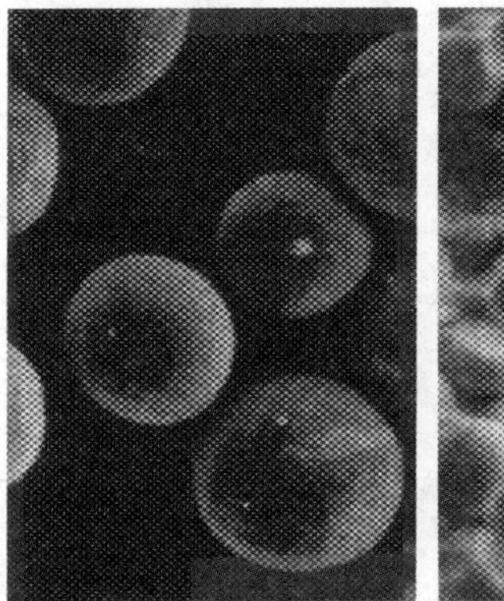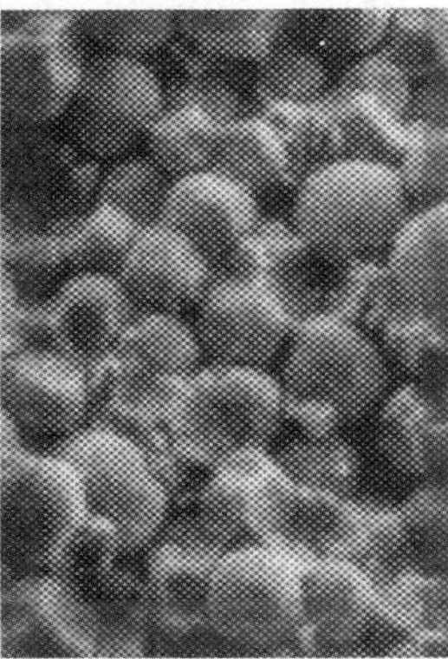

FIGURE 1. Initial morphology of the α-Al_2O_3 spherical powder (left), and assemblage of consolidated alumina particles (right).

spherical particle. In a recent paper this problem was approximated by a right circular cylinder and a uniaxial state of stress [4]. Comparison with hydrocode calculations showed that the problem was not well represented by a uniaxial state of stress, and this led to poor acceleration time estimates. In pursuit of a better analytic model, the following assumptions are made. A sphere initially at rest is impacted by a rigid plane travelling at velocity v_0. The whole particle mass is decelerated based on the force applied by the circular intersection of the sphere with the current location of the plane. This area is given by

$$A = \pi(R^2 - (L - R)^2) \qquad (1)$$

where L is the current length of the sphere, measured from the rigid plane to the top of the sphere, and R is the initial radius (there is no adjustment of the area to account for the deformation). This provides an equation of motion

$$M\frac{d\text{v}}{dt} = \pi(R^2-(L-R)^2)Y \qquad (2)$$

where M is the mass and v the center of mass velocity of the particle. Using $d\text{v}/dt = (d\text{v}/dL)\,dL/dt$ and $dL/dt = \text{v} - \text{v}_0$, Eqn. (2) can be integrated to yield

$$M\left(\text{v}_0\text{v}-\frac{1}{2}\text{v}^2\right) = -\pi Y\left(R^2L-\frac{5}{3}R^3-\frac{1}{3}(L-R)^3\right) \qquad (3)$$

giving the velocity versus current length of the sphere. With the density ρ and flow stress Y, this can be written

$$\frac{\rho}{Y}\left(\text{v}_0\text{v}-\frac{1}{2}\text{v}^2\right) = 2\left(\frac{L}{2R}\right)^3-3\left(\frac{L}{2R}\right)^2+1 \qquad (4)$$

When the particle reaches the velocity of the rigid plane, deformation ceases, and Eqn. (4) provides a final length versus initial length in terms of the ratio of the initial kinetic energy density to the flow stress:

$$\frac{\rho\text{v}_0^2}{2Y} = 2\left(\frac{L_f}{L_0}\right)^3-3\left(\frac{L_f}{L_0}\right)^2+1 \qquad (5)$$

The time dependent version of this can be used to obtain an estimate of the time it takes to accelerate the particle for the low velocity case. Writing $L = L_0(1 - w)$ we have

$$\frac{\rho}{Y}\left(\text{v}_0\text{v}-\frac{1}{2}\text{v}^2\right) = 3w^2 - 2w^3 \qquad (6)$$

Time differentiation of the definition of w gives $\text{v} - \text{v}_0 = -L_0\,dw/dt$, which gives

$$\frac{dw}{dt} = \frac{1}{L_0}\left[\text{v}_0^2-\frac{2Y}{\rho}(3w^2-2w^3)\right]^{1/2} \qquad (7)$$

The acceleration time is then given by

$$t_a = L_0\int_0^{w_f}\frac{dw}{[\text{v}_0^2-(2Y/\rho)(3w^2-2w^3)]^{1/2}} \qquad (8)$$

This can be explicitly integrated if it is assumed that w is much less than 1, so that the w^3 term can be ignored in the denominator of Eqn. (8) and when w_f is calculated in Eqn. (5). This gives the characteristic acceleration time:

$$t_a = \frac{\pi L_0}{2}\sqrt{\frac{\rho}{6Y}} \qquad (9)$$

Thus, for low impact velocities, the acceleration time is independent of the impact velocity. For a 10 μs alumina particle with density 3.98 g/cm^3 and $Y = 3.8$ GPa, $t_a = 6.6$ ns.

The characteristic time allows an examination of relative particle motion. The question is whether the time is such that small particles have the opportunity to move a great distance relative to large particles, and thus have the chance to change their relative positions, leading to additional consolidation. To consider this, we provide a formula for the minimum velocity required for a small particle to move the length of a big particle during the big particle's characteristic acceleration time:

$$\text{v} \approx \frac{2}{\pi}\sqrt{\frac{6Y}{\rho}} \qquad (10)$$

(the motion of the big particle and the acceleration of the small particle have both been neglected). Thus, for the alumina particles under consideration, a particle velocity in the compaction on the order of 1.5 km/s is required before the small particle's motion is comparable to the large particle's size. The particle velocity in most compaction experiments is less than 1 km/s, and so rearrangement is unlikely as a mechanism.

Numerical Simulations

To confirm the above model's results, the problem was examined numerically by impacting a single spherical alumina particle into a rigid anvil [4]. The numerical simulations were conducted using the two-dimensional axisymmetric option of CTH, a hydrocode developed at Sandia National Laboratories [5]. The fully dense alumina ($\rho_s = 3.98$ g/cm^3) particles were modeled with a Mie-Grueneisen equation of state and an elastic-perfectly plastic constitutive model ($Y = 3.8$ GPa [6]).

The value for the characteristic time constant from Eqn. (9) roughly agrees with the velocity dependent hydrocode values (Fig. 2). The deformed length calculated numerically is plotted in comparison with the results of Eqn. (5) in Fig. 3. The agreement between the two solutions is acceptable, and both solutions indicate that a substantial amount of deformation takes place in order to accelerate the particles within the characteristic acceleration time.

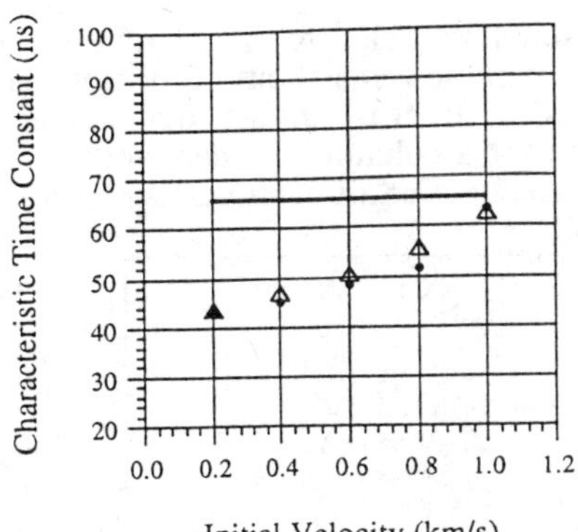

FIGURE 2. Characteristic acceleration time vs. impact velocity (for $100\,\mu m$ and scaled $10\,\mu s$ alumina spheres). Straight line is Eqn. (9).

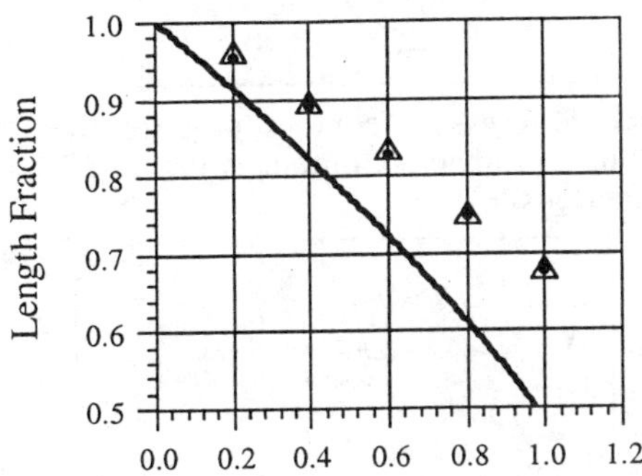

FIGURE 3. L_f/L_0 vs. velocity. Solid line is Eqn. (5).

MULTIPLE PARTICLE MODEL

A column of elastic-perfectly plastic spheres spaced Δx apart and of radius R is now considered. This elementary model is to be representative of a macroscopic assemblage of spherical alumina particles just prior to dynamic loading. Similar to the approach used by Duvall and Band to study the conceptual system of rigid beads, we examine this system through continuum methods by defining a shock velocity and a particle velocity for a column of particles driven by a moving rigid piston, as depicted from a CTH calculation in Fig. 4.

The rigid wall travels at constant velocity U_p. As the rigid wall proceeds forward it encounters the first sphere. The first sphere is deformed and is accelerated to velocity U_p. The rigid wall and sphere travel together until the next sphere is encountered. Now both the original sphere and the next sphere undergo deformation as the next sphere is accelerated to the velocity U_p. Clearly, U_p is the post-disturbance

particle velocity for the column. The velocity at which the disturbance travels through the column of spheres can be calculated by assuming the rigid wall was originally at $x=0$. Then, the location of the front just as particle $n+1$ is being struck is

$$x = nL_f + U_p t \tag{11}$$

Subtraction of this same equation for particle n gives

$$\Delta x + L_0 = L_f + U_p \Delta t \tag{12}$$

and this gives the velocity at which the disturbance is traveling as

$$U_d = U_p \frac{\Delta x + L_0}{\Delta x + L_0 - L_f} \tag{13}$$

To use the Hugoniot jump condition to calculate the pressure requires an average density. Taking the volume of the sphere divided by the volume of the cylinder gives

$$\rho = \rho_s \frac{2L_0}{3(L_0 + \Delta x)} \tag{14}$$

These can then be placed into the momentum Hugoniot jump condition ($P = \rho U_d U_p$) to give

$$\sigma = \frac{2}{3}\rho_s U_p^2 \frac{L_0}{\Delta x + L_0 - L_f} \tag{15}$$

This is a steady state axial stress behind the shock front.

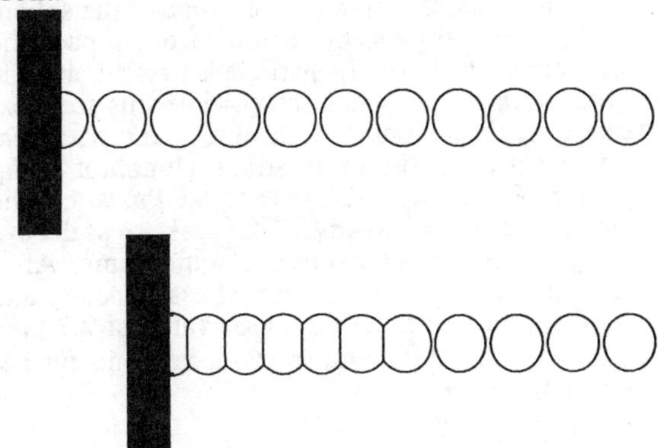

FIGURE 4. Accumulation of spheres in front of the rigid wall.

Two potential mechanisms exist for the deformation of the spheres: 1) inertial effects during acceleration of the spheres to the rigid wall velocity, and 2) pressure exerted on the spheres located between the disturbance interface and rigid wall which are travelling at velocity U_p. For a bulk solid, only mechanism (1) occurs during the passage of a shock, as the material both in front of and behind the shock front are in steady state. However, for a particle system, it has been argued that mechanism (2) is more important. We will show that for the column of particles considered here, only mechanism (1) occurs: particle deformation occurs during particle acceleration.

Numerical Simulations

A CTH simulation was comprised of 10 evenly spaced collinear 10 μm alumina spheres with a spacing of $\Delta x = 1$ μm. The first particle impacted by the piston was hemispherical in shape, with all remaining particle modeled as full spheres.

Figure 4 (above) contains an initial and a deformed geometry plot associated with the 10 collinear particles after the disturbance wave has propagated through half the spheres. The diameter of the particles is 10 μm, and the piston velocity is 0.6 km/s. The spheres behind the disturbance front are deformed with flat interfaces between the particles. The lack of deformation after passage of the disturbance front is evident from the uniformity in shape of the particles behind the disturbance front.

Figure 5 contains a contour plot of the stresses associated with the acceleration of the fourth particle, as well as the stress states in the particles that are at steady state behind it. Tensile regions can be seen in the particles, which could potentially lead to particle fracture.

A stress history in the spherical particle is given for the impact side, center, and far side of the sphere (Fig. 6). The peak pressure resulting from impact (the first peak occurs when the particle is first hit, and the second peak occurs when the particle hits the next particle in the column) is in agreement with that predicted by the uniaxial strain Hugoniot jump equation: $P = \rho_s c_0 u_p + (2/3)Y = 11.6$ GPa, where the sound speed is $c_0 = 7.6$ km/s. The peak stress decays in roughly the characteristic acceleration time. After this decay, a steady state axial stress appears, and roughly agrees with the Eqn. (15) value of 2.7 GPa ($L_f/L_0 \approx 0.75$). After the stress decays, no further deformation is seen.

SUMMARY

Analytic models and numerical simulations were used to examine the deformation of single spheres and a column of spheres. This was used to help understand the deformation mechanisms of particle compaction. It was shown that rearrangement of the particles is not likely, as the particle velocity is not high enough to allow small particles to move relative to big particles. A particle velocity cutoff for that condition was derived. It was shown that the particle deformation for a column of spheres occurs during the initial acceleration of each particle , and not at later times.

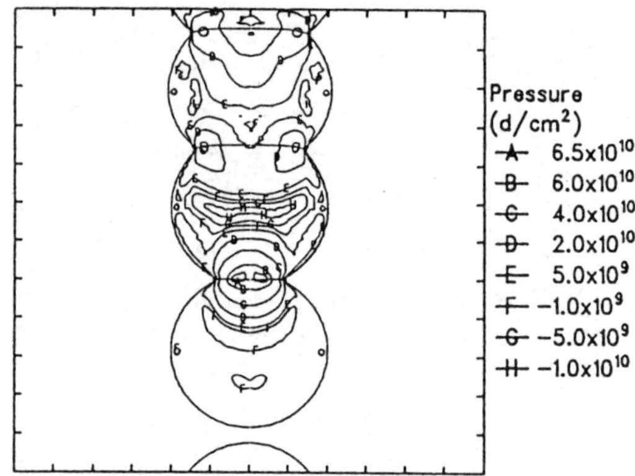

FIGURE 5. Pressure contour plot for the fourth sphere during impact. Contour A is 6.5 GPa and H is -1 GPa (tensile).

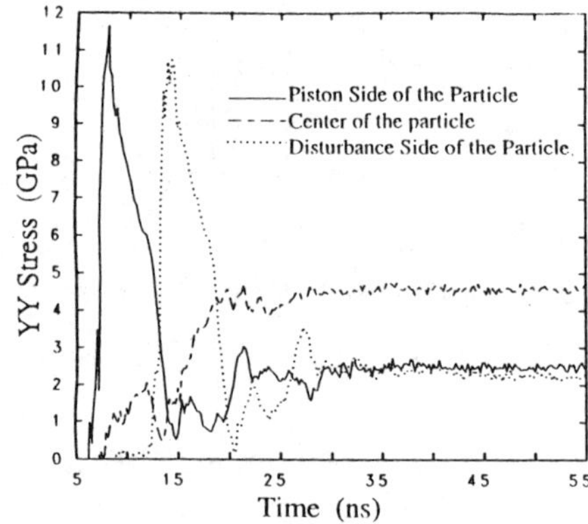

FIGURE 6. Stress history for the third sphere.

REFERENCES

1. Beauchamp, E. K., Carr, M. J., and Graham, R. A., *J. of the American Ceramic Society* **68**, 696-699, (1985).
2. Graham, R. A. and Webb, D. M., *Shock Waves in Condensed Matter-1983*, Ed. by J. R. Asay, R. A. Graham, and G. K. Straub. North-Holland, New York, 1984, pp. 211-214.
3. Davison, L., Webb, D. M., and Graham, R. A., *Shock Waves in Condensed Matter-1981*, Ed. by W. J. Nellis, L. Seaman, and R. A. Graham. American Institute of Physics, New York, 1982, pp. 67-71.
4. Young, R. D., Walker, J. D., and Beauchamp, E. K., "Intergranular Deformational Mechanisms During Dynamic Consolidation, to appear in EXPLOMET95 Proceedings (1995).
5. McGlaun, J. M., Thompson, S. L., and Elrick, M. G., *J. Impact Engrg.*, **10**, 351-360, (1990).
6. Wilkins, M. L., "Third Progress Report of the Light Armor Program," UCRL-50460, LLNL, Livermore, CA (1968)

PROBING THE SHOCK INDUCED REACTION THRESHOLD OF POWDER MIXTURES

R. D. Young

Materials and Structures Division, Southwest Research Institute, San Antonio, TX 78228

The results of LORC experiments are presented comparing the dynamic response of materials on the threshold of a shock-induced reaction. The problem is understanding the initiation criteria and chemical kinetics associated with shock induced reactions of powder mixtures and other energetic materials subjected to loading in the low pressure, long duration, and large strain regime typical of an accident scenario. Information concerning the initiation threshold and reaction conditions can be obtained using the split-Hopkinson pressure bar apparatus modified for radial specimen confinement and direct projectile impact, called the LORC apparatus. The experimental domain is 1 to 50 kbar pressure, 10 to 10,000 microseconds duration, and strain rates from 10^{-3} to 10^{-6} s^{-1}. For energetic materials, the reaction threshold and reaction kinetic information is obtained from the initial experimental conditions and diagnostic monitoring of an impulse delivered by the specimen to a transmitter bar. The diagnostic wave form differs significantly upon initiation of the reactions relative to unreacted materials. Initiation of shock-induced chemical reactions in powder mixtures can take 10's of microseconds when subjected to low pressure, long duration, moderate strain rate conditions.

INTRODUCTION

Mixtures of powdered fuels and oxiders constitute an interesting class of energetic materials which can react and perform work when stimulated by a shock wave. The products from these reactions can possess high mechanical stiffness, which theoretically enables extraction of a considerable amount of work through small volume expansions at very high pressures. This differs significantly from standard propellants and explosives, which perform most of their work on the environment through extremely large volume expansion at low pressure.

The reactant and product Hugoniots of these powder materials are such that a sustained Chapman-Jouguet point is not easily established, except under theoretical or unusual physical conditions. Conventional quantification of these reactions is difficult since a steady-state condition is rarely established. The LORC apparatus is useful for probing reaction thresholds and quantifying the amount of work associated with shock-induced solid-state reactions in energetic mixtures of powders.

SPLIT HOPKINSON PRESSURE BAR THEORY (BACKGROUND)

Hopkinson [1] was one of the first to experimentally measure momentum by propagating an elastic stress pulse into a steel bar, similar in nature to the experiments conducted herein. The lack of electronic devices limited his ability to time resolve the momentum, but his experimental technique was sufficient to discern the total magnitude of momentum delivered to the bar.

Davies [2] enabled time resolved momentum measurements by capacitively measuring the displacement at the bar's free surface perpendicular to its axis of revolution. He also defined the apparatus' useful experimental domain and limiting experimental conditions.

Kolsky [3] extensively modified this apparatus to what is presently known as the split Hopkinson pressure bar [4] (SHPB), also known as the Kolsky apparatus. The Loves Radial Constraint (LORC) apparatus is based in principle on the Split Hopkinson pressure bar, modified through mathematics first brought forth by A. E. H. Love [5].

MODIFICATIONS

The desired stimulus domain for this experiment is a pressure pulse twenty to forty kbar in amplitude, thirty to sixty microseconds in duration, at a strain rate of approximately 10^5 s^{-1}. There are three deficiencies associated with the standard Split Hopkinson pressure bar (SHPB): 1) the pressure range for low impedance porous media is limited to a few kbar [6]; 2) the induced specimen strain-rate for the SHPB is on the order of 10^3 s^{-1}; and 3) the large divergent specimen flow extrudes the specimen from between the ends of the bars resulting in significant loss of specimen materials during the test. Gorham [7] recognized that direct projectile impact could be used to overcome the first two deficiencies. Gaffney [8] et al. recognized that the third deficiency is overcome through the use of a thick wall confining ring. A problem associated with this test method is extrusion of the sample between the Hopkinson bars and confinement cylinder. Ross [9] et al. further modified the confinement cylinder to include thin interference-fit specimen covers or a hardened steel wafer/"O" ring combination as sealing methods. This arrangement was successful in stopping the flow of specimen material through the pressure-bar-to-confinement-ring interface.

THE LORC APPARATUS

Simultaneous use of Gorham's direct impact modification and Gaffney's radial confinement modification constitutes the Love Radial Constraint modification which enables: 1) achievement of higher pressures for low impedance specimens; 2) a shorter rise time to obtain peak test specimen pressure; 3) obtainment of strain rates of 10^5 s^{-1}; and 4) reduction in divergent flow of the specimen material. A conceptual schematic of the LORC apparatus is contained in Figure 1.

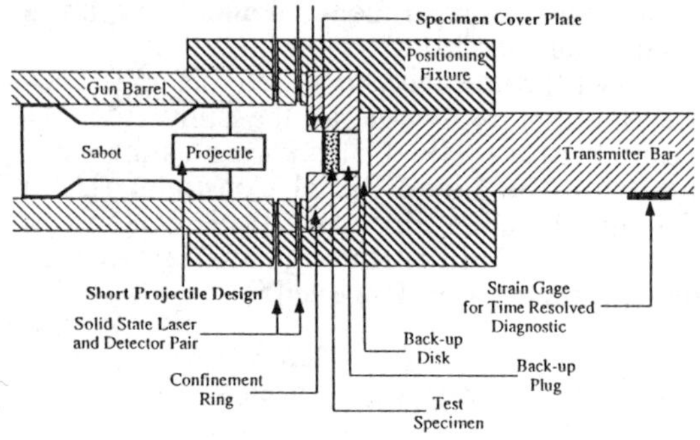

Projectile velocity resolved by solid-state laser and dector pair

FIGURE 1. Experimental configuration.

The peak pressure experimentally obtained is estimated a priori by use of the formula:

$$P = 1/2\, \rho\, C_o\, V \tag{1}$$

where P is pressure in kbar, ρ is the density of VascoMax 350 CVM Steel measured at 8.1 g/cc, C_o the steel's bulk sound speed measured at 4.851 km/s, and V is the projectile impact velocity in km/s. Although Equation (1) predicts the pressure induced when two pieces of VascoMax steel are impacted together, its use here is enabled by the small sample thicknesses and multiple pressure pulse reverberations that increase the specimen pressure to approach the upper bound values predicted by Eqn. (1). For short projectile lengths, approximately half this value is obtained experimentally due to velocity reductions in the projectile during the reverberation process. Close to full pressure values are realized for long projectiles used in these tests.

Specimen pressures from 20 to 40 kbar are achieved using unequal cross-sections of the specimen and transmitter bar. When the specimen is pressurized in excess of 22 kbar, small amounts of yielding occur on the back-up disk, a thin disk of VascoMax steel placed between the back-up plug and transmitter bar. The function of this disk is to preserve the structural integrity of the transmitter bar enabling its reuse. Small amounts of plastic deformation in the back-up disk influence the overall energy balance, but does not influence the momentum transfer in terms of first order effects.

To achieve short pressurization intervals, small specimen thicknesses with .05mm thick brass specimen cover plates are used. The standard specimen thickness is a tenth of an inch, resulting in an initial maximum pressure pulse transit time of two to three microseconds, with subsequent reduction to about a microsecond as the specimen sound speed increased with increasing pressure. As a result, complete specimen pressurization is achieved in a ten to fifteen microsecond time interval. When a specimen cover is incorporated into the design to eliminate sample extrusion around the projectile, only a very thin cover plate can be used in order to maintain short pressurization time intervals. Ross' "O" Ring wafer design possesses more mass than is contained in the test specimen used in these experiments. If used, the "O" ring design would reduce the peak achievable pressure obtainable, and thus was not adopted for use in this experiment. Acceleration of the thin brass shim stock theoretically takes place in less than a microsecond, without an observable reduction in projectile velocity.

Numerical simulations of the LORC modifications were made using Sandia's Hydrocode called CTH. The details of these numerical experiments are published [10] and will not be included herein.

DIAGNOSTICS

Two diagnostics are used in this study: projectile velocity measured prior to impact and time-resolved strain measurements on the transmitter bar. The projectile velocity is detected by sensing interruptions of the two laser beams induced as the projectile passes between the emitter and detector. The transversely oriented beams are precisely placed on the upstream side of the confinement ring such that the time differential between beam interruptions is relatable to projectile velocity. This configuration is contained in Figure 1.

The momentum delivered to the transmitter bar is diagnosed by time-resolved strain measurements. The strain gages are mounted on the bar with instrumentation capable of resolving changes in strain in the microsecond domain.

EXPERIMENTS

Known quantity of the powder are pressed into a confinement ring covered with a four mil thick brass specimen cover plate. A long and short projectile configurations are used to control the dwell duration of the experiment, which is conceptually illustrated in the schematic contained in Figure 1.

For illustrative purpose, Figure 2 contains two time-resolved pressure measurements, one for material "A" believed to be undergoing a reaction, and material "B" which shows no evidence of reaction upon post-test examination. There are two points worthy of discussion. The first is the general similarity and degree of overlap between the two time-resolved pressure measurements. This observation indicates that these two tests are fundamentally consistent. Thus, the two traces can be compared with confidence.

The second point is the departure of the two time-resolved pressure measurements which occurs at about 300 microseconds into the test. The pressure profile of the material not exhibiting post-test evidence of chemical reaction (material "B") drops significantly before the pressure profile of the material which does exhibit post-test evidence of chemical reactions (material "A").

A further indication that material "A" undergoes a chemical reaction within the experimental time domain is obtained by examining the momentum delivered to the transmitter bar. The momentum is determined by integrating the product of the time-resolved pressure and cross-sectional area of the transmitter bar over the experimental time duration. For the experimental comparison made in Figure 2, the initial projectile velocities are virtually identical. In this case the increased relative momentum

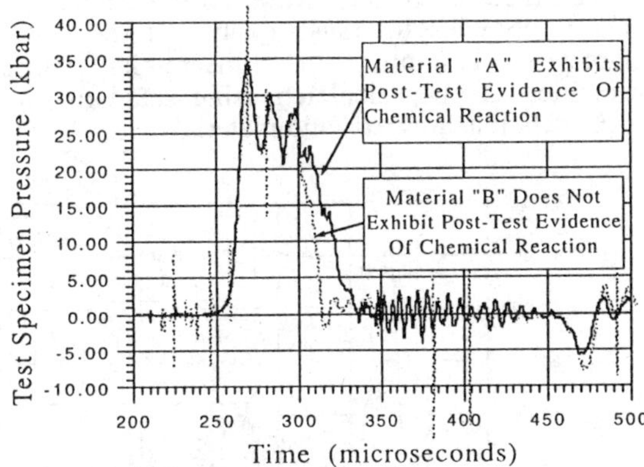

FIGURE 2. Pressure profile comparison.

delivered to the transmitter bar is proportional to the increase in area under the time-resolved pressure measurement. It is clear from the experimental traces contained in Figure 2 that the area under the pressure-history trace of material "A" is greater than that of "B". Quantification of this difference is used as an indication of the degree to which the reaction has proceeded toward completion.

It is interesting to note that the reaction in material "A" appears to be retarded by the decreasing pressure. Post-test examination reveals the abundance of reactants capable of sustaining a significantly longer reaction. In reality the reaction has been self limiting.

The same experiment was performed at a shorter dwell times (short projectile), at slightly elevated peak pressures, for which example experimental results are shown in Figure 3. Examination and comparison of the time-resolved pressure profiles do not indicate any significant departures between the two curves. In addition, the integrated momentum is about the same for these two materials. However, material "A" shows post-test evidence of reaction and the material "B" does not. The conclusion drawn is that the reaction experienced by material "A" occurred after completion of the experiment.

The following hypothesis concerning solid-state reaction kinetics can be formulated from the experimental data at hand. This experimental data is consistent with the theory that the shock wave is responsible for the mass transport which sustains the chemical reaction under shock conditions due to differential particle velocities induced by virtue of impedance differences between the fuel and oxidizer components of these energetic materials. The delay in the apparent chemical reaction could be a manifestation of the time required for the fuel and oxidizer components to translate into close proximity.

This experimentally observed time delay is consistent with the theoretical time it would take one of the thirty micron particles to translate about three particle diameters, assuming the differential particle velocities can be calculated using standard shock U_s-U_p relationships for solid materials.

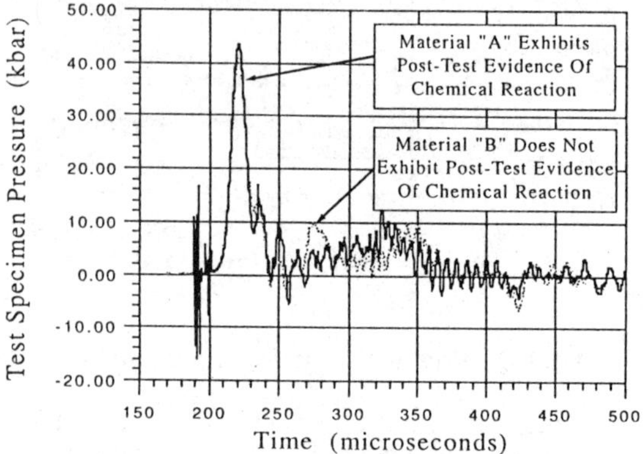

FIGURE 3. Retesting materials "A" and "B" within a different pressure time domain.

The LORC experimental results are consistent with other experimental observations that a reaction threshold exists, under which the reaction does not proceed. However, the LORC results indicate that the threshold may be related to a combination of pressure and time, which again is consistent with the theoretical mass transport mechanism.

SUMMARY

The LORC experimental technique is based upon the split Hopkinson pressure bar modified for simultaneous implementation of direct projectile impact and radial specimen confinement. This experimental technique is useful to probe the reaction kinetics associated with mixtures of fuel and oxidizer powders subjected to pressures in time domains usually associated with impact. The LORC experiment is focused upon resolving the relative ability of materials to perform work within the time frame of tens to hundreds of microseconds, stimulated by dynamic pressure pulses with magnitudes in the tens of kbars. This experimental technique is capable of relative characterization of energetic materials with respect to their ability to perform work on the environment within the time frame of interest.

The following insight into shock-induced reaction kinetics is obtained through use analysis of LORC test results. An intercomparison of test specimens which exhibit post-test evidence of chemical reactions with those which do not yielded the following conclusions:

1. The bulk mass-transport mechanism does not operate in the sub-microsecond time frame at pressures below 40 kbars.
2. Low pressure shock waves have the ability to sustain solid-state chemical reactions at pressures above 30 kbar, provided the dwell time at pressure is long enough.
3. The shock-induced chemical reactions studied showed little capability, if any, of performing work on the environment in this experimental low pressure domain.
4. The stimuli domain of 30 kbar pressure for a duration of 40 ms minimally initiates the chemical reaction, which is quenched when the pressure is reduced to ambient conditions.

REFERENCES

1. Hopkinson, B., Phil. Trans., *Royal Soc. of London*, Series A, p. 437 (1914).
2. Davies, R. M., Phil. Trans. *Royal Soc. of London*, Series A, p. 375 (1948).
3. Kolsky, H., *Proc. Phys. Soc.*, B62, p. 676 (1949).
4. Lindholm, U. S., *J. Mech. Phys. Solids*, **12**, pp. 317 (1964).
5. Love, A. E. H., New York Dover Publications (1944).
6. Nagy, A., Ko, L. W., Lindholm, U. S., *J. Cellular Plastics*, **10**, 3 (1974).
7. Gorham, D. A., Inst. Phys. Conf. Ser., 47, pp. 16 (1979).
8. Gaffney, E. S., Brown, J. A., Felice, C. W., Los Alamos Report LA-UR-84-4024 (1985).
9. Ross, C. A., Thompson, P. Y., Charlie, W. A., Doehring, D. O., Experimental Mechanics, **5**, pp. 80 (1980).
10. Young, R., Silling, S., *Proceedings for the Joint International Association for Research and Advancement of High Pressure Science and Technology and American Physical Society Topical Group on Shock Compression of Condensed Matter Conference*, pp. 1647 (1994).

APPROXIMATE ESTIMATE OF LOADING PARAMETERS IN COMPOSITES FOR THE CASE OF STRONG SHOCK WAVES

I. V. Yakovlev, V. V. Pai, and G. E. Kuz'min

Lavrentyev Institute of Hydrodynamics, Novosibirsk, Russia

The main purpose of explosive compaction of porous reinforced and nonreinforced materials is the obtaining of strong compacts with the density close to the monolithic density. As is known, only the dynamic regime of explosive compaction allows for the solution of the problem. This way, the increase in the accuracy of calculation of the pressure value in a porous material results in the possibility of a more optimal selection of the initial loading parameters to provide the dynamic compaction regime throughout the depth of the sample under consideration. We present a technique for rough estimating loading parameters in powder materials for the case of strong shock waves. The comparison of the experimental results with the calculated shape of the shock wave has shown that the discrepancy of the results does not exceed 5%.

The principal task in explosive compaction of porous composite (reinforced and nonreinforced) materials is to obtain a strong compact with the density close to that of corresponding monolith material. The result of the shock loading of a concrete sample of porous material is dictated by various conditions. Here such limiting conditions will be referred to as compactibility criteria. The most general form of such a criterion is the equation $F(\sigma_1, \sigma_2, \ldots, a_1, a_2, \ldots) = 0$ of some hypersurface in the space of dynamic parameters $\sigma_1, \sigma_2, \ldots$ The equation involves critical values $a_1, a_2, \ldots$ of dynamic parameters themselves and/or some of their combinations. The critical values are deduced from various constitutive assumptions describing behavior of the porous material under shock loading. *A priori* conclusions on the result of dynamic compaction of the material could be done if the position of the actual point $\sigma_1, \sigma_2, \ldots$ of the dynamic parameters space were known with respect to the surface $F = 0$. In the simplest case such a requirement can be presented as a set of inequalities, for example, $p \geq p_{cr}$ (i.e., the loading pressure should be not less than its specific critical value), etc. On the other hand there

are two well-known approaches for determining the position of the actual point in the dynamic parameters space depending on the initial data (powder density, detonation velocity and density of the explosive and so on) or for adjusting the initial data ensuring the proper values of the dynamic parameters. One of them is the empirical method in which the possibility or impossibility of obtaining a good compact under given initial conditions is concluded with the help of extrapolation of the available data on the basis of experience and skill of investigators. The second method is computer simulation of the equations of the porous material motion under dynamic loading based on more or less plausible constitutive models of the material and explosive with consideration for the genuine physical and chemical phenomena. Both of these lines of investigations are very interesting and important, either of them has undoubted merits. Unfortunately, both of these methods are too sophisticated and moreover by now they aren't quite reliable. We suggest a new way, sufficiently accurate and not so expensive, for estimating dynamic parameters under strong shock loading of porous composite materials and other similar substances. In this work we regard shock loading at which the compacting continues up to the stage of the monolith formation as strong loading.

Let us consider the process of compaction of porous material under loading through a metal plate. Let the process be one-dimensional and the plate be absolutely rigid. Under the plate of thickness δ_M and density ρ_M there is a semi-infinite layer of porous material (powder) of initial density ρ_{p0}. Through the powder the shock wave with coordinate $y(t)$ and velocity $U(t) = y'(t)$ propagates. Let p, u, ρ_{p1} be the shock pressure, the mass velocity (which is equal to the plate velocity) and the material density behind the wave. The motion of the material behind the shock wave is governed by the equations

$$\rho_{p0}U = \rho_{p1}(U - u),\tag{1}$$

$$p = \rho_{p0}Uu\tag{2}$$

expressing conservation of mass and momentum across the shock. Although the problem of the complete compaction of porous material is of the main interest we can consider some more common case of compression of the material to a certain preassigned density ρ_* so that $\rho_{p1} = \rho_*$.

Introduce into consideration a function $P(t)$ - a time dependence of pressure on the surface of the plate and write the equation describing combined motion of the plate and the powder in connection with the shock wave propagation. Taking account of the associated mass of the powder as, for example, in [1, 2] and using (1) the equation is as follows

$$P(t) = \eta \frac{d}{dt}\left[\left(\rho_M\delta_M + \rho_{p0}y(t)\right)y'(t)\right],\tag{3}$$

where $\eta = 1 - \rho_{p0} / \rho_{p1}$.

Integrating Eq. (3) twice with respect to time with the initial conditions $y(0) = 0, y'(0) = 0$ and solving for the shock wave coordinate $y(t)$ yields:

$$y = \frac{\rho_M\delta_M}{\rho_{p0}}\left[\sqrt{1 + \frac{2\rho_{p0}}{\eta\rho_M^2\delta_M^2}\int_0^t dt\int_0^t P(t)dt} - 1\right].\tag{4}$$

Then the shock wave velocity is found from $U(t) = y'(t)$, the mass velocity and the pressure are determined using (1), (2).

As seen from the above calculations, the shock Hugoniots of materials are not needed here, unlike in [1, 2], because the consideration is restricted to the most important case of the compaction of a porous material up to some fixed density.

Now we apply the results obtained to the two popular configurations of the explosive loading of powders.

In the case of one-dimensional impact of a plate moving with velocity v_0 against an initially resting powder the total momentum per unit area of the surface of the plate is

$$\int_0^t P(t)dt = \rho_M\delta_M v_0.$$

Substituting this to (4) and using the above consideration we can obtain the shock parameters, in particular,

$$p = \frac{\rho_{p0}v_0^2}{\eta + \dfrac{2\rho_{p0}v_0 t}{\rho_M\delta_M}}.\tag{5}$$

Let now a porous material covered by a metal plate be loaded by explosion products pressure under detonation of a charge of thickness δ_e and density ρ_e at the surface of the plate. To obtain the formula for the distribution of pressure on the plate we consider a problem on the acceleration of the plate by detonation products under sliding detonation. Considering the plate as a set of rigid material elements with mass per unit area $\rho_M\delta_M$ write the equation of motion for such an element along a curvilinear path as

$$\rho_M \delta_M D^2 \frac{d\beta}{dl} = P. \qquad (6)$$

Here D is the detonation velocity of the explosive, β is the slope of the tangent to the curve, l is the coordinate along the plate.

The expression for the angle β is selected from the results of the numerical computation of the plate acceleration [3] in the form:

$$\beta = \left(\sqrt{\frac{k+1}{k-1}} - 1 \right) \frac{r\pi\{1 - \exp[-s\lambda(r+a)]\}}{2(r+a)}.$$

Here $r = (\rho_e \delta_e) / (\rho_M \delta_M)$, $s = l / \delta_e$, k is the polytropic exponent of the detonation products, the constant λ can be determined from the pressure at the detonation front:

$$\lambda = 2 / \left[\pi(k+1)\left(\sqrt{(k+1)/(k-1)} - 1 \right) \right]$$

and a is selected by the least squares method from the experimental results and two-dimensional computations of the plate acceleration [4].

Then the physically clear asymptotics are valid: $\beta(0, s) = 0$, $\beta(r,0) = 0$, $\beta(\infty, s) = \beta_{max}$, where β_{max} is the limiting angle of the turn in the full Prandtl-Meyer flow for the given k.

If we bear in mind that $l = Dt$ the needed formula for the pressure on the surface of the plate is obtained in the form:

$$P = P_D \exp(-ct). \qquad (7)$$

where $c = \lambda(r+a)D / \delta_e$.

This pressure distribution is valid for the case of free plate acceleration by detonation products under sliding detonation over the range of $0 < r < 3$, $2 < k < 3.5$ in which it describes the numerical solution results of the two-dimensional steady problem within the 3% uncertanty. Note that based on the computations [3] the similar formula for pressure was proposed in [5].

In the case in question the plate motion can be assumed as practically one-dimensional, and from (4) and (7) we obtain the position of the shock:

$$y = \frac{\rho_M \delta_M}{\rho_{p0}} \left[\sqrt{1 + \frac{2\rho_{p0}r^2(e^{-ct} + ct - 1)}{\lambda^2 \eta(k+1)(r+a)^2 \rho_e}} - 1 \right] \qquad (8)$$

Now the shock wave parameters U, u, p can be found, for example, the pressure is:

$$p = \frac{\rho_{p0}\left[\dfrac{Dr(1 - e^{-ct})}{\lambda(k+1)(r+a)} \right]^2}{\eta + \dfrac{2\rho_{p0}r^2(e^{-ct} + ct - 1)}{\lambda^2(k+1)(r+a)^2 \rho_e}}. \qquad (9)$$

Furthermore, the elimination of the parameter t from (8) and $l = Dt$ enables one to find the theoretical shape of the shock wave in the powder.

Thus the loading parameters as functions of the strong shock propagation time through the material is obtained for the compaction of a porous material up to some fixed (for example, monolithic) density in the first shock wave. The details of these computations are in [6].

To check out the applicability of the results obtained to estimating the shock loading parameters in porous materials under strong shocks we carried out the experiment according to the scheme in Fig. 1. The shape of the shock wave in the material is determined by the resistance method [7] adapted to the case. The explosive charge 1 is placed on the surface of the plate 2 which covers the sample 3 of the material under study. The resistance transducer 4 (a straight segment of varnish-insulated nichrome wire) is tautened in the sample at some angle to the plate surface. During the experiment the shock wave changes the length of the insulated part of the transducer by transforming the powder into the monolithic material.

If the detonation velocity is known, then measuring the resistance of the indicator on time one can easily determine the profile of the shock wave 5 in the powder. The details of the measuring method is considered in [7].

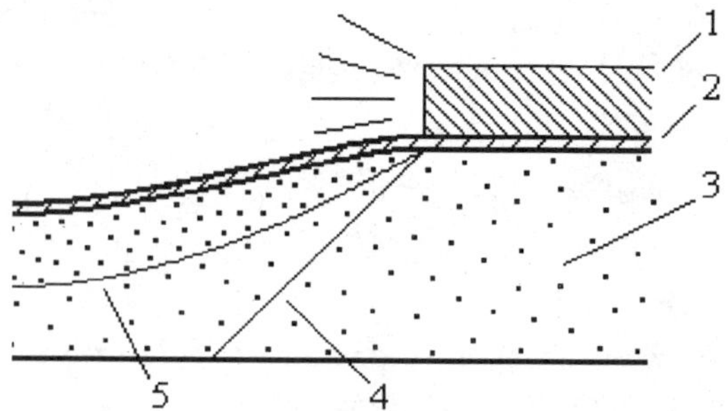

FIGURE 1. Experimental set-up. 1 - explosive, 2 - plate, 3 - powder, 4 - transducer, 5 - shock wave

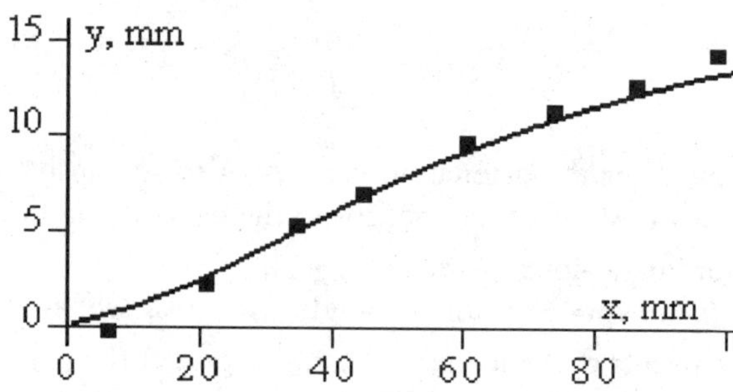

FIGURE 2. Experimental and calculated shock wave shapes

The comparison of the experimental (dots) and calculated (line) shapes of the shock wave for the one test is given in Fig. 2. In this experiment the porous material is minced brass wire (pieces with diameter 0.2 mm and lentgh 2±0.1 mm). Initial density of the material 4.53 g/cc, monolithic density 8.4 g/cc. The sample is covered by the copper plate of thickness 2 mm, width 90 mm, length 180 mm. The loading is effected by the charge of RDX ($\rho_e = 1.0$ g/cc, $\delta_e = 20$ mm, $D = 6.3$ km/s, $k = 2.8$). Figure 2 shows that the discrepancy between calculation and experimental results does not exceed 5%.

The powder counter-pressure can be included into (6) for obtaining the better accuracy. However our calculations show that the pressure on the plate surface and consequently the other results change insignificantly.

Note that though the compression waves in the detonation products are brought about by the presence of a point of inflection at the plate and lead to the increase in the pressure on the plate surface, but their influence on the shape of shock wave in the powder is weak. This is due to low intensity of the compression waves, as well as the integral dependence of the wave shape in the powder on the pressure distribution on the powder surface.

We see two main ways of using the results obtained. Provided a satisfactory compactibility criterion has been ever obtained our data will help to predict the result of a concrete experiment on the explosive loading of porous material.

Another use appears to be more promising today. Suppose that conditions for loading of the porous sample with strong shock wave are created and the sample is compacted to monolithic density. If the dynamic parameters values are met some compactibility criterion that serves in support of validity of the criterion, otherwise the latter is inadequate. In a word we have some guideline for testing criteria of compactibility.

REFERENCES

1. Shtertser, A.A., *Fizika Goreniya i Vzryva*, **18**, No. 1, 141-143 (1982).
2. Deribas, A.A., Staver, A. M., and Shtertser, A.A., "Some aspects of explosive compaction of porous layers," in: *High Energy Rate Fabrication*, 8th Intern. Conf., San Antonio, Texas, 1984, pp. 111-113, New York, 1984.
3. Deribas, A.A., Kuz'min, G.E., *J. Prikl. Mekh. Tekh. Fiz.*, No.1, 137-142 (1970).
4. Kuz'min, G.E., *Dinamica Sploshnoi Sredy*, No. 29, 137-142 (1970).
5. Vacek, J., "The acceleration of metal plates packing an explosive charge on both sides," in: *1st Intern. Sympos. Explosive Cladding*, Marianske Lazne, pp. 79-92., Pardubice- Semtin, 1970.
6. Pai, V.V., Kuz'min, G.E., and Yakovlev, I.V., *Fizika Gorenia i Vzryva*, **31**, No.3, 134-138 (1995).
7. Kuz'min, G.E., Mali, V.I., and Pai, V.V., *Fizika Goreniya i Vzryva*, **9**, No. 4, 558-562 (1973).

REACTION BEHAVIOR OF SHOCK COMPRESSED ALUMINUM AND IRON-OXIDE POWDER MIXTURES

V. Subramanian and N. N. Thadhani

School of Materials Science and Engineering, Georgia Institute of Technology, Atlanta, GA 30332-0245

Fine and coarse morphology powders of Al-Fe_2O_3 and Al-Fe_3O_4 mixed in an equivolumetric distribution were shock compressed at different packing densities and loading conditions. Compacts of powders packed at 53% TMD and shock-compressed using Momma Bear A fixture and Baratol explosive (P~16 GPa) underwent complete reaction, while no reaction was observed when the same powders were shock compressed at lower pressures (P~5GPa) using Poppa Bear fixture and Baratol explosive. The Al-Fe_2O_3 mixtures were also shock-compressed using a higher 70% TMD packing density. No reactions were observed for MBA-B (P~16 GPa) while MBA-CB (P~22 GPa) showed complete reaction. The microstructure of the cross-section of the fully reacted samples showed features of melting and solidification, with an Al-Fe intermetallic core and an aluminum oxide layer surrounding it. The unreacted compacts consisted of heavily deformed aluminum particles and iron oxide particles that were fractured (with cracks oriented along the shock direction) and the amount of fragmentation dependent on the shock pressure. In this paper, the influence of shock conditions on microstructural characteristics of the recovered reaction products and unreacted compacts will be described.

INTRODUCTION

The occurrence of shock induced reactions in powders is controlled by factors such as shock pressure, packing density, powder morphology as well as the difference between the mechanical properties of the constituents (1, 2). Past studies have also shown the influence of volumetric distribution of reactants on shock induced reactions (3). In this study, the reaction behavior of shock compressed Al-Fe_2O_3 and Al-Fe_3O_4 powders mixed in an equivolumetric distribution was examined. Both Al-Fe_2O_3 and Al-Fe_3O_4 react in an exothermic manner to form aluminum oxide and iron. The effect of shock pressure, packing density and powder morphology on the reaction behavior was also studied.

EXPERIMENT

Mixtures of Al-Fe_2O_3 and Al-Fe_3O_4 blended in an equivolumetric ratio were shock compressed at different packing densities and shock conditions. The shock conditions were varied by using different combinations of Sandia Bear series shock recovery fixtures and explosives (4, 5). The Momma Bear A fixture with Baratol explosive (MBA-B), Poppa Bear fixture with Baratol explosive (PB-B) and Momma Bear A fixture with Comp B explosive (MBA-CB) were used. The pressure and temperature obtained in these fixtures have been characterized by numerical simulation (4). The peak pressure and mean bulk temperatures computed for 54% rutile powder, in the different regions of the capsule are shown in Table 1. It should be noted that peak pressures generated

TABLE 1. Characteristics of the Bear Fixtures (5).

Fixture	Pressure, GPa		Mean Bulk Temperature, K	
	Bulk	Focus	Bulk	Edge
PB-B	5	4.5	423	348
MBA-B	16	32	498	583
MBA-CB	22	46	673	923

TABLE 2. Experimental Conditions

Fixture/explosive	Composition (morphology)	Density (%TMD)
MBA-B	Al + Fe_2O_3 (fine and coarse)	53%, 70%
	Al + Fe_2O_3 (fine*)	70%
	Al + Fe_3O_4 (fine and coarse)	53%
PB-B	Al + Fe_2O_3 (fine and fine*)	53%
	Al + Fe_3O_4 (fine)	53%
MBA-CB	Al + Fe_2O_3 (fine)	53%, 70%

*Same powders as those used in Sandia Experiments

due to radial focussing effects are independent of the equation of state of a material. The details of the experimental conditions of the present study are shown in Table 2. The recovered samples were characterized using optical and scanning electron microscopy. X-ray diffraction was used to analyze the phases formed and to calculate the residual strain in the unreacted powders. The peaks were fit with a Lorentzian function to obtain the peak width at half intensity, which after subtraction of instrumental broadening was used in Williamson-Hall plots (6) to obtain the residual strain values.

RESULTS AND DISCUSSION

The recovered samples showed occurrence of reaction as a function of the density and shock loading conditions. The MBA-B configuration showed reaction for only the 53% TMD sample. The cross-section of the reacted samples (Figure 1) showed separation of the reaction products Al_2O_3 and FeAl, due to differences in their melting temperatures. The microstructures contained features of melting and solidification as shown in

Figure 1. The MBA-CB (~22 GPa) configuration showed reaction for both 53% and 70% densities. The 53% density MBA-CB sample reacted in an explosive manner causing a hole to form at the top

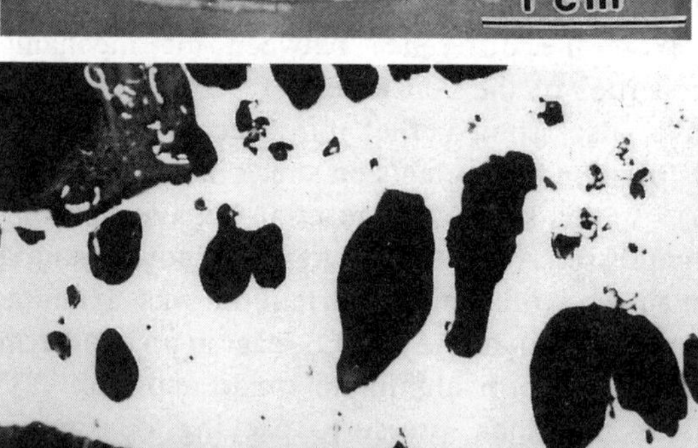

FIGURE 1. Cross-section and microstructure of reacted sample. MBA-B, Al + Fe_2O_3 (fine), 53% TMD

TABLE 3. Lattice Parameter of Reacted Samples.

Fixture	Density (%TMD)	Composition (Morphology)	Lattice Parameter ($\mathring{A}$) Al_2O_3	FeAl
MBA-B	53%	Al + Fe_2O_3 (fine)	a=4.761, c= 13.000	2.8964
	53%	Al + Fe_2O_3 (coarse)	a=4.753, c=12.962	2.903
	53%	Al + Fe_3O_4 (fine)	a=4.767, c=12.991	2.8968
	53%	Al + Fe_3O_4 (coarse)	a=4.764, c=13.081	2.9203
MBA-CB	70%	Al + Fe_2O_3 (fine)	a=4.764, c=13.018	2.8966

TABLE 4. Residual Strain in the Unreacted Al + Fe_2O_3 Samples

Fixture Density	(%TMD)	Morphology	Strain Fe_2O_3
MBA-B	70%	Fine	0.0071
	70%	Coarse	0.0066
	70%	Fine*	0.0075
PB-B	53%	Fine	0.0041
	53%	Fine*	0.0050

* Same powders as those used in Sandia Experiments

of the capsule, through which most of the sample was ejected. The PB-B (~5 GPa) configuration showed no reaction even for the 53% density. Melting of the reaction products is expected since the temperature due to the reaction for the Al-Fe_2O_3 (ΔH = -3400 kJ/kg) and Al-Fe_3O_4 (ΔH = -3061 kJ/kg) compositions is 2790 K. The exothermic heat also caused discoloration of the copper capsule. The lattice parameter of the Al_2O_3 and FeAl formed due to the reaction did not change significantly for the different conditions used, as shown in Table 3.

The unreacted samples were comprised of deformed aluminum particles and fractured Fe_2O_3 particles, with cracks oriented in the shock direction. The microstructure of an unreacted sample is shown in Figure 2 (grey Fe_2O_3 and dark Al particles). The strain in the Fe_2O_3 powders is shown in Table 4. The strain in Fe_2O_3 powder was slightly higher for the higher pressure MBA-B sample compared to the PB-B sample. The Al particles showed no strain (~2 10^{-5}), due to the annealing effects with the shock temperatures exceeding the recrystallization temperatures.

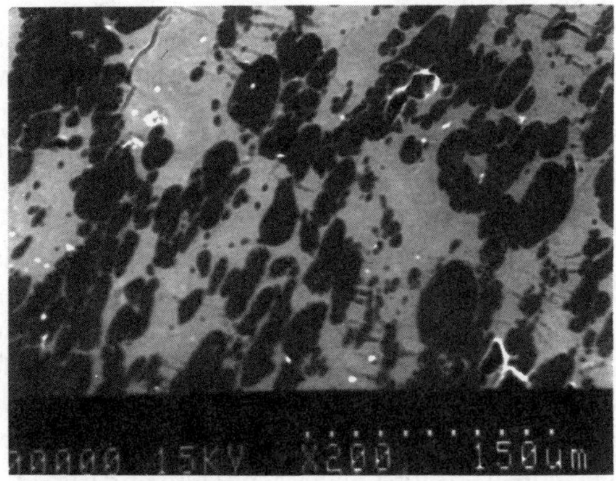

FIGURE 2. Scanning electron micrograph of an unreacted sample. MBA-B, Al + Fe_2O_3 (fine), 70% TMD

The results for Al-Fe_2O_3 are summarized in the reaction map, shown in Figure 3, as a plot of pressure and packing density. The pressures were based on the values obtained for 53% rutile powder by two dimensional numerical simulation (4).

It can be noted from Figure 3 that for packing densities of 53% TMD, reactions are observed at pressures greater than 16 GPa. With increase in

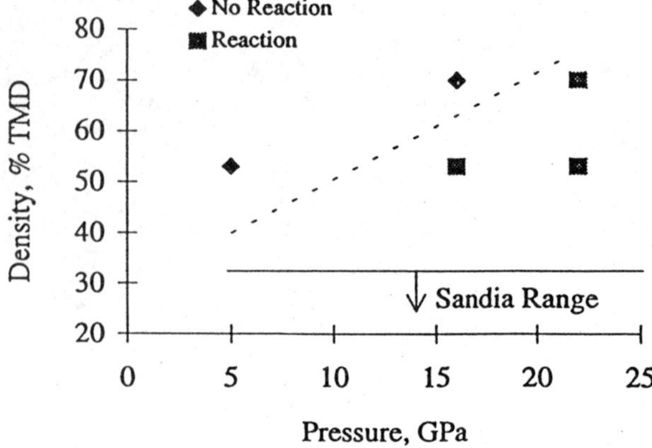

FIGURE 3. Reaction Map for Al + Fe$_2$O$_3$ (fine) as a function of pressure and density.

packing density to 70% TMD, an even higher shock pressure is required (22 GPa). This is expected since lower densities provide increased volume of voids, which permit better flow and mixing of the constituents. In addition, the residual temperatures obtained after thermal equilibrium are also higher at lower densities.

The results of the present experiments performed on non-stoichiometric but equivolumetric mixtures were compared with experiments on stoichiometric Al-Fe$_2$O$_3$ (2:1 atomic ratio) done at Sandia National Laboratories with even lower densities (<35% TMD) using MBA-CB and MBA-B fixtures (7). The range of densities in their experiments is also shown in Figure 3. It is interesting to note that although the densities of the Sandia experiments were lower than those used in the present study, for similar values of pressure, some of their samples did not react. The Sandia experiments were performed on stoichiometric mixture of Al-Fe$_2$O$_3$ in contrast to the equivolumetric mixtures used in the present work. This suggests that the volumetric distribution of the constituents, is another experimental parameter controlling the reaction behavior. The stoichiometric composition corresponds to Al-Fe$_2$O$_3$ in a 2:3 volumetric ratio. Thus with excess Fe$_2$O$_3$, the flow and deformation during shock compression, may be restricted, which can inhibit the reaction. The results of the present study thus show that equivolumetric distribution of reactants can increase the reactivity of powder mixtures, and cause reaction to occur at packing densities as high as 70% TMD.

CONCLUSIONS

Reaction between Al and iron oxide takes place depending on the initial powder packing density and shock conditions. The reaction product undergoes melting and solidification, forming an intermetallic core and an aluminum oxide layer surrounding it. The unreacted samples are comprised of deformed aluminum and fractured iron oxide particles. Comparing these results with the Sandia experiments on stoichiometric composition, shows that the volumetric distribution of the components is an important experimental factor influencing the reactivity of mixtures under shock compression. An equivolumetric distribution allows for better mixing between reactants and thus enhances the propensity for reaction initiation.

ACKNOWLEDGMENTS

Funding for this research was provided by Office of Naval Research Contract No. N00014-94-1-0169. The authors acknowledge the help of R.A. Graham and G.T. Holman for the Bear experiments which were performed at Sandia National Labs.

REFERENCES

1. Dremin, A.N. and Bruesov, O.N., *Russian Chemical reviews*, 37, 392-402, (1983)

2. Thadhani, N.N., *Progress in Materials Science*, 37, 117-226, (1993)

3. Dunbar, E., "Effect of Volumetric Distribution of Starting Powders on Shock-Induced Chemical Synthesis", Masters Thesis, New Mexico Institute of Mining and Technology (1991)

4. Graham, R.A., *High Pressure Explosive Processing of Ceramics*, Transtech, 1987, pp. 30-64

5. Graham, R.A and Webb, D.M., *Shock Waves in Condensed Matter*, Elsevier Science Publishers, 1984, pp. 211-214

6. Williamson, G.K., and Hall, W.H., *Acta. Met.*, 1, 22-31, (1953)

7. Graham, R.A., unpublished results

Shock-Induced Reaction from Mechanically Alloyed Precursor in 3Ni-Al System

Y.Asakawa[1], T.Aizawa[2], Y.Shono[3], K.Fukuoka[3] and J.Kihara[2]

1 Graduate School of Metallurgy, University of Tokyo
7-3-1 Hongo, Bunkyo-ku, Tokyo 113
2 Department of Metallurgy, University of Tokyo
3 Institute of Materials Research, Tohoku University

Constituent element powders with 3Ni-Al are pretreated by mechanical alloying machine into finely-mixed state materials and meta-stable solid solution. These are employed as precursors for shock reactive synthesis by one stage powder gun in order to understand the intrinsic process of shock chemical reactions. The recovered samples were examined by EPMA and XRD to make comparison with that synthesized from as-mixed samples. Pretreatment by mechanical alloying should be a key technique to describe the shock induced reactivity of binary system of refractory metal and aluminum.

INTRODUCTION

Shock Reactive Synthesis has been extensively investigated to propose several reaction mechanisms. [1-4]

A common physical scenario is shared among them: significant mass-mixing with severe shear deformation should be accompanied with the shock induced reaction during shock loading. Since the conventional recovery testing employs a constituent elemental powder mixture as a starting sample, however, original trace of shock reactions taking place in short duration time must be screened, or sometimes overwhelmed, by large temperaure transients and residual heating.

On the assumption that severity of mass-mixing should be positively correlated to intimacy of constituent elements leading to promotion of reaction rate, pretreatment of powder mixture into finely-mixed state must require for a smaller amount of mechanical mixing during shock loading, resulting in the reduction of ignition conditions for reaction. This possibility was positively demonstrated by the previous study[5] for 1Ni-1Al system.

In the present paper, 3Ni+Al system is employed to investigate the effect of pretreatement on the shock chemical reactivity.

EXPERIMENTAL PROCEDURE

Preparation and pretreatment of sample materials is briefly introduced with some comments on the experimental conditions.

Preparation of samples

Preparation of samples Carbonyl Ni powders with their average particle size D less than 10μ are employed as elemental constituent materials together with pure aluminum powders with D $\leq$

45μ. Powder mixture with 3Ni+Al is blended for 24 h and used as an as-mixed sample. This powder is further pretreated by using our developed Mechano-Metallurgical Processing (MMP).

In this MMP, the mechanical alloying takes place in completely different manner from the conventional milling or attriting. Through repeatedly pressing and extrusion in cycles, original powders are subjected to severe shear deformation and fragmentation, resulting in the fine-structured materials. The point to be interested is a fine controllability of the premixed state up to the prealloyed or non-equilibrium state by the numbers of cyclic loading and the promotive change of loading sequence in MMP.

As-mixed powder mixtures were pretreated by application of cyclic loading with the numbers of 200 and 1000 in the present study. These pretreated samples are called MA200 and MA1000 respectively after the number of cycles. In the diffraction pattern for MA200, although a peak intensity of Al(1 0 0) is further reduced from that of as-mixed conditions, original peaks for Ni and Al are still observed with a little broadening. While for MA1000, peaks for Al were completely diminished and an original Ni peak was shifted toward the lower angle in the diffraction pattern with a significant broadening of peaks. This reveals that both Ni and Al should be fully premixed into a Ni-Al super-saturated solid solution with the estimated average grain size of $20 \sim 30$ nm.

These as-mixed and pretreated powders were uni-axially pressed into a cylindrical disc specimen with the thickness of 5mm, the diameter of 10.3mm and the density 5.46g/cm^3(the porosity of 20%). One-stage powder gun was utilized for recovery testing with various flyer plate velocities. The whole experimental conditions are listed in Table.1.

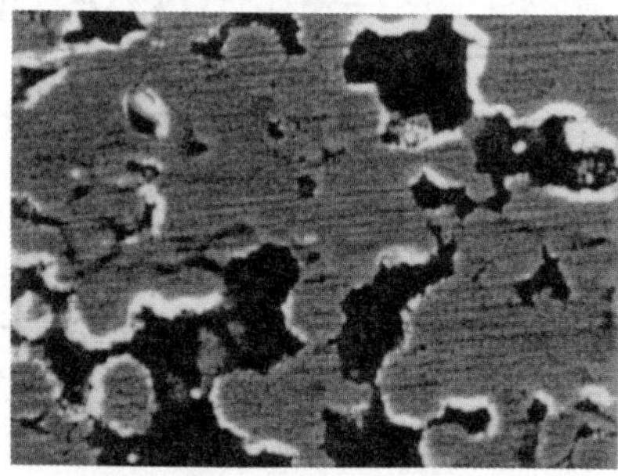

FIGURE 1.as-mixed sample

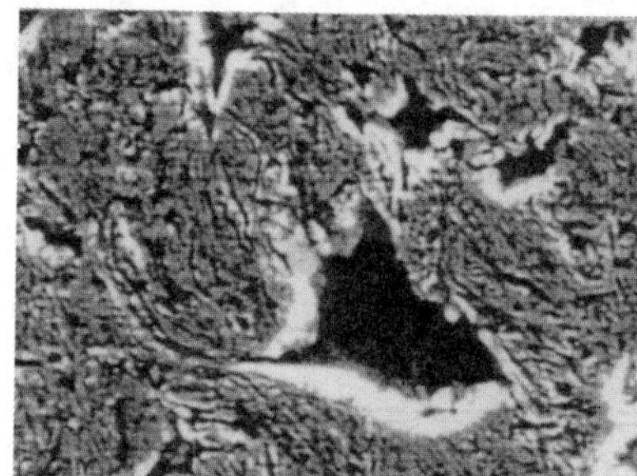

FIGURE 2.MA200

EXPERIMENTAL RESULTS

The successfully recovered samples were further worked to provide samples for XRD and EPMA. Synthesized products are listed in Table 1 together with the residual Ni and Al. In case of the shock reactive synthesis from as-mixed powders,various products were identified in the diffraction pattern with significant amount of residual Ni and Al elements. To be noted, the above identified peaks for Ni$_2$Al$_3$ or Ni$_3$Al only took place at the spots where the shock waves were thought to be converging in passage. Hence, neglecting this trace nearby no reactions should occur from the as-mixed powder. The pretreated sample into finely-mixed state or MA200 was reacted into Ni$_3$Al irrespectively of the flyer velocity. Both peak intensities for Ni$_3$Al and residual

TABLE 1. Experimental conditions and result of XRD analysis of products

flyer velocity (km/s)	sample (%)	porosity	XRD results
1.7 *	as-mixed	19.3	Ni,Al,Ni$_3$Al
1.7 *	MA200	20.2	Ni(Al),Ni$_3$Al
1.5 *	as-mixed	19.6	Ni,Al,Ni$_2$Al$_3$(t) Ni$_3$Al(t)
1.537	MA200	20.0	Ni(Al),Ni$_3$Al
1.539	MA1000	22.1	Ni(Al) Ni$_3$Al(t)
1.333	as-mixed	20.0	Ni,Al,NiAl$_3$(t)
1.322	MA200	19.7	Ni(Al),Ni$_3$Al(t)
1.338	MA1000	21.6	Ni(Al),Ni$_3$Al(t)
0.835	MA1000	21.8	Ni(Al)

* expected flyer plate velocity (t) trace

Ni in the diffraction pattern are nearly the same, and their ratio becomes invariable to the flyer velocity. No aluminum can be identified even in the trace level; aluminum is thought to be dissolved into nickel phase. Through the structural analysis, a significant shift of Ni to lower angle is detected.

Precise observation by EPMA and SEM reveals that homogeneous structure and chemical composition should be obtained, but that a small amount of residual nickel was left in material. This is because the number of cycles loadings by N = 200 must be insufficient to attain completely premixed state without any residual constituent element particles. The point to be noted with respect-to the phase and microstructure is that the morphology of post-shock materials should be nearly the same as that of starting MA200 compact. This reveals that intimately pretreated materials require for no large amount of mass-mixing and temperature transients during shock loading and its subsequence. Owing to an induced activation mechanism during the mechanical alloying, very short mass distance must be necessary to make local homogenization of phase and microstructure in shock loading. Since this local homogenization is little dependent on the shock pressure or the flyer velocity, which drives mass-mixing process in the shock reactive synthesis from powder mixture, the synthesized products from MA200 become indifferent to the flyer speed.

Since more finely-premixed state can be attained by increasing the number of cyclic loading, the shock reaction into a single-phase Ni$_3$Al might be expected to take place even by using slower velocity flyer plate. As listed in Table.1, however, in case of MA1000, only a trace of Ni$_3$Al was detected by XRD analysis. In particular, when MA1000 was shot by flyer plate velocity of 0.8km/s, MA1000 seems to be shock consolidated since the XRD profile of post-shock materials becomes just the same as that of MA1000. Even increasing the flyer plate velocity to 1.5km.s, XRD profile made a little change: sharp peak of Ni(Al) solid solution with a trace peak of Ni$_3$Al. Since a broadening width for Ni(Al) is reduced, however, grain growth must take place in shock loading and its subsequence .

DISCUSSION

An interfacial energy of grain boundaries is activated with grain size refinement or enlargement of boundary surface area during pretreatment by the mechanical alloying. Further more, each constituent element is forced to dissolve into other element phase, and the apparent diffusion rate of elements can be increased by larger numbers of dislocations induced through plastic deformation in cyclic loading. Pretreatment effect should result in a significant reduction of local mass transportation distance, and lead to shock reaction from MA200 to Ni_3Al taking place irrespectively of the applied shock pressure, below the melting temperature of Ni and products.

Drastic change observed in the shock reactions between MA200 and MA1000 suggested that a meta-stable solid solution should play a role in the control of shock induced reactions. If an original powder mixture is pretreated into fine-grain sized structures without formation of non-equilibrium phase, shock reaction must result in production of Ni_3Al. Owing to the formation of $Ni(Al)$, since a free energy barrier exists in the reaction process, no reactions are thought to take place in the shock experiments by lower velocity flyer. In other words, just like a thermal activation necessary for the conventional chemical processes such as SHS or powder metallurgy, shock activation is needed to over come the free energy barrier from the super-saturated solid solution to Ni_3Al.

CONCLUSION

Pretreatment by the mechanical alloying provides us a new point of view in the shock chemistry. In particular, as the practical point, use of pretreated precursor with finely premixed state in the binary system enables us to synthesize refractory metal aluminides even by the reduced shock conditions.

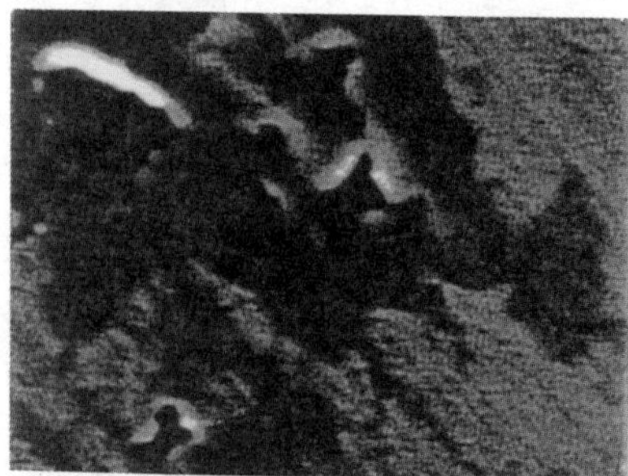

FIGURE 3.as-mixed samples recovered from shock velocity of 1.7km/s

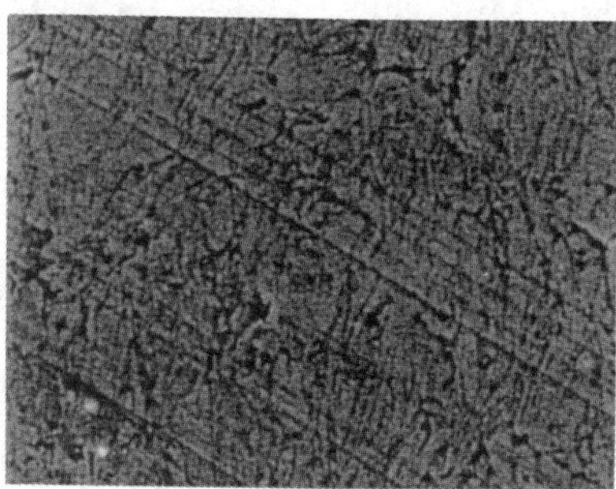

FIGURE 4.MA200 recovered from flyer plate velocity of 1.7km/s

REFERENCE
1. Thadani,N.N.et al., *J.Appl.Phys* **76**, pp.2129-2138
2. Horie,Y., *Shock Waves in Materials Science*, Tokyo, Springer-Verlag, 1993, ch.6, pp.67-100
3. Meyers,M.A., *acta.metal.mater*, 1993, **42**,. pp.+715-729
4. Aizawa,T. et al.,*J.Faculty of Engineering,*Univ.Tokyo **18**, 1995,pp.57-101.
5.T.Aizawa et al., *Annales de Chimie* , 1995, in press

SHOCK COMPRESSION OF QUARTZ AND ALUMINUM POWDER MIXTURES

V. S. Joshi[*], N. N. Thadhani[**], R. A. Graham[***] and G. T. Holman Jr.[***]

[*]*Energetic Materials Research and Testing Center, New Mexico Tech, Socorro, NM 87801, USA*
[**]*School of Materials Science and Engineering, Georgia Institute of Technology, Atlanta, GA 30332, USA*
[***]*Advanced Materials Physics and Devices, Sandia National Laboratories, Albuquerque, NM 87185, USA*

We report about the shock-compression response of highly porous (55% and 65% dense) mixtures of $4Al + 3SiO_2$ powders having shock-induced phase transitions and chemical reactions. Shock recovery experiments were performed using the CETR/Sawaoka plate-impact system (P=40 to 100 GPa) and the Sandia Momma Bear A Comp B fixture (P=22 to 45 GPa). The recovered compacts contained the high pressure stishovite phase, products of chemical reaction, as well as unreacted constituents. The reaction products formed included Al_2O_3, metallic Si (ambient and high pressure phases), SiAl intermetallic, and kyanite (Al_2SiO_5). The shock-induced chemical reaction in $4Al + 3SiO_2$ powder mixtures, appears to have been accompanied (or assisted) by the formation of stishovite, a high pressure phase of quartz.

INTRODUCTION

Polymorphic transitions in quartz have been studied both by static (1-2) as well as by dynamic high pressure methods (3-5), as have also been reviewed by Duvall and Graham (6). The continued interest in phase transformation of quartz has been partly due to lack of understanding of the mechanisms governing its behavior under shock-wave loading. Similarly, shock-induced chemical reactions(7-9), initiated in powder mixtures during propagation of shock waves, are also being studied to understand the reaction mechanisms. Aluminum+quartz mixture is of interest and is one of the weakly exothermic (ΔH_R= -6.2 kJ/cal) metal/ metal-oxide (thermite) system.

Podurets and coworkers (10) have reported about shock-induced phase transformation of quartz to stishovite in mixtures of $Al+SiO_2$ in 50:50, 60:40 and 70:30 weight proportions. The starting material had virtually no porosity, thus, the probability of a chemical reaction was minimized. Various other studies by Chao et al. (2), Wackerle (4) and McQueen et al.(5) have demonstrated that the stishovite phase is relatively stable at room temperature, and once formed can be retained. Podurets and coworkers (10) observed stishovite formation in their mixtures at pressures significantly lower than those reported by other investigators (2-5), attributing it to the increase in pressures caused by local impedance mismatch between Al and SiO_2.

Shock-induced chemical reactions have been extensively investigated in different metal/ metal-oxide and metal/ metal powder mixtures. Mixture of Aluminum and quartz represents an interesting materials system which can not only undergo chemical reactions, but also have transformation of the starting quartz material as well as of the metallic Si material (product of reaction). Thus, the objective of the research presented here was to examine the behavior of the 55% and 65% dense $4Al+3SiO_2$ powder mixtures under shock-compression.

EXPERIMENTAL

Shock recovery experiments were performed over a range of shock pressures using the well designed and calibrated 12-capsule CETR-Sawaoka system for *high shock pressures (60-100 GPa)* at two different velocities (1.9 km/s and 1.1 km/s), a modified 12-capsule system using the CETR-Holman capsule for *moderate shock pressures (40-60 GPa)* at a velocity of 1.1 km/s, and the Sandia Momma Bear A Comp B fixture for *low pressure (22-46 GPa)* experiments, using a driver plate arrangement. Details of each of these shock recovery fixtures and the corresponding two-dimensional numerical simulations have been discussed and documented elsewhere (11).

Pure aluminum powder obtained from ALCOA and quartz powder obtained from CERAC were blended in the ratio of 4:3 (stoichiometric ratio to obtain complete reaction) in a V-blender (with a high-speed intensifier bar) for 1½ hours. A characteristic scanning electron micrograph of the powder mixture is shown in Fig.1. Aluminum powder was relatively coarse (~44 microns) and appeared more rounded than the quartz powder.

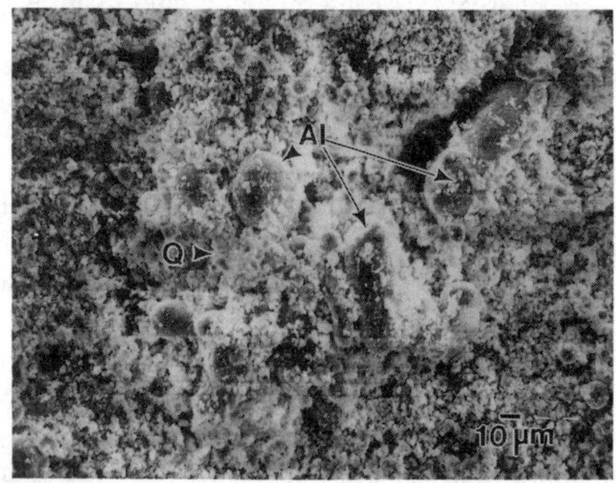

FIGURE 1. Scanning Electron Microscope image of the starting powder mixture, showing aluminum (Al) and quartz (Q) particles.

The powders were pressed in capsules at the desired packing densities, and sealed in the respective holding fixtures. Table 1 lists the type and characteristics of powders, their packing densities, and other details of experimental conditions.

RESULTS AND DISCUSSIONS

Optical and scanning electron microscopy observations as well as x-ray diffraction analysis were conducted on the longitudinal cross-sections of the recovered shocked compacts. A low magnification optical microscopy analysis of the longitudinal cross-sectional surfaces, revealed extensive reaction in the compacts of the *high velocity* Sawaoka experiment (#9404-03), partial reaction in localized areas in the *low velocity* Sawaoka (#9022-12) and Holman (#9028-09) capsule experiments, and no reaction in the *low pressure* (Momma Bear) compact. A typical optical micrograph of the cross-sections of compact, revealing partial reaction is shown in Fig. 2. The central (axial) region towards the bottom (non-impact side) of the compact has a different contrast compared to the bulk of the compact, indicating the possibility of reaction in regions dominated by radial shock-wave focussing effects.

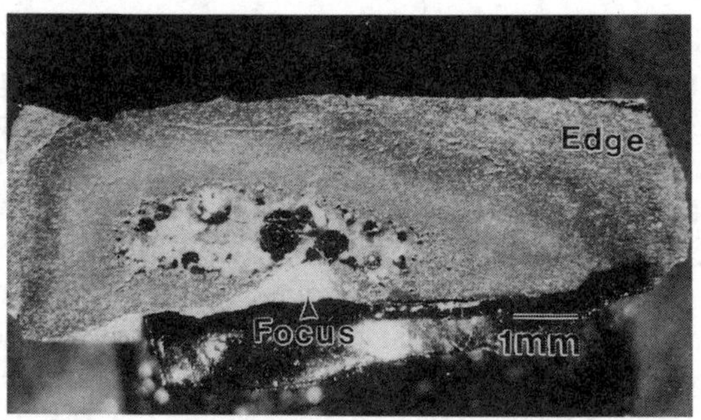

FIGURE 2. Optical Microscope image of the cross-section of the compact #9022-12, indicating different extent of reaction in various regions.

Detailed analysis of these regions at higher

TABLE 1. Experimental Conditions and Powder Characteristics.

Sample No.	Mix No.*	Fixture	Impact velocity	Packing Density	Peak Pressures
9022-12	SNL 0790-2	Sawaoka	1.10 km/s	1.53 gm/cc(57.4%)	60-100 GPa
9028-09	"	Holman	1.16 km/s	1.51 gm/cc(56.4%)	40-60 GPa
9404-03	"	Sawaoka	1.90 km/s	1.81 gm/cc(66.9%)	100 GPa
94B-12	"	Sandia	MBA-CB	1.48 gm/cc(55.1%)	22-46 GPa

*SiO$_2$ - CERAC S-1061, Lot# 10542, -325 mesh, 99.9%, Al - ALCOA 56251, Lot# 3336-21, -325 mesh, 90.6%

magnification, revealed that extensive reaction had occurred. Typical reaction areas of sample is shown in Fig. 3. The presence of spherical voids (SV), and shiny contrast of globules of metallic reaction product (MR) is an indication of chemical reactions in the central high pressure regions of the compacts.

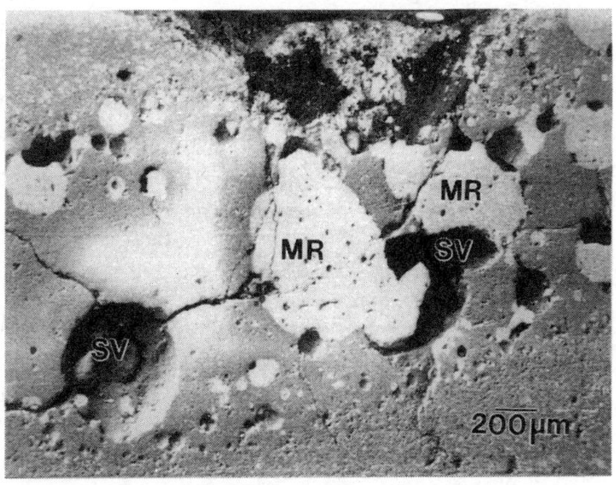

FIGURE 3. Scanning Electron Microscope image of the reacted region of sample 9022-12, showing voids (SV) and metallic reaction product (MR).

In some cases, these metallic globules were found to contain Al-Si intermetallic composition. A typical unreacted region of a compact is shown in Fig. 4, revealing the differential contrast of the starting powder mixtures.

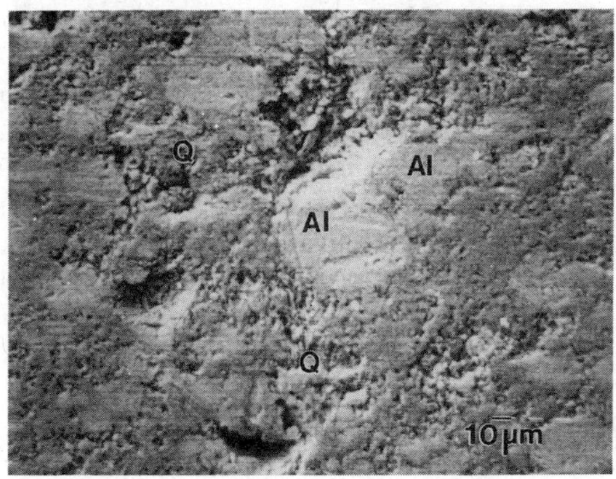

FIGURE 4. Scanning Electron Microscope image of unreacted region of sample #94B-12, showing original particles of aluminum (Al) and quartz (Q).

X-ray diffraction (XRD) analysis showed peaks for a combination of products of chemical reaction as well as peaks for materials that had undergone phase transformation. The individual XRD plots for the starting powder mixture and each of the four compacts are shown in Fig. 5 (a-e).

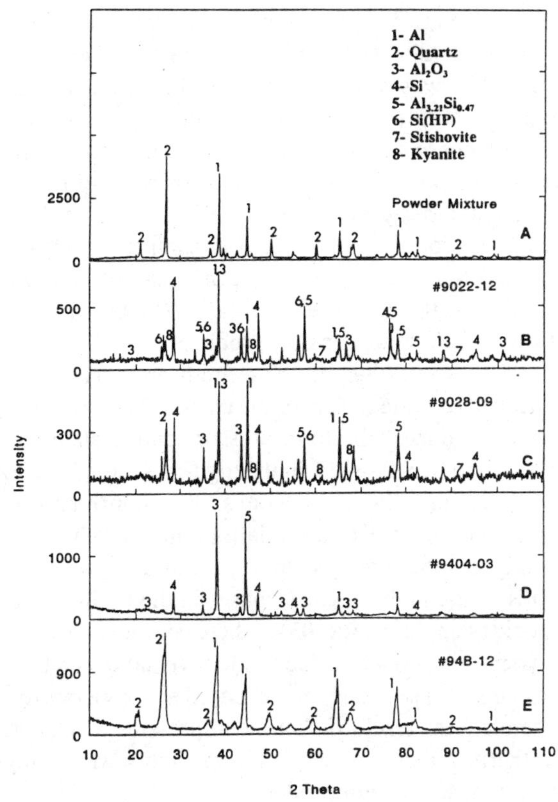

FIGURE 5. X-ray diffraction patterns of starting material and compacts; (a) starting powder mixture, (b) sample # 9022-12, (c) sample #9028-09, (d) sample #9404-03, and (e) sample #94B-12.

The low pressure MBA-CB sample (#94B-12) showed predominant peaks of unreacted powders (Al, SiO_2). The Holman capsule sample (#9028-09) also showed predominant peaks of unreacted constituents and low-intensity peaks of reaction products such as Al_2O_3 (corundum) and Si metal. Both the MBA-CB (#94B-12) and Holman sample (#9028-09) showed shifts in the d-spacings of the majority of the Al diffraction lines, indicating possibility of extensive strain retention in the shocked material. The Sawaoka capsule experiments (samples #9022-12 and 9404-03), showed prominent peaks of fully reacted material. The lower impact velocity sample (#9022) showed presence of small proportions of kyanite (Al_2SiO_5) and $Al_{3.21}Si_{0.47}$ intermetallic, in addition to the common products Al_2O_3 (corundum) and Si metal. The higher impact velocity sample (9403-03) which underwent complete reaction did not show the kyanite (Al_2SiO_5) phases. In addition, the high pressure phase of silicon and that of quartz (stishovite), observed in #9022-12, was not detected in this high impact velocity sample. Surprisingly, coesite, a moderate pressure phase of SiO_2 was not detected in any of the shock-compressed samples. The phase transformation of SiO_2 (starting material) as well as that of silicon (reaction product), as detected in sample # 9022-12 suggests that both, the reactant SiO_2 and product Si, coexisted in the high-pressure state. The coexistence of one of the reactants and products, both of which undergo transformation to the high pressure phase, indicates that at least part of the chemical reaction has happened during the high pressure shock state. In fact, it may be possible that the high-pressure phase of the reactant, SiO_2, generates the configuration necessary for reaction initiation, in an otherwise weakly exothermic system. The absence of XRD peaks of the high pressure phases in the high-impact velocity sample (#9404-03), where complete reaction had occurred, could be due to the reversal of the high pressure phases due to excessive temperatures, generated as a result of shock-compression as well as exothermic reaction, when the reactants were fully converted to reaction products.

CONCLUSION

It is shown that highly porous $4Al + 3SiO_2$ powder mixtures undergo simultaneous shock-induced chemical reaction and shock-induced phase transformation of both reactant as well as of reaction products. The coexistence of one of the reactants (quartz) in the form of a high pressure phase (stishovite) and the high pressure phase of the reaction product (Si), detected by x-ray diffraction analysis, suggests occurrence of the chemical reaction between Al and SiO_2 during the high pressure shock-loaded state. The observed high pressure phases also suggest that the shock-induced chemical reaction in the weakly exothermic metal (Al) and metal-oxide (SiO_2) system, follows a path involving the phase transformation of the SiO_2 reactant, prior to actual chemical reaction.

ACKNOWLEDGMENTS

This work was supported by Sandia National Laboratories under grant No. 42-5737, at New Mexico Tech. Extended support was also provided by EMRTC, New Mexico Tech., and in part by NSF Contract No. DMR-936132, at Georgia Tech.

REFERENCES

1. Stishov, S. M., and Popova, S.V., *Geokhimiya* **10**, 837-839 (1961).
2. Chao, E. C. T., Fahey, J. J., Littler, J., and Milton, D. J., *J. Geophys. Res.* **67**, 419-421 (1962).
3. Adadurov, P. A., Dremin, A. N., Pershin, S. V., Rodionov, V. N., and Ryabinin, Y. N., *Zh. Prikl. Mekh. Tekh. Fiz.* **2**, 81 (1962).
4. Wackerle, J., *J. Appl. Phys.* **33**, 922-934 (1962).
5. McQueen, R. G., Fritz, J. N., and Marsh, S. P., *J. Geoph. Res.* **68**, 2319-2322 (1963).
6. Duvall, G. E., and Graham, R. A., *Rev. Mod. Phys* **49**, 523-579 (1977).
7. Graham, R. A., Morosin, B., Ventruini, E. L., and Carr, M. J., *Ann. Rev. Mater. Sci.* **16**, 315-341 (1986).
8. Dremin, A. N., and Bruesov, O. N., *Russian Chemical Reviews* **37**, 392-402 (1968).
9. Thadhani, N. N., *Progress in Materials Science* **37**, 117-226 (1993).
10. Podurets, M. A., Simakov, G. V., and Trunin, R. F., *Izvestiya, Earth Physics* **24**, 267-270 (1988).
11. Thadhani, N. N., Holman, G. T., Romero, B., and Graham, R. A., "The CETR/Sawaoka 12-Capsule Plate Impact Shock Recovery Fixture: Design and Experimentation," CETR Report A-01-91 (1991).

PRESSURE MEASUREMENT IN A RECOVERY TUBE FOR SUPERHARD BN SYNTHESIZED BY EXPLOSION

S.R.Yun   Z.P.Huang   Y.F.Shun

Mechanical and engineering department, Beijing, B.I.T, P.R.China

Abstract   Double π -form gauge pair made of the manganin foil and the constantan foil, with the sensing element being 0.6mm long and 0.15mm wide, is carefully embedded in the different radius R interior BN in the recovery tube. The digitizing oscilloscope TEK 2430A is used for recording the gauge output signals supported by the constant current pulsed-power supply. the pure pressure analogue signals involved in the manganin gauge records are acquired by that the records minus the pure tensile analogue signals involved in constantan gauge records, then the pressure P can be worked out by the use of dynamic piezoresistance function and the tensile coefficient according to the above clue the P-R correlation interior BN in the recovery tube, corresponding to the different explosion-load condition, is measured and the phase transition condition of BN is also given.

INTRODUCTION

The synthesis technology by a dynamic high pressure is a field emphasized currently. The material in the recovery tube loaded by an explosion with an axial symmetry experience, which is an usual loading method. Under the condition of explosion-load the pressure and the temperature fields of the material in the recovery tube are the important parameters. This paper emphasized on the measurement technology of the pressure field.

RECOVERY DEVICE

The recovery devices are shown in Figure 1, in one case the recovery tube is in contact with the explosive, the detonation wave goes through the tube-wall and roduces the shock wave interior the material; in the other case the explosion drives the outer round tube, when the outer tube is compressed, it compacts the wall of recovery tube, and the shock wave is formed interior the material.

Under the condition of axial symmetry, the preassure P interior the materials the function of the time t, the radius R and the axial distance Z.

When shock wave travels from the wall to the center. The effection of covergence and attenuation occurs, in the place offe of axis, there occurs the regular and irregular reflections. The wave system is quite complicated, if numerical calculation is used to simulate the pressure field, the experimental data are needed to check and correct it.

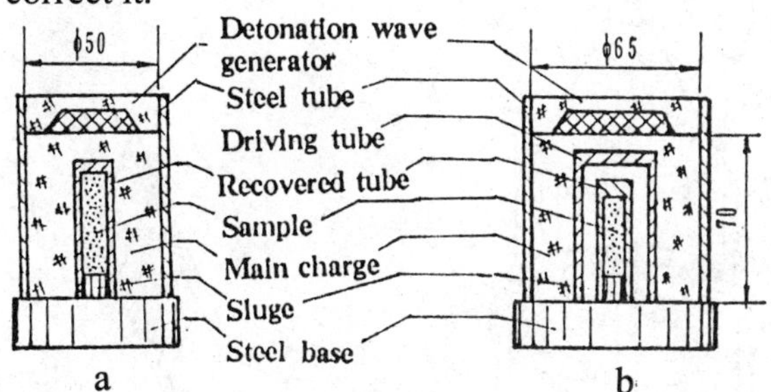

FIGURE 1. Recovery Devices

CALIBRATION OF THE MANGANIN GAUGE

Because the direction of shock wave cannot be predicted, so in this case the pressure which calculated from measured velocity[1] is not accuracy. Considering that the requirement of small size and quick response time (less than 10^{-7}S) for the gauge, double π-form gauge pair made of the manganin foil and the constantan foil, designed by us, is selected to measure the pressure.

The size of sensing element is: length, 0.6mm; width, 0.15mm; thickness, 0.02mm. The relation between the pressure P and radius R is calibrated dynamically by the electro-magnetic method. The magnetic field installation consisted of a pair of Helmhotz coil with 1 m of diameter, which is specially supplied to electricity, can form intensity of magnetigation B up to 80 mT (millitesla) in $\Phi 200 \times 200\ mm^3$ space, with its uniform of B less than 0.1%. The calibration installation is shown in Figure. 2, it is placed in to the centric part of Helmhotz coil. The plane wave initiates the main charge driving the flying plate to impact the composite gauges, so shock wave passed through it, and effective length L of the velocity gauges (made of copper foil with 0.02mm thickness) move cutting the magnetic line of force, the electromotive force ε is produced.

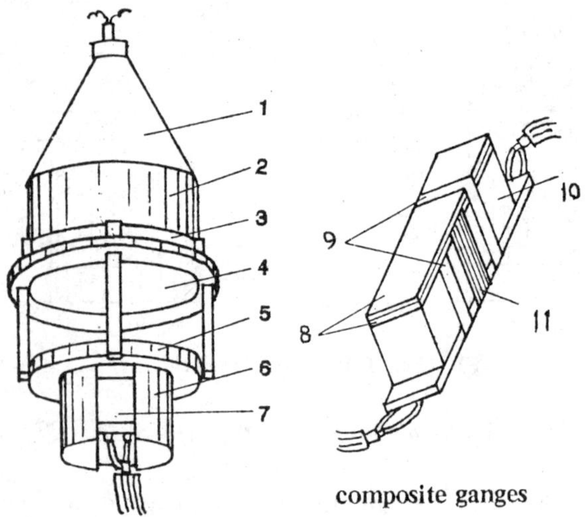

FIGURE 2. Calibration Installation
1. Plane wave generator 2. Main charge 3. Air layer 4. Flying plate 5. Sample disk 6. Sample piece 7. Composite gauges 8. Sample plate 9. Electric magnetic velocity gauge 10. Sample piece 11. Manganin gauge

ε and B are recorded by digitizing ocilloscope. According to the principle of EMVG, the particle velocity u after shock wave,

$$u = \frac{\varepsilon}{BL} \tag{1}$$

If the distance h (2-3mm) between two gauges is relatively small, the velocity of shock wave and the particle velocity all can be treated as the linear attenuation.

The manganin gauge is set up in the centric place between two velocity gauges. The corresponding pressure of shock wave P

$$P = \frac{\rho_0 h(u_1 + u_2)}{2\Delta t} \tag{2}$$

Where ρ_0 is the density of the test piece, Δt is the time which shock wave passes, h, u_1 and u_2 are particle velocities measured by two velocity gauges. When the velocity gauge's signal is recorded, meanwhile the resistance change ΔR for the manganin piezoresistance gauge, is also recorded, then P-ΔR relation can be obtained.

Because the plane wave's diameter is 50mm, much bigger than the distance between two velocity gauges (2-3mm), meet the condition of one-dimension plane and the condition of linear attenuation approximation.

In the zone of low pressure the main charge explodes in contact with the sample piece; in the zone of high pressure, the flying plate made of glass fibre reinforced plastic is used to impact the sample piece. The shock wave's velocity is within the range of 2.6-4mm/s, the time shock wave move through the manganin gauge last 5.0-7.7ns. when the response time of the manganin gauge is regarded as 4 times time length which shock wave go through the response time of the manganin gauge is within 20-50ns. shock wave's time width produced by the flying plate is over 500-1000ns, so the respond process of the manganin piezoresistance gauge do not affect on the accuracy of measurement.

For the media of sample piece, PMMA is accepted in the low pressure zone; in the high pressure zone, considering that the electric-conductive and magnetic-conductive materials cannot be used, WO_3 powders is selected. After the experimental data imitated gives the following relation

$$P = 40.4\frac{\Delta R}{R} + 0.075\,GPa$$
$$5.07 \le P \le 19 \qquad (3)$$
$$P = -2.2 + 52.6\frac{\Delta R}{R} - 15.9\left(\frac{\Delta R}{R}\right)^2 GPa$$
$$9 \le P \le 35.6\,GPa$$

For the situation that the pressure is higher than 35.6 GPa. (beyond the calibration range), the following way is used to get the estimation value. The tangent line's equation for the calibration curve at P=35.6 GPa is

$$P = 19.56\frac{\Delta R}{R} + 14.98 \qquad (4)$$

The estimation value is regarded as the average value of the result of formula (3) and formula (4).

INSTALLATION OF PRESSURE IS MEASURED

The shock wave that travels interior the material in recovery tube is not one-dimension plane wave. The method to resolve the tensile effect is to utilize constantan gauge with the same shape and size, as that of the mangantan gauge. A teflon foil separate them, which become the gauge pair. Specially be careful to make two sensing element be in the same seat. When shock-load is brought to bear on, the piezoresistance effection for the constantan gauge can be ignored, and its output is only the tensile signal $\left(\frac{\Delta R}{R}\right)_c$. The manganin gauge's output is the tensile and piezoresistance com-posite signal $\left(\frac{\Delta R}{R}\right)_m$, the pure piezoresistance signal $\frac{\Delta R}{R}$ is given as the following formula.

$$\frac{\Delta R}{R} = \left(\frac{\Delta R}{R}\right)_m - \frac{k_m}{k_c}\left(\frac{\Delta R}{R}\right)_c \qquad (5)$$

Where k_m, k_c are the coefficient of dynamic tensile strain for the manganin and the constantan, k_m and k_c are given by reference [2].

The graphite-like BN (gBN) is pressed into cylinder, which is sawed longtitudinally in some radius (R) by a thin saw with 0.14 mm of thickness, the gauge pair is binded in it. The manganin gauge and the constantan one should be strictly coincided with each other, put the whole test installation into the recovery tube, the conductor wire is led from the detonation-free end.

For assuring the gauges is set up correctly, the special tools that saw the gBN cylinder and press the gauges tightly are designed and made. The signal is recorded by 2430A digitizing ocilloscope, the classically recorded wave form is shown in Figure. 3.

The place (R) where installing the gauges is changed for each and every test (The axial place Z is not changed), the corresponding function relation between P-R in the cross section of the recovery tube under the given explosion-load condition can be obtained.

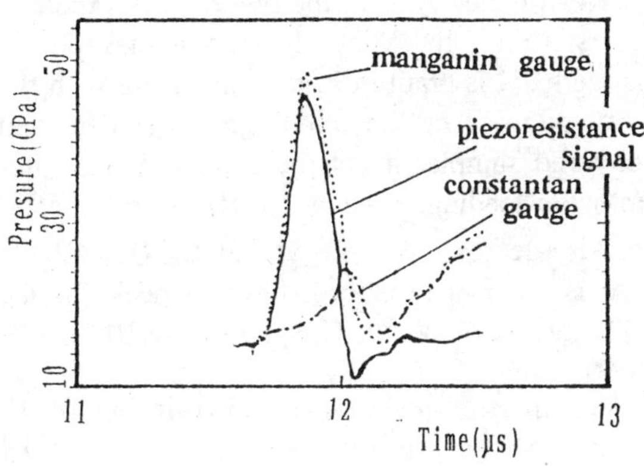

FIGURE 3. The wave form recorded

TABLE 1. Pressure distribution interior gBN cylinder in the recovery tube in Experiment 1

R(mm)	3.92	3.40	2.75	2.43	1.30	0.82	0.22
P(GPa)	10.18	12.34	14.20	16.54	57.4	66.3	73.8

TABLE 2. Pressure distribution interior the pressed BN in the recovery tube in experiment 2

R(mm)	4.1	3.1	2.02	1.52	0.94	0.55	0.26	0.15	0.1
P(GPa)	21.4	23.0	24.7	26.1	27.6	28.6	30.8	32.4	34

MEASUREMENT RESULT AND ANALYSIS ON SAMPLES RECOVERED

Among many experimental programmes, two of them are taken for example.

Experiment 1

The explosion installation shown in Figure. 1a is selected, the explosive is the mixture of RDX and plaster stone   (RDX : plaster stone = 10 : 6).

Its detonation velocity D is measured, equal to 5950m/s, and its detonation pressure P is measured too, equal to 15.86 GPa, the initial density of gBN cylinder is 1.72-1.73g/cm^3, P-R data measured in this example are shown in Table. 1.

Without the gauge pair in the gBN cylinder, but under the same condition the explosive experiment is done. The recovery tube is cut open, the centric part with 6 mm of diameter along the axial line shows darkness, it state clearly that the phase transition has taken place, but the edge part shows whiteness yet. In the place of axis there is a very thin white line, it means that phase-changed BN is graphitilized. Comparing with the X-ray pictures of the initial gBN powders and recovered sample, it can be observed that after explosion-loading,   each diffraction peaks' intensity decrease obviously, and at $2\theta = 42.9°$, there is a characteristic diffraction peak for the (002) surface of WBN. The yield of WBN is 8%.

Experiment 2

The explosion installation shown in Figure. 1b is selected. The explosive is the mixture of RDX and plaster stone (RDX : plaster stone = 10 : 5). Its D and P are measured, equal to 6200m/s and 17.48 GPa, the density of gBN sample is same as above. The pressure distribution interior the passed gBN in the recovery tube is shown in tube 2. The color of all recovered sample is nearly dark, it shows that everywhere the gBN has been transformed. The sample is purified by chemical method, and the analysis of X-ray diffraction result makes known that there is WBN left, no CBN. The yield of WBN is 30%. The averages size of WBN particles is 7 μm by SA-cp3 particle size analysis instrument.

RESULT AND DISCUSSION

1.The function relation between the pressure P-radius R is directly measured by the manganin-constantan gauge pair. The data is of the reference value for the synthesis condition of experiment and numerical calculation. Because the wave system is complicated in the recovery tube, it cannot be judged when the gauge group is out of effection, the post-wave shape recorded is difficult to represent the pressure history , but it can be act as the reference of the time the pressure exerted.

2. Under this experimental condition, above 13 GPa the phase change of gBN takes place (gBN-WBN), the pressure range of high transition rate is 20-40 GPa. The higher pressure is better for the phase change ,but the graphitilization effection is also be enhanced.

3. In view of the two explosion-load way, the recovery device   shown in Figure. 1b can produce uniform pressure, otherwise the yield of WBN is relatively high.

REFERENCES

[1] M.R.Wixom et. al. Shock Waves in Condensed Matter, P403, 1987.

[2] Shi Huan et. al. Shock Compression of Condensed Matter, P361, 1989.

STUDY ON EXPLOSIVE COMPACTION OF FINE COPPER, IRON AND NICKEL POWDERS

A.A. Kiiski, A.A. Deribas* and A.A. Shtertser*

Technical Research Centre of Finland, Manufacturing Technology, Box 17031, FIN-33101 Tampere, Finland
*Design & Technology Institute of High-Rate Hydrodynamics, Siberian branch of RAS, 29 Tereshkova Str., Novosibirsk, Russia

Explosion compaction of fine (avg. 5 μm) copper, iron and nickel powders was studied. Special explosive geometries were developed for each metal powder since different powders needed different peak pressures. The peak pressures and pressure duration times were calculated. The microstructure of the compacts was studied by both optical and SEM microscopy. Special attention was given to particle boundaries in the compacts. The consolidation mechanism was found to be different for each material. In iron, a thick layer of melt was found along boundaries between particles. In copper and nickel no visible signs of melting were observed with the SEM. Vickers hardness measurements and four point bending tests were performed to measure the mechanical properties of the compacted materials.

INTRODUCTION

Many papers have been written about the compaction of metal powders e.g. [1-2]. The bonding processes during the compaction have been studied by Morris [3]. The particle size of the powders to be compacted have normally been between 50 - 150 μm [4-5]. In this study the aim was to compact fine metallic powders with an average particle size of about 5 μm. The bonding mechanism of copper, iron and nickel powders was studied.

EXPERIMENTAL

The materials to be compacted were iron (Fe), copper (Cu) and nickel (Ni). All of the powders had an average particle size of about 5 μm. The initial densities before the compaction were 4.45 g/cm^3 for Cu, 3.83 g/cm^3 for Fe and 6.5 g/cm^3 for Ni powder. The size of the compacted samples were 50 mm in diameter and 5 mm in thickness.

The metal powder was placed into a capsule which was evacuated and heat treated at 600 °C for two hours.

The explosive geometry used is shown in Fig. 1.

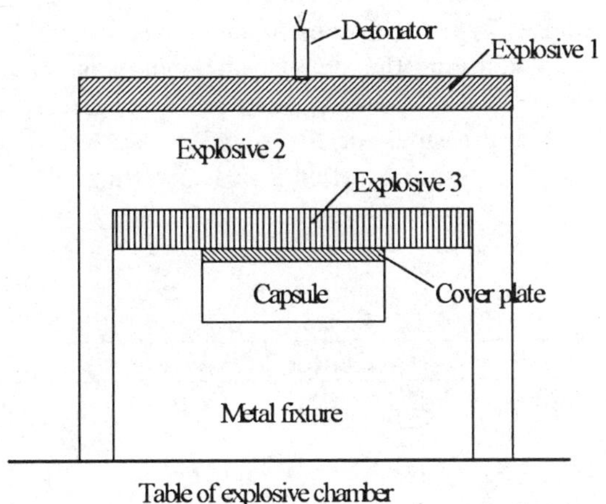

FIGURE 1. The explosive geometry used in the experiments.

For the compaction of the copper (Cu) powder for explosive 1 RDX powder was used, and for explosive 2 Ammonite N6 (80% Ammonium Nitrate and 20% TNT) were used. The RDX explosive and Ammonite N6 had detonation velocities of 6200 m/s and 4200 m/s, respectively. The thickness of the explosive 1 was 5 mm, and for the explosive 2 it was 45 mm. With the copper powder there was no explosive 3. The cover plate was of 2 mm thick steel plate.

For the compaction of iron (Fe) powder explosives 1 and 2 were Ammonite N6, and explosive 3 was RDX-based plastic explosive with a detonation velocity of 7500 m/s. The thickness of Ammonite N6 was 20 mm, and of the plastic RDX it was 18 mm.

For the compaction of nickel (Ni) powder Explosives 1 and 2 were ANFO with a detonation velocity of 2900 m/s and the thickness of 240 mm. Explosive 3 was RDX-based plastic HE with a thickness of 10 mm. The explosive experiments were done at the Design & Technology Institute of High-Rate Hydrodynamics, Russia.

The pressure profile calculations were estimated using the impedance matching technique [6] and pressure - particle velocity (P-u_p) Hugoniot diagrams. For the powders static and dynamic pore collapse of ductile porous material was used [7]. In the calculations the detonation front was moving normal to the powder surface. The direct and reflected pressures in the powders are presented. Also the pressure duration time was estimated.

Scanning electron microscope (SEM) was used in this study.

The hardness measurements were done by the Vickers-method using a 0.3 kg load. Hardness was measured along five different points (3 measurements at each point) from the center of the compacted disk to the edge of the disk.

The bending test samples were of 2 x 4 x 25 mm^3 in size. Three samples of each metal were tested.

RESULTS AND DISCUSSION

For the copper powder the pressure in direct shock was calculated to be 1.2 GPa with a duration of 7 µsec. The reflected shock wave increased the pressure to 9 GPa with a duration of 2.5 µsec.

For the iron powder the pressure in direct shock was calculated to be 10 GPa with a duration of 2.4 µsec. The reflected shock increased the pressure up to 33 GPa with a duration of 2.5 µsec.

For the nickel powder the direct shock was calculated to be 0.2 GPa with a duration of 20 µsec. The reflected shock increased the pressure to 0.7 GPa with a duration of 120 µsec.

The hardnesses of the compacted samples at a distance from the center are presented in table 1. The bending test strength σ is also presented in table 1.

TABLE 1. Hardnesses and the bending test strength of the compacted disk samples

	HV Center	HV (6 mm)	HV (12 mm)	HV (18 mm)	HV (24 mm)	σ (MPa)
Copper	135	137	132	134	106	69
Deviation	5.2	2.7	5.6	8.5	2.1	11.2
Iron	134	162	199	222	199	24
Deviation	27.1	13.5	18.4	0	1.8	5.3
Nickel	186	204	199	174	189	83
Deviation	22.1	4.9	4.7	6.5	8.5	44.4

The hardness of the copper compact is uniform except for the outer edge of the disk. The deviation of hardness at each distance is minimal. The bending strength result of 69 MPa exceeds the typical 0.2 % yield strength of 48 MPa [8] and is approximately 1/3 of the UTS 216 MPa [8] for bulk copper.

The hardness of the iron compact decreases from the outer edge to the center of the sample. The compaction of the powder is not uniform. The bending test strength of 24 MPa is low. Most probably this results from some cracking which was found inside the samples.

The hardness of the nickel compact is fairly uniform throughout the sample. The bending test strength of 83 MPa has a high deviation due to one sample with only a 30 MPa strength. This sample possibly had a crack within it.

The microstructure of compacted copper powder is shown in Fig. 2. No cracking is visible in this sample. The copper powder particles have been heavily deformed. Some porosity can be seen at the interface of copper powder particles. No melting can be detected along the particle interfaces with the SEM. With the bending test the breaking of the sample at 69 MPa is concluded to be due to the porosity.

The microstructure of compacted iron powder is shown in Fig. 3. With the SEM large cracks were observed in the sample. Along the interface of powder particles no porosity was detected. At the particle interfaces melted bands can be detected in Fig. 3. Without cracking the melted bonding of particles would have resulted in a much better bending test result. The compaction of iron powder was not successful with the explosive geometry used for copper powder so that the explosive 3 was added for better compaction. However, some cracking was found in the sample due to the high pressure of the explosive 3.

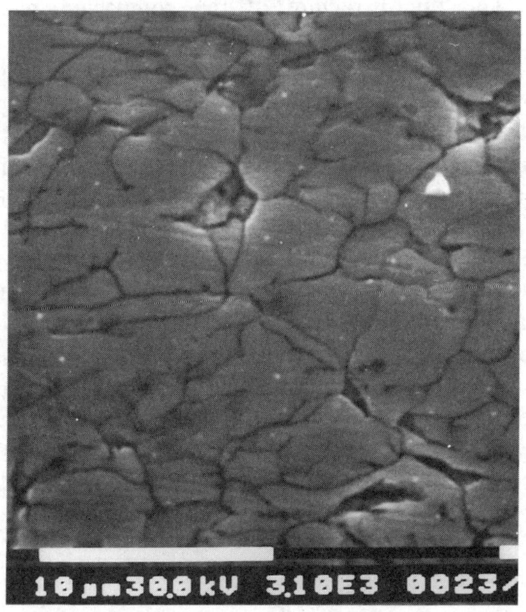

FIGURE 2. SEM micrograph of the compacted copper powder.

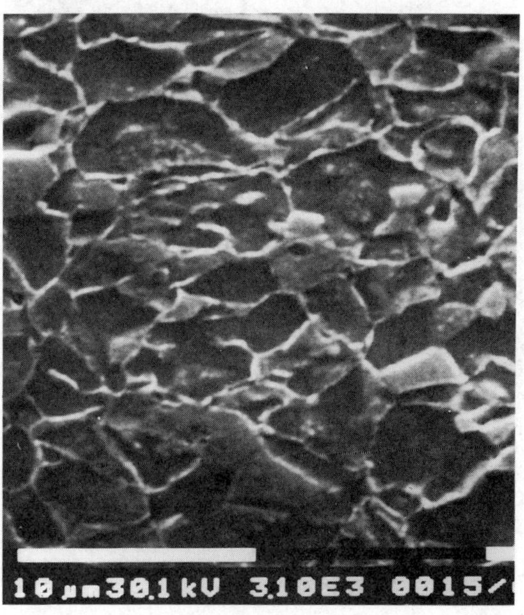

FIGURE 3. SEM micrograph of the compacted iron powder.

The microstructure of the compacted nickel powder is shown in Fig. 4. With the SEM no cracking was observed in the samples. However, the low bending test result of 30 MPa of one sample was probably due to some cracking. Some porosity can also be seen along the particle interfaces, which may reduce the strength of the compacted sample. No visible regions of melting can be seen with the SEM. The compaction of nickel powder was not successful with the explosive geometries used for copper or iron powders, for more cracking was observed for nickel than for iron when using the same geometry. For nickel powder the pressure was decreased and the pressure time duration increased significantly. These changes made possible the compaction of nickel powder with very few cracks, as seen in Fig. 4.

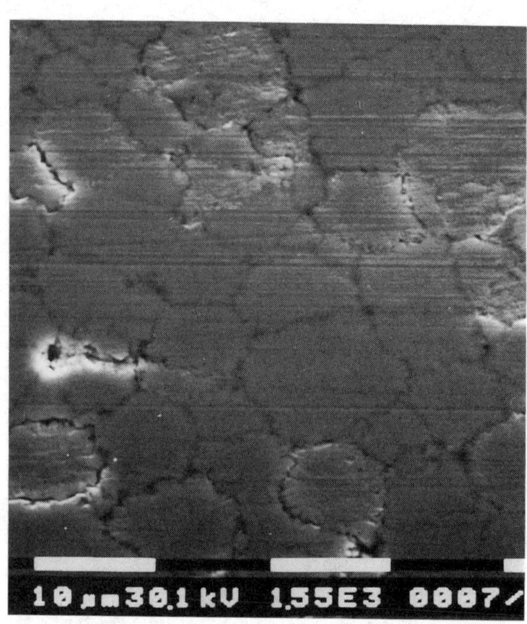

FIGURE 4. SEM micrograph of the compacted nickel powder.

CONCLUSION

Results of this study show that the compaction of fine (avg. 5 μm) metallic powders is possible with the explosive technique. Crack-free compacts of copper and nickel powders were made. No melting was observed with the SEM for the copper or nickel compacts. Iron compacts had some cracking as well as melted bonding of the particles. The compacted copper and nickel samples shows that a good strength of a compacted powder can be achieved without melting at the particle interfaces. In this study the experimental parameters were not yet optimal and the strength of the compacts can be improved. This "cold-welding" process requires a long pulse duration time. Formulas for estimation of the pressure pulse duration times to form this "cold-welding" process have been presented by A. Shtertser [9].

REFERENCES

1. Meyers M.A. and Wang S.L., *Acta Metall* **36**, 925-936 (1988).
2. Prümmer R., *Explosive welding, forming and Compaction*, (ed.) Blazynski T.Z., New York: Applied Science Publishers, 1983, ch. 10, pp. 369-391.
3. Morris D.G., *Mat. Sci. Eng.* **57**, 187-195 (1983).
4. Mammalis A.G. and Gioftsidis G.N., *J. Mat. Pross. Tech.* **23**, 333-345 (1990).
5. Gourdin W.H., *Mat. Sci. Eng.* **67**, 179-184 (1984).
6. Rice M.H., McQueen R.G. and Walsh J.M., *Solid State Physics* **6**, 1-63 (1958).
7. Carroll M.M. and Holt A.C., *J. Appl. Phys.* **43**, 1626-1636 (1972).
8. Brandes E.A. and Brook G.B (eds.), *Smithells Metals Reference Book 7th Ed.*, Oxford: Butterworth-Heinemann, 1992, ch. 22.
9. A. Shtertser, *Comb. Expl. Shock Waves*, **31**, No. 6, (1995), accepted for publication.

SHOCK COMPRESSION SYNTHESIS OF B1-TYPE TANTALUM NITRIDE

K. S. Vandersall and N. N. Thadhani

School of Materials Science and Engineering, Georgia Institute of Technology, Atlanta, GA 30332-0245

Shock compression was used to synthesize tantalum nitride with the B1-type (cubic) crystal structure. Hexagonal phase (Co-Sn Structure) tantalum nitride powder was packed with densities of ~35% and ~60% T.M.D into steel capsules, and shock loaded at 1.0 km/s impact velocity, corresponding to a 40-60 GPa calculated peak pressure. X-ray diffraction and optical and scanning electron microscopy were used to characterize the recovered compacts and starting powder. The highest yield of B1 phase was obtained from the capsule with the lower packing density in regions toward the non-impact face which correspond to the highest pressure and temperature. The lattice parameter of the shock synthesized B1 phase was determined to be 0.433 nm which coincides with a stoichiometry of approximately $TaN_{0.96}$.

INTRODUCTION

Cubic (B1-type) tantalum nitride is theoretically predicted to have high hardness and high super-conducting critical temperature in comparison with other transition metal nitrides with the same structure (1). Processing methods previously used for synthesizing B1-type tantalum nitride include self propagating high temperature synthesis (2), static high pressure (3), high temperature nitrogen atmosphere (4), plasma jet synthesis (5), and shock synthesis (6). With the exception of shock synthesis, all other processing methods give a low yield of the B1-phase or yield a product containing a mixture of off-stoichiometric phases. A bulk polycrystalline product is also not always obtained with these other techniques, and must be further processed before it can be used.

In the current study, B1-type tantalum nitride was shock synthesized from commercially available hexagonal phase tantalum nitride powder. Since shock compression results in the generation of very high pressures occurring over a microsecond duration, the possibility of nitrogen loss in the sample during the phase transformation is eliminated. Formation of the metastable phases is facilitated at high pressures due to the stability of the high density phases in the shock state.

EXPERIMENTAL PROCEDURE

Commercially available hexagonal phase tantalum nitride powder was obtained from Cerac incorporated. The Sawaoka 12-capsule plate-impact assembly (7) was used for performing the shock recovery experiments. The powder was packed into two capsules, one at tap density of ~35% theoretical maximum density (TMD) and the other capsule at a packing density of ~60% TMD using 66.75 kN (15,000 pounds) load. The two different packing densities were used to generate different shock conditions (temperature and pressure) within the sample. The Sawaoka design was slightly modified to include a cylindrical gap around the sample (referred to as the "Holman design"(8)). The "Holman Design" is utilized to reduce the radial focusing of the shock waves during the shock event which produce excessively high pressures in the axial regions of the sample (9). An impact velocity of ~1 km/s was used in the present experiment. The calculated peak pressure for the 1.0 km/s impact velocity corresponds to 40-60 GPa pressure in the compact, along the central core towards the non-impact side, decreasing to $\approx$15 GPa pressure towards the outer edges.

X-ray diffraction (XRD), optical microscopy, scanning electron microscopy (SEM), microhardness measurements, and sheet resistivity measurements were used to characterize the starting powders and shocked samples.

RESULTS/DISCUSSION

Analysis of the commercially acquired starting powder revealed that it consisted of ~5-10% of an off stoichiometric hexagonal phase $TaN_{0.8}$ (W-C structure) with the balance of hexagonal phase TaN (Co-Sn structure). The lattice parameters for the hexagonal phase starting powder were determined by XRD analysis to be a_0= 0.5201±0.0008 nm and c_0= 0.2906±0.0002 nm based on the least squares approximation of measured d-spacings. The morphology of the starting powder was observed using scanning electron microscopy to have an irregular shape with different textured surfaces. This is most likely due to fracture along cleavage planes at the surface of the powder particles during milling or a feature characteristic of gas phase powder synthesis. Figure 1 shows an SEM micrograph of the starting powder. The average size of the particles was in the range of ~10 microns.

The shock-compressed samples were recovered as 10 mm diameter by 4 mm thick compacts in the bulk polycrystalline state. Table I shows a summary of results of the yield, and different properties of the shock synthesized B1-TaN. X-ray

FIGURE 1. SEM image of starting hexagonal phase TaN powder, revealing its textured surfaces.

TABLE 1. Summary of Results

Sample	Phases Present	Peak Intensity Ratio I_a/I_b	Lattice Parameters (nm) [cubic for shocked samples]	Micro-hardness (kg/mm^2)	Sheet Resistivity (Ω/□)
Powder	Hexagonal	--	a_0= 0.5201±0.0008 c_0= 0.2906±0.0002	--	--
Shocked 1 (35% TMD)					
Impact Face	Hex. and Cubic	2.27	a_0= 0.4329±0.0005	2078.33 ± 336.8	2.735±5.938
Non-Impact	Hex. and Cubic	6.97	a_0= 0.4331±0.0001	Porous Face	--
Shocked 2 (60% TMD)					
Impact Face	Hex. and Cubic	0.30	a_0= 0.4327±0.0001	2071.78 ± 278.1	--
Non-Impact	Hex. and Cubic	0.22	a_0= 0.4329±0.0001	2023.13 ± 204.7	0.00082 ± 0.00195

diffraction patterns comparing the structural characteristics of the starting powder with each face of the two samples, are shown in Figure 2a-2e. The lattice parameters determined for the different samples and positions within each sample ranged around 0.4327-0.4331 nm and were independent of the yield of the cubic phase. The average lattice parameter of the shock synthesized B1 phase was taken as 0.433 nm which coincides with a stoichiometry of approximately $TaN_{0.96}$.

Correlation of the yield of cubic phase can be made with the starting powder packing density, and the shock loading conditions. Comparing patterns 2b and 2c to 2d and 2e reveals that both faces of the 35% TMD sample exhibit higher yield than the 60% TMD sample faces, which indicates that the yield is a function of packing density. It can also be seen that for the 35% TMD sample, the non-impact face exhibited a higher yield than the impact face. Lower temperature and pressure effects exist on the impact face than the non-impact face (due to radial focusing that occurs during the shock event); thus more transformation is observed on the face with the higher temperature. For the 60% TMD sample, the trend appears to be reversed, with a slightly higher yield visible on the impact face than the non-impact face. The higher impedance of TaN powder (60% dense) leads to more predominant one-dimensional effects, since radial focusing no longer dominates the overall loading history. Thus, the impact face subjected to the initial pressure has a higher temperature than the non-impact surface. Typical micrographs of regions with highest yield are shown in Figure 3. Thus, in general, with a lower packing density and increasing two-dimensional effects, higher temperatures are generated signifying a dominant role of thermally activated processes in the transformation mechanism.

Microhardness measurements were carried out on all faces of the compacts, except the non-impact face of the highest yield sample due to the porosity present. The impact and non-impact faces of the lower yield sample were found to have Knoop hardness values 2071.78±278.1 and 2023.13±204.71 kg/mm² respectively, and the impact face of the higher yield sample had a value of 2078.33±336.38 kg/mm². A trend of slightly increasing hardness value and standard deviation with yield is observed. The increase in standard deviation is probably due to the sampling of areas within the sample, which became more porous with increasing yield.

Sheet resistance measurements were also made on one face of each compact using the four point probe technique to correlate the electrical properties of the phase formed in the shocked state. The impact side of the higher yield sample and non-impact side of the lower yield sample were observed to have average sheet resistivity readings of 2.735±5.938 and 0.000823±0.00195 ohms/square respectively. This reveals that the higher yield sample has an increased resistivity over the reduced yield sample at room

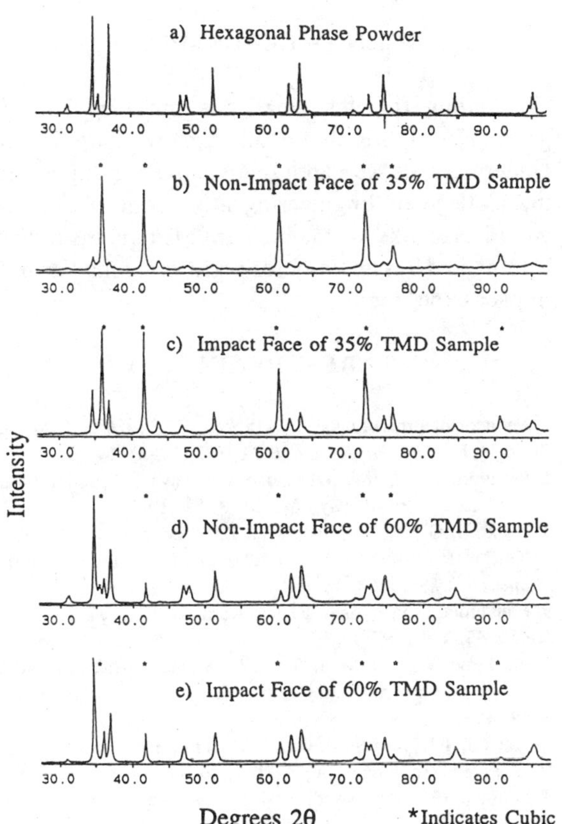

FIGURE 2. Comparison of XRD scans of Hexagonal Phase Powder and Shocked Compacts.

temperature, and it is therefore more electrically insulating at that temperature.

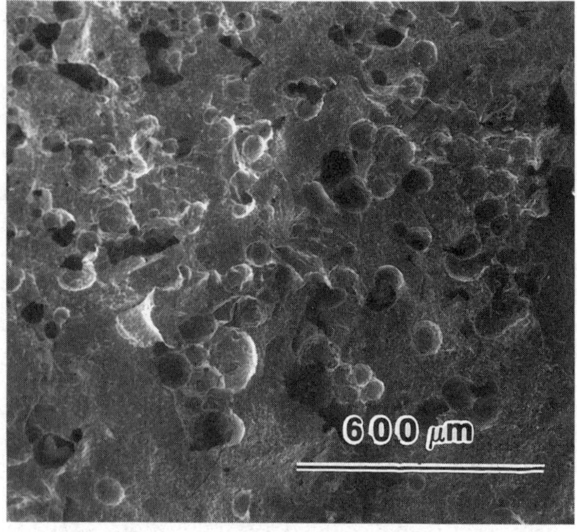

(a)

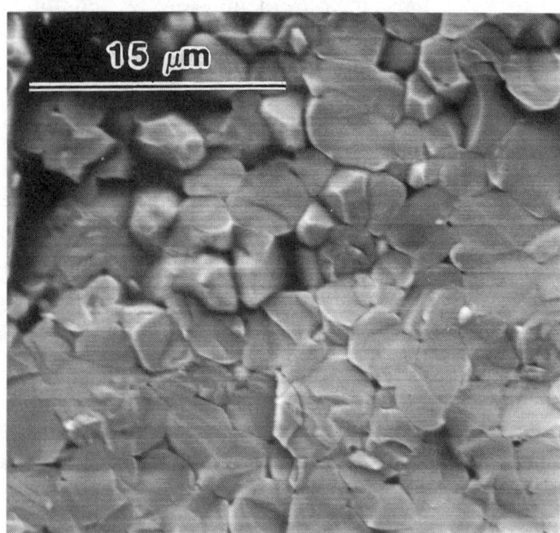

(b)

FIGURE 3. SEM micrographs of (a) porous and (b) equiaxed grain structure of shock synthesized compacts.

CONCLUSIONS

Shock compression was used to synthesize B1-type tantalum nitride in the bulk polycrystalline state. The yield was observed to be a function of packing density with the 35% TMD sample exhibiting close to 100% yield. It was found that temperature effects play a significant role in the phase transformation indicating that the transformation may involve thermally activated diffusion mechanisms. The X-ray analysis indicated that a stoichiometric TaN compound was obtained which is a distinct advantage of the shock compression process over other processing methods. This appears to be due to the high pressure and temperature effects generated in a microsecond duration of shock compression.

ACKNOWLEDGEMENTS

Funding for this work was provided by NSF grant No. DMR 9396132 and in part by the Undergraduate Research Internship Program through the College of Engineering at Georgia Tech. We would also like to thank New Mexico Institute of Mining and Technology for performing the Sawaoka impact experiments.

REFERENCES

1. Papaconstantopoulous, D.A., Pickett, W.E., Klein, B.M., and Boyer, L.L., *Physical Review B* **31** ,752 (1985).
2. Petrunin, V.F., Sorokin, N.I., Borovinskaya, I.P., and Pityulin, A.N., *Trans. Poroshkovaya Metall* **3**, 62 (1980).
3. Boiko, L.G. and Popova, S.V., *JETP Letters* **12,** 70 (1970).
4. Keiffer, R., Ettmayer, P., Freudhofmeier, M., and Gatterer, J., *Manatsch. Chem.* **102**, 483 (1971).
5. Matsumoto, O., Hyashi, E., and Kanzaki, Y., *J. Less Common Metals* **60**, 147 (1978).
6. Mashimo, T., Tashiro, S., Toya, T., Nishida, M., Yanazaki, H., Yamaya, S., Oh-Ishi, K. and Syono, Y., *J. Materials Science* **28**, 3439 (1993).
7. Akashi, T.K. and Sawaoka, A.B., "High-Density Compacts," U. S. Patent No. 4,655,830, April 7, 1987.
8. Holman, G.T., Norwood, F. N. R., and Graham, R.A., Sandia National Laboratories, unpublished results, 1991.
9. Graham, R.A., *High Pressure Explosive Processing of Ceramics*, Trans Tech, 1987, ch.2 , pp.31.

SHOCK INDUCED REACTIONS OF TITANIUM ALUMINIDES FROM MECHANICALLY ALLOYED PRECURSOR

T. Aizawa[1], Y. Kashiwabara[2], Y. Asakawa[2], K. Fukuoka[3], Y. Shono[3] and J. Kihara[1]

1   Department of Metallurgy, University of Tokyo, Hongo, Tokyo 113
2   Graduate School of Metallurgy, University of Tokyo
3   Institute of Materials Research, Tohoku University

In the planar shock reactive synthesis from element powder mixture with 1Ti + 1Al by one-stage gun, only $TiAl_3$ is synthesized with large amount of residual Ti and Al. The pretreated powders by our developed mechanical alloying system or the MA-precursors are fully reacted into two phase binary system ($TiAl + Ti_3Al$) with little amount of residual titanium and without $TiAl_3$ and residual aluminum. For varying the pretreated state, a role of this pretreatment by mechanical alloying on shock induced reaction is discussed with precise microstructural observation of synthesized materials.

INTRODUCTION

Typical two phase binary alloy of TiAl is a structural intermetallic material to be working at the elevated temperature [1]. Its specific strength, ductility and toughness are all strongly dependent on the grain size and the grain boundary characteristics [2]. For instance, TiAl can make significant superplastic deformation when the average grain size is reduced in the order of submicron [3]. In general, there are two ways to control the grain size and grain boundary characteristics: rapid solidification and mechanical alloying (MA). In particular, recent work [4] on the mechanical alloying of Ti-Al system provided effective information about the stability of synthesized non-equilibrium materials by mechanical alloying and the effect of nonstoichiometry on the chemical reaction or alloying taking place in mechanical mixing. As have been discussed in Refs. 5,6,7,8,9, the shock reactivity of pretreated materials by mechanical alloying must be different from that of constituent element powders. Since very small amount of yields is obtained by the conventional mechanical alloying, however, new MA-system is necessary to investigate the pretreatment effect on the shock reactivity. The present study concerns with the MA-pretreatment effect on the shock induced reaction of Ti-Al system by using our developed high speed mechanical alloying system [10].

PRETREATMENT BY HIGH SPEED MECHANICAL ALLOYING

The blended constituent element powder mixture is pretreated into various pre-mixed state materials by using our developed high speed mechanical alloying system. Different form the conventional milling or attriting for mechanical alloying, as depicted in Fig. 1, original powder mixture, which is pored into a closed die cavity, is cyclically subjected to high pressing and shear deformation in the prescribed pass schedule. In the present study, nearly the same loading pass schedule as shown in Fig. 1 is commonly utilized; two pressing modes and one forward extrusion mode are combined into one cycle loading. The number N of cyclic loadings are parametrically varied to obtain different premixed or pretreated state samples. The pretreated samples are called here by MA200, MA500 and MA700 after the numbers of cycles. One cycle requires for about 7 sec in effective operation of one cyclic loading. Two points to be noted here are: 1] Temperature of a die is controlled never to exceed 323 K (50 C), and 2] Liquid paraffin is used as a lubricant to reduce adhesion of mixed powders onto a die wall. Hence, raw mixed pretreated materials were subjected to debinding for elimination of carbon. As discussed later, active control of this residual carbon enables us to change the shock reactions.

Figs. 2 and 3 depict the microstructures of the blended powder mixture for 24 h with 1Ti + 1Al (as-mixed sample) and the pretreated sample by N = 200 (MA 200). In the as-mixed materials, each particle boundary of Ti and Al can be distinctly recognized. While, both Ti and Al particles were deformed and mixed in MA200. Three MA-precursors (MA200, MA500 and MA700) indicate no exothermic peaks in the DTA trace corresponding to ignition of reaction between Ti and Al.

EXPERIMENTAL PROCEDURE

To make preparations of the as-mixed and the pretreated powders, gas-atomized Ti and Al powders are commonly used with the maximum particle diameter of 45 µm and the purity of 99.7 % or more. These powder mixtures are uniaxially pressed into cylindrical disc samples with the thickness of 5 mm, the diameter of 10.3 mm and the porosity of 20 %. One-stage powder gun was utilized for recovery testing with various flyer plate velocities. The whole experimental conditions are listed in Table 1.

EXPERIMENTAL RESULTS

The successfully recovered post-shock samples were further mechanically worked to provide specimens for XRD and EPMA. Synthesized products are also summarized in Table 1 together with the residual Ti and Al; Fig. 4 shows the dependency of the shock reactivity on the flyer plate velocity both for the as-mixed and the pretreated samples. When the as-mixed sample is shot by 1.51 km/s, both Ti and Al powder particles were severely deformed and consolidated without reactions. Increase of the flyer velocity up to 1.61 km/s is required only to make partial reactions to $TiAl_3$ phase (δ-phase). As discussed in Refs. 5,7,8, nucleation and growth of embryos of local reactants with intermediate phase takes place with large mass mixing during shock loading; if these local reactants are further reacted with residual Ti and Al during residual heating, the final product must be in equilibrium phase or $TiAl_3$-phase. In other words, this partial reaction is a typical shock assisted reaction.

In case of shock reactions from MA200 and MA500, the synthesized products became nearly the same irrespectively of the applied flyer velocity; the pretreated materials were mainly reacted into TiAl with a small amount of Ti_3Al and with a trace of $TiAl_2$ and $Al_2Ti_4C_2$. No peaks for residual Ti and Al can be observed in the XRD profiles. With respect to the phase and microstructure of synthesized products, noticeable difference cannot be recognized between MA200 and MA500. The point to be interested for microstructure of synthesized sample from MA200 and MA500 is that a needlelike structure with its length of 2 - 3 µm is closely distributed in the matrix, as shown in Fig. 5. Even using the high-resolution of SEM-EDX, any significant segregation of Ti and Al can never detected. In case of MA700, although the grain size was monotonically decreased with increasing the numbers of cycles, the synthesized products became different from those of MA200 and MA500. No needlelike structures were seen in microstructure.

DISCUSSION

Since Ti and Al was mainly reacted into TiAl for MA200 and MA500 irrespectively of the applied shock pressure, large mass mixing, which must be a driving mechanism of shock reactions from as-mixed powder mixture, is thought never to take place in shock loading and its subsequence. With refinement of microstructural intimacy in pretreatment, local shock reactivity is inclined to play a role in the shock induced reaction, partially because no homogenization process of heterogeneous powder mixture becomes unnecessary to ignite chemical reactions between elements. In case of MA200 and

MA500, when structural refinement is terminated by the intermediate stage with noticeable local distribution of Ti and Al, nucleis of local reactions were first formed at the spots where both Ti and Al were premixed more finely than other parts. Homogeneous interaction of these nucleis would result in homogeneous phase of TiAl in the finally synthesized materials; however, when structural refinement is distributed in the pretreated materials, only local interactions of nucleis take place and result in needlelike structure. In fact, only homogeneous structures without needlelikes were yielded in the shock reactions of MA700, where more finely premixed state was attained by the mechanical alloying.

In our developed mechanical alloying, since liquid paraffine lubricants are added in every 10 - 20 cycles, the amount of residuals paraffine is thought to increase with the number of cycles. Although careful debinding was performed before experiments, very little but a certain amount of carbons might be left in samples. Hence, a screening effect of carbons on the shock induced reactions must be considered in the present study. In fact, little effects were observed in case of MA200 and MA500, but main reactions were screened by significant formation of carbide by $Al_2Ti_4C_2$; a nuclei of local reactant is thought to be reacted with a mobile carbon before interactions of nucleis. Advanced studies are taking place for both the lubricant-controlled and the lubricant-free mechanical alloying procedures.

CONCLUSION

A role of pretreatment in the shock induced reaction of titanium aluminides is summarized by the three points: 1) Activation of shock reactions taking place irrespectively of the applied shock pressure, 2) Microstructural modification by local shock reactions, and 3) Sensitive reaction mechanism to a small amount of carbon. With respect to 1) and 2), finely layered specimen is also utilized to investigate the effect of microstructural intimacy on the local shock reactions. Under precise control of residual carbon in the pretreated powders, the effect of non-equilibrium phase formed in pretreatment on shock reactivity is to be discussed.

ACKNOWLEDGMENTS

This study is financially supported in part by the Grand-in-Aid by the Ministry of Education, Science and Culture with the contract numbers of # 07555216 and # 07555512.

REFERENCES

1. Y.W. Kim, J. Metals 41 (1989) 24-30.
2. M. Yamaguchi Ed., Intermetallics. Proc. Int. Conf. SAMPE-3 (1993, Makuhari).
3. T. Wajata, et al., ISIJ International. 33 (1993) 884-888.
4. K. Aoki, Private Communication (1995).
5. T. Aizawa, et al., J. Faculty of Engineering, University of Tokyo. XLIII (1995) 57-101.
6. N.N. Thadhani, American Institute of Physics. 76 (1994) 2129-2138.
7. T. Aizawa, et al., J. Intermetallics. (1995) (in Press).
8. T. Aizawa, et al., Annales de Chimie (1995) (in Press).
9. Y. Asakawa et al., APS-95 (1995, August, Seattle) (to be published).
10. T. Aizawa, et al., materials Trans. JIM 36 (1995) 138-149.

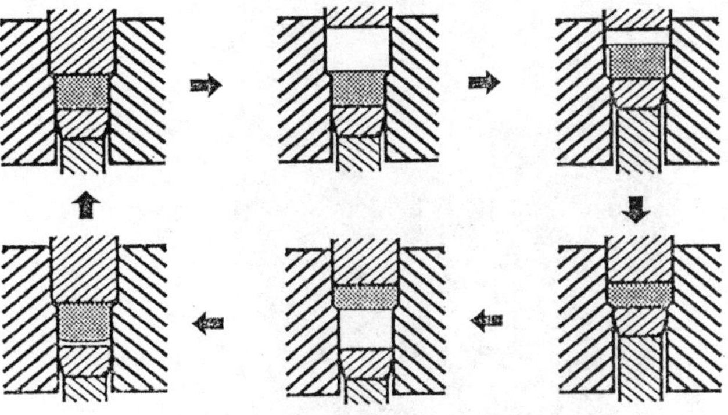

Fig. 1 Typical pass schedule in mechanical alloying.

Table 1 Experimental conditions and list of synthesized products for as-mixed, MA200, MA500 and MA700.

Flyer Velocity (km/s)	As-mixed	MA 200	MA 500	MA 700
1.6	$TiAl$, $Ti_3Al(t)$			
1.5	Ti, Al	○		
1.4			○	*
1.2		○	○	*
1.0		○		
0.8		○		

○ $TiAl$, Ti_3Al, $TiAl_2(t)$, $Al_2Ti_4C_2(t)$

* $TiAl$, $Al_2Ti_4C_2$, $Ti_{3.3}Al$, Ti_3Al

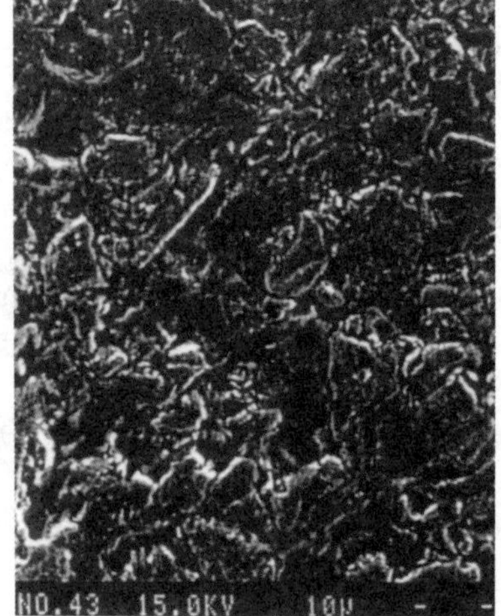

Fig. 2 Microstructure of as-mixed sampe.

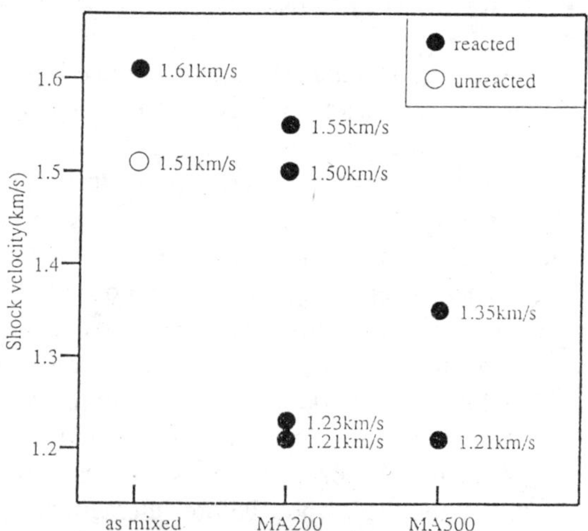

Fig. 4 Shock reactivity diagram.

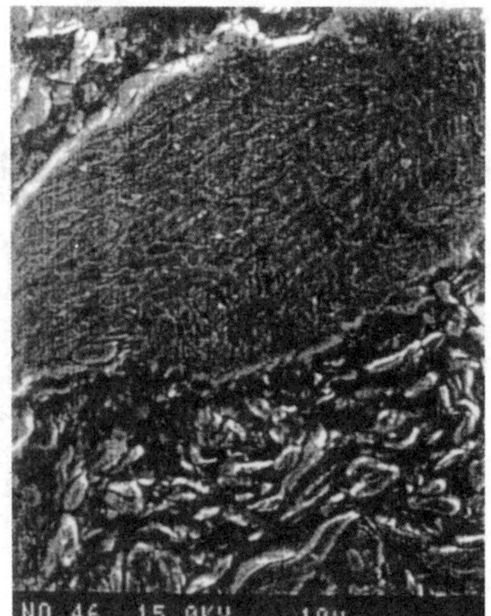

Fig. 3 Microstructure of MA200.

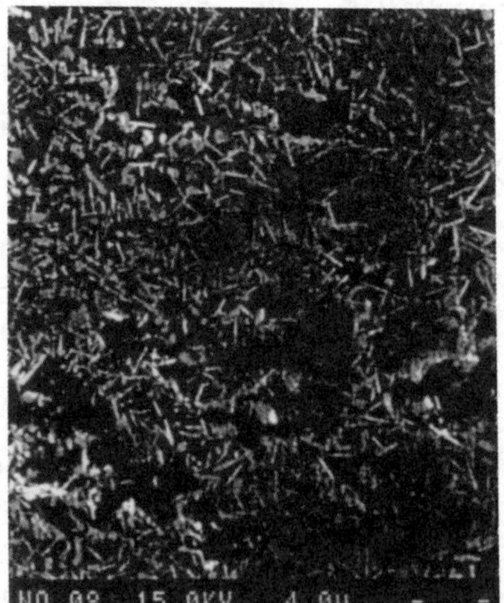

Fig. 5 Microstructure of synthesized materials from MA200.

THE EFFECT OF PULSE DURATION ON SHOCK-INDUCED CHEMICAL REACTION IN Ti-Si POWDER MIXTURES

N.N. Thadhani*, V. Subramanian*, R. Russell*, D. Savage and Y.M. Gupta****

**School of Materials Science and Engineering, Georgia Institute of Technology, Atlanta, GA 30332-0245*
***Shock Dynamics Center, Washington State University, Pullman, WA 99164-2814*

Mixtures containing Ti and Si powders in a 5:3 atomic ratio were shock-compressed using a four-cavity planar impact fixture. Copper flyer plates of 2.47 mm and 5.03 mm thickness were used to generate shock pulses of two different durations, keeping the impact velocity constant. The packing density of the powder mixtures was also varied to attain compacts of ~ 43%, 54%, 61% and 66% TMD, prior to shock compression. Microstructural characteristics of the recovered samples revealed occurrence of shock-induced reactions affected by the pulse duration and initial packing density. The reacted samples showed high levels of porosity and formation of Ti_5Si_3 compound uniformly throughout the compact cross-section. The unreacted samples showed extensive deformation of Ti particles, while silicon particles were deformed and/or fractured depending on the particle size.

INTRODUCTION

Shock compression of reactive elemental powder mixtures can lead to chemical reaction, forming compounds and rapid increases in temperature (1). The occurrence of reaction during shock compression has been investigated for many systems, and it has been suggested to depend on factors such as particle size, particle morphology, packing density as well as the peak shock pressure and temperature (2-4). The effect of **duration of the shock pressure pulse** has not been investigated in any prior studies. The reactions which occur during the high pressure state have been defined as shock induced reactions, controlled by mechanisms that prevail in the microsecond duration of high pressure shock state (5). Altering the pulse duration affects the time available for processes involved in compression of powder mixtures, and therefore, can influence the chemical reaction behavior.

In the present study, the effect of pulse duration is investigated for the Ti-Si system using medium morphology powder. Ti and Si powders in a 5:3 atomic ratio, react exothermically to form the Ti_5Si_3 intermetallic (ΔH_R = -1786 kJ/kg). Previous studies (6,7) on shock-compression of Ti-Si powders have found that the threshold conditions for reaction depend on the powder morphology and packing density. Medium morphology powders appear to be most reactive, with reactions occurring at pressures as low as 1 GPa at ~ 53 % TMD, based on results of time-resolved (6) and recovery (7) experiments.

EXPERIMENT

Medium morphology Ti and Si (amorphous) powders were mixed in the atomic ratio of 5:3. The powder mixtures were packed in a four cavity planar impact copper fixture (8). The impact experiments were done using a copper flyer plate, at an impact velocity of ~ 0.35 km/s, with the 4" diameter gas gun at WSU. The thickness of the flyer plate was varied (5mm and 2.47mm) to obtain two different pulse durations. The components of the fixture and their dimensions are shown in Figure 1. The packing densities used were 66, 61,

54 and 43% TMD. One experiment was also done with crystalline medium morphology Si powder using the thin flyer plate and packing densities of 54% and 43% TMD. Optical and scanning electron microscopy were used to analyze the recovered samples. X-ray diffraction was used to identify the phases and line broadening analysis was used to calculate the residual strain in Ti, in the unreacted samples. The peaks were fit with a Lorentzian function to obtain the peak width at half the intensity, which after subtraction of instrumental broadening was used in Williamson Hall plots to calculate the residual strain and crystallite size (9).

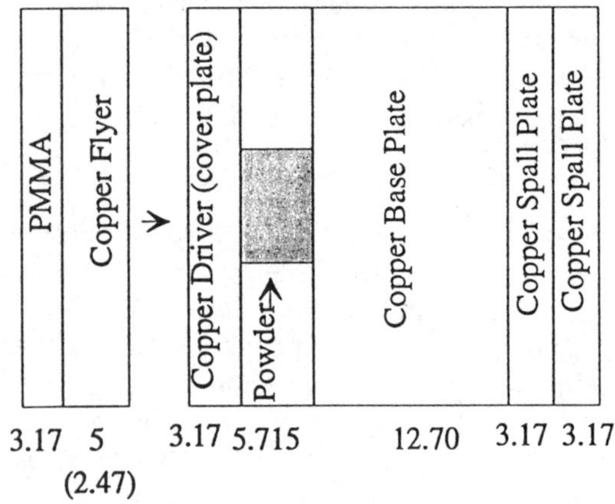

FIGURE 1. Components of the four-cavity fixture and their dimension in mm.

RESULTS AND DISCUSSION

Table 1 lists the experimental conditions. Using the impedance matching technique and the calculated Hugoniot for 5Ti+3Si powder mixture an initial pressure of ~ 6 GPa in the copper driver and 0.5 GPa in the powder was computed. The recovered samples showed the occurrence of reaction depending on the pulse duration and the packing density. For the various experiments performed, the reaction map is shown in Figure 2.

TABLE 1. Experimental Conditions (V_{imp}~.35 km/s)

Shot #	Flyer Thickness	Densities(% TMD) Reaction (Y/N)
1	5 mm	43(Y), 61(Y)
2	2.47 mm	66(N), 61(N), 54(N)
3	5 mm	66(N), 61(N), 54(Y)
4	2.47 mm	54(N), 43(Y) (amorphous & crystalline)

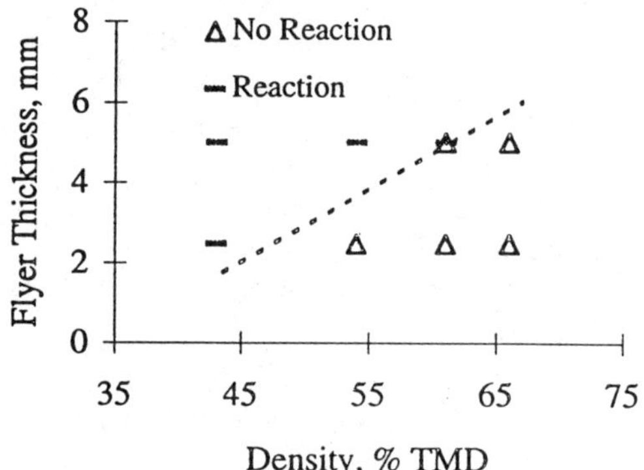

FIGURE 2. Reaction Map as a Function of Flyer Thickness and Density.

The reacted samples formed single phase Ti_5Si_3 compound with a uniform microstructure and high levels of porosity throughout the compact, as shown in Figure 3 (shot #3, 54% TMD). The lattice parameter of Ti_5Si_3 was measured and is shown in Table 2. The lattice parameter did not change significantly for the Ti_5Si_3 reaction product formed under conditions of different densities and pulse durations.

TABLE 2. Lattice Parameter of Reacted samples

Flyer Thickness	Density (%TMD)	Lattice Parameter (Å) a	c
5 mm	61	7.4511	5.1502
	54	7.4588	5.1689
	43	7.4563	5.1594
2.47 mm	43(a-Si)	7.4627	5.1540
	43(x-Si)	7.4581	5.1525

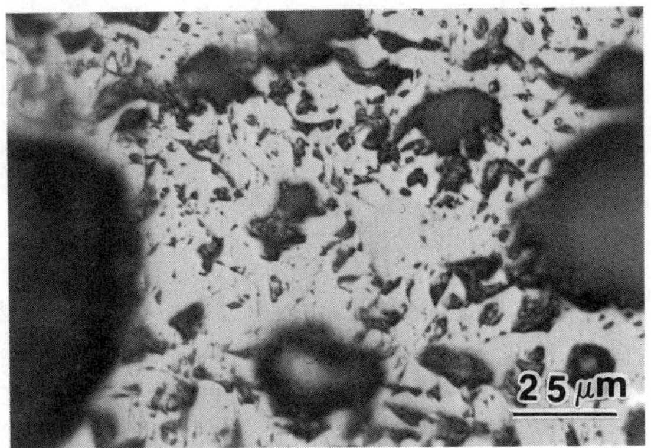

FIGURE 3. Microstructure of Reacted Sample. Flyer Thickness - 5 mm, Density - 54% TMD, shot # 3.

The unreacted samples were comprised of deformed Ti and Si particles. Some large Si particles showed cracking. A typical microstructure of regions close to the impact, center and back surface for the unreacted Ti-Si compact is shown in Figure 4 (shot# 2, 66% TMD). No obvious variation in the deformation characteristics are observed in these micrographs. In addition, there is no evidence of any partial reaction at regions of interparticle contacts between Ti and Si as shown in the higher magnification micrograph in Figure 5 (shot # 3, 61% TMD). Similar results were also observed for the other unreacted compacts. The residual strain values in the unreacted Ti are shown in Table 3. No significant change in strain was observed as a function of density and pulse duration. The strain values for Ti correspond to dislocation densities of ~ 10^{11}-10^{12} cm^{-2} based on Williamson and Smallman approach (10).

TABLE 3. Residual Strain in Unreacted Ti.

Flyer Thickness	Density (% TMD)	Strain
5 mm	66	0.0040
	61	0.0032
2.47 mm	66	0.0036
	61	0.0030
	54 (a-Si)	0.0039
	54 (x-Si)	0.0038

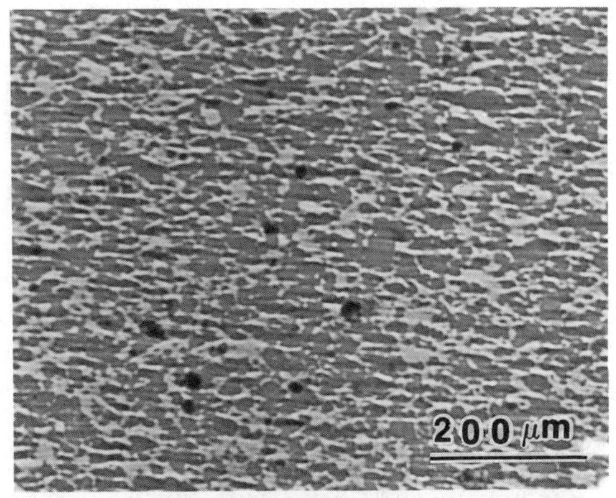

(a) Impact Surface

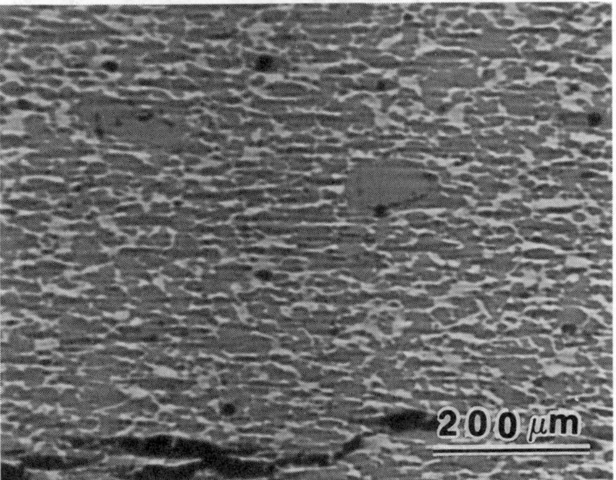

(b) Center

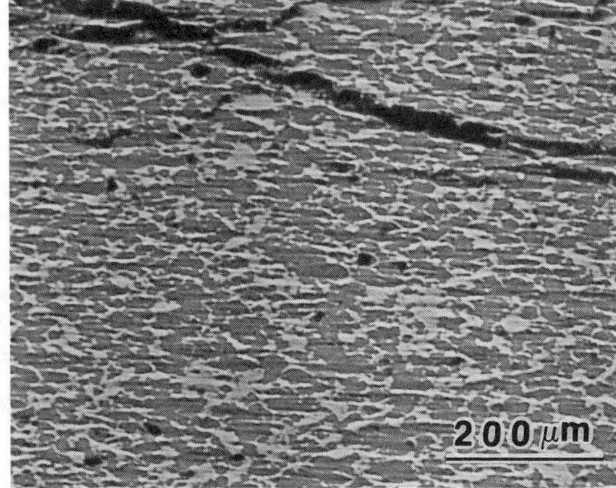

(c) Back Surface

FIGURE 4. Microstructure of Unreacted Sample at different positions. Flyer Thickness 2.47 mm, Density 66%, Shot # 2.

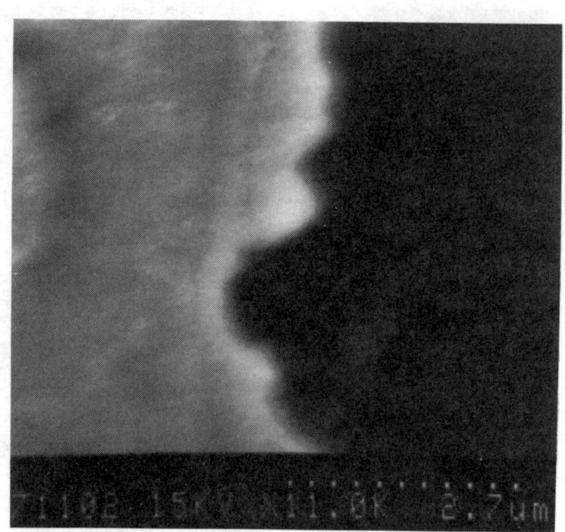

FIGURE 5. Scanning Electron Micrograph of Ti - Si particle boundary region. Flyer Thickness - 5 mm, Density - 61% TMD, Shot # 3.

Based on the observation of a uniform microstructure and occurrence of either complete or no reaction, it appears that the reaction in the Ti-Si powder mixtures maybe shock induced. Once the reaction is initiated, it becomes self-sustaining and moves through the sample thickness consuming all reactants. The final reaction product is subsequently controlled by thermal effects of the reaction and not pressure effects, which explains the uniform microstructure and lattice parameter of reaction products obtained under different densities and pulse duration.

The reduction of flyer thickness for the same packing density, reduces the pulse duration of the pressure. This reduces the time available for mechano-chemical events that may lead to reaction initiation, and therefore no reaction is seen, except at a lower density (43% TMD). Reducing the density from 53% to 43% for the lower duration experiment may increase the volume of voids present where better mixing and flow of the constituents can be attained, which results in greater propensity for reaction to occur (5).

The results described in the paper are of a preliminary nature, but reveal the important contribution of shock pulse duration on reactions occurring in powder mixtures due to shock compression. Further work is necessary to ascertain if the observed reactions are solely due to planar pressure effects of longer pulse duration or other more complex loading conditions.

CONCLUSIONS

Shock compression of Ti and Si powder mixtures results in chemical reaction, leading to the formation of Ti_5Si_3 intermetallic. The unreacted samples obtained due to loading with higher packing density or smaller pulse duration are comprised of deformed Ti and Si reactants, uniformly distributed through the sample cross-section. The results show an important effect of pulse duration on the propensity for reaction in Ti-Si powder mixtures.

ACKNOWLEDGMENTS

Funding for this research was provided by Office of Naval Research Contract No. N00014-94-1-0169.

REFERENCES

1. Dremin, A.N. and Bruesov, O.N., *Russian Chemical reviews*, **37**, 392-402, (1983)
2. Thadhani, N.N., *Progress in Materials Science*, **37**, 117-226, (1993)
3. Graham, R.A., Morosin, B., Venturini, E.L., and Carr, M.J., *Ann. Rev. Mater. Sci.*, **16**, 315, (1986)
4. Thadhani, N.N., Work, S., Graham, R.A., and Hammetter, W.F., *J. Materials Research*, **7**, 1063-1075, (1992)
5. Song, I., and Thadhani, N., N., *J. Matls. Synthesis and Processing*, **1**, 347-358, (1993)
6. Royal, T.E., "Investigation of Mechanistic Processes for Shock Induced Chemical reactions in Ti-Si, Ti-Al and Ti-B Powder Mixtures", Masters Thesis, Georgia Institute of Technology, (1994)
7. Krueger, B.R., Mutz, A.H., and Vreeland, T. Jr., *Met. Trans.*, **23A**, 55-58, (1992)
8. Korth, G.E., Flinn J.E., and Green, R.C., *Metallurgical Applications of Shock-Wave and High-Strain-rate Phenomena*, Marcel Dekker Inc, 1986, ch.6., pp. 129-147
9. Williamson, G.K., and Hall, W.H., *Acta. Met.*, **1**, 22-31, (1953)
10. Williamson, G.K., and Smallman, R.E., *Phil. Mag.*, **1**, 34-46, (1956)

CHEMICAL REACTIONS IN CONTROLLED HIGH-STRAIN-RATE SHEAR BANDS

V.F. Nesterenko[1], M.A. Meyers, H.C. Chen, and J.C. LaSalvia

Department of AMES, University of California, San Diego, CA. 92093-0411

Controlled high-strain-rate shear bands were generated in porous mixtures(Nb+Si, Ti+Si) using axially symmetric experimental configurations("Thick-Walled Cylinder") method. Shear strains up to 100 and strain rates of approximately 10^7 sec^{-1} were generated inside shear bands. Particle fracture, melting, and regions of partial reaction were observed inside shear bands for the Nb+Si system. Under the same conditions of deformation, for the Ti-Si system the reaction initiated inside shear bands and propagated through the entire sample.

INTRODUCTION

The role of intense shear in the initiation of chemical reactions and phase transformations under mechanical deformation was emphasized by Bridgman[1], Dremin and Breusov[2], Graham[3], Batsanov[4] and others. Nesterenko et al.[5,6] developed an experimental technique providing conditions for controlled high-strain-rate shear bands. In this paper this method is applied for the comparative investigation of shear induced reactions in porous Nb+Si and Ti+Si mixtures.

EXPERIMENTAL SET-UP

The thick-walled cylinder method[6] was applied for Nb-Si (68-32% by weight), and Ti-Si(74-26% by weight) systems. The powders were purchased from CERAC and had sizes of -325 mesh(<44 μm).

These two systems have widely different enthalpies of reaction ΔH:

$$5Ti + 3Si \rightarrow Ti_5Si_3 \qquad \Delta H = 580 \text{ kJ/mole,}$$

$$Nb + 2Si \rightarrow NbSi_2 \qquad \Delta H = 138 \text{ kJ/mole.}$$

This results in widely different sensitivity to initiation and ease of propagation.

Two experimental configurations, shown in Figure 1, were used. In Configuration 1 (Figure 1a) porous mixtures with a density of 2 g/cm^3 for Nb-

Si and 1.87 g/cm^3 for Ti-Si(~50% of theoretical value) were initially densified by an explosive (explosive 1, Figure 1a) with a low detonation velocities (2.5 km/sec and 3.12 km/sec, respectively) up to a density of 3.2 g/cm^3(~75% of theoretical value for Nb-Si)) and 3.4 g/cm^3(~90% of theoretical value for Ti-Si)).

In the next stage(R_0=11mm) these composite cylinders were collapsed by an explosive charge (explosive 2, Figure 1b) with a detonation velocity of ~4 km/sec, density of 1 g/cm^3, and an outer diameter of 60 mm. It produced plastic deformation highly localized in shear bands. The measurements of the segment lengths of collapsed layer indicate that shear localization started in the first stage of material movement, at overall effective strain of ~ 0.1. The values of r and r_1 (radii of porous layer after collapse) for Nb-Si mixture were equal 5.9 and 7.5 mm; initial values are 8.1(r_0) and 9.35 mm (r_{10}). For Ti-Si mixture: r = 6.05 mm(r_0 = 8 mm), r_1 = 7.5 mm(r_{10} = 9.15 mm).

Conf. 2(Fig. 1c) was used for the same powders (density 2 g/cm^3 for Nb-Si and 1.78 g/cm^3 for Ti-Si) as in Conf. 1. The inner and outer diameters of tubular cavity were 16.4 mm and 21.7 mm, respectively, R_0=11 mm. A few layers of amorphous ribbon were placed along the cylinder walls to initiate shear localization in the powder at an early stage of collapse. The average values of r and r_1 were equal to 6.5 and 8.1 mm (Nb-Si) and

---

[1] On leave from Lavrentyev Institute of Hydrodynamics, Russian Academy of Sciences, 630090, Novosibirsk, RUSSIA

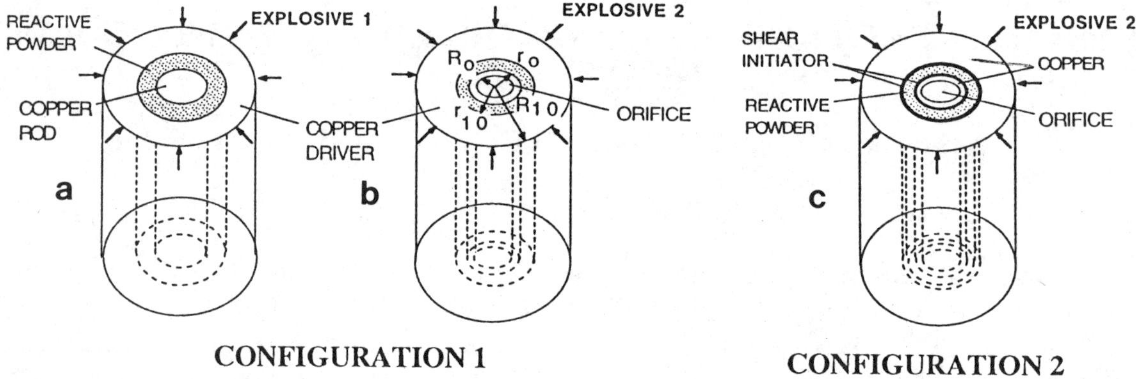

FIGURE 1. Geometry and sequence of deformation events in thick-walled cylinder method.

6.05 mm, 8.4 mm for Ti-Si mixture. The comparison between the data for the two mixtures indicates that overall strains were close.

EXPERIMENTAL RESULTS

Nb-Si mixture: Fig. 2 shows the cross-section of specimen tested in Configuration 2. The overall shear strain γ inside the shear zone(simple shear) is the ratio of the shear displacement, Δ and thickness of shear band, δ. The former is in the range Δ ~100-600 μm and δ varies between 5 and 10 μm for Conf. 1, and 10-20 μm for Conf. 2. The strains γ inside shear bands are found to be in the range 10-100 and shear strain rates $\dot{\gamma}$ are 10^6-10^7 s^{-1}[5,6].

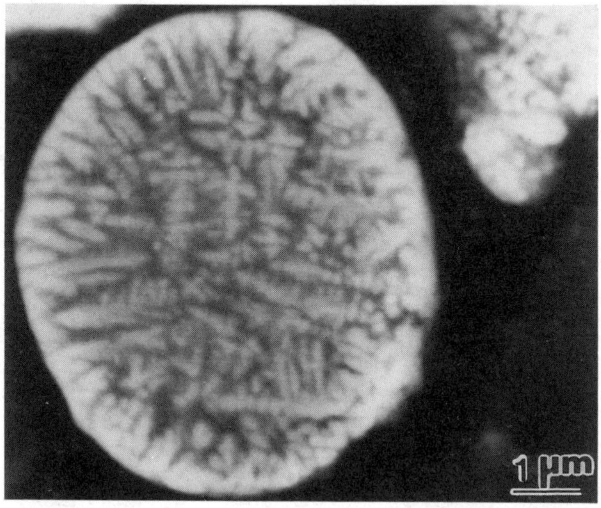

FIGURE 3. The structure of reaction product for Configuration 2, Nb-Si system.

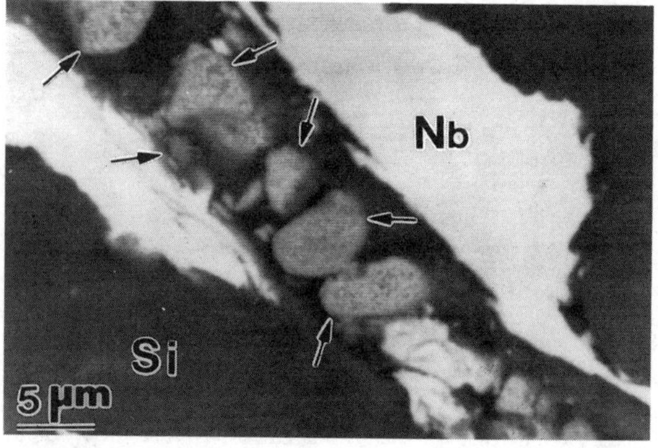

FIGURE 2. Shear band with reacted material inside(shown by arrows), Nb-Si mixture

Shear localization seems to produce partial melting of one component (Si), as in shock wave loading, as shown by Meyers et al.[7]. The multiple fracturing of the Nb particles due to shear localization, at the mesolevel, leads to a decrease of the initial particle size from 44 μm to 0.1 - 2 μm, resulting also in intense heating of the newly created thin particles. Material flow inside the shear band is unstable and results in vortex formation[5,6]. The combination of these effects (melting of Si, fracturing of Nb and

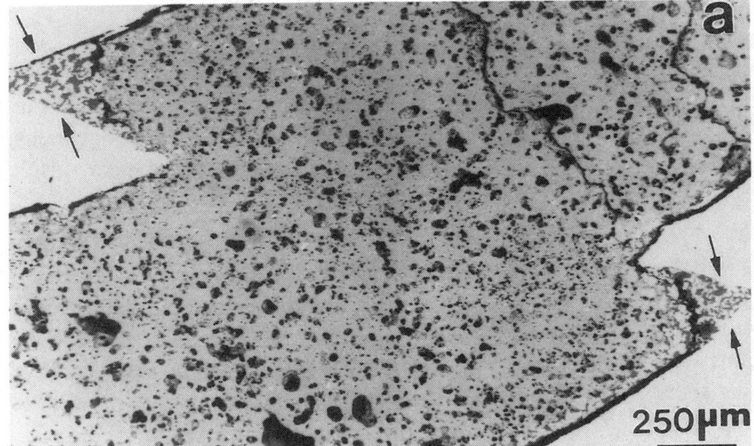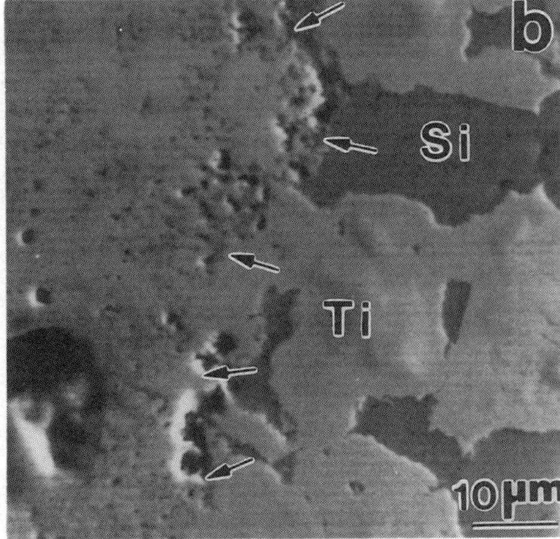

FIGURE 4. (a) - Overall view of shear band in Ti-Si mixture (Configuration 1);   (b) - structure of reacted-unreacted material interface (shown by arrows) near inner copper wall.

vorticity) provides the reaction product inside shear band with the sizes up to 8 μm (Figs. 2,3) for Configuration 2.

Ti-Si mixture: The deformation of this mixture also resulted in localized plastic flow, Fig. 4a. However, unlike in the Nb-Si mixture, the reaction propagated throughout most of the specimen. Small triangular regions, adjacent to the copper wall (shown by arrows in Fig. 4a) remained unreacted. The structure of reacted compact is depicted in the Fig. 4b. The evaluation of shock pressure in the first wave (< 2GPa) allows to conclude that shock did not play an essential role in initiation of chemical reaction[8] and that shear bands are the main reason for it.

The average values of Vickers microhardness, measured in mixtures after different stages are: 177 (Conf. 1, stage 1), 914 (Conf. 1, after collapse), and 970 (Conf. 2). They demonstrate that reaction took place for Conf. 1(after collapse) and Conf. 2.

MECHANISM OF REACTION INSIDE SHEAR BAND: COMPARISON WITH SHOCK WAVES

On the basis of the observations a mechanism of *shear-assisted* chemical reaction can be proposed. The main stages are:

(1) Nb particles are split into foils with thickness on the order of magnitude of 0.1-1 μm by localized mesoshear. They are heated as a result of intense shear deformation which precedes their formation and have "fresh" surfaces.

(2) Reaction begins due to the extensive relative flow of Nb particles and Si inside the shear band, which is accompanied by melting. The reaction proceeds along the Nb(sliver)-Si interfaces, with thickness l_r up to 1μm as mentioned by Thadhani[9].

(3) Instability of gradient flow on another, larger scale inside shear band, results in vorticity and the coiling of partially reacted Nb foils (as snow balls) together with adjacent Si into rounded particles with diameters up to 8 μm. Vorticity can also promote the collection of partially reacted small particles to the center of rotation, providing typical sizes of reaction product on the order of 1 μm.

(4) Reaction continues up to the scale length D_r in places where temperatures are sufficiently high and is then quenched by the surrounding material. This mechanism enables the evaluation of the size of final reaction product D_r [6]:

$$D_r \approx l_r \gamma. \tag{1}$$

It is important to emphasize that Equation (1) represents the connection between the size of reaction region at the microlevel and the dimensions of reacted particles at the macrolevel determined by mechanical shear movement of material.

The reaction inside shear band for Ti-Si mixture develops in the same manner. The propagation of the reaction in this case throughout the entire sample is mainly accomplished due to the differences in enthalpy of reaction.

The main differences between shear and shock assisted(induced) chemical reactions are due to essential differences in displacements of neighboring particles. Under shock-wave loading the relative displacements of particles are comparable or less than the particle sizes at normal porosities close to 50 % and cannot induce essential particle fracture, as is observed inside shear band[5,6]. This is also responsible for flow instabilities, enhancing the reactivity of mixtures.

EVALUATION OF THE REACTION TIME

The time of reaction can be evaluated from the time of shear band quenching. For the Nb-Si system reaction was quenched as a result of heat removal to the relatively cold material outside shear band. Taking the thermal diffusivity for surrounding material as k $\sim$ 0.1 cm^2/s, $\delta \approx$ 10 μm, the quenching time τ_q and $\dot{T}$ are obtained:

$$\tau_q \sim \delta^2 / \kappa \sim 10^{-5} \text{ s} \tag{2}$$

$$T \sim \Delta T/\tau_q \sim 10^3 \text{ K} / 10^{-5} \text{ s} \sim 10^8 \text{ K/s} \tag{3}$$

Reaction products can have either a microdendritic or a microcrystalline(Fig. 3) structure[5,6]. The secondary dendrite-arm spacings of the reaction product $\lambda \approx$ 0.1 μm(Fig. 3) enable independent evaluation of cooling rate. The value is of the same order of magnitude as Eqn. 3: $\dot{T} \approx 10^8$ K/s[6]. This confirms that the reaction products inside the shear bands were formed during 10^{-5} s.

It is worthwhile to mention that only the beginning of spherule formation is noticeable in Figs. 11a, 12[6], being very well developed for shock assisted chemical reactions for the same materials[7]. The mechanism of their formation can be connected with the instability of planar reaction-diffusion front[10].

CONCLUSIONS

The thick-walled cylinder method was successfully applied to generate controlled, high-rate localized reactive shear bands in heterogeneous, porous materials with overall parameters inside shear bands up to $\gamma = 100$, $\dot{\gamma} = 10^7$ sec^{-1}. A mechanism of *shear-assisted* chemical reaction, qualitatively different from the one observed for shock-wave loading, is proposed. It predicts a linear dependence of the reaction product size on shear strain. For Nb-Si mixtures, reaction was restricted to shear bands, but for Ti-Si system(with the heat of reaction three times higher) the reaction, initiated inside shear bands, propagated throughout the entire specimen, with exception of small regions, adjacent to copper driver tube, where it was arrested due to rapid heat transfer.

ACKNOWLEDGEMENTS

This research is supported by the U. S. Army Research Office, Contract DAAH 04-94-G-0314, U. S. Office of Naval Research, Contract N00014-94-1-1040, National Science Foundation, Grant MSS 90-21671, and the Institute for Mechanics and Materials, UCSD. The help of M.P. Bondar and Y.L. Lukyanov(Lavrentyev Institute of Hydrodynamics) is gratefully acknowledged.

REFERENCES

1. Bridgman, P.W., *Phys. Rev.,* **48**, 825-847 (1935).
2. Dremin, A.N., and Breusov, O.N., *Russ. Chem. Rev.,* **37**, 392-402 (1968).
3. Graham:, R.A., *Solids Under High Pressure Shock Compression: Mechanics, Physics and Chemistry,* New York: Springer-Verlag, 1993.
4. Batsanov, S.S., *Effects of Explosions on Materials,* New York: Springer-Verlag, 1993.
5. Nesterenko, V.F., Meyers, M.A., Chen, H.C., and LaSalvia, J.C., *Applied Physics Letters,* **65**, 3069-3071 (1994).
6. Nesterenko, V.F., Meyers, M.A., Chen, H.C., and LaSalvia, J.C., *Metall. and Mater. Trans.,* **26A**, (*1995*) (in press).
7. Meyers, M.A., *Dynamic Behavior of Materials* , New York: John Wiley & Sons , 1994, pp. 222, 640.
8. Chen H.C., Meyers M.A., and Nesterenko V.F., EXPLOMET-95 Proc., Elsevier (in press).
9. Thadhani,: N.N., *Journal of Applied Physics* , **76**, 2129-2138 (1994).
10. Showalter, K., *Nonlinear Science Today,* **4**, 3-10 (1995).

PLASTIC FLOW GENERATED SOLID STATE METAL/METAL REACTIONS

Diana L. Woody*, Jeffery J. Davis, J. Scott Deiter
Naval Surface Warfare Center
White Oak, Maryland 20903-5000

This paper will discuss an experimental effort to discern the basic mechanism of reactions in porous metal/metal compositions under rapid plastic flow conditions. The method will employ small scale impact tests on samples containing titanium and silicon. The unconfined samples were impacted at 17 m/s to induce a plastic flow. With the addition of aluminum and Teflon, self-propagating high temperature synthesis (SHS)-like reactions have been observed. The diagnostics consisted of a two-color infrared detector for real time emissivity measurements of the reacting materials and x-ray diffraction for recovery studies.

INTRODUCTION

A series of small scale experiments has been performed to understand the basic mechanism of reactions in porous metal/metal compositions under rapid plastic flow conditions produced by impact. This paper will discuss one of the mixtures investigated containing titanium and silicon. The unconfined samples were impacted at 17 m/s to induce a plastic flow. In these experiments, the effect of the addition of Teflon, and a metal additive, aluminum, to the Ti and Si mixtures was also measured and will be discussed. With the addition of aluminum and Teflon, self-propagating high temperature synthesis (SHS)-like reactions have been observed. A two color infrared detector was used for real time emission measurements of the reacting materials. X-ray diffraction was performed on chosen recovered samples.

EXPERIMENT

The rapid plastic flow of the samples was obtained from a drop weight impact machine. The impact machine consisted of an anvil, accelerated guided drop weight, base, and release triggering device. This setup is illustrated in Figure 1. The

impact machine is described fully in another publication.[1] Elastic shock cords were used to accelerate the drop weight to obtain impact velocities of 17 m/s. The impact of the drop weight on the anvil was planar to within 2 mrad.

The samples were 0.2 g and were in loose powder form prior to impact. Since the samples were loose powders, the porosity was not measured. The samples were opaque in both the visible and infrared ranges. Therefore, the light emanating from the impacted sample was from the edge of the sample.

The Ti and Si mixture was formed by mixing in the atomic ratio of 5 Ti to 3 Si using a slurry of Ti, Si, and 1,1,2-trichloro-1,2,2-trifluoroethane. All of the 1,1,2-trichloro-1,2,2-trifluoroethane was removed by vacuum drying prior to use. Throughout the paper, this mixture will be referred to as 5Ti + 3Si. The Aluminum and Teflon were mixed with the 5Ti + 3Si by the mortar and pestle method. The Ti, Si, and Al powders were 325 mesh from Cerac Inc. The Teflon used was $35\mu m$, 7A from Dupont.

The two color infrared detector consists of a HgCdTe element juxtaposed to an InSb element. Each element has dimensions of 0.101 cm by 0.101 cm with an active area of 0.010 cm^2. The elements were housed in a liquid nitrogen cooled Dewar and kept at an operating temperature of 77 K. The

---

*Current address: Naval Air Warfare Center, Weapons Division, China Lake, California 93555

InSb element's spectral response was from 2 μm to 5 μm. The HgCdTe element was capable of detecting wavelengths from 5 μm to 12 μm. The signal from each infrared detector element was transmitted as a voltage through an initial voltage amplifier and then transferred to a LeCroy digital oscilloscope.

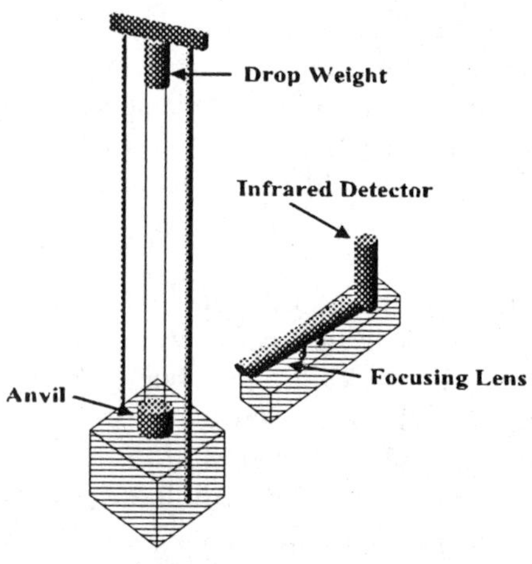

FIGURE 1. Impact Machine Apparatus

RESULTS AND DISCUSSION

Previous shock compression studies have been performed on Ti and Si mixtures at shock pressures between 0.80 and 7.5 ± 2.5 Gpa.[2,3,4] For the shock compression experiments, Ti and Si were mixed in a ratio of 5 Ti to 3 Si. The recovered product was Ti_5Si_3.

Initially, the 5Ti + 3Si system was impacted to observe if it was possible to obtain the product, Ti_5Si_3, under lower impact velocities but high plastic flow conditions.[5] As seen in Table 1, this condition did cause an emission. However, the time to peak emissions shows that this is more indicative of a heating of the metals under impact rather than a chemical reaction. The x-ray diffraction results on these samples showed no Ti_5Si_3 was formed in the recovered sample.

Extensive exothermic reaction was observed for the 5Ti + 3Si mixture containing Teflon and aluminum. Extent of reaction was quantified from the infrared emission vs. time data. The highest reactions were defined as the largest peak emissions and the smallest time to peak emission. The results from the data indicate that the 5Ti + 3Si showed a significant decrease in reaction time and increase in emission with the addition of Teflon and Al.

Qualitative signs of reactions consisted of such parameters as a strong visible light emission, a large audible signal, and observance of the recovered sample's surface. The most exothermic samples exhibited a visible flash and sustained burning upon impact up to 1 - 3 seconds in duration, and burning of the recovered sample and its holder. It was observed that the burning duration could be varied by changing the percentages of the materials in the composition. The peak emission and time to peak emission for some of the experiments are given in Table 1.

The addition of Teflon and Al to 5Ti + 3Si considerably increased the peak emissions and reduced the time to peak emission two orders of magnitude from milliseconds to microseconds. The signature of the infrared emission curve was dependent upon the relative percentages of the materials in the composition.

Changing the percentages of the aluminum and Teflon had a significant influence on the extent of the exothermic reaction. Each percentage was tested at least twice. As shown, the composites containing a greater percentage of the 5Ti + 3Si mixture showed a bright flash and sustained burning upon impact. The composites containing a greater amount of Teflon were more likely to produce little sustained burning. The sustained burning reactions continued until damped by the drop weight's rebound on the order of seconds. Also, Ti alone added to Teflon and Al produced a similar hearty reaction in the same time range as observed for the 5Ti + 3Si/Teflon/Al mixture.

X-ray diffraction was performed on the recovered sample of the impacted 5Ti + 3Si/ Teflon/Al mixtures listed in Table 1. X- ray diffraction of the recovered sample of the impacted 5Ti + 3Si revealed only Ti and Si. X-ray diffraction of the recovered sample of the impacted 30% 5Ti + 3Si, 60% Teflon, and 10% Al mixture revealed only the starting powders Ti, Si, Al and Teflon. The Teflon

TABLE 1. **List of results for some of the compositions tested.***

Material and composition	Radiation from InSb detector(mv)	Radiation from HgCdTe detector(mv)	Time to Peak Emission
5Ti+3Si	29	1.41	2.95 ms
Teflon(35µm)	17	3	1.32 ms
(50%)Teflon/(50%)Al	32	6.6	1.4 ms
(30%)5Ti+3Si (60%)Teflon (10%)Al	640	13.9	80 µs
(75%)5Ti+3Si (20%)Teflon(35µm) (5%)Al	3812	58.7	65 µs
(75)%Ti (20)%Teflon(35µm) (5%)Al	3340	40.6	70 µs

* Percentages are by mass.

was not totally consumed and there was no significant change in particle size detected. However, the recovered products of the impacted mixture containing 75% 5Ti + 3Si, 20% Teflon, and 5% Al were TiC, AlF_3, and Ti_5Si_3 . The Teflon was consumed and not present in the recovered sample. Therefore, it appears that it is possible to make the product Ti_5Si_3 under impact conditions with the proper percentages of additives.

CONCLUSIONS

The basic understanding of the initiation mechanism is ultimately crucial to the application and subsequent formulation of these and future materials. Under impact conditions discussed in this paper, the chemical reaction $5Ti + 3Si \rightarrow Ti_5Si_3$ did not occur under plastic flow conditions alone. However, when additives, aluminum and Teflon, are included in the compound, chemical reactions occur. Therefore, it appears that the interactions with the additives generate sufficient heat to allow for the $5Ti + 3Si \rightarrow Ti_5Si_3$ reaction to occur at low velocity impacts as evidenced by the results from the recovery studies. It was also shown that varying the percentages of the materials in the composites could determine whether a material is more likely to produce a sustained burning reaction. The exact chemical reaction mechanism that facilitates the formation of Ti_5Si_3

is still under investigation. The current effort is to understand SHS reactions comparing the temperature that can be produced under impact conditions with the temperature necessary for the reacton $5Ti + 3Si \rightarrow Ti_5Si_3$ to occur.

ACKNOWLEDGMENTS

This research was performed under the sponsorship of the 6.1 Independent Research program at the Naval Surface Warfare Center, Silver Spring, Md.

REFERENCES

1. Coffey, C. S., Devost, V. F.and Woody, D. L.,"Towards Developing the Capability to predict the Hazard Response of Energetic Materials Subjected to Impact," presented at the Ninth International Symposium on Detonation, Portland, OR, Sept. 1989.

2. Thadhani, N. N., Dunbar, E., and Graham, R. A., "Characteristics of shock-compressed configuration of Ti and Si powder mixtures", Proceedings of the Joint International Association for Research and Advancement of High Pressure Science and Technology and American Physical Society Topical Group on Shock Compression of Condensed Matter Conference, Colorado Springs, Colorado, June 28-July 2, 1993.

3. Dunbar, E., Graham, R. A., Holman, G. T., Anderson, M. U., and Thadhani, N. N., "Time-Resolved Pressure Measurements In Chemically Reacting Powder Mixtures,"in Proceedings of the joint International Association for Research and Advancement of High Pressure Science and Technology and American Physical Society

Topical Group on Shock Compression of Condensed Matter Conference, Colorado Springs, Colorado, June 28-July 2, 1993.
4. Krueger, B. R., Mutz, A.H., and Vreeland T., Metallurgical Transactions A, Vol. 23A, (1992).
5. Woody, D. L., Davis, J. J., and Miller, P. J., "Impact Induced Solid State Metal/Metal Reactions", in Proceedings of the Conference on JANNAF Hazards Meeting, San Diego, California, 1994.

SHOCK COMPRESSION OF ZnO+α-Fe$_2$O$_3$ POWDER MIXTURE

Hongliang He[a] Shangchun Shi[a] Hua Tan[a] Kang Xu[b]

a). *Laboratory for Shock Wave and Detonation Physics Research, Southwest Institute of Fluid Physics, P.O.Box 523, Chengdu, Sichuan 610003, P.R.China*

b).*Lanzhou Institute of Chemical Physics, Chinese Academy of Sciences, Lanzhou 730000, P.R.China*

Hugoniot data were experimentally measured for the powders of ZnO, α-Fe$_2$O$_3$, and ZnO+α-Fe$_2$O$_3$ mixture in the pressure range of 5 ~ 40GPa. Shock compressive characteristic of ZnO+α-Fe$_2$O$_3$ mixture was analyzed based upon the experimental results and a mixture model.

INTRODUCTION

To develop the ability to control and predict the chemical reactions in a shock wave process, theoretical model and numerical simulation are basically needed, for example Horie's VIR model[1]. Post-shock analysis, wave profile and shock Hugoniot data measurements are the three usual methods for experimentally study of shock-induced chemistry . Such results are also the basis for the construction of a realistic model. The reaction of mixed powders of zinc oxide (ZnO) and hematite (α-Fe$_2$O$_3$) to form a zinc ferrite (ZnFe$_2$O$_4$) is a particularly ideal system for the study of solid state reactions under shock wave loading. Graham and his coworkers have studied this system and the properties of the product in detail[2]. Herein we report an experimental measurement of Hugoniot data for the powders of ZnO, α-Fe$_2$O$_3$, and ZnO+α-Fe$_2$O$_3$ mixture.

EXPERIMENT

ZnO and α-Fe$_2$O$_3$ powders with analytically pure, and particle sizes -100 mesh were used. The mixture of ZnO + α-Fe$_2$O$_3$ was prepared by mechanical blending in 1:1 molar ratio. For shock impact measurement, powders were compacted into disc shape with densities of about 60% of crystal density, and sizes of about 6.7mm in diameter and 2.5mm in thickness. A schematic of the experimental set-up is shown in Fig.1, which is based on the shock impedance match principle. Three samples are placed at the back surface of the driver plate at 120 degrees, and both the impactor and driver plate are Ly12 aluminium. Two basic parameters have been measured during the shock loading process, (1). impact velocity; (2). shock wave velocity in the sample. Experiments were performed on a two stage light gas gun (23mm diameter), the impact velocities were measured to an accuracy of 0.1%[3]. Electrically charged pins were used to measure the shock wave velocity in the sample. For each sample, three pins (group 1) were placed at the back surface of the driver plate (i.e. the front surface of the sample), and one pin (group 2) at the back surface of the sample (see Fig.1). Shock wave velocity was then calculated from the transit time of the wave between the two group pins and the known length of the sample. Transit time was measured to ± 2ns, and sample thickness to ±0.001mm.

TABLE 1.   A Summary of The Experimental Measurements

Shot No.	Impact velocity (km/s)	Initial density (Mg/m^3)	Shock wave velocity (km/s)	Particle velocity (km/s)	Pressure (GPa)	Relative volume V/Voo
ZnO + α -Fe$_2$O$_3$, about 57.5% of crystal density						
9401	2.711	3.13	3.535	1.699	18.82	0.5196
9402	3.420	3.13	4.240	2.053	27.24	0.5158
9403	4.195	3.13	5.076	2.412	38.31	0.5249
9404	1.846	3.13	2.634	1.249	10.26	0.5273
9405	1.866	3.14	2.763	1.254	10.87	0.5462
9407	1.115	3.15	1.951	0.805	4.95	0.5871
9408	1.290	3.15	2.038	0.924	5.93	0.5467
9410	2.712	3.13	2.995	1.418	13.29	0.5264
ZnO, about 59.8% of crystal density						
9410	2.172	3.41	2.736	1.421	13.25	0.4808
9411	0.980	3.41	1.527	0.736	3.84	0.5179
9412	2.812	3.41	3.297	1.758	19.77	0.4667
9413	4.863	3.43	5.067	2.754	47.87	0.4565
α -Fe$_2$O$_3$, about 61.5% of crystal density						
9410	2.172	3.23	3.320	1.358	14.56	0.5911
9411	0.980	3.19	2.062	0.693	4.56	0.6638
9412	2.812	3.19	3.970	1.688	21.38	0.5747
9413	4.863	3.19	5.664	2.715	49.06	0.5206

Note: In all shots, both impactor and driver plate materials are Ly12 Al (similar to 2024 Al), its Hugoniot data is known from Ref.4.

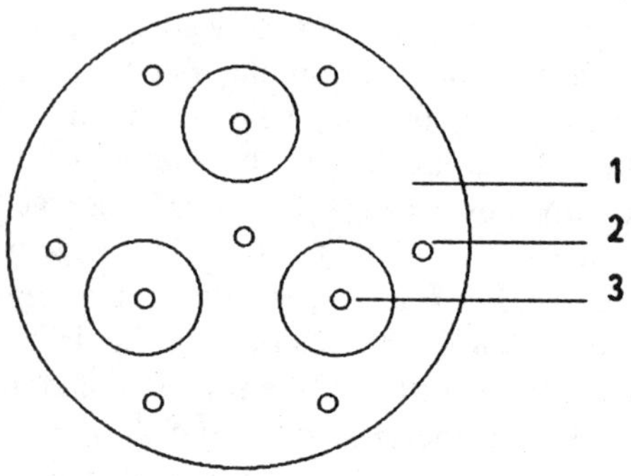

RESULTS AND DISCUSSION

Hugoniot data were measured in the pressure range of 5 ~ 40GPa. Table 1 is a list of the shock loading condition and preliminary results.

Relation of Shock Wave Velocity and Particle Velocity

Particle velocities are calculated from the impedance match solution. Figure 2 is a schematic of the shock wave velocity (D) versus particle velocity (Up). Results are fitted by the least square method, and expressed as $D = c_0 + \lambda\, Up$ (km/s) as follows:

$$ZnO :\quad D = 0.235 + 1.753\, Up \quad (1)$$
$$\alpha\text{-Fe}_2O_3 :\quad D = 0.886 + 1.777\, Up \quad (2)$$
$$ZnO + \alpha\text{-Fe}_2O_3 :\quad D = 0.262 + 1.959\, Up \quad (3)$$

FIGURE 1.   A schematic of the experimental set-up, 1--- driver plate (aluminium), 2--- electrically charged pin, 3--- sample.

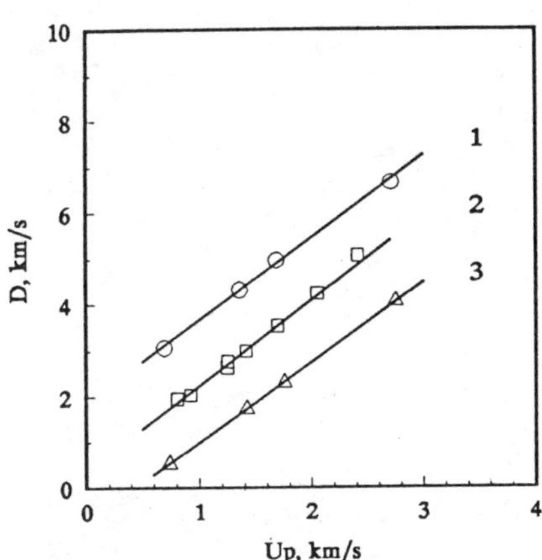

FIGURE 2. Relation of shock wave velocity and particle velocity, 1--- α-Fe2O3 (shift up 1km/s), 2---ZnO+ α-Fe2O3 , 3---ZnO (shift down 1km/s).

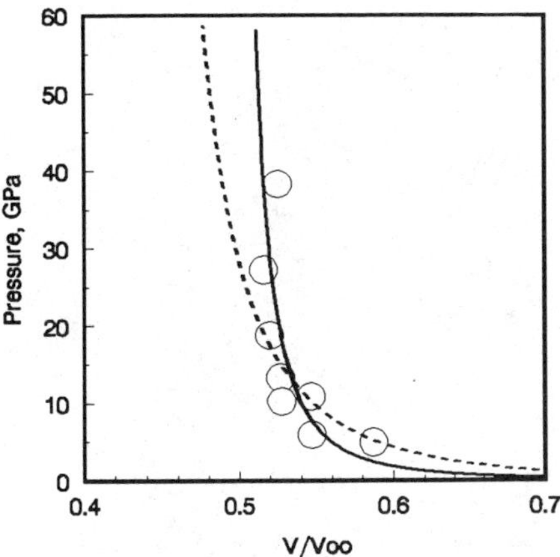

FIGURE 3. A diagram of pressure versus relative volume for ZnO + α-Fe2O3 mixture, open circle points are experimental data, solid line is fitted with Eq.(3), dashed line is calculated from mixture model.

Compressive Characteristic of ZnO+ α-Fe$_2$O$_3$ Mixture

Figure 3 is the result of pressure versus relative volume for ZnO + α-Fe$_2$O$_3$ mixture. Open circle points are experimental data, and solid line is calculated from Eq.(3). Graham and his coworkers[2] have studied the shock-synthesis of $ZnFe_3O_4$ in the pressure range of 7.5 ~ 27GPa with the shock recovery technique, and they reported that the product yield varied from 1% to 75%, correspondingly. Therefore, Eq.(3) describes a shock-induced chemistry process in ZnO + α-Fe$_2$O$_3$ mixture.

A mixture model is assumed to construct an un-reacted Hugoniot curve for ZnO+ α-Fe$_2$O$_3$ mixture from the available ZnO and α-Fe$_2$O$_3$ data. The theoretical basis of the model is the volume additivity at constant pressure[5].

Define $\quad P_m = P_1 = P_2$

$$V_m = k_1V_1 + k_2V_2 \qquad (4)$$

where P, V are pressure and specific volume, k is mass ratio of the component. Subscripts *1,2,m* refer to ZnO, α-Fe$_2$O$_3$, and ZnO+ α-Fe$_2$O$_3$ mixture, respectively. In this study, ZnO and α-Fe$_2$O$_3$ are mixed in 1:1 molar ratio, therefore k_1=0.3378, k_2=0.6622. Ignore the difference of porosity between ZnO, α-Fe$_2$O$_3$, and ZnO+ α-Fe$_2$O$_3$ mixture powders, and use Eqs.(1) and (2), a theoretical mixture Hugoniot curve is calculated,

and it is shown in Fig.3 as a dashed line. It is found that the theoretical line lies very closely to the experimental line (i.e. the solid line in Fig.3). This behavior could be a special feature of the ZnO+ α-Fe$_2$O$_3$ system. For this reaction the system has little or no reaction heat, and has little specific volume difference between the reactants and the product.

ACKNOWLEDGMENTS

This study is supported by the Chinese National Science Foundation, contract number 29273144, and Science Foundation of CAEP, contract number 9301005.

REFERENCES

1. Hwang, M., Horie, Y., and You, S., "Modeling of Shock - Induced Chemical Reactions in Powder Mixtures I: the VIR model," in *Shock Compression of Condensed Matter 1991* (S.C.Schmidt, R.D.Dick, J.W.Forbes, and D.G.Tasker, Ed.), Elsevier Science Publishers B.V., 1992, pp.597-600.

2. Graham, R.A., *Solids Under High-Pressure Shock Compression*, Springer-Verlag, 1992, pp.180-184.

3. Shi, S.C., Chen, P.S., Huang, Y., *Chinese J. High Pressure Physics* 5, 205-214(1991).

4. Marsh, S.P., *LASL Shock Hugoniot Data*, University of California, Berkeley, CA,1980.

5. McQueen, R.G., Marsh, S.P., Taylor, J.W., Fritz, J.N., and Carter, W.J., *High Velocity Impact Phenomena* (R.Kinslow, Ed.), Academic, New York, 1970, pp.293-417.

SHOCK-WAVE INDUCED FORMATION OF "DIAPLECTIC" GLASSES OF MULLITE 2Al₂O₃:1SiO₂ COMPOSITION: A TEM STUDY

W. Braue*, H. Schneider*, U. Hornemann**

(*) German Aerospace Research Establishment (DLR), Materials Research Institute,
D-51147 Cologne, Germany
(**) Fraunhofer Institute For High-Speed Dynamics, Ernst-Mach-Institute,
D-79576 Weil/Rhein, Germany

For the first time, shock-wave induced intragranular glass lamellae similar to the so-called "diaplectic glasses" reported from shock-loaded framework silicates such as SiO_2-polymorphs and feldspars have been observed by TEM for mullite $2Al_2O_3{:}1SiO_2$ polycrystals after exposure to peak pressures up to 40 GPa. A polyphase "fused-mullite" brick consisting of lath-shaped mullite and corundum grains among other constituents has been employed in this study thus offering direct comparison between the well-established shock deformation phenomena of corundum and those for mullite within the same specimen.

INTRODUCTION

Mullite, $Al_2^{VI}(Al_{2+2x}Si_{2-2x})^{IV}O_{10-x}$, has an oxygen-deficient $(0.17 < x < 0.50)$ orthorhombic structure with straight chains of edge-sharing (AlO_6) octahedra extending parallel to the c-axis which are cross-linked by doublets of $(Si/Al)O_4$ tetrahedra. With the bulk composition ranging from $3Al_2O_3{:}2SiO_2$ $(x = 0.25)$ to $2Al_2O_3{:}1SiO_2$ $(x = 0.40)$, mullite is the only stable crystalline phase in the Al_2O_3-SiO_2 system at normal conditions. Plastic deformation of mullite has been investigated so far in the high-temperature creep regime only, revealing grain-boundary sliding and/or cavitation as the major deformation mechanisms (1). Little research has yet been focused on the structural response of mullite with respect to high strain rates.

EXPERIMENTAL AND RESULTS

Due to the lack of suitable mullite single crystals, a coarse-grained 2:1 "fused-mullite" refractory brick consisting of lath-shaped mullite crystals sandwiched between corundum, a continuous silica-rich amorphous phase and interdispersed m-ZrO_2 particles (Fig. 1) was employed in this study.

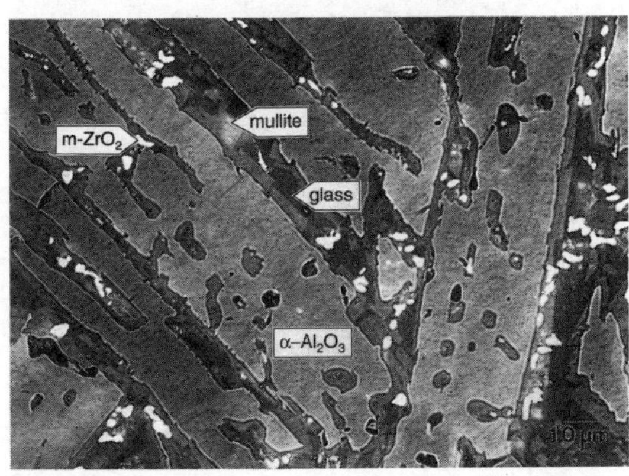

FIGURE 1. Optical micrograph of "fused-mullite" refractory brick employed in this study.

The shock loading experiments were carried out by planar impact techniques using a high explosive device (2). The mullite samples, 0.5 mm thick and 16 mm in diameter, were embedded in a cylindrical ARMCO-iron container surrounded by thick steel blocks and plates to suppress the rarefaction waves (3). The test setup is shown in Fig. 2. The peak pressures of the shock waves transmitted into the samples could be varied by using different high explosives and by varying the thickness of the flyer plate and the cover layer of the container.

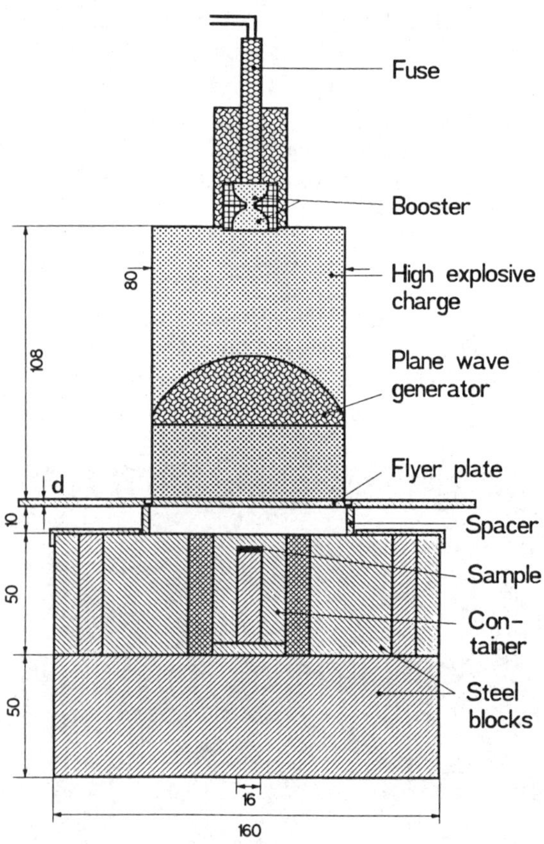

face. Peak pressures between 20 and 85 GPa can be reached. The pressure in the mullite sample was determined by impedance matching in a multiple reflection mode using the Hugoniot data of ARMCO-iron and mullite. The advantage of this method is that it is not necessary to know the exact mullite Hugoniot curve which has not been measured up to now. But thin samples had to be used. After shock loading the steel container is slightly deformed at the impact surface and the sample can be recovered completely in the original position relative to the shock wave direction.

Fragments of the specimens recovered from the shock experiments were thinned by dimpling and argon ion-beam milling and investigated in a Philips EM 430 transmission electron microscope operating at 300 kV. The surprising result of this research was the occurrence of shock-wave induced thin intragranular glass lamellae of perfect 2:1 mullite-normative composition which creates a continuous network parallel to well-defined crystallographic directions in the mullite grains, as shown in Fig. 3.

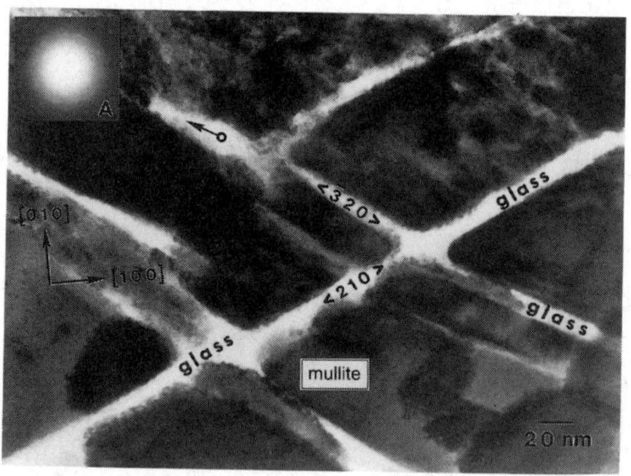

FIGURE 3. HREM image of 2:1 mullite grain (oriented // to [001]), displaying the extension of thin intragranular glass lamellae parallel to the < 210 > and < $\bar{3}$20 > directions.

The peak pressures were determined by separate calibration tests using the pin contactor technique. This technique measures the free surface velocity at the upper container-sample inter-

Their characteristics are similar to the "diaplectic glasses" reported for shock-loaded framework silicates such as quartz and feldspar (see ref. 4 for review) and it is the first time they have been described from dynamic shock experiments involving a chain silicate. The second deformation mechanism observed simultaneously (Fig. 4) involves the local decomposition of individual mullite grains to form an alumina phase (predominantly corundum) and non-crystalline silica (as determined by small-probe microanalysis (EDS, not shown)), which is actually enhanced by the effects of high pressures due to the decrease in molar volume involved.

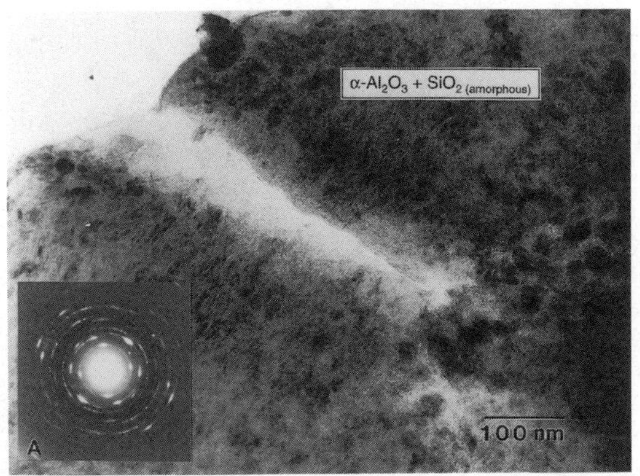

FIGURE 4. Decomposition of individual mullite grain to form corundum (see SAD) and non-crystalline silica.

This transaction preserves the shape of the parental crystal and apparently does not follow a topotactical orientation relationship. The shock-loaded alumina grains coexisting with mullite in the same specimens exhibit the typical shock lamellae consisting of basal twins and dislocation activity on the basal planes (Fig. 5), as reported elsewhere (5).

The complete transition from previously undeformed mullite grains into extended intergranular mullite-normative glass aggregates was observed in the 40 GPa shock experiment. Contrary to the homogeneous thin glass lamellae, these large areas of mullite-normative glasses tend to partially devitrify to form 2:1 mullite microcrystals (Figs. 6, 7) by taking advantage of minor constituents (e.g. m-ZrO₂) as heterogeneous nucleation sites. These microcrystals have been formed upon cool-down

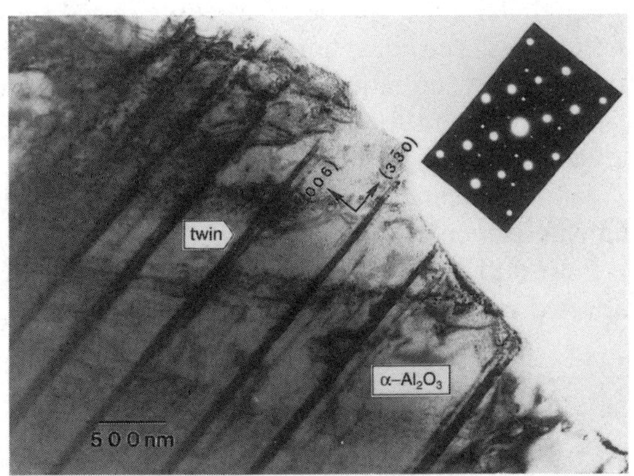

FIGURE 5. Shock-loaded corundum grain oriented parallel to [11$\bar{2}$0], emphasizing the formation of thin twin bands lying on basal planes.

after the pressure has been released from the sample. They exhibit a needle-like habit parallel [001], similar to those precipitated from alumina-rich SiO₂-Al₂O₃ melts (6).

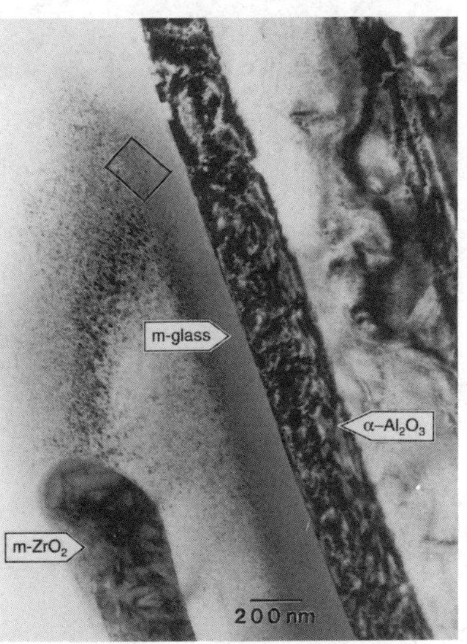

FIGURE 6. Extended area of shock-wave induced mullite glass which had replaced a former undeformed mullite crystal adjacent to a corundum grain. For details in boxed region see Fig. 7.

These preliminary results suggest that the thin intergranular glass lamellae define the initial stage of shock-induced glass formation in mullite poly-crystals which culminates in the "amorphization" of more extended mullite areas as the shock pressure is increased.

Future research will focus on the structural characterization of mullite normative glasses based on both diffraction and spectroscopic information collected with a small electron probe.

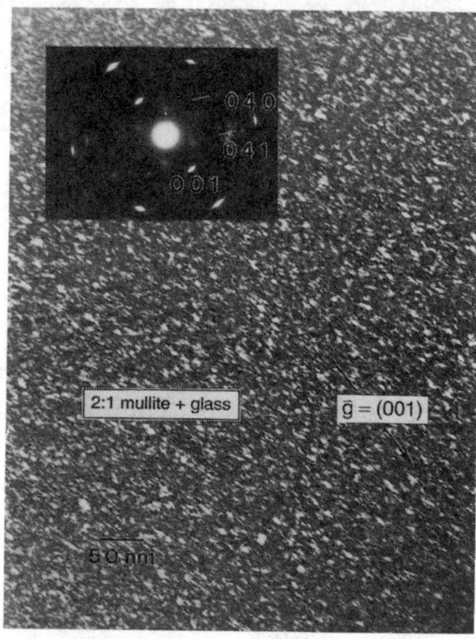

FIGURE 7. Partial crystallization of 2:1 mullite microcrystals from shock-induced mullite glass.

REFERENCES

1. Okamoto, Y. et al. J. *Europ. Ceram. Soc.* **6** 161-168 (1990).

2. Müller, W. F. and Hornemann, U., *Earth and Planetary Science Letters* **7**, 251-264 (1969).

3. Doran, D. G. and Linde, R. K., *Solid State Physics* **19**, New York: Academic Press, 1966, pp. 229-290.

4. Stoeffler, D., *J. Non-Cryst. Solids* **67**, 465-502 (1984).

5. Wang, Y. and Mikkola, D., J. *Am. Ceram. Soc.* **75**, 3252-3256 (1992).

6. Risbud, S. H. and Pask, J., *J. Am. Ceram. Soc.* **61**, 63-67 (1978).

EFFECT OF SHOCK WAVE ACTIVATION ON NITRIDING OF Sm$_2$Fe$_{17}$

Juxian Gao, Ronging Deng, Xiaohong Xu

Institute of Mechanics, Academia Sinica Beijing 100080, China

Xiaolei Rao

Institute of Physics, Academia Sinica Beijing 100080, China

Shock wave activation treatment for Sm$_2$Fe$_{17}$ under explosive loading (pressure 1.4~4 GPa) has been studied. After shock treatment the dislocation density in Sm$_2$Fe$_{17}$ was strikingly increasing, and the nitrogen absorption capacity of Sm$_2$Fe$_{17}$ heightens obviously no matter the nitriding condition is continuous rise temperature, isothermal or high pressure. Especially it can effectively restrain disproportionation of Sm$_2$Fe$_{17}$N$_y$. It makes Curie temperature slightly increase.

1. INTRODUCTION

Application of shock activation in materials science and industrial fabrication is very attractive research subject[1]. A potential application of shock activation is for Sm$_2$Fe$_{17}$N$_y$. Recently, Sm$_2$Fe$_{17}$N$_y$ has been discovered to have excellent magnetic performance such as Curie temperature T$_c$ = 470℃ , a large uniaxial anisotropy field (B$_a$ = 14T) and a respectable spontaneous magnetization (μ_0M$_s$ = 1.54T) . The theoretical upper limit on the attainable energy product is 470KJ / m^3 (59MGOe)[2]. In addition, a substantial progress has already been reported in developing coercivity in nanocrystalline powders[3]. But for the development and application of new family of rare earth iron nitrides there are still many important subjects needed to be solved, for example, the best magnetic performance of Sm$_2$Fe$_{17}$N$_y$ can be obtained when how many does y equal to; it decomposed at elevate temperature, so it can not be sintered by normal method; the disproportionation exist during the nitriding process; a long−term stability of magnetic performance in operation needs to be determined, etc. Recent research result[4] shows Sm$_2$Fe$_{17}$N$_y$ can be consolidated by explosive consolidation process. It proved that the explosive consolidation process does not change the preformed orientation of grain, initial morphology and intrinsic magnetic properties, and the rectangle of demagnetization curve is improved to be compared with glued magnet of same powder so the maximum energy product (BH)$_{max}$ is obviously increased. Therefore, magnets of Sm−Fe−N should have the potential to compete favorably with the well−established Nd−Fe−B magnets, if a processing route can be established, which is compatible with the metallurgical and thermodynamic features of the nitride system.

In this work, we have attempted to study the effect of shock activation on nitriding of Sm$_2$Fe$_{17}$.

2. EXPERIMENTAL PROCEDURES

The explosive implosive shock wave was produced by slide detonation of a cylindrical charge[5]. The mixed explosive of RDX$^{\#}$ (cyclonite) and AN$^{\#}$ (ammonium nitrate fuel mixture) in different proportions were used in our experiments. The detonation velocity of the mixed explosives was measured in the experiment, and based on this the detonation parameters[6] were calculated.

In order to study the microstructure a cylindrical bulk of Sm_2Fe_{17} was adopted instead of powder of Sm_2Fe_{17} to put into a steel capsule. The pressure distribution along the radius of cylindrical bulk was estimated to be about 1.4GPa~4GPa. Before and after explosive action some samples were taken from the bulk of Sm_2Fe_{17} to be examined by micrograph, SEM, TEM and XRD. Then the bulk Sm_2Fe_{17} was smashed into powder. A special phenomenon deserved extra attention, that is the smashing time of the bulk after shock action was obviously shortened by contrast to that of initial bulk. The both differ by ten times.

Nitriding experiments of Sm_2Fe_{17} powder before and after shock action were respectively completed under three different conditions for the sake of contrast. They are continuous rising temperature nitriding, isothermal nitriding and high−pressure nitriding. The first two nitriding experiments were conducted by means of the thermopiezic analyser. The isothermal nitriding process at 495℃ . The high pressure nitriding was in a high pressure furnace with high pure nitrogen at 0.5 MPa and 490℃ to keep 3hr. The absorbing nitrogen was determined by quantitative analysis.

3. EXPERIMENTAL RESULT

Figure 1 shows nitrogen absorption characteristics, upper line for shock activation Sm_2Fe_{17} powder (abbreviation: treatment powder) and lower line for un−shock−activation Sm_2Fe_{17} powder (abbreviation: untreatment powder). The solid line represents predetermined rising temperature line. The dotted line indicates measured rising temperature line. Table I gives experimental results about the effect of shock activation on nitriding of Sm_2Fe_{17} under three different conditions. The data shows under three different condition the nitrogen absorption of shock−treatment Sm_2Fe_{17} all are higher than that of untreatment Sm_2Fe_{17}. This indicates the shock activation can increase nitrogen absorption of Sm_2Fe_{17}, especially under continuous rising temperature the increment of nitrogen absorption is the maximum , but the y value is not up to 3. The pressure increasing is not beneficial to nitrogen

absorption. For the sake of contrast X−ray diffraction patterns of nitriding shock−treatment powder and of nitriding untreatment powder was put in one figure. Fig. 2 gives the main portion of their X−ray diffraction patterns. The contrast shows shock treatment can restrain the disproportionation of $Sm_2Fe_{17}N_y$, a−Fe in nitriding untreatment powder is obviously higher than that in nitriding shock treatment powder under the same comparing condition, a−Fe exists, which shows decomposition of $Sm_2Fe_{17}N_y$ occur. This is obviously not hoped occurrence.

Fig. 3 is in illustration of microstructure change of Sm_2Fe_{17} before and after shock action. TEM micrograph shows typical features of dislocation distribution found in untreatment Sm_2Fe_{17} (left) and in shock treatment Sm_2Fe_{17} (right). It can be seen that in untreatment Sm_2Fe_{17} there are only a few dislocations, but in shock treatment Sm_2Fe_{17} a lot of dislocations exist and cross each other. This shows shock wave action makes defect strikingly increase.

Both shock treatment Sm_2Fe_{17} and untreatment Sm_2Fe_{17} were smashed in the same

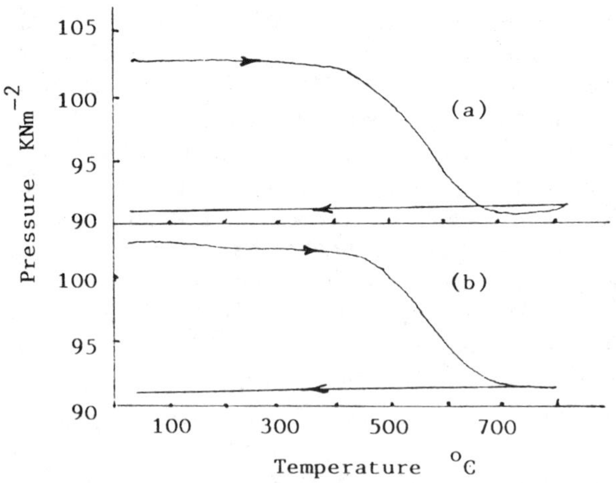

FIGURE 1. Nitrogen absorption characteristics for Sm_2Fe_{17} powders heated in the thermopiezic analyser in ~0.1MPa of N_2 (a) for shock activation Sm_2Fe_{17} powder, (b) for un−shock−activation powder

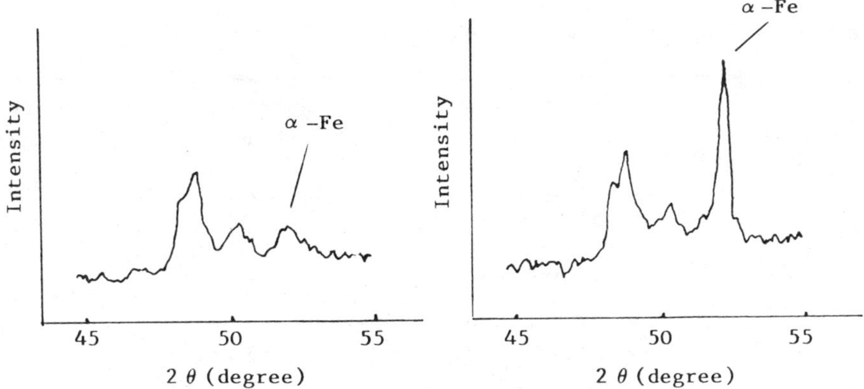

FIGURE 2. The contrast of main portion of X–ray diffraction patterns for nitriding shock treatment powder (left) and for nitriding untreatment powder (right)

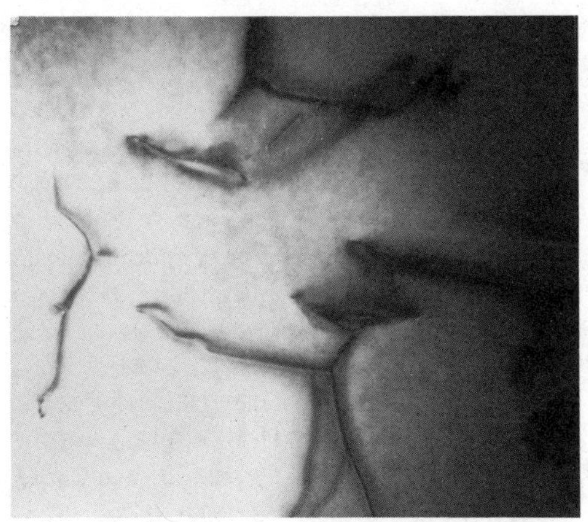

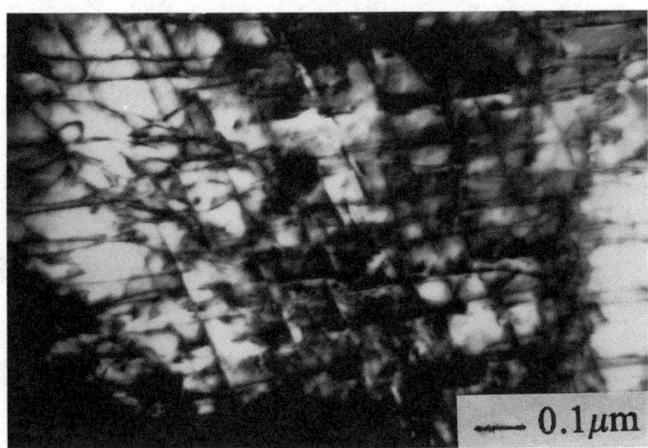

FIGURE 3. TEM bright field micrograph showing typical features of dislocation found in untreatment Sm_2Fe_{17} (left) and in shock treatment Sm_2Fe_{17} (right)

time, so the powder size of shock treatment is finer than that of untreatment. Several magnetic performance for the two powders were measured. Curie temperature of shock treatment Sm_2Fe_{17} is slightly higher than that of untreatment Sm_2Fe_{17}. Other magnetic performance do not have obvious change.

4. RESULT AND DISCUSSION

The effect of shock wave activation on nitriding of Sm_2Fe_{17} is obvious. It makes nitrogen absorption increase, no matter the nitriding condition is continuous rising temperature, isothermal or high pressure. It can restrain the disproportionation of $Sm_2Fe_{17}N_y$ and increase stability of $Sm_2Fe_{17}N_y$. It makes Curie temperature slightly increase. The microstructures observed by TEM show it makes defects strikingly increase. The density and distribution of dislocation in shock treatment Sm_2Fe_{17} is obviously different from that in untreatment Sm_2Fe_{17}. The shock wave activation is evidently related to

TABLE I. Experimental results of nitriding of Sm_2Fe_{17}

	Continuous rising temperature nitriding	Isothermal nitriding	High pressure nitriding
Experimental condition	Continuous rising temperature $\sim 0.1MPa$ of N_2	$T = 768K$ $P \sim 0.1MPa$ of N_2 3hr	$P = 0.5MPa$ $T = 763K$, 3hr
Untreatment Shock treatment	$y = 2.60$ $y = 2.92$	$y = 2.36$ $y = 2.57$	$y = 2.27$ $y = 2.44$
Nitrogen absorption increasing	12.3%	8.9%	7.5%

Note: Here y represents an amount of absorbed nitrogen atom in the structure of $Sm_2Fe_{17}N_y$ compound.

microstructure change. It is well known that the most important obstruction for application of $Sm_2Fe_{17}N_y$ is its stability (high temperature and a long period of time). The results of this study show the shock wave activation may help solve this problem.

Because smelting pure Sm_2Fe_{17} is very difficult, it always accompanies with a few a–Fe and rich samarium phase. As relative contrast the results can still improve some problems. If mixture of NH_3 and H_2 was used instead of N_2, a better result may be obtained.

ACKNOWLEDGEMENT

Support for this research was provided by the National Advance Materials Committee of China and the China National Science Foundation Committee.

REFERENCE

1. Graham, R.A., edited by Schmidt, S.C., and Holmes, N.C., "Shock compression of solids as a physical–chemical–mechanical process", *in Shock Waves in Condensed Matter 1987*, North–Holland, New York 1987, 11.

2. Coey, J.M.D., Hong Sun and Yoshichika Otani, "A new family of rare earth iron nitrides", *in Proceedings of the Sixth International Symposium on Magnetic Anisotropy and Coercivity in Rare Earth Transition Metal Alloys* (Carregie Mellon University, Mellon Institute, Pittsburgh, PA15213, USA, 1990) pp.36.

3. Schnitzke, K., Schultz, L., Wecker, J., and Katter, M., *Appl. Phys. Lett.* **57(26)** 2853 (1990).

4. Zheng, D., Cai, M., Li, D., Hu, B., Rao, Liu, G., Xu, J., and Dong, X., "A new technique of consolidation of $Sm_2Fe_{17}N_y$ magnet" *in Proceeding on the 8th National Magnetics and Magnetic Materials in China* (1993) 283(in Chinese).

5. Gao Juxian, Shao, B., and Zheng, K., *J. Appl. Phys.***69(11)**, 7547(1991).

6. Gao Juxian, Zheng, K., and Zheng, Z., *Acta Metall Sinica,***4B(3)**, 234(1991).

SHOCK-INDUCED DEFECTS IN HgO

B. Morosin, E. L. Venturini, G. T. Holman,
P. N. Newcomer, R. G. Dunn, and R. A. Graham

Sandia National Laboratories, Albuquerque, NM 87185-1421

Powder compacts of HgO have been subjected to shock-loading and preserved for post-shock analysis to understand its reactivity and stability under transient temperature-pressure excursions. Recovered samples indicate several solid state reactions which are dependent on shock conditions. Metallic Hg is recovered in small amounts in the HgO compact as well as an as-yet unidentified ferromagnetic impurity not present in the as-received HgO powder. Further, there is evidence of reaction with the copper capsule at its interface with the HgO powder, forming an intermetallic alloy.

INTRODUCTION

Our synthesis experiments on Hg-containing superconductors showed HgO, known to decompose near 500° C, could be subjected, in the presence of other oxides, to rapid temperature excursions above 800° C without significant HgO decomposition (1). Prior to employing a shock-induced chemical synthesis route similar to our previous work (2,3) for the formation of Hg-containing superconductors (for which some phases are formed only under high static pressure environments), the properties of HgO under similar conditions were considered desirable. An earlier shock compression study employed water or oil suspensions of HgO to reduce post-shock annealing effects and produced a brown material with corresponding shifts in the ir absorption bands and x-ray absorption spectrum, but not in the x-ray powder pattern (4). On some of this brown material, shoulders on neutron diffraction peaks were interpreted as possible evidence for a cubic sphalerite (a = 5.43 Å) HgO form (5). This note reports on powder compacts of HgO with no additives subjected to a variety of shock conditions considered appropriate for such superconductor synthesis.

EXPERIMENTAL

Commercially available HgO from Alfa/Aesar (reagent grade, stock #12276, lot #b15e10) was employed "as-received" in this study under careful conditions both in preparing and loading the powder compacts and in the subsequent analysis because of environmental concerns. This source of material was shown by x-ray diffraction to be the usual orthorhombic form of HgO (Space group Pnma with cell edges 6.6121, 5.5201, and 3.5213 Å). Another form isomorphic with cinnabar, HgS, is of hexagonal symmetry with a = 3.577 and c = 8.681 Å. The orthorhombic form has infinitely long zigzag but planar chains of mercury and oxygen atoms extending along the a-axis while the hexagonal form is a distortion of the NaCl arrangement and is very slightly more dense with 32.06 Å^3/molecule (verses 32.13 Å^3/molecule for the orthorhombic form). The powder is orange in color. The X-ray line profiles of this powder were shown to be slightly broadened, consistent with a crystallite domain size near 1500 Å.

The pressed (to ~55% theoretical density) compacts were subjected to controlled shock compression conditions to peak pressures in the

Sandia Bear recovery fixtures described and characterized previously (6). The shock conditions used in this study are given in Table 1. Each experiment yielded several samples, denoted as a "center" or "outer", corresponding to the location of the powder compact within the shock recovery fixture. Each sample experienced a different mean bulk temperature, as shown in the table.

X-ray diffraction data were taken using a commercial Rigaku Θ-Θ automated powder diffractometer equipped with a monochrometer on the detector set so both $CuK\alpha_1$ and $CuK\alpha_2$ radiation were obtained.

Transmission electron microscopy (TEM) specimens were prepared by dispersion of shock-loaded as well as the as-received HgO powders onto the holey carbon film of copper grids. A JEOL 1200EX TEM instrument at 120KV was employed. On most samples, both bright field imaging and dark field diffraction were carried out on selected grains.

The diamagnetism characteristic of mercury materials as well as that for superconductivity of trace amounts of metallic Hg were demonstrated by low temperature static magnetization data collected using our commercial SQUID magnetometer (Quantum Design model MPMS).

RESULTS

Experiment LE95-22 ruptured during shock-loading and did not yield any powder compact samples. The recovered sample-copper interface showed the same dull silver layer that the LE95-23 experiment also showed. The recovered powder compacts of the other two shock experiments were separated into four samples, as described below, and together with parts of the interface layer, subjected to x-ray, TEM, and magnetization studies.

The compacts were removed essentially whole, but were somewhat fragile, particularly for the extreme edge of LE95-23 (resulting in a sample designated as "extreme edge") while a thin layer of powder adjacent to the top plug of LE95-50 stuck to the copper fixture plug and was, hence, separated from the compact, gently scraped from the plug, and

TABLE 1. Schedule of Shock Experiments (a)

Experiment	Compact Density	Peak Press.	Mean Bulk Temp. (b)
	Mg/m^3 (%)	GPa	°C
LE95-22 MBA-CB	5.95 (54)	22	400(650)
LE95-23 MB-B	6.03 (54)	7.5	225(300)
LE95-50 PB-B	5.90 (53)	5.0	150(75)

(a) For more detailed descriptions of pressure and temperature, see ref 4. MBA, MB, and PB indicate the Momma Bear A, Momma Bear, and Pappa Bear fixtures while -CB and -B indicate the composition B and baratol explosives, respectively.

(b) The calculated mean bulk temperature in the center bulk and, in parenthesis, the outer edge or "ring" of the samples are shown.

designated as "top plug". The other samples resulted from separating the compact into the usual "center", "bulk", and "edge" samples. The edge samples appeared slightly more reddish-brown than the original orange color while the top plug sample appeared slightly more reddish. The center of the LE95-50 and the extreme edge of the LE95-23 samples were a bit more consolidated than the remainder of the samples and required more gentle grinding so as to pass through a 0.2 mm screen.

X-ray diffraction measurements on a specimen of the dull silver interface cut from the lid of experiment LE95-23 showed that it was the intermetallic compound $Cu_{15}Hg_{11}$. This alloy crystallizes in space group R3m with a = 13.351 and c = 16.175 Å, a rhombohedrally distorted gamma brass structure type. Note that the corresponding rhombohedral cell would be a = 9.4067 Å , α = 90.413° and, hence, this intermetallic has occasionally been reported as cubic. This intermetallic has also been confirmed by EDAX on SEM specimen.

X-ray diffraction showed only the orthorhombic form of HgO with no detectable traces of any other compound. The presence of well dispersed liquid Hg metal droplets, if present (see below) would not be detected at low concentration by the usual x-ray diffraction experiment. There were no discernable differences between either the more reddish top plug or the reddish brown extreme edge and the other samples. The x-ray line profiles were all, but with the exception of the center LE95-50 sample, slightly narrower than the initial as-received or "standard" powder employed in the experiments. The line widths for most of the shock-loaded samples are consistent with those for a well annealed ~200 nm or larger crystalline domain-sized material, essentially those obtained for our Si powder previously run on our diffractometer. The single exception had line widths essentially the same as the standard and of the order of ~140 nm. This prevents any quantitative profile analysis using these samples. It was subsequently concluded that the gentle grinding on the LE95-50 "center" sample was sufficient to reduce the grains (crystallite domain size) because of the van der Waal interactions between the HgO zigzag chains of the crystal structure.

TEM bright field imaging was used to determine the size and microstructure of the shock-loaded and as-received HgO grains. HgO proved to be very sensitive to the electron beam. All of the HgO grains observed break up into a fibrous morphology when exposed to a focused electron beam. Because of this sensitivity, detailed microstructural study is difficult; however, some changes in the shock-loaded samples were observed.

While the as-received grains were typically 100-300nm, 500-1500 nm grains for the edge sample of LE95-23, where the temperature is greatest, and 300-1000 nm for the center sample indicates grain growth, possibly occurring after the shock event. The lower temperature LE95-50 samples showed a smaller amount of grain growth. The edge sample consisted of 200-700 nm grains, with fine grains of fibrous-like clusters, less than 50nm in size found between the larger grains. The center of this sample showed a substantial amount of these fine grained fibrous-like clusters; however, recall that this sample was ground a bit more than the other samples. Occasionally, a second phase was found among these clusters which under the electron beam appeared to move. These were subsequently believed to be small droplets of metallic Hg (see below).

The magnetic properties of the recovered HgO samples were examined at temperatures between 2 and 300K in applied magnetic fields up to 5 tesla. The as-received HgO powder employed in this study contains a significant paramagnetic impurity phase. Pure HgO has a small diamagnetic susceptibility at room temperature, -2.03×10^{-7} cm^3/g. In contrast, the as-received HgO is dominated by the impurity phase. Assuming the paramagnetism arises from a spin = 1/2, g-factor = 2 ion, approximately 0.4% impurity ions per Hg are present.

In contrast, the shock-loaded samples exhibit a large positive moment at 300 K which increases dramatically with peak shock temperature and pressure, reaching $+3.3 \times 10^{-3}$ emu/g for the "outer extreme" LE95-23 sample. All four LE95-23 shock-loaded samples exhibit a roughly linear increase in moment with decreasing temperature, and the slopes are remarkably similar for the different shock conditions. The large positive moment at 300 K is attributed to an unknown ferromagnetic impurity introduced (formed) during shock loading. The nonlinear behavior of the magnetic moment verses applied field at 5 K, after the appropriate corrections for the plastic sample container and the reference as-received HgO powder, is characteristic of "hard" ferromagnetic materials where the moment "saturates" in large applied magnetic fields. The extreme outer LE95-23 has a ferromagnetic "saturation" value of 1.2(1) $\times 10^{-2}$ emu/g. If the unknown ferromagnetic impurity were magnetite, Fe_3O_4, with a saturation moment of 97 emu/g at low temperatures, this sample response would arise from an impurity level of 0.01% by weight. It is conceivable that shock loading transforms about 3% of the paramagnetic impurity in the as-received HgO to a ferromagnetic phase.

The magnetic measurements also showed evidence for a small superconducting component in the shock-loaded HgO powders. In a magnetic field of 0.5 millitesla, sufficiently small so that the contribution for the just mentioned ferromagnetic impurity is negligible, data measured with decreasing temperatures from 5 to 2 K indicated the appearance of a substantial negative magnetic moment below a transition temperature, i.e., the onset of superconductivity. This negative moment arises from the Meissner effect (expulsion of the applied field from the superconducting grains.) All of the four LE95-23 samples show a superconducting transition just below 4.2K, consistent with metallic Hg. Since the as-received HgO powder showed no evidence for superconductivity, the shock-loading converts a small quantity of the HgO to Hg. These data suggest that the amount of metallic Hg increases going from the center, bulk, edge and extreme edge samples, -2.4, -4.0, -5.0 and -31. (all x 10^{-4} emu/g), respectively, at 2 K.

CONCLUDING REMARKS

There is strong evidence that metallic Hg is formed for samples subjected to the shock conditions of experiment LE95-23. This suggests that even under pressure of 7.5 GPa, with a mean bulk temperature rise to ~225 °C, HgO is decomposing into metallic Hg and that the Hg adjacent to the Cu fixture is sufficiently reactive to form the intermetallic compound $Cu_{15}Hg_{11}$. At locations away from the Cu interface, this Hg coalesces into microdroplets.

These microdroplets reside between the grains and were disturbed by the heating effect of the TEM electron beam. The beam also frayed and defoliated the HgO grains, showing the weak van der Waal interaction between the zigzag chains in the structure of this orthorhombic form.

Further, at these low mean bulk shock-induced temperatures, the HgO is able to significantly (x5) enlargen its grain size as evidenced by TEM bright field imaging. The lack of breadth of the x-ray lines beyond that resulting from the usual experimental diffraction optics is additional evidence for annealing at these low temperatures. Together, these results rule out the possibility of a pressure induced reversible phase transition consistent with the small reported density difference of the two forms. The observed color changes without discernible x-ray pattern changes are also consistent with the previous work on HgO (4). This suggests defect, probably vacancy, formation occurs in the HgO lattice. Our lack of diffraction lines for any other phase(s) as well as published details in papers given in ref. 4 contradict the interpretation that a new HgO phase is formed under shock-loading. (5)

Also of significance were the magnetic measurements of the as-yet unidentified impurity formed from the shock environment and possibly involving the metallic Hg present. Such low level detection of phases may prove useful in other shock studies.

On a more general vein, the present materials study contributes towards resolving several shock issues concerning our Bear fixtures and TiO_2 model calculations. Although there remains ambiguities whether events occur at pressure or result from post-shock effects, the major shock effects are consistent with previous studies. That is, there is no radial focusing in the LE95-50 experiment, leading to a more uniform pressure over the entire sample in contrast to LE95-23.

ACKNOWLEDGMENTS

The assistance of Marvin Banks at the NM Tech Center for Explosives Technology Research, Socorro is gratefully acknowledged. This work supported by US DOE contract # DE-AC04-94AL85000.

REFERENCES

1. Morosin, B., Venturini, E.L., Schirber, J.E., and Newcomer, P.P., *PhysicaC* **226,** 175-183 (1994).
2. Morosin, B., Graham, R.A., Venturini, E.L., Ginley, D.S., and Hammetter, W.F., "Shock-Induced Chemical Synthesis of Phases Similar to the High Temperature Superconductor Oxides," *Proceedings of the Conference on Shock Waves in Condensed Matter, 1987,* pp 439-442.
3. Morosin, B., Graham, R.A., Venturini, E.L., and Ginley, D.S., *Synthetic Metals* **33,** 175-183 (1989).
4. Batsanov, S. S., *Effects of Explosions on Materials- Modification and Synthesis Under High-Pressure Shock Compression,* New York: Springer-Verslogen, 1994, 108.
5. Ovsyannikova, I. A., Moroz, E. M., and Platkov, A. I., *J. Eng. Phys.* **22,** 511-513 (1972).
6. Graham, R.A., *Solids Under High Pressure Shock Compression- Mechanics, Physics, and Chemistry,* New York: Springer-Verslogen, 1993, 151-159.

SHOCK-SYNTHESIS OF A METASTABLE TETRAGONAL PHASE OF EuBa$_2$Cu$_3$O$_y$

Y. Syono, H. Hikosaka, M. Kikuchi and K. Fukuoka

Institute for Materials Research, Tohoku University, Katahira, Aoba-ku, Sendai 980-77, Japan

Shock-synthesized EuBa$_2$Cu$_3$O$_y$ was found to be a metastable tetragonal phase with $c/a = 3.00$, which is only obtainable at relatively low temperatures by conventional solid state reactions. The annealing of the shock-synthesized phase above 900°C and subsequent oxidation lead to superconduction together with the increase in the c/a ratio to 3.05, presumably due to formation of perfect ordering of Eu and Ba atoms. These results suggest that no complete ordering between Eu and Ba, hence incomplete formation of CuO$_2$ plane necessary for superconduction, would be established during a short time interval of shock synthesis process.

INTRODUCTION

Extraordinarily rapid chemical reactions of powder materials in the shock process have attracted much attention of researchers and is now one of the current issues in shock wave sciences.(1) Enormously enhanced solid state reactivity under shock loading is believed to be due to the introduction of large concentration of defects. The highly nonequilibrium shock process may provide an opportunity to get a metastable phase which is hardly obtainable in other ordinary processes. Since the first discovery of high T_c oxides, many attempts have been made to synthesize cuprate superconductors by shock compression method. Noteworthy is that (La, Sr)$_2$CuO$_4$ was successfully synthesized by shock loading,(2) while no convincing result for the shock synthesis of YBa$_2$Cu$_3$O$_y$ was reported.(3)

In the present study, shock synthesis of EuBa$_2$Cu$_3$O$_y$ will be reported. We expect much easier synthesis of Eu-analogue of YBa$_2$Cu$_3$O$_y$ by shock loading, considering smaller ionic radii difference between Eu and Ba, than that of Y and Ba. Actually Eu$_{1+x}$Ba$_{2-x}$Cu$_3$O$_y$ system shows a certain solubility range between Eu and Ba ($0 \leq x \leq 0.5$), (4) while strict stoichiometry of YBa$_2$Cu$_3$O$_y$ is required as far as cation ratio is concerned. Special considerations were given on the oxygen content of the starting mixtures, because shock synthesis is conducted in a virtually closed system.

EXPERIMENTAL

Shock synthesis experiments were performed by a 25 mm bore propellant gun. The desired amount of BaO$_2$, BaCuO$_2$, Eu$_2$O$_3$ and CuO or Cu$_2$O were thoroughly mixed in a ball mill for 3 hours, and the mixture formed into a pellet of 10 mm in diameter and 1 mm in thickness was encased in a stainless steel container which was protected by momentum trap recovery assembly. The synthesis condition was controlled by peak pressure, which was estimated from the measured impactor velocity with the impedance match concept, and porosity of the starting pelletized material.

The recovered specimens were studied by X-ray powder diffraction analysis and electron microscopic observation. The metal ratio and oxygen content were measured by EPMA analysis and iodometric titration respectively. Superconducting critical temperature was determined by measuring temperature variation of DC susceptibility. Experimental details have been elucidated elsewhere.(5)

RESULTS AND DISCUSSION

X-ray powder diffraction analysis of shock-synthesized material showed formation of a tetragonal phase, similar to the oxygen defective triple perovskite $EuBa_2Cu_3O_y$, with the c/a ratio of 3.00. No single phase was achieved in the present study. Major impurities were $BaCuO_2$, Eu_2CuO_4 and also $BaCO_3$ which might be formed at ambient atmosphere from BaO residue in the shock-recovered material. Difficulty in obtaining single phase material suggests that synthesis of such compounds via solid state reaction among more than three components would be difficult.

Figure 1 shows the dependence of apparent yield of the tetragonal phase as a function of peak pressure and porosity of the starting material. The shock temperature isotherm estimated from the peak pressure and porosity is also indicated. Below 1500 °C, no reaction took place, while the yield was remarkably decreased above 4000 °C, indicating the instability of the tetragonal phase. The optimum region with the yield above 50 % was found between 2000 °C and 3000 °C. These observations suggest that partial melting of low melting components such as $BaCuO_2$ or CuO in the shocked state might play an important role for reaction enhancement.

The main difference between shock-synthesized tetragonal phase and the superconducting phase of $EuBa_2Cu_3O_y$ lies in the c-dimension, much shorter in the former resulting in the c/a ratio exactly equal to 3.00. X-ray diffraction peaks such as *001, 104* and *005*, and also the superspots of the electron diffraction pattern indicate three times period of the basic

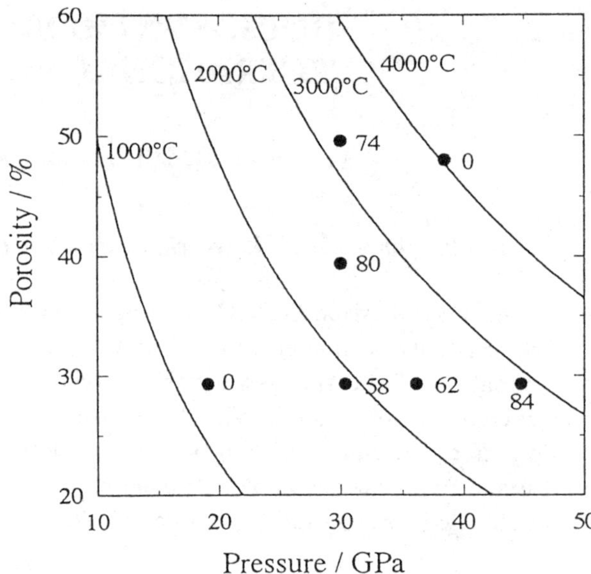

FIGURE 1. Apparent yield in shock synthesis of a metastable tetragonal $EuBa_2Cu_3O_y$ shown in a plane of peak pressure and porosity of starting material together with estimated shock temperature isotherm.

perovskite unit along one of the main axis which was clearly revealed in a high resolution electron micrograph (Fig. 2). The integer axial ratio lead to formation of unique twin pattern in which the principal axis of each domain was found to be orthogonal. These observations show that Eu and Ba atoms are ordered to some extent in the shock-synthesized specimen.

The oxygen content was preserved for $y = 6 - 6.5$, but decreased to 6.5 if it was chosen above 6.5, indicating that the oxidation state of Cu above 2 was not achieved during shock compression. No superconductivity appeared when the shock synthesized phase was oxidized at 400 °C so as to have hole carriers. This fact strongly indicates that no complete CuO_2 plane necessary for superconduction could be established probably due to incomplete Eu and Ba ordering.

In order to confirm this hypothesis, annealing experiments of the shock-synthesized specimen

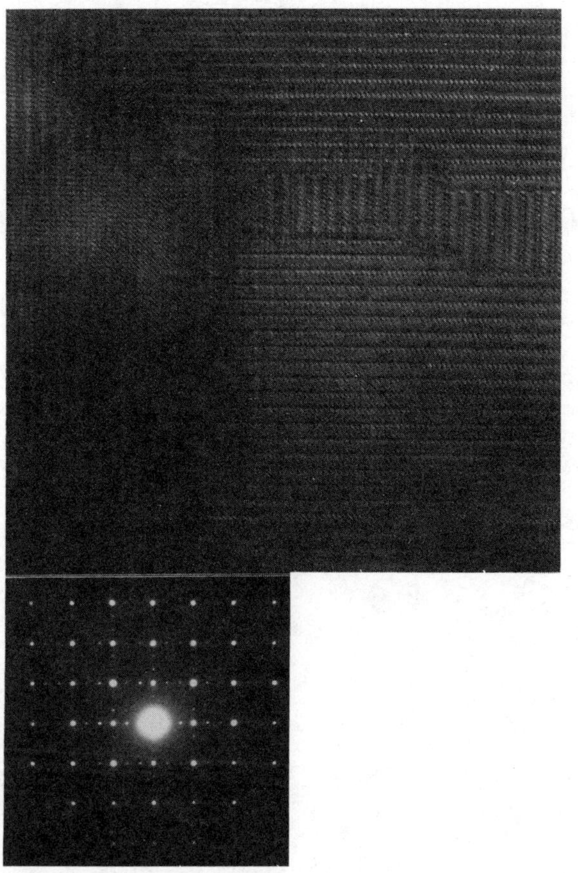

FIGURE 2. High resolution electron micrograph and electron diffraction of tetragonal EuBa$_2$Cu$_3$O$_y$ with c/a = 3.00 which was metastably formed by shock synthesis.

were carried out. It was annealed at a predetermined temperature for 2 hours in oxygen atmosphere and then quenched to liquid nitrogen temperature. The c/a ratio of the specimen annealed below 900 °C remained unchanged, but began to increase abruptly around 900 °C, approaching the c/a ratio of 3.05 of conventionally prepared phase of EuBa$_2$Cu$_3$O$_y$, as shown in Fig. 3. The increase in the c/a ratio was accompanied by the decrease in the oxygen content below 6.3. The introduction of oxygen at 400 °C in thus annealed material lead to

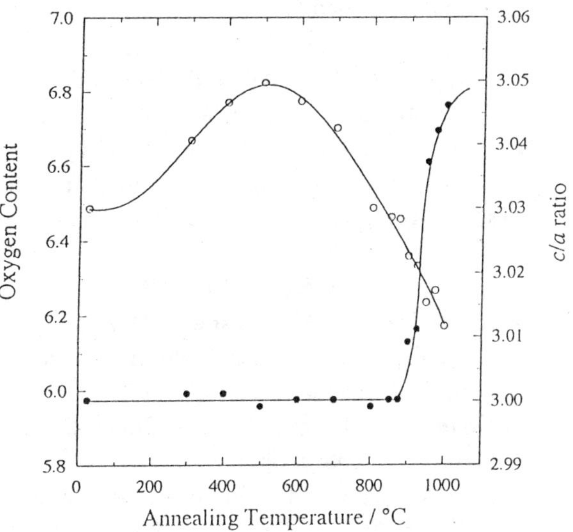

FIGURE 3. Variation of the c/a ratio and oxygen content of shock synthesized EuBa$_2$Cu$_3$O$_y$ with the annealing temperature.

superconductivity with T_c above 70 K, corroborating complete formation of CuO$_2$ plane, *i. e.*, ordering of oxygen vacancy which was promoted by remarkable improvement in Eu and Ba ordering.

No microscopic mechanism for the formation of these metastable phases has been discussed yet. However, it is to be noted that a similar tetragonal non-superconducting phase of YBa$_2$Cu$_3$O$_y$ with c/a = 3.00 was reported to be metastably prepared at relatively low temperatures around 800 °C, (6) where perfect ordering of Y and Ba might be hindered. These facts suggest difficulty in achieving perfect ordering between Eu and Ba during a short time interval of the shock synthesis process. The degree of such disordering might be taken as a useful parameter for the shock synthesis process and used as a probe of reaction kinetics of the shock synthesis, in which shock-induced or shock-assisted reaction concept has been extensively argued.(7)

ACKNOWLEDGMENT

The authors thank Prof. M. Ishikawa, Institute for Solid State Physics, University of Tokyo, and Drs. K. Kusaba and T. Atou for useful discussions.

REFERENCES

1. R. A. Graham, B. Morosin, Y. Horie, E. L. Venturini, M. Boslough, M. J. Carr and D. L. Williamson, *Shock Waves in Condensed Matter*, ed. Y. M. Gupta (Plenum, New York, 1986), pp.693-711.
2. R. A. Graham, E. L. Venturini, B. Morosin and D. S. Ginley, *Phys. Lett. A* **123**, 87-90 (1987)
3. B. Morosin, R. A. Graham, E. L. Venturini, D. S. Ginley and W. F. Hammetter, *Shock Waves in Condensed Matter 1987*, eds. S. C. Schmidt and N. C. Holmes (Elsevier, Amsterdam, 1988), pp.439-442.
4. S. Ohshima and T. Wakiyama, *Jpn. J. Appl. Phys*, **27**, 219-224 (1988).
5. H. Hikosaka, T. Atou, K. Kusaba, T. Suzuki, K. Fukuoka, M. Kikuchi and Y. Syono, *Jpn. J. Appl. Phys.*, **34**, 1506-1509 (1995).
6. Y. Nakazawa, M. Ishikawa, T. Takabatake, K. Koga and K. Terakura, *Jpn. J. Appl. Phys*, **26**, L796-L798 (1987).
7. N. N. Thadhani, *J. Appl. Phys.*, **76**, 2129-2138 (1994).

EVALUATION OF SHOCK-COMPACTED $Bi_2Sr_2CaCu_2O_{8+\delta}$ BY X-RAY DIFFRACTION AND MAGNETIC TORQUE MEASUREMENTS.

M.Kikuchi, S.Kawamata*, K.Okuda*, K.Fukuoka and Y.Syono

Institute for Materials Research, Tohoku University, Sendai 980-77, Japan
**Department of Physics and Electronics, University of Osaka Prefecture, Sakai, Osaka 593*
Japan

Shock consolidation of tap-oriented powders of $Bi_2Sr_2CaCu_2O_{8+\delta}$ along the c plane was conducted by the gun method. Optimum compaction of 95 % crystal density was achieved by 8 GPa shock-loading for 44-75 μm size powders with 67 % packing density. Degree of crystallographic orientation was evaluated by X-ray diffraction and SEM observation as well as magnetic torque measurements. High degree of orientation of tap-oriented bulk was preserved in shock loading, but post-annealing was necessary to recover superconductivity. Hysteresis in the torque curve measured below 50 K was an evidence for shock-induced pinning centers.

INTRODUCTION

Shock consolidation technique is a unique means to obtain bulk solid materials through instantaneous compaction and sintering of powders and has been expected to be promising for industrial applications. Morphology of particles is known to be an important factor for shock compaction besides shock pressure and porosity of starting materials. Most successful case has been found in shock consolidation of flaky powders of amorphous alloys.(1) The amorphous alloy powders were successfully compacted by shock process without crystallization, hence keeping the amorphous characteristics.

Several attempts have been made so far to obtaine densified bulk high T_c oxides by shock compaction.(2) Bismuth-based cuprate, $Bi_2Sr_2CaCu_2O_{8+\delta}$, among other high T_c oxides, has a very unique layered structure which leads to well developed cleavage along the basal plane, and is obtained as flaky powders. Therefore, $Bi_2Sr_2CaCu_2O_{8+\delta}$ powders were particularly suited to be effectively compacted by shock loading and oriented bulk solids could be obtained rather easily.(3,4)

Because of this peculiar two dimensional structure, unusual flux behavior in the superconducting state has been observed. However it is also known that effective pinning centers for high critical currents were hardly available in $Bi_2Sr_2CaCu_2O_{8+\delta}$. Defects introduced by shock loading are also expected to work as promising pinning centers.(5)

Evaluation of shock compacts is indispensable to achieve better superconductive characteristics. In the present study, shock compaction experiments of $Bi_2Sr_2CaCu_2O_{8+\delta}$ powders have been carried out by controlling the shock pressures and porosity of the starting materials. The degree of crystal orientation in the shock compacts was examined by SEM and XRD and also by measuring magnetic torque which could reveal bulk information about oriented grains.

EXPERIMENTAL

$Bi_2Sr_2CaCu_2O_{8+\delta}$ powders with the particle size of 44-74 μm were pre-aligned by tapping and pelletized in a disk form with the packing density varying from 50 - 75 %. Shock compaction experiments were carried out at 8 GPa by a 25 mm bore propellant gun. The shocked specimen was annealed at 800 - 875 °C in air to recover superconductivity. For the sake of reference polycrystalline pellet of $Bi_2Sr_2CaCu_2O_{8+\delta}$ synthesized by conventional solid state reaction was used.

Degree of crystallographic orientation was examined by X-ray diffraction and also by observing the pellet surface by SEM. The torque measurements were carried out by a torque magnetometer. The specimen was cooled down to the measuring temperature in zero field. The magnetic torque was measured by rotating the field direction in a plane containing the axis perpendicular to the disk plane, where the angle 0 is taken to be the field direction perpendicular to the disk plane.

RESULTS AND DISCUSSION

$Bi_2Sr_2CaCu_2O_{8+\delta}$ has been known to be rather weak against intensive shock loading.(6,7) Shock pressures as low as 10 GPa would produce remarkable degradation of superconductive fraction but no appreciable decrease in T_c, and partial decomposition took place above 20 GPa. In the present study, shock loading experiments were carried out at a peak pressure of 8 GPa with different packing density of starting materials.

The degree of crystallographic orientation of shock compacts, which is represented by the intensity ratio of I_{008}/I_{200} of X-ray diffraction peaks is shown against packing density of starting materials in Fig.1. Most efficient crystal orientation was achieved for the starting material with the packing density of 67 %. The annealing experiments of as-

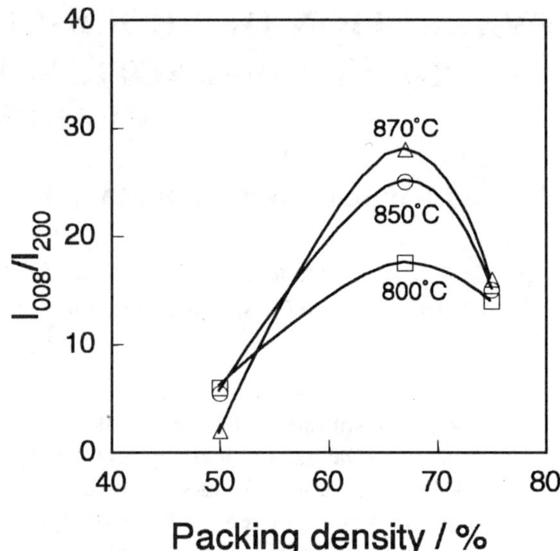

FIGURE 1   Degree of crystallographic orientation, represented by the intensity ratio of *008* and *200* reflections of the shock compact annealed at various temperatures versus packing density of starting materials.

shocked compacts to recover the superconductivity also remarkably increased the I_{008}/I_{200} ratio. This was corroborated in the SEM photograph in growth of platy crystals along the surface of the shock-compacted pellet (Fig. 2a). Interestingly completely different results were obtained for the post-annealing of the shock compact at 875 °C, in which small platelet crystals grew perpendicular to the shocked pellet (Fig. 2b). This was confirmed by the increase in the 200 diffraction peak, hence reduction in the I_{008}/I_{200} ratio. As shown in the observation of texture by SEM, the shock compact showed complete filling of voids and breakage of grains, causing the densification. The crystal density of 95 % measured by Archimedian method are consistent with the observed texture.

In order to get bulk information of crystal orientation in the shock compact, magnetic torque measurements were carried out. The magnetic torque of all the specimens increased in amplitude with the decrease in temperature and showed hysteresis below

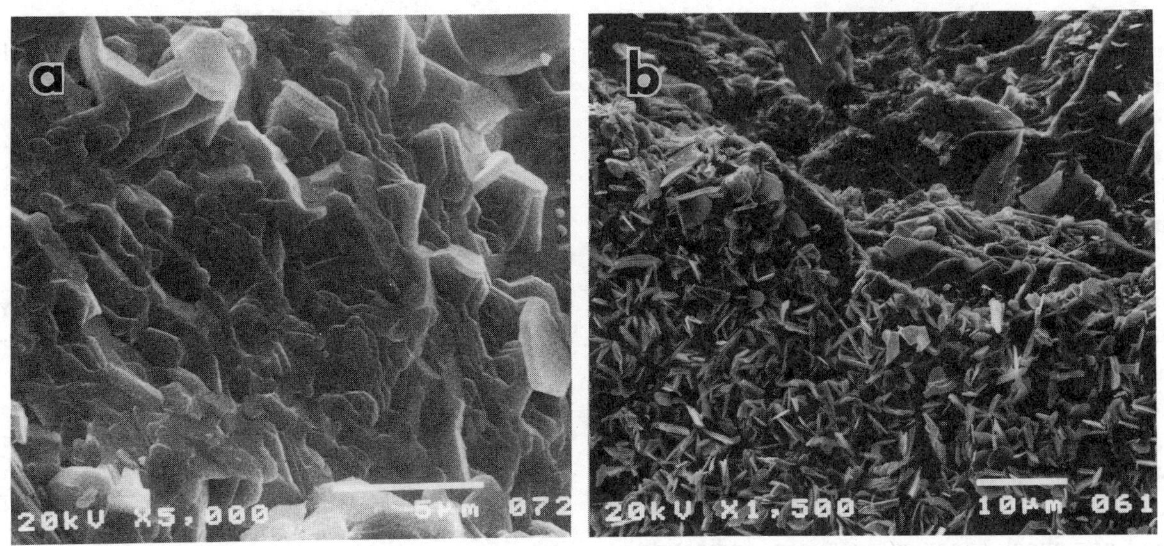

FIGURE 2 SEM photograph of the shock compact annealed at 850 °C (a) and at 875 °C (b).

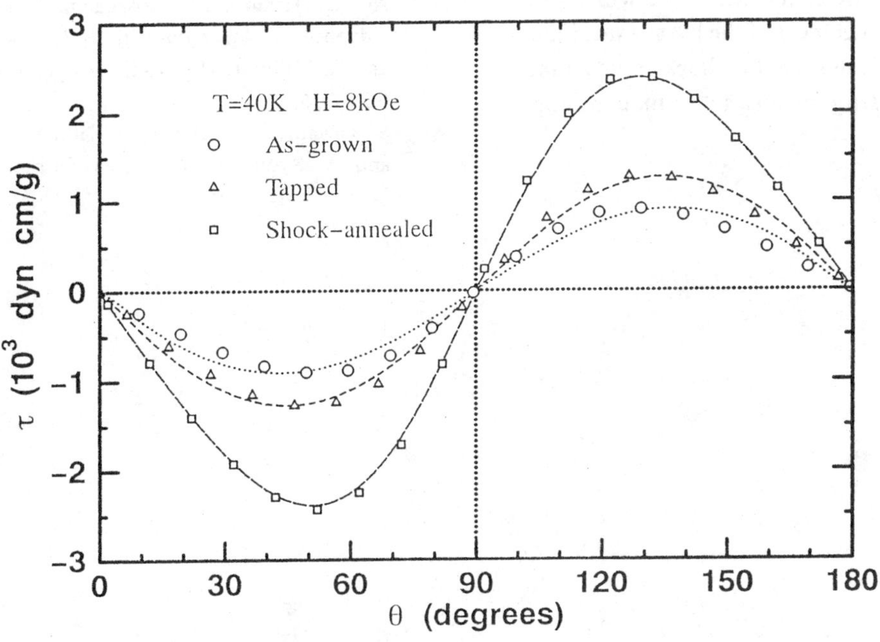

FIGURE 3   Angular dependence of reversible magnetic torque of as-grown , tapped and shock-annealed specimen. Calculated torques are shown by broken line, where the fitting parameter σ is 51.7, 43.8 and 30.2 respectively.

50 K. The reversible component obtained by averaging the torque measured with increasing and decreasing angle is shown in Fig. 3. The stable direction of each specimen is the field direction parallel to the plate plane, *i. e.* the basal plane of majority crystallites is oriented parallel to the disk plane. The amplitude of τ_{rev} increases in order of as-grown, tapped and shocked specimen, consistent with the evaluation by X-ray diffraction. (Fig. 3)

The observed angular dependence of τ_{rev} was well fitted by assuming Gaussian distribution of the direction of the crystal axis, as shown in Fig. 3. Detailed description of the model has been given elsewhere.(8) The σ value which is a measure of crystal orientation decreases in order of as-grown, tapped and shock-annealed specimen, indicating higher degree of orientation by shock compaction, although the surface orientation revealed by XRD was more pronounced than the bulk orientation obtained by torque measurements. Noteworthy is that remarkable enhancement of the hysteresis of the magnetic torque indicates that the shock compaction could be a promising means to induce pining centers.

REFERENCES

1. Y. Toda, T. Ogura,T. Masumoto, K. Fukuoka and Y. Syono, Sci.Rep.Res. Inst. Tohoku Univ.,**A 3 2**, 267-276 (1985).

2. W. J. Nellis and L. D. Woolf, MRS Bull., **1 4**,63 (1989).

3. S. T. Weir, W. J. Nellis, C. L. Seaman, E. A. Early, M. B. Maple, M. Kikuchi and Y. Syono, Physica C, **1 8 4**, 1-12 (1991).

4. M. Kikuchi, T. Atou, H. Hikosaka, K. Fukuoka, Y. Syono, N. Kobayashi, S. Kawamata and K. Okuda, Jpn. J. Appl. Phys., **3 3**, 6525-6529 (1994).

5. Y. Sakaguchi, M. Kikuchi, N. Kobayashi, K. Kusaba, K. Fukuoka, Y. Minagawa and Y.Syono, Physica C, **1 8 5 - 1 8 9**, 183-184 (1992).

6. Y. Syono, M. Nagoshi, M. Kikuchi, A. Tokiwa, E. Aoyagi, T. Suzuki, K. Kusaba and K. Fukuoka, Shock Compression of Condensed Matter-1989, eds., S. C. Schmidt et al., (North Holland, Amsterdam, 1990), pp. 579-572.

7. M. Kikuchi, Y. Syono, M. Nagoshi, A. Tokiwa, E. Aoyagi, T. Suzuki, K. Kusaba and K. Fukuoka, Advances in Superconductivity II, eds., T. Ishiguro and K. Kajimura (Springer, Tokyo, 1990), pp. 603-606.

8. S. Kawamata, K. Okuda, M. Kikuchi, H. Hikosaka and Y. Syono, J. Mater. Res., in press.

QUASI-STATIC COMPACTION STUDIES OF A POROUS PYROTECHNIC POWDER

A. I.. Atwood and P. O. Curran, *Naval Air Warfare Center, China Lake, CA 93555-6001*
C. F. Price and J. Wiknich, *COMARCO, Inc., Ridgecrest, CA 93555*

The compaction and relaxation properties of a live and an inert pyrotechnic powder simulant mixture have been evaluated under quasi-static loading conditions. The pyrotechnic powder consisted of a mixture of potassium perchlorate, magnesium-aluminum alloy, and inert binder. Potassium chloride replaced the potassium perchlorate in the inert mixture. Porous beds of powder were compacted using a double acting piston arrangement, operating at a constant loading rate of 0.11 in/min. Applied and transmitted forces were measured using either 7,500 or 20,000 lbf capacity strain gage load cells. The intragranular stress as a function of percent TMD was determined from the compaction data. The experimental intragranular stress data were further analyzed using a modified Carroll-Holt model to describe the compaction process and to allow extrapolation to a density range not achievable by experiment. The porous bed of pyrotechnic powder was much more rigid than homogeneous crystalline powders such as ammonium perchlorate (AP). Microscopic examination of the compacted material showed only light damage to the crystalline particles with little fracture. Bed relaxation resulted in a 2.8 to 4.5 percent change in bed height after compaction. These data demonstrate the presence of elastic deformation properties in a porous bed of non-viscoelastic material.

INTRODUCTION

Quasi-static compaction data were generated to determine the compaction characteristics of a pyrotechnic incendiary powder. Intragranular stress as a function of percent theoretical maximum density (TMD), bed relaxation, with and without applied load, followed by microscopic examination of the compacted powders were determined. The experimental intragranular stress data were further analyzed using a modified Carroll-Holt model to describe compaction over a range not achievable by experiment.

EXPERIMENT

Quasi-static compaction experiments were run on two powders; an incendiary consisting of 49 percent potassium perchlorate, 49 percent magnesium aluminum alloy and 2 percent calcium resonate, and an inert simulant where potassium chloride replaced the perchlorate. Pre- and post-compaction samples were examined using a Zeiss optical photo microscope under transmitted and direct illumination. Approximately 15 grams of powder were used for each of the compaction experiments.

The quasi-static compaction experiment used in this work was that of Elban and Sandusky (1-3). The material was contained in a 25 mm ID steel mold and compacted using a double acting ram technique. Applied and transmitted compressive forces were measured using either two sets of strain gage load cells, having a capacity of either 7500 lbf or 20000 lbf. The samples were compacted using a 60000 lbf screw testing machine. Top platen displacement was monitored using a linear variable differential

transformer (LVDT). Radial displacement of the steel mold was measured using constantan strain gages. The load cell, LDVT, and strain gage output were recorded on a digital oscilloscope. Load cell and LDVT output for a typical experiment are shown in Fig. 1.

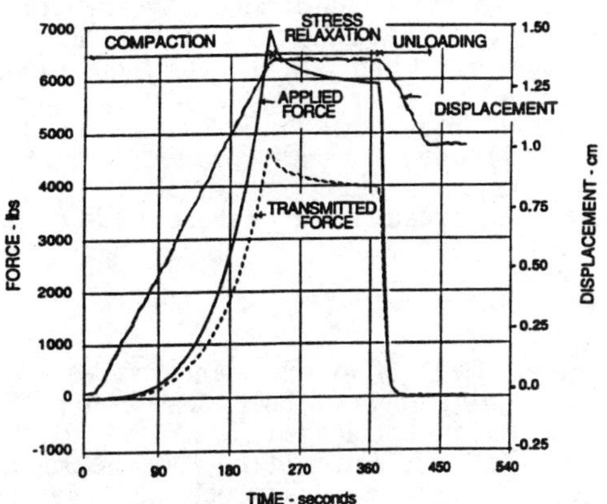

FIGURE 1. Load Cell and LVDT Output for a Typical Compaction Experiment.

Before compaction the initial percent of theoretical maximum density (TMD) was calculated. The porous bed was compacted at a constant rate of 0.11 in/min. The crosshead was stopped once the predetermined amount of compaction or the maximum load cell output occurred. The stress relaxation was monitored with the applied load, followed by an unloading of the bed at the rate of application. The inert simulant samples were loaded a second time after the unloading of the bed.

ANALYSIS

The analysis of intragranular stress follows that described by Sandusky (4). This procedure uses the method of Kuo, Moore and Yang (5-6) and considers two force balance terms. Acceleration terms are neglected, since the experiments are run quasi-statically.

Radial expansion of the mold during compaction is evaluated by converting the measured tangential strain to tangential stress, through the tensile modulus. Lame's formula for thick-walled cylinders is used to calculate the internal pressure from the tangential stress (7).

The compaction data were further evaluated using the model of Carroll and Holt (8) with the modification of Kooker and Anderson (9) using a nonlinear optimization curve fit of the data. The intergranular stress-porosity evaluation is divided into three phases of bed collapse: elastic, elastic-plastic and plastic in the analysis (11). The mathematical equations describing the analysis can be found in Ref. 12. There is a typographical error in the Mode 2 equations as published that the analyst should be aware of. The modified Carroll and Holt relations have been applied to a large variety of materials with almost uniform success, even though crystalline materials that compact by fracture are significantly different in behavior in comparison to nitrocellulose based ball propellants which compact by deformation. This analysis lends itself to incorporation into other models dealing with initiation of porous material by compressive means. The three compaction descriptors obtained from this technique are critical porosity, shear modulus, and yield stress.

RESULTS

The initial TMD of the simulant powder ranged from 47.5 to 49 percent of TMD. The initial TMD of the inert simulant ranged from 45.3 to 45.9 percent of TMD. The lower initial TMD is probably due to the difference in density between potassium perchlorate and potassium chloride (potassium perchlorate being more dense). During the acquisition of the compaction data, it was found that due to the rigidity of the bed, maximum output of the 7500 lbf load cells occurred before the desired bed displacement was achieved. The 7500 lbf load cells were replaced by 20000 lbf load cells after three experiments. While it was planned to compact the incendiary samples to 94 percent TMD or greater, that degree of compaction was not possible due to load cell limitation. A maximum of 87 percent TMD was achieved within the operating range of the 20000 lbf load cells(maximum load cell capacity available) for the incendiary powder and 89.5 percent TMD was obtained for the inert simulant. The percent TMD was calculated using a measured solid density of 2.247 gm/cm^3 for the simulant and 2.089 gm/cm^3 for the inert.

Incendiary powder intragranular stress determined from the loading portion of the force curves is plotted versus percent TMD in Fig. 2 for three, 7500 lbf experiments. A typical compaction curve for the simulant powder is compared to 200 µm AP compacted under identical conditions in Fig. 3. The AP was easily compacted to 92 percent TMD under conditions that only produced 76 to 77 percent TMD in the incendiary. Incendiary powder intragranular stress determined from the loading portion of the force curves is plotted versus percent TMD in Fig. 4 for the 20000 lbf experiments. Four compaction experiments were performed using the 20000 lbf load cells with the inert simulant powder. The intragranular stress determined from the initial loading portion of the force curves is plotted versus percent TMD in Fig. 5. Intragranular stress data for both live and inert simulant compaction experiments (20,000 lbf load cells) are compared in Fig. 6. The lower initial TMD of the inert simulant is the most remarkable difference in the data for the two powders.

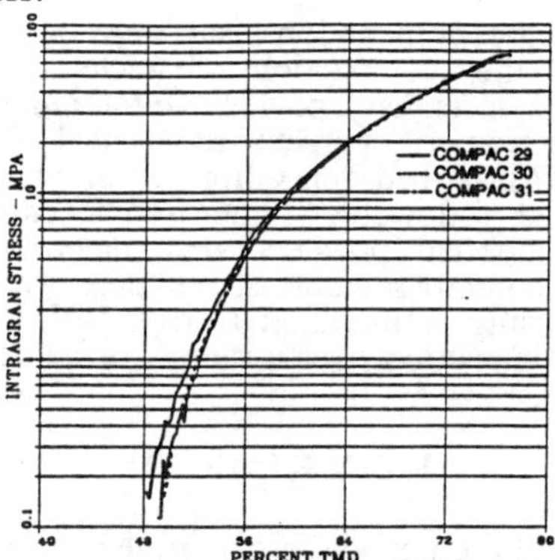

FIGURE 2. A Comparison of Three Incendiary Powder Stress Experiments Versus % TMD.

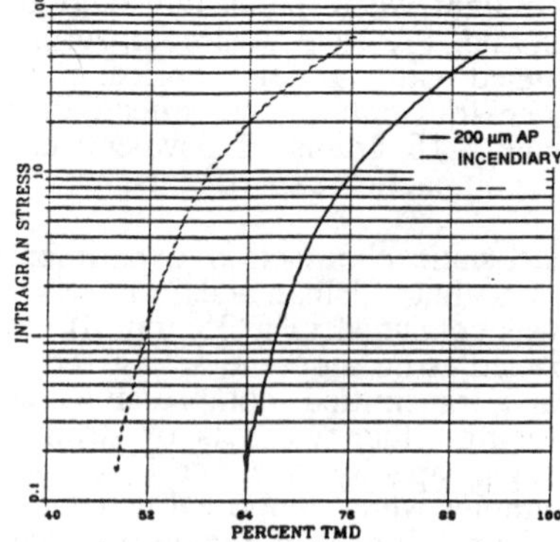

FIGURE 3. Intragranular Stress Data of Incendiary Compared to 200 µm AP.

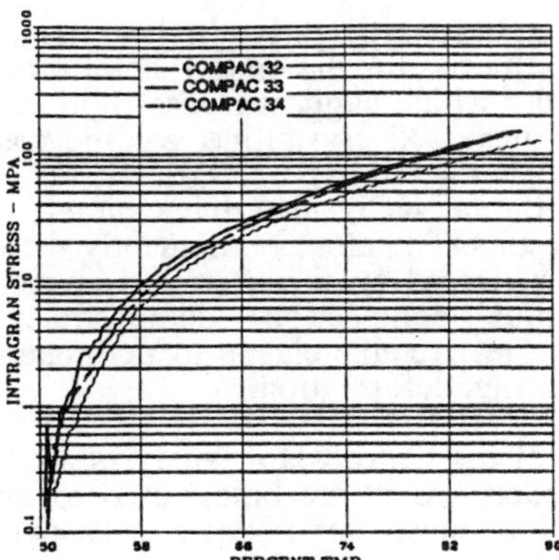

FIGURE 4. A Comparison of Incendiary Powder Intragranular Stress Data Versus % TMD.

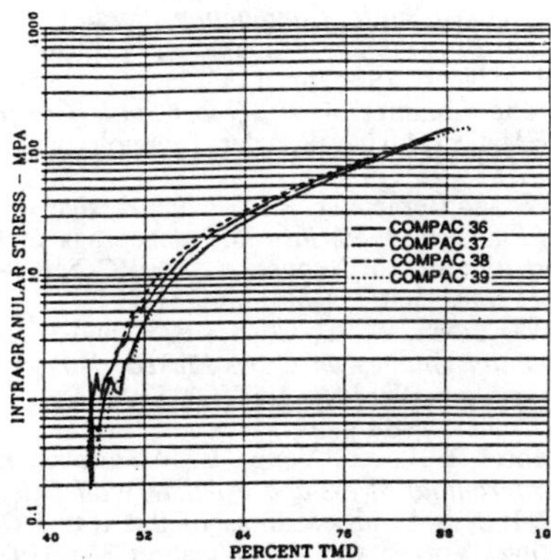

FIGURE 5. A Comparison of Intragranular Stress Versus % TMD for Four Inert Simulant Compaction Experiments.

Examination of the compacted incendiary powder by transmitted and direct illumination microscopy, showed a roughening of the perchlorate crystals with only a small amount of "fines" produced by grinding due to the compaction process. Little or no fracture of the crystals was observed. The Mg/Al appears to have experienced only a small amount of plastic deformation.

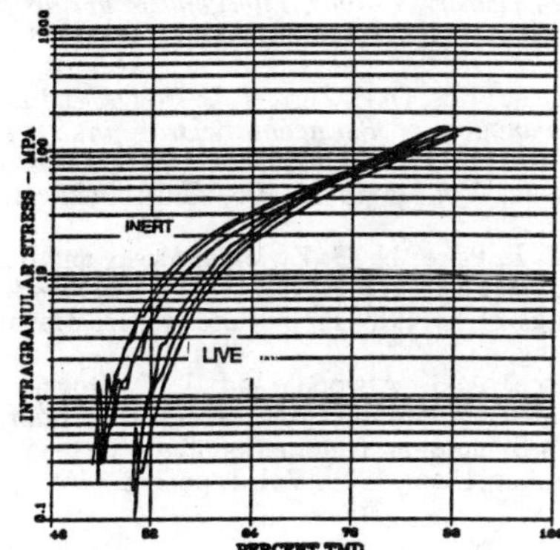

FIGURE 6. A Comparison of Live and Inert Intragranular Stress Data (20,000 lbf Gages).

Prior to removal of the sample from the apparatus, a "relaxed" bed height was measured for each material. Two weeks after the experiments were performed the height of the extracted pellets was again measured. A length increase of 2.8 to 4.5 percent was observed two weeks following compaction. An average length change between 0.5 and 2.6 percent was measured for the inert simulant two weeks after compaction. The inert bed experienced increased permanent deformation as indicated by the small changes in bed height. The increased deformation of the inert material is due to the double compaction process experienced by the inert bed. No significant changes in the radial pellet dimensions were observed for the conditions tested.

Figure 7 presents the results of the Carroll-Holt analysis of the incendiary compaction data. The data were adjusted to account for variations in initial TMD, and the density was adjusted at the higher levels of TMD to account for the bulk compressibility of the individual particles. The calculated intragranular stress curve yields the factors of Table 1.

TABLE 1. Compaction Parameters from Carroll-Holt Analysis of Compaction Data.

Parameter	Live	Inert
Sheer Modulus	$0.1563 \times 10^9 Pa$	$0.1468 \times 10^9 Pa$
Yield Stress	$0.9491 \times 10^8 Pa$	$0.1409 \times 10^9 Pa$
Bulk Modulus (fixed)	$0.6964 \times 10^{11} Pa$	$0.6964 \times 10^{11} Pa$
Critical Porosity	0.5159	0.5532

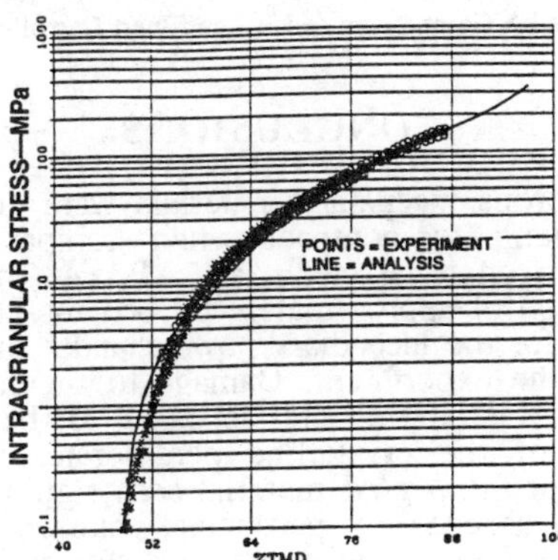

FIGURE 7. Carroll-Holt Analysis of Incendiary Intragranular Stress Data.

Figure 8 presents the results of the Carroll-Holt analysis of the inert simulant compaction data. As with the live material, these data were adjusted to account for variations in the initial TMD with the density adjusted for bulk compressibility at the high level of TMD. The calculated factors describing the inert compaction can be compared to the live incendiary powder in Table 1. Figure 9 compares the calculated intragranular stress (based on the factors of

Table 1) for the live and inert simulant powders. The two curves merge at about 76% TMD.

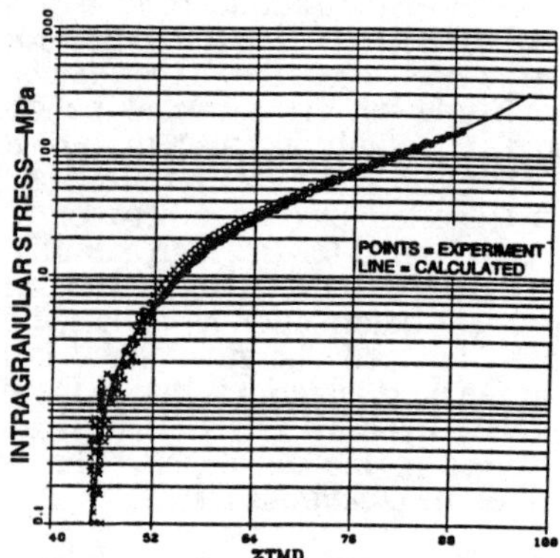

FIGURE 8. Carroll-Holt Analysis of Incendiary Intragranular Stress Data.

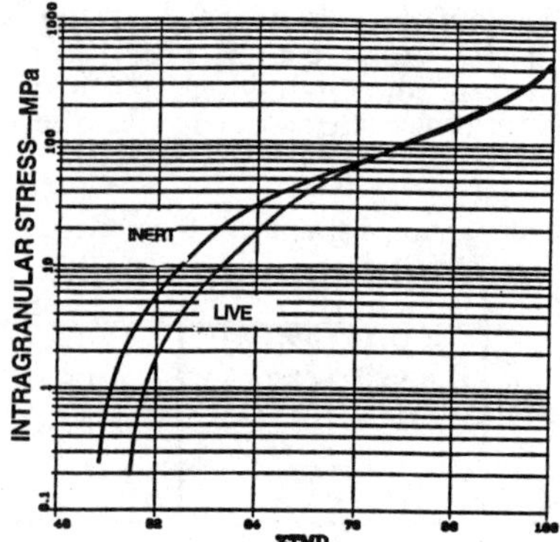

FIGURE 9. A Comparison of Live and Inert Carroll-Holt Analyses.

CONCLUSIONS

The porous incendiary materials were very rigid and difficult to compact when compared to a crystalline material such as AP. A maximum compaction of approximately 87 percent TMD was obtained for the incendiary powder under the conditions of the experiment. Damage to the individual particles was slight and appeared to be due to grinding of the crystalline perchlorate. It would appear that the Mg/Al material adds rigidity to the bed. The absence of particulate damage and the apparent bed relaxation after compaction are evidence of the presence of an elastic component in the compaction of a rigid, non- viscoelastic material.

The inert simulant has a lower powder density than that of the live incendiary due to the substitution of KCl for the $KClO_4$. The addition of an equivalent weight percent of KCl for $KClO_4$ results in a larger volume fraction of crystalline material. This is seen by a 3 percent lower initial TMD for the inert. The larger volume fraction effectively increases the crystalline solids loading of the total formulation. Slightly higher maximum TMD's were achieved for the initial compaction of the inert powder which suggest a less rigid crystal. However, the increased crystalline volume fraction and differences in the initial particle shape ("chunky" versus plate for the KCl) also have an effect on compaction results. The more permanently deformed inert bed (as evidenced by the decreased changes in bed length after compaction) demonstrates the transition from recoverable elastic to permanent, in this case fracturing, deformation.

The results of the Carroll-Holt analysis are being input into models which are being used to predict the behavior of the incendiary powder under dynamic loading conditions.

REFERENCES

1. Elban, W. L., *Quasi-Static Compaction Studies for DDT Investigations. Ball Propellants,* Propellants, Explosives, and Pyrotechnics, Vol. 9, No. 4, 1984, pp. 119-129.
2. Elban, W. L., and Chiarito, M. A., *Quasi-Static Compaction Study of Coarse HMX Explosive,* Powder Technology, Vol. 46, No. 2-3, 1986, pp. 181-193.
3. Sandusky, H. W. and Bernecker, R.R., *Compressive Reaction in Porous Beds of Energetic Materials,* in Proceedings of Eighth Symposium (International) on Detonation, NSWC MP 86-194, Naval Surface Weapons Center, 1986, pp. 881-891.
4. Sandusky, H. W. , Elban, W. L., Coyne, P. J., and Clairmont, JR., A. R., *Compaction Studies on Cross-Linked, Double-Base, High-Energy Propellant Shreds,* NSWC TR88-124, Naval Surface Warfare Center, 1988 p.46.
5. Kuo, K. K., More, B. B., and Yang, V., *Measurement and Correlation of Intragranular Stress and Particle-Wall Friction in Granular Propellant Beds,* in proceedings of the 16th JANNAF Combustion Meeting, Vol. I, CPIA Publication 308, 1979, pp 559-581.
6. Kuo, K. K., Yang, V., and Moore, B. B., *Intragranular Stress Particle-Wall Friction and Speed of sound in Granular Propellant Beds,* Journal of Ballistics, Vol. 4, No. 1, 1980, pp. 697-730.
7. Theodore Baumeister, ed. Standard Handbook for Mechanical Engineers, 7th ed New York McGraw-Hill Book Company.
8. M. M. Carroll and A. C. Holt, *Static and Dynamic Pore-Collapse Relations for Ductile Porous Materials,* Journal of Applied Physics, vol. 43, April 1972, pp. 1626-1636.
9. D. E. Kooker and R. D. Anderson, *A Mechanism for the Burning Rate of High Density, Porous, Energetic Materials,* 7th Symposium (International) on Detonation, 16-19 June 1981, pp. 198-215.
10. Atwood, A. I., C. F. Price, D. E. Zurn, T. L. Boggs, and H. P. Richter, *The Determination of Permeability/Drag in Systems Described by DDT Analysis,* 1986 Propulsion Systems Hazards Subcommittee Meeting, Vol. 1, pp 99-106, CPIA Publication, 446, March 1986.
11. Atwood, A. I. C. F. Price, N. G. Zwierzchowski and T. L. Boggs, *Gas Permeability Through Porous Beds compacted by Fracture,* Proceedings of he 1989 Propulsion Systems Hazards subcommittee Meeting,
12. Price, C. F. Price, A. I. Atwood, and T. L Boggs, *An Improved Model of the Deflagration-To-Detonation Transition in Porous Beds ,* Ninth Symposium (International) on Detonation, OCNR 113291-7 28 Aug-1Sept, 1989, Vol. 1, pp 363-376.

A STUDY OF SHOCK-INDUCED REACTIVITY IN A POROUS PYROTECHNIC POWDER MIXTURE

A. J. Lindfors, S. A. Finnegan, and J. M. Boteler

Research & Technology Division , Naval Air Warfare Center, China Lake, CA 93555-6001

Shock and reactive properties of a pressed pyrotechnic powder mixture were examined using gun-launched planar impact techniques. The pyrotechnic powder consisted of a mixture of potassium perchlorate, magnesium-aluminum alloy, and inert binder pressed to approximately 84% theoretical maximum density (TMD). Polyvinylidene fluoride (PVDF) piezoelectric polymer film shock-pressure gages were used to track the progress of the shock wave through the mixture and establish the shock Hugoniot and pressure-time trends. A comparison of experimental pressure-time trends with those obtained using a one-dimensional hydrocode shows good agreement for input pressures below 2 GPa, but increasing differences for pressures above 2.6 GPa. These differences, in the form of rising pressure levels in the region immediately behind the shock front for the experimental data, are tentatively attributed to shock-induced chemical reaction.

INTRODUCTION

The chemical reaction behavior of non-high explosive materials, particularly nonenergetic, inorganic powder mixtures and energetic, pyrotechnic powder mixtures, under high-pressure shock has been receiving increasing attention during the past two decades or so (1-3). Part of the interest in nonenergetic materials is due to the realization that shock-induced reaction synthesis has the potential for producing technologically advanced, unique or novel materials with commercial applications. The interest in pyrotechnic materials stems from the increasing use of shock as an initiation mechanism for various applications (4).

The presence of shock-induced reactivity in powder mixtures has historically been studied by post-mortem analysis of heavily confined, encapsulated samples (1-2,5). However, the development of sophisticated real-time measurement techniques has led to an increased reliance on pressure, velocity, and temperature measurements as a way to infer the presence of chemical activity (6-11). These techniques also permit the use of much simpler and less expensive test arrangements. In addition, they allow reaction times relative to the shock front to be established, thereby providing information on reaction kinetics. Finally, they provide a means for obtaining material shock property data (e.g., shock Hugoniots).

The present paper presents the results of a study, using gun-launched planar impact techniques and in-situ polyvinylidene difluoride (PVDF) shock-pressure gages (12), to obtain shock Hugoniot and chemical reaction data for a pyrotechnic powder mixture composed of potassium perchlorate (oxidizer), magnesium-aluminum alloy (fuel), and inert resin binder. The presence of chemical activity was inferred from a comparison of experimental and hydrocode-calculated pressure-time trends.

SAMPLE DESCRIPTION

The test material consisted of a 50/50 mixture of potassium perchlorate and magnesium-aluminum alloy (50/50 composition) held together with a small amount (<3%) of a resinate binder and compacted to approximately 84% (1.89 g/cc) of theoretical maximum density (TMD). Individual disc-shaped samples, compacted separately, were approximately 25 mm in diameter by 4 mm thick. Particle shapes for the two principal ingredients were quite different, with the crystalline potassium perchlorate particles being more chunky (equiaxed) and the metal alloy more flake-like. Particle sizes were similar, ranging between approximately 70-150 μm for the perchlorate and 60-125 μm for the metal. Ingredients were mixed with a twin shell blender and then compacted using a double-action press.

EXPERIMENTAL TECHNIQUE

Samples were shocked by impacting a thin gun-launched, sabot-mounted flyer plate having well characterized shock properties (13) against a driver plate of the same material upon which the sample was mounted. A 76 mm evacuated powder gun was used as the launcher, and PMMA or 2024-T4 aluminum as flyer/driver materials. There was a cutout in the sabot at the flyer plate interface so all but the outer edge of the flyer plate rear surface was free. As the flyer plate impacted the driver plate (i.e., sample holder) a compressive shock was generated in both materials and in the test sample. The subsequent reflection of the shock from the flyer rear free surface produced a rarefaction wave that propagated back to the test sample, providing the desired release behavior. Impact speeds were measured using velocity pins while impact planarity was measured with tilt pins.

Each test sample consisted of a sandwich of two 4-mm-thick discs of the pyrotechnic mixture along with two stress gage packages, one mounted on the front (impact) surface of the first (upstream) disc and the second at the interface with the second disc. Each gage package contained a 0.025-mm-thick PVDF stress transducer, K-Tech Model B-25-09, with a 3 mm square active area. The gages were covered with 0.038-mm-thick sheets of Teflon foil to provide mechanical armoring. Each gage package was bonded to the discs with Hardman D-50 polyurethane adhesive to fill voids and prevent gas pockets in the sample near the transducer. The gages were operated in the manner of Graham and Anderson of Sandia National Laboratories (14). The current mode was used. In this method, the PVDF gage has a resistor connected between the leads. The recorded voltage is used to determine the gage charge and subsequently the pressure using Graham and Anderson's calibration.

RESULTS: SHOCK PROPERTIES

A total of six tests were conducted. Experimentally-determined shock parameters for all of them are listed in Table 1. In addition, a plot of shock pressure versus specific volume (i.e., shock Hugoniot) is presented in Fig. 1. Volume was calculated using the conservation equations (15).

TABLE 1. Shock Wave Parameters

Test	Density, g/cc	Pressure, GPa	Shock Velocity, mm/µs	Particle Velocity, mm/µs
1[a]	1.886	0.38	1.00	0.20
2[a]	1.900	1.05	1.65	0.33
3[a]	1.867	2.05	1.99	0.55
4[a]	1.897	2.58	2.13	0.64
5[a]	1.880	2.70	2.21	0.65
6[b]	1.905	3.90	2.68	0.76

[a] PMMA flyer/driver material

[b] 2024-T4 flyer/driver material

A linear fit to the shock velocity-particle velocity data provided a good fit to the pressure-volume data. The observation that the fit lies near the 100% TMD point at ambient pressure (Fig. 1) also provides evidence that the sample has essentially no strength.

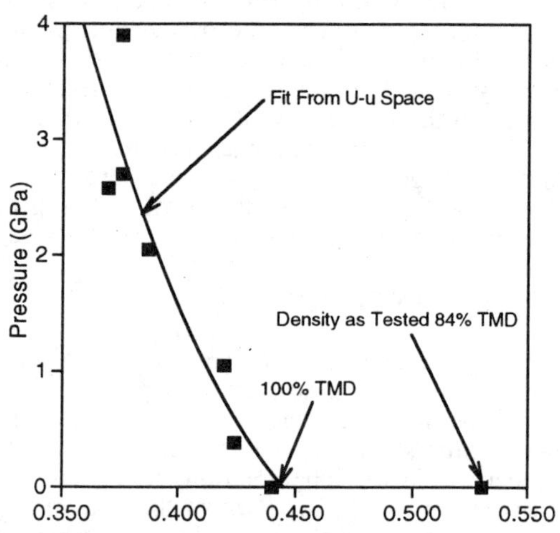

FIGURE 1. Pressure-volume (shock Hugoniot) plot.

RESULTS: PRESSURE-TIME TRENDS

Pressure-time trends for the six experiments are shown in Fig. 2. A comparison of pressure histories at the first, upstream gage plane, shows only minor changes in the shape of the input shock with increasing pressure, as expected. A similar comparison at the second, downstream gage plane shows significant changes, however. A comparison of the gage records for the three lowest input pressures, 0.38, 1.05, and 2.05 GPa, shows a gradual transition from a triangular-shaped wave form, unlike that of the input shock, to a square-topped wave, similar to the input shock. This shape change, with increasing pressure, reflects reduced attenuation and dispersion effects in the sample due to higher shock velocities and diminished rarefaction wave effects.

At the three highest input pressures, 2.58, 2.70, and 3.90 GPa, the wave form at the second gage changes again with the addition of a second pressure rise beginning approximately 200-500 ns behind the shock front. To examine the origin of this second rise, and provide confirmation of the experimental data, a simple one-dimensional, nonreactive Lagrangian hydrocode, developed for use on a personal computer (16) and using a Mie-Gruniesien equation of state (15), was used to obtain pressure histories for the same test conditions. Comparisons between measured and calculated pressure histories are shown in Figs. 3-8. Good agreement is observed between experimental and computational records at both gage planes for pressures of 2.05 GPa or less; however, at pressures of 2.58 GPa and above, the presence of the second pressure rise at the second gage plane is not observed in the computational results. This suggests that the second pressure rise is not due to mechanical processes but, instead, to phase change or chemical reaction. The estimated reaction threshold pressure (2.05-2.58) is similar to values for potassium perchlorate/zirconium (2.2 GPa) and potassium perchlorate/titanium subhydride (2.9 GPa) powder mixtures (17) which would further suggest that the second pressure rise is the result of chemical activity.

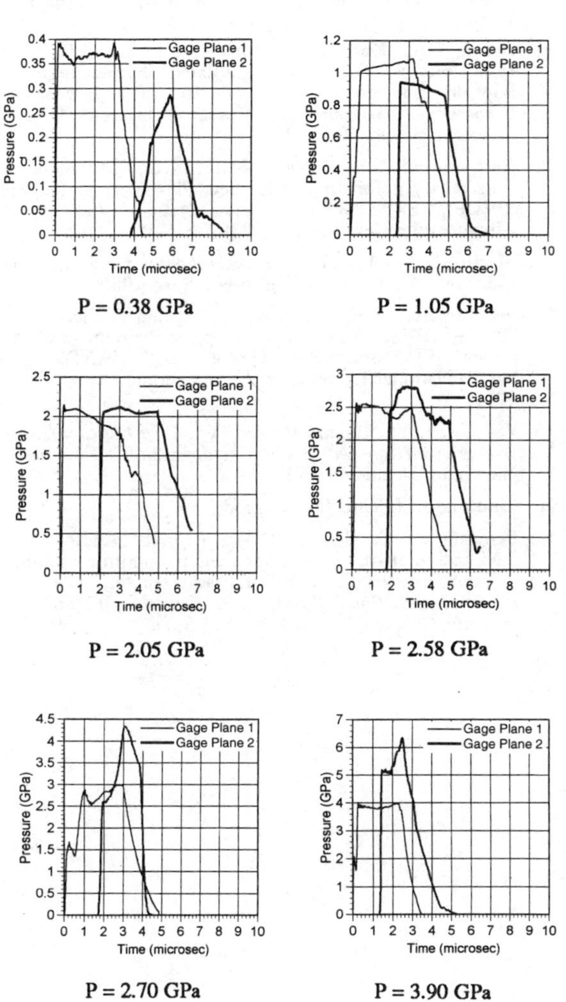

FIGURE 2. Pressure-time trends for different input pressures.

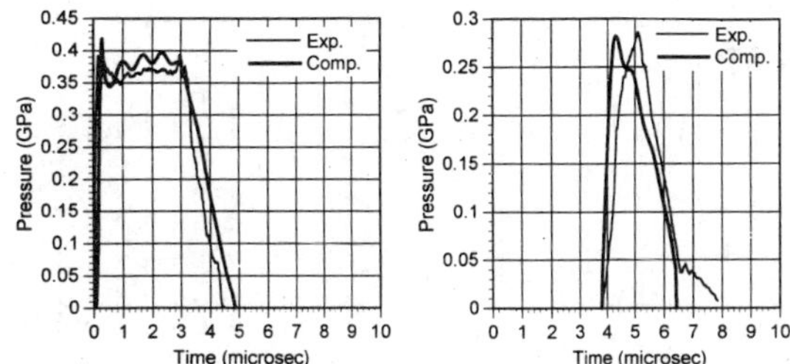

FIGURE 3. Comparison of measured and computed pressure histories for 0.38 GPa input pressure. (left - gage plane 1; right - gage plane 2)

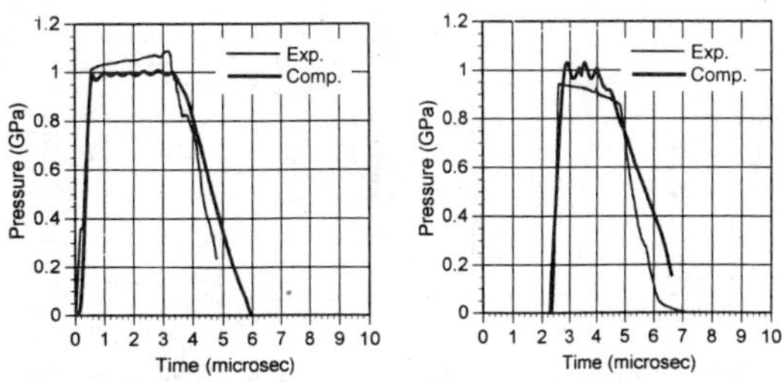

FIGURE 4. Comparison of measured and computed pressure histories for 1.05 GPa input pressure. (left - gage plane 1; right - gage plane 2)

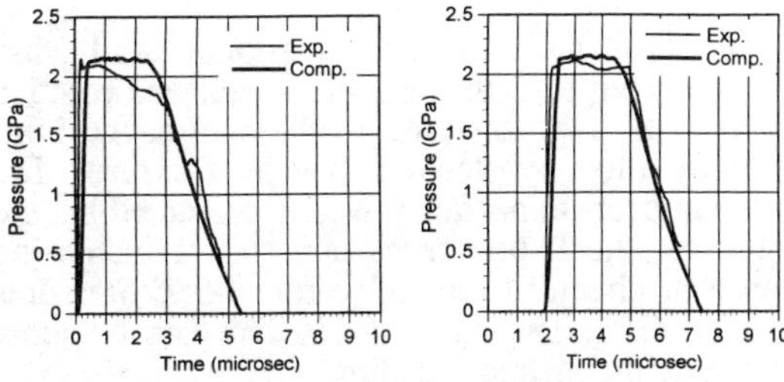

FIGURE 5. Comparison of measured and computed pressure histories for 2.05 GPa input pressure. (left - gage plane 1; right - gage plane 2)

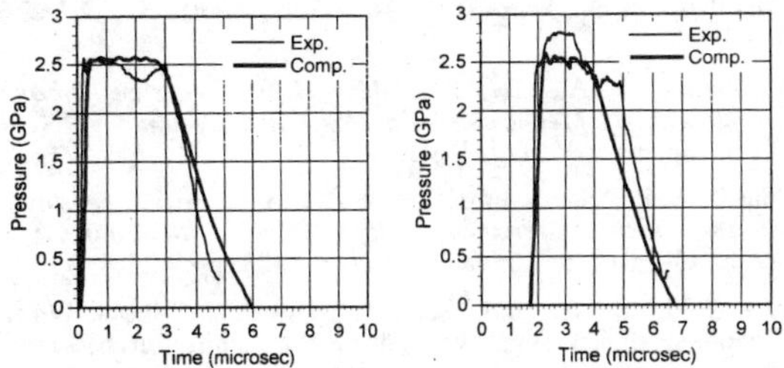

FIGURE 6. Comparison of measured and computed pressure histories for 2.58 GPa input pressure. (left - gage plane 1; right - gage plane 2)

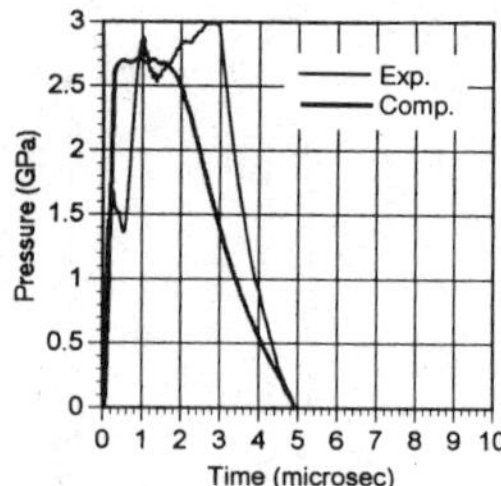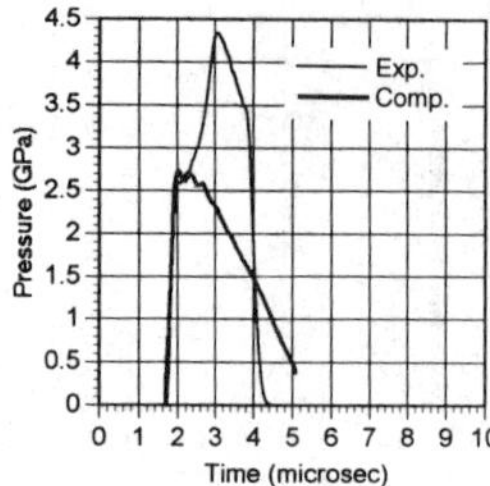

FIGURE 7. Comparison of measured and computed pressure histories for 2.70 GPa input pressure. (left - gage plane 1; right - gage plane 2)

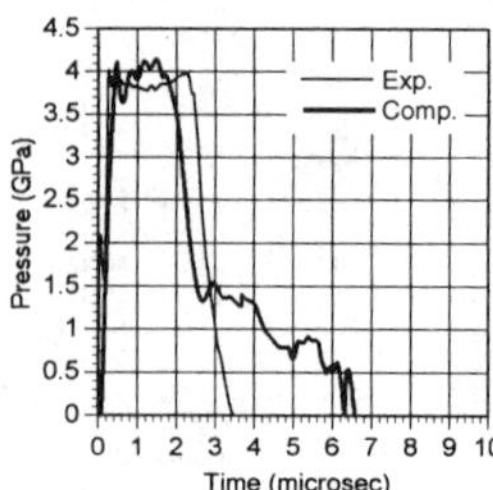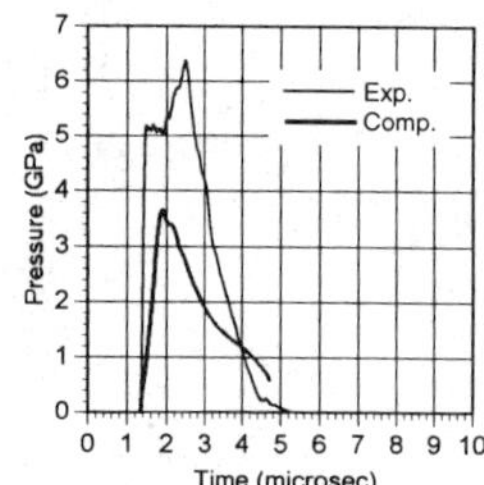

FIGURE 8. Comparison of measured and computed pressure histories for 3.90 GPa input pressure. (left - gage plane 1; right - gage plane 2)

CONCLUSIONS

An experimental and computational study has been completed to determine the shock and reactive properties of a potassium perchlorate/magnesium-aluminum alloy pyrotechnic powder mixture. The material appears to be fairly well represented by the data for its shock properties and shows increasing evidence of chemical reactivity for shock pressures above 2.6 GPa. The material also appears to show no strength under shock loading.

REFERENCES

1. Thadhani, N. N., *Progress in Materials Science* **37**, 117-226 (1993).

2. Graham, R. A., *Solids Under High Pressure Shock Compression: Mechanics, Physics, and Chemistry*, New York, Springer Verlag, 1993.

3. Hardt, A. P., "Shock Initiation of Thermite," in *Proceedings of the 13th International Pyrotechnics Seminar*, IIT Research Institute, Chicago, Illinois, 1988, pp. 425-438.

4. Sheffield, S. A. and Schwarz, A. C., "Shock-Wave Response of Titanium Subhydride-Potassium Perchlorate," in *Proceedings of the 8th International Pyrotechnics Seminar*, IIT Research Institute, Chicago, Illinois, 1982, pp. 972-990.

5. Kovalenko, A. N., Ivanov, G. V., and Usov, V. F., *Combustion, Explosion, and Shock Waves* **20**, 335-339 (1984).

6. Batsanov, S. S., Doronin, G. S., Klochkov, S. V., and Teut, A. I., *Combustion, Explosion, and Shock Waves* **22**, 765-768 (1986).

7. Batsanov, S. S., Gogulya, M. F., Brazhnikov, M. A., Simakov, G. V., and Maksimov, I. I., *Combustion, Explosion, and Shock Waves* **30**, 361-365 (1994).

8. Bennett, L. S., Sorrell, F. Y., Simonsen, I. K., Horie, Y., and Iyer, K. R., *Appl. Phys. Lett.* **61**, 520-521 (1992).

9. Boslough, M. B., "Postshock Spectral Radiance Measurements in Nickel and Nickel/Aluminum Powders," in *Shock Compression of Condensed Matter 1991*, S. C. Schmidt, R. D. Dick, J. W. Forbes, and D. G. Tasker, ed., Elsevier Science Publishers, Amsterdam, 1992, pp. 617-620.

10. Dunbar, E., Graham, R. A., Holman, G. T., Anderson, M. U., and Thadhani, N. N., "Time-Resolved Pressure Measurements in Chemically Reacting Powder Mixtures," in *High-Pressure Science and Technology - 1993*, S. C. Schmidt, J. W. Shaner, G. A. Samara, and M. Ross, ed., American Institute of Physics, New York, 1994, pp. 1303-1306.

11. Hornig, H. C., Kury, J. W., Simpson, R. L., Helm. F. H., and Von Holle W. G., "Shock Ignition of Pyrotechnic Heat Powders," in *Proceedings of the 11th International Pyrotechnics Seminar*, IIT Research Institute, Chicago, Illinois, 1986, pp. 699-719.

12. Anderson, M. U., and Wackerbarth, D. E., "Technique and Data Analysis for Impact-Loaded Piezoelectric Polymers (PVDF)," SAND88-2327, Sandia National Laboratories, Albuquerque, New Mexico, 1988.

13. Marsh, S. P., ed., *LASL Shock Hugoniot Data*, Berkeley, California, University of California Press, 1980.

14. Wackerbarth, D. E., Anderson, M. U., and Graham, R. A., "PVDFSTRESS: A PC-Based Computer Program to Reduce Bauer PVDF Stress-Rate Gauge Data," SAND92-0046, Sandia National Laboratories, Albuquerque, New Mexico, 1992.

15. Zel'dovich, Ya. B., and Raizer, Yu. P., *Physics of Shock Waves and High-Temperature Hydrodynamic Phenomena*, New York, Academic Press, 1966.

16. Lindfors, A. J., Unpublished work.

17. Schwarz, A. C., Bickes, R. W., Jr., and Headley, P. S., "Evaluation of Alternate Materials in the Ignition Column of a Shock-Initiated Through-Bulkhead Actuator," in *Proc. of the 9th International Pyrotechnics Seminar*, IIT Research Institute, Chicago, Ill, 1984, pp. 553-567.

ON THE MECHANISM OF DIAMOND FORMATION FROM EXPLOSIVES

S. V. Batalov, A. N. Averin, I. A. Batalova, B. G. Loboiko, B. V. Litvinov, V. P. Filin

Russian Federal Nuclear Center-Institute of Technical Physics
P. O. Box 245, Snezhinsk, Chelyabinsk region 456770 Russia

Experimental data on producing ultra fine diamonds from HMX-containing explosives are presented. The results of investigations into the influence of inert organic additives on the process of explosive synthesis of ultra fine diamonds from explosives with negative oxygen balance are displayed. Obtained experimental data are in agreement with the hypothesis concerning influence of the original molecular structure of explosive, organic additives, as well as loading conditions both on the dynamics of detonation diamonds synthesis, and on its properties. The experimental data on the influence of the original explosive material density on the detonation diamond synthesis are of interest in the light of the hypotheses on many-step processes of diamond formation and accompanying products of explosive material detonation transformation.

INTRODUCTION

The question of the formation of diamond particles under the explosion of the explosive compounds in the last time is vividly discussed by many authors (1-6). One of the actual directions is the study of the influence of the explosive compound internal structure on the process of generation and growth of diamond particles. The results of the investigations of explosive compounds structure influence, published in a number of recent papers (6-8), confirmed once again that the growth of carbon particles in conditions of explosive detonation is complicated, depending on many parameter processes. The investigations pursued as on aromatic, so on non -aromatic materials made evident that ultra-dispersed diamonds (or nano-scale diamonds) (UDD) explosives occurs more effectively because of lack of carbon groups in non-aromatic materials (9). Besides it the instability of aromatic molecules in combination with relatively low concentration inside them leads to the formation of carbon atoms, free or connected to one another, occurs early in the loading stage, that significantly accelerate the process of generation and growth of the carbon diamond phase. More detailed investigation of the influence of structure factors on the process of UDD synthesis under explosive   detonation has been made possible by using inert carbon-containing organic additions. The purposefulness of the insertion of organic additions during the diamond explosive synthesis is mainly determined as by "supply" of additional carbon, so, due to their negative oxygen balance, by preservation of diamond phase of carbon (11). In the paper by Gubin et.al.(13), the influence of organic additions compressibility and the velocity of their decomposition on the process of diamond particles growth is shown.

In a number of papers on investigation of diagram of carbon phase state under the high pressures (5-6) and in particular in the paper by Mal'kov of 1991 (7), the influence of the temperature in the zone of diamond synthesis on the arrangement of equilibrium diamond-carbon pressure curves in the dependence on the dimensions of forming carbon particles is represented. That is why in some papers on investigation of the factors, influencing the growing of diamond particles under the explosion, the possibility of temperature regulation in the zone of a synthesis by the insertion of organic additions (10-13) is discussed. So, for example, in the work by Mal'kov of 1995 (16), for the elevation of the environments temperature in low-temperature explosive reaction zone, the use of organic filler of low density with filled explosive is proposed. One of the important conclusions concerning a study of diamond particles generation in conditions of mixed explosive explosion is the hypotheses on the existing of diamond particles growth limitation that is connected for example with non-steady flow of detonation products because of non-uniformity of the internal structure (14). Indirect verification of this suggestion may be the results, which did not reveal the dependence between diamond crystal dimensions and such of explosive charge (8). The works concerning investigation of the influence of the regulation of internal explosive structure on the process of UDD explosive synthesis were not still performed. In this report results of experimental investigations into the influence of a number of organic additions and of internal explosive structure regulation on the process of diamond explosive synthesis are given.

PERFORMANCE OF THE RESEARCH

To provide a qualitative picture of mentioned factors influence on the process of diamond generation the HMX was chosen as an object of the investigations. It is well known that the content of carbon diamond phase in condensed explosion products of this explosive is very insignificant and comprise not more than 2% (of explosive mass) (15). However, some of its specific properties made the choice of this explosive preferable for solution of the problems set by the authors.

It is well known, that HMX forms the compositions with some of organic solvents (cyclipentane, dimethylacetamide etc.). The peculiarity of these HMX compositions besides their characteristic adsorption, crystallographic properties and relative stability is regulated internal spatial structure of alternating between themselves HMX and solvent molecules. One of these stable HMX compositions is such with dimetilamide of formic acid $(N,N$-dimetilformamid $-HCON(CH_3)_2))$ (DMFA). Having in its composition two methyl groups and one oxygen atom, DMFA was found to be a suitable material to investigate the influence of internal explosive structure on the process of explosive diamond synthesis.

HMX-DMFA composition is a regulated molecular structure, combining one DMFA molecule with one molecule of HMX. This makes the compositions an attractive structure for explosive diamond synthesis from the carbon of both constituents of composition (HMX and DMFA), because every DMFA molecule is surrounded by HMX ones and vice versa.

The experiments were carried out on 20 mm diameter samples of the composition (HMX plus DMFA) of filled density with the mass of 15g both pure one and with the addition of organic materials: butanol $(CH_3(CH_2)_2CH_2OH)$ and isopropyl alcohol $((CH_3)_2CHOH)$. HMX of fine dispersion was used to prepare the compositions. Besides that in the experiments was used the mechanic mixture of HMX and DMFA in the same proportions as in molecular compositions (the mixture has been prepared just before the experiment). To solve the questions, associated with application of utilized explosives, the pressed samples ($\varnothing$40x40 mm with density of 1,6 g/cm^3 from the composition (HMX + DMFA + binder) obtained in the result of the explosive of HMX type 90/10 (HMX -90% + binder -10%) dissolving into DMFA (A composition) were also used. All tests were performed in hermetically sealed cylindrical explosive chamber of approx. 0,9 m^3 working capacity. After blasting using the special cooling scheme, obtained condensed products of the explosion were gathered and purified to isolate the diamond-containing phase. The results of the experiments are given in a Table 1.

DISCUSSIONS ON THE RESULTS

From the experimental data (Table 1) it is evident that content of a diamond phase of carbon in the form of ultra-dispersed diamonds (UDD) in condensed products of the explosion compositions (HMX+DMFA) is nearly two times higher than UDD content obtained in the experiments with the mechanic HMX DMFA mixture as well as with A composition. This result confirms the supposition that the regulation of the internal structure of mixed explosives may considerably influence the increase of carbon diamond phase output.

The results of experiments with the mechanic mixtures of the compositions (HMX+DMFA) and organic additions also showed the UDD formation 1,2-1,9 times increase in comparison with compositions itself (HMX+DMFA). Obtained results allow to confirm the organic additions influence on diamond formation under application of low-density explosive mixtures + organic compound, in particular, as a source of additional carbon.

The most UDD output was obtained in the experiments with pressed samples from DMFA

TABLE 1. The UDD output in experiments

Compound	ρ, g/cm^3	C_{UDD},%[a]	Note
HMX + DMFA	1,1	0,5	mechanic mixture
[HMX+DMFA]$_k$[b]	1,1	1,0	powder density
[HMX + DMFA]$_k$ + butanol	1,1	1,2	mechanic mixture of composition and solvent
[HMX + DMFA]$_k$ + isopropyl alcohol	1,1	1,9	mechanic mixture of composition and solvent
Composition A	1,1	0,4	powder density
[Composition A + DMFA]$_k$	1,6	4,9	pressed samples

[a] C_{UDD}, % - mass output of UDD related to charge mass

[b] [... + ...]$_k$ - molecular composition

composition on base of A composition, which can be explained by more high density of investigated samples.

CONCLUSION

Hence, obtained experimental data allows to confirm the supposition on the efficiency of organic additions incorporated into reaction zone under the diamond explosive synthesis. Obtained UDD formation increase during the experiments may be explained for example by "delivery" of additional carbon to the zone of UDD synthesis, leading to decrease of the oxygen balance of the system explosive - organic addition. Not touching the questions of the growth of generating UDD crystals, on the base of the experiments with HMX compositions it is possible to make the suggestion that the regulation of internal explosive structure permits to expect UDD formation increase under investigation of mixed condensed explosives, as well as increase of crystal dimensions. The authors were not aimed at achievement in performed experiments the optimum UDD formation, that is the object for further investigations.

REFERENCES

1. Rudenko,A. P., Kulakov, I. I., *About Mechanism of Origination and Growth of Diamonds in the Conditions of Chemical Synthesis* Collection "Superhard Materials". Kiev, 1983.
2. Staver, A. M., Gubareva, N. V., Lyamkin, A. I., Petrov, Ye. A., *Russ. Physics of Combustion and Explosion.* **20**,100-104 (1984).
3. Lyamkin, A. I., Petrov, Ye. A., Yershov, A. P., Sakovich, G. V., Staver, A. M., Titov, V. M., *Reports to Russian Academy of Sciences* **302**,611-613 (1988).
4. Anisichkin, V. F., Mal'kov, I. Yu., Titov, V. M., *Reports to Russian Academy of Sciences* **303**, 625 (1988).
5. Volkov, K. V., Danilenko, V. V., Yelin, V. I., *Russ. Physics of Combustion and Explosion* **26**,123-125 (1990).
6. Titov, V. M., Anisichkin, V. F., Mal'kov, I. Yu., *Russ. Physics of Combustion and Explosion* **35**,117-126 (1989).
7. Mal'kov, I. Yu., *Russ. Physics of Combustion and Explosion* **27**, 136-140 (1991).
8. Anisichkin, V. F., Mal'kov, I. Yu., Titov, V. M., *Russ. Physics of Combustion and Explosion* **31**, (1995).
9. Kolomiychuk, V. N., Mal'kov, I. Yu., *Russ. Physics of Combustion and Explosion* **29**,120-128 (1993).
10. Mal'kov, I. Yu., *Russ. Physics of Combustion and Explosion* **30**, 130-132 (1994).
11. Gubin, S. A., Odintsov, V. V., Pepekin, V. I., *Russ. J. of Chemical Physics* **9**, (1990).
12. Mal'kov, I. Yu., *Russ. Physics of Combustion and Explosion* **30**, 120 (1994).
13. Gubin, S. A., Odintsov, V. V., Pepekin, V. I., *Russ. J. of Chemical Physics* **3**, 754-756 (1984).
14. Pershin, S. V., Petrov, Ye. A., Tsaplin, D. N., *Russ. Physics of Combustion and Explosion* **30**, 102-104 (1994).
15. Anisichkin, V. F., Mal'kov, I. Yu., Titov, V. M., *Reports to Russian Academy of Sciences* **303**, 625-627 (1988).
16. Mal'kov, I. Yu., *Russ. Physics of Combustion and Explosion* **30**,155 (1994).
17. Gubin, S. A., Odintsov, V. V., Pepekin, V. I., *Russ. J. of Chemical Physics* **5**,111-120 (1986).
18. Orlova, Ye. Yu., *Chemistry and Technology of Powerful Explosive Compounds,* Leningrad, Chemistry, 1973.

WERE CARBONADOS SYNTHESIZED BY AN ANCIENT IMPACT?

P.S. DeCarli

SRI International, Menlo Park, CA 94025

It has been suggested that carbonado, a type of polycrystalline cubic diamond, was formed by an ancient impact. Shock wave calculations indicate the mechanism is possible, but further studies are needed before one can say that it is plausible.

INTRODUCTION

Carbonados are polycrystalline diamonds characterized by a light carbon isotopic ratio and by the presence of inclusions of crustal, as opposed to mantle, minerals.(1,2) They are commercially mined from placer deposits in Brazil and in Central Africa; other occurences have been reported. The carbon isotopic ratio is considered evidence that the source carbon was of organic origin; the nature of the inclusions suggests that the diamonds formed at relatively shallow depths, at crustal pressures well below the diamond stability field.

Smith and Dawson suggested that carbonados were formed by early impacts of crustal rocks.(1) If this were true, one might expect carbonados to be associated with fossil impact craters, and this did not appear to be the case. Recently, Girdler et al have suggested a possible Precambrian impact origin for a ca. 800 km diameter magnetic anomaly in Central Africa.(3) At that time, Brazil and Central Africa were contiguous, and one might even hypothesize that a single giant impact could have been responsible for both Brazilian and Central African carbonados.

The present work was inspired by consideration of that wild hypothesis. However, it is independent of the resolution of the the controversy over the impact origin of the magnetic anomaly. Rather, the question I wish to address is whether impact synthesis of carbonado is plausible or even possible in a geological setting, on the basis of laboratory experience in shock synthesis of diamond. This paper is a preliminary report; the results are still inconclusive, and we emphasize that shock synthesis is only one of a variety of possible explanations for the origin of carbonados.

BACKGROUND

Laboratory studies of shock synthesis and of uncatalyzed static high pressure synthesis have shown that there are two distinct diamond synthesis regimes.(4-10)

If the source carbon is dense, well-crystallized graphite, a mixture of cubic diamond and hexagonal diamond (lonsdaleite) can be made by a quasi-diffusionless mechanism at pressures above about 15 GPa and simultaneous temperatures in the range of about 1300 K to 2000 K.

If the source carbon is a less well-ordered graphitic carbon, cubic diamond can be formed by an inferred nucleation-and-growth mechanism at pressures above about 15 GPa and temperatures above 3000 K. In this latter case, the newly formed diamond must be quenched to a lower temperature (to avoid graphitization) before the pressure is decreased below the diamond stability line.

For diamonds found in nature, the presence of lonsdaleite is considered diagnostic evidence of impact synthesis. For example, the diamonds found in some fragments of the Canyon Diablo nickel-iron meteorite contain lonsdaleite and are correlated with metallurgical evidence of strong shock.(11)

Lonsdaleite is also found in diamonds associated with impact metamorphized rocks found in the vicinity of the Popigai and Reiss impact craters. Accordingly, yakutites, lonsdaleite-bearing diamonds found in placers in Yakutia and the Ukraine are presumed to be of impact origin.(2)

It must be noted that Smith and Dawson presumed that the Yakutian diamonds were a form of carbonado; Kaminsky has pointed out that there are numerous differences between yakutites and both Brazilian and Central African carbonados.(2) Indeed, one must consider the possibility that different occurences of carbonado may have been formed by different mechanisms. Shock origin is only one posssibility; various other mechanisms have been proposed.(2,12) Finally, some of the polycrystalline diamonds recovered from kimberlite pipes, presumably of deep-seated origin, are superficially indistinguishable from carbonado. (13)

Well-crystallized graphite is found in marbles, granites, basalts, quartzites, schists, and nickel-iron meteorites. The range of shock conditions for formation and post-shock survival of lonsdaleite-diamond will depend on the matrix. The peak temperature of the graphite at pressure must be in the range of 1300 K to about 2000 K, for peak pressures above at least 15 GPa. However, the temperature of the surrounding matrix must be below 2000 K on release of pressure, if rapid graphitization is to be avoided.(14)

SIMPLE   CALCULATIONS

If one makes the simplifying assumtions that the graphite is in the form of flakes and that the shock propagation direction is normal to the plane of the flakes, simple one-dimensional calculations can provide a measure of shock pressure and temperature histories in graphite and matrix material.

The graphite in a high shock impedance matrix, such as meteoritic iron, will achieve pressure equilibrium via a series of shock reflections. In the pressure range of interest, almost all of the increase in the internal energy of the graphite occurs as a consequence of the initial shock transmitted from the higher impedance material. Subsequent shock reflections may account for more than half of the total pressure increase, but they have a negligible effect on calculated shock temperatures.

For graphite in nickel-iron meteorites, the peak temperature range of 1300 K to 2000 K will correspond to peak pressures in the range of about

60 to 120 GPa; the temperature of the iron on release from 120 GPa will be less than 1700 K.

For graphite in marble, the peak temperature range of 1300 to 2000 K corresponds to a pressure range of about 30 to 45 GPa; the temperature of the marble on release will be below 1800 K.

For graphite in granite or quartzite, the 1300 to 2000 K peak temperature range corresponds to about the 27 to 40 GPa pressure range. However, the peak pressure could be limited to less than 35 GPa by the requirement that the release temperature of the rock be less than 2000 K.

These simple calculations confirm our observations of nature; it is indeed possible to make and preserve lonsdaleite-diamond in a geologic setting. Note that the assumption of non-porous flake-like graphite in our calculations is consistent with the habit of natural graphite inclusions in the cited matrices.

The calculations of matrix temperatures on pressure release were deliberately maximized. In the case of quartzite, for example, we assumed that the release adiabat corresponded to pure stishovite.

Brazilian and Central African carbonados consist of cubic diamond; lonsdaleite has not been observed. As we have noted, the shock synthesis of cubic diamond requires high carbon temperatures, above 3000 K.

For non-porous graphite in a matched impedance matrix such as granite or quartzite, cubic diamond synthesis would require a shock pressure greater than about 70 GPa. However, the release temperature of the matrix would be well above 2000 K, and there would be no possibility of quenching the hot diamond to avoid graphitization as pressure decayed into the graphite stability field.

Calculations for graphite in other crustal rocks, including shales and carbonates, lead to the conclusion that it is highly unlikely that carbonados are shock synthesized from crystal density graphite; the release temperature of any crustal rock would be too high to quench the hot diamond.

In the commercial shock synthesis of diamond, the source carbon is porous, permitting the attainment of high shock temperatures at more modest pressures. The carbon is mixed with an appropriate (relatively incompressible) material so that the hot diamond can be quenched to lower temperature as the pressure decays.

Although carbonados have a much larger crystallite size than the usual product of laboratory shock synthesis, the ca. three order of magnitude difference may well be accounted for by the ca. six order of magnitude difference between time scales of laboratory experiments and large cratering events.

One may note that porous carbon does exist in nature; coal and charcoal are examples. We have made diamond from anthracite in laboratory shock experiments. Furthermore, the observation of polyaromatic impurities in carbonado (2) is a possible indication of a coal-like carbon source.

STATUS REPORT

Preliminary calculations indicate that shock synthesis of carbonado from porous carbon may be possible in a geological environment. More detailed shock calculations are currently on hold; the immediate problem centers on determination of the plausibility of the necessary geological environment.

At least some occurences of carbonado have been dated as Precambrian. Isotopically light (organic origin) graphite is found in rocks of this age, but coal does not appear in the geologic record until hundreds of millions of years later. On the other hand, the mineral inclusions found in carbonado are mostly crustal in nature. One would not be surprised at finding similar inclusions in a lump of coal.

If the hypothesis of shock origin of carbonado is to be viable, we must either find an appropriate Precambrian porous carbon source or show that some families of carbonados formed much later, within the past 400 million years. As we have noted, it is not necessary that the shock origin hypothesis account for all carbonados. Indeed, the range of variation in the characteristics of carbonados might be easier to explain if one assumed different origins for different "classes" of carbonado.

ACKNOWLEDGEMENTS

This investigation of the origin of carbonados is being conducted in collaboration with Judith Milledge of University College London, Ron Girdler and David Collinson of the University of Newcastle-upon-Tyne, Colin Pillinger of the Open University, and others. My collaborators have not yet commented on this paper, and I must note that models for the origin of carbonados will have to be consistent with the results of their investigations of other aspects of carbonados.

I thank Gene Shoemaker for advice and encouragement.

REFERENCES

1. Smith, J.V., and Dawson, J. B., Geology 13, 342-343 (1985)
2. Kaminsky, F.V., "Carbonado and Yakutite: Properties and Possible Genesis", Proceedings of the Fifth International Kimberlite Conference, 1991, Vol 2, pp136-143
3. Girdler, R.W., Taylor, P. T., and Frawley, J. J., Tectonophysics 212, 45-58 (1992)
4. Bundy, F.P., and Kasper, J. S., J. Chem. Phys. 46, 3437-3446 (1967)
5. DeCarli, P. S., and Jamieson, J. C., Science 133, 1821-1822 (1961)
6. DeCarli, P.S., "Shock Wave Synthesis of Diamond and Other Phases" in Proceedings of MRC Symposium on Diamond, April 1995, in press, Materials Research Society, Sept. 1995
7. Trueb, L.F., J. Appl. Phys. 19, 4707-4716 (1968)
8. Trueb, L.F., J. Appl. Phys. 42, 503-510 (1971)
9. DeCarli, P. S., U.S. Patent 3,238,019 (1966)
10. Cowan, G.R., Dunnington, B. W., and Holtzman, A. H., U. S. Patent 3,401,019 (1968)
11. Hanneman, R. E., Strong, H. M., and Bundy, F. P., Science 155, 995-997, (1967)
12. Haggerty, S. E., "Ice to Diamond: a Model for the Transformation of Clathrate to Carbonado", AGU Spring Meeting, abstract M22A-8 (1995)
13. Milledge, H.J., Private Communication, 1995
14. Davies, G., and Evans, T., Proc. Roy. Soc. London A238, 413-427 (1972)

CHAPTER X

EXPLOSIVE AND INITIATION STUDIES

SHOCK RESPONSE OF SEVERAL
PLASTIC BONDED EXPLOSIVES

G.T. Sutherland, K.D. Ashwell, J.H. O'Connor, R.N. Baker, and E.R. Lemar

White Oak Detachment, Indian Head Division, Naval Surface Warfare Center, Silver Spring, MD 20903-5640

Multiple in situ manganin gauge experiments were performed on explosive compositions IRX-1, IRX-3A, PBXN-111 (PBXW-115), and PBXN-110 (PBXW-113). For most explosives, stress-time profiles show substantial reaction. These profiles are compared to those measured in other explosives and plans are discussed for future reactive rate modeling involving these explosives.

INTRODUCTION

Our laboratory uses computer simulations to investigate the shock response behavior of various explosives in a variety of experiments. We use light gas gun experiments to obtain some parameters for the Lee-Tarver (1) reactive rate model. The growth terms for the model are best obtained from experiments that measure substantial stress growth from chemical reaction. Stress-time profiles showing high levels of stress growth are presented here for a variety of explosives. We present a discussion of the general features and trends observed in the stress-time profiles and discuss plans for future work and analysis. Stress-time profiles obtained for lower levels of reactive stress growth are given in two earlier papers (2,3) for the explosives PBXN-111 and PBXN-110.

PBXN-110 (PBXW-113) has served as our baseline ideal explosive for a variety of reasons. PBXN-110 is a two component system; the composition includes the HTPB binder system (see Table 1), and fine and coarse nitramine particles. In addition, computer simulations can be done for various shock sensitivity (4,5) and detonation performance (5,6) experiments.

PBXN-111 (PBXW-115) is our baseline nonideal explosive. PBXN-111 is similar to many other non-ideal explosives as it contains a nitramine (RDX), a fuel [aluminum (Al)], an oxidizer [ammonium perchlorate (AP)], and HTPB binder. Forbes and

TABLE 1. Explosive Composition in Volume Percent.

Ingredient	PBXN 111	PBXN 110	IRX 1	IRX 3A
Sieved HMX (60-210μm)			52.8	58.5
Coarse HMX (Class 3)		58.1		
Fine HMX (Class 2)		19.4		
Coarse RDX (Class 3)	11.9			
Fine RDX (Class 2)	7.9			
Aluminum	16.6			5.9
AP (≈200μm)	39.5			
HTPB Binder	24.1	22.5	47.2	35.6

coworkers have measured a variety of PBXN-111 detonation properties (7). Application of the Lee-Tarver model to the PBXN-111 experiments will test the suitability of this model for non ideal explosives.

We also undertook gun experiments on the research explosives (8,9) IRX-1 and IRX-3A. For these explosives, the particle size of HMX is controlled and from the same batch. These features allow better comparison of experimental data between these explosives. As before, existing (8,9)

IRX-1 and IRX-3A detonation property experiments will be simulated.

EXPERIMENT

Figure 1 displays a schematic drawing of the experiment and Table 2 displays various experimental parameters. Manganin gauges (2) were used to obtain the stress-time profiles. The diameter of the explosive samples were about 70 mm. Front cover plate material is either Teflon, copper (Cu) or aluminum (Al).

Figures 2-6 display selected stress-time profiles. The time of impact for experiment 2 is approximate, because the experiment consisted of only two gauges. In addition, Teflon replaced the explosive on the right side of the last gauge (see Figure 1) for experiments 2 and 5.

TABLE 2. Experimental Parameters.

Exp. #	Flyer	Cover Plate	Explosive Density (gm cm^{-3})	Projectile Velocity mm μsec^{-1}
1 (N110)	Cu	Teflon	1.68	0.99
2 (IRX-1)	Al	Teflon	1.43	0.91
3 (IRX-3A)	Al	Al	1.58	0.83
4 (N111)	Cu	Cu	1.79	1.00
5 (N111)	Al	Teflon	1.79	0.85

stresses because the fine HMX particles in PBXN-110 are less reactive than the coarse HMX particles. Fine particle HMX/water compositions are less reactive than coarse HMX/water compositions for input pressures of about 5 GPa (10).

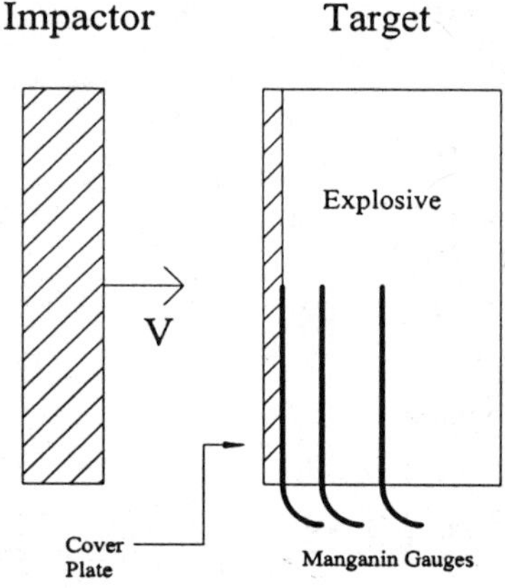

FIGURE 1. Schematic drawing experimental arrangement used in this study.

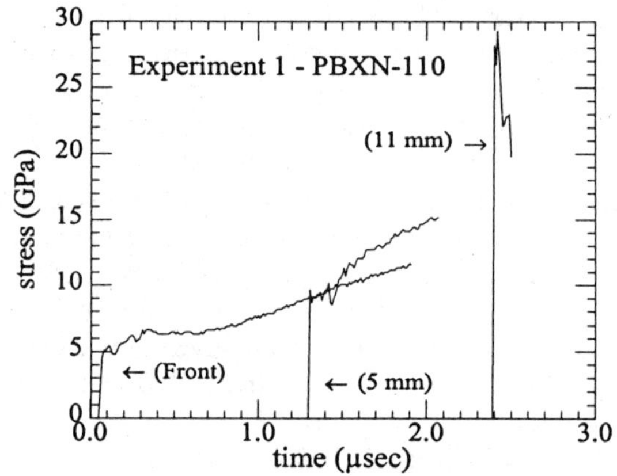

FIGURE 2. Stress-time profiles from experiment 1.

RESULTS AND DISCUSSION

For PBXN-110 (Experiment 1), reactive stress growth occurs both at the shock front and behind the shock front. Likewise, Simpson and coworkers (10) measured similar profiles for an explosive containing water and 58 volume percent HMX. This volume percentage is equal to that of the coarse HMX particles present in PBXN-110. Stress-time profiles for both materials should be similar at low input

We observed similar stress-time profiles in IRX-1 and IRX-3A. For experiments 2 and 3 the initial stresses imparted into the samples were the same. Similarities of IRX-1 and IRX-3A profiles is likely due only to HMX reactions since evidence of Al reaction is not apparent in the IRX-3A profiles. To determine the amount of reactive stress growth due to the aluminum, we plan to perform computer simulations of the IRX-1 and IRX-3A experiments. Comparison of the model parameters may allow us to gauge the extent of Al reaction.

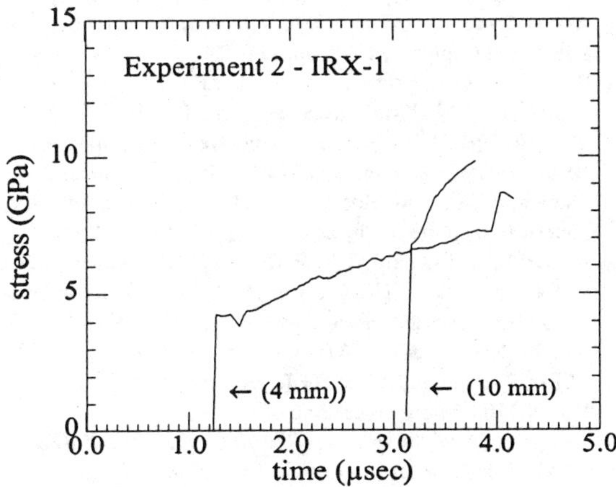

FIGURE 3. Stess-time profiles from experiment 2.

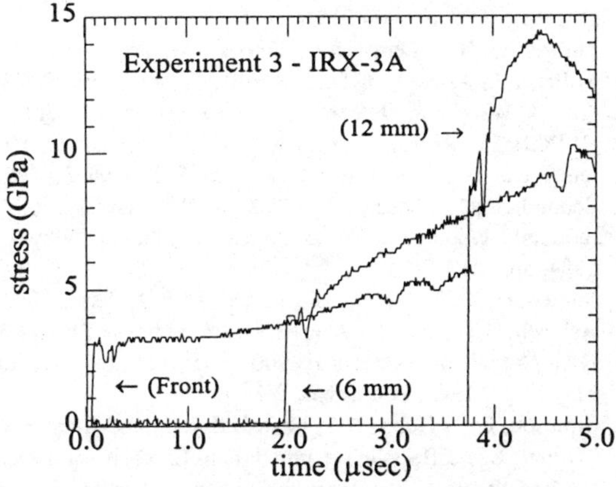

FIGURE 4. Stress-time profiles from experiment 3.

For PBXN-111, stress-time profiles in Figure 4 are similar to those observed by Tarver and Green (11) for propellant B which contains 12% HMX, Al, AP, and binder. The amount of stress growth in PBXN-111 is too great to be due to reactions of the nitramine alone (11-12); some reactive growth is due to either AP or aluminum or both. The shape of the stress-time profiles is different than those of the other explosives. We attribute this difference to the smaller volume percentage of nitramine in PBXN-111 and the different reaction rates of the Al and AP. Tarver and Green (11) generated a reactive rate model for propellant B which includes a term for the growth rate of ignited hot spots using existing HMX rates. In addition, they used a growth term for the global reaction of the Al, AP and binder. We will investigate using a similar approach for PBXN-111. We have also observed that PBXN-111 is less shock sensitive than the other explosives. Figure 5 shows an experiment performed for an impact stress of about 4 GPa; only about 1.5 GPa of reactive stress growth is observed in the profiles. This is agreement with results (4) of the modified gap test

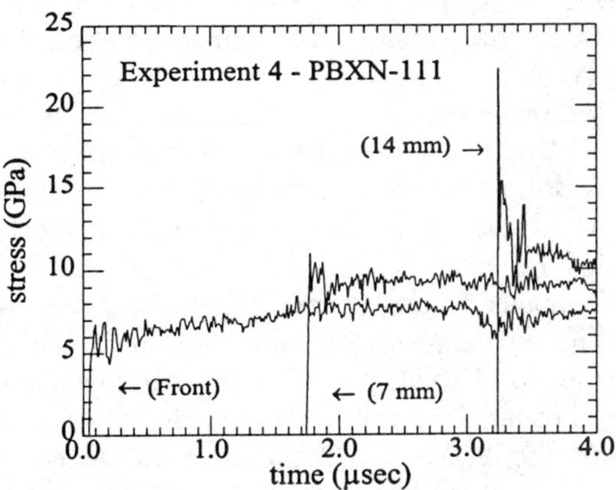

FIGURE 5. Pressure-time profile from experiment 4.

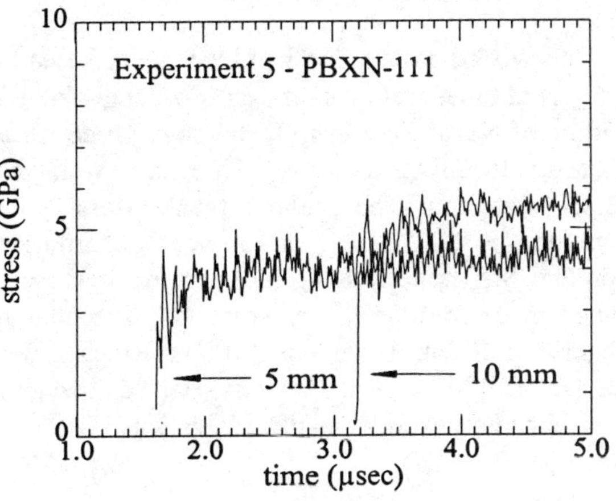

FIGURE 6. Stress-time profiles from experiment 5.

FUTURE WORK

Future work involves using the profiles from the IRX-1, IRX-3A, and PBXN-111 experiments to obtain Lee-Tarver model parameters. Comparisons of these parameters for our explosives and others will reveal relationships or trends between model parameters. Parameter trends are expected due to features such as: nitramine loading, nitramine size, and the presence of aluminum and AP.

We will perform gas gun experiments on the non-ideal research explosive IRX-4 which contains sieved HMX, Al, AP and HTPB binder. The same HMX has been used in the manufacture of all IRX explosives.

We also suggest that gun experiments be done on an HMX/AP/binder explosive. These experiments allow for a better determination of the contribution of AP to reactions occurring during a shock to detonation transition.

Computer simulations of existing shock sensitivity and detonation performance experiments will be performed. We will also use computer simulations based of the CPex (12) code to predict the gas gun profiles for PBXN-111. The reactive rate parameters for CPex were based on PBXN-111 detonation velocity decrement and wave curvature data.

ACKNOWLEDGMENTS

This work was done for the Independent Research Project at the Naval Surface Warfare Center and the Office of Naval Research (Technology Directorate) through the Explosive and Undersea Warheads Block Program. The authors thank Messrs. Pat Femiano, Brian Glancy, Carl Groves, Richard Marion and Ms. Beverly King-McLean for their assistance in performing the experiments. The authors thank Mr. Erwin Anderson, and Drs. Richard Bernecker, Phil Miller, Ruth Doherty, Harold Sandusky, and Jerry Forbes for valuable suggestions.

REFERENCES

1. Lee, E.L., and Tarver, C.M., *Phys. Fluids* **23**, 2362-2372 (1980).

2. Sutherland, G.T., "Multiple Stress-Time Profiles In The Navy Explosive PBXN-110 (PBXW-113), *Shock Compression of Condensed Mater-1989*, New York: Elsevier Science Publishing Company, 1990, pp. 721-724.

3. Sutherland, G.T., Forbes, J.W., Lemar, E.R., Ashwell, K.D., and Baker, R.N., "Multiple Stress-Time Profiles in a RDX/AP/Al/HTPB Plastic Bonded Explosive," in *High-Pressure Science and Technology-1993*, Schmidt, S.C., Shiner, J.W., Samara, G.A., and Ross, M., (editors), Woodbury, NY, American Institute of Physics, 1994, pp. 1381-1384.

4. Lemar, E.R., Liddiard, T.P., Forbes, J.W., Sutherland, G.T., and Wilson, W.H., "The Analysis of Modified Gap Test Data for Several Selected Insensitive Explosives," NAVSWC TR 89-290 (Naval Surface Warfare Center, Dalhgren, VA, 1993)

5. Miller, P.J., and Sutherland, G.T., "Reaction rate Modeling of PBXN-110," These proceedings.

6. Anderson, E. W., "Explosives," in *Tactical Missile Warheads - Vol 155 of Progress in Astronautics and Aeronautics*, Carleone, J., (editor), Washington, D.C., American Institute of Aeronautics and Astronautics, Washington, DC, pp. 81-163.

7. Forbes, J.W., Lemar, E. R, and Baker, R.N., "Detonation Wave Propagation in PBXW-115", in *Proceedings of the Ninth Symposium (International) on Detonation*, Portland, OR, Aug 1989, pp. 806-815.

8. Sutherland, G.T., Lemar, E.R., Forbes, J.W., Anderson, E.W., Miller, P.J., Ashwell, K.D., Baker, R.N., and Liddiard, T.P., "Shock Wave and Detonation Wave Response of Selected HMX Based Research Explosives with HTPB Binder Systems" in *High-Pressure Science and Technology-1993*, Schmidt, S.C., Shaner, J.W., Samara, G.A., and Ross, M., (editors), Woodbury, NY, American Institute of Physics, 1994, pp. 1413-1416.

9. Sutherland, G.T., Lemar, E.R., Forbes, J.W., Anderson, E.W., Ashwell, K.D., and Baker, R.N., in Proceedings of the JANNAF Propulsion Systems Hazards Subcommittee Meeting, May 11-13, 1993, Fort Lewis, WA.

10. Simpson, R.L., Helm, F.H., Crawford, P.C., and Kury, J.W., "Particle Size Effects in the Initiation of Explosives Containing Reactive and Non-Reactive Continuous Phases", in *Proceedings of the Ninth Symposium (International) on Detonation*, Portland, OR, Aug 1989, pp. 25-38.

11. Tarver, C.M., and Green, L.G., "Using Small Scale Tests to Estimate the Failure Diameter of a Propellant" in *Proceedings of the Ninth Symposium (International) on Detonation*, Portland, OR, Aug 1989, pp. 701-710.

12. Kennedy, D.L., and Jones, D.A., "Modelling Shock Initiation and Detonation in the Non-Ideal Explosive PBXW-115", in *Proceeding of the Tenth Symposium (International) on Detonation*, Boston, MA.

SHOCK LOADING STUDIES OF AP/AL/HTPB BASED PROPELLANTS

J. M. Boteler and A. J. Lindfors

Shock/Detonation Physics Laboratory
NAWC code 474330D, China Lake, California, 93555

The unreacted Hugoniots of three class 1.3 propellants have been investigated by shock compression experiments. Shock waves were generated by planar impact in a 75 mm single stage powder gun. Manganin and PVDF pressure gauges provided pressure-time histories to 200 kbar. The propellants were of similar formulation differing only in coarse AP particle size and the addition of a burn rate modifier (Fe_2O_3). All propellants contained 90% solids by weight and HTPB as binder. Results show negligible effect of AP particle size on shock response in contrast to the addition of Fe_2O_3 which appears to "stiffen" the host material. Within experimental error, the unreacted Hugoniots compare favorably to the solid AP Hugoniot. Five shock experiments were performed on propellant samples strained to induce in situ voids. The material state was quantified by uniaxial tension dilatometry. Experimental records suggest chemical reactivity behind the shock front. These results are discussed and compared to experimental studies reported previously .

INTRODUCTION

In recent years extensive experimental studies have generated a wealth of information regarding the inherent risk and performance tradeoffs of certain Class 1.1 propellant formulations. To date, the more common Class 1.3 propellants, have not been afforded such treatment. In contrast to the Class 1.1 propellants, 1.3 propellants are widely believed to be non-detonable and essentially unreactive at low pressure. Recent studies by Bai and Ding [1] and Huang et al. [2] on the response of several composite propellants to shock loading have indicated that AP/Al/HTPB[*] type propellants begin to decompose behind a 140 kbar shock front. They observed two distinct peaks in the pressure-time profiles, one at or near the shock front and a second occurring 1-2 μs later. They conclude that the observed increase of pressure behind the shock front (first peak) is primarily due to the shock decomposition of AP and HTPB, and the second peak is due to the reaction between these products and the Al.

The purpose of the present work is to establish the unreacted Hugoniots for three AP/AL/HTPB propellants and investigate the effect of induced damage on the shock response of these same propellant formulations.

SAMPLE PREPARATION

The propellant samples used in this study were formulated at NAWC, China Lake. The measured density, AP particle size, and percentage present are given in Table 1. All propellant samples contained 20%

weight of 29 μm aluminum, HTPB binder , and DOS (Dioctylsebaceate) plasticizer. In addition to these constituents, the PS4 samples contained a small amount (1.2%) of Fe_2O_3 as a burn rate modifier. The total solid content for each propellant was 90% by weight.

Target samples were either guillotined, or machined from bulk blocks, into 100 mm diameter slabs 6.25 mm thick (nominal). Prior to target assembly, the samples were measured for uniformity and ambient density. All samples for each propellant formulation were obtained from the same material batch in order to minimize density fluctuations.

TABLE 1. Propellant Density and AP Content.

MATERIAL	DENSITY (g/cc)	AP (μm/%wt.)
PS1	1.841	200/50 20/20
PS2	1.836	400/50 20/20
PS4	1.853	200/48.8 20/20

EXPERIMENTAL PROCEDURE

Shock experiments were performed in a single stage powder gun with a 75 mm bore. The target assembly consisted of a buffer plate (driver) and two or more propellant slabs. Teflon armored gauge packages were placed at each interface. For longitudinal stress less than 75 kbar, piezoresistive manganin gauges were used; otherwise Bauer type PVDF piezoelectric gauges were

---

[*] Ammonium Perchlorate/Aluminum/HydroxyTerminated PolyButadiene

employed. The stress gauges provided time of arrival for the shock wave in addition to pressure versus time histories. A small layer of urethane adhesive was used to fill any surface voids and bond the slabs together.

Impact configurations were planar and symmetric. Tilt was measured to be 1 miliradian or less. In order to achieve the desired stress range, three impact/buffer materials were used: PMMA, 2024T-4 aluminum (Al), and 304 stainless steel (SS). With these materials, and projectile velocities approaching 1.9 mm/μs, peak input pressures to 140 kbar were attained.

The shock experiments performed on damaged propellant samples used uniaxial tension dilatometry to quantify the state of the material prior to shock loading. This technique was developed at NAWC by Lepie [3] and Richter [4] and permits simultaneous measurement of stress-strain and void volume-strain properties. For details of the method and diagnostic instrumentation the reader is referred to the latter two references.

EXPERIMENTAL RESULTS AND ANALYSIS

A total of seventeen shock experiments were performed in the course of this study; twelve experiments to establish the unreacted Hugoniots of the three propellants, and five experiments to explore the dynamic response of pre-damaged propellant samples. On the basis of coarse AP particle size, the PS2 material was chosen for the latter set of experiments. The results and analysis for these two data sets are addressed separately.

Undamaged Hugoniot Measurements

Experimental results for the pristine samples are given in Table 2. The impact/buffer materials are indicated in parentheses following the shot number. Projectile velocity, shock velocity, particle velocity and peak pressure are given in columns 2-5 respectively. The peak pressure was measured at the buffer-sample interface. The best fit linear U_S-U_P relationship for each propellant is indicated in Figure 1. The error in shock velocity is due to the uncertainty in arrival time. The PS1 and PS2 Hugoniots include two high pressure data points provided by LLNL [5] These results compare favorably with the shock Hugoniot for pure AP (TMD=1.95 g/cc) as reported by Sandstrom et al. [6] indicated by the dotted lines in this figure. Below 50 kbar the three Hugoniots are indistinguishable from the AP Hugoniot. At higher input pressure the PS4 Hugoniot begins to diverge, becoming stiffer at higher pressure. This observation is consistent with the reported stress-strain data for this material [7]. It is unclear what the mechanism is for this anomalous behavior, however, we conjecture that the Al particles and AP crystals enhance the local binder crosslinking in a synergistic manner.

TABLE 2. Results For Undamaged Materials

SHOT #	V_P (km/s)	U_S (km/s)	U_P (km/s)	P (kbar)
2PS1 (Al)	1.37	4.11	0.93	70
3PS1 (PMMA)	1.04	3.30	0.40	27
4PS1 (PMMA)	1.41	3.59	0.56	38
1PS2 (PMMA)	1.06	3.32	0.43	26
2PS2 (Al)	1.26	4.20	0.86	66
3PS2 (Al)	1.05	3.98	0.72	52
1PS4 (PMMA)	0.76	2.82	0.33	17
2PS4 (PMMA)	1.06	3.62	0.38	26
3PS4 (Al)	1.07	3.99	0.72	54
4PS4 (Al)	1.16	3.94	0.80	59
5PS4 (SS)	1.66	4.90	1.20	109
6PS4 (SS)	1.60	5.1	1.43	135

The last shot in this series (Shot 6PS4) was designed to search for any evidence of reactivity in the pristine PS4 material. Four PVDF gauges were embedded between four PS4 samples nominally 6.25 mm thick. Figure 2 represents the pressure time histories for this experiment.

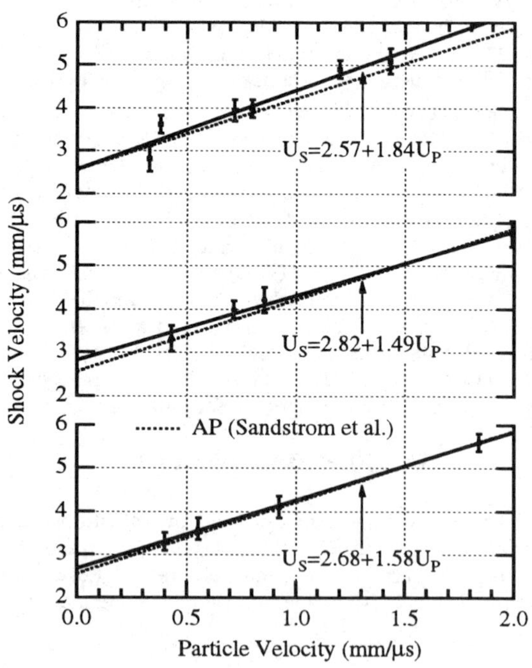

FIGURE 1. Best fit shock-particle velocity relationships for PS1 (bottom), PS2 (middle), and PS4 (top).

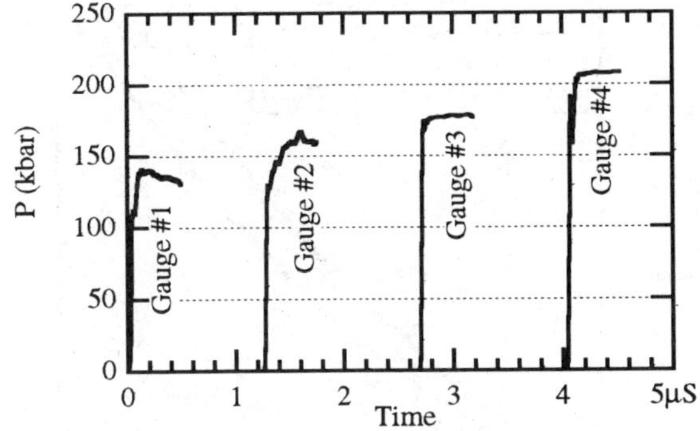

FIGURE 2. Gauge Records for Shot 6PS4 Showing Pressure Evolution. Input Pressure was 135 kbar.

For clarity, the records have been truncated to show only the first 0.5 μs of gauge response. The measured input pressure at gauge # 1 is 135 kbar and grows to 210 kbar at gauge # 4 after the shock has transited 19 mm of material. Such a pressure evolution is indicative of chemical reactivity feeding the shock front. These observations are in agreement with the findings of Bai et al.[8] for a similar propellant formulation. Bai et al. conclude that the reaction at the shock front is due primarily to the decomposition of the AP and HTPB binder.

Pre-Damaged Reactive Measurements

Five shock experiments were performed on the PS2 material for static tensile strains of 5, 10, and 15 %. The strains were applied along one dimension orthogonal to the shock front resulting in measured void volumes of 0.2, 0.5, and 1.7% respectively. The impact con-figurations and measured quantities are given in Table 3. For a given strain, the void volume (%) was determined from the uniaxial tension dilatometry results. Shock pressure, given in the last column, was measured at the first gauge plane located at the buffer-sample interface. In each case the shock experiment was performed within one hour after the static strain was applied. The records for 5% and 15% applied strain are shown in Figs. 3 and 4, respectively. In these figures the measured shock pressure is given in kbar units and time in microseconds.

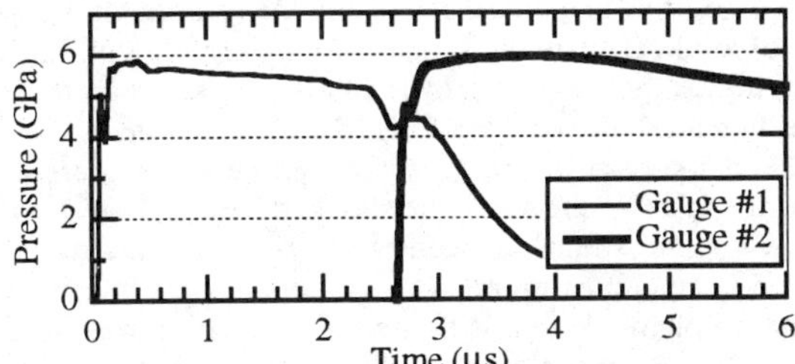

FIGURE 3. Gauge Records for 5 % Applied Strain.

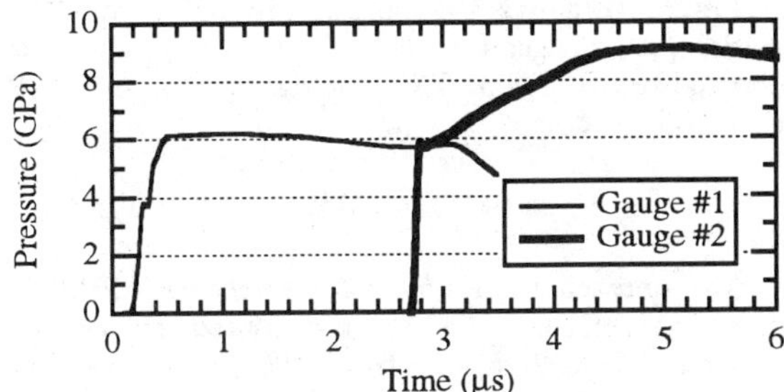

FIGURE 4. Gauge Records for 15 % Applied Strain

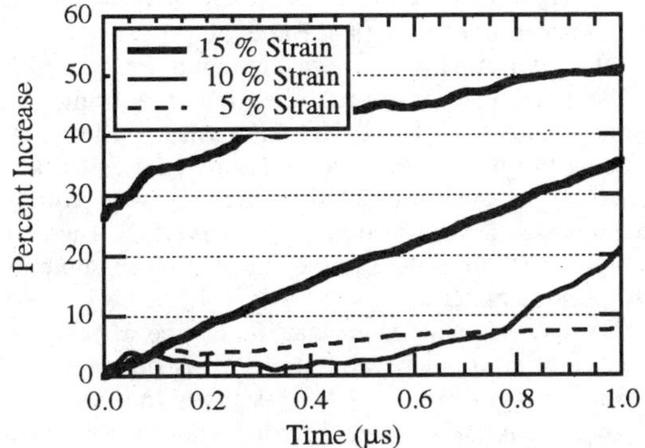

FIGURE 5. Comparison of Pressure Evolution for Three Applied Strains Indicated. Note that all records are Normalized to the Input Pressure at Gauge # 1.

TABLE 3. Results for pre-damaged PS2 samples.

SHOT	% STRAIN	% VOIDS	Vp (km/s)	U_S (km/s)	Up (km/s)	P (kbar)
1DPS2 (AL)	10.0	0.5	1.10	4.10	0.74	56.5
2DPS2 (Al)	9.78	0.5	1.11	4.00	0.71	52.5
3DPS2 (Al)	15.27	1.7	1.22	3.62	0.93	61.0
4DPS2 (Al)	5.4	0.2	1.18	3.73	0.84	57.5
5DPS2 (SS)	14.84	1.7	1.30	4.16	1.15	86.0

The time of arrival at the gauge and the sample thickness are used to calculate the shock velocity reported in Table 3. Whereas the pressure growth in Fig. 3 is small (perhaps questionable) the records of Fig. 4 at 15% strain, show a marked increase in pressure. This increase in pressure closely follows the incident shock front. In Fig. 5 the three applied strains are contrasted by normalizing the second gauge measurement to the input stress measured at gauge plane #1. In this figure the top trace corresponds to the highest input pressure of 86 kbar (Shot 5DPS2); clearly the reactive contribution to the shock front has overtaken the initial loading.

Figure 6 shows the porous Hugoniot for PS2 material pre-damaged to induce 1.7 % void volume. It is compared to the pristine PS2 material and the solid AP results of Sandstrom et al. [6].

SUMMARY

The unreacted Hugoniots for three AP/Al/HTPB propellant formulations have been measured. The change of coarse AP particle size from 200 μm to 400 μm appears to have little affect upon the dynamic response of these materials to 200 kbar. The inclusion of 1.2 % Fe_2O_3, as a burn rate modifier, alters the material response resulting in a stiffer Hugoniot. Reactive flow experiments on PS4 suggest chemical reaction near the leading edge of the shock front. These observations are consistent with the findings of Bai et al. [8].

The experiments on pre-damaged PS2 samples suggest that one (or more) chemical reaction(s) are present at the shock front. Uniaxial tension dilatometry was used to quantify the material state as a function of applied strain, and associated increase in void volume. The increase in void volume is observed to lower the reaction threshold of the sample material when compared to the pristine material.

We have demonstrated that the degree of reactivity increases with applied static strain. This observation is not surprising since it is well known that damaged energetic materials show greater shock sensitivity. However, in a real scenario the true state of a damaged propellant prior to the subsequent arrival of a shock wave remains an area of controversy. Relaxation effects must be addressed to better understand real material response.

ACKNOWLEDGMENTS

The authors gratefully wish to acknowledge the Defense Nuclear Agency for their support of this work. Mr. Scott Pockrandt is thanked for his technical support.

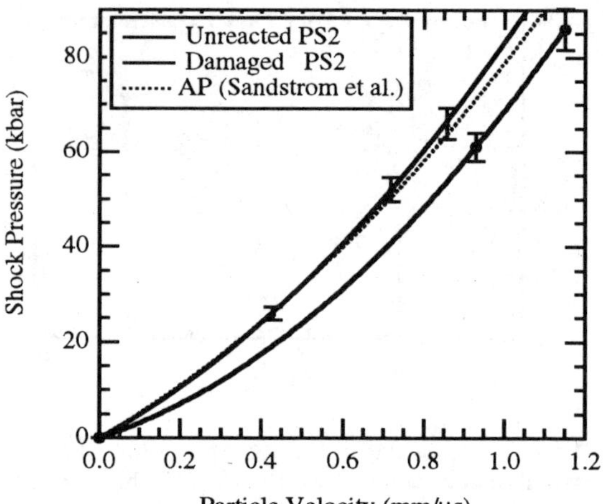

FIGURE 6. Hugoniots for pristine and pre-damaged PS2 propellant material.

REFERENCES

1. Bai, C. and J. Ding, Response of Composite Propellants to Shock Loading, in *Ninth Symposium (International) on Detonation*, 1989, Portland, Oregon:
2. Huang, F., C. Bai, and J. Ding, Mechanical Response of a Composite Propellant to Dynamic Loading, in *Shock Compression of Condensed Matter*, 1991, Williamsburg, Virginia: North Holland.
3. Lepie, A.H. and A. Adicoff, Advanced Physical Characterization of Solid Propellants. Part I: Damage, Volume Change and Dynamic Properties Under Strain, in *Proceedings of the JANNAF Structures and Mechanical Behavior Working Group*, 1977, Applied Physics Laboratory, Johns Hopknis University, Laurel Maryland: CPIA Publication 283.
4. Richter, H.P., L.R. Boyer, and A.H. Lepie, Shock Sensitivity of Damaged Energetic Materials, in *Ninth Symposium (International) on Detonation*, 1989, Portland Oregon:
5. Urtiew, P., *Private Communication*, 1995.
6. Sandstrom, F.W., P.A. Persson, and B. Olinger, Isothermal and Shock Compression of High Density Ammonium Nitrate and Ammonium Perchlorate, in *Proceedings of the 1993 Conference on Shock Compression of Condensed Matter*, 1993, Colorado Springs, Colorado: AIP Press.
7. Boteler, J.M. and A.J. Lindfors, *High Strain-Rate Mechanical Properties of PS1, PS2, and PS4 Propellants*, in *DNA Propellant Review, May 24, 1995,*
8. Bai, C., F. Huang, and J. Ding, Behavior of Ammonium Perchlorate Under Shock Loading, in *Shock Compression of Condensed Matter*, 1991, Williamsburg, Virginia: North Holland.

OBSERVATIONS OF SHOCK - INDUCED REACTION IN LIQUID BROMOFORM UP TO 11 GPa[†]

S. A. Sheffield, R. L. Gustavsen, and R. R. Alcon

Los Alamos National Laboratory, Los Alamos, NM 87545

Shock measurements on bromoform ($CHBr_3$) over the past 33 years at Los Alamos have led to speculation that this material undergoes a shock-induced reaction. Ramsay observed that it became opaque after a 1 to 2 μs induction time when shocked to pressures above 6 GPa (1). McQueen and Isaak observed that it is a strong light emitter above 25 GPa (2). Hugoniot data start to deviate from the anticipated liquid Hugoniot at pressures above 10 GPa. We have used electromagnetic particle velocity gauging to measure wave profiles in shocked liquid bromoform. At pressures below 9 GPa, there is no mechanical evidence of reaction. At a pressure slightly above 10 GPa, the observed wave profiles are similar to those observed in initiating liquid explosives such as nitromethane. Their characteristics are completely different from the two-wave structures observed in shocked liquids where the products are more dense than the reactants. As with explosives, a reaction producing products which are less dense than the reactants is indicated. BKW calculations also indicate that a detonation type reaction may be possible.

INTRODUCTION

Shock experiments on bromoform ($CHBr_3$) were done by Ramsay (1) at Los Alamos in the early 1960's. The objective of this work was to understand why some liquid explosives become opaque during shock-initiation. Nonexplosive liquids were also studied and bromoform was found to go opaque with an induction time of 1 to 2 μs when shocked above 6 GPa (1). Ramsay made Hugoniot measurements from 3 to 24 GPa, but from these no definitive reason for the material becoming opaque could be determined. He noted, however, that when compared with water, the Hugoniot had an odd shape in the shock-velocity vs. particle-velocity plane.

Experiments by McQueen and Isaak in the early 1980's showed that when bromoform is shocked to pressures above 25 GPa, the shock front emits radiation whose intensity varies with the shock pressure (2). In fact, light emission from shocked bromoform is used at Los Alamos as both a shock time-of-arrival detector and as an indicator of wave profile changes occurring in materials which are in contact with the bromoform. McQueen and Isaak's study did not lead to new information regarding a shock-induced reaction.

For some time we have been using the "universal" liquid Hugoniot developed by Woolfolk, Cowperthwaite, and Shaw (3) to estimate the Hugoniot for many liquids. Deviation from this Hugoniot often indicates the condition at which a shock-induced reaction might occur (4). When the Hugoniot data from Ramsay (1) and McQueen and Isaak (2) were plotted with the "universal" liquid Hugoniot for bromoform, deviations indicated that a reaction might be occurring at pressures as low as 10 GPa. Based on this, we have done further experiments to try to determine the shock pressure threshold and nature of the reaction.

EXPERIMENTAL DETAILS

Because bromoform has a relatively high density, 2.89 g/cm^3, pressures over 10 GPa could be obtained in single-shock experiments using our single-stage gas gun. Eight electromagnetic particle velocity gauging experiments of two different types have been completed in the pressure range of 3 to

---

[†] Work performed under the auspices of the U.S. Dept. of Energy.

10 Gpa. Parameters for these gas gun experiments are summarized in Table 1.

In the first type of experiments, called "Stirrup" experiments, magnetic "stirrup shaped" gauges at the front and back of the bromoform were used to measure the input and transmitted shock wave profiles. Stirrup experiments used a liquid cell 3 mm thick, 28.6 mm in inside diameter and 68.6 mm in outside diameter made from Kel-F plastic. A 3-mm-thick Kel-F front, and a 12-mm-thick Kel-F back plate completed the cell. The front, center ring, and back of the cell were epoxied and screwed together with nylon screws. Copper stirrup gauge elements, 5-μm thick on a 50-μm-thick Kapton substrate, were epoxied to the front and back cell pieces. The active gauge length was 9 mm, and the Kapton backing was in contact with the liquid. Five stirrup experiments were done.

The second type of experiment, called "MMG", for Multiple Magnetic Gauge experiment, consisted of a thin gauge package (with up to 10 particle velocity gauges in it) suspended at an angle in the liquid bromoform. This enabled the wave profile to be monitored at various depths in the liquid.

The MMG experiment is shown in an exploded view of Fig. 1. It consists of a two-piece PMMA body with an MMG package epoxied between the two pieces. The gauge package is on a plane at a 30 degree angle with the top of the cell. A Kel-F front completes a cell which is 40.6 mm inside diam. by 9 mm thick. The inside of the cell was lined with either Teflon or epoxy to keep the bromoform from dissolving or reacting with the PMMA. On some experiments a stirrup gauge was epoxied to the cell top as shown in Fig. 1. MMG cells were also epoxied and screwed together with nylon screws. Three of these experiments were completed.

Cells were filled just before the impact

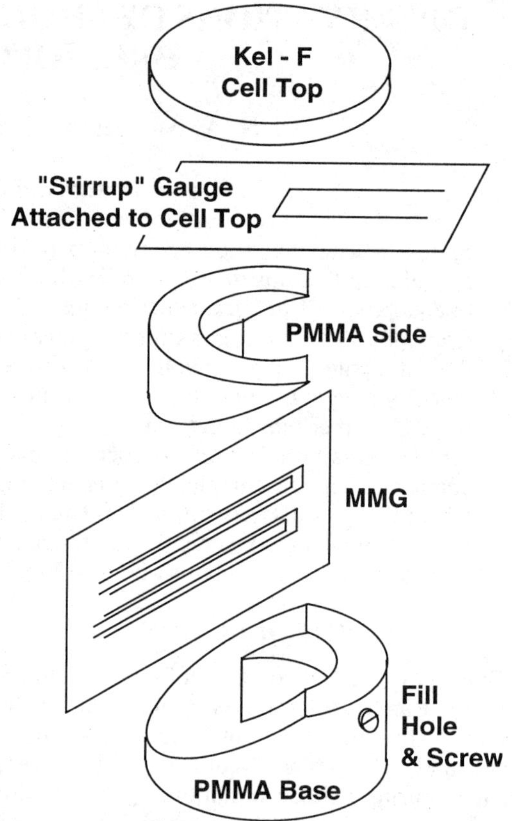

FIGURE 1. Exploded view of the MMG experiment showing the magnetic gauge and construction details.

experiment using Aldrich Chemical Co. bromoform (Aldrich #24,103-2). This bromoform is 99+ % pure, the major impurity being a small amount of ethanol stabilizer added by Aldrich.

Projectiles were made of Lexan and faced with impactors of either Vistal (pressed polycrystalline sapphire) or single crystal z-cut sapphire.

TABLE 1. Gas Gun Shot and Unreacted Hugoniot Data for Liquid Bromoform.

Shot No.	Type of Experiment	Impactor Material	Impact Velocity (mm/μs)	Particle Velocity (mm/μs)	Shock Velocity (mm/μs)	Shock Pressure (GPa)	Relative Volume (V/V$_0$)
741	Stirrup	Vistal	0.603	0.534	2.05	3.17	0.740
742	Stirrup	Vistal	0.798	0.680	2.45†	4.83	0.723
743	Stirrup	Vistal	1.000	0.840	2.58	6.26	0.674
744	Stirrup	Sapphire	(1.25)*	1.07	2.95	9.10	0.638
745	Stirrup	Sapphire	1.410	1.14‡	3.07‡	10.1‡	0.629‡
1033	MMG	Sapphire	0.964	0.83	2.542	6.09	0.674
1034	MMG	Sapphire	1.267	1.06	3.035	9.30	0.651
1035	MMG	Sapphire	1.391	1.16	3.147	10.6	0.631

† Back gauge data not good.　　* Projectile velocity estimated.　　‡ Evidence of reaction so data suspect.

RESULTS AND DISCUSSION

With these techniques, it was possible to measure the shock velocity in the bromoform quite accurately. In the stirrup experiments, the cell was rigid and the distance between gauges accurately known. Shock velocity was the distance between gauges divided by the wave transit time. In the MMG experiments, there were several gauges at fixed depths. The slope of a line fitted to the gauge depth vs. the wave arrival time gave a good shock velocity measurement. These quantities were determined for each of the experiments, even those suspected of having reaction, and are presented in Table 1. They are also plotted in Fig. 2 along with the data of Refs. 1 and 2 and the universal liquid Hugoniot for bromoform. With both the shock and particle velocity known, the mechanical state of the bromoform could be completely determined. Relevant quantities are also presented in Table 1.

Figure 2 clearly shows that Ramsay's lower pressure data are different from ours. Since his data were obtained from explosively driven experiments, at relatively low pressures, the inputs may not be accurately known. Our gun data should be more accurate because the pressure input is constant and easily controlled with the projectile velocity. That our data fall on or near the expected liquid Hugoniot is another indication of their accuracy.

Starting at pressures between 10-15 GPa the data of Refs. 1 and 2 lie below the expected liquid

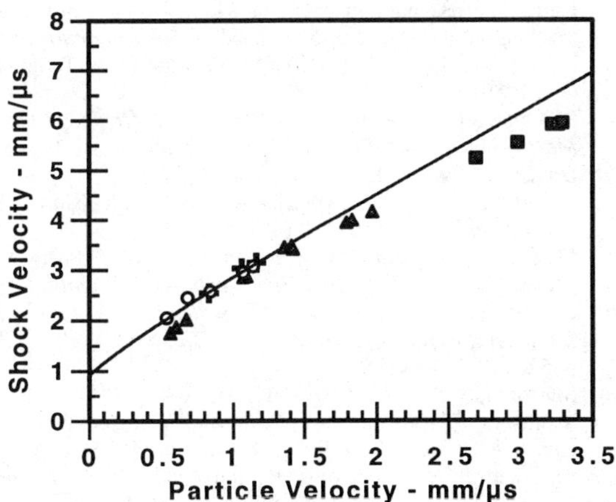

FIGURE 2. Hugoniot data for liquid bromoform. The line is the universal liquid Hugoniot using an initial condition sound speed of 0.931 mm/μs (5). Triangles are data from Ref. 1, and squares are data from Ref. 2. Data from our "stirrup" experiments are shown as circles and "MMG" data are shown as crosses.

bromoform Hugoniot. This is evidence that a reaction is occurring. Since the data are below the line, the products of the reaction are expected to be more dense than the reactants. This is similar to what has been observed in carbon disulfide (CS_2) (6), acrylonitrile (7), and other organic liquids. It is unknown whether or not this reaction causes the shocked bromoform to emit as indicated by McQueen and Isaak (2).

Ramsay states that bromoform becomes opaque at pressures above 6 GPa with an induction time of 1 to 2 μs (1). Neither the Hugoniot measurements nor the particle velocity waveforms measured in our study show any mechanical evidence of a reaction in the 6 to 9 GPa range. Particle velocity waveforms from a 9.3 GPa input MMG experiment are shown in Fig. 3a. There is no evidence in the waveforms of a chemical reaction. However, the bromoform has been held at pressure for scarcely one microsecond before the pressure is reduced by a rarefaction from the back of the impactor. It is possible that the reaction is too slow to be seen in this experiment. If a reaction does occur within one microsecond it does not result in a large enough volume change to be measurable with our particle velocity gauges. We do observe subtle waveform differences in this pressure regime but they are so small it would be unwise to interpret them as an indication of a reaction.

In contrast to the 9.3 GPa experiment of Fig. 3a very interesting waveforms were obtained at 10.6 GPa as can be seen in Fig. 3b. The four waveforms obtained from the MMG gauges in Fig. 3b are much like those obtained in homogeneous NM shock initiation experiments (8). In those experiments a reaction starts behind the shock front producing a spread out wave that then begins to move toward the shock front. As the reactive wave moves it steepens into a shock which grows in amplitude and eventually overtakes the initial shock. After overtake it has the character of a detonation wave. Analysis of the four waveforms shown in Fig. 3b indicates that bromoform is initiating in the same manner as the NM. In addition to this experiment, Shot 745 at 10.1 GPa had comparable behavior. Because there were only two gauges, one at the front and the other at the back of the bromoform, we did not understand what the waveforms meant until we saw the records obtained in Shot 1035.

Because bromoform has not been mentioned as an explosive material, these results were quite

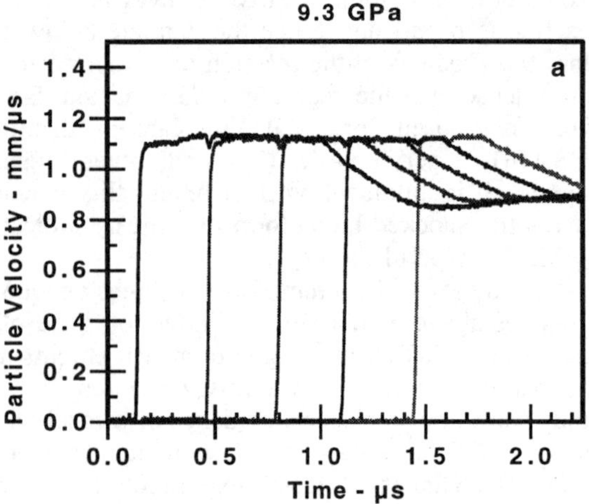

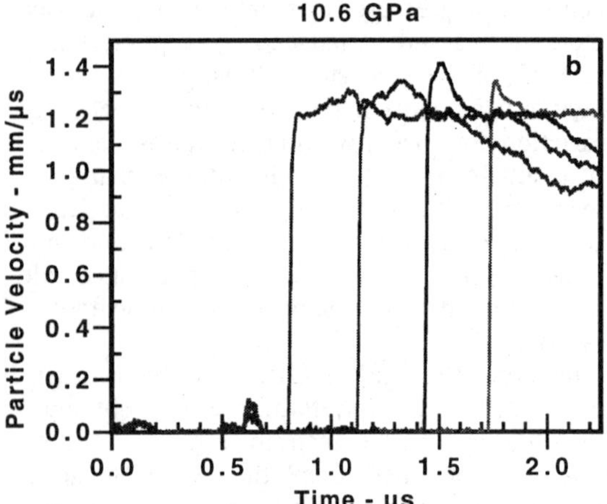

FIGURE 3. Magnetic particle velocity gauge waveforms from MMG experiments 1034 (a) and 1035 (b). Shot 1034 was at 9.3 GPa and showed no unusual behavior. Shot 1035 was at 10.6 GPa and has totally different waveforms. The two sets of waveforms are not time correlated because the gauges were at different depths.

surprising. A further evidence of bromoform's explosive behavior was that the aluminum shroud surrounding the target was expanded and cracked. This shroud protects the gun's target chamber from shrapnel originating from reacting explosive targets. It is never damaged during experiments on inert materials.

After this experiment was completed, we obtained BKW calculations on bromoform (9). These indicate that the expected reaction products are the gases HBr, Br_2, and CBr_4, and carbon as a solid. Further,

a detonation could occur with a C-J pressure of 3.2 GPa. This C-J pressure does not agree with our measurements, but it does indicate that a regime in which the products are less dense than the reactants exists and explosive initiation like waveforms are expected. It is unknown at this time whether or not bromoform would detonate in a cylinder of finite diameter. Ours are 1-D measurements and do not really indicate what may happen in 2-D geometry.

Above 15 GPa, the Hugoniot data in Refs. 1 and 2 fall below the expected liquid Hugoniot, indicating the products are more dense than the reactant. Thus, either the reaction mechanism changes at this pressure or else some of the product gases are compressed to the point they become condensed. This remains to be determined, perhaps in MMG experiments at higher pressures on our two-stage gas gun.

In summary, some very interesting reactions occur in shocked bromoform. It apparently becomes opaque beginning at about 6 GPa but either in a slow reaction or with a small volume change. At 10 GPa a detonation like reaction (products less dense than the reactant) is observed. This changes in nature somewhere above an input of 15 GPa to be a reaction in which the products are more dense than the reactants. Clearly, there is room for more research to determine the exact nature of these reactions.

REFERENCES

1. Ramsay, J. B., unpublished Los Alamos data from 1962. For Hugoniot data see p. 552, *LASL Shock Hugoniot Data*, Ed., Marsh, S. P., (University of California Press, Berkeley, CA, 1980).
2. McQueen, R. G., and Isaak, D. G., *Shock Comp. of Cond. Matter-1989*, Eds. Schmidt, S. C., Johnson, J. N., and Davison, L. W., p. 125.
3. Woolfolk, R. W., Cowperthwaite, M., and Shaw, R., *Thermochimica Acta* **5**, 409 (1973).
4. Sheffield, S. A., *Bull. Am. Phy. S.* **33**(3), 710 (1988).
5. Moses, A. J., *The Practicing Scientist's Handbook: A Guide For Physical and Terrestrial Scientists and Engineers*, (Van Nostrand Reinhold Co., New York, 1978), p. 526.
6. Sheffield, S. A., *J. Chem. Phys.* **81**, 3048 (1984).
7. Yakushev, V. V., Nabatov, S. S., and Yakusheva, O. B., *Comb. Expl. and Shock Waves* **10**(4), 509 (1975).
8. Sheffield, S. A., Engelke, R. P., and Alcon, R. R., *in Ninth Symposium (Intl.) on Detonation*, 1989, pp. 39-49.
9. Baer, M. R., and Hobbs, M. L., Sandia National Laboratories, private communication.

A STUDY OF ORGANIC REACTIONS
DRIVEN BY SHOCK WAVES IN LIQUIDS

Lloyd L. Davis and Kay R. Brower

Department of Chemistry, New Mexico Institute of Mining and Technology, Socorro, NM 87801

A method has been developed for investigating the chemical reactions of organic compounds under detonation conditions. Sealed metal capsules are placed along the axis of a cylindrical charge, which is initiated at one end. The reaction conditions are similar to those present in an explosive being shocked to detonation. The capsules allow recovery and subsequent analysis of solid, liquid, and gaseous reaction products. This method has been utilized to explore the chemical reactions of well over one hundred different compounds, covering many classes of organic compounds and most classes of explosives, using shock pressures ranging from 6 to 16 GPa. We have discovered profound effects on reaction rates and mechanisms induced by pressure, as well as some completely new reactions.

INTRODUCTION

In contrast to the extensive body of information available from the science of inorganic materials, remarkably few experimental data have been published on the organic reactions occurring under the high temperature and pressure conditions of detonation. The effect of temperature on reaction

$$k = A e^{\frac{-E_a}{RT}} \qquad (1)$$

rates is well understood, and usually expressed in the Arrhenius equation (Equation 1) where k is the rate constant, E_a the activation energy, A is the pre-exponential factor, T the temperature in Kelvin, and R the gas constant. (1,2). The effect of pressure on

$$\Delta V^{+} = -RT \left(\frac{\partial \ln k}{\partial P} \right)_T \qquad (2)$$

reaction rates, while less widely known, is also well understood (1-3). The parameter which governs the pressure effect is the activation volume ΔV^+, where P is the pressure (Equation 2). The effect of pressure is largely due to two different factors:

$$\Delta V_e = -\frac{Nq^2}{2r} \cdot \frac{1}{\varepsilon} \cdot \left(\frac{d\varepsilon}{dp} \right)_T \qquad (3)$$

reduction of free space, and electrostriction. The volume change due to electrostriction (ΔV_e) may be calculated from the Drude-Nernst equation (Equation 3) where N is Avogadro's number, q is the charge on the ion of radius r, and ε is the dielectric constant of the medium (2). While these effects have been understood and studied under less severe reaction conditions for many years, experimental difficulties have complicated work in the temperature and pressure range usually associated with detonating explosives.

EXPERIMENTAL METHOD

We have developed and characterized a method for encapsulating solutions of organic compounds, subjecting them to explosive-driven shock waves, recovering and analysing the contents (4). The reaction conditions are determined by the characteristics of the explosive used. The bulk of the sample (95%) experiences a peak pressure approximately equal to the CJ pressure (4). A small portion of the sample occupies a Mach stem as pointed out by other researchers (5). The chemical reactions occurring within the Mach stem result from general homolytic dissociation of carbon-

carbon and carbon-hydrogen bonds leading to the formation of soot, hydrogen, and methane. These reactions do not affect the bulk of the sample (4,5).

Because this is essentially a single-shock experiment it may be assumed that the sample is compressed along its unreacted Hugoniot, and that the chemical reaction in the sample occurs in a reaction zone immediately behind the shock front as in a detonating explosive. Experimental evidence that chemical reaction occurs while the sample is still in a state of shock compression has been obtained by other researchers (6).

Reaction Conditions

We have calculated the reaction conditions in these experiments by several different methods. The Lagrangian hydrodynamic analysis discussed previously (4) has yielded results consistent with all other methods used in this work as well as those obtained by other researchers (5). Various solid polymeric inserts have also been used to verify the temperature and pressure distribution present in the sample space (4,5,7). These experiments show that the amount of sample destroyed in the Mach stem is approximately 5% of the total volume.

Shock Pressure

Five different types of explosives have been used in these experiments: emulsion with an initial density of 0.90, a detonation velocity of 4.81 mm/μS and a CJ pressure of approximately 6 GPa; emulsion with $\rho_0 = 1.05$, P=8 GPa and D=5.15 mm/μs.; nitromethane sensitized with 2.5% diethylenetriamine (DETA) and containing 10% acetone, $\rho_0 = 1.095$, D=5.95 mm/μS and $P_{CJ} = 13$ GPa; nitromethane with 2.5% DETA only, $\rho_0 = 1.128$, D=6.21 mm/μS and $P_{CJ} = 15$ GPa; and a saturated solution of Pentolite in nitromethane (25% by weight Pentolite) $\rho_0 = 1.33$, D=6.41 mm/μS and P=16 GPa. Pressures were calculated by the Tiger code using a BKW-R EOS (8) from measured detonation velocities. These explosives encompass the range of shock pressures most useful for the study of organic chemistry. The low-density emulsion generates little or no reaction in most of the materials tested, and the NM-Pentolite solution

creates conditions at the threshold of solvent decomposition.

Shock Temperature

The average shock temperature in our experiments has been calculated for each type of explosive used, using chemical reactions whose activation parameters are well known, either from the literature or from measurements in our laboratory. For a reaction with an activation volume known to be nearly zero, the temperature may be calculated directly from the observed extent of reaction, the reaction time, and the relevant Arrhenius law: T= $-E_a/(R \ln(k_{obs}/A))$.

One example is the isomerization of β-pinene, which undergoes unimolecular biradical ring-opening leading to formation of limonene and myrcene. Because the activation volume of the reaction yielding limonene is zero the average temperature may be calculated from the observed yields of these two compounds if a characteristic reaction time of 1 μS is taken from the hydrodynamic analysis (4).

This method might be criticized if it were based on only one or two observations, but a more comprehensive manuscript in preparation will report temperature calculations based on 38 Arrhenius laws and their respective activation volumes, both positive and negative. For this method of temperature measurement we selected reactions known to be well-behaved under a wide range of conditions. The most useful types of reaction are the biradical isomerizations, concerted electrocyclic processes such as the Claisen rearrangement and retro-Diels-Alder reactions, and simple homolytic dissociations.

RESULTS

Most compounds exhibited an extent and type of reaction in good agreement with expectations. Temperature calculations proved to be reproducible and consistent. Solvent choice affects the temperature somewhat, as the shock Hugoniots and polarities may differ significantly. For cyclohexane the calculated temperature is always lower than for aromatic solvents. For benzene the temperature produced by the five explosives listed above in the order given is 910, 1090, 1210, 1210, and 1400 $\pm$

100 K. Methanol is known to become highly ionized (9) in explosive-driven shocks, and acid catalysis is apparent in some reactions.

Biradical Isomerizations

These reactions are initiated by homolysis of a strained-ring or π-bond. As discussed previously, the activation volumes will be small, and these reactions will be affected little by pressure. 1,2-diphenylcyclopropane undergoes cis-trans isomerization as well as a ring-opening process which yields 1,3-diphenylpropene. The pre-exponential factor for isomerization is $10^{11.2}$, and it has an activation energy of 33.5 Kcal/mole (10). In the NM / DETA / acetone explosive, the extent of reaction observed for this compound in benzene (average for 25 shots) yields a shock temperature of 1230 K. In the emulsion explosive with a density of 1.05 and a shock pressure of 8 GPa., no reaction was observed. If 5% is chosen as the maximum extent of reaction, an upper boundary is placed on the sample shock temperature generated by this explosive at 1100 K. For the saturated solution of Pentolite in nitromethane, the extent of reaction is greater than 80% and the minimum temperature is 1500 K. The reactions of β-pinene have been discussed earlier in this paper. The two major products are shown in Fig. 1 below. This compound has been used to calculate temperature in

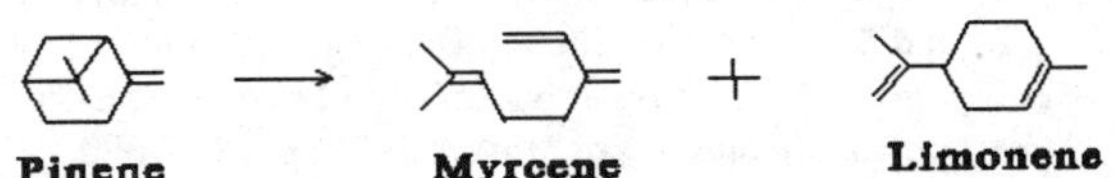

Figure 1. β-pinene undergoes two competing biradical isomerization reactions to produce myrcene and limonene in equal yields when the shock pressure is 13 GPa.

35 different experiments. In the case of rearrangement to myrcene (11), the pre-exponential factor is $10^{15.9}$ and the activation energy is 49.9 Kcal/mole. The formation of myrcene will be retarded at ~ 10 GPa by an approximate factor of 30. The rearrangement to limonene (11) has an A value of $10^{13.5}$, an activation energy of 45.1 Kcal/mole, and an activation volume of zero. In benzene or toluene solvent, it does not react when

shocked by emulsion explosive which sets an upper boundary on the temperature of 970 K. With the NM/DETA/acetone mixture the average temperature is 1190 K for limonene kinetics. When the explosive is the NM/Pentolite solution approximately 80% of the β-pinene reacts, and the average shock temperature is 1340 K for the limonene formation.

The effect of pressure on the conversion of cis-stilbene to the <u>trans</u> isomer has been discussed previously (4). This reaction has been used more than any other to determine the sample temperature for the NM/DETA/acetone explosive. For an average of many trials, the observed extent of reaction under these conditions was 33%; the known Arrhenius A value of $6 \bullet 10^{12}$ and activation energy of 42.8 Kcal/mole (1), together with an activation volume of -2mL/mole, yields a shock temperature of 1290 K in benzene.

Concerted Electrocyclic Reactions

A variety of retro-Diels-Alder reactions have been observed, including the dissociation of dicyclopentadiene into cyclopentadiene, which is trapped by a forward Diels-Alder reaction with ethyl acrylate. For norbornylene, the retro-Diels-Alder reaction is effectively irreversible. This reaction indicates shock temperatures of 1190 K for the NM/DETA/acetone solution, and 1100 K for the higher-density emulsion.

The Claisen rearrangement of allyl cresyl ether is a pressure-assisted process that competes with the homolysis of the ArO-allyl bond. The activation volumes for these two processes are approximately equal in magnitude and opposite in sign. We estimate a ten-fold acceleration of the rearrangement at high pressure. At low pressure (12) the activation volume is -16 mL but at 0.8 GPa it falls to -3 mL and will decline further at higher pressures. For the homolytic dissociation, the activation energy is simply the bond strength. For the Claisen rearrangement, we have one of the most extensive Arrhenius plots ever constructed, with reaction times ranging from six months to 0.10 seconds. In order to account for the observed product distributions, the temperature and pressure must both be close to the previously calculated values. This compound indicates a temperature of 1230 K for the 13 GPa. nitromethane solution, and 1020 for the higher-density emulsion.

Homolytic Dissociations

Nitrite esters homolyse to produce an alkoxy radical and NO. Hydrogen abstraction yields the corresponding alcohol, and N_2O by dimerization nitroxyl and dehydration of hyponitrous acid. We have used pentyl nitrite to calculate the temperature generated by the low-density emulsion at 916 K, based on an observed reaction extent of 43% and published activation parameters. N-nitropiperidine also undergoes a similar homolytic dissociation; it yielded a calculated temperature of 912 K for the same explosive.

Ionic Reactions

In the range of pressures we discuss here, the effects of electrostriction become significant whenever an ionic reaction mechanism is involved. Most of these reactions occur with strong acid or base catalysis; their observation without catalysis in shock waves is completely new and demonstrates the effect of electrostriction. As previously discussed, methanol becomes highly ionized in explosive-driven shock waves (9). In our experiments this results in formation of substantial amounts of methyl ether and water. Methanol also adds to double bonds such as in cyclohexene, forming cyclohexyl methyl ether.

The Claisen condensation usually requires a strong base catalyst, but in our experiments the hydrogen of acetone becomes sufficiently acidic to produce the enol form, which rapidly condenses to produce diacetone alcohol and 4-methyl-3-penten-2-one. Alkyl halides produce carbocations that may eliminate, rearrange, and alkylate benzene exactly as in the Friedel-Crafts reaction, albeit without Lewis acid catalysis.

Perhaps one of the most significant reactions is the ionization of nitrate esters, which was previously found at lower pressures and temperatures (12). Cyclohexyl nitrate, for example, forms the cyclohexyl carbocation in our experiments instead of cyclohexyloxy radical by homolytic dissociation of the $O-NO_2$ bond observed at low pressure. In cyclohexane, the carbocation eliminates to produce cyclohexene; in benzene the intermediate alkylates solvent to produce cyclohexyl benzene as in the Friedel-Crafts reaction. If the solvent is methanol, the product is methyl cyclohexyl ether.

CONCLUSION

The results obtained in this study provide a comprehensive basis for understanding the shock-induced reactions of organic compounds. Application of transition state theory with measured or reliably estimated activation parameters predicts reaction rates that are in close agreement with observed results. Compounds that were not expected to react did not do so. Pressure-assisted mechanisms were strongly promoted, and pressure-retarded processes were either slowed significantly or were not observed at all. The temperatures calculated from chemical kinetics agree well with values obtained from hydrodynamic codes (4,5).

REFERENCES

1. Benson, S.W., *Thermochemical Kinetics*, New York: Wiley, 1968
2. Isaacs, N.S., *Physical Organic Chemistry*, New York: Wiley, 1992
3. Asano, T., Le Noble, W.J., *Chem. Rev.* **78** (4), 407-489 (1978)
4. Davis, L.L., Brower, K.R., and Libersky, L.D., *Rev. Sci. Inst.* **66** (5), 3321-3326 (1995)
5. Morris, C.E., Loughran, E.D., Mortensen, G.F., Gray, G.T. (III), and Shaw, M.S., "Shock Induced Dissociation of Polyethylene," in *Proceedings of the Conference on Shock Compression of Condensed Matter*, 1989, pp. 687-690.
6. Joshi, V.S., Thadhani, N.N., Graham, R.A., and Holman, G.T., "Shock Compression Behavior of Quartz and Al Powder Mixtures", in these Proceedings.
7. Brower, K.R., Piunti, R., Kurniadi, W., Cash, M., and Davis, L.L., RCEM Report A-0-1-93, (1993)
8. Persson, P.A., and O'Connor, E., Private Communication.
9. David, H.G., and Hamann S.D., *Trans. Far. Soc.* **56**, 1043-1050, (1960)
10. Rodewald, L.E., and DePuy, C.H., Tetrahedron Lett., 40, 2951-2953, (1964)
11. Hawkins, J.E., and Vogh, J.W., *J. Phys. Chem*, **57**, 902-905, (1953)
12. Brower, K.R., *J. Amer. Chem. Soc.* **83**, 4370-4372, (1961)
13. Naud, D.L., and Brower, K.R., *J. Org. Chem.* **57**, 3304, (1992)

REACTION ZONE MEASUREMENTS IN DETONATING ALUMINIZED EXPLOSIVES

S. N. Lubyatinsky, B. G. Loboiko

Russian Federal Nuclear Center-Institute of Technical Physics
P. O. Box 245, Snezhinsk, Chelyabinsk region 456770 Russia

Detonation reaction zone measurements have been made on five RDX-based explosives (60 μm average particle size RDX), containing 6% polymer binder and from 0 to 19% aluminum of different particle size (from 2 μm to 20 μm). A photoelectric technique was employed to record the radiation intensity history of the shock front propagating through chloroform in contact with the charge face. The record was then translated into the explosive/chloroform interface velocity history. In all cases, the Zeldovich-von Neumann-Doering detonation wave structure was observed. Aluminum particle size was found to have no appreciable effect on the reaction zone length, which increases from 0.34 mm to 0.58 mm as aluminum content increases from 0 to 19%. Nevertheless, the reaction zone lengths of the studied explosives are less than that of RDX/TNT 50/50 (0.59 mm), which implies relatively high rate of the reaction between aluminum and RDX detonation products.

INTRODUCTION

The subjects of this study are aluminized explosives. An essential fraction of the detonation energy of these explosives is due to the reaction between the aluminum and products of the base explosive. To gain a better understanding of the mechanism of this reaction detonation reaction zone measurements have been made on five RDX-based aluminized explosives, X-0, X-9-2F, X-9-6S, X-9-20S, and X-19-6S. The first index is the percentage of aluminum. The second index is the average diameter of spherical aluminum particles (S) or the average thickness of flake aluminum particles (F) in microns. The average particle size of RDX was about 60 μm. Each explosive contained 6% of a polymer binder and was pressed to 1% porosity.

EXPERIMENTS

The reaction zone measurements were carried out by recording the radiation intensity history of the shock front in chloroform placed on the end of a detonating explosive using a photoelectric technique

(1-9) (see Fig. 1). The main advantage of the technique is the combination of high sensitivity (about 0.2% in particle velocity), high time resolution (about 5 ns), and long recording time (over 1 μs), unlimited by the perturbations at the explosive/liquid interface, which are critical for laser interferometry.

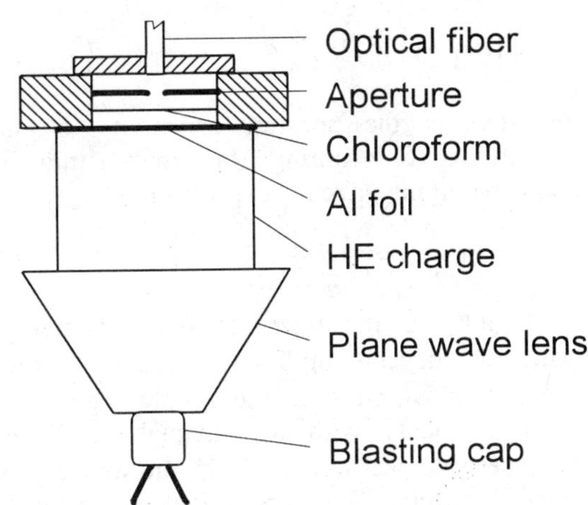

FIGURE 1. The experimental setup.

The 10 μm aluminum foil, screening detonation radiation, is attached to the face of the 40 mm diameter and 80 mm long charge with an epoxy. The 1.5 mm aperture restricts the viewed portion of the shock front to 5 mm in diameter. The shock front radiation is transmitted by a 1.5 mm plastic optical fiber to a SNFT-3 photomultiplier, whose output is recorded on a SUR-1 oscilloscope.

The shock Hugoniot for chloroform was determined by Dick (10). In the range 1.8-4.8 km/s the data were fitted to the linear relation

$$U_S = C_0 + S U_P,\qquad(1)$$

where $C_0 = 1.774$ km/s, $S = 1.367$ (9).

The relationship between the photomultiplier output I and shock parameters was derived regarding the shock front as the black body, whose radiation in the visible region is described by the Wien law

$$B_\lambda = \frac{2hc^2}{\lambda^5}\exp\left(-\frac{hcL}{\lambda RT}\right),\qquad(2)$$

where $c=2.998\times10^8$ m/s is the speed of light, $h=6.625\times10^{-34}$ Js is the Plank constant, $L=6.022\times10^{23}$ mol^{-1} is the Avogadro constant, $R=8.314$ JK^{-1}mol^{-1} is the gas constant, λ is wavelength, and T is temperature. The expression for I was simplified by introducing the effective wavelength λ_{eff}

$$I(T) = \int_0^\infty \alpha(\lambda)B_\lambda(\lambda,T)d\lambda \propto B_\lambda\left(\lambda_{eff},T\right),\qquad(3)$$

where $\alpha(\lambda)$ is the spectral sensitivity of the photomultiplier. Substituting from Eq. (2) into Eq. (3) and rearranging gave the expression

$$\ln\left(\frac{I}{I_0}\right) = \frac{hcL}{\lambda_{eff}R}\left(\frac{1}{T_0}-\frac{1}{T}\right),\qquad(4)$$

where $I_0 = I(T_0)$ is the reference point. Generally speaking, λ_{eff} depends on T but, because of strong dependence of B_λ on λ in the visible region, λ_{eff} practically coincides with the maximal wavelength of the observed spectrum $\lambda_{max}\approx625$ nm and can be considered constant. The pressure dependence of the shock temperature for chloroform was determined by Gogulya et al. (11) in the range 8.5-31.7 GPa. The data were fitted to the relation

$$T = aU_P^b,\qquad(5)$$

where $a=943$ K, $b=1.17$, and U_P is in units of km/s

(9). Substituting from Eq. (5) into Eq. (4) and rearranging gave the expression for $U_P(I)$

$$U_P = \left(U_{P0}^{-b} - A\cdot\ln\left(\frac{I}{I_0}\right)\right)^{-1/b},\qquad(6)$$

where U_{P0} is the particle velocity at the reference point, I_0 is the corresponding photomultiplier output, and $A = \dfrac{a\lambda_{eff}R}{hcL}$ is a constant. To determine A we carried out three calibration shots in which there was an aluminum or copper plate between chloroform and an explosive with well known CJ parameters. The thickness of the plate was 1 mm for an HMX-based explosive and 2 mm for an RDX-based explosive. The length of the charge was 150 mm. Shock reverberations in the plate generated a sequence of shock waves in chloroform. The particle velocities behind these shocks U_{P1}, U_{P2}, ... were calculated using the method of shock-impedance matching. The corresponding oscilloscope record looked like a staircase with the heights of the "steps" I_1, I_2, ... For each shot the value of A was found by substituting the parameters of the first two shocks in Eq. (6). In the range 1.9-2.6 km/s, covered by the measurements, the average value of A is 0.039 ± 0.001, which corresponds to $\lambda_{eff}=(595\pm15)$ nm. As expected, λ_{eff} is really very close to λ_{max}.

With characteristics method it can be shown that the metal/chloroform interface velocity at time t_i is equal to the particle velocity at the shock front at time t_f with

$$Z = \frac{dt_i}{dt_f} = 1 - \frac{U_S - U_P}{C},\qquad(7)$$

where C is the sound velocity in chloroform at the shock front. $Z=0$ at $U_P=0$. As U_P increases Z increases as well. The upper limit to Z was estimated considering chloroform as a perfect gas

$$Z_\infty = \lim_{U_P\to\infty} Z = 1 - \frac{1}{\sqrt{\delta_\infty+1}} = 0.54,\qquad(8)$$

where

$$\delta_\infty = \lim_{U_P\to\infty}\frac{U_S}{U_S - U_P} = \frac{S}{S-1} = 3.7\qquad(9)$$

is the maximum shock compression, which was calculated on the assumption of the validity of the linear U_S–U_P relation (1) when $U_P\to\infty$.

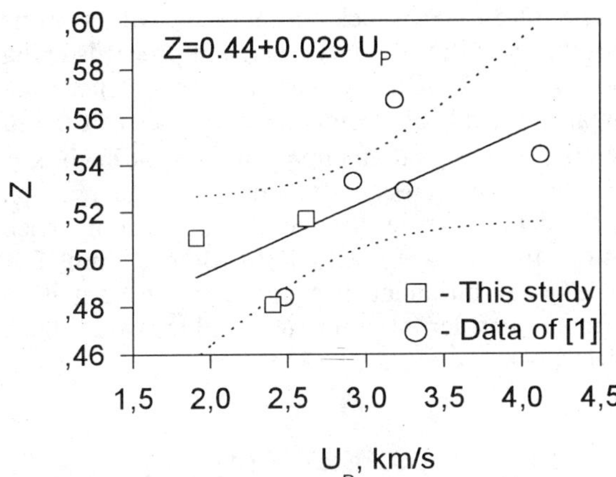

FIGURE 2. Z versus U_P. The linear fit is shown with a solid line and the 95% confidence interval is shown with dotted lines.

For each calibration shot the value of Z was found as the ratio of the shock reverberation time in the plate to the corresponding interval between the two steps in the oscilloscope record. The shock reverberation time was calculated using the experimental sound velocities in shocked metals (12). The values of Z found in this study are plotted in Fig. 2 along with the values of Z calculated using the experimental sound velocities in shocked chloroform (1). One can see that although Z tends to increase with increasing U_P, in the range 1.9-4.1 km/s this increase is negligible. This can be explained by Z almost coinciding with Z_∞, which enables us to consider Z constant and equal to its average value

$$Z = \frac{t_i}{t_f} = 0.52 \pm 0.02. \qquad (10)$$

Two shots were carried out for each of five explosives. Both oscilloscope records were digitized and averaged. In the region unaffected by the reaction zone the averaged record $I(t_f)$ was least-squares fitted to the equation

$$I = I_0 + I_1 \cdot t_f + I_2 \cdot t_f^2, \qquad (11)$$

where $I_0 = I(U_{P0})$ is the reference point. The particle velocity U_{P0} behind the shock wave produced in chloroform by the detonation products expanding from the CJ point was calculated using the CJ parameters determined earlier. The record $I(t_f)$ was then transformed into the interface velocity history $U_P(t_i)$ using Eqs. (6) and (10). A representative profile $U_P(t_i)$ is shown in Fig. 3.

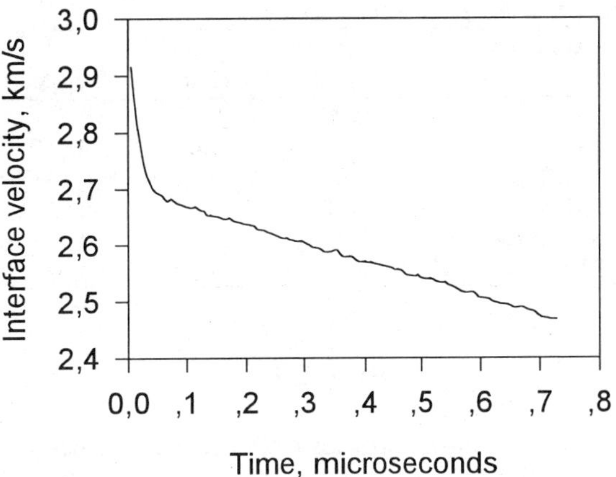

FIGURE 3. Interface velocity versus time for X-9-6S.

For all the explosives the Zeldovich-von Neumann-Doering detonation wave structure was observed. The rise time of the von Neumann spike was about 5 ns (the resolution time of the technique). The reaction started at the front without delay. The reaction rate being initially maximal decreased as the reaction proceeded. The reaction zone parameters were found as follows (9). The smoothed profile $U_P(t_i)$ was differentiated and the function $\ln(-dU_P/dt_i)$ was plotted against t_i. The plot consisted of two straight lines corresponding to different laws of shock attenuation. The point of their intersection t_J was associated with the CJ point. The reaction zone length was calculated with the formula

$$x_J = \int_0^{t_J} (D - U_P)\, dt_i, \qquad (12)$$

derived from analysis of the x-t diagram. D is the detonation velocity of the explosive. Linearity of the function $\ln(-dU_P/dt_i)$ in the reaction zone implies the exponential dependence of U_P on t_i, which was used to fit the $U_P(t_i)$ profiles,

$$U_P = U_{PI} + (U_{PS} - U_{PI})\exp(-t_i/\tau), \qquad (13)$$

where τ is the characteristic time of the reaction, U_{PS} is the initial interface velocity, and U_{PI} is a fitting parameter. For the studied explosives the standard deviations of the $U_P(t_i)$ data from the fits range from 3 to 6 m/s. The reaction zone lengths and characteristic times of reaction for the studied explosives are listed in Table 1 in comparison with those for RDX/TNT 50/50 (9).

TABLE 1. Reaction Zone Parameters.

Explosive	t_J (μs)	x_J (mm)	τ (μs)
X-0	0.06	0.34	0.011
X-9-6S	0.07	0.40	0.017
X-9-20S	0.08	0.46	0.021
X-19-6S	0.10	0.58	0.025
X-9-2F	0.11	0.63	0.027
RDX/TNT 50/50	0.12	0.59	0.029

RESULTS

The exponential profiles of the particle velocity in the reaction zone, earlier noted for a number of explosives (9), probably imply that the reaction is of the first-order, characteristic of diffusion controlled reactions.

Both 6 and 20 μm spherical aluminum particles react with RDX detonation products more readily than 2 μm flake aluminum particles.

The increase of spherical aluminum particles diameter from 6 to 20 μm results in the reaction zone length increase of only 10-20%. This can be explained on the assumption that the reaction is diffusion controlled. Then the reaction rate is inversely proportional to the characteristic heterogeneity size, which is determined by the particle size of the most coarse component. In this case, this is RDX, whose average particle size was about 60 μm. Further increase of aluminum particle size must result in more noticeable increase of the reaction zone length.

Admixing of 19% 6 μm spherical aluminum particles to the RDX-based explosive results in a twofold increase of the reaction zone length. This indicates that the reaction rate is markedly affected by the detonation front parameters, which tend to decrease as aluminum content increases. At the same time, higher CJ temperatures of the aluminized explosives do not result in higher rates of the reaction.

The reaction zone lengths of the aluminized explosives do not exceed that of RDX/TNT 50/50. That is, the rate of the reaction between aluminum and detonation products of RDX is comparable with the rate of the reaction between TNT and detonation products of RDX. Moreover, using an optical pyrometer Gogulya et al. (5, 13) showed that the mixture of 30-60 μm aluminum and 10 μm sulfur behind 33-36 GPa shock waves reacts within 50 ns with the extent to which the reaction proceeds being more than 50%. That is, reactive mixtures need not contain an explosive component to react in time of the order of 0.1 μs if an appropriate shock loading is provided. This possibly implies the same reaction mechanism for all reactive mixtures in strong shock waves. In all cases, actual diffusion is probably preceded by turbulent dispersive mixing as it is in inorganic powder mixtures subjected to strong shock waves (14).

REFERENCES

1. Gogulya, M. F., Dolgoborodov, A. Yu., *Khimicheskaya Fizika (Russian J. of Chemical Physics)* **13**, 118-127 (1994).
2. Voskoboinikov, I. M., Gogulya, M. F., Voskoboinikova, N. F., Gelfand, B. E., *DAN SSSR (the USSR Academy of Sciences Reports)* **236**, 75-78 (1977).
3. McQueen, R. G., Hopson, J. W., Fritz, L. N., *Review of Scientific Instruments* **53**, 245-250 (1982).
4. Gogulya, M. F., *FGV (Russian J. of Physics of Combustion and Explosion)* **25**, 95-104 (1989).
5. Gogulya, M. F., Voskoboinikov, I. M., Dolgoborodov, A. Yu., Dorokhov, N. S., Brazhnikov, M. A., *Khimicheskaya Fizika (Russian J. of Chemical Physics)* **11**, 244-247 (1992).
6. Voskoboinikov, I. M., Gogulya, M. F., *Khimicheskaya Fizika (Russian J. of Chemical Physics)* **3**, 1036-1039 (1984).
7. Lubyatinsky, S. N., Vorobey, V. A., "Study of detonation wave front structure in high explosives using a photoelectric technique," in *Proceedings of the Fifth All-Union Meeting on Detonation*, Krasnoyarsk, USSR, August 5-12, 1991, pp. 369-373.
8. Lubyatinsky, S. N., Vorobey, V. A., "Study of detonation front structure in high explosives using a photoelectric technique," in *Proceedings of the Second International Symposium on Intense Loading and Its Effects*," Chengdu, China, 1992.
9. Lubyatinsky, S. N., Loboiko, B. G., "Study of chemical reaction zone structure in detonating high explosives using a photoelectric technique," presented at the Symposium on Energetic Materials Technology, Pleasanton, California, May 18-25, 1994.
10. Dick, R. D., *J. of Chemical Physics* **74**, 4053-4061 (1981).
11. Gogulya, M. F., Voskoboinikov, I. M., Bulanov, I. V., *Khimicheskaya Fizika (Russian J. of Chemical Physics)* **5**, 1426-1428 (1986).
12. Altshuler, L. V., Kormer, S. B., Brazhnik, M. I., Vladimirov, L. A., Speranskaya, M. P., Funtikov, A. I., *ZhETF (Russian J. of Experimental and Theoretical Physics)* **38**, 1061-1073 (1960).
13. Gogulya, M. F., Voskoboinikov, I. M., Dolgoborodov, A. Yu., Dorokhov, N. S., Brazhnikov, M. A., *Khimicheskaya Fizika (Russian J. of Chemical Physics)* **10**, 420-422 (1991).
14. Y. Horie, M. E. Kipp, J. Applied Physics **63**, 5718-5727 (1988).

STRUCTURE AND PROPERTIES OF DETONATION SOOT PARTICLES

I.Yu. Mal'KOV and V. M. Titiov

Lavrentyev Institute of Hydrodynamics, SD RAS, Novosibirsk, 630090, Russia

The influence of TNT/RDX (50/50) detonation parameters and conservation conditions of detonation products during their expansion in hermetic detonation chamber on structure and phase composition of the detonation carbon has been considered. Systematic studies made it possible to establish the real structure of detonation carbon depending on experimental conditions. It has been shown that both during explosion in a chamber and thermal annealing in vacuum the nanoparticles of diamond have the tendency to transform not into graphite particles, as was assumed earlier, but into onion-like structures of fullerene series, composed of closed concentric carbon shells, the so-called carbon onions. The nanometer carbon particles have been obtained which comprise a diamond nucleus surrounded by a graphite-like mantle composed of quasi-spherical carbon shells which are the intermediate products of annealing of nanodiamond. The influence of initial sizes of the diamond particles and temperature on the annealing of diamond has been studied.

INTRODUCTION

Investigation of structural transformations of carbon nanoparticles is a new area in carbon physics and chemistry. In this sense detonation soot is perspective material for scientific investigations. It should be noted that in spite of the well known fact that solid carbon is one of the major detonation product (DP) of high explosives (HE) with negative oxygen balance, previously insufficient attention was paid to investigation of structure of detonation soot particles.

Necessity in such study arose as early as in 60-th, when numerical modeling of detonation was begun to develop intensively. Another stimulus to such investigation was given in 80-th after observation of ultra-disperse diamond (UDD) in detonation products of a number of HE [1-6]. At present the investigation of diamond formation under detonation conditions is being quickly developed at several scientific centers.

The main objectives which we persuaded when launching the present study were: 1) What is the real structure of detonation soot particles (both diamond and non-diamond)? 2) In what degree the structural transformations of carbon nanoparticles differ from ones of bulk carbon phases?

EXPERIMENTAL

To obtain diamond containing soot TNT/RDX 50/50 charges were detonated in 0.17 m^3 hermetic detonation chamber. Depending on objectives, the charges were confined in ice shells and/or chamber was filled up to required pressure with gaseous N_2. We used both cast and pressed charges of different density and with mass up to 100 g. Varying pressure of N_2 and density of HE made us possible to investigate the "response" of soot structure on preservation conditions and thermodynamic parameters of detonation. In more details the preservation detonation product technique is described elsewhere [6].

Obtained soot samples were studied by X-ray phase analysis (XRD), high resolution transmission microscopy (HRTEM) and small angle X-ray scattering (SAXS) technique.

RESULTS AND DISCUSSION

Investigation have shown that both preservation condition and thermodynamic parameters of detonation strongly affect the detonation soot structure.

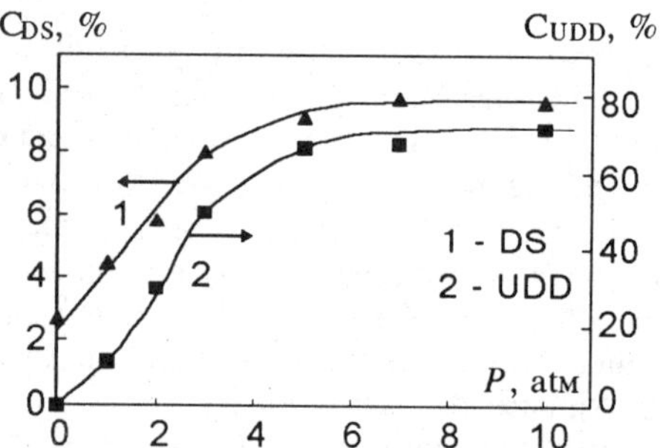

FIGURE 1. Dependence of detonation soot output and UDD content on N_2 pressure in detonation chamber. $C_{DS}=M_{DS}/M_{HE}$; $C_{UDD}=M_{UDD}/M_{DS}$.

Figure 1 demonstrates the dependence of detonation soot (DS) output and UDD content in it on N_2 pressure. When N_2 pressure is not sufficient to suppress the overheating DP during expansion in the chamber this results in partial gasification of DS and annealing of UDD.

Recently we have shown that in general case detonation carbon can contain at least 4 types of particles: 1) amorphous particles 5.0-10.0 nm in size without apparently ordered structure, 2) stretched graphite-like particles with high degree of interplane disordering up to 20.0 nm in size, 3) diamond particles 2.0-15.0 nm in size; 4) quasi-spherical particles with so-called onion-like structure 2.0-10.0 nm in size [7,8].

Figure 2 shows TEM micrographs of the detonation soot obtained under unfavorable for UDD preservation conditions. One can see partially annealed diamond particle consisting of diamond core and several curved graphite-like planes, defected onion-like carbon (OLC) particles and highly disordered carbon.
To confirm that OLC particles form due to annealing of UDD we undertook the investigation of diamond annealing in vacuum [7,8,9]. In this case isolated by the oxidative

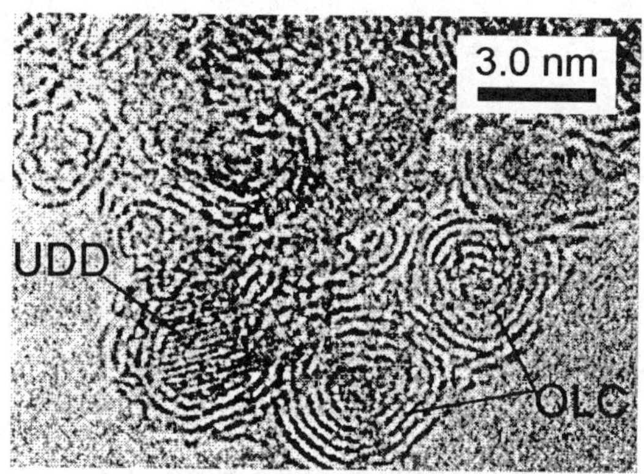

FIGURE 2. High-resolution image of detonation soot particles. Partially annealed diamond particle, defected onion-like carbon particle and highly disordered carbon are clearly seen.

removal of non-diamond carbon from the detonation soot UDD was heated in vacuum up to 1800 C.

Figure 3 a) shows medium resolution TEM image of UDD. One can see that UDD mainly consist of particles from 2,0 to 15,0 nm. Lager particles with good developed facets are clearly seen.

After annealing UDD almost completely transformed to carbon particles consisting of closed graphite-like shells (Fig. 3 b)).

Form of annealed particles may vary from near ideal spherical to polyhedral with flat layers, last been vary rare. Besides there are particles with one or several common external shells which most likely form from twinned and stuck diamond particles.

Varying the temperature and duration of heating we obtained the samples containing the intermediates of UDD transformation. Figure 4 demonstrates the pseudo-time representation of the annealing of the UDD particles. It is evident that the structural rearrangement of nanodiamond particles starts from the surface towards a crystal bulk.

When heating the diamond in the region of its thermodynamical metastability the transformation from sp^3- to sp^2- type of carbon bonds occurs, which results in transformation of diamond

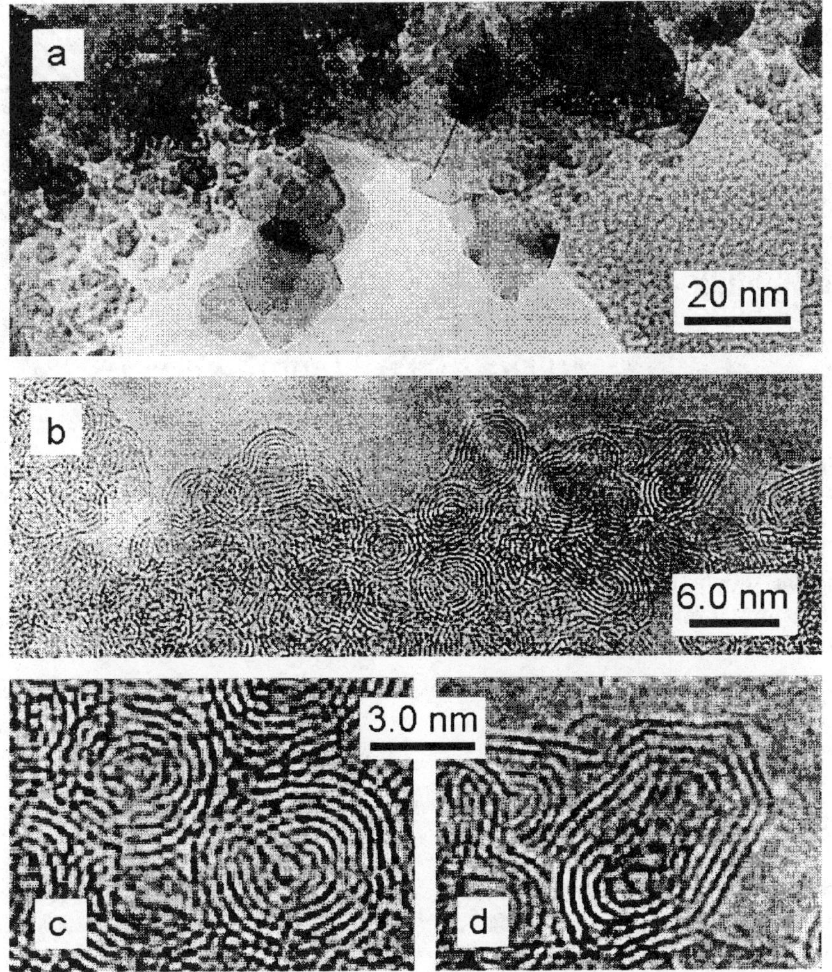

FIGURE 3. TEM images of the original ultra-disperse diamonds (a) and their annealing products ((b), (c), (d)). The relatively large diamond particles with good developed facets are clearly seen (a). It is clear that after annealing the UDD particles almost completely transform to carbon particles with closed graphite-like shells (b). Fig. 3 (c) and (d) are magnified representation of spherical and polyhedral particles respectively. The distance between the graphite-like shells ~ 0.35 nm. It is appears as if formation onion-like carbon particles proceeds trough the spiral-like network.

structure to graphite one. Physically the formation of particles composed of inserted carbon shells may be explained by the system tendency to minimize surface energy owing to the saturation of high-energy free bonds of edge atoms of formed graphite networks. In this case the tense carbon pentagons are generated.

As opposed to thermal annealing UDD in vacuum, the transformation of diamond particles during high rate expansion in explosive chamber occurs in strongly nonequilibrium condition, which in some degree may explain the more defective structure of closed particles formed under such conditions. Besides, there is large number of the atoms of oxygen, nitrogen and hydrogen which bonding with surface carbon atoms may inhibit the complete sewing forming graphite networks.

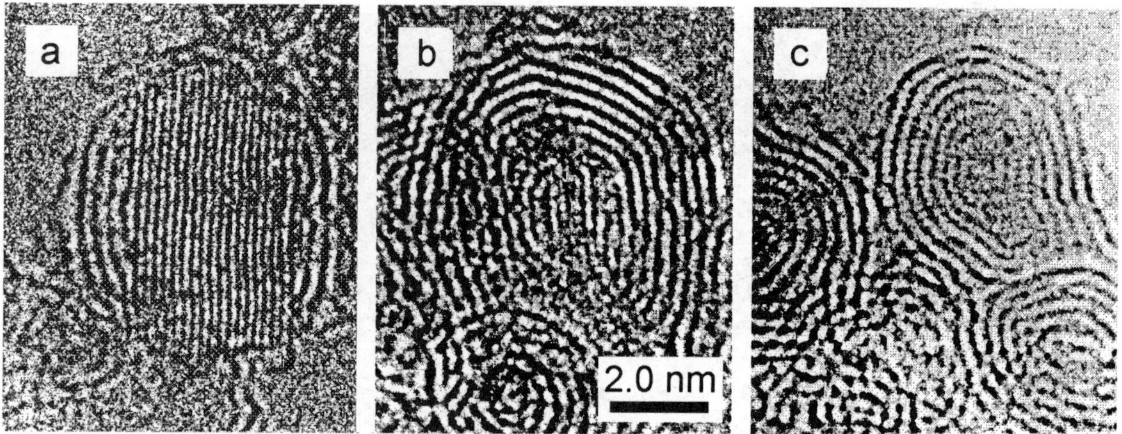

FIGURE 4. High-resolution TEM images of intermediate products of UDD annealing. a) The initial step of diamond particle transformation. b) The intermediate structure consisting of diamond core and onion-like mantle. c) Onion-like carbon. The rate of transformation of crystallographic planes with index (111) is higher than that for the plane with the large index.

Investigation has shown that temperature of the UDD-OLC transformation onset is dependent on diamond dispersivity. Thus for diamond with mean particle size of ~100, ~5.0 and ~2.5 nm it was 1950, ~1300 and less than 1200 K respectively.

When heated, first relatively small diamond particles begin to transform. Analysis of TEM images has shown that small diamond particles wherein the portion of surface atoms is the largest predominantly transform into quasi-spherical particles, while the bigger ones transform into polyhedron particles. The heating up to higher temperatures may favor to the rounding of the latter.

It should be emphasized that the formation of multi-shell carbon structure from ultra-disperse diamond is not exception. Thus the analogs of fullerene structure containing up to 70 closed graphite-like shells were obtained early in intense irradiation of soot and tubular carbon by an electron beam [10]. Hollow quasi-spherical particles were also obtained [11] by the soot annealing up to 2700 K.

Besides the multi-shell carbon particles [12] as well as ultra-disperse diamond [13] have been found in the Allende carbonaceous chondrite meteorites. So, we hope that presented here results will be useful for understanding of nature of transformation of carbon nanoparticles.

ACKNOWLEDGMENTS

The authors are thankful to A.L. Chuvilin and A. Gutakovskii for assisting with the JEM 4000 and V.L. Kuznetsov for fruitful discussions and help in diamond annealing experiments.

REFERENCES

1. Volkov, K.V., *et al., Fiz. Goreniya Vzryva*, No. 3 (1990) 123.
2. Lyamkin, A.M., *et al. Dokl. Akad. Nauk SSSR.*, 302 (1988) 611.
3. Savvakin, G. I., *Zh. Mendeleev V.Kh. O.*, 36 (1991) 141.
4. Greiner, N. R., *et al. Nature* 333 (1988) 440.
5. Titov, V.M., *et al., Prepr. Papers 9th Symp. on Detonation.*- Portland, 1989.-p.175.
6. I.Yu. Mal'kov, *Fiz. Goreniya Vzryva*, No. 5 (1993) 93.
7. I.Yu. Mal'kov, *et al., Fiz. Goreniya Vzryva*, No. 1 (1994) 130.
8. V.L. Kuznetsov, *et al. Chem.Phys.Letters* 222 (1994) 343.
9. V.L. Kuznetsov, *et al. Carbon* 32 (1994) 873.
10. Ugarte,D., *Nature* 359 (1992), 707.
11. Heer, W.A., and Ugarte,D., *Chem.Phys.Letters* 207 (1993) 480.
12. Smith, R.S., and Buseck, P.R., *Science,* 212 (1991) 322.
13. Lewis, R.S., *et al, Nature*, 326 (1987) 160.

DETONATION WAVE VELOCITY AND CURVATURE OF BRASS ENCASED PBXN-111

J. W. Forbes, and E. R. Lemar
Naval Surface Warfare Center
Indian Head Division
Silver Spring, Maryland 20903-5640

Detonation velocities and wave front curvatures were measured for PBXN-111 charges encased in 5 mm thick brass tubes. In all the experiments (charge diameters from 19 to 47 mm) the brass case affected the detonation properties of PBXN-111. Steady detonation waves propagated in brass encased charges with diameters as small as 19 mm, which is about half of the unconfined failure diameter. The radii of curvature of the detonation waves at the center of the wave fronts ranged from 52 to 141 mm for charge diameters of 25 to 47 mm. The angles between the detonation wave fronts and the brass/charge interfaces were between 72 and 74 degrees.

BACKGROUND

PBXN-111 (formerly PBXW-115) is a cast cured explosive with RDX/AP/Al/HTPB binder with 20/43/25/12 weight percent. The composition of PBXN-111 is similar to some propellants[1,2] except that PBXN-111 contains RDX instead of HMX. The initial density of the PBXN-111 used in this work was 1.79 g/cm^3. The median RDX particle size was calculated[3] to be 60 μm. The nominal sizes for AP and Al particles were 200 and 5 μm, respectively. Unconfined PBXN-111 cylinders have failure diameters of a few centimeters and curved wave fronts.[4,5]

DETONATION VELOCITY

Cylindrical PBXN-111 charges were machined to slip fit into brass tubes with 5 mm thick walls and were boosted by 50.8 mm diameter by 50.8 mm long cast pentolite plane wave boosters (PWB's) or pressed pentolite (density 1.56 g/cm^3) cylinders. A streak camera was used to record the shock velocity along the last 150 mm of the brass tube by recording changes in intensity of reflected diffuse light from a strip of teflon tape attached to the brass tube as depicted in Figure 1. This lighting technique was described previously.[4]

Since the brass cases were only a few millimeters thick, it was assumed that the shock waves in the brass were attached to the detonation wave in the explosive and traveled down the tubes with velocities equal to the detonation velocity.

Incomplete consumption of the charge, decreasing wave velocity, and lack of deformation of the brass tube near the far end of the charge were used as criteria to determine if the detonation failed.

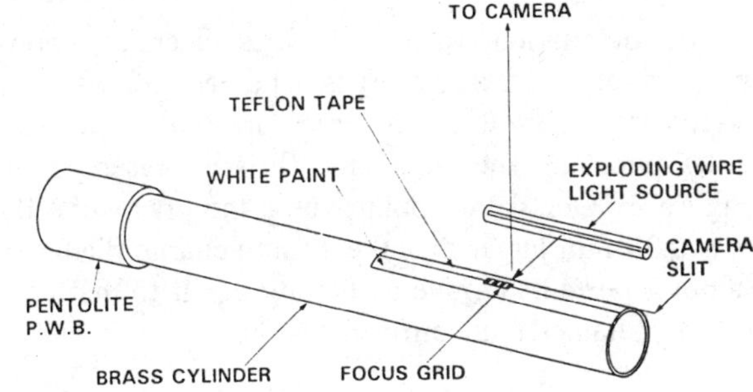

FIGURE 1. Detonation velocity test schematic of a brass confined cylindrical charge

DISCUSSION OF VELOCITY RESULTS FOR CONFINED CHARGES

The results from the experiments on confined charges are presented in Figure 2. The line through the data for confined charges is presented to assist the reader. Insufficient data exist to explore the significance of the case thickness to charge diameter ratio.

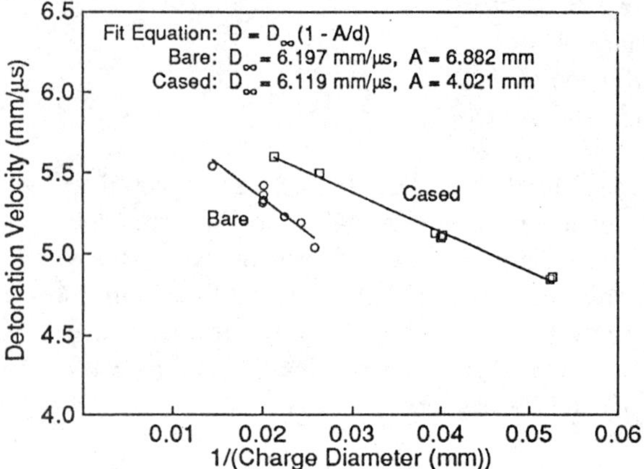

FIGURE 2. Detonation velocity of confined and unconfined cylindrical charges as a function of reciprocal diameter

Detonation waves failed to propagate in 16 mm diameter charges confined by 5 mm thick brass cases but did propagate in charges with diameters greater than 19 mm. The failure diameter for the brass confined charges is about one half that of the unconfined failure diameter.[4]

The detonation wave velocities near the non-boosted ends of the charges (the last 30-40 mm) were steady within the measurement error of ± 1.5 percent. The data and fit are presented in Figure 2. One datum point where the brass tube ID was 0.38 mm larger than the 47 mm charge diameter is not reported. It gave a velocity equal to that of a 47 mm diameter unconfined charge.

DETONATION WAVE ARRIVAL AT END OF CHARGE

A streak camera was used to record the detonation wave arrivals at the ends of cylindrical charges. The experimental setup used to obtain these breakout data is shown in Figure 3. A reflector of aluminized Mylar or Teflon tape was glued to the end of the charge. The alignment and/or reflectivity was destroyed when the shock arrived at the reflecting material. On some of the tests, lines were drawn on the Teflon tape at accurately measured distances from the edges of the charges. These marks resulted in lines on the streak camera records which showed that the breakout data extended to the edges of the charges. The experimental methods are explained in detail in reference 4.

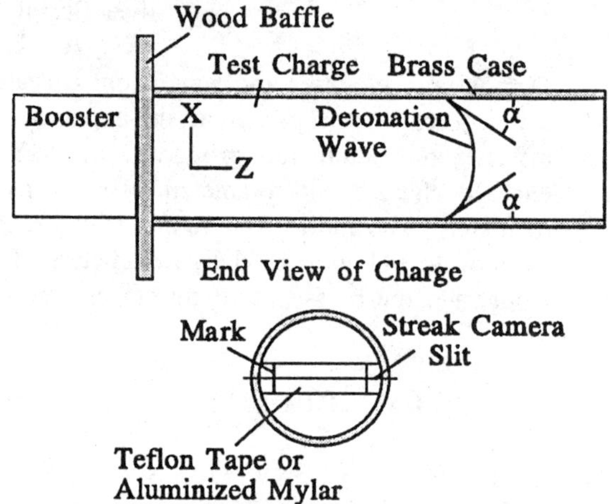

FIGURE 3. Experimental setup for breakout experiments

DATA ANALYSIS

The wave arrival time as a function of position across the charge diameter and the detonation velocity were used to calculate the wave front shape inside the charge. In the following discussion, the Z axis (wave propagation axis) is taken to be along the axis of the cylindrical charge and the X axis is taken to be along a charge diameter (see Figure 3).

FIT OVER ENTIRE DIAMETER

Chaissé and Oeconomos[6] have fitted wave front curvature data to the natural logarithm of a Bessel function. To obtain the full-diameter fit of our PBXN-111 data, we have used

$$Z = Z_o + B*\ln[J_o(A*(X-X_o))] \quad (1)$$

where J_o is the Bessel function of zero order, Z is the wave position in the direction of the charge axis at a position X across the diameter of the charge (see Figure 3), and Z_o and X_o are parameters used only to translate the origin to the center of the wave and do not depend on material properties.

Equation 1 was expanded in a series around X-X_o and this series expansion was used in a non-linear least squares fitting program to obtain material dependent parameters, A and B. In this work, terms in the expansion up to and including 14th order were used. In most cases, the data can be fit well with lower order expansions but the terms up to 14th order must be used in order for the series to approximate the Bessel function for the charge diameters in these particular tests. The values of A and B obtained from the fit are dependent on the order of the expansion used in the fit and approach limiting values as the order of the expansion is increased. If the values of A and B are used to discriminate between various models as Chaissé and Oeconomos[6] have done, care must be taken to ensure not only that the expansion fits the data well, but also that the expansion is a good approximation to Equation 1.

Parameters obtained from selected fits are presented in Table 1. Figure 4 shows the data along with the calculated fit for a PBXN-111 experiment. In all cases, the actual data are fit well. For some of the experiments, the detonation wave center differs by 1 or 2 mm from the center of the explosive cylinder. This is probably due to a slightly tilted input shock caused by the booster being at a slight angle to the axis.

TABLE 1. Parameters Obtained by Fitting Selected Record Readings for Brass Encased Charges to Equation 1

DIA. (mm)	A	B
25.2	0.0744	7.080
37.9	0.0886	2.395
47.0	0.0780	2.342

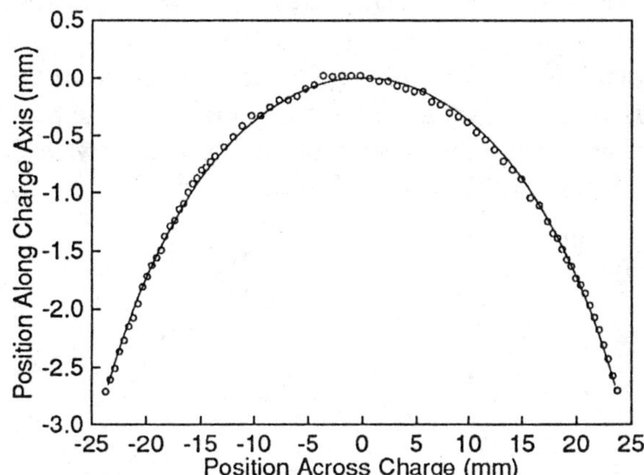

FIGURE 4. Fit to Digitized Breakout Data

WAVE FRONT ANGLE

The wave front data were used to determine the angle, α, (see Figure 3) that the detonation wave makes to the edge of the charge. The angle was obtained from the fit of Equation 1. The value for the angle at the edges of each charge is presented in Table 2.

RADIUS OF CURVATURE

The radii of curvature at the center of the charge for various diameter cased charges are presented in Table 2. The radius of curvature for 47 mm diameter cased charges is about 1.7 times that of unconfined charges of similar diameter[7].

TABLE 2. Detonation Wave Curvature for Charges in 5 mm Thick Brass Cases

DIA. (mm)	RAD. OF CURV. AT CENTER (mm)	α (deg)
25.2	52	74
37.9	106	74
47.0	141	72

The radius of curvature as a function of position was obtained from the series expansion of Equation 1. The resulting functions for three of the PBXN-111 charges are plotted in Figure 5. The peak value occurs at the center of the detonation wave.

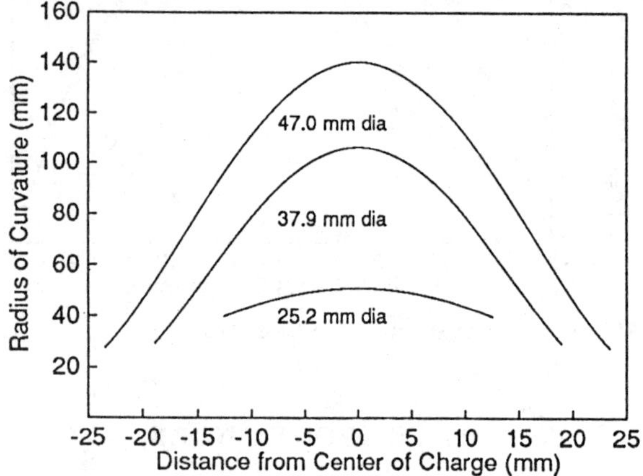

FIGURE 5. Radius of Curvature as a Function of Position across Charge

SUMMARY

All the results from the detonation wave radii of curvature and velocity experiments (diameters from 19 to 47 mm) are affected by the 5 mm thick brass confinement. The failure diameter for a brass-encased charge is about one half that of an unconfined charge. Detonation wave profiles have been fitted across charge diameters for brass encased PBXN-111. The fit equation has been used to obtain the angle of the detonation front and the radius of curvature of the wave front as a function of radial position. These fits can be used to calibrate or test the results of curved front detonation theories and hydrodynamic code calculations of PBXN-111 detonation properties.

ACKNOWLEDGEMENTS

This work was funded by the Office of Naval Research, Code 333, as part of the Explosives and Undersea Warheads Technology Block Program PE602314N. The authors wish to acknowledge technicians Dale Ashwell and Robert Baker for conducting the experiments. E. Anderson and Dr. R. Bernecker provided constructive comments.

REFERENCES

1. Dick, J. J., *Comb. and Flame*, Vol. 37, 95-99 (1980).
2. Price, D., and Clairmont, Jr., A. R., *Comb. and Flame*, Vol. 29, 87-93 (1977).
3. Forbes, J. W., and Watt, J. W., NAVSWC TR 90-168, 10 Aug 1990.
4. Forbes, J. W., Lemar, E. R., and Baker, R. N., *Proc. Ninth Symposium (International) on Detonation*, Portland, OR, Aug 1989, pp. 806-815.
5. Forbes, J. W., Lemar, E. R., Sutherland, G. T., and Baker, R. N., NSWCDD/TR-92/164, 19 March 1992.
6. Chaissé, F. and Oeconomos, J. N., *Proc. Tenth Symposium (International) on Detonation*, Boston, MA, July 1993.
7. Lemar, E. R., and Forbes, J. W., in *High-Pressure Science and Technology-1993*, Schmidt, Shaner, Samara, and Ross, (editors) AIP Conference Proceedings 309, pp.1385-1388.

DETONATION WAVE VELOCITY AND CURVATURE OF IRX-4 AND PBXN-110

E. R. Lemar, J. W. Forbes, and G. T. Sutherland
Naval Surface Warfare Center
Indian Head Division
Silver Spring, Maryland 20903-5640

Detonation velocities and wave front curvatures were measured for bare cylindrical charges of IRX-4 and PBXN-110 charges. Steady detonation waves propagated in IRX-4 charges with diameters as small as 33 mm. The failure diameter of IRX-4 is between 25 and 33 mm. A fit of detonation velocity data gives 5.83 mm/µs for IRX-4's infinite diameter velocity. Detonation wave curvature experiments have been done on 48 mm diameter cylindrical IRX-4 charges with lengths from 9 to 28 cm. The data have been fitted accurately over the entire charge diameters using the natural logarithm of a Bessel function.

MATERIALS

IRX-4[1] (HMX/AP/Al/HTPB, 30/24/16/30 wt %) and PBXN-110[2] (formerly PBXW-113, HMX/HTPB, 88/12 wt %) are cast-cured explosives. The initial densities of the IRX-4 and PBXN-110 used in this work were 1.50 and 1.68 g/cm^3 respectively.

DETONATION VELOCITY

All but one of the cylindrical charges were boosted by 50.8 mm diameter by 50.8 mm long pressed pentolite (density 1.56 g/cm^3) cylinders. The experiment for the 48.8 mm diameter IRX-4 charge was boosted by a 50.8 mm diameter by 50.8 mm long cast pentolite plane wave booster (PWB). A streak camera was used to record the shock velocity along the length of the cylindrical charges by recording changes in intensity of reflected diffuse light from a strip of teflon tape attached to the charges as depicted in Figure 1. This lighting technique has been described previously.[3,4]

DISCUSSION OF VELOCITY RESULTS

Five detonation velocity experiments were done on IRX-4 charges with diameters from 25.1 to 48.8 mm. Detonation waves failed to propagate in 25.1 mm diameter IRX-4 charges but did propagate in charges with diameters 33 mm and greater. The detonation wave velocities were measured over the last 13 cm of the charges and were steady within the measurement error of ± 1.5 percent. The data are presented in Table 1 and Figure 2.

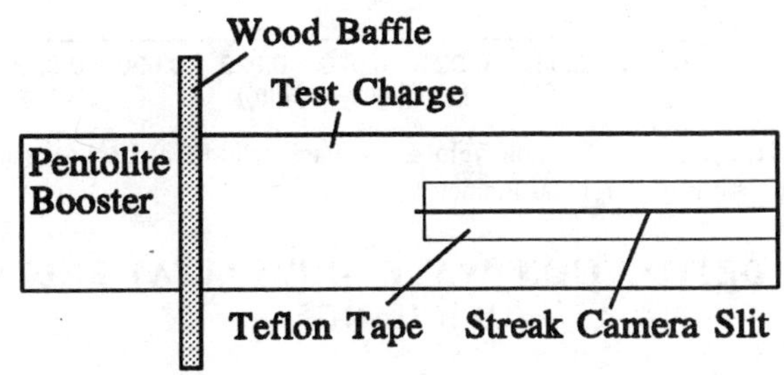

FIGURE 1. Detonation velocity test schematic

Only one PBXN-110 detonation velocity experiment was done on a cylinder from the same batch as the breakout experiment. It was done on a 49.9 mm diameter charge and gave a velocity of 8.39 mm/µs. This value was used in the analysis of the 50.0 mm diameter PBXN-110 breakout experiment.

EXPL	DIA (mm)	L (cm)	DENS (g/cm³)	DET VEL (mm/μs)
IRX-4	25.1	18.8	1.513	failed
IRX-4	33.0	19.6	1.502	5.51
IRX-4	40.6	24.4	1.504	5.60
IRX-4	48.6	27.9	1.514	5.60
IRX-4	48.8	27.9	1.499	5.62
N-110	49.9	17.8	1.68	8.39

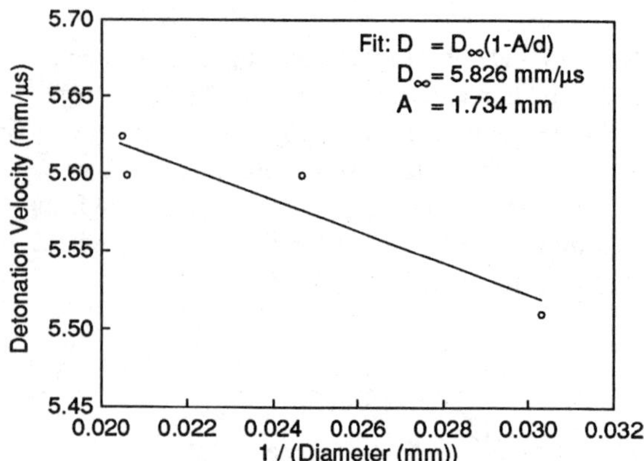

FIGURE 2. Detonation velocity of bare cylindrical IRX-4 as a function of reciprocal diameter

DETONATION WAVE ARRIVAL AT END OF CHARGE

A streak camera was used to record the detonation wave arrivals at the ends of cylindrical charges boosted by 50.8 mm diameter by 50.8 long pressed pentolite cylinders. The experimental setup used to obtain these breakout data is shown in Figure 3. A reflector of Teflon tape was glued to the end of each charge. The reflectivity was destroyed when the shock arrived at the reflecting material. The experimental methods are explained in detail in references 3 and 4.

Figure 4 gives the position-time wave breakout profiles for IRX-4 charges with the same diameter but three different lengths. The IRX-4 profiles are

nearly the same for all charge lengths studied. In addition, the wave breakout for a PBXN-110 and a pressed pentolite booster are given in the figure.

DATA ANALYSIS

The wave arrival time as a function of position across the charge diameter and the detonation velocity were used to calculate the wave front shape inside the charge. In the following discussion, the Z axis (wave propagation axis) is taken to be along the axis of the cylindrical charge and the X axis is taken to be along a charge diameter (see Figure 3).

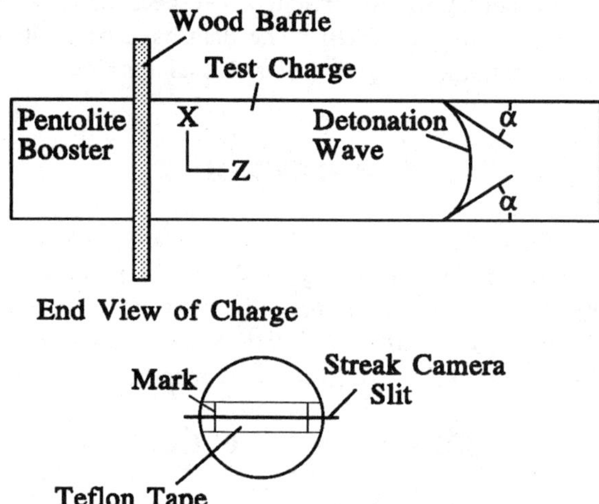

FIGURE 3. Experimental setup for breakout experiments

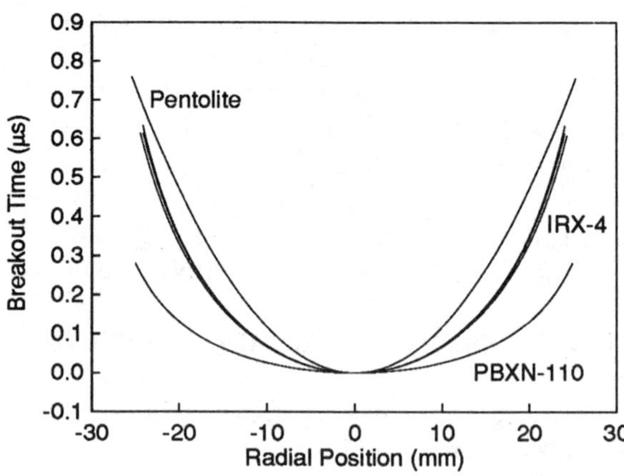

FIGURE 4. Position-time breakout profiles at the end of cylindrical charges of IRX-4, PBXN-110, and a pressed Pentolite booster

FIT OVER ENTIRE DIAMETER

The breakout data has been least squares fitted to the natural logarithm of a Bessel function as suggested by Chaissé and Oeconomos.[5] The actual fit was done to the 14th order series expansion of

$$Z = Z_o + B * \ln[J_o(A * (X - X_o))] \quad (1)$$

where J_o is the Bessel function of zero order, Z is the wave position in the direction of the charge axis at a position X. A, B, X_o and Z_o are the fit parameters. It should be noted that X_o and Z_o are used only to translate the origin to the center of the detonation wave and do not depend on material properties.

Parameters obtained from selected fits are presented in Table 2. Figure 5 shows wave breakout data along with the calculated fits for selected IRX-4 and PBXN-110 data. In all cases, the actual data are fit well. For some of the experiments, the detonation wave center differs by 1 or 2 mm from the center of the explosive cylinder. This is probably due to a slightly tilted input shock caused by the booster being at a slight angle to the axis.

TABLE 2. Parameters Obtained by Fitting Selected Record Readings to Equation 1

Expl.	Length (cm)	DIA. (mm)	A	B
IRX-4	9.1	48.2	0.0860	2.061
IRX-4	17.8	48.2	0.0891	1.879
IRX-4	27.9	48.8	0.0888	1.784
N-110	17.8	50.0	0.0904	1.008

WAVE FRONT ANGLE

The wave front data were used to determine the angle, α, (see Figure 3) that the detonation wave makes to the edge of the charge. The angle was obtained from the fit of Equation 1. The value for the angle at the edge of each charge is presented in Table 3.

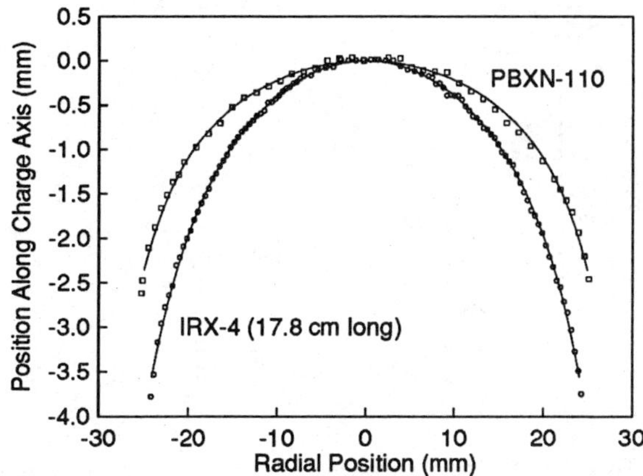

FIGURE 5. Fit to digitized breakout data

RADIUS OF CURVATURE

The radius of curvature as a function of position was obtained from the series expansion of Equation 1. The resulting function for an IRX-4 and a PBXN-110 record are plotted in Figure 6. The wave curvatures for the various lengths of IRX-4 charges are nearly the same. The radii of curvature at the center of the charge for various diameter charges are presented in Table 3.

TABLE 3. Detonation Wave Curvature and angle at charge edge for IRX-4 and PBXN-110 Charges

EXPL.	L (cm)	DIA. (mm)	RAD. OF CURV. AT CENTER (mm)	α (deg)
IRX-4	9.1	48.2	132.4	63.1
IRX-4	17.8	48.2	133.5	61.4
IRX-4	27.9	48.8	141.8	61.6
N-110	17.8	50.0	243.7	68.5

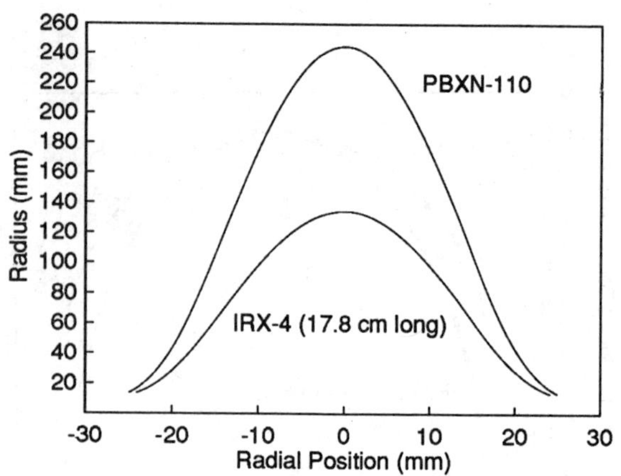

FIGURE 6. Radius of curvature as a function of position across charge

SUMMARY

Detonation velocities and wave front curvatures were measured for bare cylindrical charges of IRX-4 and PBXN-110. Steady detonation waves propagated in IRX-4 charges with diameters as small as 33 mm. The failure diameter of IRX-4 is between 25 and 33 mm. A fit of detonation velocity data gives 5.83 mm/μs for IRX-4's infinite diameter velocity. Detonation wave curvature experiments have been done on 48 mm diameter cylindrical IRX-4 charges with lengths from 9 to 28 cm.

Detonation wave profiles have been accurately fitted across the charge diameters using the natural logarithm of a Bessel function for IRX-4 and PBXN-110. The fit equation has been used to obtain the angle of the detonation front and the radius of curvature of the wave front as a function of radial position. For wave distances down the charge between 9 and 28 cm, the radius of curvature of the detonation front at the center of the charge increased slightly. These fits can be used to calibrate or test the results of curved front detonation theories and hydrodynamic code calculations of IRX-4 and PBXN-110[6].

The results of this paper add to the detonation properties data on HTPB binder explosives[1-4,6] that have been developed over the last few years.

ACKNOWLEDGEMENTS

This work was funded by the Office of Naval Research, Code 333, as part of the Explosives and Undersea Warheads Technology Block Program PE602314N. The authors wish to acknowledge technicians Dale Ashwell and Robert Baker for conducting the experiments.

REFERENCES

1. Sutherland, G. T., Lemar, E. R., Forbes, J.W. , Anderson, E., Miller, P., Ashwell, K. D., Baker, R. N. and Liddiard, T. P., "Shock Wave and Detonation Wave Response of Selected HMX Based Research Explosives with HTPB Binder Systems," in *High-Pressure Science and Technology-1993*, Eds. S. C. Schmidt, J. W. Shaner, G. A. Samara, and M. Ross, AIP Conf. Proc. 309, AIP Press, pp. 1413-1416, 1994.
2. Anderson, E. "Explosives," in *Tactical Missile Warheads-Vol 155 of Progress in Astronautics and Aeronautics,* Carleone, J., (ed.), AIAA, pp. 81-163, 1994.
3. Forbes, J. W., Lemar, E. R., and Baker, R. N., "Detonation Wave Propagation in PBXW-115," *Proc. Ninth Symposium (International) on Detonation*, Portland, OR, Aug 1989, pp. 806-815.
4. Forbes, J. W., Lemar, E. R., Sutherland, G. T., and Baker, R. N., "Detonation Wave Curvature, Corner Turning, and Unreacted Hugoniot of PBXN-111," NSWCDD/TR-92/164, 19 March 1992.
5. Chaissé, F., and Oeconomos, J. N., "The Shape Analysis of a Steady Detonation Front in Right Circular Cylinders of High Density Explosive. Some Theoretical and Numerical Aspects," *Proc. Tenth Symposium (International) on Detonation*, Boston, MA, July 1993.
6. Miller, P. J., and Sutherland, G. T., "Reaction Rate Modeling of PBXN-110," These proceedings.

DETONATION PROPERTIES OF THE NON-IDEAL EXPLOSIVE PBXW-123

W.H. Wilson, J.W. Forbes, P.K. Gustavson, E.R. Lemar, and G.T. Sutherland

Naval Surface Warfare Center, Indian Head Division, White Oak, Silver Spring, MD 20903-5640

Detonation stability and wave curvature in PBXW-123, an aluminized, non-ideal explosive, have been studied. Reaction failed very slowly in unconfined 75 mm, 100 mm, and 126 mm dia. samples. Peak output pressure was still ~28 kb after a run distance of 548 mm in a 126 mm dia. charge. Confinement had a significant effect on reaction stability. Detonation velocity was steady at ~5.5 mm/µs in 76 mm dia. samples confined in brass tubes. Reaction wavefront curvature was measured in unconfined and confined samples; detonation wavefront radius of curvature was ~140 mm at the centerline in the confined charge.

INTRODUCTION

Detonation stability in non-ideal explosives can be difficult to resolve, because sub-detonation reaction can propagate at high velocity which decreases only gradually. Small scale tests can be misleading, because vigorous, slowly decaying sub-detonation reaction can be confused for detonation. Stability of reaction and detonation in PBXW-123, a highly non-ideal explosive, have been studied in this work. Experiments were conducted to measure detonation velocity, critical diameter, shock pressure, and wave curvature.

SAMPLE MATERIAL DESCRIPTION

PBXW-123 is a cast cured formulation using aluminum and ammonium perchlorate for high heat of detonation, plus continuous phase trimethylolethane trinitrate (TMETN), an easily detonable constituent. Aluminum is a heat-producing fuel, but also increases non-ideal reaction behavior. A rubbery polycaprolactone binder is used for improved insensitivity. In the NSWC Modified Gap Test, PBXW-123 has a sustained ignition threshold[1] of 69 kb (in the explosive); its Large Scale Gap Test sensitivity[2] is ~80 kb (in the gap). The average density of the samples used in these experiments was 1.92 g/cc.

EXPERIMENTAL

Unconfined Sample Experiments

Reaction stability was studied in unconfined PBXW-123 cylinders in which overdriven detonation was initiated by Pentolite plane wave boosters (PWB). Planar shock output from the PWB was much greater than the ~69 kb needed to initiate detonation in PBXW-123. Reaction was characterized by measuring either its propagation velocity or shock pressure. Experiments were run in samples of three sizes, shown in Figure 1.

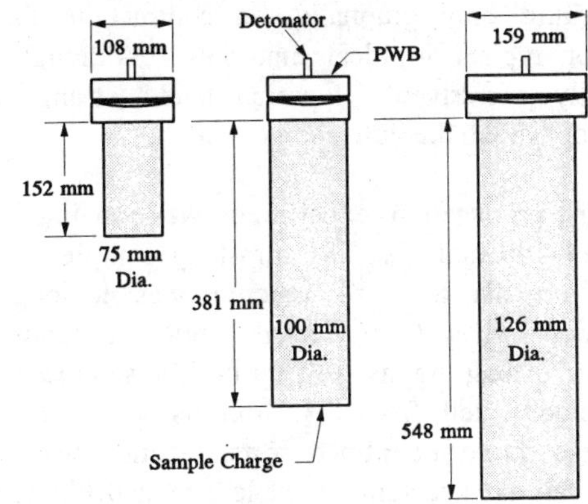

FIGURE 1. Unconfined sample and plane wave booster (PWB) configurations.

Two streak camera experiments were run in 75 mm dia. samples. Reaction progress along the final 82 mm of the charges was recorded. In both 75 mm dia. tests, the velocity observed was below the detonation velocity measured in confined PBXW-123 (described below), and was gradually slowing. At the point of first observation, 70 mm from the PWB, reaction velocity was slightly more than 5 mm/µs. The reaction propagation velocities measured in the two 75 mm dia. tests are plotted in Figure 2 as a function of distance from the PWB. It is obvious that the reaction was not stable; however, reaction

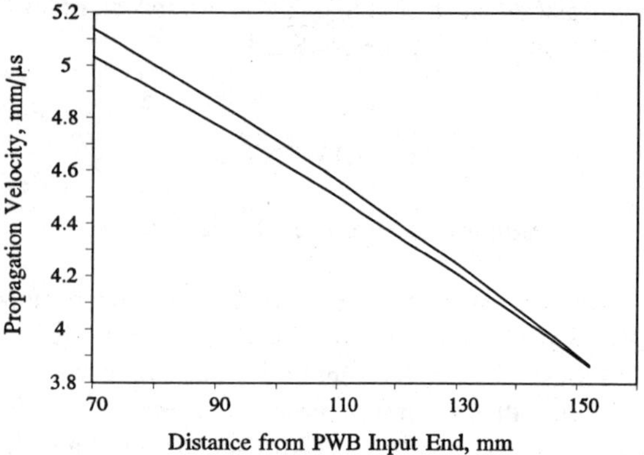

FIGURE 2. Reaction propagation velocity vs. distance from PWB, 75 mm samples.

was failing only gradually, in contrast to failure behavior typical in ideal materials. Although the originally overdriven detonation failed, strong reaction propagated the full charge length.

A third propagation experiment was run in an unconfined 100 mm dia., 381 mm-long sample. As in the 75 mm dia. tests, the reaction was no longer a detonation when first viewed, and was gradually slowing. Along the last 142 mm of the sample, reaction velocity fell from ~3.6 mm/µs to ~2 mm/µs. For these three unconfined tests, reaction was only slowly failing, decelerating at less than 0.1 mm/µs^2.

Shock output at the end of a 126 mm dia., 548 mm long unconfined charge was measured using PVDF and manganin gauges[3]. The gauges were mounted on a Teflon block attached to the output end of the charge. Two 25 µm-thick Metravib PVDF trans-

ducers, with an active area of 25 mm^2, were used along with a Dynasen manganin gauge package. Each gauge was laterally offset ~8 mm from center, so that they did not shadow each other.

One PVDF gauge was used in the current mode, in which the signal is proportional to stress rate dσ/dt. The other was used in the charge mode, in which the gauge output is integrated by a capacitor and so is proportional to stress at the gauge. The records from the three gauges are shown in Figure 3. There is good agreement among the records for the struc-

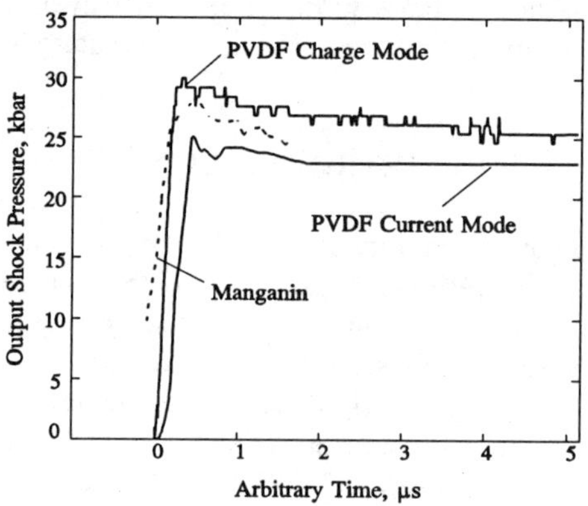

FIGURE 3. Output pressure at end of 126 mm diameter sample.

ture and magnitude of the output pressure. The two PVDF records have been time-shifted for comparison. Actually, the charge mode gauge responded ~200 ns before the current mode gauge, likely due to deviation of the curved reaction front from the charge centerline, plus any slight difference in gauge offsets. Peak stress at the charge mode gauge was ~30 kb; the numerically integrated current mode gauge peak was ~25 kb. The manganin gauge recorded a peak of ~28 kb; this gauge was closer to the charge, so it responded before the PVDF gauges. Wave curvature may also have contributed to a difference in its response compared to the PVDF gauges. Based on the moderate level of peak output stress, it is obvious that detonation also failed in the unconfined 126 mm dia. test. However, the shock pressure, after a run distance of 548 mm, was not insignificant, again implying that reaction failure in PBXW-123 is very gradual.

Confined Sample Experiments

Three reaction propagation experiments were conducted in confined samples of PBXW-123. The charge dimensions, shown in the test configuration of Figure 4, were very similar to those of the unconfined 75 mm dia. tests. The brass confinement tubes had an inner diameter of 76 mm and wall thickness of ~6 mm. The charges were cast directly in

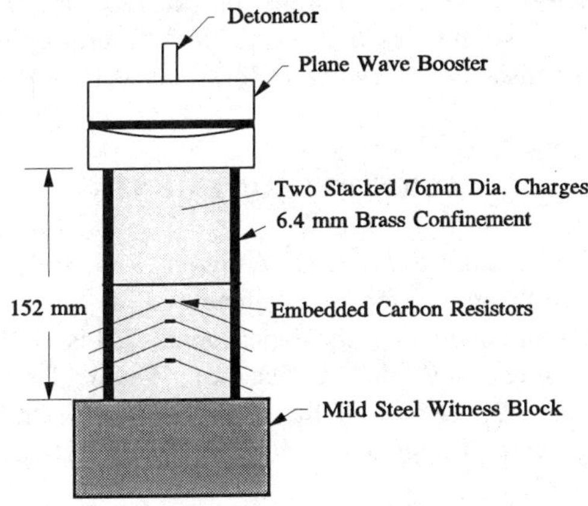

FIGURE 4. PBXW-123 confined sample test configuration.

76 mm-long tubes; 152 mm-long samples were made by stacking two filled cylinders; all interfaces were coated with a thin grease layer to fill small voids and maintain good contact. A mild-steel witness block was used, in contact with the charge output end, to gauge the strength of reaction at the end of the sample. The back of the 64 mm-thick block was spalled off and the remaining steel was shattered in all three confined sample tests, implying that detonation had propagated the full charge length.

The first confined sample experiment was uninstrumented. The last two tests used embedded carbon resistor gauges to monitor reaction front position and strength. These resistors function as time-of-arrival gauges when exposed to detonation. However, their resistance change has been calibrated[4] for pressure up to ~100 kb, so they can be used to estimate shock strength in cases such as sub-detonation reaction. Four resistors were used in the second test, all in the lower charge-half. In the last test, eight resistors, four in each charge-half, were used. In each case, all

resistor signals exceeded the 100 kb maximum calibrated response. Reaction front distance-vs.-time plots for each test, derived from gauge time-of-arrival data, were linear, implying that reaction was steady. Arrival time vs. gauge position data for the last test is plotted in Figure 5. The slope of the least squares line fit is 5.56 mm/μs. For comparison, the position-time plot for slowly failing reaction in an unconfined 75 mm dia. test is also shown.

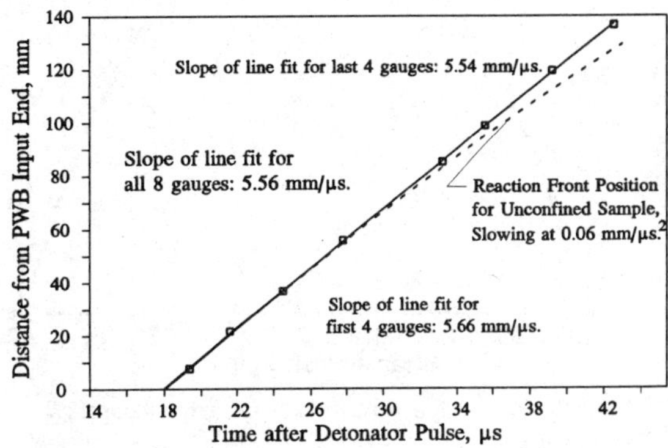

FIGURE 5. Signal arrival times at eight embedded resistor gauges, confined sample test.

Based on the response of the witness blocks, shock pressure in excess of 100 kb at all gauge locations, and steady reaction velocity, it may be concluded that detonation was stable along the full length of the confined samples.

Wave Curvature Measurements

Wave curvature was studied by observing reaction breakout at the far face of cylindrical charges. Shock arrival across a diameter was recorded by a streak camera. (See Forbes and Lemar[5], this proceedings, for a description of the experimental technique.) Curvatures of a failing reaction wave in an unconfined 100 mm dia. charge, and of a detonation wave in a 77 mm dia. confined charge were measured. Sample dimensions were essentially the same as shown in Figures 1 and 4, respectively.

In the unconfined test it was assumed that reaction

decayed to a velocity of ~2 mm/μs at the far end, based on propagation test results described above. Using that velocity, the wavefront shape was derived from the streak record. This analysis yielded a radius of curvature at the center of the wave of ~195 mm. It was also observed that the wavefront was ~2 mm off-center from the charge axis.

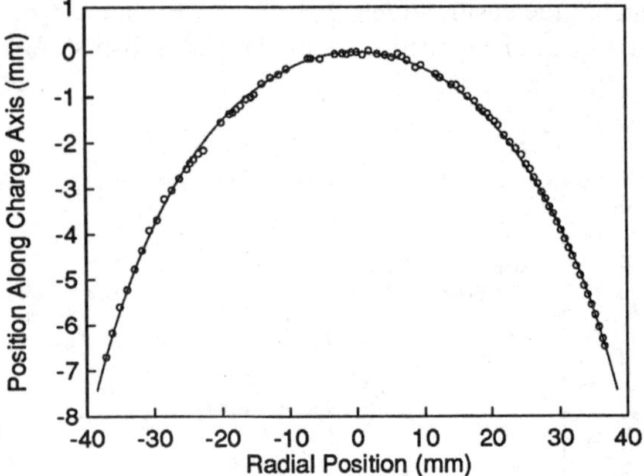

FIGURE 6. Detonation wavefront shape at breakout for 77 mm dia. confined charge.

Detonation wave shape data from the confined sample test are shown in Figure 6. These were derived from the streak record using a detonation velocity of 5.56 mm/μs. The solid line is a logarithmic Bessel function fit to the data[6]. The wavefront radius of curvature is plotted as a function of radial position in Figure 7. The radius of curvature at the center of the wave was ~140 mm.

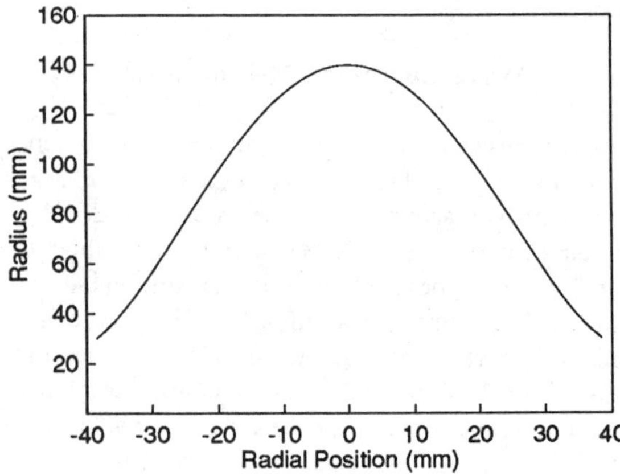

FIGURE 7. Detonation wavefront radius of curvature vs. radial position for 77 mm dia. confined charge.

CONCLUSIONS

PBXW-123 has an unconfined failure diameter greater than the largest diameter tested in this work, 126 mm. Brass confinement of thickness ~6 mm has a significant effect on detonation stability in PBXW-123. Detonation is stable in brass confined diameters as small as 76 mm, and propagates at ~5.5 mm/μs. Although detonation fails in unconfined PBXW-123 charges up to 126 mm in diameter, strong reaction continues to propagate for long distances, slowing at less than 0.1 mm/μs².

ACKNOWLEDGEMENTS

The authors thank K.D. Ashwell, R.N. Baker, and J.H. O'Connor, for their valuable skills and suggestions in conducting the experiments. This work was sponsored by the Office of Naval Research, Code 333, as part of the Explosives and Undersea Warheads Technology Block Program PE602314N.

REFERENCES

1. Lemar, E.R.; Liddiard, T.P.; Forbes, J.W.; Sutherland, G.T.; and Wilson, W.H., *The Analysis of Modified Gap Test Data for Several Selected Insensitive Explosives*, NAVSWC TR 89-290, Sep 1993, NAVSWC, Dahlgren, VA.

2. Hutcheson, R.S. Jr., unpublished test data, NAVSWC Indian Head Div. Silver Spring, MD, 25 Jun 1990.

3. Gustavson, P.K.; Tasker, D.G.; and Forbes, J.W., *PVDF Pressure Transducers for Shock Wave and Explosive Research* (U), NAVSWC TR 91-506, 1993, NAVSWC, Dahlgren, VA.

4. Watson, R.W., "Gauge for Determining Shock Pressures," *Review of Scientific Instruments*, Vol. 38, No. 7, July 1967, pp 978-980.

5. Forbes, J.W. and Lemar, E.R., "Detonation Wave Velocity and Curvature of Brass Encased PBXN-111," presented at APS Topical Conference on Shock Compression of Condensed Matter - 1995, Seattle, WA, Aug. 13-18, 1995.

6. Chaisse, F. and Oeconomos, J. N., *Proc. Tenth Symposium (International) on Detonation*, Boston, MA, July 1993.

EXPERIMENTS ON THE MODIFICATION OF THE ENERGY RELEASE RATE IN A NON-IDEAL EXPLOSIVE

S. Andrews, B. Glancy, J. Forbes, S. Collignon, and L. Hudson
Naval Surface Warfare Center
Indian Head Division
Silver Spring, Maryland 20903-5640

A non-ideal explosive was overdriven in order to increase its energy release rate. Experiments were run with a dual cylindrical ring configuration where a ring of faster detonating explosive surrounds a core of slower non-ideal explosive with a higher energy density. Detonation of the ring explosive caused a faster than normal detonation wave in the non-ideal material. Experiments were performed on the single explosives and on a composite ring charge, each in the cylinder expansion test configuration.

BACKGROUND

Computational analyses of three configurations of dual explosive charges, PBXN-110 surrounding PBXN-111 cores, were performed by Orlando Technology, Incorporated (OTI) in April 1992. The HULL Code used by OTI (1) to model these configurations uses the standard Jones-Wilkens-Lee (JWL) equation of state (EOS) to predict pressure and density information in the reacting energetic materials. Their results predict that a Mach stem, a region of very high pressure due to the convergence of shock waves, is formed from the effect of the outer explosive detonating faster than inner core. The HULL prediction of the Mach stem development is highly dependent on the correct EOS, and the JWL EOS is not accurate for the overdriven state. OTI's computations indicate the wave front within the PBXN-111 core is very angular in shape, with it significantly lagging the detonation of the PBXN-110. In one of the configurations modeled, a 25 mm core of PBXN-111 was surrounded by 120 mm outer diameter PBXN-110 charges. Maximum pressures and densities were seen to occur ~2.5 mm behind the Mach stem front, being greatest on the centerline of the charge. The other cases having larger inner core diameters had no visible Mach stem formation, yet displayed significant pressures in the compressed reaction region.

Experimental testing of the advanced insensitive high explosive (IHE) booster concept, termed a "ring booster" charge, in the configurations modeled by OTI was completed at NSWC-WO. Diagnostics included multi-slit streak photography of the breakout wave shape exiting the 102 mm diameter charges (with 25, 51, or 76 mm cores), axial streak measurements of detonation velocities, and microwave interferometric measurements of internal wave velocities. The multi-slit film records indicated much smoother detonation front profiles than previously predicted, and they gave evidence of substantial Mach stem formation in the 25 mm and 51 mm cores of the PBXN-111. Velocity measurements from streak records along the outside length of the charge and internal microwave interferometric measurements were used to verify and refine in-house computational modeling of these experiments. Steady state detonation velocities of the composite charges were more representative of the PBXN-110 outer sleeve than of the PBXN-111 inner core.

NSWC-WO computational studies of the "ring booster" concept were completed using the Lagrangian hydrocode DYNA-2D. The main findings were the smooth shape of the wave front (significantly different than OTI's predictions), and the magnitude of the Mach stem pressure along the center line of the charge. In the case of the 25 mm core of PBXN-111, the calculated peak pressure within the core was greater than 400 Kbar, as compared with the published CJ pressure of 120Kbar (2).

Further testing of the dual explosive concept was done where the downstream end of the charge was used to throw a copper plate. The plate was 6.35mm thick, 102 mm in diameter with a mass of approximately 0.45 kg. A rubber buffer was sandwiched between the charge and the plate to attenuate the shock into the plate. This test resulted in the extensive fracturing of the plate with early time fragment velocities of be 5.1 mm/µs after accelerating over a 2 mm distance. The velocity decayed to a steady ~2.8 mm/µs over the next 150 mm. These tests were followed up with cylinder expansion tests reported here to confirm previous experimental and modeling results.

ENERGETIC MATERIALS

PBXN-110 (formerly PBXW-113) is HMX/HTPB, 88/12 wt. %, and PBXN-111 (formerly PBXW-115) is RDX/AP/Al/HTPB binder with 20/43/25/12 wt. %. Both energetic materials are cast-cured explosives (2) and were manufactured at NSWC-IH at the Indian Head site. The nominal densities of PBXN-110 and PBXN-111 are 1.68 and 1.79 g/cm^3, respectively.

TEST SETUP AND DESIGN

A cylinder expansion (CYLEX) test series was performed on charges of PBXN-110, PBXN-111 and PBXN-110/PBXN-111 composite to determine the radial expansion histories of the cylinders. The intent of the composite test was to force the energy release rate of PBXN-111 to be higher than that which normally occurs in a detonation wave for this size of charge. The explosives were cast into 51mm OFHC copper tubes. For the composite arrangement, a 25 mm diameter hole was drilled through the length of a PBXN-110 loaded cylinder. PBXN-111 was then cast into the void to form the core. Each charge was boostered with two 51 mm diameter by 25 mm long Pentolite pellets. A Cordin 132 streak camera was used to obtain the radial expansion record of the cylinder, with the slit located 102 mm from the end of the charge. Back lighting was provided by xenon flash tubes behind Fresnel lenses.

The experimental arrangement was modeled using DYNA-2D prior to testing. CYLEX data for PBXN-111 were available for 102 mm diameter charges and were used in the predictions, even though PBXN-111 performance is known to be non-scalable due to its non-ideal components (3).

A copper cylinder (see Fig. 1) containing a 25mm core of PBXN-111 and a 12.7 mm outer ring of PBXN-110 was used for the composite test. The detonation velocity was measured using a pin probe circuit. Six pins at measured spacing were positioned along the length of the cylinder, separated from it by a layer of adhesive tape. As the detonation wave swept down the charge, and the cylinder expanded radially behind it, each pin was struck consecutively, electrically connecting it to the cylinder which was grounded. This action closed a simple capacitive discharge circuit connected to each pin. The output from all the circuits was monitored with a single Pearson current transformer connected to a digital oscilloscope. Charges of PBXN-111 and PBXN-110 were also fired in the CYLEX configuration to set baseline data on the individual components of the composite formulation. A schematic of the CYLEX tests can be seen in Fig. 1.

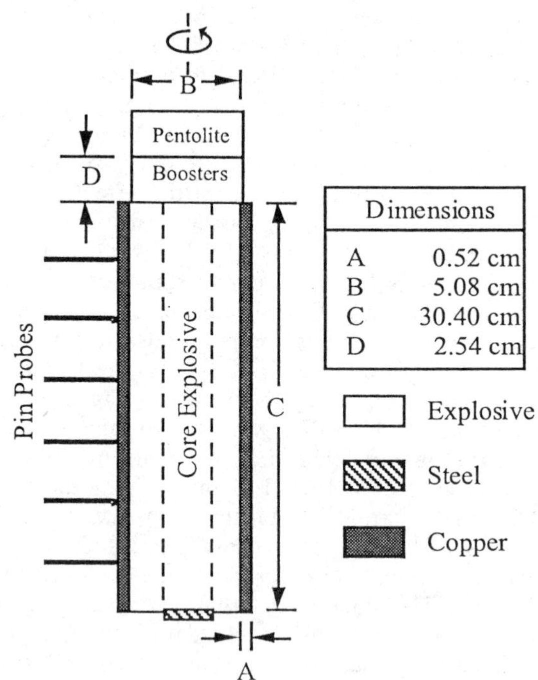

Figure 1. Schematic of CYLEX Test

In an effort to also see how the three different charges could project a fragment, a steel projectile, 25 mm in diameter and 9.53 mm thick, was placed at the end of each charge, separated by a rubber buffer. A baffle was used to allow the projectile to pass but trap the gasses from projecting forward. A Cordin 375 framing camera was used to record the projectile velocities and a steel witness block was placed downstream to stop the projectile.

ANALYSIS AND RESULTS

The streak records generated in the CYLEX tests were digitized and used as input to a software program that smoothes the radial expansion data and calculates the radial expansion velocity for each data point. The radial expansion velocity of the copper cylinder for the PBXN-111 charge (see Fig. 2) was approximately half that of the PBXN-110. The composite formulation with a PBXN-111 core performed almost as well as the PBXN-110 charge.

These experiments were modeled using DYNA-2D with standard JWL EOS's for the explosives (2). The JWL EOS for PBXN-110 was based on the copper cylinder expansion of a 25 mm diameter explosive while the JWL EOS for PBXN-111 was based on 102mm diameter explosive. Figure 2 shows a comparison of the experimental and predicted data. The predicted copper wall velocities from the PBXN-110 and the PBXN-110/PBXN-111 ring charges agree with the experimental results. However, the calculated and experimental results for the copper wall expansion from a 51 mm diameter PBXN-111 cylinder do not agree. This disagreement can be clarified by energy released behind the sonic plane, as explained in the Discussion and Conclusions section.

In each experiment, the downstream end of the charge was used to launch a single pre-formed fragment, see Fig. 1. The fragments launched from the 51 mm baseline cases (PBXN-110 and then PBXN-111) agreed with predictions whereas the composite charge completely shattered the pre-formed fragment and projected a spray at very high velocity. Unfortunately, the extensive gas products

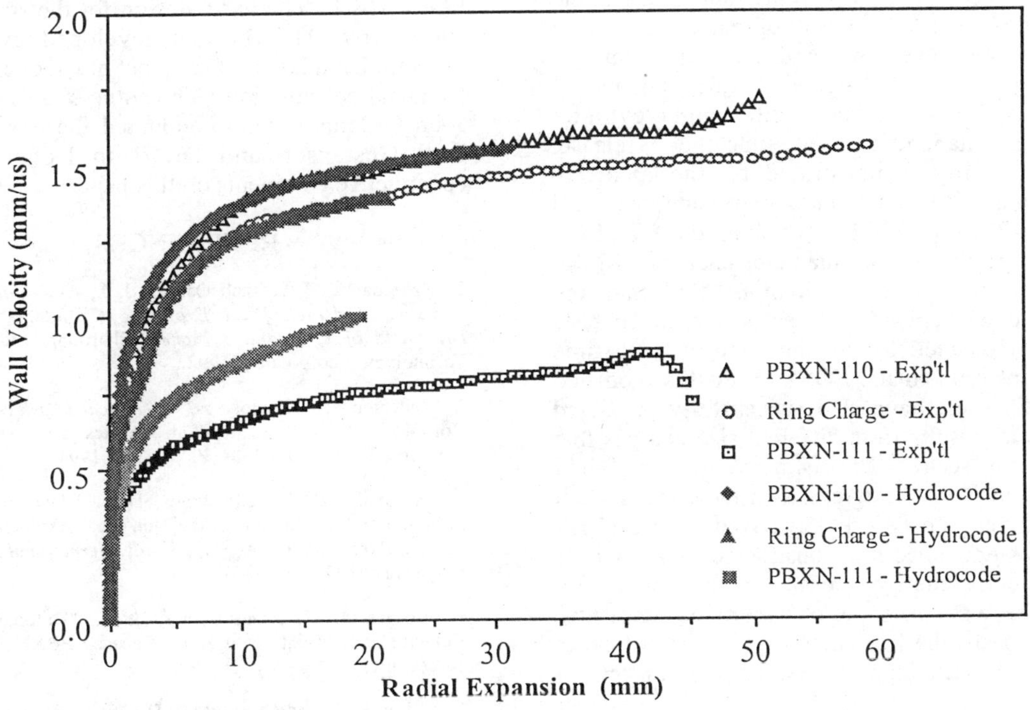

Figure 2. Experimental vs. Predicted Data

caused the film records to be unreadable; hence, the velocity measurements were not obtained.

As previously mentioned, the detonation velocity was measured in each experiment using a pin probe circuit. Table 1 lists the results. Data given for PBXN-111 are in agreement with those presented in a paper by Forbes and Lemar (4) in these proceedings. The velocity of the ring composite charge was within 2% of the PBXN-110 charge.

TABLE 1. CYLEX test Detonation velocities.

Shot #	Explosive Type	Detonation Velocity (mm/us)
1	PBXN-111	5.55
2	PBXN-110	8.43
3	Composite	8.27

DISCUSSION AND CONCLUSIONS

Energy release from a non-ideal explosive like PBXN-111 continues beyond the sonic plane of the detonation wave. The amount of reaction occurring between the wave front and the sonic plane is dependent upon the charge size (3,5). The use of the JWL EOS's obtained from copper cylinder expansion data ignore the late time energy release. This is clearly demonstrated by the incorrect predictions of 51 mm cylinder expansion velocities (see Fig. 2) for PBXN-111 which used a JWL EOS obtained from 102 mm diameter charges. Miller and Guirguis (5) have demonstrated that the discrepancies between different size cylinder tests can be accounted for by the addition of a time dependent term to the EOS. With this modified JWL EOS they have also successfully predicted underwater results from different size charges that could not be scaled. The distinctive feature of their calculation is that it agrees with the experimental data, whereas the JWL EOS derived from CYLEX data does not, since it is constructed to allow only for the early energy release. Since underwater effects have much longer characteristic times than CYLEX tests, the late energy release is critical to proper understanding of underwater performance.

The intent of the ring charge experiments was to overdrive the inner core of PBXN-111 and increase the rate of energy release over what would be expected from a 25 mm diameter charge. The agreement between calculated (using JWL EOS for 102 mm diameter PBXN-111) and experimental cylinder wall velocities for the ring charge suggests that the 25 mm diameter PBXN-111 core is releasing energy at a rate comparable to that from a 102 mm diameter cased PBXN-111 cylinder.

The rate of energy release from PBXN-111 in the ring charge was not high enough to accelerate the cylinder walls to velocities greater than that measured for the PBXN-110 cylinder. Further calculations need to be done using the modified JWL EOS of Miller and Guirguis (5) or the reaction rate model of PBXN-111 of Kennedy and Jones (3) to predict configurations that would yield increased energy release rates for PBXN-111, such that a ring charge would have wall velocities greater than the PBXN-110 cylinder.

ACKNOWLEDGMENTS

This work was funded by the Office of Naval Research, Code 333, as part of the Explosives and Undersea Warheads Technology Block Program PE602314N. The basic design for the ring charge studied by OTI (1) was developed by Patrick Spahn. Douglas Tasker is acknowledged for his technical consultation. The authors are grateful to John O'Connor who conducted the experiments. Paul Gustavson and Dr. Ruth Doherty made constructive comments on this paper.

REFERENCES

1. Szczepanski, J. F.., and Osborn, J. J., *Hull Calculational Analysis of 4-Inch Dual Explosive Cylinders with Core Diameters of 1, 2, and 3 Inches*, Shalimar, FL., Orlando Technology Incorporated, 1992.

2. Anderson, E.., "Explosives" in *Tactical Missile Warheads*, Vol. 155 of Progress in Astronautics and Aeronautics, Carleone, J., (editor), AIAA, PP. 81-163, 1994

3. Kennedy, D. L., and Jones, D. A., "Modeling Shock Initiation and Detonation in the Non-Ideal Explosive PBXW-115", in *Paper Summaries of the Tenth International Detonation Symposium*, 1993, pp. 36-39.

4. Forbes, J. W. , and Lemar, E. R., "Detonation Wave Velocity and Curvature of Brass Encased PBXN-111," these proceedings.

5. Miller, P. J., and Guirguis, R. H., "Experimental Study and Model Calculations of Metal in Al/AP Underwater Explosives", *Structure and Properties of Energetic Material*, in *Proceedings of Material Research Society*, MRS Publications, Vol. 296, Pittsburgh PA., 1993.

WEDGE TEST DATA FOR THREE NEW EXPLOSIVES: LAX112, 2,4-DNI, AND TNAZ *

L.G. Hill, W.L. Seitz, J.F. Kramer, D.M. Murk, and R.S. Medina

Los Alamos National Laboratory, Los Alamos, New Mexico 87545 USA

High pressure Pop-plots and inert Hugoniot curves have been measured for three new explosives: LAX112 (3,6-diamino-1,2,4,5-tetrazine-1,4-dioxide), 2,4-DNI (2,4-dinitroimidazole), and TNAZ (1,3,3-trinitroazetidine). LAX112 and 2,4-DNI are of interest because of their insensitivity, while TNAZ is useful for its performance and castability. The shock sensitivity of LAX112 and 2,4-DNI fall between that of pressed TNT and PBX9502, LAX112 being the less sensitive. The shock sensitivity of TNAZ falls between that of pressed PETN and PBX9501. The inert Hugoniots for all three materials are comparable to those of other explosives.

INTRODUCTION

LAX112 (3,6-diamino-1,2,4,5-tetrazine-1,4-dioxide), Fig. 1a, was developed at Los Alamos in an effort to find an insensitive high explosive with better performance than TATB. LAX112 is distinguished by its high nitrogen content and the absence of nitro groups. It has a high detonation velocity ($\approx$ 8.3 mm/μsec), but cylinder tests show that its metal pushing performance, while marginally better than TATB, is significantly below that of HMX, RDX, and PETN based explosives (1).

2,4-DNI (2,4-dinitroimidazole), Fig. 1b, is another candidate for an insensitive high explosive. Its detonation velocity ($\approx$ 7.8 mm/μsec) is slightly less than that of LAX112, while its metal pushing performance appears to be slightly better (2). Drop-weight impact tests have shown significant batch-to-batch variations in sensitivity (as much as a factor of three), and 4-nitroimidazole impurities are the suspected cause (1).

TNAZ (1,3,3-trinitroazetidine), Fig. 1c, first appeared in the open literature in 1990 (3), but was initially of little practical interest due to excessive synthesis cost. Efforts at Los Alamos and the Aerojet corporation to find alternate synthesis routes (4) have been successful, and TNAZ is, at the time of this paper, starting to be produced in quantity by Aerojet. TNAZ is very promising in

that it has a performance similar to HMX but is melt castable. Thus it is a potential replacement for octols, cyclotols, and even HMX-based PBXs in many applications.

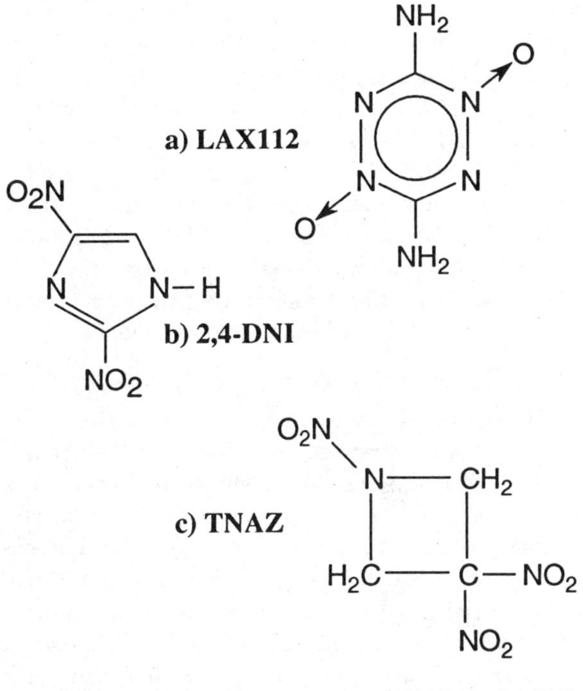

FIGURE 1. Molecular structures of LAX112, 2,4-DNI, and TNAZ.

---

*This work jointly supported by the US DoD and DOE.

EXPERIMENT
Sample Preparation

The LAX112 and 2,4-DNI materials were both plastic-bonded formulations. The samples were ram-pressed to cylindrical shape, sawn on a diagonal to form two wedges per cylinder, and finish-machined along the sawn faces.

The LAX112 samples were formulation X-0535, composed of 95 wt.% LAX112 and 5 wt.% OXY 461 (1). The molding powder was pressed at 42,000 psi and 110 C to achieve about 97.8% of the 1.829 g/cc formulation theoretical maximum density (TMD). X-0535 was found to have excellent mechanical properties— it was dimensionally stable, and pressed and machined well.

The 2,4-DNI samples were formulation X-0552, composed of 95 wt.% 2,4-DNI and 5 wt.% Estane. The molding powder was pressed at 42,000 psi and 90 C to achieve about 98.4% of the 1.720 g/cc formulation TMD (2). X-0552 pressed and machined rather poorly, and the quality of the data is correspondingly lower than for the other two materials.

TNAZ has a critical temperature far above its melting point so that it can be melt cast or hot-pressed. The TNAZ wedges were neat-pressed at 42,000 psi and 97 C directly to the final wedge shape. The densities achieved were between 99.1% and 99.4% of the 1.840 g/cc TMD, which alleviates concern about density variations near the corner opposite the pressing die (2). The sample quality using this technique was excellent.

Description of the Wedge Test

There have been many variations on the wedge test over the years; we used the so-called "mini-wedge" test of Seitz (5), as shown in Fig. 2. The sample is small (about 7 g) and so restricts the run distance to about 1 cm. But for new explosives existing only in small quantities, material minimization is critical. The driver system was a 7.8-inch diameter plane wave lens, a 2-inch thick pad of booster explosive, and up to three 0.5-inch thick attenuator plates to tailor the pressure delivered to the sample.

The diameter of the circular wedge face was 1 inch and the wedge angle was 30 degrees. This angle must be less than the "critical" value at which release waves travel into the material, so that the

propagation of the shock/detonation wave will be unaffected by that boundary. The critical angle is rarely known precisely, but 30 degrees is considered sufficiently conservative for all materials. The elliptical face of the wedge is glued to the last attenuator plate, the circular face thus serving as the observation surface. This configuration gives a slightly longer run distance (before the release wave from the opposite free boundary affects the measurement) than the reverse orientation.

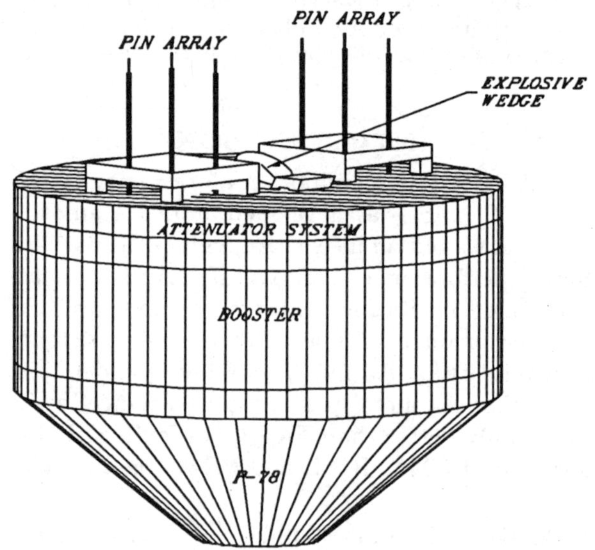

FIGURE 2. Schematic drawing of the wedge test (drawing by Herbert Harry).

The assembly in Fig. 2 is suspended upside down with the observation surface of the wedge parallel to the ground. Viewed from the side the wedge then appears as in Fig. 3a (with the attenuator plate now at 30 degrees to the ground), and viewed from below as in Fig. 3b. The image of the internal slit aperture is centered upon the wedge and, since the line of focus lies on the observation surface, there is no magnification variation or depth of field problem. The wedge is illuminated with an argon bomb, so that specularly reflected light from the observation surface is directed into the camera as in Fig. 3a. As the shock/detonation wave breaks out of the observation surface its reflectivity decreases and the light is attenuated as in Fig. 3b. Thus the wavefont appears as a curve of discontinuous exposure on the film.

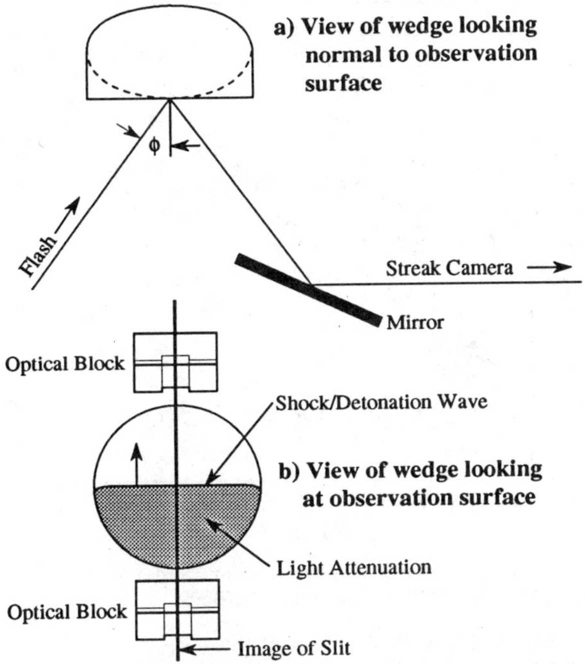

FIGURE 3. Optical configuration.

The other needed information is the free surface velocity of the final attenuator plate. Two methods were used for redundancy. In the first, two plexiglas blocks with 0.5 mm deep machined notches on one side were glued, notch side down, to the last attenuator plate, one on either side of the wedge and in the field of view of the camera slit (Fig. 3b). A thin layer of white paint was applied to the notch side of the blocks prior to gluing, so that the surface was initially reflecting to the flash light. Upon shock wave break out the attenuator reflectivity decreases. Later, the gap reflectivity decreases when impacted by the free surface. The difference between these two times is time-of-flight across the known gap width, from which the free surface velocity follows immediately. The second method involved two clusters of four piezoelectric pins spaced in increments of 0.5 mm from the plate (Fig. 2). As the free surface strikes the pins a voltage spike is produced. An x-t diagram is constructed from the known spacings and measured arrival times, and the velocity is found from the slope of a fit to the points as $x \to 0$. With good data the two methods typically agree to within a few percent.

ANALYSIS

Data was read directly from the film record by optical comparitor. By consideration of the geometry one finds that the run to detonation x^* and the time to detonation t^* are related to the respective film coordinates X^* and Y^* by

$$x^* = x_{toe} \cos\theta + \left(\frac{\sin\theta}{mag}\right) X^*, \; t^* = \frac{Y^*}{WrtSpd}, \; (1)$$

where x_{toe} is the thickness of the wedge "toe" (it is never possible to achieve a knife edge), θ is the wedge angle, mag is the magnification, and $WrtSpd$ is the camera writing speed. The shock/detonation velocity $U_{s/d}$ is related to the angle of the film trace ψ by

$$U_{s/d} = \left(\frac{WrtSpd}{mag}\right) \sin\theta \cot\psi. \qquad (2)$$

The input shock to the explosive is ideally a step rise to constant pressure, the value of which is inferred by impedance matching. The necessary ingredients are 1) the Hugoniot of the final attenuator plate, 2) the measured free-surface velocity of the final attenuator plate, 3) the measured initial shock velocity in the explosive, and 4) the initial density of the explosive. From this one can deduce the initial particle velocity in the explosive (hence the inert Hugoniot) and the initial pressure in the explosive (hence the Pop-plot). We assume that the isentrope for release wave reflected from the free surface of the final attenuator plate is the reflection of the incident shock Hugoniot (in p-u space) about its particle velocity. This is a good approximation for metal attenuators and, by comparison to more sophisticated methods, appears to be well within experimental error.

RESULTS

For a heterogeneous explosive one sees a constant initial shock velocity followed by a smooth acceleration to the detonation velocity. For a homogeneous explosive one sees a much sharper transition, followed by an overshoot in shock velocity, followed by a relaxation to the detonation velocity. LAX112 behaved like a classical heterogeneous explosive, whereas TNAZ behaved more like a homogeneous explosive. The 2,4-DNI records were, due to the aforementioned formulation properties, somewhat erratic. But otherwise, heterogeneous behavior would be expected.

Run to detonation vs. input pressure for the three explosives is shown in Fig. 4 along with four common reference explosives. The sensitivity of LAX112 and 2,4-DNI both fall between that of pressed TNT and PBX9502, LAX112 being the less sensitive. The sensitivity of TNAZ falls between that of pressed PETN and PBX9501.

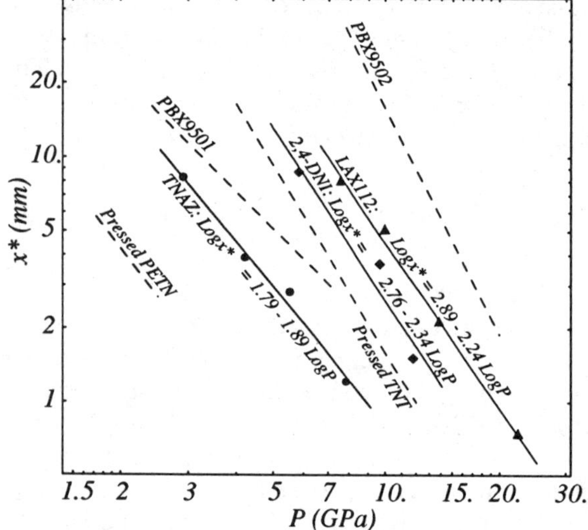

FIGURE 4. Run distance to detonation vs. input pressure (Pop-plot) points and linear fits for LAX112, 2,4-DNI, TNAZ, and selected other explosives.

The u_p-U_s inert Hugoniots for the three explosives and PBX9502 are shown in Fig. 5. The curves are similar to those of other explosives. For TNAZ the lowest velocity point agrees well with the gas gun of data of Sheffield et al. (6), indicated by square symbols. The two higher-velocity points deviate from the trend (perhaps suggesting some reaction or a phase transition) and are omitted from the fit. For the highest input pressure TNAZ case the run distance was too short to measure an accurate initial shock speed, yet the transition point could still be deciphered from the film. The Pop-plot point was therefore generated from the Hugoniot based on the other points.

The numerical values of the data points are given in Table 1. The times to detonation transition are also indicated.

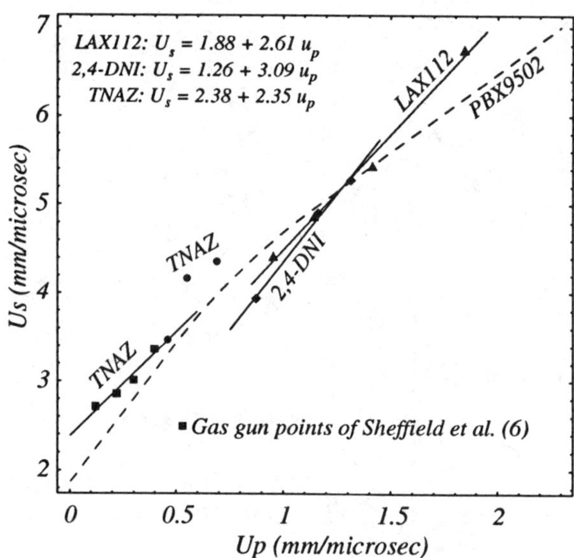

FIGURE 5. Inert Hugoniot points and linear fits for LAX112, 2,4-DNI, TNAZ, and PBX9502.

Table 1. Numerical values of the data points.

HE	ρ_0 g/cm^3	P_0 GPa	u_p km/s	U_s km/s	x^* mm	t^* μs
LAX112	1.793	7.5	0.95	4.41	8.00	1.67
	1.794	9.9	1.14	4.86	5.06	0.99
	1.794	13.7	1.41	5.43	2.12	0.36
	1.793	22.2	1.84	6.74	0.74	0.12
2,4-DNI	1.692	5.8	0.87	3.93	8.64	2.09
	1.692	9.6	1.16	4.90	3.63	0.74
	1.692	11.7	1.31	5.27	1.50	0.30
TNAZ	1.825	2.9	0.46	3.47	8.23	2.34
	1.826	4.2	0.55	4.17	3.88	1.09
	1.826	5.5	0.69	4.36	2.80	0.66
	1.828	7.8	0.85	—	1.21	0.41

REFERENCES

1. Kramer, J. F., et al. "Joint DoD/DOE Munitions Tech. Dev. Prog.", LANL Rpt. LA-12568-PR (1993).
2. Repa, J. V., et al. "Joint DoD/DOE Munitions Tech. Dev. Prog.", LANL Rpt. LA-12806-PR, V.1, (1994).
3. Archibald, T.G., et al., *J. Org. Chem.*, **55**, 2920 (1990).
4. Coburn, M. D., and Hiskey, M. A., "An Alternate Synthesis of 1,3,3-Trinitroazetidine.", LANL Rpt. LA-CP-95-145 (1995).
5. Seitz, W. L., "Short-Duration Shock Initiation of Triaminotrinitrobenzene (TATB)", in *Shock Waves in Condensed Matter*, J. R. Asay et al., eds. (1983).
6. Sheffield, S. A., et al., "Hugoniot and Initiation Measurements on TNAZ Explosive", These proceedings.

ENERGY DISSIPATION AND THE INITIATION OF EXPLOSIVES DURING PLASTIC FLOW

C.S. Coffey

White Oak Laboratory, Indian Head Division
Naval Surface Warfare Center, Silver Spring, MD 20903-5640

Energy dissipation and localization in crystalline solids during shock or impact are shear driven processes and are responsible for initiation of explosive crystals. These are extended to the macroscopic level where the applied shear stress is shown to link the initiation zone processes with the detonation shock wave.

INTRODUCTION

The central issue determining the initiation response of explosive crystals to shock or impact is the generation of the energy necessary to cause molecular dissociation. This controls the rate of release of chemical energy. The conventional theories of detonation of Chapman & Jouguet and of Zeldovich, Von Neumann and Doering, ZND, emphasize just the beginning and end states of this process and completely gloss over the processes responsible for initiation and reaction.[1] The fundamental assumptions of these approaches do not allow them access to the initiation and reaction zones. As a result, these asymptotic theories cannot address the issues of initiation sensitivity and reaction growth in terms that are known to be important such as crystal size, shape, defect content, and plastic flow. The same applies to numerous other features of detonation that relate to initiation and reaction.

In order to deal with some of these problems, several empirical, engineering, codes have been developed to treat the special case of detonations and in limited ways some of these have been quite successful.[2,3,4] However, as in all such cases, the predictions are only as good as the calibration experiments which provide the input parameters needed to run the codes. More importantly, they provide no information about the physics and chemistry of initiation and reaction.

Elsewhere, we have shown that on the microscopic level energy dissipation and localization during shock or impact induced plastic flow of crystalline explosives are responsible for initiation.[5-8] Here, these shear driven processes of plastic deformation and energy dissipation are extended to the macroscopic level and are shown to place an additional restriction on the initiation portion of a detonation wave beyond that imposed by the conventional theories for it is here that the shear must be developed and plastic deformation must begin. Two examples are given, detonation in an unconfined charge and detonation due to projectile impact.

Begin by slightly modifying the ZND assumptions. Assume that instead of a jump discontinuity the shock has a short but finite rise time in which shear and plastic flow occur. This accounts for energy dissipation, localization and initiation. The time for this to occur is less than a nanosecond depending on the number of defects present in the crystals. Initiation occurs when a critical energy density is reached within the crystals. If plastic flow is prevented, dissipation does not occur and neither does initiation. The remaining ZND assumptions need little if any modification.

The high velocity dislocations that account for the high rate plastic deformation during shock generate optical phonons that can resonantly excite the internal molecular modes and cause the rapid, multi-phonon stimulated, molecular dissociation associated with detonations. These high velocity dislocations are driven by the applied shear stress and occur in the asymptotic region where the dislocation velocity begins to approach the local sound wave speed. This is advantageous when dealing with detonations because it is only necessary to determine this critical shear stress. Mild impact and low level shock generate only low velocity dislocations and lower plastic strain rates. These also produce initiation but at a much slower rate and are not treated here.

A CYLINDRICAL CHARGE

Consider a steady state detonation wave propagating with a velocity D along the axis of an unconfined cylindrical explosive charge of radius r_0. Let the charge be composed of an aggregate of randomly oriented and sized explosive crystals held in a soft polymer binder. It will be assumed that plastic deformation can always occur. Here the focus will be on the shear due to the pressure gradient produced by the relief waves that enter the explosive from the unconfined side walls.

Rather than writing the shear stress as the deviatoric elements of a stress tensor, it is more informative to directly write the shear stress, τ, in terms of the gradient of the force, F, acting on a crystal element of length Δx, $\tau \approx (dF/dx)\Delta x/A$. Where A is the cross sectional area. While only the gradient of the forces perpendicular to the direction of wave propagation directly cause shear and plastic flow to occur it is necessary to include in the generalized shear stress the gradient of the compressive force. This can be expressed as

$$\vec{\tau} = \frac{1}{A} (\vec{dr} \cdot \vec{\nabla}) \vec{F}. \qquad (1)$$

The quantity $\vec{dr}$ is the differential length over which plastic deformation occurs. When plastic flow occurs the components of the force are coupled to each other which allows the simplification

$$\vec{\tau} \approx (\vec{dr} \cdot \vec{\nabla}) \vec{P} \qquad (2)$$

where $\vec{P} = \vec{P}(r,t)$ is the pressure. If the pressure is written as $\vec{P} = P\hat{u}$ where $\hat{u}$ is any unit vector and P is the pressure amplitude, then the amplitude of the shear stress becomes

$$\tau = (\vec{dr} \cdot \vec{\nabla}) P \qquad (3)$$

In order for a steady state detonation to develop, molecular dissociation must proceed at a constant rate on the surface of the initiation portion of the detonation wave. To achieve this constant rate of molecular excitation and dissociation requires that the shear stress be a constant over the entire surface of the initiation front so that $d\tau = (\vec{dr}\cdot\vec{\nabla})\tau = 0$. Combining these, gives the condition for a steady state detonation wave as the Laplacian, $\nabla^2 P = 0$. For a cylindrical charge this has the solution

$$P(r,z) = P_0 J_0(\alpha r) e^{\alpha z} \qquad (4)$$

where $J_0(\alpha r)$ is the zero order Bessel function and P_0 is the on axis pressure of the detonation wave. The magnitude of the shear stress is

$$\tau = \alpha P_0 e^{\alpha z} (J_1(\alpha r)^2 + J_0(\alpha r)^2)^{1/2} dr \qquad (5)$$

For an aggregate of randomly oriented explosives crystals it is only necessary that this magnitude be a constant on the initiation surface at the front of the steady state detonation wave. Thus, on the axis of the charge, r = 0, z = 0, and at any other point (r,z) on the constant shear/initiation surface, $\tau(r=0,z=0) = \tau(r,z)$, or

$$e^{-\alpha z} = (J_1(\alpha r)^2 + J_0(\alpha r)^2)^{1/2}$$

(6)

This determines the initiation front. The distance separating the initiation front, eqn. (6), from the peak pressure profile, eqn. (4), on the axis of the charge is the steady state reaction zone thickness, δ. The critical shear stress for initiation is obtained for r = 0, z = 0 from eqn. (5), $\tau_c \approx \alpha P_0 \ell$, where ℓ is the average particle size of the explosive crystals. Figures (1) and (2) show the radial and axial pressure profiles. There is good agreement with experimental data.[9]

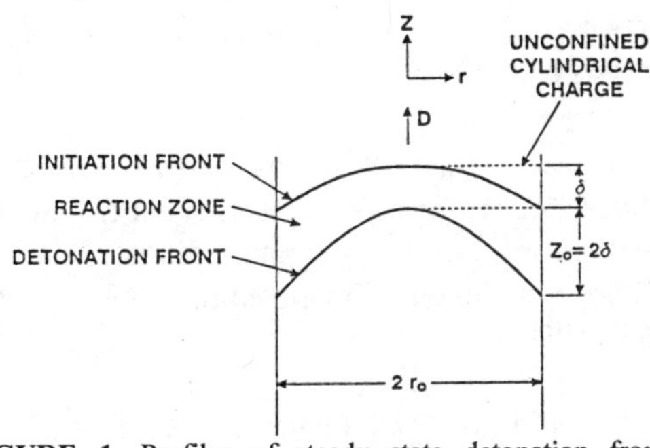

FIGURE 1. Profiles of steady state detonation front and constant amplitude shear stress/initiation front.

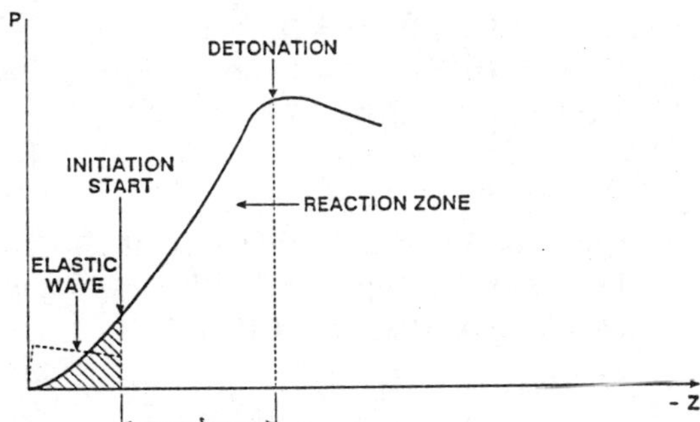

FIGURE 2. Pressure profile of detonation wave along the z axis. Cross hatched area prior to initiation is the region of plastic work needed to establish energy density for initiation.

Detonation Due to Projectile Impact

Treating the energy dissipated during plastic deformation as the source of initiation allows impact to be treated in much the same manner as shock. Consider briefly the problem of a spherical projectile impacting an explosive charge. Assume that to first order the flow around the projectile resembles the classical flow of an incompressible fluid around a sphere. The flow velocity is

$$V = -V_0 (1 - \frac{a^3}{r^3}) \cos\phi\, u_r + V_0 (1 + \frac{a^3}{2 r^3}) \sin\phi\, u_\phi$$

(7)

A mass element located at a radius r and angle ϕ from the z axis of a sphere of radius a will experience a time varying velocity gradient and consequently a force gradient and a resultant shear due to the passage of the projectile, $\tau = (m/A)d/dt(dr \cdot v)V$. If the velocity of the projectile is nearly constant, the time differential transforms as $d/dt \rightarrow V_0 d/d\ell$ where $d\ell$ is the distance a mass element travels in a time dt. From this the critical shear stress for initiation of a steady state detonation is

$$\tau_c = \frac{\rho (a + r_f) \ell V_0^2}{4 \pi a^2} [\quad]^{1/2} .$$

(8)

where r_f is the failure radius of the explosive material and the brackets enclose the angular terms.

Equation (8) can be used with experiment to determine the shear stress, τ_c, required for the initiation of a steady state detonation. Listed below are the experimental conditions required to achieve steady state detonation for three different explosives along with the calculated value for the critical shear stress.[10] The crystal size, ℓ, was not well known in these experiments. An average particle size of $\ell = 10^{-4}$ m was assumed. The angular terms in (8) were set equal to unity and the projectile radius was 6.35 mm.

Explosive	r_f(mm)	V_0(km/s)	ρ(kg/m^3)	τ_c(Pa)
PBXN-110	3.	1.987	1.66×10^3	$.12 \times 10^8$
PBXN-111	18.	2.9	1.79×10^3	$.72 \times 10^8$
Comp B	2.15	1.64	1.7×10^3	$.08 \times 10^8$

From eqn. (5) the critical shear stress was determined to be $.29 \times 10^8$ Pa, $.48 \times 10^8$ Pa and $.29 \times 10^8$ Pa for PBXN-110, PBXN-111 and Composition B respectively. In a different set of experiments the stress on crystals of the explosive HMX at the onset of detonation was found to be about 10^7 to 10^8 Pa depending on particle size.[11] This are also the approximate stress levels at which rapid plastic flow occurs as it should be.[11]

CONCLUSIONS

At the microscopic level energy dissipation and localization due to plastic deformation are responsible for the initiation of detonation. Here, this has been extended to the macroscopic scale where shear has been identified as the important driver of plastic deformation. This specifies an initial condition for the initiation portion of the detonation wave since this shear stress must be supplied by the shock supported by the detonation. This couples the energy dissipation processes in the initiation zone with the detonation shock wave. It is impossible to achieve this with the classical asymptotic theories of detonation. The present development specifies the form of the detonation wave, critical shear level for initiation and other observables. It avoids the artificial separation of excitation by shock and by impact, they differ only in the rate of energy dissipation and, consequently, in the rate of molecular dissociation.

ACKNOWLEDGEMENTS

This work was supported by the Office of Naval Research and the Naval Surface Warfare Center Independent Research Program.

REFERENCES

1. Fickett, W. and Davis, W. C., *Detonation*, Berkeley: University of California Press, 1979, ch. 2, p 42.

2. Lee, E. L. and Tarver, C. M., *Phys. of Fluids* **23**, 2362 (1980).

3. Johnson, J., Tang, P and Forest, C. A., *J. Appl. Phys.* **57**, 4323 (1985).

4. Bdzil, J. B. and Stewart, D. S., *Phys. of Fluids* A, **1** (7), 1261 (1989).

5. Coffey, C. S., *Phys. Rev.* B **24**, 6984 (1981).

6. Coffey, C. S., *Phys. Rev.* B **32**, 5335 (1984).

7. Coffey, C. S., *J. Appl. Phys.* **70** (8), 4248 (1991).

8. Coffey, C. S., in *Structure and Properties of Energetic Materials*, D. H. Liebenberg, R. W. Armstrong and J. J. Gilman, editors., Material Research Society Symposium, Volume 296, 63 (1993).

9. Forbes, J. W., and Lemar, E. R., This Symposium.

10. Liddiard, T. P., and Roslund, L. A., "Projectile/Fragment Impact Sensitivity of Explosives", NSWC TR 89-184, p. 28.

11. Coyne, P. J., Elban, W. L., and Chiarito,"Strain Rate Behavior of HMX Porous Bed Compaction", *8th Symposium on Detonation*, 1985, pp. 645-657.

NATURE OF IGNITION SITES AND HOT SPOTS, STUDIED BY USING AN ATOMIC FORCE MICROSCOPE

J. SHARMA and C.S. COFFEY

NAVAL SURFACE WARFARE CENTER, CARDEROCK DIVISION
NAVAL SURFACE WARFARE CENTER, INDIAN HEAD DIVISION
SILVER SPRING, MD 20903-5640

The morphology of ignition sites and hot spots, with structures ranging from ten microns to the molecular size, has been studied by using an atomic force microscope. Sub-ignited and ignited samples of RDX from drop hammer tests have been investigated. In the central (caked) area, the particles get crushed but they still retain crystal edges and their height is comparable to their lateral dimensions. Particles from outer streaked area are more flat, rather rounded and they do not show crystal edges. Shear bands and dislocations are seen on them. Two kinds of reaction sites have been identified. One kind are volcano-like hillocks, ranging from 100-1000 nm in size, occasionally with holes. The shape gives evidence of melting and they seem to arise from the pressure of trapped gases generated in deep lying sites. The other reaction sites consist of hemispherical craters 20-100 nm in size. They appear to arise from shear forces and generation of dislocations. Molecularly resolved images give evidence of very high density ($5X10^{12}$ /cm^2) of dislocations and crystallographic disarrays.

INTRODUCTION

Understanding the structure and formation of hot spots and ignition sites in energetic materials is important as it can supply information on the safety aspects of explosive handling. The idea of hot spots has its origin in the work of Bowden and Yoffe (1) who proposed a size of 0.1-10 μm for them. The advent of the atomic force microscope (AFM), gives us the opportunity to investigate the hot spots and reaction sites down to the molecular scale. In the present work such a study has been carried out in impacted RDX.

An AFM (atomic force microscope) yields images from the excursions of a sharp tip mounted on a tiny cantilever, when it is made to raster scan a given sample (2). It has become a powerful technique and can reveal topographical structures such as crystal defects, dislocations, shear bands, craters and hillocks, ranging in size from nanometers to microns, including molecular or atomic arrays.

In the present work, control and impacted samples of RDX have been investigated. For the latter, sugar like powder of RDX crystals, 100-300 μm in size, sandwiched between sheets of heat sensitive films or of cleaved mica, was impacted. Specimen were picked up at different distances from the center of the impact pattern (3), and were imaged by the AFM. It is found that in the center, a cake-like layer is formed, composed of crystals crushed to micron size, which still retain crystal edges and shapes. As the particles are displaced radially outward, they become rounded like pebbles and lose their crystal appearance. At larger radius the pebbles get more and more flattened. At the end region of their travel, where ignition usually starts, the surface of the particles is rough, isolated and groups of uplifted molecules are seen having jogs and dislocations. The number density of dislocations found is rather high-$5X10^{12}$ /cm^2. It has been found that in the dislocations, not only the molecules are displaced but they also suffer rotation. This has occurred because RDX is a molecular crystal, for an atomic solid this question does not arise. Up and down undulations are shown by rows of molecules. On larger scale, evidence of melting, formation of volcanoes, micron size Munroe jets, hillocks arising from internal reactions caused by adiabatic compression and heating, have been observed. Crater like hemispherical reaction sites were also observed. The smallest reaction site was 20 nm in size which translates into an action spot of mere 10,000 molecules releasing only 10^{-14} J of energy. This may represent the smallest hot spot which could survive thermal quenching, by the lattice.

EXPERIMENTAL

A Digital Instruments Nanoscope II, Scanning Probe Microscope was used in the present work. The measurements were carried out in air at room temperature using a 100 µm v-shaped cantilever with a 3 µm oxygen sharpened silicon nitride tip. The cantilever was used in repulsive mode to avoid confusion from surface contamination, and therefore a high force of 10^{-7}N was used. The images were obtained in height mode.

RESULTS AND DISCUSSION

In the center of the impact pattern, the crystals were found to have been crushed to microns size and formed a thin cake, but they still showed crystal faces and edges. As the crystals move out radially they become rounded like pebbles and get decreased in size to 100-300 nm. Simultaneously the ratio of height to lateral dimension progressively changes from a value of 1:2 in the center to a value of 1:20 near the end of the track. In other words the particles are getting flattened. The surfaces of the particles do not end in flat crystallographic planes. During the high speed radial motion (v≈150 m/s), deformation of the particles takes place. Friction and cold working causes melting in some areas. Since the melting and decomposition temperatures are very close in RDX, simultaneous decomposition must be taking place and products such as nitroso derivative of RDX could be produced.

Figure 1 shows the formation of volcano like reaction sites of 200-1000 nm size, complete with craters through which gases might have escaped. The surrounding area shows evidence of melting and re-solidification. Fig. 2 shows a Munroe jet formed on a cylindrical surface of mica, which gives the tracks a curved appearance. The squirted liquid tracks are about ten microns in length and have left 300 nm wide tracks. This is the smallest Munroe jets ever reported. In some areas a large number of hillocks are produced ranging in size from 200-1000 nm. From their shape it appears that they are formed by reaction taking place in the interior, leading to melting and ballooning (Fig. 3). Probably these hot spots arise from adiabatic compression and heating of trapped gases as suggested by Bowden and Yoffe (1).

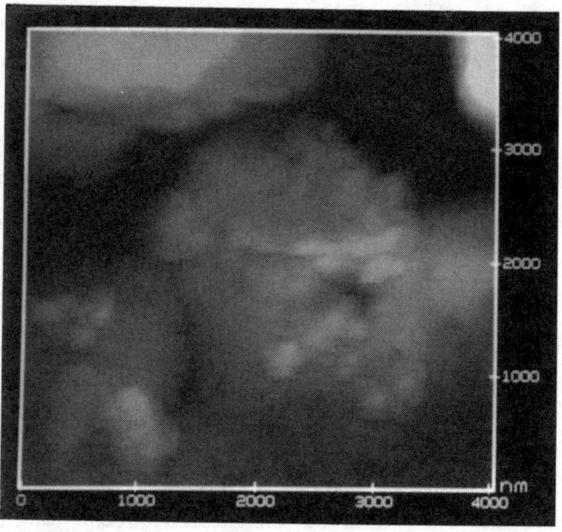

FIGURE 1. An AFM image of two reaction sites, looking like volcanoes, in impacted RDX is shown. The structure indicates that melting and gas ejection has taken place.

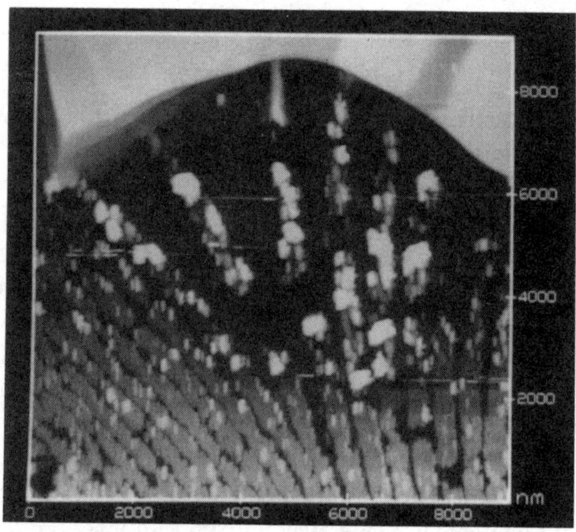

FIGURE 2. Micro Munroe jets approximately 10 microns long and 300 nm wide are shown by this AFM image, giving direct first time proof of Roth's suggestion.(Ref.6).

Another kind of reaction site observed consists of hemispherical craters, as shown in Fig. 4. Material has reacted and disappeared. The craters range over 20-300 nm in diameter. In some areas, concentration as high as 10^9 /cm² of such craters has been seen, obviously the critical density needed to consume the entire portion of the crystal would be higher. Among these craters, we find the smallest reaction site to be 20 nm in diameter, which would encompass about 10,000 molecules and energy evolution of 2×10^{-14} J. This result agrees well with the calculations of Armstrong et al (3), who arrived at a number of 3500 molecules for the smallest hot spot, with 40 ns as characteristic time for growth.

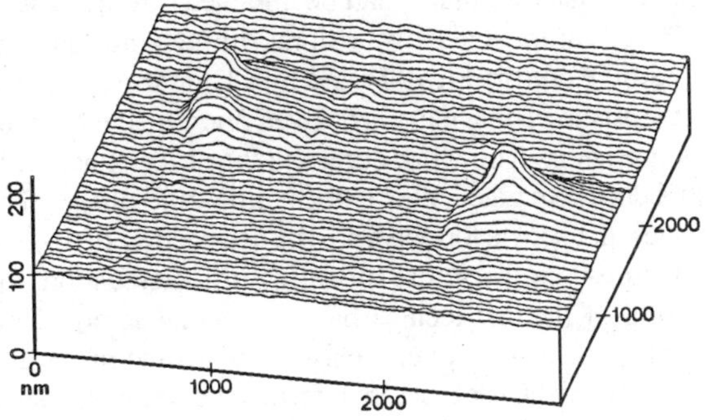

FIGURE 3. A line plot image (AFM) of hillock-like reaction sites is shown, in which melting and ballooning is evident, due to action in the interior.

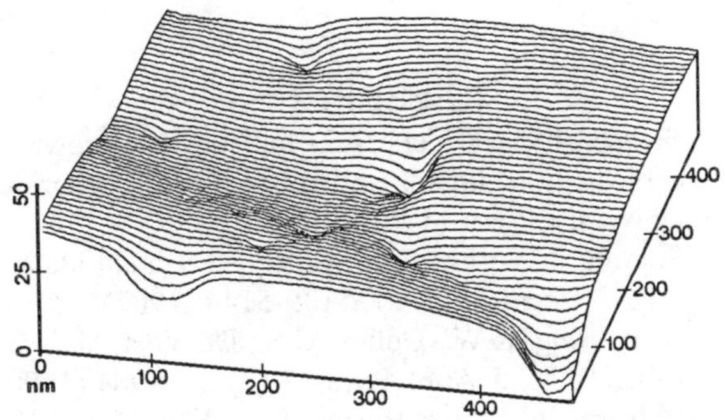

FIGURE 4. Hemispherical craters arising from loss of material are shown. The smallest crater found is 20 nm in diameter.

In molecularly resolved images a large number of molecular disarrays, jogs and dislocations can be seen in impacted samples. Fig. 5 shows image of a control single crystal RDX. In contrast to the control crystal, the surface of an impacted sample, as shown in Fig.6, does not end in a crystallographical flat plane. Some molecules arise out of the plane and sit at higher level in an isolated way. Repeated jogs in the molecular rows are seen as evidence of dislocations. Shear bands of raised molecules are frequent. Alignment in molecular rows, is often violated. In a 10nmX10nm area, quite a few jogs can be seen. This means that dislocation density as high as $5 \times 10^{12}/cm^2$ is possible.

FIGURE 5. Molecularly resolved image of a control RDX crystal (001) plane, is shown giving regular arrays. Each blob represents a molecule (2.5X2.5 nm image).

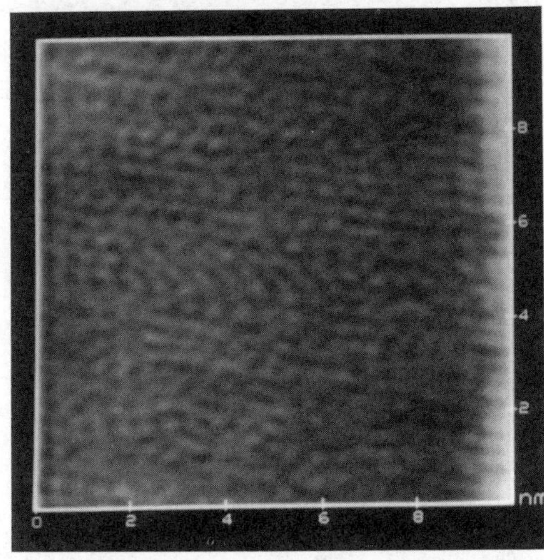

FIGURE 6. Image of a particle of RDX, which has been shot out radially during impact. Molecular disarrays and jogs from generations of dislocations are exhibited.

Some rows of molecules show gradual undulations in the z-direction with a periodicity of 5-8, as shown in Fig. 7. This can arise from residual strain in the crystal. Else, due to chemical alterations to nitroso derivative (4,5), the size of the molecules would slightly decrease and this can produce up and down undulations.

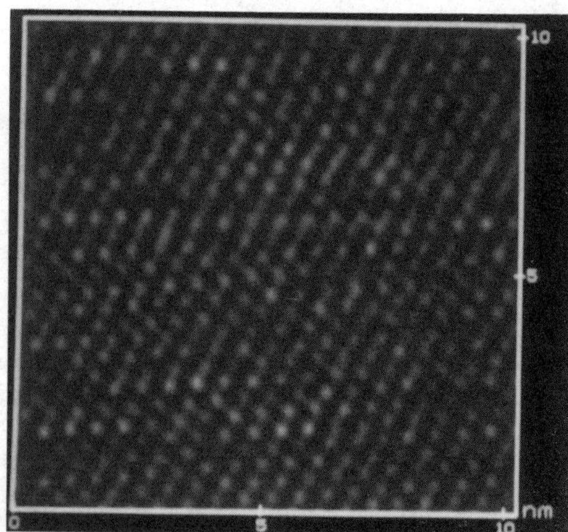

FIGURE 7. A 10X10 nm image of radially driven particle is shown. The variations in the size of the blobs indicate fluctuations in the height and orientations of the molecules.

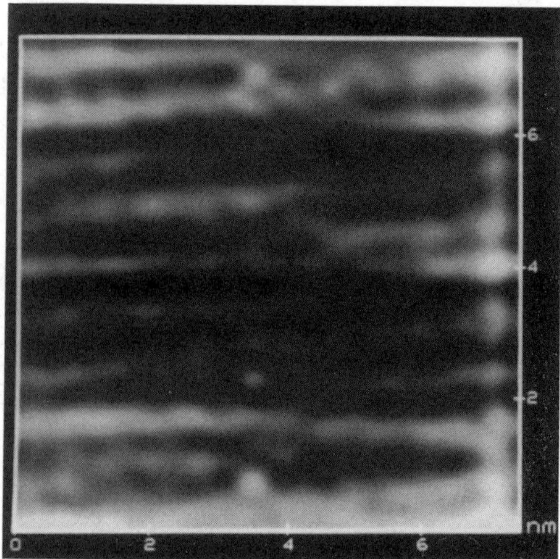

FIGURE 8. A region of effected RDX showing a slip plane (vertical) and a few dislocations in horizontal rows.

Figure 8 shows a region where edge dislocations can be seen. The positions of the molecules are changed in a way commensurate with an edge dislocation, but it appears that in the new sites the molecules with larger space around them undergo re-orientation. Since RDX molecules are not spherical, it is not surprising that orientation would change. This aspect of molecular reorientation in the dislocations has so far not been taken into consideration in the theoretical model. Now that it has been seen, it is obvious that in the region of a dislocation the molecules should adjust not only their positions but also their orientations.

CONCLUSIONS

The present work shows that by using an atomic force microscope on variously effected and partially damaged explosives, a large amount of information on the processes that take place, can be obtained. In the case of impacted samples it shows that most of the mechanisms listed by Roth(6), Field(7) and by Chaudhri (8) for hot spot formation come into play. Even micron size Munroe jets are seen. The work confirms that a lot of action in RDX does take place after it goes into liquid phase. The smallest hot spot is formed from the destruction of 10,000 molecules. In the future modeling of a dislocation, the rotation of the molecules has to be considered. This would apply not only to explosives, but also to molecular solids in general. The fact that particles of explosives are found in various stages of reaction, explains that during impact the hot spots develop independently at numerous sites till enough gas is released to cause a shock wave and general explosion. This also explains the existence of many thermal spikes observed in the BIC tests of Coffey (9).

REFERENCES

1. Bowden, F.P. and Yoffe, A.D. Initiation and growth of explosions in liquids and solids, U.K., Cambridge University Press, (1952), 26, 65.
2. Sharma, J., Forbes, J.W., Coffey, C.S. and Liddiard, T.P., J. Chem. Phys. 91, 5139-5144 (1987).
3. Armstrong, R.W., Coffey, C.S., De Vost, V.F. and Elban,W.L., J. Appl. Phys. 68 (3), 979-984 (1990).
4. Behrens, Jr, R. and Bulusu, S., J.Phys. Chem. 95, 5838 (1991).
5. Sharma, J., Hoffsommer, J.C., Glover, D.J., Coffey, C.S., Forbes, J.W., Liddiard, T.P., Elban, W.L. and Santiago, F.in Eighth Symposium (International) on Detonation. Albuquerque, NM. (1985), 725.
6. Roth, J.,Encyclopedia Explos.7,Ed. Fedroff, B.T. and Sheffield, O.E.(1975), H170.
7. Field, J.E., Bourne,N.K., Palmer, S.J.P. and Walley, S.M., Eds.Field, J.E. and Gray, P., Energetic Materials,London, The Royal Society, (1992), 270.
8. Chaudhri, M.M., in Ninth Symposium (International) on Detonation,857, (1989).
9. Coffey, C.S., De Vost, V.F., and Woody, D.L. in Ninth Symposium (International) on Detonation, 1243, (1989).

ORIENTATION-DEPENDENT SHOCK RESPONSE OF EXPLOSIVE CRYSTALS *

J. J. DICK

Group DX-1, MS P952, Los Alamos National Laboratory, Los Alamos, New Mexico 87545 USA

Some orientations of PETN crystals have anomalously high shock initiation sensitivity around 4 to 5 GPa. Results of a series of laser interferometry experiments at 4.2 GPa show that this is associated with an elastic-plastic, two-wave structure with large elastic precursors. Implications for the initiation mechanism in single crystals is discussed. Initial work on beta phase, monoclinic HMX is also described.

INTRODUCTION

Anomalous luminescent emission and initiation of detonation have been observed for two orientations of single crystals of pentaerythritol tetranitrate (PETN) in shock experiments near 4 GPa.(1) The crystals were more sensitive at 4.2 GPa than at 8.5 Gpa. From the data available it was not clear what was responsible for this anomaly. In addition to the sensitivity anomaly observed in wedge experiments, there was an unusual intermediate velocity transition between the initial shock velocity and the final detonation velocity in a wedge experiment on a [110] crystal. After consideration of these results it seemed that measuring time-resolved histories at several thicknesses through the initiation regime would be very helpful in clarifying the nature of the anomaly. Therefore a series of measurements of particle velocity vs time at several thicknesses through the initiation regime was undertaken using velocity interferometry. The results indicate that the anomaly is associated with separated elastic and plastic waves with large elastic precursors. In addition to the [110] experiments, experiments were performed at 4.2 GPa on [100] and [001] orientations as well. The records show orientation dependence in accord with previous luminescent emission experiments and a model of orientation dependence of shock sensitivity based on steric hindrance to shear.(2,1) Interferometry experiments were performed on [110] crystals at stresses ranging from 4.0 to 7.2 GPa in order to look at the variation in material response from the anomalous regime to the higher stress regime. The records show a continuous variation from one type of history to another. At higher stresses with a single shock, the initiating flow peaks further behind the shock wave. This results in slower shock growth at 6 and 7.2 GPa than at 4 to 5 GPa.

EXPERIMENTAL TECHNIQUE

PETN crystals were subjected to shocks using a light-gas gun. Particle velocity vs time histories were recorded at the PETN/PMMA window interface using a velocity interferometer. Projectiles made of 2024 aluminum were impacted on Kel-F (polytrifluorochloroethylene) discs 50 mm in diameter and 5 mm thick. The PETN crystals were mounted on the Kel-F discs with a silicone elastomer. The crystals had typical lateral dimensions of 15 mm.

The measurement system used was a dual, push-pull, VISAR system.(3) The dual VISAR with different fringe constants removes ambiguity in determining the particle-velocity jump at the shock when extra fringes must be added. The light was transported from the argon-ion laser to the target and thence to the interferometer table with fiber optics.

EXPERIMENTAL RESULTS

In previous work(1) anomalous detonation was observed in a [110] PETN crystal in a wedge experiment at about 4.26 GPa. The run distance to detonation in wedge experiments was shorter at

*Work performed under the auspices of the U. S. Department of Energy

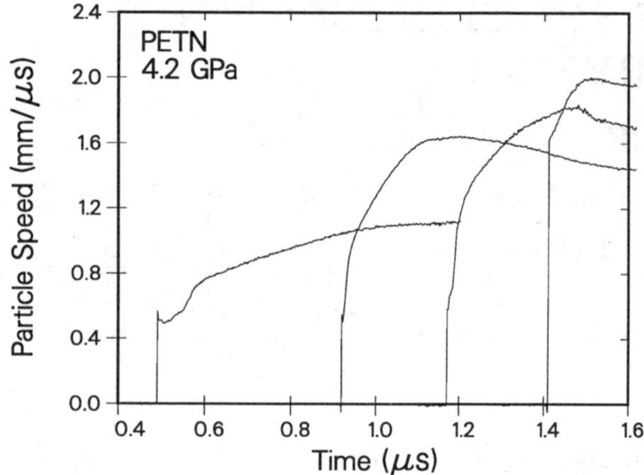

FIGURE 1. Particle vs time histories at the PETN/PMMA interface for a 4.15 GPa input shock at 1.825, 3.47, 4.44, and 5.55 mm PETN thicknesses of [110] orientation

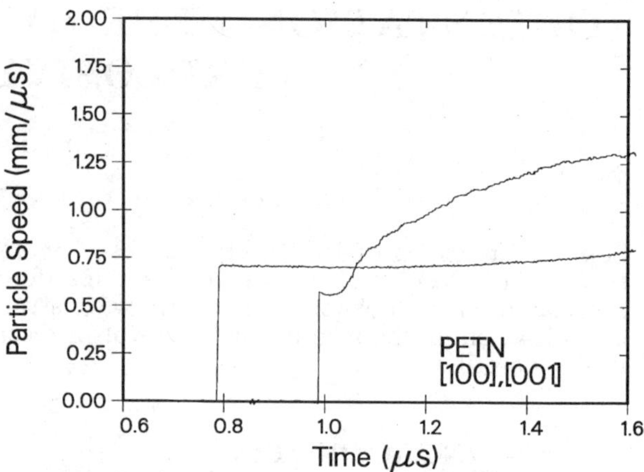

FIGURE 2. Particle vs time histories at the PETN/PMMA interface for a computed 4.15 GPa input shock strength for [001] and [100] orientations. The [001] crystal thickness was 3.79 mm and the [100] crystal thickness was 2.90 mm.

4.26 GPa than at 8.5 GPa. The run distance normally would increase and the sensitivity decrease with decreasing input shock stress. Furthermore, the run distance was the same at 4.2 Gpa and at 9.2 Gpa, a double-valued behavior. More experimental information was needed to clarify the behavior. In order to observe the behavior behind the leading shock wave, a series of four VISAR experiments was performed at 4.15±0.01 GPa on [110] crystals of different thicknesses in order to obtain particle vs time histories through the initiation regime. The first two with crystal thicknesses of 1.825, and 3.47 mm were in the region of constant initial shock velocity in the wedge experiment. The third at 4.44 mm thickness was at the onset of the intemediate velocity transition. The fourth experiment with a crystal thickness of 5.55 mm was in the region of the intermediate velocity transition. The particle velocity vs time histories obtained at the interface are shown in Fig. 1. It was unexpected to see a two-wave structure. In Ref.(1) the leading wave was thought to be the bulk or plastic wave to the final shock state.

The two-wave structure recorded at 1.825 mm is a large elastic shock followed by a more-dispersed plastic wave. The elastic wave amplitude is 2.74 GPa in PETN. This precursor strength is much larger than those seen for input shock strengths of 1.14 GPa. There the elastic precursor shock strength for [110] crys-

tals was 1.0 GPa. Dependence of elastic precursor strength on input shock strength in [110] and [001] PETN crystals was noted in earlier work for shock strengths up to 2.7 GPa.(4) Elastic precursor strengths were as strong as 2.0 GPa after 5 mm of wave propagation in that work.

The profile behind the plastic wave is not flat and steady as would be expected in an inert material. Instead, there is evidence of exothermic initiation chemistry causing increasing particle velocity immediately behind the plastic wave. The initiating flow accelerates the second wave so that it completely overtakes the elastic shock by about 4.6 mm causing the intermediate velocity transition. The detonation transition was at 6.6±0.2 mm in the wedge experiment.

In Fig. 2 particle velocity vs time histories for [001] and [100] orientations for the same input stress are displayed. For the [001] orientation an elastic-plastic, two-wave structure is displayed similar to that observed in [110] orientation. The elastic precursor strength is 3.15 GPa, larger than observed in [110] orientation. However, the initiating wave is weaker than in [110] at that thickness. In contrast, the [100] crystal displays a single wave to the final state followed by a nearly constant particle velocity indicative of essentially inert behavior. These behaviors correlate with the relative steric hindrance to shear.

DISCUSSION OF THE RESULTS

Elastic-Plastic Wave Structure

In the [110] and [001] orientations there is an elastic-plastic wave structure in the region of the low-shock-stress sensitivity anomaly observed for [110] crystals. Initiation begins in or immediately behind the plastic wave. This is consistent with our model of steric hindrance to shear.(1,2) In the model the endothermic first step in explosive decomposition is chemical bond breaking in the sterically hindered shear flow in the plastic wave or shock. This leads to the exothermic decomposition steps on the way to initiation of detonation, especially at low stresses. Our previous geometric analysis of steric hindrance for rigid molecules found [110] and [001] orientations to be hindered and [100] and [101] orientations to be relatively unhindered. These results were corroborated by molecular mechanics analysis of deformable molecules for the cases considered, [100], [101], and [110]. For the [100] orientation there is a single wave with a flat following flow indicative of no initiation response. This is consistent with the minimal steric hindrance for this case. The small elastic precursor(2) has been overdriven by the plastic wave at this level of shock strength; i.e., the wave speed on the plastic Hugoniot is faster than the wave speed on the elastic Hugoniot for the input particle velocity of 0.616 mm/μs.

The two-wave structure explains another feature noted in earlier work.(1) From photodiode records of the luminescent emission it was inferred that there was an absorbing or dark zone behind the leading shock. This is consistent with the emission coming from the region of the plastic wave, not the leading elastic wave. The interpretation is that the peak in the photodiode signal and the subsequent fall in signal level is due to quenching of the emission in the crystal by the rarefaction from the free surface after arrival of the plastic wave. In Fig. 3 particle velocity records are shown of the elastic-plastic wave structure for [110] and [001] orientations. There are arrows on each record marking the time at which the photodiode peak would be based on data presented in our 1991 article.

In order to determine the position of the photodiode peak for the sample thicknesses correspond-

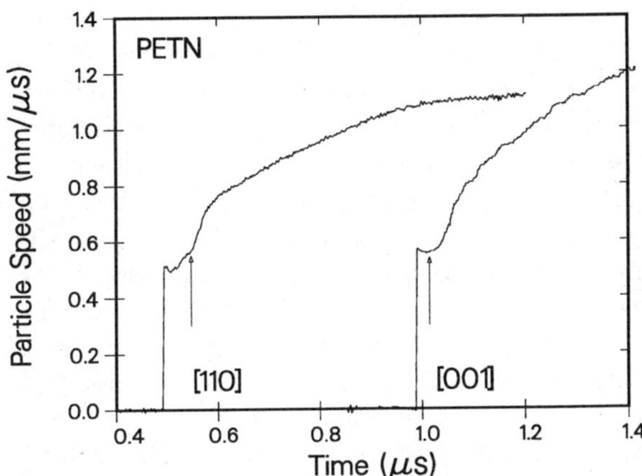

FIGURE 3. Particle vs time history at the PETN/PMMA interface for [110] and [001] crystals shocked to about 4.16 GPa. An elastic-plastic,two-wave structure is displayed by both records. The plastic wave is followed by increasing particle velocity due to exothermic initiation processes. In each case the arrow indicates the inferred position of the peak in emission as determined from a photodiode record in earlier work. The beginning of the fall in emission intensity coincides roughly with the beginning of the plastic wave. This indicates that the luminescent emission begins at the base of the plastic wave.

ing to the VISAR experiments, the following analysis was performed. The photodiode records were obtained for crystal thicknesses different from those used in the VISAR experiments. For [110] orientation the photodiode record was for a crystal 2.79 mm thick vs 1.825 mm for the VISAR record. For [001] orientation the photodiode record was for a crystal 3.94 mm thick vs 3.79 mm for the VISAR record. The input shock stresses were were equal within 0.22 GPa for [110] orientation and within 0.11 for the [001] orientation. An analysis was performed in the position-time plane to determine the arrival time for the event associated with the photodiode peak at the sample thicknesses of the particle velocity records assuming a constant velocity for the disturbance. Account was taken of the particle velocity of the PETN/PMMA interface in the VISAR experiment and PETN free surface velocity in the emission experiment, but wave interactions were ignored. The disturbance arrival time t is given by:

$$t = \frac{t_0 - \frac{x_0}{u_i}}{1 - \frac{U_{pk}}{u_i}} \qquad (1)$$

where t_0 is the elastic wave transit time in the VISAR experiment, u_i is the velocity of the PETN/PMMA interface in the VISAR experiment, and U_{pk} is the apparent velocity of the photodiode peak from the photodiode experiment.

The striking result as seen in Fig. 3 is that quenching of the emission begins as soon as the initial portion of the plastic wave arrives at the free surface, at least within the 10-20 ns accuracy of the analysis. This implies that the emission originates from the entire plastic wave not just behind it. It suggests that onset of emission coincides with the onset of sterically hindered shear. From time-resolved spectral measurements this emission was interpreted as due to excited electronic states of NO_2.(1) This raises the possibility that the emission is due to direct nonequilibrium excitation by the sterically hindered shear. As suggested in an earlier article(2) the endothermic first step in initiation may involve nonequilibrium excitation of molecules on a femtosecond time scale caused by a mechanical process, sterically hindered shear occuring in the plastic flow associated with the uniaxial strain in a plane shock. In the molecular mechanics calculations in that article the dihedral angle changed by up to 60°, a much larger change than that caused by thermal motion. Also, the calculations indicated that significant bond angle strain occurred in PETN for the most hindered cases. Ref. (5) suggests ways in which bond angle distortion can drastically change the electronic state of a molecule.

It is worth mentioning that the calculated homogeneous temperature rise at 4.2 GPa is about 100 °C. The peak in the spectral data corresponds to 5000 to 6000 K by Wien's law, an unreasonable heterogeneous temperature, much higher than detonation temperature. Furthermore the spectral curves do not fit those of a gray body with constant emissivity. Rather, the spectra have the character of a chemiluminescent edge on the blue side. This result substantiates the previous conclusion that the observed emission is due to luminescence from excited electronic states.(1) While we consider the nature and timing of the emission to be evidence for a mechanoluminescent mechanism, the possibility that the sterically hindered shear causes vibronic uppumping followed by bond breaking, and that the electronic excited states are due to subsequent chemical reactions on a nanosecond time scale cannot be ruled out.

HMX STUDIES

Work has begun on studying the unit cell of this monoclinic crystal. The space group is $P2_1/c$. Possible slip systems are being studied for relative steric hindrance for different possible shock orientations. Because of the reduced symmetry of the unit cell, there are many more cases to consider than for PETN. The known slip systems of anthracene and other molecular crystals of the same space group have been studied for possible guidance. The importance of twinning in deformation of HMX is another complication.(6,7) Crystals of 110 and 011 orientations in $P2_1/n$ have been cut into slabs in preparation for VISAR experiments.

REFERENCES

1. J. J. Dick, R. N. Mulford, W. J. Spencer, D. R. Pettit, E. Garcia, and D. C. Shaw, J. Appl. Phys. **70**, 3572 (1991).

2. J. J. Dick and J. P. Ritchie, J. Appl. Phys. **76**, 2726 (1994).

3. Willard F. Hemsing, Rev. Sci. Instrum. **50**, 73 (1979).

4. P. M. Halleck and Jerry Wackerle, J. Appl. Phys. **47**, 976 (1976).

5. J. J. Gilman, Phil. Mag. B **71**, 1057 (1995); in *High-Pressure Science and Technology-1993-Part 2, Amer. Inst. Phys. Conf. Proc. 309*, edited by S. C. Schmidt, J. W. Shaner, G. A. Samara, and M. Ross, p. 1349, New York, 1994, AIP Press.

6. S. J. P. Palmer and J. E. Field., Proc. R. Soc. Lond. A. **383**, 399 (1982).

7. R. W. Armstrong, H. L. Ammon, Z. Y. Du, W. L. Elban, and X. J. Zhang, in *Structure and Properties of Energetic Materials, MRS Symposium Proceedings, Vol. 296*, edited by D. H. Liebenberg, R. W. Armstrong, and J. J. Gilman, p. 227, Pittsburgh, 1993, Mat. Res. Soc. Press.

PARALLEL/OBLIQUE IMPACT ON THIN EXPLOSIVE SAMPLES

Vincent M. Boyle, Robert B. Frey, and Alfred L. Bines

U.S. Army Research Laboratory, Aberdeen Proving Ground, Maryland 21005-5066

Parallel/oblique impact experiments were conducted on thin (0.6 mm–1.0 mm) explosive samples in order to induce a well-defined state of combined pressure and shear in the sample. A compressed gas gun accelerated a projectile into a thin explosive sample over a 130-mm diameter. The impacted explosive was viewed by a high-speed framing camera through a transparent anvil in order to detect evidence of reaction. Simplified assumptions were made in order to calculate the pressure and the strain rate; the pressure range of these experiments was from 0.33 GPa to 1.31 GPa, and the strain rate across the sample was 50,000 per second assuming a viscosity of 50,000 poise. No explosive reaction was detected for the 15-µs duration of these tests. The temperature due to viscoplastic heating of the explosive was calculated for several explosive thicknesses, viscosities, yield strengths, and projectile impact velocities. Using reasonable values of viscosity and yield strength, the maximum temperature increase for these tests was calculated to be 111° C.

INTRODUCTION

The shearing of explosive materials under pressure is an effective way to produce localized heating by viscoplastic work concentrated in a small region of the deforming explosive. This localized heating can cause the explosive to react, releasing additional heat to accelerate the reaction. In the experiments reported here, we have attempted to study the ignition of several explosives as they were impacted under conditions that would cause the explosive sample to shear in a known manner under the high pressure of the impact. A maximum pressure of 1.3 GPa was reached with a strain rate of about 50,000 per second over an explosive layer 0.6 mm thick.

EXPERIMENTAL APPROACH

In order to obtain well-defined conditions for pressure-shear impact on explosive, we adapted a technique described by Abou-Sayed, Clifton, and Hermann (1), Kim and Clifton (2), and Li and Clifton (3). In this technique, one-dimensional combined pressure shear waves were generated in a flat target plate by the impact of a flat, high acoustic impedance flyer plate; both the flyer plate and the target plate were inclined at an angle to the velocity vector of the flyer plate in order to produce a shear component of particle velocity in the impacted target. The impact occurred simultaneously at all points of the flyer-target interface. In addition, a high acoustic impedance anvil supported the target plate. The flyer plate and anvil have higher acoustic impedance than the target plate in order to prevent unloading of the target by reflection of waves at the target interfaces. This arrangement is illustrated in Figure 1 for the flyer plate impacting a target at 30° obliquity. A gas gun was used to accelerate the flyer plate. The explosive layer was observed with a framing camera. Upon impact, the normal component of the flyer plate velocity generates a stress wave in the target; this stress wave is reflected between the high impedance boundaries several times until the target reaches a state of uniform stress determined by the flyer plate velocity and the material properties of the flyer plate, target plate, and anvil. Likewise, the parallel

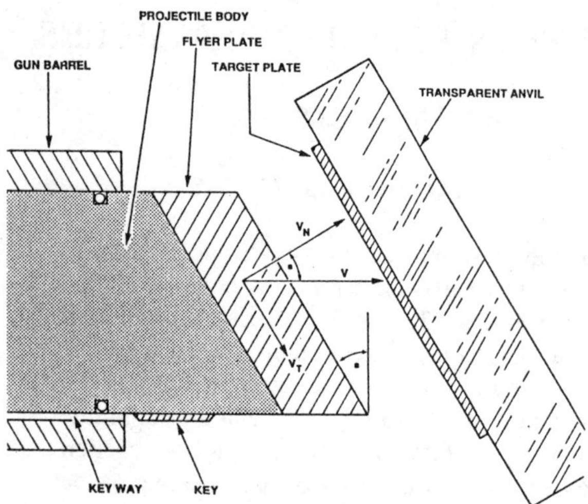

FIGURE 1. Parallel/oblique impact of a flyer plate on a thin explosive sample. The normal and tangential components of the flyer plate before impact are illustrated.

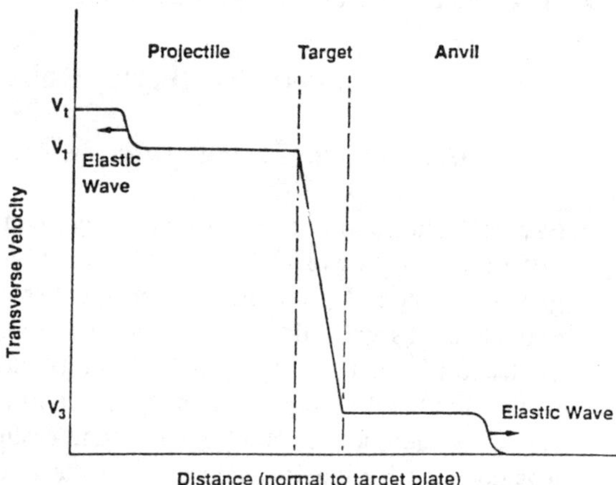

FIGURE 2. The transverse velocity (the component parallel to the interfaces) in the flyer plate, explosive target, and the anvil after impact.

component of flyer plate velocity, by its traction with the target surface, produces a shear wave in the target which, after several reverberations, induces a state of uniform shear. The strain rate associated with this shear is determined by the parallel component of flyer plate velocity; the material properties of the flyer plate, target plate, and anvil; and the thickness of the target plate.

In the experiments reported here, the strain rate in the explosive target can be calculated as

$$(V_{tp} - V_{ta}) / x , \qquad (1)$$

where V_{tp} and V_{ta} are the components of the projectile velocity and the anvil velocity parallel to the interface after impact and x is the original thickness of the explosive sample. The calculated strain rate depends on the values assumed for explosive viscosity and yield strength and the elastic properties of the flyer plate and anvil. Figure 2 illustrates these transverse components in the flyer plate, explosive target, and anvil after impact.

In order to calculate the stress in the explosive sample, we assumed that the flyer plate and the anvil remained elastic during the impact and, after several reverberations, the explosive attained a stress level equal to what would be achieved by

the impact of the flyer plate directly on the anvil. With these assumptions and the requirement that the particle velocity and pressure remain equal at the flyer plate-anvil interface, we were able to use the interface equation to calculate the stress in the explosive.

RESULTS

The following summary is the data obtained from a total of 30 tests. For flat impacts the pressure in the sample varied from 0.39 to 1.43 GPa. For oblique impacts it ranged from 0.33 to 1.31 GPa, and the shear strain rate varied from 9,600 s^{-1} to 49,000 s^{-1}. The impact simultaneity over a 130-mm diameter ranged from 0 to 20 μs for the 15 tests that were observable by the framing camera. The average nonsimultaneity was 7 μs. For the impacts that we were able to observe, we did not see any obvious sign of explosive reaction such as light emission or the expulsion of reaction products from the region of impact. In all cases, the explosive in the impacted region became darker in about 4 to 6 μs; after this, the darkness did not appear to increase during the available time of observation, about 15 μs. However, the darkness did appear to increase with the impact pressure.

We were able to compare several shear and nonshear tests at pressures which were nearly equal; the presence or absence of shear did not appear to have an effect on explosive darkening.

For one test, the explosive target consisted of a 0.5-mm cast sheet of Pentolite explosive in which the grain boundaries were very prominent. Upon impact at 0.68 GPa, the grain boundaries were noticeably darker than the rest of the explosive for several microseconds, and then the entire impacted region became uniformly dark.

Since we were not able to tell if the explosive darkening meant that reaction was occurring, we tried to detect infrared (IR) radiation by using a photovoltaic silicon photodiode that was sensitive to wavelengths from the visible to the near IR (300 nm to 1,100 nm). The photodetector viewed a small region on the edge of the impact area. For one test, the argon bombs did not function and the photodiode did not detect any signal during 6 ms of observation. For a second test, the argon bombs functioned and the photodiode detected a signal, but it corresponded to the turn on of light from the argon bombs before the flyer plate even impacted the explosive target.

Several tests for which the rear surface (the surface facing the camera) of the explosive was marked beforehand with fine lines using a permanent marker were done. The lines appeared to remain undistorted during the time of observation, even though the impacted area of the explosive became dark.

Examination of the debris recovered after the tests did not reveal any evidence of explosive reaction having occurred.

DISCUSSION

We were surprised that we were unable to detect any obvious sign of explosive reaction for Detasheet since, as reported previously (4), we were able to cause Detasheet to react under what appeared to be a milder stimulus of a 0.2-GPa pressure and a 60-m/s shear velocity. The duration of those tests was about 500 μs, whereas the tests reported here would be terminated when release waves originating at the boundary of the flyer plate reached the axis, a time of about 15 μs. The longer duration of those earlier tests may have allowed the explosive to reach temperatures required for reaction.

In these tests, if the strain rate is uniform across the target plate, the temperature increase in the target plate can be expressed by the formula,

$$\Delta T = (\nu \ [d\varepsilon/dt]^2 + Y \ [d\varepsilon/dt] \ ) \ t/\rho c \ , \qquad (2)$$

where

ΔT = temperature increase (°C)
ν = viscosity (poise)
$d\varepsilon/dt$ = strain rate (1/s)
t = time duration (s)
ρ = density (g/cm^3)
c = specific heat (ergs/g-°C)
Y = yield stress in shear (dynes/cm^2).

For the experiments reported here, the strain rate of the explosive is a function of its thickness, viscosity, and yield strength, as well as the component of the flyer plate velocity parallel to the explosive surface and the material properties of the flyer plate and anvil. We computed the strain rates corresponding to a range of explosive viscosities and yield strengths for a 30° angle steel flyer plate impacting 0.6-mm-thick Detasheet explosive at a velocity of 153 m/s. We then used equation 2 to calculate the corresponding temperature increase, assuming a time duration of 15 μs, an explosive density of 1.48 g/cm^3, and a specific heat of 1.25×10^7 ergs/g-°C. Figure 3 shows the temperature increase in the explosive target as a function of its viscosity and yield strength. It can be seen that the calculated temperature increase, over a wide range of viscosity and yield strength, is no greater than 116° C. We would not expect to see evidence of explosive reaction in our experiment at such a low temperature.

We can use Frank-Kamentskii's equation for the adiabatic explosion time (5) to calculate the temperature required to produce a thermal explosion in 15 μs. Using parameters for PETN explosive (6), the calculated temperature for a

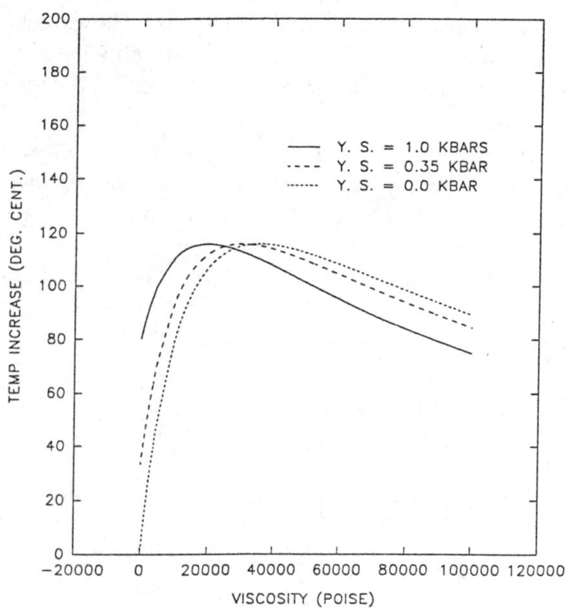

FIGURE 3. The temperature increase in an explosive target plotted as a function of viscosity and yield strength of the explosive. The sample thickness is 0.06 cm, and the transverse component of the flyer plate velocity is 7,650 cm/s.

thermal explosion time of 15 μs is 818 K, which corresponds to a temperature increase of 525° C. This temperature increase is much higher than those calculated for the parallel/oblique experiments.

CONCLUSIONS

We were not able to detect explosive reaction under conditions of combined pressure/shear. Using our experimental data and reasonable values for explosive yield strength and viscosity, we calculated a temperature increase of about 111° C in the impacted explosive. This increase in temperature is much too small to cause a thermal explosion in the time of our experiment, approximately 15 μs, and is also probably too small to generate observable reaction products.

REFERENCES

1. Abou-Sayed, A. S., Clifton, R. J., and Hermann, L., *Experimental Mechanics*, 127–132 (1976).
2. Kim, K. S., and Clifton, R. J., *Journal of Applied Mechanics*, **47**, 11–16 (1980).
3. Li, C. H., and Clifton, R. J., "Dynamic Stress-Strain Curves at Plastic Shear Strain Rates of 10^5 sec^{-1}," *Shock Waves in Condensed Matter*, AIP Conference Proceedings No. 78, edited by W. J. Nellis et al. Brown University, Providence, RI, 1981.
4. Boyle, V., Frey, R., and Blake, O., "Combined Pressure Shear Ignition of Explosives." *Ninth Symposium (International) on Detonation*, OCNR 113291-7, vol. 1, Portland, OR, Aug 28–Sep 1, 1989.
5. U.S. Army Materiel Command, *Engineering Design Handbook - Principles of Explosive Behavior*, AMCP 706-180, Alexandria, VA, 1972.
6. Rogers, R. N., *Thermochimica Acta*, **2** (1975).

PROJECTILE IMPACT INITIATION OF A HOMOGENEOUS EXPLOSIVE

M D Cook and P J Haskins

Defence Research Agency, Fort Halstead, Sevenoaks, Kent TN14 7BP, England

In this paper, we report results of projectile impact experiments into both pure, and amine-sensitised nitromethane. A variety of projectile geometries were used to impact charges covered by aluminium barriers of varying thickness. From these experiments, threshold conditions for detonation were determined. These results have been interpreted in terms of both a classical thermal explosion model, and by use of a 2D hydrocode incorporating an Arrhenius burn routine. The values of the Arrhenius parameters used to fit the results are discussed in the light of possible decomposition schemes for nitromethane.

INTRODUCTION

It is widely accepted that projectile impact is one of the major hazards which explosively filled stores may encounter. Despite the vast amount of work carried out in this area of research, there is still much to understand about the underlying processes involved. Furthermore, a comprehensive capability to predict the response of a munition to projectile impact is still lacking. Indeed, even prompt shock initiation, which is relatively well understood, still evades complete prediction. A number of empirical methods exist which can predict the prompt shock response of an explosive covered by various barrier plates, thicknesses and materials. For example, the method developed by James(1) based on a critical energy criterion. However, these methods really apply only to certain geometries, and cannot predict prompt shock responses in all cases.

Over the last few years, we have carried out a systematic series of experiments aimed at generating data on projectile impact phenomena on a variety of different types of explosives. In previous papers(2,3), we have reported projectile impact results on RDX/TNT 60:40, a plastic explosive containing 88% RDX with 12% grease, and more

recently, a study of XDT in both melt cast and PBX explosives(4). In this paper, we report the results of projectile impact experiments on both pure and amine-sensitised nitromethane.

Nitromethane is a valuable energetic material with which to investigate explosive phenomena. It has the advantage of being a clear liquid, which allows high speed photography of the onset and build-up of reaction. It is also homogeneous, which allows considerable simplifications to be made when attempting to interpret experimental results in terms of new theoretical models. An additional advantage of this material is that its sensitivity can be drastically increased by the addition of quite small quantities of amines. Since this is a chemical phenomenon, it was hoped that a study of both pure and sensitised nitromethane might give valuable insight into the chemistry involved in the initiation and growth of reaction in this material.

We also report, in this paper, an outline of a new explosive burn routine that we have developed and implemented in the DYNA2D hydrocode. The nitromethane projectile impact data have been modelled using this burn routine, and the data used are consistent with those derived from a thermal explosion model incorporating a single step Arrhenius reaction.

EXPERIMENTAL

Analar grade nitromethane was used in all experiments. Sensitised nitromethane was made-up fresh before each firing. The nitromethane was sensitised with 5% v/v diethylene triamine.

The liquid explosive was poured into either 9mm thick steel-walled or 3mm thick PMMA-walled containers having internal dimensions of 57mm in diameter and 100mm in length. The front face of the container, which was subject to impact, was made of aluminium plate of various thicknesses, and was glued to the body of the vessel. For the PMMA containers, a PMMA base-plate was also glued in place. A hole in the top of the cylindrical body allowed the liquid explosive to be poured into the vessel just prior to firing.

The projectiles used were flat-nosed steel rods. Two diameters of projectile were employed, 13.15mm and 20mm, that had the same mass. A few experiments were also carried out using a 13.15mm diameter projectile that had a nose angle of 165°. The projectiles were housed in a nylon sabot and fired from a 30mm RARDEN gun.

Each experiment was back-lit by flash bulbs and filmed using a quarter height fastax camera framing at ca. 30,000 fps. The film record was used to calculate the projectile velocity, observe the event, and check for normality of projectile strike. The estimated error in measuring the projectile velocities was no more than 2.5%.

PURE NITROMETHANE EXPERIMENTS

Experiments were carried out in both steel and PMMA containers. Figure 1 shows the results obtained for pure nitromethane. The error bars indicate the difference between the lowest velocity experiment which detonated, and the highest velocity that did not. The threshold plots obtained for both diameters of projectile are curved, and show a steady increase in threshold velocity with increasing barrier thickness. Analysis of the film records revealed that prompt detonations (the build-up to detonation occurring within 30µs), were always observed with

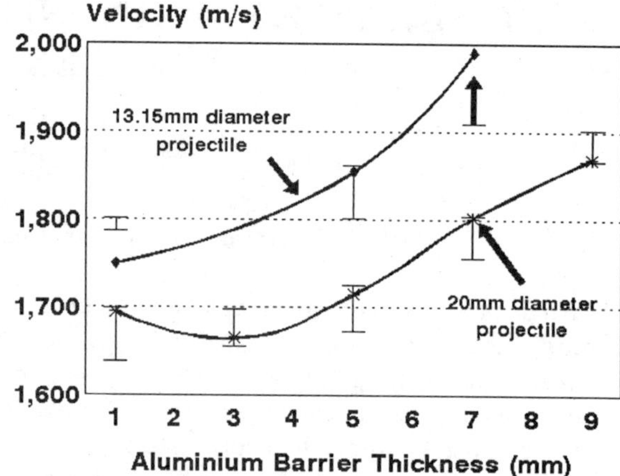

Figure1. Projectile impact results for pure nitromethane for both 13.15mm and 20mm diameter flat-nosed mild steel rods. The solid lines are the theoretical prediction from the DRA Explosive Burn Model.

experiments which were carried out using PMMA vessels. Steel containers also gave prompt detonation events, but occasionally delayed detonations (occurring at times in excess of 30 µ s) were observed. The reason for these delayed detonations is not clear, but might be due to shock-focusing or cavitation effects. Only the prompt events are shown in Fig. 1, and discussed here.

SENSITISED NITROMETHANE EXPERIMENTS

In addition to the flat-nosed rods, a conical projectile, having an included angle of 165° was also used. The results for these experiments are shown in Fig. 2. The sensitised nitromethane results gave lower threshold values than the equivalent pure nitromethane experiments, as would be expected. A detonation threshold curve, for aluminium barrier thicknesses of between 1 and 9mm, was obtained using a 13.15mm diameter projectile. From Fig. 2 it can be seen that an 'S' shaped response curve was obtained for this projectile. The significance of the shape of this response curve especially in relation to the pure nitromethane results is not clear. Results for the 20mm diameter projectile impacting sensitised nitromethane, covered with 1mm of aluminium, show a significantly lower threshold velocity than

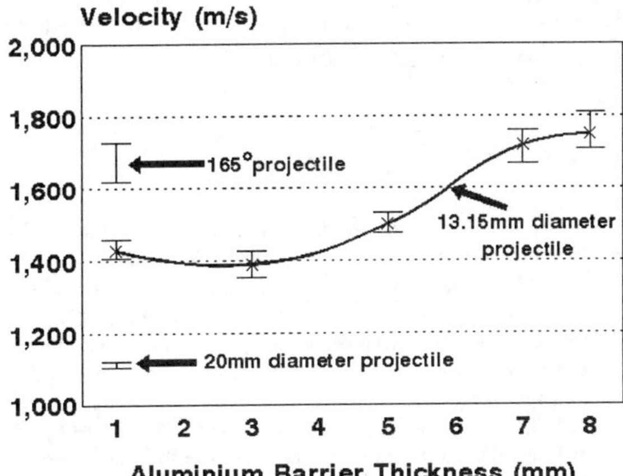

Figure 2. Projectile impact results for sensitised nitromethane using 13.15mm (both flat-nosed and 165° conical) and 20mm diameter projectiles. The nitromethane was sensitised with 5%v/v of diethylene triamine. For clarity, the experimental data is also shown as a solid line.

for the equivalent experiment for pure nitromethane. A conical-nosed (165°) projectile required a much higher velocity to achieve detonation in sensitised nitromethane protected with a 1mm aluminium barrier. As with pure nitromethane, experiments carried out in steel vessels occasionally gave delayed detonations.

DRA EXPLOSIVE BURN ROUTINE

An explosive burn routine, based on coupled Arrhenius chemical kinetics, has been developed, and implemented as a separation equation of state within the DYNA2D hydrocode. The Arrhenius burn model considers the growth of an explosive burn as a series of coupled kinetic steps. The rate of reaction of each step is a function of the concentrations of the constituents, the temperature and the pressure. Each reaction releases energy which fuels the burning process; if the release is sufficiently fast, the burn will run-away to detonation. The model is an attempt to consider the chemistry of the explosive realistically, and does not require arbitrary parameters. The model basically describes the behaviour of a homogeneous energetic material. The model requires both un-reacted and product hugoniots, the heat capacity at constant volume for the unreacted material, and activation energies and pre-exponential factors for a one-, two- or three-step Arrhenius kinetic scheme. The model is active throughout the calculation and, requires no arbitrary ignition or growth terms.

Using this basic model, we have attempted to parameterise the nitromethane Arrhenius kinetics using chemically sensible values, and to test the model against the projectile impact experimental data for nitromethane presented in the earlier sections of this paper.

Initially, a single-step exothermic reaction scheme was parameterised. An initial guess of both the activation energy and the frequency factor were obtained from thermal explosion theory in which a single exothermic Arrhenius term had been employed.

It has been demonstrated by Hubbard and Johnson(5) that, for a single step first order reaction obeying Arrhenius kinetics, the time for a fraction (f) of the original material to have reacted is given by:

$$t = v^{-1} \int_0^f (1-f)^{-1} \exp\left(\omega / (E_o + Qf)\right) df$$

In this expression, v is the frequency factor, Eo is the shock energy, Q the exothermicity of the reaction, and ω is given by:

$$\omega = C_v E_a / R$$

where C_v is the heat capacity, which is assumed to be constant, E_a is the activation energy, and R is the molar gas constant. We have used the above expression to obtain a fit to experimental aluminium flyer impact data on pure nitromethane(6). The calculated induction times that we present in the paper assume that $f = 0.9$ which is equivalent to 90% reaction of the explosive. The values used are shown in table 1.

TABLE 1. Thermal Explosion Theory fit to pure nitromethane aluminium flyer impact data.

E_a (J mol^{-1})	92296
v	2.68 x 1^8
C_v (J kg^{-1} K^{-1})	1720
Q (MJ Kg^{-1})	5.35

The data from the thermal explosion fit was used in the explosive burn model to attempt to reproduce the pure nitromethane projectile impact data. The standard DYNA2D fluid material model 9 was used for pure nitromethane. The viscosity was obtained from the CRC data handbook(7), and the specific heat data from Ref. 8. Whilst the values given in table 1 did not provide an exact fit, they were very close to the final optimised values, shown in table 2.

TABLE 2. Pure nitromethane data fitted to the projectile impact results using the explosive burn model

Density	1280 kg m^{-3}		
Viscosity coefficient	0.63 x10^{-3} Pa s (@ 25 °C)		
R	8.31 J kg^{-1} mol^{-1}	C_v	1720 J kg^{-1} K^{-1}
Murnaghan EOS Data		**JWL Data**	
A	1560 m s^{-1}	A_J	2.0925 x 10^{11} Nm^{-2}
B	1.721	B_J	5.69 x 10^9
Arrhenius Kinetics Data		R_1	4.4
ν	4.08 x 10^9	R_2	1.2
E_a	114000 J mol^{-1}	W	0.3
Q	5.35 MJ kg^{-1}	E_{0V}	5.1 x 10^9

The actual fit obtained from the burn model in DYNA2D is shown as the solid lines in Fig. 1. It can be seen that the fit is good apart from the result for the 13.15mm diameter projectile impacting through a 1mm thick aluminium plate. The discrepancy here is just within the experimental error.

DISCUSSION AND CONCLUSIONS

In this paper we have presented projectile impact data on both pure and sensitised nitromethane. We have outlined the principles behind a new homogeneous burn model that has been implemented in the DYNA2D hydrocode, and shown that we can reproduce the pure nitromethane experimental data using a chemically sensible single-step exothermic Arrhenius term. The fact that the experimental data can be fitted with a single exothermic step implies that it is this step which, to a very large extent, controls the behaviour in the prompt shock regime. Currently, this work is being extended to add an initial endothermic step to the pure nitromethane explosive burn parameterisation. Preliminary work suggests that the endothermic term is more consistent with a bi-molecular first step in the decomposition of pure nitromethane than with a unimolecular process In the near future, it is hoped to parmeterise the burn model for the sensitised material.

REFERENCES

1. James, H.R., "Critical Energy Criterion for the Shock Initiation of Heterogeneous Explosives" *Propellants, Explosives and Pyrotechnics,* **13**, 35, (1988).
2. Cook, M.D., Haskins, P.J., and James, H.R., "Projectile Impact Initiation of Explosive Charges" *in the Proceedings of The Ninth Symposium (International) on Detonation,* 1989, pp 1441-1450.
3. Cook, M.D., Haskins ,P.J., and James ,H.R., "An Investigation of Projectile and Barrier Effects on Impact Initiation of a Secondary Explosive", *in the Proceedings of the APS Topical Conference on Shock Compression of Condensed Matter,* 1991, pp. 675-678.
4. Haskins, P.J., Cook, M.D., "An Investigation of XDT Events in the Projectile Impact of a Secondary Explosive" *presented at The Tenth International Detonation Symposium,* July 12-16, 1993.
5. Hubbard, H.W., Johnson, M.H., J. Appl. Phys. **30,** pp. 765-769 (1958).
6. de Longueille, Y., Fauguignon, C., and Moulard, H., *in the Proceedings of The Sixth Detonation Symposium,* 1976, pp 105.
7. Lide, D.R., *CRC Handbook of Chemistry and Physics,* CRC Press Inc, 1994, p 6-241 and p
8. Jones., W.M. and Giauque., W.F., *J. Am. Chem. Soc.,* **69**, 938 (1947).

EXTINCTION AND INITIATION OF DETONATION OF NM-PMMA-GMB MIXTURES

J. C. Gois, J. Campos and R. Mendes

Lab. of Energetics and Detonics
Mech. Eng. Dep. - Fac. of Sciences and Technology
Universidade de Coimbra, 3000 Coimbra, Portugal

The influence of the addition of a small amount of glass microballoons (GMB) on the extinction and initiation of NM-PMMA mixtures is investigated. Numerical simulation using DYNA 3D code shows the collapse of small GMB, under NM detonation. GMB of a mean particle diameter of 45μm were added into NM-PMMA explosive mixture. Polymethylmetacrylate (PMMA) was added to nitromethane (NM) to avoid buoyancy movements of GMB, increasing the viscosity of the original NM. Different failure tests were done initially proving the low values of critical extinction diameter and thickness. Accurate failure thickness experiments were performed using a prismatic configuration with a small opening angle of 2.9°. A double resistive wire technique and an aluminium witness bar were used to determine its critical thickness, as a function of GMB mass concentration. This double resistive wire technique shows, in real time, the induction time and distance of initiation. The failure thickness was correlated with those previously obtained with cylindrical critical configuration. Measured values are almost exactly the half values of critical cylindrical diameters.

INTRODUCTION

The addition of small amounts of inert particles to liquid nitromethane (NM) changes drastically its detonation properties decreasing abruptly its critical diameter and the minimum energy to initiation (1). Glass microballoons (GMB) were used with success to sensitise emulsion explosives (2). The same effect was observed by Presles et al. (3) and Gois et al. (4) for nitromethane-polymetylmetacrylate (NM-PMMA) mixture. The sensitivity of NM-PMMA-GMB was found to arise with a GMB mass fraction increasing and a particle diameter decreasing (5). The failure or critical diameter (d_c), by definition, determines the charge size threshold limit for the propagation of detonation under steady conditions. However, at d_c, it is observed a complex initiation phenomena, principally with heterogeneous explosives, showing detonation velocity values much less than its characteristic detonation value.

The decreasing of critical diameter, due to the presence of GMB has generally been attributed to an increasing in the energy release rate, caused by particle induced hot-spots. As the particle concentration is increased, the density of re-initiating sites grows. However, it is not very clear the cooperative mechanisms involved in hot-spot generation.

The deformation of GMB, as a function of its size, under a NM detonation, is evaluated using DYNA3D code (6) by Cortez (7). The material assumed for GMB shell was SiO_2 defined in (8). The evolution of $\delta t/\tau$, as a function of inverse particle diameter $1/d$, is shown in Fig. 1 (being δt the time duration between the instant when detonation front touch the GMB shell and the instant when the deformed GMB shell arrives at the centre of particle, and τ is the time duration takes by the detonation front to cover the same distance, which is obtained dividing the particle radius by the detonation velocity). A minimum limit value of

$\delta t/\tau$=1.5, for very large particles, corresponding to jet formation, is in a good agreement with open literature. Small particles show clearly a pure collapse situation.

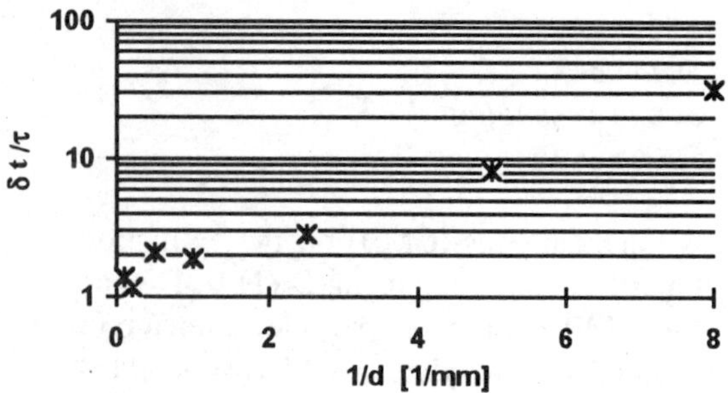

FIGURE 1 - Evolution of $\delta t/\tau$ as a function of 1/d.

Corner turning experiments allow to detect the existence of failure and re-initiation. Experiments by Ramsay (9) with PBX 9502, using a prism test, based on a wedge test, have shown that the failure thickness decreases when the thickness of confinement and its impedance are increasing. The failure measured thickness is one-half of the failure diameter measured in long cylindrical charges.

This paper reports the last recent results of our continuous study on extinction and initiation process of NM-PMMA-GMB mixtures. Experiments have been done with PMMA confinement with the aim to compare failure thickness results with those obtained in a previous paper for long cylindrical charges (4).

EXPERIMENTS AND RESULTS

Mixtures

GMB QCel 520 FPS, supplied by Akzo Inc., have a wall thickness of about 1 μm and diameters in range 16-79 μm. The mean diameter obtained by laser diffraction spectrometry is 45$\pm$1 μm. The effective density measured by helium picnometric is 220$\pm$3 kg.m^{-3}. In order to increase NM viscosity and avoid buoyancy movements of GMB inside NM, 4% (by weight) of polymetylmetacrylate (PMMA) was added. Table 1 shows the initial density of NM-PMMA-GMB mixtures as a function of GMB mass fraction, X_m.

The experiments were performed at an initial temperature of mixtures in the range 12-16 °C. The detonation velocity was measured for different cylindrical charges and extrapolated to an infinite charge diameter (vd. Table 1). The detonation pressure was measured by the shock induced polarisation (SIP) of a 1 mm thickness PMMA plate, by impedance mismatch method (10).

TABLE 1. Density, detonation velocity and pressure of NM-PMMA-GMB mixtures as a function of GMB mass fraction.

X_m [%]	0	1	2	3
ρ [kg.m^{-3}]	1149	1090	1034	987
D_∞ [m.s^{-1}]	6360	6120	5340	5062
P [GPa]	12.6	11.6	10.7	9.8

Extinction thickness

Corner turning propagation

Corner turning propagation experiments have been performed for NM-PMMA-GMB mixtures using brass confinement. The re-initiation or not at the channel, in a perpendicular position to the main channel, was detected by double resistive wire technique (10) and by the crush on a witness aluminium plate placed under the channel T tube. The results show at the corner, an extinction followed by re-initiation into the perpendicular channels. For 1% (by weight) of GMB concentration in explosive mixture, re-initiation was observed in perpendicular channels with brass confinement of 4 mm thickness section. For higher GMB concentration this critical thickness decreases drastically, generating many problems to obtain a precise experimental set-up. Consequently these very small values of extinction thickness led to more accurate failure thickness experiments, where the extinction zone could be clearly observed and measured.

Prismatic configuration

The experimental prism test apparatus, used to measure the failure thickness of NM-PMMA-GMB mixtures, consists of a prism of the explosive, initiated along its large end base by a 200 mm long plastic explosive donor, based in RDX, delivered by Portuguese Explosives Society (SPEL), with $D_\infty = 6200$ m.s^{-1} (vd. Fig. 2). The prism lateral confinement was made in PMMA (thickness 5 mm) with a length of 150 mm, an opening angle of $2.9°$ and 50 mm of wide. Two foils of copper close the prismatic NM-PMMA-GMB acceptor.

In order to verify the initiation and the detonation velocity of the explosives (donor and acceptor) two sets of a double thin resistive wire (resistivity of 177 W.m^{-1} and diameter of 100 mm) were isolated and fixed. One between the plastic explosive booster and prismatic acceptor and the other between the acceptor and the witness aluminium plate, along the axis of charges. The witness plate was 25 mm thick.

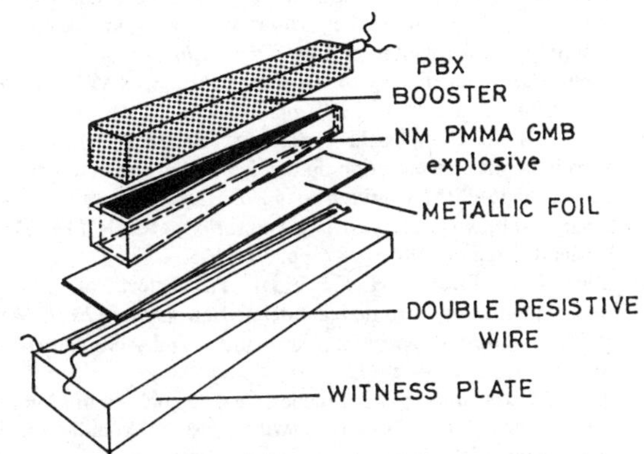

FIGURE 2. Experimental prism test apparatus

Both resistive wires were connected, as a voltage divider (with a variable resistance formed by double resistive wire and a constant resistance R_c) to a voltage source with $V_{ref} = 5$ V. When the explosive mixture is initiated, the shock front on copper sheet is enough to short-circuit the resistive wire, generating a variation of voltage that allows to obtain its position along the wires (vd. Fig. 3).

This position, as a function of time, can be obtained from V_{out} by the following equation:

$$x = \frac{\dfrac{V_{ref}R_c}{V_{out}} - 2\rho L_0 - R_o - R_c}{\dfrac{R_1 - R_o}{L_0} - 2\rho} \qquad (1)$$

where V_{out} represents the measured voltage at the constant resistance R_c, x is the distance measured from the taper and R_o and R_1 are respectively the measured internal resistance of voltage source at initial and at maximum output voltage (vd. Fig. 3).

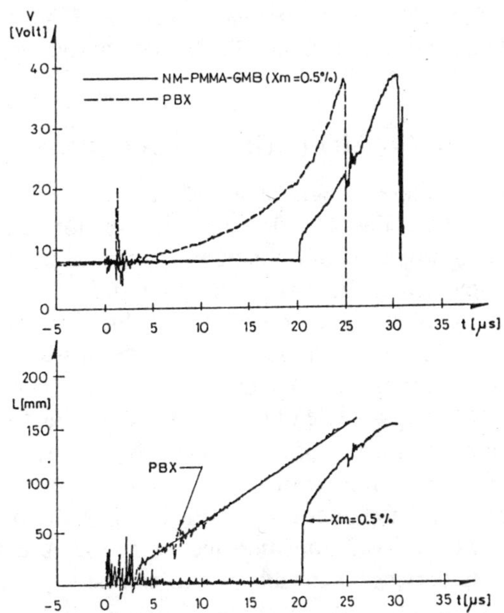

FIGURE 3. Voltage record and calculated position, as a function of time, of the two double resistive wire systems.

In Figure 4 the recorded signals, are shown as a function of GMB mass fraction X_m.

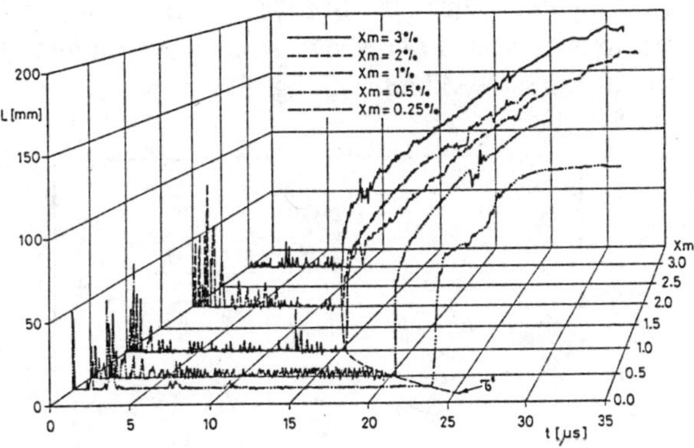

FIGURE 4. Calculated position as a function of time, for differents GMB mass fraction.

The signal of upper double resistive wire is always the same for all the experiments, being the detonation velocity of donor given by the slope of the signal. The signal of lower double resistive wire, for each explosive mixture, shows clearly the location of initiation transition in the explosive acceptor, by the variation of voltage. The NM-PMMA mixture, without GMB, does not initiate.

DISCUSSION AND CONCLUSIONS

Polynomial regression of shock transition records (vd. Fig. 4), after transition section, allows the determination of transition to initiation position. Its velocity is increasing towards the detonation velocity of PBX donor explosive (which drives the progression of the upper layer of the NM-PMMA-GMB acceptor). The values obtained are dependent on the opening angle of our prismatic configuration. In our case the evolution of time delay τ' (measured between the initial instant t_o, where the plastic explosive detonation touches the taper of the prism and the observed transition section in the acceptor) and its correspondent distance x, can be expressed, as a function of GMB mass fraction X_m, by $\tau'=\exp(3.1741 - 0.3546\ X_m)$ and $x=\exp(4.3739 - 0.4705\ X_m)$, for X_m between 0.25 and 3%. The thickness value, in the prismatic configuration, at the initiation position, is defined as G_{II}. The correlation between $2*G_{II}$ and the critical diameter d_c for the same NM-PMMA-GMB mixtures, previous presented (4), shows similar evolution and almost the same values (vd. Fig. 5).

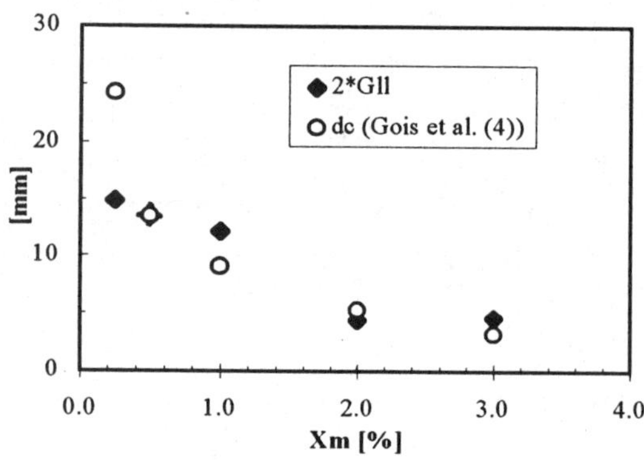

FIGURE 5. Double G_{II} and d_c evolution as a function of X_m.

Presented results demonstrate the validity of experiments and show the drastic reduction of detonation extinction thickness with small GMB concentration in NM-PMMA explosive mixtures.

ACKNOWLEDGEMENTS

We are very grateful to P. Cortez by the numerical DYNA3D calculations.

REFERENCES

1. Engelke R., Effect of a physical inhomogeneity on steady-state detonation velocity, *Physics. Fluids*, 22, nº 9, 1979, pp. 1623-1630.
2. Yoshida M., Ilda M., Tanaka K., Fujiwara S., Kusakabe M., Shiino K., Detonation behaviour of emulsion explosives containing microballoons, in *Proceedings of the 8th Symposium (International) on Detonation*, NSWC MP 86-194, Albuquerque, USA, 1985, pp. 993-1000.
3. Presles H.N., Campos J., Heuzé O., Bauer P.., Effects of microballoons concentration on the detonation characteristics of nitromethane-PMMA mixtures, in *Proceedings of the 9th Symposium (International) on Detonation*, OCNR 113291-7, Portland, Oregon, USA, 1989, pp. 362-365.
4. Gois J.C., Presles H.N., Vidal P., Effect of hollow heterogeneities on nitromethane detonation, *Dynamics Aspects of Detonations*, Progress of Astronautics and Aeronautics ed, vol. 153, 1993, pp 462-470.
5. Gois J.C., Campos J., Mendes, R., Shock initiation of nitromethane-PMMA mixtures with glass microballoons, in *Summaries 10th Symposium (International) on Detonation*, Boston, Massachusetts 02117, USA, and accepted for publication in the Proceedings of this Symposium, 1993.
6. Halquist, J., *Theoretical Manual for Dyna 3D*, University of California, Lawrence Livermore Laboratory, Rept. UCID-19401.
7. Cortez, P., Hollow glass sphere test, in *Internal Report*, Lab. Energética e Detónica, 1994.
8. Marsh, S., *LASL Shock Hugoniot Data,*. Eds. University of California Press, 1987.
9. Ramsay J., Effect of confinement on failure in 95 TATB/5 Kel-F, in *Proceedings of the 8th Symposium (International) on Detonation*, NSWC MP 86-194, Albuquerque, USA, 1985, pp. 372-377.
10. Mendes R., Campos J., Gois J.C., Moutinho C., Shock initiation and detonation stability of industrial explosives, in *Proceedings of the 24th International Conference of ICT*, Karlsruhe, Germany, 1993, pp. 29.1-29.13.

SHOCK COMPRESSION OF LIQUID HYDRAZINE[†]

Benjamin O. Garcia

Los Alamos National Laboratory, Los Alamos, New Mexico 87545

David J. Chavez

Rockwell, White Sands Test Facility, Las Cruces, New Mexico 88004

Liquid hydrazine (N_2H_4) is a propellant used for aerospace propulsion and power systems. Because the propellant modules can be subject to debris impacts during their use, the shock states that can occur in the hydrazine need to be characterized to safely predict its response. Several shock compression experiments have been conducted to investigate the shock detonability of liquid hydrazine; however, the experiments' results disagree. Therefore, in this study, we reproduced each experiment numerically to evaluate in detail the shock wave profiles generated in the liquid hydrazine. This paper presents the results of each numerical simulation and compares the results to those obtained in experiment.

BACKGROUND

In the past, the White Sands Test Facility (WSTF) has performed tests that attempted to shock initiate liquid hydrazine (N_2H_4). No evidence of hydrazine reaction was found (1, 2). The most recent test of the shock detonability of liquid hydrazine was conducted by Science Applications International Corporation (SAIC) for Eglin Air Force Base (3). The results of this test suggested some evidence of hydrazine reaction, although the details of the results were poorly substantiated.

PURPOSE

Given the differences in these results, we conducted this study in an attempt to investigate in detail the shock stimuli that would be necessary to achieve an appreciable hydrazine reaction. Specifically, this study determines the minimum power density needed for the detonation of homogeneous liquid hydrazine. The term "power density" is meant to suggest the pressure delivered by a projectile per unit time. The minimum power density will generate sufficient heat for attaining the critical temperature for a detonation condition. Based on the results of this analysis, we propose an experiment aimed at re-producing the most realistic situation in which a hypervelocity projectile impact might initiate homogeneous liquid hydrazine.

APPROACH

1. Determine the unreacted equation of state (EOS) of liquid hydrazine.

Accurate determination of the thermodynamic variables of the shock states of any substance usually requires that substance's EOS. However, the EOS of liquid hydrazine is not currently known. For the numerical simulations performed in this study, we assumed that the EOS of liquid hydrazine is similar to the EOS of water, based on the fact that the density, boiling point, melting point, and critical temperature of liquid hydrazine are within 2% of the values for water. Therefore, we assumed that the EOS of liquid hydrazine can be taken as

$$U_s = 1.5 + 1.5\,U_p. \tag{1}$$

With this EOS, we used a code called SEQS (Solid Equation of State) to compute single shock hugoniots in the temperature vs. specific volume plane, pressure vs. particle velocity plane, and the shock velocity vs. particle velocity plane.

[†] This work was performed under the auspices of the National Aeronautics and Space Administration (NASA).

2. Determine the reactive hugoniot equation of state for the detonation products.

To determine the C-J state and a description of the expansion isentrope, we used the BKW (Becker-Kistiakowski-Wilson) code. The C-J parameters determined from this computation are as follows:

C-J pressure	139 kbar
C-J velocity	7750 m/s
gamma	3.36

The expansion isentrope was determined based on the following chemical equilibrium at high pressure:

$$N_2H_4 \rightarrow N_2 + 2H_2 \qquad (2)$$

3. Perform thermal stability experiments.

Thermal stability experiments are needed to determine the chemical reaction rate, products, critical temperatures, and activation energy for liquid hydrazine. These quantities are constants in the Arrhenius rate law for burn, which is needed to perform hydrodynamic calculations using SIN (a one-dimensional Lagrangian, reactive hydrodynamics code) and TDL (a two-dimensional Lagrangian, reactive hydrodynamic code). A series of small-scale isothermal kinetic experiments was performed at the New Mexico Institute for Mining and Technology (NMT) to validate existing kinetic parameters for liquid hydrazine. The results of these experiments are shown in Table 1.

The numerical models generated in this study used an activation energy (E_a) of 17 kcal/mol and a frequency factor (A) of 3.61×10^6 s^{-1}.

4. Determine the minimum power density for steady-state detonation.

The tests were modeled with the following assumptions. A shock travels into the hydrazine, compressing and heating the propellant. The shock heating results in chemical decomposition that accelerates exponentially. The reaction begins at the rear boundary, because it has been hot the longest, and a detonation wave propagates at the C-J state of the shock propellant. This type of bulk heating or thermal initiation analysis has been successfully performed by C. L. Mader for homogeneous energetic materials using the Frank-Kamenetskii equation, with the Arrhenius kinetic parameters determined from laboratory thermal stability experiments (7, 8).

We used SIN as the primary workhorse for modeling the tests and in determining the detonability of liquid hydrazine for the proposed experiment. After determining the power density requirement for attaining a steady-state detonation, we used TDL to determine the effects of geometry on the release wave attenuation. The two-dimensional calculations were augmented by calculations using the ZEUS code, which provided a check on the shock pressures generated in the liquid hydrazine.

RESULTS

Condensed Phase Detonation Studies (WSTF #1)

In this test, 887 g of C-4 was detonated on top of a stainless steel tube (4" diam x 10" long) filled with liquid hydrazine. The calculation indicated that the shock wave generated by the C-4 explosive was not sufficient to generate a hydrazine reaction. The shock pressure generated in the hydrazine was 150 kbar, for a duration of ~2 μs.

Demonstration of Hazardous Hypervelocity Test Capability (WSTF #2)

In this test, a spherical aluminum projectile (1/8" diam) was fired with a velocity of 6.1 km/s at a cylindrical, 300-ml stainless steel vessel filled with liquid hydrazine. ZEUS calculations were per-

TABLE 1. Arrhenius Parameters of N_2H_4 Decomposition

	Isothermal (this work)	DSC[a]	ARC[b]	Isothermal[c]
E_a (kcal/mol)	26.8	17	23.4	20.5
A (s^{-1})	6.51×10^6	3.5×10^6		
Temperature Range (°C)	184-314	202-317		

[a]Differential scanning calorimetry (4)
[b]Accelerating rate calorimeter (5)
[c](6)

formed to provide a better representation of the shock wave generated by the sphere impacting the cylindrical vessel. This model showed that the release waves caused the shock wave to attenuate very quickly, rendering the shock wave ineffective. The pressure and duration, as calculated by ZEUS, were 83 kbar and ~0.8 μs, in agreement with the TDL calculation. The TDL calculation revealed that the pressure generated by the projectile was not sufficient to generate a hydrazine reaction. Figure 1 shows the initial projectile impact.

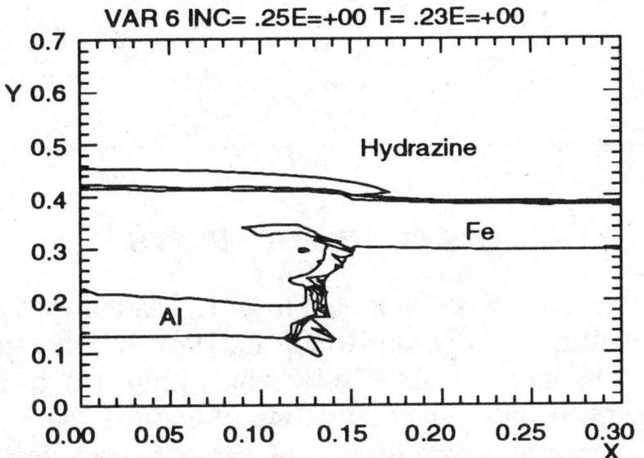

FIGURE 1. TDL calculation of projectile impacting the cylindrical vessel for WSTF #2.

Fuel Tank Explosion Lethality (SAIC)

In this test, a 100-g cylindrical aluminum projectile was fired with a velocity of 5.0 km/s at a spherical (100-mm-diam) aluminum vessel filled with liquid hydrazine. The SIN code calculation revealed that the shock wave generated by the projectile was sufficient to cause some decomposition of the hydrazine. One reason for the higher shock pressure generated in this experiment was the impedance matching of materials adjacent to the liquid hydrazine. Figures 2 and 3 provide a graphical representation of the shock impedance matching of this test and of WSTF #1, respectively. The pressure and duration as calculated by SIN were 285 kbar and ~5 μs. The pressure and duration as calculated by ZEUS were 267 kbar and ~1.5 μs, in agreement with the SIN calculations. Another reason that there was more hydrazine decomposition in this test as compared to the other tests is that the projectile geometry sustained the shock pressure for a longer duration before the release wave reduced its magnitude.

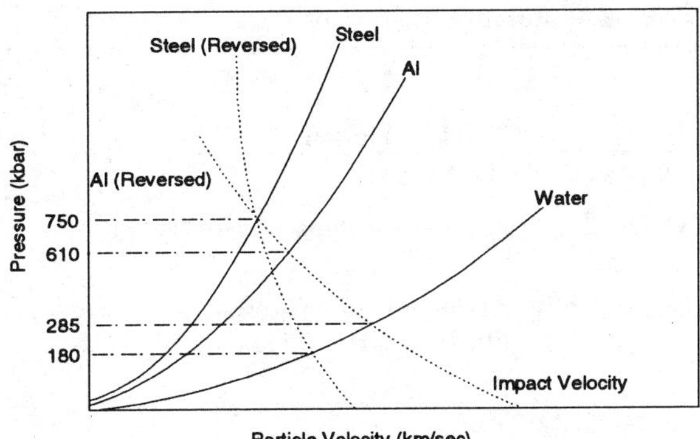

FIGURE 2. Shock-matching curves for the SAIC test.

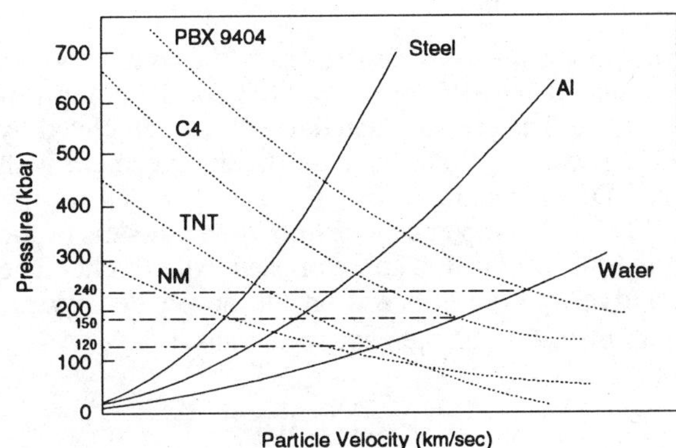

FIGURE 3. Shock-matching curves for WSTF #1.

Proposed Experiment (TITANK)

The goal of our study was to use the analyses of the previous tests to define a test situation that replicates the minimum power density conditions that might occur in a real-life scenario. Our numerical model of such a situation simulated a steel slug impacting a titanium vessel filled with liquid hydrazine. A cylindrical projectile was chosen to provide the sustained pressure necessary to generate enough shock heating. The projectile velocities were chosen such that they could be achieved using WSTF's 1-in. light gas gun. The desired projectile velocity was between 7-7.5 km/s.

The result of the SIN and TDL models revealed that the shock wave generated by the steel slug was sufficient to achieve a substantial hydrazine reaction.

TABLE 2. Reactive Analysis of Tests

Test	Test Description	SIN (1-D)			ZEUS		Power Density (kbar^2-s)
		Pressure (kbar)	Duration (μs)	Mass Fraction	Pressure (kbar)	Time (μs)	
WSTF #1	C-4 on N_2H_4.	150	0.7	.99	--	--	0.016
WSTF #2	Al projectile impacting on N_2H_4 @ 6 km/s.	--	--	--	83 62.7	T_0 0.8	0.0024
SAIC	Al slug impacting Al sphere filled w/ N_2H_4 @ 5 km/s.	285	5	0.81	267 267 178 178	T_0 1.5 1.8 2.4	0.137
TITANK	Steel slug impacting Ti tank filled w/ N_2H_4 @ 7.5 km/s.	600	2.4	0.58	600 444 300 178	T_0 1 1.5 2	0.398

The shock pressure generated in the hydrazine, as calculated by SIN, was 600 kbar for a duration of ~2.4 μs. The pressure and duration, as calculated by ZEUS, were 600 kbar and ~1 μs, in agreement with the TDL calculations.

Table 2 compares the results of the models of the previous tests and of the proposed experiment. The relative ranking is based on the power density requirement for homogeneous materials, $P^2\tau$.

CONCLUSIONS

Numerical reactive models of the WSTF and SAIC tests successfully reproduced their respective test results. Those tests emphasized that, under particular shock loading conditions, a minimum power density is required to achieve a hydrazine reaction. The models suggest that shock heating increases as a function of the power density applied to the liquid hydrazine.

Future work should involve tests in which many of the parameters used in reactive modeling are unknown for propellants of interest. This should include a program for determining the EOSs of such propellants, and a program for enlarging the data base of kinetic parameters for these propellants, specifically for use in reactive hydrodynamic models.

Finally, the proposed experiment should be carried out to verify the results of the our modeling effort. The experiment must employ enough detailed instruments to provide sufficient information on the extent of the hydrazine reaction and to verify model parameters for future tests.

ACKNOWLEDGMENTS

The authors are indebted to C. L. Mader of Mader Consulting Company, Honolulu, Hawaii, who spent many hours of his time in sharing with us his knowledge of the numerical modeling of explosives.

The authors would also like to acknowledge the following people from NMT: Dr. Jimmie Oxley and her students, Dr. Per-Anders Persson, and Marvin Banks.

REFERENCES

1. Rathgeber, K., and Radel, B., "Condensed Phase Detonation Studies," White Sands Test Facility special test data report WSTF #90-24354, September 28, 1990.

2. Rucker, M. A., Beeson, H., Stoltzfus, J. M., and Benz, F. J., "Demonstration of Hazardous Hypervelocity Test Capability," White Sands Test Facility test report TR-692-001, September 24, 1991.

3. Wilson, C. W., Warne, D., and Chatfield, M. D., "Fuel Tank Explosion Lethality," SAIC technical report SAIC 91-5425-SH, Shalimar, FL, April 1991.

4. Bishop, C. V., Miller, E. L., and Benz, F. J., "Liquid Propellant Thermal Hazards Estimation Using Differential Scanning Calorimetry," in *JANNAF Safety and Environmental Protection Subcommittee Meeting*, 1983.

5. Welich, R. C., and Davis, D. D., *Thermochimica Acta*, **171**, 1-13 (1990).

6. Anworthy, A. E., Sullivan, J. M., Cohy, S., and Wely, E., "Research on Hydrazine Decomposition," final report AFRPL-TR-69-146, Rocketdyne, Canoga Park, CA, July 1969.

7. Mader, C. L., "Numerical Modeling of Impact Involving Energetic Materials," in *High Velocity Impact Dynamics* by Jonas A. Zukas, 1990.

8. Mader, C. L, *Numerical Modeling of Detonations*, Los Alamos, NM:Los Alamos National Laboratory, 1979.

Shock Reactivity Experiments on a Porous Composite Propellant

H. W. Sandusky and R. R. Bernecker

Naval Surface Warfare Center, Indian Head Division, Silver Spring, MD 20903-5640

Porous beds of a propellant containing ammonium perchlorate and aluminum in an inert binder were subjected to two-dimensional (2-D) shocks from 95.2 mm diameter, cylindrical donors. The 50.6 mm diameter beds consisted of 3.2 mm cubes packed at a solids fraction of 59%. One bed was lightly confined by a plastic tube and the others were heavily confined by steel tubes with 17.6 mm walls. For a shock pressure in the gap (P_G) of 7.53 GPa, the reactive shock velocity (U) rapidly declined for the lightly confined bed. For heavily confined beds with the same P_G, U was nearly constant for each bed but less for the longer bed; when P_G was reduced for the same bed length, U was lower. Furthermore, dent depth in the witness block declined as P_G was reduced from >12.0 to 6.08 GPa. While steel tube fragments and witness block dent indicated a detonating event at the highest P_G, it was unlikely to have been steady detonation. The data suggest that the critical diameter for steady detonation in a lightly confined bed would be considerably in excess of 50.6 mm. While previous 2-D shock loading experiments with smaller donors and beds of another composite propellant also indicated a detonating event, based on punching of the witness plate and steady values of U, it is likely that steady detonation was not achieved there in view of the present experiments.

INTRODUCTION

Composite propellants typically consist of ammonium perchlorate (AP) and aluminum (Al) in an inert binder. In their as-manufactured state and without other energetic components, such as nitramines, they are classified as Hazard Division 1.3 or mass deflagrating because of their insensitivity to shock initiation and their large critical diameter (d_c) compared to explosives. Large d_c indicates a slow rate of shock reactivity that allows rarefactions to quench the reaction. This was verified (1) by Green et al. with manganin gauges inserted into TPH-1123 (86% AP and Al in inert binder). They did not record reaction within the ~4 μs recording time of the gauges for shock pressures of 8.8 and 14.4 GPa.

Accident scenarios often involve multiple insults to energetic materials. Damaging the material during an early phase of the scenario makes it more likely to violently react or transit to detonation if subsequently shocked. A characteristic of AP formulations is that d_c decreases with the introduction of porosity.(2) Price and co-workers measured detonation velocities in simple composite propellant models consisting of

25 μm AP and 125 μm carnauba wax.(3) Mixtures of AP with 0, 10 and 20% wax were pressed over a range of densities into 50.8 mm diameter by 203 mm long cylinders. When the charges were shock loaded by large scale gap test (LSGT) donors, each composition exhibited a maximum density at which detonation would no longer propagate. Detonation failed at 72.3% of theoretical maximum density (TMD) for 100/0 AP/wax and at 85.2% TMD for 90/10 AP/wax, and was near failure at 86.8% TMD for 80/20 AP/wax.

EXPERIMENTAL ARRANGEMENTS

Gap experiments were conducted on 50.6 mm diameter porous beds of 3.2 mm cubes of a composite propellant packed at 1.045 g/cc, which is 58.9% of TMD. The porous beds were confined by two types of tubes. In one experiment, light confinement was provided by a plastic tube (polymethylmethacrylate, PMMA) with a wall thickness of 6.4 mm. This transparent tube permitted the reaction front to be directly observed with a streak camera. In the other experiments, the beds were heavily confined by a steel tube with a 17.6 mm wall thickness. Ionization probes

(Dynasen, Inc. model CA-1040) were inserted through the steel tube wall and projected into the bed by ~1 mm for tracking reaction front, and a witness block provided a relative measure of the energy in the front (reactive shock or detonation) at the far end of the bed.

The arrangement for the steel tube experiments is shown in Fig. 1. The same shock donor was used for the lightly confined porous bed. The donor consists of 1.56 g/cc pellets of pressed pentolite, and is a composite of the LSGT donor and one of the two pellets from the NSWC expanded large scale gap test (ELSGT). The shock pulse from the pentolite is attenuated by a 95.2 mm diameter PMMA gap, as in the ELSGT. When maximum shock pressure in the porous bed was desired, a 2.5 mm thick gap was employed to avoid direct impingement of the pentolite detonation products onto the porous bed. The calibration of peak shock pressure in the gap (P_G) versus gap thickness was taken (4) to be the same as for the ELSGT donor, since the shock from both pentolite pellet combinations attenuate at the same in water.

EXPERIMENTAL RESULTS AND DISCUSSION

The initial conditions and results are summarized in Table 1. The reactive front velocity (U) in the PMMA tube (Shot DNASR-2) declined from 6.8 to 2.8 mm/µs. The final value of U is lower than in a somewhat longer bed receiving the same shock (P_G = 7.53 GPa) in steel

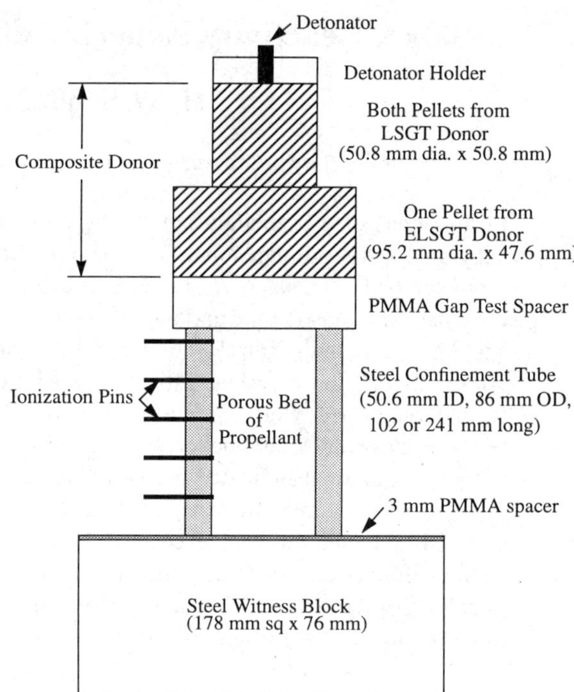

FIGURE 1. Arrangement for composite donor gap experiments.

tube confinement (Shot DNASR-3), thus confinement is important at this bed diameter. In a similar arrangement with light confinement, the explosives PBXW-115 (AP, Al, RDX, inert binder) and

TABLE 1. Summary of Composite Donor Gap Tests

Shot DNASR-	Confining Tube* Mat'l	Confining Tube* OD (mm)	Bed^ L (mm)	PMMA Gap Spacer Gap (mm)	PMMA Gap Spacer P_G (GPa)[#]	U[+] (mm/µs)	Witness Dent (mm)
2	PMMA	63.4	76.1	25.4	7.53	6.8 - 2.8	-
1	Steel	85.9	101.7	2.5	16.4	-	12.3
5	Steel	85.9	101.4	11.4	10.24	-	11.1
3	Steel	86.0	101.7	25.4	7.53	3.45	9.3
6	Steel	86.0	241.4	25.4	7.53	3.38**	8.4
4	Steel	86.0	101.8	38.0	6.08	3.11	7.2

* ID of confining tube and bed diameter = 50.6 mm

^ Loading density = 1.045 g/cc, 58.9% TMD; L = bed length

[#] P_G = Peak shock pressure in the gap

[+] U = Velocity of reactive shock in porous bed

** Declining velocity from 3.49 to 3.27 mm/µs when pin data fit with a quadratic

PBXN-103 (AP, Al, energetic binder) did not transit to detonation (5); therefore, it was unlikely that the less energetic composite propellant would detonate. These plastic-bonded explosives did appear to detonate in the higher confinement of steel tubes. At the highest P_G in steel tube confinement (Shot DNASR-1), the plastic deformation of the tube fragments and the dent in the witness block suggested a detonating event. However, dent depth and U both declined with decreasing values of P_G for the four steel tube experiments with 102 mm long beds (Shots DNASR 1,3,4,5), indicating that steady-state detonation was not achieved. Furthermore, there was never any indication of a buildup to detonation, as would be expected as P_G is decreased. Whatever reaction was initiated by the donor shock appeared to propagate at a near steady velocity.

In a final test (Shot DNASR-6), the tube length was increased to 241 mm to determine if a failing reaction front could be observed in a longer bed. Both U and dent depth were less than that measured for the same P_G in the shorter tube, verifying that the reaction front was failing. As shown in Fig. 2, the ionization probe data could be described by a constant U of 3.28 mm/μs that fitted all but the first and last probe responses. Despite the precision of the pin locations and response times, there is enough variation in the propagation of the reaction front through the bed of cubes to mask a slow decline in U. This is seen by the improved fit to the data in Fig. 2 obtained with a quadratic equation, which results in U declining from 3.49 to 3.27 mm/μs. The 3.46 mm/μs velocity at ~50 mm from the gap is essentially the average value of U measured for the 102 mm bed in Shot DNASR-3.

Dent profiles from the steel tube experiments are shown in Fig. 3. Dent depth decreases monotonically for decreasing values of U as well as P_G. The somewhat higher 3.27 mm/μs value of U at the end of the bed in Shot DNASR-6 versus the average 3.11 mm/μs in Shot DNASR-4 resulted in a deeper dent (8.4 versus 7.2 mm). For Shot DNASR-1, there were pock marks within the dent that were somewhat larger than the cubes, as if individual cubes violently reacted on the witness block. The depth of the pock marks declined along with the lessening dent as U decreased.

If a 19 mm thick ELSGT witness plate had been used instead of the block, it would probably had been punched in Shots DNASR-1,3,5 by the slowly failing reaction. Steady detonation may have occurred for

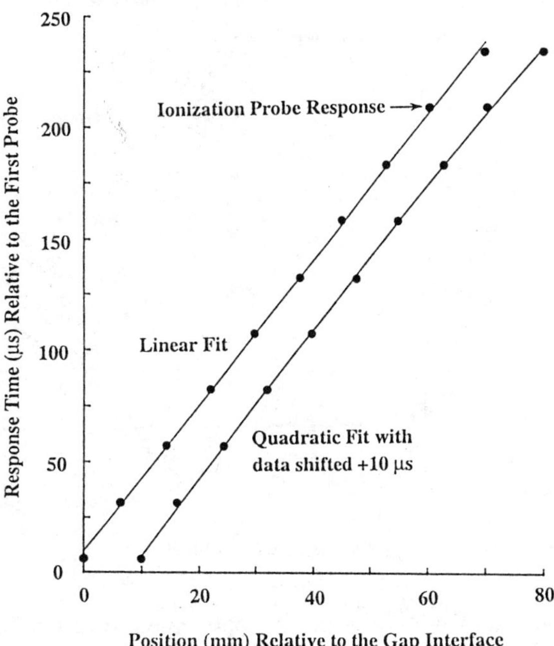

FIGURE 2. Ionization probe responses to reactive shock in Shot DNASR-6.

smaller particles. Particle size and shape (2.1 mm cubes versus 2.5 mm wide by 0.79 mm thick shreds from lathe turnings) significantly affected shock reactivity of PBXN-109 (RDX, Al, inert binder).(6) The comparatively small (25 μm) AP particle size in the work of Price et al. in the LSGT arrangement may have involved detonation or, as observed in the current experiments, only failing reaction that was vigorous enough to punch a plate. Increased diagnostics, as in the current experiments, would be needed to determine that.

SUMMARY AND CONCLUSIONS

Gap experiments, using 95.2 mm diameter donors, were conducted on 3.2 mm cubes of a composite propellant packed at 59% TMD into 50.6 mm diameter beds. These beds simulated a uniform state of damage, but not necessarily the extent of damage that would occur in an accident scenario. High-speed photography of a bed lightly confined in a plastic tube showed a rapidly failing reaction front, whereas in heavy confinement, the reaction fronts were nearly steady.

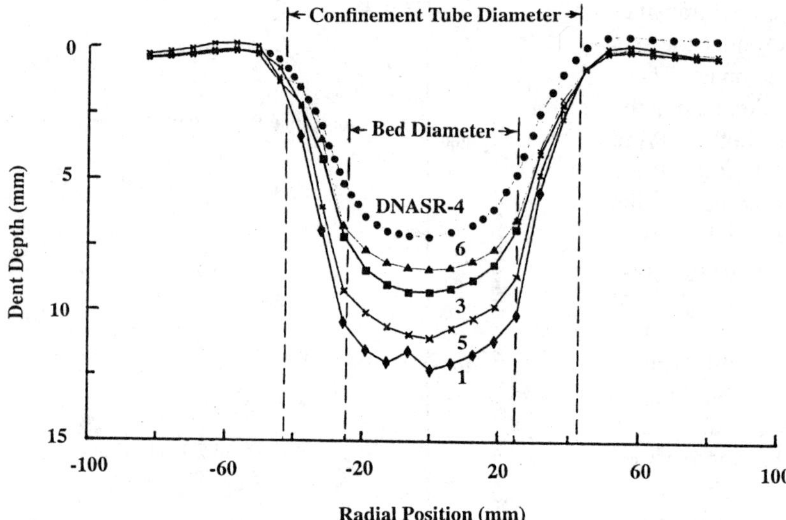

FIGURE 3. Dent profiles in the witness block.

For those beds with heavy confinement receiving the strongest shocks (P_G > 10 GPa), the steel tube fragments and witness block suggested that detonation had occurred and would have punched a standard witness plate. Such limited diagnostics in a standard test would not have shown that the reaction front was slowly failing, as measured by both declining front velocity and dent depth in the witness block with decreasing P_G. It was concluded that steady-state detonation was not achieved for the particle dimensions and bed diameter that were examined. Either reduced particle dimensions or increased bed diameter, while maintaining high confinement, may have produced a detonable system. Since U rapidly declined in lightly confined beds, the critical diameter for an unconfined bed would be considerably in excess of 50.6 mm.

REFERENCES

1. Green, L., James, E., and Lee, E., "The Detonability of Composite Propellants," in Proceedings of 1986 JANNAF Propulsion Systems Hazards Subcommittee Meeting, CPIA Publ. 446, Vol. I, Mar. 1986, pp.13-19.
2. Zerilli, F. J., ed., "Notes from Lectures on Detonation Physics," NSWC MP 81-339, Naval Surface Warfare Center, Oct 1981.
3. Price, D., Clairmont, A. R., Jr., and Erkman, J. O., "Explosive Behavior of a Simple Composite Propellant Model," Combustion and Flame, Vol. 17, 1971, pp. 323-336.
4. Sandusky, H. W. and Bernecker, R. R., "Influence of Fresh Damage on the Shock Reactivity and Sensitivity of Several Energetic Materials," in Proceedings of Tenth Symposium (International) on Detonation, to be published.
5. Bernecker, R. R. and Clairmont, A. R., Jr., unpublished data.
6. Bernecker, R. R. and Clairmont, A. R., Jr., "Shock Initiation Studies of Cast, Damaged, and Granulated PBXs," in Proceedings of Tenth Symposium (International) on Detonation, to be published.

INVESTIGATION OF EXPLOSIVES MECHANIC IMPACT SENSITIVITY ON THE SAMPLES

**B. G. Loboyko, A. V. Alekseev, B. V. Litvinov, V. D. Sumin,
N. P. Taybinov, V. P. Filin**

*Russian Federal Nuclear Center-Institute of Technical Physics
P. O. Box 245, Snezhinsk, Chelyabinsk region 456770 Russia*

Several results of investigation into HMX-based explosive compound sensitivity to mechanic impact on the samples are presented. Mechanic loading of samples was effected by dynamic insertion of a pin. Alternation of physical state of explosive compound on account of preliminary thermal treatment or destruction of samples increased their sensitivity considerably.

INTRODUCTION

The investigation of the explosives sensivity to the mechanic impacts always received, and keep on now, a great attention (1,2,3,4,5,6). By sensivity it is meant the explosive capability to respond to mechanic impact by the occurence of the explosive transformation in the form of burst, combustion or detonation. Explosive sensivity is determined by its chemical structure and physical parameters as well as by external conditions. A good number of experimental and theoretical works were devoted to investigation of explosives chemical structure influence on their sensivity (7,8,9,10,11,12,13,14). A great deal of structural parameters, influencing on the explosive sensivity were revealed. In our opinin, the influence of explosives physical state on their sensivity is investigated more poorly. In some extent the influence of such factors on the sensivity of the explosives as the size and degree of perfection of crystal, dispersity (15,16), temperature, presence of inert additions (17) is studied.

As a rule, the experimental checking of factors mentioned above influence was performed on small weighed portions of powder explosive, using the drop techniques (2,3,6). However, the conditions of drop tests don't reflect all diversity of mechanic impacts, occuring in practice. Moreover, the explosives are applied not only in the form of a powder, but in pressed and cast samples (charges). That is why the results of the

investigations of drop techniques on small powder-like weighed portions of explosive are very conditional and don't permit to judge about specific factor influence on the degree of hazard of one or another explosive in real conditions of explosive charges applications.

To extend the knowledge about the nature of explosive sensivity, about the degree of explosive hazard in real conditions one started to use the techniques, which differ from drop ones as by the nature of mechanic impact, so by the mass of investigated explosive. In USA the technique "Susan-test" (18) has received wide acceptance; in this technique the pressed explosive samples 51mm diameter and 102mm high (with mass of 0,4kg) are used.

In this paper the experimental results of the investigation of the influence of mechanic physical state of pressed explosive samples on the base of HMX (60mm diameter, 50mm high, approx.0,3kg mass) on their sensivity are represented.

INVESTIGATION TECHNIQUE

The variation in physical state of the explosive in the sample is usually carried out by such ways as change of initial explosive dispersity, modification of the degree of samples pressing etc. In the present work the another approach was used.

From the explosive materials based on HMX, using the pressing technique, the samples of 60x50mm diameter were produced. One third of the samples was held during 4 hours at the temperature of 180°C. In the result of thermal procedure the shape of the samples was preserved, but some increase of geometrical dimensons, also mass and density decrease by 0,3% and 13% respectively was occured. The net of fractures up to 1,5mm of width was formed. These samples we'll indicate by "T" index. The second third was exposed to low-velocity mechanic impact as a result of which on the side surface of the samples the hollow with the depth up to 5mm and approx.14 mm in diameter, a number of radial fractures and small particles in the form of dust and fragments of 1...5mm size were formed. These samples we'll indicate by "M" index. All other samples were not exposed to any impact and they were the reference ones ("K" index).

The investigating of the samples sensitivity to the mechanic impacts was performed according to the technique, described in paper (19); this one permits to realize the states (inside the investigated sample) close to three-axial compression, pure shear, to simulate the bar insertion or the gas compression into the cavities close to the explosive etc. In the work the experimental device was used for this technique in which the sample under investigation was exposed to bar insertion. The simplified scheme of device is given in Fig. 1.

Investigated explosive compound sample was put into thick-walled steel housing. Mechanical loading was effected by steel load, accelarated by the products of additional charge explosion; the velocity of the load was registrated by the electric contact gages system. Using the way of change the additional charge mass one can vary the piston velocity from some to 70μs.

The fact of explosion and intensity of explosive transformation in the sample are controlled visually by the charcter of housing destruction. Sensitivity is characterized by critical velocity of load. As critical velocity (V_{cr}) half of a sum of minimum velocity under which explosion of investigated explosive sample occurred and maximum velocity under which the explosion was not observed.

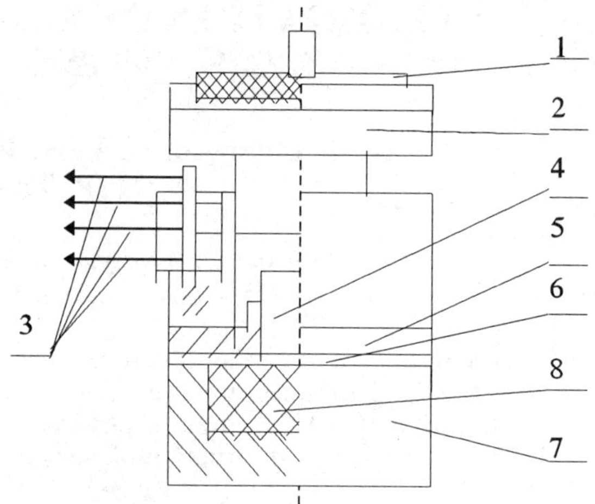

1 - additional charge;
2 - load q = 2,5(kg) ;
3 - system of electric contact gages;
4 - pin $\varnothing$ 12 (mm) ;
5 - directing ring ;
6 - aluminum 0.3 mm plate ;
7 - housing ;
8 - investigated charge.

FIGURE 1. Scheme of experimental device.

RESULTS AND DISCUSSION

According to a technique, described above, the investigations of the sensitivity of "T", "M", "K" samples were carried out.

Generalized experimental results of the investigations are given in the Table 1.

The Table 1 shows that the sensitivity of the samples, in advance exposed to thermal and mechanic impact is considerably higher than one of reference ones.

It may be explained in the following way. The explosives initiation process may be conventionally divided into two stages: the stage of "hot points" formation (heating-up and chemical decomposition centers) and the stage of the explosive decomposition reactions development up to the explosion (19). In general, the possibility of chemical decomposition centers formation and reaction development before the explosion is determined as by physic mechanical characteristics so by the chemical structure of the explosive. And as for the passing of the decomposition of reaction the parameters, associated with explosive reactionability, are influencing.

Now we'll analyze the process going on in the investigated sample during the experiment performance. The analysis will be performed from the stand-point of possible influence of these processes on each of the stages of explosion initiation. Bar insertion into the explosives is one of the most widespread and intense impact types under which the explosives flowing and destruction inside impact zone is provided. In doing so the benificial conditions for the stage of the formation of explosive, decomposition centers are created practically under any physical state of the explosive in the sample. That is why the limiting stage of the explosion process excitement in this case is the stage of development of explosive decomposition reaction i.e. the second stage.

In the samples "T" and "M" the decomposition reaction development to an explosion is simplified in comparison with the reference samples because of the porosity increase and of the presence of fractures which, as it is known (20), contribute to decomposition center development, its transformation to the convective combustion and then to explosion.

From the experiments results, realized in the given studies, it is possible to conclude that the preliminary mechanic destruction ("M" samples) effects some more influence on the sensitivity, than preliminary thermal explosive decomposition ("T" samples). It may be explained in our opinion by availability of dust and small particles of compound, which not only facilitate reaction development in decomposition centers up to exlosion (second stage), but also contribute to generation of "hot points" (first stage).

Obtained results is well comparable with data of those few investigations of physical factors influence on the pressed samples sensivity.

TABLE 1. Experimental Results of the Investigation of the Sensitivity of the Explosive Samples, Based on HMX.

Tempera-ture of sample, °C	Index of sample	Number of tests	V_{cr}, µs
20	K	10	15,1
	T	10	12,5
	M	10	10

For example, significant sensivity increase of mechanically damaged samples from LX-10 (octogen-94%, viton A-6%) was observed in the work of Chidester et al.(21). The value of threshold (critical) loading velocity for the samples with fractures is approx. by 33% less than for non-damaged ones.

CONCLUSION

The results of the investigation of explosive sensivity to the mechanic impact on the pressed samples are presented.

It is shown that the alternation of physical state of explosive is achieved on account of preliminary thermal treatment and mechanic impact, significantly increase the sensitivity of the samples to mechanic impact.

The interpretation of the experimental data was performed in the frame of the hypothesis of two-stage process of the explosives initiation, which permited to explaine the obtained results.

REFERENCES

1. Kast, G., *Explosive Compounds and Means of Ignition,*. Moscow: Goskhimtekhizdat, 1942, pp. 448.
2. Bouden, F. P., Ioffe, A. D., *Excitement and Development of Explosion in Solid and Liquid Explosive Compounds,* Moscow: Inostrannaya Literatura, 1955, pp. 120.
3. Afanas'ev, G. T., Bobolev, V. K, *Shock Initiating of Solid Explosive Compounds,* Moscow: Nauka, 1968.
4. Bouden, F. P., Ioffe, A. D., *Rapid Reactions in Solid Compounds,* Moscow: Inostrannaya Literatura, 1962, pp. 244.
5. Kholevo, N. A., *Shock Sensitivity of Explosive Compounds,* Moscow: Mashinostrojenie, 1974.
6. Kamlet, M. J., "Association between Shock Sensitivity and Structure of Polynitroalyphatic Organic Explosive Compounds." Collection *Detonation and Explosive Compounds,* Moscow: Mir, 1981, pp. 142-159.
7. Pivina, T. S., Petrov, E. A., Agranov, G. A., Shlyapochnikov V. A., "Evaluation of Explosive Compound Shock Sensitivity on

Molecular Level. Problems of Combustion and Explosion" in *Proceedings of IX Simposium on Combustion and Explosion,* Chernogolovka, 1989, pp. 89-93.

8. Mullay, J., *Propelants, Explosives, Pyrotechnics* **12**, 60-63 (1987).

9. Litvinov, B. V., Garmasheva, N. V., Filin, V. P., Loboyko, B. G., "Relationship between impact sensitivity and molecule structure of organic high explosives" in *Proceedings of the 2nd International Symposium on Intense Dynamic Loading and Its Effects,* Chengdu, China, June 9-12, 1992.

10. Delpuech, A., *Propellant and Explosive* **V3,** N6, 169-175, (1978).

11. Owens, F. J., *Jour. of Mol. Struct. (teochem)* **121,** 213-220 (1985).

12. Xiaa He Ming. *Acta Chimica Sinica,* **43**, 14-18 (1985).

13. Belik, A. V., Potemkin, V. A., .Zefirov, N. S., *Reports at Academy of Sciences* **V308**, N4, 882-885 (1989).

14. Veller, L., Ventselberg, O., "New Data on Explosive compounds Shock Sensitivity" in: *Initiating Explosive Compounds* Moscow: ONTI, 1935.

15. Tailor, W., Will, A.,. *Mechanism of Excitement and Spread of Detonation in Solid Explosive Compounds.* in: "Initiating Explosive Compounds", Moscow: ONTI, 1935.

16. Brunswig, G., *Theory of Explosive Compounds,* Moscow-Leningrad: Goskhimtekhizdat, 1932.

17. Belyaev, A. F., *Condensed Systems Combustion Transfer into Explosion.* Moscow: Nauka, 1973.

18. Dobratz, B., M., *LLNL Explosives Handbook Properties of Chemical Explosive Simulants,* LLNL, 1981.

19. Taibinov, N. P., Litvinov, B. V., Loboyko, B. G., Filin, V. P., Alekceev, A. V., Sumin, V. D., Garmasheva, N. V., "Investigation of Explosive Material Sensitivity to Mechanical Loading," in *Proceedings of the 2nd International Symposium on Intense Dinamic Loading and Its Effects,* Chengdu, China, 1992, pp. 96-98.

20. Chidester, S. K., Green, L. G. and Lee, C. G. "A Frictional Work Predictive Method for the Initiation of Solid High Explosives from Low-Presure Impacts," presented at the Tenth International Detonational Symposium, Boston, 1993.

ON THE MECHANISM OF INFLUENCE OF EXPLOSIVE COMPOUNDS: DESTRUCTION PROCESS ON SENSITIVITY OF THESE COMPOUNDS TO MECHANIC IMPACTS

V. P. Filin, B. G. Loboyko, A. N. Averin, B. V. Litvinov, I. G. Korotkikh, A. V. Alekseev, Yu. A. Belenovsky, N. P. Taibinov.

Russian Federal Nuclear Center-Institute of Technical Physics
P. O. Box 245, Snezhinsk, Chelyabinsk region 456770 Russia

The results of investigations into sensitivity of the HMX-based explosive compound samples to mechanic stimuli are shown in the presented report. As a result of experimental studies it was illustrated, that explosives deformation and destruction processes under mechanical stimuli are accompanied by occurrence of different electric phenomena. The hypothesis on possible influence of electric phenomena occurring under deformation and destruction on the mechanism of formation of zones with high density of energy is discussed in the report.

INTRODUCTION

The development of explosive composition, which have reduced sensitivity to mechanic impacts, is approached in several directions. One direction is the development and synthesis of specific explosive compounds, having reduced sensitivity to mechanical impacts. Another direction is the selection of efficient deterrents, compatible with explosive compound to decrease the compound sensitivity.

For solution of these questions, knowledge about the properties of explosive compound and binder, influencing explosive sensitivity, are necessary. The quest for such knowledge is effected by our concepts about the process and mechanism of explosion initiating under mechanical impacts. It also is necessary to mention here, that hypothesis about this or that mechanism determines the direction of research, including recorded parameters, investigation techniques and construction of corresponding theories.

In numerous studies (1-7) it is established, that the process of explosive compound initiating under external impacts consists of several stages, the main of which are the following:

1. The stage of formation of zones with high density of energy. The main questions, requiring understanding in this stage are:
- mechanism of external energy concentration in the zones;
- mechanism of its transformation into energy, received by each molecule of explosive compound;
- mechanism of this energy consumption by the molecule.

2. The stage of explosive compound chemical decomposition beginning in the zones of energy concentration. In this stage it is required to understand the mechanisms of energy concentration on the separate chemical bondings of explosive compounds molecules and the mechanism of molecular decomposition with energy output and formation of final products.

3. The stage of explosive compound chemical decomposition transition from the zones to the whole mass of compound. In this stage it is required to understand the mechanisms of energy transition from decomposed molecules to neighboring molecules of explosive compound and mechanisms of interaction of primary decomposition products with molecules of original explosive compound.

Depending on explosive compound state, conditions and parameters of external impacts, the critical contribution to sensitivity may be done by each of these stages.

In the current presentation we shall stop only at the first stage, taking into account, that the rest stages will also contribute their shares in probability of explosion.

Currently researchers support mainly the opinion about thermal mechanism of energy concentrators formation (1-8). In this case heating zones formation occurs on account of:
- rapid adiabatic compression of air inclusions, contained in explosive compound charge;
- ductile heating under rapid friction;
- annihilation of dislocations under shift deformations;

- destruction of explosive compound crystals with formation of new surface, having increased reactivity (destruction of explosive compound crystal is accompanied by different electric and emission effects, leading to local heating of newly formed surface of explosive compound).

One way of corroboration of this or that mechanism of heating zones formation is expansion of investigations(temperature range and comparison of experimental results of sensitivity alternation in this process with predicted by this or that mechanism.

For this purpose authors have conducted investigations into influence of temperature on HMX-based explosive compound. The experimental technique and results are described below.

EXPERIMENTS AND RESULTS

For experimental works we used the technique of investigation into explosive compound sensitivity to mechanical impacts, described in paper (8). This technique allows to test explosive samples of different sizes under different loading conditions. In the current paper we used experimental device, the scheme of which is given in Fig. 1.

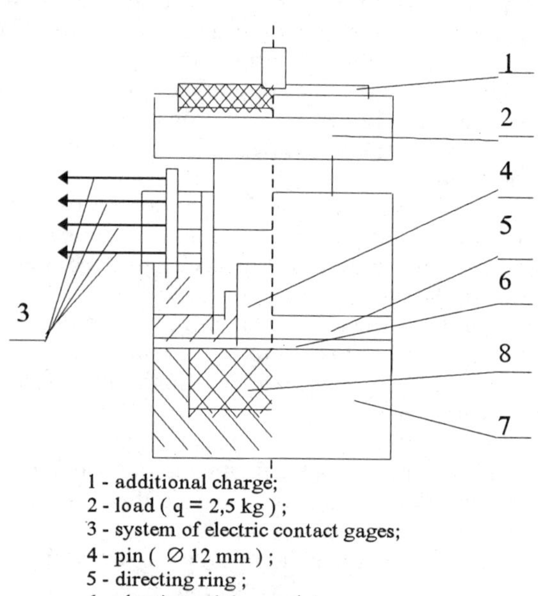

1 - additional charge;
2 - load (q = 2,5 kg) ;
3 - system of electric contact gages;
4 - pin (∅ 12 mm) ;
5 - directing ring ;
6 - aluminum 0.3 mm plate ;
7 - housing ;
8 - investigated charge.

FIGURE 1. Scheme of experimental device

Mechanical loading (insertion of steel pin with diameter 12 mm into investigated sample) was conducted by steel load with the mass of 2.5 kg. The load was accelerated by explosion products of additional charge and the load velocity was measured by system of electric contact gages.

By adjustment of the additional charge mass it is possible to vary pin velocity from several meters per second to 70 meters per second. The sensitivity characteristic was assumed to be the critical velocity, the value of which was determined according to the formula:

$$V_{cr} = \frac{V_{min}^{+} + V_{max}^{-}}{2} \qquad (1)$$

where V_{min}^{+} is minimum velocity of load, under which explosion of investigated sample occurred, and V_{max}^{-} is the maximum velocity, under which explosion wasn't observed. In the experiments under this scheme the HMX-based explosive compounds samples ∅ 60 by 50 mm were investigated at temperatures +20°C and -196°C. Experimental device cooling to the temperature of -196°C was carried out by liquid nitrogen.

The average (related to 5 experiments) values of V_{cr} are given in the Table 1.

TABLE 1. Experimental Results of Investigation into Samples Sensitivity.

Temperature of sample, °C	Critical velocity, m/s
+20	15.1 ± 0,5
−196	13.5 ± 0,5

It is obvious from the analysis of results, given in the Table 1, that explosive compound sensitivity to mechanical impacts at the temperature -196°C does not decrease, as it should follow from some initiating mechanisms. These mechanisms are associated with transformation of mechanic energy into thermal energy on account of collapse of air inclusions, contained in explosive charge, or on account of ductile heating of explosive compound under different types of flow and friction of explosive, as a matter of fact this velocity even a little increases.

This fact, in our opinion, may be explained from the standpoint of mechanism of energy concentration zones formation due to electric charges, occurring in explosive compound crystals under deformation and destruction (5).

The essence of mechanism is following. During loading of explosive compound crystal on its opposite facets appear electric charges, equal by value, but different by sign. And the more is the force, compressing or stretching the crystal, the more is relative value of charges. And in crystal destruction the growing fracture is a rapidly separating condenser, in which under specific conditions occurs electric discharge. The temperature in the discharge zone may reach tens thousand °C. And this energy is sufficient for beginning of decomposition of particular explosive compound molecules. Proceeding from the most common explosives activation energy (20-60 Kcal/mole) the energy of decomposition beginning corresponds to formal temperature of 10000-30000K. It is practically impossible to obtain such concentration of energy for separate explosive molecule by non-electric mechanisms. Alternation of investigated sample initial temperature (by 200°C or by 2%) under such concentrations of energy practically must not influence the beginning of decomposition. In the given case, the influence of temperature on explosive compound sensitivity will be affected through temperature influence on fragile destruction of explosive compound crystal. The explosive compound crystal tendency to fragile destruction increases alongside with temperature decrease. The indicated alteration of crystals properties does not contradict with observed in experiments sensitivity increase under mechanical impacts.

For realization of this effect necessary (but not sufficient) condition is separation of electric charges under crystal of explosive compound destruction (9). One of the reasons, leading to separation of charges is associated with piezo-effect availability in explosive compound crystals (10). For establishing of piezo-effect availability in explosive compound crystals authors conducted investigations into HMX mono-crystals behavior under mechanical loading.

For this purpose authors used the technique, described in the papers (11, 12). The scheme of experimental device is given in Fig. 2.

HMX mono-crystal 5 mm length, 5 mm width and 3 mm height was put in steel housing-electrode. Mechanical loading of crystal was affected by steel roller-electrode, on which the load was freely falling. In this case between electrodes occurs the difference of potentials, recorded by oscillograph. The typical oscillogram of experiments described above is given in Fig. 3.

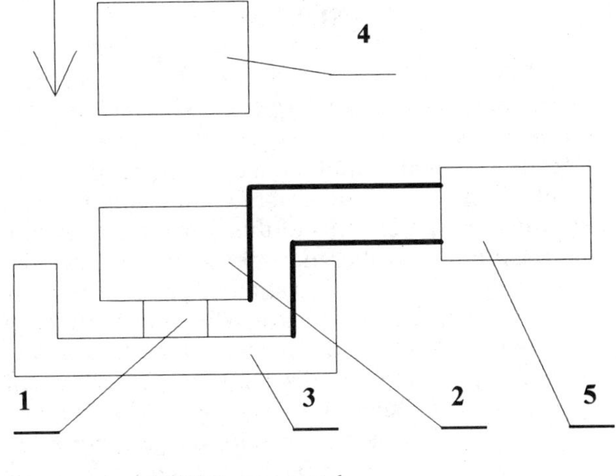

1- HMX mono-crystal;
2- steel roller-electrode;
3- steel housing-electrode;
4 - load;
5 - oscillograph.

FIGURE 2. Scheme of experimental device.

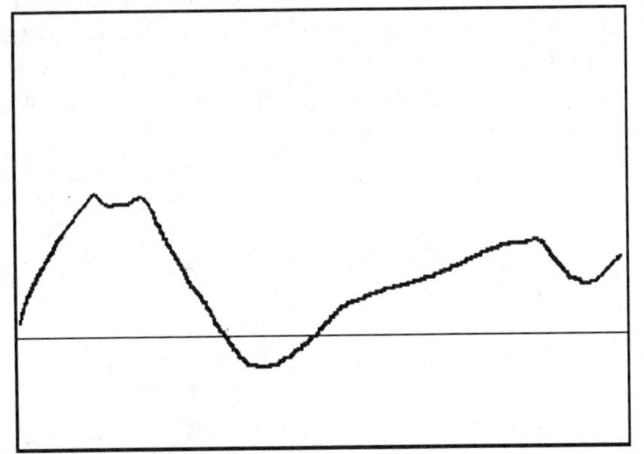

FIGURE 3. Oscillogram of experiment.

As a result of the conducted experiments it was shown, that under impact loading of HMX crystals between opposite facets occurs a difference of potentials. This fact corroborates the possibility of realization of mechanism of energy concentration zones occurrence on account of electric discharge. For a final solution of the issue of mechanism of energy concentrators formation, additional studies of electrophysical processes, going under deformation and destruction of crystals are necessary.

CONCLUSION

Results obtained in this paper do not contradict with mechanism of high density of energy concentrators occurrence on account of electrophysical phenomena, occurring under explosive compound crystals deformation and destruction. As benefits of this mechanism, in case of its future corroboration, we may consider the following. From consideration of physical process of initiating, it is obvious that we may influence explosive compound sensitivity, varying the probability of electric discharge in the charge of explosive compound. For example, sensitivity may be reduced on account of electric voltage decrease on free facets of crystal. In case of new explosives synthesis it is necessary to design a molecule in such a way, that crystals of this explosive compound would have a minimum piezoelectric effect. In our opinion, this mechanism of explosive compound initiating may by applied for consideration of other types of external impact (heat, light, etc.) on explosive compound. Therefore, proceeding from stated above it is considered to be purposeful to conduct studies in electrophysic processes, going in explosive compound and factors, influencing these processes.

REFERENCES

1. Kholevo, N. A., Shock *Sensitivity of Explosive Compounds*, M., Mashinostrojenie, 1974.
2. Andreev, K. K., Belyaev, A. F., *Theory of Explosive Compounds*, M., Oborongiz, 1960.
3. Stanyukovich, K. P., *Physics of Explosion*, M., Nauka, 1975.
4. Afanas'ev, G. T., Bobolev, V. K., *Shock Initiating of Solid Explosive Compounds*, M., Nauka, 1968.
5. Varentsov, Ye. A., Khrustalev, Yu. A., Khrapal', V. N., *Russian Journal of Physical Chemistry*, **LXII**, 2197-2204 (1988)
6. Bouden, F., Ioffe, A., *Excitement and Development of Exploion in Solid and Liquid Compounds* Moscow: Inostrannaya Literatura, 1955.
7. Dubnov, L. V., Sukhikh, V. A., Tomashevich, I. I., *Collection "Explosion Matters"*, **71/28**, (1972).
8. Taibinov, N. P., Litvinov, B. V., Loboyko, B. G., Filin, V. P., Alekceev, A. V., Sumin, V. D., Garmasheva, N. V., "Investigation of Explosive Material Sensitivity to Mechanical Loading," in *Proceedings of the 2nd International Symposium on Intense Dinamic Loading and Its Effects*, Chengdu, China, 1992, pp. 96-98.
9. Miles, M. H., Dickenson,. T., Jensen, L. C., *J. Appl. Phys.*, **57**, 5048-5055 (1985).
10. Deryagin, B. V., Krotova, N. A., Smilga, V. P., *Adhesion of Solid Bodies* Moscow: Nauka, 1973.
11. Plonski, A. F., *Piezoelectricity*. State Publishing House of Engineering-Theoretical Literature. M.,1956.
12. Zheludev, I. S., *Physics of Crystal Dielectrics*. M., Nauka., 1968.

INVESTIGATION INTO LOW-TEMPERATURES INFLUENCE ON HIGH EXPLOSIVE COMPOUNDS SENSITIVITY TO SHOCK-WAVE IMPACTS

A. N. Averin, A. V. Alekseev, S. V. Batalov, B. G. Loboiko, B. V. Litvinov,
V. D. Sumin, V. P. Filin, A. N. Yagnakov

Russian Federal Nuclear Center-Institute of Technical Physics
P. O. Box 245, Snezhinsk, Chelyabinsk region 456770 Russia

Study of shock-wave sensitivity of explosives under various temperatures is of great significance for correct analysis of safe application of different industrial processes, technologies, as well as for correct understanding of explosion initiation mechanism in (explosives). Currently, the influence of low, (-100°C...-200°C) temperatures on explosive sensitivity to weak shock waves is poorly studied. This paper gives experimental results on the influence of low temperatures on the sensitivity of HMX - based explosives to weak shock-waves. In the present paper an attempt is made to experimentally determine dependence of HMX - based explosive sensitivity to weak shock waves on temperatures. The original technique of the experiment is presented in the report.

INTRODUCTION

Explosive research involves many operations and applications where safety is a concern, for example, techniques of explosive cutting. In addition to safety, the desire to understand the peculiarities of initiation in condensed explosives has given great significance to investigations of shock-wave sensitivity of explosives at different temperatures. The interest in the influence of temperatures on the critical parameters of detonation appeared long ago. In particular, classic investigations into the temperature-dependence of critical diameter of detonation in trinitrotoluene powder in the range -240∞C...60∞C show a linear decrease of critical diameter with temperature (1,7). Investigations into the sensitivity response of explosive compounds to different impacts under heating were conducted in a considerably wide range of positive temperatures (1,3). For example, results of recent investigations into the influence of temperature on shock-wave sensitivity of samples, based on such explosive with low sensitivity as TATB, heated to the temperature of 250∞C, showed increase of shock-wave sensitivity to the level of HMX compositions of VBX-9404 type (2). Behavior of explosive sensitivity to weak shock-wave impacts under low temperatures (-100∞C ... -200∞C) currently is poorly investigated.

In the current paper experimental results of investigations into the influence of low temperatures (-170∞C) on HMX-based explosive sensitivity to weak shock-wave impacts are reported.

TECHNIQUE OF EXPERIMENT

Explosive sensitivity to weak shock waves (hereinafter WSW) was investigated according to experimental scheme, given in Fig. 1.

When the active charge triggers, the weak shock wave of required intensity (it was varied by adjustment of inert Plexiglas pad of different thickness) entered the investigated sample of explosive and caused there detonation in some depth with delay or the shock wave faded.

Explosive sensitivity to WSW was estimated regarding detonation delay time t_d.. Detonation delay time meant the difference between time interval between shock wave entrance into sample and the moment of shock (detonation) wave arrival to the opposite end of the sample (t_{exper}) and calculated time of charge detonation with stationary rate (t_{st}).

Value of detonation delay time (t_d) of investigated samples was determined using formula:

$$t_d = t_{exper} - H_{expl}/D_{st},\qquad(1)$$

where:

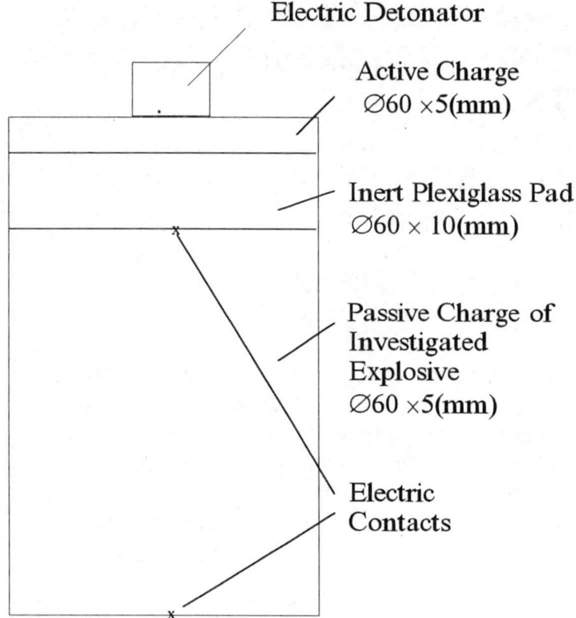

FIGURE 1. The experimental scheme.

t_{exper} - time, measured in experiment, microseconds;
H_{expl} - height of investigated explosive, millimeters;
D_{st} - stationary detonation rate for investigated explosive for given density of sample and its geometrical sizes, millimeters per microseconds (was determined in series of experiments at prescribed temperature using experimental unit (Fig. 1) without inert pad).

It is also necessary to mention here, that values of initiating shock wave P, entering investigated sample, was also determined in modeling experiments (using manganin gages) under prescribed temperatures, varying thickness of inert Plexiglas pad.

The time t_{exper} was measured using electric contacts technique applying ionization type electric contacts. The first gage was located between Plexiglas inert pad and passive charge, the second one on the bottom end of investigated explosive sample.

As investigated, explosive compound samples were pressed discs of technological density with porosity ~ 2% ∅40 × 40 mm made of HMX-containing explosives with inert (composition A) and TNT-containing (composition B) binders. Investigated samples in the experimental unit were slowly cooled in special thermostat to -170°C. In

TABLE 1. Experimental Data.

Explo-sive	Experimental temperature, °C	Number of experi-ments	Δt_d, μs
Compo-sition A	25	5	2.26+0.1
	-170	5	2.13+0.1
Compo-sition B	25	5	2.58+0.1
	-170	5	2.76+0.1

this case the uniform cooling of samples volume, excluding cracks formation, was achieved.

DISCUSSION OF RESULTS

It is possible to make a conclusion from obtained experimental data, that cooling of investigated TNT-containing explosive samples of Composition B to the temperature -170°C resulted in some decrease of sensitivity to weak shock wave impact. So, sensitivity of cooled pressed samples with such composition (composition B) in case of shock wave impact with an amplitude of P≈30 Kbar reduced approximately by 8%. The explanation of sensitivity decrease as a consequence of alternation (increase) of density, resulting from samples cooling to -170°C is poorly convincing,; change of sample density is not accompanied by alternation of non-uniformity degree of explosive sample for volume share of pores remains the same. More acceptable, in our opinion, is explanation of this event starting from the general principles in the framework of "hot points" theory of detonation occurrence, connected with increase of reaction rate alongside with the increase of temperature of explosive compound. Obtained data is in correspondence with results of other papers, particularly, with results of paper (4), where increase of pre-detonation area for TATB is registered under temperature decrease up to -70°C.

From another side, the obtained experimental data for composition A samples showed some absolute increase of sensitivity, though taking into account the deviation of obtained numerical results, the statement would be more correct, that sensitivity of samples to weak shock waves impact practically did not change. This corroborates given above suggestion, that alternation of density does not play critical role in the process of alternation of explosive compound sensitivity to weak shock waves. It looks like the more important role is played by the state of grains separation. Therefore, the general decrease of initial temperature is compensated by possible change of prevailing mechanism of explosive transformation in the second stage of the process which is the stream character of the interaction between hot gas under pore collapse.

A general result is that explosive compound temperature decrease to -170°C does not result in uniform change of their sensitivity to weak shock wave impact. The tests do not allow us to make uniform conclusions about influence of low temperatures on explosive compounds sensitivity to weak shock waves and about the very internal factors, influencing its behavior. Nevertheless, the difference of (t_d) for investigated HMX explosive compounds with different types of binders, require, in our opinion, further investigation . As internal factors, the character of alternation of which in case of cooling to low temperatures may considerably influence the alternation of explosive compounds sensitivity to weak shock waves, we may list, according to our opinion and opinion of several researchers (5,6), such as type of binder, porosity, size of grains.

CONCLUSION

In such a way, given in presentation experimental data of investigation into shock wave sensitivity of cooled up to low temperatures (-170°C) pressed HMX-based explosive compounds samples, allow the suggestion, that development of explosive transformation in explosive compound under such types of impact in pressure range less than ~30Kbar is determined by the process of transformation in the "hot" points. Further development of these works in the direction of investigation into low temperatures influence on sensitivity of explosive compounds with different types of binder, different porosity or different sizes of grain to shock wave impact of different amplitude and duration, will allow to create a more full qualitative picture of occurrence and development of explosive transformation in mixtures of condensed explosive compounds.

REFERENCES

1. Yuhanson, K., Person, P., *Detonation of Explosive Compounds*, Moscow: Mir, 1973, pp.52-64

2. Urtiew, P., Maienschein, J. and Tarver, C., "Shock Sensitivity Dependence on Initial Temperature," presented at the Symposium on Energetic Materials Technology, Pleasanton, California, May 18-25, 1994.

3. *Theory of Explosive Compounds*, Collection of Articles, Moscow: Oborongiz , 1963.

4. Jackson, R. , "Regularities of Initiating and Distribution of Detonation in TATB." *Detonation and Explosive Compounds,* Collection of Articles, Moscow: Mir, 1981.

5. Stresso, P., Cennedy, J.. "Critical conditions of Detonation Shock-Wave Initiating in Explosive Compounds of Practical Use." *Detonation and Explosive Compounds*, Collection of Articles, Moscow: Mir, 1981.

6. Mouland, H., "Particular Aspects of the Explosive Particle Size Effect on Shock Sensitivity of Cast HMX Formulations" in *Proceedings of the 9th Symposium on Detonation*, 1989.

7. Belyaev, A. F., *Combustion, Detonation and Explosion Work of Condensed Systems,* Moscow: Nauka, 1968.

LOW PRESSURE SHOCK INITIATION OF POROUS HMX FOR TWO GRAIN SIZE DISTRIBUTIONS AND TWO DENSITIES[†]

R. L. Gustavsen, S. A. Sheffield, and R. R. Alcon

Los Alamos National Laboratory, Los Alamos, NM 87545

Shock initiation measurements have been made on granular HMX (octotetramethylene tetranitramine) for two particle size distributions and two densities. Samples were pressed to either 65% or 73% of crystal density from fine (≈ 10 μm grain size) and coarse (broad distribution of grain sizes peaking at ≈ 150 μm) powders. Planar shocks of 0.2 - 1 GPa were generated by impacting gas gun driven projectiles on plastic targets containing the HMX. Wave profiles were measured at the input and output of the ≈ 3.9 mm thick HMX layer using electromagnetic particle velocity gauges. The initiation behavior for the two particle size distributions was very different. The coarse HMX began initiating at input pressures as low as 0.5 GPa. Transmitted wave profiles showed relatively slow reaction with most of the buildup occurring at the shock front. In contrast, the fine particle HMX did not begin to initiate at pressures below 0.9 GPa. When the fine powder did react, however, it did so much faster than the coarse HMX. These observations are consistent with commonly held ideas about burn rates being correlated to surface area, and initiation thresholds being correlated with the size and temperature of the hot spots created by shock passage. For each grain size, the higher density pressings were less sensitive than the lower density pressings.

INTRODUCTION

The present work is a continuation of our efforts to develop an understanding of the low pressure shock compaction and initiation of highly porous HMX (1-3). References contained in our previous work (1,2) and others (4-7) indicate that the initiation sensitivity for porous explosives is a complex function of density (porosity), particle size, pulse duration, and input pressure. In an effort to determine how these parameters affect the initiation of HMX, we prepared samples with densities of 65% and 73% of TMD, grain sizes varying by more than an order of magnitude (from ≈ 10 μm to ≈ 150 μm) and used input pressures varying from 0.2 - 1 GPa. Sustained pulses were used and the response of the explosive was recorded using particle velocity gauges.

EXPERIMENTAL DETAILS

Description of HMX Powders

Two different lots of HMX powder with two different particle size distributions were used in this series of experiments. One powder was composed of "coarse" particles which had the appearance both to the naked eye and under a microscope of granulated sugar. This HMX was made by Holston (Lot HOL-920-32) and had a bulk or pour density of ≈ 1.16 g/cm^3 (8). The material was screened to eliminate agglomerates and a few of the largest particles. Sieve analysis of the powder done by Dick (8) is given in Table 1 and shows a broad particle size distribution with a peak near 150 μm. All the crystals have sharp corners and edges.

TABLE 1. Particle Size Distribution for "Coarse" HMX, Holston Lot 920-832.

Sieve Opening μm	500	350	250	177	125	88	62	44	Subsieve
Weight % Retained on Sieve	1.3	4.0	15.6	18.2	27.8	11.8	12.1	4.9	4.3

[†] Work performed under the auspices of the U.S. Dept. of Energy.

The second powder, whose size distribution is shown in Table 2, was composed of "fine" particles and had the appearance of powdered sugar. This HMX was also manufactured by Holston (Lot HOL-83F-300-023) and also had a bulk or pour density of ≈ 1.16 g/cm^3. The particle size at the peak of the distribution is about 10 µm. Particle sizes were determined by Microtrac analysis. The mean particle size of the coarse and fine HMX is different by a factor of more than 10. Photographs show that this powder also contains an occasional large particle with a diameter of ≈ 50 µm. The rounded appearance of the particles indicates that this material was probably prepared by milling.

Gas Gun Experimental Setup

The experimental setup for the initiation experiments is shown in Figure 1. Experiments used gas gun driven projectiles to obtain sustained-shock input conditions. HMX powder was confined in sample cells which had a polychlorotrifluoro-ethylene (Kel-F) front face and a poly 4-methyl-1-pentene (TPX) or polymethylmethacrylate (PMMA) cylindrical plug back. The front face was attached with screws to a Kel-F confining cylinder with an outside diameter of 68.6 mm and an inside diameter of 40.6 mm. The pressed HMX (between the Kel-F and TPX) was ≈ 3.9 mm thick. The back plug was pressed into the Kel-F confining cylinder and held in place with an interference fit. Projectiles faced with Kel-F impacted on the Kel-F target face.

Magnetic particle velocity gauges were located on the front and back surfaces of the HMX. These were constructed of a 5 µm thick aluminum "stirrup" shaped gauge on a 12 µm thick FEP Teflon sheet. The active region of the gauge was 10 mm long. Particle-velocity histories were measured at both the front and back of the HMX sample. The gauge at the interface of the Kel-F front disk and HMX gives the input or loading profile. The gauge at the interface of the back plug and HMX gives the transmitted wave profile. The transmitted wave profile is not equivalent to what would be observed if the gauge was suspended in the HMX powder because of the impedance mismatch between the HMX and the plastic back plug. However, it is representative of the transmitted wave profile and gives a reasonable

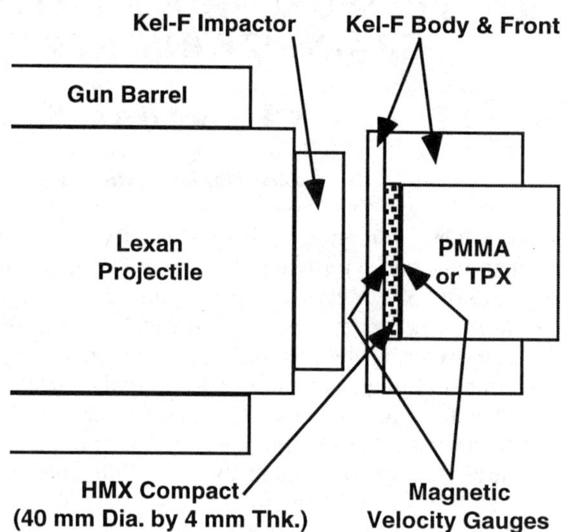

FIGURE 1. Cross section view of the projectile and target.

estimate of the rise time. Wave profiles were recorded on fast digitizing oscilloscopes.

RESULTS

A total of sixteen experiments were performed; four each for each of the two particle size distributions and for each of the two nominal densities. The nominal densities used were 1.24 g/cm^3 or 65% TMD (35 % porous) and 1.40 g/cm^3 or 73% TMD (27 % porous).

Figure 2 shows wave profiles for four experiments. The projectile velocities on these experiments were very close to the same at ≈ 0.6 mm/µs, resulting in an input to the HMX of ≈ 0.72 GPa. Complete results for the entire series of experiments will be presented elsewhere. With this input, the coarse HMX (Fig. 2a and 2c) begins to react as soon as the wave passes the front gauge and enters the powder. The front particle velocity decreases because the reacting HMX is decelerating the cell front where the gauge is located. Stress measurements show the stress at this interface increasing (1,2). The transmitted wave is growing and steepening up considerably. By the time the wave reaches the back of the HMX, the particle velocity has doubled. There appears to be a little

TABLE 2. Particle Size Distribution for "Fine" HMX, Holston Lot 83F-200-023

Particle Diameter µm	> 45	25.1	17.7	12.5	8.9	6.3	4.4	3.1	2.2	1.3	0.8	0.6
Weight % of Particles	6	8.3	10.3	11	15.8	12.5	11.3	9.5	5.8	5.0	3.0	1.5

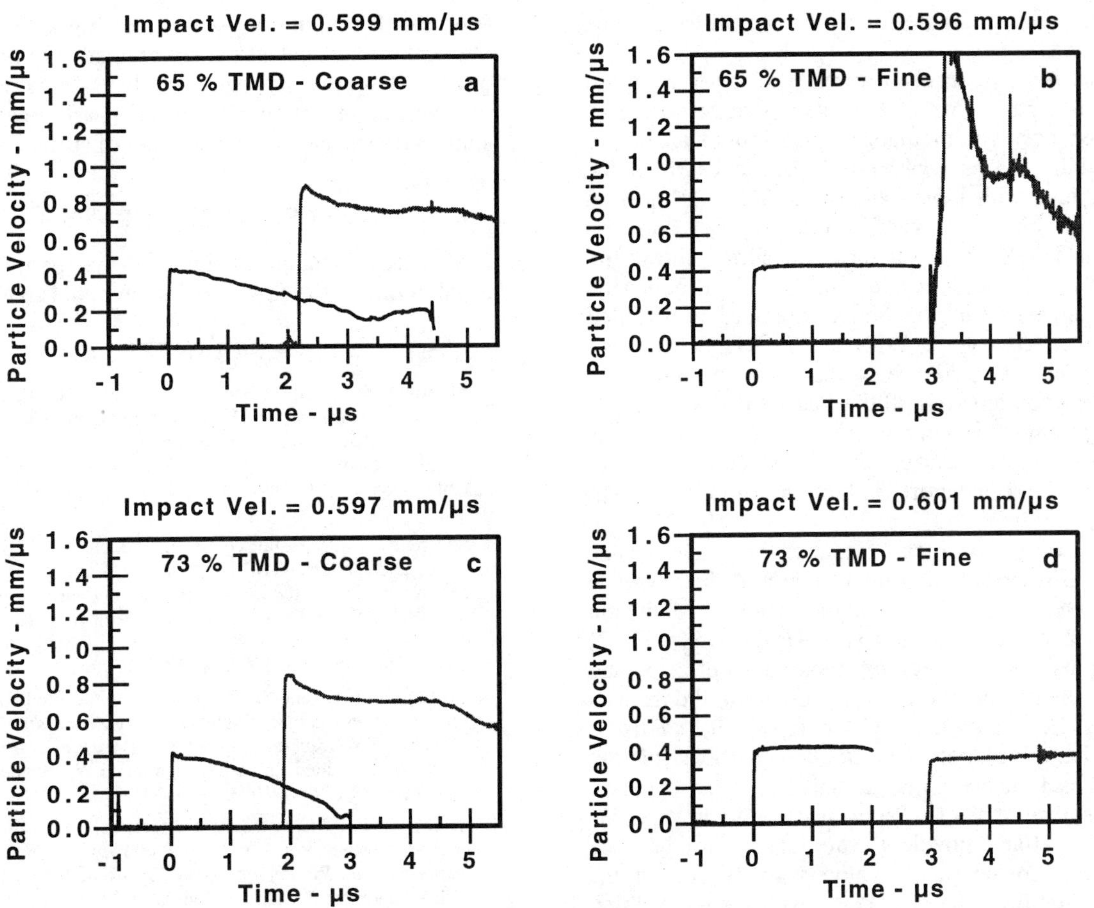

FIGURE 2. Input and transmitted particle velocities in porous HMX compacts. All the samples were about 3.9 mm thick. (a) 1.24 g/cm^3 (65% TMD), "coarse" particle powder. This experiment had a TPX back disk. (b) 1.24 g/cm^3 (65% TMD), "fine" particle powder. This and items c and d shown in this panel had a PMMA back disk. (c) 1.40 g/cm^3 (74% TMD), "coarse" particle powder. (d) 1.40 g/cm^3 (74% TMD), "fine" particle powder.

more reaction in the 1.24 g/cm^3 (Fig. 2a) than the 1.40 g/cm^3 (Fig. 2c) coarse material.

In this regime of 0.5-1 GPa, the reactivity of the fine particle HMX differs greatly from that of the coarse particle HMX. In these ≈ 0.72 GPa input experiments in particular, there is no reaction evident in the front gauge profile at either density for the fine HMX. (See Figs. 2b and 2d.) The transmitted wave profile in the 1.40 g/cm^3 fine powder (Fig. 2d) also shows no reaction.

The transmitted wave profile in the 1.24 g/cm^3 fine particle powder (Fig. 2b), however, shows a great deal of reaction. After the wave reflects off the back PMMA disk, the initial ≈ 0.72 GPa pressure in the HMX is approximately doubled. There is no

reaction for several tens of nanoseconds. Then reaction begins and proceeds very rapidly. We have estimated this reaction to be more than 10 times more rapid than that shown by the input gauge in the coarse HMX.

DISCUSSION

From the four experiments shown in Figure 2 and others like it but at different inputs, we have drawn the following conclusions. Reaction (reactivity) depends slightly on density for both the fine and coarse particle HMX. In general the higher density HMX seems to be less sensitive. This is what might be expected from looking at the energy deposited during compaction. Less energy is deposited in the higher density material for a given input pressure.

Reactivity depends a great deal on the initial particle size. For the coarse particle material, reaction occurs immediately when inputs are above 0.7 GPa. We have observed reaction begin at the front gauge with inputs as low as 0.5 GPa (after an induction time of several hundred ns). The fine particle HMX, by contrast, does not show any reaction at inputs less than ≈ 0.72 GPa at either the front gauges or in the higher pressure transmitted and reflected waves. With ≈ 0.72 GPa inputs and above, there is evidence of reaction, but only after the wave has reflected off the back PMMA disk and the pressure is approximately doubled. This reaction, which occurred only in the lower density powder, had a short induction time and was extremely rapid.

In addition to these features, the wave profile characteristics of initiating coarse and fine particle explosives are different. Jerry Dick's Manganin gauge measurements on coarse HMX (65 % TMD) for inputs of 0.81 GPa and thicknesses of 2, 3, and 4 mm, clearly show the wave growing in the front as the wave traverses the HMX compact (9). This is seen also in our more than doubled particle velocity at the back gauge (Figs. 2a and 2c). By contrast, run distance to detonation measurements in very fine particle HNS powders (nominal particle size 1-2 μm) showed strong velocity overshoots at the onset of detonation (4). These results suggested that a reactive wave developed well behind the shock front and caught up during the transition to detonation. This is similar to the mechanism by which a homogeneous explosive builds up to detonation (10). Thus, coarse particle explosives have a growing reactive wave at the shock front while fine particle explosives likely have a growing reactive wave behind and eventually overtaking the shock front.

These observations are generally explained in the following way. Most hot spot reaction theories indicate that the size of a hot spot is very nearly the size of a particle or of a void. Large particles thus lead to large hot spots. The initial temperature of the hot spot is scaled by the shock pressure. Large hot spots would cool slowly enough that they could begin reacting, even if the hot spot temperature was fairly low. The following reaction is relatively slow because the large particles don't have much surface area. By contrast, the small particles lead to small hot spots. These small hot spots cool more rapidly.

Thus it takes higher pressures and higher hot spot temperatures to get the fine grained explosive to ignite before the hot spot cools. Once ignited, however, the reaction is relatively fast because the small particles have a large amount of surface area.

ACKNOWLEDGMENTS

Mr. J.G. Archuleta obtained the particle size distribution for the fine HMX shown in Table 2.

REFERENCES

1. Sheffield, S. A., Gustavsen, R. L., Alcon, R. R., Graham, R. A., and Anderson, M. U., "Shock Initiation Studies of Low Density HMX Using Electromagnetic Particle Velocity and PVDF Stress Gauges," presented at the Tenth Symposium (International) on Detonation, July 12-16, 1993, Boston, Mass.

2. Sheffield, S. A., Gustavsen, R. L., Alcon, R. R., Graham, R. A., and Anderson, M. U., "Particle Velocity and Stress Measurements in Low Density HMX," in *Proceeding of joint AIRAPT/APS Conference on High Pressure Science and Technology*, Edited by S.C. Schmidt, J.W. Shaner, G.A. Samara, and M. Ross, 1993, pp. 1377-1380.

3. Gustavsen, R. L., and Sheffield, S. A., "Unreacted Hugoniots for Porous and Liquid Explosives," in *Proceeding of joint AIRAPT/APS Conference on High Pressure Science and Technology*, Edited by S.C. Schmidt, J.W. Shaner, G.A. Samara, and M. Ross, 1993, pp. 1393-1396,

4. Setchell, R.E., "Microstructural Effects in Shock Initiation of Granular Explosives," *in Proceedings of the International Symposium on Pyrotechnics and Explosives*, Ding Jing, Ed. China Academic Publishers, Beijing, 1987, pp. 635-643.

5. Hayes, D.B., "Shock Induced Hot-Spot Formation and Subsequent Decomposition in Granular, Porous HNS Explosive," in *Shock Waves, Explosions, and Detonations*, edited by J.R. Bowen, N. Manson, A.K. Oppenheim, and R.I, Soloukhin, Vol. 87 of Progress in Astronautics and Aeronautics, 1983 pp. 445-467.

6. Moulard, H., "Particular Aspect of the Explosive Particle Size Effect on Shock Sensitivity of Cast PBX Explosives," *Ninth Symposium (Intl.) on Detonation*, Office of the Chief of Naval Research Report OCNR-113291-7, Arlington, VA, 1989, pp. 18-24.

7. Simpson, R. L., Helms, F. H., Crawford, P. C., and Kury, J. W., "Particle Size Effects in the Initiation of Explosives Containing Reactive and Non-Reactive Continuous Phases," *Ninth Symposium (Intl.) on Detonation*, 1989, pp. 25-38.

8. Dick, J. J., *Combustion and Flame* **54**, 121-129 (1983).

9. Dick, J. J., *Combustion and Flame* **69**, 257-262 (1987).

10. Sheffield, S. A., Engelke, R. P., and Alcon, R. R., "In-Situ Study of the Chemically Driven Flow Fields in Initiating Homogeneous and Heterogeneous Nitromethane Explosives," *Ninth Symposium (Intl.) on Detonation*, 1989, pp. 39-49.

SHOCK INITIATION OF PBX-9502
AT ELEVATED TEMPERATURES

R. N. Mulford and R. R. Alcon

Los Alamos National Laboratory, Los Alamos, New Mexico 87544

The shock sensitivity of PBX-9502 is known to change with temperature. Both volume expansion and increased internal energy may contribute to this phenomenon. PBX-9502 was heated and its initiation and detonation behavior was examined, using MIV and shock tracker gauging on a light gas gun. Sensitivity and reactive wave profiles were measured. Complementary experiments were done on PBX-9502 made to undergo "ratchet growth", or non- reversible anisotropic thermal expansion, under carefully controlled thermal cycling. This process causes noticeable size changes and significant changes in sensitivity. Sensitivity and reactive wave profiles are discussed in terms of density and microscopic material morphology.

INTRODUCTION

TATB-based explosive PBX-9502 exhibits interesting thermal behaviors that may be expected to influence the sensitivity of the material at elevated temperatures. Both morphological and chemical changes may be important in understanding the response of PBX-9502 at temperatures above ambient.

Morphological changes include alteration in size and distribution of voids and a decrease in bulk density with increasing temperature. On heating, TATB-based explosives undergo "ratchet growth," or non-reversible thermal expansion, as a result of grossly anisotropic thermal expansion of TATB crystallites in the material. The phenomenon is complicated, with the material exhibiting several different growth regimes at different temperatures, (1) due to interactions of crystallites and to binder behavior. As the crystallites and binder are expanded and redistributed, intercrystalline porosities are believed to decrease, while intracrystalline porosities are believed to increase, (2) both of which may be expected to influence initiation properties.

Dependence of reaction rate on temperature almost certainly increases the reaction of the TATB at elevated temperatures, although extensive experiments by Buntain (3) have shown the increase in reactivity with increasing temperature to be small.

EXPERIMENTAL

Experiments are done on a single stage light gas gun, using in-material magnetic (MIV) gauging (4,5,6) in a target specially modified to allow controlled heating and temperature monitoring, shown in Figure 1.

The target is heated front and back using silicone rubber flexible heaters (7) rated to 45W and 220°C. The entire target was insulated with household fiberglass insulation. Temperature was monitored at 4 locations including the gauge plane with both wire and fine foil (8) copper constantan thermocouples. Thermocouple mortality was high during heating, possibly due to expansion of the PBX material onto which the TC was fastened. One minute before firing, the front heater and insulation was jettisoned using a spring assembly, to provide an unimpeded planar surface for projectile impact. Heating rates and temperature uniformity within the target were thoroughly studied, as was cooling after ejection of the front heater. Heating rate was between 0.5 and 1°C/minute, to prevent thermal stresses, distortion, or cracking in the PBX material.

The gas gun projectile impact generates reliably well-supported shock waves with a well-characterized wave shape. The square pressure pulse simplifies consideration of the time-dependent behavior of the growth of the reactive wave. The gas gun can reach projectile velocities of up to 1.4 mm/μsec, corresponding to pressures of up to about 10.5 GPa in full-density PBX materials when single crystal sapphire impactors are used. This maximum pressure is insufficient to detonate cold PBX-9502 within the observable time, but if the material is hot or has undergone non-reversible thermal expansion to a lower density, then the run to detonation will be short enough to provide good data. Particle velocity, u_p, was measured directly using ten nested magnetic gauges, and shock velocity U_S was obtained from time of arrival at the different gauges and from a shock tracker gauge. (6)

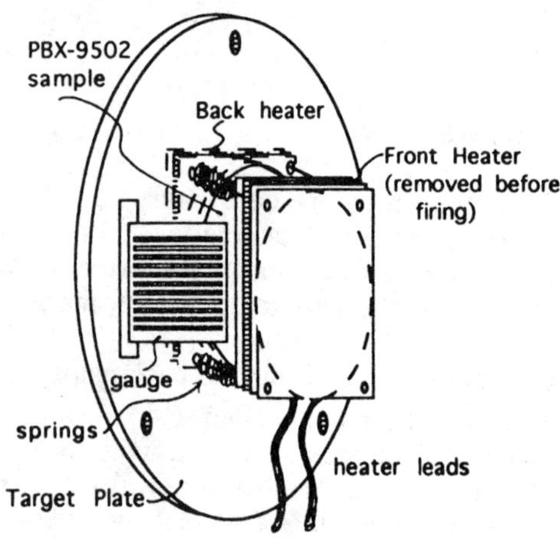

Figure 1. The hot MIV target

DISCUSSION

Heated material exhibits quite different reactive behavior from other detonable (HMX-based) PBX materials studied at these pressures. Waveforms observed in PBX-9502 at 80°C are shown in Figure 2. The mechanism for increased reactivity at high temperatures consists of two factors, temperature dependence of the reaction, and material morphology. These data confirm that significant changes in material morphologies occur between ambient and 100°C, as has been proposed to explain anomalous thermal expansion and contraction data. (1)

When heated between 25°C and the pressing temperature of the material, TATB crystallites undergo thermal expansion to reoccupy the voids created when they cooled and contracted after pressing. Pressing temperature varies between 80°C and 100°C. This process produces some net volume

expansion, but is generally recognized (2) to result in a decrease in porosity of the PBX material. Above the pressing temperature, further expansion of TATB crystallites pushes them apart. This "graphitic" growth along with increased flow of the binder increase the void fraction as the net volume increases rapidly. Porosity discussed here refers only to intercrystalline voids. It is anticipated that intracrystalline voids increase uniformly with temperature, but data is inconclusive.

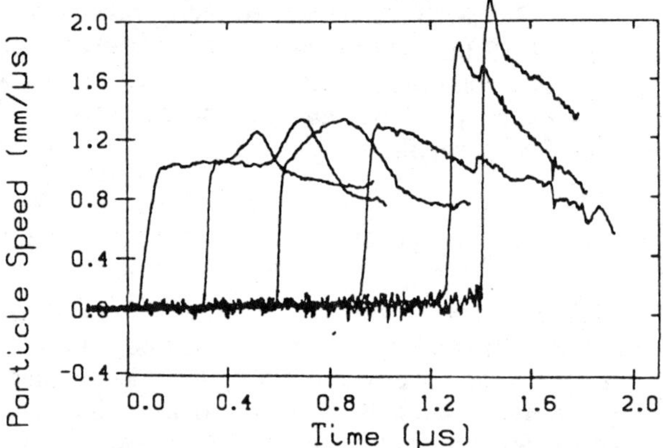

Figure 2. MIV gauge records for PBX-9502 reacting at 80°C. Baseline drift due to an electronic artifact has been corrected.

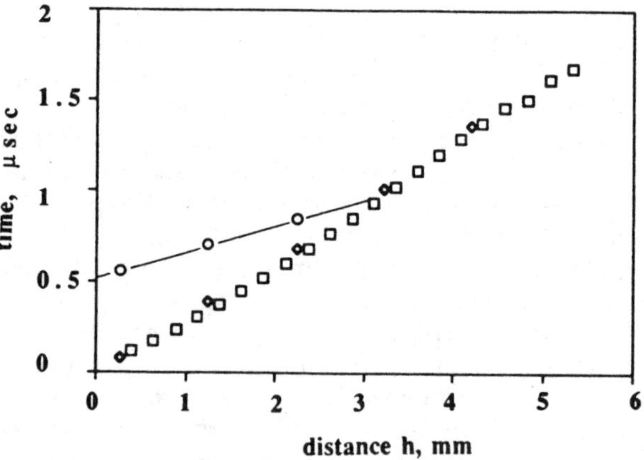

Figure 3. The x-t diagram drawn from experimental data shows the overtake of the shock (◇, □) by the reactive wave. (○) The ◇ points are u_p gauges, and □ points are from the shock tracker gauge.

These few data may be sufficiently consistent with extant ideas to suggest possible descriptions of PBX 9502 behavior at elevated temperatures.

Material near the pressing temperature can thus be expected to have quite low intercrystalline porosity, resembling a very dense solid or a liquid. The development of detonation in the absence of hot spots relies on homogeneous initiation, (9) in which the reactive wave arises in the hot shocked material after the shock front has passed, and then accelerates to join the shock front as a full detonation wave. The process is chemical, rather than mechanical. Here the reactive wave, marked "A" in the u_p records shown in Figure 2, cannot develop into a superdetonation before overtaking the shock. Its trajectory is shown in the x-t diagram in Figure 3. This is similar behavior to that seen in sensitized nitromethane and other liquids. (10,11)

At higher temperatures, the (intercrystalline) void fraction increases, restoring the material to a typical porous PBX, and the initiation to a heterogeneous mechanism. The data shown in Figure 4 was taken at 100°C, and shows u_p profiles typical of heterogeneous initiation, comparable with that of room temperature HMX-based PBX materials.

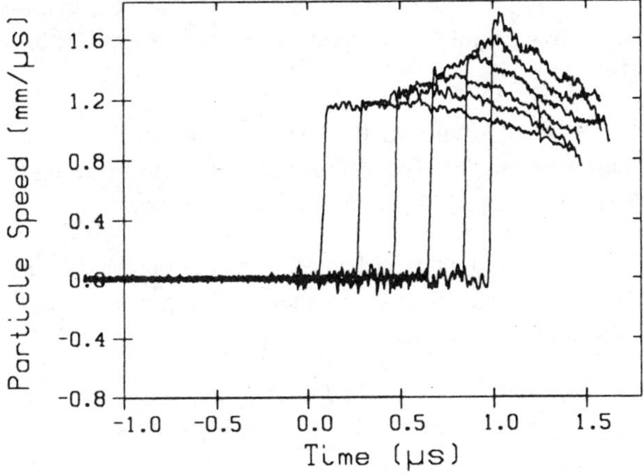

Figure 4. Detonation behavior of PBX-9502 at 100°C.

These records are consistent with the expectation that the material morphology is altered in different ways as the material passes through different temperature regimes, and that detonation behavior depends strongly on the material morphology, specifically on intercrystalline voids.

An alternate explanation should be considered for these data. The detonation behavior of TATB-based materials has been suggested to be homogeneous, because the void size is small relative to the reaction zone of the TATB. In this case, the 80°C data may be viewed as an extension of the room temperature behavior. The records obtained at 100°C may be anomalous, resulting from microcracking during heating. This kind of microcracking has been seen (12) in a few of many identical PBX-9502 samples subjected to the same very controlled heating. Microcracking has a marked effect on the growth of the reactive wave. A larger number of experiments will assist in distinguishing between these possibilities.

The apparent irrelevance of intracrystalline porosity to the initiation mechanism lends some support to the assertion that small voids do not assist in the hot spot initiation of the PBX-9502. However, the true behavior of intracrystalline voids with temperature is not well characterized.

The chemical contribution to the increase in reactivity at these elevated temperatures may be evaluated by comparison of the run distance for PBX-9502 at reduced density with the run distance vs. run time points for these heated samples. Plots are shown in Figure 5.

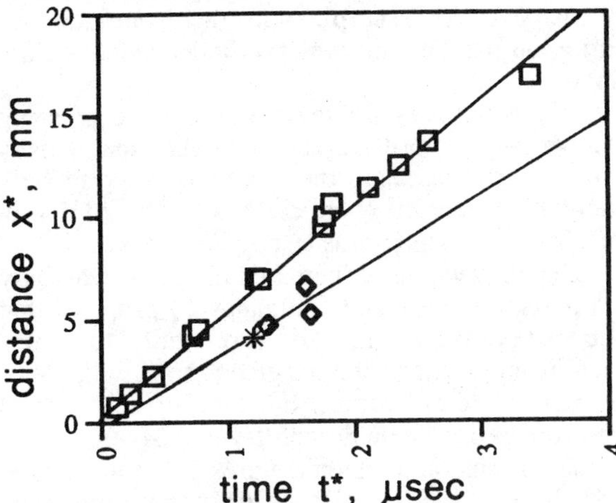

Figure 5. The run distance for the heated PBX-9502 sample (✳) is comparable to the run distance for PBX-9502 at initial density of 1.825 (◆), but differs from the run for cold PBX-9502 (□) from reference (13).

The points for reduced density material were obtained from shots done on cold PBX-9502 that had been subjected to ratchet growth by repeated heating in an oven to 211°C, cooling very slowly to room temperature in between cycles. The resulting samples were then machined into MIV wedges and fired cold on the gas gun. Wackerle and Dallman (2) show that the sensitivity of material expanded in this way does not differ from material pressed to the same reduced density, ruling out chemical decomposition during the heating cycles.

The density of the 80°C sample can be estimated (1,2) to be 1.848 g/cm^3, compared with 1.825 g/cm^3 for the expanded material, allowing for the possibility that its reaction may be slightly accelerated by temperature in order for its run distance to fall with that of lower density material. Points for both hot and expanded PBX-9502 fall on the Hugoniot given by Dallman and Wackerle (2) for PBX-9502 at 75°C.

CONCLUSION

Based on the limited data we have obtained, we suggest the following.

The rate of chemical reaction of TATB is increased with temperature, since the 80°C sample exhibits a shortened run distance despite the decreased porosity of the sample and the homogeneous initiation mechanism, which is independent of hot spots.

In the temperature regime above the pressing temperature, porosity increases and hot spot density and activity dominate the reaction, masking bulk thermal effects on chemical rate. The initiation mechanism in this case is heterogeneous.

The observation of increased chemical reactivity supports Wackerle and Dallman's (2) proposal that those hot spots that are too small to cause significant reaction behind a given shock will, above a certain temperature, become effective and contribute to the initiation of the explosive.

The limits on the temperatures and porosities at which PBX-9502 will exhibit homogeneous initiation behavior will contribute interesting data to the understanding of hotspot reaction, yielding data on the interaction of chemical reaction rate and hotspot effectiveness.

1. Howard Cady, Los Alamos National Laboratory, M-1, informal report written to J. Dallman concerning high temperature materials property measurements on PBX-9502 and LX-17, May 1993.

2. J. Dallman and J. Wackerle, Tenth Symposium (International) on Detonation, #110, July 12-16, 1993.

3. G. A. Buntain, Los Alamos National Laboratory, M-1, informal report written to J. Dallman, M-1, concerning the impact sensitivity of TATB powders at ambient, 250°C, and 300°C, June 1990.

4. R. Mulford, S. Sheffield, and R. Alcon, in "High Pressure Science and Technology," Colorado Springs, 1993, p. 1405.

5. S. Sheffield and R. Alcon, in "High Pressure Science and Technology," Colorado Springs, 1993, p. 1405.

6. R. Mulford and R. Alcon, "Shock Tracker Configuration of In-material Gauge," this volume.

7. Watlow Corporation, 003030C1.

8. RDF Corporation, Hudson, New Hampshire.

9. A. W. Campbell, W. C. Davis, and J. R. Travis, Phys. Fluids 4, 498 (1961).

10. S. A. Sheffield, Ray Engelke, and R. R. Alcon, Ninth Symposium (International) on Detonation, p. 39 (1989).

11. S. A. Sheffield, R. L. Gustavsen, and R. R. Alcon, "Observations of Shock-Induced Reaction in Liquid Bromoform up to 11 GPa," these proceedings.

12. R. Mulford, unpublished MIV records obtained from PBX-9502 subjected to thermal expansion and contraction.

13. J.J. Dick, C.A. Forest, J.B. Ramsay, and W.L. Seitz, J. Appl. Phys., 63, 4881-4888 (1988).

Gun-Launched Impact Initiation of the Composite Explosive PBXN-103

H. W. Sandusky and R. R. Bernecker

Naval Surface Warfare Center, Indian Head Division, Silver Spring, MD 20903-5640

A well-instrumented variation of the Susan test has been developed for simultaneously quantifying the extent and rate of deformation, the time for onset of reaction, and the magnitude of reaction occurring from the impact of energetic materials onto a flat plate. Unconfined, 40 mm diameter samples are mounted onto a 50.8 mm diameter aft body for launching from a powder gun. The mass of the aluminum aft body extends the time for sample deformation at a uniform strain rate, without crushing the sample against the target plate as can occur with the steel aft body in the Susan test. Initial experiments were conducted on the composite explosive PBXN-103, consisting of ammonium perchlorate and aluminum in an energetic binder. There was no reaction at an impact velocity of 46 m/s, mild reactions at 67 and 78 m/s, and deflagrations at ~95 m/s. Times to reaction from 220 to 1400 µs were based on the high-speed photographs of an air blast wave, which appeared prior to luminous reaction. Reactions occurred as a result of PBXN-103 deformation, not from pinching the explosive between the projectile and target.

INTRODUCTION

The impact sensitivity of energetic materials (EMs) is often determined for small (~5 mm diameter by 1 mm) samples by dropping a weight on them and for larger samples ($\geq$17 mm diameter) by propelling them from a gun into a steel plate or vice versa. In the Susan test, a 50.8 mm diameter by 101.6 mm sample, that is lightly confined by an aluminum cup and attached to a steel aft body, is propelled from a 76 mm powder gun into a steel plate. The threshold and violence of reaction are determined by measurements of blast overpressure versus impact velocity. While the large sample size is of interest, the Susan test is complicated by the possibility of ignition by pinching the sample with the aluminum cup as if deforms just after impact or by pinching an otherwise unreacting sample between the steel plate and the heavy aft body. These concerns have been addressed with a somewhat smaller projectile that has an unconfined sample attached to an aluminum aft body. A preliminary series of experiments were conducted on samples of the underwater explosive PBXN-103, which consists of ammonium perchlorate (AP) and aluminum in an energetic binder. Impacted EMs which contain AP as a major component often exhibit a relatively low threshold for reaction and violently react for velocities just in excess of the threshold.

EXPERIMENTAL ARRANGEMENT

A schematic of the arrangement is shown in Fig. 1. An unconfined 40 mm diameter by 75 mm long sample is bonded into a 15.0 mm deep recess in an aluminum aft body. The sample with its aft body is propelled from a 50.8 mm powder gun into a 178 mm square by 25 mm thick plate of hardened 4340 steel with a 0.4 µm surface finish. The target plate is clamped to four, 76 mm thick steel blocks for increased inertia. The arrangement is aligned by a laser beam along the gun axis that is reflected off the target and back onto itself, thus assuring perpendicularity of impact. The support for the target assembly does not affect the first blast wave reaching nearby overpressure transducers.

Blast overpressures are recorded by piezoelectric transducers at the locations shown in Fig. 1. Transducer #1 (PCB 113A23) near the gun muzzle records muzzle blast and is far enough from the target that compressive waves from sample reaction have coalesced. Two PCB 102A02 transducers (200 psig range) are located close to the target for observing deflagrating events. Transducer #2 is 254 mm before the target and 254 mm from the gun axis. Transducer #3 is just 180 mm above the impact point, which is close enough to distinguish the time frame for sample reaction. Muzzle blast did not interfere with overpressure measurements of sample reaction.

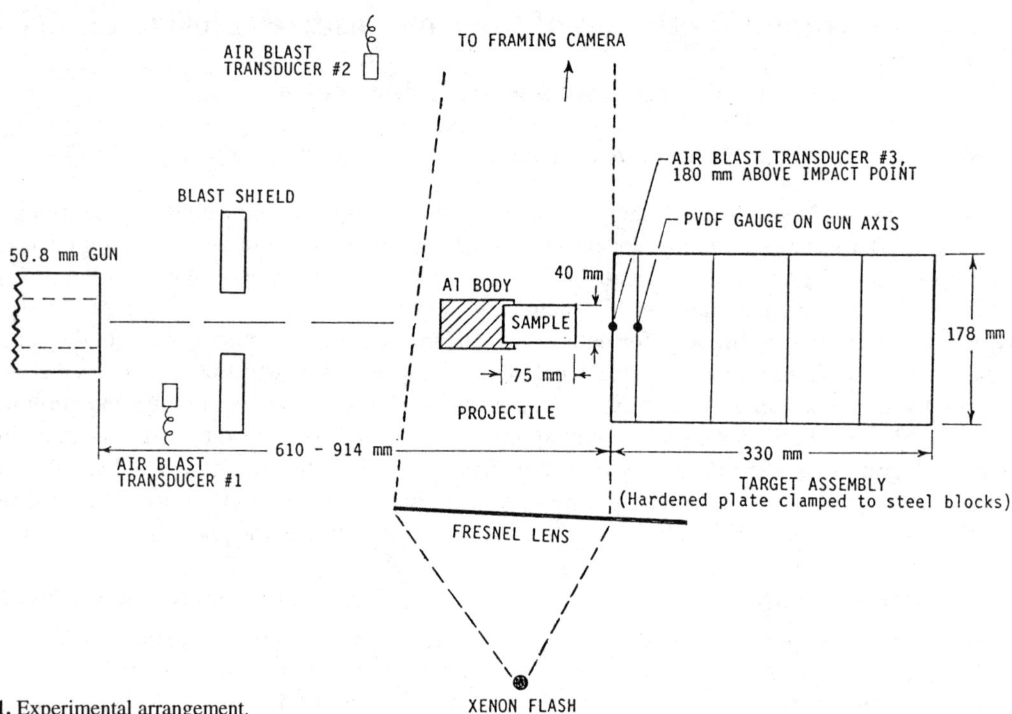

FIGURE 1. Experimental arrangement.

Sample velocity before impact, impact flatness from one perspective, deformation following impact, time for onset of reaction, and some indication of the extent of reaction are recorded either at 20 or 25 µs intervals with a continuous access framing camera. The field of view was backlit with a xenon flash and Fresnel lens. Another xenon tube is located in the field of view to be pulsed as a time fiducial for the other instrumentation. A blast shield near the gun muzzle prevents the products of the propelling charge from disrupting the field of view. Shorting switches, which contacted the shoulder of the projectile as it exits the muzzle and blast shield, verify projectile velocity and trigger the electronic flashes and recording instrumentation. A polyvinylidene fluoride (PVDF) gauge is sandwiched between the impact plate and first steel block in the target assembly to record the impact time.

EXPERIMENTAL RESULTS

The results are summarized in Table 1 in the order of increasing impact velocity (V_i). Two batches of PBXN-103 were used; Mix No. 1 was prepared for these experiments, while Mix No. 2 was prepared several years prior. Tabulated overpressures are the first peak from transducer #2. Traces from transducers #2 and #3 from Shot APIE-7 are shown in Fig. 2. The large second peak for transducer #2 was unusual. Since transducer #3 is closer to the impact point, the amplitude of the signal is lower because compressive waves generated over a time frame of >100 µs had not yet coalesced.

Increasing impact velocity did not necessarily result in higher overpressures, as seen in comparing the 67.4 m/s impact in Shot APIE-6 with the 77.7 m/s impact in Shot APIE-8. In Shot APIE-6 there was a a long delay prior to reaction. Since the front of the aft body was only 20 mm from the target when reaction began, much of the sample had already been severely deformed or broken up. This extensive damage permitted the higher overpressure than in the test with the next higher impact velocity, Shot APIE-8.

The times of reaction listed in Table 1 are the first appearance of an air shock on the high-speed camera film relative to time of impact. Even weak waves are clearly visible when backlighting the field of view with a lens that converges the light from a point source. Appearance of an air shock was coincident with the deformed region of the sample being cloudy versus

TABLE 1. Summary of Gun-Launched Impact Experiments on Unconfined PBXN-103*

SHOT APIE-	MIX NO.	V_i (m/s)	P_b (psig)	Δt (μs)	COMMENTS
5	2	46.0	0.1	f	Sample unevenly deformed and projectile rotated after planar impact
6	2	67.4	12.7	1400	Extensive deformation prior to reaction
8	1	77.7	0.5	-	No framing photography
4	1	93.7	>29	<40[#], 220	Luminous reaction at 220 μs
7	2	97.6	67.8	<40[#], 260	Luminous reaction at 260 μs
3	1	~95	-	-	No backlighting and overpressure measurements, luminous reaction

* MIX NO. = PBXN-103 samples from two mixes

V_i = Sample impact velocity

P_b = Peak overpressure at transducer #2 in Figure 1

Δt = time between impact and detection of reaction

f = no reaction

[#] = weak reaction which extinguished

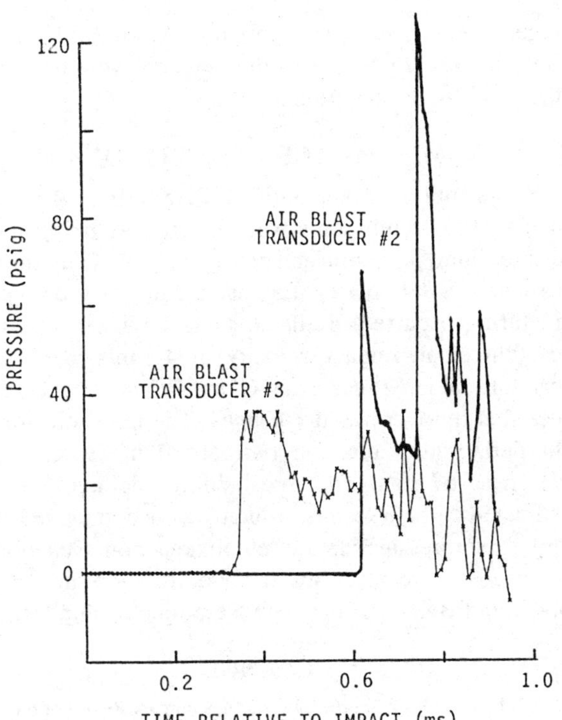

FIGURE 2. Overpressures from the 97.6 m/s impact, Shot APIE-7.

having a distinct outline. Selected frames from the high-speed photographs from Shot APIE-4 are shown in Fig. 3, with times relative to that at impact. The target plate is the dark zone at the right edge of the field of view. The clamping bars for the target assembly, which appear above and below the impact point, are not as close to the edges of the incoming sample as it appears; however, the clamping arrangement was subsequently changed to avoid interference with a highly deformed sample. There are several regions within the field of view, especially at the left edge and in the center, which appear to be clouded by gases but are actually due to nonuniform backlighting. At -10 μs there is a very narrow but uniform line of backlighting between the sample and target, indicating the flatness of impact from the perspective of the camera. By +40 μs (not shown), an air shock emerges from the impact region. At +70 μs, both an opaque cloud (debris and gases) and a spherically diverging air shock are seen near the target. The next frame shows a slowing of the opaque cloud and a weakening of the air shock, indicating a failing reaction. The projectile had not perceptibly slowed prior to the onset of luminous reaction at +220 μs within the opaque cloud. The +230 μs frame shows an air shock from this new reaction. In subsequent frames, the luminosity and velocity of the cloud increased.

There was no evidence of any aft body impacting the target, and except for the more violent reaction in Shot APIE-3, the aft bodies were recovered undamaged. In Shots APIE-6 and -8 with V_i < 95 m/s, much of the PBXN-103 was recovered. The largest fragments were several millimeter thick layers from the impacting end of the sample and the outer circumference, while the core was broken into many small pieces with millimeter dimensions.

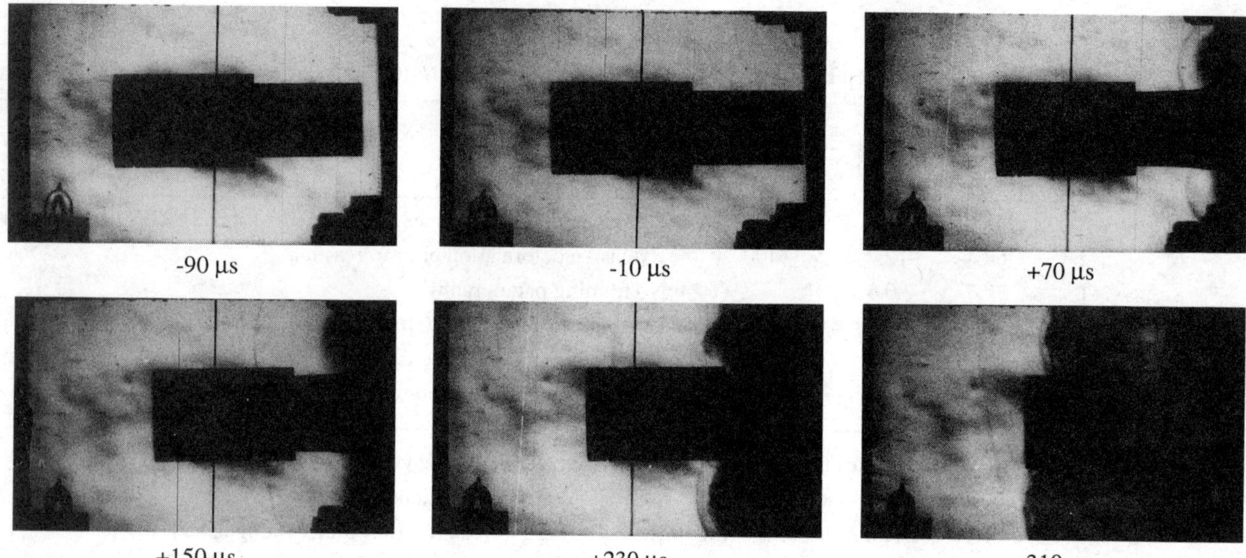

FIGURE 3. Photographs from the 93.7 m/s impact, Shot APIE-4.

DISCUSSION

Upon impact, an on-axis length of sample approximately equal to its radius is weakly shocked. For the ~95 m/s impacts, the first appearance of reaction at <40 μs probably occurred as a result of limited, but rapid, deformation from the initial 4.4 kbar shock. This pressure is about half of the 8.9 kbar threshold for burning of PBXN-103 in the underwater sensitivity test (UST).(1) The lower reaction threshold for an impacted sample may be due to its unrestricted radial deformation relative to confinement by water in the UST and because the reaction was unsustained.

Once the initial shock attenuates, the sample deforms more slowly from dissipation of its kinetic energy and that of the aft body. Sample pressure may be limited to <1 kbar, depending on the ability of the material to deform and flow. Reaction will probably be initiated in regions of relatively low pressure and high strain rate.(2) Recovery of the end layer of the samples in tests with $V_i < 95$ m/s indicates that the PBXN-103 failed near the edges as material behind the impacting surface was forced toward the target plate.

The threshold for a sub-deflagration reaction (without flame spreading) was near 67 m/s considering the long (1400 μs) delay prior to reaction. While the luminous reaction associated with ~95 m/s impacts were deflagrations because flame had spread through the debris cloud, the overpressures did not displace the target assembly in two of those tests (Shots APIE-4,7). This suggests that the ~95 m/s impacts were near the threshold for deflagration.

SUMMARY AND CONCLUSIONS

Six samples of unconfined PBXN-103, 40 mm diameter by 75 mm long, were attached to the front of an aluminum projectile and propelled from a gun into a hardened steel plate. Extensive deformation and fracturing occurred in all the samples. There was no reaction at an impact velocity of 46 m/s and mild reactions, at late times, from 67 and 78 m/s impacts, as recorded by overpressure gauges. The threshold for a sub-deflagration reaction (without flame spreading) was near 67 m/s. The threshold for deflagration is estimated to be ~95 m/s, which was accompanied by larger overpressures and luminous reaction. Reactions occurred as a result of material deformation, not from pinching PBXN-103 between the projectile and target.

REFERENCES

1. Liddiard, T. P. and Forbes, J. W., "A Summary Report of the Modified Gap Test and the Underwater Sensitivity Test," NSWC TR 86-350, Naval Surface Warfare Center, Mar 1987.
2. Sandusky, H. W., Coffey, C. S., and Liddiard, T. P., "Rate of Deformation as a Measure of Reaction Threshold in Energetic Materials," Mechanical Properties of Materials at High Rates of Strain, Conference Series No. 70, The Institute of Physics, Bristol and London, 1984, pp. 373-380.

REACTIVE WAVE GROWTH IN SHOCK-COMPRESSED THERMALLY DEGRADED HIGH EXPLOSIVES

Anita M. Renlund

Sandia National Laboratories, P. O. Box 5800, MS 1454, Albuquerque, NM 87185-1454

We have performed experiments to study the effect of thermal degradation on shock sensitivity and growth to detonation of several high-density plastic bonded explosives, confined in stainless steel cells. Assemblies were heated *in situ* in the target chamber of a light-gas gun. Confinement was varied to allow, in some cases, for thermal expansion of the explosive, and in other cases to vent the decomposition gases. Particle velocity profiles were measured using VISAR at a LiF window interface. Results for the IHE PBX-9502 showed that its sensitivity to shock initiation could be dramatically increased or decreased depending on the confinement conditions during heating. Effects were much less pronounced for PBX-9404 and PBX-9501.

INTRODUCTION

Shock initiation sensitivity of heated explosives, particularly insensitive high explosives (IHEs) containing TATB (triaminotrinitrobenzene), has received much attention in recent years. (1,2) The susceptibility of thermally degraded high explosives (HEs) and IHEs to violent reaction is the focus of studies to guarantee safety of explosive systems. To predict DDT in HEs it is necessary to know the shock initiation behavior of the material. While it was clear that heated IHEs were sensitized to shock initiation, the controlling parameters had not been identified. We chose to focus on the effect of confinement of heated HEs subjected to shocks. In brief, we found that the sensitivity of the heated system could be either increased or decreased relative to ambient IHE, depending on the heating conditions. Much less effect was observed for heated HMX- (tetranitro tetraazacyclooctane) based HEs.

It is reasonable to expect temperature to affect shock sensitivity, but it may do so via different mechanisms, and the effects can act in opposite directions. Chemical reaction rates increase with temperature. Any decomposition products may be more or less sensitive than the pristine material.

But there are also physical effects that accompany heating: phase transitions, thermal expansion (affecting porosity), trapped gases inside voids, and fractures that affect particle size. Confinement conditions may play a crucial role in the changes that accompany heating by affecting both chemical and physical processes. Confinement allows the decomposition products to remain in contact with the bulk material, perhaps enhancing autocatalytic processes. Formation of low-volume products may be favored at equilibrium. Also, dissolution of products into the lattice may alter fracture patterns, and confinement can limit thermal expansion.

EXPERIMENTAL

We conducted a series of gas-gun shots on heated PBX 9404, PBX 9501 and PBX 9502 using different physical confinement conditions. The explosive pellets (0.75-in diameter) were made at Pantex by hot-pressing the molding powders. Our experimental arrangement is shown schematically in Fig. 1. Briefly, an explosive pellet was heated inside a stainless steel cell with at least 0.5-in wall thickness to provide strong confinement. LiF was chosen as the window to allow acquisition of

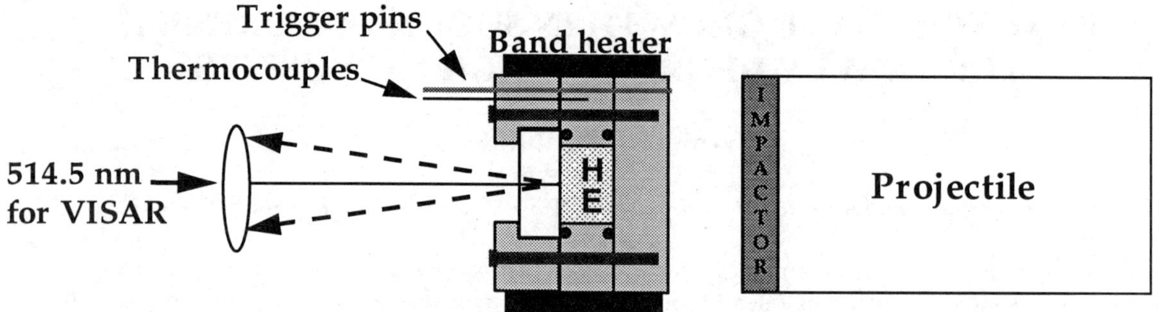

FIGURE 1. Schematic of experimental arrangement.

VISAR signal because it could be heated to the temperatures of interest. The cell was sealed with silicone or Viton o-rings, but in some cases the o-rings were cut to allow escape of the decomposition gases. Different cell thicknesses were used: some were matched to the HE pellet thickness, others were oversized to allow the pellet to expand during heating.

The target assembly was heated *in situ* inside the target chamber of a 63-mm light-gas gun. A band heater around the cell was used to heat the HE assembly at a 4°C/min ramp to a final temperature. In some experiments we allowed the HE to soak at the temperature for up to an hour. Three thermocouples were used to monitor the temperature at various positions in the cell. These temperatures agreed to within 4°C. We performed an initial series of shots on heated inert targets to check for target tilt introduced via heating, and observed no significant deviation from ambient shots. Tables 1 and 2 provide descriptions of the various shot assemblies and conditions for PBX

9502 and PBX 9501. The LiF window included a 0.020-in-thick LiF buffer to protect the diffuse VISAR reflector. We used a dual-delay leg VISAR system to record the particle velocity in the LiF.

RESULTS AND DISCUSSION

Results of 7 shots on PBX-9502 are shown in Table 1. In all cases the impact velocity was between 1.98 and 2.01 km/s, giving an initial shock to the PBX 9502 of approximately 8.6 GPa. If there was no growth in the shock wave due to reaction, the reshock from the LiF window raised the pressure in the HE to about 11.5 GPa. In shots 1, 3 and 5 there was no evidence of reactive wave growth; the HE behaved much like an inert material. Shot #2 showed wave growth after the reshock from the window, but not sufficient to lead to detonation. The target for shot #6 was heated to 240°C for one hour, then cooled overnight before firing the shot. The measured particle velocity was

TABLE 1. Summary of shot parameters and results for PBX 9502 (ρ=1.89 g cm^{-3}), P=8.6 GPa (PBX 9502 -- TATB/Kel-F 800 95/5 wt. %)

Shot #	Temperature (°C)	HE Thickness (in)	Cell Thickness (in)	Soak time (min)	o-rings	Result (mm/µs)
1	150	0.196	0.196	30	yes	0.65
2	150	0.197	0.221	60	yes	0.65 - 0.95
3	240	0.196	0.195	60	yes	0.65
4	240	0.197	0.227	60	yes	Detonation
5	240	0.197	0.195	30	cut	0.65
6	240 / 20	0.197	0.196	60/cool	cut	0.72 - 0.8
7	150	0.393	0.413	30	yes	Detonation

TABLE 2. Summary of Shot parameters and results for PBX 9501 (ρ=1.84 g cm^{-3}, 0.197 in thick)
(PBX 9501 -- HMX/estane/BDNPA(F)   95/2.5/2.5 wt. %)

Shot #	Temperature (°C)	Soak time (min)	Impact Velocity (mm/µs)	Shock arrival time (µs)	Result
1	20	NA	0.9	1.12	Detonation
2	200	30 min (vent)	0.9	0.96	Detonation
3	200	8 min	0.9	1.00	Detonation
4	200	20 min	0.9	1.35	Detonation
5	200	8 min	0.75	1.65	Detonation on reshock
6	200	30 min	0.75	2.0	Detonation on reshock

slightly higher than others measured at elevated temperatures and did show some growth after reshock. The two samples that detonated had some additional space to allow for expansion of the pellet. These results are in rough agreement with run-distances to detonation measured in previous work of Dallman and Wackerle(1) and Urtiew, et al. (2) for heated IHEs. In those experiments the materials were not confined, allowing the material to expand and showing increased shock sensitivity at elevated temperature. These present results indicate that an important mechanism may be the change in density. Indeed, we do not expect significant thermal decomposition to occur at the temperatures and soak times of these studies. Any increase in sensitivity due to enhanced reaction rates at elevated temperatures appear to be effectively countered by the strong confinement limiting expansion. Concerns of increased sensitivity need to be seen in light of the whole system, because strong confinement can mitigate the potential sensitization.

The results of experiments on PBX 9501 are presented in Table 2. Impact velocities of 0.9 and 0.75 km/s correspond to initial shocks in the HE of 5.7 and 4.5 GPa, respectively. All targets were made with cells and pellets of matched thicknesses, and full o-rings, usually silicone, were used to seal the cells. The target on shot #2 was assembled with Viton o-rings that allowed venting of gases that had built up as a result of decomposition.

PBX 9501 showed less dramatic changes than were observed for PBX 9502. The arrival time of the shock/detonation wave at the window interface, measured referenced to an arbitrary but uniform time, t=0, was noticeably affected due to heating in the fashion described. If the arrival times corresponded to a difference in the rate of shock build-up to detonation, then it appeared as though heating for longer times acted to retard the reactive wave growth. Unlike the PBX 9502 experiments, the temperature of these shots did allow for substantial decomposition prior to the shot. Indeed, we had two experiments that cooked off before the gun shot. It is therefore reasonable to expect high gas pressures in the cell, especially in shots 4 and 6. Also note that the shot temperature was above that for the solid - solid phase transition temperature to form δ-HMX.

Five experiments were also performed on PBX 9404 (HMX/NC/CEF/DPA   94/3/3/0.1 wt %) at temperatures from 140 to 200°C and impact velocities from 0.7 to 1.2 km/s. All samples had grown to detonation within the 0.197-in thickness of the pellet. We had no uniform reference time on those shots, so relative differences between the shots are not instructive. At best we can say that no dramatic changes occurred for this HE due to heating.

Any detailed understanding of the effects of temperature on shock sensitivity ultimately requires a complete description of the thermally degraded HE material. In particular we would like to know, as functions of temperature, soak time, and confinement, parameters such as the extent of decomposition, transient and final decomposition products and their sensitivities, evolved gas pressures, and porosity (both closed and connected). To aid in understanding the results of our gas-gun shots, we performed preliminary experiments aimed at measuring the pressure of evolved gases as a function of temperature and soak time. We assembled cells with explosive samples just as we had for the gun shots, mounted the cell in such a way as to transfer force to a cooled load cell, and then heated the explosive assembly while monitoring the force measured by the load cell. Much of the force was generated due to thermal stresses and thermal expansion, but we hoped to derive gas pressures by comparing forces from cells heated that were fully sealed with those that were heated with cut o-rings. Unfortunately, the analysis proved much more difficult because for PBX 9502 higher forces were measured from cells with cut o-rings than for fully sealed cells. Clearly the details of chemical decomposition are greatly influenced by how a sample is sealed and heated.

We are now investigating the state of thermally degraded HEs in more sophisticated experiments that reduce the effects of thermal stresses. We are also using complete 3-dimensional modeling and postmortem analysis of the explosive samples to develop a better understanding of degraded materials and how changes in chemical and physical parameters may lead to changes in shock sensitivity. We hope that these insights may be coupled with detailed reaction kinetics to develop a predictive understanding of the sensitivity of thermally degraded energetic materials.

ACKNOWLEDGMENTS

This work was supported by the United States Department of Energy under Contract No. DE-94AL85000. The author gratefully acknowledges the excellent technical assistance of Jill Miller, who contributed to all phases of these experiments, and M. J. Navarro and H. M. Anderson, Ktech, for gun operation and target preparation.

REFERENCES

1. Dallman, J. C. and Wackerle, J., "Temperature-dependent shock initiation of TATB-based high explosives", in Proceedings of the 10th Symposium (International) on Detonation, 1993, in press.
2. Urtiew, P. A., Cook, T. M., Maienschein, J. L., and Tarver, C. M., "Shock Sensitivity of IHE at Elevated Temperatures", in Proceedings of the 10th Symposium (International) on Detonation, 1993, in press.

ANALYTICAL INITIATION CRITERIA OF HIGH EXPLOSIVES AT DIFFERENT PROJECTILE OR JET DENSITIES

M. Held

TDW Gesellschaft für verteidigungstechnische Wirksysteme mbH
86523 Schrobenhausen

Test results of initiation threshold values with tungsten spheres of different diameters d and Lexan spheres of 76 mm diameter can be very well described by the u^2d-Held-criterion where u is the cratering velocity or $v/(1 + \sqrt{\rho_t/\rho_p})$. For covered composition B the initiation threshold value is 23 mm^3/us^2. This value fits also shaped charge tests where the high explosive charge is also in a covered arrangement.

INTRODUCTION

The initiation criterea of high explosive charges with projectiles or shaped charge jets can be very well described with the v^2d-criterion which was the first time described for shaped charge jets by the author in 1968 [1]. A summary of results which confirm this criterion for a large number of high explosive charges and load conditions - shaped charge jets, projectiles, flying foils with a L/D larger than 0.2 and of numerical calculations - are published in [2]. All these tests have been generally done with copper jets and with steel fragments, therefore with similar densities.

The question arises how this formula has to be modified if the density of the projectile is strongly changed. Mader has used in his numerical calculations a ρv^2d-criterium where ρ is the projectile density ([3] and [4]). Chick et al. [5] have given a relation of $\sqrt{\rho} \cdot v^2d$ from experimental results with copper and aluminium jets. The author has considered in [6] as a rough rule the stagnation pressure for the initiation of high explosive charges and has used a u^2d criterion where u is the cratering velocity in a high explosive charge.

Very recently test results of the initiation of composition B by tungsten spheres of different diameters and plastic spheres with 76 mm diameter were published. These initiation tests have used a magnitude of different projectile densities from tungsten with 17.77 g/cm^3 to the plastic - Lexan - with the density of 1.2 g/cm^3. With these results it can be very well proved the influence of the density ρ of the projectiles to the initiation criterea of high explosives. The reaction behaviour was also modelled in detail in <7> by numerical calculation with the ignition and growth model.

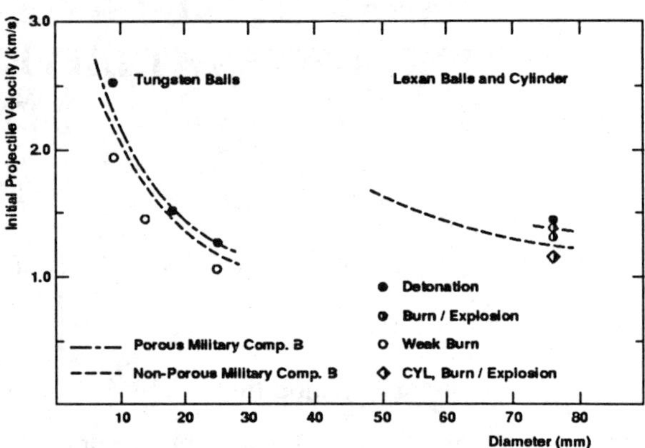

FIGURE 1. Experimental detonation threshold velocities and calculated data <7>

TEST RESULTS

Fig. 1, a direct copy of Fig. 4.07 of Reference <7>, shows the results of the initiation tests with cast composition B of 1.63 g/cm^3 density with tungsten balls of 9, 14, 18 and 25 mm diameter and the density of 17.77 g/cm^3, resp. with the Lexan sphere of 76 mm diameter and a density of 1.2 g/cm^3. The tests were conducted by the Naval Research Laboratory (NRL) and are in detail described in Reference <7>. Calculations have been made with this type of Comp. B with a density of 1.63 g/cm^3 (porous Military Comp. B) and a pressed material with higher density (1.712 g/cm^3) which is called "Non-Porous Military Comp. B". According to the standards Comp. B should be cast, never pressed. But in experiments not so well educated theoretical people often give "new definitions".

All the experimental results with the large variety of diameters, ranging from 9 mm to 76 mm, and densities, ranging from 1.2 g/cm^3 to 17.77 g/cm^3 can be very well described by the Held-equation <6>

$$I_{Cr} = u^2 d = 23 \text{ mm}^3/\text{us}^2 \qquad (1)$$

The cratering velocity is given by the Bernoulli-equation to

$$u = v/(1 + \sqrt{\rho_t/\rho_P}). \qquad (2)$$

This gives

$$I_{Cr} = u^2 d = \frac{v_{Cr}^2}{(1 + \sqrt{\rho_t/\rho_P})^2} d \qquad (3)$$

The projectile velocity v_{Cr} can now be calculated if the above equation is solved to v_{Cr}

$$v_{Cr} = \sqrt{I_{Cr}/d} \cdot (1 + \sqrt{\rho_t/\rho_P}) \qquad (4)$$

For the tungsten spheres with the density 17.77 g/cm^3, resp. Lexan sphere with the density 1.20 g/cm^3 we get the following equations for the critical velocities with respect to the diameter d

$$v_W = 1,303\sqrt{23/d} = 6.25/\sqrt{d} \quad \text{resp.} \quad (5)$$

$$v_{Le} = 2,165\sqrt{23/d} = 10.39/\sqrt{d} \quad (6)$$

With the initiation threshold value of 23 mm^3/us^2 all the experiments in <7> can be very well described resp. predicted (Fig. 2).

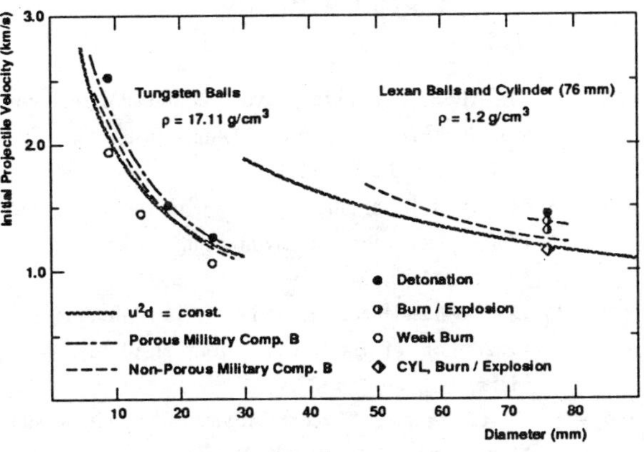

FIGURE 2. Comparison of values with $u^2d = 23$ mm^3/us^2 with the detonation threshold velocities of Fig. 1

The numerically calculated detonation threshold velocities for the tungsten balls are also very good represented. There is maybe a small difference in the trends between the theoretical prediction to the u^2d criterion for the Lexan balls, but this is marginal.

By using 23 mm^3/us^2 threshold value for a copper jet we get the equation

$$v_{Cu} = 1,428\cdot\sqrt{23/d} = 6.85/\sqrt{d} \quad (7)$$

If we give this equation in the v^2d-diagram of Reference <2> we get the strong marked line in the log-log-diagram (Fig. 3). This confirms full the Comp. B data of Chick & Hatt <5> where the high explosive charge was covered. All the tests under <7> are made under covered conditions.

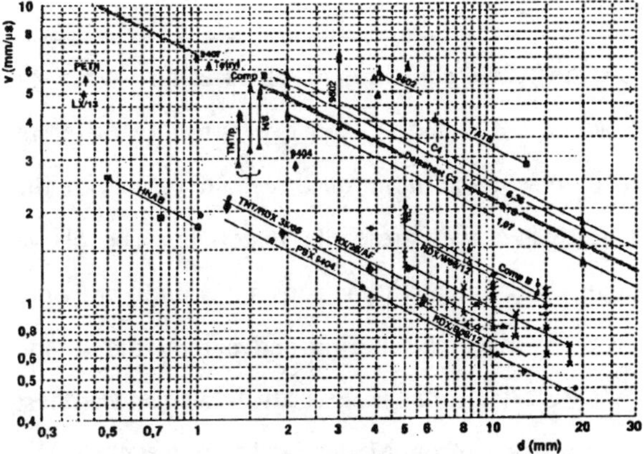

FIGURE 3. $u^2d = 23$ mm^3/us resp. $v_{Cu} = 6.85/\sqrt{d}$ for a copper projectile or jet for covered composition B

By comparing the constants for critical velocities for tungsten and copper it is very remarkable, that they are nearly equal. The value from tungsten with 6.25 is only about 10 % different to copper with 6.85.

TABLE

Initiation Criteria				$u^2 \cdot d = \dfrac{v^2}{(1 + \sqrt{\rho_t / \rho_p})^2} \cdot d$	$\sqrt{\rho} \cdot v^2 \cdot d$	$\rho \cdot v^2 \cdot d$
Projectiles				Held (1987)	Chick (1986)	Mader (1983 / 1986)
Material	ρ (g/cm^3)	d (mm)	v (km/s)	mm^3 / μs^2	$\sqrt{\text{g/cm}^3} \cdot$ mm^3/μs^2	(g/cm^3) $\cdot$ mm^3/μs^2
Lexan	1,20	76	1,19	23	118	129
Tungsten	17,77	25	1,25	23	165	694

Obviously the equations of Chick et all. <5> or Mader (<3> and <4>) give strongly different constants of the initiation threshold values for the investigated two projectile materials. The above table shows clearly that constant values are only given by the u^2d-criterion and not by the square root $\sqrt{\rho} \cdot v^2$d from Chick et all. <5> and at least not for $\rho \cdot v^2 \cdot$d from Mader (<3> and <4>).

CONCLUSION

The experimental as numerical results of the initiation treshold values of tungsten balls with different diameters and plastic balls - Lexan - with very strong different densities can be very well described with the Held-criterion u^2d where u is the cratering velocity and d the diameter of the projectile. The test results for Comp. B with 23 mm^3/us^2 are in full agreement with shaped charge jet results, if the covered test results are used.

REFERENCES

<1> M. Held, "Initiierung von Sprengstoffen, ein vielschichtiges Problem der Detonationsphysik" Explosivstoffe, 5 2-17, 1968

<2> M. Held, "Initiation Phenomena with Shaped Charge Jets", 9th Int. Symposium on Detonation, Vol. II 1416-1426, 1989

<3> C. L. Mader and G. H. Pimbley, "Jet Initiation and Penetration of Explosives", Journal of Energetic Materials 1, 3-44, 1983

<4> C. L. Mader, "Recent Advances in Numerical Modelling of Detonations", Propellants, Explosives, Pyrotechniques, 11, 163-166, 1986

<5> M. C. Chick, T. Bussell, R. B. Frey and G. Boyce, "Initiation of Munitions by Shaped Charge Jets", 9th Int. Symposium on Ballistics, 421-430, 1986

<6> M. Held, "Discussion of the Experimental Findings from the Initiation of Covered but Unconfined High Explosive Charges with Shaped Charge Jets", Propellants, Explosives, Pyrotechniques 12, 167-174, 1987

<7> M. J. Murphy, E. L. Lee, A. M. Weston and A. E. Williams, "Composition B Shock Initiation Report (U), UCRL-ID-118300, 1994

SHOCK INDUCED SUB-DETONATION CHEMICAL REACTIONS IN 1,3,5-TRIAMINO-2,4,6-TRINITROBENZENE

Henric Östmark

National Defence Research Establishment, S-17290 Stockholm, SWEDEN

The technique of combining slapper ignition with fast (12 µs/scan) time-of-flight mass spectroscopy has been used for studying the early shock-induced decomposition of TATB (1,3,5-triamino-2,4,6-trinitrobenzene). By varying the slapper energy, and hence the shock intensity, it was possible to vary the degree of reaction from a weak detonation to full detonation. As the slapper energy was lowered, several larger mass fragments were detected, e.g. m/z 242. This indicates that it is possible to detect early decomposition products using this technique. The same set of experiments was also conducted on isotopically labelled TATB (TATB-$^{15}N_6$) where the corresponding peak occurred at m/z 248. From this it was concluded that the peak resulted from a shock-induced elimination of oxygen, and hence that the early decomposition is a conversion of a nitro group to a nitroso group.

INTRODUCTION

The low temperature (100-500°C) thermal decomposition mechanisms of the most commonly used high explosives, e.g. RDX (1-5) or HMX (1,5,6) are fairly well known, or at least well examined. In contrast to this, reported experimental measurements of shock-induced decomposition mechanisms and chemical reactions of high explosives, at the higher temperatures (500-2000°C) and higher pressures (200-400 kbar) characteristic of shock and detonation states, are very rare and often incomplete. A common approach to this problem has been to use the thermal decomposition mechanism also for shock ignition modelling. This is not necessarily a valid application of the thermal decomposition mechanism. Indeed for HMX it has been shown that the solid phase decomposition rate varies greatly with pressure (7). A more fundamental knowledge of shock-induced decomposition and reaction should lead to a deeper understanding of the chemical processes occurring at shock ignition. It would also be of great value to the development of new high explosives, if the data shows a way of correlating the shock sensitivity and possibly the non-ideal behavior of a high explosive to its chemical structure and decomposition path. This knowledge must of course also be combined with an understanding of the mechanical properties of the high explosives at high deformation rates.

This paper describes how the technique and equipment developed by Blaise et al. (8), consisting of electrical gun (9) ignition ("slapper ignition") combined with a very fast time-of-flight mass spectrometer, can be used for studying early shock-induced reactions (sub-detonation chemical reactions) in TATB (1,3,5-triamino-2,4,6-trinitrobenzene) in "real time". It also proves that the technique of combining slapper ignition with mass spectroscopic detection is feasible for ignition studies (by varying the slapper energy), and shows a novel decomposition path for TATB. All experimental work for this paper was carried out at Los Alamos National Laboratory, which is gratefully acknowledged.

THERMAL DECOMPOSITION

In order to facilitate the interpretation of data acquired by the combination of mass spectroscopy with slapper ignition, it is of great importance to have some understanding of which reactions can occur in the sub-detonation/detonation zone. To this end, a literature survey was conducted. This gave as a result at least two fundamentally different decomposition paths for TATB. One is from an early, purely mass spectroscopic study (10) which is characterised by an early breakdown of the aromatic ring. The other is a thermal, UV, impact and shock decomposition path (11-14) proposed by Sharma *et al.* The latter decomposition path, which appears to be more accepted today, is characterised by elimination of one or more water molecules from TATB, forming a series of furoxanes and furazanes (14, see Fig. 1). A third decomposition path which has been proposed is the elimination of a nitro group from the aromatic ring.

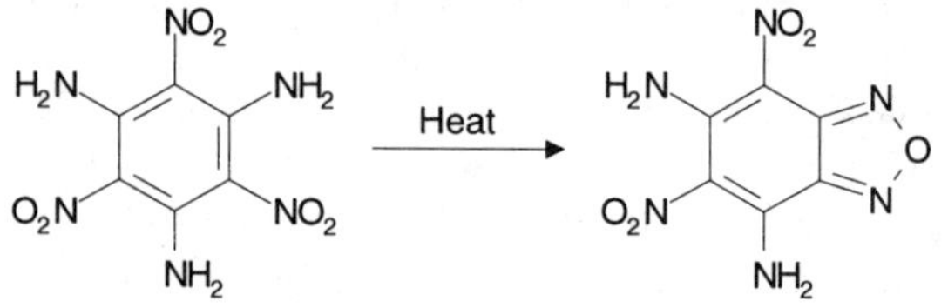

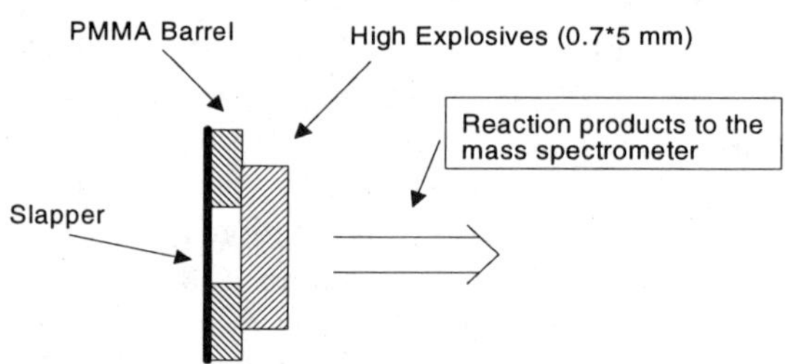

FIGURE 1. TATB thermal decomposition according to Sharma *et al.* (14).

EXPERIMENTAL

The equipment used was the one described by Blais *et al.* (8) and consisted of a combination of slapper ignition with very fast time-of-flight mass spectroscopic detection of the detonation products. The equipment allowed up to 95 scans to be acquired, each with a mass range of 1-300 *m/z*. The time for each scan was 12 μs. The explosive set-up is shown in Fig. 2. The experiments were conducted in the following way: a sample was placed in a high vacuum chamber and ignited with an electrically accelerated flyer plate (slapper). The reaction products then expanded into the vacuum and most of the reactions were thereby frozen. After expansion, the reaction products were analysed with a time-of-

flight mass spectrometer. A procedure earlier applied to thermal decomposition (4) may be adapted for use in shock ignition. For thermal decomposition, the reaction was started with a laser and the products were analysed with a quadropole mass spectrometer. It was shown to be possible to follow the transition from a weak decomposition to a self-sustained deflagration by varying the laser power. The analogous method was applied to shock-induced reactions/detonations by substituting the slapper for the laser. The reactions could then be studied by varying the shock pressure (by varying the slapper energy) and possibly the shock length (by varying the flyer thickness or the acoustical impedance), and then analysing the reaction products with the mass spectrometer. When using high shock amplitudes, the reaction should preferably be terminated before a full detonation has developed. This was achieved by using thin samples. The occurrence of the various species was then measured as a function of shock pressure and time.

The TATB samples used in this experiment were 5 mm in diameter and 0.7 mm thick (thinnest possible obtainable by conventional pressing methods).

FIGURE 2. Explosive set-up (schematical).

The main objective of these measurements was to find out whether different species can be observed at different shock pressures, and how the relative amounts of these reaction products vary with the shock pressure. This data would form the basis for trying to conclude the shock decomposition path and to determine which reactions take place in the shock ignition zone. No equipment for measuring the slapper velocity was available at the time of measurement, so any presented correlation between shock pressure and chemical reaction is qualitative only.

RESULTS

A series of experiments on TATB was carried out, where the slapper velocity was varied by varying the high voltage applied to the slapper between 4-6.5 kV, and the mass data was acquired for the m/z range 1-300 with a scan time of 12 µs. The overall reactions were studied by plotting selected masses as a function of time (SIM or single ion monitoring). Typical SIM spectra can be seen in Figs. 3 and 4.

These figures show that, as slapper energy was reduced, the time during which reaction products could be observed was greatly increased from about 75 µs to about 400 µs as the ignition threshold was approached. The increased observation time might be explained by assuming that when the shock pressure induced by the flyer is lowered, a steady detonation is not achieved. The detonation velocity was in this case estimated to about 1200 m/s. By varying the slapper energy, and hence the shock intensity, it is thus possible to vary the degree of reaction from a weak detonation to a full detonation. The very late (>700 µs) peaks in the SIM spectrum (Fig. 4) are probably from small TATB particles which have been thrown out unreacted. The spectra for those particles agree very well with a TATB spectrum as measured by conventional mass spectroscopy. This spectrum is used as reference spectrum for unreacted TATB.

Figure 5 shows the high mass region (m/z 235-265) for TATB and for a sub-detonation decomposition. Note particularly the peak with m/z 242 (=258-16) which is greatly enhanced as the threshold of a detectable reaction is approached, and which indicates that loss of a fragment with m/z 16 occurs in the early shock-induced decomposition. Two possible fragments with this mass are easily recognised: NH_2 and O.

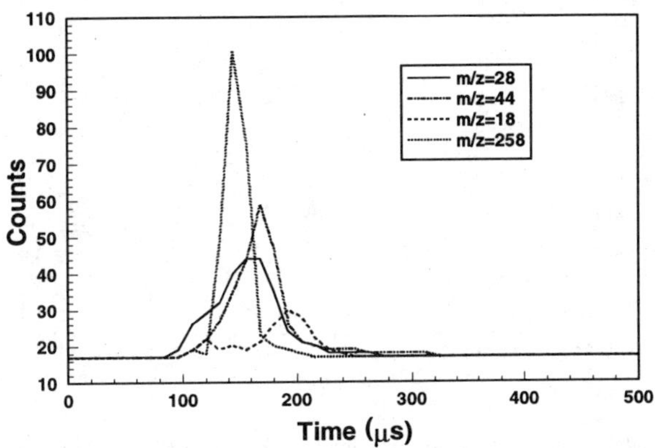

FIGURE 3. TATB SIM spectrum at high shock pressure (full detonation).

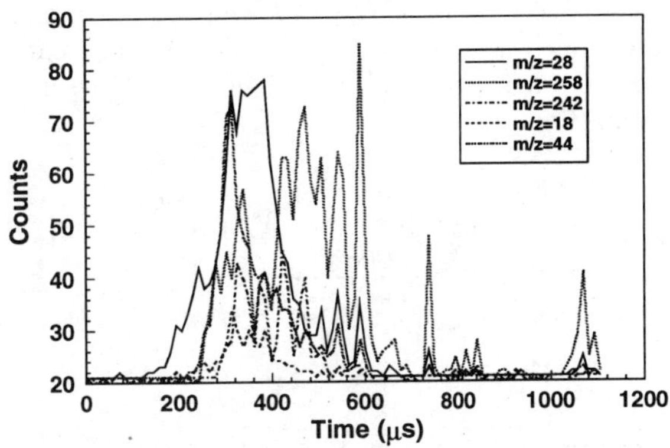

FIGURE 4. TATB SIM spectrum for near threshold ignition.

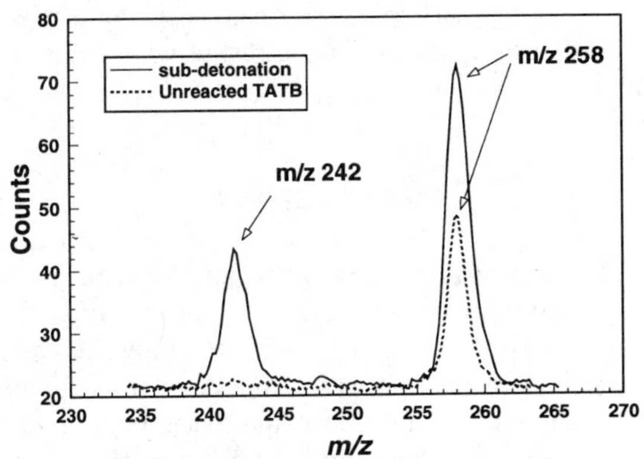

FIGURE 5. Mass spectra of shock-induced reaction in TATB (high mass region).

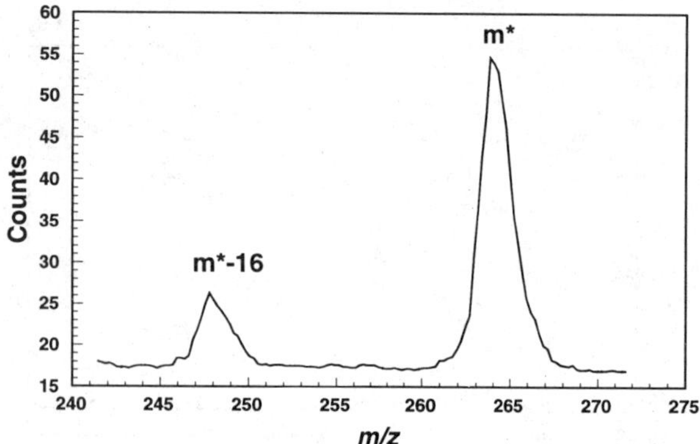

FIGURE 6. Mass spectrum for TATB-^{15}N$_6$ (high mass region) for a sub-detonation.

The same set of experiments was conducted using labelled TATB (TATB-^{15}N$_6$) showing the occurrence of a peak (Fig. 6) with the same mass difference, *m/z* 248 (=264-16). This indicates that the lost fragment is an oxygen atom from the nitro group (Fig. 7).

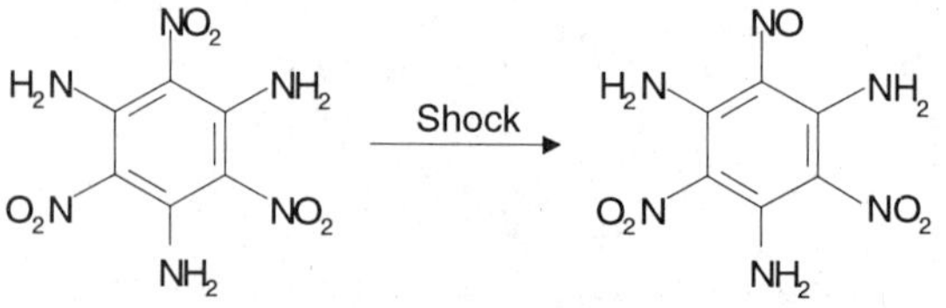

FIGURE 7. Proposed TATB shock decomposition.

This is a very interesting result in that it indicates that "non-thermal" processes can occur in shock-induced decomposition, in contrast to what has been suggested.

CONCLUSIONS

The combination of slapper ignition and mass spectroscopy, and the technique of varying the slapper energy, makes it possible to study the early shock-induced decomposition in a detonation. Using this technique on TATB, it has been proven that TATB exhibits a shock-induced decomposition path which has not been observed for thermal decomposition, i.e. the conversion of a nitro group to a nitroso group by shock-induced elimination of oxygen. During the course of these experiments some problems and inconsistencies were identified which need to be addressed in future work: A true one-dimensional experimental set-up is desirable, a method for producing thinner explosive samples should be found, and measurements of slapper velocity should be performed.

ACKNOWLEDGEMENTS

I would like to thank Roy Greiner and Herbert Fry at Los Alamos National Laboratory for allowing me to use their equipment and for all their help during my sabbatical stay at LANL. I would also like to thank Jimmie Oxley at New Mexico Tech for the nitrogen-15 labelled TATB.

REFERENCES

1. Robertson, A.J.B., *Transactions of the Faraday Society*, **45**, 85-93(1949).

2. Zhao, X., Hintsa, E.J., and Lee, Y.T., *J. Chem. Phys.*, **88,** 801-810(1988).

3. Faber, M.,, and Srivastava, R.D., *Chem. Phys. Lett.*, **64,** 307-310(1979).

4. Östmark, H., Bergman, H., and Ekvall, K., *J. Analytical and Applied Pyrolysis*, **24,** 163-178(1992).

5. Bulusu, S., *et al.*, *J. Phys. Chem.*, **90,** 4121-4126(1986).

6. Shaw, R., and Walker, F.E., *J. Phys. Chem.*, **81,** 2572-2576(1977).

7. Piermarini, G.J., Block, S., and Miller, P.J., *J. Phys. Chem.*, **91,** 3872-3878(1987).

8. Blais, N.C., Fry, H.A., Greiner, N.R, *Rev. Sci. Instr.,* **64,** 174-184(1993)

9. Chau, H.H., *et al.*, *Rev. Sci. Instr.*, **51**, 1676-1681(1980).

10. Faber, M. and Srivastava, R.D., *Combust. Flame*, **42,** 165-171(1981).

11. Sharma, J. and Owens, F.J., *Chem. Phys. Lett.*, **61,** 280-282(1979).

12. Sharma, J., *et al.*, *J. Phys. Chem.*, **86,** 1657-1661(1982).

13. Sharma, J., *et al.*, "Comparative Study of Molecular Fragmentation in Sub-Ignited TATB Caused by Impact, UV, Heat and Electron Beams", in *Shock Waves in Condensed Matter - 1983*, J.R. Asay, R.A. Graham, and G.K. Straub, Editors. 1983, Elsevier Science Publishers B.V. p. 543-546.

14. Sharma, J., *et al.*, *J. Phys. Chem.*, **91,** 5139-5144(1987).

EVOLUTION OF EXPLOSION IN TATB HE IN THE PROCESS OF ITS EXPANSION INTO A FREE SPACE FOLLOWED BY IMPACT AGAINST HARD BARRIER

I.Ye.Plaksin, V.I.Shutov, V.M.Gerasimov, V.F.Gerasimenko, V.G.Morozov, I.I.Karpenko, S.S. Sokolov

Russian Federal Nuclear Center VNIIEF, Arzamas-16, Russia.

The poorly studied behaviour of high-resistant heterogeneous TATB HE, that is its sensitivity increase during double shock wave compression with inter-load relief is being considered.

In experiments with weak shock-wave loading of thin HE samples successively followed by their relief, expansion into the gap and repeated shock loading caused by deceleration against the rigid barrier, there is recorded either initiation and evolution of explosive transformation ($\Delta > \Delta^*$) in the damaged HE or shock initiated reaction decay at $\Delta < \Delta^*$, which is dictated by Δ value (the gap between HE and the barrier).

Analysis of experimental data and results of other works allows to conclude that after the first shock and expansion into the gap have occurred, HE structure experiences irreversible changes, that is there is an increase in porosity, in TATB crystal defects. The result of rheological changes is explosion of the damaged HE at pressures of the repeated shock, which are lower than the initiation threshold of the starting HE.

For the numerical analysis of this phenomenon good use was made of the improved phenomenological model of HE decomposition kinetics, which is based on the concepts of "hot spots" evolution. The model includes the preshock HE density change effect on the shock-wave sensitivity. This model was used before to describe sensitivity decrease of HE previously subjected to shock compression.

INTRODUCTION

It is well known (see, for example, Mader, 1970 (1)) that HE shock compressed with underthreshold intensity becomes less sensitive to succeeding shock-wave loadings. HE shock desensitization is principally caused by the pore reduction and disappearance, and heterogeneity lessening (Makarov Ju., et al, 1992 (2), Plaksin I., et al, 1992 (3), Morozov V. et al, 1993 (4)).

Later experiments (5) also made it possible to reveal another no less remarkable feature of these high explosives, namely their aftershock sensitization. When HE unloads and expands into a free space after having been shock compressed, its shock-wave sensitivity rises again and becomes higher compared to that of the starting HE.

This HE feature relates to one of the poorly investigated phenomena. Effects of initiation threshold reducing at the initial phase of HE relief, which were recorded by x-ray pulse technique, are considered in the latest work of Yu.M. Makarov (6). In a special purpose works of V.G.Morozov et.al. (7,8) these processes are discussed in terms of their numerical simulation ability.

This HE phenomenon can potentially result in different abnormal environments and produce numerous effects.

In this connection there was posed the problem of analytical and numerical simulation of the damaged HE behaviour. Its solution would make predictions of a complicated HE behaviour in abnormal environments more precise and raise the safety level of HE-containing facilities.

EXPERIMENT AND RESULTS

Three runs of experiments were carried out, wherein three thin HE samples (TATB having 10% of plastic binder, (ρ_0=1.88-1.89g/cm3, theoretical maximum density 1.93g/cm3) were accelerated onto barriers using shock-wave generators. The barrieres were made of materials having different shock impedance such as magnesium, aluminum and steel. HE initiation pressure threshold was 8GPa.

Experimental schemes are shown in Figures 1-3. In Fig.1 a thin HE sample was explosion loaded through a attenuate spacer to the pressure of P=6GPa, whereupon it relieved into the air. Crushed HE fragments (granules) having traveled (1m impacted against the magnesium barrier and exploded forming craters 1 - 7 mm in diameter.

In experiments schematically shown in Fig.2 HE impact against the aluminun barrier was studied. Two cases of HE response to 5GPa shock-wave loading were compared which differ only in boundary conditions. In scheme 1 HE (Fig.2.1) or its inert simulator of polytetrafluoroethylene (PTFE, Fig.2.2) was placed between the loading generator and aluminum barrier free of gaps. In scheme 2 the aluminum barrier was placed 15 mm away from HE (Fig.2.3).

Experimental results are presented in Fig.2 and Table 1. From comparison of HE reactions to the loading device, it can be concluded that in the case of no gap between the

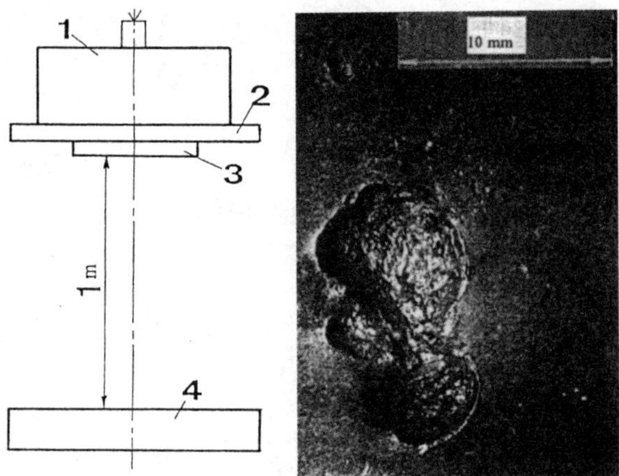

FIGURE 1. HE throwing onto magnesium barrier. Experimental setup and photograph of the craters formed in Mg-barrier as result of explosion: 1, 2 - shock-wave generator: 1 - HE-donor; 2 - HE-attenuator; 3 - HE-acceptor (TATB-composition); 4 - magnesium barrier.

Scheme of the HE-acceptor shock loading	Experimental setup view after the shot	Crater depth, H, in Al-barrier, mm	Residuel thickness of Cu-plate, Δ, mm
1. HE acceptor without gaps (Fig. 2.1)		19	6,9
2. "Cold HE" simulator without gaps (Fig. 2.2)		20	6,6
3. Experiment with HE-acceptor gap placing (Fig. 2.3)		26	≈0 break-through

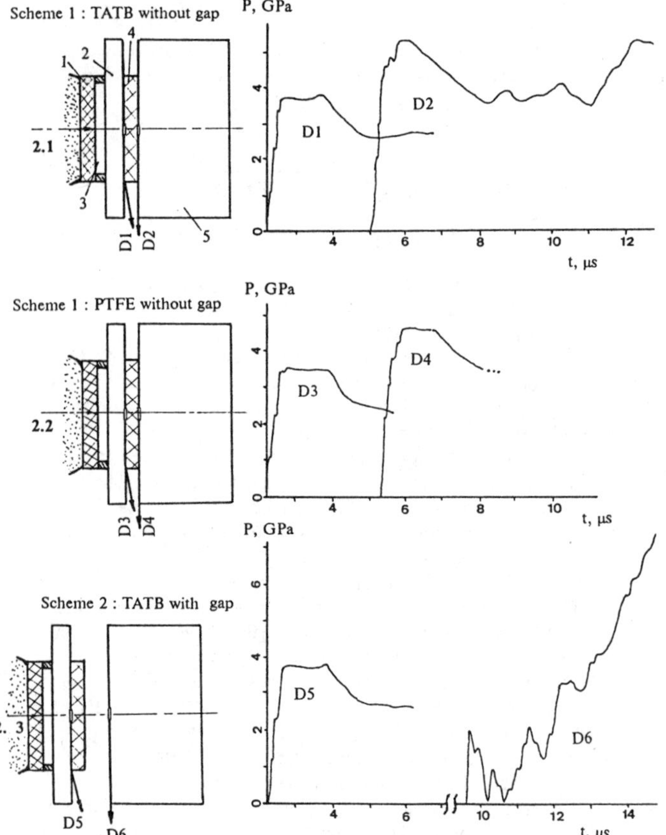

FIGURE 2. HE impact against the aluminium barrier.
1-3 - shock wave generator: 1-HE-donor; 2 - copper barrier; 3 - air gap; 4 - HE-acceptor; 5 - aluminium barrier, D1-D6 - manganin gauges.

HE and aluminun barrier its shock-wave loading results in no explosion. In this case HE effect is actually no different from "zero" response of the "cold" simulator. HE remains were crushed granules having small burning traces along the edges. But, in the second experiment with HE gap placing the sample of the same HE explodes when impacting against aluminum. As manganin gauges indicate, the rate its of energy release rises within 3 - 5 μs after its coming into collision with the barrier.

In experiments of Fig.3 the dynamics of explosive transformation during steel barrier impact of HE was recorded. Records were made of P(t) pressures and the amount of the barrier extracted energy during HE reaction.

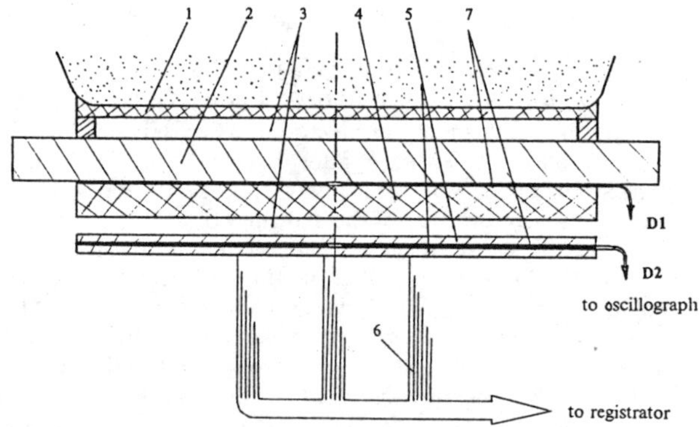

1-3-shock wave generator: 1-HE-donor; 2-copper barrier; 3-air gap; 4-HE-acceptor; 5-steel barier; 6-(x-t)-diagram registrator; 7-manganin gauges

FIGURE 3. Recording of the dynamics of explosive transformation during steel barrier deceleration of HE. Experimental setup.

The value Δ of the gap between HE and the barrier was varied. (x-t) diagram of shock waves and steel barrier (shot with gap value (Δ=7.5 mm) is shown in Fig.4. V(t) barrier velocity dependence on the time is given in Fig.5. Fig.6 shows the time dependence of pressure P(t) at the middle of the

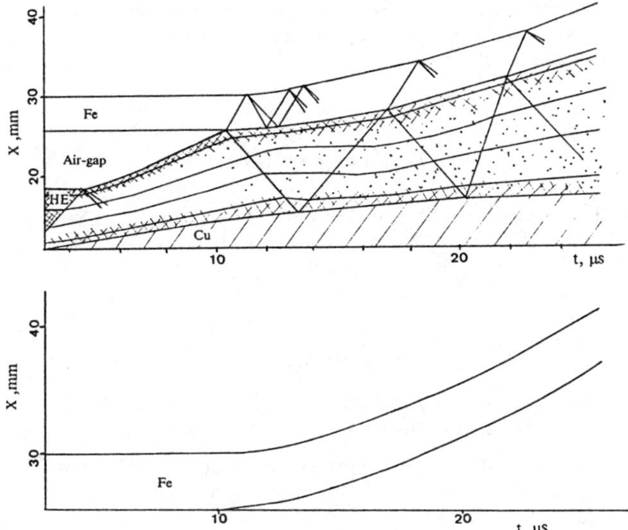

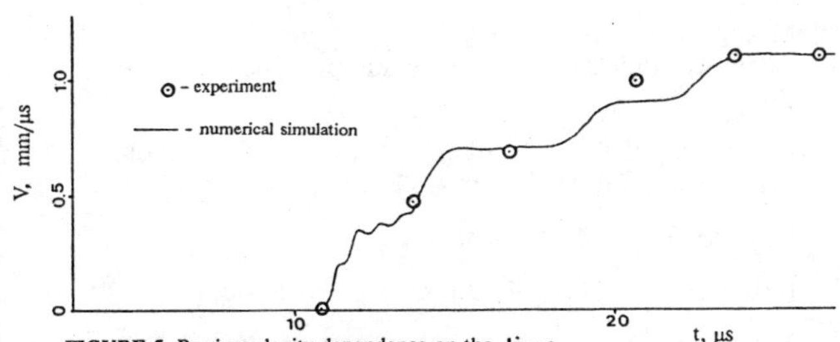

FIGURE 4. (x-t)-diagram of shock waves and steel barrier

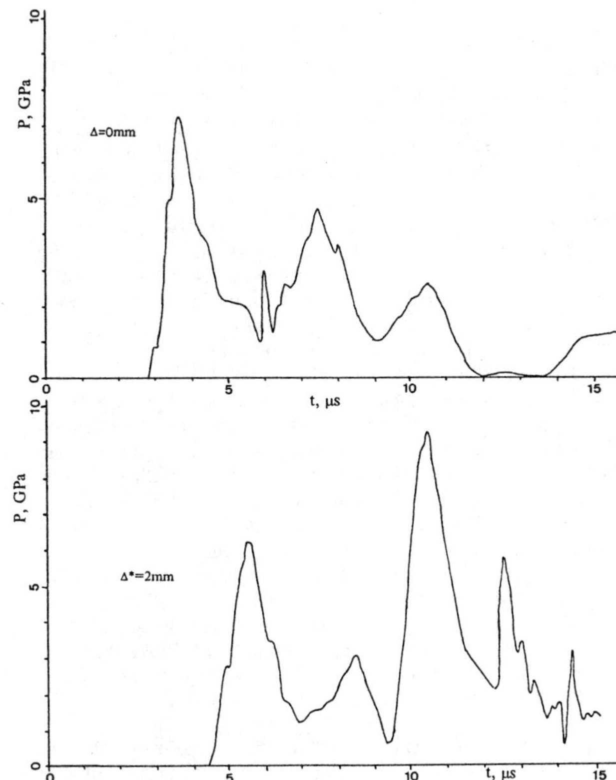

FIGURE 6. Time dependence of pressure P(t) which was recorded by manganin gauges embeded in steel barrier.

FIGURE 5. Barrier velocity dependence on the time.

barrier section which was recorded by manganin gauges built therein.

DISCUSSION AND NUMERICAL SIMULATION

In experiments with HE impacting against magnesium barrier (Fig.1) a remarkable phenomenon was detected: the diameter of exploded HE granules was less than the starting HE failure diameter. At the same time the pressure amplitude of shocked HE was 6 GPa, which is lower than HE initiation threshold (8 GPa). In its impacting against aluminum and steel the pressure amplitudes were still lower:

$$P_{Al}=3.5 \text{ GPa}, \ P_{Fe}=4 \text{ GPa}$$

Evolution of explosion initiated by pulses of such amplitude means that during relief and expansion the starting low-sensitive HE changes its structure greatly and approximates to RDX and HMX compositions in sensitivity.

From the experiments of Fig.2 and Fig.3 it was found out that the damaged HE sensitivity increased as a function of the gap between HE and aluminum or steel barriers . In Fig.7 the relative energy of the accelerated steel barrier is plotted as a function of the gap value: $E/Eid=(V/Vid)^2=f(\Delta)$ ("id" symbol applies to the case of ideal detonation of HE, when it impacts against barrier and decelerates).

For this HE type the gap value $\Delta^*=2$mm is critical. At $\Delta<2$ mm the exothermic reactions initiated in the

damaged HE die down after repeated shock. This is evidenced by pressure P(t) relations recorded by manganin gauges. The amplitude at the front of the shock wave circulating between the generator and the barrier decrease.

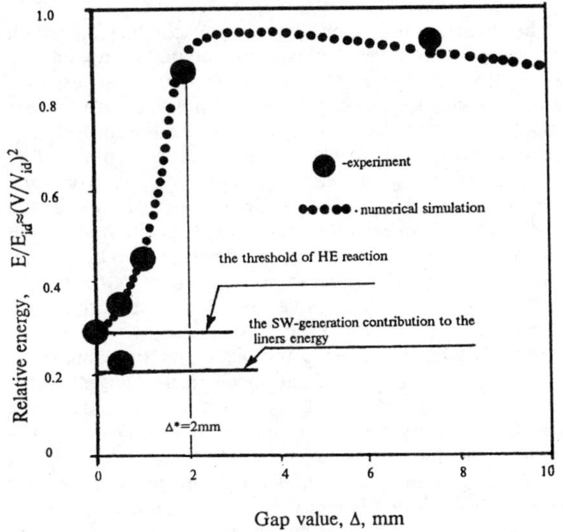

FIGURE 7. Relative energy of the accelerated steel barrier as a function of the gap value.

Both rheological factors, that is porosity of the explosive (which increases during its expansion into the gap) and TATB crystal defects arising under initial shock-wave explosive loading are decisive in changing of the explosive sensitivity to repeat impact (J.Sharma et al., 1987 (9), (Plaksin I. et al, 1995 (10)). Experimental estimation of the role played by both rheological factors, i.e. porosity and inner crystal defects, in the increase of the explosive sensitivity to repeat shock-wave loading, has been made by I.Plaksin, et.al., 1995 (10). The obtained results show that isotropic porosity (uniformly distributed microscopic pores at TATB crystal boundaries) is more important for evolution of the explosive decomposition reaction than crystal defects.

If the path lenth and the time of HE expansion ($\Delta>2$mm) rise, the increased porosity will lead the postinitiation mechanism. As a result a self-sustained explosive transformation is initiated in HE after its impacting against the barrier. The recorded P(t) relations in the shot with gap value $\Delta=2$ mm prove evolution of the damaged HE explosion. Explosive transformation of HE occurs during circulation of shock waves between the barrier and the generator plate.

In terms of concepts of "hot spots" (see Mader C.,1979 (1), Lee, E.L. and Tarver, C.M.,1980 (11)) a phenomenological model of HE decomposition kinetics has been suggested to describe the explosive behaviour. Previously this model was used to describe the sensitivity decrease of shock compressed HE (Morozov V.,1993 (4)).

The novelty included is a "hot spot" evolution analysis based on estimations and approximated solutions of the equations describing the "hot spot" reaction behaviour in terms of heat conduction and hydrodynamics considerations.

The proposed model capabilities are illustrated in Fig.7, where experimental relations are given to be compared with the numerical ones.

SUMMARY

The experiments were staged and conducted, which made it possible to reveal and describe the poorly investigated low-sensitive TATB HE composition feature: when expanding into a free space HE shock compressed with under-threshold intensity becomes much less explosion resistant. The results obtained should be taken into account while developing the models of detonation initiation and evolution.

For the numerical and analytical simulation of the complicated HE behaviour a phenomenological model of HE decomposition kinetics has been developed. The model includes the preshock HE density change effect on the shock-wave sensitivity. The model gives an adequate description of HE response to the repeated shock-wave loading.

There is a need to make a thorough structural consideration of HE, which experienced rheological changes during shock compression followed by relief, to elucidate the mechanism of initiation and evolution of explosive transformation. Estimating the presence of defects increase in TATB crystals, phase state, heat conduction, strength and viscosity changes of HE, which is repeatedly shock compressed under intermediate relief seems to be the immediate investigation aim in this area.

Analytical investigations and simulations of the sensitivity of relieved explosive involving the transition to spall fractures are to be pursued.

ACKNOWLEDGMENTS

The authors thanks Eng. Formin A.I., Mr. Bragin A.Yu. and Eng. Zimalin V.S. for their great contribution into preparation and performance of the experiments.

REFERENCE

1. Mader.C.,1979. Numerical Modeling of Detonations, University of California Press.

2. Макаров Ю.М. и др., 1991 Пятое Всесоюзное Совещание по детонации, Сборник докладов. Том 2. Красноярск, 5-12 августа 1991г.

3. Герасимов В.М., Плаксин И.Е. и др., 1992. "Химическая физика", Том 12., N5.

4. Morozov V., et.al., 1993. Combustion and Internal Chamber Processes. Nova Scientific Publisher. New York.

5. Plaksin I., et.al.,1994. Energetic Materials Technology Symposium May 18-25, 1994. Pleasanton, California.

6. Makarov Yu.,et.al., 1994. Ibid.

7. Morozov V., et.al.,1994. Ibid.

8. Morozov V., et.al., 1995. "Химическая физика", том 14, N2-3.

9. Sharma J., et.al., 1987. J.Phys.Chem. 1987, **91**, p.p.5139-5144.

10. Plaksin I., et.al.,1995. On Sensitivity of TATB-Based Shock Wave Damaged Explosive.
" Detonica " Bull. of AP 3 E. Portugal. Vol. III , N 10, 1995.

11. Lee.E.L., and Tarver,C.M., Phys. Fluids, **23**, 2362 (1980).

HUGONIOT AND INITIATION MEASUREMENTS ON TNAZ EXPLOSIVE[†]

S. A. Sheffield, R. L. Gustavsen, and R. R. Alcon

Los Alamos National Laboratory, Los Alamos, NM 87545

Particle velocity measurements have been made on samples of TNAZ (1,3,3-trinitroazetidine) explosive pressed to 98 - 99% of theoretical maximum density. Measurements were made with magnetic particle velocity gauges and a VISAR interferometer. Stirrup shaped magnetic particle velocity gauges were mounted on the front and back of the TNAZ pressing. The back gauge was located at the interface of the TNAZ and a PMMA window and was also used as the diffuse reflector for the VISAR measurement. This allowed the simultaneous measurement of particle velocity by both a magnetic gauge and a VISAR. Well defined inputs to the TNAZ, ranging from 0.6 to 2.4 GPa, were produced by gas gun projectile impact. Unreacted Hugoniot data were obtained from the front gauge measurement and shock transit times through the TNAZ. A linear shock velocity vs. particle velocity fit of $U_s = 2.38 + 2.33u_p$ mm/µs was obtained for the unreacted Hugoniot and should be accurate to at least 3.0 GPa. An elastic-plastic transmitted wave, similar to that which has been seen in other explosive materials, was observed in the 0.6 GPa input experiment. Considerable amounts of reaction were observed in experiments with inputs of 1.6 and 2.4 GPa.

INTRODUCTION

TNAZ (1,3,3-trinitroazetidine, see Fig. 1) is a relatively new explosive that has an output similar to that of HMX (octahydro-1,3,5,7-tetranitro-1,3,5,7-tetrazocine) based plastic bonded explosives, such as PBX9404, but has a relatively low melting point near 100 °C. Experiments have been done on TNAZ at both Los Alamos and LLNL to characterize the initiation behavior and the unreacted Hugoniot. Wedge experiments were completed by Hill et al. (1) of Los Alamos and Manganin pressure gauge measurements were made at LLNL (2). These studies indicate that TNAZ is slightly more sensitive to sustained shock initiation than PBX 9404.

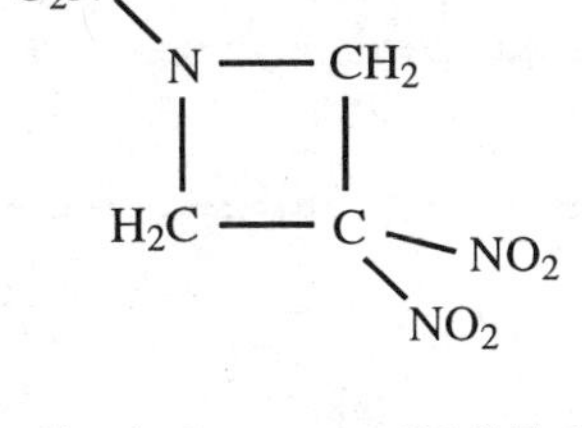

Because there was considerable variation in the unreacted Hugoniot data measured in Refs. (1) and (2), we made measurements to characterize the shock and initiation properties of this material up to 2.4 GPa. Particle velocity waveforms were recorded as a function of time at both the input and output faces of the TNAZ. This allowed us to obtain initiation profiles on the higher pressure shots at the same time we were measuring Hugoniot points.

EXPERIMENTAL DETAILS

TNAZ samples used in these experiments had densities of 1.81 - 1.82 g/cm^3, or 98 - 99% of the 1.84 g/cm^3 theoretical maximum density (TMD). Density measurements were made on each sample.

Impact experiments were performed using a gas gun to provide well controlled inputs to the TNAZ. A cross-section view of the experimental setup is shown in Fig. 2. A Lexan projectile, faced with a single-crystal z-cut sapphire impactor, strikes a target comprised of the TNAZ sample (25.4 mm diam. by 7.8 mm thick), a Kel-F confinement ring, and a PMMA back window. Stirrup shaped

FIGURE 1. Chemical structure of TNAZ, $C_3H_4N_4O_6$.

[†] Work performed under the auspices of the U.S. Dept. of Energy.

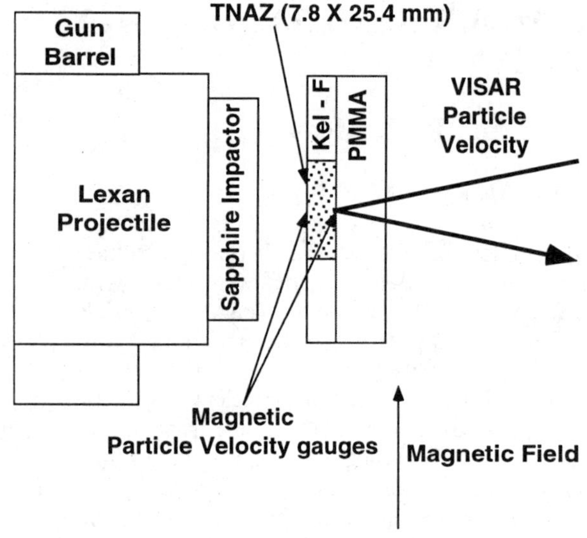

Figure 2. Experimental setup for magnetic gauge and VISAR particle velocity measurements in TNAZ.

magnetic particle velocity gauges, fabricated from 5 μm thick aluminum foil, were located on the impact surface and at the interface of the TNAZ and PMMA window. The front gauge was insulated on both sides with 12 μm of Kapton film. The back gauge was insulated on the side next to the TNAZ with 12 μm of FEP Teflon. The active element of these gauges is 10 mm long. The center of the back gauge was used as a diffuse mirror for VISAR measurements (3).

Because of the aspect ratio of the TNAZ samples, we were concerned about the length of time the 10 mm long back stirrup gauge would remain in one dimensional strain. (The Kel-F confinement ring was one method we used to help maintain 1-D strain.) The small VISAR measurement area (≈ 100 μm diam.) on the axis of the target should be in 1-D strain for ≈ 1 μs longer than the edges of the stirrup gauge. Differences in the stirrup gauge and VISAR measurements indicate that the back stirrup gauge is no longer in a state of 1-D strain.

Because the PMMA window has a lower shock impedance than TNAZ, a small rarefaction is sent back into the TNAZ when the shock wave reaches this interface. The transmitted wave profile is thus perturbed by a small amount, but is still representative of the transmitted wave.

RESULTS AND DISCUSSION

Four experiments, covering an input stress range of 0.6 to 2.4 GPa, were completed. Shot data, including unreacted Hugoniot points, are summarized in Table 1. All the waveforms from the magnetic gauges and the VISAR were successfully obtained in each experiment.

Particle velocities for the Hugoniot measurements were obtained from the front gauge record. Shock velocities were obtained by dividing the TNAZ thickness by the shock transit time. Unreacted Hugoniot data for each experiment are presented in Table 1 and are plotted in Fig 3.

In Shot 1030, with an input of 2.4 GPa, there was considerable reaction in the wave as it traveled through the TNAZ. Analysis of the back gauge record indicates that most of the reactive growth is behind the shock front. The reactive wave had not quite caught up to the shock front at the time it reached the gauge plane. This means the shock velocity should be reasonably accurate despite the reactivity. However, since the shock front has grown a little bit, a slight error in shock velocity (on the high side) would be expected.

The unreacted Hugoniot data from this study are plotted in the shock-velocity vs. particle-velocity plane in Fig. 3. The data fit a linear relationship with $U_s = 2.387 + 2.319u_p$ where U_s is the shock velocity and u_p is the particle velocity. Also shown

TABLE 1. Gas Gun Shot and Unreacted Hugoniot Data for TNAZ.

Shot No.	Impactor Material	Impact Velocity (mm/μs)	Initial TNAZ Density (g/cm³)	Particle Velocity (mm/μs)	Shock Velocity (mm/μs)	Shock Pressure (GPa)	Relative Volume (V/V₀)
1028	Sapphire	0.134	1.82	0.121	2.716†	0.60	0.9555
	(Elastic wave data)‡			0.058	2.774	0.29	0.9791
1027	Sapphire	0.247	1.81	0.221	2.86	1.14	0.9227
1029	Sapphire	0.336	1.81	0.300	3.016	1.63	0.9007
1030	Sapphire	0.454	1.81	0.397	3.363	2.41	0.8820

† Shock velocity obtained by using the time to 1/2 maximum particle velocity on back gauge waveform.

‡ Elastic wave data obtained by using the first part of the back gauge waveform to determine both U_s and u_p.

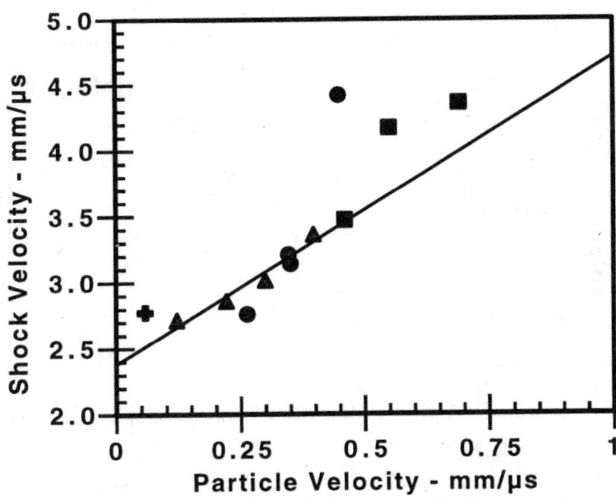

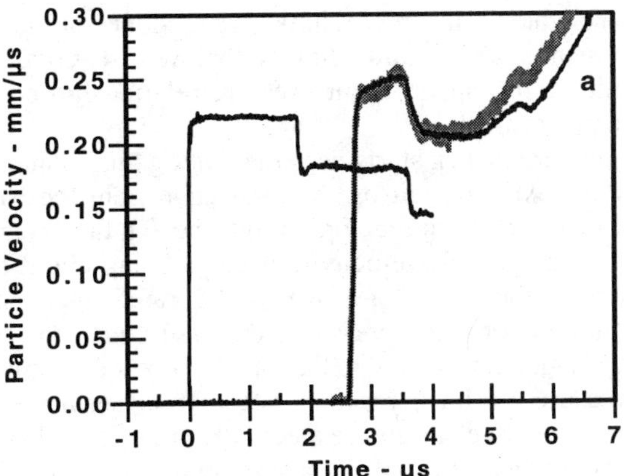

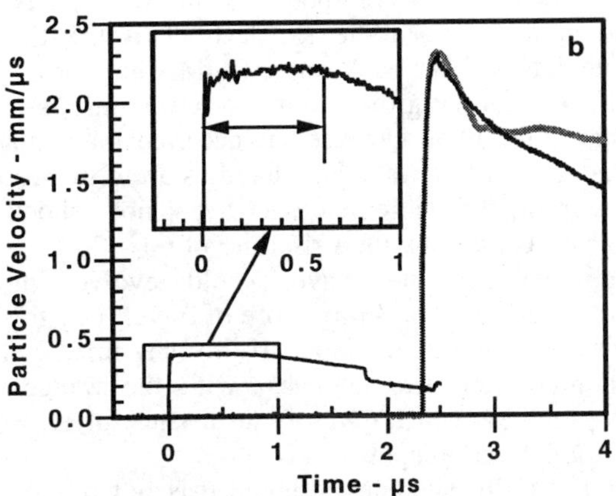

FIGURE 3. Unreacted TNAZ Hugoniot plot. Data from this study are the triangles, Hill et al's. wedge test data (1) are the squares, and data from the LLNL Manganin gauge experiments (2) are the circles. The cross is the point for the elastic wave.

are the earlier data from Los Alamos (1) and LLNL (2). Our data are in good agreement with the lowest pressure Los Alamos wedge test data point and with two of the points from LLNL. If we discard the low pressure LLNL point (because of its poor resolution and large error bars) as well as the high pressure LLNL point and the two highest pressure Los Alamos points, the fit is $U_s = 2.38 + 2.33u_p$. This should be considered a good unreacted Hugoniot for TNAZ of density 1.81 - 1.82 g/cm^3 for inputs below 3.0 Gpa. We cannot at present explain the differences between our fit and the higher pressure data, allthough it appears to be systematic. It may be due to a transition to a lower density phase, such as a liquid.

Particle velocity waveforms for two of the experiments are shown in Fig. 4a and 4b, and cover the regime from very little reaction to a great deal of reaction. Front and back particle velocity waveforms for Shot 1027, which had an input of 1.14 GPa, are shown in Fig. 4a. The front gauge shows no evidence of reaction in this shot. (The rarefaction which appears at ≈ 2 μs comes from the back of the 10-mm-thick sapphire impactor.)

The back gauge waveforms obtained from the VISAR (light line) and the magnetic gauge are essentially identical up to ≈ 4 μs, at which time the magnetic gauge record gets increasingly lower than the VISAR record. This indicates that the stirrup gauge is experiencing 2-D strain. The drop in particle velocity which occurs at a time of ≈ 3.5 μs is due to the rarefaction from the back of the sapphire

FIGURE 4. Particle velocity waveforms obtained from Shot 1027 with a 1.14 GPa input are shown in (a). Those from Shot 1030, with a 2.41 GPa input, are shown in (b). The dark line is the magnetic gauge measurement and the light line is the VISAR measurement.

impactor. The particle velocity measured by the VISAR and the back magnetic gauge were very close at early times in all four experiments. This agreement gives us confidence in the accuracy of both measurements.

No reaction was observed in the front gauge records for the shots with inputs of 0.6, 1.14, and 1.63 GPa. Shot 1030, with an input of 2.41 GPa (shown in Fig. 4b) showed evidence of reaction at the front gauge after an induction time of about 0.6 μs. This is shown by the particle velocity decrease starting at ≈ 0.6 μs and is clearly seen in the inset in Fig. 4b. The particle velocity decreases because the TNAZ is reacting, causing the pressure to increase and the impact interface (gauge plane) to decelerate. If other shots were done at input pressures above and slightly below 2.4 GPa, an induction time vs. input pressure relationship could be determined.

After reaction starts near the front gauge plane, it is not extinguished by the rarefaction from the back of the sapphire impactor. Evidence for this comes from the particle velocity/time slope being about the same before and after arrival of the rarefaction. An estimate of the pressure decrease due to the rarefaction is $\approx$ 0.6 GPa or 25% of the initial pressure. With a reaction rate that is sensitive to pressure one might expect the rate to change dramatically due to this decrease in pressure.

As mentioned earlier, for this experiment there is a shock front followed by a large reactive wave (with a particle velocity of about 2.3 mm/µs) that is just about to overtake the shock front at the time it interacts with the back gauge/PMMA interface. The shock front has grown from 0.4 to 0.9 mm/µs. That this much of an increase has occurred may indicate that the reactive wave has already started to overtake the front. The large reactive wave is not a shock but has a steep front with a risetime of 60 - 70 ns. We estimate that the wave would evolve into a detonation in 2 or 3 mm more of travel, i.e., the run distance would be about 10 to 11 mm. This estimate compares favorably with the wedge data Pop-plot (1) which gives a run distance of 12.4 mm for a 2.41 GPa input.

From this single experiment it is not possible to determine if the initiation is more homogeneous than heterogeneous in character. Because the reactive wave is very large behind the shock front, and the shock front amplitude has increased very little, we think the initiation is behaving more homogeneously than heterogeneously. Multiple embedded gauge experiments would be needed to verify this.

Shot 1029 with an input of 1.63 GPa showed no reaction at the front gauge or in the shock front as it moved through the sample. Reaction did begin after the shock interacted with the PMMA window. This is puzzling because the PMMA has a lower impedance than the TNAZ and interaction of the wave with this interface reduces the pressure. An estimate of the reaction rate at the back gauge is $\approx$ 0.4 µs^{-1}.

An interesting material response was observed in Shot 1028, the lowest input experiment at 0.6 GPa. Figure 5 shows the transmitted wave profile. The wave is composed of a shock with a sharp jump up

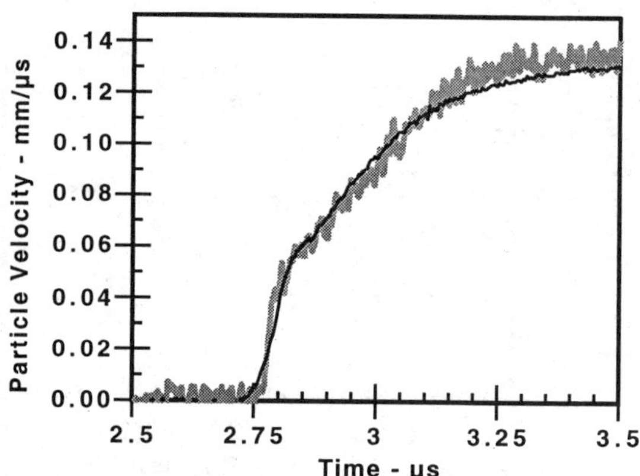

FIGURE 5. Transmitted wave profile for Shot 1028 showing elastic-plastic behavior. The dark line is the magnetic gauge measurement and the light line is the VISAR measurement.

to 0.04 mm/µs followed by a disperse wave spread out over $\approx$ 0.5 µs. We think this is elastic-plastic behavior, similar to that which has been seen in other explosives by Lemar et al. (4), Dick et al. (5), and Wasley and Walker (6). Using the initial jump, the elastic wave in TNAZ has an amplitude of about 0.29 GPa. This can be compared to the estimate of 0.14 GPa for Comp B-3 (4). The elastic wave would be overdriven by a wave with a particle velocity greater than 0.2 mm/µs.

ACKNOWLEDGMENT

We thank John Kramer of DX-16 at LANL for providing the TNAZ samples.

REFERENCES

1. Hill, L. G., Seitz, W. L., Kramer, J. F., and Murk, D. M., "Wedge Test Data for Three New Explosives: LAX-112, 2-4-DNI and TNAZ," These proceedings.
2. Lawrence Livermore National Laboratory, unpublished experimental data on TNAZ, 1995.
3. Barker, L. M. and Hollenbach, R. E., *J. Appl. Phys.* **41**, 4208–4226, (1970).
4. Lemar, E. R., Forbes, J. W., Watt, J. W., Elban, W. L., *J. Appl. Phys.* **59**, 3404-3408, (1985).
5. Dick, J. J., Forest, C. A., Ramsay, J. B., Seitz, W. L., *J. Appl. Phys.* **63**, 4881-4888, (1988)
6. Wasley, R. J., and Walker, F. E., *J. Appl. Phys.* **40**, 2639-2648, (1969).

SHOCK INITIATION OF 1,3,3-TRINITROAZETIDINE (TNAZ)

R. L. Simpson, P. A. Urtiew and C. M. Tarver

Lawrence Livermore National Laboratory,
P.O. Box 808, L-282, Livermore, CA 94551

The shock sensitivity of the pressed solid explosive 1,3,3-trinitroazetidine (TNAZ) was determined using the embedded manganin pressure gauge technique. At an initial pressure of 1.3 GPa, pressure buildup (exothermic reaction) was observed after ten μs. At 2 GPa, TNAZ reacted rapidly and transitioned to detonation in approximately 13 mm. At 3.6 GPa, detonation occurred in less than 6 mm of shock propagation. Thus, pure TNAZ is more shock sensitive than HMX-based explosives but less shock sensitive than PETN-based explosives. The shocked TNAZ exhibited little reaction directly behind the shock front, followed by an extremely rapid reaction. This reaction caused both a detonation wave and a retonation wave in the partially decomposed TNAZ. An Ignition and Growth reactive flow model for TNAZ was developed to help explain this complex initiation phenomenon.

INTRODUCTION

The four membered ring explosive 1,3,3-trinitroazetidine (TNAZ) was first synthesized by Archibald (1). This explosive is of interest because of its high energy density and its melting point of 100°C, which means that melt casting may be possible. The small scale safety properties, calorimetric heat of detonation, and thermal explosion behavior of TNAZ were reported by Simpson et al. (2). In this paper, the shock sensitivity of TNAZ was measured at three shock pressures using embedded manganin gauge techniques (3) and calculated using the Ignition and Growth reactive flow model (4) for shock initiation in the DYNA2D hydrodynamic code (5).

EXPERIMENTAL

The experimental geometry is shown in Fig. 1. A 12.7 mm thick, 80 mm diameter aluminum flyer plate impacted a target consisting of a 6 mm thick, 90 mm diameter aluminum buffer plate and a 25 mm thick, 50.8 mm diameter TNAZ charge. The TNAZ charge was held in place by a 5 mm thick Lexan ring. Six 0.3 mm thick Teflon-insulated manganin gauges were placed in pairs along the center line of the TNAZ charge at distances of 0, 6, and 12 mm. Three experiments were fired in the 100 mm powder gun with aluminum flyer velocities of 0.352 mm/μs, 0.483 mm/μs and 0.684 mm/μs, producing initial shock pressures of approximately 1.3 GPa, 2 GPa and 3.6 GPa, respectively. The measured changes in the resistance of manganin were converted to pressure histories and compared to Ignition and Growth reactive flow calculations.

REACTIVE FLOW MODELING

The Ignition and Growth reactive flow model (4) uses two Jones-Wilkins-Lee (JWL) equations of state, one for the unreacted explosive and another one

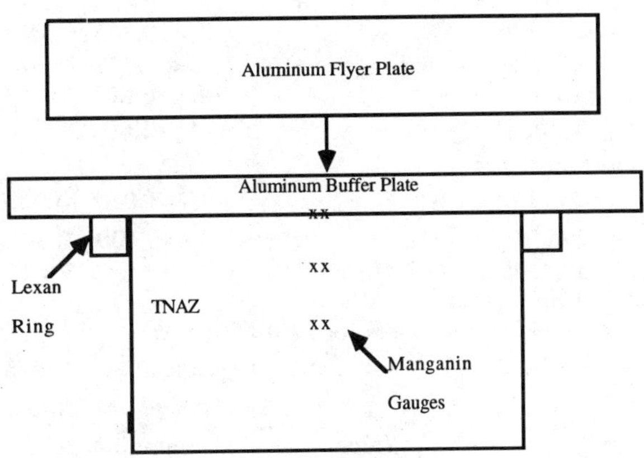

FIGURE 1. Experimental geometry for the shock initiation of TNAZ

for the reaction products, in the temperature dependent form:

$$p = A\, e^{-R_1 V} + B\, e^{-R_2 V} + \omega C_v T \qquad (1)$$

where p is pressure in Megabars, V is relative volume, T is temperature, ω is the Gruneisen coefficient, C_v is the average heat capacity, and A, B, R_1 and R_2 are constants. The equations of state are fitted to the available shock Hugoniot data. The reaction rate law is:

$$dF/dt = I(1-F)^b(\rho/\rho_0-1-a)^x + G_1(1-F)^c F^d p^y$$
$$0<F<\text{Figmax} \qquad 0<F<\text{FG1max}$$
$$+ G_2(1-F)^e F^g p^z \qquad (2)$$
$$\text{FG2min}<F<1$$

where F is the fraction reacted, t is time, ρ is the current density, ρ_0 is the initial density, p is pressure in Megabars, and I, G_1, G_2, a, b, c, d, e, g, x, y, and z are constants. This reaction rate law models the three stages of reaction generally observed during shock initiation of heterogeneous solid explosives. The equation of state parameters for TNAZ, aluminum, and Teflon, and the Ignition and Growth reaction rate law parameters are listed in Table 1.

COMPARISON OF RESULTS

Figure 2 compares the experimental pressure histories (solid lines) with those predicted by the Ignition and Growth model (dashed lines) for the TNAZ experiment with the intermediate aluminum flyer velocity of 0.483 mm/μs, which propagates a 2 GPa shock pressure into the TNAZ charge. The 0 mm gauges measured very little growth of reaction for the first 2 μs after impact, and then the pressure rose slowly to 5 GPa over the next 4 μs. An instantaneous jump in pressure then occurred at 7 μs. The 6 mm deep gauges recorded a relatively slow pressure rise to 8 GPa over the first 3 μs, followed by an instantaneous jump to over 20 GPa at 6 μs. The 12 mm deep gauges recorded a rapid increase in pressure to over 20 GPa just behind the shock front, indicating that the transition to detonation had nearly occurred at this distance into the TNAZ charge. The instantaneous pressure increases at the 0 and 6 mm gauge locations at later times than the rapid reaction at the 12 mm gauges indicate that the rapid transition to detonation at approximately 13 mm into the TNAZ charge also resulted in a rearward moving retonation wave in the partially reacted TNAZ that was recorded by all of the gauges at the 0 and 6 mm depths.

TABLE 1. Equation of State and Reaction Rate Parameters

1. Ignition and Growth Model Parameters for TNAZ

Unreacted JWL	Product JWL	Reaction Rate Parameters	
ρ_0 =1.83 g/cm^3			
A=3300 Mbar	A=10.32518 Mbar	I=4.0e+13	G$_2$=3000
B=-0.0452515 Mbar	B=0.9057014 Mbar	a=0.0	e=0.667
R$_1$=13.5	R$_1$=6.00	b=0.667	g=0.667
R$_2$=1.35	R$_2$=2.60	x=20.0	z=3.0
ω=0.8695	ω=0.57	G$_1$=230	F$_{igmax}$=0.02
C$_v$=2.7814e-5 Mbar/K	C$_v$=1.0e-5 Mbar/K	y=2.0	FG1max=0.35
T$_0$=298°K	E$_0$=0.100 Mbar	c=0.667	FG2min=0.35
Shear Modulus=0.03 Mbar		d=0.333	
Yield Strength=0.001 Mbar			

2. Gruneisen Parameters for Inert Materials

$$p = \rho_0 c^2 \mu[1+(1-\gamma_0/2)\mu-a/2\mu^2] / [1-(S_1-1)\mu-S_2\mu^2/(\mu+1)-S_3\mu^3/(\mu+1)^2]^2+(\gamma_0 + a\mu)E$$

$$\text{where } \mu=\rho/\rho_0-1 \text{ and } E \text{ is thermal energy}$$

Inert	ρ_0(g/cm^3)	c(mm/ms)	S$_1$	S$_2$	S$_3$	γ_0	a
6061-T6 Al	2.703	5.24	1.4	0.0	0.0	1.97	0.48
Lexan	1.193	1.933	2.04	0.0	0.0	0.61	0.0
Teflon	2.15	1.68	1.123	3.98	-5.8	0.59	0.0

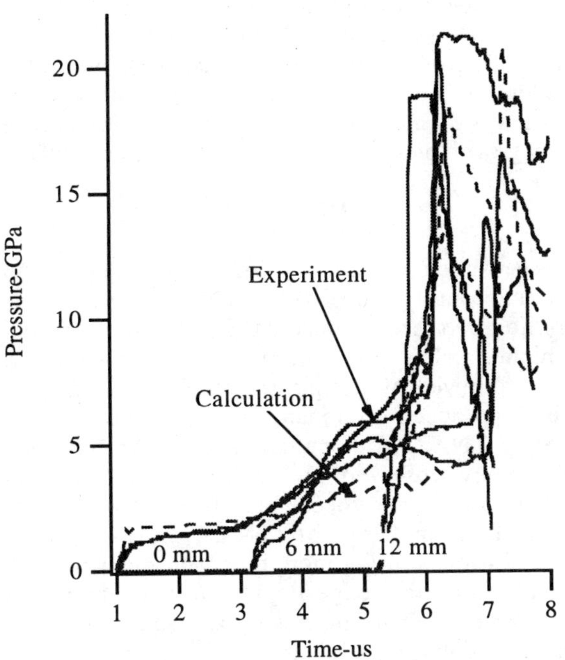

FIGURE 2. Pressure histories for TNAZ impacted at 0.483 mm/µs by an aluminum flyer plate

Other solid explosives exhibit some degree of shock front growth, followed by a continuously increasing growth of reaction behind the shock front. Detonation occurs when the resulting pressure pulse overtakes the shock front. No retonation is observed, because the explosive has been entirely reacted during the buildup process. Homogeneous liquid explosives and void free solid explosives exhibit a different shock initiation pattern, in which a rapid reaction is initiated at or near the rear boundary of the explosive charge after a time delay proportional to the initial shock pressure. This rapid reaction creates a "super" detonation wave in the precompressed material, which then overtakes the initial shock front and decays to a Chapman-Jouguet (CJ) detonation. Thus the TNAZ initiation in Fig. 2 seems to be a combination of the two limiting cases.

To account for this initiation process, the Ignition and Growth reactive flow model for TNAZ ignites less than 2% of the TNAZ near the shock front in the first rate of Eq. (2), allows a relatively slow growth of reaction to 35% reacted using the second rate in Eq. (2), and then rapidly consumes the remainder of the explosive using the third rate in Eq. (2). The calculated pressure histories agree well with the manganin records in Fig. 2, particularly with regard to the pressures and times of the transition to detonation and the arrival of the retonation wave at the 0 and 6 mm depths. The retonation velocity between the 0 and 6 mm gauges is 7.4 mm/µs.

The other two experiments measure the relatively slow growth of reaction behind the shock front and the rapid transition to detonation at high pressure. Figure 3 contains the experimental and calculated pressure histories for the lowest impact velocity experiment, 0.352 mm/µs, which yields an initial shock pressure of 1.3 GPa. After a delay of 5 µs, the pressure grows slowly over the next 5 - 6 µs, reaching 5 - 8 GPa, as in Fig. 3. The pressures then increase rapidly near the end of the records at 13 - 14 µs. The Ignition and Growth pressure histories also reach 5 - 8 GPa, corresponding to 35% reaction for these TNAZ equations of state. The calculation predicts a transition to detonation near the end of the 25 mm thick TNAZ charge, and the resulting retonation wave then reaches the gauge positions at 12 - 14 µs. Two-dimensional calculations of this experiment predict that the gauges are still in the one-dimensional flow region, but the sound velocity in such a rapidly reacting mixture is uncertain. So it

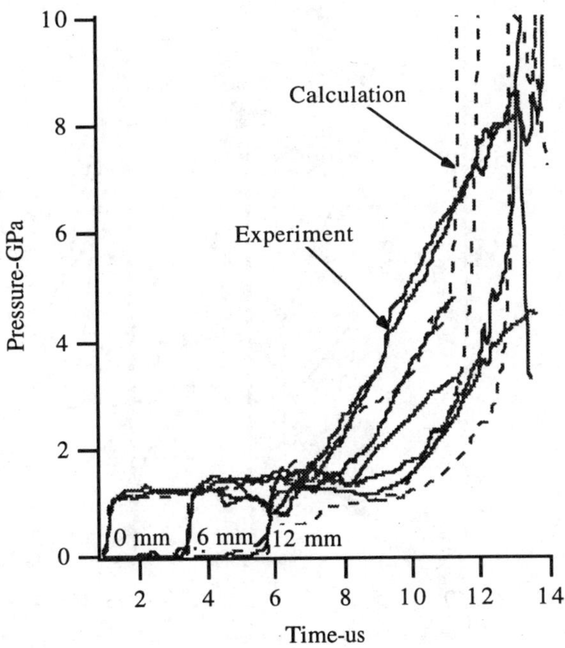

FIGURE 3. Pressure histories for TNAZ impacted at 0.352 mm/µs by an aluminum flyer plate

is not certain that the experimental pressure increases at late times in Fig. 3 measure the arrival of the retonation wave. However, the earlier portions of the gauge records and calculations do illustrate the slow growth part of the TNAZ reactive flow at 1.3 GPa. To account the large differences in TNAZ reactivity with respect to the initial shock pressure, the growth rate has a p^2 dependence in Eq. (2) and therefore decreases sharply when the rarefaction wave from the rear of the aluminum flyer plate reaches the gauge location. This results in lower pressures than the gauges in the 7 - 12 μs time frame of Fig. 3.

The experimental and calculated results for the highest aluminum flyer velocity experiment, 0.684 mm/μs, producing a 3.6 GPa shock in TNAZ, are shown in Fig. 4. After 0.6 μs, the 0 mm gauges record a rapid growth reaction to 14 GPa. The 6 and 12 mm gauges record high pressures characteristic of a detonation wave with a velocity of 8.24 mm/μs between gauge packages. Since the steady state CJ detonation velocity of TNAZ is 8.7 mm/μs, the buildup process is not complete at the 6 mm gauges, as evidenced by the lower peak pressures and the excess transit times to the 12 mm gauges. The Ignition and Growth calculations predict detonation

at 3 mm and a peak pressure of 15 GPa at the 0 mm gauge. However, calculation underestimates the initial portion of the pressure rise at 0 mm. No evidence of a retonation wave at 0 mm is observed. Thus, at 3.6 GPa, TNAZ appears to pass through the slow growth process rapidly and detonate promptly.

SUMMARY

The shock sensitivity of TNAZ has been determined using embedded manganin gauges and the Ignition and Growth reactive flow model. TNAZ exhibited a slow growth of reaction behind the shock front followed by a rapid transition to detonation, which produced a retonation wave in the partially reacted TNAZ. Retonation waves have not been previously observed in the heterogeneous solid explosives. This process may be related to time-dependent melting of TNAZ during shock compression, causing more homogeneous behavior. Shock initiation experiments at various initial densities and initial temperatures are required to better understand the shock initiation of TNAZ.

The molecular formula of TNAZ is $C_3H_4N_4O_6$, which results in an oxygen balance greater than HMX and less than PETN. This paper shows that TNAZ is also intermediate between HMX and PETN in terms of its shock sensitivity.

ACKNOWLEDGEMENTS

The TNAZ was kindly provided by Dr. Gary Parsons of Eglin Air Force Base. The authors would like to thank Frank Garcia for firing the gun shots and Leona Meegan for assembling the targets. This work was performed under the auspices of the U.S. Dept. of Energy by Lawrence Livermore National Laboratory under contract No. W-7405-ENG-48.

REFERENCES

1. Archibald, T. G., *Organic Chemistry* **55,** 2920-2924 (1990).
2. Simpson, R. L., Garza, R. G., Foltz, M. F., Ornellas, D. L., and Urtiew P. A., "Characterization of TNAZ," Lawrence Livermore National Laboratory Report UCRL-JC-ID-119672, December 1994.
3. Urtiew, P. A., Erickson, L. M., Hayes, B., and Parker, N. L., *Combustion, Explosion and Shock Waves* **22,** 597-614 (1986).
4. Tarver, C. M., Hallquist, J. O., and Erickson, L. M., in Eighth Symposium (International) on Detonation, Naval Surface Weapons Center NSWC 86-194, Albuquerque, NM, 1985, pp. 951-961.
5. Whirley, R. G., Engelmann, B. E., and Hallquist, J. O., "DYNA2D User Manual," Lawrence Livermore National Laboratory Report UCRL-MA-110630, April 1992.

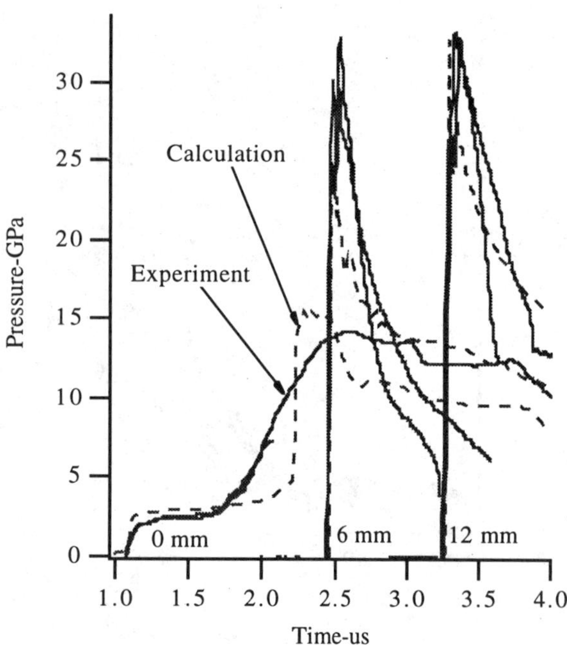

FIGURE 4. Pressure histories for TNAZ impacted at 0.684 mm/μs by an aluminum flyer plate

SHOCK INITIATION OF 2,4-DINITROIMIDAZOLE (2,4-DNI)

P. A. Urtiew, C. M. Tarver and R. L. Simpson

Lawrence Livermore National Laboratory,
P.O. Box 808, L-282, Livermore, CA 94551

The shock sensitivity of the pressed solid explosive 2,4-dinitroimidazole (2,4-DNI) was determined using the embedded manganin pressure gauge technique. At an initial shock pressure of 2 GPa, several microseconds were required before any exothermic reaction was observed. At 4 GPa, 2,4-DNI reacted more rapidly but did not transition to detonation at the 12 mm deep gauge position. At 6 GPa, detonation occurred in less than 6 mm of shock propagation. Thus, 2,4-DNI is more shock sensitive than TATB-based explosives but is considerably less shock sensitive than HMX-based explosives. An Ignition and Growth reactive flow model for 2,4-DNI based on these gauge records showed that 2,4-DNI exhibits shock initiation characteristics similar to TATB but reacts faster. The chemical structure of 2,4-DNI suggests that it may exhibit thermal decomposition reactions similar to nitroguanine and explosives with similar ring structures, such as ANTA and NTO.

INTRODUCTION

The five membered ring explosive 2,4-dinitroimidazole (2,4-DNI) was first synthesized by Lancini et al. (1) by nitrating 2-nitroimidazole. It was later obtained by the thermal rearrangement of 1,4-DNI (2). Currently 2,4-DNI is being made from 4-nitroimidazole, which is commercially available (3). This allows 2,4-DNI to be produced in large quantities in a cost effective manner. The goal of the synthesis project is to produce a relatively inexpensive explosive that has a higher energy density than TATB and TNT, yet is still insensitive. The small scale safety properties, thermal explosion behavior, and detonation velocity versus charge density of 2,4-DNI were reported by Jayasuriya et al. (3). In this paper, the shock sensitivity of 2,4-DNI was measured at three shock pressures using embedded manganin gauge techniques (4) and calculated using the Ignition and Growth reactive flow model (5) for shock initiation and detonation in the DYNA2D hydrodynamic code (6).

EXPERIMENTAL

The experimental geometry is shown in Fig. 1. A 12.7 mm thick, 60 mm diameter Lexan flyer plate impacted a target consisting of a 6 mm thick, 90 mm diameter Teflon buffer plate and a 25 mm thick,

50.8 mm diameter 2,4-DNI charge. The 2,4-DNI charge was held in place by a 3 mm thick Lexan ring. Six 0.3 mm thick Teflon-insulated manganin gauges were placed in pairs along the center line of the 2,4-DNI charge at distances of 0, 6, and 12 mm. Three experiments were fired in the 100 mm powder gun with Lexan flyer velocities of 0.979 mm/μs, 1.518 mm/μs and 2.273 mm/μs, producing initial shock pressures of approximately 2, 4, and 6 GPa, respectively. The measured changes in resistance of the manganin gauge elements were converted to

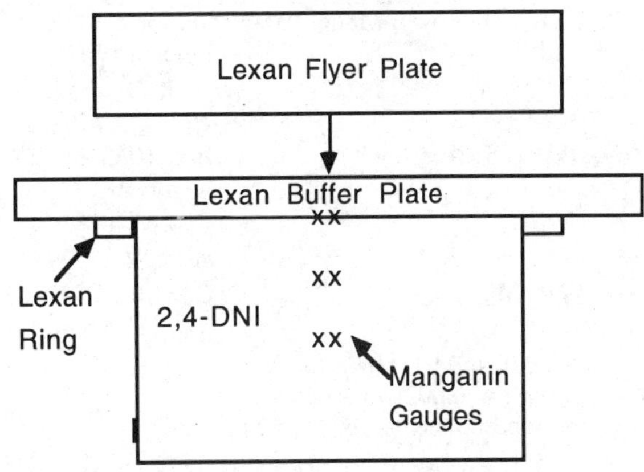

FIGURE 1. Experimental geometry for the shock initiation experiments on 2,4-DNI

pressure histories and compared to Ignition and Growth reactive flow calculations.

REACTIVE FLOW MODELING

The Ignition and Growth reactive flow model (5) has been incorporated into several hydrodynamic codes and used to solve many explosive and propellant safety and performance problems (7). The model uses two Jones-Wilkins-Lee (JWL) equations of state, one for the unreacted explosive and another one for the reaction products, in the temperature dependent form:

$$p = A\,e^{-R_1 V} + B\,e^{-R_2 V} + \omega C_v T \qquad (1)$$

where p is pressure in Megabars, V is relative volume, T is temperature, ω is the Gruneisen coefficient, C_v is the average heat capacity, and A, B, R_1 and R_2 are constants. The equations of state are fitted to the available shock Hugoniot data. The reaction rate law is:

$$dF/dt = I(1-F)^b(\rho/\rho_0 - 1 - a)^x + G_1(1-F)^c F^d p^y$$
$$0 < F < Figmax \qquad 0 < F < FG1max$$

$$+\; G_2(1-F)^e F^g p^z \qquad (2)$$
$$FG2min < F < 1$$

where F is the fraction reacted, t is time, ρ is the current density, ρ_0 is the initial density, p is pressure in Megabars, and I, G_1, G_2, a, b, c, d, e, g, x, y, and z are constants. As explained in previous papers (4), this three term reaction rate law models the three stages of reaction generally observed during shock initiation of heterogeneous solid explosives. The first term ignites some of the solid explosive as it is compressed by a shock or compression wave creating heated areas (hot spots) as the voids in the material collapse. Generally the amount of explosive ignited by a strong shock wave is approximately equal to the original void volume. The second term in Eq. (2) represents the relatively slow growth of reaction from the hot spots into the surrounding solid in a deflagration-type process. The third term in Eq. (2) describes the rapid transition to detonation observed when the growing hot spots begin to coalesce and transfer large amounts of heat to the remaining unreacted particles, causing them to react very quickly and to create a high pressure pulse which overtakes the leading shock front and thus causes detonation. The equation of state parameters for 2,4-DNI, Lexan, and Teflon, and the Ignition and Growth rate law parameters used in the reactive flow calculations are listed in Table 1. The measured pressure histories and those predicted by the Ignition and Growth model are compared in the next section.

TABLE 1. Equation of state and reaction rate parameters

1. Ignition and Growth Model Parameters for 2,4-DNI

Unreacted JWL	Product JWL	Reaction Rate Parameters	
ρ_0 =1.67 g/cm^3			
A=2700 Mbar	A=6.113 Mbar	I=2.0e+8	G_2=25
B=-0.0519165 Mbar	B=0.1065 Mbar	a=0.0	e=0.667
R_1=13.0	R_1=4.40	b=0.667	g=0.667
R_2=1.30	R_2=1.20	x=15.0	z=2.0
ω=0.9	ω=0.32	G_1=9.5	Figmax=0.3
C_v=3.0e-5 Mbar/K	C_v=1.0e-5 Mbar/K	y=1.0	FG1max=1.0
T_0=298°K	E_0=0.089 Mbar	c=0.667	FG2min=0.3
Shear Modulus=0.035Mbar		d=0.667	
Yield Strength=0.002 Mbar			

2. Gruneisen Parameters for Inert Materials

$$p = \rho_0 c^2 \mu[1+(1-\gamma_0/2)\mu - a/2\mu^2] \,/\, [1-(S_1-1)\mu - S_2\mu^2/(\mu+1) - S_3\mu^3/(\mu+1)^2]^2 + (\gamma_0 + a\mu)E$$

where $\mu = \rho/\rho_0 - 1$ and E is thermal energy

Inert	ρ_0(g/cm^3)	c(mm/ms)	S_1	S_2	S_3	γ_0	a
6061-T6 Al	2.703	5.24	1.4	0.0	0.0	1.97	0.48
Lexan	1.193	1.933	2.04	0.0	0.0	0.61	0.0
Teflon	2.15	1.68	1.123	3.98	-5.8	0.59	0.0

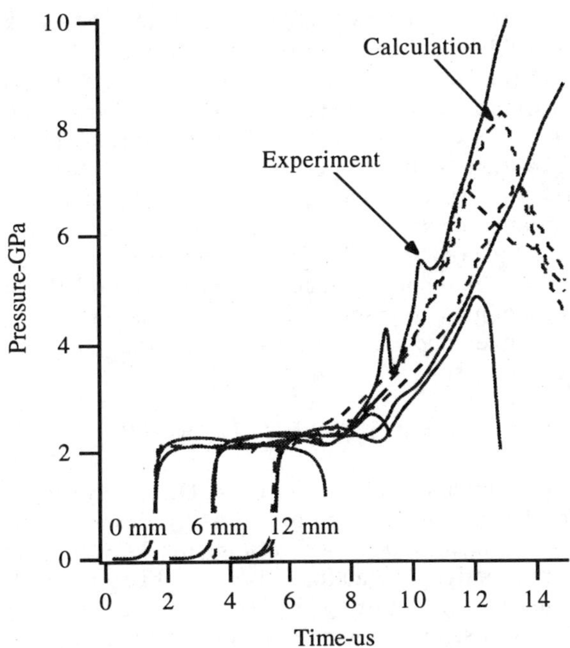

FIGURE 2. Pressure histories for 2,4-DNI impacted at 0.979 mm/µs by a Lexan flyer plate

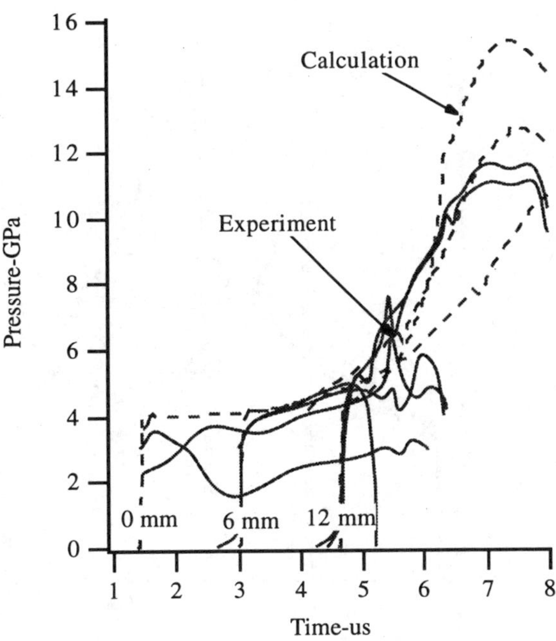

FIGURE 3. Pressure histories in 2,4-DNI impacted at 1.518 mm/µs by a Lexan flyer plate

COMPARISON OF RESULTS

Figure 2 compares the experimental pressure histories (solid lines) with those predicted by the Ignition and Growth model (dashed lines) for the lowest pressure 2,4-DNI experiment with a Lexan flyer velocity of 0.979 mm/µs, which imparts a 2 GPa shock pressure into the 2,4-DNI charge. The 0 mm gauges measured very little growth of reaction for the first 6 µs after impact, and then the pressure rose slowly to 3 GPa over the next µs before the gauges failed. The 6 mm deep gauges recorded no pressure increase for the first 3 µs, followed by a continuous growth to over 10 GPa in the next 5 µs. The 12 mm deep gauges recorded similar pressure histories. The Ignition and Growth model pressure histories exhibit similar continuous energy releases up to about 8 to 9 GPa, corresponding to approximately 50% reaction. At this shock pressure, TATB-based explosives do not react (8), and HMX-based explosives exhibit a faster growth of reaction behind the shock front than does 2,4-DNI (9).

Figure 3 shows the experimental and calculated pressure histories for 2,4-DNI impacted by Lexan at 1.518 mm/µs, creating a 4 GPa shock. At this shock amplitude, TATB-based explosives do not react, while HMX-based explosives transition to detonation at run distances of less than 10 mm (10). The gauge records in Fig. 3 clearly show that 2,4-DNI is not close to detonating at the 12 mm gauge position. Although the 0 mm gauge records are not very good, the other 4 gauges clearly show some shock front amplitude increase at the 6 and 12 mm depths, followed by a pressure growth to 11 GPa over the next 3 µs at the 12 mm gauges. The Ignition and Growth model calculates these increases accurately by allowing about 3% reaction during shock compression (ignition) and a subsequent reaction rate with a p^1 dependence.

The experimental and calculated results for the highest Lexan flyer velocity experiment, 2.273 mm/µs, producing a 6 GPa shock are shown in Fig. 4. The 0 mm gauges record a rapid growth reaction to 12 GPa over 2 µs. The 6 and 12 mm gauges record high pressures characteristic of a detonation wave. The Ignition and Growth calculations agree closely with the gauge records and predict that detonation occurs just before the reactive shock reaches the 6 mm gauge. At a shock pressure of 6 GPa, HMX-based explosives detonate in about four

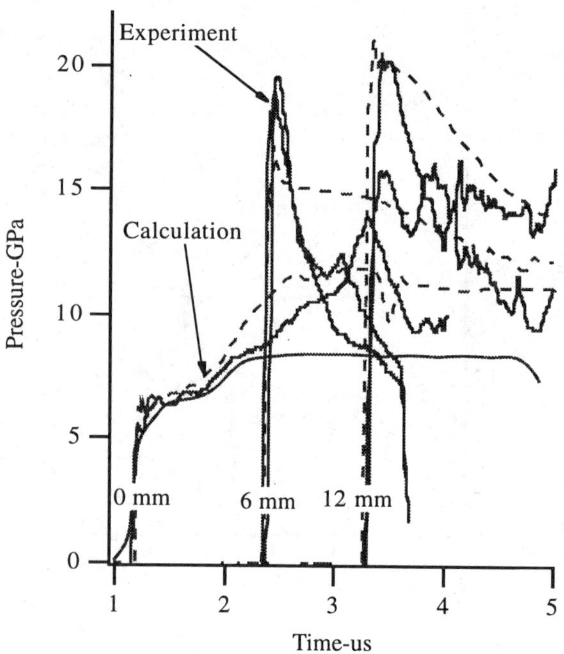

FIGURE 4. Pressure histories for 2,4-DNI impacted at 2.273 mm/μs by a Lexan flyer plate

mm, while TATB-based explosives do not decompose to any significant degree. Therefore, 2,4-DNI is much more reactive than TATB at 6 GPa, but but not as reactive as most explosive molecules.

SUMMARY

The shock sensitivity of 2,4-DNI has been determined using embedded manganin pressure gauges and the Ignition and Growth reactive flow model. Its shock initiation characteristics have been shown to be similar to TATB-based explosives in that a shock of sufficient strength to ignite a few percent of the charge then exhibits some amplitude growth as it propagates through the charge, followed by a growth of reaction that can be modeled with a linear pressure dependence. The transition to detonation then occurs in the usual fashion when the growing pulse pulse overtakes the leading shock front. Since it reacts relatively slowly at 2 and 4 GPa shock pressures, 2,4-DNI is definitely less shock sensitive than HMX. Since it reacts slowly at 2 and 4 GPa and detonates within 6 mm at 6 GPa, 2,4-DNI is more shock sensitive than TATB. However, since most hazard scenarios involve shock

pressures well below 6 GPa, 2,4-DNI may be shock insensitive enough for many applications.

Williams et al. (11) have studied the thermal decomposition of several explosives with ring structures similar to that of 2,4-DNI, such as 3-nitro-1,2,4-triazol-5-one (NTO) and 3-amino-5-nitro-1,2,4-triazole (ANTA), and found that these compounds partially decompose forming a polymer-like residue called melon. Thus 2,4-DNI should decompose in a similar manner. Since at least one of these compounds, nitroguanidine, exhibits Group 2 explosive behavior (12), 2,4-DNI may also fall into this category.

ACKNOWLEDGEMENTS

The authors would like to thank Dr. K. Jayasuriya and Dr. R. Damauarapu of the Army Research and Development Center for furnishing the 2,4-DNI used in this study. The authors also would like to thank Frank Garcia for firing the 100 mm gun shots and Leona Meegan for assembling the embedded gauge targets. This work was performed under the auspices of the U.S. Dept. of Energy by Lawrence Livermore National Laboratory under contract No. W-7405-ENG-48.

REFERENCES

1. Lancini, G. C., Maggi, N., and Sensi, P., *Farmaco (pavia), Ed. Sci.*, **18**, 390-395 (1963).
2. Sharnin, G. P., Fassakhov, R. Kh., and Orlov, P. P., USSR Patient 458553 (1975).
3. Jayasuriya, K., Damauarapu, R., Simpson, R. L., Coon, C. L., and Colburn, M. D., "2,4-Dinitroimidazole: A Practical Insensitive High Explosive," Lawrence Livermore National Laboratory Report UCRL-JC-ID-113364, March 1993.
4. Urtiew, P. A., Erickson, L. M., Hayes, B., and Parker, N. L., *Combustion, Explosion and Shock Waves* **22,** 597-614 (1986).
5. Tarver, C. M., Hallquist, J. O., and Erickson, L. M., in Eighth Symposium (International) on Detonation, Naval Surface Weapons Center NSWC 86-194, Albuquerque, NM, 1985, pp. 951 -961.
6. Whirley, R. G., Engelmann, B. E., and Hallquist, J. O., "DYNA2D User Manual," Lawrence Livermore National Laboratory Report UCRL-MA-110630, April 1992.
7. Tarver, C. M., Urtiew, P. A., Chidester, S. K., and Green, L. G., *Propellants, Explosives, Pyrotechnics* **18**, 117-127 (1993).
8. Tarver, C. M., *Propellants, Explosives, Pyrotechnics* **15**, 132-142 (1990).
9. Lee, E. L. and Tarver, C. M., *Phys. Fluids* **23,** 2362-2372 (1980).
10. Dobratz, B. M. and Crawford, P. C., "LLNL Explosives Handbook," Lawrence Livermore National Laboratory Report UCRL-52997, January 1985.
11. Williams, G. K., Palopolli S. F., and Brill, T. B., *Combustion and Flame* **98**, 197-204 (1994).
12. Price, D., Clairmont, A. R., and Erkman, J. O., *Combustion and Flame* **17,** 323-336 (1971).

SHOCK INITIATION OF AN ε-CL-20-ESTANE FORMULATION

C. M. Tarver, R. L. Simpson and P. A. Urtiew

Lawrence Livermore National Laboratory,
P.O. Box 808, L-282, Livermore, CA 94551

The shock sensitivity of a pressed solid explosive formulation, LX-19, containing 95.2% by weight epsilon phase 2,4,6,8,10,12-hexanitrohexaazaisowurtzitane (HNIW) and 4.8% Estane binder, was determined using the wedge test and embedded manganin pressure gauge techniques. This formulation was shown to be slightly more sensitive than LX-14, which contains 95.5% HMX and 4.5% Estane binder. The measured pressure histories for LX-19 were very similar to those obtained using several HMX-inert binder formulations. An Ignition and Growth reactive flow model for LX-19 was developed which differed from those for HMX-inert binder formulations only by a 25% higher hot spot growth rate.

INTRODUCTION

2,4,6,8,10,12-Hexanitrohexaazaisowurtzitane (HNIW) was synthesized by Nielsen (1) and called CL-20. CL-20 has at least 5 different crystalline structures (2). The epsilon phase of CL-20 is used with 4.8% by weight Estane binder to produce a formulation, LX-19 (formerly called RX-39-AB and RX-39-AC), that is an equivolume of binder analog of LX-14, which contains 95.5% HMX and 4.5% Estane. The shock sensitivity of LX-19 was measured at three shock pressures using the wedge test and the embedded manganin gauge techniques (3). One manganin gauge shot was fired using PBXC-19, a 95% ε-Cl-20 in an ethyl vinyl acetate (EVA) binder formulation (4). PBXC-19 has been tested by Wilson et al. (2) in the Modified Gap Test. An Ignition and Growth reactive flow model for LX-19 and PBXC-19 was developed, based on previous work on LX-14 and LX-10, which contains 94.5% HMX and 5.5% inert higher density binder Viton (5).

EXPERIMENTAL

The experimental geometry for the LX-19 wedge tests is shown in Fig. 1. A 100 mm powder gun was used to fire 19 mm thick aluminum discs into 6.35 mm thick aluminum buffer plates on which 30 mm thick LX-19 wedges were mounted. Shock transit times were measured using 30 piezoelectric transducers at precisely measured depths. The aluminum flyer velocities were 0.3335, 0.5804, and 0.8004 mm/μs, producing initial shock pressures of 1.4, 2.7 and 3.9 GPa, respectively. The initial shock velocities are used to determine states on the unreacted Hugoniot. The distances and times when the shock velocities accelerate represent detonation.

The experimental geometry for the embedded gauge experiments is shown in Fig. 2. A 12.7 mm thick, 60 mm diameter Lexan flyer plate impacted a target consisting of a 6 mm thick, 90 mm diameter Lexan buffer plate and a 25 mm thick, 50.8 mm diameter LX-19 charge. This charge was held in place by a 3 mm thick Lexan ring. Six 0.3 mm thick Teflon-

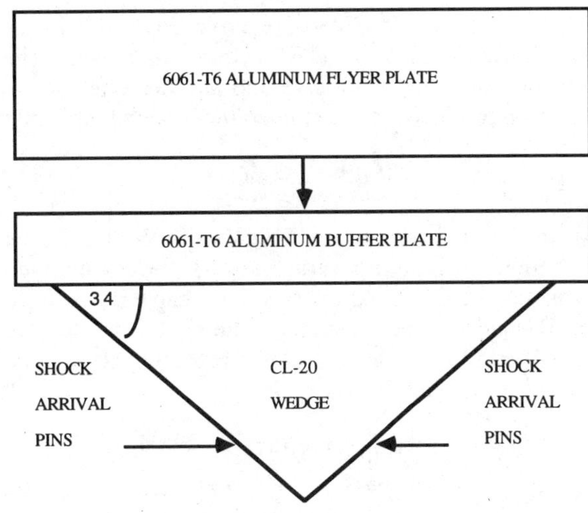

FIGURE 1. Experimental geometry for the wedge test

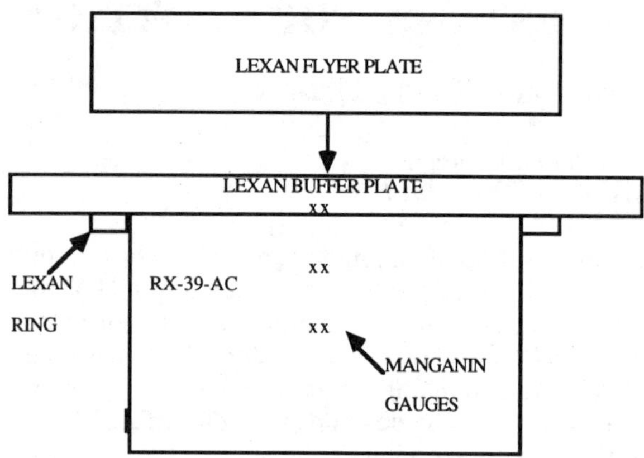

FIGURE 2. Geometry of the embedded manganin pressure gauge experiments

insulated manganin gauges were placed in pairs along the center line of the charge at distances of 0, 6, and 12 mm. Three experiments were fired in a 100 mm powder gun with Lexan flyer velocities of 0.54, 0.56, and 0.851 mm/µs, producing initial shock pressures of ~1.1, 1.3, and 2 GPa, respectively. One experiment was fired using PBXC-19 and a Lexan flyer velocity of 0.867 mm/µs, producing ~2.1 GPa. The measured pressure histories were compared to reactive flow calculations.

REACTIVE FLOW MODELING

The Ignition and Growth reactive flow model uses two Jones-Wilkins-Lee (JWL) equations of state, one for the unreacted explosive and another one for the reaction products, in the temperature dependent form:

$$p = A\, e^{-R_1 V} + B\, e^{-R_2 V} + \omega C_v T \qquad (1)$$

where p is pressure in Megabars, V is relative volume, T is temperature, ω is the Gruneisen coefficient, C_v is the average heat capacity, and A, B, R_1 and R_2 are constants. The equations of state are fitted to the available shock Hugoniot data. The reaction rate law is:

$$dF/dt = I(1-F)^b(\rho/\rho_0-1-a)^x + G_1(1-F)^c F^d p^y$$
$$0<F<Figmax \qquad 0<F<FG1max$$

$$+\, G_2(1-F)^e F^g p^z \qquad (2)$$
$$FG2min<F<1$$

where F is the fraction reacted, t is time, ρ is the

current density, ρ_0 is the initial density, p is pressure in Mbars, and I, G_1, G_2, a, b, c, d, e, g, x, y, and z are constants. This three term reaction rate law models the three stages of reaction generally observed during shock initiation of pressed solid explosives (4). The equation of state parameters for LX-19, aluminum, Lexan, and Teflon, and the Ignition and Growth rate law parameters are listed in Table 1. The reaction rates are the same as those used for LX-10 (5), except that the hot spot growth rate parameter G_1 in Eq. (2) has been increased from 120 to 150.

COMPARISON OF RESULTS

Table 2 contains the experimental and calculated run distances and times to detonation for the three LX-19 wedge tests. The transition to detonation was very rapid in both the experiments and calculations. The Ignition and Growth unreacted Hugoniot yielded excellent agreement with the three measured shock velocities preceding the transitions to detonation. The reaction rate law with G_1=150 yielded excellent run distances and times to detonation for LX-19.

A more difficult test of a reactive flow model is the calculation of measured pressure histories during a shock initiation experiment. Figure 3 shows the

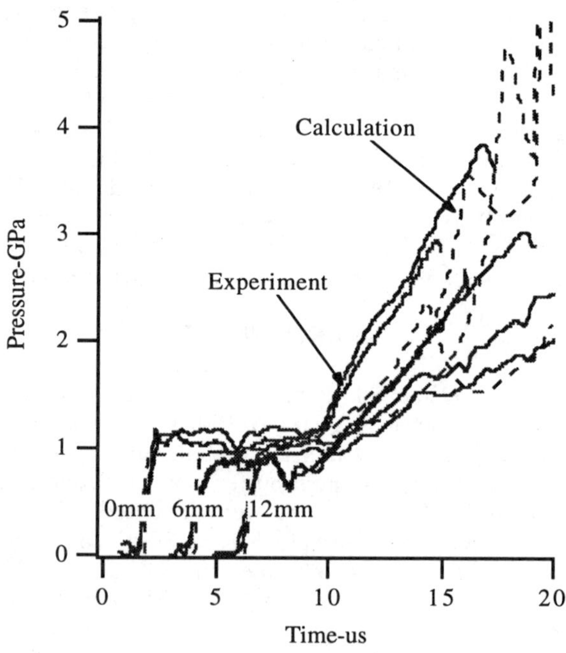

FIGURE 3. Pressure histories for LX-19 impacted at 0.54 mm/µs by a Lexan flyer plate

TABLE 1. Equation of State and Reaction Rate Parameters

1. Ignition and Growth Model Parameters for LX-19

Unreacted JWL	Product JWL	Reaction Rate Parameters	
ρ_0 =1.942 g/cm^3			
A=4444 Mbar	A=16.379 Mbar	I=7.43e+11	G_2=400
B=-0.0513014Mbar	B=1.8629 Mbar	a=0.0	e=0.333
R_1=13.5	R_1=6.50	b=0.667	g=1.0
R_2=1.35	R_2=2.70	x=20.0	z=2.0
ω=0.8695	ω=0.55	G_1=150	F_{igmax}=0.3
C_v=2.7814e-5 Mbar/K	C_v=1.0e-5 Mbar/K	y=2.0	F_{G1max}=0.5
T_0=298°K	E_0=0.115 Mbar	c=0.667	F_{G2min}=0.5
Shear Modulus=0.0454 Mbar		d=0.333	
Yield Strength=0.002 Mbar			

2. Gruneisen Parameters for Inert Materials

$$p = \rho_0 c^2\mu[1+(1-\gamma_0/2)\mu-a/2\mu^2] / [1-(S_1-1)\mu-S_2\mu^2/(\mu+1)-S_3\mu^3/(\mu+1)^2]^2+(\gamma_0 + a\mu)E$$

where $\mu=\rho/\rho_0-1$ and E is thermal energy

Inert	ρ_0(g/cm^3)	c(mm/μs)	S_1	S_2	S_3	γ_0	a
6061-T6 Al	2.703	5.24	1.4	0.0	0.0	1.97	0.48
Lexan	1.193	1.933	2.04	0.0	0.0	0.61	0.0
Teflon	2.15	1.68	1.123	3.98	-5.8	0.59	0.0

TABLE 2. Experimental and calculated wedge test results

Flyer Velocity (mm/μs)	Pressure (GPa)	Experimental results		Calculated results	
		Distance(mm)	Time(μs)	Distance(mm)	Time(μs)
0.3335	1.437	>29	9.4	>29	9.4
0.5804	2.663	13.17	3.38	13.3	3.4
0.8004	3.924	8.89	2.47	8.8	2.4

comparison between the measured and calculated pressure histories for the lowest impact velocity on LX-19, 0.54 mm/μs, which produces a ~1.1 GPa shock. The gauges and calculations both show a slow growth of pressure for 6 to 8 μs, followed by a rapid increase over the next 5 to 7 μs. This growth pattern is nearly identical to that observed for LX-10 and LX-14 (5). Figure 4 contains records for a Lexan flyer velocity of 0.56 mm/μs, which results in a shock of ~1.3 GPa. The measured pressures at the two 0 mm gauges may be in error, because they increase sooner than those of the other four gauges and the six gauges in Fig. 3. The calculated pressures show fast reaction to 15 GPa at 15 to 18 μs, at which time the 6 mm and 12 mm deep gauges may be stretched by the radical rarefaction wave.

At higher shock pressures, the gauge records and calculated pressure histories agree more closely. Figure 5 shows the pressure histories resulting from a Lexan flyer velocity of 0.851 mm/μs, which shocks LX-19 to ~2 GPa. The agreement is excellent, even during the rapid pressure increase

from 7 to 20 GPa in 2 μs. The LX-19 pressures reach 20 GPa at about 11 μs, whereas the pressures in a similar experiment on LX-14 reached 15 GPa in 12 μs with similar rates of pressure increase. Figure 6 contains the pressure histories for an experiment using PBXC-19 impacted by a Lexan flyer at 0.867 mm/μs, imparting ~2.1 GPa to the explosive. The Ignition and Growth model parameters were not changed, except for the initial density of PBXC-19, which is 1.896 g/cm^3. The calculated pressure histories are very close to the experimental records at this shock pressure, which causes rapid reaction and transition to detonation just beyond the last gauge.

SUMMARY

The shock sensitivity of two ε-CL-20 formulations has been quantitatively demonstrated to be slightly greater than that of similar formulations based on HMX. The Ignition and Growth reaction rates for ε–CL-20 formulations are the same as those

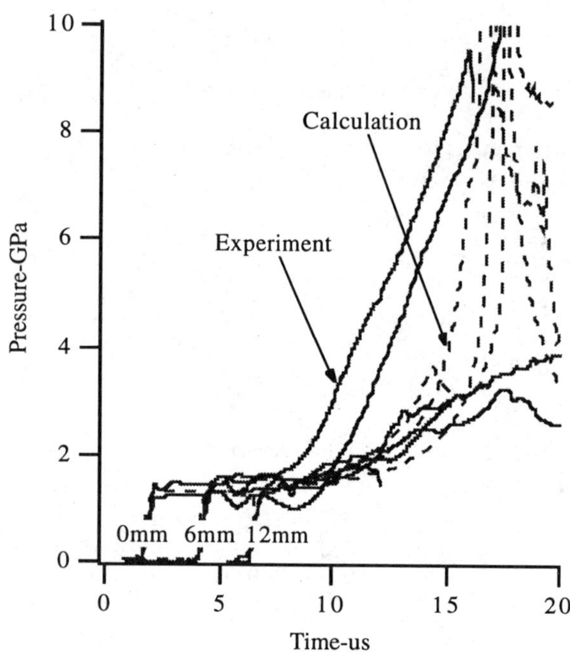

FIGURE 4. Pressure histories for LX-19 impacted at 0.56 mm/μs by a Lexan flyer plate

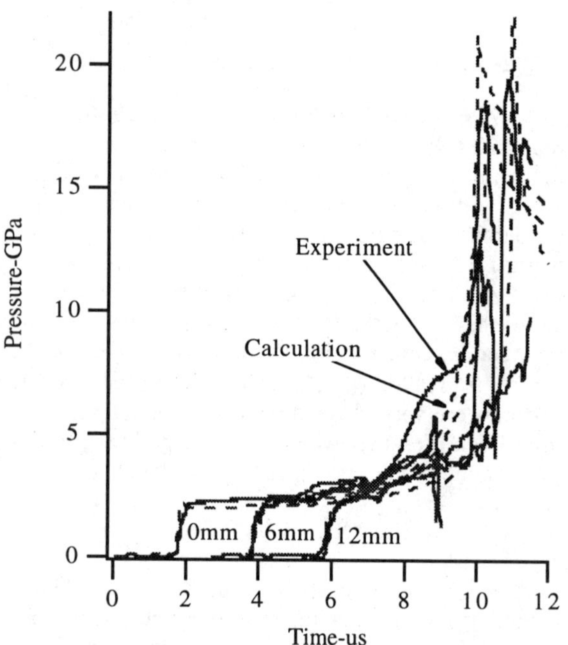

FIGURE 5. Pressure histories for LX-19 impacted at 0.85 mm/μs by a Lexan flyer plate

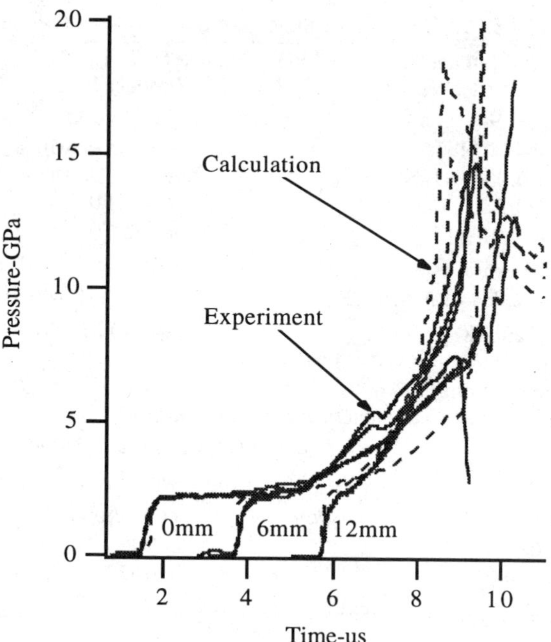

FIGURE 6. Pressure histories for PBXC-19 impacted at 0.867 mm/μs by a Lexan flyer plate

previously developed for similar HMX-based formulations, except for a 25% increase in the hot spot growth coefficient G_1 from 120 to 150, and accurately calculate the measured pressure histories.

ACKNOWLEDGEMENTS

The authors would like to thank Frank Garcia for firing the 100 mm gun shots and Leona Meegan for assembling the embedded gauge targets. This work was performed under the auspices of the U.S. Dept. of Energy by Lawrence Livermore National Laboratory under contract No. W-7405-ENG-48.

REFERENCES

1. Nielsen, A., "Polycyclic Amine Chemistry," in *Chemistry of Energetic Materials,* Olah, G. A. and Squires, D. R., eds., Academic Press, San Diego, 1991, pp. 95 -124.
2. Wilson, W. H., Forbes, J. W., Liddiard, T. P., and Doherty, R. M., in *High -Pressure Science and Technology-1993,* Schmidt, S. C., Shaner, J. W., Samara, G. A., and Ross, M., eds., AIP Press, New York, 1993, pp. 1401-1404.
3. Urtiew, P. A., Erickson, L. M., Hayes, B., and Parker, N. L., *Combustion, Explosion and Shock Waves* 22, 597-614 (1986).
4. Bui, Q. T., Naval Air Warfare Center, China Lake, formulated and kindly provided the PBXC-19.
5. Tarver, C. M., Urtiew, P. A., Chidester, S. K., and Green, L. G., *Propellants, Explosives, Pyrotechnics* 18, 117-127 (1993).

CHAPTER XI

ELECTRICAL, MAGNETIC, AND OPTICAL STUDIES

INVESTIGATION OF RUBY LUMINESCENCE R-LINE BY ISENTROPIC MEGABAR PRESSURE

A.I.Pavlovskii, O.M.Tatsenko, V.V.Platonov, A.A.Volkov, I.M. Markevtsev

Russian Federal Nuclear Center VNIIEF, Arzamas-16, Russia

This work is devoted to investigation of ruby R-line luminescence for pressure measurements under isentropic material compression.
Compression was carried out by ultrahigh magnetic field. Influence of non-hydrostatic compression on ruby luminescence spectrum was studied. Paper describes experimental set-up and experimental results of ruby R-line shift up to 1 Mbar.
The problem of background radiation, bounding the upper limit of recording, is described.

Over the last years the method of isentropic material compression by MC-generator ultrahigh magnetic field pressure has been developed [1]. One of the problems arising from analysis of material equation of state in megabar pressure range is the lack of reliable method for pulsed pressure measurement. To determine the static pressure, produced in diamond anvils, the shift of ruby luminescence R-line length is used [2]. The works analyzing the ruby optical spectrum by shock- wave compression became available [3]. The present work considers the possible use of ruby pickup to measure the pulsed pressure, generated by ultrahigh magnetic field in MC-1 generator. Furthermore, spectroscopic studies carried out in conditions of isentropic compression are of interest to understand the effect of dynamic compression on material properties at the atomic- molecular level.

In reference [4] the spectrum of ruby luminescence upon its compression in MC-1 generator was investigated. It was found that R-line luminescence vanishes and the absorption R-line appears against the background of arising glow. To judge by estimation results, its shift is detected up to ~ 200 GPa pressures, (Figure 1).

However, instability of absorption R-line appearance with every experiment caused one to give consideration to more profound investigation of ruby luminescence R-line. Apparatus for material compression by magnetic field pressure comprises MC-generator and compression chamber - a copper tube, screening the compressed material with ruby pickup inside from magnetic field.

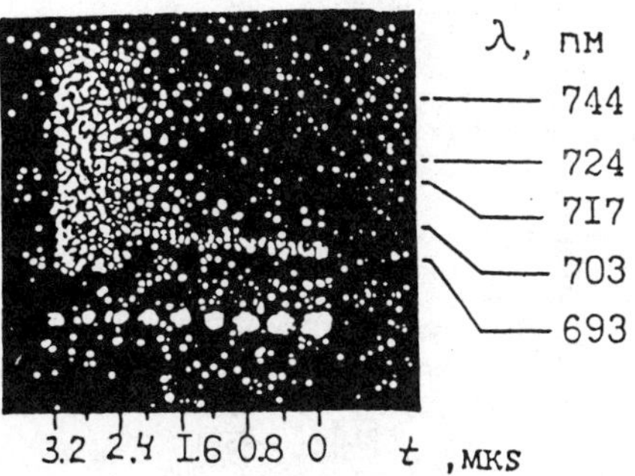

FIGURE 1. R-line luminescence and absorbtion of Ruby.

Owing to this, satisfactory (up to 1%) coincidence of pressure values in ruby pickup and material around it is possible with materials whose density is $\rho < 1.0 g/cm^3$. Therefore aluminum as a compressed material [4] was changed by a softer polyethylene material. Within polyethylene a sapphire-ruby-sapphire core ($\varnothing$=3 mm, l=110 mm) grown by united process is placed. The ruby sample was 40 mm long and Cr^{3+} concentration ranged from 0.03 to 0.05 %. Sapphire cores were used to introduce the ruby laser pump radiation and remove the ruby luminescence through the boundary regions of nonuniform pressure. Fig.1 presents the detection scheme of ruby luminescence optical spectrum in experiment with MC-1 generator.

To generate luminescence the ruby laser was used. Since the time of luminescence is $\tau \sim 3 \times 10^{-3}$ μs and the process duration is $\sim 10^{-5}$ μs, ruby luminescence may be considered as the continuous source of radiation, transmitted through the guide and dividing prism on two photomultipliers and into spectrograph slit with 70 A/mm dispersion. To prevent exposure of PEM (photoelectric multiplier) and EOT (electrooptical transducer) photocathodes during ruby laser operation electromechanical shutter was used. The spectrum in 690-810 nm range was registered by electron- optical camera. Electron-optical camera operated in the mode of photochronographic recording. The scanning factor was 100 ns/mm.

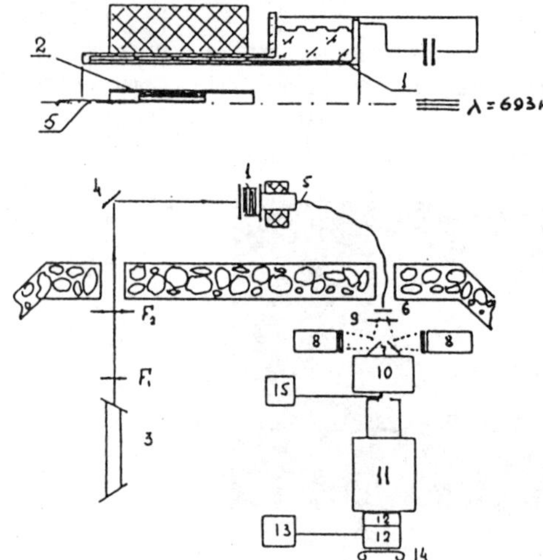

FIGURE 2. Optical spectrum measuring scheme 1 - generator MC -1; 2 - camera of compression; 3 - laser λ=694.3 nm; 4 - mirror; 5 - fiberoptic; 6 - filters; 7 - mirror; 8 - photo receiver; 9 - lens; 10 - spectrograph; 11,12,13 - streak camera; 14 - photo film.

The total transformation coefficient of measuring path, the time, space and spectral resolutions were ~10^{+6}, ~8×10^{-8} s, ~ 10lp/mm, ~ 3 nm respectively that didn't permit ruby lines R_1, R_2 resolution. Calibration of luminescence R-line shift was performed by helium-neon tube spectrum. While the procedure was being developed, direct relation between the level of ruby pump and duration of the recorded line shift was observed. As a result the ruby was pumped to saturation with luminescence output on the guide end ~0.5W. Fig.3 presents a photograph of luminescence R-line shift during ruby compression.

It is seen that as the pressure grows R-line shift into the long-wave region. Profile densitometry of R-lines on photographic film from EOT at different compression moments showed the intensity drop of R-lines and their merging with background illumination. Decrease of luminescence amplitude

may be attributed to the loss of sapphire windows transparence, amorphicity of crystalline structure, thermal heating of ruby. Though the ruby subjected to isentropic compression gets heat far less than under shock compression [5] (Fig.4).

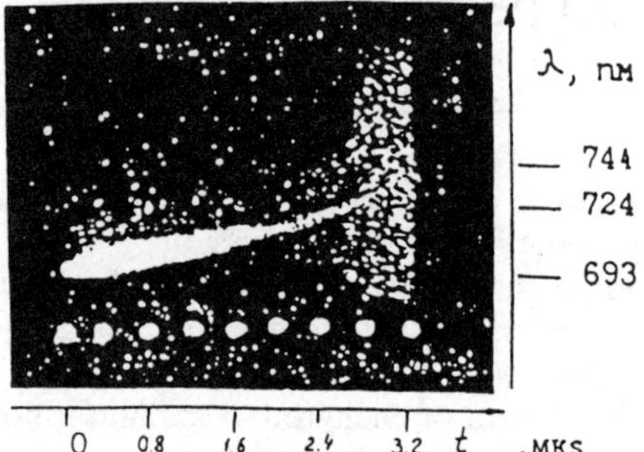

FIGURE 3. Photograph of ruby luminescence R-lines shift. Calibration spectrum of helium-neon tube is applied next to.

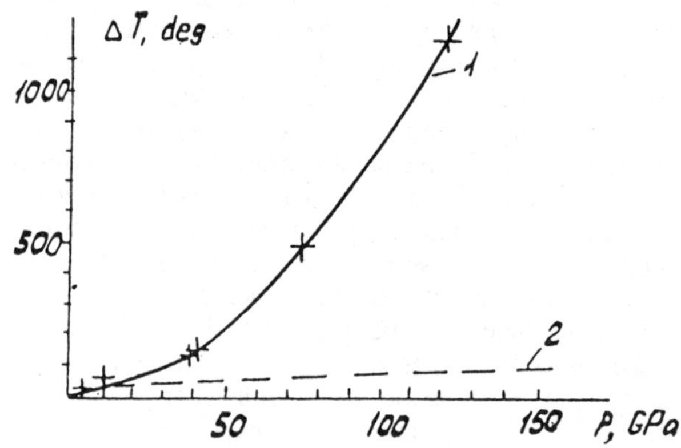

FIGURE 4. Dependence of sapphire heating on pressure upon shock compression
(curve 1) and by isentropic compression (curve 2).

Therefore R-lines may be observed up to p~70GPa when shock compression occurs and in case of isentropic compression they are seen at P up to 500GPa. Another factor presenting from recording R-lines shift in high pressures region is the occurrence of background illumination most frequently of a band-type structure that did not depend on pressure value. In Fig.4 the photograph of sapphire's own glow under compression in MC-1 generator is presented. The glow character like this may be put down to occurrence of shift zones by plastic flow.

The spectrum obtained is of a definite interest by itself because the glow has a structure typical of a line spectrum, and the power of some bands comes up to about 1 Watt that is considerably different from experimental results produced by shock compression. Treatment of experimental results showed that continuous character of R lines is observed prior to shift $\Delta\lambda=302$ A. The subsequent recording of shift up to $\Delta\lambda=490$ A is indicated by station density of individual points. Masmuch as R_1 and R_2 lines of luminescence shift with pressure by different functional dependencies the spatial and spectral resolvability of apparatus should be increased upon subsequent improvement of recording procedure, that will make possible to lower the estimation error of $\Delta\lambda=\pm15$ A.

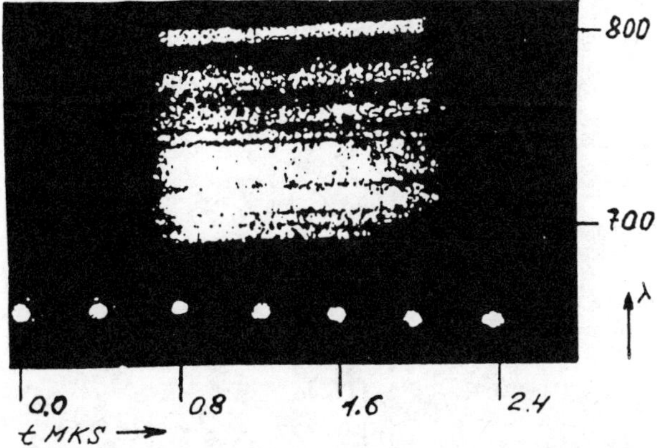

FIGURE 5. Photograph of sapphire core's deformation glow.

To estimate the pressure value by $\Delta\lambda=f(t)$ shift one can make use of hydrostatic relation [2], since above Hugoniot limit the pressure components are equalized, the plastic flow of crystal initiates and all-sided compression approximates to hydrostatic one. The difference from hydrostatics consists in slight ruby heating, Fig.4. Then the total shift of R-line may be represented by $\Delta\lambda=\Delta\lambda_P+\Delta\lambda_T$, where $\Delta\lambda_P$-hydrostatic relation [2] and $\Delta\lambda_T=0.065\Delta T$ ($\Delta T=T-T_o$). Having derived the calculated values of ΔT from Fig.3, one can present $\Delta\lambda=f(P)$ relation as it is shown in Fig.6.

The true pressure value may be resulted only from calibration.

Thus there is a real possibility to measure pressure up to 90-150 GPa by ruby R-line luminescence shift under isentropic compression of materials with $\rho \le 1.0$ g/cm^3. Stability of results obtained points to the fact that preference should be given to measurement by ruby luminescence R-lines, but not by absorption R-line.

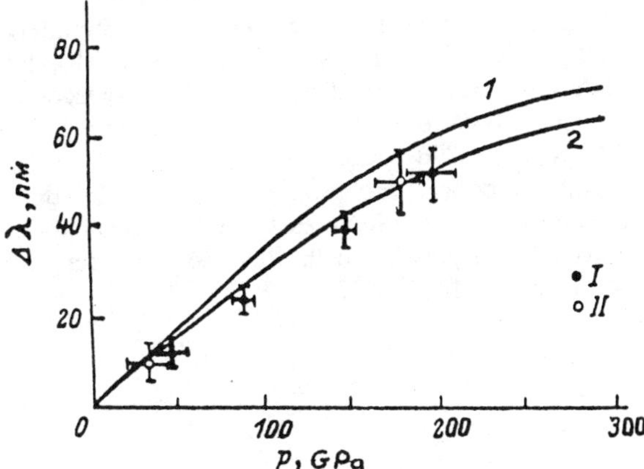

FIGURE 6. Shift R-line of ruby from pressure
1- $\Delta\lambda_T = 0.364\cdot p - 3.5\cdot10^{-6}p^2$ - isothermal regime [6,7]
2- Isentropic regime - calculation (solid line)
I - the experimental data,   II - the X ray data of compression Al

REFERENCES

1. A. I. Pavlovskii, N. P. Kolokolchikov, O. M. Tatsenko, Reproducible generation of multimegagauss magnetic fields. Megagauss Physics and Technology, Washington, New York, (1979). P. 627.
2. Xu Ji-an. Mao Ho-kuang, Peter Bell, The pressure calibration up to Mbar and the achievement of 5.5 Mbar under hydrostatic and nonhydrostatic condition. Acta Physica Sinica, Vol. 36, N 4, (1987). P. 501.
3. P. D. Horn, Y. M. Gupta, Wavelength shift of the ruby luminescence R-lines under shick compression. Appl. Phys. Lett., Vol. 49, N 14, (1986). P. 856.
4. А. И. Павловский, О. М. Таценко, В. В. Дружинин, А. А. Волков, М.В.Добруник, В. В. В. Платонов, П. В. Соснин, Люминесценция и антилюминесценция рубина при импульсном сжатии магнитным полем до 2 Мбар. Письма в ЖЭТФ, т. 12, в. 22, (1986).

5. R. Hawke, D. E. Duerre, J.G. Huebel, R. N. Keeler, W. C. Wallace, Electrical properties of Al_2O_3 under isentropic compression up to 500 GPa. J. Appl. Phys., Vol. 49, N6, (1978). p 3298.

6. Mao H.K., Bell P.M., Shaner J.W., Steinberg D.J. Specific volume measurements of Cu, Mo, Pd and Ag and calibration of the ruby R_1 fluorescence gauge from 0.06 to 1 Mbar. - J. Appl. Phys., 1978, v. 49, N 6, p. 3276-3283.

7. K.A. Goettel, Ho-kwang Mao, P.M. Bell, Generation of static pressures above 2.5 megabars in diamond-anvil pressure cell. Rev. Sci. Instrum. 56 (1985), No. 7, 1420-1427.

OPTICAL STUDIES ON Rh_2O_3

J. GHOSE and A. ROY

Department of Chemistry, Indian Institute of Technology, Kharagpur - 721 302, INDIA.

IR and reflectance spectra of Rh_2O_3 have been observed at room temperature. X-ray diffraction pattern show that the sample has the corundum structure. The IR spectra shows the presence of 5 peaks at $706,647,597,456$ and 266 cm^{-1}, corresponding to the corundum structure. Optical constants (n,k) were determined by a Kramers-Kroning analysis from the reflectance spectra recorded in the 200-1200 nm wavelength region following a procedure involving a straight line approximation of the reflection attenuation. Rh_2O_3 shows a direct band gap of 3.4 eV and an indirect band gap of 1.22 eV

INTRODUCTION

Rh_2O_3 appears in 3-crystallographic modifications, Rh_2O_3-I, space group symmetry R3c, Rh_2O_3-II, Pbna, the high pressure form and Rh_2O_3-III Pbca the high temperature form (1). Earlier studies on Rh_2O_3-III have shown that it is Pauli paramagnetic. Its electrical behaviour is characterised by a small activation evergy, and the Seebeck coefficient indicates p-type conduction (2). The present work was taken up to study the characteristics of Rh_2O_3-I and compare with those of Rh_2O_3-III and also to determine the optical band gap of this oxide.

EXPERIMENTAL AND RESULTS

The sample Rh_2O_3 used was supplied by Johnson and Mathey U. K. (Specpure). X-ray diffractogram of the sample was recorded by Philips X-ray Diffraction Unit (Model PW 1710) using CuK$_\alpha$ (λ=1.5418) radiation with Ni-filter. IR spectra of the sample was recorded on a Perkin-Elmer 883 model Spectrophotometer in KBr pellet in the range 4000-200 cm^{-1}. The lines

of the X-ray diffractogram of the sample compare very well with the lines of Rh_2O_3-I published in the ASTM cards [25-707]. No extra lines were observed, indicating that the sample is single phase Rh_2O_3-I. The IR spectra shows the presence of 5 peaks at 706, 647,597,456 and 266 cm^{-1}. The positions of the peaks match with the published IR spectra of Rh_2O_3-I. (1).

Resistivity and thermoelectric power measurements of the sample was carried out in air following the procedure described elsewhere (3,4). Magnetic succeptibility was measured by the Gouy method at room temperature. Hg Co(CNS)$_4$ was used for calibration. The temperature variation of the electrical resistivity of Rh_2O_3-I is shown in Fig.1. Figure 1 shows that the resistivity decreases linearly with increasing temperature and the activation energy (Ea) calculated from the slope is 0.1 eV. The thermoelectric power measurements show that conduction in Rh_2O_3-I is p-type and the room temperature value of the Seebeck coefficient is $+60\mu V/^oC$. The magnetic succeptibility measurements show that at room temperature Rh_2O_3-I is paramagnetic with X_g=1.29 X 10^{-6} emu/g.

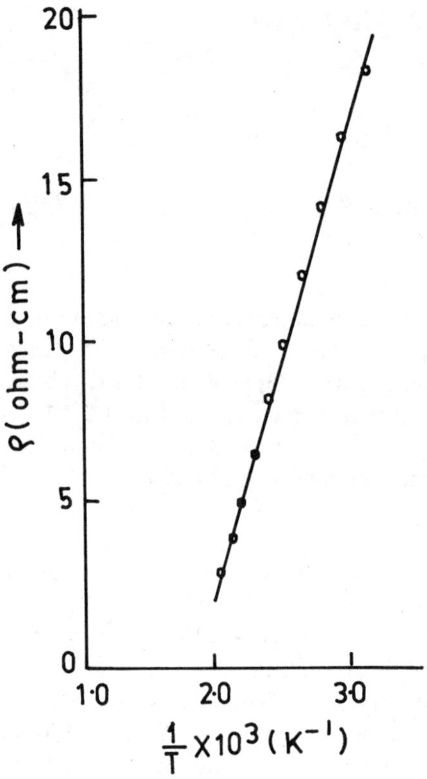

FIGURE1. ρ vs 1/T plot of Rh₂O₃-1 in air.

Reflectance spectra of the sample was taken in the 1200-200 nm wave length region with a Shimadzu UV-3100 spectrophotometer. The reflectance spectrum (%R vs λ) of Rh_2O_3-I is shown in Fig.2. The spectrum indicates strong transition at 350 nm (3.54 eV). In addition, a small kink at 830 nm (1.49 eV) is observed. The optical constants - n (refractive index), k (extinction coefficient) and α (absorption coefficient) for Rh_2O_3-I were determined in the 1200-200 nm wave length region by a Kramers-Kronig analysis of reflectance data (5,6). The plot of α vs hυ is shown in Fig.3. To ascertain the nature of the optical transition use of Tauc plots were made i.e., $(\alpha h\upsilon)^{1/2}$ vs $h\nu$, α^2 vs hν and $(\alpha h\nu)^{3/2}$ vs hν. Only the plot of α^2 vs hν is found to be linear and is shown in Fig.4a for the transition at 350 nm. A value of 3.4 eV is obtained by extrapolating α^2=0 whoch gives the optical band gap (Eg). For the transition at 830 nm the $(\alpha h\nu)^{1/2}$ vs hν plot is linear and is shown in Fig.4b. A value of 1.22 eV is obtained by extrapolating $(\alpha h\nu)^{1/2}$=0 which gives the optical band gap (Eg) for this transition.

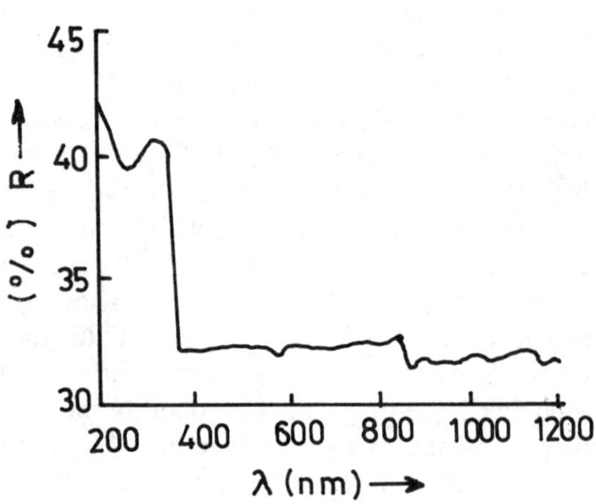

FIGURE2. Reflectance spectrum of Rh₂O₃-1.

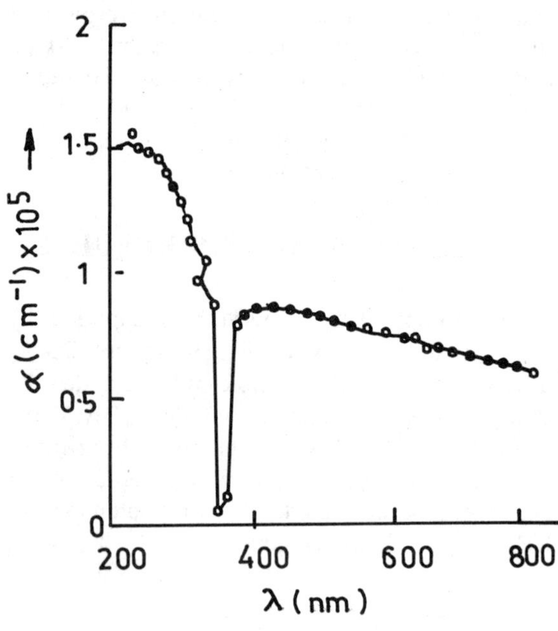

FIGURE3. α vs λ plot of Rh₂O₃

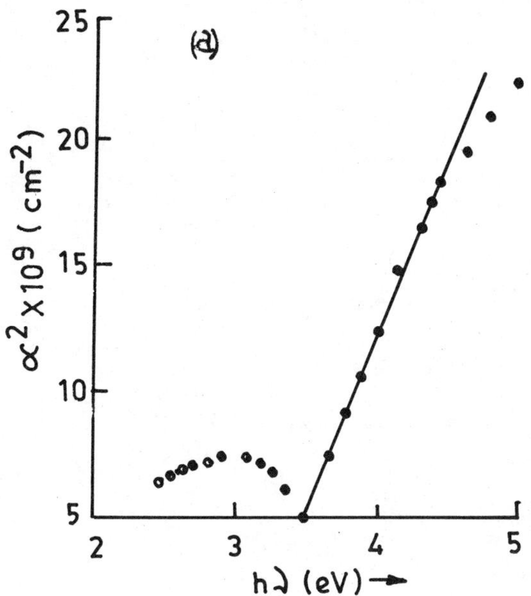

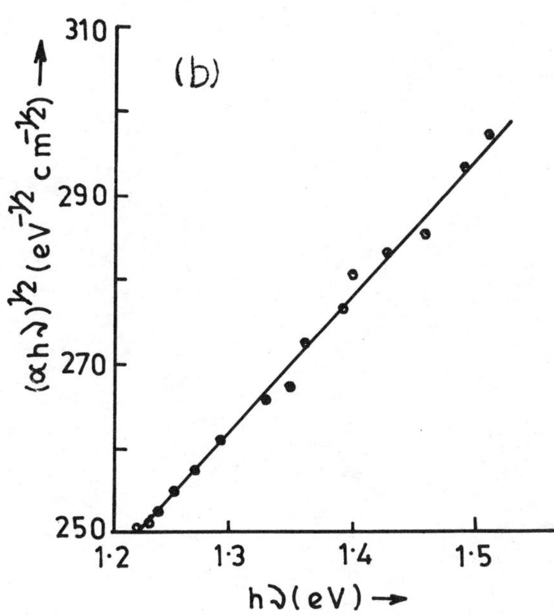

FIGURE4(a). α^2 vs hν plot of Rh$_2$O$_3$-1 at 350nm. (b). $(\alpha h\nu)^{1/2}$ vs hν plot of Rh$_2$O$_3$-1 at 830nm.

DISCUSSION

The result of X-ray analysis show that the compound has a corundum structure. The IR studies show that the spectrum of the sample is similar to the published IR spectrum of Rh$_2$O$_3$-I (1). Thermal analysis has shown that the compound is stable upto 400°C. These results clearly indicate that the compound investigated is Rh$_2$O$_3$-I, which is the low temperature, low pressure form of rhodium oxide. The result in Fig.1 shows that ρ vs 1/T is linear as is usually observed in semiconductors which show a decrease in resistivity with increase in temperature., following the relation ρ=A exp (Ea/KT). Leiva et al (2) have reported that Rh$_2$O$_3$-III may be a semimetal. But the results show that Rh$_2$O$_3$-I is probably a semiconductor. The room temperature resistivity of Rh$_2$O$_3$-I is 18.7 ohm cm, which is higher than that of Rh$_2$O$_3$-III (4.0 ohm cm). The thermal activation energy is also higher, i.e., 0.10 eV compared to 0.05 eV, as is the Seebeck coefficient, +60.0 and +37.0 μV/°C respectively. However, the magnetic susceptibility of both the oxides are of the same order, i.e. 1.29 x 10^{-6} and 0.59 x 10^{-6} emu/g respectively.

In order to determine the nature of the optical transition and the optical band gap of this semiconductor, reflectance spectra of the sample was taken and the data was analysed for n (refractive index), k (extinction coefficient) and α (absorption coefficient). The results show that a strong transition occurs at ~350 nm and a weak one at ~830 nm. Fig.3 shows that a is quite large at ~350 nm. A large absorption coefficient has been observed when there are two overlapping transitions and the overlapping process is a charge transfer process involving metal to metal bond (7). Thus, the high absorption coefficient suggests that the two transitions in Rh$_2$O$_3$-I are overlapping each other, and that a metal to metal bond is present. In Rh$_2$O$_3$-III, presence of metal-metal bond was suggested by Leiva et al (2), and this oxide was considered to be a semimetal.

The Tauc plots in Fig.4 clearly show that the

nature of the optical transition at ~350 nm is direct with a band gap of 3.4 eV and the weak transition is indirect with a band gap of 1.22 eV.

In V_2O_3 which is considered to be a semimetal (8), the optical transitions at lower energy range is direct in nature. The results on Rh_2O_3-I however show that the lower energy range transition is indirect in nature. Thus it appears that Rh_2O_3-I may not be a semimetal and as indicated in the resistivity measurements, is a semiconductor.

CONCLUSION

From these results it may be concluded that Rh_2O_3-I behaves like a p-type semiconductor and has two optical transitions one direct, with a large absorption coefficient and another weak indirect transition.

REFERENCES

1. Poeppelmeier K. R., Newsam J. M. and Brown J. M., J. Solid State Chem. <u>60</u> (1), 68 (1985).
2. Leiva H., Kershaw R., Dwight K. and Wold A., Mat. Res. Bull. <u>17</u>, 1539 (1982).
3. De K. S., Ghose J., Murthy K.S.R.C., J. Solid State Chem. 43, 261 (1982).
4. Padmanaban N., Avasthi B. N., Ghose J., J. Solid State Chem. 81, 250 (1989).
5. Kramers H. A., Phys. Z. 30, 522 (1929).
6. Kronig R De L., Phys. Rev. 30, 521 (1929).
7. Kennedy J. H. and Fress K. W., J. Jr., J. Electrochem. Soc. 125, 713 (1978).
8. Valiev K. A., Kopaev Yu. V., Mokerov V. G. and Rakov A. V., Sov. Phys. JEPT 33, 1170 (1971).

ULTRAFAST SPECTROSCOPY OF THE FIRST NANOSECOND

I-Y. Sandy Lee* , David E. Hare, Jeffrey R. Hill, Jens Franken,
Honoh Suzuki[+] and Dana D. Dlott[†]

School of Chemical Sciences, University of Illinois at Urbana Champaign, Urbana, IL 61801

Bruce J. Baer and Eric L. Chronister

Department of Chemistry, University of California, Riverside, CA 92521

Studies of the first nanosecond of shock compression require picosecond temporal resolution and sub micrometer spatial resolution. A series of experiments is described, where laser-generated shock waves are produced in microscale or nanoscale thin layers. The molecular response in these layers is probed by optical absorbance or coherent Raman spectroscopy. In nanolayer experiments, shock risetimes and molecular energy transfer processes are seen to occur on the tens of ps timescale. In microlayer experiments, subnanosecond shock-induced temperature and pressure jumps are seen in grains of an energetic material, TATB (triamino-trinitro benzene).

INTRODUCTION

This article describes optical experiments to study the first nanosecond of shock compression of molecular materials. In the first ns, shock fronts propagate a few μm. Direct sub-ns resolution of shock-wave induced effects requires a probing technique with sub-μm spatial resolution, which is problematic, and picosecond temporal resolution, which is straightforward. Although ps lasers have been used in the past for many types of shock measurements, our experiments are the first to provide the necessary spatial, spectral and temporal resolution to probe the detailed behavior of molecular materials, including high explosives, immediately behind a solid state shock front.

All experiments are performed using high repetition rate picosecond lasers and microfabricated shock target arrays with thousands or millions of shock elements. The arrays are scanned through intersections of several laser pulses so that a fresh target element is presented upon every laser shot. The experimental details are found in refs. 1-5.

OPTICAL NANOGAUGE EXPERIMENTS

The optical nanogauge experiment is diagrammed in Fig. 1. The multilayer target is fabricated by sputtering and spin coating (4). A ps pulse from a Nd:YAG laser produces a shock wave in the Al layer, which propagates into the (transparent) polymer, PMMA. For the first few ns, this shock is planar and it propagates at a constant velocity (see Fig. 4). A ps white light pulse enters the PMMA, and it passes twice through the nanogauge before exiting into an optical spectrograph, where the instantaneous absorption spectrum is obtained.

*Present address, Jet Propulsion Laboratory, California Institute of Technology, Pasadena, CA 91109.
[+]Present address, Department of Chemistry, Kyushu University, Fukuoka 812, Japan.
[†]Author to whom correspondence should be addressed.

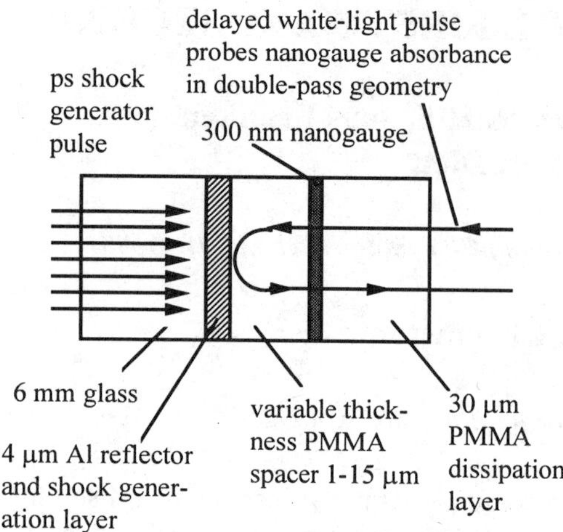

FIGURE 1. Schematic diagram of nanogauge experiment, from ref. 4. The gauge consists of R640 dye in PMMA, which has an intense absorbance near 560 nm.

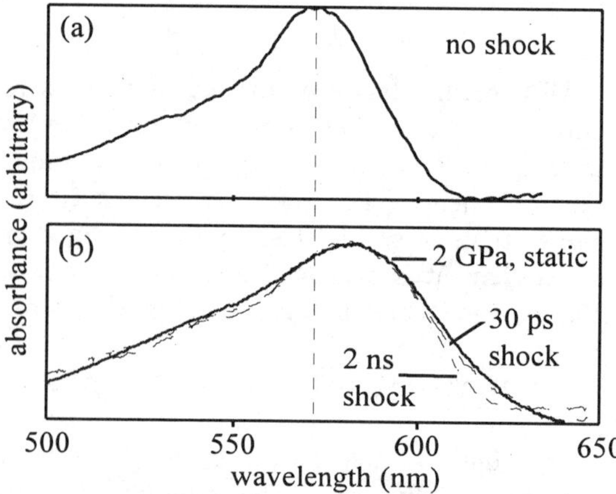

FIGURE 2. (a) Absorption spectrum of dye in nanogauge, from ref. 4. (b) Comparison of a static 2 GPa spectrum taken in a diamond anvil apparatus, to ps time resolved spectra obtained within a few tens of ps of shock passage through the nanogauge. The dynamic and static spectra have the same peak redshift. The shocked spectrum shows 100 ps time scale transients on the red absorption edge.

The nanogauge is an ~300 nm thick PMMA layer containing R640 dye (4). In independent experiments, we determined that R640 absorption *redshifts* upon *increasing pressure*. In addition, R640 develops an *increased red edge absorption* upon *heating* due to hot-band transitions (4). Figure 2(a) shows the spectrum of the dye under ambient

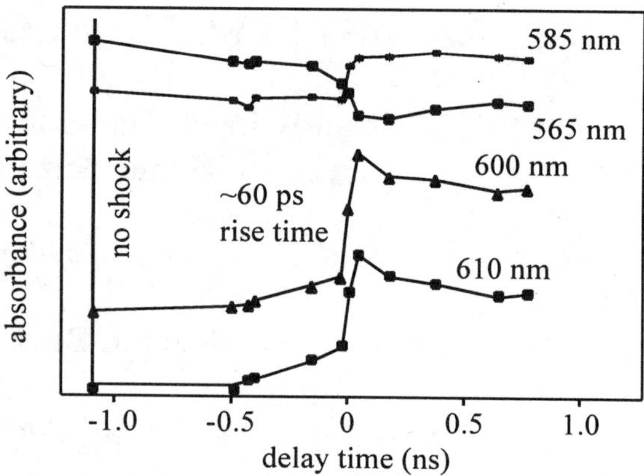

FIGURE 3. Time dependence of dye absorption at different wavelengths (see Fig. 2a spectrum). An ~60 ps risetime is observed. Adapted from ref. 4.

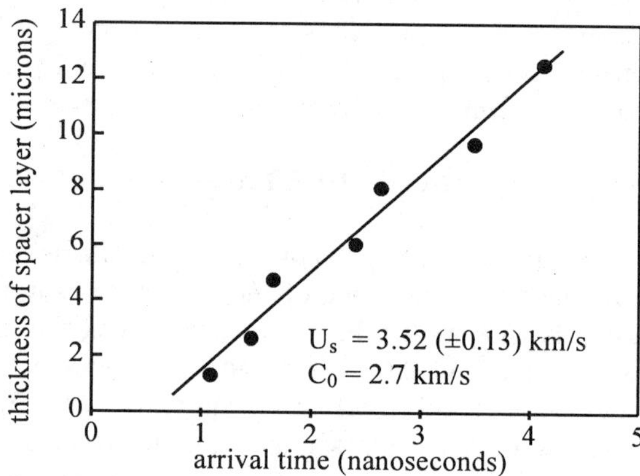

FIGURE 4. Velocity of shock through the target array remains constant for the first few ns, indicating a planar shock whose velocity U_s is about 20% greater than the acoustic velocity C_0. In PMMA, this velocity indicates a shock pressure of P = 2.1 GPa. From ref. 4.

conditions. Figure 3 shows the time dependence of the spectrum at different wavelengths. The data show a fast (~60 ps) transient, whose midpoint is denoted t = 0. This time is thought to represent the shock front midway through the nanogauge. The decrease at 565 nm and increase at 585 nm, which are respectively on the blue and red side of the absorbance peak, indicate an ultrafast peak redshift due to the pressure jump at the front. The risetime of the data appears to be limited by the time

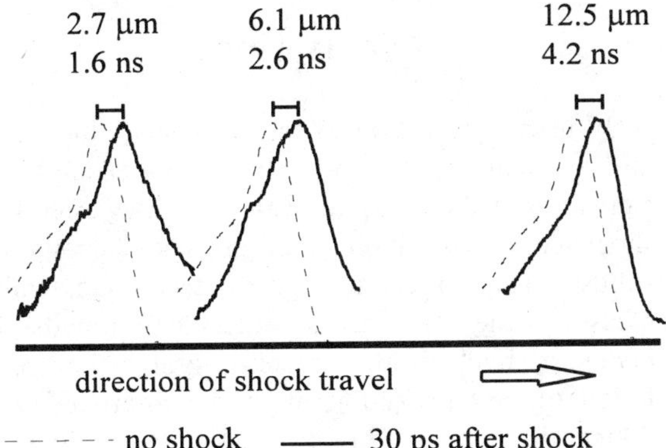

FIGURE 5. Comparison of nanogauge spectrum with no shock, to dynamic spectra taken 30 ps after the passage of the shock front, after the shock propagates 2.7, 6.1 and 12.5 μm through PMMA. For each spectrum, the red edge is to the right.

required for the shock front to pass through the 300 nm gauge layer (4).

We made a series of targets with different thickness spacer layers (see Fig. 1), and determined the arrival time of the front at the gauge layer, as shown in Fig. 4. The slope of Fig. 4 gives the shock velocity, $U_s = 3.5$ km/s. According to reference data (6), this shock velocity in PMMA corresponds to a pressure of 2.1 GPa.

Figure 2(b) compares the spectrum of the nanogauge, taken in a static high pressure diamond anvil apparatus at the University of California, to dynamic measurements under shock compression. The peak redshift of the nanogauge spectra at 30 ps and 2 ns agree very well with a static 2 GPa spectrum, showing the peak redshift can be used as an instantaneous (60 ps risetime) optical probe of the shock pressure (4).

There are some very interesting time dependent effects occurring on the 100 ps time scale, visible on the red absorption edge of the dye in the shocked nanogauge (Fig. 2b). These are thought to arise from shock-induced heating of the dye and surrounding polymer, in which different degrees of freedom are heated at different rates (7,8) during the first 100 ps.

We also see some changes in the response of the nanogauge as the shock front propagates different distances through the PMMA. Figure 5 shows nanogauge spectra taken at different points along the shock path through the PMMA. At all three points, the dye redshift is equal to the value for 2 GPa, but the red edge absorption effect decreases with increasing distance of propagation. This effect is consistent with what we expect (4,8) for a shock front whose risetime is increasing as it propagates, since a steeper shock front should pump the lower energy dye vibrations more efficiently than a broader shock front.

PICOSECOND CARS EXPERIMENTS

We now describe ps CARS experiments used to study ultrafast shock compression of energetic materials. In these experiments, the spatial resolution is obtained by using very small (<1 μm) grains of material, here the insensitive explosive TATB. With ps CARS, we can obtain an instantaneous high resolution vibrational spectrum of the material. This work is described in more detail in ref. 5. The idea is shown in Fig. 6. A sample is fabricated on a glass substrate, consisting of TATB grains suspended in a layer of PMMA containing IR-165 dye (1). A ps pulse from a Nd:YAG laser suddenly heats the PMMA-dye layer initially at (T_0, P_0), but does not heat the TATB. Pressure builds up to P_1 in the hot PMMA layer because it is heated to T_1 by the ps pulses faster than it can undergo thermal expansion. The increased pressure P_1 in the PMMA causes shock compresion of the TATB.

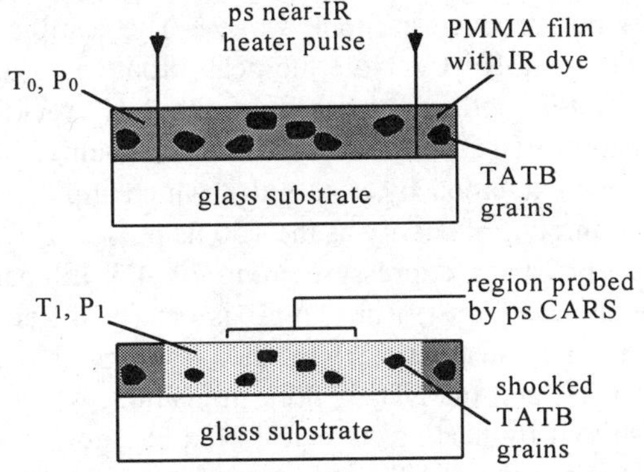

FIGURE 6. A ps pulse is used to suddenly heat a PMMA layer doped with near-IR dye. Pressure build-up in the rapidly heated PMMA shock compresses embedded grains of TATB, until the PMMA layer unloads a few ns later. The vibrational spectra of the polymer and the explosive are monitored by ps CARS.

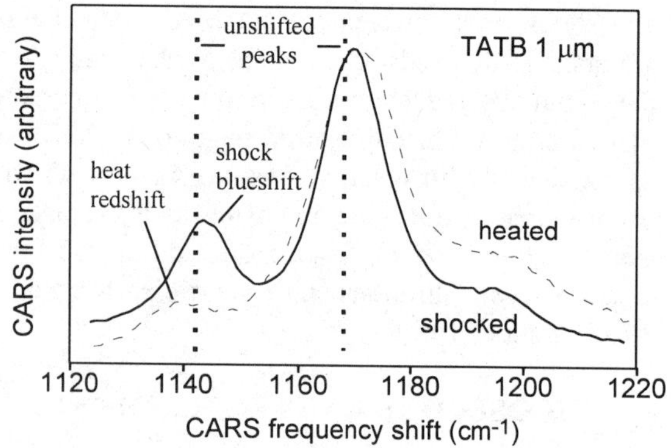

FIGURE 7. Comparison of TATB CARS spectrum immediately after a longer duration pulse which heats to 210°C, and a shorter duration 100 ps pulse which produces a 4.3 kilobar shock wave. Taken from ref. 5. The vertical dashed lines indicate locations of TATB peaks under ambient conditions.

Shock compression occurs far faster than the TATB can be heated by thermal conduction from the hot PMMA (5). The TATB remains shock compressed for a few ns until the PMMA layer unloads.

In previous work (2), we have shown we can use ps CARS to measure the temperature and pressure in the PMMA layer during shock loading and unloading, so we can determine the shock pressure in the TATB with good accuracy. Figure 7 shows the ps CARS spectra of two TATB vibrational transitions, an NH_2 rocking mode at ~1140 cm^{-1} and a C-NH_2 stretch at ~1170 cm^{-1}. The "heated" spectrum is obtained using an 150 ns duration pulse which does not produce a shock wave. The sample is heated to ~210 °C. Note the peak broadening and the heating induced redshift of the NH_2 rocking transition. The "shocked" spectrum is obtained 300 ps after excitation by a ps pulse which inputs the same amount of energy as the 150 ns pulse. The ps pulse produces a pressure jump of 4.3 kilobars. The shock effect (9) on TATB is clearly different from the heating effect--the transitions are narrower, and the NH_2 rocking transition, which is redshifted by heating, is blueshifted by shock. By calibrating the TATB vibrational peak shifts in static heating and compression experiments, it should be possible to instantaneously determine the temperature induced by shock compression. Ultimately we hope to monitor the onset of shock-induced chemical reactions in high explosives.

SUMMARY

By microfabrication of target arrays which contain elements of sub micrometer dimensions, we can observe the ps time scale dynamic effects of shock compression on amorphous and crystalline solids. The optical nanogauge technique, which measures the spectra of molecules immediately behind a shock front, has demonstrated time resolution of ~60 ps and it might be improved with thinner layers. The ps CARS technique is especially exciting, as it promises to provide instantaneous measurements of the temperature, pressure, and composition of crystalline materials tens of ps after shock loading.

ACKNOWLEDGMENT

Work at Illinois was supported by Air Force Office of Scientific Research contract F49620-94-1-0108, Army Research Office contract DAAH 04-93-G-0016, and National Science Foundation grant DMR 94-04806. Work at the University of California was supported by National Science Foundation grant CHE-94-00542. H. S. thanks the Taro Yamashita Foundation for financial support in 1994.

REFERENCES

1. David E. Hare and Dana D. Dlott, Appl. Phys. Lett., 64, pp. 715-717 (1994).
2. David E. Hare, Jens Franken and Dana D. Dlott, J. Appl. Phys. 77, pp. 5950-5960 (1995).
3. I-Yin Sandy Lee, Jeffrey R. Hill and Dana D. Dlott, J. Appl. Phys. 75, pp. 4975-4983 (1994).
4. I-Yin Sandy Lee, Jeffrey R. Hill, Honoh Suzuki, Bruce J. Baer and Eric L. Chronister, and Dana D. Dlott, J. Chem. Phys (in press).
5. David E. Hare, Jens Franken and Dana D. Dlott, Chem. Phys. Lett, (in press).
6. LASL Shock Hugoniot Data, S. P. Marsh, Ed. (University of California, Berkeley, 1980).
7. Xiaoning Wen, William A. Tolbert and Dana D. Dlott, J. Chem. Phys. 99, pp. 4140-4151 (1993).
8. D. D. Dlott and M. D. Fayer, J. Chem. Phys. 92, 3798 (1990). A. Tokmakoff, M. D. Fayer and D. D. Dlott, J. Phys. Chem. 97, 1901 (1993).
9. W. M. Trott and A. M. Renlund, J. Phys. Chem. 92, 5921 (1988). S. K. Satija, B. Swanson, J. Eckert and J. A. Goldstone, J. Phys. Chem. 95, 10103 (1991).

TIME RESOLVED EMISSION STUDIES OF ALUMINUM/WATER COMBUSTION

C. A. Brown and T. P. Russell

Chemistry Division, Code 6110, Naval Research Laboratory, Washington DC 20375

A series of single-shot, time-resolved emission experiments have been performed to probe the reactions of laser heated aluminum in both air and water. The major species observed are atomic Al and molecular AlO. The temporal behavior of observed species is being used to gain insight into reactions that lead to the formation of Al_2O_3. These experiments are being extended to studies of Al/H_2O reactions in gem anvil cells as a function of pressure.

INTRODUCTION

Composite materials with metal additives have vital applications in the defense and aerospace industries. Despite their importance, basic understanding of the underlying chemistry of metallized systems has yet to be realized. The high temperatures and pressures obtained in these reactions add to the complexity of experimentally probing these systems. We have undertaken a series of novel time-resolved experiments to probe simple reactions of a common metal additive, aluminum, with oxygen and water. Since water is an end product of many energetic material reactions, Al/water reactions should have relevance to the late time energy release in composite explosions.

There are numerous studies of Al reactions with oxidizers, including water. Early spectroscopic studies involved metal foil flashbulbs.(1) More recent examples include studies of the dynamics of gas phase Al and oxygen reactions.(2-4) The experiments most similar to those described here are those which monitor emission from exploding wire reactions of Al and water.(5-7) The highest published pressure obtained was 446 kPa with no time resolution of emission.(5) The Al and water spectroscopic experiments described herein are unique in both the time regimes studied and the pressures obtained.

EXPERIMENTAL

In these experiments, a sample is placed in the focal plane of a dye laser and laser heated with a single pulse. The sample consists of either aluminum foils or particles. The foils, purchased from Goodfellow Corporation, are 0.8 microns thick and have purities > 99.1 %. The Al particles are ~5 microns in diameter and are also purchased from Goodfellow.

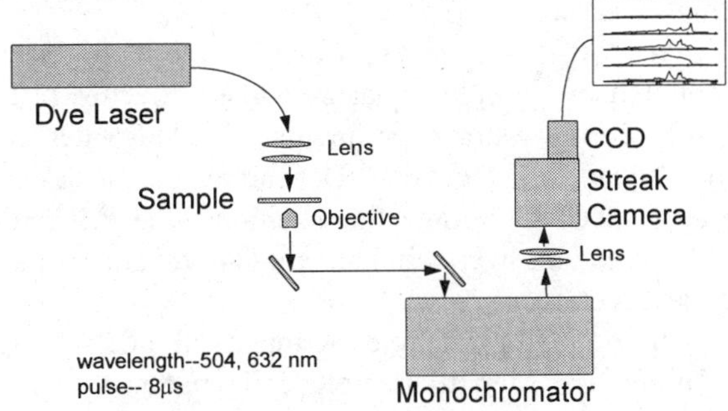

Figure 1. Experimental design.

The experimental technique employed has been described in detail elsewhere,(8,9) therefore only a brief description will be provided. A schematic of the design is shown in Figure 1. A single pulse (8 μs) from a Cynosure flashlamp pumped dye laser is used to heat the sample. Typical laser energies delivered to the sample are

30-100 mJ. Since the light collection occurs on the side opposite that of heating, an amount of energy able to produce a 'window' in the material must be deposited on the sample in order to detect emission.

Upon reaction, the collected emission is dispersed by a Spex 500M monochromator. Before each set of experiments, the monochromator is calibrated in the wavelength region being monitored. After leaving the monochromator, a Hamamatsu C2830 Streak Camera streaks the signal in time. The time resolution can be as high as 10 ns for studying early reaction sequences, or as slow as hundreds of microseconds to obtain global time information. A liquid nitrogen cooled Spex Spectrum 1 CCD camera images the signal from the streak camera. The timing of the laser pulse and streak camera is controlled by a Stanford Research Systems model DG535 pulse/delay generator and is monitored with a Tektronix DSA 602A.

The three types of experiments performed are: Al foil reactions in air (ambient pressure), Al foil reactions in water (ambient pressure) and Al reactions in a cubic zirconia anvil cell, CZAC, (high pressure).

In the ambient experiments, the foils are mounted inside either quartz or glass cuvettes (10 mm). In the water experiments, distilled water is also added to the cuvette. Depending on the laser energy used to initiate the reactions, the Al and H_2O reactions were violent enough to crack the glass cuvette.

In the high pressure experiments, the CZAC is a miniature Merrill Bassett cell made of 301 stainless that is designed for 180° transmission measurements.(10) The gasket is made of Inconel 600 with a gasket-hole that has dimensions of 250 microns (diameter) by ~200 microns (depth). The cubic zirconia anvils can yield pressures of > 10 GPa.(11) Pressure inside the CZAC is measured with the ruby fluorescence measurement technique and is calculated on a peak shift model.(12,13) To load the cells, a small ruby sphere < 15 μm in diameter is placed in the gasket-hole, followed by aluminum (<25% of the volume), and then the remaining volume is filled with the pressure transmitting material, water.

RESULTS AND ANALYSIS

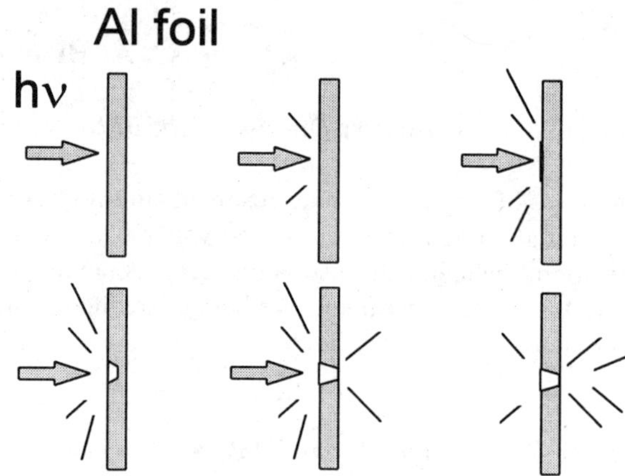

Figure 2. Scheme illustrating the laser/foil interaction.

Figure 2 gives a diagram of our present understanding of the laser ablation process. Initially, part of the laser pulse is reflected from the face of the foil due to the high reflectivity of Al. Next the layer of Al_2O_3 covering the surface is ablated off the foil. The laser then starts drilling a hole in the foil with the heating and evaporation of the Al metal. Finally, a hole is drilled through the foil and emission can be detected from the opposite side. The emission continues after the laser pulse until the reaction cools. The two major species observed in these reactions are atomic Al and molecular AlO. The bands detected are the Al doublet at 394 and 396 nm and at least 4 of the AlO (B→X) vibrational bandheads: (2,0) at 447 nm, (1,0) at 465 nm, (0,0) at 484, and (0,1) at 508 nm.

Representative spectra of the Al and air reactions are shown in Figure 3. The excitation wavelength in both cases is 632 nm and a laser energy of 30 mJ. The timing for each experiment is identical with each track or spectrum separated by 2 μs.

The left half of Figure 3 shows the atomic Al emission. While the Al atomic doublet at 394 and 396 nm is unresolved in this spectrum, experiments that are performed with a higher order grating have resolved the doublet. While the most intense track shows Al emission, the other tracks exhibit self-reversal of the Al emission. This indicates that a layer of cooler atomic Al exists which absorbs the light produced in the reaction. The Al self-reversal continues for microseconds after its appearance.

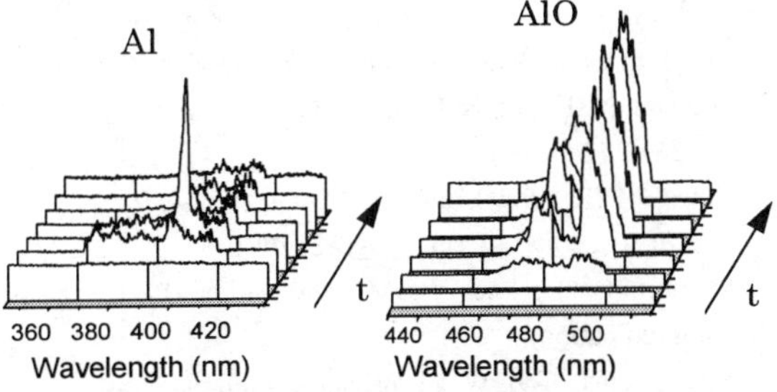

Figure 3. Spectra obtained from Al/air reactions with 2 μs time resolution. The Al emission and self-reversal are shown on the left, while the AlO bands at 465 and 484 nm are shown on the right.

The (1,0) and (0,0) vibrational bandheads of AlO emission are shown in the right half of Figure 3. These features increase in intensity on the time scale shown. The vibrational manifold in these bands can also be observed. The AlO emission continues for microseconds after its initial appearance.

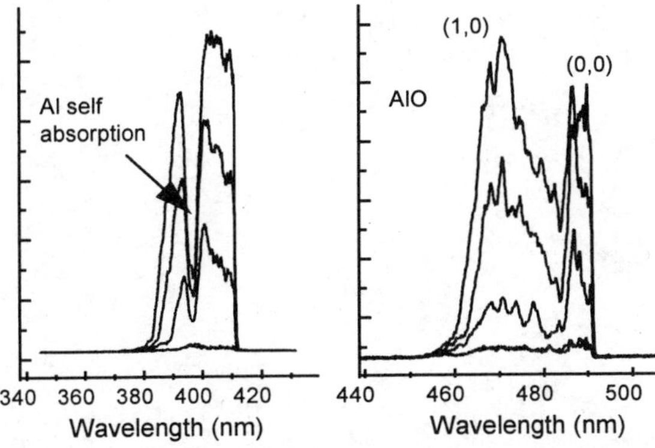

Figure 4. Spectra obtained from Al/water reactions. The spectra, which are separated by 400 ns, are overlaid for clarity. The Al self-absorption is shown on the left, while the AlO bands at 465 and 484 nm are shown on the right. These spectra occur after the emission has peaked and the intensity is decreasing in time.

The spectra of Al foil reactions with water using the same laser excitation (632 nm) and laser energy (30 mJ) are shown in Figure 4. The time resolution is faster than shown previously for Al/air

reactions, with each track or spectrum separated in time by 400 ns. The spectra are overlaid for clarity.

The spectra on the left illustrate the features due to atomic Al. In contrast to the Al/air reactions, a large background continuum emission is present in the Al/H_2O reactions. This results in the self-absorption of Al, appearing as a dip in the continuum background. In the experiments in water, Al is not observed in emission, it is only detected in absorption. The light produced in the reaction absorbed by a layer of cooler atomic Al.

The AlO (1,0) and (0,0) bandheads are shown on the right. While there are contributions from the continuum, it is not as pronounced in this region. The band shapes and relative intensities of the AlO bands are significantly different in the emission obtained from Al/water reactions compared to the emission obtained from Al/air reactions. These spectral differences are currently being explored in detail.

The ambient pressure reactions of Al in air and water aid in the description of Al particle reactions with water at high pressure. Experiments with pressures up to 0.8 GPa have been performed.(9) Preliminary experiments yield Al, AlO, a broad continuum background, and other species. Additional experiments are in progress to probe these species further.

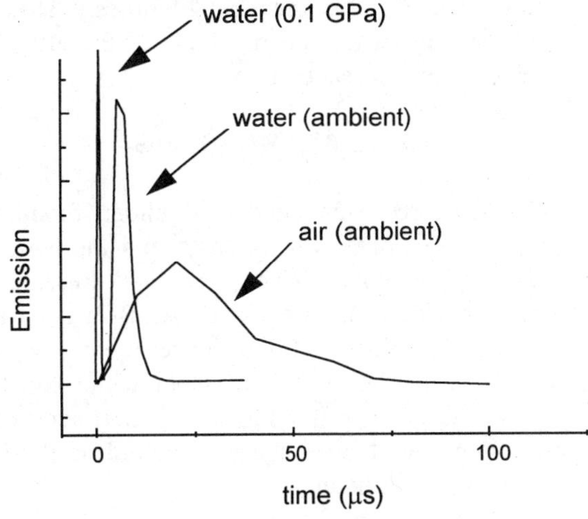

Figure 5. Plot of the intensities of AlO emission from Al reactions in air, water, and in water at pressure (<0.1 GPa). For comparison, the plots are referenced to the first appearance of emission (t=0).

An example of the information obtained is the appearance and disappearance of intermediates (chemical lifetimes). A comparison of the AlO lifetimes is shown in Figure 5. The Al lifetimes (not shown) also show the same trends. Note that the air and water spectra have the same experimental conditions and are thus directly comparable. The high pressure data was performed under different conditions, and while not directly comparable, qualitative trends can be observed.

The AlO emission from Al and air reactions lasts the longest, with a lifetime > 75 μs. The AlO emission from reactions in water appears and disappears much faster. A chemical lifetime of 15 μs is obtained. At high pressure, the observed AlO emission has an even shorter lifetime, 1μs. This behavior can be explained by the expected increase in the collisional frequency of intermediates in the reaction with increasing effective pressure.

CONCLUSION

In conclusion, these single-shot time-resolved emission studies are providing insight into Al reactions in air and water. The major intermediates observed in emission are Al and AlO. The temporal information available from these experiments is being used to probe chemical reaction sequence and chemical lifetimes. These experiments also allow for the study of the effects of high pressure on reaction.

ACKNOWLEDGMENTS

The research was conducted under funding from the Office of Naval Research and the Naval Research Laboratory. The National Research Council provided funding for C. A. Brown as a post-doctoral fellow. Some experiments were conducted in the lab of Y. M. Gupta at the Shock Dynamics Center in the Physics Department of Washington State University with the aid of T. M. Allen and G. Pangilinan.

REFERENCES

1. Brzustowski, T. A. and Glassman, I., *Heterogeneous Combustion, Progress in Astronautics and Aeronautics Series*, **15**, Wolfhard, I, Glassman, I, and Green, L. eds., New York: Academic Press, 1965, pp. 41-73.

2. Dagdigian, P. J., Cruse, H. W., and Zare, R. N., *J. Chem. Phys.*, **62**, pp. 1824-1831 (1975).

3. Fontijn, A., Felder, W., and Houghton., P. J., *International Symposium on Combustion*, **16**, Pittsburg:Combustion Institute, 1977, pp. 871-880.

4. Garland, N. L. and Nelson, H. H., *Chem. Phys. Lett*, **191**, pp. 269-272 (1992).

5. Jones, M. R., and Brewster, M. Q., *J. Quant. Spectroc. Radiat. Transfer*, **46**, pp. 109-118, 1991.

6. Stanton, C. T., Lee, W. M., and Miller, P. J. (private communication).

7. Frank, A. M., Tao, W. C. (private communication).

8. Russell, T. P., Allen, T. M., Rice, J. K., and Gupta, Y. M., *J. Physique* (in press).

9. Russell, T. P., Allen, T. M., and Gupta, Y. M., *Chem. Phys. Lett.* (submitted).

10. Merrill, L. and Bassett, W. A., *Rev. Sci. Instr.*, **45**, pp. 290-294 (1983).

11. Russell, T. P. and Piermarini, G. J., (manuscript in preparation).

12. Block, S. and Piermarini, G. J., *Phys. Today*, **29**, 44-55 (1976).

13. Piermarini, G. J., Block, S., Barnett, J. D., and Forman, R. A., *J. Appl. Phys.*, **46**, pp. 2774-2780 (1975).

TIME-RESOLVED TEMPERATURES OF SHOCKED AND DETONATING ENERGETIC MATERIALS

C. S. Yoo, N. C. Holmes, and P. C. Souers

Lawrence Livermore National Laboratory, Livermore, California 94551

Chemical processes occurring in shock-compressed and detonating high explosives have been studied using fast time-resolved emission spectroscopy and a two-stage gas-gun. The spectral characteristics of emission from shock-compressed nitromethane, tetranitromethane and single crystals of pentaerythritol tetranitrate are typically very broad and structureless, likely representing thermal emission. Assuming the thermal emission from a gray-body, the emission intensity can be correlated to the temperature changes in shock-compressed and detonating high explosives. We report Chapman-Jouguet temperatures of 3800 K for nitromethane, 2950 K for tetranitromethane, and 4100 K for pentaerythritol tetranitrate. In this paper we also compare the data with the chemical equilibrium models.

INTRODUCTION

The shock temperatures of high explosives (HE's) are important for understanding shock-initiation and detonation processes; however, it has been one of less understood thermodynamic variables. On the other hand, unlike other variables like pressure or energy, the temperature is very sensitive to chemical equilibrium models which require experimental verification. In our previous study[1] we demonstrated a method to measure the time-resolved thermal emission from shocked and detonating nitromethane (NM). In this paper, we explain how the time-resolved thermal emission data can be correlated to the temperatures at various thermodynamic states such as the Chapman-Jouguet (*CJ*) condition. We will also present the results of other homogeneous high explosives like tetranitromethane (TNM) and pentaerythritol tetranitrate (PETN) single crystals.[2] Homogeneous high explosives

are used to simplify the problem by separating other complicated mechanical issues associated with hot spots, non-ideal density, grain boundaries, etc.

EXPERIMENTS

Figure 1 shows the experimental setup to measure time-resolved shock temperatures of high explosives. The sample is loaded in

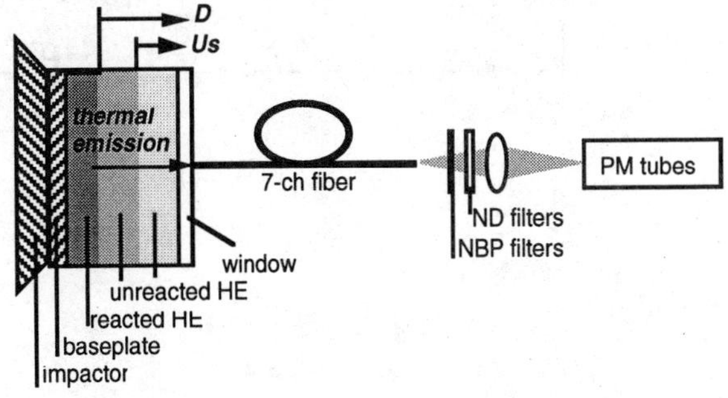

FIGURE 1. The experimental setup for time-resolved shock temperature measurements of transparent homogeneous high explosives.

"""

6-8 mm cavity between a baseplate and a transparent window and is shock-compressed by using a two-stage gas-gun which can accelerate an impactor up to 8 km/s.

The thermal emission from shock-compressed high explosives is time-resolved by using six photomultiplier tubes in the spectral range between 350 and 700 nm.[3] In this setup, thermal emission is collected from the central 3 mm area of HE, limited by the acceptance cone of an optical fiber bundle. This results in fast time resolution, a few ns, primarily depending on the rise time of recording electronics. The temperature of HE is then obtained by fitting this time-resolved thermal emission data to a gray body Plank function at 1 ns intervals during the event of interest, typically 1 μs.

SHOCK-INITIATION AND DETONATION

In homogeneous HE's the chemical reaction is believed to initiate behind the shock front near the baseplate/HE interface as shown in Fig 1.[4-7] This means that the detonation initially occurs initially from the shock-compressed HE (superdetonation), but later from the unshocked HE (detonation) as the detonation front catches the shock front. Figure 2 illustrates the thermodynamic states through which homogeneous HE molecules are evolved under shock compression.

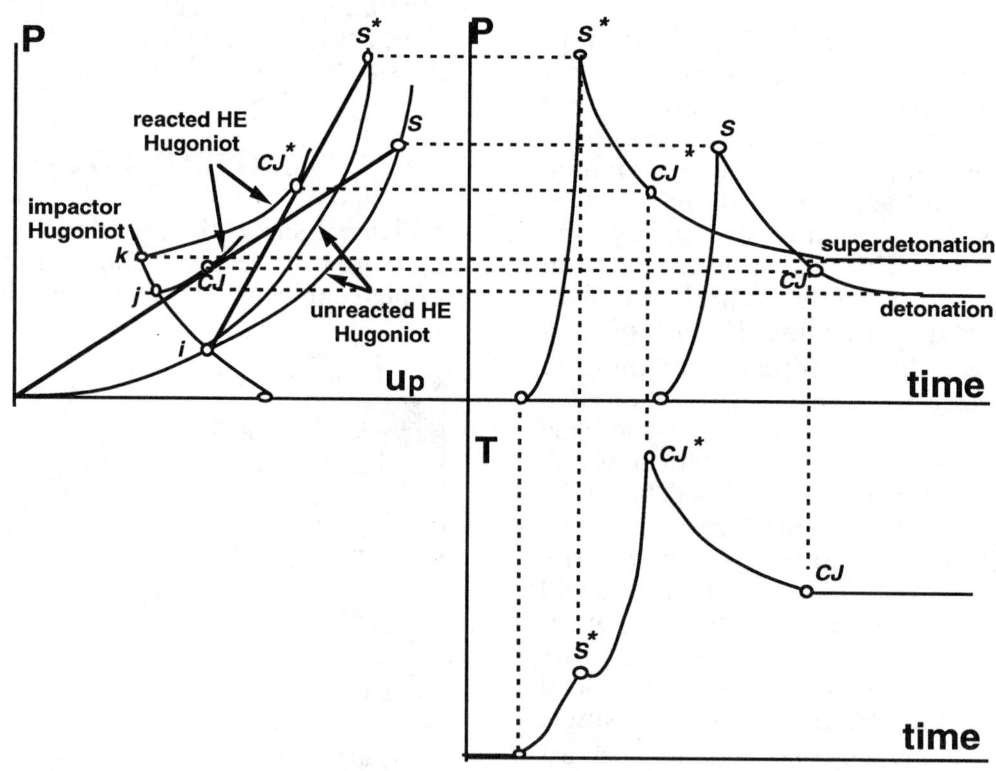

FIGURE 2. The thermodynamic states of shock-compressed and detonating homogeneous HE and the reaction products, together with the pressure and temperature changes at the reaction front.

The initial shock condition, i, is defined by the impedance match between an impactor and unreacted HE. The pressure of this shocked but unreacted HE increases to the von-Neumann spike, S^*, by chemical energy behind the shock front. The chemical reaction then initiate at S^*, and the products eventually evolve to the CJ^* state as the reaction is fully developed. These detonation products are then expanded isentropically to the condition matching the impedance of the impactor k, as long as the reaction front remains behind the shock front. In the later time, however, the detonation front catches the shock front, and the detonation occurs directly from unshocked HE at S, and the products evolve to the CJ and to the impedance match condition, j, between the isentropically expanded HE products and the impactor.

The temperature, on the other hand, increases as the reaction progresses and peaks at the CJ. The opacity of the reaction products is substantially higher than that of unreacted HE, due to formation of graphite and other hydrocarbons at high temperatures. Under these conditions the thermal emission measured across the reaction zone (Fig 1) is mostly from the reaction front and can be correlated to the temperatures at various thermodynamic states discussed in Fig 2. For example, the measured temperature will initially increase to that at S^* prior to shock-initiation and to the CJ^* temperature of shock compressed HE products. Then, the CJ^* temperature decays to the CJ temperature, as the detonation front catches the shock front. The measured temperature should then remain at the CJ, as long as the impact condition is same or below the CJ condition.

TIME-RESOLVED TEMPERATURE

Figure 3 shows the time-resolved temperatures of NM, TNM, and (110) PETN single crystal shocked to 10.7, 12.9 and 13.3 GPa, respectively. The temporal profiles are similar to one discussed in Fig 2 and can be characterized into the pre-detonation, superdetonation and normal detonation zones. The shock-initiation is evident from the initial increase in temperature and the subsequent small decrease in the pre-detonation zone. The induction times of the reactions are: 50 ns in NM and PETN and 200 ns in TNM. The second strong temperature jump indicates the detonation of these shock-compressed and reacting high explosives. The transitions to the CJ^* states occur in 20-50 ns in TNM and PETN but in 200 ns in NM. The peak temperatures then decay to steady-state values for the next 100-400 ns, representing the transition to the detonation of uncompressed HE. Notice that, unlike TNM and NM, PETN detonates at substantially lower pressures than the CJ pressure near 31 GPa, consistent with the previous observation that PETN is sensitive along the (110) direction.[8]

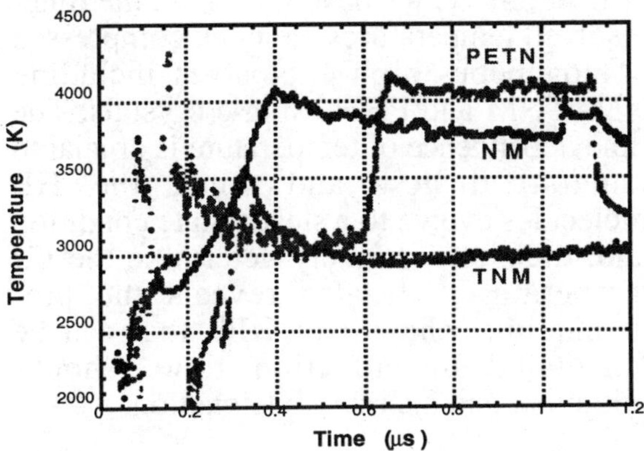

FIGURE 3. Time-resolved shock temperatures of selected homogeneous HE's.

The *CJ* temperatures of these explosives can be obtained from the steady-state values shown in Fig 3. The results are summarized in Table I, together with the calculated *CJ* temperatures using a chemical equilibrium code for comparison.[9] The *CJ* temperature of PETN is higher than other two due to the abundance of oxygen. The *CJ* temperature TNM is substantially lower than that of NM, likely due to the absence of hydrogen in TNM.

TABLE I. The steady-state temperatures of shock compressed homogeneous high explosives in comparison with the calculated *CJ* temperatures.

HE	P (GPa)		T (K)	
	unreacted	reacted	Steady-state	*CJ cal*
NM	107	128	3800	3750
TNM[1]	129	137	2950	2640
PETN	133	191	4100	4430

*1   98 % pure TNM was used in experiments and the calculation was also done based on 98 % TNM mixed with 2 % NM.

In summary, we have presented the time-resolved temperatures of shock compressed homogeneous high explosives including NM, TNM and PETN single crystals. The time dependence of temperature is crucial to understand how and when the HE molecules evolve to a steady state condition and, thus, to accurately determine the *CJ* temperature. It also reveals the pre-detonation behavior of HE's that can be correlated to induction time, shock-initiation, and the rate of detonation.

ACKNOWLEDGMENTS

We thank Ervin See, Jim Crawford and Bruce Morgan at LLNL for the technical assistance. We also appreciate J. Dick at LANL for providing the PETN single crystals and for comments valuable to the study. The discussions with S. Sheffield at LANL and B. Nellis, D. Erskine, F. Ree and M. van Thiel at LLNL were very useful for the study. This work was performed under auspices of the U.S. DOE by the LLNL.

REFERENCES

1. C.S. Yoo and N.C. Holmes, Shock Initiation of Nitromethane in Shock Compression of Condensed Matter - 1993, edited by S. Schmidt et. al. (AIP, 1994).
2. PETN single crystals were oriented along (110) direction and was obtained from J. Dick at the LANL.
3. N.C. Holmes, Rev. Sci. Instru. 66, 2615 (1995).
4. A.W. Campbell, W.C. Davis, J.B. Ramsey, and T.R. Travis, Phys. Fluid 4, 511 (1961).
5. A.N. Dremin, S. Savrov, and A.N. Amdrievskii, Comb. Expl. and Shock Waves, vol.1, 1965, pp 1.
6. S.A. Sheffield, R. Engelke, and R.R. Alcon, In-situ study of the chemically driven flow field in initiating homogeneous and heterogeneous nitromethane explosives in Proceedings of the Ninth International Symposium on Detonation, pp39, Portland, Oregon (1989).
7. D.R. Hardesty, Combustion and Flame 27, 229 (1976).
8. J.J. Dick, R.N. Mulford, W.J. Spencer, D.R. Pettit, E. Garcia, D.C. Shaw, J. Appl. Phys. 70, 3572 (1991).
9. M. van Thiel, F.H. Ree, and L.C. Haselman, Accurate Determination of Pair Potentials for a CwHxNyOz system of Molecules: a Semiempirical Method, UCRL-ID-120096, LLNL (March, 1995).

APPARENT SPECTRAL RADIANCE AT SHOCKED METAL/WINDOW INTERFACE

Wenhui Tang[*+], **Ruoqi Zhang**[*], **Fuqian Jing**[+], **Jinbiao Hu**[+]

* *Department of Applied Physics, National University of Defense Technology, Changsha, Hunan, 410073, P. R. China*
\+ *Laboratory for Shock Wave and Detonation Physics Research, Southwest Institute of Fluid Physics, P. O. Box 523—61, Chengdu, Sichuan, 610003, P. R. China*

When a shock wave passes through an ideal metal/window interface, the interface temperature is time independent, accordingly, the apparent spectral radiance at the interface is also regarded as time independent. However, every material has certain transparency, so the apparent spectral radiance at the interface must be the sum of the radiance of a thin layer rather than the surface. In this paper, the relationship among optical absorption coefficient of a metal and its pressure and temperature is studied based on classical electromagnetics. It is obtained that the absorption coefficient decreases and the optical depth increases as shock pressure increases. By using radiative transport equation, we also calculate the apparent spectral radiance at an ideal iron/sapphire interface following shock compression. It is showed that the apparent spectral radiance is time dependent, although the interface temperature is time independent. As an example, the apparent spectral radiance history at iron/sapphire interface under shock pressure of about 230 GPa is presented.

INTRODUCTION

Measurements of shock temperatures of metals are usually performed by observing the spectral radiance emitted from a sample through a window. The basic method is to deduce the interface temperature by fitting the measured spectral radiance to the Planck greybody model, then, shock temperature of the metal is calculated by using the result of thermal conduction (Grover and Urtiew [1]) and the Grüneisen equation of state [2]. When a shock wave passes metal/window interface, the thermal contact resistance [3], gaps [4] and the absorption of the window [5] can all lead to a time dependent behaviour of the interface temperature and its corresponding spectral radiance. However, for an ideal interface, the interface temperature is time independent [1]. If the window is fully transparent, the apparent spectral radiance from the interface is also regarded as time independent.

It is well known that the spectral radiance of an actual material at high temperature can be expressed as the Planck greybody function

$$I_P(\varepsilon, \lambda, T) = \frac{\varepsilon C_1}{\pi \lambda^5} \frac{1}{\exp(\frac{C_2}{\lambda T}) - 1} \quad (1)$$

where λ is the wavelength, ε, generally a wavelength independent constant, is the emissivity, T is the surface temperature of the material, and $C_1 = 3.7418 \times 10^{-16} \, \mathrm{Wm^2/Sr}$ and $C_2 = 1.4388 \times 10^{-2} \, \mathrm{mK}$ are the first and the second radiation constants, respectively. For an idealopaque body, $I_P(\varepsilon, \lambda, T)$ is the surface spectral radiance. For a real material, however, it always has certain transparency, the observed spectral radiance is actually the sum of the radiance of all the layer in the optical depth. Thus, the spectral radiance does not completely represent the interface temperature. One can see that to determine the shock temperature of a metal, it is important to understand the physical implication and the picture of the spectral radiance emitted from the shocked metal.

In this paper, the relationship among the optical absorption coefficient of a metal and its pressure and temperature is studied. Then, the spectral radiance history from an ideal metal/window interface following shock compression is calculated by using radiative transport equation.

ESTIMATION OF THE OPTICAL DEPTH Of A METAL

When light passing through a distance x in a medium whose absorption coefficient is μ , the attenuation of the spectral intensity can be expressed as

$$I_\lambda(x) = I_\lambda(0)e^{-\mu x} \qquad (2)$$

where $I_\lambda(0)$ is the intensity at $x=0$. μ is determined by the extinction coefficient κ and the wavelength of the light in vacuum λ:

$$\mu = \frac{4\pi\kappa}{\lambda} \qquad (3)$$

To describe the transport character of a material, a common used parameter is the optical depth δ, which represents the propagation distance when the light intensity attenuates to $1/e$ of the initial intensity, and equals reciprocal of the absorption coefficient:

$$\delta = \frac{1}{\mu} = \frac{\lambda}{4\pi\kappa} \qquad (4)$$

From the electromagnetics theory, the relation between κ and electromagnetic parameters can be described as

$$\sigma = 2n\kappa\omega \in_0 \qquad (5)$$

where σ is the electrical conductivity, $\in_0 = 8.8541872 \times 10^{-12}$ F/m is the vacuum dielectric constant, $\omega = 2\pi c/\lambda$ is the angular frequency of the light wave, c is the speed of light, and n is the index of the medium. For iron, as $\lambda = 0.58$ μm, under standard conditions, $\kappa = 3.85, n = 3.36$[6], so the value of σ_0 at $\lambda = 0.58$ μm is $7.44 \times 10^5 \ \Omega^{-1} \ m^{-1}$, and $\delta_0 = 0.012$ μm , where the subscript 0 denotes values under standard conditions.

In addition, when $T > 0.5\Theta$, where T is temperature and Θ is the Debye temperature, respectively, the electric conductivity of metal can be simplified from the Bloch—Grüneisen formula as [7]

$$\sigma = \frac{4A\Theta^2}{BT} \qquad (6)$$

where A is the atomic weight, B is a material parameter. Using the definition of the Grüneisen parameter:

$$\gamma(v) = -\frac{\mathrm{d}\ln\Theta}{\mathrm{d}\ln v} \qquad (7)$$

the relation between σ and T and specific volume v can be obtained as

$$\frac{\sigma}{\sigma_0} = \frac{T_0}{T}\exp[2\gamma_0(1 - \frac{v}{v_0})] \qquad (8)$$

here, we used the empirical formula $\gamma/v = \gamma_0/v_0$.

Finally, combining equations $(3) - (5)$ and (8), the optical depth of a metal under high temperature and high pressure can be obtained as

$$\frac{\delta}{\delta_0} = \frac{T}{T_0}\exp[2\gamma_0(\frac{v}{v_0} - 1)] \qquad (9)$$

APPARENT SPECTRAL RADIANCE AT IDEAL INTERFACE

Assuming an ideal metal/window interface contact, when a shock wave passes the interface, the temperature distribution in the metal can be written as[1]

$$T(x,t) = T_R - \frac{T_R - T_W}{1 + \alpha}\mathrm{erfc}(\frac{|x|}{2\sqrt{Dt}}) \quad (10)$$

where T_R and T_W are the release temperature of the metal and the shock temperature of the window, respectively, D is the thermal diffusivity of the metal. x is the spatial coordinate, origin set at the interface, $x<0$ in metal, t is time, started at the instant that the shock arrivals at the interface. α is a parameter connected with the thermal conductivity K, the specific heat at constant pressure C_p, and the density ρ,

$$\alpha = \left[\frac{(k\rho C_p)_M}{(k\rho C_p)_W}\right]^{1/2} \qquad (11)$$

where the subscripts M and W represent the metal and the window, respectively. When an iron flyer impacts on iron/sapphire window assembly at velocity of about 6.5 km/s, it generates a shock pressure of about 230 GPa in iron. Using the shock wave jump conditions and the Reimann integral equation , the specific volume of iron v_R , which is at partially release state , is 0.086 cm³ / g , the release temperature T_R equals 5600 K , and the shock temperature of sapphire, T_W, is 1500 K. Here, we used the following material parameters: for iron, $\rho_0 = 7.85$ g/cm^3, $U_s = 3.955 + 1.58u_p$(km/s), $\gamma_0 = 1.9$; for sapphire, $\rho_0 = 3.977$ g/cm^3, $U_s = 8.724 + 0.975u_p$(km/s), $\gamma_0 = 1.32$ (from Ref. 8), where U_s and u_p are shock wave and particle velocities, respectively. According to Tan and Ahrens[8], $\alpha = 4.5$, D can be calculated to be about 0.023 μm^2/ns. So, the temperature distribution in iron under above shock loading is

$$T(x,t) = 5600 - 745 \text{erfc}\left(\frac{|x|}{0.305\sqrt{t}}\right) \quad (12)$$

the units in above equation are: $[T] = $K, $[x] = \mu$m, $[t] = $ns.

Using equation (9) by taking value of T_0 as 300 K, we have $\delta_R/\delta_0 = 5.43$. Thus, $\delta_R = 0.065$ μm, which is 4 times more than δ_0. This result indicates that the optical depth increases when a metal under shock compression.

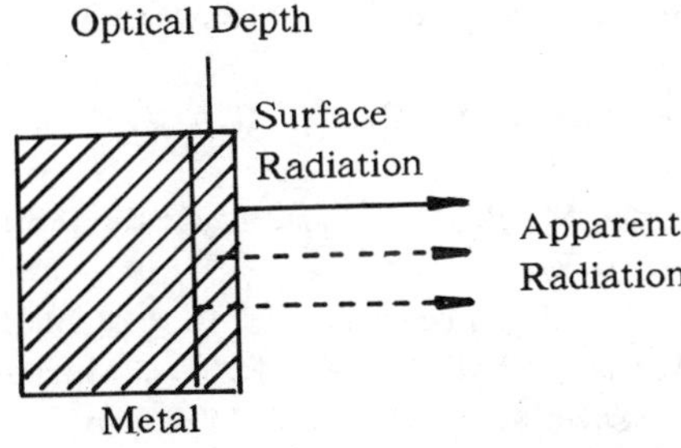

FIGURE 1. Schematic of radiation of an actual radiating body.

Since iron has some transparency under shock loading, the observed spectral radiance from a window is not only the contribution of the interface temperature but the total radiative effective of all the layer in the optical depth. Figure 1 is a schematic of radiation of an actual radiating body. Therefore, it is necessary to solve the radiative transport equation for getting the apparent spectral radiance. As shown by equation (12), the temperature field is nonsteady, but, the characteristic time of the temperature occurring distinct variation is much longer than the time that light travels a free path, the radiation field can be regarded as quasi-steady, which is suitable to the instantaneous distribution of the temperature at any time. So, the radiative transport equation in a one dimensional plate can be expressed as

$$\frac{dI(x)}{dx} = \mu[I_P(T(x,t)) - I(x)] \quad (13)$$

where $I_P(T(x,t))$ is the Planck spectral radiance, $I(x)$ is the total spectral radiance at position x. Combining equations (1), (4), (9), (10), and (13), we can get the apparent spectral radiance at iron/sapphire interface versus time as shown in Fig. 2. From Fig. 2, we can see that the spectral radiance starts from an initial spike, then attenuates to a stable value within about 10 ns. The calculation result indicates that the initial spike of the spectral history corresponds to the release temperature of iron, T_R, while the stable value refers to the interface temperature T_I[1]:

$$T_I = T_R - \frac{T_R - T_W}{1 + \alpha} \quad (14)$$

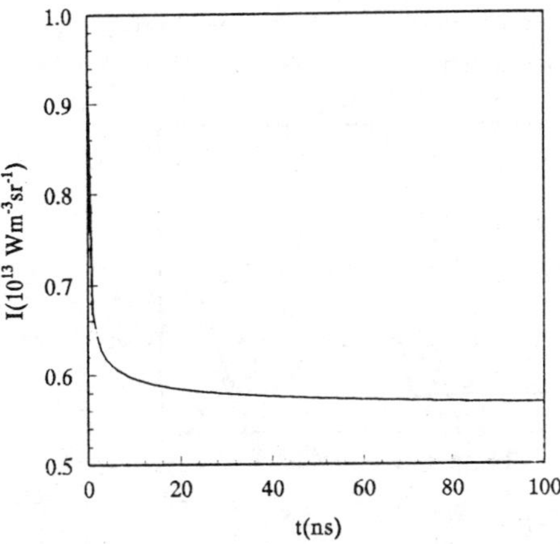

FIGURE 2. The apparent spectral radiance emitted from the shocked iron/sapphire interface versus time ($\lambda = 0.58$ μm).

DISCUSSION AND CONCLUSION

In the measurement of shock temperature of a metal, nonideal thermal contact can lead to a long thermal relaxation process at metal/window interface. Thus, to record the spectral radiance history presented in Fig. 2, the interface should be nearly perfect, and the time resolution of the measurement system should be higher than 1 ns.

So the experimental verification is very difficult due to the harsh experimental conditions. Figure 3 is a typical spectral history from shocked iron — film/sapphire interface obtained by Bass et al. [2]. One can see that the radiance history after about 10 ns in Fig. 2 is in good agreement with the experimental result. Of course, the spectral histories within the initial 10 ns, which are shown in Fig. 2 and Fig. 3, respectively, are not comparable due to the time resolution of the measurement system, which is about 10 ns. However, with the raise of the time resolution of the pyrometer system, it is possible to record the spectral history presented in Fig. 2. In this case, the release temperature of the metal can be measured directly. This is of significance to increase the measurement precision of the shock temperature of the metal.

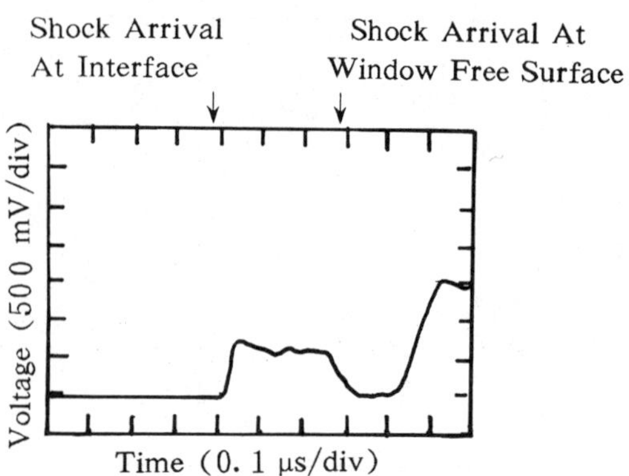

FIGURE 3. Reproduction of oscillogram of the spectral radiance from iron/ sapphire interface following shock compression, where the shock pressure of iron is 225 GPa, the interface pressure is 177 GPa($\lambda = 0.45\,\mu$m, from Ref. 2)

To sum up, when a metal under shock loading, the absorption coefficient decreases, accordingly, the optical depth increases. For an ideal metal/ window interface, the apparent spectral radiance is time dependent due to the transport of the light even though the interface temperature is time independent . Because the initial value of the apparent spectral history refers to the release temperature of the metal, this result provides a theoretical evidence for measuring the release temperature directly. Of course, the direct measurement of the release temperature presupposes the raising the time resolution of the pyrometer.

REFERENCES

1. Grover,R. , and Urtiew,P. A. , *J. Appl. Phys.* **45**,146—152(1974).
2. Bass,J. D. ,Svendsen,B. , and Ahrens,T. J. , in *High Pressure Research in Mineral Physics*,edited by Manghnani,M. H. ,and Syono,Y. ,Tokyo: Terra Press, 1987,pp. 393—402.
3. Tang, W. , Zhang, R. , Jing, F. , and Hu,J. , Restudy on the Thermal Relaxation at Interfaces Following Shock Compression, Presented at the Conference on *Shock Compression of Condensed Matter*—1995,Seattle, Washington, August 13 — 18, 1995, in this proceedings.
4. Urtiew,P. A. , and Grover,R. , *J. Appl. Phys.* **45**,140—145(1974).
5. Boslough, M. B. , *J. Appl. Phys.* **58**, 3394—3401(1985).
6. Billings,B. H. , Frederikse,H. P. R. , Bleil,D. F. , Lindsay,R. B. , Cook,R. K. , Marion,J. B. , Crosswhite,H. M. , and Zemansky,M. W. (ed.), in *American Institute of Physics Handbook*, Third Edition, New York: McGraw — Hill Press,1972,pp. 6—140.
7. Gerritsen, A. N. , *Metallic Conductivity*, *Handbuch der Physik*, Berlin: Springer — Verlay Press, 1956,pp. 137.
8. Tan. H. , and Ahrens, T. J. , *High Pressure Research* **2**,159—182(1990).

HIGH-SPEED OPTICAL STUDIES OF THE DRIVING PLASMA IN LASER ACCELERATION OF FLYER PLATES*

Wayne M. Trott

Sandia National Laboratories, Albuquerque, NM 87185-0834

High-speed optical methods have been used to study several characteristics of the high-temperature driving plasma in laser acceleration of flyers with optical fiber coupling. Flyer materials examined include pure aluminum and composite targets containing an insulating layer of Al_2O_3. Results from fast-gated spectroscopy show that silicon appears early in the plasma spectrum, indicating that the coupling fiber material plays an active role in flyer acceleration. Plasma emission persists longer and pressure broadening of discrete lines decays more slowly in the launch of composite flyers. Fast-gated spectroscopy has also been used in conjunction with very thin ("trace") metal layers deposited in the target fabrication process to examine the depth of ablation by the driving plasma.

INTRODUCTION

Laser-driven acceleration of flyer plates affords a promising approach to well-controlled, short-pulse shock compression of condensed phase materials. With this application in mind, the generation and characterization of laser-accelerated flyers have been subjects of active investigation (1-8). The use of step-index, multimode optical fibers for coupling laser energy to the flyer target offers several advantages in optimization of this experimental technique (1) and has been exploited in detailed studies of flyer velocity-time histories as a function of incident fluence, optical pulse duration, target area and thickness, etc. (3). These experiments in combination with parallel theoretical studies (4,5) have shown that a modest laser driver can deliver predictable and precisely variable flyer impactor velocities over a wide range of experimental parameters. For many practical operating conditions, simple scaling laws adequately describe the dependence of final flyer velocity on several of these parameters, including incident fluence, target mass, etc.

The dynamic physical behavior of laser-driven flyers is an additional, crucial factor in determining their utility in shock compression studies. Tests with pulsed-laser stereo photography (6) and other high-speed imaging techniques (6,7) have shown that excellent overall flyer planarity can generally be obtained; however, the acceleration process can give rise to a number of complex, multidimensional effects that are evident on a small scale. For example, very early in the launch process, the flyer surface typically exhibits a fine-structure nonplanarity that is consistent with the apparent spatial frequency of modal noise in the optical intensity distribution at the fiber output (7). Reflecting changes in spatial properties of the evolving plasma, these small-scale perturbations grow significantly with time. Rapid heat transfer from the high-temperature plasma to the accelerating flyer can also lead to melt and vaporization of a significant fraction of the plate material (9,10). The vital role of the expanding plasma in laser acceleration of flyers motivates direct studies of this aspect of the launch process.

In this paper, we describe recent observations of the driving plasma, including broadband measurements of the time-dependent intensity of plasma emission in combination with spectral data obtained with a fast-gated optical multichannel analyzer (OMA) system. Characteristics of the plasma generated in the launch of pure aluminum flyers are compared to those observed with composite materials

*This work performed at Sandia National Laboratories supported by the U. S. Department of Energy under contract DE-AC04-94AL85000.

containing an insulating layer of Al_2O_3. This comparison is of interest in view of the well-established (2,8) superior performance of composite flyers/impactors (i.e., consistently higher peak velocity vs. incident optical fluence and larger amplitude and duration of pressure pulses delivered to an acceptor). Fast-gated spectroscopy has also been used to examine the plasma generated in acceleration of targets containing very thin layers of different ("trace") metals deposited with the usual materials. These tests suggest a useful approach for examination of the depth of flyer ablation by the plasma.

EXPERIMENTAL

Important elements of the experimental design used in this work are shown schematically in Fig. 1. The setup for flyer generation using optical fiber coupling has been described in detail previously (1,3). The driving laser was a Q-switched Nd:Glass oscillator (Lasermetrics Model 9380). The laser output ($\lambda = 1.054$ μm) was horizontally polarized and multimode. Laser energy was focused by a 100-mm focal length lens and coupled into the large-diameter (1000 μm) end of a fiber taper that was positioned so that the diverging beam slightly underfilled the input face. This taper concentrated the optical energy into a 400-μm-diameter fiber section. From this element, the energy was proximity coupled either to a thin (0.5 mm) fused silica substrate coated (at the output face) with the flyer target material or to a 1-m-long fiber prepared in a similar manner. The different target samples (described below) were fabricated by physical vapor deposition. Absorbing neutral density filters were used to vary the energy incident on the flyer targets. All tests were performed with laser pulse durations near 18 ns (FWHM).

As shown in Fig. 1, the optical diagnostics were configured to examine plasma emission collected and transmitted in the "backward" direction (i.e., opposite to that of the driving laser propagation path) by the fiber elements. In addition to providing some degree of spatial resolution (defined by the numerical aperture of the final fiber), this sampling scheme favored the detection of fiber material (e.g., Si, O) in the plasma. Direct evidence of such species was sought in order to test computational results which indicate that the fiber compound plays a substantial

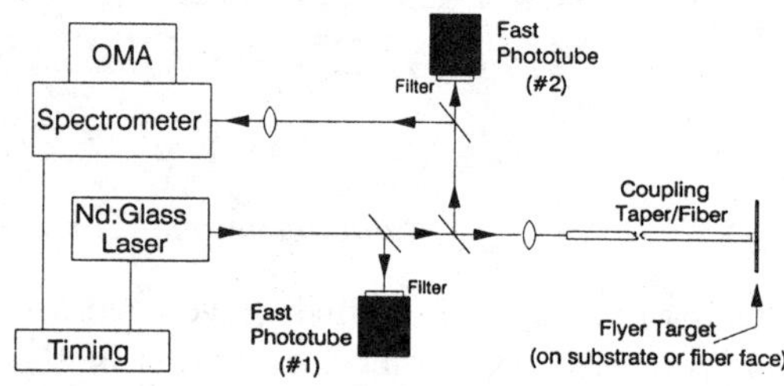

FIGURE 1. Schematic diagram of experimental design for optical studies of the driving plasma in laser acceleration of flyers.

role in the ablation process driving the flyer target (10). UV-transmitting optics were used to facilitate detection of strong Si emission features in the 2500-3000 Å region. Plasma light was directed to the entrance slit of a 0.33-m spectrometer. The dispersed spectrum was viewed by a silicon target vidicon coupled to a gateable proximity focused channel intensifier tube and the signal was processed by the OMA. The detector was gated on by a voltage pulse, which was triggered at variable delay with respect to the driving laser pulse by using digital delay generators. Gates as short as 10 ns could be obtained; however, data collection over 60-ns intervals was sufficient to track the important changes in the plasma spectrum. The temporal profile of plasma intensity was measured by a fast photodetector (time constant < 0.4 ns). For this diagnostic, appropriate broadband filters were used to isolate desired regions of the spectrum and to discriminate against reflected laser light.

RESULTS AND DISCUSSION

Figure 2 presents phototube records of visible plasma light generated in the launch of two different flyer targets: (a) an 8-μm-thick film of pure Al and (b) an 8-μm-thick composite film with an embedded 0.25-μm-thick layer of Al_2O_3 placed 0.25 μm from the film/fiber interface. In both cases, the emission intensity was observed to rise abruptly near the time corresponding to the onset of flyer motion. Comparable peak intensities were obtained with the two materials but emission persisted for a significantly longer time in acceleration of the composite film. Higher peak intensities and shorter emission "pulses" were seen in tests on very thin (2-μm) Al and composite films. Again, a longer pulse was ob-

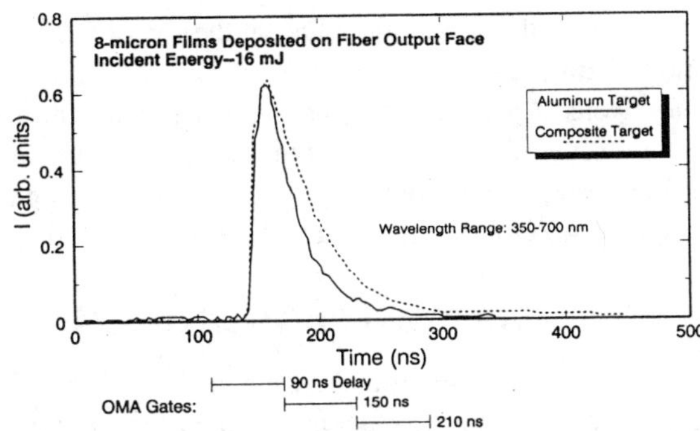

FIGURE 2. Temporal profile of plasma emission intensity observed in the acceleration of two different flyer materials. OMA gates (60-ns intervals) used in spectral studies are shown in relationship to the emission records.

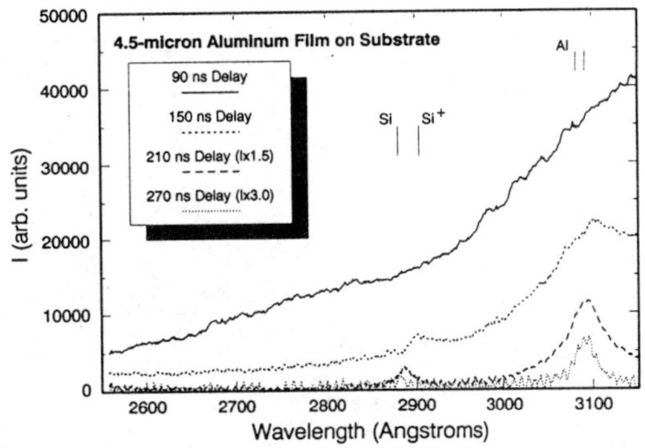

FIGURE 3. Spectra of emission from plasma driving an Al film deposited on a fused silica substrate; 15 mJ incident energy. Each curve represent s the sum of six individual shots.

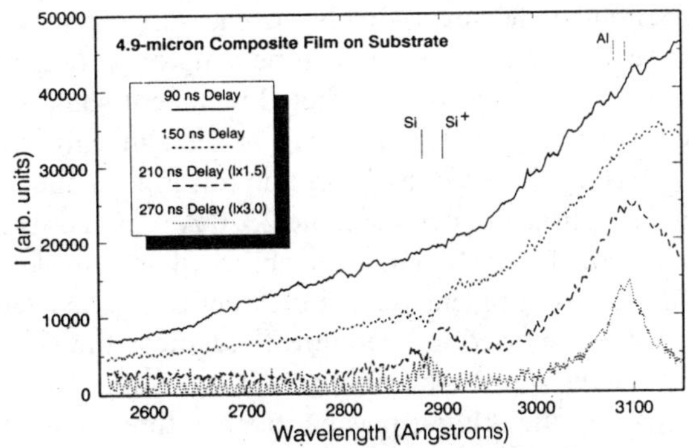

FIGURE 4. Spectra of emission from plasma driving a composite film deposited on a fused silica substrate; 15 mJ incident energy. Each curve represents the sum of six individual shots.

served in the latter case. Relatively long-lived emission appears to be a characteristic of composite flyer acceleration since similar results were obtained in other wavelength bands as well as with different flyer thicknesses. Definitive interpretation of this phenomenon is difficult due to the limited spatial resolution and wavelength bandpass of the data, uncertain optical thickness of the plasma, etc. It is likely that the thermal confinement of the insulating Al_2O_3 layer results in the plasma remaining hotter for a slightly longer time. Also, interactions giving rise to pressure broadening occur over a longer time in the composite flyer case (as demonstrated below).

Changes in the spectral properties (2550-3150 Å region) of plasma driving Al and composite films are illustrated in Figs. 3 and 4, respectively. The relationship of the nominal delay times cited in these figures to the time-dependent emission intensity is shown in Fig. 2 (e.g., the gate designated "90 ns delay" captured light in the first ~30 ns of flyer acceleration, etc.). For both flyer types, the early spectrum is dominated by continuum emission, a reasonable result in view of the high particle densities and field strengths in the plasma during this time. Discrete features from vaporized and ionized material appear at later intervals, as the plasma rapidly expands. Identified features include an Al multiplet (3082 Å, 3093 Å), a Si line at 2882 Å, and a Si$^+$ line at 2906 Å (11,12). Similar features have been seen in other wavelength regions; e.g., a Si multiplet near 2520 Å, a Si$^+$ line at 4131 Å, and strong Al lines at 3944 Å and 3962 Å.

In accordance with the phototube data, the overall intensity of plasma emission falls off more quickly in the launch of a pure Al flyer. Pressure broadening of the Al lines clearly decays faster in this event, as well. This result suggests that material associated with the plasma remains more dense and compact during acceleration of a composite target, an interpretation consistent with other data on the visual appearance of the region behind the flyer in both cases, as observed by back lit shadowgraphy (7) and an image motion camera technique (8).

Of particular interest is the definite identification of silicon features, indicating substantial vaporization and ionization of the fiber material. During the earliest stages of acceleration, the high particle density, etc. in the plasma seems to preclude direct

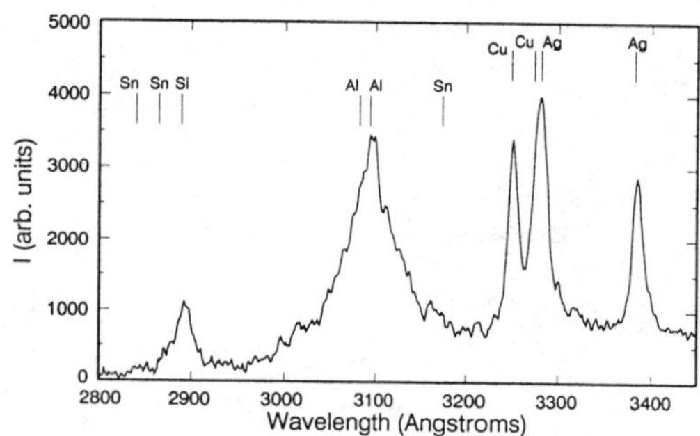

FIGURE 5. Late-time spectrum of emission from plasma driving an Al flyer target containing embedded 0.01-μm-thick layers of copper, silver and tin; 15 mJ incident energy.

resolution of these features. Their relatively early appearance, however, lends strong support to the concept that the fiber core can be a major participant in pushing the flyer, as predicted by recent computational work (10). Initially, Si^+ is seen in emission while Si is observed in absorption (cf. Figs. 3 and 4). Si absorption also occurs in the 2520 Å region during this time. Later, Si alone is observed in emission. This time-dependent behavior is entirely consistent with the expected temperature fields near the fiber/target interface. As the fiber compound begins to participate in the driving plasma, temperature increases from the "cold" fiber to a region of vaporized material, with still higher temperatures occurring in the ionized regime. At late times, expansion cools the plasma so that only neutrals remain in emission.

The utility of thin layers of different metals as a diagnostic approach for the study of flyer ablation and thinning is demonstrated in Fig. 5. The prepared target (4-μm total thickness) consisted of aluminum embedded with 0.01-μm layers of copper, silver, and tin. These layers were positioned at 0.10 μm, 0.21 μm, and 0.32 μm from the fiber/target interface, respectively. Prominent lines of all relevant materials appear in the 2800-3400 Å region, facilitating acquisition of a useful spectrum on a single shot. Important features include two Cu lines (3248 Å, 3274 Å), two Ag lines (3281 Å, 3383 Å) and three Sn lines (2840 Å, 2863 Å, 3175 Å). The Cu and Ag features are very strong, implying vaporization to a depth of at least 0.22 μm. The relatively weak Sn lines may reflect less vaporization at 0.33 μm; however, lower intensities for these features are also consistent with published line strengths (11). Spectra from samples

with layers of copper and silver at depths of 0.25 μm and 0.5 μm, respectively, exhibit strong Cu lines and very weak Ag features. This suggests that ablation can involve >0.5 μm of the flyer target. The substantial vaporization depths observed here are once again in good agreement with computational results (10).

ACKNOWLEDGMENTS

The excellent technical assistance of Jaime N. Castañeda is gratefully acknowledged. Flyer targets were skillfully prepared by Catharine Sifford (Thin Film & Brazing Team, SNL).

REFERENCES

1. Trott, W. M., and Meeks, K. D., "Acceleration of Thin Foil Targets Using Fiber-Coupled Optical Pulses," in *Shock Waves in Condensed Matter--1989*, eds. S. C. Schmidt, et al., New York: Elsevier Science Publishers, 1990, pp. 997-1000.

2. D. L. Paisley, "Laser-Driven Miniature Flyer Plates for Shock Initiation of Secondary Explosives," in *Shock Waves in Condensed Matter--1989*, eds. S. C. Schmidt, et al., New York: Elsevier Science Publishers, 1990, pp. 733-736.

3. Trott, W. M., "Studies of Laser-Driven Flyer Acceleration Using Optical Fiber Coupling," in *Shock Waves in Condensed Matter--1991*, eds. S. C. Schmidt, et al., New York: Elsevier Science Publishers, 1992, pp. 829-832.

4. Farnsworth, A. V., and Lawrence, R. J., "Numerical and Analytical Analysis of Thin Laser-Driven Flyer Plates," in *Shock Waves in Condensed Matter--1991*, eds. S. C. Schmidt, et al., New York: Elsevier Science Publishers, 1992, pp. 821-824.

5. Lawrence, R. J., and Trott, W. M., "Theoretical Analysis of a Pulsed-Laser-Driven Hypervelocity Flyer Launcher," *Int. J. Impact Engineering* **14**, 439-449 (1993).

6. Paisley, D. L., Montoya, N. I., Stahl, D. B., and Garcia, I. A., "Interferometry, Streak Photography, and Stereo Photography of Laser-Driven Miniature Flyer Plates," in *SPIE Proceedings No. 1358*, Cambridge, UK, 1990, pp. 760-765.

7. Frank, A. M., and Trott, W. M., "Stop Motion Microphotography of Laser Driven Plates," in *SPIE Proceedings No. 2273*, San Diego, CA, 1994, pp. 196-206.

8. Trott, W. M., "Investigation of the Dynamic Behavior of Laser-Driven Flyers," in *High-Pressure Science and Technology--1993*, eds. S. C. Schmidt, et al., New York: AIP Press, 1994, pp. 1655-1658.

9. Frank, A. M., and Trott, W. M., "Investigation of Thin Laser-Driven Flyer Plates Using Streak Imaging and Stop-Motion Microphotography," (this volume).

10. Farnsworth, Jr., A. V., "Laser Acceleration of Thin Flyers" (this volume).

11. Meggers, W. F., Corliss, C. H., and Scribner, B. F., *Tables of Spectral-Line Intensities, 2nd Edition*, Washington: National Bureau of Standards, 1975.

12. Lanz, T., and Artru, M.-C., *Physica Scripta* **32**, 115-124 (1985).

MEASUREMENT OF STRAIN AND TEMPERATURE FIELDS DURING DYNAMIC SHEAR OF EXPLOSIVES

Blaine W. Asay, Gary W. Laabs, Paul D. Peterson,
David J. Funk

Los Alamos National Laboratory, Los Alamos, NM 87545

We have used dynamic speckle photography to spatially resolve the displacement field in a precracked specimen of the plastic bonded explosive PBX 9501 during asymmetric impact. We have also used a linear array of InSb detectors to acquire preliminary data which show a temperature rise in the explosive upon impact. Although the work is in its initial stages, these results show the potential applicability of these methods to processes which include high rate deformation of reactive materials.

INTRODUCTION

Laser methods have been used for many years to obtain in-plane displacement fields on the surface of materials during quasistatic and dynamic deformation.[1] These methods include laser speckle interferometry, laser and white light speckle photography, moirè photography and many others. Huntley[2] has shown how laser speckle photography can be performed at high rates using inert materials. These methods have also been applied to low rate[3] experiments involving high explosives and high rate processes in inert materials.[2,4] Other methods are capable of determining out-of-plane displacement.[5]

Zehnder and Rosakis[6] have demonstrated high-resolution temperature measurements at the tip of propagating cracks in inert materials using precisely aligned InSb diodes. Others[7] have used visible light in an attempt to determine temperatures of explosives during initiation.

We are beginning to apply white light and laser speckle photography and IR radiometric techniques to determine displacement and temperature fields in explosives during moderate- and high rate deformation and penetration. These data are essential to corroborate hydrocode calculations which predict damage and, in some cases, initiation of the explosive. The data are also useful in the construction of constitutive models. Visualization (by means of spatially and temporally resolved temperature and strain measurements) of the energy localization (e.g., shear bands) that occurs during dynamic deformation of explosives is a great aid to a better understanding of initiation mechanisms.

SPECKLE PHOTOGRAPHY

When coherent light strikes a rough surface the reflected light constructively and destructively interferes. When imaged, this interference gives rise to a speckle pattern. This pattern is a unique function of the surface, and as the surface is displaced, the speckle pattern also moves proportionally. In conventional laser speckle photography, the speckle pattern from the undisplaced surface is first recorded. The surface is then deformed, and another photograph is taken on the same film, creating a double exposure. This image is then analyzed either with a laser probe[8] or digitally.[9] The size of the speckle is controlled by limiting stop of the system and the wavelength of the laser used. There is an optimum speckle size which maximizes resolution.

Alternatively, a surface can be painted to achieve a speckle-like appearance and a noncoherent source can be used for illumination. These artificial speckles (sometimes called "white light speckles") then serve the same purpose as laser-induced speckles, and the analysis is the same. The white light speckle method is much more robust and not as sensitive to decorrelation arising from out-of-plane motion of the specimen. However, it is also not as accurate in part, because the speckle size cannot be as accurately determined.

DYNAMIC DISPLACEMENT IN PBX 9501

We used white light speckle photography in the experiment reported here. This was done to maximize the correlation in the region of large out-of-plane motion in the vicinity of the projectile. Figure 1 shows a schematic of the projectile impacting the explosive sample. The PBX 9501 (25 x 15 x 5 mm) was prepared by lightly spraying the surface with alternating black and white paint. The sample was unconfined and attached to the stand with epoxy. A small notch (0.2 x 4 mm) was cut in the specimen 12.5 mm from the bottom. The brass projectile was a 10 x 17 cm right circular cylinder, and was fired from a gas gun.

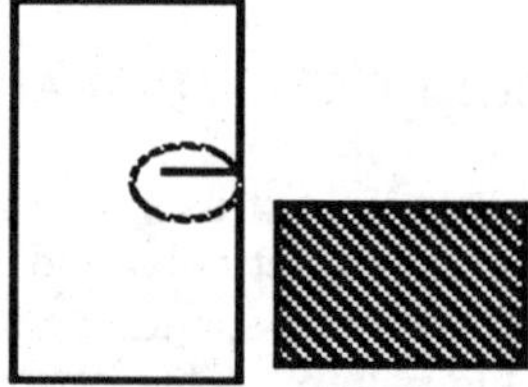

Figure 1. Schematic showing the spatial relationship of the projectile to the explosive sample. The crack is shown as a horizontal line. The circle refers to the location of the data acquisition shown in Figs. 2 and 3.

The front face of the explosive was illuminated with a Nd:YAG laser (New Wave Research Inc., Mini/Lase) operating at 532 nm with 15 mJ/pulse and a pulse length of 6-7 ns. The flash lamp and Q switch were independently triggered. The flash lamp was triggered when the projectile crossed a laser beam placed in its path at a known distance from the gun, and the Q switch was triggered (after an appropriate delay) by a PVDF gauge (0.228 mm- thick) mounted on the impact surface of the explosive. The gauge was covered by a 2 mm-thick layer of neoprene.

A coherent illumination source was used in this experiment to achieve very rapid shuttering and sufficient exposure. However, special care was taken to reduce the size (inasmuch as possible) of the laser speckle. This was done by using an f/2 lens (Pentax, 50 mm) and a magnification of 0.922. Examination of static images showed that the laser speckle was negligible. The explosive surface was imaged onto a CCD camera (Kodak, Megaplus 4.2), comprised of an array of 2048 x 2029 square pixels, 9 microns on each side. An image was recorded before the experiment. This was then combined with the image taken after projectile impact. The images were then analyzed using two forward fast Fourier transforms[9] with autocorrelation halo subtraction.[8]

Figures 2 and 3 show displacement contours in the region identified in Figure 1, 6 μs after impact. The impact velocity was 91.1 m/s. One dimensional shock calculations predict a shock velocity in the explosive of 2.75 mm/μs and a shock pressure of 0.2 GPa. These values are well below the initiation threshold of PBX 9501. There is a large gradient in displacement (strain) in both the x and y directions in the vicinity of the crack, showing the effects that small inclusions can have on the flow field during penetration of projectiles.

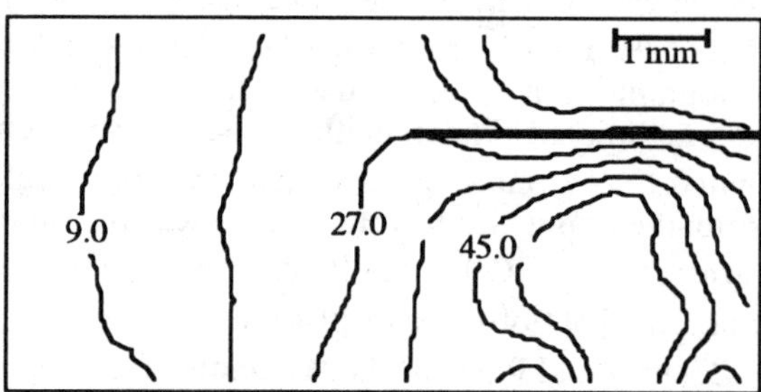

Figure 2. Contours of displacement in the vertical direction, 6 μs after impact. Contours are spaced 9 μm apart. The horizontal line shows the approximate location of the crack.

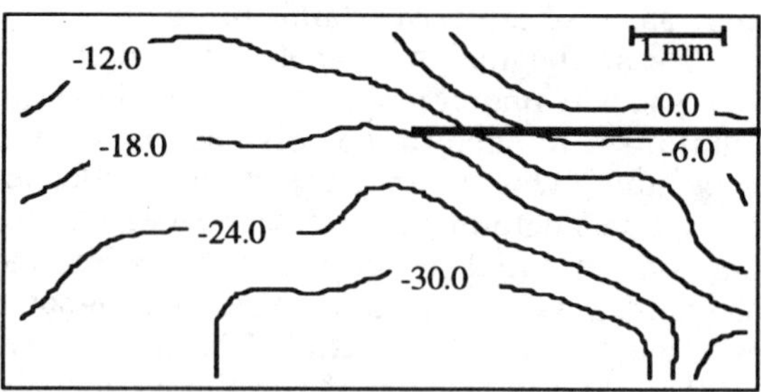

Figure 3. Contours of displacement in the horizontal direction, 6 μs after impact. Contours are spaced 6 μm apart. The horizontal line shows the approximate location of the crack.

Figure 4 shows the same region of the specimen 15 μs after impact in a separate experiment. The impact velocity was slightly lower (87.0 m/s) but the records can be compared qualitatively. Only horizontal displacement is shown, because of space limitations. Note the larger overall displacement as well as the increased gradients that occur at later times.

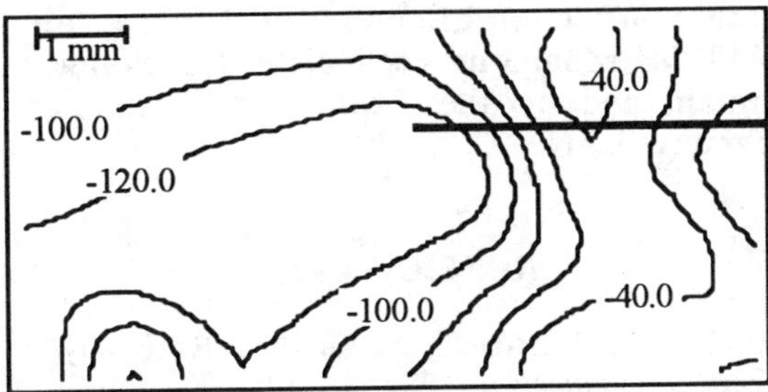

Figure 4. Contours of displacement in the horizontal direction., 15 μs after impact. Contours are spaced 20 μm apart. The horizontal line shows the approximate location of the crack.

TEMPERATURE MEASUREMENTS

Zehnder and Rosakis[6] have demonstrated the utility of using a linear array of InSb diodes to measure temperature at the tip of propagating cracks in metals. We have used a similar device in an attempt to measure temperatures in HE undergoing impact. Because the mechanical strength of an explosive is so much less than that of a metal, the temperatures we expect, particularly in unconfined HE undergoing deformation at a moderate rate, will be significantly lower. However, relatively low temperatures (i.e., 200 C), when localized, can initiate reaction which may lead to a detonation.

Our camera (Cincinnati Electronics, model SDD-066) consists of two sets of eight diodes, with each set spaced 5 mm apart. The individual diodes are 81 μm square spaced on 103 μm centers. The frequency response is 2 MHz, and the spectral sensitivity is from 1 to 5.5 μm. Digitizing modules (LeCroy Model 8818 and Jorger Model TR200) sampled the diode output at 160 ns/point. An f/4 Ge/Si optimized lens was used to image the target at a magnification of 1.0

These experiments used a symmetric impact instead of the asymmetric impact used to measure displacement (see Fig. 5). The first linear array was imaged onto a region 3 mm from the front surface, and the second array was thus 8 mm from the surface. The impact velocity was 167 m/s and the digitizers were triggered by a PVDF gauge with no neoprene cover.

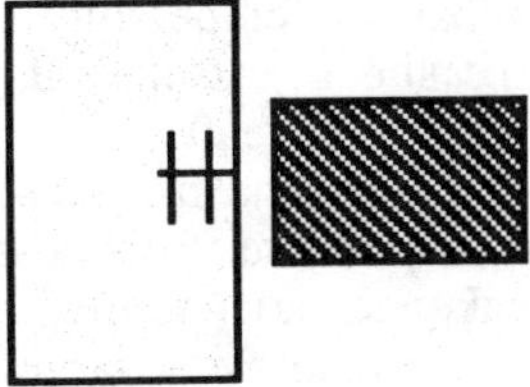

Figure 5. Experimental configuration used to measure temperature during impact. Horizontal line represents crack, while two vertical lines represent orientation of two linear InSb arrays (not to scale).

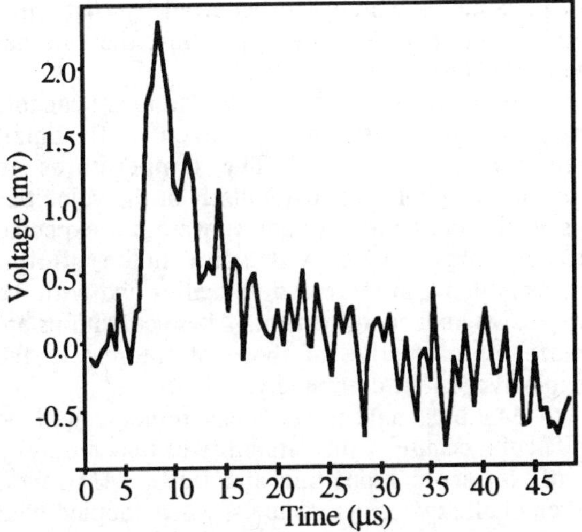

Figure 6. Voltage record obtained from a single diode in the first linear InSb array after projectile impact. Zero time corresponds to impact.

The record obtained from a single diode in the first array is shown in Fig. 6. Records from the other diodes in that array were similar. One dimensional shock calculations predict a shock velocity in the explosive of 2.81 mm/μs and a pressure of 0.3 GPa. Thus, 5 μs after impact, assuming one dimensional conditions and no attenuation, the shock would reflect from the rear surface.

Based on measurements of a calibrated black body source and calculations, we estimate that a 2 mv signal corresponds to a temperature rise of from 5-50 C. The uncertainty arises because the emissivity of the explosive is unknown and varies across the surface. We will be making measurements to better quantify this value in the near future.

It appears that the temperature rise occurred shortly after the shock passed, but that cooling occurred very quickly thereafter. The second array positioned 8 mm from the front surface did not report a rise in temperature. It is possible that the shock had weakened sufficiently, and that the multidimensional effects were large enough that any temperature rise that did occur was below our detection limits. Also, triboluminescence effects have not been ruled out.

At times exceeding 100 μs, other voltage peaks of somewhat increased magnitude were noted with both arrays. However, because of the large displacements which had occurred by that time, and the large uncertainty in position, they are not reported here.

These data show that the detectors are capable of making measurements, even with small temperature changes. The temperatures in unconfined explosive, particularly at the velocities used in this preliminary study, were not expected to be large. We will soon make similar measurements at increased velocities and with the explosive under confinement. These conditions are more representative of those of interest to the explosives safety community.

Making quantitative measurements will be difficult because of the variability of the emissivity with surface composition and finish. However, even qualitative measurements, when coupled with the displacement/strain fields, will be useful as we model the penetration/impact process.

CONCLUSIONS

We have shown that speckle photography and linear arrays of InSb detectors are capable of providing information regarding the dynamic behavior of explosives during deformations at moderate strain rates. Small diode signals were observed. However, energy localization (and thus temperature) will be greatly increased when confined explosives are used.

ACKNOWLEDGMENTS

Special thanks are given to P. Dickson, T. Goldrein, and J. Huntley for many helpful discussions. This work was supported by the Department of Energy and the Department of Defense/Office of Munitions under the Joint DoD/DOE Munitions Technology Development Program and by the Explosives Technology Program at LANL.

REFERENCES

1. Chiang, F. P.; Adachi, J.; Anastasi, R.; Beatty, J. *Optical Engineering* **1982**, *21*, 379-390.
2. Huntley, J. M.; Field, J. E. In *Prococeedings of the Second International Conference on Photomechanics and Speckle Metrology*; San Diego, CA, 1991.
3. Goldrein, H. T.; Huntley, J. M. In *Tenth Symposium (Int.) on Detonation*; Boston, MA, 1993.
4. Vogel, D.; Michel, B.; Totzauer, W.; Schreppel, U.; Clos, R. In *SPIE Vol. 1554A Speckle Techniques, Birefringence Methods, and Applications to Solid Mechanics*; 1991; pp 262-274.
5. Mason, J. J.; Rosakis, A. J.; Ravichandran, G. *J. Mech. Phys. Solids* **1994**, *42*, 1679-1697.
6. Zehnder, A. T.; Rosakis, A. J. In *Experimental Techniques in Fracture*; J. S. Epstein, Ed.; VCH: ; pp 125-169.
7. Setchel, R. E. In *Shock Waves in Condensed Matter*; Elsevier Science: 1987; pp 553-556.
8. Huntley, J. M. *J. Phys. E:Sci. Instrum* **1986**, *19*, 43-48.
9. Chen, D. J.; Chiang, F. P. *Experimental Mechanics* **1992**, 145-153.

THEORY OF SHOCK MAGNETIZATION OF ASTEROIDS GASPRA AND IDA

George Q. Chen, Thomas J. Ahrens

Lindhurst Laboratory of Experimental Geophysics, Seismological Laboratory
California Institute of Technology, Pasadena, CA 91125

and Raymond Hide

Department of Physics and Earth Sciences, University of Oxford
Oxford OX1 3PU, England, U.K.

The observed magnetism of asteroids such as Gaspra and Ida (and other small bodies in the solar system including the Moon and meteorites) may have resulted from an impact-induced shock wave producing a thermodynamic state in which iron-nickel alloy, dispersed in a silicate matrix, is driven from the usual low-temperature, low-pressure, α, kaemacite, phase to the paramagnetic, ϵ (hcp), phase. The magnetization was acquired upon rarefaction and reentry into the ferromagnetic, α, structure. The degree of re-magnetization depends on the strength of the ambient field, which may have been associated with a solar-system-wide magnetic field. A transient field induced by the impact event itself may have resulted in a significant, or possibly, even a dominant contribution, as well. The scaling law for catastrophic asteroid impact disaggregation imposes a constraint on the degree to which small planetary bodies may be magnetized and yet survive fragmentation by the same event. Our modeling results show it is possible Ida was magnetized when a large impact fractured a 125±22 km-radius proto-asteroid to form the Koronis family. Similarly, we calculate that Gaspra could be a magnetized fragment of a 45±15 km-radius proto-asteroid.

INTRODUCTION

Magnetism of the Moon and other small bodies in the solar system has been a controversial topic (see, *e.g.* (1, 2)), and has only become more interesting since the recent flybys of the asteroid 951 Gaspra and the larger asteroid 243 Ida by the Galileo spacecraft, which have found that both of them may be sufficiently electrically conducting so as to perturb the interplanetary magnetic field, or they are magnetic (3). Ida is a member of the Koronis family, a group of asteroids with similar eccentricities and inclinations which are thought to all be the post-collision fragments of a single proto-asteroid. Here we present a quantitative model evaluating the extent of magnetization by hypervelocity impacts–one of a few magnetizing mechanisms previously suggested–using phase diagrams of magnetic minerals, shock and post-shock temperature calculations, and a fracturing model by Housen *et al.* (4). We conclude pressure-induced structural changes are responsible for magnetization of low-porosity rocks; Impacts are generally incapable of magnetizing a planetary body throughout, but impact magnetization may offer a valid explanation for small magnetic asteroids like Gaspra or Ida which are thought to be impact fragments of larger bodies.

SHOCK-INDUCED MAGNETIZATION

We first study metallic iron embedded in a silicate (lunar rock, as described in (5)) matrix. Shock temperature calculations are shown with iron's phase diagram in Fig. 1. Three distinct magnetization mechanisms are possible in different shock pressure-temperature regimes:

1. If the Hugoniot in *P-T* plane crosses the Curie point at pressures between 0 and about 1.75 GPa (Fig. 1), natural Curie-point writing occurs during or after being shock-heated to above the Curie temperature (1043K for pure iron at 1bar). The phase change is second order. This mechanism requires intensive shock heating and only occurs upon shocking silicate iron-bearing rocks that are less than ~40% of crystal density.

2. If the Hugoniot crosses the phase boundary between (1.75 GPa, 1043 K) and the α-ϵ-γ triple point at (11.0 GPa, 750 K), iron undergoes a first order phase transformation from ferromagnetic body-center-cubic (bcc) structure (α phase) to paramagnetic face-center-cubic (fcc) structure (γ phase) (6, 7). When on the release of pressure the system returns through the phase boundary, the reverse transition occurs and the material becomes stably magnetized. Silicate rocks with between 40 to 80% of crystal density containing kaemacite can be magnetized via this method.

3. Shocked silicate rock with greater than $\sim$ 80% crystal density may be magnetized upon the crossing of the Hugoniot with the α-ϵ phase boundary (between the α-ϵ-γ triple point and (273 K, 14 GPa)). The high pressure ϵ phase has hcp structure and is paramagnetic. The transition pressure is slightly temperature dependent (from about 14 GPa at room temperature to about 11 GPa at the triple point), but can be taken to be approximately 13GPa.

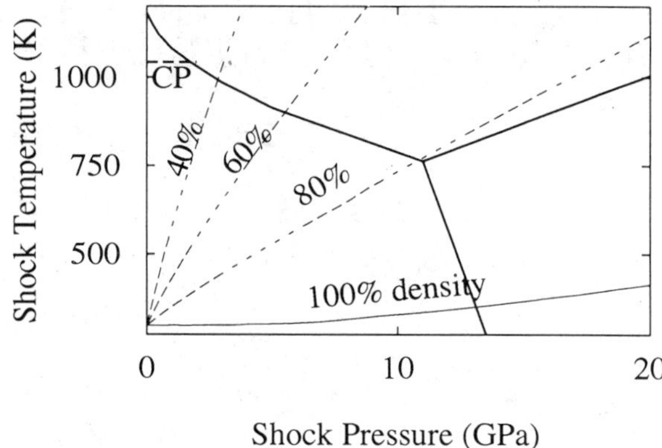

FIGURE 1. Shock temperatures vs. shock pressure of gabbroic anorthosite of different porosity values. "100% density" is 2.936 g/cm^3. Iron phase diagram is superimposed to demonstrate different transitions at different porosities. The dashed line with the label "CP" is the Curie temperature of iron.

Similar calculations have been conducted for more realistic magnetic carriers, *e.g.*, kaemacite (FeNi) and magnetite (8). Although they have different Curie temperatures and phase diagrams than those of iron, the conclusion remains that phase changes at relatively low pressures (<20 GPa) are a major shock magnetization mechanism.

From the Holsapple-Schmidt scaling of planetary impacts (9), the radius inside which the target is shocked above the threshold pressure (hereafter called magnetization radius) can be obtained for various impact conditions.

FRAGMENTATION OF ASTEROIDS

An important question is whether the proto-asteroid can remain largely integral and yet be driven to a sufficient shock pressure when it is shock-magnetized. Housen *et al.* (4) developed a catastrophic fragmentation (CF, defined as when the largest fragment mass is equal to one-half of that of the original target) threshold based on dimensional analysis and laboratory fragmentation experiments. The ratio of the largest fragment mass (M_{L}) to the total proto-asteroid mass (M) is given by:

$$\frac{M_{\mathrm{L}}}{M} = F'\left(\frac{Y_{\mathrm{t}} + \sigma_{\mathrm{G}}}{\sigma_{\mathrm{I}}}\right) \qquad (1)$$

where Y_{t} is the material fracture strength, σ_{G} is the lithostatic stress, and σ_{I} is impact-induced tension. The above equation suggests lithostatic stress has the effect of strengthening the target, which was demonstrated in Housen *et al.*'s hydrostatically loaded fragmentation experiments (4).

The function $F'(x)$ has the form:

$$F'(x) = 1 - 2^{3\mu/2-1}K'x^{-3\mu/2} \qquad (2)$$

where μ is a measure of shock wave attenuation in the target material ($\mu=0.4$ for sand, 0.55–0.6 for rock), K' is an experimentally determined constant ($\sim 2.4 \times 10^{-3}$).

CONCLUSION

At a given impact velocity and target size, there is a maximum impactor size above which the proto-asteroid is fragmented catastrophically. The radius of magnetization at this impactor size is the limit of magnetization for the target, if it survives the impact. This limit (*vs.* impact velocity) is plotted in Fig. 2. It can be seen from the figure that it is very unlikely or impossible to magnetize an asteroid by hypervelocity impact without severely fracturing it. On the other hand, impact-induced magnetization on an unfragmented body (like the Moon) must be limited to the vicinity of impact center, and if it has been under multiple impacts, its magnetic field should have a "patchy" characteristic.

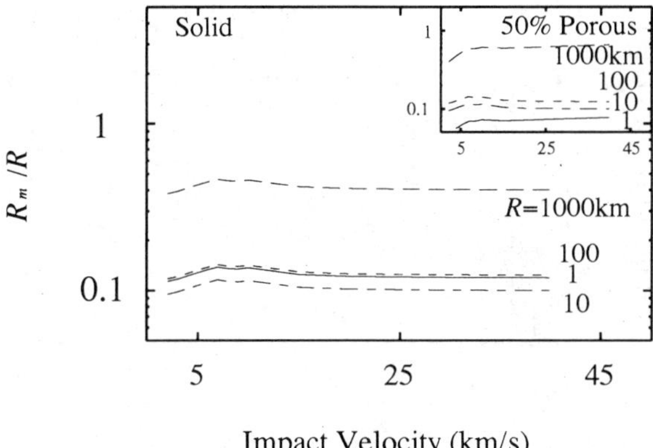

FIGURE 2. Ratio of magnetization radius (R_m) over proto-asteroid radius (R) at catastrophic fragmentation threshold for 100% density rock. Inset: same calculation for 50%-porosity rock. Calculations are done for difference proto-asteroid sizes, the radii are labeled next to the curves. All curves are below $R_m/R=0.9$, suggesting that the proto-asteroid is fragmented before it can be completely magnetized.

DISCUSSIONS

Assuming both Gaspra and Ida were completely impact-magnetized, we can obtain a constraint on the minimum sizes of the impactors. Then, requiring the largest fragments (from the same impact) be larger than Gaspra or Ida, lower limits on the pre-impact asteroid sizes can be set using Equation 1 (Fig. 3). At 5 km/s impact velocity, which is about the most probable in the asteroid belt, we obtained that the proto-Gaspra body was at least 45 ± 15 km and the impactor at least 7.6 ± 0.8 km in radius; For Ida, the minimum radii for parent body and impactor are 125 ± 22 and 27 ± 2 km respectively.

Based on geometrical considerations, the estimated minimum radius of the parent body of the Koronis family (of which Ida is a member) is 45 km to 56 km (10, 11). Considering the numerous uncertainties, especially the importance of frag-

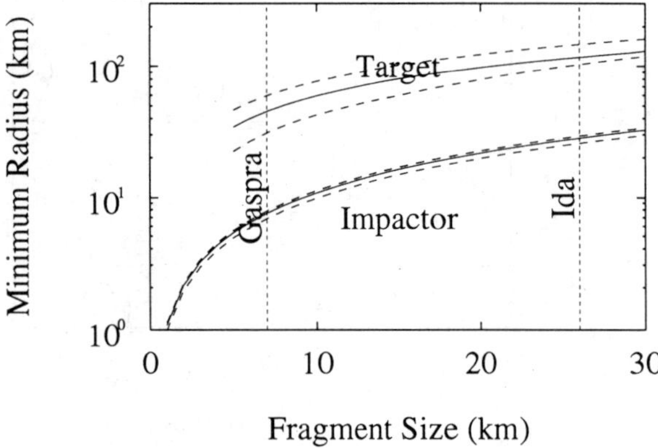

FIGURE 3. Dependence of minimum radii of impactor and pre-impact body on size of the final magnetized fragment, for a given impact velocity of 5 km/s. The dashed lines are obtained by varying target strength and density to determine uncertainty of the model. The minimum radius of impactor is calculated such that R_m is twice the fragment radius.

ment reaccumulation after break-up, we suggest that the present analysis allows, but does not prove, Ida could have been magnetized when a large impact fragmented a proto-asteroid to form the Koronis family.

ACKNOWLEDGMENTS

Research supported by NASA Grant NAGW-1941. Contribution #5592, Division of Geological and Planetary Sciences, California Institute of Technology.

REFERENCES

1. Hood, L. L., and Huang, Z., *J. Geophy. Res.* **96**, 9837–9846 (1991).
2. Collinson, D. W., *Surveys in Geophysics* 14,89–118 (1993).
3. Kivelson, M. G., Bargatze, L. F., Khurana, K. K., Southwood, D. J., Walker, R. J., and Coleman, Jr., P. J., *Science* **261**, 331–334 (1993).
4. Housen, K. R., Schmidt, R. M., and Holsapple, K. A., *Icarus* **94**, 180–190 (1991).
5. Ahrens, T. J., and O'Keefe, J. D., "Equation of state and impact-induced shock-wave attenuation on the moon," in *Impact and Explosion Cratering*, 1977, pp. 639–656.
6. Bundy, F. P., *J. Appl. Phys.* **36**, 616–620 (1965).
7. Liu, L., and Bassett, W. A., *Elements, oxides, and silicates : High-pressure phases with implications for the Earth's interior*, New York: Oxford University Press, 1986, pp. 52–54.
8. Chen, G., Ahrens, T. J., and Hide, R., *Icarus* **115**, 86–96 (1995).
9. Holsapple, K. A., and Schmidt, R. M., *J. Geophys. Res.* **92**, 6350–6376 (1987).
10. Gradie, J. C., Chapman, C. R., and Williams, J. G., "Families of minor planets," in *Asteroids*, 1979, pp. 359–390.
11. Fujiwara, A., *Icarus* **31**, 277–288 (1982).

SHOCK-INDUCED CONDUCTION WAVES IN SOLIDS AND THEIR APPLICATIONS IN HIGH POWER SYSTEMS

S.D.Gilev and A.M.Trubachev

Lavrentyev Institute of Hydrodynamics, Lavrentyev Prosp. 15, Novosibirsk 630090, Russia

Physical processes in the shock wave accompanied by the electroconductivity changes have been investigated theoretically and experimentally. The analysis based on the classical electrodynamics is performed for the dielectric-metal and metal-metal transitions. This has permitted us to develop the new techniques for experimental investigation of condensed matter in shock waves. Allowance for non-stationary electromagnetic processes in a specially designed measuring cell is the distinctive feature of these techniques. The conductivity value and its dependence on the intensity and the type of loading, the temporal characteristics of the transitions, the behavior under the release and other data are obtained in experiments with the silicon, the ytterbium, the metallic powders and foams. The conclusions have been made regarding the character and the mechanism of the shock-induced phase transitions. The shock-induced conduction waves have been applied to switching the high electric currents and producing the ultrahigh magnetic fields.

INTRODUCTION

Compression of the condensed matter in the shock wave (SW) is accompanied by changes of their electrophysical properties. We shall use the term "shock-induced conduction waves" (SICW) if the conduction wave is generated by the shock compression of a matter. The varied conductivity region is formed as SW is traveling through a sample. If the sample is connected to the emf source or it is placed in the outer magnetic field the interaction between SW and the electromagnetic field causes the induced current wave which propagates through the sample.

The analysis of the physical processes in the shock wave accompanied by the conductivity changes in the electromagnetic field is a poorly explored field of the shock wave physics. The study of both of the matter mechanical characteristics and the electromagnetic properties changing in SW provides the new class of physical phenomena. The problem is of both fundamental and applied importance. The fundamental importance of SICW

research is apparent from the fact that the effect of the conductivity change is usual for all materials. The analysis of the shock-wave electrodynamics proves to be highly profitable. Firstly, it enables one to clear up the peculiarities of the electromagnetic picture in the matter compressed by SW and to interpret the results of a variety of experiments. Secondly, it allows us to develop some new methods of the shock wave diagnostics and measurements of different physical properties of matter. Thirdly, SICW can be usable in the field of physics and technology of the high-powered systems.

Briefly reviewed in the present paper are the results of theoretical and experimental investigations of SICW. They are obtained at the Lavrentyev Institute of Hydrodynamics.

ELECTRODYNAMICS OF SICW

The analysis of the electromagnetic processes in a matter under the shock compression can be made on the basis of the macroscopic approach in the framework of the classical electrodynamics. Mathematically, the problem is reduced to the solution of a boundary-value problem for the equation of the magnetic field diffusion in the region with movable boundaries, in particular, the SW front. Two most important cases have been carefully investigated: 1) a medium is initially nonconductive and it becomes well conductive under SW action (the dielectric-metal transition), 2) a medium is initially conductive and changes its conductivity in SW (the metal-metal, metal-dielectric transitions). The character of solutions obtained differs considerably from the classical problems of the electromagnetic field diffusion.

The physical generalities of SICW electrodynamics are stipulated by two principal peculiarities of shock compression. Firstly, there are two different mechanisms of changing electromagnetic field: diffusion and convection. Secondly, SW divides the matter into two regions: the first one ahead of the shock front and the second one behind the shock front with different properties. Dimensions of these regions are ever changing when SW moves through the matter.

The electromagnetic processes in a conductive matter depend on two main physical times: the relaxation time of the electromagnetic field in the matter θ and the SW propagation time through the matter t. The ratio of the times θ/t defines the principal parameter for the class of electrodynamics problems with shock front motion. The ratio θ/t is $R = \mu_0 \sigma (D-U)^2 t$ for the dielectric-metal transition and $Re_m = \mu_0 \sigma x_0 D$ for the metal-metal transition (σ is the electroconductivity, D and U are the shock and mass velocities, x_0 is the thickness of a conductive sample).

If $\theta/t \ll 1$, then the electromagnetic state of the compressed matter is equilibrium one, the current density is uniform over the sample. In this case, the electrical engineering approach can be used. If $\theta/t \gg 1$, then the electromagnetic state of the sample is highly nonequilibrium. The mutual effects of the electromagnetic diffusion, the shock

front motion and the matter compression cause the qualitative changes in the electromagnetic picture into the sample. The electromagnetic field distributions have the surprising and, at first glance, paradoxical character. A number of unusual effects occurs, for example, the separation between the current "wave" and the shock front or the rise of the induced anti-currents in the surface layers of the sample (1,2).

METHODS OF INVESTIGATION OF SICW

The analysis performed has permitted us to develop the new methods for experimental investigation of electrophysical properties of a condensed matter in SW.

The nonstationary electromagnetic interaction between the compressed matter and the electromagnetic field is the distinctive feature of the methods proposed (3). The experimental record of the voltage from the sample surface or from the special measuring loop is used to determine the matter properties. The realization of the method is essentially based on the following fundamental steps: the analysis of the matter shock wave electrodynamics, the practical construction of a measuring cell, the development of an effective software.

Statement of a realistic model of the electromagnetic interaction between the conductor and the electromagnetic field is the governing factor. The problem of reconstructing the matter property from the data of the conductive region surface is the inverse problem with respect to the above problem of the magnetic field diffusion. From mathematical viewpoint, this problem is ill-posed and has no single solution. It can be solved if one specifies the physical nature of the transition and necessary efficiency of the approach.

The special efforts are taken to develop the measuring techniques of the conductivity behind the shock front (4-7). The essence of the procedure proposed consists in recording the voltage at the boundaries of the conducting material in the specially designed measuring cell filled by the tested material (8). When the plane wave propagates along the cell, the compressed material changes its conductivity. This results in the changes of the current distribution in the cell and produces the recording voltages. We have elaborated the full technology of conductivity reconstruction providing the solution of the inverse problem of the current diffusion.

The methods elaborated allow one to improve the accuracy and the time resolution by one-two orders as compared to the known technique.

EXPERIMENTAL INVESTIGATION OF SICW

The use of the new methods made it possible to investigate experimentally the electrophysical properties of the various materials in SW: the porous and monocrystal silicon (4-7) (the semiconductor-metal transition), the metallic powders (7,9) (the transformation into a conductive state), the high porous metallic foam (10) (the metal-metal transition), the ytterbium (6,7) (the metal-semiconductor transition). The features of SICW are investigated: the conductivity value and its dependence on the intensity and the type of loading, the time characteristics of the transitions, the matter behavior under the release etc. The main results can be summarized as follows.

The porous silicon. The advent of the conductivity corresponds with the time precision $10\,ns$ to the moment of the SW arrival at the sample. The silicon conductivity in the shocked state remains practically constant. The silicon conductivity σ increases as the shock pressure P increases too. The dependence of $\sigma(P)$ tends to a saturation for $P > 10\,GPa$. The limit conductivity is about $10^6\,Ohm^{-1}m^{-1}$. The conductivity character is metallic in the process. The silicon conductivity is of thermal nature, the advent of the high conductivity is related to the silicon melting in SW.

The monocrystal silicon. The SW pressure dependence of the conductivity consists of two parts: the sharp rise and the plateau. The break in the dependence $\sigma(P)$ is fixed at $P \approx 12\,GPa$. The conductivity of the metallic silicon at $P > 12\,GPa$ is about $4 \cdot 10^6\,Ohm^{-1}m^{-1}$. In dynamic condition the transformation to the conductive state occurs at lower pressures as compared to static condition and it is softer. The transition is reversible under stress release. The back transition has significant hysteresis of the conductivity $\sigma(P)$.

The aluminum powder. The pressure dependence of the conductivity has two parts: the sharp rise and the relatively weak decrease. The first part is due to the occurrence of electrical contacts between grains laminated initially by nonconductive films. The second part corresponds to the conductivity decrease due intensive heating matter in SW.

The high porous nickel foam. The electrical resistance of the nickel foam increases by several tens of times during the shock compression of the sample. The experimental data can be described by the step-like conductivity change model. Under multiple compression and following release the resistance has complex changes. Tests show the high spatial inhomogeneity of the material after a shock front and the thermal nonequilibrium of the metallic skeleton and the gas located in pores.

The ytterbium. The pressure dependence of the resistivity has four characteristic regions: the resistivity maximum, the minimum, the increase, the plateau. The first maximum corresponds to the ytterbium transition to the semiconductive state. The experiments at the normal and cryogenic temperatures permit us to estimate the energetic gap of the semiconductive state $\Delta E_g \approx 0.04\,eV$.

APPLICATION OF SICW FOR HIGH POWER SYSTEMS

The active experimental studies showed that under shock loading and following release the conductivity of some materials may change by several orders increasing or decreasing in the times $\approx 10\,ns$. This effect can be used for transmittance and transformation of the electromagnetic energy in high-powered systems. The new principles of formation and control by dynamic conduction have been used for switching the high electric currents and for producing the ultrahigh magnetic fields.

The problem of short time commutator is one of the central ones of the high current pulsed technique. The features of a commutator on conduction waves (6,7) are as follows: 1) the current is broken without physical rupture of the circuit, so that conditions for arcing do not arise, 2) the elimination of air gaps next to the current conductors ensures high electric strength on switching, 3) the switching time is determined by the passage time of the wave through the sample and can be quite short (tens of nanoseconds), 4) the switch can integrate switch on operation with switch off operation. The model experiments conducted on different transitions in waves of compression and release have demonstrated the possibility of designing the current switch based on SICW. The current is up to $10^5 A$ in the tests. For special composition of the metallic foam with saturation by high explosive we have registered the commutator resistance change $R/R_0 \approx 10^2$ for times $\approx 300 ns$.

The closed configuration of SICW converging to one point is used to obtain ultrahigh magnetic fields (11-15). This approach to solving magnetic cumulation problems has advantages as compared to the traditional methods. One can more easily introduce the initial magnetic flux into the compression region, reduce heating of current layer and increase the compression stability. The fundamental difference between shock-wave compression from classical magnetic cumulation by metallic liner lies in the fact that when matter is compressed a significant fraction of the magnetic flux remains frozen in the conductor formed. In our small scale experiments the 90-fold field amplification with $4 MGs$ magnetic field have been reached. It corresponds to the energy density of electromagnetic field about $6 \cdot 10^{10} J/m^3$ and the magnetic pressure about $60 GPa$. The most efficiency for producing field is demonstrated by the high porous aluminum powder. The shock-wave method of producing ultrahigh magnetic fields opens up a number of new possibilities for obtaining extreme energy densities and related physical experiments. As shown by numerical investigations, the record magnetic energy density can be achieved by optimal design of the shock-wave magnetic field generator.

ACKNOWLEDGMENTS

The authors would like to thank E.I. Bichenkov for helpful discussions.

The research is supported in part by Grant No. 94-02-04022 of the Russian Foundation for Fundamental Research and by Grant No.RB0000 of the International Science Foundation.

REFERENCES

1. Gilev S.D., in *Dinamika Sploshnoi Sredy* **88**, pp. 31-46 (1990). /In Russian/
2. Gilev S.D., *Combust., Explos. Shock Waves*, **31**, N4 (1995).
3. Gilev S.D., "Electromagnetic methods for investigation of chemical and phase transformations of solids in a shock wave", presented at the Intern. Conf. on Metall. and Mater. Appl. of Shock-Wave and High-Strain-Rate Phenom., El Paso, USA, August 6-10, 1995.
4. Gilev S.D., and Trubachev A.M., *J. Appl. Mech. Tech. Phys.*, **29**, 818-824 (1988).
5. Gilev S.D., and Trubachev A.M., "Method for measurement of matter electroconductivity in shock waves", in *Proceedings of the Fourth All-Union Conference on Detonation*, 1988, p.2, pp. 8-12. /In Russian/
6. Bichenkov E.I., Gilev S.D., and Trubachev A.M., *J. Appl. . Mech. Tech. Phys.*, **30**, 291-303 (1989).
7. Gilev S.D., *Shock-induced conduction waves*, Ph. D. thesis, Novosibirsk, 1990. /In Russian/
8. Gilev S.D., *Combust., Explos. Shock Waves*, **30**, pp. 204-208 (1994).
9. Gilev S.D., in *Dinamika Sploshnoi Sredy* **99**, pp. 105-109 (1990). /In Russian/
10. Gilev S.D., *Technical Physics*, **65**, (1995) in print.
11. Gilev S.D., and Trubachev A.M., *Sov. Tech. Phys. Lett.*, **8** (1982).
12. Gilev S.D., and Trubachev A.M., *J. Appl. Mech. Tech. Phys.*, **24**, 639-643 (1983).
13. Bichenkov E.I., Gilev S.D., and Trubachev A.M., "Shock-wave magnetic cumulation generators," in *Ultrahigh Magnetic Fields: Proceedings of Third Intern. Conf. on Megagauss Magn. Field Generation and Rel. Topics*, Moscow: Nauka, 1984, pp. 88-93.
14. Bichenkov E.I., Gilev S.D., Riabchun A.M., and Trubachev A.M., "Shock wave method for generation of megagauss magnetic fields," in *Megagauss Technology and Pulsed Power Application: Proceedings of Fourth Intern. Conf. on Megagauss Magn. Field Generation and Rel. Topics*, New York: Plenum Press, 1987, pp. 89-105.
15. Trubachev A.M., in *Dinamika Sploshnoi Sredy* **88**, pp. 132-147 (1990). /In Russian/

INVESTIGATION OF HEAVY CURRENT DISCHARGES WITH HIGH INITIAL GAS DENSITY

A. Budin, A. Bogomaz, V. Kolikov, A. Kuprin, V. Leontiev, Ph. Rutberg and N.Shirokov

Institute of Problems of Electrophysics of Russian Academy of Sciences, Dvortsovayanab.,18, St. Petersburg, 191065 Russia

Piezoelectric pressure transducers, with noise immunity and time resolution of 0,5 μs were used to measure pulse pressures of 430 MPa along the axis of an electrical discharge channel. Initial concentration of He was $2,7\cdot10^{21}$ cm^{-3}, dI/dt=$6\cdot10^{11}$A/s, and I_{max}=560 kA. Shock waves with amplitudes exceeding the pressure along the axis, were detected by a pressure transducer on the wall of the discharge chamber. Typical shock velocities were 2·4 km/s. Average pressure measurements along the discharge axis at different radii were used to estimate the current density distribution along the canal radius. The presence of the shock waves, promoting the additional hydrogen heating in the discharge chamber, has been registered during the discharge in hydrogen for I_{max}~1 MA and an initial concentration of 10^{21} cm^{-3}.

INTRODUCTION

High current discharge in dense gases has been used for gas heating in light gas accelerators [1,2]. Such discharges can also act as a source of ultra violet and soft X-ray radiation [3]. (Stability of the discharge, being radiation source, and effective gas heating in the whole volume of the discharge chamber (DC) can hardly be achieved simultaneously.) So these two tasks were set up in series of experiments to form a discharge with predicted properties. In this series pulse pressure was measured by a pressure transducer of original design, described below. Simultaneous current and voltage measurement were also made.

EXPERIMENTAL RESULTS

Discharges in air at atmospheric pressure and in helium and hydrogen with initial density of 1-$4\cdot10^{21}$ cm^{-3} [4] were studied. Pulse current 600 kA, current rise time of 1.5-2 μs and rate, dI/dt, of $6\cdot10^{11}$ A/s for air and helium DC design is given in Fig. 1. The discharge was initiated along the anode axis by a supplementary electrode in the case of air, and by plasma jet in the case of helium [4]. In the case of hydrogen, the pulse current was of 300-1000 kA with rise times of 100- 150 μs and dI/dt of $1.6\cdot10^{10}$ A/s. This discharge was initiated by a wire explosion using either a configuration like that

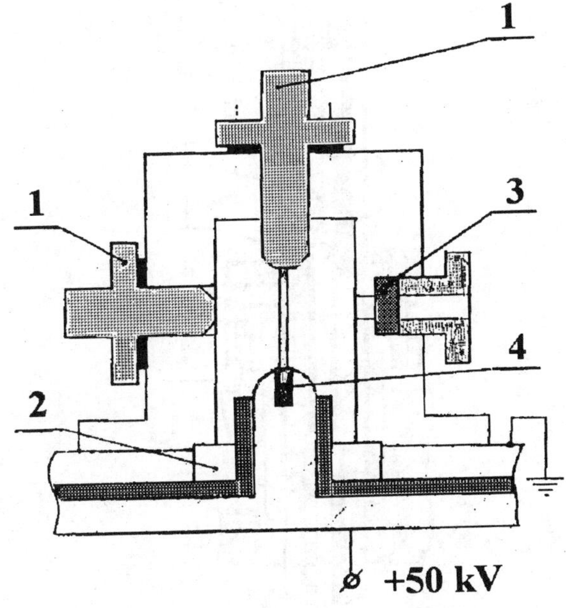

FIGURE 1. Discharge chamber
1. Piezoelectric transduser
2. Rogovsky coil
3. Transparent window
4. Plasma injector

shown in Fig. 1 or with the igniting wire position between a conic cathode and the DC wall [1,2]. Current, voltage and pressure were acquired together with high speed photography and time integrated spectra of the discharge radiation. The pressure was measured by a piezoelectric transducer placed one end of a piston (Fig 2). Its design differs from earlier designs [5] by full length rubber sealing of the piston and providing sound absorption at the back edge. Absorption was achieved by using materials with gradually decreasing density and increasing section with constant acoustic resistance. This enabled us to weaken the multitude of pressure pulses, passing by the piston as a result of reflections at the edges. This way the piezoelement strain force was reduced and the transducer registration time limit increased. Rise time of the transducer signal, determined by sound dispertion in the front piston and thickness of the piezoelement, was less than 0.5 μs, its delay due to the pulse passing by the piston was about 15 μs. To measure the pressure radius distribution, the front edge of the transducer was screened by diaphragms with opening radius r_1 less than that of the arc r_0 (Fig. 3).

DISCUSSIONS

The magnetic pressure associated with the constant current density discharge, P(r) is:

$$P(r) = P_{max} - \mu j^2 r^2 / 4 \ , \tag{1}$$

where

$$P_{max} = \mu I^2 / 4\pi^2 r_o^2 \tag{2}$$

The average pressure over the diaphragm opening is measured by the pressure transducer. This pressure:

$$\bar{P}(r_1) = 1/(\pi r_1^2) \int_0^{r_1} P(r) 2\pi r \, dr = $$
$$= P_{max}(1 - r_1^2 / 2r_0^2) \tag{3}$$

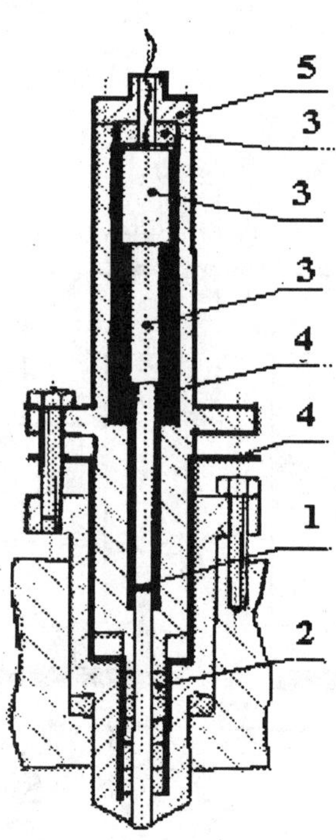

FIGURE 2. Piezoelectric transducer.
1. Piezoelement (quartz, turmalin)
2. Rubber sealing
3. Sound absorbtion unit
4. Insulation
5. Cable connection

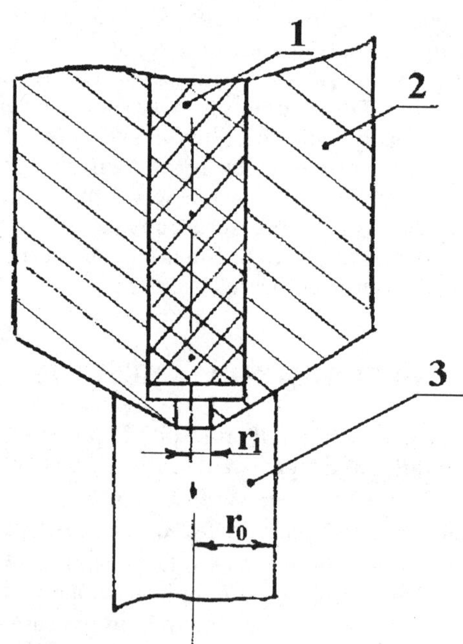

FIGURE 3. Scheme of pressure measurements along the discharge axis.
1. Ceramic pivot
2. Measurement tip, working as a cathode
3. Arc

Where I is the maximum current amplitude which corresponds to the discharge radius of r_0. The ratio of two average pressures at radii r_1 and r_2 enables us to get the arc radius.

$$\frac{\bar{P}(r_1)}{\bar{P}(r_2)} = \frac{2r_0^2 - r_1^2}{2r_0^2 - r_2^2} \qquad (4)$$

For the discharge in air with I_{max}=144 kA; r_1=0.10 cm, r_2=0.20 cm, $P(r_1)/P(r_2)$=1.6 we get r_0=0.21 cm which is close to the value of 0.25 cm taken from the high speed photography. For the discharge with I_{max}=290 kA, $P(r_1)/P(r_2)$ >2 which is impossible force constant current density along the radius. This indicates a pressure rise in the axis zone. This is confirmed by observations of a uniform melting zone over the half-spherical cathode following experiments using a 0.2 cm piston and currents of 500 kA. For current of 290 kA, with r_1-0.1 cm the pressure was $P(r_1)$=171 MPa. For the discharge in air, with similar conditions, the magnetic pressure was 160 MPa [6]. For the discharge in helium with an initial concentration of $2.7 \cdot 10^{21}$ cm and I_{max}=560kA, we got $P(r_2)$=436 MPa which is close to that calculated for a constant which current density and r_0=0.5cm. This radius is close to that taken from the high speed photography. Furthermore, the discharge channel extends with a speed of 2-4 km/s. The pressure transducer, placed at the DC wall, registered shock waves, produced by the discharge. Their amplitude was the same value as the pressure at the discharge axis. According to the calculations, the shock waves heat the gas in the DC. For the discharge in hydrogen with initial density of 10^{21} cm^{-3}, energy of 1 MJ and rate of current rise of $1.6 \cdot 10^{10}$ A/s with axial electrode position the discharge was held by its own magnetic field. For the coaxial electrode design, typical of light gas launchers [2], shock waves caused by secondary breakdowns promoting heat transfer were also detected.

CONCLUSIONS

The stage of equilibrium of magnetic and gaskinetic pressures of high current discharges in high density media was investigated by means of specially designed pressure transducers. For the discharge in air with currents of 290-500 kA the current density rise at the axis was detected in this stage. For the discharge in helium the stage of equilibrium is possible in the assumption of constant current density in the channel. For the discharge in hydrogen with 1 MJ energy stored this stage was detected by the pause in the current channel extension.

ACKNOWLEDGMENTS

This work was done under the financial support of the Russian Foundation of Basic Research (project 43-02-1742).

REFERENCES

1. A.A. Bogomaz, A.V. Budin, A.Kolikov, Ph.G.Rutberg, *IJIE*, V17, pp.94-98 (1995)
2. V.A. Kolikov, A.A. Bogomaz, A.V. Budin, A.G. Kuprin, Ph.G. Rutberg, "Investigation of Processes which take Place at Hydrogen of HIgh Initial Density Heated in Powerful Electric Discharge Launchers", in *Proceedings of the 5-th Europe an Symposium on Electromagnetic Launch Technology, Toulouse, France*, 1995
3. Arapchuk l.E., Bogolubsky S.P., Telkovskaya O.B., *Journal of Technical Physics*, V55, N11, pp. 2222-2224 (1985)
4. D.A. Andreev, A.A. Bogomaz, Ph.G. Rutberg, A.M. Shakirov, *Journal of Technical Physics*, V62, N6, pp.74-82 (1992)
5. T.J. Jones and C.C. Vhases, *RSI*, V38, N8, pp. 1038-1042 (1967)
6. Skowronek M., Romeas P., Vu-Tien-Jia, *Rev. Phys. Appl.*, V12, N10, pp.1723-1727 (1977)

CHAPTER XII

EXPERIMENTAL TECHNIQUES

EOS IMPEDANCE MATCHING EXPERIMENTS AT HIGH PRESSURE WITH SMOOTHED LASER BEAM

B. Faral[1], M. Koenig[1], J.M. Boudenne[1], D. Batani[2], A. Benuzzi[2], S. Bossi[2], M. Temporal[3], S. Atzeni[4], Th. Löwer[5].

1 LULI, Ecole Polytechnique, Palaiseau, FRANCE, 2 Dep. Fis.Univ. Milano, Milan, ITALY, 3 INFN, Legnaro, ITALY, 4 ENEA, Frascati, ITALY, 5 MPQ, Garching, GERMANY

High quality shock waves in the range of 10 to 50 Mbars have been obtained in laser experiments with pulse energy $\approx$ 100 J. Laser beams smoothed by Phase Zone Plates produced high quality, planar shock waves in two-step, two-material targets, allowing the simultaneous measurements of the shock velocities in the two materials. By the use of the impedance-matching technique, the relative consistency of the equations of state of these materials can be tested, or relative equation of state data can be measured. Pressures higher than 35 Mbar were achieved in gold. EOS data for copper, relative to aluminium EOS, at pressures up to 34 Mbar, have been obtained. Comparisons with preliminary Hohlraum experiments in the same range of pressure have been made.

INTRODUCTION

A knowledge of the equation of state (EOS) of materials at pressures over 10 Mbar is important in several branches of physics, including astrophysics and inertial confinement fusion research.

Pressures above a few Mbar can only be achieved with shocks, and above 10 Mbars can only be produced by nuclear weapon driven experiments or high power lasers. Most of the time, the value of the pressure is determined indirectly by the measurement of the shock velocity D, assuming known the EOS of the shocked material. A direct measurement of an EOS point, instead, requires the experimental determination of an additional parameter of the shocked material (1). The impedance match technique (1) is intermediate between the usual indirect pressure determination and the direct one. This technique allows for testing the relative consistency of the EOS of the two materials, or, alternatively, for measuring an EOS point, using the EOS of an other material as a reference.

We report the first demonstration of such a technique at pressure P > 10 Mbar (actually up to 35 Mbar in gold), using direct-drive laser irradiation, and two-step, two material targets. We used this technique for checking the consistency of the couple Al-Au (2), and for the obtention of EOS points for copper, relative to aluminum. The experiment was performed at the Laboratoire pour l'Utilisation des Lasers Intenses (LULI), Ecole Polytechnique. Instrumental to the success of the present experiment was the production of a smooth, nearly one-dimensional shock wave (3), obtained by the use of the beam smoothing technique based on the so-called phase-zone plates (PZP) (4). As a result of the good coupling of laser energy to matter and of the reduce level of preheating (as compared to indirect drive), it has been possible for the first time to perform EOS studies at pressure P = 10–35 Mbar, and to obtain EOS data for copper, at pressures two times higher than the higest data available for copper up to now, by using laser pulses of relatively modest energy (E $\approx$ 100 J at λ = 0.53 µm).

We have used double-step targets, made of a "base" foil of a material A, which is irradiated by the laser on one side, and supports, on the opposite side, two steps made, respectively, of the same material A, and of a different one B. Using rear-face, time-resolved imaging we experimentally determine the velocity of the shock propagating through the two steps, D_A and D_B (corresponding to particle velocities U_A and U_B), respectively. Consistency of the EOS's of the two materials requires the coincidence of the experimental point (P_B, U_B) and the one (P'_B, U'_B) deduced from (P_A, U_A) by using impedance-matching equations associated to EOS tables.

The error on the shock velocity must not exceed 5%, since the pressure deviations between different theoretical models do not exceed 10%. In our case there are three main sources of possible errors in the determination of D: the quality of the shock itself (requiring flatness over a wide region), the rate (ps/mm) of the streak camera, and the knowledge of the step thicknesses.

Experimental Set-Up

In our experiment the luminosity of the shock emerging at the rear face of the target was imaged by a photographic objective onto the slit of a visible streak camera with 5 ps time resolution. We performed the calibration of the streak sweep speeds with an etalon made up with a serie of short laser pulses (FWHM= 100 ps). The relative error in the speed used for our experiments was lower than 1%. The system magnification was M = 22, allowing a 5 μm spatial resolution, which was checked by imaging a suitable grid. Three of the six beams of the LULI laser (frequency doubled, $\lambda = 0.53$ μm) with total laser energy $E_{2\omega} \approx 100$ J were focused onto the same focal spot. The laser pulse was a 600 ps FWHM Gaussian. A fourth beam, also converted to 0.53 μm, was used as a temporal fiducial. Each beam had a 90 mm diameter and was focused on target with an f = 500 mm lens.

Phase Zone Plates Technique

We have used the optical smoothing technique of phase zone plates (PZP), which is built on the association of the main lens with a Fresnel lens array having a randomly distributed phase shift (0 or π). Each lens of the array has consecutive elements whose phase shift differs by π. The PZP's allow a flat top distribution of laser intensity to be produced at a given distance from the focal plane of the main focusing lens (4). Note that, even if we were able to produce very high quality shock waves (3), our experimental conditions were not optimal for the application of the PZP technique due to the long focal length of the focusing lens (500 mm) and to the small laser beam diameter (90 mm). Indeed our PZP's design is a compromise between the laser intensities required to achieve the desired shock pressure (P $\geq$ 10 Mbar in aluminum) and the size and flatness of the focal spots. In our case we had only five elements per Fresnel lens, which is not enough to have very sharp edge width. We were able to produce a measured focal spot consisting of a 200 μm top-hat intensity distribution with gaussian edges. The total focal spot FWHM was 400 μm, which corresponds to a laser intensity $I_L \leq 10^{14}$ W/cm^2.

EXPERIMENTS AND RESULTS

Accuracy of Measurements

In order to reduce one of the possible source of experimental errors, one needs high quality, well characterized, structured targets as described above. In addition to the accurate knowledge of the step thicknesses, sharp step edges are required. Also, the spacing between the two steps must be small compared to the flat top-hat portion of the laser focal spot. We used electron-gun deposition, since it can produces sharp edges. The target is made in three stages: first, the base material is deposited; a mask is then applied to this base in order to deposite the first step. Then, a second mask is applied, which is mechanically and optically guided, to ensure that the steps do not overlap and that their separation is limited to no more than 50 μm. The second step is then deposited. The overall quality of the targets was checked by electron microscopy.

The step heights were determined within an absolute error lower than 0.05 μm (Dektak measurements). Since the thicknesses of the aluminum and gold steps were 4 and 2 μm respectively, this ensured a relative error of about 1% for the aluminum step and about 2 % for the gold step. The EOS's of these materials have been intensively studied and differences between either various EOS calculations or experimental results are within a few percent (5). This allows us to check the overall sensitivity of the method. Figure 1 shows a typical streak image of a two-step target.

Here we can clearly determine the shock velocities in the two materials, the flat part of the focal spot being $\approx$ 200 μm while the spacing between the two steps is $\approx$ 45 μm. In order to check that the shock pressure is constant inside each steps we used targets with different thicknesses of the base. 1-D simulations, performed with the hydrodynamic code FILM developed at Ecole Polytechnique, have shown that in our conditions, the base must be thicker than $\approx$ 8 μm.

We also checked (3) the quality of the shock wave generated with the PZP smoothed beams using simple foil targets ; we found typical

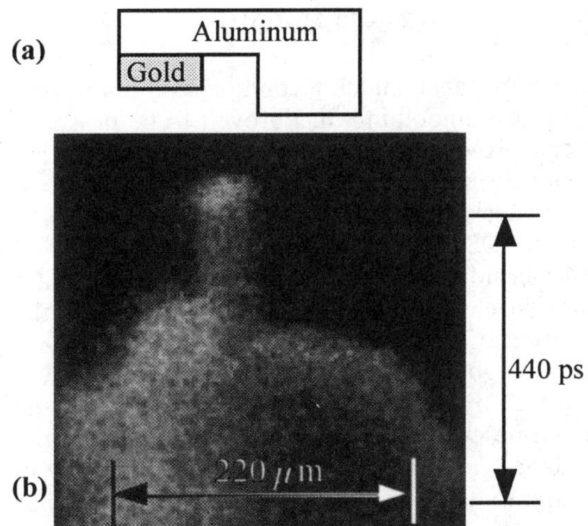

FIGURE 1. Experiment on a double step target. Optical smoothing is realized with PZPs giving a measured focal FWHM spot diameter of 400 μm. (a) sketch of the target; the aluminum base thickness is 9.5 μm, the steps of aluminum and gold are 4.25 μm and 1.85 μm respectively. (b) Streak camera record of visible light emitted by the rear side of the target. The aluminum shock velocity is $D_A \approx 24.5$ μm/ns (corresponding to $P \approx 10$ Mbar), and the gold shock velocity is $D_B \approx 13.9$ μm/ns (corresponding to $P \approx 22$ Mbar).

variations of ± 5 ps for the shock breakthrough time across the 200 μm flat region of the focal spot. In order to ensure that 2-D effects were quite negligible here, the experimental results and the FILM simulations were compared with simulations made with the 2-D hydrodynamic Lagrangian code DUED (6), developed at ENEA Frascati. With all the above errors taken into account, the shock velocities are determined within a maximum error of ± 5% for aluminum and ± 6.5% for gold.

A summary of the experimental data is presented in Fig. 2, where the shock velocity in gold D_B is plotted versus the aluminum shock velocity D_A. The corresponding pressures range from 4.5 to 16.5 Mbars in aluminum, and from 9.5 to 37 Mbars in gold, depending on the input laser energy. The data are in good agreement with the SESAME EOS (solid line) with most points lying in a 3% wide error strip around the theoretical prediction. The EOS of the two materials used have been widely studied in this range of pressure and being thus relatively well known, these results validate the whole technique used, allowing to apply it for the measure of less known EOS materials.

In this range of pressure, copper data are rare, although some results have been recently shown (7).

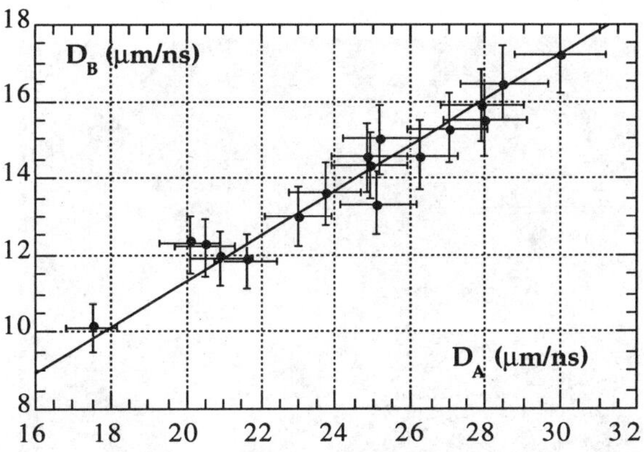

FIGURE 2. Experimental results : shock velocity in gold D_B versus aluminum shock velocity D_A. The straight line corresponds to SESAME EOS.

So we have made experiments with the couple of materials aluminium-copper, and this, in a higher range of pressure, since we used for some experiments the iodine laser of the MPQ Garching Institut, which may delivers 300 J in the blue light domain (λ=0.43 μm, frequency tripled).

This energy has been used in two distincts ways : direct irradiation of the target, coupled with PZP smoothing (laser beam diameter : 300 mm); or indirect irradiation (Hohlraum). In this last technique, a gold target is directly irradiated by the laser, producing a strong X-ray emission, which induces the shock in the two material target. Without going further in this technique, note that the impedance matching experiments require a plane shock, whatever the technique used to produce this shock, and that the indirect drive technique is known for producing high quality shocks i.e. flat ones (8) (at the price, however, of a low ratio energy on target/laser energy). We see on Fig. 3 the breakthrough of a shock generated by indirect irradiation at the rear of a two step Al-Cu target. Note the flatness of the shock over the two steps region.

Figure 4 shows the different results that were available on copper and the four ones we obtained : the lowest one has been obtained at LULI (with PZP smoothing), the two intermediate ones have been obtained by indirect irradiation (Garching), and the higher has been obtained at Garching with direct PZP smoothed irradiation. Note that the PZP technique has allowed a better use of the energy available on target, and thus, has generated pressures significantly higher than the ones

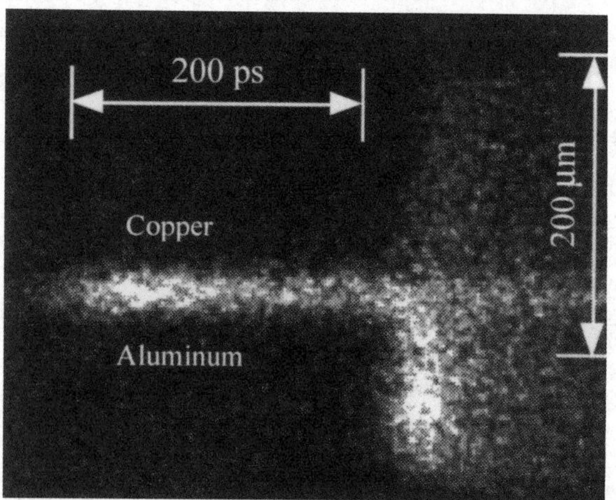

FIGURE 3 . Streak camera record of visible light emitted by the rear side of a Al-Cu double step target. Shock has been generated by indirect irradiation. The aluminum base thickness is 11 μm, the steps of aluminum and copper are 5.7 μm and 4.9 μm respectively. Shock velocity in Al is ≈ 24 μm/ns (P≈10 Mbar), shock velocity in Cu is ≈ 20 μm/ns (P≈19 Mbar).

obtained with indirect irradiation. The access to these higher pressures is seen clearly on Fig. 4 : the maximum pressure of previously available data was of the order of that we obtained in indirect irradiation ≈ 19 Mbar, while the one obtained with direct irradiation (and PZP) is ≈ 34 Mbar.

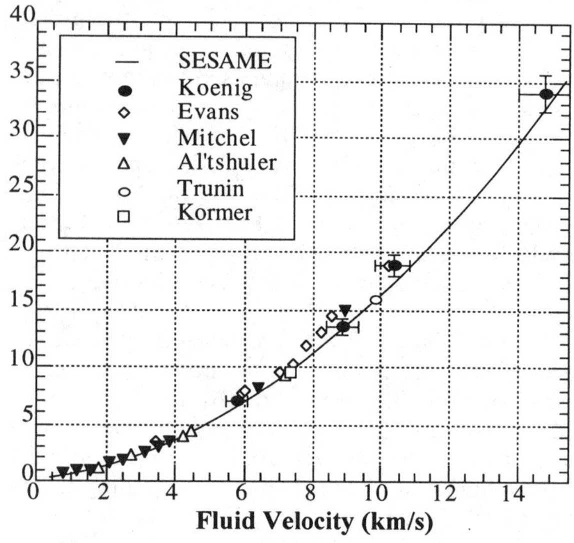

FIGURE 4. Pressure versus shock fluid velocity for copper. Data with error bars correspond to our results, obtained by direct and indirect irradiation.

CONCLUSIONS

The impedance match technique, combined with PZP optical smoothing, has proved to be powerful and can provide enough accurate data by paying attention to each part of this technique. It can be used to check the consistency of two given EOS, or to obtain EOS data if a reference EOS is known.

Furthermore, by its principle itself, the increase of pressure in the second material allows the generation of pressures which would be otherwise difficult to obtain. The poor flatness of shock waves generated by direct laser irradiation was the touchstone which prohibited the use of the impedance match technique for accurate EOS measurements. The PZP optical smoothing has removed this obstacle.

After having validated the technique on the couple Al-Au, we have obtained EOS data for a less known material, copper, with the LULI laser and the MPQ-Garching one. The first results, where we compare shocks induced by direct and indirect irradiation are very encouraging, and show that the same laser, can reach, with direct PZP smoothed irradiation, pressures two times greater than the ones obtained by indirect irradiation.

The accuracy can be enhanced to ± 2% by improving the target fabrication, by using a higher resolution streak camera and by enhancing the step heights. In this last case, a laser 2-3 times more powerful, and such lasers already exist in Europe (MPQ, RAL, CEA-Limeil), would be sufficient to ensure constant shock speeds, opening then the way to EOS data acquisition in high pressures domain.

This work was supported by EU Human Capital and Mobility program under contracts CHRX-CT93-0377, CHRX-CT93-0338, ERB-CH-CT92-0006 and ERB-CHGE-CT93-0046.

REFERENCES

1. Ya. B. Zeldovich and Yu. P. Raizer, *Physics of Shock Waves and High Temperature Hydrodynamic Phenomena* (Academic Press, New York, 1967).
2. M. Koenig et al., Physical Review Letters, **74**, 2260 (1995).
3. M. Koenig et al., Phys. Rev. E, Rapid Comm., **50** , R3314, (1994)
4. R. M. Stevenson et al., Optics Letters **19**, 363 (1994).
5. L.V. Al'tshuler et al., J. Appl. Mech. Tech. Phys. **22**, 145 (1981).
6. S. Atzeni, Comput. Phys. Comm. **43**, 107 (1986); Plasma Phys. Controll. Fusion **31**, 2137 (1989).
7. A.M. Evans et al., "*Feasability of Hugoniot EOS measurements on Helen*", 23rd ECLIM, Oxford, U.K., 19-23 Sep. 1994.
8. Th. Lower et al., Phys. Rev. Lett. **72**, 3186 (1994).

EXPERIMENTAL TECHNIQUE TO MEASURE TENSILE IMPEDANCE OF A MATERIAL UNDER PLANE SHOCK WAVE PROPAGATION

D. P. Dandekar

Army Research Laboratory - Materials Directorate
MS: AMSRL-MA-PD
Aberdeen Proving Grounds, MD 21005

This paper describes the basis of an experimental technique to measure tensile impedance of a material under plane shock wave propagation. The technique involves simultaneous measurements of free surface velocity in a conventional symmetric impact shock wave spall experiment and rear surface particle velocity in a companion experiment where the sample is backed by a well characterized low impedance material. This technique has been used to measure tensile impedances of a tungsten alloy above its Hugoniot Elastic Limit (HEL) and of silicon carbide below its HEL.

INTRODUCTION

In a conventional plane shock wave experiment, spall strength of a solid is determined by impacting a flat plate of a stationary material with a thinner plate of the same material. As a result of the impact, shock waves propagate in the stationary material and in the impacting plate. The propagating shock waves reflect from the respective free surfaces of the colliding plates and travel back in to these plates as rarefaction waves and meet in the stationary plate. As a result of their interaction tensile waves propagate towards the impact and free surfaces of the stationary plate. In other words, a tension is generated at the region of collision of the two rarefaction waves. If the tensile stress generated is of magnitude and duration such as to cause material separation in this region the tensile stress is relaxed to zero. A similar set of events occur if the stationary plate is impacted by a plate made of a material with impedance lower than that of the stationary material, provided its thickness is such that the rarefaction waves originating at the two surfaces of the stationary plate meet inside it.

In these experiments, particle velocity profile is monitored at the free surface of the stationary plate to obtain information about the spall threshold of the material. In some instances the particle velocity profile is monitored at the stationary plate - buffer/window material interface. The buffer material also has a lower impedance than the stationary material. In either case, it is necessary to assume a value for the effective tensile impedance to calculate the spall threshold stress of the material. The assumed values are generally chosen on the basis of compression and release behavior of the material under plane shock wave loading and unloading. The present work shows that it is possible to determine the value of spall threshold stress of a material without making any assumption about the tensile impedance of the material. This in turn would permit an evaluation of the correctness of the assumption made pertaining to the assumed value of the tensile impedance. In addition, it would facilitate development of a better material model for use in the numerical simulations.

DESIGN OF EXPERIMENT

The design of experiment is described in terms of symmetric impact of two flat plates made of the material of interest. A non-symmetric impact experiment does not affect the results of the experiment to be described.

A general configuration of the experiment is shown in Figure 1. The experiment consists of simultaneously impacting two sets of stationary plates with thinner plates of the same material, as the stationary plates, at a given impact velocity. The particle velocity is recorded at the free surface of the stationary plate [Fig. 1 (a)]. And, the particle velocity for the other stationary plate is monitored at

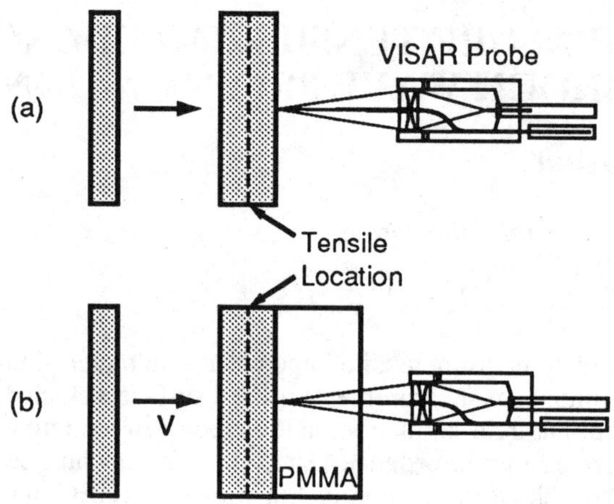

Figure 1. A general configuration of experiment to determine tensile impedance of a solid.

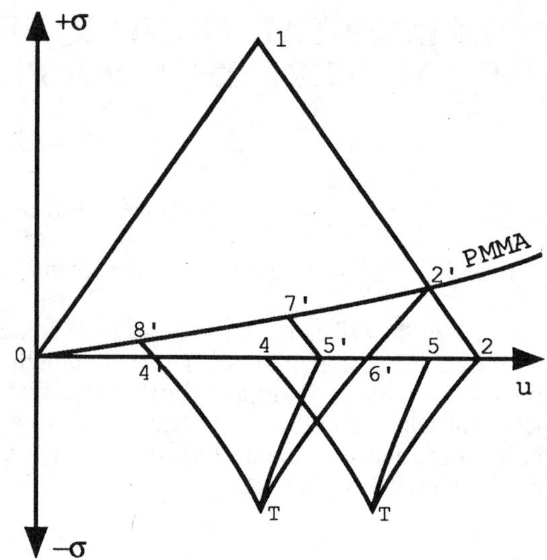

Figure 2. Stress- particle velocity diagram for the experiment to determine tensile impedance of a solid.

the plate-window interface [Fig. 1 (b)]. The window is transparent and has a lower impedance than the plate material. In addition, the design of the experiment requires that the tensile stress generated in both stationary plates be such as to exceed the spall threshold of the plate material. Further, it is assumed that shock, release and reshock response of the window material is completely known.

ANALYSIS OF THE EXPERIMENT

The stress-particle velocity diagram for the experiment described above is shown in Figure 2. Upon impact, stress-particle velocity coordinates defining the shock state of the material are (σ_1, u_1). For the configuration Fig.1(a), free surface velocities measured correspond to 2, 4, and 5. In linear elastic case free surface velocities corresponding to 2 and 5, are the same.   For the configuration Fig.1(b), the measured particle velocities correspond to 2', 8' and 7'. These correspond to initial compression, release from the initial compressed state, and re-shock from the release state of the window material. Similarly, in the case of the experiment configuration [Fig.1(b)] the points 5' and 6' will be coincident for a linearly elastic material.   In the analysis, we further assume that the properties of the window material are time independent.

Then for configuration Fig.1 (b):

$$\sigma_T = Z_T(u_T - u_{2'}) + \sigma_{2'}, \quad \text{for T6'2'} \tag{1}$$

$$\sigma_T = - Z_T(u_T - u_{8'}) + \sigma_{8'}, \quad \text{for T4'8'} \tag{2}$$

where Z_T is the tensile impedance, σ_T and u_T are the stress and particle velocity for the tensile state. For configuration Fig.1 (a)

$$\sigma_T = 0.5 x Z_T (u_4 - u_2). \tag{3}$$

σ_T is also equal to

$$\sigma_T = 0.5 x Z_T (u_{4'} - u_{6'}). \tag{4}$$

Eqs. (1), (2), and (3) yield

$$2\sigma_T = Z_T(u_{8'} - u_{2'}) + \sigma_{8'} + \sigma_{2'} = Z_T(u_4 - u_2),$$

which yields

$$Z_T = (\sigma_{2'} + \sigma_{8'})/(u_4 - u_2 + u_{2'} - u_{8'}) \tag{5}$$

In case of either non-linear elastic material or an inelastic material T-5, and T-5' provide the impedance for the shock state attained following the spallation of the material. Moreover, in the case of an elastic-plastic solid the impact velocity need not be equal to u_2 as shown in Figure 2, and release may involve both elastic and plastic releases with the tensile path dependent on the impact stress generated in the solid.

In the present work PMMA was used as the window material. The Hugoniot determined by Barker and Hollenbach[1] is used to calculate the tensile impedance.

EXPERIMENT AND MATERIALS

Spall experiments were performed on a 10 cm diameter light gas gun. A four beam VISAR was used to record the free surface velocity and particle velocity at the material-PMMA interface, simultaneously. The uncertainty of the velocity measurement is ± 1%. Impact velocity was measured by means of shorting four charged pins located around 20 mm ahead of the target. The measured impact velocities have an uncertainty of 0.5 %. The tilt angle of impact was of the order of 0.5 mrad.

Spall experiments were conducted on polycrystalline silicon carbide and polycrystalline tungsten alloy.

Silicon carbide was a hot pressed material and has a designation SiC-B. Its density is 3.221 ± 0.001 Mg/m³. The measured values of longitudinal and bulk sound wave velocities are 12.22 ± 0.05, and 9.32 ± 0.06 Km/s, respectively. The Hugoniot Elastic Limit of this material has been reported to be 11.7 GPa[2]. A Spall experiment was performed at an impact stress of 11.8 ± 0.2 GPa to determine its tensile impedance. A detailed paper dealing with the spallation in this material is presented in a companion paper at this conference[3].

Polycrystalline tungsten alloy (93 W) is nominally composed of 93 % tungsten, 5 % nickel, and 2 % iron by weight. Its density is 17.75 ± 0.05 Mg/m³. The measured values of longitudinal and shear wave velocities are 5.18 ± 0.05, and 2.81 ± 0.04 Km/s, respectively. The HEL of the alloy has been determined to be 2.55 ± 0.34 GPa. This alloy appears to deform like an elastic- plastic solid under plane shock wave compression. The above set of information is obtained from an unpublished investigation[4]. A spall experiment on this alloy was performed at 15.8 GPa.

RESULTS

The values of free surface velocities and particle velocities at the specimen-PMMA interface recorded in the experiments performed on silicon carbide and tungsten alloy are given in Table 1. The recorded wave profiles in these experiments are shown in Figures 3, and 4.

TABLE 1. Measured values of free surface velocities and particle velocities at the specimen-PMMA interface.

Material	u_2	u_4	$u_{2'}$	$u_{8'}$
SiC	0.6059	0.5615	0.5383	0.4015
93 W	0.3946	0.3378	0.3693	0.2747

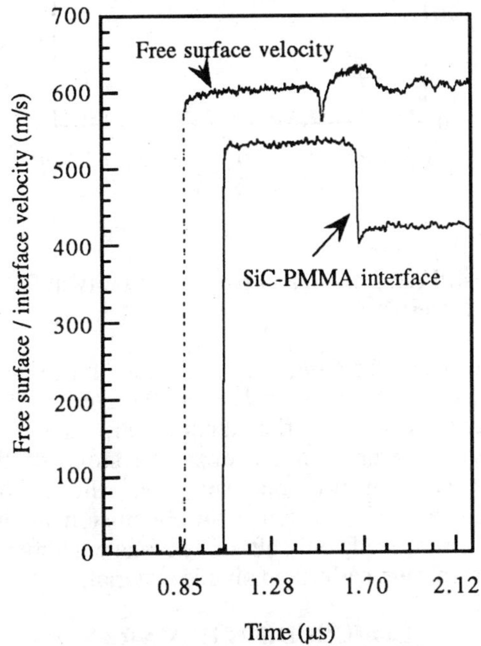

Figure 3. Free surface and particle velocity at SiC-PMMA interface.

Tensile impedance of silicon carbide

The value of tensile impedance calculated from the above sets of measurements and the Hugoniot of PMMA is 40.0± 1.1 Gg/m².s. The value of the elastic impedance of silicon carbide is 39.3 Gg/m².s. Thus, the measured value of the tensile impedance is consistent with expected response of silicon carbide up to HEL.

Tensile impedance of 93 W

The calculated value of the tensile impedance of

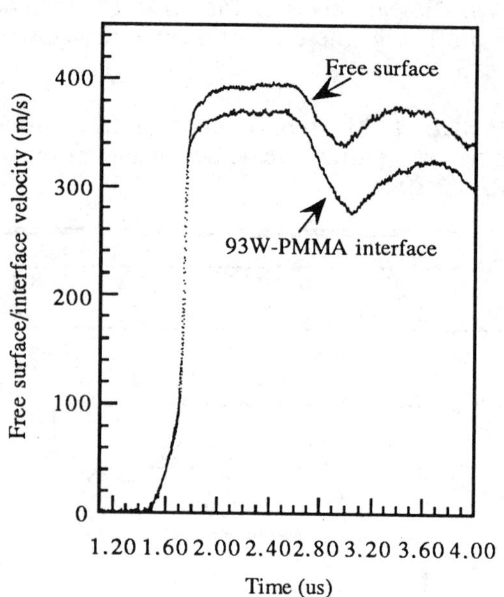

Figure 4. Free surface and particle velocity at 93W-PMMA interface.

93 W is 64.5 ± 2.5 Gg/m^2.s. The value of the plastic impedance of this alloy is 71.6 ± 2.0 Gg/m^2.s. The measured value of the tensile impedance is somewhat lower than the expected slope of the tensile path in this material, i.e., the plastic impedance. However, a firm conclusion in this regard can be made only after an adequate number of experiments are performed on this material.

CONCLUDING REMARKS

It is shown that the technique described in the present work to measure tensile impedance of a solid yields results consistent with the expected response of elastic and elastic-plastic solids. More experiments need to be conducted on these and other materials like single crystals of sapphire, and metallic material to validate this proposed technique to measure tensile impedance. Work is under progress to measure tensile impedance 93W at other impact stresses. In addition, we have begun to conduct experiments to measure the tensile impedances of single crystals Z-cut sapphire, and Z-cut quartz.

ACKNOWLEDGMENT

Author is thankful to P. Bartkowski for his valuable assistance in the preparation of figures. Author acknowledegs the careful review and suggestions made by Y. M. Gupta which led to improvement of the manuscript.

REFERENCES

1. Barker, L. M. and Hollenbach, R. E., Journal of Applied Physics **41**, 4208-4226 (1970).
2. Feng, R., Raiser, G. F. and Gupta, Y. M., Unpublished.
3. Bartkowski, P. and Dandekar, D. P., A companion conference paper.
4. Dandekar, D. P. and Weisgerber, W., Unpublished.

EXPERIMENTAL METHODS OF INVESTIGATING COMPOSITE STRUCTURES STRENGTH AGAINST UNSTATIONARY LOADING EFFECTS.

V. P. Petrovsky, A. V. Ostrik

Central Institute of Physics And Technology
Russian Federation Ministry of Defence, Sergiyev Posad

We bring forward experimental techniques to determine composite structures strength to nonstationary loadings effects. Particular attention is paid to the studies spearheaded to the impact on the functioning solid-fuel missile engines (SFME).

Composite structures nowadays are more and more widely used in aerospace and aviation equipment. When operating on the ground and in flight the structures elements are affected by numerous nonstationary loadings of shock and vibration types caused by rocket motors functioning mode changes, wind gusts, shock wave, explosion fragments, collisions of aircraft with birds, as well as by energy directed fluxes impact [1-4]. The duration of said loadings varies from 10^{-6} to $10^{-3}/s$ depending on the cause of their emergence and characteristic values of pressure pulses vary from 0.1 kPa$\cdot$s to 5 kPa$\cdot$s. As the experience of designing and exploiting the structures made of carbon-, glass-, and organoplastics shows in many cases, important for practice, just the strength to nonstationary loadings having the abovesaid characteristics determines the degree of the risk connected with using composite materials in bearing elements of aerospace vehicles.

We bring forward experimental techniques to determine composite structures strength to nonstationary loadings effects. Particular attention is paid to the studies spearheaded to the impact on the functioning solid-fuel missile engines (SFME).

1. LOADINGS GENERATING EXPLOSIVE DEVICES

At present the most widely used method to generate nonstationary loadings on structures of large dimensions is the hydrodynamic method of local loading. The technique ensures the loading generation by shock wave of elastic high explosive burst. The explosion is spatially distributed on the surface of the object being tested. The data on explosive devices generating nonstationary loadings that affect functioning SFMEs are given in table 1 and described in detail in [5].

2. MEASURED PARAMETERS

In the course of investigating nonstationary processes emerging in functioning SFMEs under the effects of lateral short - time loadings two parameter groups were measured. The first one characterises an engine response during the impact action. This group includes nonstationary deformations of a SFME casing and overloadings on a SFME vital elements. The second parameters group characterises the state of the object after the loadings impact.

TABLE 1 Parameters of hydrodynamic devices generating nonstationary loadings

Item Number	Device Designation	Parameters	
		Duration, μs	Pressure Pulse, I_d, kPa·s
1	Contact sector charge (CSC)	1 ... 2	0.8 ... 5
2	CSC with the porous rubber spacer	5 ... 10	0.1 ... 2
3	Equidistant-surface charge(ESC)	10 ... 100	0.3 ... 3

The major parameters of this group to be measured were chosen as follows: internal pressure in combustor, an engine thrust, quasistatic deformations, and casing temperature.

The nonstationary deformations measurings were made using electric strain-measuring techniques in the balanced single bridge circuit scheme. The KB - 10 - 200 strain transducers were used that allow to measure relative deformation values of $\varepsilon_{max} \approx 4$ % with a maximum error less than 15%.

To measure the overloadings the АДП - 10 - 01 piezoelectric converter was used. The gauge ensures high transducing coefficient. The converter is an HF, vibration-proof device that has the capability of measuring acceleration values up to 10^4 g.

3. TEST RESULTS

At first, working SFME was tested for attack of a load created by ESC with the value of $I_d = 0.3$ kPa· s and $\tau_1 = 100$ μs. According to the calculations, which have been done by means of numerical method [6], such load would not destroy SFME. The effect, which was exerted on the engine approximately in the middle of the time of the engine work, did not have an effect on it. Only insignificant variations in magnitudes of internal pressure and thrust were observed. It should be noted, that in the center of the load area a nonstationary longitudinal tension deformation superimposed on a static one ($\varepsilon_{max} \approx 1.3\%$)

and reached the value of $\varepsilon_{max} \approx 1.6\%$, which is close to critical. Significant overloads on combustor body and at the place of the nozzle unit fixing were observed. The spectral analysis of the process of the evolution overloads shows that a spectrum of vibrations of the body-charge system has a pronounced peak at a freguency of 600 Hz and lesser peaks at the frequencies of 1.6 and 2.6 kHz.

In this way, the tested engine had sufficient strength to withstand the load with $I_d = 0.3$ kPa·s and $\tau_1 = 100$ μs. At the same time the action of the CSC pressure pulse with $I_d = 2$ kPa·s and $\tau_1 = 10$ μs on the SFME through a rubber leads to the destruction of the damaged body by internal pressure after the time of the order of 900 μs from the beginning of the initiation of the charge. At the moment of the loss of carrying capacity the longitudinal deformations reach the value of $\varepsilon \approx 2\%$. The elevation of internal pressure after the action wasn't observed, and the main cause of the break - down of the engine was a decrease in the combustor body strength as a result of a pulse load.

Further tests were carried out with less level of pressure pulse ($I_d = 0.5$ kPa·s and $\tau_1 = 10$ μs) in order to determine other causes of destructions of SFME which are different from the body damages. But the action of the lesser pulse on the engine at the end of its work also have led to the loss of a body leak-proofness without preliminary increase of pressure.

The most interesting was a case of action by the same pressure pulse, but in the middle of the engine work, when there was enough fuel in the combustor to dampen partially a nonstationary load and not to allow catastrophic damages of the body. At the moment of the loading the pressure in the combustor was equal to 4.8 MPa and then it was increasing to 7.6 MPa with a succeeding 0.03 s period without significant increase lasting till body destruction. (See Table 2)

The total time of SFME work after the action was equal to $t \approx 0.093$ s. The significant increase of pressure by 60% (Fig. 1) apparently might have occured either because of increase of a burning surface with chipping-of charge damages or because of breakdown of a normal combustion products outflow as a result of shocking of the throat of a nozzle by chipped fuel pieces. Similar situation took place during the tests of SFME SL-3 [7] when some pieces of charge broke away and were carried off with gas stream through the nozzle that was a cause of sharp increase of pressure and destruction of the engine.

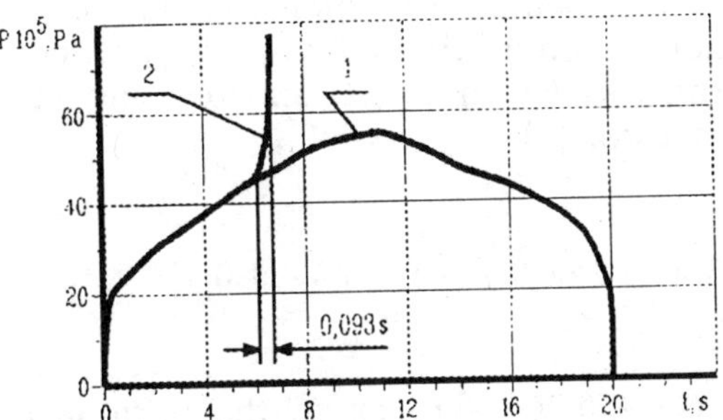

FIGURE 1. Pressure in the combustion chamber as a function of time (1 - normal burning; 2 - experimental dependence)

In order to evaluate the time of shocking of critical section after the action we shall proceed from a fact that a time of chipping-off formation is short, a form of chipped-off pieces shapes are spherical (with diameter comparable to the size of critical section R), and the velocity of gas flow is equal to its medium value $\overline{V}$ along the length of the canal. From the equation of piece motion along the canal and the expression for the force acting from the side of the flow (f)

$$f = c_x \rho_g \, (\, \overline{v}\text{-}v \,)^2 \cdot \pi r^2 / 2 \qquad (1)$$

($c_x \approx 0.5$ - aerodynamic resistance coefficient at a subsonic flow; ρ_g - gas density; V - piece velocity) we can obtain the time of chocking (t_{ch})

$$t_{ch} = 4 \, [\, (\, \rho_s / \rho_g \,) \cdot L \, R / 3 \, c_x \, v^2 \,]^{1/2} \qquad (2)$$

(ρ_s - fuel density; L - distance between the central part of the engine and the critical section).

Calculations using the equation (2) show that $t_{ch} < t*$ and, consequently, nozzle shocking by chipped-off fuel pieces may take place but increase in burning surface with a damage of the charge may be not less

TABLE 2 Internal Chamber Pressure Change

t, s	P₁,MPa	P₂,MPa	P₃,MPa	P₄,MPa
0.2	2.19	2.20	2.04	2.14
2.0	3.17	3.10	3.11	3.14
5.84	4.76	4.69	4.71	4.72
6.013	4.76	4.73	4.71	4.76
6.022	4.92	4.92	4.88	4.90
6.040	6.04	5.92	5.90	5.90
6.050	6.83	6.71	6.77	6.73
6.060	7.30	7.17	-	7.18
6.070	7.50	7.33	-	7.36
6.080	7.53	7.40	-	7.36
6.093	7.53	7.40	-	7.36

probable cause of increase in internal pressure. From the relationship between a burning surface area s(t) and pressure in the combustor [2]

$$\rho(t)/\rho(0) = [s(t)/s(0)]^{1/(1-\nu)} \qquad (3)$$

(ν = 0.266 - index of the burning area dependence on pressure) one can see that increase in pressure till value of 7.68 MPA is realized because of increase of area ΔS in the value of $\Delta S \approx 1800$ cm^2 (data on burning surface area at the normal work of SFME are represented in Table 3). The close value of increase of the area may be obtained if we assume that the form of destruction along the whole combustor is like a circle, taking into consideration peculiarities of chipped-off destructions in cylindrical charges [8, 9] and

$$\Delta S \approx \pi R(2L) \approx 1880 \text{ cm}^2 \qquad (4)$$

TABLE 3. Fuel Burning Surface Area
Depending on Vault Thickness (1)

1, sm	S, sm^2	1, sm	S, sm^2
0	2360	8.6	4825
1	2776	9	4817
2	3180	10	4600
3	3558	11	4216
4	3840	12	4080
5	4157	13	3923
6	4000	14	3935
7	4626	15	3527
8	4810	16.2	2325

Thus, carried out evaluations demonstrate possibility of destruction of the engine because of fuel chipped-off damages made by shock wave moving from nonstationary side load.

REFERENCES

1. Sinyukov, A.M., Volkov, L.U., L'vov, A.I., Shishkevich, A.M., *Solid Fuel Ballistic Mussile,* Moscow: USSR MoD Printing Office, 1972.
2. Feodosyev, V.I., *The Basics of Missile Flight,* Moscow: Nauka, 1979.
3. Karmishin, A.V., Skurlatov, E.D., Starsev, V.G., Fel'dshtein, V.A., *Nostationary Aeroelasticity of Thinwalled Structures,* Moscow: Mashinostroyeniye, 1982.
4. Velikhov, Ye.P., Sagdeyev, V.Z., Kokoshin, A.A., *Space Weapons: Safety Dilemma,* Moscow: Mir, 1986.
5. Ostrik, A.V, Petrovsky, V.P., *Chemical Physics,* **1**, 11-17, (1995).
6. Ostrik, A.V., Petrov, I.B., Petrovsky, V.P., *Mathematical Modelling,* **8**, 51-59 (1990)
7. *Missile/Space Daily,* **9** (1987).
8. Ostrik, A.V., Petrovsky, V.P., "Destruction of Cylindrical Bodies as a Result of Focusing and Interference of Stress Waves", *in Proceedings of the 8-th International Conference on Equations of State,* 1992, p. 32.
9. Ostrik, A.V., Petrovsky, V. P., *Applied Mechanics and Technical Physics,* **1**, 133-137, (1993)

PRECISE METHOD TO DETERMINE POINTS ON ISENTROPIC RELEASE CURVE ON COPPER

C. REMIOT - J.M. MEXMAIN - *L.BONNET

Commissariat à l'Energie Atomique - CEV-M - B.P. n° 7 - 77181 COURTRY - FRANCE
*CVa - 21120 IS sur TILLE - FRANCE

When a higher shock impedance foil (with several hundreds of µm in thickness) is set on the studied material surface, the release phase occurs by steps, whose duration of each plateau corresponds to a go and return of the shock wave in the foil. Step velocity levels can be easily measured by D.L.I. technique. The intermediate velocity values, connected with the knowledge of the foil Hugoniot, allow us to determine a few points on the isentropic release curve. The experiments have been achieved on a two stage light gas gun with a projectile velocity varying from 1400 to 3000 m/s. The caliber of the launcher is 30 mm. For this study concerning copper, the target is composed of a 2 mm thickness copper transmitter on which the sample is mechanically held. The tungsten (W) thick foil is, under pressure, sticked on the sample with UV stick-cord. The free surface velocity measurement accuracy of the tungsten foil is 0.4 % for values between 1500 to 3500 m/s. The first shock in the sample is varying from 40 to 120 GPa and the mass velocity from 800 to 2000 m/s. By impedance matching between the copper sample and the tungsten thick foil, we deduce for each experiment three points on the copper isentropic release curve and the final free surface velocity. The accuracy we obtain is in order of 0.4 GPa. for the pressure and 10 m/s for the mass velocity.

INTRODUCTION

Isentropic release curve of shocked materials is generally not well known. In this paper we propose an experimental approach which provids a few experimental points on a such curve : the thick foil method. So we present experimental results obtained with this method on copper.

1. EXPERIMENTAL APPROACH : THE THICK FOIL METHOD

This approach was already studied by J.M. LEZAUD and al. /1/ /2/.

When a higher shock impedance foil (with several hundreds of µm of thickness) is set on the studied material surface, the release phase occurs by steps (fig. 1), whose duration of each plateau corresponds to a go and return of the shock wave in the foil. Step velocity levels can be easily measured by DLI technique (Doppler Laser Interferometry) (fig. 2).

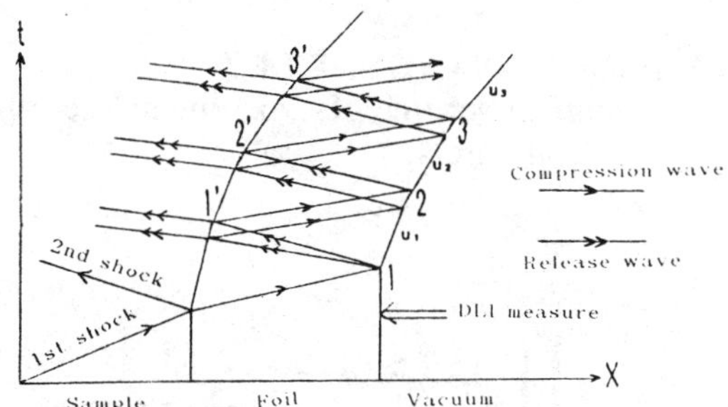

FIGURE 1. X-t diagram

FIGURE 2. Free surface velocity versus time

The intermediate velocity values, connected with the knowledge of the foil Hugoniot, allow us to determine some points on the isentropic release curve (fig. 3).

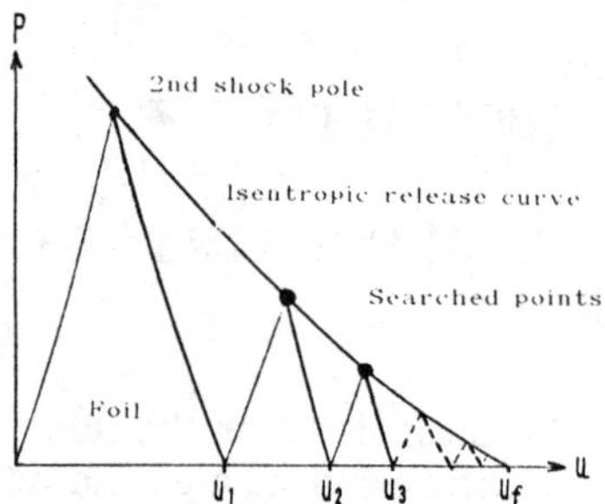

FIGURE 3. P-u diagram

2. SURVEY OF OUR FACILITY

We use a two stage light gas gun (T.S.L.G.G.) : first powder stage and second gas stage.. The characteristics are following :
Powder mass : 100 to 425 g
Projectile mass : 50 g
Projectile velocity : 1400 to 3000 m.s^{-1}
Projectile diameter : 30 mm

3. EXPERIMENTAL SET UP

Simplified scheme of experimental set up is presented on figure 4.

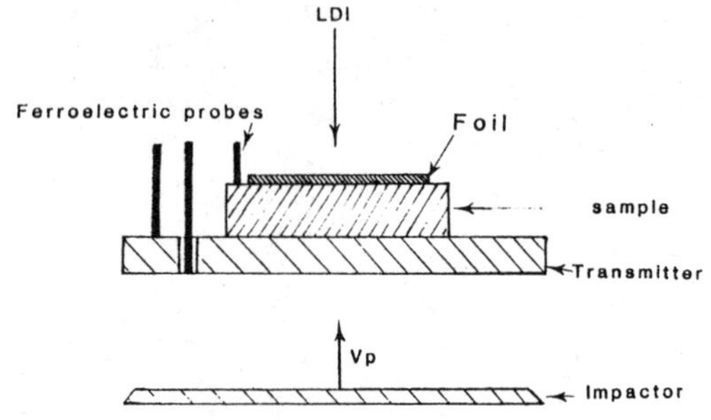

FIGURE 4. Experimental set-up

The target consists of a 2 mm thickness copper transmitter on which the copper sample is mechanically held (2 mm thickness). The tungsten thick foil is under pressure sticked on copper with UV stick-cord.

4. OBTAINED RESULTS

Two experiments have been achieved with Echo launcher.

Table 1 summarizes the experimental conditions of these two shots.

Tungsten foil Hugoniot parameters A and B (corresponding to the classic relation D = A + Bu, D : shock velocity, u : mass velocity) which have been used for data processing are :

r_0 = 19249 kg.m^{-3} (our measure)

A = 4.022 ± 0.020 km/s

from /3/

B = 1.260 ± 0.011

We obtain tungsten free surface velocity versus time. The accuracy of these measurements is less than 0.4 % (fig. 5)

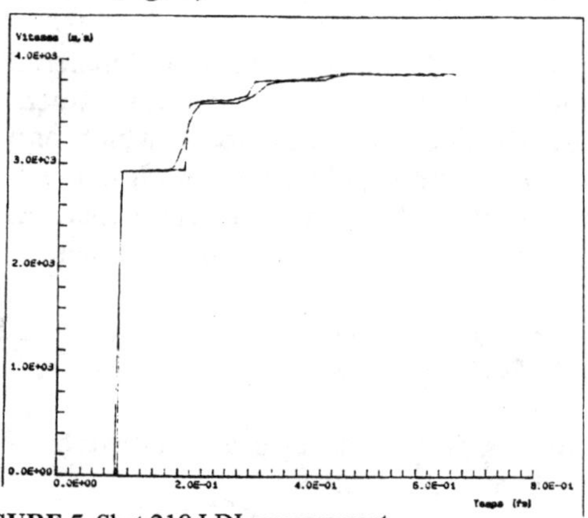

FIGURE 5. Shot 218 LDI measurement

	SHOT 218	SHOT 226
Impactor thickness W	1.72 mm	1.72 mm
Impact velocity (m/s)	3090 ± 10	1326 ± 4
Target thickness	2.1 mm	1.96 mm
Incident shock pressure in target (GPa)	119.5	40,7
Foil thickness W	0.27 mm	0.26 mm

Table 1

Table 2 presents the measured intermediate velocity levels and the obtained points (in pressure/mass velocity diagram) of the isentropic release curve of copper.

5. DISCUSSION ON RESULTS

Figure 6 shows a comparison between our experimental results (P,u points) and the mirror equation deduced from the Copper Hugoniot.

These coefficients are :

A = 3941 ± 21 m/s
B = 1.497 ± 0.025
r_0 = 8930 ± 5 kg.m^{-3}

from Bernier experimental results /4/.

Main assumption :

we suppose that the second shock Hugoniot is the same as the first one for W
-it is the pressure of incident shock that drives the choice of Copper Hugoniot and the Copper Hugoniot is adjusted on the first shock point.

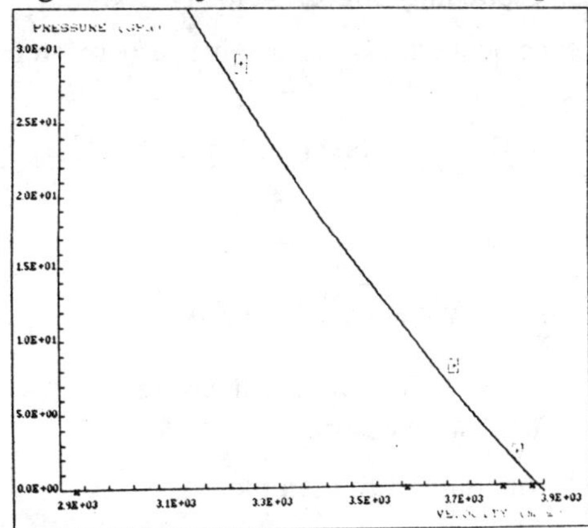

FIGURE 6. Comparison experimentation-calculation

			SHOT 218	SHOT 226
		u_{w_1}	2933 ± 11	1200 ± 6
W foil velocity steps		u_{w_2}	3613 ± 11	1528 ± 7
(m/s)		u_{w_3}	3817 ± 12	1657 ± 8
FINAL VALUE (Free Surface Velocity)		u_{w_4}	3875 ± 11	1717 ± 7
	P_{c_1} (GPa)		29,1 ± 0.7	13.3 ± 0.4
	u_{c_1} (m/s)		3273 ± 11	1364 ± 6
Deduced points	$\dfrac{\Delta u_{c_1}}{2}$ (m/s)		340 ± 6	164 ± 4
on	P_{c_2} (GPa)		8.2 ± 0.5	5.1 ± 0.4
the isentropic	u_{c_2} (m/s)		3715 ± 11	1593 ± 8
release	$\dfrac{\Delta u_{c_2}}{2}$ (m/s)		102 ± 6	65 ± 5
curve	P_{c_3} (GPa)		2.3 ± 0.5	2.3 ± 0.4
	u_{c_3} (m/s)		3846 ± 11	1687 ± 8
	$\dfrac{\Delta u_{c_3}}{2}$ (m/s)		30 ± 5	30 ± 5

$$u_{c_i} = \frac{u_{w_i} + u_{w_{i+1}}}{2}$$

Table 2. $\Delta u_{c_i} = u_{w_{i+1}} - u_{w_i}$

Measurement accuracy :

The accuracy of deduced pressure is calculated as follows.

Let us named u_{w_i} the W step velocity and u_{c_i} the corresponding copper velocity and Δu_{c_i} the difference between two successive velocity steps.

Typically $\Delta u_{c_i} = u_{w_{i+1}} - u_{w_i}$

u_{w_i} and Du_{c_i} are taken in table 2.

$$P_{c_i} = r_0(A + B\frac{\Delta u_{c_i}}{2})\frac{\Delta u_{c_i}}{2}$$

$\quad$ A $\qquad$ W values from LANL
$\quad$ B
$\quad$ r_0 $\qquad$ W density (our measure)
$\quad$ Δu_{c_i} our measure

and P_{c_i} pressure in the sample

$$P_{c_i} = \rho_o A \frac{\Delta u_{c_i}}{2} + \rho_o B \left(\frac{\Delta u_{c_i}}{2}\right)^2$$

$$dP_{c_i} = A\left(\frac{\Delta u_{c_i}}{2}\right)d\rho_o + \rho_o\left(\frac{\Delta u_{c_i}}{2}\right)dA + \rho_o A d\left(\frac{\Delta u_{c_i}}{2}\right)$$

$$+ B\left(\frac{\Delta u_{c_i}}{2}\right)^2 d\rho_o + \rho_o\left(\frac{\Delta u_{c_i}}{2}\right)^2 dB + \rho_o B d\left(\frac{\Delta u_{c_i}}{2}\right)\Delta u_{c_i}$$

$$\Delta u_{c_i} = u_{ci+1} - u_{ci} \qquad d\left(\frac{\Delta u_{c_i}}{2}\right) = du_{c_i}$$

$$\text{So} \quad dP = d\rho_o\left[A\left(\frac{\Delta u_{c_i}}{2}\right) + B\left(\frac{\Delta u_{c_i}}{2}\right)^2\right] + \frac{\Delta u_{c_i}}{2}\rho_O dA$$

$$+ \rho_O\left(\frac{\Delta u_{c_i}}{2}\right)^2 dB + \left[\rho_o A + \rho_o B \Delta u_{c_i}\right]d\left(\frac{\Delta u_{c_i}}{2}\right)$$

taking $\qquad dr_0 \approx 10 \text{ kg.m}^{-3} \qquad dA = 20 \text{ m.s}^{-1}$
$dB = 0.011$.

$$d\left(\frac{\Delta u_{c_i}}{2}\right) \approx 5 \text{ m.s}^{-1} \text{ (our classical accuracy)}$$

and experimental values of A, B and r_0.

dP becomes equation $\quad$ (1) $\quad dP = 0.12 + 0.085$

$$d\left(\frac{\Delta u_{c_i}}{2}\right) \text{ in GPa.}$$

$d(Du_i)$ is in the order of 5 m.s^{-1}. So dP is in order of 0.4 GPa.

CONCLUSION

This method of determining the isentropic release curve of materials is very satisfactory because of a very good accuracy on tungsten free surface velocity measurement due to very good DLI results and loads to 3 isentropic points measured <u>on one shot</u> with a Two Stage Light Gas Gun.

ACKNOWLEDGEMENT

These experiments were realized in collaboration with J.C. MIRABEN, P. PICARD, M. SIMONOT and J. MATHIAS (CEA Centre d'Etudes de Valduc). Acknowledgement for their contribution for training T.S.L.G.G.

BIBLIOGRAPHY

/1/ $\qquad$ J.M. LEZAUD and al.
$\qquad$ Revue de Phys. Appl. 20 (1985) 51-62.

/2/ $\qquad$ J.M. LEZAUD and al.
$\qquad$ J. Phys. III France 3(1993) 207-222.

/3/ $\qquad$ R.S. HIXSON - J.N. FRITZ
$\qquad$ J. Appl. Phys. 71 (4) 15 february 1992.

/4/ $\qquad$ CEA Internal report.

INCREASE OF THE DYNAMIC RANGE OF CATCHUP EXPERIMENTS BY HIGH-PASS FILTERING

David J. Erskine

Lawrence Livermore National Laboratory, Livermore, CA 94551

The release-catchup shock experiment is an important tool for measuring the speed of sound in compressed matter. The catchup of the release wave to the leading shock is sensitively detected optically, through an indicating fluid which produces light approximately to the 4th power of the shock pressure. However, this sensitivity demands a dynamic range which exceeds the capabilities of our digitizer. The catchup signature lies at the top of a flat pulse, thus any signal clipping is a catastrophic loss of data. We have invented a simple and accurate method for recording the catchup signature that is insensitive to signal clipping. A high pass circuit prior to the digitizer is used with post experiment integration. The insensitivity to clipping allows recording the catchup signature at higher gain, and thus with an improved signal to noise ratio.

INTRODUCTION

An important kind of shock experiment is the "release-catchup" experiment promoted by McQueen et al.[1-3], used to measure the speed of sound in compressed materials. In this experiment, a thin impactor generates a shock and release wave both traveling forward through the sample. The depth at which the release catches up to the leading shock is a measure of the sound speed in the compressed sample. The moment of catchup is detected most sensitively optically, from behind the target. Behind the sample is a fluid which produces light under shock. When the shock enters the fluid the light intensity jumps immediately to a nearly constant level (Fig. 1).

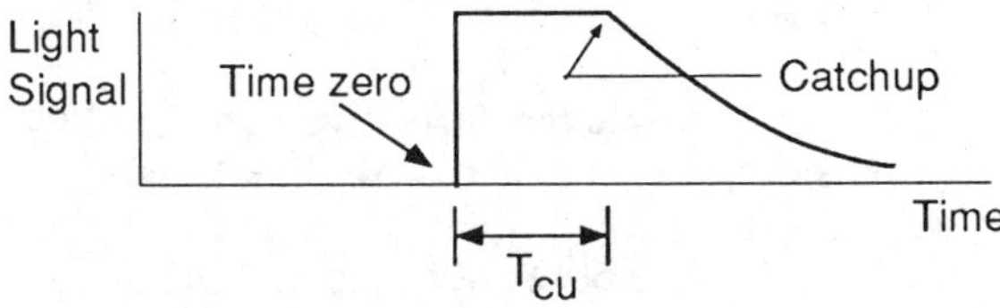

Figure 1. Shock entering liquid generates light related to pressure by $I \sim P^4$. When the rarefaction wave catches up to leading shock the intensity changes slope. Time from start of light to catchup (T_{cu}) is the measurement of interest.

When the release wave catches up to the leading shock in the fluid, the light intensity drops sharply, producing a change in slope denoted the catchup point. Measuring the time difference T_{cu} between the start of the light pulse and the catchup point is the goal of the experiment.

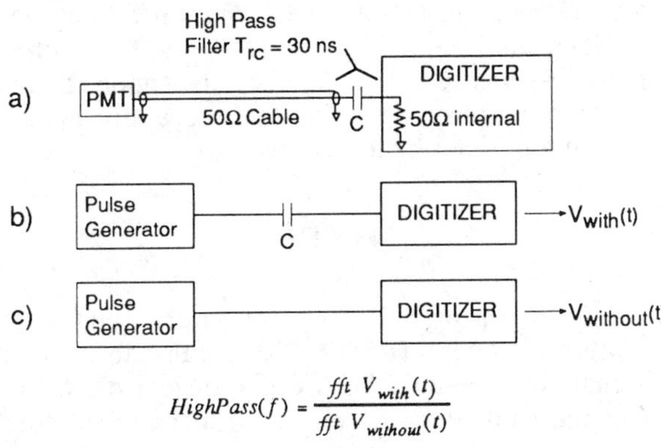

$$HighPass(f) = \frac{fft \; V_{with}(t)}{fft \; V_{without}(t)}$$

Figure 2. a) Electrical configuration. Photomultiplier (PMT) generates signal from shock generated light. Capacitor and internal 50Ω impedances of recorder and cable form a high pass filter with time constant T_{rc} = 30 ns. Transfer function of filter *HighPass(f)* measured by applying test pulse with b) and without c) filter and dividing their Fourier transforms (*fft*).

McQueen et al report[2] that the light intensity I varies with the impactor velocity U_f approximately

$$I \sim U_f^8. \tag{1}$$

This is consistent with $I \sim T^4$ by the Stefan-Boltzmann law, and typically observed shock behaviors $T \sim P$, and $P \sim U_f^2$, where T is temperature, and P is the shock pressure. We accelerate our projectiles with a two stage gas gun, initially propelled by gunpowder. The variation in powder burning produces a final projectile velocity variation of 10% in some cases. Because of the high exponent of Eq. (1), this yields a variability in signal intensity of ~2:1. Unknown sample equation of state may also contribute a significant uncertainty through the variation in P, through $I \sim T^4 \sim P^4$. Because the useful portion of the signal lies at the top of a pulse, any overloading is a catastrophic loss of the catchup signature. This forces us to be especially conservative in choosing the nominal vertical gain of our digitizing recorder. However, the 8 bit vertical resolution is insufficient at the reduced nominal level to determine the catchup time to the desired precision.

As a solution, we have invented[4] a simple method to eliminate the catastrophic sensitivity to clipping. The signal is recorded through a high pass circuit. The inverse operation is applied numerically during data analysis. We will demonstrate that the method is robust to clipping and preserves the high frequency information needed to determine the change of slope at the catchup point. As a generic method, it can be applied to any pulse experiment where the pertinent information lies on top of a roughly flat pulse.

METHOD

Our optical signal is detected with photomultipliers (PMT). Coaxial cables lead signals to Tektronix DSA 602 transient digitizers. Coaxial cable delays are used to increase the total cable length to 300 ns to make the appearance of spurious echos obvious. The vertical resolution of the digitizers is 8 bits, or 256 vertical points. The sampling interval is 2 ns.

The traditional way to handle signals of high dynamic range is to use redundant channels with different vertical gain settings. We assign 2 PMTs to observe each of 6 target locations, with a high and low optical attention preceding each PMT tube. We must be mindful that in spite of the increased dynamic range of the digitizer using this method, the PMTs have their own dynamic range limitation. Excessive light will cause PMT saturation. However, this saturation is more forgiving than the

clipping of the digitizer. The change in slope at the catchup point can still be observed in the saturated signal, albeit weaker.

Our method[4] is based on the idea that the catchup point is a change of slope, from a slope that is initially small. Therefore, if one records the derivative of the signal, the signal immediately preceding the catchup will always be near zero, and this can be placed at the center of the digitizing range for any gain.

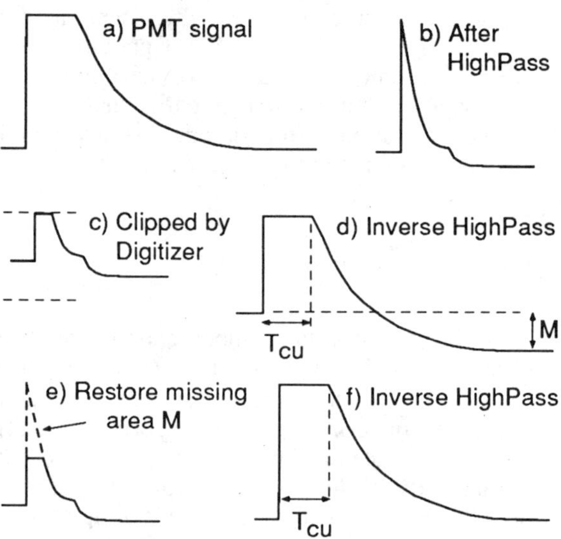

Figure 3. Steps in recording the experimental signal. a) Photomultiplier signal. b) After high pass circuit. c) Digitizer clips signal. d) After numerically applying inverse of high pass filter. Catchup time T_{cu} can be obtained from this signal. Optionally, further processing will restore proper shape to tail. Baseline offset M is the area that needs to be restored to derivative signal e). f) final result after applying inverse of high pass filter to e).

We can generate the derivative-like signal by a high pass circuit. This is trivially implemented by inserting a small capacitor in series at the digitizer input (Fig. 2). The 50Ω impedances of the digitizer input and cable with the capacitor form a high pass network with a time constant $T_{rc} \approx 30$ ns. Due to stray inductances, the behavior of the filter is more complex than a simple RC circuit. We do not attempt to model its effect by circuit theory. Instead, we measure the actual transfer function for the filter of each channel. This is done both accurately and conveniently by fast Fourier transforms of the digitized signals.

An electrical test pulse is applied to each channel, with and without the high pass filter and recorded by the digitizer as $V_{with}(t)$ and $V_{without}(t)$, respectively. The Fourier transforms (*fft*) are divided to produce the frequency response of the filter

$$HighPass(f) = \frac{fft\ V_{with}(t)}{fft\ V_{without}(t)}. \qquad (2)$$

If $V_{PMT}(t)$ is the PMT signal to be found, and $V_{rcrd}(t)$ is the recorded digitizer signal, then without clipping the PMT signal is obtained

$$fft\ V_{PMT}(t) = \frac{fft\ V_{rcrd}(t)}{HighPass(f)}. \qquad (3)$$

The steps in the recording process are illustrated in Fig. 3. The source PMT signal a) is converted to a derivative-like signal b) by the high pass circuit. This signal suffers clipping, because a high gain is deliberately used to improve the signal to noise ratio around the catchup point. We have determined that large amounts of clipping can be tolerated by the process without detrimenting the catchup signature, as long as the catchup point is not closer than about T_{rc} to the pulse start.

After the experiment, the recorded derivative-like signal is "integrated" by dividing its Fourier transform by *HighPass(f)* by Eq. (3). This signal (Fig. 3d) is sufficient to determine the catchup point, since all the high frequency information, except under the clipped region, is present. However, for mostly cosmetic reasons, the signal can be further processed to bring a closer resemblance to the original in low frequencies as well. The difference in baselines (Fig. 3d) is defined as M. A triangle of area M is added to the clipped signal (Fig. 3e). When this is integrated via Eq. (3), the result Fig. 3f is very close to the original in both low and high frequency information.

The only portion of the signal significantly altered by the clipping is directly under the clipped region, which is approximately a time constant T_{rc} from the pulse start. We have numerically tested the ability to resolve the catchup point close to the pulse start on the severity of clipping. For the PMT pulse, we used an experimental signal (Fig. 4a, bold) measured previously with direct coupling between the PMT and digitizer. The simulated derivative-like signal (Fig. 4b, bold) was obtained by applying Eq. (3), and artificially clipped at either of two levels. This was then integrated via Eq. (3) and shown as light curves against the original curve in Fig. 4a).

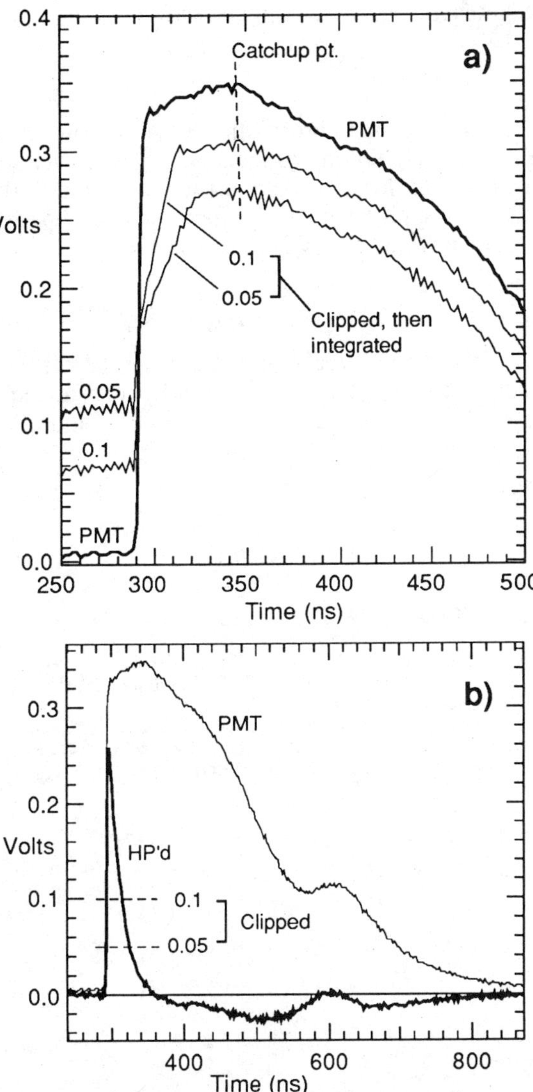

Figure 4. Effect of clipping on reconstituted signal is tested numerically. a) Bold curve: source PMT signal. b) Bold: simulated highpassed signal obtained by multiplying Fourier transform of PMT by *HighPass(f)*, which was measured from an actual filter. Signal was then clipped at either level 0.1 or 0.05 and integrated by division of its Fourier transform by *HighPass(f)*. These reconstituted curves are light curves in a). Curves shifted vertically for clarity. Data for a short time after shock is altered, but change in slope at catchup point remains.

The results show that this method is very robust to severe clipping. Only the region directly clipped was affected. All other regions retained the high frequency information content needed for detecting the change in slope.

We observe that our digitizer does not appear to be "blinded" by overloading; it recovers promptly.

CONCLUSION

This method improves the signal to noise ratio of the catchup portion of the signal. By preventing catastrophic loss of data through clipping, the nominal signal level can be increased to a greater level than would otherwise be advisable.

ACKNOWLEDGMENT

This research was performed under the auspices of the U.S. Department of Energy by the Lawrence Livermore National Laboratory under contract W-7405-Eng-48.

REFERENCES

1. R. G. McQueen, J. W. Hopson, and J. N. Fritz, Rev. Sci. Instsrum. 53, 245 (1982)
2. R. G. McQueen, D. G. Isaak, in "Shock Compression of Condensed Matter-1989", Ed. S. C. Schmidt et al., (Elsevier Science Pub., NY 1990), p. 125.
3. J. N. Fritz, C. E. Morris, R. S. Hixson, and R. G. McQueen, in "High-Pressure Science and Technology-1993", Ed. S. C. Schmidt et al., (AIP Press, NY 1994), p. 149.
4. D. J. Erskine, Rev. Sci. Instr. Sept. 1995.

A SIMPLE METHOD OF DETERMINING THE ATTENUATION OF THE SHOCK PRESSURE IN POROUS MATERIALS

S. S. Feng and H. F. Wang

Department of Mechanics and Engineering, Beijing Institute of Technology, Beijing 100081, P.R.C.

A simple method of determining the attenuation of the shock pressure in porous materials is proposed. The proposed method is based on a convenient experiment carried out to measure the shock wave velocity and a simple porous Hugoniot independent of the effective Mie-Gruneisen function γ. By combining the experimental results and the calculated Hugoniots, the attenuations of shock pressures in porous iron with several initial densities are obtained.

INTRODUCTION

The porous material has a larger initial specific volume compared with the corresponding solid material. When they are compressed to the same specific volume, the porous material will need a higher shock pressure. It means that the porous material has the better capability of attenuating the shock pressure or absorbing the shock energy compared with the corresponding solid material. This property of the porous material mainly depends on its shock Hugoniot in the complete compaction regime. The porous Hugoniot in the complete compaction regime is extrapolated usually from that of the corresponding solid material. The extrapolation has been done most frequently by using the Mie-Gruneisen equation of state written as follows [1]

$$(\frac{\partial P}{\partial E})_V = \frac{\gamma}{V} \qquad (1)$$

where γ is an effective Gruneisen function and P, E, and V are pressure, specific internal energy, and specific volume, respectively. Although this extrapolated porous Hugoniot is used frequently, the behavior of γ at high pressures and high temperatures has not been well understood. The extrapolation through equation (1) is only good within a limited range of states adjacent to the reference state. In order to extrapolate the porous Hugoniot within a wide range of states, an approximate empirical equation of state is suggested as follows [2] :

$$(\frac{\partial E}{\partial V})_P = -(\frac{\partial E}{\partial V})_H \qquad (2)$$

The extrapolation through equation (2) along the constant specific internal energy in a $P - E$ plane will lessen the extrapolated deviation, however, there is an inherent error in the Eq. (2), and it is quite complex to solve for the porous Hugonoit by combining eight equations. In this paper , a simple porous Hugoniot independent of γ is used to calculate the relationship between the shock wave velocity and the particle velocity. A convenient experiment is carried out to measure the shock wave velocity. By combining the experimental results and the calculated Hugoniots , the attenuations of shock pressures in porous iron with several initial densities are obtained.

EXPERIMENTS AND RESULTS

Experimental Arrangement and Principle

Figure 1 shows the experimental arrangement. In our experiment, the planar shock wave generator diameter is 50 mm. The TNT charge density, length and diameter are 1.57 g /cm³, 50 mm and 45 mm , respectively. The cylindrical porous iron plates thickness and diameter are 5 mm and 40 mm , respectively. Both the copper foid and the insulating

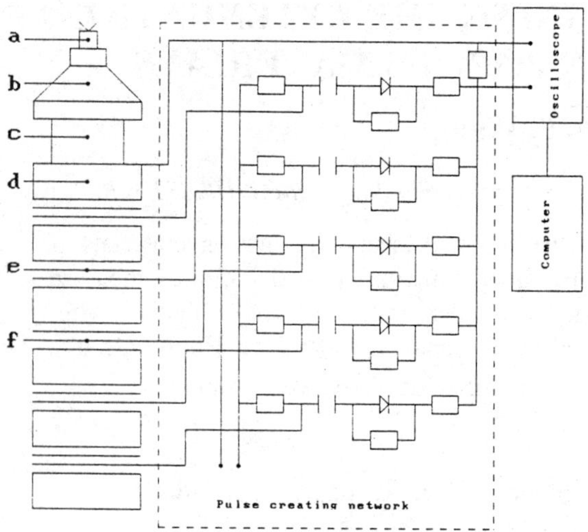

FIGURE 1. Experimental arrangement for measuring shock velocity. Labels (a) , (b) , (c) , (d) , (e) . and (f), refer, respectively, to   detonator, planar   wave generator, TNT charge, cylindrical plate , insulating paper. copper foid.

respectively. Both the copper foid and the insulating paper (polytetrafluoroethylene) thickness are 0.1 mm. The export pulse width of the pulse creating network is about 0.5 μs.

When the planar shock wave propagates to the interfaces of the cylindrical plates, the insulating papers between two adjacent plates are broken down in turn to result in the circuit shorting out. The capacitance   discharge   pulses   obtained   by   the matching resistance are imported into the Textronix 2440 oscilloscope. Thus, the pulse singals with a certain interval time are obtained. By processing the experimental data , the relationship between the shock wave velocity and the propagation distance can be obtained.

Experimental Results

Figure 2 shows the experimental results for porous iron. The experimental data are regressed by a fourth order polynomial written as follows:

$$X = a_1 t + a_2 t^2 + a_3 t^3 + a_4 t^4 \qquad (3)$$

Differentiating equation (3) , we obtain

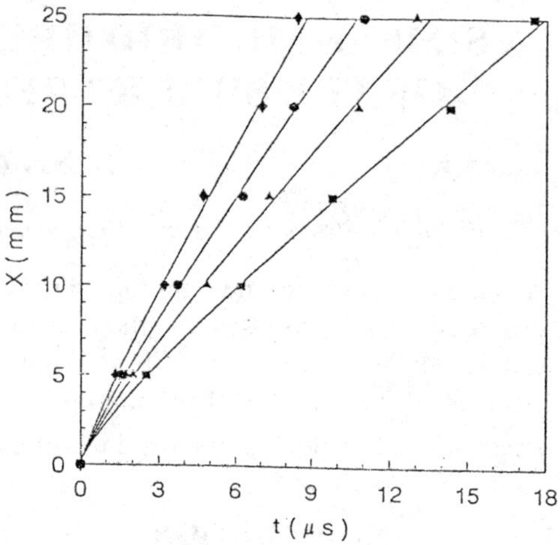

FIGURE 2. Experimental results for porous iron illustrated in a $x - t$ plane. The experimental data points ● , ◆ , ▲ ,and ■ refer, respectively, to porous iron with initial densities 80% , 70 % , 60 % ,and 50 % densities of the corresponding solid initial density.

$$D = a_1 + 2a_2 t + 3t^2 + 4t^3 \qquad (4)$$

where a_1 , a_2 , a_3 , and a_4 are the regression coefficients shown in Table 1. By combining Eqs. (3) and (4), the relationship between the shock wave velocity D and the propagation distance X can be obtained . Figure 3 shows the shock velocity versus the propagation distance for porous iron .

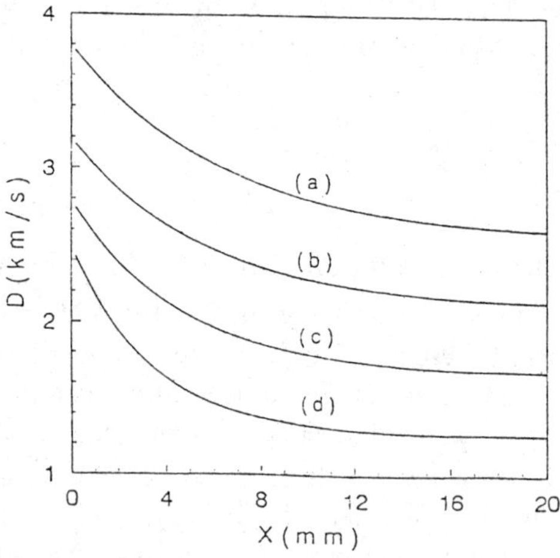

FIGURE 3. Shock velocity D versus distance X obtained by combining Eqs. (3) and (4) for porous iron . Curves (a) . (b), (c).and (d) refer. respectively. to the porous iron with initial densities 80% , 70 % , 60 % ,and   50 % densities of the corresponding solid initial density.

964

TABLE 1. Regression Coefficients in Eqation (3)

ρ_{00}/ρ_0	0.5	0.6	0.7	0.8
a_1	2.046	2.547	3.042	3.665
a_2	- 0.112	- 0.144	- 0. 171	- 0.231
a_3	0.0071	0.0107	0.0154	0.0235
a_4	- 0.00015	- 0.0003	- 0.00051	- 0.00094

$D-U$ HUGONIOT RELATION

In general, experimental Hugoniot data are obtained under the normal atmospheric temperature and pressure as the initial state. If various temperatures are chosen as initial states under the normal atmospheric pressure, a group of shock Hugoniots will be obtained [3]. Figure 4 shows three Hugoniots . Suppose the specific internal energy to be same (E_H) at the state point $H(P_H, V_H)$ for both the porous Hugoniot and the dotted Hugoniot , the initial specific internal energy for the dotted Hugoniot to be $E_0 + Q$, the corresponding state point in the $P-V$ plane to be $E(0, V_*)$, the dotted Hugoniot to be written still in the following $P-V$ Hugoniot form[1] known for solid materials:

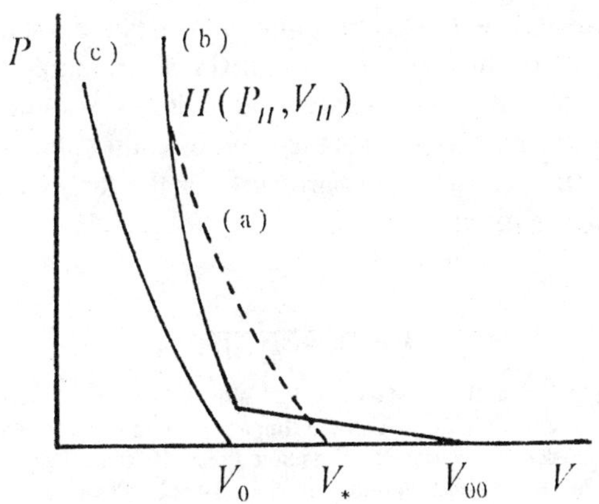

FIGURE 4. Shock Hugoniots in a *P-V* plane. The dotted cuve (a) is one of the solid shock Hugoniots with various initial temperatrues , and curves (b) and (c) are shock Hugoniots for the porous material and corresponding solid material, respectively.

$$P = \frac{a^2(V_* - V)}{[V_* - b(V_* - V)]^2} \qquad (5)$$

By using the thermal expansion equation, V_* can be written in the following expression :

$$V_* = V_0(1 + \alpha \frac{Q}{C_p}) \qquad (6)$$

where a and b are the material constants , and V_0 , E_0 , α , and C_p are the initial specific volume, initial specific internal energy, volume expansion coefficient ,and constant pressure specific heat of the corresponding solid material, respectively. Q is the heat energy absorbed by the corresponding solid material from state point $A(0, V_0)$ to state point $E(0, V_*)$ under the normal atmospheric pressure.

Based on above assumptions [3], a simple porous Hugoniot in a $P-V$ plane is obtained as follows:

$$V_H = \frac{V_0}{K}(1 - \frac{Z - \sqrt{Z^2 - 1}}{b}) \qquad (7)$$

$$K = \frac{1 + \dfrac{P_H}{2}\dfrac{\alpha}{C_p}V_0}{1 + \dfrac{P_H}{2}\dfrac{\alpha}{C_p}V_{00}} \qquad (8)$$

$$Z = 1 + \frac{a^2 K}{2V_0 b P_H} \qquad (9)$$

where V_{00} is the initial specific volume of the porous material. By combining Eqs. (7) , (8) , (9) and the planar shock wave jump equations written , in a material that is initially stationary, as follows:

$$VD = V_{00}(D-U) \qquad (10)$$

$$V_{00}(P - P_0) = DU \qquad (11)$$

the $D-U$ Hugoiot can be obtained. Figure 5 shows the calculated $D-U$ Hugoniots for porous iron .

ATTENUATIONS OF SHOCK PRESSURES

Figure 6 shows the shock pressure P versus propagation distance X obtained by combining Eqs. (3) , (4) , (7) , (8) , (9) , and (11) . Curves (a) , (b) , (c), and (d) are for porous material with initial densities 80% , 70% , 60% , and 50% densities of the corresponding solid initial density, respectively. It can been from Fig. 6 that the initial shock pressures will increase while increasing initial densities. For the same propagation distance within the range of experiments, the shock pressure for the larger initial density is higher.

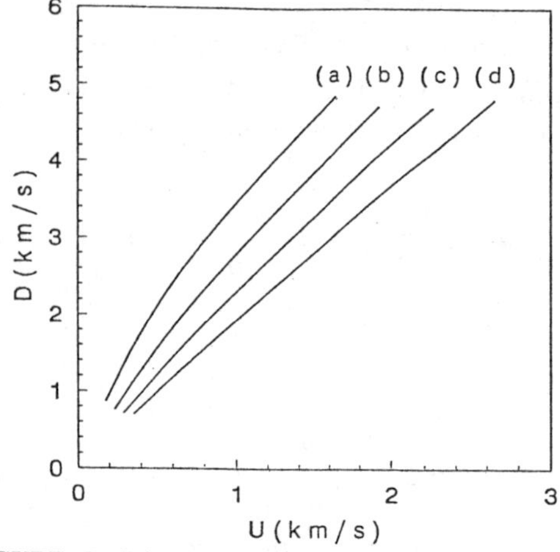

FIGURE 5. Calculated Hugoniots for porous iron illustrated in a *D-U* plane. Curves (a) , (b) , (c), and (d) are for initial densities 80% , 70% , 60% , and 50% densities of the corresponding solid initial density, respectively. $\alpha = 3.51 \times 10^{-5}$ (1 / k)^4 , C_p (300 k ~ 1000 k) = 724 (J / kg . k)^4 , b=1.92, a = 3.57 (k m / s)^1, V_0 = 1/ 7.85 (cm^3/ g) .

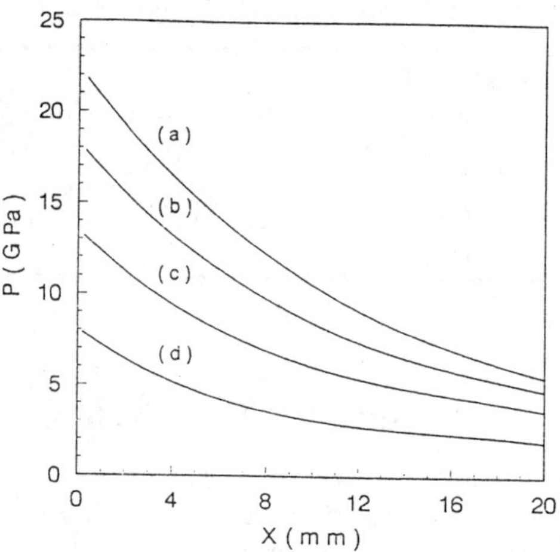

FIGURE 6. Curves of shock pressure P versus propagation distance X for porous iron with initial densities 80% , 70% , 60% , and 50% densities of the corresponding solid density, respectively.

CONCLUSIONS

A simple experiment is carried out to measure the shock wave velocity in inert solid materials accurately. The shock wave velocity versus the propagation distance for porous iron with several initial densities are obtainad. A simple porous Hugoniot is proposed to calculated the relationship between the shock wave velocity and the particle velocity. The attenuations of shock pressures in porous iron with several initial densities are obtained. the results show that the initial shock pressures will increase while increasing initial densities. For the same propagation distance within the range of experiments, the shock pressure for the larger initial density is higher.

REFERENCES

1. McQueen, R, G., Marsh, S. P., Talor, J. W., Fritz, J. N., and Carter, W. J., High Velocity Impact Phenomena, edited by Kinslow R., Orlando: Academic Press, 1970, pp.297.
2. Oh, K. H., and Perrson , P. A., J. Appl . Phys. 65 , 3852 (1989).
3. Li , X. J., J. High Pressure Phys. (CHINA) 5 (4) , 301 (1991).
4. Scheider , P. J. , Hand book of Heat Transfer Fundamentals, edited by Rohsenow, W. M., McGraw-Hill Book Company, 1985.

NEW EXPERIMENTAL APPROACH OF IMPACT ON SILICON CARBIDE

P. Riou, C.E. Cottenot and M. Boussuge*

DGA/CREA/MCS, 16 bis Avenue Prieur de la Côte d'or, 94114 Arcueil, FRANCE

**Ecole des Mines de Paris, Centre des Matériaux, 91003 Evry, FRANCE*

This paper deals with a new experimental approach of impact on ceramics. Beam target geometry allows real time data during SiC/projectile interaction to be obtained. A comparison between experimental results and numerical simulations is performed. The capability of a simple brittle criterion to describe the SiC damage is discussed.

INTRODUCTION

A lot of works point out the interest to use ceramics in armor structure (1). Considering their specific mechanical properties (2), ceramics are often employed in bi–layers concepts where:

– the first layer (ceramic) breaks or erodes the projectile and spreads the impact pressure on a large area on the second layer, by the creation of a conical crack.
– the second layer (steel or composite), which ensures a structural function, holds back the ceramic fragments and stores the impact energy by plastic deformation or delamination.

Although very important, the choice of the ceramic is often guided by empirical criteria, since it is very difficult to reach the interesting area (cracked cone), during impact, in order to visualize and understand the chronology of fracture. Thus, in this study, we have chosen to investigate the impact behavior of Silicon Carbide which appears to be one of the best materials for ballistic applications (2).

The aim of this paper is therefore to determine the fracture phenomenology within the cracked cone. In order to reach this goal, a beam shaped target geometry has been defined. The advantage of this kind of specimen resides in the possibility to investigate the conical crack area evolution in real time (3). After the experimental aspect, the second part of this work concerns the numerical simulation of these impacts.

EXPERIMENTAL ASPECT

Test system and associated measurements

Experimental set–up is presented in Fig. 1. The beam target can be used confined or not.

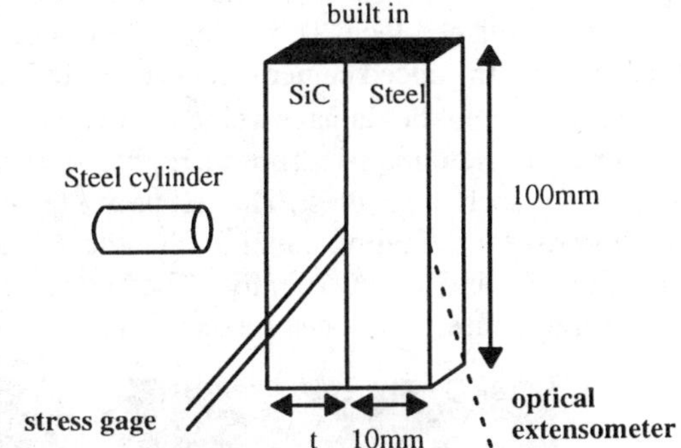

FIGURE 1. Experimental set–up.

The ballistic test configuration is the following : a steel cylinder (l=20mm, ϕ=11mm) impacts with a velocity range between 100 to 350 m/s a SiC beam. This latter, the dimension of which are 100x10xt mm^3 where the thickness t is equal to 10, 15 or 20mm, is fixed at its ends.

Some physical properties of SiC are given below :
. density = 3.15 g/cm^3
. porosity = 1.8 %
. Young's modulus = 404 GPa
. Hardness (Vickers) = 20.8 GPa
. Fracture stress 3pts bend. = 370 MPa
. Fracture Toughness = 3.2 MPa.m$^{1/2}$
· Longitudinal wave velocity = 11800 m/s
. Tranversal wave velocity = 7600 m/s

This configuration allows two kinds of results to be collected :

– firstly, to obtain real time data during impact. Thus, a stress gage is placed at the SiC/Steel interface to follow the evolution of the axial stress during impact. Moreover, an optical extensometer is employed to follow the displacement of the rear side of the target, during the interaction.
– secondly, to understand the phenomenology of the fracture. For this, pictures of impact are taken. Hence, we have the possibility to visualize the damage kinetics.

Results

Considering the second point, the obtained results presented in Fig. 2, 3 and 4, are very attractive. This technical approach is based on the work of Strassburger and al.(4) and adapted to this experimental set–up (5). The ceramic target presents one mirror–polished face. This face is illuminated by a flash and a ultra–speed camera receives the reflected light. Considering the damage velocity, a gap of 1 μs between each photographs is chosen. Eight pictures are obtained for each impact test. The test presented here, corresponds to a 20mm ceramic thickness and a projectile velocity of 203m/s. The three more interesting pictures have been selected.

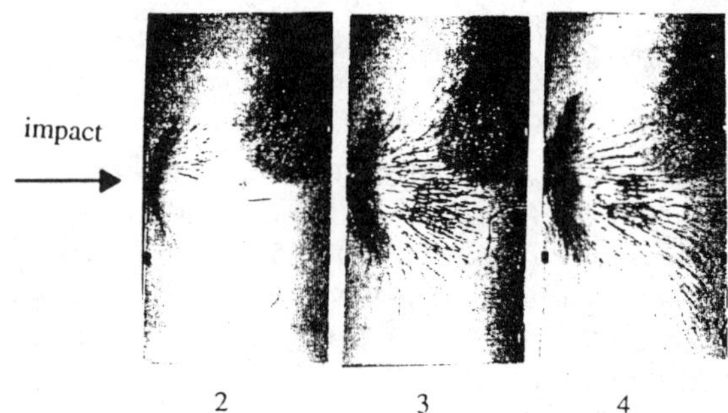

FIGURES 2, 3 et 4. Crack zone extension during impact. Each photo is separated by 1 μs.

The first picture, which is taken 1.9 μs after impact, presents already some cracks. No damage being visible at 0.9 μs after impact, cracking should start between 0.9 and 1.9 μs, under the projectile. Taking into account photos 2 and 3, it is possible to calculate a damage expansion velocity of roughly 8200m/s. This result can be compared with (4). Namely, they obtain, for the same projectile velocity and although the experimental set–up is lightly different, about 9000m/s. With regard to the accuracy of the measurements, both values can be considered as similar. If we consider our value for damage expansion, we can determine that the first crack occurs 1 μs after impact.

The study of the damage expansion suggests some remarks :

– the field of cracks appears complex and not really symmetric. The general form is circular and it is obvious that the damage expansion is slower than the longitudinal wave velocity ($c_L = 11800$m/s).
– some cracks do not initiate from the impact area. These cracks are nucleated on the material surface, far from the impact : hence, in addition to the damage generated by crack propagation, some cracks nucleate on defaults because of the tensile stress field that follows the first compression wave.
– on these pictures and the following, we do not notice any rear face damage. It could be interesting to check whether this observation is confirmed with a confined beam.

NUMERICAL SIMULATION
Elasticity

Previous works on microstructural observations of impacted targets (3) point out that the SiC failure, in this configuration, is transgranular and induced by a cleavage mechanism. Therefore, due to the brittleness of this ceramic, an elastic model with a simple failure criterion could be used to model the impact behavior of SiC.

Considering the particular geometry (for impact test) of the beam specimen, a three dimensional code: EFHYD3D, an explicit finite element code from ESI, has been chosen. The behavior of the steel is considered as elastic–plastic using Von Mises yield criterion.

The adopted procedure is the following :

– firstly, a purely elastic model is used to validate the first microseconds of the numerical simulation.
– secondly, the fracture problem is approached, considering a simple brittle modeling.

To validate the first part of the numerical simulation, a comparison is made between the experimental curve of the axial stress at SiC/steel interface, obtained by a Manganin gage, and the simulation (Fig. 5). The results exhibit a good agreement during the first microseconds corresponding to the time during which the ceramic is not too much damaged. Then, the apparition of the damage makes impossible any comparison, using this modeling, based on a simple elastic behavior of the ceramic.

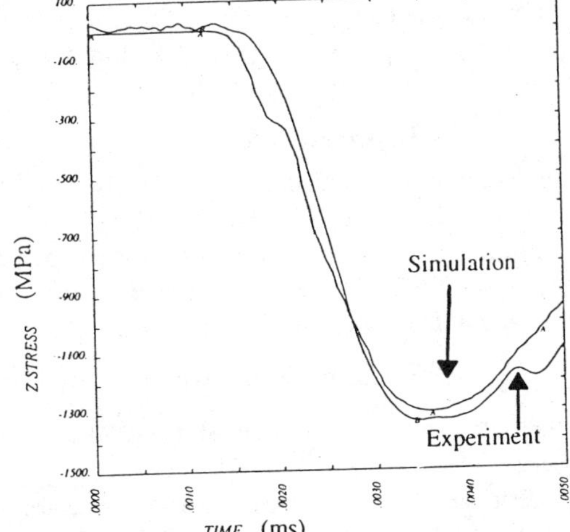

FIGURE 5. Comparison between the experimental axial stress at SiC/steel interface and the numerical simulation.

Moreover, a second comparison based on the rear surface displacement of the target can be used to complete the correlation. Results presented on Fig. 6 confirm the good correlation between numerical simulation and experiment.

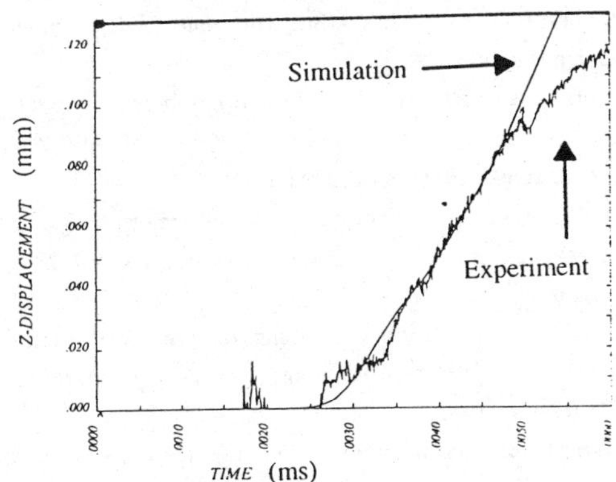

FIGURE 6. Comparison between the experimental rear target displacement and the numerical simulation.

Fracture

Up to now, to describe the damage and the fracture of the ceramic, a brittle modeling is often employed, where a threshold fracture criterion is applied. The most often used criteria are critical strain or stress (6). Some more complicated models have been used like (7), generalizing the Griffith criterion.

For this work, we have therefore decided to use a critical principal stress criterion. The choice of the threshold is very important and can be guided by two values :

– the bending strength, obtained by three points bending tests in quasi–static loading. For the SiC, this value is 370 MPa (8).
– the spall strength, obtained by plate impact. This value is obtained under dynamic loading and is also the result of a fracture under tensile stress. For the SiC, in our stress range, a threshold of 500 MPa seems realistic (9).

Taking into account numerical aspects, we have to consider that when the larger principal stress reaches the threshold, the cell is damaged. In order to obtain a good numerical stability, we have to use a progressive damage. Therefore, the concerned cell is broken, and has no more stiffness, after a number of cycles that the users can choose. A compromise between an immediate fracture and a progressive damage must be found. In our case, we used 10 cycles to reach the fracture of the cell, which seems acceptable considering the small time step. The tests made with different number of cycles produce the same results.

Figure 7 presents the state of damage of a ceramic beam 2 μs after the impact at 200m/s. The threshold is set to 370 MPa, which is the strength in a quasi–static case. The shape of the damaged zone is not very well described when compared with Fig. 2, 3 and 4. Moreover, the fracture chronology is to fast (the start of the cracking calculated is about 0.5 μs and the damage expansion velocity about 10000 m/s compared to 1 μs and 8200 m/s for the experiment). The criterion seems therefore too much severe.

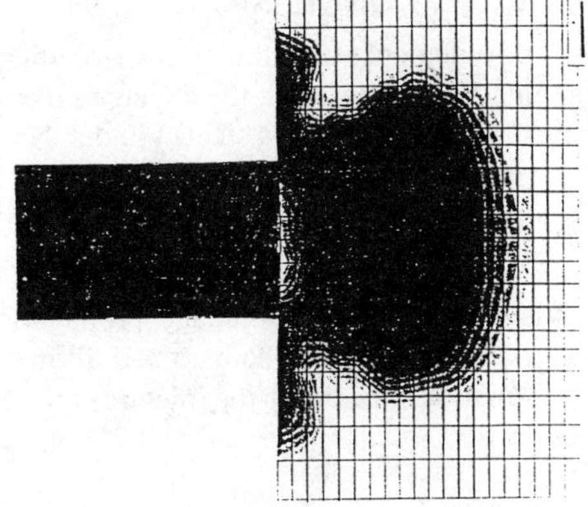

FIGURE 7. Damaged zone calculated with a threshold of 370 MPa at 2μs. Comparison with Figure 2.

The damaged zone represented in Figure 8 has been calculated for a threshold of 500 MPa. The general damage shape is less good than the previous simulation. This means that it seems difficult to improve the correlation by simply increasing the fracture threshold.

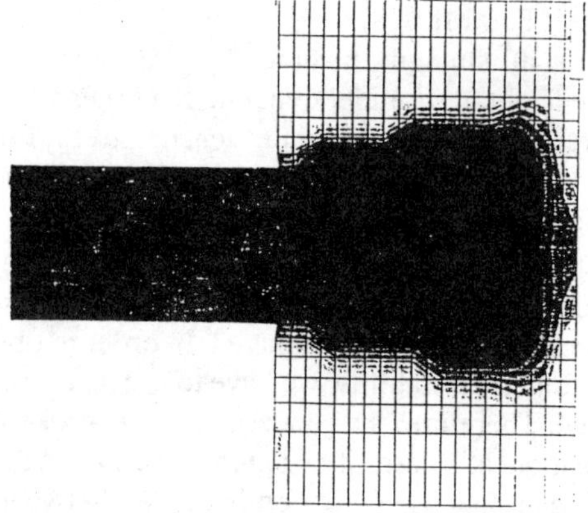

FIGURE 8. Damaged zone calculated with a threshold of 500 MPa at $2\mu s$. Comparison with Figure 2.

Two other brittle models (with a pressure cut–off and (7)) have been tested. It appears that a simple brittle criterion is unable to reproduce the experimentally observed damage. This will incite us to use a damage model.

In the case of a damage model, the high stress levels induce by the impact are responsible of an accumulation of damage leading finally to a catastrophic failure.TCK (Taylor Chen Kuszmaul) (10) or Rajendran (11) models are interesting to be tested to try to improve the agreement between modeling and experience. .

CONCLUSION

This study demonstrates the interest of this new target geometry to understand the damage kinetics of a ceramic target impacted by a steel cylinder. Namely, the beam geometry allows observations of the conical cracked area which plays a main role in the impact resistance.

The visualization of the damage propagation is possible, with a ultra–speed camera and allows very useful information concerning the fracture chronology to be obtained.

The use of a stress gage and of an optical extensometer, in spite of technical difficulties, is very precious to validate the numerical simulation. Nevertheless, the usual approach of fracture using a simple brittle criterion is not completely satisfying. A damage model appears necessary to obtain a good modeling of the damaged zone shape.

ACKNOWLEDGMENTS

The research program described in this paper has been supported by DGA/DRET/STRDT (General Delegation for Armament, Direction of Research and Technology, Technical Service of Research and Technological Development).

REFERENCES

1. den Reijer P.C., "Impact on ceramic faced armour", Thesis, Delft, 28 Novembre 1991.

2. Cottenot, C. and Riou, P., "Caractérisation de céramiques pour blindages légers", Rapport ETCA n° 93R035, Avril 1993.

3. Riou, P., Beylat., L, Cottenot, C.E.,and Derep, J–L.,"Impact damage on Silicon Carbide",, supplément au Journal de Physique III, Volume 4, pp. 281–287, Septembre 1994.

4. Strassburger, E., Senf, H., and Rothenhäusler, H., "Fracture propagation during impact in three types of ceramics", ibid 3, pp.653–658.

5. Chaillot, S., "Visualisation de l'endommagement à l'impact du SiC", Rapport ETCA n° 95R065, Stage de Maîtrise de l'Université de Valenciennes, Juin 1995.

6. Rajendran, A.M., Cook, W.H, "A comprehensive review of modeling of impact damage in ceramics", Report of University of Dayton Research Institute, December 1988.

7. Margolin, L.G.,"A generalized Griffith criterion for crack propagation", *Engineering Fracture mechanics*, Vol. 19, N°3, pp 539–543, 1984.

8. Denoual, C., and Riou, P.,"Comportement à l'impact de céramiques techniques pour blindages légers", Rapport ETCA n°95R005, Février 1995.

9. Cagnoux, J., and Chambaudie, S., "Comportement sous choc de TiB_2, B_4C et SiC", Rapport du Centre d'Etudes de Gramat, 1993.

10. Taylor, L.M., Chen, E.P., and Kuszmaul, J.S.,"Microcrack–induced damage accumulation in brittle rock under dynamic loading", *Computer methods in applied mechanics and engineering*, pp 301–320, 1986.

11. Rajendran, A.M.,"Modeling the impact behavior of AD85 ceramic under Multi–axial loading", Report of ARL, May 1993.

HEAD-ON COLLISION OF A PLANAR SHOCK WAVE WITH A GRANULAR LAYER

A. Britan, T. Elperin, O. Igra and J. P. Jiang

Ben-Gurion University of the Negev, Beer-Sheva, Israel

An experimental study was conducted in a vertical, 31×31 mm^2 cross-section, shock tube. The investigation included measurements of the gas pressure and the compressive stress; based on the observed measurements the trajectories of the principal disturbances inside the granular layer were reconstructed. Results of this study show that two different effects, gas filtration and mechanical response of a granular layer to the weak shock wave loading, are of paramount importance and depend on the layer characteristics.

INTRODUCTION

In a weak shock wave impacts the granular layer compaction effect strongly depends on material characteristics. In [1] pressure histories were recorded at the shock tube end wall which was covered with various types of granular material for incident shock wave Mach number M_s=1.36. The unsteady pressure spike at the end wall was observed; its amplitude sometimes exceeds the equilibrium pressure level and depends on the layer thickness (h) and the diameter of solid particles (d_p). Pressure spike of about 6.2 MPa was observed in [2] for nylon powder ($d_p = 5\mu$, $h = 1cm$, $M_s = 1.7$). In such materials the mechanical response and gas filtration through the layer are of importance. Contrary to these studies, only filtration pressure was registered in [3] inside a sample consisting of sand particles with diameter $d_p = 340\,\mu m$. In order to prevent compaction, particles were fixed at the contact points with an epoxy resin. Gas flow pictures and pressure profiles inside an array of cylinders were obtained in [4] for $M_s = 2$. Rigid porous materials were also used for pressure measurements in [5] and obviously, in this cases the skeleton compression can be neglected and the most important effect is due to gas filtration.

The main goal of the present study is to investigate the coupling between granular layer compaction during the impact and unsteady waves interaction inside the material. Experiments were performed in a vertical small-size shock tube with a square (31×31 mm^2) cross section, 1.5m long channel and 0.8m long driver [6]. In all tests the thickness of a granular layer was $h = 138\,mm$

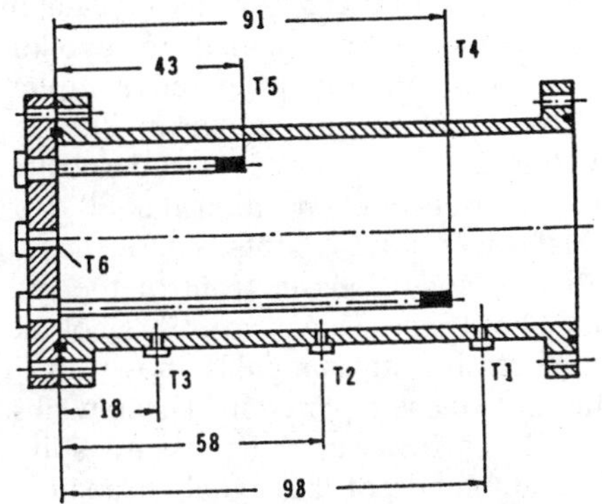

Fig. 1 Schematic of shock tube test section

Transducers T1 -T3 (Suntzov 5136) at the test section (see Fig. 1) were protected from having direct contact with the solid particles by a screen permeable to gas in order to measure the local gas pressure. Transducers T4 - T6 (Kistler 601 H) were used without a protected screen to enable measuring stresses inside the granular layer. All recorded data were stored using a data acquisition system sampling at 500 KHz per channel on IBM -486 computer.

Since previous studies [1], [2] confirmed that the magnitude of M_s does not affect the obtained results, in the present experiments a single value $M_s = 1.33$ was employed.

EXPERIMENTAL RESULTS

When choosing a pressure transducer for stress recording inside a granular medium conflicting requirements have to be met. On the one hand it must be as small as possible in order to measure stress at the desirable point inside the layer; on the other hand it should be significantly larger than a diameter of the particles d_p in order to exclude effects of particle orientation and quality of their contacts with the transducers. In Table 1 characteristics of the investigated material, i.e., particle density ρ_p, material density ρ_s and bulk densities ρ_c, relative density $v = \rho_c / \rho_s$, porosity $\varepsilon = 1 - v$, d_p, and the Young's moduli for material E_s, are given. Since the diameter of the pressure transducer used is 5.5 mm, the stress measured by transducers T4 - T6 may depend on the above mentioned size effect (at least for materials N1 - N3). In order to verify repeatability of the obtained results, pressure signals for two kinds of materials with identically prepared samples and different characteristics are shown in Fig. 2. Also shown in Fig. 2 (at the top) is a superposition of signals from transducer G5 positioned at 80 mm from the granular layer. These plots demonstrate good repeatability of conditions in front of the granular layer in the shock tube. However, this is not the case for the compressive stress signals shown in Fig. 2, which, similar to those reported in [1], show unsteady pressure spikes. However the pressure spike and quasisteady amplitude of the signals in material N1 varied significantly from run to run. Identical signals in this figure are observed only after the strong rarefaction wave, reflected from the driver's end wall, reaches the granular layer-gas interface. During the test time period the amplitudes of signals change from the gas pressure up to the compressive stress in the granular material. Minimum scattering was recorded in powder with a small particle diameter. In material N5 (see Fig. 2), it did not exceed 18%.

Compressive stress recorded by transducer T4 inside this material is two times higher than the gas pressure in front of the granular layer (for initial pressure spike this pressure ratio is about 2.6) and it decreases quickly along the layer due to friction losses. In order to demonstrate the effect of filtration flow on the compression stress developed in the granular material, two experiments were conducted. In the first, a "standard" granular layer (see signals (a) in Fig. 3) was placed on the shock tube end wall. In the second, the top of an identical layer of granular material was covered by thin (10 μm) polyethylene film (marked (b) in Fig. 3). This film prevents gas filtration through the layer during the impact. At the top of Fig. 3 are also shown gas pressure signals obtained for transducer G5 placed in front of the layer. These signals exhibit a slight rise in time and a time delay due to time elapsed until the reflected shock wave reaches transducer G5. Such unsteady pressure behavior recorded ahead of the granular layer is more pronounced in materials having larger particles and, obviously, is not observed in experiments with a covered granular layer. Notably, compressive stress signals, (a) and (b) in Fig. 3 have approximately the same shape for the initial unsteady spike, smooth leading front due to the attenuation and dispersion of transmitted and compaction waves. In some cases, stress signals demonstrate an initial slowly rising precursor, which is clearly visible for signals T5 and T6. Analysis of experimental results shows that the characteristic points at the stress signal (where the signal sharply changes its behavior) are well reproduced and can by used as reference points for measuring velocities: of precursor c_1, compaction wave v_s, and the leading edge velocity of stress reduction at the spike back side, v_r. In order to measure the transmitted wave velocity, v_g, and its trajectory inside the granular layer, gas pressure signals from transducers T1 - T3 were used. All time of arrival values were determined

TABLE 1

N	Material	ρ_p, g/cc	ρ_s, g/cc	ρ_c, g/cc	v	ε	d_p, mm	E_s, kg/mm^2
1	PVC	1.4	1.400	0.856	0.611	0.389	3.33	300
2	Al$_2$O$_3$	0.940	3.288	0.511	0.544	0.456	1.67	38000
3	Fe	7.414	7.414	4.499	0.607	0.393	1.04	20000
4	Fe	7.414	7.414	4.457	0.601	0.399	0.45	20000
5	Potash	1.905	2.050	1.089	0.572	0.428	0.45	6000

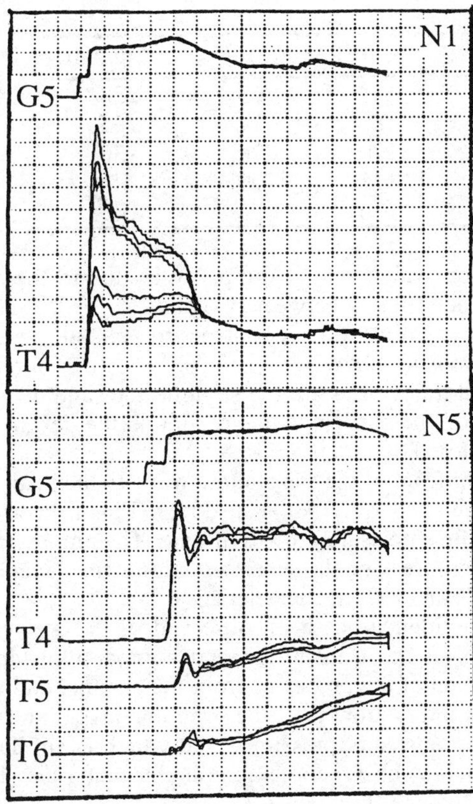

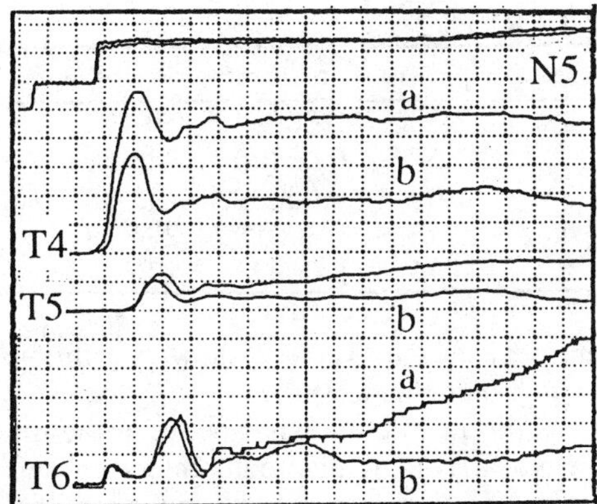

Fig.3 Compressive stress signals for standard (a) and protected (b) granular layers. Vertical scale 1 bar / div for transducers G5, T4, T5 and 0.1 bar / div for T6; horizontal scale 0.5 ms / div

Fig.2 Repeatability of compressive stress signals inside the granular materials. Vertical scale 1 bar / div, horizontal scale 1 ms / div (material N1), 0.5 ms / div (material N5)

from the accumulated data by averaging results of 5 - 9 experiments which were conducted at the same conditions. When the leading edge of a disturbance was ill defined the detectable minimum amplitude for this measurements was 10 KPa.

DISCUSSION

In the present experiments the test section was subjected to a weak shaking during its filling with granular material. Sound velocity inside a layer is given by the following formula [8]:

$$c_0 = 2\sqrt{\frac{E_s}{\rho_s}}\, v\, k^{-2} \left(v_0 - v\, k^{-1} \right) \qquad (1)$$

where $k = v_1/v$ is the packing coefficient, v_1 is a ratio of bulk densities after and prior to

shaking, v_0 is the current relative density [8]. For estimating the permeability coefficient f the well known Carman-Kozeny formula [9] was used. Two values of c_0 (for $v_0 = v$ and for $v_0 = v_1$) calculated using Eq. (1) and the permeability coefficients f are given in Table 2 along with measured precursor velocities v_1, transmitted pressure wave velocity v_g, compaction wave velocity v_s and velocity of the leading edge of the stress reduction v_r.

Since the granular layer was subjected to shaking, measured values c_1 are close to the maximum calculated sound velocity c_0. Material N1 has the lowest Young's modulus, and its precursor velocity is too low to overcome a transmitted wave; no elastic precursor was observed in this material.

Velocities of the transmitted waves increase with f, at least inside material N1 - N3, and calculated value of f can be used as characteristic of gas filtration when $v_g > v_s$.

Materials N4 and N5 have extremely low f values and a transmitted wave reflects from the end wall after the compaction wave because in this case $v_g < v_s$. This results in additional reduction of f in vicinity of the channel end wall, where transmitted waves trajectories change their slope. In a flexible foams compaction wave velocity is determined by gas filtration and consequently, depends on f [7].

TABLE 2

N	d_p, mm	f, mm^2	c_0, m/s	c_1, m/s	v_g, m/s	v_s, m/s	v_r, m/s	a_0, m/s[10]
1	3.33	0.00954	50-102	-	287	192	363	233
2	1.67	0.00496	246-513	418	276	189	314	165
3	1.04	0.00099	253-532	531	226	170	270	130
4	0.45	0.00018	233-481	493	143	175	196	55
5	0.45	0.00027	176-376	418	144	236	255	55

The data presented in Table 2 show that this is not the case in granular materials.

In foam a time delay exists between the time of shock wave arrival at the foam interface and the appearance of a compaction wave in the material [7]. In the present study a delay time of about $150\mu m$ was observed only in materials N4 and N5 and the time did not change significantly with varying particle density. For all other materials the transmitted and compaction waves start at the same point.

In contrast to a compaction wave, the magnitude of v_r decreases when the permeability coefficient f decreases, and it depends on the friction and heat transfer between the two phases. The present results were compared with those of [10] for equal d_p. Contrary to conditions in experiments reported in [10], in the present study the gaseous phase was compressed by a compaction wave and/or a transmitted wave, which affects the rarefaction wave speed v_r. The possible reason for the appearance of this rarefaction wave may be refraction of the wave reflected from the channel end wall at the granular layer-gas interface [7]. In the present study, the first reflected wave in materials N2 - N5 is a precursor while for material N1, similarly to flexible foam, it is a transmitted wave. If this explanation is correct, then the wave velocity after its reflection from the channel end wall must be from 1.5 to 5.5 times higher than that before reflection.

CONCLUSIONS

A detailed experimental study of a weak shock wave impact on a granular layer has been conducted. It is shown that separate, discrete waves pass through the material skeleton and through the gas filling the pores between the granules. The wave succession depends solely on two processes. The first is a complex interactions of a compaction wave with the granular material and transition of compaction due to the stress between particles. The second process is gas filtration which acts at the particles due to the drag forces and friction between two phases.

Wave diagrams constructed based on the measurements show:
1. Most trajectories are the straight lines and depend solely on the particle diameter, not on the density of the material.
2. Sharp front of the transmitted wave (which can be distinguished only for transducer T1), follows by continuous pressure growth in time. The amplitude of gas pressure signals decreases along the layer and depends on material characteristics.
3. No evidence has been observed of the reflection process of the transmitted and compaction waves. Reflection of the precursor from the channel end wall and its interaction with the granular layer interface result in unsteady pressure spike which is similar to the previous observations in granular materials [1, 2] and in foams [7].

REFERENCES

1. Gelfand B.E., Medvedev S.P., Borisov A.A., Polenov A. N., Combustion **9**, 153-165 (1989).
2. Sakakita H., Hayashi A.K., "Study on interaction between a powder layer and a shock wave," presented at the 18 ISSW Symposium, Kagoshima, Japan, May 18 - 22, 1992.
3. van der Grinten J.,G., Sniekers R., W., van Dongen M. E. H., "An experimental study of shock- induced wave propagation in dry, water-saturated porous media,"presented at the ISBN Symposium, Rotterdam, Holland,1988.
4. Rogg B., Herman D., Adomeit G., Int. J. Heat Mass Transfer **28,** 2285 - 2297 (1985).
5. Levy A., Ben-Dor G., Skews B., Sorek S., Experiments in Fluids **15**, 183 - 190 (1993).
6. Britan A., Elperin T., Igra O. and Jiang J. P., "Acceleration of a sphere behind planar shock wave,"presented at the 20th ISSW Symposium, Pasadena, USA, July 13 - 18, 1995.
7. Skews B., W., Atkins M., D., Seitz M., W. J., Fluid Mech. **253**, 245 - 265 (1993).
8. Gorobcov V. G., Mirilenko A. P., Pikus I. M., Combustion, Explosion and Shock Waves, **23,** 54- 57 (1987).
9. Dullen F.A., Porous Media, New York, Academic Press, 1992.
10. Gelfand B. E., Medvedev S. P,. Borisov B.E., Polenov A. N.,Timofeev E. I., Frolov S. M., Tsuganov S. A., Journal of Applied Mechanics and Technical Physics **27** 127 - 130 (1986).

HUGONIOT MEASUREMENTS OF ZINC SULFIDE

M. UCHINO and T. MASHIMO

High Energy Rate Laboratory, Kumamoto University, Kumamoto 860, Japan

Hugoniot-measurements experiments were performed on a high-density zinc sulfide (ZnS) polycrystal to study the elastoplastic and structural phase transitions under shock compression. The Hugoniot parameters were measured by means of the inclined-mirror method combined with a powder gun. As a result, the structural phase transition was observed in addition to the elastoplastic transition.

INTRODUCTION

Zinc sulfide (ZnS) is used as a pressure calibration material in static high-pressure researches. ZnS has a cubic zinc-blende structure at ambient state. It was reported that this material transformed to NaCl-type structure under static compression [1-6]. Balchan and Drickamer reported that the phase transition occurred at 24.0 GPa in 1961 using the fixed point scale (FPS-1) based on an extrapolation of the Bridgman's data [1]. In 1970, Drickamer revised the scale downward changing the transition pressure to 18.5 GPa, using of the x-ray diffraction work based on the shock data of Rice *et al*. [2]. Piermarini and Block reported that the phase transition occurred at 15.0 ± 0.5 GPa using the ruby fluorescence line-shift based on the Decker equation of state for NaCl [3]. In addition, Le Neindre *et al*., Yu *et al*., and Zhou *et al*. reported that the phase transition also occurred at about 15 GPa by the electrical resistance [4] and x-ray diffraction measurements [5,6].

We can directly determine pressure-density relation by the measurement of the Hugoniot parameters (shock velocity and particle velocity). Gust reported that the phase transition occurred at 17.4 GPa under shock compression using the inclined-prism technique [7].

In this study, the Hugoniot-measurement experiment was performed on a high-density ZnS polycrystal to study the elastoplastic and structural phase transitions under shock compression. The Hugoniot parameters were measured by means of the inclined-mirror method combined with a powder gun, using a new high-speed streak camera and a xenon flash lamp [8].

EXPERIMENTAL PROCEDURE

The plate-shaped polycrystals of about 4 mm in thickness and about 19x19 mm in width were prepared by chemical vapor deposition technique. The mean bulk density was measured to be 4.086 g/cm^3 by the Archimedean method, which was in accord with x-ray density of 4.086 g/cm^3. The grain sizes were 2-5 μm. The plane shock waves were generated by the high-velocity impact method. Tungsten or copper plates of 1.5-2.0 mm and about 1 mm in thickness were used as the impact plate and the driver plate, respectively. These plates and the specimens were ground parallel and rapped to an accuracy of 2-3 μm. In this method, the mirror-rotating type streak camera whose maximum streak rate to film was faster than 10 mm/μs and a strong xenon-flash lamp were used [8]. The impact velocities were measured by the electromagnetic method within an accuracy of 0.2 % [9]. The shock velocities of the second and third waves were analyzed considering the interactions of shock waves and rarefaction waves [10]. The particle velocities of the elastic wave and the second wave were determined by the free-surface approximation, and that of the third wave was determined by the impedance matching method. The Hugoniot data of the impact and driver plates used

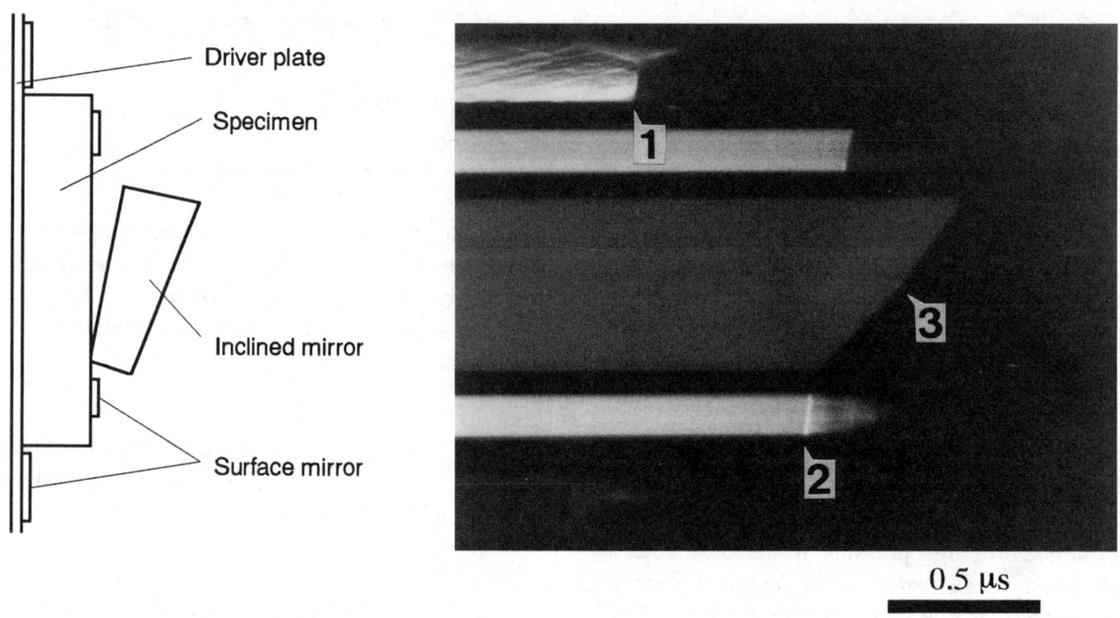

FIGURE 1. Typical streak photograph of the ZnS by the inclined-mirror method, when the impact velocity (tungsten impact plate) was 1.682 km/s

were from Ref. 11.

RESULTS AND DISCUSSION

Figure 1 shows a typical streak photograph by the inclined-mirror method, when the impact velocity was 1.682 km/s. At points 1 and 2, the elastic wave arrived at the rear surface of the driver plate and specimen, respectively. At point 3, a clear kink due to the structural phase transition is seen. The kink of elastoplastic transition is also observed at the beginning of the inclined mirror image, but can not be well identified by this photograph, because the particle velocity of the elastic wave is very small and the difference between the elastic shock velocity and plastic one is small in this case.

In this study, the Hugoniot-elastic limit (HEL) and structural phase transition stresses were determined to be within 1.5-3.0 GPa and 15-16 GPa, respectively. The present phase transition pressure was lower than that by Gust [7], but, was consistent with the recent static one [3-6].

REFERENCES

1. Balchan, A. S., and Drickamer, H. G., Rev. Sci. Instrum., **32**, 308-313 (1961).
2. Drickamer, H. G., Rev. Sci. Instrum., **41**, 1667-1668 (1970).
3. Piermarini, G. J., and Block, S., Rev. Sci. Instrum., **46**, 973-979 (1975).
4. Le Neindre, B., Suito, K., and Kawai, N., High Temp.-High Pressures, **8**, 1-20 (1976).
5. Yu, S. C., Spain, I. L., and Skelton, E. F., Solid State Commun., **25**, 49-52 (1978).
6. Zhou, Y., Campbell, A. J., and Heinz, D. L., J. Phys. Chem. Solids, **52**, 821-825 (1991).
7. Gust, W. H., J. Appl. Phys., **53**, 4843-4846 (1982).
8. Mashimo, T., Nakamura, A., and Hamada, Y., "A compact high-speed streak camera for impact-shock study of solids," in *Proceedings of 20th Int. Cong. High-Speed Photography and Photonics*, SPIE-1801, 1992, pp. 170-175.
9. Mashimo, T., *Shock Waves in Material Science*, Tokyo: Springer-Verlag, 1993, ch. 6, pp. 113-144.
10. Ahrens, T. J., Gust, W. H., and Royce, E. B., J. Appl. Phys., **39**, 4610-4616 (1968).
11. McQueen, R. G., Marsh, S. P., Taylor, J. W., Fritz, J. N., and Carter, W. J., *High-Velocity Impact Phenomena*, New York: Academic Press, 1970, ch. 7, pp. 293-417.

GENERATION OF CYLINDRICAL IMPLODING SHOCKS IN SOLID USING A HIGH EXPLOSIVE

T. Hiroe*, H. Matsuo*, K. Fujiwara*, S. Sakai*, T. Abe*
M. Yoshida and S. Fujiwara****

**Department of Materials Science & Resource Eng., Kumamoto University, Kumamoto 860, Japan*
***Department of Advanced Chemical Technology, High Energy Density Laboratory,*
National Institute of Materials and Chemical Research, Ibaraki 305, Japan

Strong cylindrical imploding shock waves have been generated in PMMA by the simultaneous detonation of high explosive PBX shells. Surface initiation of the PBX shell begins at the collision of a cylindrical thin flyer driven by the detonation of a low-density PETN shell, which is initiated by the explosion of fine copper wire-rows using a high voltage impulsive current. As preliminary studies, shock initiation of the PBX, wire-row conditions for the PETN detonation and cylindrical flyer acceleration systems were investigated. Observed shadowgraphs of imploding shocks in PMMA cylinder (40mmϕ dia.) show that cylindrical shock waves are focusing with a fairly good axisymmetry, but the comparison with the numerical results indicates that the actual steady state C-J pressure of the PBX seems to be on the way of build-up in this study.

INTRODUCTION

The one-dimensional implosion of strong cylindrical shock waves in solids or powders generates the extremely high pressure near the focusing axis without heavy facilities and it can be applied to the investigations on high pressure properties and syntheses of materials (1). The authors have been making progress to produce such strong converging shocks in solid, using the wire-rows explosion technique. First, the conditions of wire-rows and the pentaerithritoltetranitrate (PETN) density were investigated (2) to produce the wide and flat plane detonation, initiating the powder PETN by the simultaneous explosion of fine copper wire-rows with use of the impulsive electric current from the capacitor bank. Next, this experimental technique was utilized to initiate a cylindrical shell of the low-density PETN at the periphery and the symmetrical imploding shock wave has been produced in a polymethylmethacrylate (PMMA) cylinder placed inside of the PETN shell (3). However, in case of the charge assembly of PETN and PETN/silicon rubber (SR), the observed imploding shock trajectory has indicated that the generated shock pressure is not so strong as expected because the Chapman-Jouguet (C-J) detonation is not built-up in the main explosive PETN/SR shell.

In this study, a high explosive PBX is employed, of which shock initiation characteristics is not so clear. Then the initiation method of the PBX is developed by detonating the PETN. An advanced explosive system for generation of the strong imploding shocks in PMMA is developed based on the preliminary studies. The observed shock trajectories are compared with numerical results by the random choice method (RCM) code.

PRELIMINARY STUDIES
Detonation of PETN Shells

In the previous study (3), the conditions of wire-rows were investigated for the simultaneous detonation of long cylindrical PETN shells. The low density powder PETN (0.6 g/cc) was charged into a cylindrically-shaped shell with the thickness d of 6 mm. The rows of 100 μm-dia. copper wires are cemented circumferentially with the intervals W of 6 or 12 mm over the inner surface (100 mm dia.) of the outer cylinder of PETN container. The wires (the number of rows N : 2 - 22) are connected to the capacitor bank of 12.5 μF, max. 40kV as an impulsive electric current supplier and the detonation arrival at the inner surface of the explosive shell was recorded with a streak camera using inclined mirror system. The records indicate that the time jitter be-

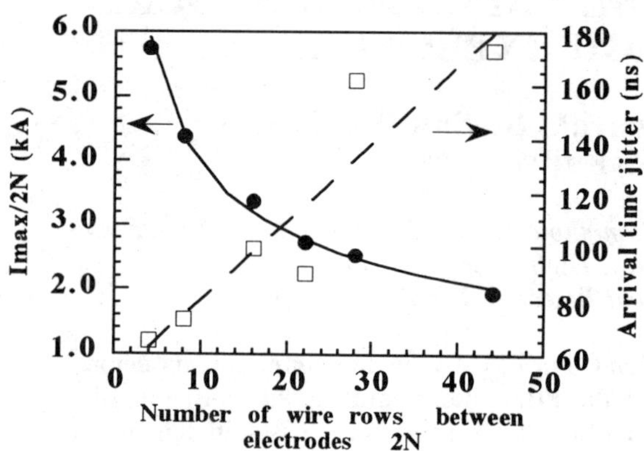

FIGURE 1. Relations of number of wire-rows, peak current per wire and arrival time jitter of detonation front for producing cylindrical detonation for PETN by wire-explosion.

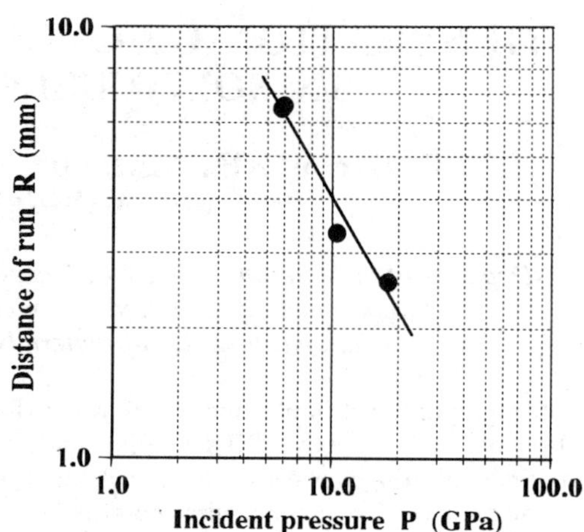

FIGURE 2. Distance to the C-J velocity vs. incident shock pressure for PBX 80U.

tween neighboring wires become negligibly small for W/d = 1 or less because of the interferences of the initial detonation waves. Figure 1 shows the relations of wire-row number 2N between electrodes, peak current per a wire Imax/2N and the arrival time jitter of detonation. The time jitter becomes large as the number of wire-rows is increased because the current value per a wire decreases.

Shock Initiation of the PBX

The wedge-shaped samples of PBX 80U (contents: HMX wt. 80 %, active binder wt. 20 %, density: 1.75 g/cc, C-J velocity: 8.3 km/s) were provided and shocked by the system composed of a plane wave generator, a booster explosive and a driver plate. The light emission from the slant surface of the wedges has been recorded by a streak camera to measure the run distance, and the emission from the gap assembly placed on the driver plate was also recorded to estimate the incident pressure by using the impedance-match technique. Figure 2 shows the relation between the incident pressure and the distance of run for the PBX. This figure suggests that the detonation pressure of low-density PETN is too low for the direct contact system to detonate the PBX within a small distance.

In order to achieve the higher incident pressure, a flyer impact system is examined experimentally, where a thin plastic plate (Mylar or Teflon, 0.1 - 0.3 mm thick) has been accelerated up to 3 km/s by the explosion of the PETN. As the results, the PBX incident pressures for the cases of the PETN-flyer-PBX system and the PETN-PBX direct contact system are estimated using cross curves with the PBX

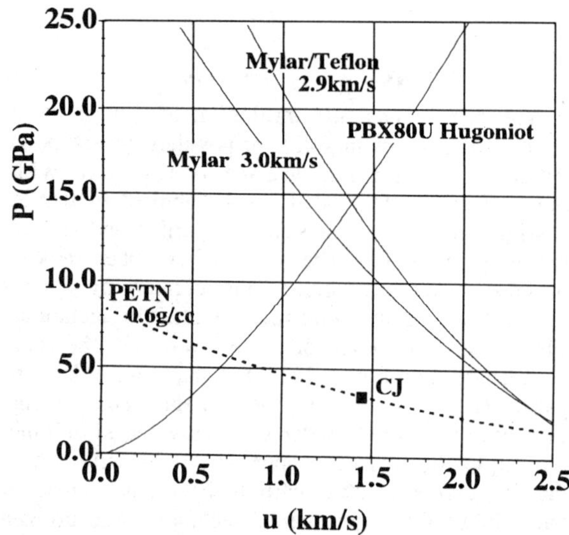

FIGURE 3. The estimation of the incident pressure from the Hugoniot curves of the PBX, PETN and the flyer materials.

Hugoniot curve as shown in Fig. 3. In the former case, the incident pressure and the run distance are to be 13 - 15 GPa and 2 - 3 mm respectively.

GENERATION OF IMPLODING SHOCKS
Experimental

The experimental setup for the generation of cylindrical imploding shocks in PMMA has been developed as shown in Fig. 4. These dimensions are

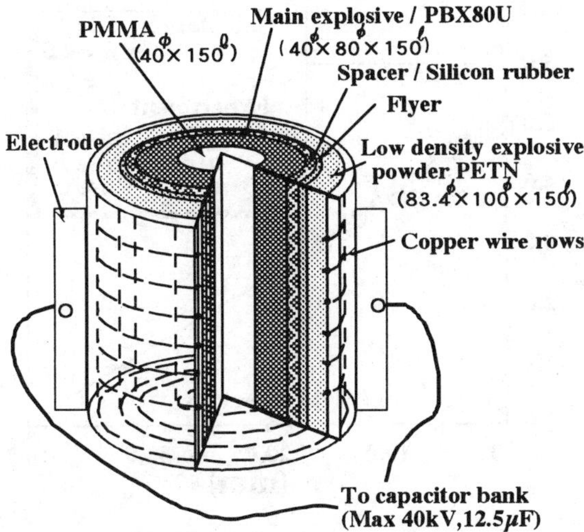

FIGURE 4. Sketch of the experimental setup for generation of cylindrical imploding shocks in PMMA.

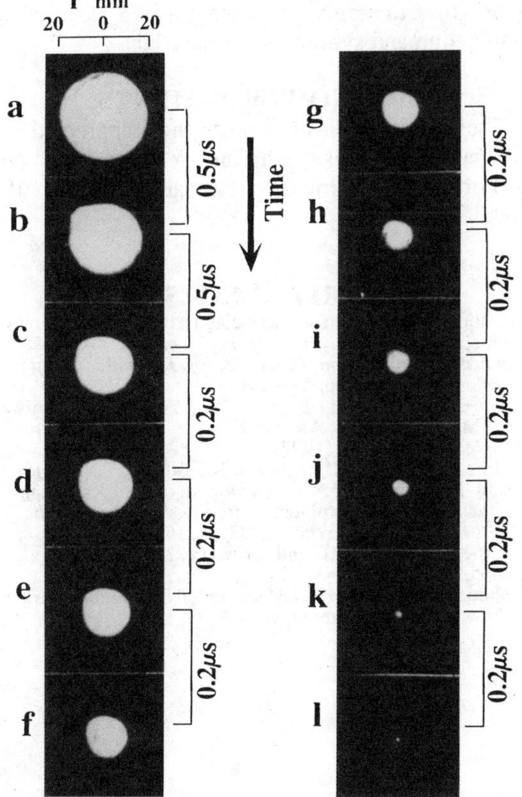

FIGURE 5. Typical series of sequential framing camera shadowgraphs taken at intervals of 0.5 and 0.2 μs for case B.

TABLE 1. Test Cases and Conditions for Implosion Test

Case	Number of wire-rows N	Flyer materials (mm)
A	8	Mylar(0.1) + Teflon(0.1)
B	12	ditto
C	22	Mylar(0.2)

based on the initiation and detonation data of the PETN and the PBX and the flyer acceleration data by the PETN explosion. This cylindrical testing assembly consists of outermost wire-rows (N :8, 12, 22, W :6 mm), the PETN, the thin plastic flyer, a spacer sheet, the PBX and innermost PMMA. A cylindrical plastic flyer of 0.2 mm thickness is accelerated into the holes (4 mm dia., 6 mm pitch) of a silicon rubber spacer of 1.5 mm thickness and impacts the outer surface of the PBX. The shadowgraphy of imploding shock waves in PMMA is taken by framing and streak cameras with an argon flash. Table 1 shows the test cases. In cases A and B, the axial length covered by wire-rows is not sufficient to produce strictly one-dimensional implosion. Nevertheless, since the PETN, PBX and PMMA cylinders have enough axial lengths, the influence of the lateral unloading on the imploding flow in the middle section can be neglected.

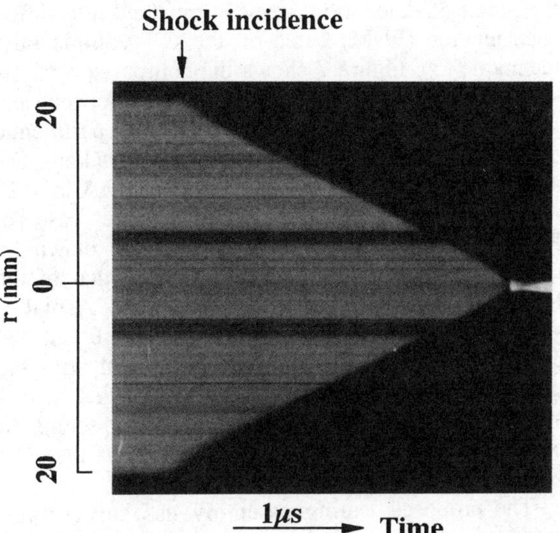

FIGURE 6. Streak camera record of shadowgraph for case A.

Results and Discussions

Typical series of sequential framing camera shadowgraphs are shown in Fig. 5 for case B. The initial slight distortion of the shock front does not affect the convergence of the shock up to the focusing stage. It can be said that the imploding shock in solid seems to be rather stable. Simultaneously, streak camera photographs are also taken using a beam splitter, one of which is shown in Fig. 6 for case A. Two symmetrical shock trajectories are recorded. Although the incident shock velocity is slightly decreased by the expansion behind the shock, it is soon increased by the self-amplifying effect of pressure. Especially near the center, the acceleration of imploding shocks can be seen. The strong emission light is also recorded after focusing. The shock pressure of 250 - 300 GPa is estimated near the focusing point from the shock trajectories. Experimental results have shown that the discrepancies in shock trajectories for cases A, B and C are small, and it suggests that the one-dimensional imploding shocks have been generated in a middle section of the PMMA cylinder.

The numerical simulations are performed using the one-dimensional RCM code for solids and gases (4). It is assumed that the C-J propagation velocity suddenly appear on the PBX surface at the collision of the flyer, and the combustion gas of the PBX can be treated as a perfect gas with the ratio of specific heat γ. The numerical analysis by the RCM has started just after the detonation wave has arrived at the contact surface of the PMMA and PBX, using the result of detonation analysis by the finite difference method (FDM) based on the C-J volume burn technique (5). Figure 7 shows the comparison of the shock trajectories for cases A, B and C between experimental and numerical results. In experimental data, the average radii of shock fronts is taken. The average imploding shock velocity in PMMA is 1.25 - 1.30 times larger than that in the case of PETN/SR (3). Numerical trajectories are drawn by connecting the results of the FDM with that of the RCM, for two cases of γ = 3.0 or 4.5. Actual γ value of the PBX is estimated as 2.7 - 3.0 but the experimental results show good agreement with the numerical results for γ = 4.5. It indicates that in this study the stable C-J pressure of the PBX seems to be on the way of build-up (5) though the C-J velocity is attained in the initial stage.

The proposed testing assembly has successfully produced strong cylindrical imploding shocks in solid using a high explosive. Moreover an advanced

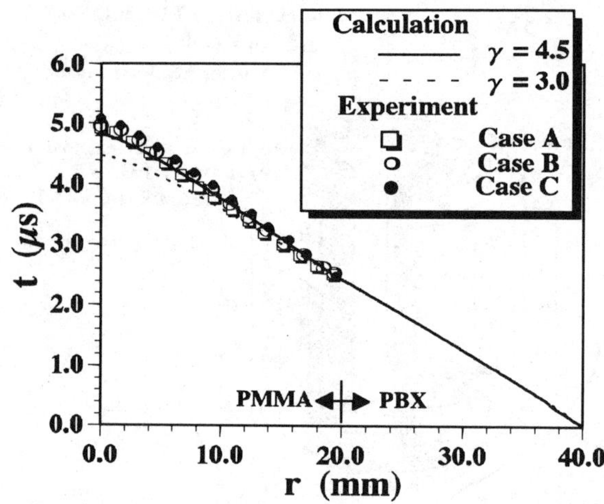

FIGURE 7. Comparison of experimental and numerical results for shock trajectories in the PBX and PMMA.

system for recovery is needed and its improvement is in progress to apply the assembly to the dynamic compaction and syntheses of materials.

ACKNOWLEDGMENTS

The authors wish to express their appreciation to K. Murata, K. Takahashi and Y. Katoh of NOF Corporation. for providing the samples of the PBX 80U.

REFERENCES

1. Yoshida, M., "Cylindrical imploding fixtures for high pressure generation and recovery", in *Proc. of 2nd Workshop on Industrial Application Feasibility of Dynamic Compaction Technology*, 1988, pp. 263-264.
2. Hiroe, T., Matsuo, H., Fujiwara, K., Yoshida, M., Fujiwara, S., Miyata, M. and Akazawa, T., *J. Industrial Explosive Society* **53**, 219-226 (1992).
3. Hiroe, T., Matsuo, H., Fujiwara, K., Tanoue, T., Yoshida, M. and Fujiwara, S., "A production of cylindrical imploding shocks in solid by exploding wire rows", in *Proc. of the joint AIRAPT/APS Conference*, 1993, pp. 1667-1670.
4. Hiroe, T., Matsuo, H. and Fujiwara, K., *J. Appl. Phys.* **72**, 2605-2611 (1992).
5. Maider, C. C., *Numerical Modeling of Detonation*, Berkeley: University of California press, 1979.

NON-ONE-DIMENSIONAL AND QUASI-SPHERICAL LOADING OF METAL BALLS BY SHOCK-WAVES HAVING UP TO 3 Mbar PRESSURE, WITH THE INVESTIGATION OF THE PRESERVED SAMPLES

B. V. Litvinov[*], V. I. Zel'dovich[**], N. P. Purygin[*], I. V. Khomskaya[**], V. I. Buzanov[*], A. E. Kheifetz[**], O. S. Rinkevich[**], N. Yu. Frolova[**], G. A. Sobyanina[**]

[*] *Russian Federal Nuclear Center-Institute of Technical Physics, Snezhinsk, Chelyabinsk region, Russia*
[*] *Metal Physics Institute, RAS Ural Divizion, Ekaterinburg, Russia*

Solid steel, brass and duralumin balls with 60 and 40 mm in diameter were exposed to explosion of spherical explosive charge 10 and 20 mm thick respectively. The explosion was initiated on the charge surface simultaneously in 2, 4 and 12 points, distributed uniformly over the sphere. To preserve the balls, massive case was used in test. In the diametral plane of section there were found either cracks, distributing mainly along the shock-waves collision lines, or the cavity. In most cases, the cavity was surrounded by the metal layer, having dendritic structure. In balls in case of explosion initiation in 12 points the metal motion was quasi-spherical (one-dimensional), in case of initiation in 2 points - two-dimensional and in 4 points - three-dimensional. The pressure evaluations, base on the convergent motion, as well as on the value of the residual temperature on the boundary of dendrits zone, showed, that inside crystallization zone the pressure exceede 2 Mbar.

The phenomenon of energy cumulation in spherically convergent shock waves was investigated by E.I.Zababakhin (1). But the cumulation phenomenon can be observed in a convergence of the shock front of complicated configuration, occurring as a result of the explosion of explosive compound in some initiation points. In this case the important part can be played by the effects of interaction of waves, coming from different initiation points.

Solid ball samples made of steel, brass and duralumin of 60 or 40 mm diameter were exposed to the explosion impact of the spherical charge 10 and 20 mm thick respectively. The explosion was initiated on the charge surface simultaneously in 2,4 and 12 points, uniformly distributed over the sphere. To preserve the samples, in tests the massive case was used, which permits to decelerate the unloading process and, by this, to decrease the tension stresses, occurring in the sample. The asynchronism of explosion initiation in different points did not exceed 10^{-7} sec.

Depending on a number of initiation points (2,4,12) it was supposed that the material movement in the sample will be two-dimensional, three-dimensional and quasi-spheric respectively.

After a loading, on the balls surface one can see the lines of detonation waves interaction, forming regular geometric shapes. For the case with 2,4 and 12 points there were circumference, triangles and pentagons respectively. Besides that, as a result of explosion impact, the balls were slightly deformed. The balls were looking like ellipsoid, tetrahedron or dodecahedron by deformation character and by the pattern of shape on the samples surface.

To investigate the structure and measure the microstructure the balls were cut in two parts, according to diametral plane, coming along the projection of initiation points. In the central part of the samples, as a result of intreacion of

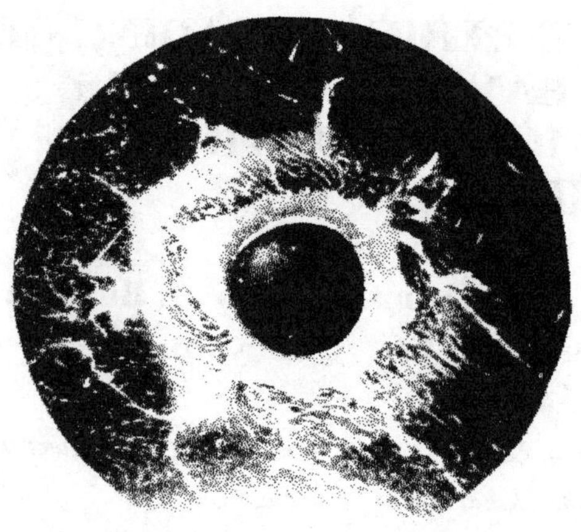

FIGURE 1. Center part of diametral cross-section of steel ball 40 mm in diameter, 12 initiation points.

compression wave with unloading one, coming from the surface (2) in the dependence of a number of initiation points, either cracks or cavity were occured. In the most of cases, the cavity was surrounded by the metal layer, having dendrictic structure. Dendrite presence indicates that in case of shock waves focusing, metal melted, cristallizing under subsequent cooling.

In Fig.1 the photo of central cross-section part of steel sample (steel with 0,37wt.% of C and 1,1wt.% of Cr) 40 mm in diameter, subjected to quasi-spherical loading, is represented. There are clearly seen: the central cavity approx. 5 mm in diameter, surrounded it dendrites zone and more extended annular zone in which the transformation $\alpha \to \gamma \to \alpha$ cycle took place, resulting the generation of beinite structure (light etching, see Fig.1).

Spherical cavity shape and such of it surrounding dendrites area makes evident the one-dimensional character of the movement near the ball center. However, from the measurements of microhardness and kind of sample microstructure, it follows that the forming of spherical convergent movement is a complicated process. In Fig.2 the scheme is given, demonstrating the change of quasi-spheric (i.e., as a matter of fact, tree-dimensional) front shape as far as it moves to the center of the ball. Head waves of Mach trishocks (BC segment on the scheme), developing in the places of collision of divergent shock waves, moving from the neighbour initiation points, have more elevated velocity (3,4), that leads

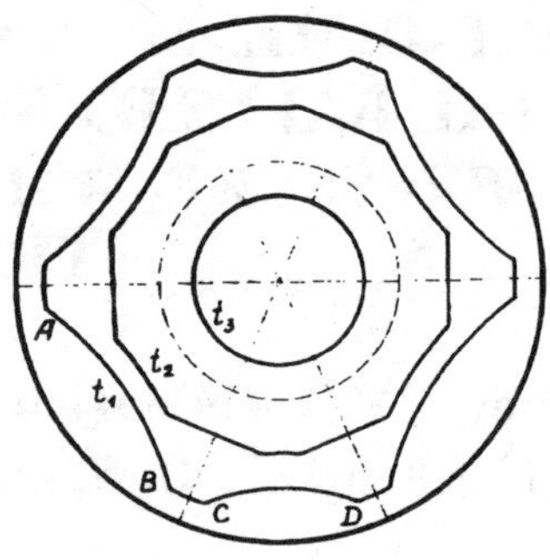

FIGURE 2. Scheme of shock waves propagation in the ball under quasi-spheric loading in time moment t_1, t_2, t_3; $t_1 < t_2 < t_3$. AB and CD are fronts of waves, coming from neighbor initiation points; BC is head wave front.

to front smoothing and, finally, to one-dimensional movement character.

On the basis of the comparison of the results on isentropic unloading calculation, made according to data (5), with dynamic T-P iron diagramm the estimations of pressures, observed in steel ball of 40 mm diameter were made. This estimations showed that on the external boundary of dendrits zone the pressure reached 2MBar and, besides that, they permited to calculate approximately the value of "k" index in the law of pressure increase $P = P_0 (r_0/r)^k$ in cumulation process (1). Obtained value k=0,8-0,9 is in a good agreement with accumulated before the experimental data and with well known theoretical concepts.

The picture of shock front propagation similar to observed in steel ball 40 mm in diameter takes place in brass ball 40 mm in diameter and in duraluminium one of 60 mm diameter, subjected to the loading, according to the same 12-pointed scheme. Individual peculiarities of physical characteristics of investigated materials lead to the distinctions in cavity and melting zone dimensions, in the degree of residual deformation and in the range of observed phase transformations; but in kind, the impact results are identical.

In our opinion, the studying of duralumin sample, loaded by the scheme with 4 initiation points (quasi-tetrahedron) is of great interest. In the center of a ball there is a cavity of irregular shape (Fig.3), that makes evident the preservation of three-dimentional character of the movement during the all impact time.

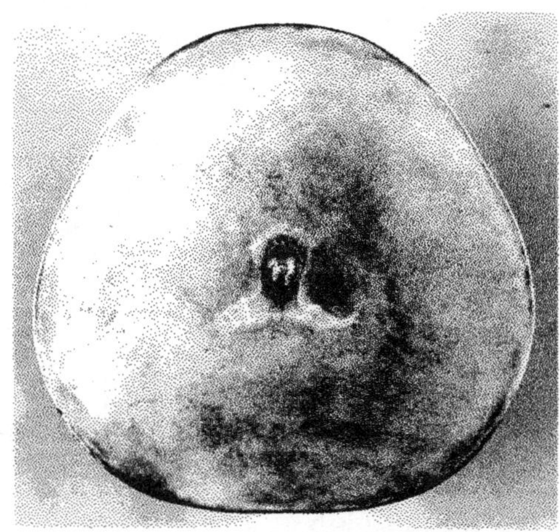

FIGURE 3. Diametral cross-section of duralumin ball 60 mm in diameter, 4 initiation points. In the center is a cavity and it surrounded melting zone.

However, the presence of extended melting area determines the cumulation process presence. On the basis of mentionned one can conclude that in the given sample the irregular (of Mach) process of shock waves interaction takes place.

In the most of cases the high-intensive pulsed impact and the subsequent unloading cause the formation of cracks, propagating predominantly along the shock waves interaction lines. So, in steel sample, loaded by two-dimensional scheme (Fig.4), the extended crack is situated in the diametral plane which is perpendiculare to the line, connecting the initiation points projections. Microhardness measurements and the results of microstructural investigation showed that with movement from the ball surface to the central crack, the material riveting on firstly become less, then it increases. More high riveting on nearby the sample surface makes evident, that shock waves attenuation, and microhardness increase in the central part in this case can be the result of the impact of waves, reflected from the collision plane. Immediately nearby the crack, the shock impact intensity was enough to perform the phase $\alpha \rightarrow \gamma$ transfer, that is indicated by the martensite structure, observed in this area.

It is necessery to note, that in a number of cases the direction of cracks propagation is determined by the   texture of the sample in initial state. For example, in steel ball 40 mm in diameter (fig.1) the

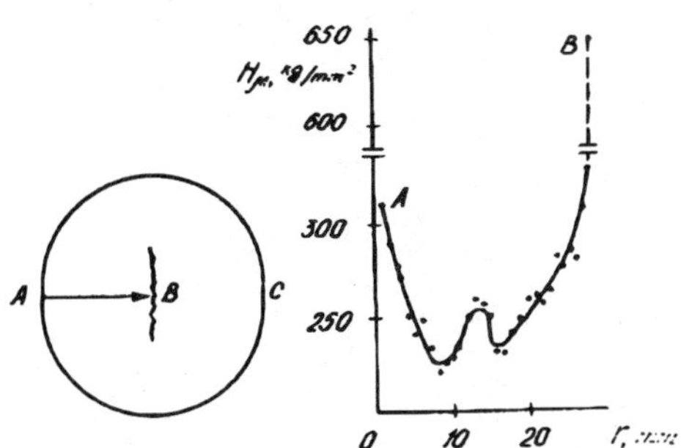

FIGURE 4. Steel ball 60 mm in diameter, 2 initiation points.

radially located fractures have tendency to bend along the direction of the rolling of blank from which the sample was cut. So, the material structure anisotropy leads to additional microscopic defect formation in the process of shock impact.

In the most of investigated samples the effects of shock waves interaction lead to the localised deformation areas and the adiabatic shear bands formation. Observed deformated structure illustrates the process character and shock impact intensity.

Thus the investigation of preserved samples permits to observe the large phenomena class, these phenomena were determined by the action on different materials (metals) of the superposition of shock waves with complicated configuration of the front. Given investigation confirmed, in whole, the original knowledges on the movement character and, besides this, it permited to find out some properties of convergent front formation in diffefent materials.

The present work was performed under the financial support of the Russian Fund of Fundamental Investigations (project code 93-02-2762).

REFERENCES

1. Zababakhin, E. I., Zababakhin, I. E., "Phenomenon of Unlimited Cumulation," M.: Nauka, 1988, p.p. 56-66.
2. Zeldovich, Ya. B., Rayzer, Yu., P., "Physics of Shock Waves and High-Temperature Hydrodinamic Phenomena," M.; 1963, p.p. 513-518.
3. Curant G., Fridrix C., "Supersonic Flow and Shock waves," M., Izdatelstvo Inostrannoi Literatury, 1950.
4. Wisem, J., "Linear and Non-Linear Waves," M., Mir, 1977.
5. Properties of Condensed Compounds under High Pressures and Temperatures. Ed. Trunin, R. F., Arzamas-16,1992, p.p. 36-48.

INTERFACE INSTABILITIES BETWEEN SHOCK-COMPRESSED METAL LAYERS

A.L.Mikhailov, R.S.Osipov

Russian Federal Nuclear Center VNIIEF, Arzamas-16, Russia

Instability of the interface between shock-compressed metal layers, with the shock wave normally incident to this, has been investigated. Experimental setups and test data are described. The instability development is considered in terms of its relationship with shock pressure, initial perturbation amplitudes, and postshock strength and melting behaviour of metals.

INTRODUCTION

Phenomena of liquid or gas boundary instability during acceleration of them perpendicularly to the boundary are well known. It is so called Taylor instability (1). If we have acceleration caused by shock wave effect, we deal with Richtmyer-Meshkov instability (2, 3).

When we deal with solid mediums, meaning of strength becomes important. It is able in some cases, for example, when strength is higher than forces of "gravitational or inertial local disbalances" on interface to stop development of instability decidedly. We observe this phenomenon usually during explosion acceleration and deceleration of solid bodies.

But, obviously, the situation may be changed if relation between these forces is opposite. It will happen, when we have greatly high accelerations, in the other words, in certain relation between amplitude of shock wave, characteristics of materials and parameters of initial disturbances either on the wave front or on surface of the boundary.

In this case, when we deal with instability of Richtmyer-Meshkov type and shock wave acceleration of solid bodies, shock wave heating of materials becomes also important. At sufficient amplitude of shock wave phase transitions and among them melting and vaporization even are possible. Here the importance of strength may be lessened considerably.

In this paper we represent some results of "metal-metal" boundary instability study, when compression of them took place by shock wave - moving perpendicularly to the boundary.

1.Experimental Procedure

Generally, development of boundary instabilities is studied using either optical or pulse radiography methods.

Obviously, optical methods are not useful for study of "metal-metal" boundary instability. It seems to be possible to use pulse radiography with high penetrability. But considerable procedure difficulties, connected with extraction of small-contrast disturbed boundaries, take place.

Finally, we developed method of shock wave loading of studied samples, packed tightly in hermetically sealed and survivable during explosion container of "flat" geometry, with following metallographic analysis of container contents (4).

Drawing of container is depicted in Figure1; photograph of container and its separate details is presented in Fig.2.

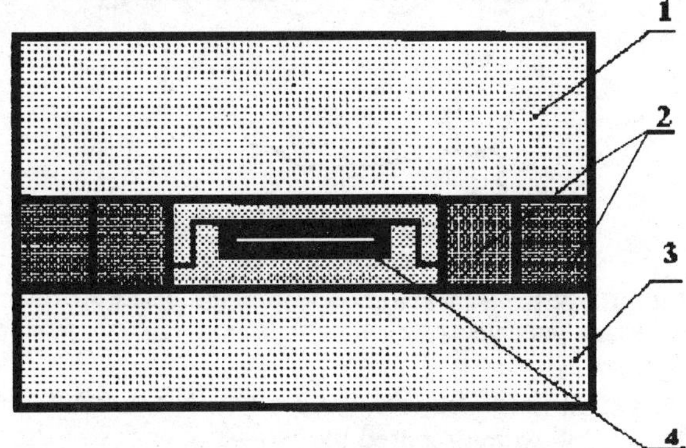

FIGURE 1. "Plane" geometry container scheme: 1-shock-wave plane geometry generator; 2, 3-passive shield; 4-inner germetisation shell.

Sample in the shape of two tightly retained discs from different metals contacted each other along pressed in inner contour of hermetic sealing 4 of container (Fig.1).

FIGURE 2 ."Plane" geometry container.

Loading device 1 based on charge of solid explosive provides sample with shock wave having flat front and amplitude up to 100 Gpa. We have developed several loading devices, which deal with wide range of shock wave pulses having well-defined shape, amplitude and duration. Time of keeping of sample in compressed state up to arrival of rarefaction wave is from 1 to $5\mu s$ in dependence on applied loading device and amplitude of shock wave.

Geometry of shock wave front (its flatness) and shape of pulse have been certified using photo-chronography method of "flash gaps" and manganine pressure gauge respectively. Typical profiles P(t) of shock wave recorded by manganine gauge at entrance in sample for two types of loading devices are depicted in Fig.3.

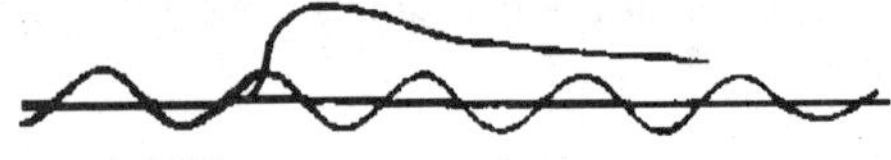

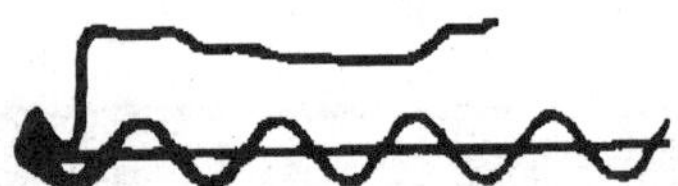

FIGURE 3. Experimental shock-wave form before specimen in "Plane" containers.

Initial disturbances, starting development of instability, were prepared by two ways:

1. geometry (a)-by "sinusoidal" concentrical distortions in boundary between two metals, constituted the sample (Fig.4);

2. geometry (b)-by regular distortions of shock wave front falling on flat boundary. For this purpose "lens element", constituted from two plates of different metals contacting each other along distorted boundary, was placed before sample (where shock wave entered him) (Fig.5).

In our experiments we studied samples constituted by the following metal pairs: Sn-Bi, Pb-Bi, Cd-Bi, Zn-Bi, Bi-Zn, Sn-Pb, Sn-Cd.

In these pairs the first places are given to metals, which are the first on the way of shock wave.

2.Results of experiments and their discussion

Results of the experiments are given as photographs of characteristic sections in Fig.4, 5 and summarized in Table 1.

a)

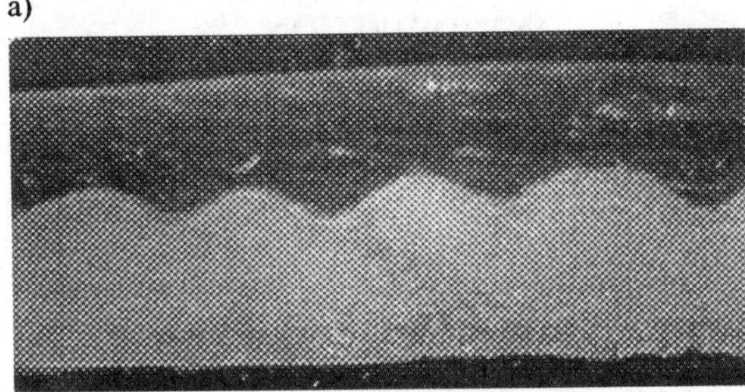

b)

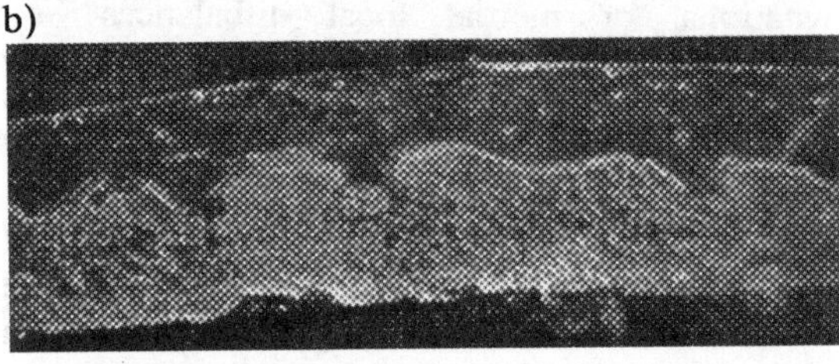

c)

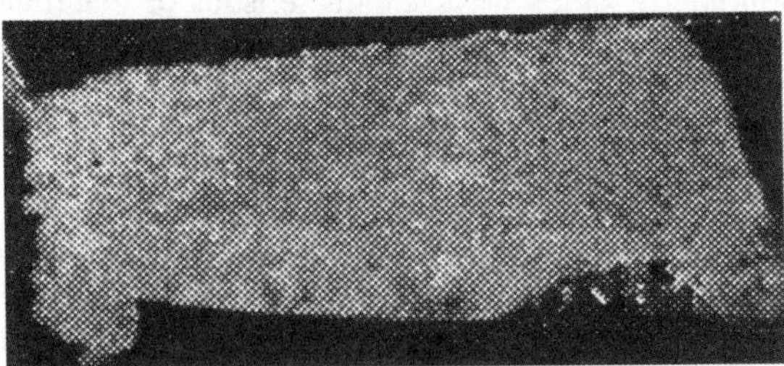

FIGURE 4. Sn-Cd, λ_0=1.6 mm, a_0=0.2 mm

These photographs illustrate development of disturbances on boundary and mutual mixing of metals in dependence on increase of shock wave pulse amplitude. (As this took place, pulse duration decreased as velocity of shock wave grew.)

Designations in the table: λ_0, a_0 - wavelength and amplitude of initial disturbances of boundary or shock wave front;

Ps-pressure of sample shock wave compression, GPa;

Pc-"critical" minimum pressure, when disturbance of boundary stability has been recocrded;

Pi-interval of shock wave compression pressures, when we managed to record development of instability and difference of metal mix extent.

It has been recocrded in experiments that critical amplitude of shock wave Pc exists and, when it takes place, stability of boundary is broken. To say more precisely, it looses its initial shape and metals of sample start to mix in volume of hermetic sealing inner contour at Ps>Pc.

a)

b)

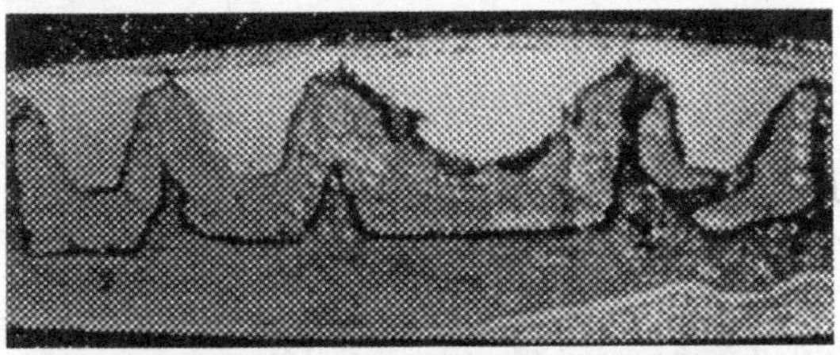

c)

FIGURE 5. Pb-Bi, λ_0=1.1 mm, a_0=0.2 mm

Value of Pc depends on metal pair, which forms the sample and also on amplitude of initial disturbances. (It was recorded, when we dealt with pairs Sn-Bi, Sn-Cd.) As we decreased amplitude of initial disturbances a_0, Pc increased slightly. We can explain this phenomenon by reducing of local overfalls of pressure, in the other words, by difference in dynamics on the boundary between places, where passed and reflected shock waves and rarefaction waves, which cause development of disturbances, converge and diverge.

In experiments, performed in geometry of Fig.5 with pair Pb-Bi, we managed to record gradual increase of disturbances amplitude, formed on originally flat boundary (Bi enters Pb) as amplitude of shock wave increases (see photograph).

When pressures were higher than upper value of interval Pi, metals mixed homogeneously in volume without noticeable difference of section kind because of shock wave pressure. Width of this interval, recorded in our experiments, depends not only on physics of phenomenon, but on discrete character of pressure interval (pressure step), initiated by loading device, as well.

Existence of shock wave critical amplitude Pc and its comparatively weak dependence on amplitude of initial disturbances a_0, in our opinion, indicate that there are quality changes of system strength characteristics. We may try to connect this fact with melting of metals, which constitute samples, behind shock wave front and this results in corresponding fall of shear strength, responsible for "blockage" of disturbance development.

We prepared estimations of parameters of shock wave, which is able to initiate melting of studied metals, using Simon formula (5).

Results of these estimatioms as well as values Pc from table 1 are represented in Table 2.

Analysis of these two tables reveales some correlation between of boundary instability and shock wave melting interval co-ordinates of more refractory metal from pair. This would confirm mentioned supposition.

If so, dependence Pc on amplitude a_0 of initial disturbances, revealed in our experiments, may, probably, be used for more precise determination of melting at extrapolation $a_0 \to 0$.

These results are in relatively reasonable agreement with represented in paper (6) fall of lead dynamic yield point at shock wave amplitudes 22-26 GPa, which was registered by the mean of main strains difference method and associated by authors with shock melting of lead.

Generally speaking, supposition about less strict connection Pc with Pm is reasonable, if we take into consideration that, as amplitude of shock wave

TABLE 1

Sample	Geometry Fig 4	λ_0, cm	a_0, cm	Ps in shots, GPa	Pc, GPa	Pi, GPa
Sn-Bi	a	0.16	0.05	11, 14, 16, 18, 21.5		18 - 21.5
	a	0.16	0.02	18, 18.5, 20, 21.5,	20 - 21.5	20 - 21.5
	b	0.32	0.02	20, 21.5		20 - 21.5
Pb-Bi	b	0.11	0.02	14, 18, 18.5, 20, 21, 21.5, 22, 22.5, 23, 25, 26.5, 28	$\cong$18.5	18.5 - 22
Cd-Bi	a	0.16	0.05	21.5, 25, 28	25 - 28	25 - 28
Sn-Pb	a	0.16	0.05	18, 21.5, 23	23	21.5 - 23
Sn-Cd	a	0.16	0.02	23.5, 25, 27, 28, 29, 30	29 - 30	29 - 30
	a	0.16	0.05	21.5, 28	$\cong$28	-
Zn-Bi	a	0.16	0.05	28, 30	>30	>30
Bi-Zn	a	0.16	0.05	28, 30	>30	>30

grows, local disturbances of pressure on its front shaped eighter by "lens element' or distortions of boundaries, can finally overcome dynamic yield point of metals. This dynamic yield point can have tendency to decrease as shock wave heating of metals increases already before metals will melt (7).

Here we speak about "correlation" Pc and Pm taking into our consideration also that metals with normal shape of melting curve Pm(T) melt in rarefaction wave, followed the front of shock wave up to the moment, when amplitude of shock wave reaches values, which are sufficient for melting just behind the shock front.

Some correlation occurs between Pc and static shear moduli G of investigated metals, exept Pb (see Table 2). Here we must take into account also the pressure dependence of these moduli.

TABLE 2

Metal	Pm,GPa	Pc, GPa	G, GPa
Bi	16.0	-	12.0
Sn	32.0	20-21.5	11.0-15.0
Pb	34.0	18.5-22	7.5
Cd	47.0	25-30	23
Zn	59	>30	35

The next note is connected with presence of reflected shock waves in sample and its smooth aftershock compression behind the front of the first wave because different shock impedances of metals of sample and container material where thickness of sample is limited.

Experiments with fusible metals located in container made of strong material with close impedances or experiments with containers of "cylindrical geometry", where flat shock wave exists in Mach stem, are of specific interest from this point of view.

Liquidation of reflected waves using such methods, in our opinion, would give opportunity to study dynamics of disturbances growth as well.

ACKNOWLEDGMENTS

We wish to thank the Program Committee of the 1995 APS Seattle Conference for the given opportunity to present this paper and the US Department of Energy for financial support of our participation in this Symposium.

REFERENCES

1. Taylor J., Proc.Roy.Soc., v.201, N1065, 192 (1950).
2. Richtmayer R.D.,Comm. on pure and appl.math.,v.XIII, 297 (1960).
3. Е.Е.Мешков, Механика жидкости и газа, т.5, 151 (1968).
4. О.Б..Дреннов, А.Л.Михайлов, Р.С.Осипов, А.В.Федоров, "Проблемы прочности", N10, 120, 1989
5. S.E.Babb, Rev.Mod.Phys., v.35, N2, 400 (1963).
6. Ю.В.Батьков и др. "Плавление свинца при ударном сжатии", ПМТФ, N1, 149 (1988).
7. Ю.В.Батьков, Б.Л.Глушак, С.А.Новиков, "Сопротивление материалов пластической деформации при высокоскоростном деформировании в ударных волнах

PLASMA PRODUCTION FROM SHOCK COMPRESSION OF CONDENSED MATTER

Tai-Ho Tan and Stanley P. Marsh

DX-15, Los Alamos National Laboratory, Los Alamos, NM, 87545

The experimental investigation of HE-driven, phased, cylindrical, SS liner implosion has yielded many interesting results. Plasma and radiation are found to be copiously produced. Plasmas with velocity up to 17 cm/μs are observed. The temperature in the expansion surface reaches 8 - 10 eV and stays hot for tens of microseconds. The signatures of plasma interactions with the imploding wall and the glass port are clearly identified. Finally, a cluster of cooler but still self-luminous, high-density debris is observed to travel at 1.8 cm/μs. Additional experiments were carried out to study the plasma flow and reconvergence inside the liner cavity by inserting a diverting disk along the axis of implosion. Significant emission of vuv and soft x rays is detected. All the experiments are guided by the calculations using the MESA 2D hydrocode and the results agree with many of the predictions.

INTRODUCTION

The study of plasma and high-frequency radiation production from a purely hydrodynamic-driven shock compression of solids is difficult because the pressure needed to reach the ionized states requires very high impacting velocity that is not easily attainable. Furthermore, when plasma is produced it is often constrained by the configuration from being observed. From our investigation of HE-driven, phased, cylindrical liner implosion we are able to measure copious production of plasma and radiation. The study is motivated by the results of hydrocode calculations showing that when a hollow-core HE cylinder is initiated to drive a thin ss tubing, the convergence velocity can reach above 1 cm/μs, thus producing a 15-Mb pressure as the density is shock compressed to 19.5 g/cm^3. The temperature from the heating rises rapidly above 8 eV and is followed by a combination of radiation and plasma emissions. The process continues for many microseconds as the phased implosion proceeds along the axis, and the evolving dynamics can be observed from the open end of the cylinder. The presence of structures in the expansion front and the maintenance of high temperature during the expansion are predicted, too. By inserting a small diameter disk along the phased implosion axis, simulation shows that the plasma flow can be perturbed and redirected to converge toward the axis. Cusps and hot spots are formed in the plume with temperature in excess of 100 eV. These are intriguing predictions that if proven valid will provide us with a means to investigate a variety of topics in interesting dynamic regimes.

In these exploratory experiments, the HE initiation simultaneity and liner integrity are closely monitored to ensure that the implosion is symmetrical and free from evidence of material failure prompted by instabilities. Radiographs of implosions are compared with those from numerical simulations. Finally, multiple diagnostics that include an extensive array of time and spatially resolved instruments are employed to measure the history and profiles of plasma expansion and radiation temperature. Some of the results have been reported elsewhere.[1,2] In this paper we include samples from code calculations and a discussion of the device and experimental setup, diagnostic instrumentation, data and code comparisons.

SAMPLE CALCULATIONS

The 2D MESA Eulerian hydrocode calculation is used to guide the design and execution of the experiments. A few examples from the implosion of a SS liner are provided here to facilitate the discussion. Similar calculations have been extended to other materials for comparison but will not be presented. The EOS input and zone size play a critical role in the outcome of the simulations. These EOS values are often not totally derived from measurement, but rather from calculations based on a complicated set of physical models to cover the entire domain from solid to partially ionized gas. The end cap shown is incorporated to seal the cavity under vacuum during intense dynamic loading, a feature that makes it possible to transport plasma and soft x rays unimpeded.

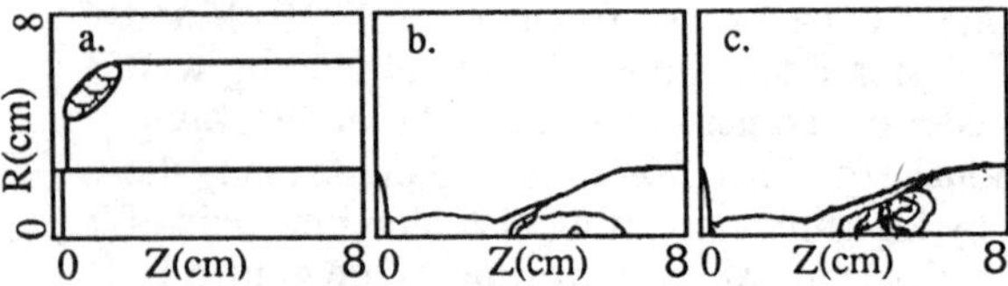

FIGURE 1. Samples of 2D hydrocode calculation.

Snapshots from two different calculations are presented in Fig. 1. Time starts at zero when the corner of the HE cylinder is initiated. Relative to firing unit discharge, an additional 15.2 μs burn time must be added. Figure 1a shows the device setup and the progression of the HE detonation wave at 2 μs after the corner initiation. The liner collapses on axis at 10.3 μs after reaching a velocity of 1 cm/μs. The maximum compressed density is calculated to be 19.5 g/cm^3, while maximum pressure and temperature are 15 Mb and 8 eV. At this temperature the generated hot gas is partially ionized. The simulated event at 12.0 μs is shown in Fig. 1b, where the expanding plasma density envelope of ~ 10^{-3} g/cm^3 is moving at about 6 cm/μs. The fastest velocity for the very low-density component can exceed 14 cm/μs. Closer examination also reveals a wavy structure in the expansion front, suggesting the presence of instability. Because many details are clearly visible the dynamic profile at this time is usually chosen to compare with the radiograph. The hot gas continues to expand and manifest sign of instability at the expansion front. The temperature remains high, but may be a numerical artifact since the code does not account for heat loss. At later time, a high-density cluster of around 3 g/cm^3 liner debris begins to be formed and travels down the axis at about 1.8 cm/μs, which eventually impacts the glass port to generate a 25 eV temperature. Figure 1c shows the convergence of plasma flow in an arrangement where a 1.25 cm diam x 0.1-cm-thick SS disk is inserted at a distance of 5 cm. Cusps and hot spots are developed in the flow, and temperature at the convergence can reach as high as 100 eV.

EXPERIMENTS

Device Description and Typical Experimental Setup

The device description and arrangement of a typical experiment are illustrated in Fig. 2. The device, which consists of a simple, hollow core, PBX-9501 HE cylinder (13.5 cm o.d x 5.08 cm i.d. x 13.5 cm long), is corner initiated to implode a 5.08 cm o.d., 0.081-cm thick, seamless 304 SS tubing. The ring initiation is accomplished with a single detonator by means of a thin-wall conical structure fabricated from a C2 explosives. A 0.48-`cm-thick SS sealing disk is welded to one end of the liner. The open end is glued to an expansion chamber to facilitate the fielding of diagnostics. The chamber is made either from plastic or aluminum and may be evacuated or backfilled to any level of pressure. To measure the plasma expansion and propagation of uv and higher frequency radiation the chamber is evacuated. A plastic chamber is employed when a vacuum above a few millitorrs is adequate and it enables cameras to view sideways through the chamber. A metal chamber is used when higher vacuum is needed. The insert at the lower left corner shows the arrangement where a SS disk is placed along the axis to divert the plasma flow. In this configuration the HE cylinder is shortened to 6.25 cm in length.

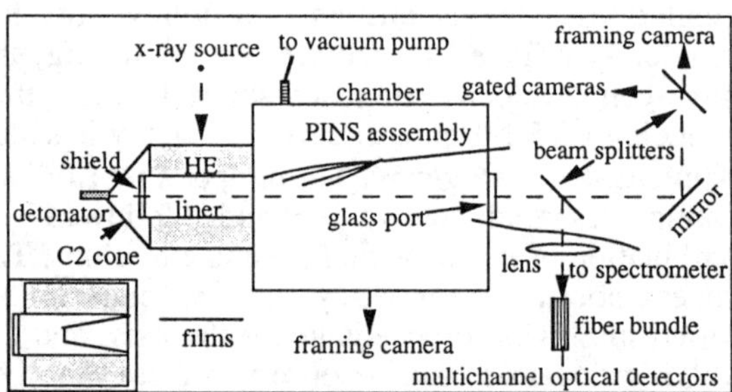

FIGURE 2. Schematic of device and experimental setup.

Itemized Diagnostics and Functions

An extensive array of diagnostic instrumentation is fielded in each shot, and is designed to assess the performance and to reveal temporal and spatial information that can be correlated with the simulations. Some are not shown and other are not fielded in every shot. Often, 8 electrical pins are placed in a collar around the HE to measure the initiation simultaneity. A radiograph is taken to assess the implosion symmetry and overall dynamic behavior. A time and spectrally resolved multichannel optical fiber detector system is employed to record light emission history inside the imploding cavity. These detectors with their color or neutral density filters are calibrated and arranged to cover four orders of magnitude in intensity. Sharing the view along the axis with the aid of a beam splitter are several time-resolved cameras. A gated 1024-channel spectrometer is fielded via an optical fiber to grab an interval of light emission for absolute spectrum determination, and it provides an in-situ calibration for the fiber detectors. An IMACON framing camera viewing sideways through the plastic chamber is included to record the expansion of plasma and liner debris. Optical pins were introduced once to measure the plasma expansion velocity, but tend to be intrusive to the experiment. Photoelectric diodes (XRD) are employed to measure the uv, vuv and soft x-ray emissions.

Results and Comparisons with Code Predictions

In all the experiments the electrical pin data have consistently shown that the initiation around the corner of the HE cylinder is simultaneous to within 25.5 ns and starts at 15.2 μs after the firing discharge. All the experimental times are measured from the discharge signal. A radiograph is shown in Fig. 3. Aside from showing implosion symmetry, it also reveals a wealth of dynamic information after the initial axial collapse, which includes the size and shape of the imploding liner over its entire length, signatures of higher density plasma and jet emissions, shock reflection into the expanding HE gas, and behavior of the sealing cap. The profile from the calculation overlaps remarkably well with the observed features. Similar results are obtained from exposures where a disk is inserted, and shows that the disk remains in place before collision with the imploding liner.

Measurement of light emission from the evacuated cavity provides the history of the entire dynamic process for more than 20 μs. Signal of a typical fiber optic channel is shown in Fig. 4. The start at 25.5 μs corresponds exactly to the calculated time of collapse. It rises steeply to reach a peak in

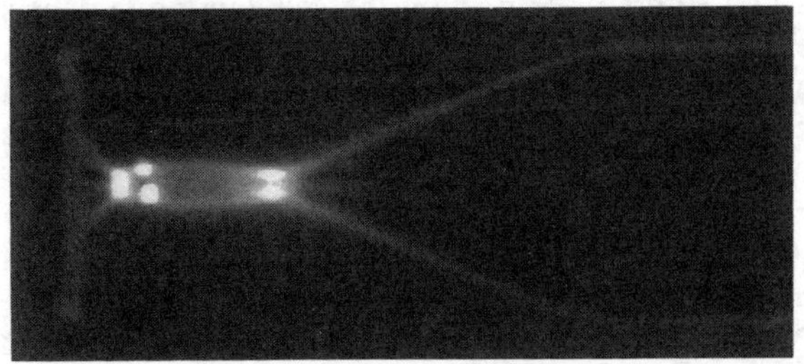

FIGURE 3. An x-ray radiograph taken at 27μs.

about 1 μs and may be explained by plasma expansion into the imploding wall. The temperature from 26.25 to 27.25 μs is estimated to be 8 eV from intensity calibration, which is consistent with the fit to a blackbody-like spectrum from the spectrometer and in agreement with the prediction. Continuous intensity increase before the second peak may be attributed to emissions from the expanding plasma surface, hot spots generated by instability and continuous implosion. The calculation does predict maximum temperature to remain between 8-10 eV during the expansion. The location of the second peak at 32 μs coincides with the calculated arrival of higher density plasma at the glass port. The third peak at 36.6 μs correlates well with the predicted final impact by the massive cluster. This is supported by the observations from the side viewing IMACON and the axially imaging rotating mirror camera. The absence of light earlier than 25.51 μs suggests that the vacuum has no leak.

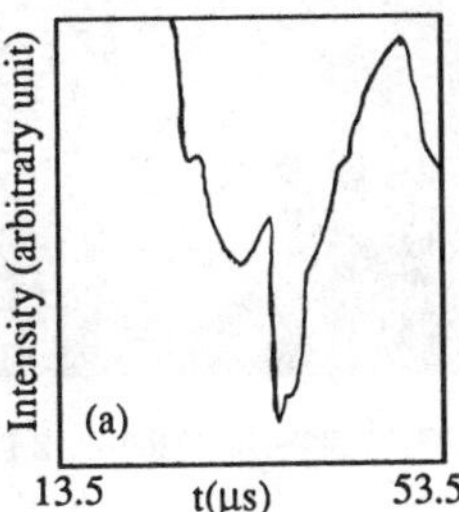

FIGURE 4. Light emission history during implosion.

The IMACON images from the side through the plastic chamber wall were recorded at 0.5 μs intervals. Using a 1-ND filter, the early unsaturated records reveal the impact at the window by the transparent plasma. The fastest plasma velocity is thus estimated to be between 12-17 cm/μs, consistent with the calculation and supported by the optical pin measurements. Those images taken with

2-ND filters clearly show the self-luminous debris entering the plastic chamber at 29.5 μs, advancing at a speed of about 1.9 cm/μs, and producing the third peak in Fig. 4, all tending to confirm the predictions.

The usefulness of ultrafast framing photography of the imploding liner cavity is compromised by its inability to accommodate the rapid variation in the object location and light emission intensity. In one setting the IMACON images were saturated because of insufficient attenuation. In another setting, a rotating mirror camera focusing on the expansion chamber yielded no information on the collapsing cavity. However, the images of a cloud-like expansion over several microseconds corresponds closely to the debrisobserved in the IMACON. The punch-through at the port is clearly visible. Finally, a well focused snapshot taken along the axis by a fast-gated electronic camera at 1.1 μs after the liner collapsed is shown in Fig. 5. The appearance of structures in the plasma expansion implies nonuniform density and/or temperature distribution, strongly indicating the presence of turbulence and hot spots, as suggested by the simulation. Further study of this structure may shed light toward understanding the nature of plasma behaviors in a phased implosion process.

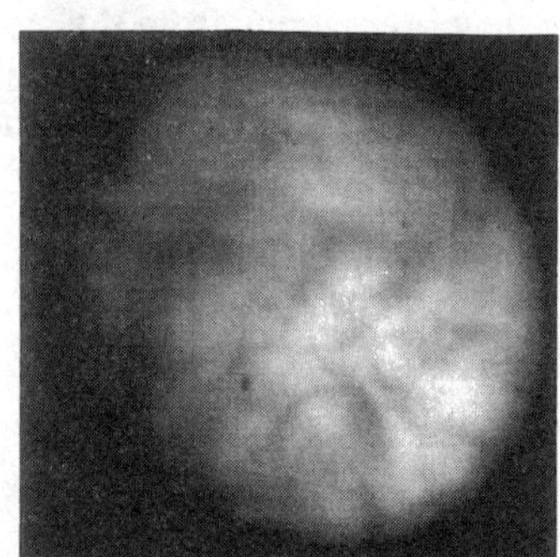

FIGURE 5. Axial view of expanding plasma.

In the two experiments where a disk is inserted to divert the plasma flow, a large amount of soft x rays is detected. The unfiltered channels with a threshold of 16 eV are near saturation. However, the signal from the high-energy, carbon filtered channel is small, and the ratio suggests that the integrated temperature measured over the entire plasma plume is only about 10 eV. This is to be expected if the very high temperature hot spots represent only a very small fraction of the total area of emission. Three gated cameras triggered at different times yielded information on plasma before, during, and after flow around the disk. Interesting structures are seen in the expansion, but all the photographs were taken before the plasma plume reconverged on axis.

CONCLUSIONS

We have demonstrated that a very thin SS cylindrical liner can be driven by HE to implode symmetrically and intact to very high velocity, with a substantial plasma and radiation yield. We are pleasantly surprised to find that the data are in reasonable agreement with many of the predictions from the numerical simulation. It appears that we may have found a powerful tool to study interesting ultra-high pressure and temperature phenomena in the regimes that are normally investigated only with electromagnetically driven, pulsed power devices.

ACKNOWLEDGEMENTS

We are indebted to C. Findley, N. Gray, H. Oona, G. Idzorek, P. Rodriquez, S. Sterbenz, and L. Veeser for their contributions.

Research supported by DOE.

REFERENCES

1. T. H. Tan and S. P. Marsh, Proceedings of the 4th International Symposium on High Pressure Dynamics, Tours, France, 1995, p. 373.

2. T. H. Tan and S. P. Marsh, Proceedings of the International Symposium on Shock Waves, Pasadena, California, USA, 1995 , in print.

A STRONG SHOCK TUBE PROBLEM CALCULATED BY DIFFERENT NUMERICAL SCHEMES

Wen Ho Lee and Sean P. Clancy

University of California
Los Alamos National Laboratory
Los Alamos, NM 87545 USA

Calculated results are presented for the solution of a very strong shock tube problem on a coarse mesh using (1) MESA code, (2) UNICORN code, (3) Schulz hydro, and (4) modified TVD scheme. The first two codes are written in Eulerian coordinates, whereas methods (3) and (4) are in Lagrangian coordinates. MESA and UNICORN codes are both of second order and use different monotonic advection method to avoid the Gibbs phenomena. Code (3) uses typical artificial viscosity for inviscid flow, whereas code (4) uses a modified TVD scheme. The test problem is a strong shock tube problem with a pressure ratio of 10^9 and density ratio of 10^3 in an ideal gas. For no mass-matching case, Schulz hydro is better than TVD scheme. In the case of mass-matching, there is no difference between them. MESA and UNICORN results are nearly the same. However, the computed positions such as the contact discontinuity (i.e. the material interface) are not as accurate as the Lagrangian methods.

INTRODUCTION

For many years, hydrodynamic code developers have been using different test problems to check the new algorithms or methods in their codes. This process is gradually becoming a standard procedure to achieve the quality assurance for both Eulerian and Lagrangian hydro codes. There are two major difficulties in computing multi-material flow problems. These are the material interface reconstruction and the shock front tracking (or any discontinuity other than density). Most of the hydro codes use the SLIC [1] or Particle-in-Cell [2] methods to define the material interfaces, and in general, the results are quite acceptable. For shock front discontinuities, artificial viscosities are used. However, to find a right formula of the artificial viscosity for a particular state is not easy. Most of the 2-D hydro codes tend to use too large an artificial viscosity (to avoid the code crash) and consequently, over-dissipate the kinetic energy. For 1-D ideal gas problems, the Godunov scheme seems to posses the best formulation of viscous effect necessary to smear the shock front. Nevertheless, at least three problems exist for the Godunov scheme in solving a real material flow problems with two or three spacial dimensions. The first is the coupling of the real equation of state (most of the EOS are discrete data), the second is the treating of the free surface boundary, and the third is the construction of the Riemann solver system with elasticity and plasticity. In this paper, the calculated results are presented for the solution of a very strong shock tube problem on a coarse mesh using (1) MESA code, (2) UNICORN code, (3) Schulz hydro, and (4) modified TVD scheme.

PROBLEM DEFINITION

The initial conditions of the strong shock tube test problem are described in Fig. 1-a. The total length of the tube is 9 cm. Material 1 of density = 1 gm/cm^3 and internal energy = 1000 J/gm occupies the left 3 cm, whereas material 2 of density = 0.001 gm/cm^3 and internal energy = 0.001 J/gm is on the right 6 cm. Both materials are assumed to be ideal gas with $\gamma = 5/3$. There are 90 uniform zones for the whole tube with 30 zones for material 1 and 60 zones for material 2. Since this is a computational simulation, no real diaphragm to separate the materials is necessary. A mass-matching initial configuration is shown in Fig. 1-b. Mass matching means that the masses in zone 30 (the most rightward zone in material 1) and zone 31 (the most leftward zone in material 2) are approximately equal with the maximum mass ratio of 2 (usually about 1.2). The distribution of the mass ratio of neighboring zones inside the material 1 will be similar to a sine function. The initial velocities for both materials are set to zero. The computation is run to time = 0.06 μsec. The density distributions for the analytical solution at time = 0.06 μsec are given in the upper part of Fig. 2.

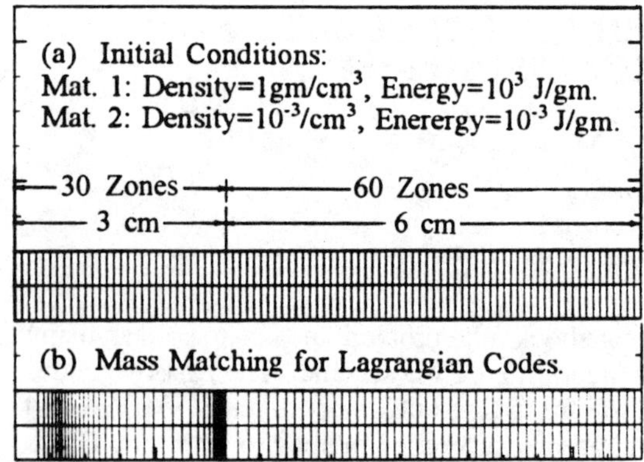

FIGURE 1. (a) The initial conditions of the shock tube. (b) Mas-matching initial configuration.

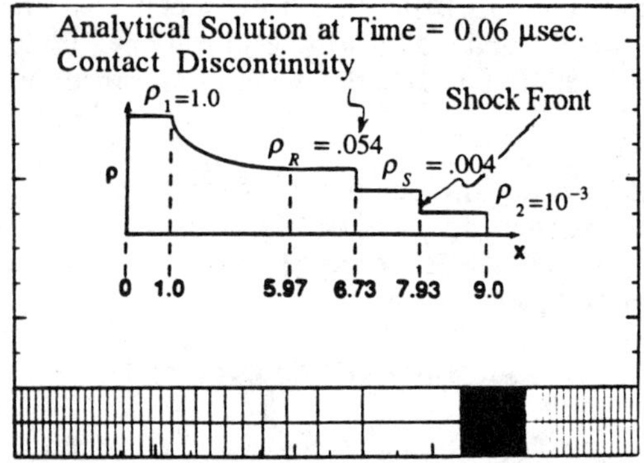

FIGURE 2. The density distributions for the analytical solution at time = 0.06 μsec (upper) and the mass matched mesh at time = 0.06 μsec (lower).

NUMERICAL SCHEMES

The conservation equations of mass, momentum, and internal energy for an inviscid fluid are solved in planar/cylindrical coordinate systems. For MESA, UNICORN, and Schulz hydro, a staggered mesh is used. The pressure p, specific   internal energy E, and density ρ are defined at cell center along with the velocity vector **u** (where **u** $= u$ **i** + v **j** with u , v the velocity components in r, z directions respectively) and coordinates r and z are defined at the vertices. However, the modified TVD scheme defines all variables at cell center. For all of the four schemes, explicit finite differences are used.

MESA and UNICORN are two-dimensional Eulerian codes. In the first phase, they solve the Lagrangian part in both r and z directions simultaneously. The momentum and energy equations solved in Lagrangian phase are

$$\frac{\partial u}{\partial t} = -\frac{1}{\rho}\frac{\partial p}{\partial r} \ , \tag{1}$$

$$\frac{\partial v}{\partial t} = -\frac{1}{\rho}\frac{\partial p}{\partial z} \ , \tag{2}$$

$$\frac{\partial E}{\partial t} = -\frac{p}{\rho r}\frac{\partial(ru)}{\partial r} - \frac{p}{\rho}\frac{\partial v}{\partial z}. \tag{3}$$

At the end of phase one, we have the temporary values of velocity $\tilde{u}$ and $\tilde{v}$, energy $\tilde{E}$, momentum $M\tilde{u}$ and $M\tilde{v}$ (where M is the mass in the cell), and total internal energy in the cell $M\tilde{E}$. In phase two or remap phase, the average velocities $\bar{u}$ and $\bar{v}$ are calculated by

$$\bar{u} = \frac{1}{2}(u + \tilde{u}) \ , \tag{4}$$

$$\bar{v} = \frac{1}{2}(v + \tilde{v}), \tag{5}$$

where u and v are the old velocities. These average velocities are used for computing the momentum and energy flux across the cell boundaries. The governing equations are split in radial and axial directions. Then calculation is performed in each separated direction. The order of this calculation is alternated for each time advancement to maintain the accuracy of one-dimensional procedure. MESA uses the monotonic advection methods of Van Leer [3] to avoid the non-physical oscillations and negative densities. UNICORN code uses LeBlanc limiting [4] for the remap phase which is a second order monotonic interpolation scheme. To describe the method, let us start with the radial mass conservation equation

$$\frac{\partial \rho}{\partial t} = -\frac{1}{r}\frac{\partial(r\rho\tilde{u})}{\partial r} \tag{6}$$

The finite difference of Eq. (6) is

$$\rho^{n+1}_{j+\frac{1}{2}} = \rho^n_{j+\frac{1}{2}} - \frac{\Delta t^{n+\frac{1}{2}}}{r_{j+\frac{1}{2}}\Delta r_{j+\frac{1}{2}}}[(rF)^{n+\frac{1}{2}}_{j+1} - (rF)^{n+\frac{1}{2}}_{j}] \tag{7}$$

Where n is the discrete time level and n+1/2 means $n + \Delta t / 2$, j is the zone index in r direction, and F_j the mass flux across interface j. Here $F_j = \rho^*_j\tilde{u}_j$. UNICORN method uses slope S_L, S_R, and S_C to obtain the optimum value of ρ^*. The expressions of S_L, S_R , and S_C are

$$S_L = \frac{\rho_{j+\frac{1}{2}} - \rho_{j-\frac{1}{2}}}{0.5\Delta r_{j+\frac{1}{2}}} \ , \tag{8}$$

$$S_R = \frac{\rho_{j+\frac{3}{2}} - \rho_{j+\frac{1}{2}}}{0.5\Delta r_{j+\frac{1}{2}}} \ , \tag{9}$$

and

$$S_C = \frac{1}{\Delta r_{j+\frac{1}{2}}} \left(\frac{\rho_{j+\frac{1}{2}}\Delta r_{j+\frac{3}{2}} + \rho_{j+\frac{3}{2}}\Delta r_{j+\frac{1}{2}}}{\Delta r_{j+\frac{3}{2}} + \Delta r_{j+\frac{1}{2}}} \right.$$

$$\left. - \frac{\rho_{j-\frac{1}{2}}\Delta r_{j+\frac{1}{2}} + \rho_{j+\frac{1}{2}}\Delta r_{j-\frac{1}{2}}}{\Delta r_{j+\frac{1}{2}} + \Delta r_{j-\frac{1}{2}}} \right). \tag{10}$$

The minimum value of the slope S is

$$S = \min\{|S_L|,|S_C|,|S_R|\}. \tag{11}$$

If the slope S_L has an opposite sign of the slope S_R, then $S = 0$. Finally, we interpolate backward along the slope S by a distance $\frac{1}{2}\Delta t^{n+\frac{1}{2}}\tilde{u}_j$ from the zone boundary j. This location gives the desired value of the interpolated mass density ρ^*. UNICORN method is second order accurate, and it greatly reduces the numerical diffusion.

The Schulz hydro uses Schulz's two-dimensional pure hydrodynamic code as a framework [5]. Modifications to the artificial viscosity are made so that the acceleration from the artificial viscosity will project onto the unit vector in the direction of local acceleration. This projection helps maintain a stable computational grid. The new momentum equations are

$$u_t = -\frac{1}{\rho j}(z_l \frac{\Delta p}{\Delta k} - z_k \frac{\Delta p}{\Delta l})$$

$$-\frac{1}{M}\frac{\Delta}{\Delta k}(rq_A \frac{z_l u_k - r_l v_k}{u_k^2 + v_k^2}u_k)$$

$$-\frac{1}{M}\frac{\Delta}{\Delta l}(rq_B \frac{z_k u_l - r_k v_l}{u_l^2 + v_l^2}u_l) \ , \tag{12}$$

$$v_t = \frac{1}{\rho j}(r_l \frac{\Delta p}{\Delta k} - r_k \frac{\Delta p}{\Delta l})$$

$$+\frac{1}{M}\frac{\Delta}{\Delta k}(rq_A \frac{z_l u_k - r_l v_k}{u_k^2 + v_k^2}v_k)$$

$$-\frac{1}{M}\frac{\Delta}{\Delta l}(rq_B \frac{z_k u_l - r_k v_l}{u_l^2 + v_l^2}v_l) \ , \tag{13}$$

where k and l are Lagrangian coordinates, j is the Jacobian of transformation, q_A is the artificial viscosity associated with differencing in the l direction, q_B is that in k direction, the subscripts k and l indicate the derivatives, Δ means finite differencing, and M is the cell mass.

The modified TVD scheme [6] solves the momentum and energy equations written

$$\frac{d\bar{u}}{dt} = -\frac{1}{\rho}\nabla p \ , \tag{14}$$

$$\frac{dp}{dt} = -\rho c^2 \nabla \cdot \bar{u} \ . \tag{15}$$

where c is the sound speed. The discrete form that emerges, after identifying the limiter ϕ_1 and ϕ_2 is

$$u^* = u_\circ^* - \phi_1\left[\frac{\Delta x \nabla p}{(\rho c)_+ + (\rho c)_-}\right]^n - (1-\phi_1)(\frac{\Delta t \nabla p}{2\rho})^n \tag{16}$$

$$p^* = p_\circ^* - \phi_2\left[\frac{\Delta x \nabla \cdot \bar{u}}{(\rho c)_+ + (\rho c)_-}\right]^n$$

$$-(1-\phi_2)(\frac{\Delta t \rho c^2 \nabla \cdot \bar{u}}{2})^n \ . \tag{17}$$

where u^* and p^* are derived from the method of characteristics and the subscripts plus and minus refer to the cell-centered values on opposite sides of the face for which the flux is desired. The limiter is required at the cell face and is a function of the cell-face velocity divergence.

RESULTS AND CONCLUSIONS

The computed results of the density along the axial
direction are presented with the analytical solutions.
Fig. 3 shows the density distributions as simulated
by MESA and UNICORN codes. The UNICORN
calculations for the shock front and contact
discontinuity are better than those of MESA, but
overall they are almost the same. The densities
versus the axial direction for uniform zoning are
shown in Fig. 4 for both Schulz hydro and
modified TVD. In this case, the location of the
contact discontinuity as calculated by Schulz hydro
is much closer to the analytical one as compared
with those of the TVD scheme. The TVD scheme
also shows an undershoot at the contact
discontinuity location. Good results are obtained for
both Schulz hydro and the TVD scheme when
masses matching zones are used as shown in Fig. 5.
The final mesh of this case for Schulz hydro is
shown in the lower part of Fig. 2. However, both
schemes compute the shock front positions ahead of
the analytical one and both give a slight undershoot
in density at the contact discontinuity location. In
conclusion, the Lagrangian codes are better than the
Eulerian codes for this problem. Uniform zoning in
the Lagrangian codes makes the scheme first order,
whereas mass matching makes them second order
accurate.

REFERENCES

[1] Noh, W.F., and Woodward, P., SLIC UCRL-
    77651 (1976)

[2] Lee, W.H., and Kwak, D., Computer Phys.
    Commu., **48**,pp11-16 (1988)

[3] Van Leer, B., J. Comput. Phys., **23**,
    pp276-299 (1977)

[4] Clancy, S.P., Ph.D. Thesis, Univ. Texas,
    Austin (1989)

[5] Schulz, W.D., Method in Computatational
    Physics, Vol.**3**,pp1-45 (1964)

[6] Kashiwa, B., and Lee, W.H., Lecture Notes in
    Phys., No.**395**,pp277-288 (1990)

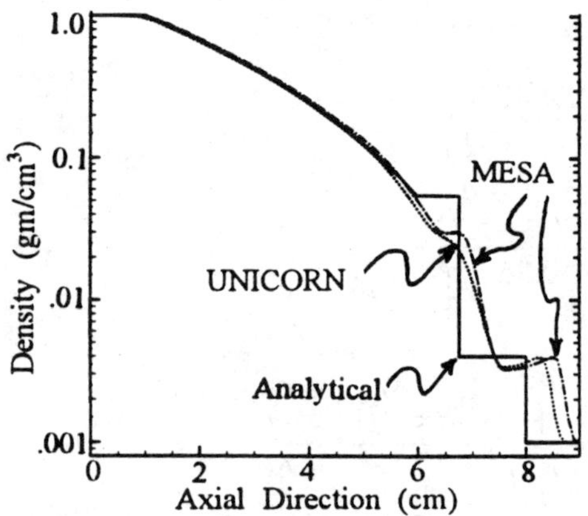

FIGURE 3. The density distributions as calculated by
MESA and UNICORN codes.

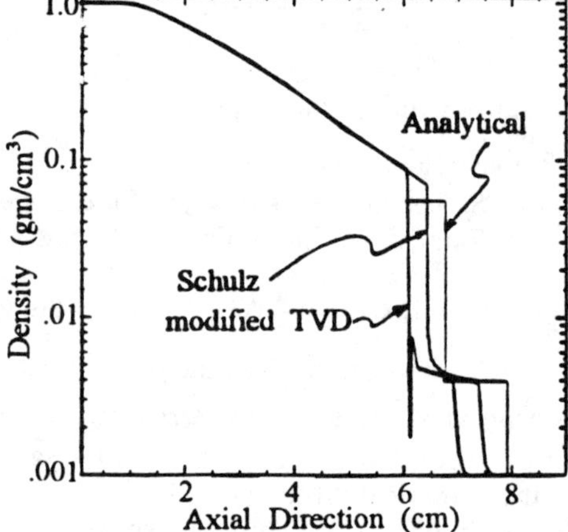

FIGURE 4. The density distributions as calculated by
Schulz hydro and TVD with uniform zoning.

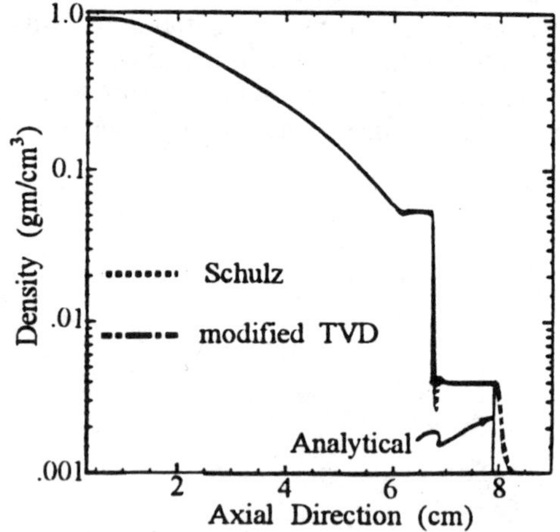

FIGURE 5. The density distributions as calculated
by Schulz hydro and TVD with mass matching.

IN SITU X-RAY DIFFRACTION FROM UNIFORM RADIATION DRIVEN SHOCKS IN CRYSTALLINE SOLIDS

N.C. Woolsey, R. Cauble, R.W. Lee, & J.S. Wark*

Lawrence Livermore National Laboratory, University of California, P.O. Box 808, Livermore, CA 94550
**Department of Physics, Clarendon Laboratory, University of Oxford , Parks Rd, Oxford, OX1 3PU, UK*

An experiment to create highly uniform shocks in crystalline solids suitable for studying the thickness of shock fronts by *in situ* x-ray diffraction is described. A laser is used to create a soft x-ray radiation source. This radiation source heats a thin layer of the silicon crystal; the ablation of this layer produces a uniform near-one-megabar shock. We discuss how x-ray diffraction can be used to measure the shock front thickness from a uniformly shock compressed solid. The x-ray diffraction technique is described through previously measured experimental data and by comparison to simulated diffraction records.

INTRODUCTION

The investigation of shocked solids by the diffraction of short, intense pulses of x-rays has been an area of experimental study for over a quarter of a century (1). The attractiveness of the technique lies in its potential to give information on the shock process on the lattice level. As a crystal is compressed or rarefied, the angle of diffraction, ϑ_B, of monochromatic x-radiation, λ, changes. This can be seen by differentiating Bragg's law $n\lambda = 2d_H \sin(\vartheta_B)$ to give:

$$\frac{\Delta 2d}{2d} = -\Delta\vartheta \cot \vartheta_B$$

which is valid in the limit of small changes and where d_H is the lattice spacing for the set of planes with Miller indices *H*.

By recording the diffraction angle, direct information concerning the degree of elastic strain, that is $\Delta 2d/2d$, within the crystal can be obtained. Further, as x-ray diffraction is a bulk phenomenon, a diffraction record also contains depth dependent information.

Over the past few years we have been using high power lasers to launch compression pulses of nanosecond duration into materials and to create separate short duration x-ray sources for flash diffraction from the compressed crystals (2-4). The technique has been developed to the level such that it is now possible to record a time-resolved diffraction pattern from the shocked crystal and with a temporal resolution of 10 picoseconds, with the total duration of the diffracted pulse of order a nanosecond. In principle the temporal development of the diffraction pattern gives information on the time and depth-dependent strain profiles within the shocked crystal.

A limitation to the work accomplished so far is the spatial uniformity of the shock launched into the solid. It is well known that high energy laser systems have non-uniform beam intensity spatial profiles. Typically, we have worked in the near field of the laser beam (as large focal spots are necessary) where the beam uniformity is superior to the far field. However, the RMS intensity across the near field may be up to 30% (5) leading to significant 2-dimensional effects. This problem has been addressed by inertially confining the ablating plasma that generates the shock with a layer of transparent plastic (2,6,7). This technique improved the uniformity of compression but non-uniformities are still present denying, for example, the possibility of making accurate measurements of the shock front thickness.

A superior way to create uniform planar shocks is to use a soft x-ray source to thermally heat a thin layer of the material under test (8-10), At sufficiently high temperatures (10 eV and above), the heated layer ablates off the surface of the material and a compression wave is launched into the cold solid by momentum conservation. Such a soft x-ray heating source can be created by focussing a high power laser onto a thin gold x-ray conversion foil. X-rays from the rear side of this foil, which approximates a 70 eV blackbody, impinges on the front surface of the crystal. The experimental geometry is illustrated in Figure 1.

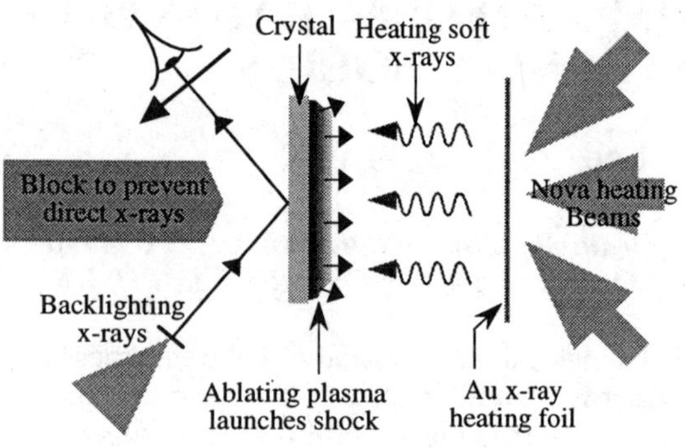

FIGURE 1. Schematic illustration of the proposed experiment.

EXPERIMENTAL

Nova, a multiple beam inertial confinement fusion laser at LLNL is used to create a soft x-ray heating source to drive a shock wave in the solid and to create an independent but synchronized backlighting x-ray source for diffraction measurements. Five beams of Nova are focussed into a 1-mm diameter spot on a 2000-Å thick free standing gold foil. A 2-ns square laser pulse with up to 10 kJ of energy in the 3rd harmonic of Nd:glass (0.35 μm) is used to irradiate the foil. At these irradiances approximately 10% of the laser energy is converted into soft x-rays (9). X-rays from the backside of this conversion foil volumetrically heat the front layer of a silicon crystal.

Using the published spectral data (9, 10) and published cold mass absorption coefficients for silicon (11) we determine that 95% of the energy produced by the conversion foil is absorbed within the first 5 μm of the silicon crystal. Temperatures of approximately 15 eV are reached in the first 2 μm of the silicon generating pressures approaching 0.5 Mbar. Simulations indicate shock pressures approaching 1 Mbar over a 3-mm diameter area are possible using energies available on Nova and this technique.

The shock takes ~4 ns to traverse a typical crystal thickness of 40 μm. When the shock is within 2 or 3 absorption depths of the rear surface an additional set of synchronized Nova beams are used to create a vanadium plasma that emits efficiently the helium-like resonance lines. This diffraction beam consists of a series of lines: the resonance line $1s^2\text{-}1s^12p^1P_1$ (2.392 Å), the intercombination line $1s^2\text{-}2p^1P_3$ (2.393 Å) and a series of unresolved lithium-like dielectronic satellites centred at 2.4 Å (12). This pseudo-monochromatic radiation is diffracted from the crystal onto both spatially resolving x-ray film and a temporally resolving x-ray streak camera.

The availability of materials and the geometry of the Nova experimental chamber dictated the choice of silicon (111) crystals and vanadium backlighters.

In silicon the 1/e absorption depth of 2.3 Å radiation measured normal to the diffracting surface is approximately 7 μm. Assuming a typical dynamical range of an x-ray streak camera of 100 we can see the same feature up to 15 μm below the crystal surface, noting that the beam must enter and exit the crystal, before and after shock breakout.

EXAMPLE DATA

An example of x-ray diffraction experimental data is shown in Figure 2. This data was taken using the VULCAN laser system at the Rutherford Appleton Laboratory, U.K. (4). Here a short approximately 60 ps Gaussian laser pulse was used to drive an approximately 100 kbar compression wave into a single crystal of silicon.

This data will be discussed in more detail in the section below by comparison with computer simulations. The discussion will compare simulation with experiment and then centre on how shock front thickness data can be extracted from a diffraction record, highlighting the advantages gained by using a large laser system like Nova.

SIMULATION

Simulations are required to aid detailed interpretation of time-resolved diffraction patterns such as Figure 2. The simulation procedure neatly divides into two parts. First a hydrodynamics model is used to calculate the shock generating mechanism, leading to the time development of the shock and strain profiles in the solid. Once the strain-depth profiles are calculated each time-step is post-processed with an x-ray diffraction model to produce a time dependent diffraction record. The instrument response is then folded into the simulation by matching the spectral and temporal resolution to the experiment to yield a simulation of experimental data. Note that the simulations assume a mono-chromatic radiation source; the experimental data is generated with an x-ray source consisting of 3 closely spaced x-ray lines.

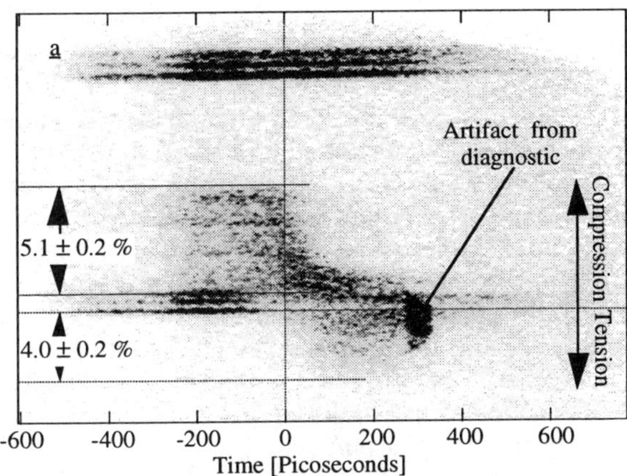

FIGURE 2. X-ray diffraction record of laser shocked crystal of silicon (111). Trace <u>a</u> is of the incident probe radiation from a spectrometer. The shock driving laser beam pulse was ~60 ps FWHM Gaussian. Time t=0 indicates shock breakout.

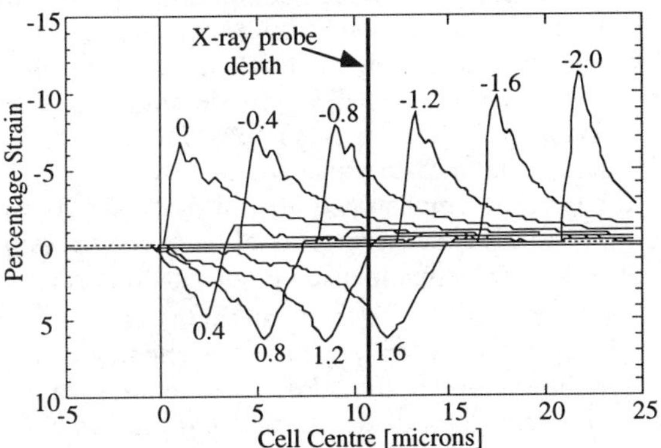

FIGURE 3. Time step of the hydrodynamics used to simulate the data shown in Figure 2. The numbers above the traces indicate approximate times before and after shock breakout.

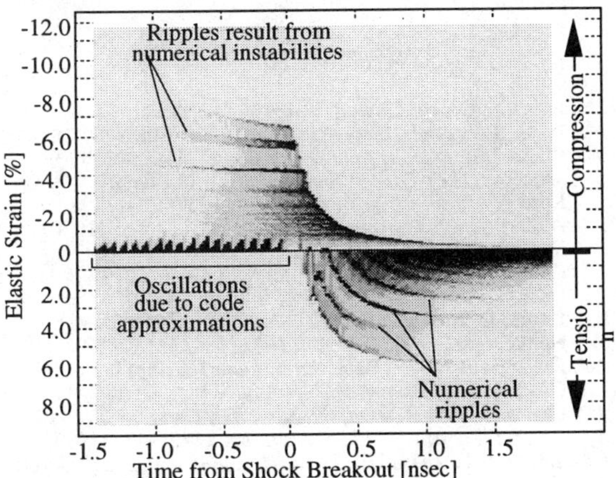

FIGURE 4. Simulation of Figure 2. Time t=0 indicates shock breakout.

A hydrodynamic time history is shown in Figure 3. These curves were calculated using MEDUSA (14) and modified to allow a Mie-Grüneisen silicon equation of state to be used. The simulated data, shown in Figure 4, was calculated using an x-ray diffraction model similar to that described by Erola *et al.* (15). The ripples in Figure 3 result in the bands of intense diffraction in Figure 4, the ripples result from numerical instabilities and are not real.

DISCUSSION

A comparison of Figure 2 with the simulation in Figure 4 indicates that the essential features of the experimental data are reproduced. Firstly, both the simulated and experimental data show diffraction simultaneously at the Bragg angle and at angles higher the than Bragg angle. Here we define the Bragg angle as the diffraction angle from the unperturbed solid only. Diffraction at greater angles correspond to diffraction from a compressed lattice. Secondly, as time evolves the diffraction profile at high angles rapidly shifts back towards the Bragg angle and decays to near zero compression 1 ns later. The backlighter has turned off in Figure 2 by this time. A few tens of picoseconds after the compression amplitude starts to decrease, we see that diffraction is now observed at angles lower than the Bragg angle. This indicates some part of the probed solid is in tension. As the compression decays the tensile portion of the diffraction record increases until it plateaus at approximately 1 ns.

The gross features of the diffraction records are readily understood using Figure 3. As mentioned above the diffracting x-ray beam and streak camera combination has a finite depth resolution. In this case the probe depth normal to the diffracting surface is approximately 11 μm. At early times, *e.g.*, t=-0.4 ns in Figure 3, the diffracting x-rays probe predominantly unperturbed crystal; beneath the crystal surface the compression wave is traveling towards the rear surface. Diffraction is principally from the unperturbed solid with a component strongly attenuated by photoelectric absorption at high angles. The high angle component gets brighter as the compression wave advances towards the rear surface since the material in front of the shock is being reduced.

At shock breakout the entire volume being probed is strained, the unperturbed signal disappears; see t=0 in Figure 3. Immediately following this a rarefaction wave forms and travels back into the solid. Tensile states are created as this rarefaction crosses a second rarefaction created at the front of the crystal as the pressure drive stops. Depth-dependent strain information is clearly demonstrated just after shock breakout when both tensile and compressive strains are observed. Here the two rarefaction waves have crossed, as illustrated at t=0.4 in Figure 3. The tensile state continues to evolve as the rarefaction wave travels further back into the crystal and passes the tail of the second rarefaction. At this point the tensile strain peaks.

The fracture strength of a brittle solid is a function of strain-rate (16), experimentally observed peak tensile strengths may be lower than the ideal fracture strength of the solid due to spall.

It is interesting to note that the strain rates measured in these experiments are between 10^8 and 10^{10} s^{-1} and are several orders of magnitude higher than the strain rates reached by other techniques. The brittle fracture model of Grady (17) indicates the ideal strength of silicon should be reached at strain-rates above 10^9 s^{-1}. Analysis of Figure 2, where the strain rates are 1×10^9 s^{-1}, indicates the peak tensile strain reached is 4.0%. This implies a fracture strength of ~75 kbars, which is close to the experimental fracture strength recorded from high quality single crystal whiskers(18).

Information on the shock front can be extracted from the diffraction data in Figure 2. At shock breakout, t=0 in Figures 2 and 4, there is a delay between the immediate onset of the decay in the peak compression and development of a tensile strain of less than 100 ps (note the streak camera temporal resolution was approximately 50 ps). The simulations in Figures 3 and 4 are useful as an indicator of the shock front, they do not provide shock thicknesses that can be compared to experiment as the shock front is a function of the hydrodynamic mesh size and artificial viscosity.

Since we are in the weak shock regime, we can conclude that the elastic shock front thickness is less than 1 μm. This conclusion is supported by the 10 ps temporal resolution experimental data.

The method we plan to use at Nova to investigate the shock front thickness involves creating a long duration compression pulse. In this case the pressure source remains on for 2 ns ensuring the rarefaction wave, which is formed as the pressure source is turned off, is many diffracting beam absorption depths behind the advancing shock front. Thus, the diffracting beam interacts with only the shock front so that any diffraction signal recorded between zero and peak compression must have originated from the shock front. The ratio of the intensity of this signal to the incident diffracting beam is directly related to the shock front thickness. It should be possible to measure shock front thicknesses less than 1000 Å thick using this technique. There are, of course, limitations on these measurements. As examples, errors will arise due to the crystals, that is in ensuring both faces are precisely parallel and aligned to the crystallographic axis. Further the difficulty in producing a noise-free enviroment on the high energy laser system will limit experimental accuracy.

CONCLUSIONS

X-ray diffraction has been shown to be a powerful tool for investigating shock compression processes. Ideas on how the technique can be used to measurement shock front thicknesses have been presented. Experiments to produce highly uniform shocks up to 1 Mbar in pressure which are suitable for shock front measurement are underway.

It has been demonstrated how diffraction data can be used to give measurements of peak compression and tension, and that measurements at strain rates of 10^9 s^{-1} are possible. Fracture strengths approaching the ideal strength of silicon have been achieved.

We have shown that the experimental data is readily understood using a combination of a hydrodynamics simulation code and an x-ray diffraction model. Simulated time-resolved diffraction spectra show remarkable, qualitative agreement with experiment.

REFERENCES

1. Johnson, Q., *et al.. Nature* **213**, 1114-1115 (1967).
2. Wark, J.S., *et al.. Phys. Rev.* **B35**, 9391-9394 (1987).
3. Wark, J.S., *et al..J. Appl. Phys.* **68**, 4531-4534 (1990).
4. Woolsey, N.C. *D.Phil Thesis* (Oxford University, 1994).
5. Kato, Y., *et al. Phys. Rev. Lett.* **53**, 1057-1060 (1984).
6. Anderholm, N.C. *Appl. Phys. Lett.* **16**, 113-115 (1970).
7. Yang, L.C. *J. Appl. Phys.* **45**, 2601-2608 (1974).
8. Cauble, R., *et al. Phys. Rev. Lett.* **74**, 3816-3819 (1995).
9. Back, C.A., *et al. JQSRT* **51**, 19-25 (1994).
10. Kania, D.R. *et al. Phys. Rev.* **A46** 7853-7868 *(1992)*.
11. Henke, B.L., *et al.. A D & N D Tables* **27**, 1 (1982).
12. Kelly, R.L. *Phys. Chem. Refer. Data* **16**, Suppl. 1 (1987)
13. Gust, W.H. & Royce, E.B. *J. Appl. Phys.* **42**, 1897 (1972).
14. Christiansen, J.P., *et al.. Comp. Phys. Comm.* **7**,272 (1974).
15. Erola, E., *et al.. J. Appl. Crystallog.* **22**, 35-42 (1990).
16. Grady, D.E. & Lipkin, J. *Geophys. Res. Let.* **7** 255 (1980).
17. Grady, D.E. *J. Mech. Solids* **36**, 353-384 (1988).
18. Eisner, R.L. *Acta Mett.* **3**, 414 (1955).

CHAPTER XIII

INTERFEROMETRY

WHITE LIGHT VELOCITY INTERFEROMETRY

David J. Erskine and Neil C. Holmes

Lawrence Livermore National Laboratory, Livermore, CA 94551

We describe a generic method for using broadband and incoherent light in velocity interferometry. Single frequency lasers are no longer necessary. Compact, powerful and inexpensive light sources previously prohibited due to their incoherence are now suitable for Doppler velocimetry of remote objects through air, including arc lamps, flash lamps, light from detonations, pulsed lasers, chirped frequency lasers and lasers operated simultaneously in several lines. These powerful sources should allow practical line and areal velocimetry. In our technique, the light source is imprinted with a coherent echo having a delay matching the delay in an analyzing interferometer. The technique is generic to all wave phenomena (i.e. radar, ultrasound).

INTRODUCTION

The VISAR[1-4] velocimeter (velocity interferometer system for any reflector) is an optical interferometer having a fixed delay τ between its arms. This delay converts small Doppler shifts into fringe shifts in the interferometer output. Let the delay be specified by the distance $(c\tau)$ light travels in that time, where c is the velocity of light in vacuum. The idealized velocity per fringe proportionality[1,2] (η) of a VISAR is

$$\eta = \frac{c\,\lambda}{2(c\tau)} \qquad (1)$$

where λ is the average wavelength of light. Equation (1) neglects dispersion[2] in the glass optics inside the interferometer and assumes $v/c \ll 1$, where v is target velocity.

In previous VISARs, the coherence length (Λ) of the illumination must be as large as $c\tau$ in order to produce fringes with significant visibility. This severely restricted the kind of light source which could be used. The Λ of white light ($\sim 1.5\ \mu m$) was insufficient. Previously, lasers were the only light sources used in VISARs because their coherence length could be made sufficiently long when operated in a single frequency mode. However, in this mode the output power is low. Typical laboratory measurements in shock physics were limited to measurement of velocity at a single point on the target due to low laser power. Measurement of velocity over a surface, or of a remote object through a telescope in the field demands orders of magnitude more power. Optical amplification of the single frequency source has been used to construct a line-VISAR[5]. However, this is expensive and cumbersome.

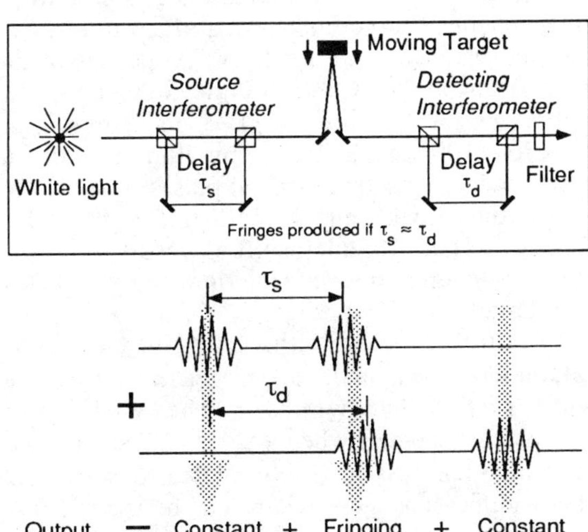

Figure 1. White light velocimeter concept. Two interferometers before and after target have similar delays τ_s and τ_d. White light is a series of independent wave packets. The first interferometer splits each wave packet into two identical packets-- the second does the same to create a total of four. Two of those packets will overlap if $\tau_s \approx \tau_d$, producing fringes. The other packets contribute constant intensity. Target velocity will change the apparent τ_s due to the Doppler effect, causing a fringe shift. The wave packet length is inversely proportional to system bandwidth, represented by the filter.

METHOD

"""

We have invented[6] a simple and generic method (Fig. 1) of preparing any wave source, regardless of its initial coherence, to illuminate a velocity interferometric experiment. A coherent echo is imprinted on the illuminating light by an interferometer (denoted the source interferometer) having a delay τ_s. The reflected light from the target is observed through a second interferometer (denoted the detecting interferometer) having delay τ_d. Partial fringes result when $c\tau_s$ and $c\tau_d$ are within a coherence length of each other. Target velocity causes the apparent value of τ_s to change due to the slight Doppler scaling of the spectrum reflected from the target, producing a fringe shift.

Double interferometer arrangements using short coherence length light have been used previously in communication[7] and to multiplex serial fiber optic sensors[8], and to measure target motion where the target is internal to one of the interferometers[9-11]. However, we believe we are the first to apply this concept to remote targets external to an interferometer. This allows the fringe phase to be independent of target distance, surface roughness, and the scattering and dispersive properties of the interposed medium. A single interferometer technique by Geindre et al.[12] measures high velocity (100 km/s) plasma by illuminating the plasma with two subpicosecond pulses separated by ~1 ps, then dispersing the reflected light with a grating. Their Doppler shift is resolvable by a grating rather than a second interferometer because it is so large.

Figure 2 shows the fringing behavior calculated[6] for a stationary target for narrow and wide system bandwidths. The white light velocimeter operates where $c\tau_s \approx c\tau_d$. The widths of the fringe bands is inversely related to the system bandwidth. Thus these widths can be increased by insertion of a bandpass color filter.

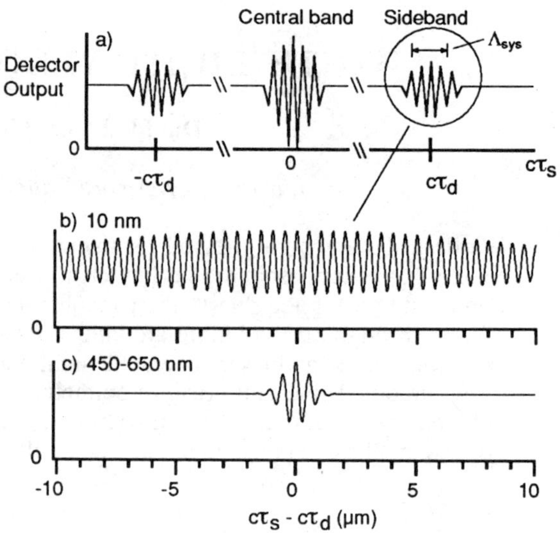

Figure 2. White light velocimeter output versus source interferometer delay length ($c\tau_s$), for a stationary target. Both source and detecting interferometers are the Michelson type. A moving target Doppler shifts apparent $c\tau_s$ by $\Delta(c\tau_s)=-(2v/c)$ ($c\tau_s$). a) Schematic of global behavior: sidebands have the same shape and width as the central band, but lower amplitude. Their width, the system coherence length (Λ_{sys}), is inversely proportional to the bandwidth of the system. b) Calculated detail for 495-505 nm bandwidth. c) For 450 - 650 nm bandwidth.

A moving target Doppler shifts apparent $c\tau_s$ by $\Delta(c\tau_s)=-(2v/c)$ ($c\tau_s$). If λ is the average wavelength, then the fluctuating part of the intensity varies locally approximately as

$$\Delta I \propto \cos\left[\left(\frac{2\pi}{\lambda}\right)(c\tau)\left(\frac{2v}{c}\right)+\phi_0\right] \qquad (2)$$

where ϕ_0 is some constant. From this we obtain Eq. (1).

The fringe visibility is defined ($I_{max}-I_{min})/(I_{max}+I_{min})$, where I_{max} and I_{min} are local maximum and minimum output intensities. Figure 2 shows the fringe visibility in the sidebands is less than unity. This is not a practical difficulty. Since the two outputs of an interferometer are of opposite phase, they can be subtracted numerically to cancel the incoherent signal portion. Also, using a Fabry-Perot interferometer as the source interferometer can increase fringe visibility arbitrarily at the expense of absolute signal strength[6].

Since a conventional VISAR uses monochromatic illumination, a discontinuous velocity jump will cause an ambiguity in the fringe shift to an integer. This because fringes are periodic. However in the white light velocimeter, by recording fringe shifts for different colors separately, the velocity can be uniquely determined. One implementation of this is to disperse the velocimeter output by a grating and record the spectrum versus time by a multi-channel detector or streak camera. Figure 3 shows a calculated streak pattern for a hypothetical velocity history. Because of Eq. (2), the output intensity varies sinusoidally versus $1/\lambda$ as well as with velocity. The number of fringes across the streak record in the wavelength direction scales with the velocity, and therefore the velocity skip across the shock is determinate.

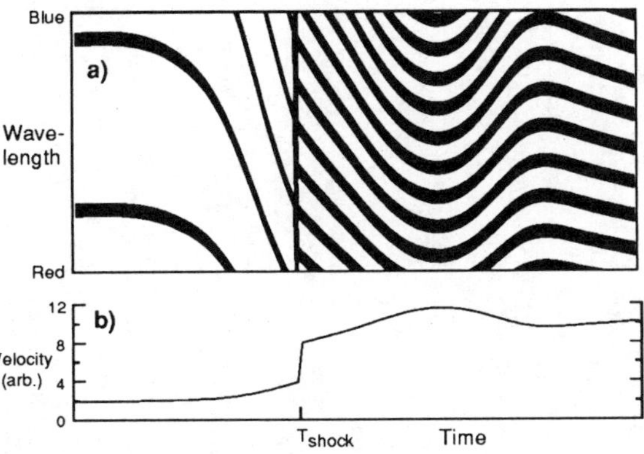

Figure 3. Multi-color fringes can be used to determine velocity absolutely. a) Output intensity map versus wavelength and time, from Eq. (2) using hypothetical velocity history b). Dark bands represent light minima (fringes). This simulates dispersing the light from the target with a grating and recording it by a streak camera or multi-channel detector. The number of fringes along the wavelength dimension scales with velocity, allowing unambiguous determination of the velocity after discontinuous shock at T_{shock}.

This type of streak record may remind readers of the use of a Fabry-Perot interferometer in conjunction with a streak camera. In that system, because the illumination is quasi-monochromatic, the number of fringes across the record in the wavelength direction is constant, so there is a fringe ambiguity across the shock.

CONCLUSION

The ability to use broadband and incoherent light sources is a great practical advantage of the white light velocity interferometer. These sources can be inexpensive, compact and powerful. Line and areal velocimetry should now be practical.

ACKNOWLEDGMENT

This research was performed under the auspices of the U.S. Department of Energy by the Lawrence Livermore National Laboratory under contract W-7405-Eng-48.

REFERENCES

1. Barker, L.M., & Hollenbach, R.E., J. Appl. Phys. 43, 4669-4675 (1972).
2. Barker, L.M., & Schuler, K.W., J. Appl. Phys. 45, 3692-3696 (1974).
3. Hemsing, W.F., Rev. Sci. Instrum. 50, 73-78 (1979).
4. Sweatt, W., Rev. Sci. Instrum. 63, 2945-2949 (1992).
5. Hemsing, W.F., in *Shock Compression of Condensed Matter-1991*, edited by S.C. Schmidt et al., (North-Holland, Amsterdam, 1992), p. 767-770.
6. Our report to be published by Nature.
7. Delisle, C., & Cielo, P., Can. J. Phys., 53, 1047-1053 (1975).
8. Brooks, J., Wentworth, H., Youngquist, M., Kim, B.Y., Shaw, H., J. Lightwave Tech., LT-3, 1062-1071 (1985).
9. Gusmeroli, V., & Martinelli, M., Opt. Lett. 16, 1358-1359 (1991).
10. Ribeiro, A., Santos, J., Jackson, D., Rev. Sci. Instr. 63, 3586-3589 (1992).
11. Harvey, D., McBride, R., Barton, J., & Jones, J., Meas. Sci. Technol. 3, 1077-1083 (1992).
12. Geindre, J.; Audebert, P., Rousee, A.; Falliès, F., Gauthier, J.; Mysyrowiczk, A.; Dos Santos, A.; Hamoniaux, G.; & Antonetti, A., Opt. Lett. 19, 1997-1999 (1994).

INCREASE IN VELOCIMETER DEPTH OF FOCUS THROUGH ASTIGMATISM

David J. Erskine

Lawrence Livermore National Laboratory, Livermore, CA 94551

Frequently, velocimeter targets are illuminated by a laser beam passing through a hole in a mirror. This mirror is responsible for diverting returning light from a target lens to a velocity interferometer system for any reflector (VISAR). This mirror is often a significant distance from the target lens. Consequently, at certain target focus positions the returning light is strongly vignetted by the hole, causing a loss of signal. We find that we can prevent loss of signal and greatly increase the useful depth of focus by attaching a cylindrical lens to the target lens.

INTRODUCTION

The motion of targets impacted by projectiles is frequently measured by a velocity interferometer system for any reflector (VISAR)[1-4]. The targets are located in a tank to contain debris and are optically interrogated remotely, keeping expensive optics outside the tank. Quite often the target is illuminated by a laser passing through a hole in a mirror, with the reflected light from the target returning nearly along the same path. The light not passing through the hole is diverted to the interferometer where the velocity is determined from the Doppler shift of light. Reference [4] gives an excellent review of several VISAR designs and relationships of important design parameters. Some of these relationships are derived by considering the vignetting of the beam by the diameters of optical components. However, the reference does not discuss the vignetting that can occur from the hole in the mirror. That is the subject of this report.

METHOD

Figure 1 shows our arrangement of optics coupling light to and from the target. A f/1.8 50 mm focal length camera lens (L_1) focuses the laser illumination and semi-collimates the reflected light. Mirror M_2 separates returning light from the incoming laser beam by a small hole which allows the laser beam to pass. Because of the significant distance between the L_1 and M_2, for certain focus positions the returning light is imaged into the

hole, eliminating or greatly reducing the signal reaching the interferometer (Fig. 2). We call this focus configuration *dead center*.

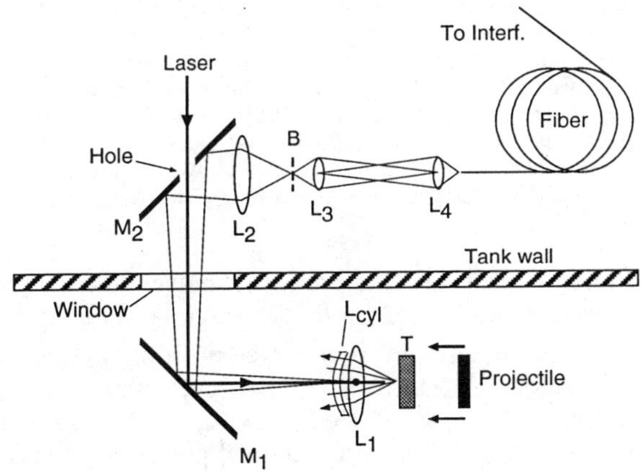

Figure 1. Target interface optics. Target (T) contained in tank is impacted by projectile from a gas gun[5]. Target is illuminated by an argon ion laser and the reflected light returned to the interferometer via optical fiber, 600 μm core diameter. A 3 mm hole in mirror M_2 separates laser and reflected light beams. L_1, L_2: f/1.8 50 mm focal length camera lenses; M_1 mirror; L_3, L_4: 10x microscope objectives. Telescope formed by L_2 and L_3 images aperture of L_1 to aperture of L_4 through intermediate image B. L_4 images aperture of L_3 onto fiber diameter. L_3-L_4 separation 22 cm. L_1-M_2 separation 110 cm. M_2-L_2 separation 8 cm. Cylindrical lens L_{cyl} (focal length -66 cm) is glued to front of L_1 to ameliorate vignetting by hole in mirror.

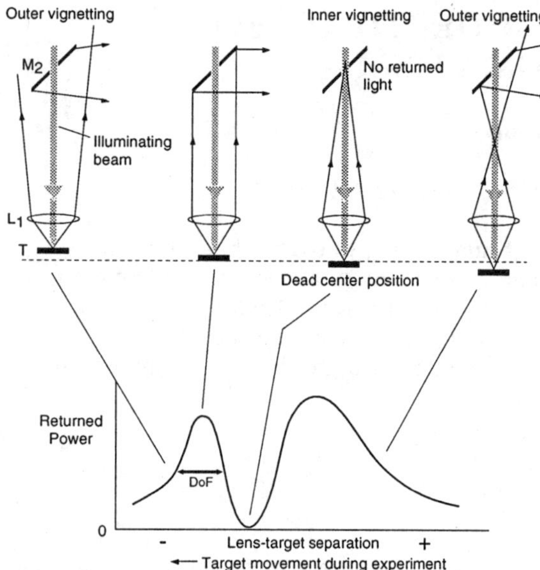

Figure 2. Vignetting during target travel, without cylindrical lens. Mirror (M$_2$) with hole collects light reflected from target (T), illuminated by laser beam passing through hole. Plot of returned power vs. target lens focus is suggestive only. As target moves toward lens (L$_1$), there is a position (dead center) where returned light is imaged into the hole and little is reflected by the mirror. To avoid this during the impact experiment, the initial separation must be set to the inside of dead center, reducing the useful depth of focus (DoF). The range reduction is greater than a factor of two because the returned power falls off more slowly on the outside of dead center.

Since impact by the projectile moves the target toward L$_1$, to avoid passing through dead center in the experiment the initial target position is set to the inside of dead center. However, this greatly reduces the depth of focus (DoF), defined as the range of travel where the returned power is at least 50% of maximum.

We discovered that attaching a simple cylindrical lens to the front of the target lens eliminates the loss of signal at dead center. Secondly, judicious choice of cylindrical focal length can produce a roughly uniform returned light power relationship with target focus. The combination of these two greatly extends the depth of focus. The reason is illustrated in Fig. 3, which diagrams the cross-section of the beam where it intersects M$_2$, when a cylindrical lens (L$_{cyl}$) is used. Only light in an annulus outside the hole and inside some effective vignetting diameter will pass on to the interferometer. Without L$_{cyl}$, the beam diameter at dead center is smaller than the hole, causing complete loss of signal. With L$_{cyl}$, the beam cross-

section is generally elliptical, except for the dead center position where it is circular with a diameter exceeding the hole. Since the average diameter never falls below the hole diameter, the signal is not completely lost at dead center.

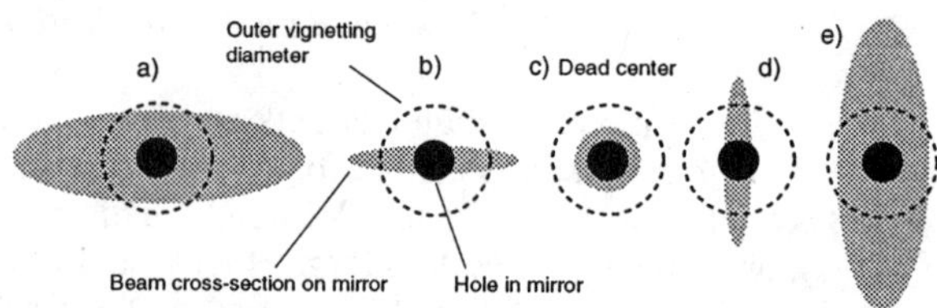

Figure 3. Returning beam cross-section on mirror M$_2$ when a cylindrical lens L$_{cyl}$ is used, for several target lens focus positions. Dark circle is hole in mirror. Dashed circle is effective aperture beyond which diameter vignetting occurs. Thus, only the portion of returned light falling on the annulus between these two circles will enter interferometer. Without L$_{cyl}$ (this case not shown), the beam cross-section at dead center is smaller than hole. With L$_{cyl}$, the average cross-section diameter is never less than the hole diameter, preventing loss of signal at dead center. Judicious choice of cylinder focal length can produce an intersection between the annulus and the beam cross-section which is approximately independent of target focus, creating uniform returned power.

Figure 4 is a measurement of the returned power versus target lens focus, achieved by twisting the camera lens (L$_1$) focusing ring. The target was semi-polished stainless steel, which was the witness plate for an equation of state experiment to be performed. Without L$_{cyl}$, the light drops to zero at one position. After gluing the cylindrical lens to the front of the camera lens we repeated the measurement. No drop in power was observed at the previous dead center position. Secondly, for a -66 cm cylindrical focal length found empirically, the power was roughly uniform for the entire range of focus accessible by twisting the focusing ring. Apparently, the cross-section of the beam overlapping with the accepting annulus of M$_2$ was roughly constant. Such a uniformity had never been achieved with our target optics without L$_{cyl}$.

In VISAR experiments the velocity is determined by counting fringe shifts from an interferometer output. If there is a break in the data, these shifts become ambiguous to an integer number of fringes. To avoid such a break, the target position must start inside the dead center position, since it will be pushed toward the lens by the impact. In Fig. 4 this would correspond to a position $\approx$2 mm. Since the power drops by to 50% of local maximum at the 0 mm mark, the depth of focus would be 2 mm. With L$_{cyl}$, the data of Fig. 4

indicate the depth of focus is beyond 6 mm, and quite likely as great as 10 to 12 mm. The average power is lower in the L_{cyl} case, but only by a factor of two. The lack of fluctuation in the power is more important for good recording than its absolute value.

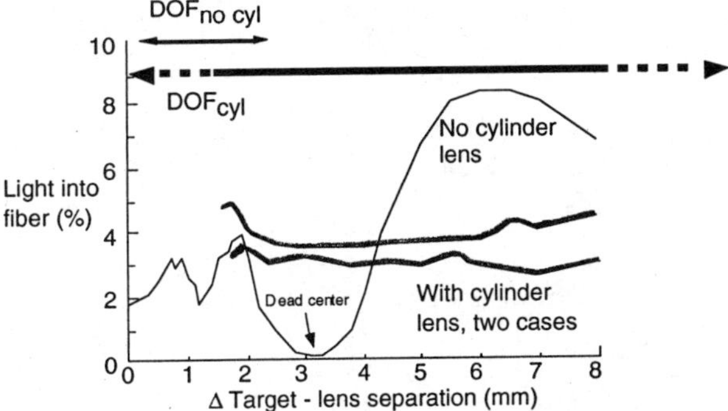

Figure 4. Measured returned light power versus target lens focus position, for the case of no cylinder lens L_{cyl} (thin curve), and two cases with L_{cyl} (bold curves). The horizontal axis is the increase in camera lens (L_1) distance from target (by twisting its focusing ring). Power out of interferometer fiber was divided by power entering tank window. Target was semi-polished stainless steel. When L_{cyl} was glued to camera lens front, it restricted focusing ring movement to >1.7 mm. Fluctuations in signal for <2 mm are caused by growing image of surface scratches as lens approaches ∞:1 conjugate ratio. Double arrowed bars indicate practical depth of focus ranges (DoF) for the cylinder and non-cylinder cases. Dashed portions are estimated. $DoF_{no\ cyl}$ must be on inside of dead center to avoid loss of signal as target moves toward lens after impact.

We note that the holed mirrors discussed in Ref. [4] are positioned much closer to the target lens than in our configuration. This reduces the focus dependence of hole vignetting for a diffusively scattering target (Fig. 5a). In our arrangement, where the holed mirror is outside of a tank, a system of relay lenses could be used to image the target lens closer to the holed mirror.

However, we prefer to use specularly reflective targets to increase the returned light power. Consequently, a short distance between the holed mirror and target lens is a disadvantage because for specular targets normal to the beam, the beam returns along the same path and is strongly vignetted by the hole (Fig. 5b). Insertion of a cylindrical lens would not significantly help in this case. Thus, we prefer to use a larger target-mirror separation and the use of a cylindrical lens to lessen hole vignetting.

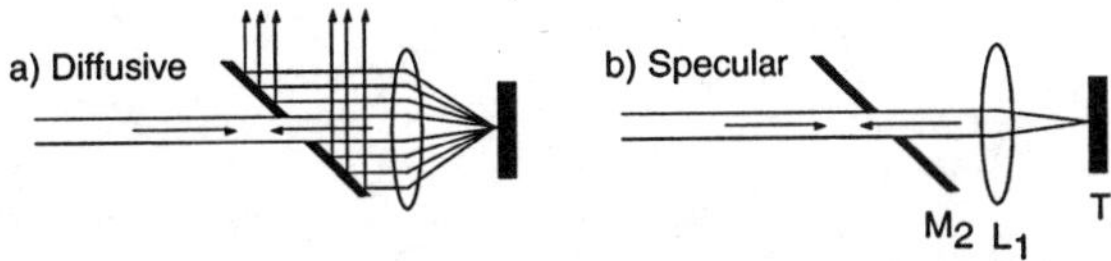

Figure 5. Hole vignetting when mirror (M_2) is close to lens (L_1) of target (T). a) For a diffusively scattering target the vignetting is not substantial and is roughly independent of target position. b) For a specularly reflective target oriented normal to the illuminating beam the vignetting is complete when the target is at the focal point of the lens. An astigmatic lens would not significantly reduce hole vignetting when the mirror is close to the lens.

ACKNOWLEDGMENT

This research was performed under the auspices of the U.S. Department of Energy by the Lawrence Livermore National Laboratory under contract W-7405-Eng-48.

REFERENCES

1. L. M. Barker and R. E. Hollenbach, J. Appl. Phys. **43**, 4669 (1972).
2. W. F. Hemsing, Rev. Sci. Instrum. **50**, 73 (1979).
3. D. J. Erskine, submitted to Rev. Sci. Instrum. 4/95
4. W. Seatt, Rev. Sci. Instrum. **63**, 2945 (1992).
5. A.H. Jones, W.M. Isbell, and C.J. Maiden, J. Appl. Phys. **37**, 3493 (1966).

A DESENSITIZED DISPLACEMENT INTERFEROMETER APPLIED TO FAST MOVING PARTICLES IN A CONTINUUM

H.D. Espinosa[†], M.Mello[‡], and Y.Xu[†]

† *School of Aeronautics and Astronautics, Purdue University, West Lafayette, IN 47907*
‡ *Division of Engineering, Brown University, Providence, RI 02912*

The present paper introduces a variable sensitivity displacement interferometer (VSDI) used to monitor both normal and in-plane particle displacements in wave propagation experiments. The general system consists of two interferometers working in tandem. Normally reflected light is interfered with each of two symmetrically diffracted light beams generated by the specimen rear surface. In cases where the surface motion simultaneously exhibits both in-plane and normal displacements, the fringes represent a linear combination of the longitudinal and transverse components of motion. Decoupling of the normal and in-plane displacement histories can be achieved through a linear combination of the two VSDI records. The sensitivity of the interferometer is shown to depend on the angle θ or equivalently, the frequency σ of a grating manufactured at the observation point and the order n of the diffracted beams. The variable sensitivity feature of the VSDI greatly desensitizes normal displacement measurements and is particularly well suited for wave propagation studies in which normal particle velocities in excess of 100 m/s are generated. Experimental results are presented which demonstrate the application of this technique to monitoring particle motion histories in plate impact recovery experiments.

INTRODUCTION

In wave propagation experiments two interferometers are commonly used in the measurement of normal particle velocities. In the case of low velocity impact experiments, a normal displacement interferometer (NDI), having a fixed sensitivity of $\lambda/2\,[mm/fringe]$ where λ represents the laser light wavelength, is frequently used because this interferometer can resolve slow changes in velocity with very high accuracy and hence provide very valuable information for damage identification. Nonetheless, the extreme sensitivity of this interferometer severely limits its application in wave propagation experiments due to the inordinately high signal frequencies which may be generated. It is reasonable to regard a particle velocity of 0.1 $mm/\mu s$ as an upper bound for the NDI in wave propagation experiments. In order to overcome this limitation, Barker and Hollenbach[1] developed a normal velocity interferometer (NVI). The NVI has a variable sensitivity given by $\lambda/(2\tau(1+\delta))\,[mm/\mu s - /fringe]$ whereby τ represents a time delay between the interfering light beams intro-

duced by an air delay leg or etalon in the interferometer. However, during an initial time period τ, an NVI is functioning as an NDI since the delayed light arriving at the detector from the delay leg or etalon is reflected from a stationary target. As a result, in the low velocity regime (0.1-0.3 $mm/\mu s$), values of τ in the neighborhood of 10 nsec or more are required in order to obtain records with sufficient sensitivity to changes in particle velocity. This in turn leads to a greater averaging of the velocity measurements. Furthermore, elastic precursors causing velocity jumps of more than 0.1 $mm/\mu s$ in a time less than τ cannot be detected because the early time NDI signal frequency may exceed the frequency response of the light detection system. The variable sensitivity displacement interferometer (VSDI) presented here has been designed to overcome the inherent frequency limitations of the NDI and NVI within the noted velocity gap of interest. The sensitivity of such an interferometer is fully variable; thus, it can be adjusted to operate over a wide range of particle velocities without exceeding the frequency response of the light detection system.

"

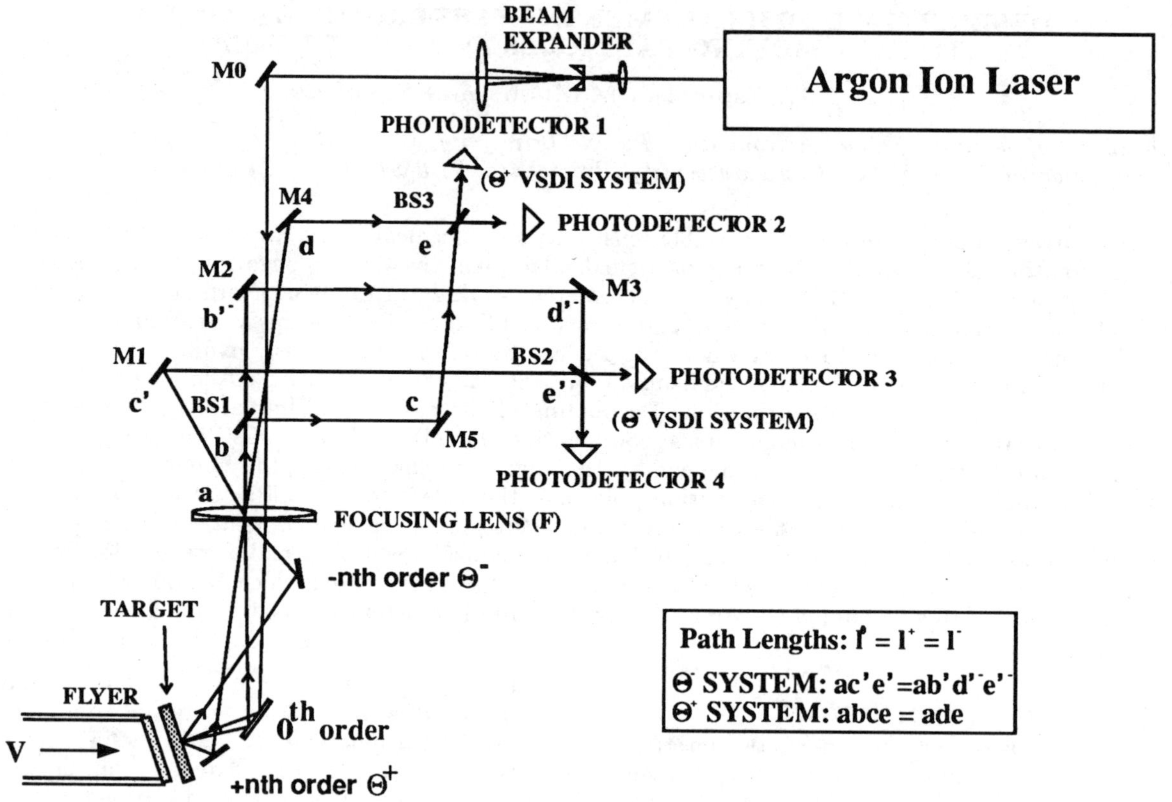

FIGURE 1. Schematic of VSDI system.

VSDI SYSTEM

To examine the governing equation for a variable sensitivity displacement interferometer, consider the effect of interfering the normally reflected beam with a beam diffracted at an angle $\theta^\pm$ with respect to the specimen normal as shown in Figure 1. The normally reflected beam is split at beamsplitter BS1. Each half of the normal beam is then made to interfere with one of the diffracted beams via beamsplitters BS2 and BS3. The resulting signals generated by each interfering beam pair are monitored by photodetectors. The combined field for either pair of interfering plane waves leads to a classical interference expression of the form

$$I(t) \propto I^0 + I^\pm + 2\sqrt{(I^0 I^\pm)}$$
$$cos(\beta)cos\Psi^\pm(t). \qquad (1)$$

Here, I^0 and $I^\pm$ represent the time averaged intensities of the respective contributing light fields and β is the angle between their respective polarization vectors. Normal and transverse particle motion introduce frequency modulation through the time varying phase term, namely,

$$\Psi^\pm(t) = \frac{2\pi}{\lambda}[2U(t - \frac{l^0}{c}) - U(t - \frac{l^\pm}{c})$$
$$(1 + cos\theta^\pm) \mp V(t - \frac{l^\pm}{c})\sin\theta^\pm \qquad (2)$$
$$+ (l^0 - l^\pm)] + \phi^0 - \phi^+,$$

where λ is the wavelength of the laser source, $U(t)$ and $V(t)$ represent the normal and in-plane displacements of the point of observation from its position at time t=0, ϕ^+, ϕ^- represent constant arbitrary phase terms, and l^+, l^-, and l^0 represent the fixed initial path lengths traversed from target to detector by the $\theta^{\pm}$ diffracted beams and the normally reflected beam respectively. Observe that the transverse motion phase term is subtracted for the case where the θ^+ beam is employed and otherwise added when interfering with the θ^- beam. Next, setting $l^0=l^+=l^-$ leads to a more simplified and useful form, i.e.,

$$\Psi^{\pm}(t) = \frac{2\pi}{\lambda}[U(t-\frac{l}{c})(1-cos\theta^{\pm})\mp$$
$$V(t-\frac{l}{c})\sin\theta^{\pm}]+\phi^0-\phi^+. \tag{3}$$

When the material point exhibits both normal and in-plane components of displacement, frequency modulation results through a linear combination of both motion components as dictated by Eq. (3). A full fringe shift associated with simultaneous displacements δV and δU corresponds to

$$\Delta\Psi(t)^{\pm} = \frac{2\pi}{\lambda}[\delta U(t-\frac{l}{c})](1-cos\theta^{\pm})$$
$$\mp \delta V(t-\frac{l}{c})\sin\theta^{\pm}] = 2\pi. \tag{4}$$

Rearranging Eq. (4) and solving for δU yields a generalized normal displacement fringe sensitivity expression given by

$$[\frac{normal\ displacement}{fringe}] =$$
$$\frac{\lambda}{1-cos\theta^{\pm}} \pm \frac{\delta V \sin\theta^{\pm}}{1-cos\theta^{\pm}}. \tag{5}$$

In the case where the θ^+ beam is interfered (Θ^+ VSDI system), the combined presence of normal and transverse motion serves to increase the value of the fringe constant, i.e., to further desensitize the normal displacement measurement. The opposite situation arises when the θ^- beam is employed (Θ^- VSDI system). In this case, the transverse motion term is subtracted which serves to decrease the value of the normal displacement fringe constant, i.e., to increase the sensitivity of the normal displacement measurement. Thus each VSDI system will generate a different signal frequency when used to monitor the same given combined state of motion. This effect is confirmed in pressure shear experiments

at the time corresponding to the arrival of the shear wave at the target rear surface. Up until this point in time, each VSDI system functions as a DNDI having a fringe sensitivity given by the first term of Eq. (5). With the delayed arrival of the shear wave, the signal now represents a linear combination of the two motion components. Consequently, the fringe sensitivity is altered as indicated by Eq. (5) and the signal frequency of each VSDI system exhibits a sudden increase or decrease depending upon whether the θ^+ or θ^- beam is being interfered.

VSDI APPLIED TO PRESSURE-SHEAR RECOVERY EXPERIMENTS

The pressure-shear soft-recovery experiment (Espinosa et al.[2]) is particularly suitable to investigate the applicability of the variable sensitivity displacement interferometer (VSDI). The configuration is shown in Fig. 2. In the present study the multi-plate impactor and target plates were made of speed-star steel. The specimen consisted of a glass fiber-reinforced polymer. Complete details of the experimental procedure are described in Espinosa et al.[2]

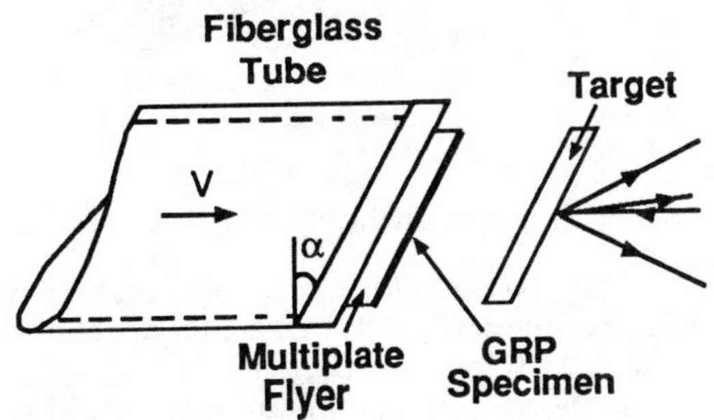

FIGURE 2. Schematic of high strain rate pressure-shear configuration.

A combined Θ^- VSDI-TDI system was utilized to monitor the normal and transverse velocities at the target free surface. A grating with 1000 lines/mm was manufactured in the target plate back surface. The zero order reflected beam was combined with the negative first order diffracted beam to produce a Θ^- VSDI. A TDI interferometer (Kim at al.[3]) was obtained by mixing the two first order diffracted beams.

The normal velocity-time profiles obtained from the experiment and the amplitude corrected signal are plotted in Fig. 3. Upon arrival of the normal wave, fringes are produce in the Θ^- VSDI channel. A significant frequency increase is observed after arrival of the shear wave in accordance with Eq. (4). A jump in velocity to a level of approximately 50 m/sec is followed by a reduction and increase in velocity due to wave reverberations in the thin polymer layer used in the multi-plate flyer. An approximately constant velocity of 95 m/sec is monitored in the next 1.8 μsec which is in agreement with the elastic prediction.

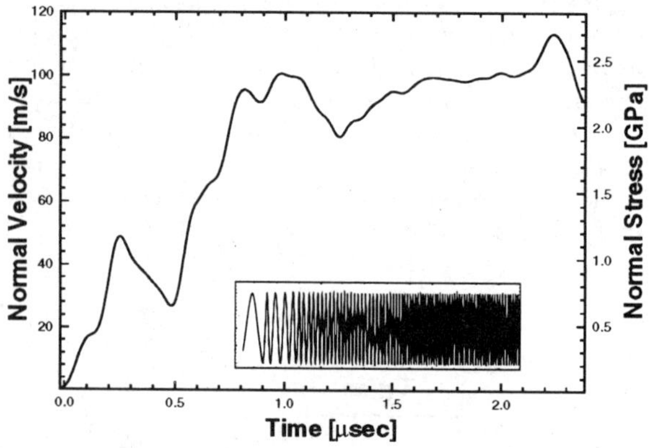

FIGURE 3. Normal velocity-time profile.

The transverse velocity history together with the amplitude corrected TDI signal are shown in Fig. 4. The transverse particle velocity progressively increases to a maximum value of 15 m/sec corresponding to a shear stress of 200 MPa.

CONCLUDING REMARKS

In principle, the VSDI interferometer presented here can be used in wave propagation experiments conducted on metallic and nonmetallic materials in a variety of impact configurations including those where an optical window is employed. It is also possible to extend the technique to the case where scattered light from a preferentially roughened surface is collected by 3 fiber optic probes. In this case the light is transported to the interferometer through optical fibers thereby increasing the versatility of the

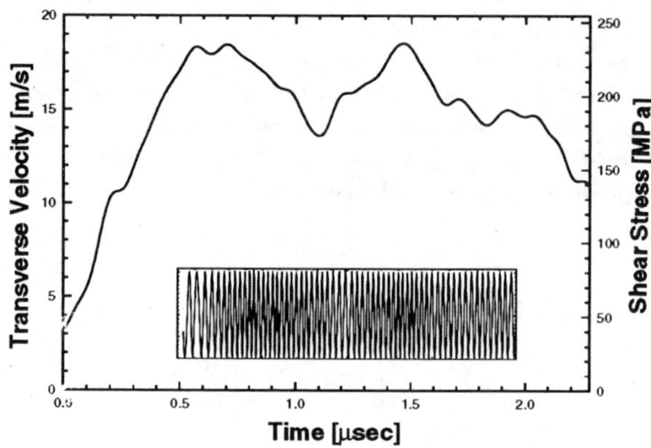

FIGURE 4. Transverse velocity-time profile.

system. Hence, it appears that the VSDI exhibits promise as a new tool for monitoring particle displacements in wave propagation experiments, particularly within the previously noted problematic velocity gap spanning 0.1 - 0.3 *mm/μs*.

ACKNOWLEDGMENTS

This research was supported by the Army Research Office through grants No. DAAH04-95-1-0385 and DAAH04-95-1-0168, and by the National Science Foundation through Grants No. MSS-9309006 and MSS-9311289.

REFERENCES

1 Barker, L.M. and Hollenbach, R.E., "Interferometer Technique for Measuring the Dynamic Mechanical Properties of Materials", *Rev. Sci. Instru.* 36, p.1617 (1965).
2 Espinosa, H.D., Emore G., and Xu Y., "High Strain Rate Behavior of Composites with Continuous Fibers", to appear in: J. Vinson editor, *High Strain Rate Effects on Polymer Metal and Ceramic Matrix Composites*, ASME Winter Annual Meeting, San Francisco (1995).
3 Kim, K.S., Clifton, R.J., and Kumar, P., "A Combined Normal and Transverse Displacement Interferometer with an Application to Impact of Y-Cut Quartz", *J. Appl. Phys.*, 48(10), pp. 4132-4139 (1977).
4 Espinosa, H.D., Mello, M, and Xu, Y, "A Variable Sensitivity Displacement Interferometer with Application to Wave Propagation Experiments", Submitted to J. App. Mech. (1995).

HYDRODYNAMIC PROTON BEAM-TARGET INTERACTION EXPERIMENTS USING AN IMPROVED LINE-IMAGING VELOCIMETER.

K. Baumung, J. Singer, S.V. Razorenov[+], and A. V. Utkin[+]

Forschungszentrum Karlsruhe, P.O. Box 3640, D-76021 Karlsruhe, Germany
[+]Russian Academy of Sciences, Institute of Chemical Physics (Chernogolovka),
Chernogolovka, 142432 Russia

The hydrodynamic response of 10- to 100-μm-thick target foils exposed to a high-power proton beam is investigated by laser Doppler velocimetry of the rear surface. For the first time, the ablative acceleration and the ablation pressure have been measured spatially resolved by use of an improved line imaging velocimeter. The instrument was operated in both the VISAR and the ORVIS mode, and takes advantage of the spatial localization of the interference fringes in the mirror planes for an intermediate imaging. This allows us to obtain optimum contrast, high spatial resolution of ≤ 10 μm, and to vary the magnification in a wide range.

INTRODUCTION

High-power light ion beams generated by pulsed power accelerators offer an interesting potential as drivers for inertial confinement fusion (ICF) because of their low cost, good efficiency, and effective energy deposition in the target. At the Karlsruhe Light Ion Facility KALIF, a 1.7 MV, 600 kA, 50 ns fwhm generator, experiments on physics of ion beam generation and focusing, and on beam-target interaction phenomena are under way.

Actually, at peak power densities of 1 TW/cm^2, up to 40 kJ of proton beam energy can be delivered to a 6- to 10-mm-diameter (fwhm) focal spot.[1] With proton ranges in condensed matter of 10 to 20 μm, energy densities of several MJ/g are realized leading to peak pressures in the 10- to 100-GPa-range, and to temperatures of 10-20 eV. The EOS of the cold, dense plasma that forms by rapid ablation of the energy deposition zone is difficult to describe theoretically, and relies on scarce experimental data only. Actually, one principal goal of the target experiments performed on KALIF is to provide empirical EOS-data for the ablation plasma. As a first step, hydrodynamic experiments

employing laser Doppler velocimetry have been performed to measure the ablation pressure and the ablative acceleration of 20- to 100-μm-thick aluminum foils.[2] This method turned out to be a useful tool for the diagnostics of beam parameters, too.[3] Finally, KALIF has been used as a generator of short, intense pressure pulses for dynamic tensile strength measurements at strain rates up to $>10^7$ s^{-1} and nanosecond load durations.[4]

For all these application, and, moreover, for the ICF-relevant issue of the development of Rayleigh-Taylor instabilities in thin ablatively accelerated layers of condensed matter, it seemed desirable to extend the velocimeter to one-dimensional spatial resolution. In the following, hydrodynamic beam target interaction experiments are reported in which, for the first time, an improved line-imaging instrument was applied that was operated in both the VISAR [5] and ORVIS [6] mode.

EXPERIMENTAL SETUP

At KALIF, the target is placed in the focal plane of the ion source inside a vacuum vessel ~100 mm away from an optical window through which the

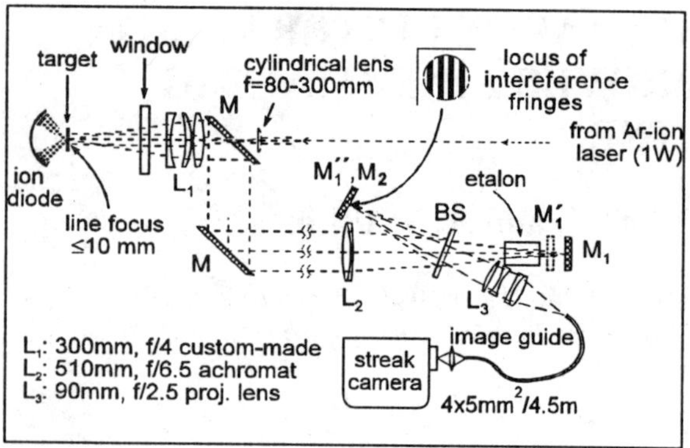

FIGURE 1. Schematic of the improved line imaging VISAR/ORVIS velocimeter with intermediate imaging onto the mirror planes M_2 and M_1''.

rear surface, opposite to the ion beam, can be observed. The line-imaging VISAR[7] described by Hemsing uses a slow collecting lens to directly image the target onto the slit of the streak camera. The big *f*-number provides a large depth of field but collects a very small part of the reflected light only. To supply sufficient intensity to the camera, an expensive amplifier is required to boost the single-frequency light of an argon ion laser to ~400 W. In addition, changing the imaging scale in practice strongly affects the effective *f*-number.

The conditions at KALIF allow us to use a 300 mm focal length *f*/4 collecting lens (L_1 in Fig. 1). This provides reasonable intensities but leads to a significant reduction of the depth of field and, as a result, to low contrast of the interference fringes if direct imaging to the camera slit is applied. Taking into account that the interference fringes in a Michelson interferometer are localized[8] in the virtual air gap between the mirror planes M_2 and M_1'' we make an intermediate image imaging into this plane. Looking from the interferometer output, M_1'' is the image, seen in the beam splitter BS, of mirror M_1 that is shifted by the etalon to the apparent position M_1'. For a maximum field of view of ~8 mm, the intermediate image is realized with lenses L_1 and L_2. For fields of view of 1-2 mm, L_1 alone is used. To provide maximum resolution, L_1 is a custom-made triplet optimized for imaging across the 10-mm-thick window of the vacuum vessel. A fast projection lens L_3 is used to project the intermediate image on the input of an 4.5-m-long image guide. For optimum overall resolution, i.e. taking into account the limited performance of the streak camera, high magnification is needed. In this case, a telescope instead of L_3 is used. In this configuration, a spatial resolution of <10 μm with a total field of view of ~200 μm has been realized. Disregarding a 50%-reduction of the intensity, the image guide allows us to position the interferometer just in front of the KALIF vacuum vessel and thus to minimize vignetting, and to place the streak camera behind a concrete shielding wall. In addition, focusing of the system and adjustment of the interferometer is greatly facilitated by connecting a reticle to the interferometer instead of the image guide.

A cylindrical lens is used to expand the laser beam vertically which then is focused by L_1 to illuminate a ≤10-mm-long and ~200-μm-wide line on the target. To fully utilize the laser power of 1W, a beam expander is employed to convert the ~1-mm-diameter Gaussian profile into a "hollow" beam with an annular intensity profile of ~4 mm outer diameter. This gives a rather uniform illumination along the line on the target. For very small fields of view, the 250-μm-diameter focus of the original laser beam is used.

The trigger pulse for the streak camera is deduced from the KALIF pulse line. This involves a jitter of < ±7 ns with regard to the events on target.

EXPERIMENTS AND RESULTS

In previous experiments using a normal ORVIS we occasionally observed an early loss of intensity that could be attributed to a loss of reflectivity of the surface, the development of hydrodynamic instabilities, or a tilt of the target. Therefore, with the first spatially resolving experiments we investigated the ablative acceleration of thin aluminum foils. Figure 2 depicts the interferogram obtained with a 50-μm-thick target in the VISAR mode: the fringes indicate curves of equal velocity, the velocity change between two fringes, the "fringe constant", being 608 m/s. Figure 3 shows the evaluation of the velocity profile for a 50 ns period.

The KALIF voltage pulse looks like a somewhat flattop Gaussian of 50 ns fwhm. The protons are accelerated and focused along a ~10-mm-distance close to the ion source and then propagate ballistically towards the focus. On their 130-mm-trajectory to the target faster protons accelerated later into the pulse can catch up or even

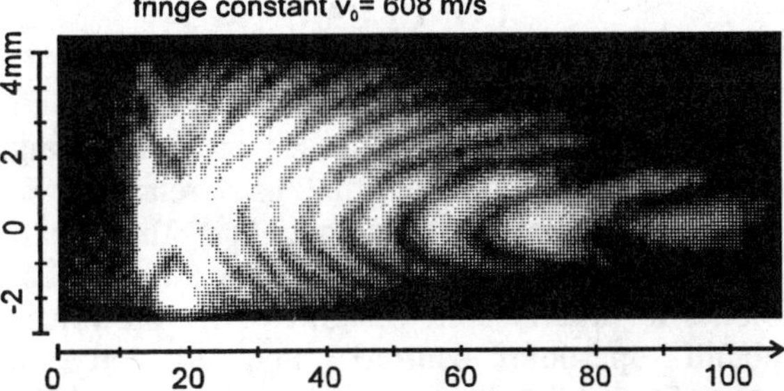

FIGURE 2. Line VISAR interferogram of an ablatively accelerated aluminum foil of 50 μm thickness. The initial jump is ~4.5 fringes.

overtake slower ones started earlier with a lower energy. The resulting time-of-flight compression or "bunching" of the ion beam front causes a step increase of the power density at the beginning of the interaction. This leads to a rapid pressurization of the energy deposition zone. A pressure wave propagates into the adjacent condensed material. When it reaches the rear free surface the decompression causes a fast acceleration. In Fig. 2, there is a jump corresponding to ~4.5 lost fringes in the center. The subsequent pressure evolution in the deposition zone is a result of both further energy input and increase of the compressibility of the material due to heat up and ablation. Therefore, the velocity history just after the initial jump strongly

depends on the actual power trace. In the present example, the velocity reduces for a few nanoseconds until the release wave that starts from the rear free surface, after having been reflected from the deposition zone boundary as a pressure wave, again reaches the rear surface. Then, as the pressure in the ablation zone drops, the acceleration reduces. Since the ablation pressure p scales with the power density I as $p\sim I^{0.7}$, the velocity across the 8 mm fwhm of the focus varies much less than a factor of two. However, it leads to an increasing curvature of the foil such that the reflected light of the regions out of the axis gets progressively vignetted. This explains the early loss of intensity in off axis positions. From Fig. 2 also follows that the position of peak intensity is slightly shifted during the pulse

To determine the ablation pressure we have employed 33-μm-thick aluminum foils glued on a 5-mm-thick window if LiF single crystal. The target and the window having nearly identical Hugoniot EOS, the interface between the two materials (with glue thicknesses <1 μm) is no discontinuity for the pressure wave propagation. This allows us, through the window, to directly measure the particle velocity u_p very close (<10 μm) to the energy deposition zone without perturbation due to wave reverberations from the rear free surface, and without significant distortion of the wave front due to steepening over long travels. Figure 4 shows the VISAR interferogram of the particle velocity. In this experiment the number of lost fringes can be clearly determined to be 7 because the final velocity is zero so that the fringes are closed curves.

The velocity profiles presented here are nearly monotonic and can be evaluated without ambiguity. For more complicated cases simultaneous recording of 2 to 4 interferograms is applied to obtain, from quadrature coding components, information on the phase of the fringes.[7] Another solution for this problem is to record "line ORVIS" interferograms as shown in Fig. 5. Here, the ordinate is a space coordinate, too. For each position, the velocity is determined by evaluating, like for VISAR interferograms, the intensity modulation along a line parallel to the time axis. However, in contrast to the VISAR records, the sign of the velocity change is clear. Of course, the spatial resolution is higher in the VISAR mode, but ORVIS interferograms provide a simple means to measure complex profiles definitely.

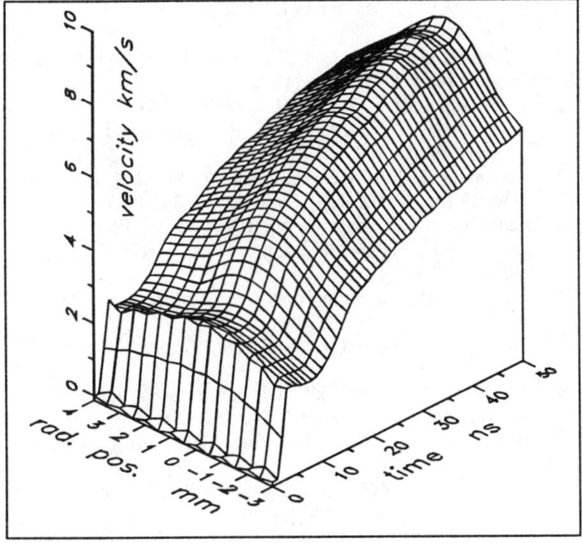

FIGURE 3. Free surface velocity profile vs. time evaluated for a 50 ns period from raw data shown in Fig. 2 .

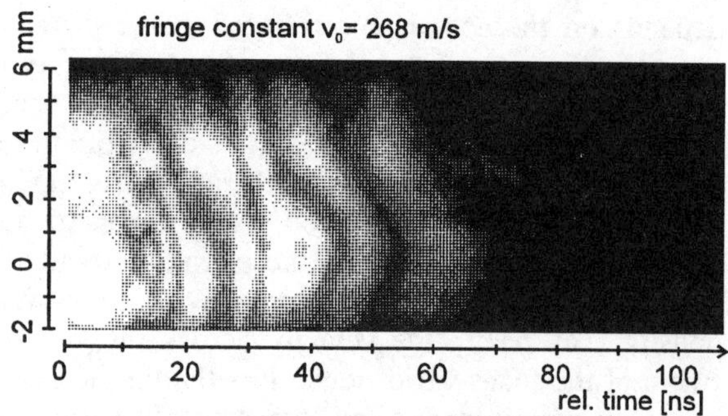

FIGURE 4. Line VISAR interferogram of the particle velocity close (<10 μm) to the ablation zone measured with a 33-μm-Al / 5-mm-LiF composite target).

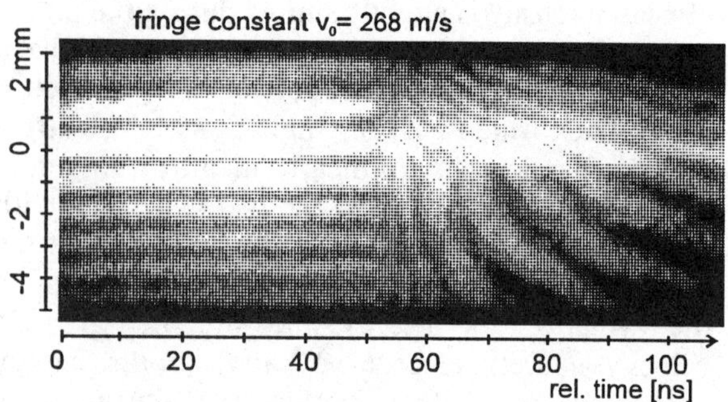

FIGURE 5. Line ORVIS interferogram of the particle velocity close (<10 μm) to the ablation zone measured with a 33-μm-Al / 5-mm-LiF composite target.

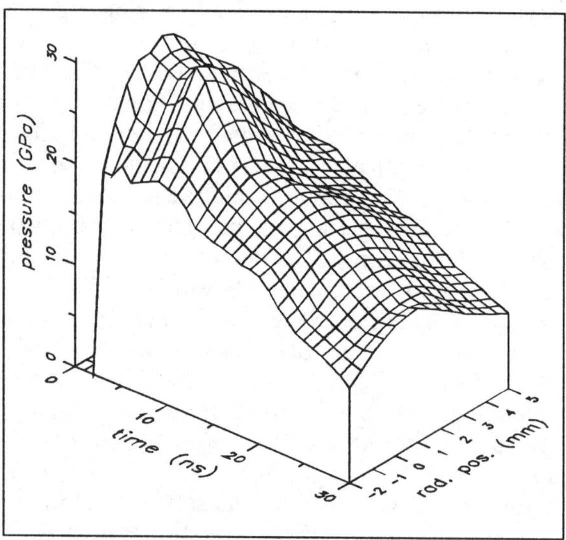

FIGURE 6. Ablation pressure in an aluminum target as a function of radial position and time, evaluated from interferogram Fig. 4.

CONCLUSION

We have presented an improved line-imaging laser Doppler velocimeter based on a wide-angle Michelson interferometer that uses relatively fast optics and takes advantage of the spatial localization of the interference fringes in the mirror planes for intermediate imaging. This allows us to obtain optimum contrast, and to vary the magnification in a wide range. A custom-made lens especially designed for imaging through the window of the vacuum vessel not only provides a high spatial resolution of ~5μm but, due to the comparatively low f-number of 4 allows us to operate the instrument at laser powers < 1W common for usual velocimeters. For the first time, we have measured the ablation pressure and the ablative acceleration of thin foils along a ~8-mm-line across the ion beam focus as a function of time.

ACKNOWLEDGMENT

This work was partly supported by the German-Russian Scientific and Technical Co-Operation Program and by the NATO scientific Programme.

REFERENCES

1. Bluhm H. J., Hoppé P. J. W., Laqua H .P., and Rusch D., *Proc. of the IEEE* **80**, 995 (1992).
2. K. Baumung, H. U. Karow, H. J. Bluhm, P. Hoppé, D. Rusch, G. I. Kanel, A. V. Utkin, S. V. Razorenov, and V. Licht: *Nuovo Cimento* **106 A**, 1771 (1993).
3. Baumung K., Karow H. U., Rusch D., Kanel G. I., Utkin A. V, and Licht, V. , *J. Appl. Phys.* **75**, 7633 (1994).
4. Kanel, G. I. Razorenov S. V., Utkin A. V., Fortov V. E., Baumung K,. Karow H. U, Rusch D., and Licht V., *J. Appl. Phys.* **74**, 7162 (1993).
5. Barker L. M. and Hollenbach R. E., *J. Appl. Phys.* **43**, 4669 (1972).
6. Bloomquist D. D. and Sheffield S. A., *J. Appl. Phys.* **54**, 1717 (1983).
7. Hemsing W. F. et al., *Ultrahigh- and High-Speed Photography, Videography, Photonics, and Velocimetry*, edited by Shaw L. L. et al. , Proc. SPIE **1346** (1990), pp. 133-140.
8. Born. M and Wolf E., *Principles of Optics*, 6th ed. (Pergamon, Oxford, 1980), p.301.

ANALYSIS OF A HIGH INTENSITY X-RAY SOURCE USING A SPECIALIZED DOPPLER INTERFEROMETER SYSTEM

K. J. Fleming

Explosive Projects & Diagnostics, Dept 2554, Sandia National Laboratories, Albuquerque, NM, USA 87123

The Saturn accelerator at Sandia National Laboratories is a high power, variable-spectrum, x-ray source capable of simulating radiation effects of nuclear countermeasures on electronic and material components of space systems. It can also function as a pulsed-power and radiation source, and as a diagnostic test bed for a variety of applications. Obtaining highly accurate measurements of the emission spectra is difficult because the high intensity x-rays and MegaAmpere levels of current inside the experiment chamber can damage or destroy electronic measurement devices. For these reasons, an optical based measurement system has been designed, developed and successfully tested in the Saturn accelerator. The system uses fiber optic coupled sensor(s) connected to a specialized Doppler interferometer system which analyzes the shock wave imparted into a target material. This paper describes the optical system, its related components, and material response data of polymethyl methacrylate.

INTRODUCTION

X-ray emissions produced by the Saturn accelerator have an important role in radiation effects simulation, especially since the future of underground nuclear testing is uncertain. Accurate modeling and testing must rely on precise measurements of the response of materials subjected to the x-ray radiation. For this reason, a "fidelity measurement series" is underway. This series of experiments uses different types of devices that measure material response to intense. short pulse, x-ray radiation. All of these measurement devices are simultaneously exposed to the same radiation source.

This test series uses the Saturn accelerator in the **Plasma Radiation Source** configuration which has an argon line energy of 3.2 keV. This scheme ascertains the most accurate measurements possible since other (different) devices such as quartz and PVDF gauges are simultaneously measuring emission output. The VISAR (Velocity Interferometer System for Any Reflector) is the only system in this series which uses optical methods for analyzing material response to x-ray emissions. An optically transparent material with known shock properties is exposed to the x-ray source. The energy is absorbed by the material causing a shock wave which is analyzed. This paper will describe the method of particle velocity measurement, the sensor design, the complete system configuration, and the data obtained.

BACKGROUND

The VISAR is a unique measurement tool in that its measurement method is based on optical, not electronic phenomena, which gives it immunity from electrical interference. The bandwidth of the system is only limited by the detectors and recording instrumentation. Because VISAR uses laser light for measurement, it is non-intrusive. The VISAR was developed by Barker and Hollenbach to perform shock wave studies on various materials. In its original configuration, the test material was impacted with a projectile and the resulting shock wave was analyzed by the VISAR (*1*). In the Saturn experiment, similar measurement techniques are used but the shock wave energy of the source is replaced by the intense x-ray radiation.

THEORY OF OPERATION

The primary components of VISAR are a single frequency laser, a modified Michelson interferometer cavity, high speed photomultiplier tubes and a recording device such as a digitizing oscilloscope. The laser used for this experiment is an argon-ion with a 1 Watt continuous wave (cw) output at 514 nm wavelength. The material exposed to the x-rays is polymethyl methacrylate (PMMA), specified as *Polycast UVB with a density of 1.184 g/cm³*. A 50-mm diameter, 1 mm thick PMMA disk is cemented to a 50-mm diameter, 12.5 mm thick PMMA disk with a thin (25 μm) aluminum foil sandwiched between the two disks. The foil and disks are bonded together under pressure with a very thin layer of epoxy glue (This assembly is referred to as a **"target"**). A special fiber optic coupled sensor (*2*) is attached to the target assembly such that the laser light is focused through the 12.5 mm thick PMMA disk onto a point on the foil (**Fig. 1**). The diffuse surface of the foil reflects the light which is collected by the sensor and focused into another fiber optic which is connected to the VISAR interferometer cavity.

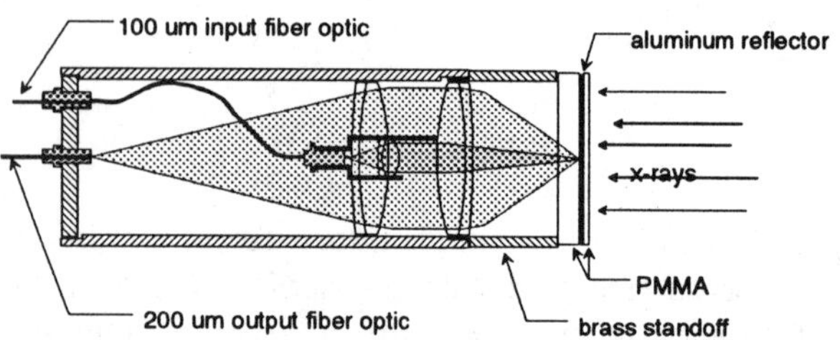

FIGURE 1. Sensor configuration. for x-ray analysis. Source is approximately 100mm from target sample.

The VISAR cavity is a modified, unequal leg Michelson interferometer. The sensitive components in the cavity are cemented together to significantly reduce the need for alignment (**Fig. 2**). The light from the output fiber optic, connected to the sensor, is collimated and sent through the cavity. The beamsplitting coating reflects one beam through the glass "delay bar" and the other beam through air and a 1/8 wave retarder. The beams are recombined to form two sets of interference patterns which have simultaneous intensities 180 degrees out of phase with each other. The interfered light is routed to optical detectors which convert optical signals to electric signals that are then transmitted to the recording device(s). The inverted-phase light is electronically inverted and added to the normal phase light which doubles the signal intensity and cancels out most of the self light generated in the irradiated fiber optics (*3*). A 20 angstrom bandpass filter placed in front of the return fiber optic further filters the laser light to keep the signal-to-noise ratio high.

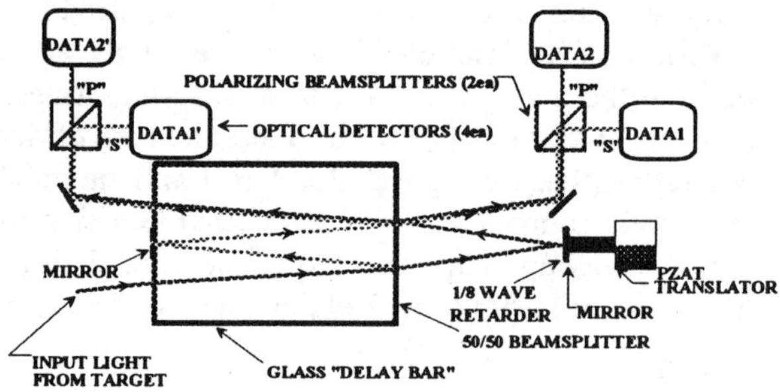

Figure 2. Drawing of "Fixed Cavity" interferometer. "DATA X" are optical detectors. The PZAT Translator simulates the return signal and fine adjusts the interferometer.

When the x-ray radiation exposes the PMMA, it generates a shock wave near the front surface that travels into the sample. This compressive wave propagates through the aluminum foil and into the rear PMMA disk. The resulting instantaneous particle velocity (PV) at the foil/PMMA interface causes a Doppler shift in the light: Since the length of the legs of the interferometer cavity are different, the Doppler shift arrives at the beam recombination point **at different times.** This relationship is defined by:

$$arrival\ time = h(1-1/n) \tag{1}$$

where h is the delay leg length and n is the index of refraction. The distance the light travels in the delay leg is longer than in the reference leg **and** the speed of light is slower in glass than in air. Using the relationship from equation (1), the delay time τ is :

$$\tau = 2h/c = \frac{2h(1-1/n)}{c} \tag{2}$$

where c is the speed of light in a vacuum. Using these relationships, the fringe count $F(t)$ relates to target velocity $u(t-\tau/2)$ as:

$$u(t - t/2) = \frac{\lambda F(t)}{2\tau(1 + \Delta v/v)} \cdot \frac{1}{1+\delta} \qquad (3)$$

in which λ is the wavelength of the laser light, $\Delta v/v$ is an index of refraction correction factor for the window, δ is a correction factor with respect to wavelength for dispersion in the delay bar. Equation (3) is manipulated to obtain the **velocity-per-fringe** (VPF) constant for an interferometer which is:

$$VPF = \frac{\lambda}{2\tau(1 + \Delta \upsilon/v)} \cdot \frac{1}{1+\delta} \qquad (4)$$

EXPERIMENT DESCRIPTION

The VISAR for the Saturn experiments is a dual-leg system which consists of two interferometer cavities with different VPFs. Leg 1 has a VPF of 38 m/s, and Leg 2 has a VPF of 108 m/s. The laser light collected from the sample is split and routed to the two interferometer cavities (**Fig.3**). This dual leg technique gives a broader measurement range for unknown velocities, and can catch very fast shock jumps that may be missed with a single leg system.

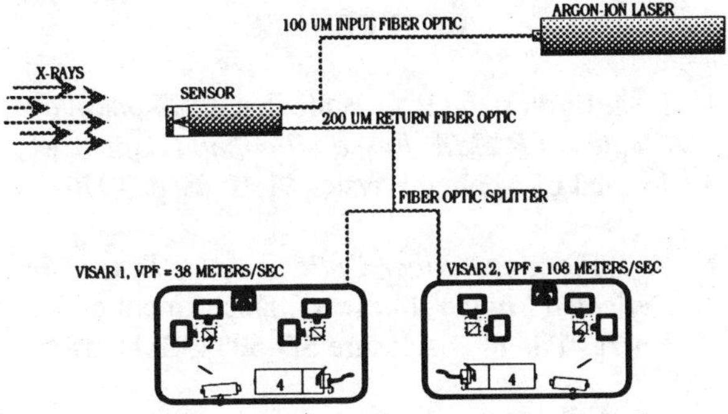

Figure 3. Diagram of Saturn VISAR experiment using the dual interferometer cavity configuration. The sensor assembly is 50mm diameter, 200mm long. The fiber optic length is 100 meters.

The digitizing oscilloscopes have a 2 Gigasample rate. A total of 32,000 points per trace were recorded, although most of the useful data are contained in about 5000 points. The data reduction software converts the sinusoidal traces to a polar plot, then to a velocity as a function of time plot.

DATA REDUCTION

The Hugoniot of PMMA is the relationship between the shock and particle velocities of the material and is expressed as :

$$U_s = C_o + SU_p \qquad (5)$$

where:
U_s is the shock velocity in mm/μs
C_o is the initial bulk sound velocity in mm/μs
S is a dimensionless empirical parameter
U_p is the particle velocity in mm/μs

For PMMA, equation 5 has the form (4):

$$U_s = 2.49 + 1.69 U_p \qquad (6)$$

RESULTS

Figure 4 on the next page shows the particle velocity data from one of the Saturn shots. Both VISAR legs have been simultaneously plotted to show any measurement differences in the velocity history. In this case, the peak velocity is 86 meters-per-second with a measurement tolerance of 4%. The high bandwidth of VISAR is valuable in tracking the fast ($\approx$10 ns) compression and release components of the shock wave. Both VISAR legs tracked the entire shock history with little differences in the measured particle velocities.

The measured particle velocity can be plotted as a function of shock velocity. By knowing the values of shock velocity, particle velocity, and density, the pressure can be calculated using conservation of momentum:

$$P - P_o = U_s U_p \rho_o \qquad (7)$$

where:

P is the final pressure in GPa
P_o is the initial pressure , which is equal to zero.
ρ_o is the initial density in g/cc (in this case = 1.85).

During shock loading, the stress at any time in the event can now be calculated. For this example, the peak pressure is .314 GPa. It is important to note that equation 7 should not be used in unloading because the relationship between U_s and U_p differs from that of equation 6 and is not currently documented.

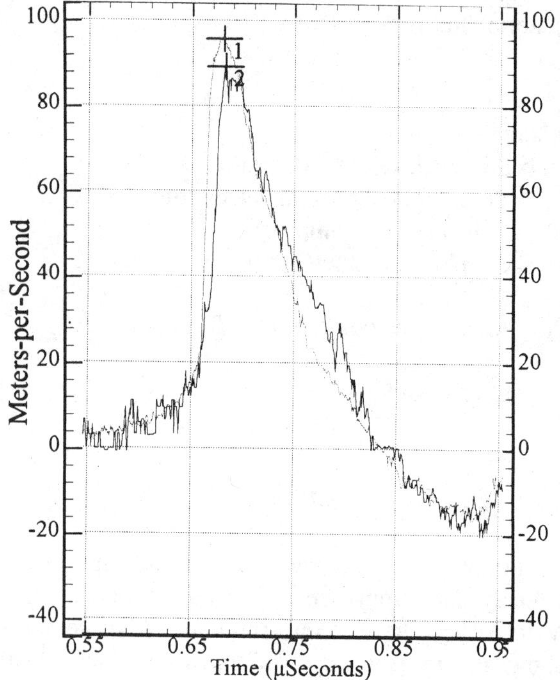

Figure 4. Particle velocity in the PMMA sample. Peak pressure is calculated to be .314 GPa.

CONCLUSION

The VISAR coupled sensors worked well, performing with the predicted immunity from electrical interference. The radiation induced self-light was blocked out by using the inherent light filtering nature of VISAR and narrow optical bandpass filters. Radiation induced fiber darkening was only detrimental after three or more exposures to x-ray radiation, so an efficient method of fiber optic replacement was developed.

The unique measurement technology of VISAR has brought material response analysis to a new level, since particle velocity can be measured with a high level of detail and accuracy. The measurements are repeatable with results closely matching predicted pressures in PMMA and measurements using piezoelectric gauges. This technology is in the early stages of development and more tests using different materials are planned for the future.

ACKNOWLEDGMENTS

This work was performed at Sandia National Laboratories for the US Department of Energy under contract DE-AC04-94Alb5000.

Developing new diagnostic techniques in a limited time frame is a very challenging goal requiring hard work and highly competent support personnel. The author gratefully thanks the following people for sharing their talent and hard work to make this a successful venture.

Ed Vieth--*Fiber optic fabrication, diagnostics.*
Richard Pepping--*Technical consulting,.*
Bill Barrett--*Testing coordination.*
Robin BroylesJohn Heise--*Data acquisition.*
Bill Brigham--*VISAR diagnostics, testing.*
O.B. Crump Jr.--*VISAR cavity assembly.*
Dan Sanchez, Theresa Broyles--*Testing support.*
Richard Wickstrom--*Data reduction.*
Mark Heddeman--*Financial support.*
Lloyd Bonzon--*Management Support.*

REFERENCES

1. L.M. Barker and R.E. Hollenbach, *"Shock Wave Studies of PMMA, Fused Silica, and Sapphire"*, Journal of Applied Physics 41:10 (Sept 1970).

2. K. J. Fleming, *"Fiber Optic Coupled Sensor for Reflected Photon Analysis"*, Department of Energy Patent Disclosure SD-5034, S-74, 181.

3. W. F. Hemsing, *"Velocity Sensing Interferometer (VISAR) Modification"*, Review of Scientific Instruments, 50:1, (Jan 1979)

4. J. D. Matthews, L. J. Weirick, *"A Hugoniot Study on PMMA Manufactured by Polycast Technology Corp."*, SAND90-2402, Sandia National Laboratories, Albuquerque, NM, (Feb 1991).

CHAPTER XIV

GAUGES AND APPLICATIONS

RETURN TO THE SHORTED AND SHUNTED QUARTZ GAUGE PROBLEM: ANALYSIS WITH THE SUBWAY CODE[*]

S. T. Montgomery, R. A. Graham, and M.U. Anderson

Sandia National Laboratories, Albuquerque, NM 87185

Simulations with finite element models of well controlled impact experiments with x-cut quartz gauges have been performed with the transient electromechanics code SUBWAY. Comparisons of measured gauge output current with calculated output current were made for four fully-electroded gauge configurations, involving two different can spacings and potting materials. The observed good agreement between measured and calculated currents provides a basis for confidence in the basic capabilities of the code.

INTRODUCTION

Quartz gauges have the ability to provide high fidelity measurements of impulsive stress pulses in materials with resolution of a few nanoseconds. Although the conditions under which the gauges can provide well defined measurements were carefully delineated in early work[1], requirements for measurements in difficult environments have led investigators to use gauge configurations which invoke two-dimensional mechanical and electrical conditions in the gauges. With the development of the transient electrodynamics code SUBWAY these conditions can now be accurately simulated and used to identify critical features of gauge responses.

Of particular interest at present is the measurement of the time-resolved stress states produced in materials by intense, short duration, soft x-rays produced by the Saturn accelerator at Sandia National Laboratories. (See paper by Barrett, et. al., present proceedings.) Three gauge configurations, as indicated in Fig. 1, are being investigated including the shunted guard ring gauge (A), the fully-electroded gauge (B), and the shorted-guard ring gauge (C).

To better understand the responses of fully electroded and shorted guard ring gauges we used the SUBWAY code with finite element models to simulate responses for several well controlled experiments.

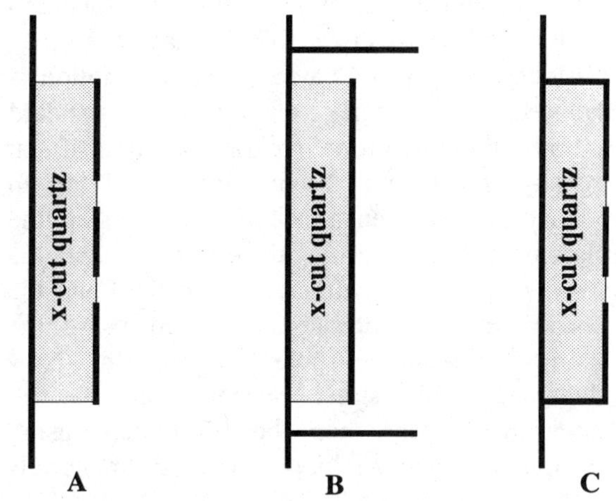

FIGURE 1. Typical Quartz Gauge Configurations.

EXPERIMENTAL

Controlled, precisely characterized impact loading experiments were carried out on the Sandia Precision Impact Loading Facility. As has been the case in prior work, the well-aligned

[*] This work supported by the United States Department of Energy under contract DE-AC04-94AL8500.

impact of X-cut quartz disks at impact velocities known to 0.1% provided the loadings. The tilt was typically less than 200 μradian. Short circuited current pulses were measured with low ohmic value current shunts connected to high frequency digitizers with low loss cables. Disks were all 37 mm in diameter and 6.35 mm thick.

COMPUTATIONAL

SUBWAY is a transient electromechanics code developed at Sandia National Laboratories[#]. The code uses a finite element description to model the motion and electric displacement in charge-free materials, as governed by Cauchy's equation of motion and Gauss's law. Using the electric field-electric potential relation, $\mathbf{E} = -\text{grad }\phi$, and the constitutive relation for the electric displacement in Gauss's law provides a Poisson equation governing the electric potential. Here $\mathbf{E}$ is the electric field and ϕ is the electric potential.

A complete description of assumptions and solution method used in SUBWAY has been given by Montgomery and Chavez.[2] Material motion is advanced over a time step Δt using standard methods[3] of explicit time integration with artificial viscosity to smooth discontinuities. Prior to updating the stress tensor at end of a time step, the Poisson equation governing the electric potential is solved using a conjugate gradient method and the electric field in materials calculated using the electric field-electric potential relation. New values for the stress tensor and electric displacement can then be evaluated using constitutive relations appropriate to materials involved in the simulation.

SUBWAY has the capability to treat general electrical loading conditions on specified surfaces. In particular, the integral form of Gauss's law is used to calculate the electric charge, Q, on a specified electrode or conducting surface at each

---

[#] SUBWAY code was developed primarily by S.T. Montgomery, P. F. Chavez, and L. M. Taylor at Sandia National Laboratories. It is not generally available.

time step of the simulation. The electric current, I, flowing from the electrode can then be calculated using the finite difference relation,

$$I = -(Q^{n+1} - Q^{n})/\Delta t^{n}.$$

Quartz has one threefold axis with two twofold axes normal it . One of the twofold axes corresponds to the cylindrical axis about which a disk of x-cut quartz is fabricated. Linear piezoelectric constitutive relations for x-cut quartz are given by Tiersten.[4] Graham[5] has shown that an accurate description of impacting x-cut quartz requires addition of nonlinear terms to the linear piezoelectric constitutive relations. In this study, selected higher order terms are included in the constitutive relations for the normal component of stress in the x-direction T_{xx} and the electric displacement vector $\mathbf{D}$. We take

$$T_{XX} = \tilde{T}_{XX} + \frac{1}{2}c^{E}_{111}S^{2}_{XX} + \frac{1}{6}c^{E}_{1111}S^{3}_{XX} - \frac{1}{2}e'_{111}E_{X}$$

$$D_X = \tilde{D}_X + \frac{1}{2}e'_{111}S^{2}_{XX} + \varepsilon'_{XX}\ \text{tr(S)}\ E_X$$

where $\tilde{T}_{XX}$ and $\tilde{D}_X$ are linear piezoelectric stress and electric displacement components, respectively, while S_{XX} and tr(S) are the normal component of the strain in the x-direction and trace of the strain tensor, respectively. The constants c^{E}_{111}, c^{E}_{1111}, e'_{111}, and ε'_{XX}, are associated with nonlinear material response and have been discussed by Graham. The dependence of dielectric constant on material density through the trace of the strain tensor is also incorporated into the Y and Z components of the electric displacement.

RESULTS

We performed simulations of experiments on several fully-electroded gauge configurations. The configuration used in the simulations is shown in Fig. 2. The figure shows a quarter section of the impactor, the gauge, and the backing aluminum tamper surrounded by potting and an aluminum can. The simulation shows electric field distributions about 0.7 µsecond after impact. The one-dimensional conditions in the center of the gauge disk, and the two-dimensional field distributions in the outer portion of the disk are clearly indicated. The can is spaced at L/2, where L is the gauge thickness.

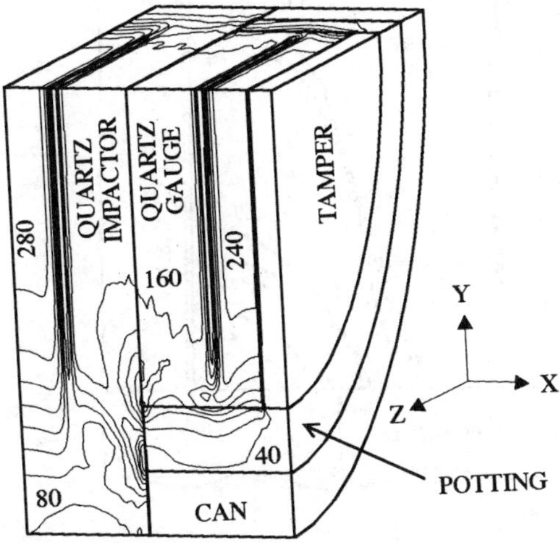

FIGURE 2. A quarter section of the finite element model for a fully electroded quartz gauge with can spacing L/2 approximately 0.7 µsecond after impact. The component materials of the gauge and impactor are labeled, with Hysol epoxy being used for the potting material in the simulation. Contours of electric field magnitude are labeled from 40 kV/cm to 280 kV/cm.

A sample comparison of measured and calculated gauge current for a fully electroded gauge loaded at a stress of about 10 kbar is shown in Fig. 3. The observed current pulse typically shows more noise than seen in prior work due to improved higher frequency recording.

As can be seen, the calculated gauge current agrees very well with the measure current until near the end of the pulse. The rounding of the calculated current is due to wave dispersion caused by the finite element discretization and artificial viscosity used in SUBWAY. As a result the wave reflection from the tamper causes the computed current to be lower than observed in the actual shock event. The observed behavior is representative of all the numerical simulations of gauge response conducted.

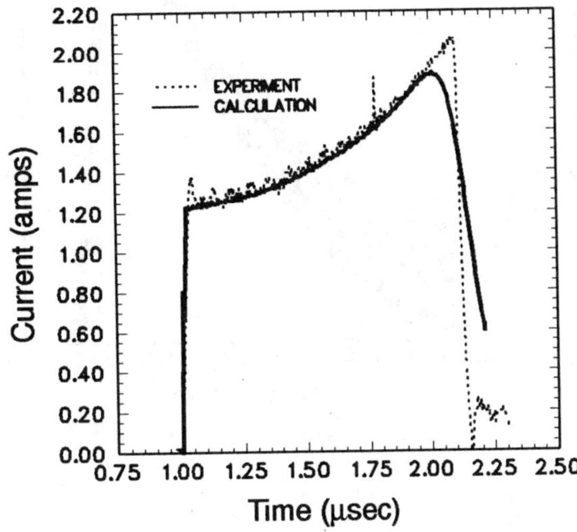

FIGURE 3. Experimental and calculated current pulses are compared for an experiment at about 10 kbar. Generally good agreements are observed, but certain features, particularly at late time, require more refined calculations.

Simulations of current pulses were carried out for four different fully electrode gauge configurations. At the same stress of 10 kbar, current pulses, stress distributions and electric field distributions were determined for two can spacing distances (L and L/2) and for two different epoxies (Hysol and alumina-loaded epoxy). The computed current pulses are shown in Fig. 4 in which the plotted currents are restricted to the values that show the initial current jump and the final values of the current at wave transit time.

The effects of both the epoxy and can spacing are readily apparent. The epoxy effect is a direct result of differences in dielectric constant. The can spacing effect is a result of electric field fringing within the gauge resulting from the proximity of the can to the edge of the disk.

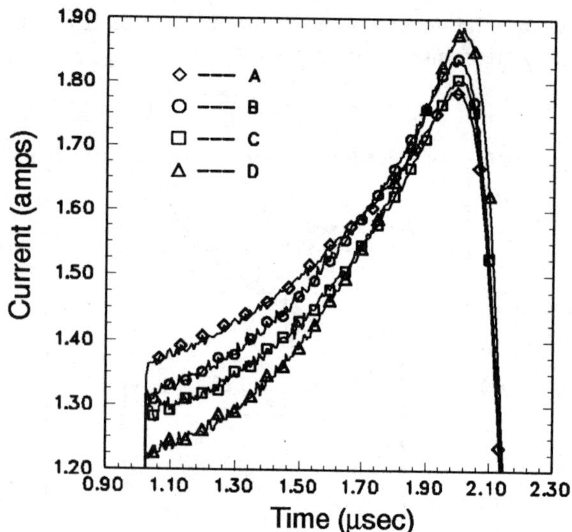

FIGURE 4. Calculated gauge currents for the fully electroded gauge configuration. The plot is restricted to the values that show the initial jump and final calculated values of the quartz gauge current. The curve labeled A corresponds to alumina loaded epoxy potting and a can spacing of L. The curve labeled B corresponds to alumina loaded epoxy potting and a can spacing of L/2. The curve labeled C corresponds to Hysol epoxy potting and a can spacing of L. The last curve, labeled D, corresponds to a calculation with Hysol epoxy potting and a can spacing of L/2.

As can be seen in this figure, the initial current jump depends on can spacing and potting material. For fixed can spacing the simulations show that the initial current jump is higher for the Hysol epoxy potting. Since the dielectric constant for the Hysol epoxy is lower than that for the alumina loaded epoxy these results indicate that the initial current jump decreases with increasing dielectric constant of the potting material. This result is a consequence of larger electric field fringing resulting from higher dielectric constant.

Figure 5 shows an enlargement of the electric field distribution in the outer region of the disk. Three different lines of constant field are shown

for can spacings of L, L/2 and L/4. There is a clearly discernible effect on field distribution caused by the can spacings. The field distribution is consistent with the current pulse predictions.

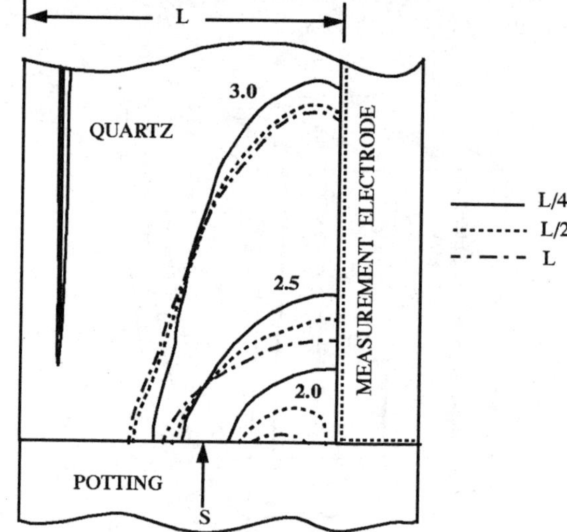

FIGURE 5. Contours of the electric field magnitude in a section of the quartz gauge, approximately 0.7 μsecond after impact.

CONCLUSIONS

The present work has shown that the SUBWAY code can be used as a quantitative tool for study of the physics and phenomenology of piezoelectric gauges. The code will prove particularly useful in study of the shorted guard ring gauges.

REFERENCES

1. Graham, R. A., Neilson, F. W., and Benedick, W. B., J. Appl. Phys. **36**, 1775-1783 (1965).
2. Montgomery, S.T. and Chavez, P. F., Basic Equations and Solution Method for the Calculation of the Transient Electromechanical Response of Dielectric Devices, Sandia National Laboratories Report, SAND86-0755, June 1986.
3.Taylor, L.M. and Flanagan, D.P., PRONTO 3D: A Three-Dimensional Transient Solid Dynamics Program, Sandia National Laboratories Report, SAND87-1912, March 1989.
4. Tiersten, H.F., Linear Piezoelectric Plate Vibrations, New York: Plenum Press, 1969, ch. 7, pp. 51-61.
5. Graham, R.A., Physical Review, **6**, 4779-4793 (1972).

NOISE MEASUREMENTS IN SHUNTED, SHORTED, AND FULLY ELECTRODED QUARTZ GAUGES IN THE SATURN PLASMA RADIATION SOURCE X-RAY SIMULATOR[1]

W. H. Barrett, J. I. Greenwoll, C. W. Smith

Sandia National Laboratories, Albuquerque, NM 87185-1159

D. E. Johnson, C. F. De La Cruz

Ktech Corporation, 901 Pennsylvania NE, Albuquerque, NM 87110

This paper describes recent work to improve the measurement of the stress response of materials to intense, short pulses of radiation. When Saturn fires, large prompt electrical noise pulses are induced in stress measurement circuits. The conventional wisdom has been that the shorted guard ring quartz gauge was the only configuration with acceptable prompt signal-to-noise characteristics for stress measurements in this pulsed radiation environment. However, because of abnormal signal distortion, the shorted guard ring gauge is restricted to a maximum stress of about 8 kbars. Below this level, the normal, quantified signal distortion is correctable with analytical deconvolution techniques. The shunted guard ring gauge is acceptable for high fidelity measurements to about 25 kbars with negligible signal distortion. Experiments were conducted on the Saturn soft x-ray source which show that higher fidelity shunted guard ring gauges can successfully measure stress with acceptable induced noise. We also found that a 50-ohm impedance matching resistor at the gauge reduced the prompt noise amplitude and improved the baseline quality of the measurement prior to shock wave arrival.

INTRODUCTION

The Saturn accelerator at the Sandia National Laboratories is a 36 module, high power, variable-spectrum, x-ray simulation source that can be operated in the bremsstrahlung or plasma radiation source (PRS) modes. For these experiments, Saturn was configured in the PRS mode utilizing argon as the source gas. The total x-ray yield from this source is approximately 300 kjoules. Approximately 40 kjoules of this is K-shell atomic line radiation near 3.2 keV in a 20 ns FWHM pulse.(1), (2) Thin filters (1.2 µm-thick polycarbonate with 200 Å of Al) were used to eliminate the UV and soft x-ray components below 1 keV.

Stress measurements which faithfully reproduce the state of stress in material samples exposed to this intense, short duration, soft x-ray source are needed to understand material behavior and determine material properties. Prompt electrical noise pulses of several hundred volts are induced in the measurement circuits. Signals of a few volts are typical amplitudes for our experiments. The stress wave transit time through the material separates the prompt noise from the stress signal. However, significant baseline disruption frequently occurs causing uncertainty in arrival times and stress amplitudes. The conventional wisdom has been that the shorted guard ring quartz gauge was the only configuration with acceptable prompt signal-to-noise characteristics for stress measurements at Saturn. (See paper by Montgomery, Graham, and Anderson, present proceedings, for a description of the different quartz gauge configurations used in these experiments.) However, because of abnormal signal distortion, the shorted guard ring gauge is restricted to a maximum

[1]Work supported by the USDOE under contract DE-AC04-94AL85000

stress of about 8 kbars. Below this level, the normal, quantified signal distortion is correctable with analytical deconvolution techniques. The shunted guard ring gauge is valid for high fidelity measurements to about 25 kbars with negligible signal distortion. The fully electroded gauge was proposed for manufacturability and economic reasons.

DESCRIPTION OF EXPERIMENT

Three series of Saturn shots, beginning in August 1994, were undertaken to determine the feasibility of using shunted and fully electroded gauges in Saturn. In Test Series 1, we exposed twelve, 0.76-mm thick PMMA (polymethyl methacrylate, Polycast UVB) samples mounted on 31.75-mm-diameter by 6.35-mm-thick quartz gauges. The active area, encompassed by the center electrode, was 12.7 mm in diameter for the shorted and shunted gauges. The active area for the fully electroded gauge was, of course, the full 31.75 mm. Two identical quartz gauges each of the three designs (shorted guard ring, shunted guard ring, and fully electroded) were fielded on each of two Saturn shots. All gauges were at the same distance from the PRS source. The nominal fluence level was 5 cal/cm^2. Therefore, four nominally identical measurements were obtained for each gauge design.

A current viewing resistor (CVR) was installed between the center electrode and the negative (front face) electrode. We chose the resistance value to obtain equal stress signal amplitudes independent of the gauge type. This allowed a direct comparison of signal-to-noise ratio of the different gauges. The CVR value for the shorted and shunted gauges was 360 ohm; for the fully electroded, it was 8.2 ohm.

The feasibility of using shunted or shorted gauges was successfully demonstrated in Test Series 1. Two follow-on series of shots, conducted in October and December 1994, were designed to explore the effects on the prompt noise of the choice of the CVR value. In earlier experiments in underground nuclear weapon effects tests, Sandia had demonstrated that the lowest noise was achieved using an impedance matching resistor at the gauge. A 50-ohm resistor at the gauge and at the recording instrument matched the 50-ohm characteristic impedance of the long (500 to 2000 ft) coaxial signal cable runs. This design minimized reflections of large noise spikes at connectors, attenuators, equalization networks, and the gauge, although the technique reduces the signal by a factor of two. We decided to try the same approach at Saturn. For comparison we also fielded gauges with the CVR value used in the August shots and with no CVR in the gauge, the open or infinite resistance configuration. The latter has been the standard method used to date at Saturn.

On Test Series 2 and 3, we used the same three quartz gauge designs. On Test Series 2, we used identical PMMA material samples. On Test Series 3, we increased the thickness of the PMMA sample to 1.27 mm in order to increase the time between the prompt noise and the stress signal. For the shunted guard ring gauge, there were a total of seven 50-ohm, three 360-ohm, and five open CVR gauges. For the fully electroded gauge, there were a total of five 50-ohm, three 8.2-ohm and five open CVR gauges. For the shorted guard ring gauges, there were only four 50-ohm gauges.

The data acquisition system consisted of Cujac 141 coax cable from the gauge to the Saturn diode vacuum feedthru port (11- to 14-ns long), RF-214 coax cables from the vacuum feedthru port to the J-Box (about 30 ns-long), the permanent ½" heliax cable plant from the J-box to the screen room wall (72-ns long), the recording instruments in the screen room, and the VAX computer controller. We used two each Tektronix SCD-1000 flash converters (with 3/8 " heliax cables, 58-ns long) on each gauge to cover the amplitude dynamic range of both the hundreds of volts prompt noise pulse and the several volt stress signal. We also used two each Hewlett-Packard HP-54524A digitizers (with 1/4" heliax cables, 41-ns long) to cover a 4-ms recording window in order to record several shock reverberations through the samples. All recorders were terminated in 50 ohm at the input.

RESULTS

Figure 1 shows typical examples of the prompt noise pulses we saw in Test Series 1. The solid line is from the shorted guard ring gauge, the dashed line is from the shunted guard ring gauge, and the dotted line is from the fully electroded gauge. The shorted gauge typically had the lowest amplitude pulse as previous experience had shown. The interesting result is that the shunted gauge pulse was less than 40% greater, on the average, than the shorted gauge pulse. The prompt noise pulse for the fully electroded gauge was more than three times greater than the pulse for the shorted gauge.

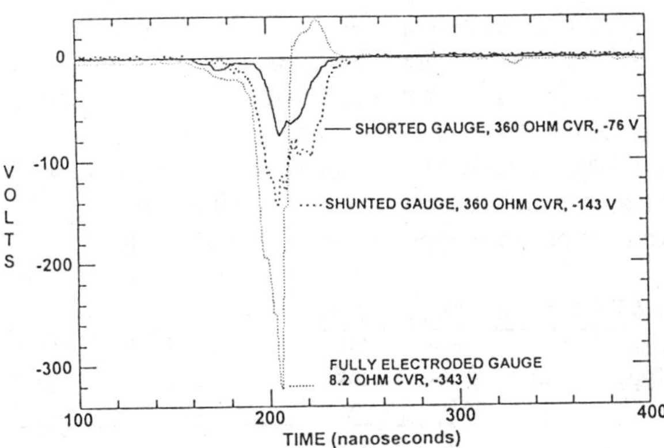

FIGURE 1. Prompt noise comparison, test series 1.

Figure 2 shows the stress waveforms from the recording instruments set for optimum resolution of the stress pulse. From top to bottom, the traces are from the shorted, shunted, and fully electroded gauges. Three typical characteristics of the baselines prior to shock arrival can be seen. The three peaks immediately following the negative prompt noise pulse are visible in all three traces to a different degree. These are probably due to reflections of the prompt noise pulse from small electrical impedance mismatches at connections in or very near the gauge housing. The trace does not return to the baseline prior to shock arrival, probably due to slow amplifier recovery from the three noise peaks. There is a pulse at about 440 ns (negative for the shorted and shunted gauges, 360 ohm CVR, and positive for the fully electroded gauge, 8.2 ohm CVR). One of the authors, Smith, traced this to small electrical impedance mismatches at the screen room wall which reflect a small part of the prompt noise pulse. This reflection pulse is then reflected from the non-50 ohm gauge and mixes with the stress signal. A mismatch of a few percent can result in the reflection of a few volts if the noise pulse is a few hundred volts, as is the case here.

The response of the shorted and shunted gauges were similar, although the shorted gauges usually had better baseline quality. While results for the fully electroded gauge did not appear very promising we decided to proceed with a more thorough investigation of the low noise quartz gauge design for Saturn PRS experiments.

Figure 3 presents typical examples from Test Series 2 and 3 of 50-ohm versus open CVR for the shunted and

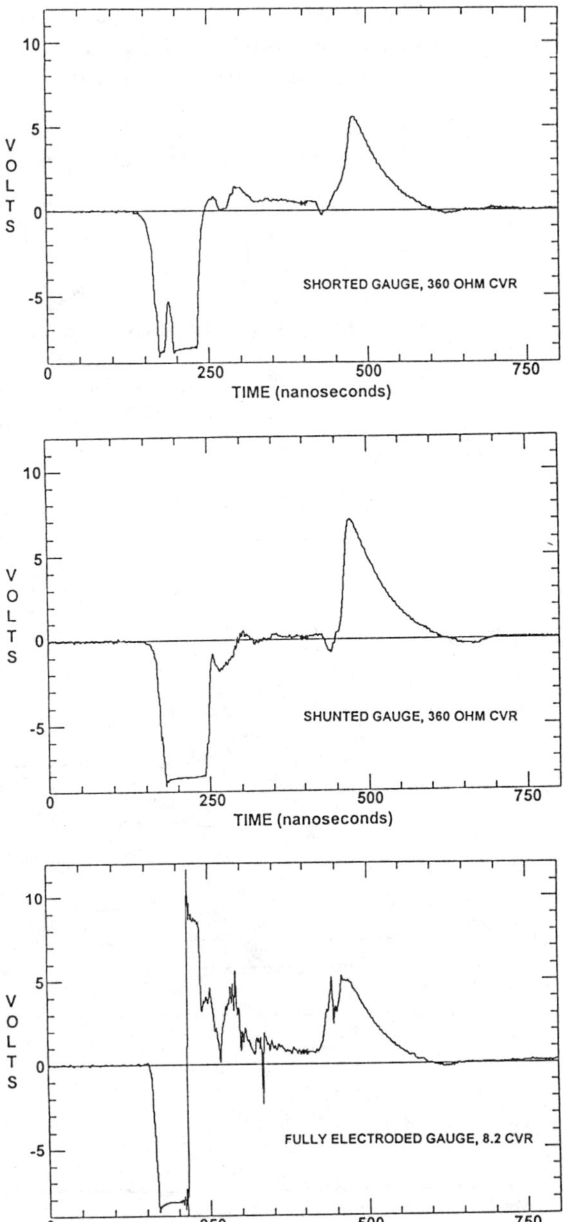

FIGURE 2. Baseline and stress waveform comparison, series 1: top, shorted gauge, 360 ohm; middle, shunted gauge, 360 ohm; bottom, fully electroded gauge, 8.2 ohm.

fully electroded gauges and 50-ohm versus 360-ohms CVR for the shorted gauge. The improvement is obvious. The reflections at the toe of the stress pulse do not appear in the 50-ohm records. The "3" noise peaks are reduced, although not eliminated. The baseline offset is less for most, but not all of the 50-ohm gauges. The longer baseline of the 50-ohm shorted gauge results from using a thicker PMMA sample on that shot.

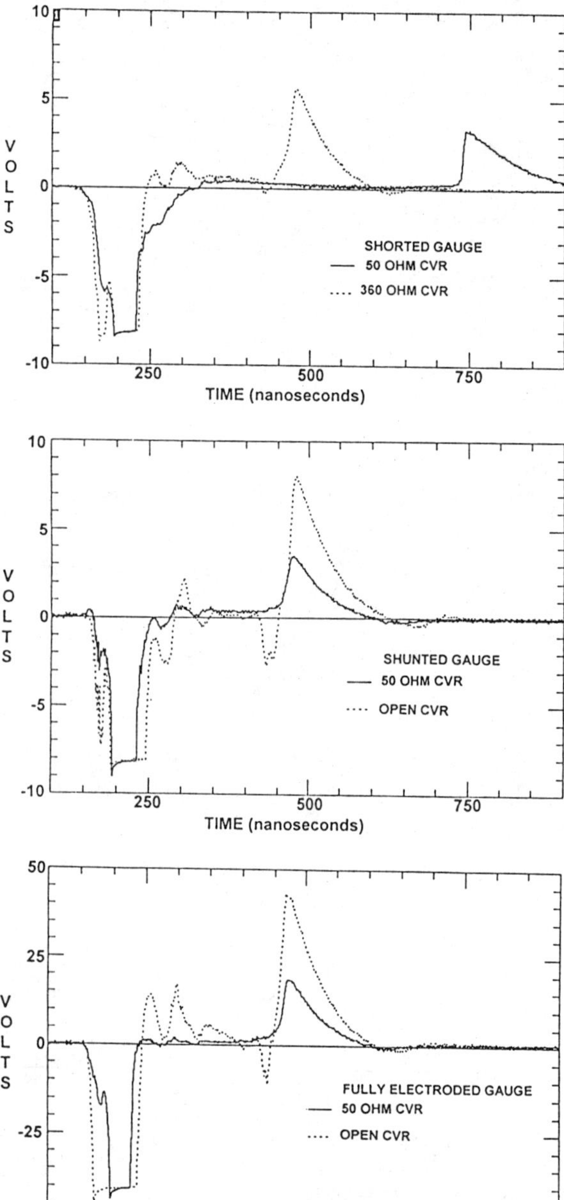

FIGURE 3. High versus 50 ohm CVR comparison: top, shorted gauge, 360 vs 50 ohm; middle, shunted, open vs 50 ohm; bottom, fully electroded, open vs 50 ohm.

Table 1 is a summary of the results of all three test series. For the prompt noise and the stress pulse, the numbers given are averages of all the good measurements for that gauge type. The numbers given for the three baseline characteristics are the average of the absolute values for all gauges of that design. A good measurement was when Saturn produced the nominal yield of between approximately 35 and 55 kjoules.

TABLE 1. Summary of results.

Gauge Type	CVR Resist. Ohms	Prompt Peak V	"3" Peaks V	Baseline at 400ns V	Reflec- tion V	Peak Signal V
Shorted	360	-82	1.16	0.19	0.46	5.2
	50	-50	0.88	0.10	0.00	3.4
Shunted	None	-156	2.04	0.21	2.04	8.1
	360	-113	1.19	0.28	0.76	6.5
	50	-131	1.11	0.18	0.00	3.8
Fully	None	-623	29.20	4.32	20.40	43.2
Electro-	50	-319	3.08	0.96	0.00	19.6
ded	8.2	-262	29.93	0.62	2.33	4.8

CONCLUSIONS

We have drawn two major conclusions from our recent quartz gauge work at Saturn. One, the shunted guard ring gauge, which has the highest fidelity, can be used at Saturn with acceptable, although slightly higher, prompt noise response than the normally used shorted guard ring gauge. Two, using a 50-ohm termination resistor at the gauge reduces the prompt noise pulse and, more importantly, significantly reduces the later baseline distortions which can cause serious uncertainty in the stress measurement.

ACKNOWLEDGMENTS

These experiments could not have been done without the dedication and professional efforts of the Saturn crew.

REFERENCES

1. SAND 92-2157, "Sandia National Laboratories Radiation Facilities Brochure", UC-9000, August 1993.
2. Spielman, R. B., et. al., "Z-Pinch Experiments on Saturn at 30 TW", in *AIP Conference Proceedings No. 195 on Dense Z-Pinches*, 1989, pp. 3-16.

QUARTZ GAUGE RESPONSE IN ION RADIATION

P.E. Taylor[*], P.H. Gilbert[*], C. Kernthaler[*], L.M. Lee[**],
E.A. Smith[**], S.T. Reeder[**], M. U. Anderson[***]

*Atomic Weapons Establishment, Aldermaston, Reading, Berkshire, UK, RG7 4PR
** Ktech Corporation, 901 Pennsylvania NE, Albuquerque, NM 87110
*** Sandia National Laboratories, P. O. Box 5800, Albuquerque, NM 87185

This paper describes recent work to make high quality quartz gauge (temporal and spatial) shock wave measurements in a pulsed ion beam environment. Intense ion beam radiation, nominally 1 MeV protons, was deposited into material samples instrumented with shunted quartz gauges adjacent to the ion deposition zone. Fluence levels were chosen to excite three fundamentally different material response modes (1) strong vapor, (2) combined vapor and melt phase and (3) thermoelastic material response. A unique quartz gauge design was utilized that employed printed circuit board (PCB) technology to facilitate electrical shielding, ruggedness, and fabrication while meeting the essential one dimensional requirements of the characterized Sandia shunted quartz gauge. Shock loading and unloading experiments were conducted to evaluate the piezoelectric response of the coupled quartz gauge/PCB transducer. High fidelity shock wave profiles were recorded at the three ion fluence levels providing dynamic material response data for vapor, melt and solid material phases.

INTRODUCTION

Development of multi-phase material constitutive models requires time resolved physical properties data i.e., stress, impulse, velocity, mass, energy, etc. Intense pulsed radiation sources, such as the GAMBLE II accelerator, located at the U. S. Naval Research Laboratory, can be used to generate material response data in solid, melt and vapor phases, which are required for model development. Due to the 10-60 ns GAMBLE II ion beam pulse duration, 0.5 Ghz transducers and recording systems must be used to acquire high fidelity data. Also, the transducer and signal cable must be configured/shielded to reject electrical noise pickup/generation at accelerator discharge. Unique quartz gauges (designated AWEQG) were used in this program to acquire aluminum material response data for multi-phase states. This paper presents ion beam test results as well as quartz gauge characterization data.

QUARTZ GAUGE DESIGN AND CHARACTERIZATION

The one-dimensional shunted x-cut quartz gauge has been a standard for stress measurement in impact physics work since 1965 when Graham characterized its response to planar shock loading (1). This original work described the shock induced piezoelectric response of a precisely defined quartz disk (designated the Sandia Quartz Gauge (SQG) Figure 1) based on crystalographic orientation, disk dimensions, electrode configuration, guard ring width, etc. Since the original quartz characterization work, other quartz gauge configurations have been utilized in an attempt to improve certain gauge features i.e., increased signal level, longer recording time, improved electrical shielding and ruggedness (2, 3). Each of the changes made to the SQG design produced a deviation in the quartz piezoelectric shock response when compared to the 1965 characterization study, however.

Two of the most difficult steps in producing a high quality SQG is soldering electrical leads to the thin

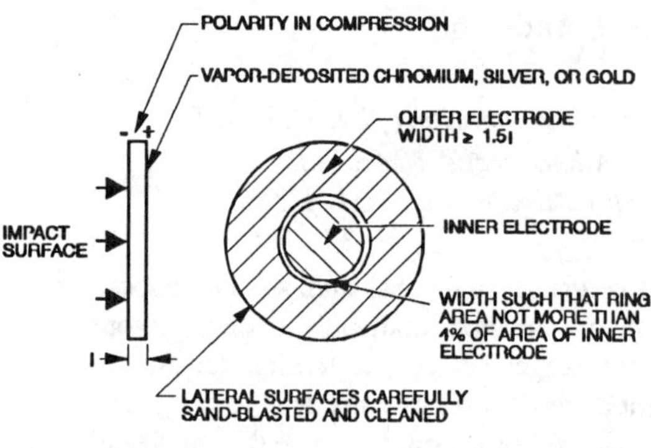

FIGURE 1. Sandia x-cut quartz guard-ring configuration to obtain one-dimensional conditions.

metal electrode and cutting the precise guard ring that separates the active and guard ring areas. For this ion beam application, a different approach (AWEQG) was taken to achieve both of these functions using printed circuit board (PCB) technology, in an attempt to ease fabrication and improve ruggedness. Details of the PCB technique are not presented here (patent pending). All other SQG criteria defined in Ref. 1 were adhered to in the AWEQG.

Shock loading characterization of the AWEQG configuration included gas shock tube testing and flat plate impact testing. Shock tube test results (Figure 2) confirmed overall gauge operation, with recorded gauge outputs agreeing with predicted signal levels based on Ref. 1 SQG characterization. Gauge rise time in the shock tube testing was primarily controlled by the precision of gauge-gas shock front alignment, and does not represent a gauge limitation. Shock tube test data taken before and after shipment from the UK to the US confirmed AWEQG ruggedness.

The shock tube results demostrated AWEQG response but only at low stress (0.2 MPa). Plate impact experiments were performed to evaluate AWEQG piezoelectric response at 0.5 and 1.0 GPa peak stress and to assess any electrical breakdown problems resulting from use of the PCB technology. Precision impact experiments were conducted using the same Sandia compressed gas gun facility used in Ref. 1. The AWEQG disk was impacted with a

precisely aligned x-cut quartz disk used as a projectile facing, whose velocity at impact was accurately measured to ±0.5%.

Table 1 summarizes AWEQG characterization results for three gas gun shots. Initial current jump and subsequent wave shape were evaluated for (a) long pulse loading at 0.5 and 1.0 GPa (2 experiments) and (b) short pulse loading in Figure 3. The initial current jump and current at complete gauge transit were in close agreement with Ref. 1 data (Table 1). The third experiment was designed to assess the short pulse (pulse duration less than quartz gauge transit time) anomaly in shock loaded quartz (4). The anomaly, resulting from shock induced conductivity in the shocked and then unloaded quartz material, manifests

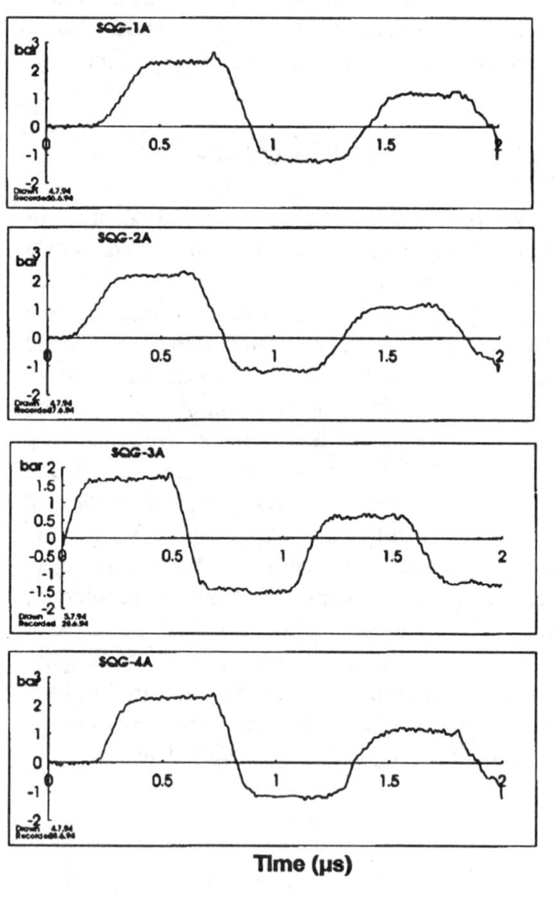

FIGURE 2. Representative AWEQG shock tube results (4 gauges).

TABLE 1. AWEQG Gas Gun Experimental Results Summary.

Shot Number	u^a_p	σ^b	k^c	$i_t/i_i{}^d$	Tilt ^{e}	Configuration ^{f}
2611	0.066	1.000	2.18	1.05	134	A
2612	0.0335	0.508	2.09	1.02	62	A
2618	0.0662	1.004	2.18	---	262	B

a. u_p is quartz particle velocity at impact surface taken as one-half the impact velocity (mm/µs).

b. σ (GPa) is impact stress calculated from ($\Delta u_p U_s \rho_o$) where U_s=5.72mm/µs and ρ_o=2.65 g/cm^3.

c.   k is the current coefficient calculated from $(i_i l)/\sigma A U_s$, where i_i is the initial current jump, l is quartz thickness, A is the active area.

d. i_i is initial current jump and i_t is current at quartz transit time, with i_t/i_i being a measure of current increase during wave-transit time.

e.   Tilt (micro rad) was calcuated using quartz output signal rise time, active area diameter, impact velocity and assuming impact of two perfectly flat plates.

f.   Configuration A was 50.8 diameter by 5.08 mm thick x-cut quartz impactor for long pulse loading and configuration B was 50.8 diameter by 0.635 mm thick x-cut quartz impactor (free rear surface) for short pulse loading.

itself as an apparent charge output when there should be no output. Figure 3(b) shows the short pulse experimental data (Shot 2618) with no anomalous current output, i.e., after shock unloading the signal returns to the base line and remains there until complete transit of the gauge. Shot 2618 was designed to have normal response, based on Ref. 4 data, but was located adjacent to the normal/anomalous response boundary at 1.0 GPa. Rise time differences shown in Figure 3 were attributed to non-planarity or tilt at impact, for Shots 2611 and 2612. This high degree of planarity, a unique capability of the gun facility used, was required to define meaningful AWEQG gauge characterization data. The asymmetric rise recorded for Shot 2618 was attributed by deformation in the thin (0.635 mm) quartz impactor due to launch loads.

ION BEAM RESULTS

The GAMBLE II accelerator produces nominally 30 kJ of 1 MeV protons, producing a 50 cm diameter beam 1m from the anode. Fluence levels can be varied from 0.1 to 100 cal/cm^2 and are a function of distance from the anode. Ion pulse width is also dependent on distance from the anode and is between 10 and 60 ns FWHM.

Thirty one samples were irradiated in 10 ion beam

experiments, with high fidelity quartz gauge data being obtained on all samples. Representative

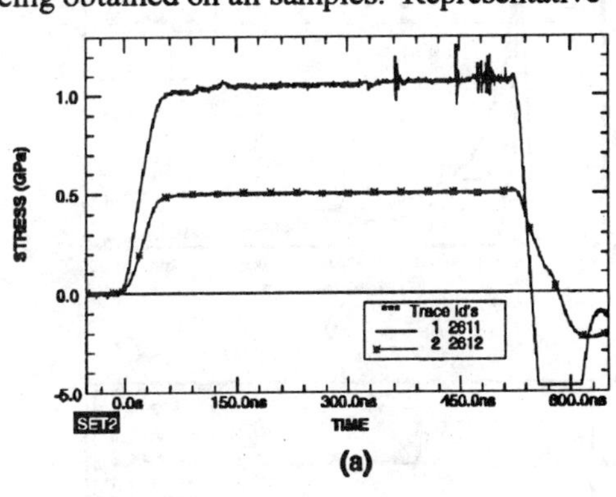

(a)

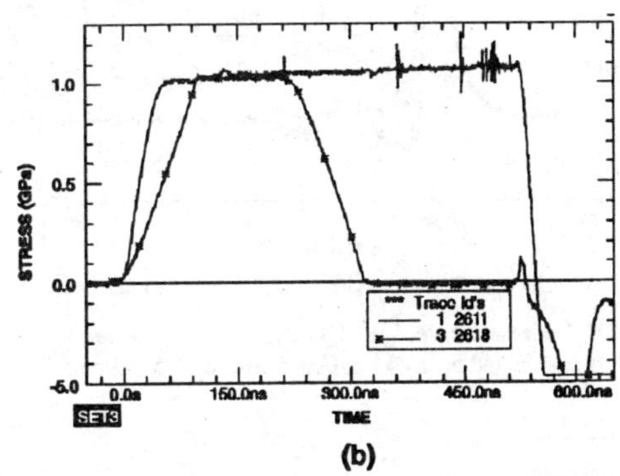

(b)

FIGURE 3. AWEQG data for (a) 0.5 and 1.0 GPa long pulse loading and (b) 1.0 GPa long and short pulse loading.

AWEQG data are shown in Figure 4 for aluminum samples exposed at 15 and 1.5 cal/cm^2. Measurable differences in stress and wave shape are apparent for the two fluence levels shown. The 15 cal/cm^2 exposures produced fast rise shock fronts followed by gradual release representing vapor dominated response. The 1.5 cal/cm^2 exposures, which produced thermoelastic responses, obviously generated lower peak stress with gradual rise followed by rapid stress release (compared to 15 cal/cm^2). Even tensile pulses are evident for the low fluence cases showing the results of elastic wave reflections in the aluminum sample.

All AWEQG gauges were insensitive to accelerator noise generation as evidenced by the signal baseline prior to shock arrival. Also the three records at 1.5 cal/cm^2, which were from a single ion experiment, show close agreement in basic wave shape.

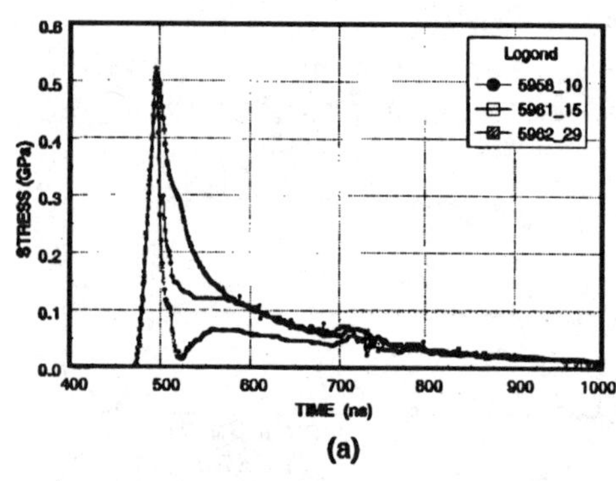

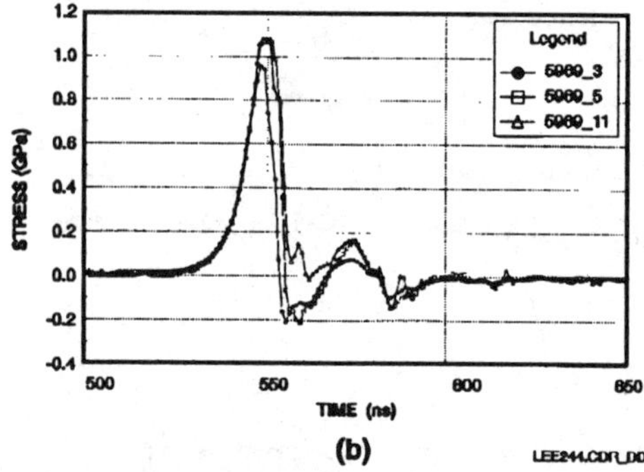

FIGURE 4. Representative AWEQG data from ion beam experiments conducted at (a) 15 cal/cm^2 and (b) 1.5 cal/cm^2

SUMMARY

The AWEQG stress gauges, based on SQG one-dimensional criteria and employing a PCB technique, have shock induced piezoelectric characteristics (loading and unloading) identical to the SQG (Ref 1), up to 1 GPa. Also, the AWEQG did not show the anomolous current response for short pulse loading at 1 GPa.. The AWEQG gauge design has excellent ruggedness and minimizes electrical noise pickup in harsh accelerator environments. In the ion beam application the quartz gauges provided high resolution stress-time data that can be used to develop multi-phase material constitutive models spanning the solid through vapor phases.

REFERENCES

1. Graham, R. A., Neilson, F. W. and W. B. Benedick, "Piezoelectric Current from Shock Loaded Quartz-A Submicrosecond Stress Gauge," *J. of Applied Physics*, Vol. *36*, No. 5, May 1965, pp. 1775-1783.

2. Graham, R. A., "Piezoelectric Current from Shunted and Shorted Guard-Ring Quartz Gauges," *SAND78-1911*, Sandia National Laboratories, NM, December 1978, pp. 89-100.

3. Reed, R. P., "The Sandia Field Test Quartz Gage, Its Characteristics and Data Reduction," *SAND 78-1911*, Sandia National Laboratories, NM, December 1978, pp. 149-176.

4. Graham, R. A., and C. E. Ingram, "Piezoelectric Current from X-Cut Quartz Subjected to Short-Duration Shock-Wave Loading," *SAND 78-1911*, Sandia National Laboratories, NM, December 1978, pp. 61-72.

CARBON PIEZORESISTIVE GAUGES IN ONE-DIMENSIONAL STRESS MEASUREMENTS

C. Hari Manoj Simha, S. J. Bless

Institute for Advanced Technology, The University of Texas at Austin.
4030-2 W. Braker Lane, Austin, TX 78759

Long rod impact experiments on 2024-T351 aluminum are presented using carbon piezoresistive gauges. The data is reduced using the one-dimensional strain calibration for carbon gauges. From the agreement between the experiments and finite-difference simulations we conclude that the one-dimensional strain calibration for carbon gauges can be used in reducing one-dimensional stress data.

INTRODUCTION

The reduction of voltage-time data obtained from piezoresistive stress gauges (carbon, manganin, and ytterbium) is by means of a calibration curve, either supplied by the manufacturer or available in the literature. The calibration curve is determined by a series of plate impact experiments, a one-dimensional (1D) strain state. This curve is well known for carbon gauges (1-3). Rosenberg et al. (4) showed that for manganin gauges the 1D strain calibration could be used for bar impact experiments which is a 1D stress experiment. Use of carbon gauges in rod impact experiments at low stresses (<20 kbar) is also highly desirable since these gauges are more responsive than manganin gauges and the noise in the measurements is much lower, or absent.

EXPERIMENTS

The experiments are performed on the 56 mm bore single-stage light gas gun (SLGG) at the IAT Hypervelocity Launch Facility (HLF) in Leander, Texas. The experiment configuration and geometry is as shown in Fig. 1. The 4340 steel plate impactor is impacted against the 2024-T351 aluminum rod instrumented with the carbon gauge. We choose this aluminum as its material properties and dynamic behavior are well known. The gauge used is the Dynasen Inc. FC 300-50-EKRTE, 2.5 mm X 1.25

mm, 0.5 mil thick. The gauge is emplaced between the rod and the backing piece using epoxy and the epoxy is squeezed out by clamping the assembly in a special fixture. The length L1 is selected such that L1 > 6D where D is the diameter of the rod. It has

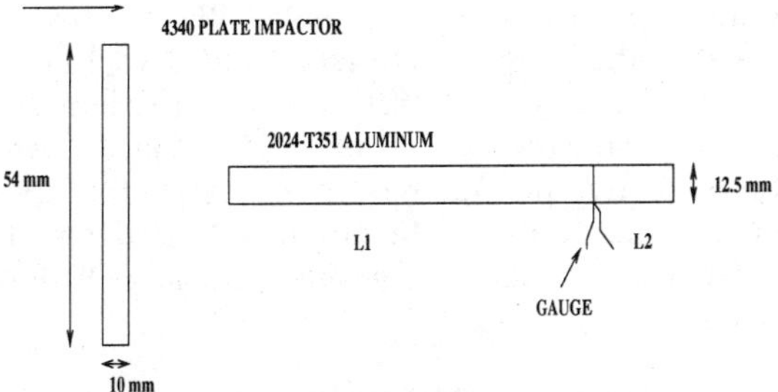

Figure 1. Experiment configuration.

been shown in (5) that this ensures a one-dimensional stress state. The length L2 is varied for each experiment in order to vary the pulse width recorded by the gauge. Impact velocity is measured by an optical technique described in Simha (6). The gauge is a part of a Wheatstone bridge, one other arm of which is a compensating gauge. The compensating gauge serves to eliminate effects due to heating. The bridge is pulsed by using a piezoresistive power supply: Dynasen CK1-50-300. Voltage-time data is acquired using a digital oscilloscope (LeCroy 9310L 300MHz). The data is reduced using the one-dimensional strain calibration (3), given by

$$\sigma = 0.255\left(\frac{\Delta R}{R_0}\right) + 0.00075\left(\frac{\Delta R}{R_0}\right)^{2.2} + 0.0000021\left(\frac{\Delta R}{R_0}\right)^{4}. \qquad (1)$$

where σ is in kbar and $\Delta R/R_0$ is a %. The results are shown in Table 1.

Table 1. Experiments on 2024-T351

Test ID	Velocity m/s	L1 mm	L2 mm
IAT001	266	125	19
IAT004	250	125	35.7
IAT007	267	125	19

SIMULATIONS

The simulations are performed using the Lagrangian processor of AUTODYN-2D. We use 8 cells across the radius of the target rod. Lynch and Simha (7) have shown that this is sufficient to guarantee convergence in this code. An elastic-perfectly plastic model is used for the aluminum and steel, a Mie-Grüneisen equation of state (EOS) for the aluminum and a linear EOS for the steel.

Table 2. Constants for 2024-T351 (8)

Density gm/cc	2.785
Bulk Sound Speed km/s	5.32
Parameter S	1.33
Grüneisen Coefficient G	1.33
Shear Modulus GPa	27.6
Yield Stress GPa	0.34

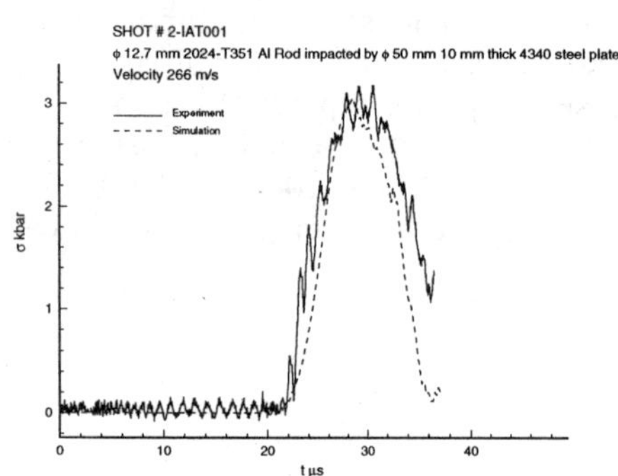

Figure 2. Gauge trace and simulation shot IAT001

These simple models are sufficient for the present case as we are not interested in the flow behavior at the impact plane. The rear portion of the rod remains elastic and a wave with an amplitude equal to the yield stress of the target rod propagates down the rod. Due to dispersion there is some decay in this amplitude. This is not of great concern as a good code should reproduce this effect. The constants used in the simulations (8) are shown in Tables 2 and 3. Figures 2, 3 and 4 show the experimental and simulation results. The agreement between simulations and experiments is excellent.

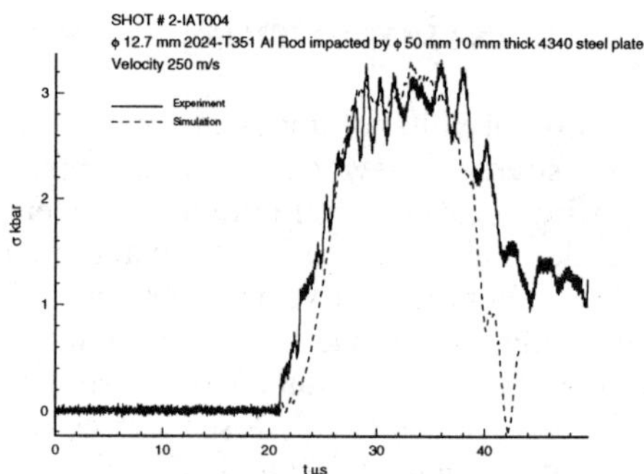

Figure 3. Gauge trace and simulation shot IAT004

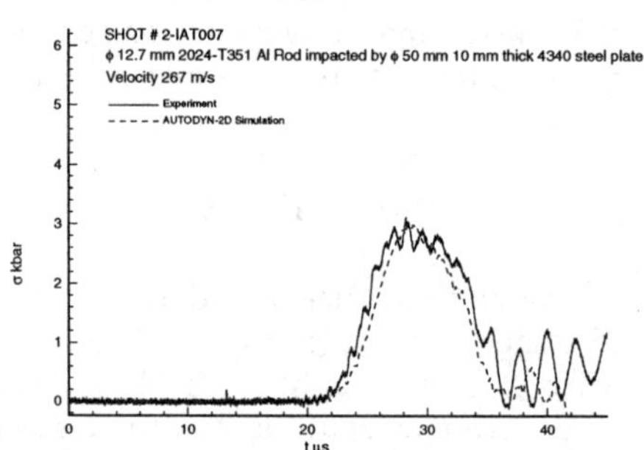

Figure 4. Gauge trace and simulation shot IAT007

Table 3. Constants for 4340 (8)

Density gm/cc	7.83
Bulk Modulus GPa	159
Shear Modulus GPa	81.8
Yield Stress GPa	0.792

The pulse width of the stress depends on the length of the backing piece L2, since this controls the time taken by the wave to reach the free surface at the rear of the rod and return as an unloading wave. This is seen in shot IAT004 where a wider pulse is obtained as a longer backing piece was used.

CONCLUSIONS

From the above we conclude that the 1D strain calibration for carbon gauges can be used to reduce 1D stress data.

ACKNOWLEDGMENTS

This work was performed under US Army Research Laboratory Contract DAAA21-93-C-0101.

REFERENCES

1. Horning, R. R. and Isbell, W.M., *Rev. Sci. Instrum.* **46**, 1372-1379 (1975)
2. Perez, M. and Chartagnac, P., *Rev. Sci. Instrum.* **51**, 921-925 (1980).
3. Charest, J. Dynasen Inc. Private Communication.
4. Rosenberg, Z. et al., *Journal of Applied Mechanics* **51**, 202-204 (1984).
5. Habberstaad, J. L., et al., *Journal of Applied Mechanics* **39**, 367-371 (1972).
6. Simha, C. H., *Institute for Advanced Technology Technical Note* **IAT.TN0054**, (1995).
7. Lynch, N. J. and Simha, C. H., *Institute for Advanced Technology Technical Note* **IAT.TN0045**, (1994).
8. Steinberg, D. J., **UCRL-MA-106439** (1991).

CALIBRATION AND USE OF A RUGGED NEW PIEZORESISTIVE PRESSURE TRANSDUCER*

R. A. Lucht and J. A. Charest

Los Alamos National Laboratory, MS C920, Los Alamos, NM 87545
Dynasen, Inc., 20 Arnold Place, Goleta, CA 93017

A new 50-ohm piezoresistive pressure gauge has been developed and calibrated in the range 0 to 4.0 GPa. This "pinducer" consists of one half of a 100 ohm, one quarter watt, carbon composition resistor mounted coaxially at the end of a small brass tube. Three techniques have been used to calibrate this new gauge. Good agreement is found between all calibration data, and a smooth curve is fit through all resistance change versus pressure data up to 1.5 GPa. The gauges exhibit rise times of about 0.5 µs. They offer advantages in ruggedness, cost, and flexibility of application. The pinducer can be successfully used in divergent flows, harsh environments, and positions where lead protection would be impossible with thin-film gauges. A unique application is demonstrated.

INTRODUCTION

With the end of the cold war, emphasis shifted from the design of new weapons and munitions to the safety of the existing stockpile. This change has renewed interest in existing long-term programs and produced several new programs to study delayed transition-to-detonation (XDT) and deflagration-to-detonation transition (DDT). XDT has been observed in propellants in shotgun and card gap tests (1, 2), and DDT in propellants (3), and granulated explosives (4). In many cases, XDT involves an insult with damage to the reactive material, resultant heating, ignition of burning, and DDT.

Whether pure DDT or XDT, the initiation mechanisms often involve relatively slow buildup of pressure from burning followed by the coalition of stress waves into a shock wave that can eventually run to detonation. Theoretical understanding of the details of these processes and accurate modeling of accident scenarios leading to XDT or DDT, requires measurement of the temporal change of pressure over a variety of time scales and at numerous positions. The measurements are complicated by the fact that the pressures often start quasi-statically and transition to waves with highly multi-dimensional flow. Thus, thin-film gauges such as carbon or manganin cannot be used because of the complicated flow and because they and their leads are not robust enough to survive the duration and violence of the experiments. While gauges such as piezoelectric quartz gauges could be used in these applications, they often have prohibitively large (and thus slow) active regions and they are quite expensive for one-time use. These needs have driven the development of the "pinducer".

DESIGN

The term pinducer implies the combination of devices known as pressure transducers and time of arrival pins such as capped or ionization pins. Pinducers work like pressure transducers, but are shaped like and have the convenience features of pins. The pressure sensing element is one half of a 100 Ω, 1/4-watt resistor, manufactured by Mouser Corp. The details of its construction are shown in Fig. 1. The lead wire from the carbon composition resistor half is connected to the center conductor of the coaxial pin arrangement. Copper is vapor plated

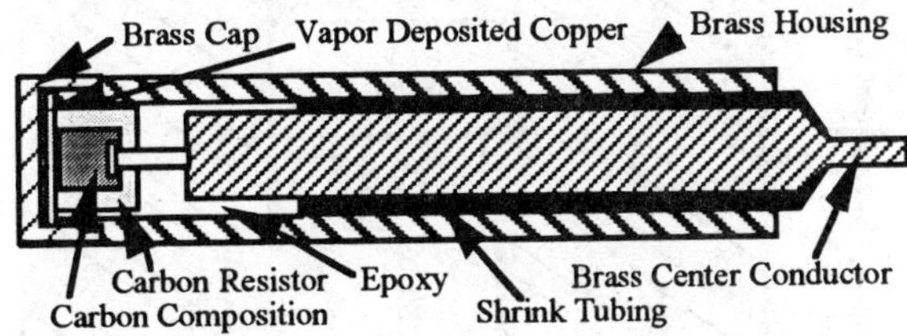

FIGURE 1. Pinducer construction. Outer diameter is 0.125 in.

*This work supported by the US Department of Energy.

out on the cut surface of the resistor and provides a connection to the outer conductor, a brass tube housing, of the coax. This design builds upon the use of carbon composition resistors by Ginsberg and Asay (5) to measure pressure in multi-dimensional flows. Like other carbon gauges, its resistance decreases as the pressure increases. Because the active element is basically a small, right-circular cylinder of length about equal to diameter, it is quite insensitive to flow direction and planarity in shock pressure detection. However, its size does limit the time resolution to that required for it to reverberate to pressure equilibrium with its surroundings.

CALIBRATIONS

Typically, the development of a new gauge such as carbon film or manganin (6) has taken many years. Pinducers have only been available for about one year and some improvements could probably be made; however, a reasonable calibration curve has already been achieved using three very different types of apparatus.

The first apparatus generated a quasi-static pressure in the 0 to 0.2 GPa range and is shown schematically in Fig. 2. It consisted of a heavy wall steel cylinder, a steel piston with two circumferential o-rings, and two gauge ports. One port fit a PCB Piezotronics Co. transducer with a 0.07315 GPa/volt calibration. The other port had 1/4-in. pipe threads to fit a 1/8-in. Swagelok fitting. The pinducer was held in this with nylon ferrules.

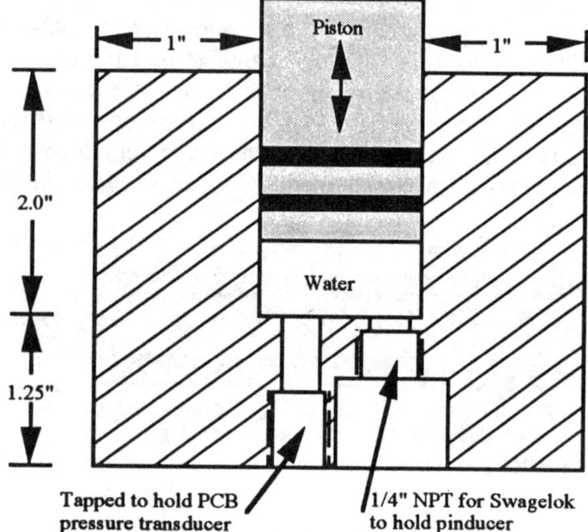

FIGURE 2. Quasi-static piston-driven calibration device.

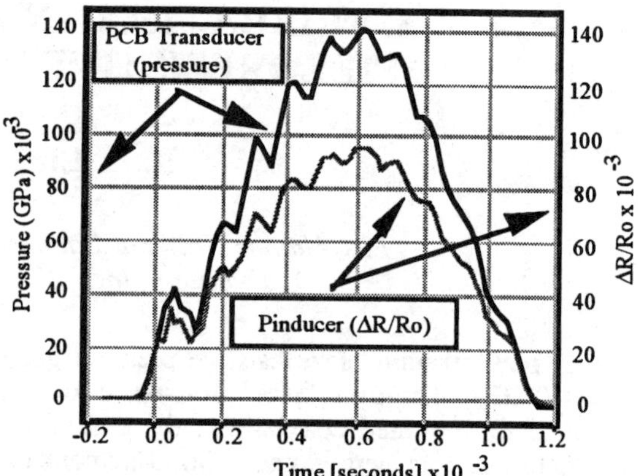

FIGURE 3. Comparison of the gauge wave profiles from a pinducer and a PCB Inc. transducer in the piston-driven quasi-static calibration test.

The piston was struck with a heavy hammer and the gauge's outputs were recorded on a LeCroy 9400 digital oscilloscope. Typically a standard pulsed-Wheatstone bridge-power supply would be used with a thin film, 50-Ω gauge in a shock experiment. Because the piston-driven system produced gauge wave profiles of 1 ms or longer, a 9v direct-current power supply with a 50 Ω viewing resistor was used with 1 MΩ termination at the oscilloscope. An example of the results is shown in Fig. 3. After each test the resistance of the pinducer was measured, and if it had not changed, an additional test was run. Each pinducer was typically capable of withstanding many such tests before damage and permanent resistance change occurred, an example of the pinducer's robust construction. The data produced in this calibration consist of 58 points from 4 different pinducers. A quadratic least squares fit (LSF) with a zero intercept yields:

$$P(GPa) == 1.318(\Delta R/R_0) + 2.202(\Delta R/R_0)^2 \quad (1)$$

The scatter in the data is remarkably small and show that at least in this regime, the pinducer response is quite reproducible.

Four calibration points were generated with the gas gun at Dynasen in the 0.2 to 0.8 GPa range. The calibration system used a 1/2-in.-thick PMMA flyer impacting a 1-in.-thick PMMA target with two embedded pinducers. The top of the pinducers were flush with the PMMA impact interface, and two carbon-film gauges were placed on this interface to produce reference peak pressures and wave forms. In addition, several different makes and sizes of

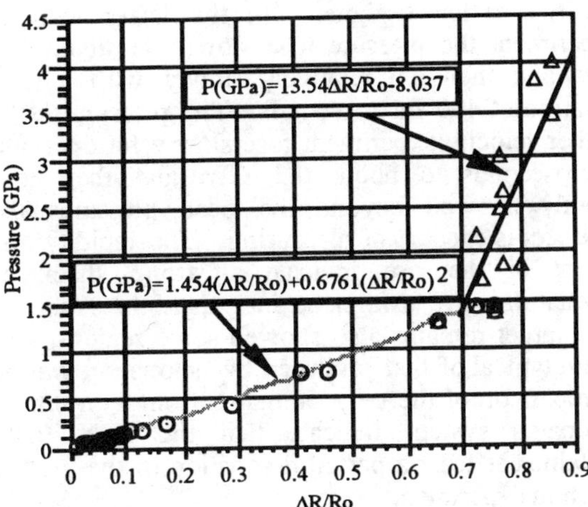

$$P(GPa)=13.54\Delta R/Ro-8.037$$

$$P(GPa)=1.454(\Delta R/Ro)+0.6761(\Delta R/Ro)^2$$

FIGURE 4. Pressure versus time plot of all calibration data. The quadratic LSF is to all data below 1.5 GPa, and the linear LSF is to all data above 1 GPa.

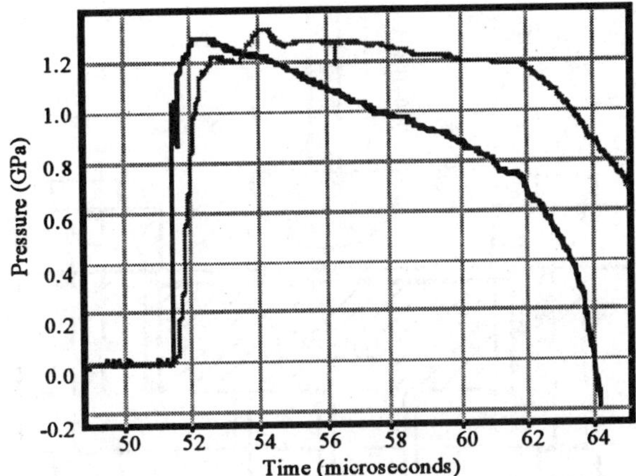

FIGURE 5. Comparison of the shock response of a carbon-film gauge and a pinducer. The experiment used a NQ plane-wave lens to drive a low-pressure shock into PMMA.

resistors were tried in the pinducers. It was concluded that 1/8-watt resistors did not reproduce the waveform adequately, and 1/4-watt Mouser brand resistors produced slightly better results than Allen Bradley.

The third set of calibration data were generated using explosive plane-wave generated shock waves. Five experiments produced 16 calibration points in the range 0.7 to 4.0 GPa. Four of the shots used Detasheet™-nitroguanidine (NQ) plane-wave lenses to drive shocks into PMMA-copper, aluminum-polymethyl methacrylate (PMMA), aluminum-copper, and PMMA-PMMA stacks of material. Reference gauges were carbon film in all cases. Some measurements were made at the interfaces of different materials, while others were made between layers of identical material. The fifth shot used a calcitol- PBX 9501 plane-wave lens, NQ driver, and an aluminum-PMMA-PBX9501 material stack. Measurements were made between the PMMA and PBX 9501. All the data from the three sources are shown in Fig. 4.

At the higher shock pressures there is evidence of ring up in the pinducers to come to the equilibrium pressure with its surroundings. An example of this can be seen in Fig. 5 where the response of a pinducer is compared to that of a carbon-film gauge. This is a result from the experiment where a NQ plane wave was used to drive a low-level shock into PMMA. It is also evident from Fig. 5 that the rise time of the pinducer ($\sim$0.5 μs) is slower than that of the carbon film. The response to the trailing rarefaction is also somewhat different, with the carbon film response falling off faster than that of the

pinducer. Because calibration of pressure drops after shocks is so difficult, it is impossible to estimate which, if either, gauge response is correct in this regime.

It is clear from Fig. 4 that pinducers are reliable from 0 GPa to about 1.0 to 1.5 GPa. At pressures above 1.5 GPa, there is very little change in resistance for significant changes in pressure. Thus, it is difficult to reliably measure the difference between a 2.0-GPa shock and a 4.0-GPa shock. Up to 1.5 GPa, there is very little scatter in the data. The LSF of all this data predicts a pressure only 3% different from that predicted by the LSF in Equation (1) at $\Delta R/Ro=0.12$, the upper extent of this data. Thus, the data from the three types of calibration are consistent up to 1.5 GPa.

APPLICATION AND RESULTS

The pressure regime of interest in XDT or DDT processes is usually below 1.0 GPa. In these experiments, pressures often rise quasi-statically before generating a shock that can build and run to detonation. Thus, pinducers are ideal for these types of applications.

A set of experiments has been designed and executed at LANL to study the XDT potential of PBX 9501. The experiments were similar to those done by Chidester et al. (7). A schematic of the set-up is shown in Fig. 6. A one piece stainless-steel holder with a 1.125-in.-deep, 6-in.-diameter hole milled in it is used to hold a 1-in.-thick, 5.8-in.-diameter piece of PBX 9501. The base of the holder was 0.5-in.-thick stainless and had 0.125-in. holes

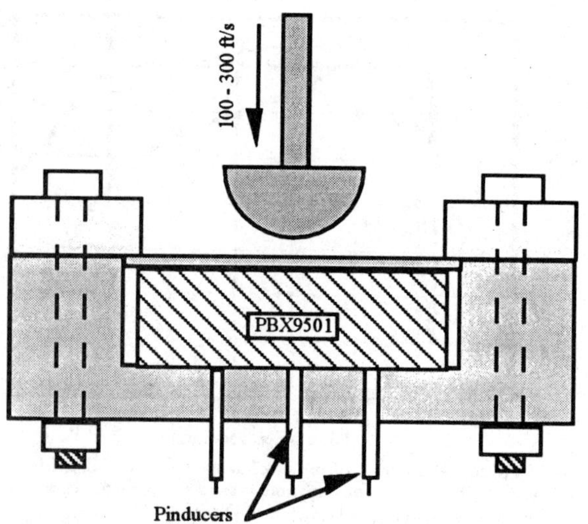

FIGURE 6. Schematic of spigot gun experiments. The PBX 9501 is held in a stainless-steel holder and is impacted by the mild steel projectile fired from the spigot gun.

drilled through it to accommodate the installation of pinducers. Above the PBX 9501 a 0.125-in.-thick stainless-steel cover plate was held in position with a heavy retaining ring and eight bolts as shown. Several shots were fired without instrumentation in the targets to determine that the velocity threshold for violent reaction was in the 250-275-ft/s region.

In two recent experiments with instrumention, one with a projectile velocity of 235 ft/s showed no significant reaction, while the other at 265 ft/s reacted violently. Pressure versus time profiles are shown in Fig. 7 for the same pinducer position in

the two different shots. In the lower velocity experiment the pressure rose slowly to about 0.1 GPa and then fell relatively slowly with a total duration of 400 microseconds. The pressure in the higher velocity experiment rose slowly for over 500 microseconds to about 0.2 GPa and then rose rapidly to well beyond the peak pressures the pinducer is capable of measuring. The rapidity and extent of the rise indicate a strong shock or detonation wave. Magnetic analysis of the stainless-steel target remains also showed some regions with strain typical of that produced by detonation waves. Comparison of the target remains from a promptly detonated system indicate that the PBX 9501 sustained at least a partial detonation in the violent reaction experiment.

CONCLUSIONS

The carbon composition resistor based pinducer has been shown to be capable of accurate pressure measurements in both the quasi-static and shock regimes up to pressures of 1.5 GPa. It is particularly useful for the hostile environments and long time regime experiments experienced in XDT and DDT processes.

REFERENCES

1. Jensen, R. C., Blommer, E. J., and Brown, B., "An Instrumented Shotgun Facility to Study Impact Initiated Explosive Reaction," in the Seventh Symposium on Detonation, 1981, pp. 299-307.
2. Keefe, R. L., "Delayed Detonation in Card Gap Tests," in the Seventh Symposium on Detonation, 1981, pp. 265-272.
3. Bernecker, R. R., Sandusky, H. W., and Clairmont, A. R. Jr., "Deflagration-to-Detonation Transition (DDT) Studies of a Double-Base Propellant," in the Eight Symposium on Detonation, 1985, pp. 658-668.
4. McAfee, J. M., Asay, B. W, Campbell, A. W., and Ramsay, J. B., "Deflagration to Detonation in Granular HMX," in the Ninth Symposium on Detonation, 1989, pp. 265-279.
5. Ginsberg, M. J., and Asay, B. W, "Commercial Carbon Composition Resistors as Dynamic Stress Gauges in Difficult Environments," Rev. Sci. Instrum. 62 (9), 2218-2227 (1991).
6. Rosenberg, T., Yaziv, D., and Partom, Y., "Calibration of Foil-Like Manganin Gauges in Planar Shock Wave Experiments," J. Appl. Phys. 51 (7), 3702-3705 (1980).
7. Chidester, S. K., Green, L. G., and Lee, C. G., "A Frictional Work Predictive Method for the Initiation of Solid High Explosives from Low-Pressure Impacts," Lawrence Livermore National Laboratory UCRL-JC-114186 (1993).

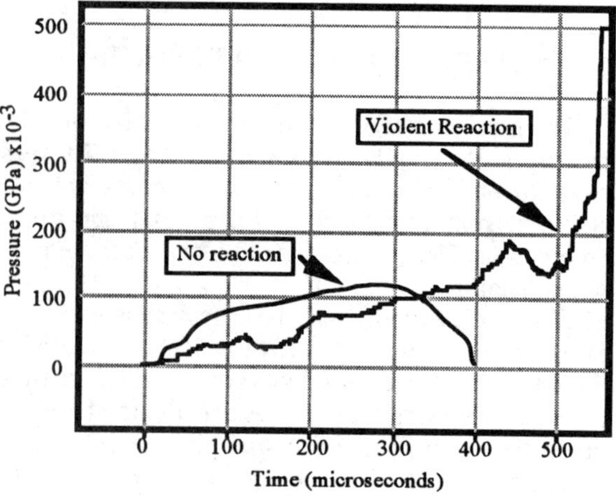

FIGURE 7. Wave profiles produced by pinducers from two experiments at slightly different impact velocities.

IMPROVED ANALYTICAL MODEL FOR THE PIEZORESISTIVE RESPONSE OF LATERAL MANGANIN STRESS GAUGES

Wang Wu, Jin Xiaogang and Zvi Rosenberg*

Southwest Institute of Fluid Physics, P. O. Box 523-61
Chengdu, Sichuan, 610003, P. R. China
** Rafael, P.O.Box 2250 Haifa, Israel*

A modified analytical model, based on Z. Rosenberg's by incorporating work-hardening in the piezoresistance analysis, for the response of lateral manganin gauges is presented. And the method to derive the model parameters by using dynamic calibration curve of the longitudinal manganin gauges is provided, too. To check its validity, planar impact experiments with aluminum alloy specimen are conducted, in which longitudinal and lateral manganin gauges are embedded in the target to measure the stresses in two directions at the same time . The longitudinal stresses in the experiments are in range of 0.87-8.2Gpa

INTRODUCTION

One of the most important issues in the field of dynamic response of materials is the exact determination of the lateral stresses in shock loaded solids. The only direct technique to measure these stresses is by using piezoresistance stress gauges, as was first done by Bernstein *et al*[1]. These gauges are embedded in a plane parallel to the shock wave propagation direction, as shown schematically in Fig. 1. They analyzed their data with the aid of the

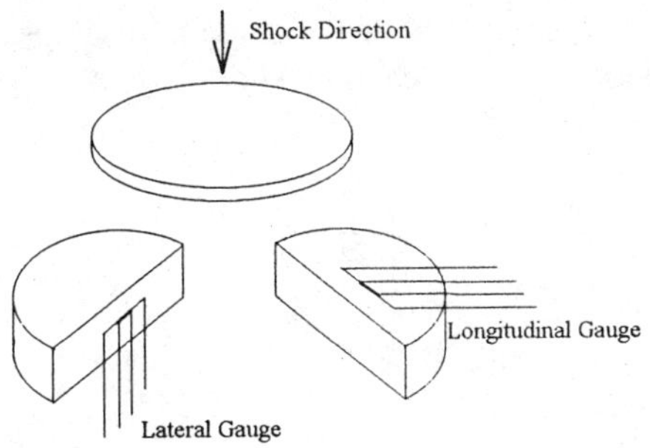

FIGURE 1. Schematic representation of longitudinal and lateral stress measurement

gauge calibration curve, in terms of relative resistance change versus shock pressure, which was determined for the longitudinal gauge configuration. However, the response of piezoresistance gauges in lateral configuration can be very different from that in the longitudinal configuration as was first demonstrated in Ref. 2 for foil gauges embedded in PMMA. In Ref. 3 a relatively simple and straightforward analytical model is presented by Z. Rosenberg *et al* to account for the different response of the gauge in these two configurations. The model is based on a few assumptions concerning the stress and strain states of the gauge material. However, the assumption of small strain for the model results in that the model can only be used for the analysis of the gauges in relatively very hard matrix, such as glass and ceramics, at low stresses.

In this paper a modified analytical model for manganin gauge is proposed based on the Rosenberg's by incorporating the work hardening properties of the gauge. We use the modified model to account for the experimental data of the manganin gauges in LY-12Al matrix loaded by planar impacting. The results verify validity of our model.

ANALYTICAL MODEL FOR MANGANIN GAUGE

In Ref. 3, the authors divide the response of piezoresistance gauge into two ranges — elastic and plastic ranges. They point out that piezoresistance response in the elastic range of the gauge material depends on the matrix material, and that, on the other hand, these gauges have a unique calibration curve, both for loading and unloading within their plastic range. They have developed an analytical model for the responses of longitudinal and lateral manganin gauges (for small strains) in plastic range and the expressions of the model is

$$\left(\frac{\Delta R}{R_0}\right)_{long.} = \frac{K}{K_g}(\sigma_x^m - \frac{2}{3}Y_g) \qquad (1.1)$$

$$\left(\frac{\Delta R}{R_0}\right)_{lat.} = \frac{K}{K_g}(\sigma_y^m + \frac{1}{3}Y_g) \qquad (1.2)$$

where ΔR is the resistance change of the gauge, R_0 is the initial resistance of the gauge, K_g is the bulk module of gauge material (manganin), K is the dynamic piezoresistance coefficient relating the relative resistance change of the gauge to its volume strain, Y_g is the yield strength of the gauge, and σ_x^m and σ_y^m are longitudinal and lateral stresses in the matrix, respectively.

We assume Y_g is pressure dependent. According to the experimental results in Ref. 4, the yield strength and can be expressed approximately as

$$Y_g = \beta \cdot \sigma + Y_{g_0} \qquad (2)$$

where σ is stress, β is assumed hardening coefficient and Y_{g_0} is initial or static yield strength of the gauge. Replacing the ratio K/K_g in equation (1.1) and (1.2) with α and by a similar elasto-plastic analysis in Ref. 3, we finally obtain the modified model for manganin gauges:

$$\left(\frac{\Delta R}{R_0}\right)_{long.} = \frac{\alpha(3-2\beta)}{3}(\sigma_x - \frac{2}{(3-2\beta)}Y_{g_0}) \qquad (3.1)$$

$$\left(\frac{\Delta R}{R_0}\right)_{lat.} = \frac{\alpha(3-2\beta)}{3(1-\beta)}(\sigma_y + \frac{1}{(3-2\beta)}Y_{g_0}) \qquad (3.2)$$

where the superscript 'm's in stress terms are omitted. The response of longitudinal gauge in elastic region can be expressed as

$$\left(\frac{\Delta R}{R_0}\right)_{long.} = \frac{\alpha(1+v_g)}{3(1-v_g)}\sigma_x \qquad (3.3)$$

where v_g is Poison's ratio which approximately equals 1/3 for metals. Since the whole experimental data of the lateral gauges in our research are in plastic range, we won't discuss the expression of the lateral gauges' response in their elastic range here.

EXPERIMENT

The planar impact experiments were conducted on our 57-mm gas gun in Lab for Shack Wave and Detonation Physics Research, Southwest Institute of Fluid Physics. The commercial foil manganin gauges we use are about 0.15 mm thick after encapsulating. The schematic view of the experimental set-up is shown in Fig. 2. The flyers and targets are made of the same material—LY-12 aluminum. In each shot, two gauges,

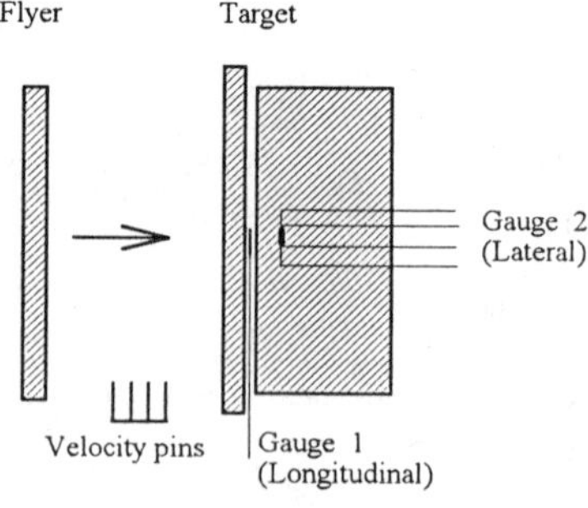

FIGURE 2. Experimental configuration for planar impacting

longitudinal one and lateral one, are embedded in the combined target.

Nine shots were conducted with impact velocities between 117m/s to 993m/s. The correspondent longitudinal stresses in the targets are between 0.87Gpa to 8.2Gpa, which are calculated through the following well-known shock-wave relations

$$u = w / 2 \qquad (4.1)$$

$$D = C_0 + \lambda \cdot u \qquad (4.2)$$

$$\sigma_x = \rho_0 \cdot D \cdot u \qquad (4.3)$$

where u, w, D and ρ_0 are particle velocity, impacting velocity, shock wave velocity and initial density, respectively, C_0 and λ are Hugoniot coefficients. The material constants, used in calculations, for LY-12Al are: ρ_0=2.78 × 10^{3}m/s, C_0=5.25 × 10^{3}m/s and λ=1.39.

RESULTS AND DISCUSSIONS

The calibration curve for manganin gauges in longitudinal configuration is shown in Fig. 3. The experimental data are fitted with two straight lines that respectively represent the elastic region and the plastic region of the piezoresistance responses of these gauges, and the critical stress marking the onset of yielding of the gauges is 1.5Gpa. Obviously, the gauges behave normally and reasonably compared with those in Ref. 5. The two straight lines in Fig. 3 can be expressed as

$$\left(\frac{\Delta R}{R_0} \right)_{long.} = 0.0212\sigma_x \qquad (0 < \sigma_x \leq 1.5GPa)$$

$$(5.1)$$

$$\left(\frac{\Delta R}{R_0} \right)_{long.} = 0.0281\sigma_x - 0.0122 \quad (\sigma_x > 1.5GPa)$$

$$(5.2)$$

The agreement between the fitting lines and the

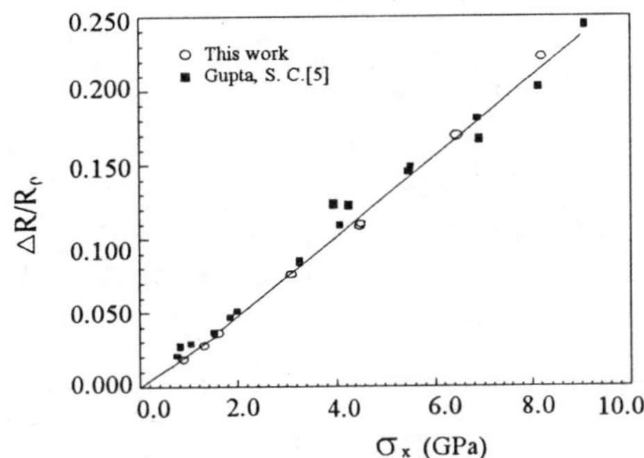

FIGURE 3. Calibration curve of manganin gauge in longitudinal configuration

experimental data is very good, as one can see from the figure. Comparing above two equations with (3.1) and (3.3), we can determine the parameters in these equations as: α=0.03186(1/GPa), β=0.1770, and Y_{g_0} =0.574(GPa).

Finally, the calibration relationship of the lateral manganin gauges in their plastic region can be expressed as

$$\left(\frac{\Delta R}{R_0} \right)_{lat.} = 0.0341\sigma_y + 0.0074 \qquad (5.3)$$

while the Rosenberg's original model for lateral manganin gauge can be expressed as

$$\left(\frac{\Delta R}{R_0} \right)_{lat.} = 0.0281\sigma_y + 0.0061 \qquad (5.4)$$

We can now use equations (5.3) and (5.4) together with the lateral gauge records in experiments to calculate the lateral stress σ_y in the matrix, separately. Assuming a Von-Misses (or Tresca) yield criterion, we thus can deduce the dynamic yield strength of the target material using following expression:

$$Y = \sigma_x - \sigma_y \qquad (6)$$

The calculation results are drawn in Fig. 4. To

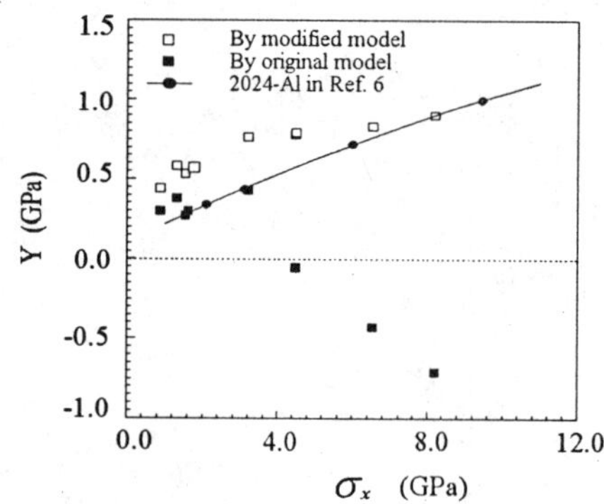

FIGURE 4. Variation of the yield strength with shock stress for LY-12Al

check these data, the yield strength curve of 2024Al, which is the counterpart of the LY-12Al, is shown in the figure, too. The curve is determined by way of measuring the elastic unloading of shock wave with two manganin gauges both in longitudinal configuration (see Ref. 6 for details). We can see, from the figure, that the data of yield strength Y calculated by the original model are reasonable only in the low stress region, and that the model becomes completely invalid with the increase of longitudinal stress. On the other hand, the data calculated by the modified model are relatively reasonable in the whole experimental stress range. We have also noticed the deviation of the data by modified model in the low stress region from the solid line. We think the deviation is due to the influence of the geometry of the gauges on their piezoresistance responses. Specifically, it is due to the deformation of the sensitive element of the gauge along current direction. We have carried out a rigorous calculation for modification about the contribution in the upper section of the stresses but an approximate one at low stresses because of limited experimental data. We will discuss this problem in detail in another paper. As a summary, we believe that the work-hardening included model is of potential values for the engineering use, although more experiments concerning various matrix materials and in a wider stress range are needed to prove its validity or to improve it.

ACKNOWLEDGMENTS

This work is supported by the sci & tech funds of CAEP.

REFERENCE

1. Bernstein, D. , *et al, Behavior of Dense Media under High Dynamic Pressures,* New York: Gorden and Breech, 1968, 461-468
2. Rosenberg, Z., Partom, Y. and Yaziv, D., *J. Appl. Phys.,* 52, 755-758(1981).
3. Rosenberg, Z. and Brar, N. S. , *Applied Acoustics,* 41, 377-386(1994).
4. Rosenberg, Z. and Partom, Y. , *J. Appl. Phys.,* 57, 5084-5088(1985).
5. Gupta, S. C. and Gupta, Y. M. , *J. Appl. Phys.,* 62, 2603-2609(1987),
6. Rosenberg, Z. and Partom, Y. , and Yaziv, D., *J. Appl. Phys.,* 56, 143-146(1984).

ON THE SOURCE FOR FAILURE
OF COMMERCIAL MANGANIN GAUGES
AT HIGH SHOCK PRESSURES

Z. Rosenberg and A. Ginzburg

RAFAEL, P.O. Box 2250, Haifa, Israel

It is a well known fact that commercial Manganin gauges fail at shock stresses near 20 GPa because of the short circuiting of their plastic encapsulation at these stresses. We present experimental data which show that these gauges can withstand much higher stresses if loaded by a multiple series of reverberation (when embedded in different materials). These experiments support a different interpretation by which the short-circuiting of the gauge at high shock amplitude has to do with the temperature rise, rather than stress (or pressure), in the epoxy layers surrounding the gauge.

INTRODUCTION

Piezo-resistive stress gauges are being used for over 30 years as a main diagnostic tool in shock wave experiments. Since the pioneering works of Fuller and Price [1] and Keough [2] on Manganin gauges two other materials became quite useful (especially in the low stress range) Ytterbium and carbon. Manganin is the most popular gauge material mainly because of its large range of applicability, which extends to 150 GPa according to [3]. Most commercial gauges are foils with various grid designs which are encapsulated between thin plastic sheets with an overall thickness of 50-100 μm. The gauge material is usually 5-10 μm thick and the plastics used for encapsulation are made of fiber-reinforced epoxy resins or Kapton. These gauges are commonly used to record stress time histories in the 1.0-20 GPa range (Manganin) and for the 0-2 GPa range (Ytterbium and carbon).

Many experimentalists in this field have realized that the epoxy and kapton encapsulated Manganin gauges fail at around 20 GPa by short circuiting (at least partially). In order to extend the measurement range, new gauges have been designed. The work of Vantine, Erickson and their colleagues [4,5] is based on the earlier findings of Champion [6] who realized that most plastics become electrically conductive at around 20 GPa. Teflon (Polytetrafluoroethylene) is an exception, as it remains a good electrical insulator up to pressures in the 100 GPa range. Thus, in order to measure pressures in the 30-50 GPa range (detonation pressures, for example) the Lawrence Livermore group encapsulated their Manganin foils in between two teflon sheets using a softer teflon as an adhesive and a hot press to make sure that the package is flat and thin enough. Stresses over 20 GPa are routinely measured with these gauges and the common assumption is that their success is due to the absence of epoxy or any other plastic near the gauge foil (except for teflon).

The purpose of the present paper is to describe several experimental results with commercial Manganin gauges, encapsulated between epoxy sheets, in which we have measured stresses in the 25-30 GPa range. These records should lead to new interpretations for the cause of failure of gauges at high shock stresses which may be linked to the actual stress history experienced by the gauge rather than the final stress.

EXPERIMENTAL TECHNIQUES AND RESULTS

In order to achieve high shock stresses we used the experimental configuration shown schematically in Figure 1. Basically, a plane detonation wave is initiated in a cylinder of HMX-type high explosive which is transmitted through various buffer materials to the gauge location. The gauges we used are manufactured by micro-measurements (type LM-SS-210FD-50). These are 50 Ω grid-like gauges about 5×6 mm wide attached to a glass reinforced epoxy sheet. The sheet thickness is about 40 µm and the gauge is 10 µm thick. Adhesion of the gauge and the different disks was achieved with a 24-hour epoxy resin (Hysol 0151 clear) under pressure to avoid air bubbles on and near the gauge.

The first experiment we performed included four gauges embedded in different materials. A buffer disk of aluminum was sandwiched between the explosive and the gauges in order to reduce the stress level to around 20 GPa. The four embedded materials were teflon, polypropylene, plexiglass (PMMA) and a sandwich of teflon (first disk) and plexiglass (second disk). In all of these arrangements the thickness of the first disk was 2 mm.

Figure 2 shows the four gauge records from this experiment where it is clearly seen that the gauge embedded in teflon measures a stress of 25 GPa with a relatively long duration of measurement, while the gauge embedded in PMMA shows irregular features and short circuits after a short time. The other gauges measure stress around 17 GPa for a long time as expected.

The record of the gauge in plexiglass shows a piezoelectric signal which the gauge picks up when the shock wave reaches the first PMMA disk. All other traces have a small disturbance at the same time which is the result of the same piezoelectric phenomenon being picked up by these gauges. The overshoot of the gauge, at the moment the shock wave enters its location, is probably also due to this piezoelectricity. However, the most interesting feature is the short-circuiting of the gauge after about 1 µs. This short measurement time is typical for gauges embedded in plexiglass or epoxy when shocked to stresses near 20 GPa and, as discussed above, this was attributed to the high pressure conductivity of these materials. Considering the gauge embedded in teflon, which experiences a much higher stress level (25 GPa as compared to 17 GPa), we find that this is probably not the right explanation.

One has to consider the fact that the gauge in teflon is also surrounded by epoxy and is actually encapsulated in epoxy rather than teflon. The only difference between the two embeddments (besides the amplitude of the signal) is the fact that the gauge in teflon reaches its peak stress level via a series of reverberations while the gauge in PMMA experiences a single shock to 17 GPa. Thus, the temperatures of the gauge encapsulation should be different in the two cases with a much lower temperature for the gauge in teflon.

In order to further substantiate this line of thought, we performed a second experiment in which the gauge was sandwiched between two teflon disks. This time we attached a copper disk on each side of the sandwich so that a stress increase is expected to reach the gauge location from the backing copper disk. Figure 3 shows the gauge record. As is clearly seen, the gauge record survives the high stress of around 30 GPa. Thus, gauges embedded in teflon, even if they are encapsulated in epoxy, do not short circuit at these high stresses. Consequently, the assumption that epoxy at about 20 GPa short circuits the gauge has to be reconsidered. It seems that this short-circuiting is a more complex phenomenon and has to do with the detailed stress history experienced by the gauge and its encapsulation

A possible explanation can rely on the high temperatures which are attained in these plastic materials when shock loaded to stresses near 20 GPa. A rough estimate yields a temperature of near 1500°C for plexiglass (or epoxy) when shocked to 20 GPa by a single shock. On the other hand, multiple shock loading is a quasi-isentropic process, resulting in a much lower temperature. We propose this mechanism for the short circuiting of gauges at high shock stresses and suggest that more experiments be performed to check this assumption. For example static high pressure loading of plexiglass (or epoxy) should confirm our assumption if no reduction in electrical resistivity is observed at 20 GPa.

CONCLUSIONS

Experiments were performed in order to investigate the source for short-circuiting phenomenon experienced by piezo-resistive gauges when shocked to about 20 GPa. These experiments indicate that it is not the stress amplitude which causes gauge failure, but rather the detailed stress history of its loading. When embedded between teflon disks, gauges withstand shock stress near 30 GPa without short circuiting, while when embedded in plexiglass they failed near 20 GPa. These findings led us to conclude that the epoxy encapsulation fails (by short circuiting) because of the high temperatures which are produced during a single shock of these amplitudes. Much lower temperatures are anticipated when gauges are embedded between teflon disks because of the impedance mismatch between teflon and epoxy. More experiments, on these lines, are needed to fully understand the reasons for gauge failure, by short circuiting, at high shock pressures.

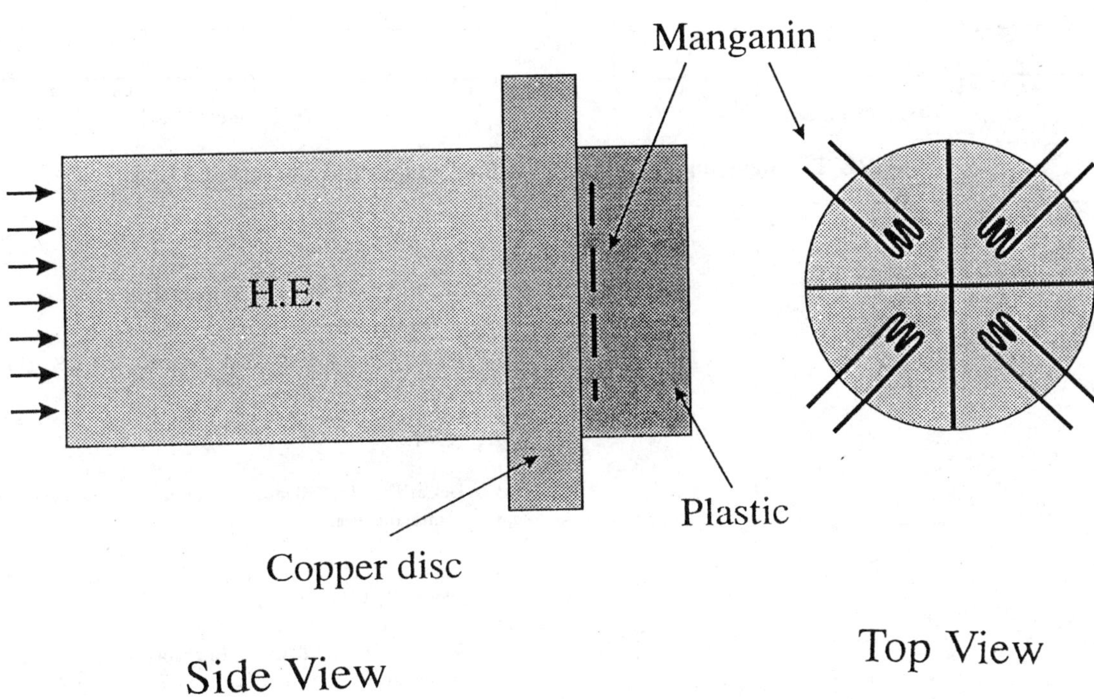

FIGURE 1. Schematics of experimental configuration.

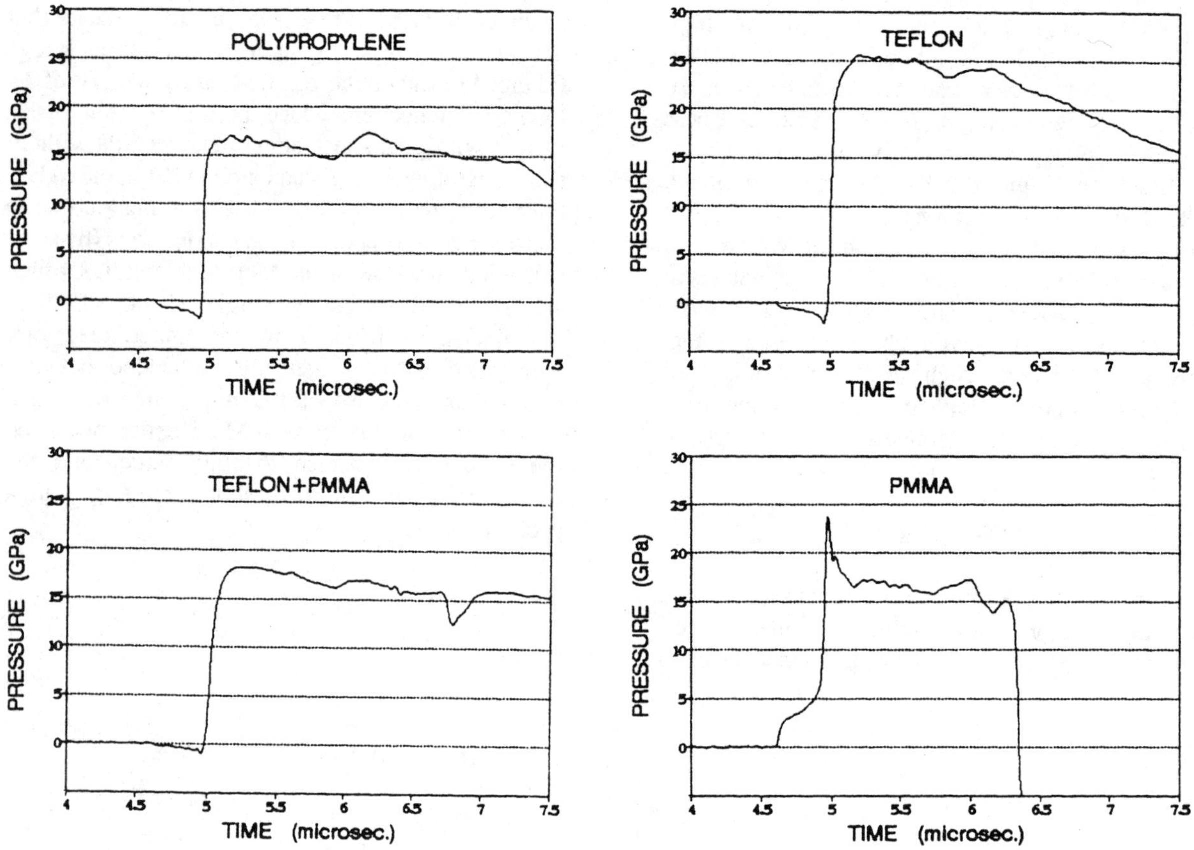

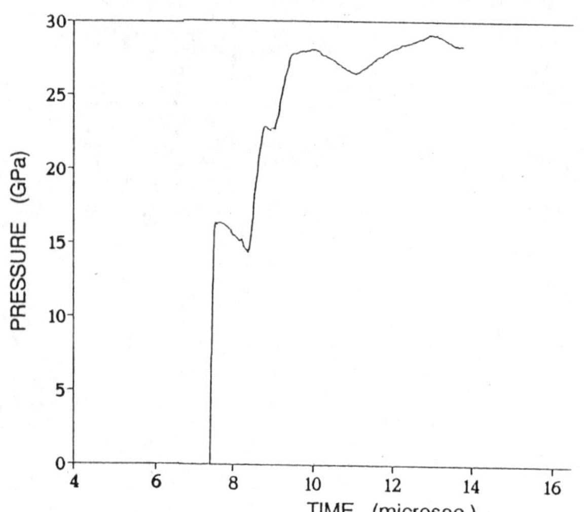

FIGURE 2. Stress time records of the four gauges in Experiment #1.

FIGURE 3. Gauge record in Experiment #2.

REFERENCES

1. Fuller, P.J.A. and Price, J.H., *Nature*, **193**, 262, **(1962)**.

2. Bernstein, D. and Keough, D.D., *J. Appl. Phys.*, **35**, 1471, (1964).

3. DeCarli, P., Stanford Research Institute, private communication.

4. Vantine, H.C., Erickson, L.M. and Janzen, J.A., *J. Appl. Phys.*, **51**, 1957, (1980).

5. Vantine, H.C., Chan. J., Erickson, L.M., Janzen, J. and Weingart, R., *Rev. Sci. Instrum.*, **51**, 116, (1980).

6. Champion, J., *J. Appl. Phys.*, **43**, 2216, (1972).

FRACTOEMISSION AND ITS EFFECT UPON NOISE IN GAUGES PLACED NEAR CERAMIC INTERFACES

N.K. BOURNE AND *Z. ROSENBERG

*Shock Physics, Cavendish Labs, Cambridge, UK. *RAFAEL, PO Box 2250, Haifa, Israel.*

It has been noticed in several works that a gauge mounted close to a metal surface shows some electrical ringing when loaded rapidly by a shock wave. This effect may be explained in terms of the capacitance introduced between gauge and target. It is not expected that such effects will be observed when gauges are placed onto insulators such as ceramics. However, a similar trace was observed in which ringing at the top of the fast-rising elastic wave was seen. We present gauge data to illustrate this effect and show a simple solution to suppress this noise with an assessment of its limitations. High speed photographs and fast emission measurements are used as an aid in explaining the phenomenon.

INTRODUCTION

The use of gauges in shock loading experiments is still widespread despite modern advances in other diagnostic instrumentation (1). Of particular value in plate impact experiments are manganin gauges (2) with fast response and well established calibration. However, with the advent of cheaper storage oscilloscopes operating at GHz it has been possible to observe additional features on the gauge traces not before commented upon. It has been noted before (3) that the traces obtained for metals showed characteristic dips before the rapid rise on shock arrival. Additionally, it is possible to see ringing on the tops of these pulses. Similar features had not previously been observed in ceramics such as alumina which are insulators. When we consistently found overshoots and ringing in alumina experiments it was decided to adopt a parallel strategy to explain these features. One part of our work consisted in rebuilding our differential power supply in order to minimise ringing in the bridge circuit. The second constituted an experimental investigation using high-speed photography and fast photodiodes to identify parts of the break-up process which could give rise to the observed phenomena. The power supply improvements reduced electrical noise by a small amount. The results of the second investigation are presented below.

EXPERIMENTAL

The results presented were collected from instrumented plate-impact experiments carried out on the single-stage gas gun at the Cavendish laboratory (50 mm bore, 1 km s^{-1} maximum impact velocity). Uniaxial strain compressive shock and release waves were recorded after travel through tiles of target materials backed with 12 mm thick PMMA blocks in order that the processes occurring at the interface where the gauge was placed could be observed. The longitudinal stress normal to the planar wave fronts was recorded using piezoresistive stress gauges. In some experiments the gauge was embedded 1 mm away from the ceramic/PMMA interface in the backing block.

The PMMA backing was chosen because it acoustically matches both the gauge's backing material and the epoxy used to assemble the sample. The signal recorded thus followed the form of the stress history accurately and with minimal smoothing of discontinuous stress rises (signal rise times of *ca.* 30 ns were typical on elastic waves). The gauge was bonded to the sample with a low viscosity epoxy and was separated from it by a 25 μm mylar sheet. The gauges used were MicroMeasurements manganin gauges (LM-SS-125CH-048) and the calibration data of Rosenberg *et al.* (4) were used in reducing the voltage data collected.

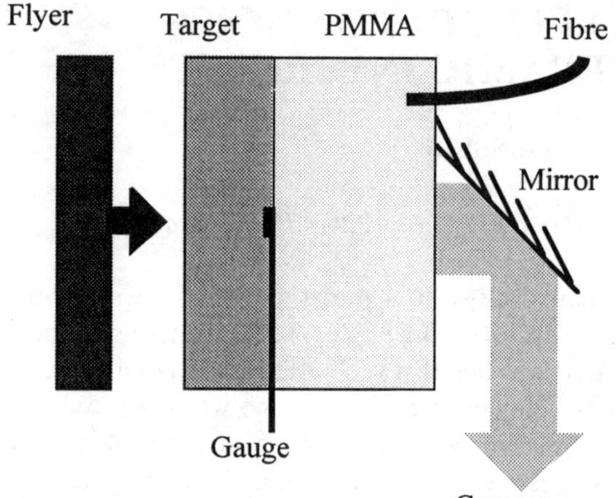

FIGURE 1. Experimental arrangement for high-speed photography experiments.

The signals were recorded using a fast (1 GS s^{-1}) digital storage oscilloscope and transferred onto a micro-computer for data reduction. Impact velocity was measured to an accuracy of 0.5% using a sequential pin-shorting method and tilt was adjusted to be less than 1 mrad by means of an adjustable specimen mount. Impactor plates were made from lapped copper and aluminium discs and were mounted onto a polycarbonate sabot with a relieved front surface in order that the rear of the flyer plate remained unconfined.

The rear of the PMMA plate contained an optical fibre fed out of the specimen tank to a fast (GHz) photodiode with maximum sensitivity in the blue, the output of which was displayed simultaneously with the gauge trace. A mirror was bonded at 45° on the rear of the target so that the backsurface could be monitored with a high-speed camera. A schematic of the arrangement is shown in FIG 1. The camera used was Ultranac FS501 (IMCO Electro-optics) programmable image converter camera taking individual frames every 200 ns with an exposure time of 160 ns. No ambient light was used to illuminate the sample so that only emission from the target was observed. A synchronisation pulse fed from the camera to a free channel on the scope to which the gauge was connected allowed synchronisation of the camera with the stress history.

RESULTS

Figure 2 shows typical traces obtained for two conducting ceramics (SiC and TiB$_2$) and for a half-hard copper. The targets were 10 mm thick and were impacted with copper flyers at varying velocities. There are several features of the traces common to each of the materials. Firstly there is an initial dip seen at the moment of arrival of the elastic shock at the gauge. Secondly there is Gibbsian overshoot and ringing at the top of the fastest rising part of the trace. This feature appears at the Hugoniot elastic limit (HEL) where it has been exceeded (*eg*. for SiC) and on the top of the trace for purely elastic shocks. In the copper trace shown it is interesting to note that there is both an overshoot at the HEL and a ringing on the top of the trace.

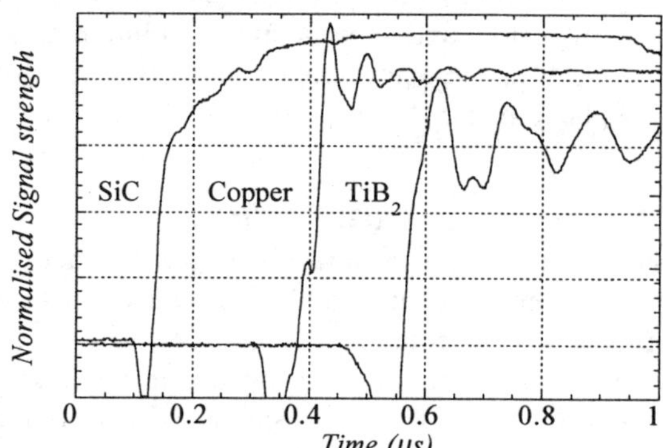

FIGURE 2. Stress pulses in several conducting materials shocked to different levels showing initial dip followed by ringing at HEL and on top of trace. Copper is shocked to *ca*. 100 kbar, SiC shocked to *ca*. 175 kbar, TiB$_2$ shocked to below first cusp. Note the normalised stress axis.

It is important to note that these phenomena are only present when the pulse rises rapidly, judged here to be in times <50 ns. As the rise time of the pulse increases the amplitude of the ringing decreases. These effects are almost never observed with targets that mechanically ramp the wave (such as polymers or glasses).

Figure 3 shows two traces taken by gauges placed on the interface between a 97.5% alumina target and a PMMA backing block and displaced 1 mm into the PMMA away from the interface. These traces are found to be very characteristic of the observed behaviour of aluminas of varying purity (see for example ref. 5 in these proceedings). The principle features of note are the lack of any dip preceding the rise at the elastic shock, and a pronounced ringing at the HEL present over the rest of the trace. It can be seen that with the gauge mounted 1 mm away into the PMMA there is no ringing giving clean traces comparable with VISAR results (see 6 for example).

Separate impact testing of 1 mm sheets of PMMA showed no significant ramping of the wave for stresses of <15 kbar. Here the HEL as measured in the PMMA is *ca.* 10 kbar and the viscoelastic response at this level is negligible so that the measured values at the interface and 1 mm in remain the same.

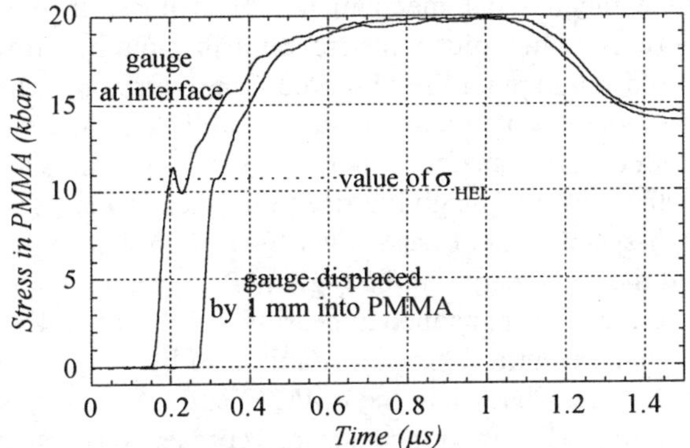

FIGURE 3. Variation of the form of the trace recorded in an alumina with different gauge mounting geometries. Alumina shocked with 3 mm copper flyer at *ca.* 530 m s⁻¹.

For aluminas with much higher HEL this is not the case and significant increases in the HEL are observed (5). This effect will be the subject of future publications.

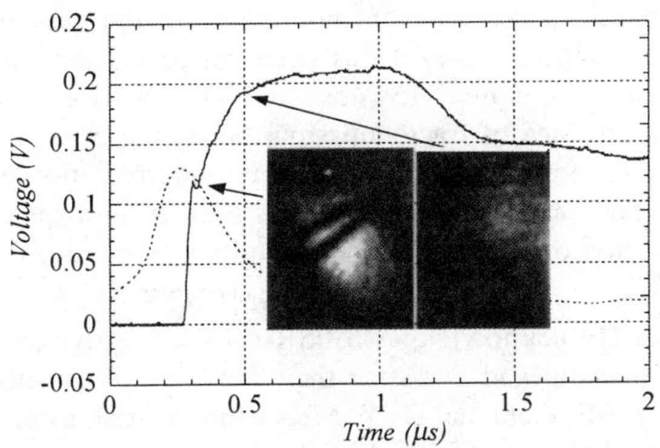

FIGURE 4. Alumina impacted with copper flyer above its HEL. The gauge trace is the full line whilst the photodiode output is shown as a dotted line. Two frames from the simultaneous high-speed record are shown with arrows pointing to the times at which they were taken.

Figure 4 shows frames from a high-speed sequence taken simultaneously with a gauge record and a photodiode trace. The material impacted is a 8 mm thick 97.5% alumina with the gauge mounted 1 mm into PMMA. The light trace rises as the shock arrives at the rear surface of the target. Simultaneously light is recorded in the high-speed camera sequence at this time. The first frame of the

sequence clearly shows the legs of the gauge crossing the centre of the target. The next frame shows light emitted towards the periphery of the tile. This is presumably due to the shock lagging here due to lateral release. There was no light observed for impacts *below* the HEL using these techniques.

DISCUSSION AND ANALYSIS

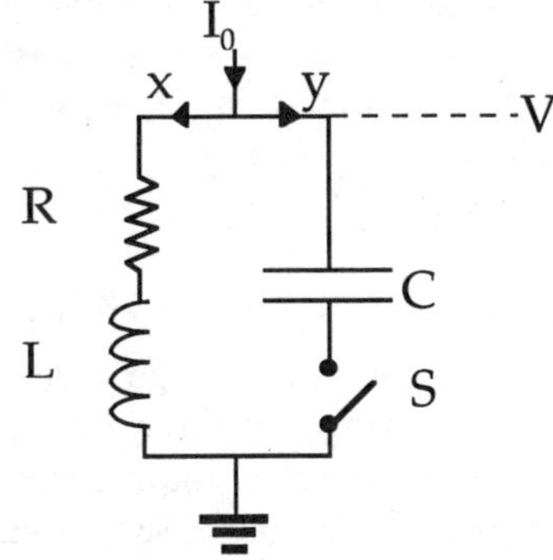

FIGURE 5. Equivalent circuit for analysis of gauge pickup.

The observed behaviour of the gauge can be explained in terms of the capacitive linkage of the gauge into the power supply circuit. The power supply is an initially balanced Wheatstone bridge through which a constant current is supplied for the duration of the experiment by the triggered discharge of a capacitor. The gauge is assumed to be purely resistive so that changes of resistance due to pressure are recorded at the scope. In fact the gauge has a small finite inductance (*ca.* 1 μH) and is linked to the power supply down coaxial cables which have both resistance and capacitance. However, the effect of cabling was shown to be negligible if the cable lengths were kept less than 2 m. Mismatches in the circuitry exist since by definition the gauge resistance changes. However, these small changes do not account for the magnitude of the observed ringing. It was thus hypothesised that the effects were due to a capacitive linkage between the gauge, its legs and the specimen on which it sits separated by a mylar sheet. The capacitance of this combination can be calculated using the well known formula

$$C = \varepsilon A / d ,\qquad (1)$$

where d is the separation between gauge and plate, A is the area of the gauge and legs and e is the permittivity of the medium (*ca.* twice the permittivity of free space for mylar). This gives a capacitance of *ca.* 150 pF for the gauge. Following the schematic of FIG. 5 it is assumed that the arrival of the shock at the gauge accelerates the target towards the gauge pseudoinstantaneously changing the capacitance C. This has the effect of switching S and pulling current onto the charging capacitor C. If the inductance is assumed negligible, the current in the two legs of the circuit is represented by x and y, and the charge on the capacitor at time t is Q, then the following relations can be applied

$$I_0 = x + y, \quad xR = \frac{Q}{C} \text{ so that } R\frac{dx}{dt} = \frac{y}{C} = \frac{(I_0 - x)}{C}.$$

$$\sin ce \ x = 0 \text{ at } t = 0 \quad V = I_0 R (1 - e^{-1/RC}). \tag{2}$$

It can be seen that the voltage V across the gauge dips, not recovering until the capacitor is fully charged. The time constant for this process is *ca.* 8 ns which agrees well with the observed value from the traces of FIG. 2. To explain the ringing oscillations a full solution for the equivalent circuit of FIG. 5 can be obtained by incorporating the gauge inductance L thus:

$$I_0 = x + y, \quad xR + L\frac{dx}{dt} = \frac{Q}{C},$$

$$\Rightarrow L\frac{d^2x}{dt^2} + R\frac{dx}{dt} = \frac{y}{C} = \frac{(I_0 - x)}{C}.$$

$$\text{So that } \frac{d^2x}{dt^2} + \frac{R}{L}\frac{dx}{dt} + \frac{x}{LC} = \frac{I_0}{LC}. \tag{3}$$

This equation has a solution of the form

$$x = I_0 \left\{ 1 - e^{-\frac{R}{2L}t} (\gamma \sin \beta t + \cos \beta t) \right\} \text{ as } x = \frac{dx}{dt} = 0 \text{ at } t = 0.$$

$$\text{where } \beta = \sqrt{\frac{1}{LC} - \frac{R^2}{4L^2}}, \text{ and } \gamma = \frac{R}{2L\beta}. \tag{4}$$

Once suitable values for R, L and C are substituted into (4) a damped oscillatory solution results with the period of oscillation *ca.* 80 ns controlled by the L and C and time constant for the damping being *ca.* 40 ns. Again this agrees with the observed behaviour. It can be seen that the capacitance of the target/gauge interface should be minimised to avoid electrical pickup. The gauge leg material gives rise to the majority of the gauge area and this may be reduced to almost zero by either using thin wires or taking the legs directly out backwards through the PMMA backing. This procedure works well; an example is given in these proceedings (7). A second strategy is to reduce the capacitance by moving the gauge away from the interface back into the PMMA so increasing d by *ca.* 40 times and thus reducing the capacitance by the same amount. This can be seen to be successful by the traces of FIG. 3.

The behaviour observed in aluminas can only be explained if the capacitance appears at the HEL by some mechanical mechanism. It is possible that there is some piezoelectric contribution but this would not explain the observed light emission. The phenomenon of fractoemission is well known (8) to produce free surface charge on newly fractured surfaces as well as giving rise to light emission. The high-speed camera has shown the presence of light, and the free charge in the region of the gauge would give rise to capacitance appearing only at the HEL. Thus the electrical ringing (minimised by movement of the gauge back into the PMMA) strongly suggests that the HEL in this material represents massive *fracture* of the target.

CONCLUSIONS

The observed form of manganin gauge traces in metals and conducting or insulating ceramics has been explained in terms of capacitive linkage between the gauge and the target surface through the dielectric sheet used in mounting. In the case of insulating ceramics it has been suggested that the apparent conductivity observed at the HEL is a consequence of fractoemission during the break-up of the ceramic. This confirms the interpretation of the HEL as a fracture threshold. Further high-speed spectroscopy will be used to quantify the emissions.

ACKNOWLEDGMENTS

NKB acknowledges EPSRC for an Advanced Fellowship and DRA for their funding. We thank Prof. J.E. Field and Dr Y. Mebar for discussions and Mr D.L.A. Cross for technical support.

REFERENCES

1. Field J.E., Walley S.M., Bourne N.K., Huntley J.M., J. de Physique II, Colloq. C8, 3-23, (1994).
2. Rosenberg Z., in APS proceedings of Shock Waves in Condensed Matter 1991, eds. Schmidt S.C., Dick R.D., Forbes J.W., Tasker D.G., Elsevier, 439-446, (1992).
3. Brar N.S. personal communication, (1992).
4. Rosenberg Z, Yaziv, D., and Partom Y., J. Appl. Phys. **51**(7), 3702-3705, (1980).
5. Murray N.H., Bourne N.K., Rosenberg Z., in APS proceedings Shock Waves in Condensed Matter 1995 (1995).
6. Kipp M.E., and Grady D.E., in APS proceedings of Shock Waves in Condensed Matter 1989, Elsevier, 377-380, (1990).
7. Galbraith S.D., Rosenberg Z., and Bourne N.K., in APS proceedings Shock Waves in Condensed Matter 1995 (1995).
8. Dickinson J.T., Donaldson E.E., and Park M.K., J. Mater. Sci., **16**, 2897-2908, (1981).

SHOCK TRACKER CONFIGURATION
OF IN-MATERIAL GAUGE

R. R. Alcon and R. N. Mulford

Los Alamos National Laboratory, Los Alamos, New Mexico 87544

A special configuration of electromagnetic in-material gauge enables measurement of shock time of arrival at up to forty-one points in 10 mm of run distance. This measurement clearly defines the shock line for Lagrangian analysis, and provides sufficient shock velocity data to determine time of turnover to detonation. The measurement mimics a wedge test, with the advantage that data is obtained inside the material, rather than at a surface. The technique has been applied to the reactive systems X-0407, PBX-9404, and thermally damaged PBX 9502. Response to multiple shock inputs is linear, and shock tracker data can be seen to reflect the velocities of the separate shocks. Several configurations will be discussed.

INTRODUCTION

Use of embedded magnetic gauges provides unique measurements in the Lagrangian frame of the time evolution of the shocks. Use of multiple gauges gives independent measurements of particle velocity u_p, shock velocity U_s, and, from impulse records, P. The precision of the gauges is better than 2%.

The shock tracker gauge supplements the MIV configuration, providing a set of shock arrival times at 0.25 mm intervals in the material. Data from MIV records is complemented by the detailed shock velocity (U_s) data provided by the shock tracker.

Lagrange analysis of MIV gauge data[1] depends for its success on definition of a shock line for the experiment. Deriving this shock line solely from ten available gauge records yields an inadequately defined x-t locus. In particular, the slope, and hence the velocities, at the end points are ill-defined. This shock locus is used to give projectile velocity u_p and pressure P in between gauges. It can also define the shock jump at the gauge itself, when the gauge record falls off the line defined by other gauges. This can result in a "kink" in the surface constructed around the line, a discontinuity which is amplified as the surface is expanded away from the shock line. A well defined shock line smooths the surface to more nearly resemble the physical case, and permits more accurate determinations of parameters derived from derivatives along the line or surface.

The shock tracker provides a measurement of acceleration and velocity analogous to the record obtained in a wedge shot, with the added benefits of a reliable, supported, reproducible input plane wave and data at each 1/4 mm. Unlike a wedge shot, the shock tracker gauge is inside the material, and is not susceptible to critical angle or other surface effects.

EXPERIMENTAL

MIV gauges (Magnetic Impulse and Velocity) gauges have been used for many years to directly measure particle velocity in materials accelerated by shock waves. The gauge consists of a set of fine wire loops, actually a 0.2 mil thick aluminum pattern etched onto a 1 mil plastic substrate.[3] The entire shock experiment is conducted in a uniform magnetic field generated by large fixed magnets, with the active region of the gauge perpendicular to the field. When the material under study is accelerated, the MIV loops yield a current as they are carried through this external magnetic field at the particle velocity of the material. The gauge is glued onto a precisely machined planar surface interior to the material to be studied, at a specified angle as shown in Figure 1.

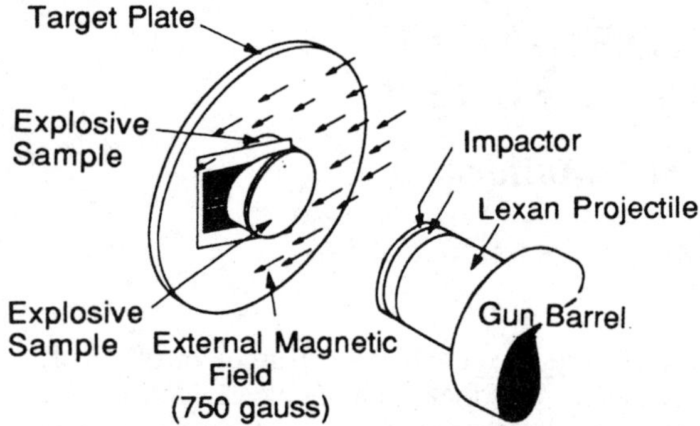

Figure 1. The MIV gauge measures particle velocity as it is propelled through an external magnetic field.

The shock tracker configuration supplements these configurations, as shown in Figure 2. The original design for the shock trackers was developed and tested by John Vorthman.[2]

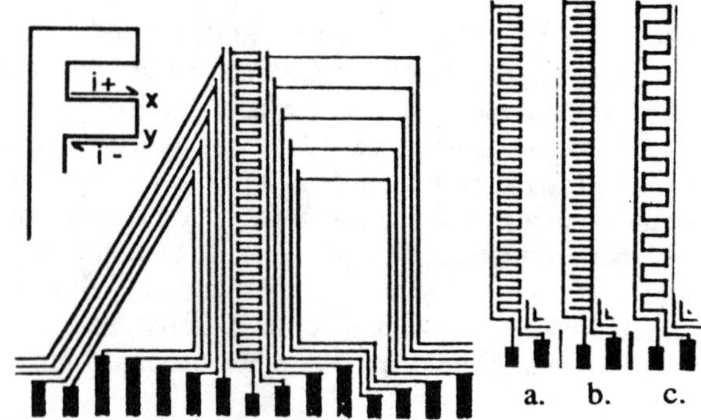

Figure 2. The shock tracker is located in the middle of the MIV gauge. Configurations a), b), and c) have been used.

The shock tracker consists of a large number of accurately spaced elements perpendicular to the magnetic field direction. Different configurations are shown in Figure 2. A closed loop is required, as configuration b was proven to give no response to the shock. Right- and left-going bars on each loop of the tracker generate positive and negative current in response to the velocity jump, as the active loop of the tracker is propelled across the flux lines of the magnetic field. The minimum interelement spacing used is 0.5 mm. The gauge is employed at an angle to prevent each incremental element from perturbing the flow at the next succeeding element. At the usual angle of 30⁰, the 0.5 mm spacing yields data every 0.25 mm.

Experiments described here are done on a single stage light gas gun. Waves generated are one-dimensional for about 4 μsec.

RESULTS

Examples of tracker traces are shown in Figures 3, 5, and 8. Figures 3 and 4 are data from detonating X-0407, Figures 5 and 7 are data from PBX 9502 at 80⁰C before detonation is reached, and Figure 8 shows the tracker response to multiple shocks.

The X-0407 sample shown in Figure 3 demonstrates how the frequency of the center crossings reflects the shock velocity. Both frequency and particle velocity increase smoothly as initiation and growth to detonation occur. Center crossings are measured and plotted against the location of each shock tracker loop in the material, to generate a graph of the shock line, analogous to a wedge record. Velocity data is summarized in the graph in figure 4, showing the turnover to detonation at 1.5 μsec.

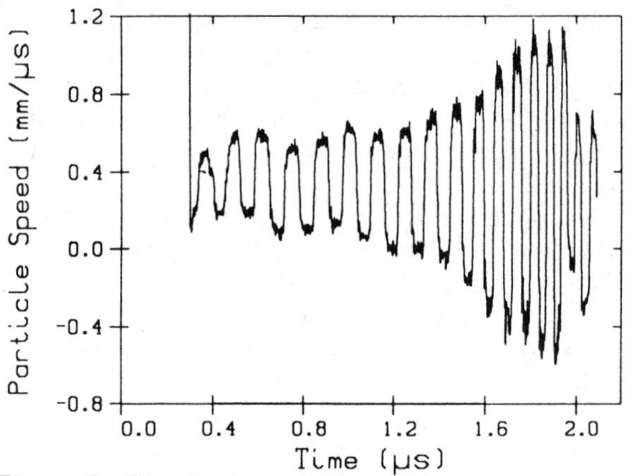

Figure 3. The shock tracker responds to each traverse of a horizontal element by the shock.

The increase evident in the positive and negative u_p amplitude is due to the acceleration of the shock as it traverses a single loop, as shown in Figure 2. The current in leg y will cancel that generated by leg x, unless acceleration of y relative to x enables it to traverse more flux lines than x, in which case the negative current contribution will more than cancel the positive current contribution, and the current excursions will increase in amplitude.

At around 1.9 μsec, the velocity becomes constant at the detonation velocity and the particle velocity returns to its steady state amplitude, reflecting zero acceleration.

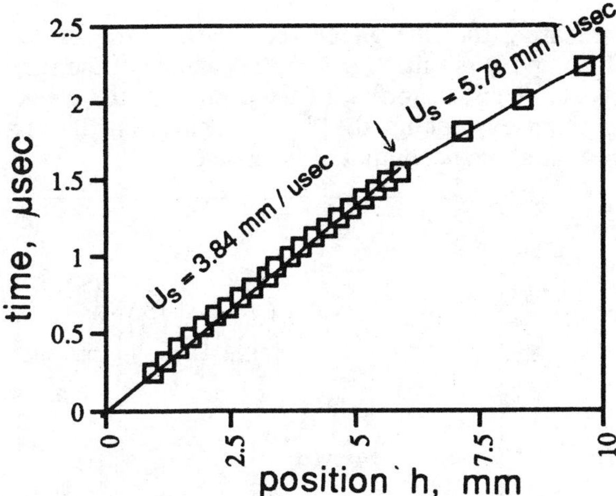

Figure 4. The shock line of the reacting X-0407 is defined by the zero crossings of the shock tracker record. It shows the turnover to detonation much as a wedge record does.

Shock velocities are easily obtained from the shock tracker data. The shock line gives a much higher degree of statistical precision than can be obtained from a few gauge positions, particularly when the shock velocity is not constant.

PBX 9502 subjected to ratchet growth before installation of the gauge yielded the record shown in Figure 5. Abrupt velocity changes due to density discontinuities or cracks are clearly evident in the tracker record and in the shock line, Figure 6, generated from the tracker data. This behavior is also evident in the u_p gauge records.

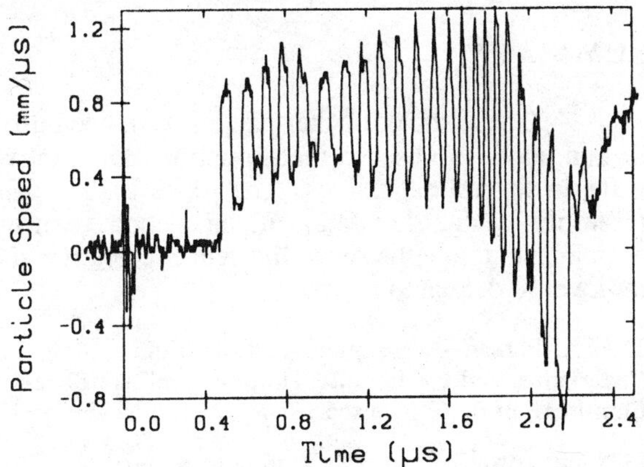

Figure 5. The shock tracker record of PBX 9502 that has undergone irreversible thermal expansion, or "ratchet growth".

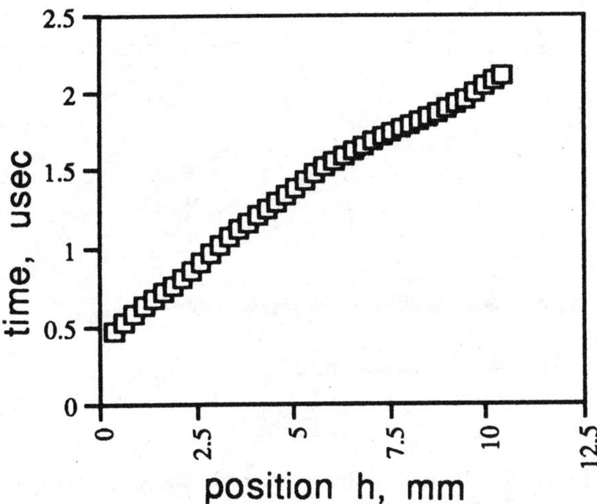

Figure 6. The shock line of PBX 9502 that has undergone irreversable thermal expansion, or "ratchet growth".

Hot PBX 9502 running to detonation generates the example shown in Figure 7. In this material, the shock tracker indicates inert or constant velocity behavior, and even a slight deceleration of the input shock. Comparison with u_p gauge records shown in Figure 8 shows that homogeneous reaction is occurring, with a reactive wave running behind the initial shock. Thus the two regions of the shock tracker are being propelled at different rates, creating a more complex output than arises from simple heterogeneous detonation, one that reflects acceleration of the material behind the shock front.

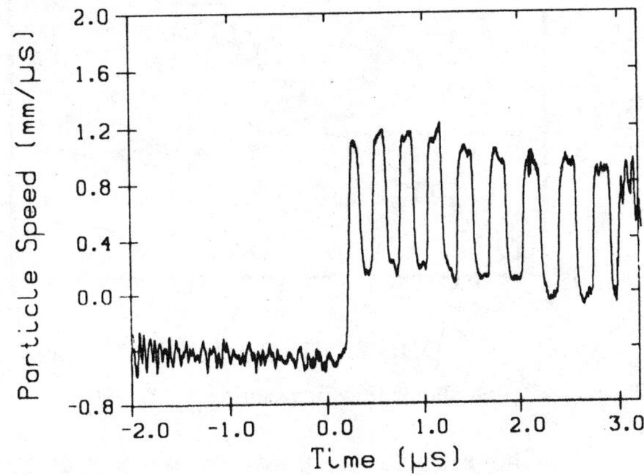

Figure 7. The shock tracker record of PBX 9502

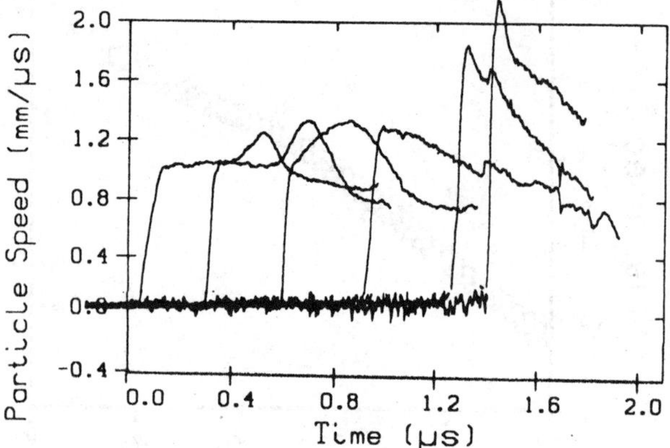

Figure 8. Particle velocity records showing reaction in PBX 9502.

Comparison of shock tracker and particle velocity gauge data in Figs. 8 and 9 indicates that the tracker yields an incomplete description of the reaction in the material, since tracker data does not reflect acceleration behind the shock front. By analogy, it may be that standard wedge experiments yield the same incomplete picture. Use of the shock tracker coupled with the particle velocity gauges is clearly important in accurately describing the turnover to detonation in materials such as hot PBX 9502.

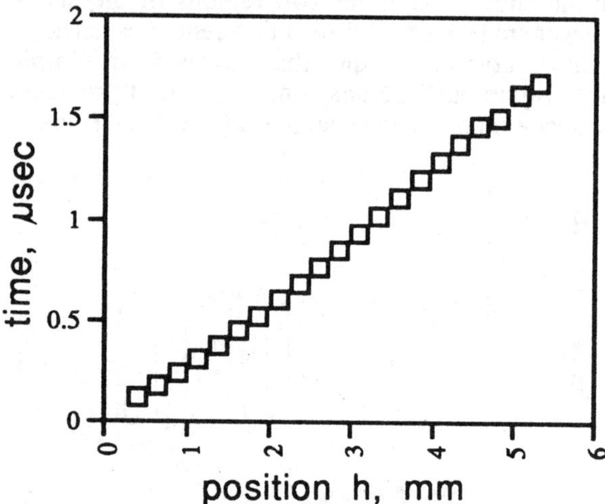

Figure 9. The shock line of reacting PBX 9502

Shock tracker response to two successive square shocks is shown in Figure 10. The data from this shot is well modeled by summing the response of the tracker to the two individual shocks,

indicating that the gauge retains its shape in the flow, and is not damaged by the passage of the first shock. Thus, distortion of the portion of the gauge in the flow behind the front is not contributing current to the net output of the gauge.

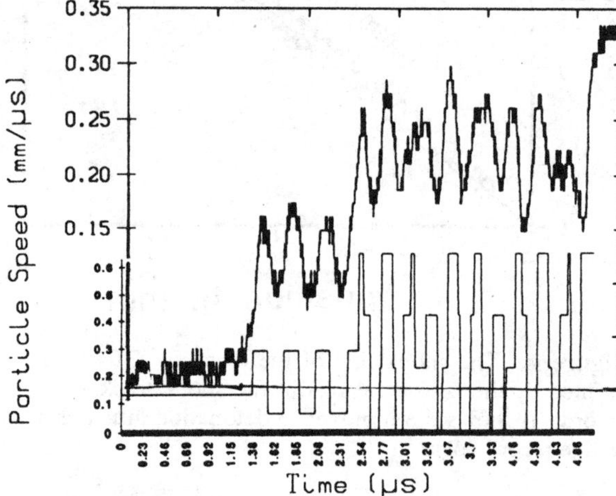

Figure 10. The shock tracker record of PBX 9502 responding to two square waves is a sum of responses to each wave.

Limitations on the accuracy of the gauge are precision of the construction and measurement of the location of the elements, and the risetime of the signal as the shock traverses a given element.

Risetimes of the individual cross bar response signals are slightly longer than the traversal time of the gauge thickness by the shock wave. The ratios of observed risetime to calculated traversal time in two X-0407 experiments are 1.76 and 1.78.

SUMMARY

These shock tracker gauges have been used in gun shots to observe the growth of the reactive wave in several explosives. The shock tracker data resembles wedge shot data, with an adequate density of points to define the shock line and characterize the turnover to detonation.

1. C.A. Forest, "Lagrangian Analysis, Data Covariance, and the Impulse Time Integral", in Shock Compression of Condensed Matter, 191, Elsevier 1992.

2. John Vorthman, private communication

3. RDF Corporation proprietary method. RDF Corporation, Hudson, New Hampshire.

CALIBRATION OF THIN-FOIL MANGANIN GAUGE IN ALOX MATERIAL

R. A. Benham

Explosive Components Department, Sandia National Laboratories, Albuquerque, NM 87185

L. J. Weirick

Explosive Projects and Diagnostics Department, Sandia National Laboratories, Albuquerque, NM 87185

L. M. Lee

Ktech Corp., 901 Pennsylvania NE, Albuquerque, NM 87110

The purpose of this program was to develop a calibration curve (stress as a function of change in gauge resistance/gauge resistance) and to obtain gauge repeatability data for Micro-Measurements stripped manganin thin-foiled gauges up to 6.1 GPa in ALOX (42% by volume alumina in Epon 828 epoxy) material. A light-gas gun was used to drive an ALOX impactor into the ALOX target containing four gauges in a centered diamond arrangement. Tilt and velocity of the impactor were measured along with the gauge outputs. Impact stresses from 0.5 to 6.1 GPa were selected in increments of 0.7 GPa with duplicate tests done at 0.5, 3.3 and 6.1 GPa. A total of twelve tests was conducted using ALOX. Three initial tests were done using polymethyl methacrylate (PMMA) as the impactor and target at an impact pressure of 3.0 GPa for comparison of gauge output with analysis and literature values. The installed gauge, stripped of its backing, has a nominal thickness of 5 μm. The thin gauge and high speed instrumentation allowed higher time resolution measurements than can be obtained with manganin wire.

INTRODUCTION

Manganin alloy has been used widely as an in-material stress gauge in planar shock wave experiments. Gauges of the wire design have been studied by Lee[1] and have been used in numerous physics and engineering applications. Because of the physical size of the manganin wire, 76 μm diameter, this design responds relatively slowly to an input shock and the gauge output is dependent on the material in which the gauge is mounted. The thin foil manganin gauge construction, as reported by Rosenberg,[2] is much thinner, 5 μm, and provides a better temporal representation of the shock in materials. This design, with gauge backing in place, is reported[2] to be insensitive to the target material. Rosenberg[2] reported that the calibration curve for the thin foil manganin gauge showed distinct elasto-plastic behavior with a linear part from 0 to 1.5 GPa. The slope of the elastic part of the calibration was ~50 GPa/(Ω/Ω). The higher stresses were represented by a fourth-order polynomial fit. The target and impactor materials used in the Rosenberg work were PMMA, copper, magnesium and aluminum, all homogeneous in nature.

The work presented here develops a shock response calibration curve for commercial foil gauges that have the customary Kapton® backing removed and the gauges imbedded in ALOX. Four gauges were installed in each target to obtain redundant gauge outputs from the same test input.

EXPERIMENTAL TECHNIQUE

The plane impact experiments were conducted in the Sandia National Laboratories 63-mm diameter light gas gun described by Sheffield.[3] Impact velocity was measured using five coaxial pins that were shorted by a metallic ring around the projectile. The pins were separated by 10 mm in the axial direction and the last pin was 25 mm in front of the target assembly face. The accuracy of the impact velocity was $\pm$ .5%. The impactor tilt was inferred using the output from the four manganin gauges, equally positioned around the target axis. The acceptable tilt of the impactor was established a priori to be less than 100 ns closure across the gauge element (3.175 mm).

Figure 1 is a schematic of the projectile and target used in this calibration program. The impactor material was PMMA made by Rohm and Haas for the three preliminary tests and ALOX made at Sandia's plastic shop for the twelve calibration tests. The impactor was nominally 2.5 mm thick and was backed by a 5-mm thick disk of carbon foam which had a nominal density of 0.2 g/cm.[3] This backing material reflects a very small percentage (~4%) of the shock wave reaching it from the impactor. The sabot to carry the impactor was made of either aluminum or syntactic foam filled nylon. Using the same material for the impactor and target produced a symmetric impact with the resultant particle velocity in the material equal to one-half the projectile velocity. The stress level in the impacted material was determined from the known Hugoniot curves for ALOX[4] and the measured impact velocity by the impedance-match technique. The manganin gauges were sandwiched between two pieces of PMMA or ALOX, respectively. The first piece, the buffer, was nominally 5 mm thick and the second, the target, was 25.4 mm. The polycast backer was used to support the instrumentation cable connection.

Commercial manganin gauges (Micro-Measurements Model No. VM-SS-110FB-048, Part No. C-941028-A) were used in this study. The

gauges are etched foil with 48 Ω nominal resistance. The foil was 5.0 μm thick and the Kapton® backing was removed after the gauges were glued to the respective target. The total thickness of the gauge installation between the buffer and the target was nominally twice the gauge thickness. The manganin gauge had two thin ribbon extensions for attachment to the instrumentation cable which were coated with a thin layer of copper (~5 μm thick) to reduce output caused by resistance change of the leads during the impact.

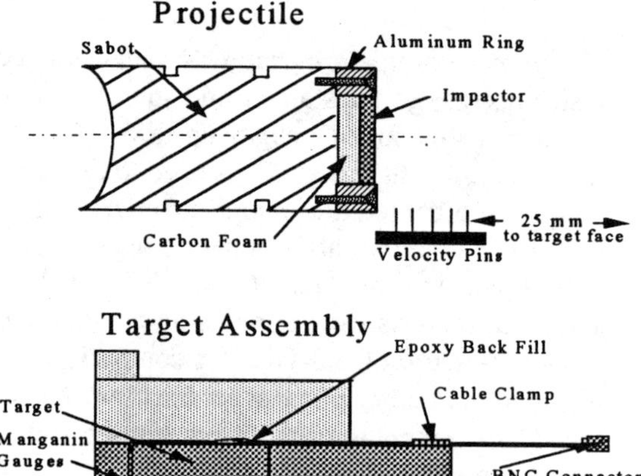

FIGURE 1. Schematic of Test Hardware

PRELIMINARY TESTS

A series of three tests was conducted at the beginning of this study to validate the experimental process. The material, PMMA, which has been characterized for shock loading[5] was used in a symmetrical impact to produce 3.0 GPa shock loading. The manganin gauges were stripped of the backing and were installed in the same manner as for the ALOX experiments.

The preliminary experiments were conducted to provide data for comparison with calculated gauge response to assess overall experiment accuracy. A calculated stress-time history in the PMMA target was directly compared with the 5 μm thick manganin output to determine the correlation. The calculations used measured impact velocity in conjunction with a rate dependent PMMA model,[5] while the manganin resistance change data were converted to stress using Rosenberg's equation.[2] Comparison of calculations and experiment are shown in Fig. 2. A high frequency oscillation was evident on the leading of the experimental data, which was attributed to manganin/PMMA equilibration. Recording system frequency response (~20 MHz) negated the ability to directly track manganin response. The 5 μm thick manganin foil was not included in the calculations.

were run at 0.7 GPa increments between these values. Four gauges were recorded at each level for a total of 48 gauges. Forty-eight measurements were obtained with two gauges losing leads early in the response time. Figure 3 shows gauge output as a function of stress level. A third order, least squares fit to these data is shown. The equation that describes the change in gauge resistance (dR) divided by the initial gauge resistance (R) is $P(GPa) = 5.5027 * (dR/R) - 0.2063 * (dR/R)^2 + 0.0061 * (dR/R)^3$. The correlation factor (R^2) is 0.9986 for this data fit. This calibration is valid for pressures between 0.5 and 6.1 GPa.

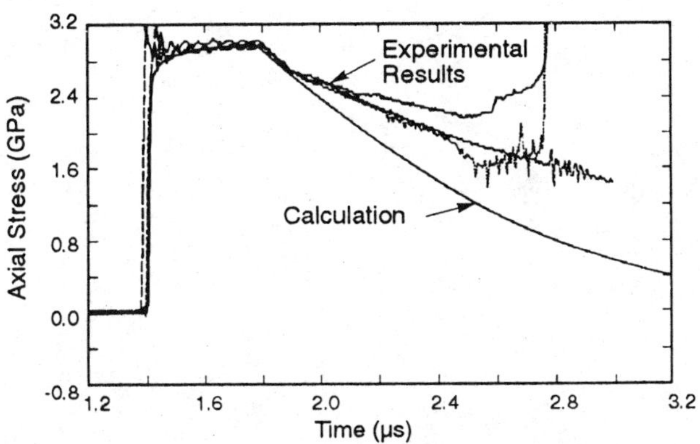

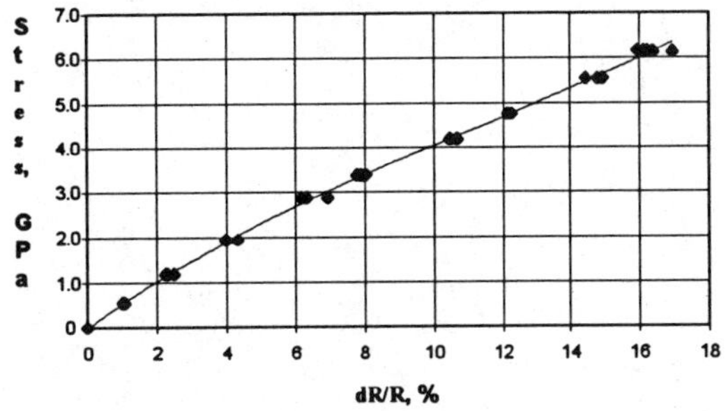

FIGURE 3. Results of the manganin foil gauge embedded in ALOX

FIGURE 2. Experimental results and calculational comparison for manganin foil gauge embedded in PMMA showing overall agreement until 1.9 μs

Overall agreement was observed between gauge response and calculations based on shock wave rise, amplitude and release. Manganin unloading response deviated from the one-dimensional calculations at about 1.9 μs. The shock wave arrived at the gauge connection to the instrumentation cable at approximately 2.8 μs.

RESULTS

Twelve tests were conducted with ALOX material for the impactor and target. Duplicate tests were run at 0.5, 3.3 and 6.1 GPa. Single tests

DISCUSSION

The calibration shows similar characteristic elastic response up to about 1.5 GPa as reported by Rosenberg.[2] The elasto-plastic response is also similar to Rosenberg data up to 3.3 GPa. At stress greater than 3.3 GPa the gauge in ALOX shows a larger change in resistance than seen in the homogeneous materials of Rosenberg's work. At 6.1 GPa the difference is around 15%. This difference is attributed to non-homogeneous response of the aluminum oxide filled epoxy at these higher stresses. The gauge repeatability was excellent; note that there are eight data points at 0.5 and 3.3 GPa and seven points at 6.1 GPa.

ACKNOWLEDGMENTS

The authors wish to acknowledge the contribution of the following people to this program: Heidi Anderson, Ktech Corp., who assembled the projectiles and targets; Frank Horine, Ktech Corp., who procured the ALOX material; Rick Saxton, Ktech Corp., who set up the data recording system; Ben Duggins, SNL 2553, who conducted the data recording; and Mike Navarro, Ktech Corp., who conducted the gas gun tests.

This work was supported by the United States Department of Energy under Contract DE-AC04-94AL85000.

REFERENCES

1. Lee, L. M., J. Appl. Phys. Vol. 44, Sept. 1973, pp 4017-4022.

2. Rosenberg, Z., et al, J. Appl. Phys. Vol. 51 (7), July 1980, pp 3702-3705.

3. Sheffield, S. A. and Dugan, D. W., "Description of a New 63-mm Diameter Gas Gun Facility," *Shock Waves in Condensed Matter*, ed. Y. M. Gupta, Plenum Press (1986).

4. Lee, L. M., "Alumina-Filled Epoxy Shock Response as a Function of Temperature," Air Force Weapons Laboratory Report, AFWL-TR-87-133 Sept. 1988.

5. Schuler, K. W. and Nunziato, J. W., "The Unloading and Reloading Behavior of Shock Compressed Polymethyl Methacrylate," J. Appl. Phys. Vol. 47, (7), July 1976, pp 2995-2998.

LAGRANGIAN ANALYSIS OF THE PVDF SHOCK SENSOR

H.Moulard and F.Bauer

French-German research institute of Saint-Louis (ISL), Saint-Louis ,FRANCE

The electrical response of the PVDF shock sensor is here revisited. An analysis with a lagrangian hydrocode shows that the electrical charge liberated during the 1D deformation of the PVDF film is strictly proportional to the true strain of the PVDF film. On the experimentally studied range (until 30 GPa), the electrical response of the PVDF gauge is linear

INTRODUCTION

This paper presents an analysis with the lagrangian hydrocode DYNA of the electrical response of the PVDF gauge during a dynamical test. This analysis enables us to establish a new correlation between the variation of the electrical quantities measured during the dynamic test and the variation of the mechanical quantities computed with the hydrocode during the numerical simulation of this dynamic test.

A SIMPLE LAGRANGIAN ANALYSIS

The Symetrical Impact Test

The dynamic test submitted to this lagrangian analysis is the symetrical impact test, routinely used for the PVDF gauge calibration. In this test (fig.1), projectile and target material are the same.

The PVDF gauge, 25 µm thick, is glued on the target impact surface, to assure the electrical insulation of the two electrodes, two 110 µm thick Kel-F sheets are glued on both sides of the PVDF film. At impact, due to impedance mismatch between PVDF and copper, a shock wave reverberation occurs in the gage. So submitted to stress and strain, the piezoelectric PVDF gage release electrical charges in a short circuit made from a CVR (fig.1.). A transient digitizer measures directly the current passing through the CVR, every current peak (fig. 2.) corresponds to the successive shock waves travelling through the PVDF film. Then, by a numerical integration of the current profile (1,2), the quantity of electrical charges released by the PVDF gauge versus time is deduced.

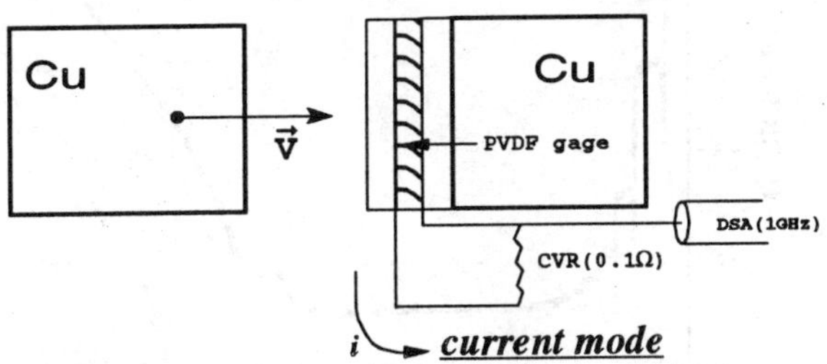

FIGURE 1. Symetric Impact test design.

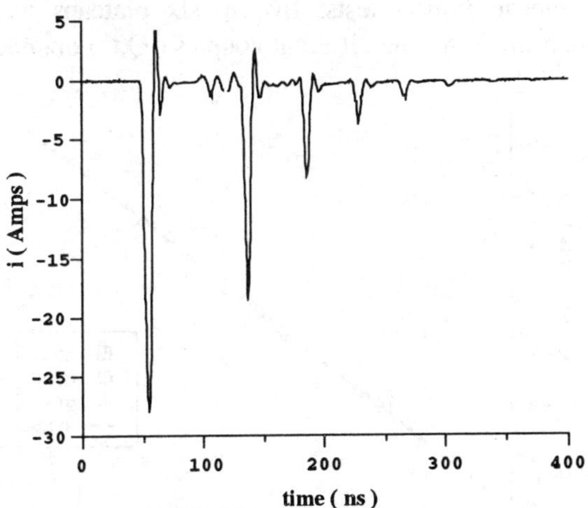

FIGURE 2. Measured current profile

Linear Electromechanical Coupling

Stimulated by the high quality level of the electric measurements obtained by the last advance in the PVDF technology (3), we set to ourselves the problem to find a phenomenological modeling, able to calculate the observed current profile from the mechanical quantities given by the numerical simulation of this symetric impact test. After testing different possible modeling, we achieved success with the following simple rule: the electrical charge Q, released by the piezoelectric PVDF film is proportional to the global 1-D true strain ε of the PVDF film:

$$Q(t) = k \cdot \varepsilon(t),$$

where k= 12.218 µC/cm2 and

$$\varepsilon(t) = \ln\left(\frac{e(t)}{e(t_0)}\right) = \ln\left(\frac{v(t)}{v(t_0)}\right) \quad (1)$$

where e = PVDF film thickness; v = PVDF specific volume and t = time. To check this linear law, it is enough in a unique symetric impact test, to measure the electrical charge Q released on every successive plateau and to compute with the hydrocode the corresponding ε value. In our symetric impact tests, five to six plateaus were measured. All the different couples (Q,ε) obtained

in three distinct symetric impact test are reported on the fig. 3: all the points are accurately aligned along an unique straight line passing through the origin. This simple linear law represents the intrinsic calibration law of the PVDF gage and has immediate consequence for the use of the PVDF gage as a lagrangian gage.

APPLICATIONS

Lagrangian Strain Gage

The first important application of this linear electromechanical coupling is to enable to compute the theoretical current or charge profile for a given dynamical test. So, the computed current profile represents the ideal current profile that the transient recorder should measure if the measurement conditions or gage design are optimum. Figure 4 shows the computation - experiment comparison for the electrical charge liberated by a PVDF gage, with a 3×3 mm2 active area in a symetric impact test where a 13 mm diameter copper projectile, 2 mm thick impacts at 655 m/s a 3 mm thick copper target. A good agreement of the two profiles is observed, except the existence of experimental ponctual overcurrents due to strong inductive effect: the initial shock wave generates intense variation of current (20 A in 2 ns!) in the 3×3 mm^2 PVDF active area. The current in the gage being directly proportional to the PVDF active area, to limit the

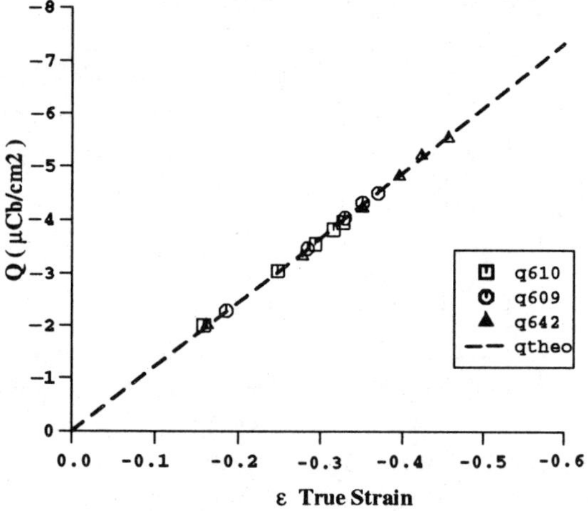

FIGURE 3. Experimental linear law of calibration obtained from three different symetric impact tests (q610,q609,q642)

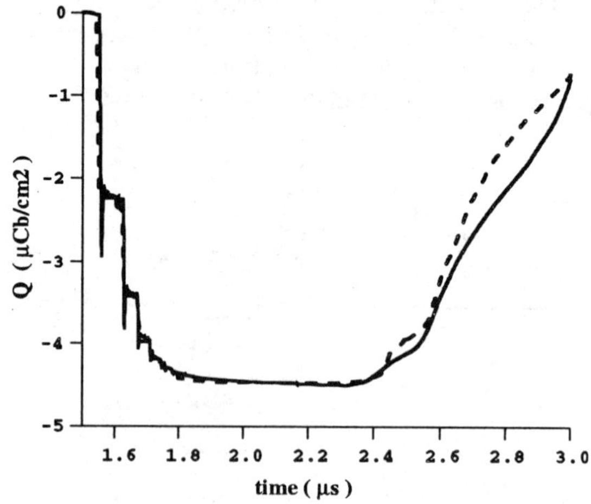

FIGURE 4. Comparison between measured (——) and computed (- - -) electrical charge.

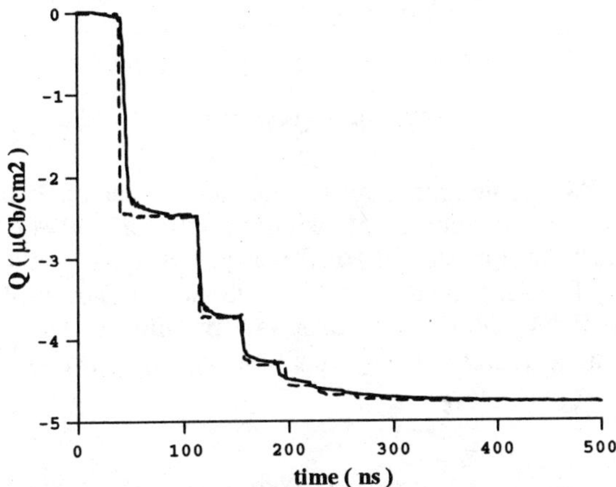

FIGURE 5. Comparison between measured (———) and computed (- - -) electrical charge for the small PVDF gage.

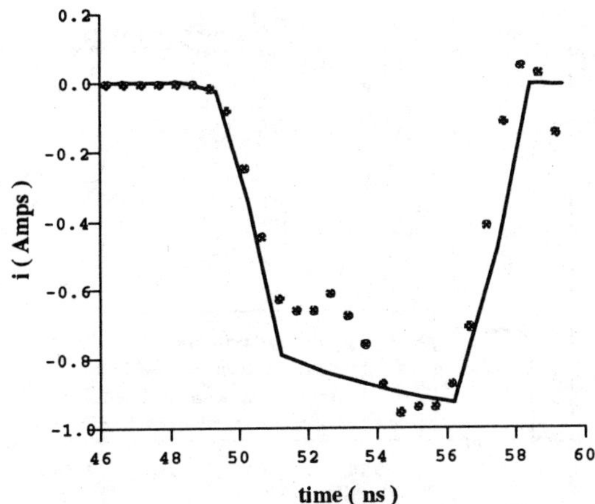

FIGURE 6. Comparison experiment (. . .) and computation (———) for the first current peak measured with the small PVDF gage.

level of this indesirable inductive effect, the same impact experiment has been done again with a smaller active area (0.5×0.5 mm^2) for the PVDF gage. This small gauge (fig. 5) enables to suppress this parasite inductive effect and simultaneously to increase the time resolution: On fig.6, the markers along the experimental curve correspond to the experimental measurement sampling interval (0.5 ns). These results show also clearly that the PVDF gage behaves as a lagrangian true stain sensor.

Lagrangian Stress gage.

At last, we set to ourselves to find an analytical solution to compute the stress σ undergone by the PVDF gage from the measured electrical charge Q. The following relationships will be used:
- linear electromechanical coupling:

$$Q = K \times \varepsilon \qquad (2)$$

$$\varepsilon = \ln\left(\frac{v}{v_0}\right) \qquad (3)$$

- Hugoniot dynamic adiabat:

$$U = C + S.U_p$$

$$\mu = \frac{v_0}{v} - 1$$

$$\sigma_h = \frac{\rho_0 . C^2 . \mu (1+\mu)}{\left(1 - \mu(S-1)\right)^2} \qquad (4)$$

- Mie-Gruneisen equation of state:

$$P(v,E) = P_h(v) + \Gamma_0 \left(E - E_h\right) \qquad (5)$$

where Γ_0 = Gruneisen constant for PVDF;
E = internal energy; P = pressure; subscript h deals with a state on the hugoniot adiabat. From these relations, two analytical solutions are now proposed.

Approximate Explicit Analytical solution

If we assume in eq (5), that PVDF internal energy keeps approximatively the same value of internal energy as on the hugoniot adiabat at the same compression ratio, then, from (2), (3), (4) and (5) so simplified, we obtain easily an explicit analytical formula giving the stress directly from the measured electrical charge. But, it is only an approximate solution. (fig.7).

Exact Iterative Lagrangian Calculus.

Here, we calculate between the time step n and n-1, the PVDF internal energy variation due mainly

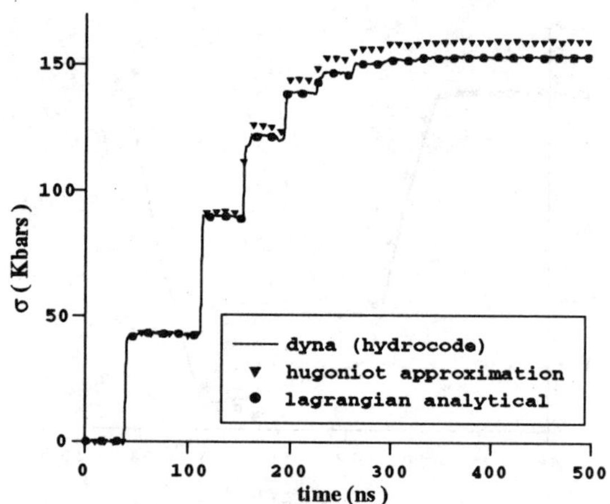

FIGURE 7. Comparison of the two methods for calculating the stress from the measured electrical charge.
(——): theoretical stress;
(▼) : approximate explicite formula
(• •) : exact iterative lagrangian calculus

to the compression work by means of the following discretization formula:

$$E(v_n) = E(v_{n-1}) - \Delta w$$

$$\Delta w = \frac{\left[P_h(v_n) + P(v_{n-1})\right]}{2}\left[v(n) - v(n-1)\right]$$

From the Mie-gruneisen equation of state, the pressure at time step n is then calculated:

$$P(v_n) = P_h(v_n) + \Gamma_0\left[E(v_n) - E_h(v_n)\right]$$

The fig.7 shows the calculus, with these two methods, of the stress undergone by the PVDF gage in the symetric impact test, the experimental electrical measurements of which have been previously shown on fig.5 and 6. By comparison to the theoretical stress profile computed with the hydrocode, we observe that the approximate analytical solution overestimates systematically the stress values, whereas the step-by-step analytical lagrangian calculus gives the correct stress profile.

CONCLUSION

This numerical analysis enables to establish a new electromechanical coupling for the PVDF shock sensor: the electrical charge released by the PVDF varies linearly with the global true strain of the PVDF film. This linear law permits to define more accurately the gage calibration and its use as a lagrangian gage.

REFERENCES

1. Bauer F. and al.,"Piezoelectric response of precisely poled PVDF to shock compression greater than 10 GPa",in *Proceedings of the International Symposium on the applications of ferroelectrics (ISAF92)*, 1992, pp 273-276.

2. Graham R.A. and al.,"Piezoelectric polarization of the ferroelectric polymer PVDF from 10 MPa to 10 GPa. studies of loading-path dependence", in *proceedings of the American Physical Society Topical Conference on shock compression of condensed matter*, Williamsburgh, 17-20 june 1991, pp 883_886.

3. Bauer and al.,"Piezoelectric response of ferroelectric polymers under shock loading: nanosecond PVDF gauge",in *proceedings of the American Physical Society Topical Conference on shock compression of condensed matter*, Seattle, 13-18 August 1995.

ELECTRICAL EFFECTS IN POLARIZED POLYMER FILMS UNDER HIGH VELOCITY IMPACT: APPLICATION TO IMPACT GAUGES.

Vladimir V.Yakushev

Institute of Chemical Physics in Chernogolovka RAS, Moscow Region, 142432, Russia

A new method of detection of high velocity bodies is presented. The method is based on the depolarization effect in polarized polymeric film as a result of the local failure at impact and perforation. In the experiments electrical signals were registered across a load resistor connecting the metal coating surfaces of the film. It is shown that the signals are determined by the velocity, form, and size of impactors. Such gauges keep efficiency even in the case of impact of many elements and have opportunity for yielding information about explosive fragmentation of metallic covers, shaped charge phenomena, impact initiation of detonation and others.

INTRODUCTION

Knowledge of velocity vectors and dimensions of rapidly flying elements is important for different technical applications, such as the explosive acceleration of solid elements and their groups, explosive fragmentation, high velocity impact, shaped charge phenomena. As a rule, experiments of this type are conducted using an optical and x-ray diagnostics. A very useful addition to these methods may be a system of detecting of multiple impact events with the indication of these events parameters.

In our opinion such system can be designed on the basis of electrically polarized thin polymeric films as impact sensors using the phenomenon of their quick depolarization induced by shock wave or high velocity impact (1).

The sensors should have the following features:

- A functional dependence of the output signal on element dimensions, velocity, and so on;

- The time resolution of the order of several microseconds;

- An ability to indicate each impact event in series of them;

- Operability with a long coaxial cable;

- High signal magnitude;

- Low price, the possibility to obtain sensors with a surface area of several square dm;

- Good keeping qualities and other necessary service characteristics.

The aim of the work reported here was to study the electrical response of polarized polyvinyl chloride (PVC) and polyvinylidene fluoride (PVF_2) films to high velocity impact. The films are the most promising as the sensor material.

MATERIALS

The measurements were made on vacuum aluminized polymeric films. Two kinds of the PVC films were used, both of which contained 93 mass % of the base polymer. The first was a commercial grade 0.45 mm thick film (PVC-1). The second was a custom-made high quality calendered 0.2 mm thick film (PVC-2). The films were thermally poled at temperatures from 85 to 92^0 C. After polarizing for some minutes at the desired temperature, the samples were cooled to room temperature under the applying of field. The total time of the polarization was adjusted to be about 1 h in each case.

Remanent polarization P_r of the films was obtained by integration of the thermally stimulated depolarization current with respect to time. The polarization of the PVC-1 and PVC-2 films was equal to 0.2 and 0.72 μC/cm^2 respectively.

A commercial uniaxially stretched aluminum coated and polarized PVF$_2$ film (30 μm thick) was supplied by the Plastpolymer Okhta Research and Production Association. The features of the film were described by Sherman and all.(2).

IMPACT TESTS

The impact experiments were conducted with a powder gun. The projectiles were steel balls of 7 mm diameter. The velocity of the projectile was ranged between 0.4 and 1.4 km/sec.

The impact gauge was manufactured as a capacitor. The polarized films about 3x5 cm^2 were dielectric of the capacitor. The PVF$_2$ target assembly was made of the film sheet as a sensing element which was bonded to a 1 mm thick textolite holder with epoxy resin. The electrodes of the gauges were connected to a load resistor R_e=75 Ω for PVC-1, 50 Ω for PVC-2 and 25 Ω for PVF$_2$ films. The ball velocity vector was perpendicular to the gauge plane.

During the experiment the current from the gauges was measured through recording of the voltage U across R_e using a digital oscilloscope which digitized at a rate of 50 nsec and had a recording window of 50 μsec. The capacitance of PVC and PVF$_2$ gauges were 0.2 and 5 nF correspondingly.

RESULTS AND DISCUSSIONS

Figure 1 shows typical current pulses from the PVC-2 specimens. As seen, the maximum current rises with the ball velocity. The signal sign indicates that the polarization of the film reduces in the course of its perforation.

It should be pointed out that the maximum voltage U_m across R_e was reasonably high. For example U_m was equal to about 17 V at the ball velocity and the load resistor were equal to 1.36 km/sec and 50 Ω respectively.

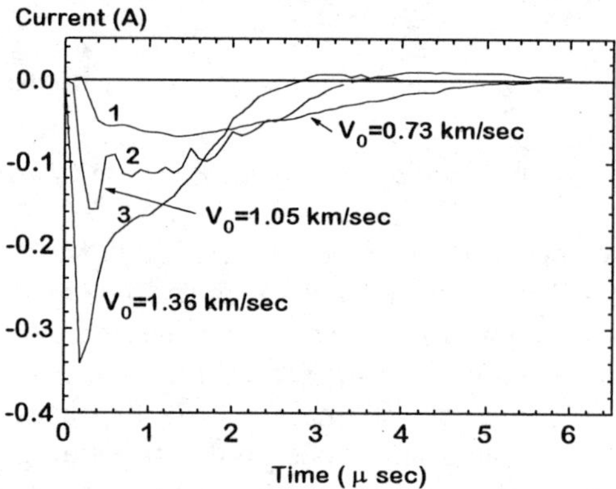

FIGURE 1. Typical current records of PVC-2 gauges.

The electric response of PVC-1 film has been investigated in our early work (1). Voltage signals across a 75 Ω resistor arising during perforation of the film by steel balls of 4 to 8 mm diameter were recorded in the experiments. Some experimental dependencies of the maximum voltage U_m on the ball size and velocity V_0 are presented in Fig. 2.

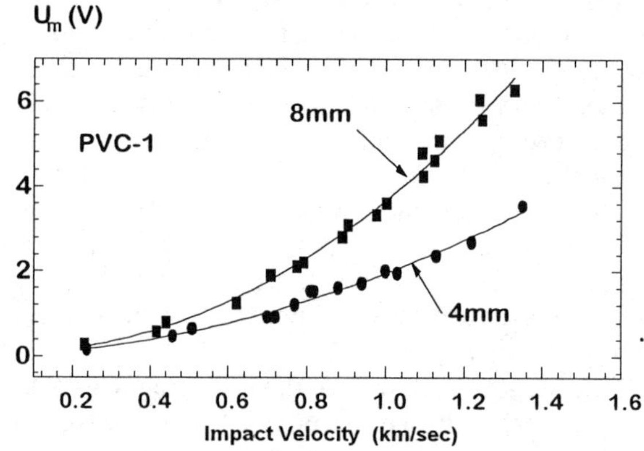

FIGURE 2. The electric signal amplitude versus impactor diameter and velocity.

Typical PVF$_2$ current records during the tests are shown in Fig. 3. The first short positive pulse of the signals appears at the moment of the ball contact with the film. It corresponds to the ordinary piezoelectricity of the polarized film. It can be very clearly seen that the first pulse undoubtedly is followed by the high negative one. This pulse uniquely testifies that a fast depolarization of the PVF_2 film arises. It is interesting that its duration is considerably less than the film perforation time in contrast to PVC films. Apparently this is due to increase of the piezoelectric current contribution to the total signal at final stages of the perforation.

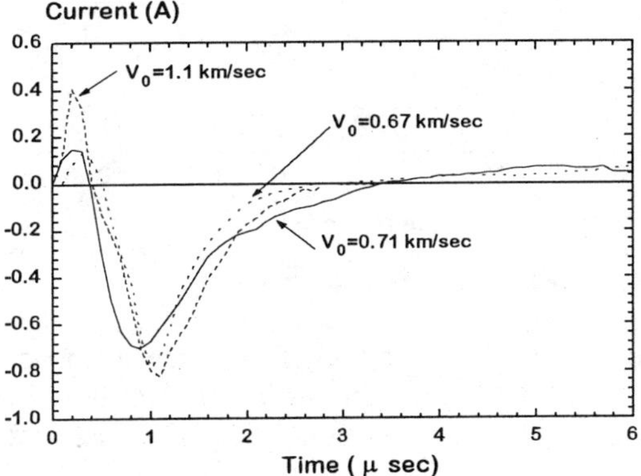

FIGURE 3. Typical current records of PVF_2 gauge.

Let us concentrate our attention on the dimensionless specific charge $q_s = Q_s/P_r$ evolved into the external load circuit during the perforation of the PVC films. The value of Q_s can be obtained by integration of the current pulse and dividing the integral by the area of the ball middle-section.

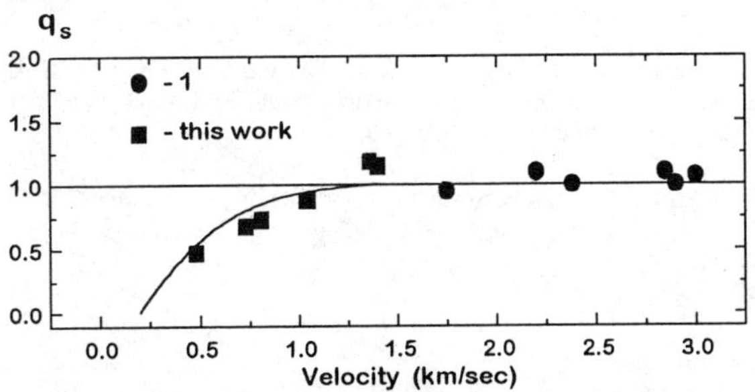

FIGURE 4. Plot dimensionless specific charge vs. velocity: 1 - the experiments by A.L.Nemanikhin and V.V.Yakushev on a light gas gun for PVC-1 with the ball diameter of 9.6 mm.

Figure 4 is a plot of q_s versus V_0 for the PVC-1 and PVC-2 films. The line represents the results of (2) for PVC-1. The increase of V_0 at first leads to increasing q_s. Then, starting approximately with $V_0 = 1$ km/sec, variation of the specific charge becomes slower and q_s tends asymptotically to 1.

By this means we can infer that at V_0 more than 1.2 km/sec and for a spherical impactor the PVC films are fully depolarized in the impingement region.

It should be noted that in the velocity range covered the depolarization degree of the PVF_2 is markedly less than unity.

As $q_s=1$, in order to calculate the electrical response of the gauge we suppose the total film depolarization arises instantly upon its contact with the surface of an impactor. In other words the gauge generates an electrical charge which is proportional to the current destruction spot area. Value of this charge Q(t) is determined by the expression

$$Q(t) = P_r\, S(t), \qquad (1)$$

where S(t) is the instantaneous contact surface area between an impactor and the film.

If the decay time constant of the measuring circuit is considerably less then the perforation time, the electrical current

$$I = dQ/dt = P_r\, dS/dt \qquad (2)$$

If the impactor has a spherical shape, then

$$I = 2\pi P_r\, V_0\, r_0\, (1 - t / t_0), \qquad (3)$$

where r_0 is the sphere radius, $t_0 = r_0 / V_0$.

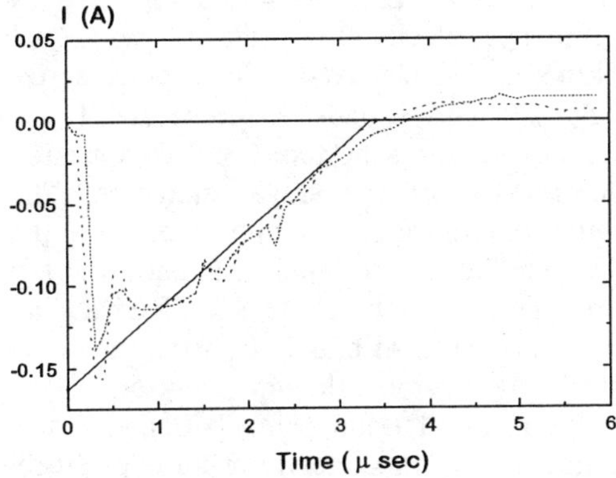

FIGURE 5. Experimental (dotted lines) and calculated (solid line) dependencies current vs. time of PVC-2. Velocity of the ball - 1.05 km/sec.

Comparison of some experimental records with calculations from Eq. 3 are shown in Fig. 5.

Figure 5 shows a good fit of the records to the equation excepting their initial parts. So the gauge current signal reflects the dynamics of the polarized film destroyed zone increasing. Therefore this signal contains information about the velocity, form, and size of the elements which act on the gauge.

As far as we know, this method of high speed impact parameters determination is unique by its potentials (time resolution, convenience of application and others). As an example we shall present two experimental oscillograms. The first oscilloscopic trace (Fig. 6) demonstrates the possibility of multiple impact event registration.

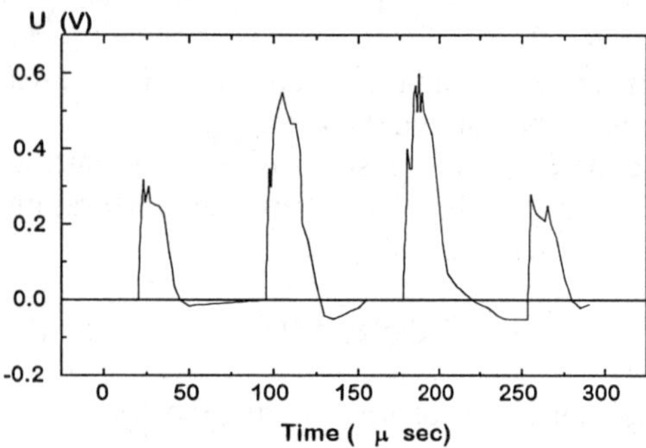

FIGURE 6. Electrical response of PVC-1 gauge on multiple impact

A group of 6 small (0.4 g) spheres made of a tungsten-nickel-iron alloy was launched by a powder gun. The velocity of the elements was equal to 0.36 km/sec. The observed voltage pulse series (R_e is 50 Ω) corresponds to impacts of the elements. The first and the fourth pulses appeared from the impacts of the single elements. The second and the third pulses appeared from the double events, that is two elements impacted at the same time. Therefore, the charges evolved during the second and the third pulses are twice as much as that during the first and the fourth pulses.

The second oscillogram (Fig. 7) demonstrates the determination of the velocity of an explosively formed projectile which is flying inside a debris cloud. In the trace one can see the pulses of each impact event both of small particles of the cloud and the main element - projectile. For the latter

case the oscilloscope amplifier was saturated. The projectile velocity can be obviously determined with the use of this oscillogram.

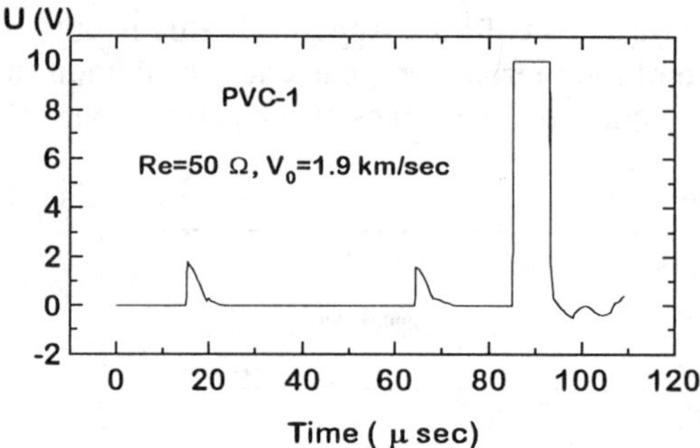

FIGURE 7. Voltage versus time.

We hope that the method of high speed impact parameters determination will be useful in conducting of many dynamic experiments. and tests.

ACKNOWLEDGMENTS

This work has been funded by Russian Foundation for Basic Research under grant No 95-02-05353-a.

REFERENCES

1. Alekseev E.M., Antipenko A.G., Dremin A.N., Krenev S.A., Kurto A.P., Nemanihin A.L., Sidorov V.A., and Yakushev V.V. *Sov.J.Chem.Phys.***11**, 330-335 (1992).

2. Sherman M.Ya., Lesnykh O.D., Vlader N.B., Artem'ev B.A., Myasnikov G.D., Lobanov A.M., Zolotova V.I. *Plastmassi*, No.10, 46-48 (1990).

PIEZOELECTRIC RESPONSE OF FERROELECTRIC POLYMERS UNDER SHOCK LOADING : NANOSECOND PIEZOELECTRIC PVDF GAUGE.

F. Bauer*, H. Moulard*, R.A. Graham**

**Institut Franco-Allemand de Recherches de Saint-Louis, (ISL), Saint-Louis, FRANCE*
***Sandia National Laboratories, Albuquerque, NM 87185-1421*

The most common piezoelectric polymer is PVDF, based on the monomer CH_2-CF_2. Electrical processing with the Bauer cyclic poling method can routinely produce individual samples with a range of remanent polarizations up to 9 mC/cm². The behavior of PVDF has been studied over a wide range of pressures with the destructive, but precise, method of very-high-pressure shock loading. It appears that low inductance electrode lead designs prepared via a new poling procedure improve significantly the precision of the piezoelectric response of the PVDF gauges under shock loading. In particular piezoelectric current response of shock compressed PVDF film is of the order of one nanosecond. Studies to pressures of 30 GPa are available which show that the piezoelectric behavior is linearly dependent on volumetric strain to a close approximation as described in a companion paper. Anomalous response of PVDF observed are identified: solutions are given. The first record of detonation profile is presented.

INTRODUCTION

Piezoelectric materials are widely used in modern technology. Early work by Bauer (1) which explored the behavior of polyvinylidene fluoride (PVDF) for high pressure applications, led to recognition of the need for such highly reproducible properties for PVDF.

The availability of highly reproducible samples provides an opportunity to study the materials under the destructive, very high pressure conditions achieved in controlled shock loading. Thus the combination of precise pressure conditions and highly reproducible samples provides an unprecedented opportunity to describe the piezoelectric response of PVDF to high pressure and large volume compression. Previous observations (4) of the PVDF piezoelectric charge response data to 30 GPa show significant deviations from idealized, continuous behavior between about 12 and 20 GPa. It appears that low inductance electrode lead designs prepared via a new poling procedure (6) improve significantly the precision of the piezoelectric response of the PVDF

gauges under shock loading as well as the rise time which attains one nanosecond. The piezoelectric behavior is observed to be linearly dependent on volumetric strain to a close approximation as described in a companion paper (8). Anomalous response of PVDF are identified : a solution is proposed. The first record of detonation profile is presented. The present paper will briefly summarize the status of such experimental work.

FERROELECTRIC POLYMERS : PIEZOELECTRIC PVDF GAUGES

PVDF is a partially crystalline linear polymer, in which the β phase or ferroelectric phase is typically achieved from biaxially stretched films. Under appropriate electrical poling PVDF in the ferroelectric phase with thickness of 25 micron exhibit well defined remanent polarization (4,5,6). With the Bauer cyclic poling process, reproducible remanent polarizations as large as 9 μC/cm² are routinely achieved in commercial processes. A higher degree of reproducibility is achieved when the maximum displacement current at the coercive

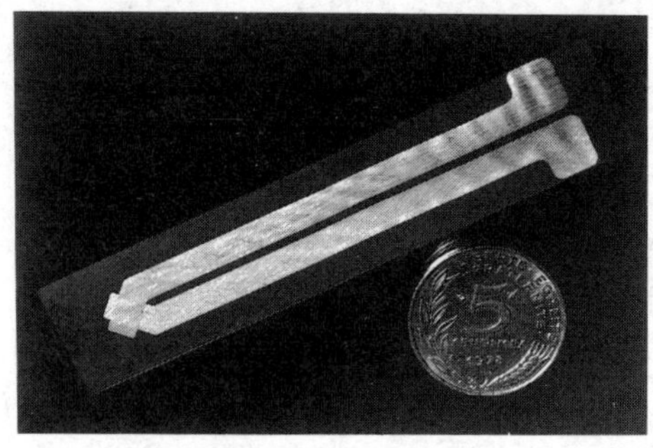

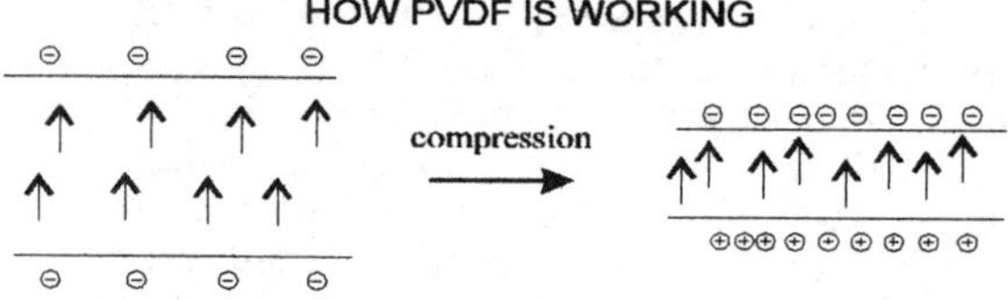

FIGURE 1. PVDF gauge : electrical configuration and "How PVDF is working".

field is stabilized. Each sample fabricated in this process is characterized with an individual poling history with well defined electrical properties which can be reproduced at will. PVDF samples are prepared from biaxially stretched film which have been poled to 9.0 μC/cm^2 through electrodes which are sputtered on the film or via a new process, through contact electrodes which allows the electrode deposition into a low inductance configuration after poling (6). Fig. 1 shows the PVDF gauge and one electrical configuration and "How PVDF is working". Low inductance PVDF gauges will be minimized for gauges having electrical leads with separation equal to zero.

Hypothesis "How is PVDF working", Fig. 1, show that low inductance and high precision gauge with fast response requires a low RC time, (R : resistance of the leads and of the current viewing resistor, C : capacitance value of the gauge), control of the maximum displacement current and remanent polarization value. Small active areas of 0.25 mm^2 have also been prepared via this new process.

EXPERIMENTATION HIGH PRESSURE APPARATUS : PRECISE IMPACT LOADING

Impact loading is produced by controlled impact in a powder gun (caliber 20 mm) at ISL, or compressed gas gun (SNLA). The symmetrical impact of an impactor and target copper provides the loading. The gauge element is placed on the impact surface of the target material. The PVDF gauge is insulated on both sides with a Kel-F film of 110 μm in thickness. Kel-F matches the shock impedance of the PVDF. In this configuration the initial stress wave produced by impact is that typical of the impact of the copper impactor on the PVDF gauge. This wave then, reverberates between the impactor and target until stress is reached equal to that for the standard impactor and target. Low loss coaxial cables and 1 GHz digitizers provide the high-frequency recording capability required to properly interpret the sensor responses. Typically, the current is recorded on two independent digitizers of different sensitivities to broaden the dynamic range of the recording. The electrical charge is determined by numerical integration of the recorded current. With the use of copper, pressures can be routinely achieved up to about 30 GPa with uncertainties of about 3%. The time scale of the event is controlled by stress waves which typically propagate through a PVDF sample in times of about 10 nanoseconds. The short duration of the loading creates conditions in which no heat is exchanged. Under this "Hugoniot" condition, the mechanical work results in specific internal energy which raises the sample temperature 10 to 1,000 °C.

RESULTS

At the same initial stress, the records observed, Fig. 2, indicate a rise time of 3.5 nanoseconds for a standard gauge (7) and of 2 ns for a low inductance and low RC time gauge. Fig. 3 shows the electrical charge versus time measured after equilibrium. The electrical charge is a continuous function of the shock pressure until 30 GPa, Fig. 4. The continuous curve is the computed response (8). The behavior indicates a strongly nonlinear character. Even though the piezoelectric polarization is observed to be nonlinear with pressure, the strongly nonlinear compressibility of

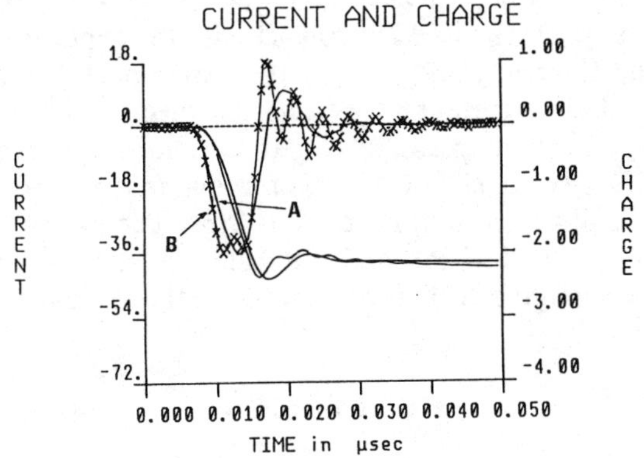

FIGURE 2. Initial stress (Current and Charge)
A: standard gauge B: low inductance gauge
Initial Shock pressure: 3.5 GPa

PVDF results in a behavior in which the observed polarization with volume compression is approximately linear with pressure (2,8).

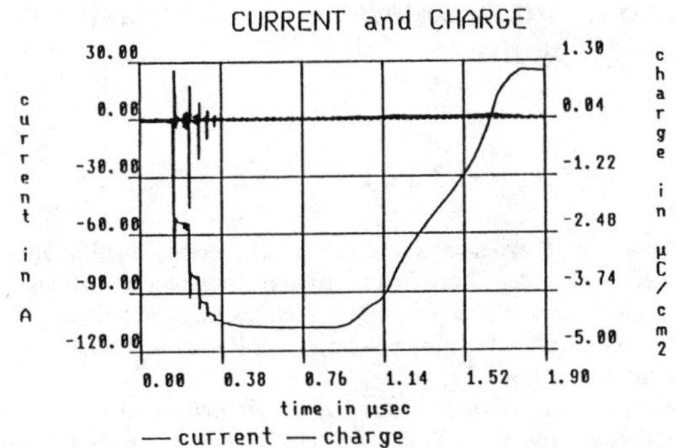

FIGURE 3. Experimental record (Shock pressure: 12.95 GPa)

There are no indications of dielectric relaxation or electrical conduction phenomenon over the indicated range. The behavior appears to be fully described by piezoelectric effects with no permanent changes in remanent polarization indicated. For low inductance gauges with smaller active areas of 0.25 mm^2 the ring up record indicate a rise time of 1 ns for an initial stress of 3.5 GPa induced into the PVDF as shown on Fig.5. The theoretical response computed by Moulard (8) is the continuous line. The agreement is noticeable.

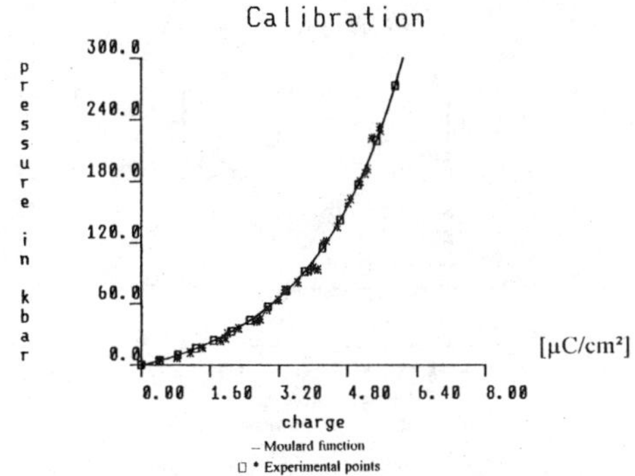

FIGURE 4. Pressure [kbar] versus charge [µC/cm^2].

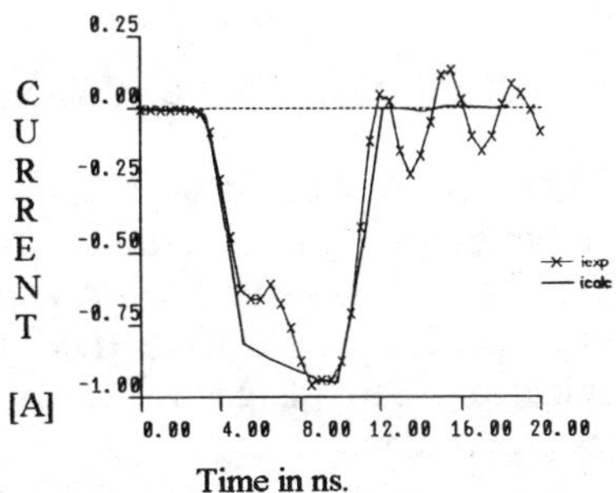

Time in ns.

FIGURE 5. Nanosecond response of a low inductance gauge with an active area of 0.25 mm^2.

ANOMALOUS RESPONSE : SOLUTION

Previous studies of the piezoelectric response of standard PVDF gauges placed on sapphire and subjected to sapphire impact (7) have shown a light increase of the electrical charge during the duration of the application of the compression. Fundamental considerations have led to the supposition that a reorientation of the small amount of the α phase into β phase could occur in the unpoled film covered by the electrodes and subjected to the shock wave. It should be recalled that the poled part of PVDF is only in β phase. PVDF gauges have been prepared with other electrical leads placed on the opposite face of the film as represented on Fig. 6. The same shock experimentation on such a gauge placed on sapphire has been repeated. As we can see, on Fig. 6, the PVDF reproduces in a good manner the step shock compression.

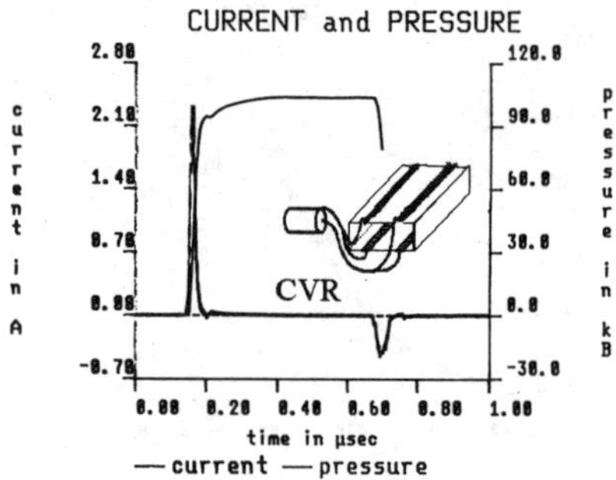

FIGURE 6. Response of PVDF with the new leads configuration to a step shock compression.

DETONATION PROFILE MEASUREMENT

It should be recalled that PVDF works differently in an powder gun experiment where it is rung-up to the final pressure, to an explosive single-shock experiment. We have tried to use « ad hoc shielded » new gauges in an explosive single-shock experiment.

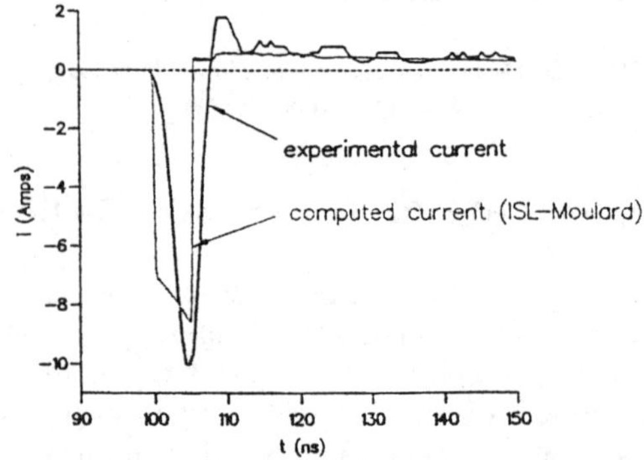

FIGURE 7. Detonation pressure profile (current).

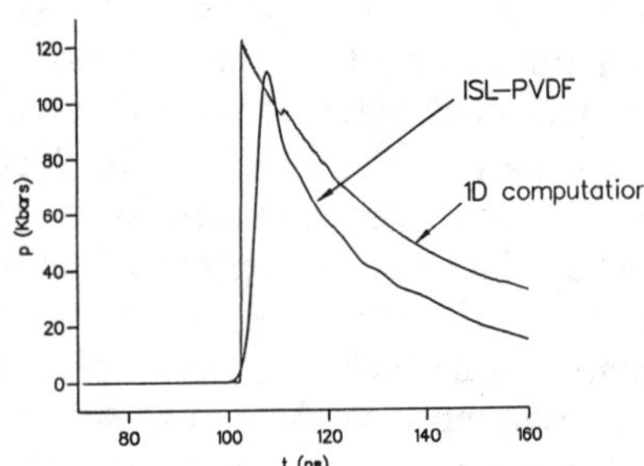

FIGURE 8. Detonation pressure profile (stress)

The explosive is a mixture of hexogen and wax 70/30 in weight percent. The gauge with Kel-F insulator was placed in contact with the explosive. The figures 7 and 8, give the experimental and theoretical current as well as the theoretical and experimental pressure in the vicinity of the explosive. It should be pointed out that the rise time of the current is equal to 2 ns. The pressure attains 10.5 GPa and indicates a detonation pressure greater than 20 GPa inside the explosive.

CONCLUSION

Some high pressure applications of ferroelectric polymers has been presented. PVDF stress rate gauges show continuous response with high reproducibility to pressure approaching 30 GPa. It appears that low inductance electrode lead designs and reproducible remanent polarization are (significantly) improving the precision of the piezoelectric response of the PVDF gauges under shock loading. After ten years studies, we were able to record the detonation pressure profile of a specific explosive.

REFERENCES

1. Bauer, F., Piezoelectric and Electrical Properties of PVDF Polymers under Shock Wave Action: Application to Shock Transducers, in *Proceedings of the American Physical SocietyTopical Conference on Shock Waves in Condensed Matter* 1983, pp. 225-228.
2. Graham, R.A., *Solids under High Pressure Shock Compression*, New York: Springer Verlag, 1993, ch 5.2, pp 103-113.
3. Bauer, F., *Ferroelectrics*, 49, 231 (1983), 115, (1991).
4. Bauer, F., Graham, R.A., Anderson, M.U., Lefebvre, H., Lee, L.M., and Reed, R.P., "Piezoelectric Response of Precisely Poled PVDF to Shock Compression greater than 10 GPa", in *Proceedings of the ISAF 92*, 1992, pp. 273-276.
5. Bauer, F., French Patent 82 21025., US Patents 4,611,260 and 4,684,337.
6. Bauer, F., French Patents 93 00064 and 93 00061
7. Graham, R.A., Anderson, M.U., Bauer, F., Setchell, R.E., "Piezoelectric Polarization of the Ferroelectric Polymer PVDF from 10 MPa to 10 GPa: Studies of Loading - Path Dependence," in *Shock Waves in Condensed Matter-1991*, 1992, pp. 883-886.
8. Moulard, H., Bauer, F., « Lagragian Analysis of the PVDF Shock Sensor « in *Proceedings of the American Physical Society Topical Conference on Shock Compression of Condensed Matter* Seattle 1995, 13-18 August

PIEZOELECTRIC P(VDF-TrFE) THICK COPOLYMERS : RESPONSE IN CURRENT MODE UNDER SHOCK LOADING.

F. Bauer*, P. Isner-Brom*, H. Moulard*, S. Couturier**, T. De Resseguier**, M. Boustie**

*Institut Franco-Allemand de Recherches de Saint-Louis, (ISL), Saint-Louis, FRANCE
**L.E.D., ENSMA Futuroscope Poitiers, FRANCE

The most common piezoelectric polymers are PVDF, and copolymers PVDF with C_2F_3H. A unique application of these polymers are time-resolved dynamic stress gauges based on PVDF studies under shock loading. P(VDF-TrFE) copolymers exhibit unique piezoelectric properties over a wide range of temperature depending on the compositions. The materials may be formed directly from a melt, but their piezoelectric properties are strongly dependent on the annealing technique as well as on the poling process. Copolymer plates of 240, 500 µm in thickness have been fabricated. On the plates, small areas have been poled using the ISL cyclic process. Under shock loading up to 14 GPa, unique piezoelectric response is observed. Experimental results obtained in current mode are presented and discussed.

INTRODUCTION

The polar polymer polyvinylidene fluoride (PVDF) and its copolymers with 20 mol % - 50 mol % trifluoroethylene (TrFE) possess a variety of scientifically interesting and technologically important properties which have made them among the most widely investigated polymers (1,2,3,4,5). They exhibit large dielectric, piezoelectric and pyroelectric constants and were the first known ferroelectric polymers. PVDF is a partially crystalline linear polymer with a carbon backbone in which each monomer $[-CH_2-CF_2-]$ unit has two dipole moments, one associated with CF_2 and the other with CH_2. In the crystalline phase, PVDF exhibits a variety of molecular conformations and crystal structures depending on the method of preparation (5). By stretching, the α phase transforms to the β phase in which the molecular conformation is in the all-trans (TT) planar zigzag with the dipole moments perpendicular to the chain axis. This β form exhibits reversible spontaneous polarization and is therefore the most useful ferroelectric and strongly piezoelectric phase (3,5). It is known, that vinylidene fluoride (CH_2-CF_2) copolymers with trifluoroethylene $(CHFCF_2)$ are yielding polymers (VF_2/VF_3) in which comonomer units are distributed randomly in the chains. Each

monomer, like PVDF, shows a dipole moment. For each cocentration of trifluoroethylene, the VF_2/VF_3 copolymer shows different characteristics, such as the transition temperature FE (ferroelectric) $\rightarrow$ P (paraelectric) and the melting temperature (5). Poling of the copolymers in the ferroelectric (FE) phase results in well ordered molecular conformation of the crystalline phase with a well defined remanent polarization (7). PVDF and its copolymers are useful in a variety of transducer and detector applications. One of the most unique of these applications is their use as the active piezoelectric elements in self-powered, nanosecond time resolved, dynamic stress gauges for the study of shock-wave compression phenomena. Made of very thin (~ 25 µm thick) films (1,2), such gauges are nonintrusive and are ideal for many shock-wave applications (1,6,10). A remarkable feature of these results is that the electrical output signal of properly-prepared gauges is a continuous function of shock pressure up to hundreds of kilobars (6). In addition to extending the direct pressure range of piezoelectric shock gauges by over an order of magnitude, this finding is remarkable since, at these high dynamic pressures, gauge temperatures can be expected to be well over 1000 K. These temperatures are, of course well above the normal melting temperatures (~440 K) of PVDF and its copolymers. The shock results raise important questions about the mechanism responsible for the shock induced electrical output from these polymers. An other feature is that films of VF_2/VF_3 copolymers can be prepared by compression molding of the melts (resin was available from Solvay). Samples of 200 up to 500 μm in thickness can also be pressed. The dynamic piezoelectric response of thick copolymers subjected to shock wave actions became an up-to-date subject for experimental investigations (2). The present paper will briefly summarize the status of such experimental work.

FERROELECTRIC COPOLYMERS

VDF/TrFE copolymers : The structure, ferroelectric properties and melting temperature of VDF/TrFE copolymers are known to depend on composition (3,5). We have chosen for the present work the $P(VDF_{0.7}-TrFE_{0.3})$ composition and the $P(VDF_{0.77}-TrFE_{0.23})$ composition because (i) they have the same crystalline β phase as PVDF at room temperature, (ii) they exhibit well defined and widely separated T_c and T_m and (iii) they have potential for applications. The resins were available from Solvay through Piezotech S.A.. The samples of 240 μm have been extruded by Solvay and the samples of 0.5mm in thickness were prepared by compression molding of the melt followed by quenching into either water or air. Samples were annealed at 413 K in order to enhance the crystallization; other samples were only quenched into cold water in order to increase the ratio of the amorphous phase to the crystallite phase.

Poling process : In the preparation of piezoelectric copolymers it is in general necessary to apply a high poling electric field to an essentially insulating material. The poling process is the same as that established for PVDF (7,8), and provides a mean for achieving a homogeneous polarization at a predetermined level. Each sample fabricated in this process is characterized with an individual poling history with well defined electrical properties which can be reproduced at will (7,8). High voltage up to 60 kV are used to pole the thicker samples. Hysteresis loops on 450 μm thick P(VDF-TrFE) (70/30) film is given Fig. 1. For each copolymer gauge, we measure the remanent polarization as well as the displacement current.

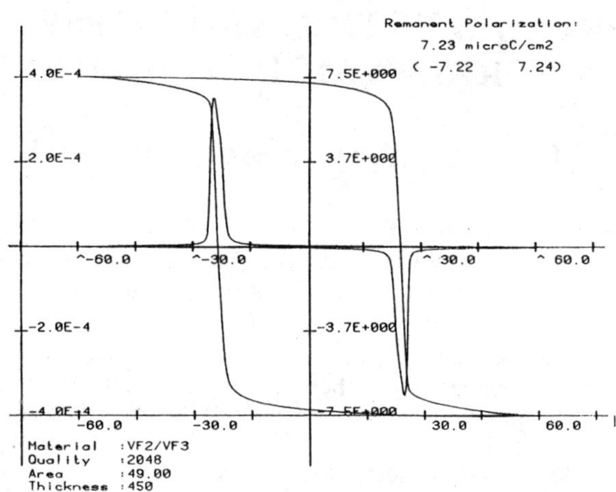

FIGURE 1. Hysteresis curve versus voltage of a 450 μm thick P(VDF-TrFE) 70/30 sample.

All gauges with low crystallinity show a remanent polarization value close to 4 μC/cm². All gauges with high crystallinity show a remanent polarization value close to 7.2 μC/cm². The spatial distribution of the polarization is homogenous. The configuration of the sputtered electrodes on the copolymer gauge is based on that established by Graham (2) called « Sandia guard ring configuration », as shown in Fig 2.

EXPERIMENTATION

Impact loading is produced by controlled impact in a powder gun (caliber 20 mm) at ISL (6). The symmetrical impact of an impactor and target copper provides the loading. The gauge element of 240 μm in thickness is placed on the impact

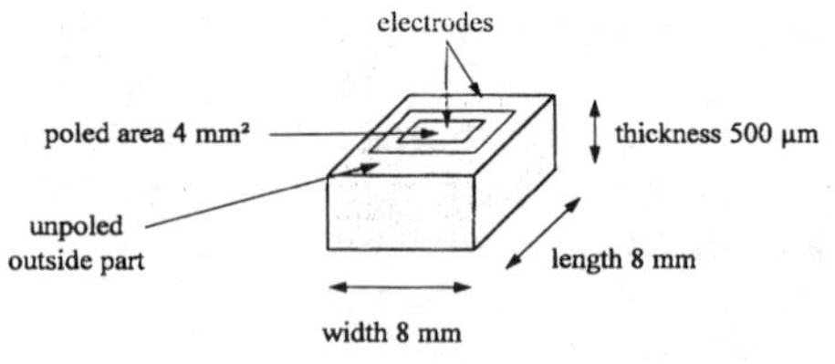

FIGURE 2. Electrical configuration of a copolymer gauge

surface of the target material. The copolymer gauge is insulated on both sides with a Kel-F film of 110 μm in thickness. Kel-F matches well the shock impedance of the copolymer. In this configuration the initial stress wave produced by impact is that typical of the impact of the copper impactor on the copolymer gauge. This wave then reverberates between the impactor and target until stress is reached equal to that for the standard impactor and target. Low loss coaxial cables and 1 GHz digitizers provide the high-frequency recording capability required to properly interpret the sensor responses. With the use of copper, pressures can be routinely achieved up to about 30 GPa with uncertainties of about 3%. The short duration of the loading creates conditions in which no heat is exchanged. For 500 μm thick copolymers, the gauge element is placed behind a 3 mm thick target material for technical reasons. In the latter case, an induced shock of 10 Gpa can be achieved but is completely dependent on the visco-plastic behavior of the copper. Nevertheless, the shock induced can be computed.

RESULTS

Response of 240 *μm* copolymer under shock compression

In using the high pressure tool described above, we were able to study the piezoelectric response of a 240 μm copolymer in thickness obtained under shock compression with the same experimental set-up used for PVDF (6). The active poled area is equal to 4 mm^2 . Typical current records are shown on Fig.3. The final pressure after equilibrium

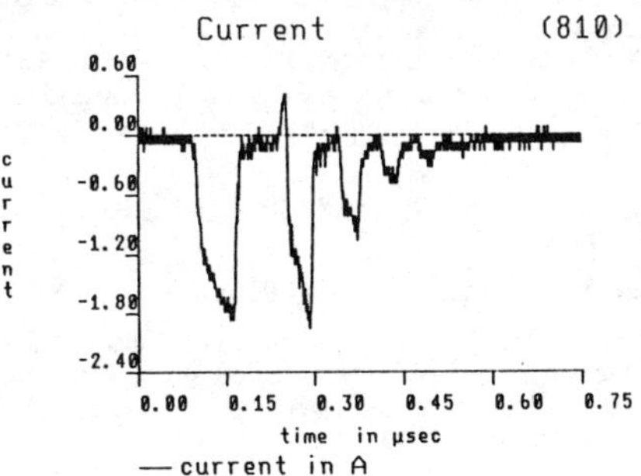

FIGURE 3. Current response of a 240 μm thick copolymer.

attains 14 GPa. The distinctive current pulses are corresponding to the waves reverberations. The first current pulse is observed to increase as the initial wave propagates through the 240 micron sample. When the wave reverberates, we observe also the same current behavior. There are no indications of dielectric relaxation or electrical conduction phenomenon over the indicated range. The behavior appears to be fully described by piezoelectric effects with no permanent changes in remanent polarization indicated.

Thick Copolymer study

ISL samples of 450 μm in thickness, were prepared as described above, with sputtered elctrodes in a guard ring configuration. They have been bonded behind a plate of copper of 3 mm in thickness placed at the muzzle of the powder gun. The level of the shock wave induced into the copolymer was chosen in the range of 1 to 4 GPa. One example of current record is given Fig.4 at a stress level of 3.4 GPa. The record obtained shows deviations from the expected signal which, theoretically, corresponds to a current jump followed by a current slightly increasing and coming back to zero when the wave is passed. As we can see on Fig. 4, the current is jumping and begins to slightly increase in time and is drastically distorted.

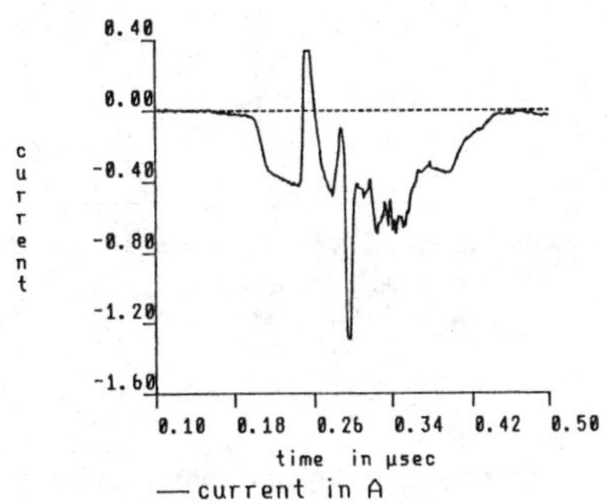

FIGURE 4. Anomalous current response of a 450 μm thick copolymer;

The control of the homogeneity of the polarization in the copolymer does not reveal any anomalies (1). After many non destructive controls of the preparation process of the samples, as well as of the annealing process, electron microscope analysis revealed defects corresponding to microscopic cracks in the bulk of the copolymer during the annealing process. In the meantime, because of the lack of resin of $P(VDF_{0.70}$ -$TrFE_{0.30})$ in mole %, we have decided to study and to optimize the annealing process on the available resin i.e. $P(VDF_{0.77}$ -$TrFE_{0.23})$ in mole % whose composition is close to the first composition studied. 500 micron samples were prepared by compression molding of the melts and were quenched into water. The crystallinity was lowered in this case and after poling at a level of 4 µC/cm², no crack was detectable.

We have repeated the shock experiments with the same set-up. Typical current records are shown on Fig. 5. One should contrast this record with that shown in Fig.4 to appreciate the distortion obtained with the high crystallized copolymer. The current from the $P(VDF_{0.77}$ -$TrFE_{0.23})$ in mole % is observed to increase as the wave propagates through the 500 micron sample. The extent of the effect depends largely on the stress amplitudes as we can see on the record obtained at 9.4 GPa.

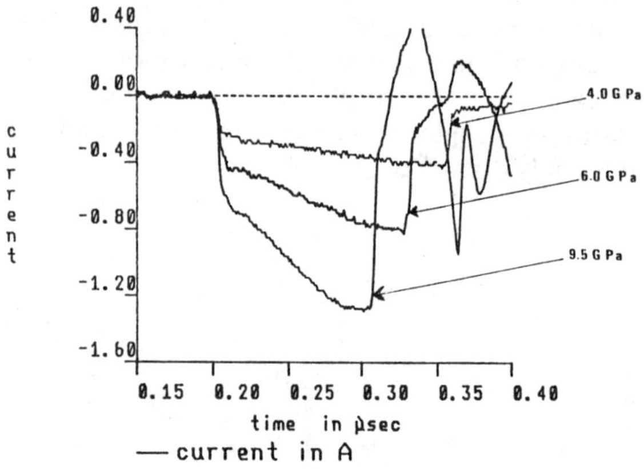

FIGURE 5. Experimental current records at different shock pressure levels

We also can note a rise time of a few nanoseconds for the first current jump. The non linear behavior of the observed records is noticeable, and has to be analyzed with the same approach used for PVDF as described in one companion paper (10).

CONCLUSION

The first experimental study of the piezoelectric response of thick piezoelectric copolymers was realized. First use of such copolymer for measuring short stress pulses (25 ns) induced by laser in a aluminum target was initiated by M. Boustie, T. de Resseguier et al. in Poitiers (9). They have observed the same anomalous response of thick copolymer at shock level higher than 3 GPa. We have determined the origin of such anomalous response and consequently defined the process of preparation as well as the poling process of a copolymer which can respond to shock level of 10 GPa in a single shock compression. A definition of a standard copolymer matched for this application is actually in progress. Lagrangian analysis (10) will be made in order to identify the dependences of the current with the strain induced.

REFERENCES

1. Bauer, F., Piezoelectric and Electrical Properties of PVDF Polymers under Shock Wave Action: Application to Shock Transducers, in *Proceedings of the American Physical SocietyTopical Conference on Shock Waves in Condensed Matter* 1983, pp. 225-228.

2. Graham, R.A., *Solids under High Pressure Shock Compression*, New York: Springer Verlag, 1993, ch 4, ch 5, pp. 71-113.

3. *Ferroelectrics* 32 Nos 1-4 (1981) and 115 N°4 (1991).

4. Samara, G A., Bauer, F., « Hydrostatic Pressure Studies of Polyvinylidene Fluoride (PVDF) and its Copolymers » in *Shock Waves in Condensed Matter-1987*, 1988, pp. 611-614

5. Lovinger, A.J., *Jap J. Appl. Phys.* 24, Supplement 24-2, 18 (1985).

6. Bauer, F., Moulard, H., Graham, R.A.,« Piezoelectric Response of Ferroelectric Polymers under Shock Loading : Nanosecond Piezoelectric PVDF » in *Proceedings of the American Physical Society Topical Conference on Shock Compression of Condensed Matter* Seattle 1995, 13-18 August.

7. Bauer, F., French Patent 82 21025., US Patents 4,611,260 and 4,684,337.

8. Bauer, F., French Patents 93 00064 and 93 00061.

9. Romain, J P., Bauer, F,.Zagouri, D., Boustié, M., . « Measurements of Laser Induced Shock Pressures using PVDF Gauges », in *Shock Waves in Condensed Matter-1993*, 1994, pp. 1915-1918.

10. Moulard, H., Bauer, F., « Lagragian Analysis of the PVDF Shock Sensor « in *Proceedings of the American Physical Society Topical Conference on Shock Compression of Condensed Matter* Seattle 1995, 13-18 August

CHARACTERIZATION AND REDUCTION OF PROMPT ELECTRICAL NOISE ON THE SATURN PRS X-RAY SIMULATOR

D.E. Johnson[*], F.W. Davies[*], and M.A. Hedemann[**]

[*] *Ktech Corporation, 901 Pennsylvania NE, Albuquerque, NM 87110*
[**] *Sandia National Laboratories, P.O. Box 5800, Albuquerque, NM 87185*

Experiments have been conducted to identify and characterize the source of prompt electrical noise observed on piezoelectric sensor data traces recorded at the Saturn soft x-ray source. The amplitude of this noise spike often exceeded the amplitude of the signal of interest. Although the noise duration was short (50 to 100 ns), it interfered with many measurements of soft x-ray induced material response detected with piezoelectric gauges. The noise spike can be due to a combination of sources, including bremsstrahlung radiation. However, it was determined that high-energy electrons generated about 90 per cent of the recorded signal in the near-source region. Fixtures employing permanent magnets reduced the noise spike amplitude by more than a factor of ten, permitting sensitive, high-fidelity, material response measurements to be made with greater accuracy and confidence. Furthermore the test environment is more accurately controlled and defined.

INTRODUCTION

The Saturn x-ray simulator [1] at Sandia National Laboratories, New Mexico can be configured as a Plasma Radiation Source (PRS) to generate soft x-rays of 1 to 5 keV energy [2]. In this configuration, Saturn is used to determine material or component response to pulsed x-ray exposure of 0.1 to 50 cal/cm^2. Material response is determined primarily by measuring the x-ray generated stress pulse that propagates through the sample using piezoelectric stress gauges [3] or VISAR. Until recently, the use of piezoelectric stress gauges has been complicated by the presence of a large noise spike that occurs coincident with x-ray generation. As Figure 1 shows, the amplitude of this noise spike often exceeded that of the signal of interest, especially when the machine used an argon gas puff as the source. The high-amplitude noise spike, though brief in duration and proceeding the stress pulse, often drove the recording amplifiers into saturation and either obscured the signal of interest or shifted the

baseline enough to prevent accurate measurement. It is also important to note that the noise spike was only large when the machine used a gas puff, such as argon, as the source. The use of wire arrays, such as aluminum, resulted in much lower noise spike amplitudes that seldom interfered with material response measurement.

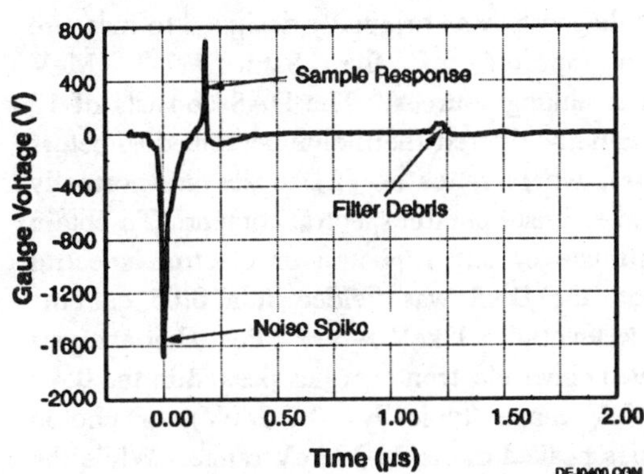

FIGURE 1. The noise spike amplitude often interfered with the signal of interest.

Work supported by Sandia National Laboratories, Albuquerque, NM, under contract AH-8570

Since Saturn in the mode of interest delivers a pulse of about 400 kJ of electrical energy to the diode at 2 MV and 8 MA, electrical noise in not unexpected. Potential noise sources include electromagnetic pulse (EMP), electrons, and bremsstrahlung radiation. A variety of techniques, including shielding, grounding, baffles, and permanent magnets, were used separately and in combination to reduce the noise spike amplitude to an acceptable level. The single most effective tool was the use of permanent magnets to sweep out high-energy electrons. These electrons were found to generate 90 to 95 percent of the observed noise signal on x-cut quartz gauges.

NOISE REDUCTION TECHNIQUES

Noise reduction techniques commonly used in pulsed power environments are routinely applied at Saturn. These include various forms of shielding, grounding, and baffles. Recently, 0.1 to 0.4 T permanent magnets have been added to the test fixtures and proven extremely effective at reducing the noise spike amplitude. The following paragraphs describe several experiments that lead us to conclude that high-energy electrons were generating most of the electrical noise.

Differential Absorption Spectrometer (DAS)

This diagnostic was originally designed to measure photon spectra of the Saturn 1-2 MeV bremsstrahlung sources. The DAS consists of 13 differentially filtered thermoluninescent dosimeters (TLDs), where ratios of TLD doses are normally used to estimate photon spectral content. To obtain an estimate of either photon or electron spectral content, the DAS was fielded at a 0.05 cal/cm^2 fluence level of 3.1 keV x-rays. Post-shot analysis indicated either electron energies peaked in the 0.4 - 2.0 MeV range (typically ~0.7 MeV), or photon energies peaked in the 20-30 keV range. While the DAS could not be used a priori to differentiate between photons and electrons, the measured TLD doses (~400 Gy) were inconsistent with source-strength capabilities in the 20-30 keV photon energy range. These results therefore strongly indicated a large electron-based component to the noise.

Shielding

Enclosing the sensor in metal shielding serves two functions: capacitive, magnetic, and radiated electrical noise are blocked or reduced by the Faraday enclosure; and noise induced by soft x-rays and electrons can be reduced. The effectiveness of the shielding depends primarily on the material properties, atomic number, conductivity, permeability, and density, and on the thickness of material used.

An experiment was conducted at Saturn using two shielding materials, aluminum and stainless steel, in several different thicknesses. Nominally identical x-cut quartz gauges housed in aluminum canisters were fielded with the sample shielding material bonded onto the front surface of the quartz element. The gauges were placed such that the sample was normal to the radiation path and directly between the source and the gauge. All samples were attached with conductive silver epoxy around the edges forming a complete Faraday enclosure.

As Figure 2 shows, the noise spike amplitude decreased in a roughly linear manner as the areal density of the sample was increased. Also, an equal thickness of stainless steel was about six times more effective than aluminum, even though aluminum is a much better electrical conductor. The dependence on areal density, rather than electrical conductivity, indicates that the electrical noise spike is probably generated by penetrating electrons and not by pulsed power noise that is electrically coupled into the sensor.

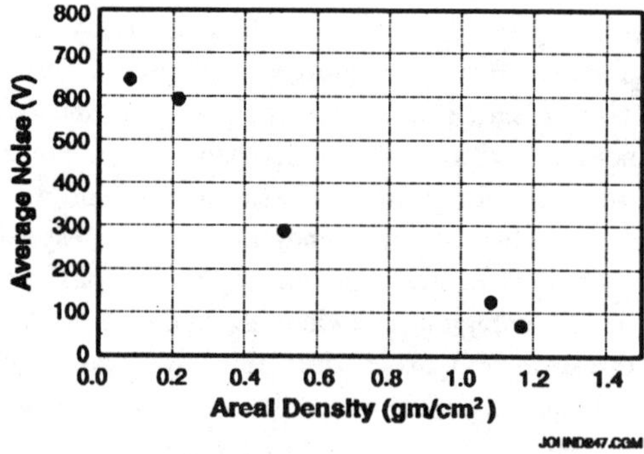

FIGURE 2. Saturn noise spike amplitude vs areal density of shielding.

The maximum penetration range, R in gm/cm^2, of electrons in both aluminum and stainless steel depends linearly on the electron energy, E in MeV [4]:

$$R = 0.542\,E - 0.133 \qquad (1)$$

This equation is accurate for electron energies from about 700 keV to 3 MeV. Below 700 keV it begins to overestimate the range, but still provides a reasonable approximation.

If we assume that the noise spike amplitude was proportional to the number of electrons that penetrated the shield layer, we can use Equation 1 to estimate the electron energies. As shown in Figure 2, the noise amplitude was reduced 60 per cent by 0.55 gm/cm^2 of aluminum, so many of the electrons were of energy 1.2 MeV or less. Significant noise levels were still observed with 1.18 gm/cm^2 of material, so some electron energies may have been as high as 2.4 MeV. This range of energies is consistent with observation based on the DAS analysis, assuming the source of the gauge noise and of the DAS activation energy was both electrons.

Orientation

The penetrating radiation or electrons that were hypothesized as the noise source appeared to originate from the plasma radiation source (PRS) region or very close to it. The shielding samples used in the experiment just described were placed between the x-ray source and the sensor. The samples were just large enough to cover the crystal and epoxy potting compound and would therefore block only x-rays and electrons that originated at the source.

In a second experiment, gauges were oriented such that the sensors were pointed away from the source. These gauges had much lower noise spike amplitudes than identical gauges pointed at the source. This second experiment supported the hypothesis that the noise was generated by x-rays or electrons that originated within or from very near the PRS.

Permanent Magnets

None of the previously described experiments could definitively differentiate between electrons and x-rays as the source of the noise spike energy, so an experiment using permanent magnets was designed. X-rays are unaffected by magnetic fields, but if the noise source was electrons, the magnets would sweep out or trap many of the electrons and reduce the noise amplitude. The permanent magnets were arranged to produce a field transverse to the electron flight path. As the electrons traveled through the

magnetic field, the Lorentz force would deflect the electrons or trap them in spiral orbits, depending on the electron energy (velocity) and the field strength. Figure 3 shows this arrangement schematically. Three fixtures were installed for each test. One fixture generated a field of approximately 0.1 T, the second fixture generated a 0.4 T field, and the third, with no magnetic field, was used as a control. Calculations showed that the 0.1 T field would trap electrons of energy 100 keV or less and sweep electrons of energy 1 MeV or less. The 0.4T field would trap electrons of energy up to 1 MeV and sweep electrons of energy 10 MeV or less. Each of the three fixtures placed the gauges at the same range from the source and included a stainless steel aperture such that the gauges were exposed only to radiation and electrons emitted directly from the source region.

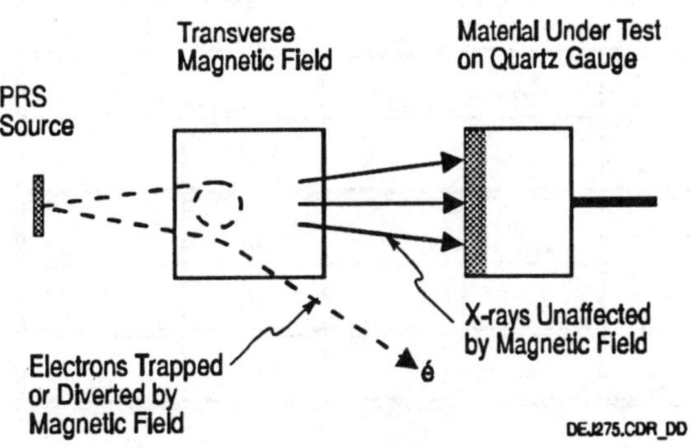

FIGURE 3. Permanent magnet fixtures for trapping or sweeping electrons.

Figure 4 shows the gauge records from one of two shots using the permanent magnet test fixtures. The noise spike amplitude was much lower on the two gauges fielded behind the magnetic fields. The test was repeated on a second shot with the same results as listed in Table 1.

The noise amplitude without magnets varied from 265 to 500 V. The 0.1 T field reduced the noise amplitude by a factor of 10 to an average value of 32 V. The 0.4 T field was even more effective, reducing the noise amplitude by 27 times to an average of 14 V.

The strong effect observed with the magnetic fields confirms that the noise spike is caused primarily by high-energy electrons produced at the source. The small remaining noise level may be due to bremsstrahlung radiation.

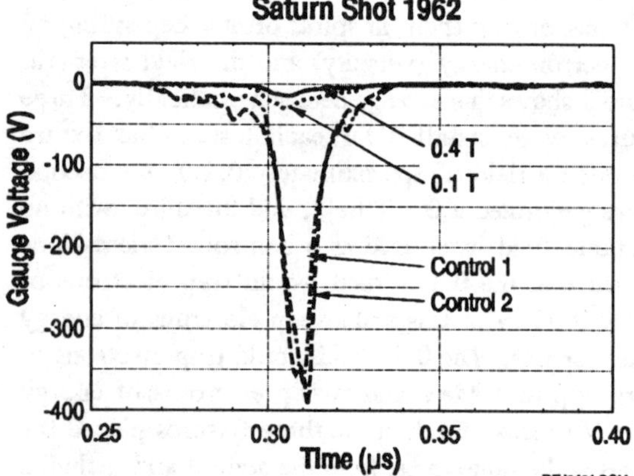

FIGURE 4. Noise spike records from Saturn tests with permanent magnets.

TABLE 1. Noise Amplitude Comparison with Magnetic Fields (Noise Peak Amplitude in Volts)			
	No Field Control		Magnetic Field	
Shot	1	2	0.1 T	0.4 T
1961	265	500	34	12
1962	394	362	30	15
Avg:	380		32	14

DISCUSSION OF RESULTS

The results of the experiments described here lead us to conclude that the noise spike observed at Saturn in gas-puff PRS mode is due almost entirely to energetic electrons emitted from the source region. These electrons range in energy from a few hundred keV up to about 2 MeV. Magnetic fields of 0.1 to 0.4 T over a volume of a few cubic centimeters are effective at trapping or sweeping out these electrons and reducing the noise spike amplitude by 10 to 30 times.

The relative absence of the noise spike when using a wire source array provides some clues as to the origin of the noise-generating electrons. We hypothesize that the electrons are generated in the gas puff source just before the gap breaks down to form the plasma channel that becomes the x-ray source. At this instant, before the gap breaks down and current flow begins, the gap voltage is several megavolts and the electric fields are oriented such that electrons would tend to be accelerated radially outward towards the gauges. In contrast, the electric fields are lower in a wire source because conduction begins immediately and is only limited by the source inductance and the relatively low resistance of the source wire array.

CONCLUSIONS

Permanent magnets can be used in Saturn and other PRS simulators to sweep out electrons and provide a test environment that is electrically much quieter. More sensitive and accurate material response measurements can be made and the test environment is better defined. Effective shielding design for both electrical and fiber-optic experiments has been simplified by this noise source identification and characterization.

REFERENCES

[1]   Choate, L. M., *The Saturn X-Ray Simulation Facility, A Technical and Logistical Information Guide for Users*, Albuquerque, Sandia National Laboratories, 1989.

[2]   Spielman, R. B., "Z-Pinched Experiments on Saturn at 30 TW", *AIP Conference Proceedings No. 195*, Dense Z-Pinches, 1989, pp 3-16.

[3]   Johnson, D. E., Lee, L.M., Hedemann, M. A., and F. Bauer, "PVDF Measurement of Soft X-Ray Induced Shock and Filter Debris Impulse," *Proceedings of the Conference on Shock Compression of Condensed Matter*, 1993, pp. 1911-1914.

[4]   B. T. Price, C. C. Horton, and K. T. Spinney, *Radiation Shielding*, Pergamon Press, 1957, pp. 71-72.

ACKNOWLEDGEMENTS

The authors are especially thankful for the technical assistance provided by C. DeLaCruz, R. Klingler, S. Heyborne, and M. Rice in performing these experiments and preparing the manuscript.

HIGH-SPEED SPHERICAL SOLID REGISTRATION BY THE USE OF PIEZOELECTRIC COMPOSITE FILMS

V.M.Shunin, S.S.Nabatov, V.V.Yakushev,

Institute of Chemical Physics in Chernogolovka, Moscow reg., 142432 Russia,
&
A.P.Volkov

Institute of Mechanization, Krasnoarmeisk, Moscow reg., 141260 Russia.

Polarized composite polymer-ceramic films as electric pulses detector of high-speed solids were designed and studied. Films were made of the silicon rubber loaded with modifying agents and piezoelectric ceramic powder. The electrical film responses to impacts of 4-10 mm steel balls accelerated from 0.5 to 2.0 km/s were measured. The gauge generated pulses with rather high amplitude (tens of volts on 50 Ohm resistance) and high time resolution (~1 µs) between impacts of solids. The pulse amplitude depends on impactor velocity and its diameter. The gauges are suitable for operating under high temperature.

INTRODUCTION

In several areas of explosive physics there is a necessity for registration of high-speed solid parameters. This is particular true for explosive acceleration of solids, explosive fragmentation of metallic covers and behavior of ammunition under the impact of fragments. The piezoelectric or pyroelectric films which are able to generate electrical signals under the solid impacts hold the greatest promise for realizing such gauges.

For the first time the application of ferroelectric films as impact gauges was suggested in 1984 in connection with the registration of cosmic dust particles within the framework of "Venus-Galley" program [1,2]. The gauge sensor was a polyvinylidene fluoride (PVDF) film polarized by an electric field. The other polymer material which may be used as a sensor is polarized polyvinyl chloride (PVC). The phenomenon of depolarization of PVC film under the action of high speed solids was used as the basis for making of the depolarized impact gauges [3].

However, the application fields of PVDF and PVC films are limited by virtue of the low temperature stability. So it is appropriate to investigate the possibility of the use of piezoelectric composite polymer-ceramic materials as the gauge sensor. Such composites are designed to combine the high and stable (under temperatures) piezoelectric activities of ceramics with the easy processibility of polymers. Selecting the suitable components we can vary in a wide range both the mechanical and electrophysical properties of these materials [4-6].

In this paper results of measurements of electric responses of flexible piezoelectric composite films to the high speed impact of steel balls will be described. The measurements were performed with high-temperature composite materials on the base of silicon rubber filled by a ceramic powder from the lead zirconate titanate family.

MATERIAL

Composite samples were prepared in the form of thin films (100-300 µm) by a solvent cast technique or piezoelectric "paint" [7,8]. The material consists of a small piezoelectric ceramic particles dispersed in an elastomer-based matrix. The matrix used was polydimethylsiloxane rubber with modifying agents to improve the rheologic and adhesive properties and temperature stability. Butyl acetate as a rule was used as a solvent. The loading of the ceramic powder (mark of TsTS-19, Curie point is about 290 ^{0}C) was 20-40% by volume. The average particle size was about 5 μm. The mixture was treated by the ultrasonic disperser and then spreaded on 7 μm aluminum foil to make the film. The polymerization process has been finished for 48 h. The solvent traces was removed by vacuum drying at 120 ^{0}C for 24 h.

The film were poled at 120-140 ^{0}C under the constant electric field 5-8 kV/mm for 40 minute. To avoid non-controlled charge relaxation which may proceed after high-temperature polarization during tens days, the samples were cured at 150 ^{0}C for 1 h. The capacities and the dielectric loss tanδ of the composites were measured at 1 kHz frequency at the room temperature. The relative dielectric constant ε_{33} was calculated from the capacitance. Piezoelectric constants d_{33} were measured with static method.

The high-speed impact results reported in this paper were obtained with the flexible composite film in which the real loading of the ceramic powder was about 30% by volume (the design proportion of the components was equal 35%). The average density of the specimen was determined by Archimedes method and turned out to be 3.65 g/cm. .The porosity amounted to about 7%. For this sample the average value of dielectric constants ε_{33}= 11.2, tanδ= 0.01, and piezoelectric constant d_{33} = 10.0 pC/N.

EXPERIMENTAL

A schematic of the experimental arrangement is shown in Fig.1. The gauge consists of the film composite sensor with two 7 μm aluminum electrodes. One of these electrodes was used in the manufacture of composite. The second electrode was cemented to other side of composite film with the same polymer which stood duty as matrix. The electrode measures 150 mm x 300 mm. The gauge thickness was equal 0.29 mm. The terminal-revets were served as the current collectors. The measured

gauge capacitance was 5.8 nF (involving adhesive layer between electrode and composite).

The steel balls with diameters 4-10 mm were used for a high speed solids. The balls were accelerated to speeds from 0.5 km/s up to 2.0 km/s by a powder-gun. A ball strikes the electrode along the normal to it. The vectors of film remnant polarization $\mathbf{P_r}$ and ball velocity $\mathbf{w}$ had the same direction in all experiments.

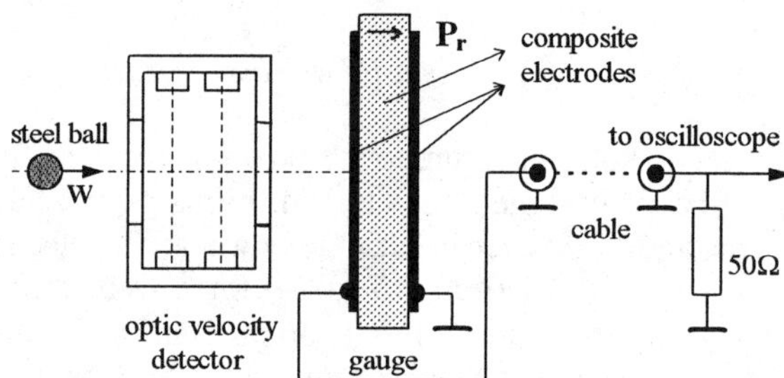

FIGURE 1. Experimental arrangement used to simultaneously measure the gauge pulse and the ball velocity.

Such experimental arrangement measures the ball velocity, starts an oscilloscope for recording of gauge response, and determines the moments of touching the ball with the gauge electrode. During the interaction between the ball and the gauge the current from the gauge is measured by recording the voltage across a 50 Ohm resistor using a oscilloscope with 50 ns/point resolution. The resistance of 50 Ohm terminates the cable in its characteristic impedance, eliminating cable reflections.

It was carried out the high temperature testing of these gauges by means of repeated many times heating from room temperature up to temperature of the matrix stability (200 ^{0}C) and shot-time (few minutes) heating up to Curie point of the ceramic. The repeated heating tests of gauge were performed in an oven at a normal pressure. The short-time heating up to 300 ^{0}C was carried out at the pressure 5 Torr (600 Pa).

RESULTS AND DISCUSSION

Figure 2 shows a record of the gauge response to impact of a steel ball of the diameter 7 μm and velocity 1.1 km/s. Maximal voltage V_{max} is reached for time 0.3 μs and then decreases practically to zero.

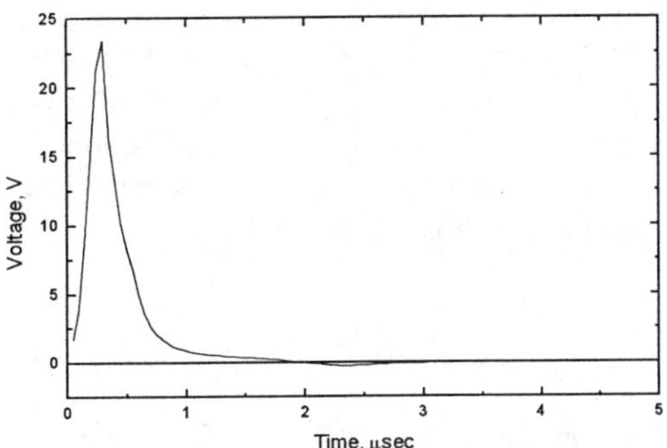

FIGURE 2. Flexible composite gauge voltage output.

The value of V_{max} depending on the diameter and velocity of the boll is presented in Fig.3. It is seen that the electrical response amplitude has a rather large values and depends on the high-speed parameters (velocity and cross-section of the boll) within the considered velocity range. V_{max} increases with increasing diameter and velocity of the solid. It permits one not only to record the impact of solids but also to connect the data with parameters of solids.

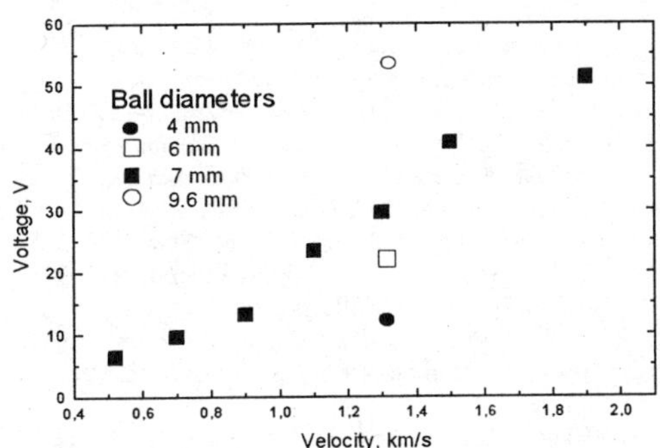

FIGURE 3. Velocity dependence of peak voltage V_{max}.

The duration of the pulse usually equals 1-1.5 μs, which is markedly smaller than the time of the ball piercing through the film (3-5 ms). Keeping in mind that the signal voltage arises at the moment of touching the ball with the film, we can suppose that the response is determined by a change of the remnant polarization behind of the shock wave front. At the same time for the polarized polymer

PVC [3] and PVDF films the high-speed solid impact results in their depolarization at the expense of plastic behaviour during penetration of the solid, with the response duration being several times longer than that for composite films. In addition, the spurious voltage impulses are absent on the oscillograms of piezoelectric composite films. These impulses, as known, appear in polymer films and last some time after the penetration of the solid.

So the composite gauges generate a short and unipolar impulses convenient for automatic registration. Such a response of composite piezofilms seems to be connected with their mechanical properties (flexible, easy compressibility of polymer matrix and others). For example, composite gauge reinforced by fiber glass fabric generates the signal with a small negative impulse (see Fig.4). It is not inconceivable that this short and unipolar signal may be also connected with the electric conductivity in the intensive plastic behaviour zone or even behind the shock wave. In any case these gauges as electronic pulse detectors of high-speed solids can reach 1 μs time resolution at best.

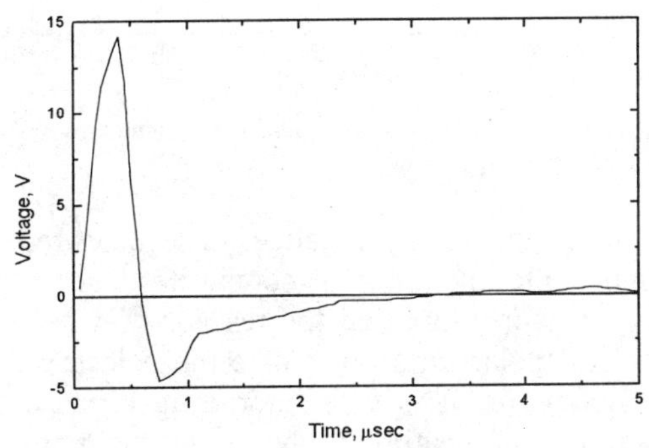

FIGURE 4. Reinforced composite gauge voltage output.

The composite piezofilm electric response to the impact of solids may be resulted from piezoelectric or depolarization effects of ceramic particles. In tests to determine piezoelectric constants d and in the high speed impact experiments signs of the electrical responses are the same. It might point to the fact that the response is mainly due to piezoelectric effect. However, it is worth bearing in mind that piezoelectric and depolarization effects in composites have the signals of the like polarity.

The same effect was observed in composites under the plane shock waves [9]. As for PVC piezofilms [3,10] the signal signs of piezoelectric and depolarization effects are opposite. In the present time the nature of the composite response is not uniquely determined.

In the work it has been performed a set of experiments to check the temperature stability of composite films. Repeated cyclic heating with various rates from the ambient temperature to 200 ^{0}C showed that d_{33} and the electrical response to solid impact do not change practically. The short-time heating up to 300 ^{0}C has no effect on the piezoelectric properties too. Results of this experiment are shown in Fig.5.

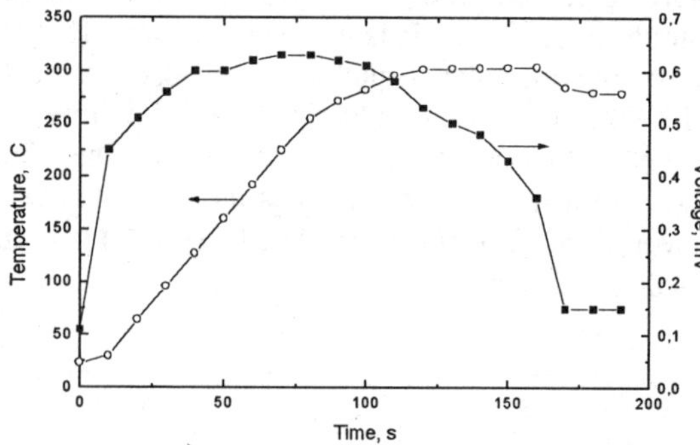

FIGURE 5. Time dependence of the temperature and the voltage generated by gauge.

The temperature of gauge was measured by thermocouple under the of composite. The output voltage was measured on the resistor of 1 MOhm. As seen depolarization of gauge piezoelectric film does not occur. It is very interesting because the temterature of heating higher then the point of Curie (290 ^{0}C) At the same time the heating of ferroelectric polymers higher than the Curie point results to disappear there remnant polarisation. This fact indicates that the composite films are a suitable material for gauge operating under high temperature.

SUMMARY

The results of the study have shown that piezoelectric composite films can serve as a sensor of high speed solid impacts gauges. These gauges have a high sensibility (tens of volts on 50 Ohm resistance) and the highest time resolution (1 μs) among of all the known impact gauges. They can operate at rather high temperature.

ACKNOWLEDGMENTS

The authors would like to thank Mr. V.E.Korolev and Mrs L.A.Smetanina for their experimental assistance, and SIC "Molniya" for their support of this work.

REFERENCES

1. Simpson J.A., Tuzzolino A.J., *Nuclear Instruments and Methods in Physics Research* **A236**, 187-202 (1985).
2. Perkins M.A. ., Simpson J.A., Tuzzolino A.J. *Nuclear Instruments and Methods in Physics Research* **A239**, 310-315 (1985).
3. Alekceev E.M., Antipenko A.G., Dremin A.N., Krenev S.A., Kurto A.P., Nemanichin A.L., Sidorov V.A., Yakushev V.V., *Chimicheskaya phizika* **11**, 235-239 (1992).
4. Newnham R.E., Safari A., Giniewicz J., and Fox B.H., *Ferroelectrics* **60**, 15-21 (1984).
5. Ting Robert Y., *Ferroelectrics* **67**, 143-157 (1986). .
6. Andrianov E.G and Shunin V.M.,*Technology of making composite piezoelectric materials and their properties,* Chernogolovka: OIChP AS USSR, 1986, pp.1-24.
7. Shunin V.M., Nebogatov V.E, Sharapov V.L., and Yakushev V.V., "High-temperature composite piezoelectric material", presented at *the International Scintific-Practice Conference,* S.-Petersburg, January 16-17, 1992 in book Piezoengineering-92, 1992, pp. 32-33.
8. Hanner K.A., Safari A., Newnham R.E., and Runt J., *Ferroelectrics* **100** 255-260 (1989).
9. Shunin V.M., Antipenko A.G., Dremin A.N., Yakushev V.V., Andrianov E.G., Klimov V.V., and Stets R.G., *Phizika goreniya i vzryva* **22**, 118-121 (1986).
10. Antipenko. A.G., Kurto A.P, Yakushev V.V., "Electrical effects of piezoelectric polymer under shock compression", presented at the *Workshop of Detonation,* Chernogolovka, 1981, in Proceedings of the Workshop on Detonation, 1981, v.2, pp.118-123.

PVDF TRANSDUCER RESPONSE FROM AN ELECTRICAL DISCHARGE IN AN ALUMINIZED SOLID

Richard J. Lee and Jerry W. Forbes

Naval Surface Warfare Center, Indian Head Division, White Oak, Silver Spring, MD 20903-5640

The electrical responses of PVDF transducers were measured for cylindrical stress waves generated from expanding electrical discharge channels in an aluminized solid. Electrical discharges with risetimes of 15 ns and durations varying from 100 to 400 ns were used to create the initial stress wave in the solid. Each electrical discharge was formed across two thin electrodes sandwiched between the test sample and a sapphire disk. The transducer response was not corrected for strain effects.

INTRODUCTION

PVDF transducers were used in a high voltage, electrical discharge environment. A specialized packaging was used to provide a relatively non-intrusive gauge capable of recording stress profiles induced by an evolving arc channel in the surrounding material. The rapid expansion of the arc channel acted as a piston establishing a shock wave in the surrounding medium. The goal is to develop a model for arc channel expansion in solids for various discharge parameters (e.g., current,[1] deposition energy,[2] deposition rate, and deposition time). Since the electrical arc is idealized as an expanding cylindrical channel and is typically small in diameter, e.g., 0.2 to 0.4 mm, the mechanical disturbance reaching the gauge is inherently divergent. Hence PVDF transducers were subjected to non-uniaxial stress and strain. Strain is known to distort the stress signature,[3] but no attempt was made to correct the PVDF transducer signals for non-uniaxial strain in the gauge.

EXPERIMENTAL

Cylindrical samples, 12.7 mm in diameter and either 3.18 mm or 6.35 mm thick, were tested in an enclosed test cell, see Figure 1. The electrical discharge was formed between two thin electrodes along the interface between one surface of the test sample and a sapphire disk, 15.9 mm in diameter and 12.7 mm thick. The electrodes were brass foils, 12.7 mm wide and 25.4 μm thick, cut to a point at a 30° angle. The gap spacing between the electrode tips was 6.3 mm. Each electrode was held in place on the sapphire disk by a strip of 63.5 μm thick adhesive-backed Kapton tape. The sapphire disk offered a surface which was resistant to arc erosion, allowing the use of the same surface and electrode configuration through the course of a given experimental series. The sapphire disk was held in place by a PMMA retaining ring which fit over two locating pins pressed into the base of the test fixture. The test samples were circumferentially confined by an RTV washer which fit in a PMMA containment cylinder with an inner diameter of 38 mm. The containment cylinder was secured in a groove cut in the top surface of the retaining ring.

The PVDF gauge was inserted through a slit in the containment cylinder located 90° from the electrode access slits and at an appropriate height from the sapphire surface. The active area of the transducer was situated directly over the center of the discharge gap. The distance between the discharge gap and the transducer was defined by the thickness of the test sample. Most experiments used a 3.18 mm thick test sample with another 6.35 mm sample and RTV washer set behind the gauge for confinement. Departures from this configuration are mentioned in the Results section. Vacuum grease was applied at all interfaces, except at the discharge gap, to exclude air at these surfaces. The discharge

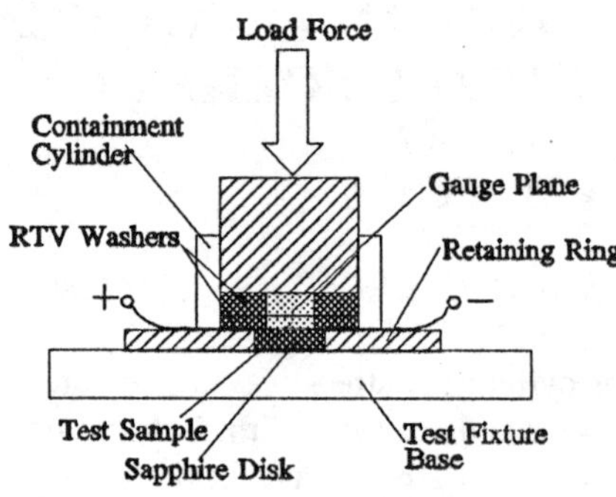

FIGURE 1. Experimental Arrangement

gap was left clear of vacuum grease so as not to impede the dielectric breakdown process.

A force of 1.8 kN, was applied to the system by a 90 mm long PMMA rod and bolt supported at the top of the test fixture to ensure intimate contact at the various interfaces.

Experimental Sample Material

An inert propellant simulant was used for test samples. This material consists of calcium chloride crystals, fine spherical aluminum particles and a Hydroxyterminated Polybutadiene (HTPB) binder system. The density of the material was 1810 kg/m³. The presence of aluminum facilitated dielectric failure of the test samples, allowing arc channels to be formed at moderate voltages across the 6.35 mm electrode gap.

Discharge Circuit

A charged cable pulser was used to provide fast 15 ns risetime, 100 to 400 ns duration electrical discharges. Electrical energy was stored on a 50 Ω co-axial cable and then rapidly transferred to the electrode gap via a closing switch. This technique typically offers fast developing rectangular current discharges if they are operated into a load matching the characteristic impedance of the cable. The current lasts for as long as it takes an electromagnetic wave to travel down the cable and back again. For this study, 10, 20, 30, and 40 m long cables were

used to provide nominal discharge pulse lengths of 100, 200, 300, and 400 ns, respectively. The dynamic resistance associated with the electrical discharge did not provide an ideal matched load. Hence, the current and voltage would ramp up and down respectively following a sharp rise (15 ns) to an initial magnitude. Despite this non-ideal behavior, the power pulse was nearly rectangular and hence defined by a single magnitude.

Electrical Diagnostics

The voltage and the current were measured with current transformers (Pearson Electronics Model 2877, featuring a 2 ns risetime). The voltage probe consisted of a Pearson coil measuring the current through a liquid resistor connected directly across the electrodes to minimize errors due to inductance. These data were used to make calculations of the electrical power profile and the energy deposited for each experiment.[4]

PVDF Transducer Packaging

The gauge package was based on the standardized PVDF transducer element marketed by K-tech Corporation using the patented Bauer poling process. The uniaxial stress is obtained from a stress versus charge density calibration.[5] Note that strain effects are also important, as described later. The charge was measured using a 1 nF integrating capacitor. A 50 Ω resistor was placed in series between the integrating capacitor and the cable for electrical impedance matching. An active area, nominally 9 mm², was selected because it provided a reasonable signal to noise ratio with the 1 nF integrating capacitor. Smaller gauges, requiring a smaller capacitance, suffered from premature signal attenuation due to a significant RC droop.

Gauges were shielded on both sides by an unbroken sheet of 25.4 μm thick copper foil folded over the transducer at the end closest to the active area. The negative lead was directly bonded to the shield while the positive lead was insulated from the shield by a 25.4 μm thick sheet of Mylar. Earlier gauges, that were shielded only over the negative lead, developed a noticeable base line shift. The double shielded gauges provided better electrical noise rejection, eliminating the base line shift, and hence allowing a closer placement of the gauge to the discharge gap.

Each gauge was electrically insulated from the high voltage area by placing a 25.4 μm thick disk of Kapton, 38 mm in diameter, between the gauge and the test sample. For experiments where the voltage exceeded 30 kV, a 127 μm Kapton disk was used. Attempts at gluing sheets of insulation to the gauge proved ineffective. The insulation had to be closely cropped to the gauge to allow access through the containment cylinder. The glue bond often failed, allowing arcing to the gauge through the test sample. The large insulating disk offered better dielectric coverage but had to be mounted separately through the top of the containment cylinder.

RESULTS

Three different studies were performed where the electrical discharge time, energy deposition rate, i.e., the power, and energy deposition were consecutively held constant. The signal shown in Figure 2 was from a transducer located 3.18 mm from the discharge gap where 514 mJ was deposited. The signal peak was 1.03 V (5.2 MPa). Signal magnitudes are given in volts with the equivalent stress (ignoring strain) in parenthesis. The risetime of this signal was 430 ns as defined between 10% and 90% of full magnitude and its duration was 300 ns as defined between the 90% points on either side of the peak.

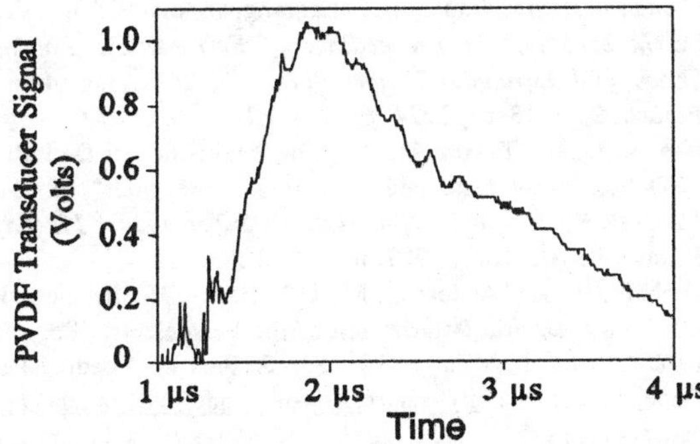

FIGURE 2. Transducer response at 3.18 mm

Discharge Energy and Power Effects

Two series of experiments were performed to investigate the effect of holding the discharge time constant (100 and 300 ns) while varying the power and energy associated with the discharge. This was accomplished by using the same cable for a given series and increasing the initial charging voltage for successive experiments. These early studies departed from the usual arrangement in that a 6.35 mm test sample was used in conjunction with a 6.35 mm PMMA disk, behind the gauge, for confinement. The data indicated that differences between discharge parameters, i.e., power and energy, could be differentiated by the PVDF gauge response.

Discharge Time and Energy Effects

Three experiments were performed where the near rectangular power profile was maintained around 1.6 MW for successive increases in deposition energy. This was accomplished by charging different lengths of cable to 20 kV. The different discharge times provided the change in energy. The gauge was placed at a distance of 3.18 mm from the discharge gap. The data show increasing transducer response with increasing energy deposition. The peak signals from the transducer 0.33 V (1.8 MPa), 1.28 V (4.9 MPa), and 1.65 V (6.3 MPa) correspond to energy depositions of 141 mJ/100 ns, 519 mJ/300 ns, and 693 mJ/400 ns respectively.

Discharge Time and Power Effects

Experiments were performed using nominal discharge times of 100, 200, 300, and 400 ns, where the energy deposition was maintained between 470 and 516 mJ. It was difficult to achieve identical energy depositions between experiments because subtle variations in the discharge can affect these values. These experiments were conducted by charging different lengths of cable to selected voltages to yield similar deposition energies. The data indicate similar transducer responses at 3.18 mm from the discharge gap for different discharge times and powers when the electrical energy deposition is the same. The initial parameters, characterized by the charging voltage and main discharge time, were 37 kV/100 ns,

25 kV/200 ns, 20 kV/300 ns, and 17 kV/400 ns which corresponded to electrical powers of 5.19, 2.57 and 2.35, 1.53, and 1.0 MW respectively. The peak signals from the transducer were 1.05 V (5.34 MPa), 1.15 V (5.6 MPa) and 1.11 V (5.6 MPa), 1.07 V (5.4 MPa), and 1.03 V (5.2 MPa) respectively. Risetimes and signal durations were nearly identical.

DISCUSSION

Wave Curvature Effects

The curvature of the wave coming from the arc channel affects the gauge response. These additions to the response are due to a poorly defined affected area over the transducer element, non-uniaxial strain and gauge/matrix coupling.[6]

The area of the sensor affected by any disturbance must be known in order to obtain an accurate measurement of the stress. This becomes problematical for a divergent wave since the entire active area will not be uniformly affected. This mostly affects the perceived risetime of the pressure front as the diverging wave initially impinges on the gauge. The risetimes resulting from this effect are expected to be 146 ns and 76 ns at 3.18 mm and 6.35 mm from the discharge gap respectively. These risetimes were calculated given the 9 mm^2 active area and assuming a cylindrical shock originating at the center of the arc channel with a wave speed of 2.3 mm/μs. This wave speed corresponds to the transit time between gauges placed at the two distances. These calculated risetimes are much smaller than those observed, 430 ns and 550 ns at the respective distances. Hence the area of the gauge does not explain the long rise times observed.

It is clear that the gauge response cannot be interpreted using an uniaxial compression analysis. The observed signal is likely affected by non-uniaxial strain and coupling effects between the gauge and the material it is included in. Future work will deal with these concerns.

Electrical Discharge Effects

These preliminary results suggest that the mechanical disturbance established by the arc channel is determined by the magnitude of the energy deposition, not the deposition rate, for short duration electrical pulses. However, all the effects of short duration discharge times (100 to 400 ns) on the mechanical response of the material can not be completely resolved with the observed signal risetimes greater than 400 ns. Better resolution of the effects of short duration discharges will require closer placement of the gauge and proper strain compensation.

CONCLUSIONS

A thin film PVDF gauge, operating in the charge mode, has been successfully shielded and insulated to produce useful data in a high voltage, electrical discharge environment. Preliminary results indicate a dependence of mechanical reponse on the electrical energy deposition, not the deposition rate, for short duration discharges.

ACKNOWLEDGEMENTS

Thanks are due to R. Main, P. Kelly and J. Volk for machining the parts required for this study. Thanks are also given to P. Miller for providing a suitable building to conduct these experiments, and D. G. Tasker for his constructive comments.

REFERENCES

1. Braginskii, S. I., "Theory of the Development of a Spark Channel," JETP, **34**(7), No.6, p1068 (1958).
2. Drabkina, S. I., J. Exptl., Theoret. Phys. (U.S.S.R.) **21**, p473, (1951).
3. Moulard, H. and Bauer, F., "Lagrangian Analysis of PVDF Shock Sensors," *in Proceedings of Ferroelectric Polymer Shock and Dynamic Sensor Workshop*, ISL, Saint-Louis, France, Sept. 13-15, 1994, pp. 4.3-4.18.
4. Lee, R. J., and Tasker, D. G., "The Acquisition of Definitive ESD Sensitivity Data and a New Test Method," in *1987 JANNAF Propulsion Systems Hazards Subcommittee Meeting*, Huntsville AL, Mar. 1987, pp. 319-325.
5. Graham, R. A., Anderson, M. U., Bauer, F., Setchell, R. E., " Piezoelectric Polarization of the Ferroelectric Polymer PVDF From 10 MPa to 10 GPa: Studies of Loading-Path Dependence," <u>Shock Compression of Condensed Matter-1991</u>, proceedings of *The APS Topical Conference*, Williamsburg VA, June 17-20, 1991, pp. 883-886.
6. Gupta, Y. M., "Stress Measurements Using Piezoresitance Gauges: Modeling the Gauge as an Elastic-plastic inclusion," J. Appl. Phys., **54**(11), Nov., pp. 6256-6266.

DEVELOPMENT OF PRESSURE GAUGE USING PVDF COPOLYMER FOR UNDERWATER SHOCK MEASUREMENT

Kenji MURATA*, Katsuhiko TAKAHASHI*, Yukio KATO* and Koichi MURAI**

**Chemicals and Explosives Divisions, Aichi Works, Taketoyo-Plant, NOF Corporation*
6 1 -1 Kitakomatsudani Taketoyo-cho, Chita-gun, Aichi 470-23, JAPAN
***Department of Materials Engineering and Applied Chemistry*
Mining Collage, Akita University
1-1 Gakuen-cho, Tegata, Akita-city, Akita 010, JAPAN

In order to realize pressure gauge which can sustain underwater shock pressure higher than 100MPa, we developed new type of pressure gauge using two materials as sensing element, PVDF and PVDF copolymer. Within 1.8 to 6.7 of the scaled distance, measurements of underwater shock profile were performed using new type of pressure gauge and compared with the results obtained by pressure gauge using tourmaline. Measured peak pressure using new type of pressure gauge presents good agreement with that using tourmaline.

As the results of development of the sensor system with buffer amplifier, underwater shock wave profile was successfully measured using new gauge with PVDF copolymer. Experimental results using PVDF copolymer presents good agreement with that using tourmaline.

On the other hand, measured duration time of underwater shock wave using PVDF is shorter than that using tourmaline because discharge time of PVDF is very short compared with time scale of underwater shock wave.

INTRODUCTION

Pressure gauge using tourmaline crystal is very widely used to measure underwater shock wave pressure [1] [2]. However, maximum measurable pressure is limited to be 20-30MPa because tourmaline crystal is destroyed by the high pressure generated due to impedance mismatch with water.

On the other hand, pressure gauge made by PVDF is used to measure very high shock pressure (>GPa) in solid [3][4].

Since shock impedance of fluoropolymer such as PVDF is very close to that of water, pressure gauge using PVDF potentially has the possibility to measure high underwater shock wave pressure (>100MPa). P.K.Gustavson et.al. [5] measured pressure profile (>4GPa) at the water/PMMA interface and Ohwa et.al. [6] measured only peak pressure of underwater shock wave.

However, there is few pressure gauge using PVDF to measure underwater shock wave profile accurately. In order to realize pressure gauge which can sustain underwater shock pressure higher than 100MPa, we developed new type of pressure gauge using two materials as sensing element, PVDF and PVDF copolymer.

DESIGN OF PRESSURE GAUGE

The piezoelectric material generates charge $Q(C)$ under the condition of stress σ (N/m^2).

$$Q = d \cdot \sigma \cdot S \tag{1}$$

Here, $d(C/N)$ and $S(m^2)$ is piezoelectric modulus

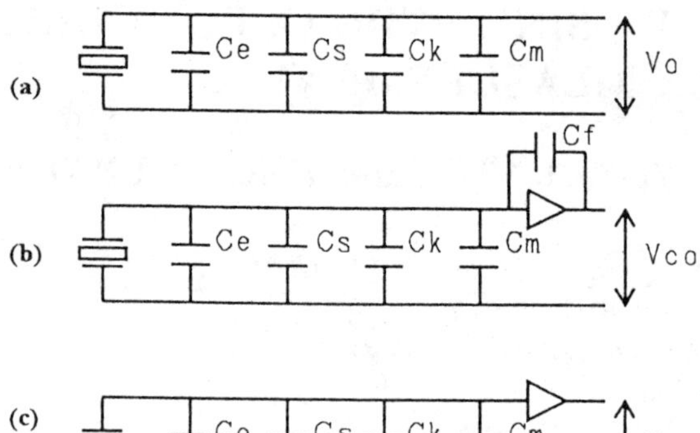

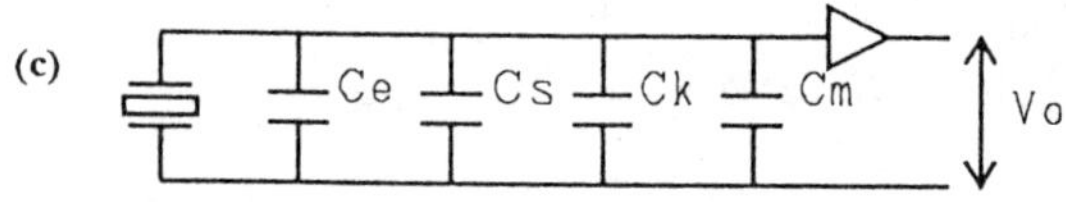

FIGURE 1. Equivalent circuit of sensor system
  (a)Without amplifier   (b)With charge amplifier
  (c)With buffer amplifier

and area of sensing element respectively.

In the case of sensor system without charge amplifier, whose equivalent circuit is shown in Figure 1(a), signal level Vo(V) at input terminal is expressed by eq.(2).

$$Vo=Q/(Ce+Cs+Ck+Cm) \qquad (2)$$

Here, Ce(F), Cs(F), Ck(F) and Cm(F) shows capacitance due to sensor element, gauge structure, cable and input terminal. In underwater explosion, cable length is usually larger than ten meter. Since Ck easily becomes a few thousands pF and very large compare to the sum of Ce,Cs and Cm,Vo mainly depends on cable length. In order to neglect the effect of cable length on Ck, charge amplifier is usually adopted for gauge using tourmaline.

In the case of sensor system with charge amplifier, whose equivalent circuit is shown in Fig.1(b),signal level Vco(V) is expressed by eq.(3).

$$Vco=Q/Cf \qquad (3)$$

Here,Cf(C) is capacitance of feedback condenser of charge amplifier. When the same sensor system with charge amplifier such as gauge using tourmaline was adopted for gauge using PVDF and PVDF copolymer, severe noise problem occurred. Because Ce of PVDF and PVDF copolymer is

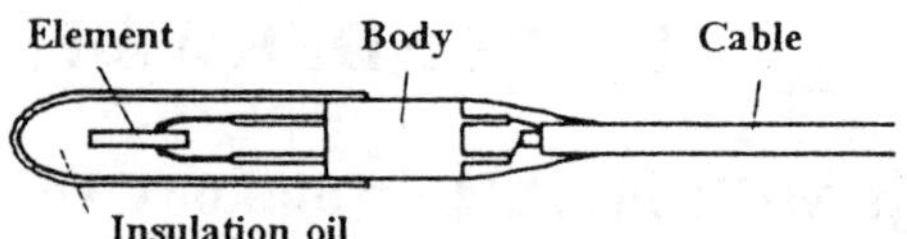

FIGURE 2. Schematic of gauge using PVDF and PVDF copolymer

TABLE 1. Physical property of PVDF, PVDF copolymer and tourmaline

Piezoelectric materials	PVDF	PVDF copolymer	Tourmaline
Piezoelectric modules d33(pC/N)	4to15	8to10	2
Dielectric constant ratio $\varepsilon/\varepsilon_o$	4to13	10	1
Density (g/cm3)	1.8	1.9	6.0
Sound velocity (km/sec)	1.5to2.3	2.4	3.1
Acoustic impedance (Mkg/m2·sec)	3.4	4.5	18.6

d33; Piezoelectric modulus of d33 axis
ε_o; Dielectric constant under vacuum

about one hundred times larger than that of tourmaline.

$$Vn=k(Ce+Cs+Ck+Cm)/Cf \qquad (4)$$

Here,Vn(V) and k(V) is noise level at input terminal and constant respectively. For the sensor system using PVDF and PVDF copolymer, buffer amplifier was adopted as shown in Fig.1(C). Finally PVDF and PVDF copolymer with properties as shown in Table 1 was used as sensing element and element size of 5mm in square and 50μ m in thickness was adopted.

Our new pressure gauge using PVDF and PVDF copolymer is shown in Figure 2. Pressure gauge is filled in insulation oil, whose shock impedance is very close to that of water.

EXPERIMENTS

Calibration of pressure gauge was performed using the oil pressed calibration apparatus(PCB Model 913A02) as shown in Figure 3. In order to generate semi-dynamic pressure in the range of 2 to 40MPa, impacter was dropped on the piston from the height of 10 to 50cm. For calibration of new gauges using

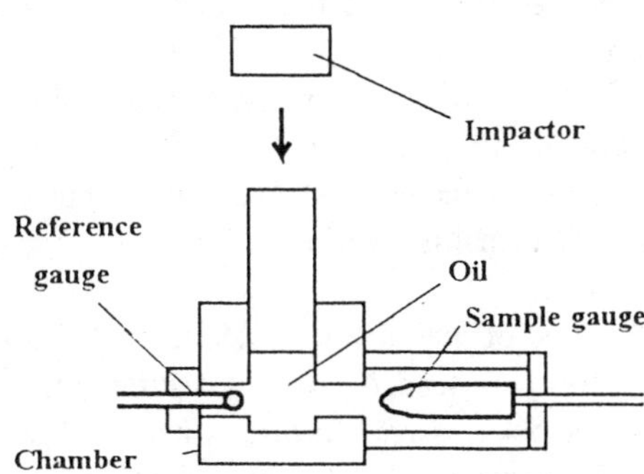

FIGURE 3. Schematic of calibration apparatus

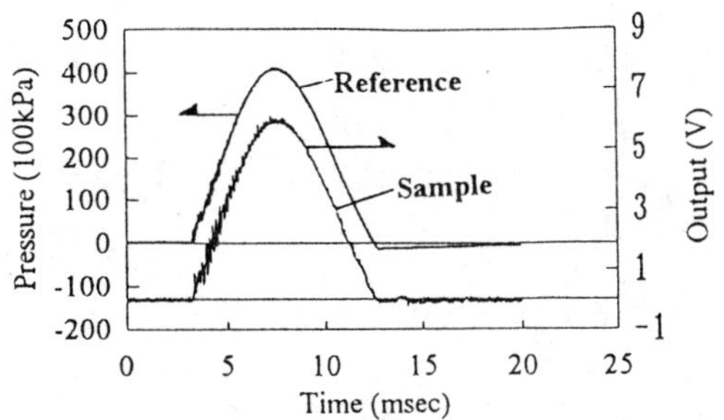

FIGURE 4 Example of measured pressure profile

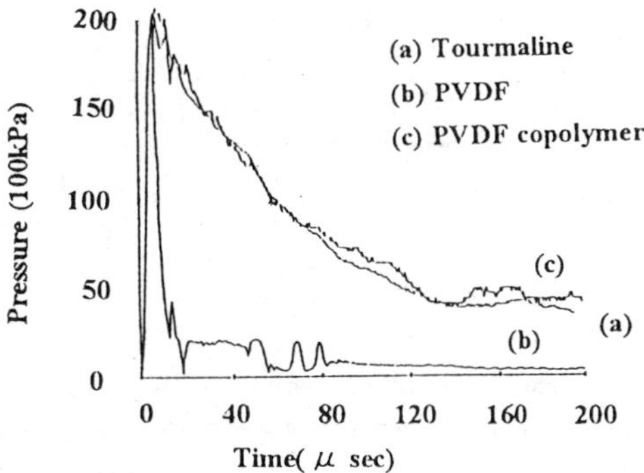

FIGURE 5 Example of measured shock wave profile

PVDF and PVDF copolymer, sample gauge using PVDF or PVDF copolymer and reference gauge using tourmaline (PCB Model 136A) was set to the chamber as shown in Fig.3. And then, measurement of pressure profile was performed using sample gauge and compared with the results obtained by reference gauge. Sensitivity coefficient of new pressure gauges were calculated by peak pressure measured by the reference gauge and maximum output voltage obtained by new pressure gauges.

In order to evaluate the performance of new pressure gauge, underwater explosion test was performed. Our testing tank is 36m in diameter and 8m in depth, and emulsion explosive (density;1.1g/cm3, detonation velocity;3.2km/s) was used for generation of underwater shock wave.

Using sensor system with charge amplifier for gauge using tourmaline and buffer amplifier for new gauges using PVDF and PVDF copolymer, measurements of underwater shock wave profile were performed. The results obtained by new pressure gauge using PVDF and PVDF copolymer is compared with the results obtained by pressure gauge using tourmaline. And also, shock wave energy (Es) and bubble energy (Eb) was calculated according to eq.(1) and (2) in reference[7] [8].

RESULTS AND DISCUSSION

As the results of calibration of new gauges, the sensitivity coefficient of PVDF and PVDF copolymer was 70.0MPa/V and 11.2MPa/V respectively. In the calibration process, measured ressure profile using PVDF copolymer presents

good agreement with that using tourmaline as shown in Figure 4. In Figure 5, it is shown that the example of measured shock wave profile by new gauges and tourmaline gauge. From experimental results, it is found that measured peak pressure using new type of pressure gauges presents good agreement with that using tourmaline. However, difficulty is to measure shock wave profile accurately, because discharge time of fluoropolymer such as PVDF is potentially very short compared with time scale of underwater shock wave. As described the above, as the results of development of the sensor system with buffer amplifier,we succeeded in measurement of underwater shock wave profile using PVDF copolymer. Experimental results using PVDF copolymer presents good agreement with that using tourmaline. On the other hands, measured duration time of underwater shock wave using PVDF is shorter than that using tourmaline. It is considered that this difference between PVDF and PVDF

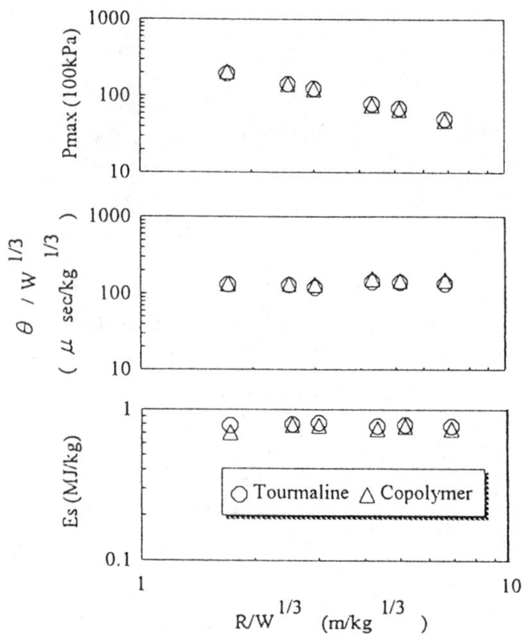

FIGURE 6. Variation of shock wave energy, scaled characteristic time, and maximum pressure with scaled distance

copolymer to underwater shock wave, is based on chemical structure, degree of crystallization, surface structure of PVDF and PVDF copolymer on discharge property. These properties will be studied in our futures' work.

From these results, PVDF copolymer was selected for underwater shock wave measurement.

And also, in order to confirm performance of gauge using PVDF copolymer, variation of Es, scaled characteristic time and peak pressure with scaled distance is summarized in Figure 6. From experimental results,it is found that results of Es, scaled characteristic time and peak pressure obtained by gauge using PVDF copolymer presents good agreement with those obtained by gauge using tourmaline within 1.7 to 6.8 of scaled distance. Therefore, it is suggested that our new gauge using PVDF copolymer is very attractive for measurement of underwater shock wave profile over 100MPa, which is our goal.

CONCLUSIONS

For underwater shock wave measurements, new type of pressure gauges using PVDF and PVDF copolymer was developed.

Measurements of underwater shock wave profile were performed using new type of pressure gauges and compared with the results obtained by pressure gauge using tourmaline within 1.7 to 6.8 scaled distance.

As the results of the sensor system with buffer amplifier, we succeeded in measurement of underwater shock wave profile using PVDF copolymer. Underwater shock wave profile obtained using PVDF copolymer was good agreement with that using tourmaline .

On the other hands, measured peak pressure using PVDF presents good agreement with that using tourmaline. However, measured duration time of underwater shock wave using PVDF is shorter than that using tourmaline,because discharge time of PVDF is potentially very short compared with time scale of underwater shock wave profile.

REFERENCES

1. J. Roth," Underwater explosives", in Encyclopedia of Explosives and Related Items, 10, pp.U38-U81, Dover , New Jersey (1983)

2. R. H. Cole," Underwater Explosion", Dover Publications, New York (1948)

3. F. Bauer , A. Lichtenberger, "Use of PVF2 hock gauges for stress measurements in hopkinson bar", in Shock compression of condensed matter, (1987)   pp. 631-634

4. F. Bauer," Behavior of ferroelectric ceramics and PVDF2 polymers under shock loading", ibdi, (1991) pp.251-262

5. P. K. Gustavson et. al, "Underwater shock wave measurements using PVDF Transducers", ibdi, (1991) pp.905-908

6. T. Ohwa et.al, "Devvelopment pyrotechnical device for extracorproreal microexplosive lithotripsy", Kogyo kayaku, 50, pp16-22 (1989)

7. K.Murata, K.Takahashi and Y.kato,"Effect of metal confinement on underwater explosion of explosives", Proceeding of 18th International symposium on shock wave, p947 (199 1)

8 K. Takahashi, K. Murata and Y. Kato, "Underwater shock enhancement by metal confinement", Proceeding of 13th International symposium on Ballistics, p531 (1992)

LASER DRIVEN SHOCK PRESSURE MEASUREMENTS BY VF_2/VF_3 AND PVDF GAGES FOR PULSES OF 2.5ns UP TO 10^{12} W/cm^2

S. Couturier, M. Boustie, T. de Résséguier, M. Hallouin, J.P. Romain

Laboratoire de Combustion et de Détonique (U.P.R. au C.N.R.S. n° 9028)
E.N.S.M.A. - B.P. 109 - 86960 Futuroscope Cedex (France)

Thick 450μm piezoelectric VF_2/VF_3 copolymer and thin 25μm PVDF gages have been used to measure the induced pressure history at the back face of aluminum and copper targets irradiated by infra-red laser pulses[†] of 2.5ns with intensities up to 7.10^{11} W/cm^2. The measured pressures in the gages infer pressures up to 200 kbar on the front face of the target. The whole shock pressure temporal pressure profile applied on the front face of the targets is determined roughly by using the laser matter interaction hydro-code FILM. The modifications to bring to this profile in order to fit the experimental record via the simulation of the propagation of the applied profile into the set-up are very limited. The comparison of the applied peak pressure given by simulation of laser matter interaction and the experiment deduced one is also rather good. These results give the peak pressure versus the incident intensity under these laser irradiation conditions and assess the possibility to use this kind of gages as a measurement device for high amplitude with rapid evolution shocks.

INTRODUCTION

High power pulsed lasers have been widely used as shock generators in recent years. The field of laser-driven shock measurements has also expanded, particularly for short time pulses which allow us to reach very high pressure (the megabar regime). Yet, before the laser becomes a reliable laboratory tool, a better knowledge of the temporal shape of the shock induced by a laser irradiation has to be got.

Previous works have demonstrated that Poly(vinilydene-fluoride) films and its copolymer (VF_2/VF_3) present strong piezoelectric effects when stretched bi-axially poled according to the Bauer process (1,2,3,4). A piezoelectric film converts mechanical energy into electrical energy with a response time of a few nanoseconds and therefore, it can be used to measure very short pulses. This study presents an experimental method using this diagnostic to measure the shock profile induced by laser irradiation of 2.5ns duration. The measurements made at the back of the targets are related by numerical simulation of waves propagation to the front pressure loading. The resulting shock profile is compared with laser-matter interaction computations and scaling laws.

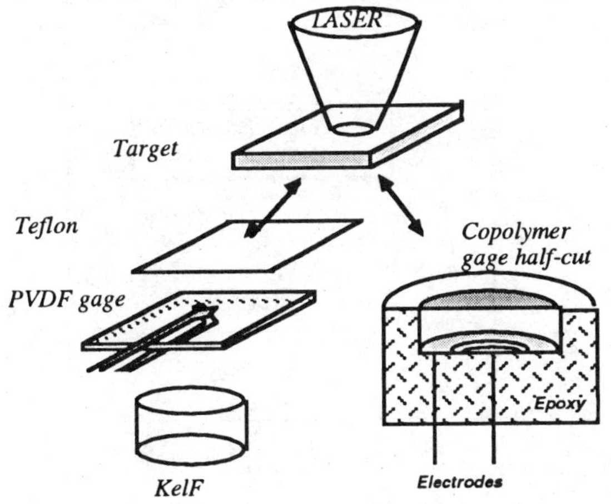

FIGURE 1. Sketch of the experimental set-up

---

† Experiments carried out at LULI (Ecole Polytechnique)

EXPERIMENTAL SET-UP

The experimental set-up is shown on Fig. 1. The gages have different shapes and we are presently making detailed studies of the 25μm PVDF and the 450μm thick copolymer sensor. The thick gage is stuck onto the rear face of an aluminium or copper target with a resin which has the same acoustic impedance as the gages one. For the thin gage, there is a 25μm teflon insulator between the target and the PVDF and an anvil of KelF at its back.

Laser energy at λ=1.06μm and τ=2.5ns (FWMH) is deposited onto the front face of the target with different intensities. The focal spot is larger than the size of the active area (typically 3mm^2) and than the thickness of target with the gage in order to remain in 1D propagation conditions. A shock propagates from the irradiated surface into the target. When it reaches the interface, the gage delivers an electrical signal recorded on a 2GHz sampling oscilloscope. This signal is correlated to the stress history of the interface (target/gage).

PVDF STATE OF THE ART

Up to now, the usual interpretation given to PVDF signal was made according to a linear relationship between the difference of stress applied on the electrodes and the current density. This piezoelectricity basic law requires to fulfill several mechanical assumptions, among which some are not really valid in this case, such as an elastic constitutive law for PVDF or the steadiness of waves propagation throughout the gage (5).

Recently, Moulard and Bauer have experimentally demonstrated that the charge delivered by the gage is a linear function of the true strain (6) :

$$\frac{Q(t)}{A} = K \ln \frac{e_0}{e(t)} \quad \text{and} \quad i(t) = \frac{dQ(t)}{dt} \quad (1)$$

Q : delivered charge , A : active area
e_0 : initial gage thickness, i : delivered current
e : gage thickness
K = 12 μC/cm^2 for 25μm gage
K = 13 μC/cm^2 for 450μm gage

NUMERICAL SIMULATION OF THE EXPERIMENTAL SIGNALS

The experimental signals recovered with a thin 25μm PVDF gage or a thick 450μm copolymer one are related to the initial loading applied at the front face of the target by the mechanisms of shock waves propagation through the structure. In order to correlate this measurement to this investigated loading, we have used a 1D shock waves propagation code SHYLAC (7). As soon as an applied pressure profile is given as input, we can compute the shock and release waves propagation and calculate the gage thickness e(t) at each time step, yielding to the current intensity

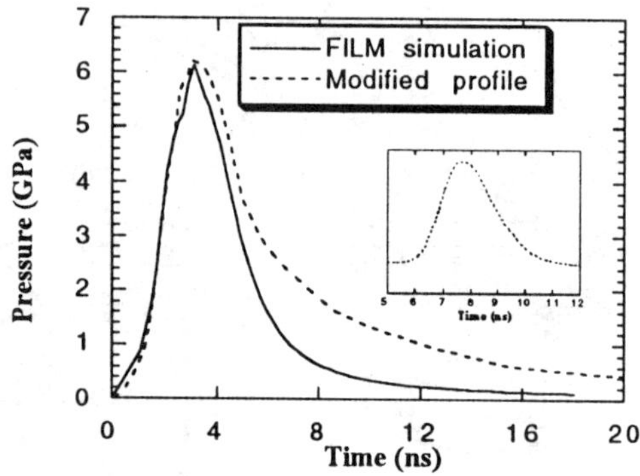

FIGURE 2-a. Comparison of the applied pressure profile simulated by laser-matter interaction code FILM and the modified profile leading to the simulation of Fig. 2-b for an irradiation of Φ=166 GW/cm^2, τ=2.5ns, λ=1.06μm. Inlaid is the laser energy pulse shape.

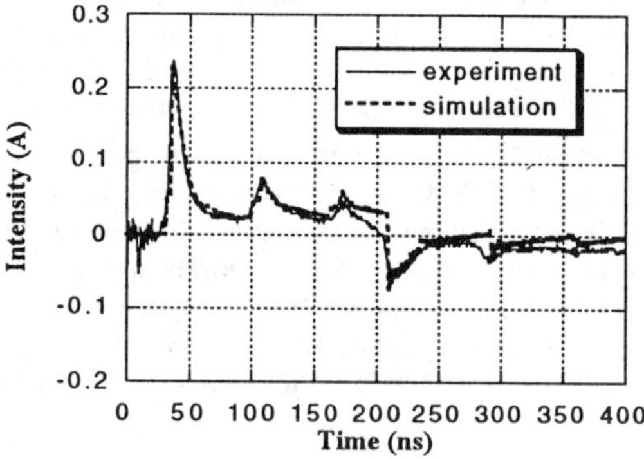

FIGURE 2-b. Experimental recorded signal by a 450μm copolymer gage at the back of a 250μm aluminum target compared with the simulated signal with SHYLAC, taking as input pressure loading the modified profile of Fig. 2-a. Laser conditions : Φ=166 GW/cm^2, τ=2.5ns, λ=1.06μm.

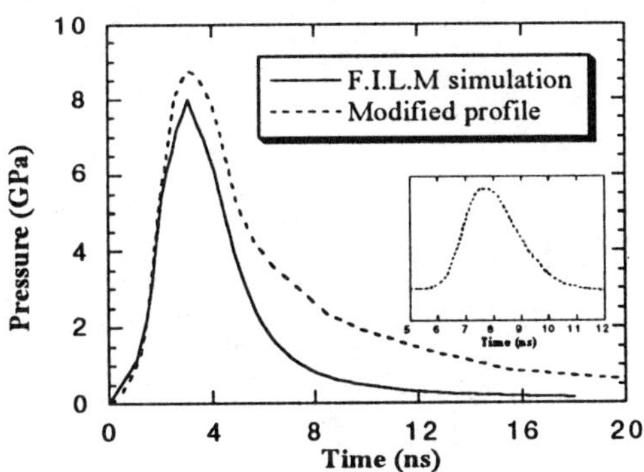

FIGURE 3-a. Comparison of the applied pressure profile simulated by laser-matter interaction code FILM and the modified profile leading to the simulation of Fig. 3-b for an irradiation of $\Phi=216GW/cm^2$, $\tau=2.5ns$, $\lambda=1.06\mu m$.

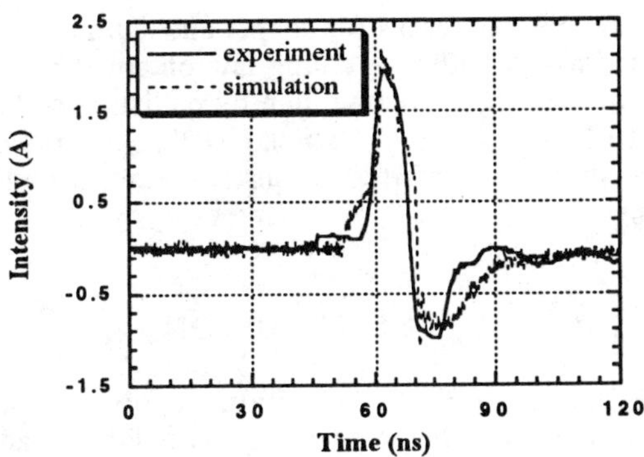

FIGURE 3-b. Experimental recorded signal by a 25μm PVDF gage at the back of a 250μm copper target compared with the simulated signal with SHYLAC, taking as input pressure loading the modified profile of Fig. 3-a. Laser conditions : $\Phi=216GW/cm^2$, $\tau=2.5ns$, $\lambda=1.06\mu m$.

according to equation 1. This computed function i(t) is compared with the experimental record and the applied pressure profile is corrected until simulation fits well with the experiment. The basis for the pressure profile induced by a given laser irradiation is obtained from the laser matter interaction code FILM (8) (see Fig. 2-a). This code assumes a gaussian temporal light deposition, what is not experimentally fully exact, since the release time is a bit longer than the rise time of energy deposition (see Fig. 2-a). Taking into account this consideration, the release time and duration of the profile given by FILM has been increased up to the modified profile of Fig. 2-a, leading to the very good fit of the experimental signal by the simulation (see Fig. 2-b). The same processes have been followed for thin PVDF recordings. An example is shown in Fig. 3, where the Fig. 3-a shows the front pressure loading modified from the one given by FILM and leading to the comparison of simulation and experiment on Fig. 3-b.

The very good agreement shown on one example of each gage let us believe that the temporal applied loading is the right one. If the amplitude of this latter was higher, we would have to narrow its duration so that the attenuation be stronger in order to fit the experimental signal. Thus, even if the amplitude of the signal could be right, the temporal shape would not follow. On the other hand, according to the same principle, by having a lower input maximum pressure, we should enlarge the duration to have the right pressure amplitude in the gage ; but this would lead again to a discrepancy concerning the chronology of the signals.

SCALING LAW BETWEEN PEAK INDUCED PRESSURE AND LASER INTENSITY AT 2.5ns DURATION

By determining the applied peak pressure profile for a given irradiation leading to the best fit with the experimental signal as described previously, we can fit the simulated peak pressure versus the laser intensity, either for the thick copolymer or thin PVDF. These measurements are compared with the results of laser matter interaction with FILM code and they lie along the same curve (see Fig. 4). For such short pulse duration, FILM's simulations lead

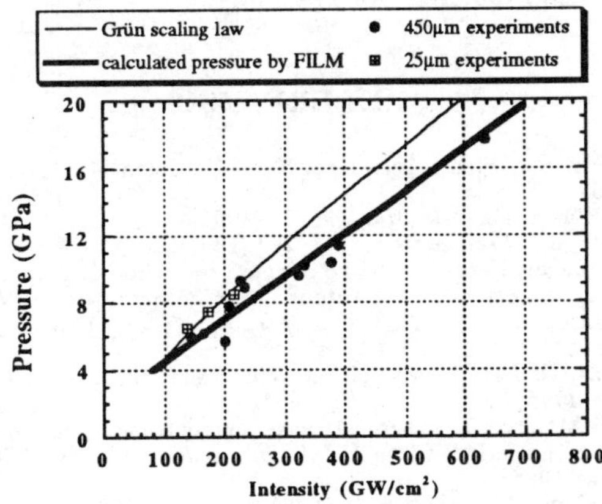

FIGURE 4. Peak pressure induced for a given laser intensity with $\tau=2.5ns$, $\lambda=1.06\mu m$. Comparison of experiments deduced results, laser-matter interaction simulation with FILM and Grun's scaling law.

to comparable results for copper and aluminum. The comparison with the scaling law obtained by Grün by measuring the momentum by ballistic pendulum with laser shocks of 4ns shows that the pressure measurements inferred in our study are a realistic range.

CONCLUSION

Thin 25μm PVDF and 450μm copolymer gauges have been used here under very short shock loading. The signals reveal a very good time resolution of the physical events occuring during waves propagation. By using a linear relationship between the true strain and the delivered charge, numerical simulations of waves propagation provides us with the shock profile leading to the best fit with the experiments. The profiles found match quite well with the results given by a laser-matter interaction code. A reference curve, relating the peak induced pressure to the laser intensity at $\lambda=1.06\mu m$ and $\tau=2.5ns$ is thereby determined on the range 100-700 GW/cm^2.

These kinds of measurements will be kept on with shorter pulse durations and higher amplitudes. Efforts will be made in parallel to get a better characterisation of the PVDF copolymer.

ACKNOWLEDGEMENTS

We would like to thank F. Bauer and his team from Institut Saint-Louis for providing us with necessary materials for this study.

REFERENCES

1. F. Bauer, French patent 822102S, US patent 4611260 and 4684337.
2. F. Bauer, *Ferroelectrics*, **49**, 231 (1983).
3. F. Bauer, *Ferroelectrics*, **115**, 247 (1991).
4. F. Bauer, Henry Moulard, "Réponse sous choc de polymeres ferroelectriques: capteur piezoélectrique PVDF nanoseconde", in *H.D.P. proceedings* (Tours, France, june 1995).
5. R.A. Graham, F. W. Neilson, W.B. Benedick, *J. Appl. Phys.* **36**, 1775, 1965
6. H. Moulard, F. Bauer, "Analyse lagrangienne de la réponse de la jauge PVDF", in *H.D.P. proceedings* (Tours, France, june 1995).
7. F. Cottet, M. Boustie, J. Appl. Phys. **66**, 4067, 1989
8. S. Atzeni, *Plasma Physics and Controlled Fusion*, **29**, 11,1987

THE NEW SIMULTANEOUS PVDF/VISAR MEASUREMENT TECHNIQUE: APPLICATIONS TO HIGHLY POROUS HMX

M. U. Anderson and R. A. Graham

Sandia National Laboratories, Albuquerque, NM 87185-1421

Time-resolved pressure and particle velocity measurements provide the foundation for modern descriptions of rate-dependent materials deformation processes in condensed matter subjected to high pressure shock loading. Much of that foundation has been based on the VISAR, nanosecond time-resolved particle velocity technique. It has been known for some time that full description of the rate-dependent processes requires both particle velocity and stress data, but stress sensors with nanosecond resolution have not been previously available for general usage. More recently, the piezoelectric polymer PVDF has been used for stress and stress-rate measurements in rate dependent materials response experiments. Neither method, by itself, gives unique material response models, particularly in the case of highly porous materials. The first simultaneous, time-resolved measurements with VISAR and PVDF have now been carried out. Measurements were taken on highly porous HMX that clearly show the multi-dimensional nature of HMX viscous compression behavior, followed by the onset of chemical reaction. This technique promises to provide a qualitative improvement in our ability to develop rate-dependent material descriptions of all solids including highly porous solids. Precise use of this technique will require new window materials.

INTRODUCTION

A new technique has been developed to study materials under shock compression by simultaneous measurement of material response in both the stress and volume planes. These simultaneous measurements allow unique observations of time-dependent material response under shock compression. Time-dependent material response occurs during shock compression of materials in the inert and reactive states, and is particularly apparent in the case of highly porous materials.

Highly porous materials are expected to have time-dependent behavior since a typical sample initially consists of particles in contact with other particles at a finite number of locations, while the rest of the surface area is surrounded by voids. A shock wave propagating into this highly porous matrix causes localized stresses at particle contact points, resulting in large shear deformation, and local regions of zero stress from the unsupported particle faces. As the material begins to consolidate toward solid density, each particle is subject to deformation in the form of fracture, shear, and plastic flow. The portions of each particle that are unsupported complicate this deformation process due to the zero stress regions. Thus, many localized deformation fronts are in constant communication with localized voids, and the material deformation that results has material, morphology and density-dependent time scales over which mechanical equilibrium is achieved. Prior work with PVDF measurements has provided clear evidence for the rate-dependent behaviors of a number of powder samples. (1)

EXPERIMENTAL TECHNIQUE

The simultaneous PVDF/VISAR measurement technique developed in the present study was carried out on the powder samples subjected to controlled shock-loading from a compressed-gas gun. The 4mm thick sample is encased in a polymer capsule with PVDF gauge packages at the impact surface, and at sample input and propagated locations as in Figure 1. The propagated-location PVDF gauge package is mounted directly on the VISAR reflector which is sputtered to a PMMA window.

The piezoelectric polymer PVDF is a 25 μm thick film that generates a stress-rate dependent current. (2) The current is recorded electronically, then numerically integrated to give the piezoelectric charge-versus-time profile. (3) The shock response of PVDF is used to convert the charge-versus-time to stress-versus-time. The particle velocity measurements of the present study use the VISAR interferometer to measure mirror motion-versus-time as evidenced by the Doppler shift of the laser light returned from the mirror through reference and delayed optical paths. Two separate interferometers having complementary delay constants were used in each experiment for enhanced accuracy.

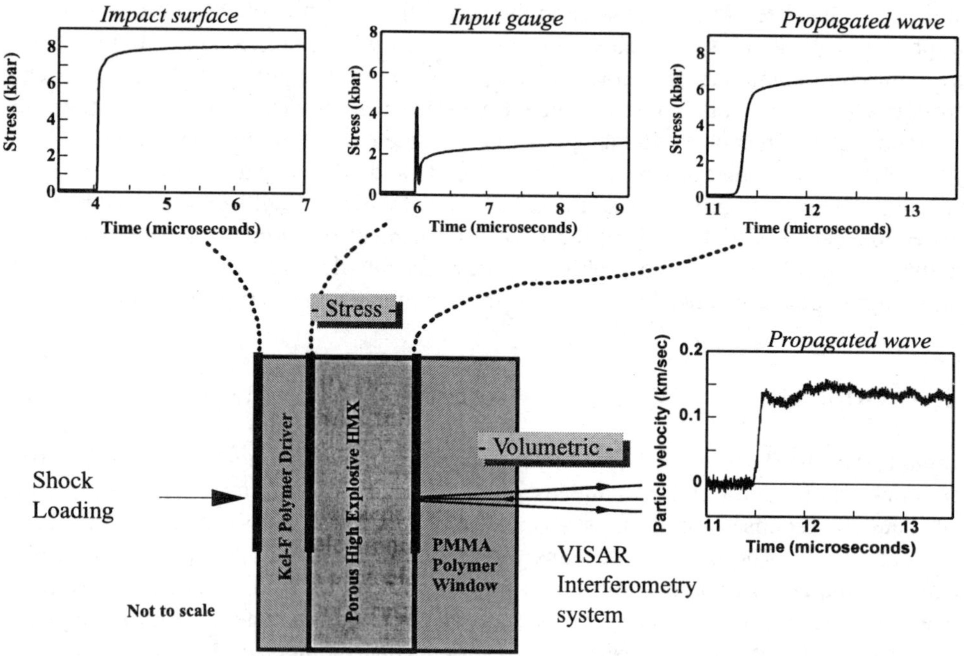

FIGURE 1. Simultaneous measurements in stress and volume planes are made with PVDF/VISAR technique. The experiment includes impact surface stress measurements with PVDF as well as PVDF measurements of time-resolved pressure of both the input and propagated stress pulses in the porous HMX high explosive sample. The propagated-location PVDF gauge package is mounted directly on the VISAR reflector which is sputtered to a PMMA window. Resulting stress-versus-particle velocity data gives unique material description of complex two-dimensional dynamic deformation of the highly porous materials

RESULTS

Figure 1 shows the three PVDF measurements and the VISAR measurements on the porous HMX sample. The 64% dense HMX sample is as described by Gustavsen. (4) The impact surface PVDF gauge provides a detailed record of the impact stress while the input gauge to the powder sample provides the initial stress signature input to the sample. This record clearly shows the importance of the viscoelastic behavior of the Kel F driver. The early time "spike" is a result of the gauge signal prior to the arrival of the wave reflected from the powder interface. The propagated wave PVDF record shows the arrival of a dispersed wave with a continuously rising stress history. The VISAR record for the propagated wave also shows a dispersed wave with relaxation.

The individual measurements of either stress or particle velocity-versus-time show time-dependent material response with significant rate dependence, but the records show that neither is particularly distinguished. The individual records show no explicit detailed features which can be connected to the material deformation. Nevertheless, when the measurements are synchronized in time, combined and plotted in the stress-versus-particle velocity plane, as shown in Figure 2, distinctive features of the material response can be identified.

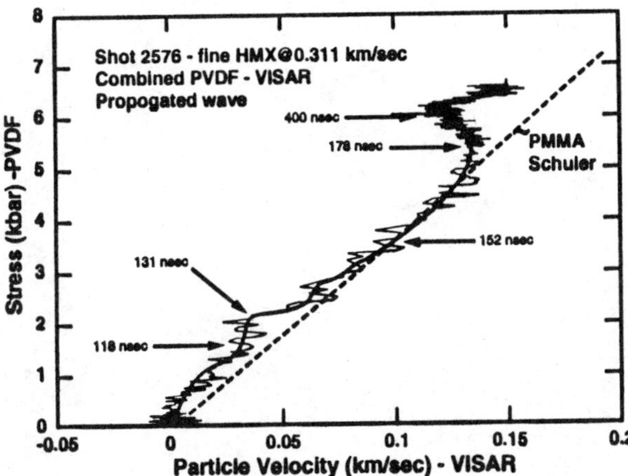

Figure 2. The time-dependent stress-versus-particle velocity relationship determined from simultaneous PVDF and VISAR measurements show detail on the deformation of the porous HMX sample as it deforms from the distended to the fully dense state. The data are shown by the discontinuous line which is fit by the solid line. The various times noted indicate the relationship at that particular time. The records indicate an approach to solid density at about 140 nsec followed by a quiescent period and an apparent chemical reaction.

The stress-versus-particle velocity record shows a behavior for the first approximately 140 nanoseconds of response which lies above the relationship of the PMMA. Since the powder and PMMA are in contact, they must have equal states. States off the PMMA one -dimensional behavior are clear indications of the two-dimensional deformation of the powder sample. Such behavior is expected for heterogeneous, multidimensional material deformation.

Prior simultaneous PVDF/VISAR measurements have demonstrated that time-dependent material behavior can be identified as soon as mechanical equilibrium is acheived at the PVDF gauge/VISAR mirror interface, typically 10 - 20 nanoseconds. (2)

After this early period the response moves to the PMMA response relationship for one-dimensional conditions. This observation indicates that the powder is now fully compacted to densities approaching solid density. After a delay time to approximately 160 nanoseconds, the upward excursion of the relationship is consistent with the onset of limited chemical reaction. Thus, the measurements appear to provide a detailed picture of the deformation process from the initial state to solid density.

Mechanical impedance mismatches between the PVDF gauge package and the PMMA cause a small effect of wave reverberation for the first 20 nanoseconds. This difficulty can be overcome with the use of amorphous Kel F windows. Experiments to describe the index of refraction for the windows are in progress.

CONCLUSIONS

The present work has shown the first simultaneous measurements of particle velocity and stress with time resolutions of a few nanoseconds.

The technique should prove generally useful for observations on a wide variety of solids in the solid state and in porous states. Most existing descriptions of rate-dependent materials responses to shock loading are typically based on VISAR measurements alone. Although this approach has proven successful in advancing the materials descriptions, the materials models are ambiguous due to reliance on particle velocity data alone. The particle-velocity-based models are essentially constructed on the volumetric characteristics of the measurements. It has been widely recognized that accurate description of the rate-dependent behaviors require both stress and particle velocity.

The ability to describe both particle velocity and stress with nanosecond time resolution will provide a much more rigorous description for models of rate-dependent materials responses to high pressure shock compression. This capability should remove ambiguity from existing materials models.

ACKNOWLEDGMENTS

The authors are pleased to recognize the collaboration with Steve Sheffield and Rick Gustavsen at Los Alamos National Laboratory DX-10, and thank them for their sample preparation and discussions concerning their observations of HMX high explosive behavior. Mel Baer's insight and numerical simulation capabilities have been vital to our progress in developing this new technique. Work supported by the USDOE under contract DE-AC04-94AL85000.

REFERENCES

1 M. U. Anderson, R. A. Graham, and G. T. Holman, "Time-Resolved Shock Compression of Porous Rutile: Wave Dispersion in Porous Solids", *High-Pressure Science and Technology - 1993*, eds. S. C. Schmidt, J. W. Shaner, G. A. Samara and M. Ross, AIP Press (1994) pp 1111-1114.

2 R. A. Graham, M. U. Anderson, F. Bauer, and R. E. Setchell, "Piezoelectric Polarization of the Ferroelectric Polymer PVDF from 10 MPa to 10 GPa: Studies of Loading-Path Dependence", *Shock Compression of Condensed Matter - 1991*, eds. S. C. Schmidt, R. D. Dick, J. W. Forbes, D. G. Tasker, North-Holland (1992) pp 883-886.

3 D. E. Wackerbarth, M. U. Anderson, R. A. Graham, SAND92-0046, February, 1992, Sandia National Laboratories.

4 S. A. Sheffield, R. L. Gustavsen, R. R. Alcon, R. A. Graham, and M. U. Anderson, "Particle Velocity and Stress Measurements in Low Density HMX", *High-Pressure Science and Technology - 1993*, eds. S. C. Schmidt, J. W. Shaner, G. A. Samara, M. Ross, AIP Press (1994) pp 1377-1380.

IMPROVED SHOCK-DETECTING PIN ARRANGEMENT

David J. Erskine

Lawrence Livermore National Laboratory, Livermore, CA 94551

Shockwave speeds are often measured by comparing arrival times at the tips of electrical shorting pins in a hexagonal array over two elevations (called up and down). In the conventional arrangement, the center pin is solely responsible for measuring the curvature of the wavefront. Without this datum the shock speed cannot be precisely determined. In some experiments this pin fail frequently enough to be a problem. We report a simple rearrangement between up and down designated pins which eliminates the critical reliance on a single pin.

INTRODUCTION

An important equation of state measurement is the velocity of a shockwave generated in a sample by the impact from a high velocity (several km/s) disk projectile. Such experiments[1-5] are vitally important in physics for establishing high pressure material behavior to high precision, including materials used by researchers as standards. In many cases, shock wave experiments are the only method available to attain the necessary pressure and temperature, or to perform an accurate measurement over a sufficiently large sample volume. The shock velocity experiment is the preferred method to determine the equation of state of standards because the developed shock parameters depend through simple relations[3] on only four values which can potentially be measured very precisely: projectile velocity and initial density, sample initial density, and the measured shock velocity (U_S).

Experimental techniques have been developed[1,2] to accurately measure these parameters using a two stage gas gun[6]. These account for the bowing distortion the projectile suffers from the acceleration of launch. In experiments involving thin samples and high projectile velocities the shock transit times are short; thus, it is increasingly important to precisely account for the projectile distortion. We report[7] a simple modification in the target design which will significantly reduce the uncertainty in the shock speed determination.

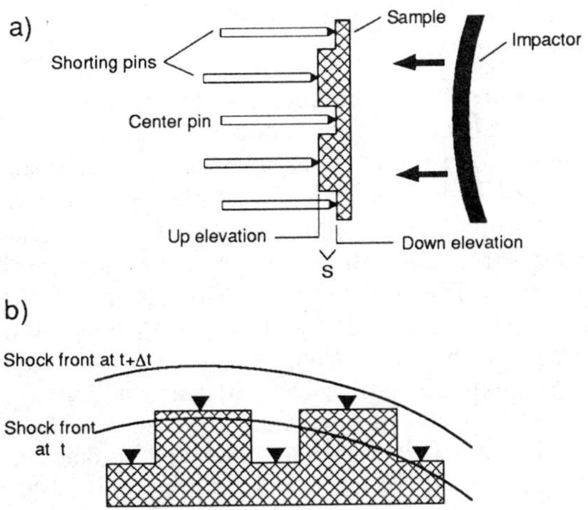

Figure 1. Target for measuring shock speed, side view. a) Sample is a "tophat" having elevations at two levels "up" and "down" differing by height S. Other supporting structures not shown. Impactor is a metal disk, which generally has a bow and tilt due to acceleration of launch. Electrical shorting pins pressed against elevation surfaces detect arrival of shock wave. b) Detail of shock front in relation to pin tips (triangles). Vertical dimension is exaggerated. The shock front preserves its shape as it moves perpendicular to target face. Shock front shown at two moments separated by time interval Δt.

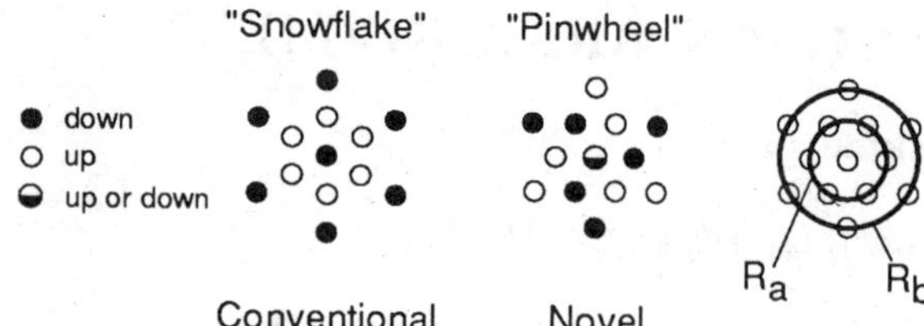

Figure 2. Arrangement of "up" and "down" designated electrical shorting pins on the sample, projected on the target face. The inner six pins are on a diameter of radius R_a, outer six on R_b. Both the snowflake and pinwheel arrangements have 13 pins, split 7/6 between down and up elevations. However, the pinwheel has pins of both kinds on each radius. The center pin in the pinwheel can be up or down. The snowflake center pin should be the same kind as the outer pins to best resolve the bowing. The up/down roles as groups can be interchanged.

METHOD

Figure 1 is a schematic of the target used to measure shock speeds. A metal disk projectile 1 - 3 mm thick is accelerated up to 8 km/s by a two stage gas gun[1,6]. Impact with the sample creates a shock wave propagating toward the sample rear. Passage of the shock through specific locations around the sample is observed by a set of electrical shorting pins. They are placed in two planes (elevations) parallel to the face, called "up" and "down", which are separated by a distance S (the step height). For samples 15 - 20 in mm diameter, S is 1 - 3 mm.

The pins generate an electrical pulse upon passage of the shock which is sent via cable to recording electronics. The shock arrival times at the target can be determined to ~0.5 ns overall precision[1], including cable length uncertainties. (At 10 km/s, this time precision corresponds to measuring pin elevations to 5 μm precision.) In the measurement of shock speed in a stiff material such as sapphire[7], the shock transit time can be as low as 100 ns. Thus this subnanosecond time precision is important.

There is an uncertainty in the shock arrival time pulse generated by a pin, due to variations in the thickness or shape of the tip, or the presence of foreign matter under the tip, all which change the effective elevation. The pins must be pressed securely against the sample by springs to ensure their location at the measured elevation surface. In some pin types, vibration from handling the target can rub off the metal film at the tip, causing the pin to fail. For these reasons, targets have as many pins as possible for redundancy. Due to the width

of pins and springs, it is difficult to space pins closer together than ~ 3 mm center to center. Avoidance of the release waves from sample corners reduces the available area on the up elevation. Previous[5,6] researchers using larger targets have used pins side by side in a line. For our smaller samples this is not practical. Instead, a hexagonal pin arrangement provides the most efficient packing.

Figure 2 diagrams the arrangement of pins. Each pin is either on the up or down elevation. There are two configurations, labeled "snowflake" and "pinwheel". Both use 13 pins divided into 6 down, 6 up and one center pin. The snowflake pattern is the conventional configuration used in some previous equation of state studies[1-3].

Figure 3 shows how the transit time (Δt) is found by interpenetrating the up and down pin data sets along the time axis. This is similar to the 1-d interpenetration algorithm of Holmes[8], but done in an additional dimension. Fig. 3b shows that without the center pin, Δt cannot be determined because the amount of radial bowing is indeterminate. Neglecting the effect of bowing[1,2,7] can alter Δt by up to ±15 ns for impactor velocities up to 8 km/s.

Furthermore, even with valid center pin data, the net transit time uncertainty cannot be less than ~75% of the center pin uncertainty. Thus, there is essentially no benefit from the 12 other pins, contrary to intuition. This because the net transit time uncertainty[9] is given by

$\sim\sqrt{\delta t^2/N + 0.75^2\, \delta t_c^2}$, where δt is a typical individual pin uncertainty, and δt_c is the center pin uncertainty (which is similar). The number of up and down pins is assumed to be the same and given by N.

The alternative configuration ("pinwheel") is introduced as a solution to this problem. In this arrangement the center pin is not needed (Fig. 3c) to accurately determine the shock transit time. Fundamentally, this is because there are both up and down pins on the same radius R_a or R_b.

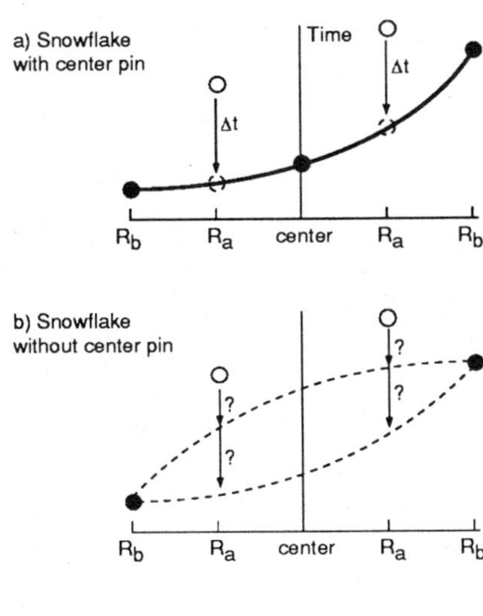

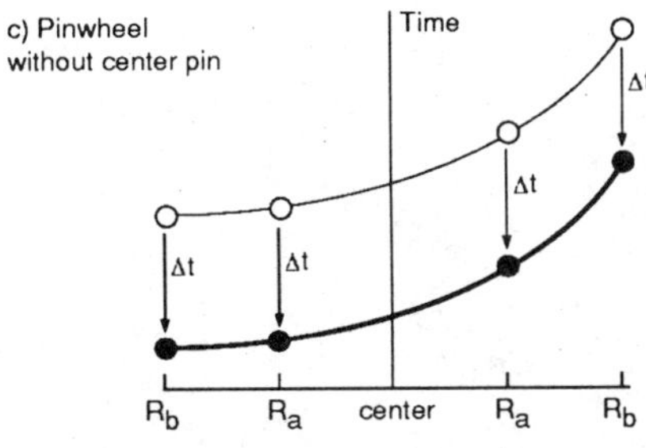

Figure 3. Finding the transit time through interpenetration of data. Arrival times of shocks at pins are plotted versus 2-dimensional pin location across the face. Only one dimension is shown on horizontal axis. Open and closed circles represent up and down pin data respectively. Here, the center pin is a member of the down set. The up data are translated in time by Δt to form a mutual best fit surface. This yields the shock speed $U_s = S/\Delta t$, where S is step height. a) Pins in snowflake arrangement. b) If center pin missing in snowflake arrangement, Δt cannot be determined because curvature of shock surface is indeterminate. Fundamentally, this is because R_a and R_b contain only up and down pins respectively. c) In pinwheel arrangement, lack of center pin does not prevent accurate determination of Δt because the circles R_a and R_b each contain both up and down pins.

For the pinwheel, we expect the net transit time uncertainty to be approximately $\sim\sqrt{2}\,\delta t/\sqrt{N}$. This estimate is appropriate for a combination of $N + N$ pins fitting two best fit surfaces with standard deviation δt.

CONCLUSION

We believe the pinwheel pattern is the most robust to pin loss, has the least uncertainty in transit time, and most space efficient arrangement of 13 pins for shock wave transit time experiments. This is important in experiments where the degree of radial bowing is significant to the shock transit time. Such is the case in stiff materials having high shock and sound speeds, since the transit times are short, the sample dimensions need to be small, and the projectile distortion significant and not reproducible from shot to shot.

ACKNOWLEDGMENT

This research was performed under the auspices of the U.S. Department of Energy by the Lawrence Livermore National Laboratory under contract W-7405-Eng-48.

REFERENCES

1. A.C. Mitchell, and W.J. Nellis, Rev. Sci. Instrum. **52**, 347 (1981).
2. A.C. Mitchell, and W.J. Nellis, J. Appl. Phys. **52**, 3363 (1981).
3. N.C. Holmes, J.A. Moriarty, G.R. Gathers, and W.J. Nellis, J. Appl. Phys. 66, 2962 (1989).
4. M. Van Thiel, L.B. Hord, W.H. Gust, A.C. Mitchell, M. D'Addario, K. Boutwell, E. Wilbarger and B. Barrett, Phys. Earth and Plan. Interiors **9**, 57 (1974).
5. A.H. Jones, W.M. Isbell, and C.J. Maiden, J. Appl. Phys. **37**, 3493 (1966).
6. W.D. Grozier and W. Hume, J. Appl. Phys. **28**, 892 (1957).
7. D.J. Erskine, to be published by Rev. Sci. Instr. 1995.
8. N. C. Holmes, Rev. Sci. Instrum. **62**, 1990 (1991).
9. The net uncertainty is analyzed in Ref. 2. However, their formulas Eqns. 26 and 28 are incorrect because they treat two correlated quantities (δt_1, δt_c, Eq. 6) as uncorrelated. The correct expression for the net uncertainty at the single standard deviation level is

$$\delta(\Delta t) = \sqrt{\frac{\sigma_d^2}{N_d}\left(\frac{R_a^2}{R_b^2}\right)^2 + \frac{\sigma_u^2}{N_u} + \delta t_c^2\left(1 - \frac{R_a^2}{R_b^2}\right)^2}$$

where σ_d and σ_u are the standard deviations of the best fit planes to the up and down pins, and N_u and N_d are the number of valid up and down pins, and δt_c is the uncertainty of the center pin.

SHOCK CHARACTERIZATION OF TOAD PINS

Lawrence J. Weirick

Explosive Projects and Diagnostics Department, 2554, Sandia National Laboratories 87185

Michael J. Navarro

Ktech Corp., 901 Pennsylvania N.E., Albuquerque, NM 87110

The purpose of this program was to characterize Time Of Arrival Detectors (TOAD) pins response to shock loading with respect to risetime, amplitude, repeatability and consistency. TOAD pins were subjected to impacts of 35 to 420 kilobars amplitude and approximately 1 ms pulse width to investigate the timing spread of four pins and the voltage output profile of the individual pins. Sets of pins were also aged at 45°, 60° and 80° C for approximately nine weeks before shock testing at 315 kilobars impact stress. Four sets of pins were heated to 50.2° C (125° F) for approximately two hours and then impacted at either 50 or 315 kilobars. Also, four sets of pins were aged at 60° C for nine weeks and then heated to 50.2° C before shock testing at 50 and 315 kilobars impact stress, respectively. Particle velocity measurements at the contact point between the stainless steel targets and TOAD pins were made using a Velocity Interferometer System for Any Reflector (VISAR) to monitor both the amplitude and profile of the shock waves.

INTRODUCTION

The TOAD pin (Time Of Arrival Detector) is currently under consideration for use as a diagnostic for accepting electroexplosive devices. Although the TOAD pin has been in use for many years, it has not been extensively studied to determine its behavior under abnormal conditions or over a range of stress inputs. This report describes a gas gun test program designed to fully characterize a specific TOAD pin (Dynasen Model CA-1135). The gas gun is employed because it provides reliable and well-measured stress levels over a wide range of amplitudes.

EXPERIMENTAL TECHNIQUE

The intent of this program is to perform tests in which four pins will be loaded simultaneously by the identical stress pulse, in conjunction with a VISAR (Velocity Interferometer for Any Reflector) measurement to provide accurate waveshape information. The results gained during the experiments allow achieving the five objectives of the TOAD characterization program: 1. Determine the response of the pin (risetime and pulse shape) over the range of stress 35-420 kbar. 2. Evaluate pin-to-pin consistency. 3. Assess the behavior of the pin when initially heated to 50.2° C. 4. Artificially age the pin and study possible degradation in performance. 5. Artificially age the pin, heat to 50.2° C and assess the pin behavior.

The experimental configuration used for this program is shown in Figure 1. The impactor material (typically stainless steel) is mounted onto the sabot and is chosen to give the desired stress level in conjunction with the stainless steel target material and the desired impact velocity. The stress wave from the stainless steel target is transmitted into the TOAD pins. The pins are held in a Vespel™ fixture. The diameter of the pin placement was approximately 0.30" which minimized the uncertainty in the stress wave arrival time. Resistance measurements were made both after assembly of the target and as mounted before a test to assure intimate contact of the pins and stainless steel targets. In this manner, the incident shock wave should contact all four pins at nearly the same instant (within 6 ns across the 0.3" diameter of the four pins). A polymethyl methacrylate (PMMA) window allows the laser beam in the VISAR system to be placed slightly offcenter, providing room for the Vespel fixture.

TABLE Shot Matrix and Results

Shot Number	Configuration	Impact Stress kbar	Particle Velocity mm/µs	Comments	Spread at 10V ns
559	SS/SS	35	0.169	Baseline	26
565	SS/SS	50	0.246	Baseline	12
562	SS/SS	160	0.703	Baseline	11
561		167	0.729		10
563	SS/SS	315	1.24	Baseline	6
564					12
576	WC/SS	420	1.56	Baseline	7
567	SS/SS	315	1.24	Aged at 45° C	10
566	SS/SS	315	1.24	Aged at 60° C	7
570					12
568	SS/SS	315	1.24	Aged at 80° C	5 (33)
571					7
572	SS/SS	50	0.246	Heated to	44
573				50.2° C	51
574	SS/SS	315	1.24	Heated to	12
575				50.2° C	16
592	SS/SS	50	0.246	Aged and	5 (25)
593				Heated	15
590	SS/SS	315	1.24	Aged and	35
591				Heated	45

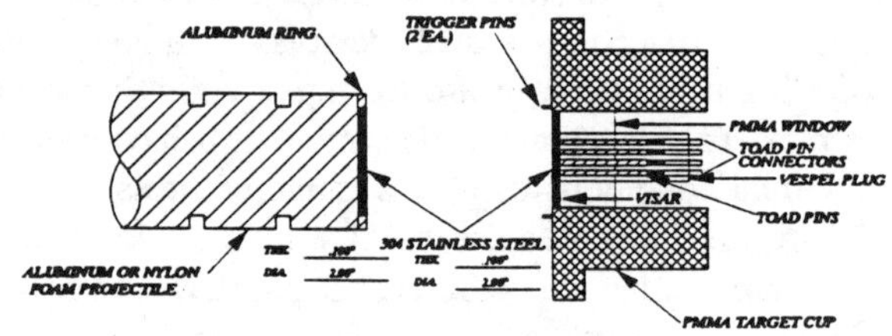

FIGURE 1. Schematic of the Projectile/ Target Configuration

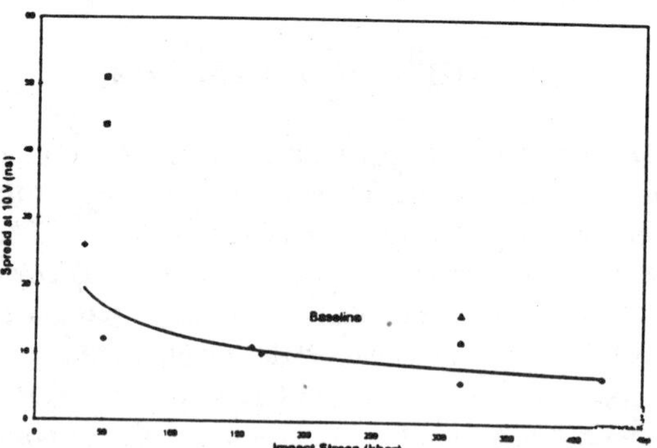

FIGURE 2. Pin-Response Spread at 10V versus Impact Stress

RESULTS AND ANALYSIS

Table 1 lists the conditions for the test matrix along with the results of the TOAD pin response for this program. The test conditions listed include the shot number, the test configuration which includes the impactor and target materials, the impact stress, the measured particle velocity in the PMMA window and a comment on the category of the test. For these "baseline" tests, the impactor material chosen was primarily 304 stainless steel with the exception of the highest impact stress shot where tungsten carbide was used. The impact stress values ranged from a low of 35 kbar to a high of 420 kbar.

Figure 2 plots the spread in the time response of the TOAD pins at increasing impact stress. The spread decreased with increased impact stress for all conditions. The spread in the response was smallest for the baseline condition and increased with aging, heating, and aging and heating.

Figure 3 plots the voltage-response time to reach 10 volts at increasing impact stress. It was observed that the time became shorter as the impact stress increased. This result can also be described in the terms of risetime. The higher the applied impact stress, the faster was the resulting risetime.

The trend in voltage amplitude response was that the higher the impact stress the lower the maximum amplitude and the shorter the response width. Figure 4 plots the maximum amplitude of the voltage response of the TOAD pins at increasing impact stress. The digitizers were set for a maximum voltage output of 70 V in order to maximize the observation of response spread and risetime. Thus, all voltage responses above this value were "clipped". The voltage response at impact stresses below 315 kbar were clipped. However, values were obtained at 315 and 420 kbar impacts.

DISCUSSION

Baseline Tests

The results from these tests found that the voltage response had a faster risetime with increased impact stress level. These results reflect the loading curve profiles measured by VISAR. This loading curve profile is analogous to that seen by the TOAD pins in the intended application, in that both have stainless steel parts butted against the TOAD pins. The Hugoniot Elastic Limit (HEL) of 304 stainless steel has been reported [5] to be 2.3 kbar (0.066 mm/µs particle velocity in PMMA window). This value corresponds in amplitude with the inflection points in the loading curves shown in Figure 2. Therefore, it is probably the elastic/plastic nature of 304 stainless steel which is producing the loading ramp behavior.

Inspite of the loading ramp behavior, impacts of TOAD pins at 50 to 420 kbar impact levels produced voltage signal spreads at 10 V outputs of 12 ns or less. These results are very acceptable for the intended use of the TOAD pins. Therefore, the TOAD pins as currently designed and built satisfy the needed requirements.

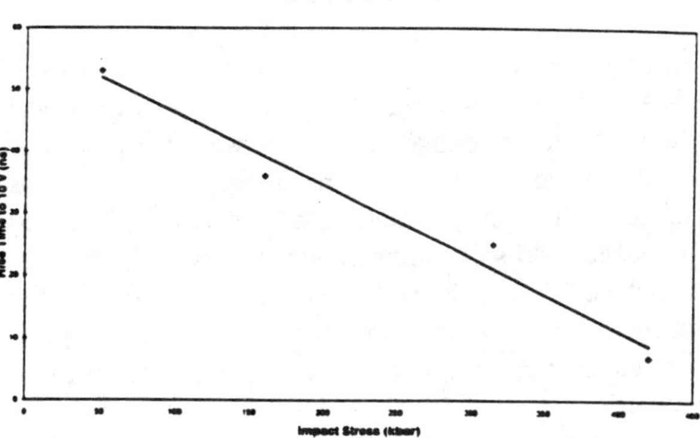

FIGURE 3. Voltage Risetime at 10V versus Impact Stress

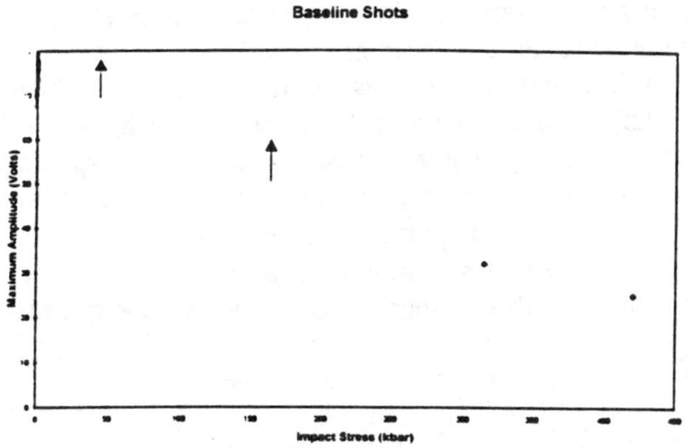

FIGURE 4. Maximum Voltage Amplitude versus Impact Stress

Aged Tests

The actual temperature used for the aging process had no experimentally measurable affect on the voltage response of the pins. Also, the process of aging the pins did not affect the spread of the voltage response between the four pins in a set. The process of aging the pins, however, did effect the voltage response by producing pins which responded with faster risetimes. The aging process had affected some physical property of the TOAD pin assembly. What exactly has changed is not known. However, it is unlikely that the piezoelectric crystal itself has changed at such low relative temperatures. The most likely change is within the conductive-epoxy potting compound. It could have further cured or perhaps outgassed. Again, the resultant spread of voltage response was not changed but the voltage-response profile was improved.

Heated Tests

Heating the pins at 50.2° C for two hours before impacting at a stress level of 315 kbar slightly increased the spread of the voltage response between the four pins, at least for one set with the other giving the usual 12 ns spread. The pins, when used in their intended application, do see a temperature history of 50.2 C for anywhere from 2 hours to four days. Therefore, the results from the present program may explain the good success of the TOAD-pin results in the application scenario.

Heating the pins at 50.2° C for two hours before impacting at a stress level of 50 kbar significantly increased the spread of the voltage response between the four pins in both sets. The reason for this result is not fully understood. There are most likely at least two contributing factors to this result. First, the loading history at these lower stress level tests is extended which decreased the risetime and could extend the response spread. Second, raising the temperature of the target to 50.2° C could change the alighnment of the target (not measureable at this point in the procedure) and increase the tilt angle thereby increasing the pin-response spread.

Aged and Heated Tests

The voltage response of the aged and heated pins was essentially the same as that response from the aged or heated pins. The same general conclusions can be made. The voltage-response profiles are sharpened (faster riseteimes) compared to the results from baseline tests. The pin-response spread is not significantly differenct from that of the baseline for the 315 kbar impacts. The spread increased for the 50 kbar impacts. The same reasoning applied to the aged or heated tests should apply to these tests on aged and heated pins.

ACKNOWLEDGEMENTS

The authors wish to acknowledge the contribution of the following people to this program. Heidi Anderson, K-Tech Corp., who assembled the projectiles and targets; Frank Horine, K-Tech Corp., who assisted on the TOAD pin assembly techniques; Bill Tarbell, SNL 2654, who interfaced with the customer and initiated this program; Jay Huttenhow, SNL 12367, the cusomer who encouraged and funded this program.

REFERENCES

1. Sheffield, S. A. and Dugan, D. W. "Description of a New 63-mm Diameter Gas Gun Facility," *Shock Waves in Condensed Matter*, ed. Y. M. Gupta, Plenum Press (1986).

2. Barker, .L. M. and Hollenbach, R. E. "Laser Interferometer for Measuring High Velocities of Any Reflecting Surface," J. Appl. Phys. 43:11 (1972).

3. Hemsing, W. F. "Velocity Sensing Interferometer (VISAR) Modification," Rev. Sci. Instrum. 50:1 (1979).

4. Marsh, S. P. ed., *LASL Shock Hugoniot Data*, University of California Press, Berkeley, CA (1980).

5. Kinslow, R. ed., *High-Velocity Impact Phenomena*, Academic Press, New York (1970).

HYPERVELOCITY IMPACT
TESTS OF OPTICAL SENSORS

J. S. Browning and J. L. Montoya

Sandia National Laboratories, Albuquerque, New Mexico 87185

Hypervelocity tests of spacecraft optical sensors were conducted to determine if the optical signature from an impact inside the sunshade resembled signals observed on-orbit. Impact tests were conducted in darkness and with the ejected debris illuminated.

INTRODUCTION

This paper contains the findings of hyper-velocity impact tests on spacecraft optical sensors conducted at the Johnson Space Center Hypervelocity Impact Test Facility (JSC HIT-F) located near Houston, Texas, in December 1994. The work was performed to observe the spacecraft optical sensor's response to impacts inside the sensor's sunshade.[*]

The JSC HIT-F was chosen as the test site because the projectiles and velocities available at the facility are most representative of the hypervelocity particles thought to be responsible for an anomalous optical sensor response observed on-orbit. The projectiles are a few micrograms, slightly more massive than the microgram particles thought to be responsible for the signal source. The test velocities are typically 7.3 km/s, which are somewhat slower than typical space particles.

BACKGROUND

History

The advent of orbiting space platforms in the early 1960's opened a whole new field of observing, monitoring and studying the Earth. Sandia National Laboratories have developed optical sensors to detect nuclear bursts. The optical sensors are capable of observing light emitted from lightning, meteorites, and perhaps other unknown processes. The primary focus has been on deciding whether a recorded optical event is evidence of a nuclear burst. What other information may be present in the signals of unknown origin has essentially remained a matter of speculation.

Theory

In January of 1994 Craig Fox and Bob Wiley (Science Applications International Corporation) suggested that a group of the unknown signals might be caused by hypervelocity microparticle impacts on the inside of the optical sensor sunshade (1). The frequency of the occurrence of the signals seemed to agree with the statistics associated with microgram-size orbital debris. However, the signal duration of several milliseconds and other characteristics were inconsistent with the optical signatures reported in the literature in a loosely defined field called "Cosmic Dust." The laboratory experiments indicated that the plasma from hypervelocity impacts lasts only tens of microseconds. The theory of Fox and Wiley was discounted by some because it was thought that the amount of light from a hypervelocity impact would not be detected by the optical sensor because the light from the impact would be too dim to be detected, or that the baffles in the sunshade would block the light from reaching the optical detector.

[*] This work performed at Sandia National Laboratories was supported by the U. S. Department of Energy under contract DE-AC04-94AL85000.

Charlie Vittitoe (Sandia) analyzed the impact scenario proposed by Fox and Wiley, and suggested the idea that the signal from the optical detector may not be caused by light emission from the expanding plasma cloud, but instead may be the result of sunlit ejecta. Clues included a characteristic inverse square time decay at late times in some signals and the signal magnitudes. Based upon these considerations a testing sequence for the purpose of observing the spacecraft optical sensor response to hypervelocity impacts on the sunshade was developed. The following were established as the test conditions:

- line-of-sight view of the impact site by the spacecraft optical sensor
- impacts inside the sunshade in darkness
- impacts inside the sunshade with the ejected debris illuminated

TEST EXECUTION

Light Gas Gun Operation

The operation of the gun begins with the percussion firing of a few grams of powder in a shotgun shell. The burning powder propels a plastic piston that compresses hydrogen gas. The piston then reaches a stop where an aluminized mylar membrane is located. The membrane bursts, and the compressed hydrogen gas is released into the rifle barrel. The compressed gas propels a sabot down the barrel. The sabot is machined from a cylindrical piece of nylatron that is cut along its axis into four wedge-shape sections. An aluminum projectile, with a mass of a few micrograms, is contained in a small cavity drilled into the axis of the sabot. The rifle barrel imparts a spin to the sabot. When the sabot reaches the end of the barrel, the four sections separate and impact an aluminum plate called a "sabot catcher." The aluminum projectile passes unimpeded through a small hole in the sabot catcher and enters the test chamber. The muzzle velocity of the aluminum projectile is typically 7.3 km/s. The diameter of the projectile is 131 micrometers, about a fifth of the length of the diagonal of a grain of salt.

Preparation

In Figure 1 the setup for line-of-sight tests and the dark and illuminated debris tests is shown. All of the tests were conducted in a large, steel vacuum chamber 60 inches in length with a diameter of 44 inches. The chamber has a large rear door and two smaller doors on each side that permit access to the chamber. The chamber vacuum of about 400 microtorr was achieved by a large piston pump.

For the dark and debris illuminated tests the spacecraft optical sensor, with the sunshade attached, was located approximately in the center of the test chamber. The spacecraft optical sensor ("BDY Sensor") was tilted at an angle of 34 degrees from the floor of the chamber, viewed a black cloth background, and did not have a line-of-sight view into the projectile entrance port. Dark green blotter paper was taped on the chamber walls to minimize any reflections of light emitted by projectile impacts, sabot impacts or from the muzzle. The projectile target was the sunshade. The projectile impacted on a baffle inside the sunshade. In this setup the spacecraft optical sensor did not have a line-of-sight view of the impact site. A second optical detector ("BDY Diode") was located near the rim of the sunshade to provide a line-of-sight view of the impact site.

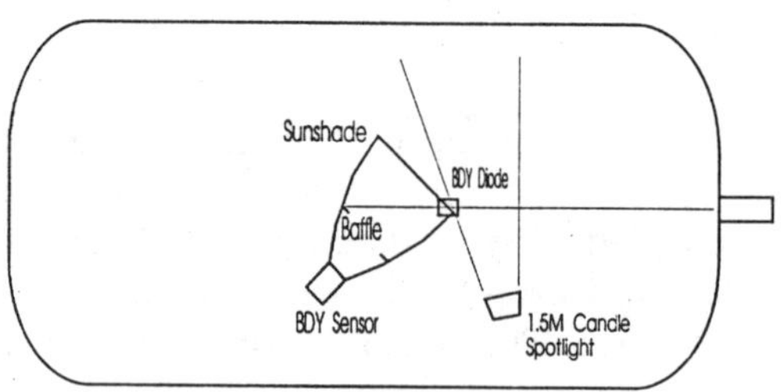

FIGURE 1. Setup for Dark and Illuminated Debris Tests. (The BDY Diode is located off axis near the rim of the sunshade and has a line-of-sight view of the impact site.)

A 1.5 million candle spotlight, with a halogen bulb, was located on the floor of the chamber to illuminate debris ejected out of the sunshade. The spotlight was powered by a rechargeable lead-acid battery. All unused ports to the chamber were sealed to block out light from the outside of the chamber. A set of three baffles with 1 inch diameter apertures was positioned between the sabot catcher and the projectile entrance port to minimize the light entering the chamber from the gas gun. For the dark tests, referred to in this report as "tests in darkness," the spotlight was turned off. For the illuminated debris tests the spotlight was turned on.

Instrumentation

The optical detector, used in all of the tests, was the silicon photodiode and the signal conditioning portion of a spacecraft optical sensor. The optical detector uses a silicon photodiode which detects light between the wavelengths of 0.4 and 1.1 microns with a peak sensitivity at 0.9 microns. The fast response of the photodiode is amplified by a chain of amplifiers. The desired gain may selected by tapping into the appropriate signal node along the chain. The slow response of the photodiode is canceled by a feedback circuit which uses a time average of the photodiode response.

Prior to using the optical detectors at the JSC HIT-F, the detectors were tested at Sandia with various light sources in order to understand the limitations and characteristics. The performance of the detectors was compared to other available optical detectors to examine bandwidth characteristics (i.e., risetime and falltime, linearity, sensitivity and saturation characteristics). From the recorded data, of significance was the fact that the satellite optical sensor had a risetime (and falltime) of 18 microseconds.

RESULTS

In Figure 2 the responses from three tests of the spacecraft optical sensor (line-of-sight, impacts in darkness, and illuminated debris) are compared with selected on-orbit signals. The on-orbit signals are shown in the top graphs and the test data from the JSC HIT-F are shown in the bottom graphs. The bottom graph is the optical response in volts measured at the output of the amplifier chain of the sensor electronics, plotted in "LD Units.*" The graphs are semilog up to 500 microseconds, and then convert to a log-log scale.

In the line-of-sight case the selected on-orbit signal compares very well with the test data. Ignoring the response due to muzzle and sabot impact light, the peak response is qualitatively very similar to the on-orbit peak response in the semi-log portion of the graphs. Unfortunately data out to several milliseconds was not acquired for comparison, but the slope of the decay in the log-log portion of the graph suggests that an optical sensor response would have been present up to 10 milliseconds after the impact. In the case of the impact in darkness the spacecraft optical sensor response is also qualitatively very similar to the selected on-orbit signal.

The spacecraft optical sensor response to illuminated debris is compared with a long duration on-orbit signal (1 second) to suggest that the duration of the signal is determined by how long illuminated debris stays within the field of view of the optical sensor. In the HIT-F tests the field of view is limited by the confines of the test chamber and by the spotlight beam. In the on-orbit case, the cone of the field of view is not truncated and the light scattered by the debris would be detected by the optical sensor until the debris sufficiently disperses.

CONCLUSIONS

Hypervelocity impact tests on optical sensors were performed at the Johnson Space Center Hyper-velocity Impact Test Facility. A two-stage gas gun was used to propel microgram-size aluminum spheres to velocities of 7.3 km/s. Spacecraft optical sensor responses were recorded from line-of-sight impacts on sunshade material, from impacts inside the sunshade in darkness, and from impacts inside the sunshade with the ejected debris illuminated.

---

*$LD = 1 + 25 \log_{10}(I/I_o)$, where I is the optical intensity and I_o is a reference intensity.

It was observed that all of the spacecraft optical
sensor responses resembled on-orbit signals. The
test results, however, do not rule out the possibility
that other processes may be responsible for the on-
orbit signals. Factors not taken into account by this
work are:

- The vacuum levels achieved during the tests are
  not equivalent to the hard vacuum of space.
  The vacuum has an effect on the formation of
  the impact plasma and the cooling and
  dispersion of the impact debris.
- The sunshade of the spacecraft optical sensor is
  likely to be electrically charged on-orbit. The
  redistribution of this charge during a
  hypervelocity impact on the sunshade may
  affect the optical sensor response.
- Variation in particle velocity, mass and impact
  site inside the sunshade will affect results.

ACKNOWLEDGMENTS

Sandia personnel who contributed to this work
were R. Spalding, N. Blocker, G. Scrivner, C.
Vittitoe, G. Christiansen, P. Gibson, and T. Welton.
The HIT-F is managed by Jeanne Crews.

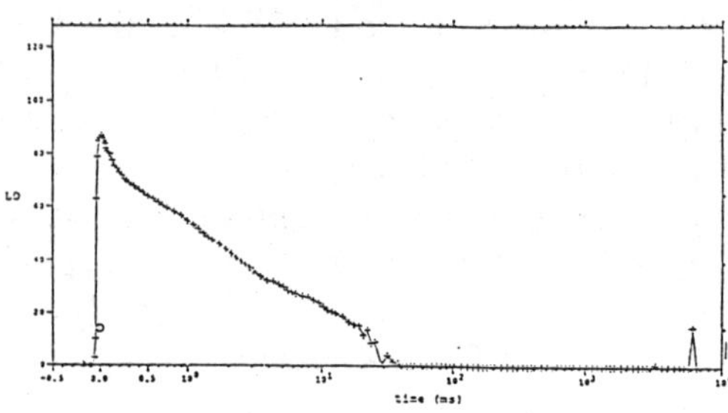

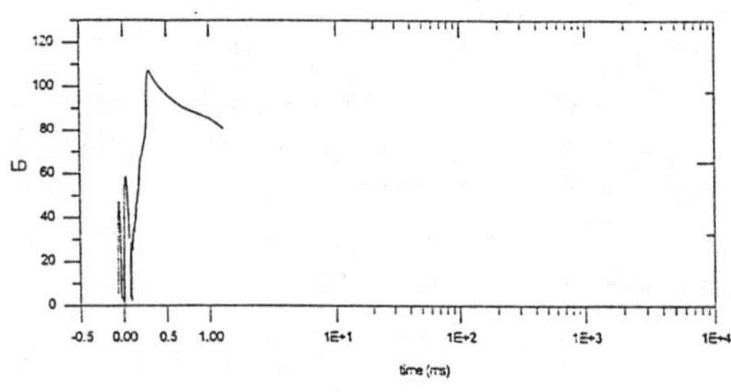

a. line-of-sight

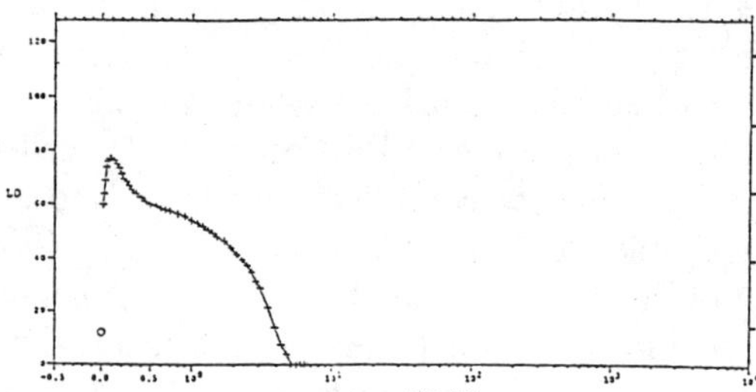

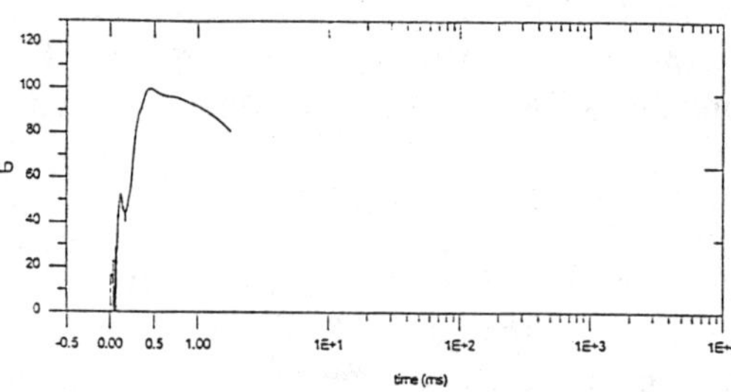

b. impact in darkness

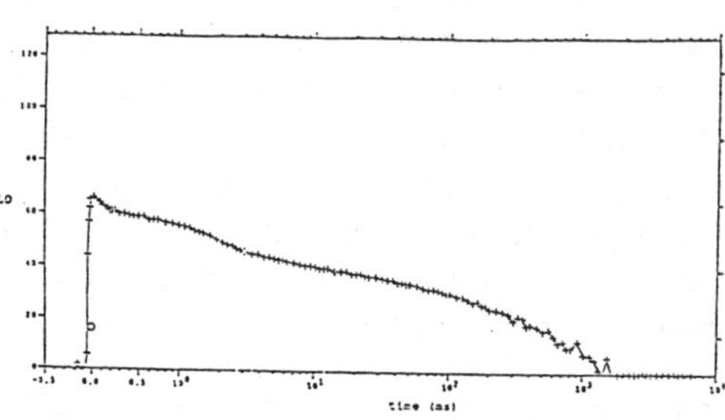

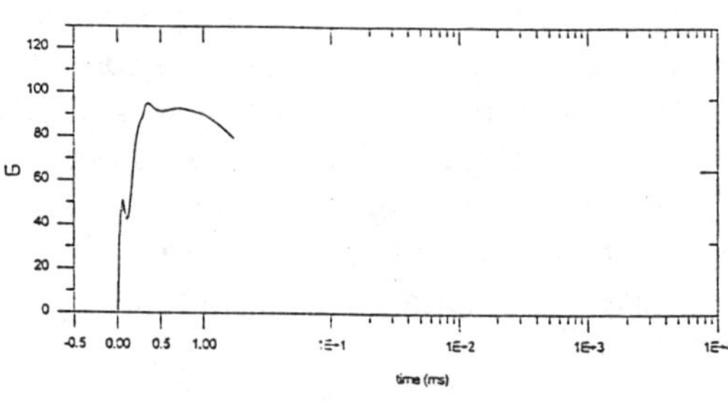

c. illuminated debris

FIGURE 3. Comparison of Test Results to On-Orbit Signals.

REFERENCE

1. C. Fox and R. Wiley, "Hypervelocity Microgram Particle
Impact (HMPI) Study," prepared for AFTAC/DOS, Patrick Air
Force Base, Florida, Contract Number F08650-93-D-0050,
October 31, 1994.

CHAPTER XV

IMPACT PHENOMENA AND HYPERVELOCITY STUDIES

PENETRATION MECHANICS
OF NON-CIRCULAR RODS

S. J. Bless

Institute for Advanced Technology, The University of Texas at Austin
4030-2 West Braker Lane, Austin, TX 78759

Penetrators that contain planar elements, such as rods with cruciform or triform cross sections, are analyzed for the limiting case of plane symmetry. Results are compared to conventional rods of axial symmetry. It turns out that rigid plane penetrators are more effective than axially symmetric penetrators below a critical velocity. Eroding plane penetrators are expected to always be more efficient than cylinderical penetrators. However, in a test calculation for an finite width eroding plate penetrator, the expected advantage of plane symmetry was not observed.

The cross sections of modern anti-armor penetrators are usually circular. Cross sections of other types of penetrators, such as arrow points or nails, however, often deviate substantially from circular symmetry. Indeed, even for anti-armor applications, there may be several advantages to non-circular cross sections. For example, they provide enhanced stiffness for the same length and mass, which may aid launchability, flight stability, and ability to penetrate certain asymmetric targets. They may provide reduced mass for the same length and major diameter, which may make higher velocities possible. Non-circular penetrators may encompass planar elements, and that is the motivation for the study reported here.

Penetration mechanics of non-circular penetrators has received little attention. Solutions are available for some earth indentation problems [1]. Cherepanov has published some interesting speculations concerning super deep penetration caused by wedge penetrators that penetrate in a cleavage mode [2]. We have recently completed a numerical study of several non-circular cross sections for penetrators that provide enhanced stiffness [3]. A more complete description of that work is available in [4].

It is relatively easy to compare the limiting cases of slender-pointed linear vs. circular penetrators for an elastic-plastic material. Penetration of circular penetrators can be treated by cavity expansion theory. The elastic-plastic solution for a cylindrical cavity was given by Bishop, Hill, and Mott [5] as:

$$R_t = \beta Y (1 + \ell n \frac{G}{\beta Y}) \qquad (1)$$

where $\beta = \sqrt{3}/3 = 0.577$, and G and Y are shear modulus and yield strength, respectively. We use the expression R_t for cavity expansion stress because this symbol is conventionally used to represent the stress that the target exerts on a penetrator during penetration [6]. The energy required to penetrate is R_t times the cavity volume. For armor steel, G = 80 GPa, Y = 0.8 GPa, giving $R_t/Y = 3.55$. Thus, the cavity expansion stress is about 3.5 times the yield stress.

Imagine instead that the penetrator mass is redistributed in the shape of a wedge of essentially infinite lateral extent and slender point. Then as the wedge indents, the stress required to open the cavity is equal to the stress required to open the crevice. The deformation in the target is one-dimensional strain. Hence, the conventional shock wave analysis applies. If the wedge tip arrives at a location at t = 0, the half width of the crevice, a, is equal to $V_c t$, where V_c is the cavity expansion velocity. For a wedge or cone of half angle ϕ, $V_c = u \tan \phi$, where u is the penetration velocity. There will be an elastic and a plastic front that move into the target with velocities C_L (the longitudinal sound speed) and C_0 (the bulk sound speed = $\sqrt{K/\rho}$). The elastic wave will initially be a ramp, but given time it will steepen to a plateau equal to $[(1-\nu)/(1-2\nu)]Y$ which is equal to about 2Y.

The cavity expansion stress will be ZV_c, where Z is the shock impedance. For steel we take this as ρC_0. The cavity expansion velocity at which axisymmetric penetration requires less stress than planar penetration is:

$$V_c = 3.55Y / \rho C_0 \qquad (2)$$

For steel, C_0 is about 4.56 km/s. Therefore, the critical value of V_c is equal to 80 m/s.

If we imagine a cone and a wedge rigid penetrator which have the same included angled, then we would expect that for low speed penetration, the force required to advance the penetrator will be less for the wedge shape. Rigid body penetration with hard penetrators has been shown to be possible in steel up to about u = 1.5 km/s. Assuming a cone half angle of 15°, the maximum impact velocity at which a wedge might out perform a conical point is about 300 m/s from Equation (2). It is important to remember, however, that this analysis ignores friction, which in the absence of melting or cavitation, may be very substantial for slender penetrators.

Eroding armor penetrators differ from the scenario just described in that most of the cavity expansion takes place at the nose of the penetrator, where the interface of the penetrator and the target is approximately hemispherical. Thus, for eroding penetrators, a better approach is to calculate the cavity expansion pressure for an axisymmetric rod from the spherical cavity expansion solution, which was also provided by Bishop, Hill, and Mott [5], and can be written as:

$$R_t = \frac{2Y}{3}\left(1 + \ell n \frac{2G}{Y}\right) \qquad (3)$$

which is equal to 4.19Y for steel. For this case, a planar penetrator would be best treated by the cylindrical cavity expansion model given above, according to which R_t equals 3.55Y. Thus, for the case of a spherical cross section penetrator nose, a planar penetrator always penetrates the target with about 18 percent less resistance than a circular penetrator. Over the range of impact velocity 1.5 to 2.5 km/s, this gives a change in the penetrator efficiency, s = $\Delta P/\Delta L$, of about 15 percent. (See Partom [7] for a discussion of formulas relating s to R_t.)

The finite difference code AUTODYNE was also used to compare axisymmetric with planar penetration. In the simulations a steel target was used, and projectiles were steel or tungsten alloy penetrators with L/D=10. Targets were steel Y=1 GPa and G=80 GPa. The tungsten was modeled as Y=2 GPa, G=140 GPa. Other parameters were standard, and are given in [7, 8]. Circular penetrators were computed with axial symmetry, D/2=5 mm, L=100 mm. Planar penetrators were computed as plane strain problems, also with D/2=5 mm, and L=100 mm. Comparisons between planar and axial symmetry were made at 1.5, 2.0, and 2.5 km/s for steel projectiles, and 1.5 km/s for tungsten projectiles.

Figure 1 shows the results of these calculations for final penetration depth. The numbers next to the plot are the ratios of target diameter to penetrator diameter (D/d). This had to be increased for the planar rods because they caused too much target flow. It can be seen that in all cases the planar penetrator penetrated more, yielding about 25 percent higher penetration at 1.5 km/s. However, the advantage of the non-circular shape decreased at higher velocity. This calculation may be affected by the relatively large volume of the planar penetrator. In the circular case, the ratio of target to penetrator volume was 400, whereas for the planar case it was 56 or less.

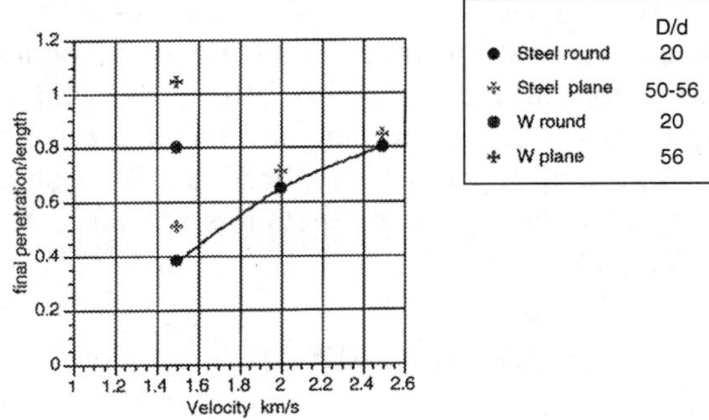

FIGURE 1. Comparison of tungsten alloy and steel plane and axisymmetric penetrators into armor steel: total penetration depth from [9].

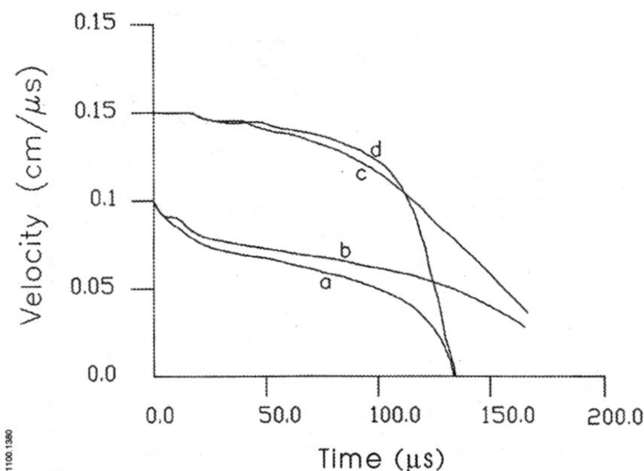

FIGURE 2. Comparison of tungsten alloy plane and an axisymmetric penetrator into armor steel at 1.5 km/s impact velocity. Lower curves are penetration velocity of circular (a) and plane (b); upper are tail velocities of circular (d) and plane (c) from [9].

For the case of greatest practical interest, tungsten penetrating steel, velocity of the interfaces is shown in Figure 2. It can be seen that as predicted, the plane penetrator penetrates faster than the round penetrator. The difference in penetration at later times is exaggerated by the rapid deceleration of the round rod, which may be an artifact of the calculation (since the target bulged at late times for the plane penetration). We can calculate penetrator efficiency as:

$$s = \frac{\Delta P}{\Delta L} = \frac{u}{v - u} \qquad (4)$$

where u is penetration velocity and v is tail velocity. At 50 μs, the penetrator efficiency of the planar tungsten rod is about 20 percent greater than of the circular cross section rod. This is fairly good agreement with the expectation from the cavity expansion solutions.

Littlefield and Anderson, in [4] recently conducted numerical simulations with the CTH code to compare the penetration of L/D=5 tungsten alloy round rods with an equal mass and equal length rectangular cross section rods. The aspect ratio of the non-circular rod was four. Two velocities were considered, 1.8 and 2.6 km/s. These calculations were very similar to the ones described in [3], where the zoning and material models are given.

In this simulation, the cylinder diameter was 5.67 mm, and the plane penetrator had the same area, but a longest side length of 11.24 mm. The calculations were both run 3-D. They were continued until penetration became steady state. Results of the calculations for 2.6 km/s are shown in Figures 3 and 4. It can be seen that the penetration velocity is about 5 percent less for the plate penetrator. The cavity is oval in shape. The penetrator efficiency values are summarized below:

cross section	1.8 km/s	2.6 km/s
1:4 rectangle	s=0.94	s=1.35
circular	s=1.08	s=1.46

where s was computed from $u/(v_0 - u)$, with u taken at the beginning of the steady state phase, and v_0 taken as the impact velocity. In these calculations the non-circular rod is less efficient than the circular one, by 17 percent at 1.5 km/s, and 8 percent at 2.6 km/s. This differs both in absolute amount and in velocity trend from the preceding analysis.

Lastly, we note that in the analysis of static penetration of clay of [1], the resistance to cone penetration was about 15 percent higher than the resistance to plane-strain wedge penetration.

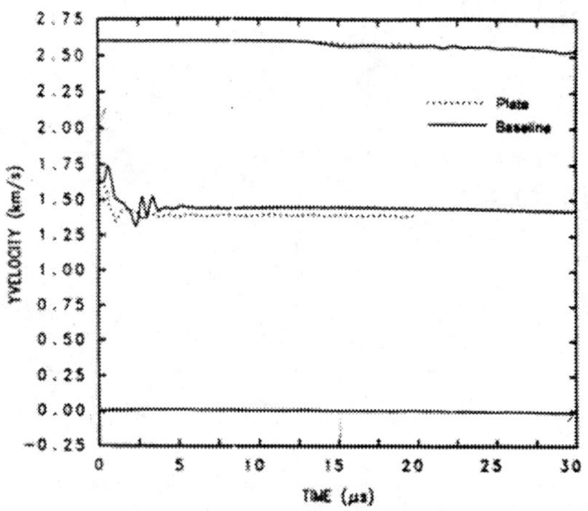

FIGURE 3. Penetration velocity and tail velocity for circular rods and rectangular cross section rods, from [4].

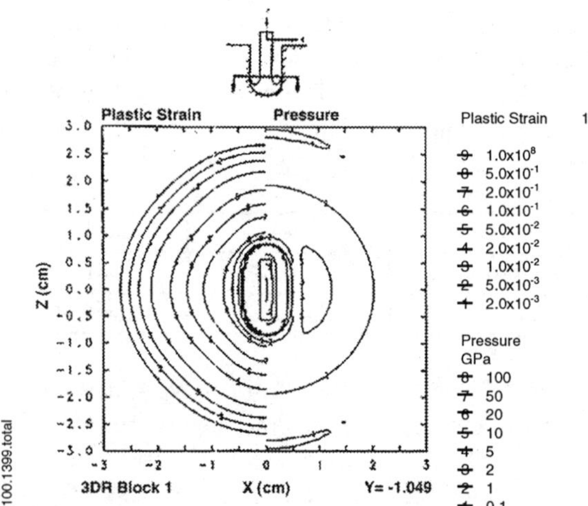

FIGURE 4. Crater profiles in the steady state behind the penetration front, from [4].

We tentatively reconcile these results by concluding that the inherent advantage of plane penetrators may be offset by effects associated with edges. At higher velocities the edge effects become less important. Implications for design of practical non-circular penetrators remain to be developed.

ACKNOWLEDGMENTS

This work was supported by the U.S. Army Research, Laboratory (ARL) under contract DAAA21-93-C-0101.

REFERENCES

1. Baligh, M. A. and Scott, R. F., "Analysis of Wedge Penetration of Clay," Géotechnique **26**, 185-208, 1976.

2. Cherepanov, G. P., "Super-Deep Penetration," *Eng. Fracture Mech.*, **47**, 691-713, 1994.

3. Bless, S. J., Littlefield, D. L., Anderson Jr., C. E., and Brar, N. S., "The Penetration of Non-Circular Cross-Section Penetrators," *Proc. 15th Int'l Symp. Ballistics*, Jerusalem, 21-24 May, 1995.

4. Littlefield, D. L., and Anderson Jr., C. E., "A Numerical Study of the Penetration Physics of Asymmetric Long-Rod Projectiles at 1.8 and 2.6 km/s," Southwest Research Institute Report 07-6716, submitted to Institute for Advanced Technology, January 1995.

5. Bishop, R. F., Hill, R., and Mott, N. F., "The Theory of Indentation and Hardness Tests," *Proc. Phys. Soc.*, **57**, Part 3, pp. 147-159, 1945.

6. Tate, A., "Further Results in the Theory of Long Rod Penetration," *J. Mech. Phys. Solids*, 17:141, 1969.

7. Partom, Y., "The Optimal Velocity of Constant Kinetic Energy Constant L/D Long Rod Projectiles." presented at the Hypervelocity Impact Symp., Santa Fe, October 1994.

8. Partom, Y., "Comparison of Penetration Efficiency in Axial and Planar Symmetries," Institute for Advanced Technology Report # IAT.R 0033, November, 1993.

9. Partom, Y., "Comparison of Penetration Efficiency in Axial and Planar Symmetries," Proc. EXPLOMET 95, El Paso, TX, August 1995.

ACCOUNTING FOR THE
PENETRATION ENTRANCE EFFECT

Yehuda Partom

Institute for Advanced Technology, The University of Texas at Austin,
4030-2 W. Braker Lane, Austin, Texas 78759 USA
Permanent Address: Rafael Ballistics Center, P. O. Box 2250, Rafael, Haifa 31021 ISRAEL

When a long rod penetrates a thick target, its penetration rate is initially higher than during the quasi steady-state phase. This "entrance effect" extends typically to a depth of 1-2 rod diameters into the target. One explanation of the entrance effect is that initially the rate equals the 1D strain particle velocity. But it has been shown that the 1D strain state dissolves at about 0.2d into the target. Another explanation has been, that as at the entrance boundary the target is less confined, and can flow backwards to form a lip, the penetrating rod sees initially less resistance. To investigate the origin of the entrance effect we performed computer simulations with AUTODYNE-2D. By constraining the backward motion of the boundary, we show explicitly that the entrance effect is not an outcome of the partial confinement. Using a non steady-state penetration model derived by Walker and Anderson we show that the entrance effect results from the initial 1D strain penetration rate, and the subsequent non steady-state penetration.

INTRODUCTION

Running a computer simulation of a long rod penetration event, one can see clearly an entrance boundary effect. Figure 1 is from an AUTODYN/ EULER run of a L/D = 10 tungsten alloy rod at V = 1.5 km/s into a RHA steel target. We see that initially, penetration velocity is significantly higher than during the quasi steady-state phase. This behavior is typical of long rod penetration events, and we refer to it as the "entrance effect." In contrast to what we get from a simulation, there is no entrance effect in a Tate model run.

In Figure 2, we show a typical plot from an AUTO-DYN simulation of material boundaries and velocity arrows at about 9 µs (~1/2D) into the penetration process. We see that a typical lip is formed on the target front surface around the crater. Because of this lip, and the backward target motion that creates it, it has been proposed that target resistance is lower near the entrance boundary, leading to the entrance effect. This view is not shared by some investigators who claim that the entrance effect results from the initial shock phase and the high (1D strain) particle velocity associated with it.

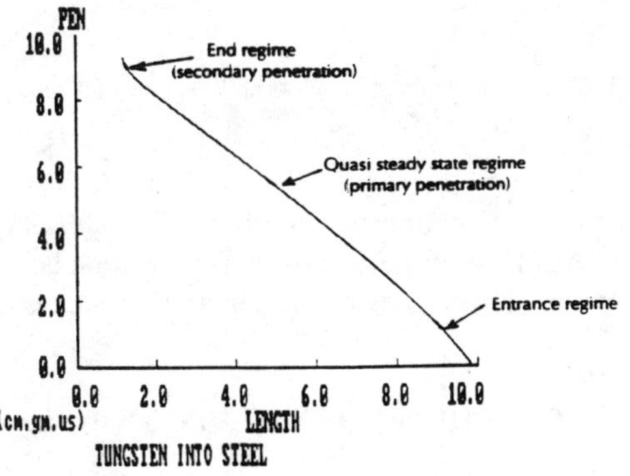

FIGURE 1. Penetration-erosion curve from an AUTODYN run of a L/D=10 tungsten alloy rod penetrating RHA steel at 1.5 km/s.

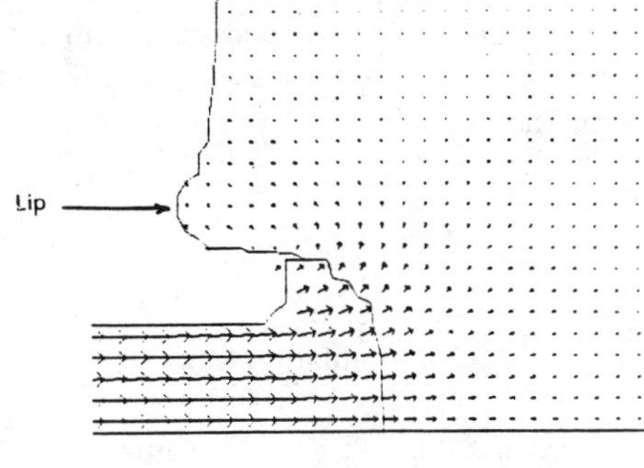

FIGURE 2. Material boundaries and velocity arrows plot from an AUTODYN simulation at about 1/2D penetration.

In what follows, we investigate the origin of the entrance effect by means of AUTODYN simulations. We do this by inhibiting the backward motion of a part of the entrance boundary. We find that the entrance effect is not affected by this inhibition. Recently Walker and Anderson [1] proposed a time-dependent model for long rod penetration, so-called WA model. It is based on integration of the momentum equation along the projectile-target center line. Among other things, Walker and Anderson show that their model captures the entrance effect.

In what follows, we use the WA approach to formulate what can be called a time dependent Tate model. It is a much simplified version of the WA model. Running this model, we show that it too is able to capture the entrance effect.

SIMULATIONS

We are using AUTODYN/EULER version 2.65. The tungsten alloy (projectile) and RHA steel (target) material models and parameters are conventional and the same as in [2] and [3]. The projectile radius is D/2 = 5 mm, and the Euler cell size is 1 x 1 mm so that there are five cells across the radius. This may not be enough to obtain convergent results, as shown in [4]. But as our purpose is to investigate the origin of the entrance effect, five cells across the radius seem satisfactory. The projectile aspect ratio is L/D = 10, and the impact velocity V = 1.5 km/s. We terminated our runs after 400 cycles (approximately 40 µs) and plotted velocity vectors plots every 100 cycles.

In all runs the entrance boundary is stress free on a ring $D/2 < r \leq D/2 + \Delta r$, and constrained in the axial direction elsewhere ($r > D/2 + \Delta r$). The different runs are listed in Table 1.

Table 1

List of Simulation Runs

No.	Δr mm
1	1
2	2
3	5
4	10
5	∞

Figure 3 shows velocity vectors plots at cycle 300 for runs 1 to 4. We see that when Δr = 1 and 2 mm, the target material is not-able to flow through the gap. For Δr = 5 and 10 mm, target material flows through the gap and a lip is formed. But in all cases, the penetration is the same, and the mushroom configuration is the same. The amount of backward flow of the target boundary does not seem to influence the penetration process.

We conclude from the simulations that the entrance effect is not caused by low target resistance near the entrance boundary.

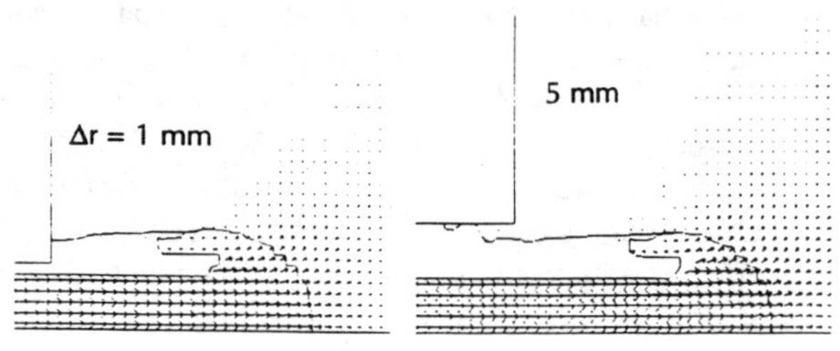

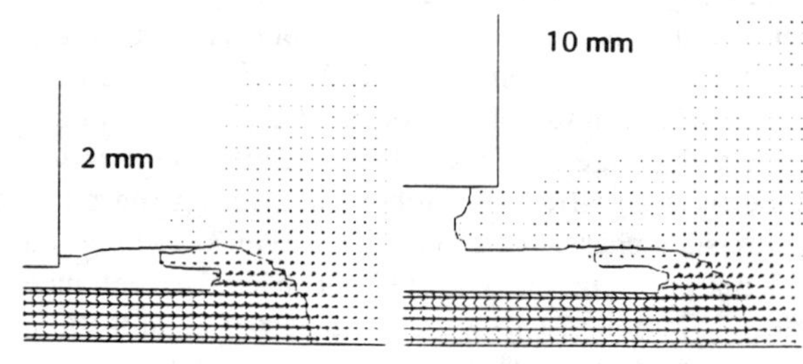

FIGURE 3. Velocity vectors plots at 300 cycles for runs 1 to 4.

In the next section we propose, as did Walker and Anderson [1], that the entrance effect can be accounted for by time dependency of the penetration process.

TIME DEPENDENT TATE MODEL

Walker and Anderson (WA) [1] have shown that by integrating the axial momentum equation along the projectile-target center line, one gets the following relation:

$$\rho_p \int_{x_p}^{x_i} \frac{\partial u}{\partial t} dx + \rho_t \int_{x_i}^{x_t} \frac{\partial u}{\partial t} dx +$$

$$\frac{1}{2} \rho_p \left\langle [u(x_i)]^2 - [u(x_p)]^2 \right\rangle +$$

$$\frac{1}{2} \rho_t \left\langle [u(x_t)]^2 - [u(x_i)]^2 \right\rangle$$

$$- \left[\sigma_{xx}(x_t) - \sigma_{xx}(x_p) \right] - \tag{1}$$

$$2 \int_{x_p}^{x_i} \frac{\partial \sigma_{xy}}{\partial y} dx - 2 \int_{x_i}^{x_t} \frac{\partial \sigma_{xy}}{\partial y} dx = 0$$

where x = axial direction, y = transverse direction, u = axial particle velocity, t = time, σ_{ij} = cartezian stress components, ρ_p = projectile density (assumed constant), ρ_t = target density (assumed constant), $x_p \leq x \leq x_t$ is the integration range along the center line ($y = 0$). Unlike WA, we choose x_p to be the elastic-plastic boundary in the projectile, and x_t the elastic-plastic boundary in the target. x_i is the projectile-target interface. We use $u(x_p) = V$ (projectile tail velocity), $u(x_i) = U$ (penetration velocity), and assume: $u(x_t) \cong 0$.

WA went on to evaluate the last two integrals in Eq. (1) by assuming a specific form for the flow field $u(x, y)$. We are using a simpler approach. Recognizing, as did WA, that:

$$-\sigma_{xx}(x_t) - 2 \int_{x_i}^{x_t} \frac{\partial \sigma_{xy}}{\partial y} dx = R_t \tag{2}$$

$$\sigma_{xx}(x_p) - 2 \int_{x_p}^{x_i} \frac{\partial \sigma_{xy}}{\partial y} dx = Y_p \tag{3}$$

where R_t and Y_p are the Tate model target resistance and projectile strength, respectively. We get:

$$\rho_t \int_{x_i}^{x_t} \frac{\partial u}{\partial t} dx - \frac{1}{2} \rho_t U^2 + R_t =$$

$$-\rho_p \int_{x_p}^{x_i} \frac{\partial u}{\partial t} dx - \frac{1}{2} \rho_p \left(U^2 - V^2 \right) + Y_p \tag{4}$$

To evaluate the integrals we assume $x_i - x_p = \text{const.} = S_p$, $x_t - x_i = \text{const.} = S_t$, $u = U + (V-U) f_p(x)$ in the pro-

jectile, and $u = U f_t(x)$ in the target, where the functions f_p and f_t are advected without change at a velocity U, as shown in Figure 4.

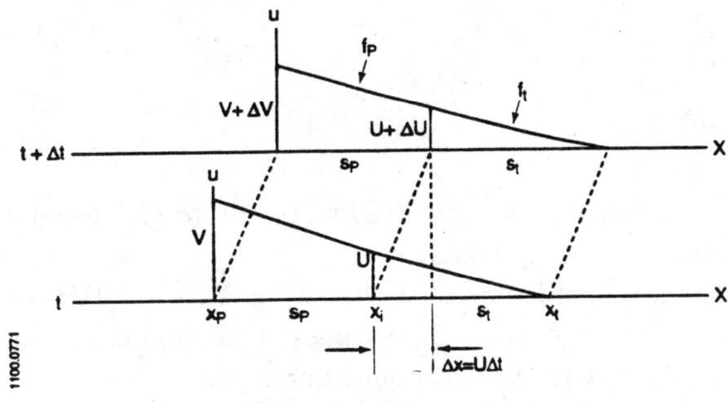

FIGURE 4. Simplified flow field on center line.

The advection assumption leads to:

$$\frac{\partial f_p}{\partial t} = -U \frac{\partial f_p}{\partial x} \; ; \qquad \frac{\partial f_t}{\partial t} = -U \frac{\partial f_t}{\partial x} \tag{5}$$

so that in the projectile:

$$\frac{\partial u}{\partial t} = \frac{dU}{dt} + \left(\frac{dV}{dt} - \frac{dU}{dt} \right) f_p - U(V-U) \frac{\partial f_p}{\partial x} \tag{6}$$

and in the target:

$$\frac{\partial u}{\partial t} = \frac{dU}{dt} f_t - U^2 \frac{\partial f_t}{\partial x} \tag{7}$$

Evaluating the integrals we get after some rearrangements and substitutions:

$$R_t + \frac{1}{2} \rho_t U^2 + \frac{dU}{dt} +$$
$$\left[\rho_p S_p (1 - \kappa_p) + \rho_t S_t \kappa_t \right] = \tag{8}$$
$$Y_p + \frac{1}{2} \rho_p (V-U)^2 - \rho_p S_p \kappa_p \frac{dV}{dt}$$

Where κ_p, κ_t are shape factors. For f_p, f_t linear, $\kappa_p = \kappa_t = 1/2$. For f_p, f_t concave 2nd degree parabolas, $\kappa_p = \kappa_t = 1/3$.

We see that Eq. (8) collapses to Tate's equation when $dU/dt = dV/dt = 0$ (steady-state assumption). We therefore refer to Eq. (3) as the "time dependent Tate equation." Using Eq. (8) together with the kinematic relations of Tate's model we thus define a time dependent Tate model, as follows:

$$\frac{dU}{dt} = F \left(U, V, \frac{dV}{dt} \right) \tag{9}$$

as defined by Eq. (8).

$$\frac{dV}{dt} = -\frac{Y_p}{\rho_p \left(L - S_p\right)} \qquad (10)$$

$$\frac{dL}{dt} = -(V - U) \qquad (11)$$

$$\frac{dp}{dt} = U \qquad (12)$$

We integrate the four ODEs Eqs. (9) to (12) using a standard ODE system solver.

At the end phase of the integration, $|dV/dt|$ becomes large as $L - S_p$ becomes small, and using Eq. (8) to evaluate dU/dt becomes unreliable.

We therefore switch to a more robust way to compute dU/dt as follows:

We compute $\left(\dfrac{dU}{dt}\right)_{Tate}$ by:

$$\left(\frac{dU}{dt}\right)_{Tate} = \frac{\rho_p\,(V-U)}{\rho_t U + \rho_p\,(V-U)}\,\frac{dV}{dt} \qquad (13)$$

as obtained from differentiating Tate's equation and get:

$$\left(-\frac{dU}{dt}\right) = \max\left[\left(-\frac{dU}{dt}\right)_{Eq.\,8}, \left(-\frac{dU}{dt}\right)_{Tate}\right] \qquad (14)$$

To see how the time dependent Tate model performs, we compare it to the regular Tate model and to simulation results. The example we show is for an $L/D = 10$ tungsten alloy rod penetrating an RHA steel target. The model parameters are:
$\rho_p = 17.3$ g/cc, $Y_p = 2$ GPa, $S_p = 14$mm, $\kappa_p = 0.5$, $\rho_t = 7.85$ g/cc, $R_t = 6$ GPa, $S_t = 65$ mm, $\kappa_t = 0.3$, $V_o = 1.5$ km/s, $U_o = 1$ km/s, $L_o = 100$ mm.

In Figure 5, we show penetration velocity plots from the time dependent Tate model, from the regular Tate mode, and from the simulation.

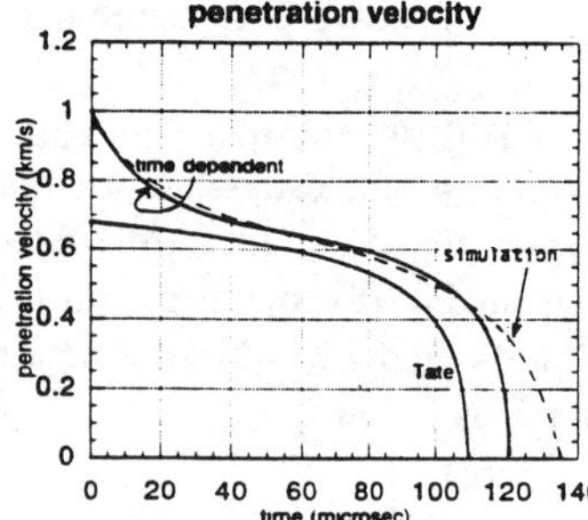

FIGURE 5. Penetration velocity histories.

We see the following:

- The entrance effect is captured quite well by the time dependent Tate model.

- As expected, the time dependent Tate does not do well at the end phase.

CONCLUSIONS

We ran computer simulations with AUTODYN to investigate the origin of the entrance effect in long rod penetration. By constraining the backward motion of the entrance boundary, we show that the entrance effect does not depend on the backward motion of the entrance boundary.

Based on Walker and Anderson's approach [1] of integrating the momentum equation along the axis, we propose a time dependent Tate model. Running the model and comparing to simulation results we show that the model captures quite well the entrance effect.

We conclude that the entrance effect results from the initial high (1D strain) penetration velocity, and from the time dependency of the penetration process.

ACKNOWLEDGMENTS

This work was supported by the U.S. Army Research, Laboratory (ARL) under contract DAAA21-93-C-0101.

REFERENCES

[1] James D. Walker and Charles E. Anderson, "A Time-Dependent Model for Long-Rod Penetration," *Int. J. Impact Engng.*, 16(1), pp. 19-48, 1995.

[2] Yehuda Partom, "Projectile-Flow Effect for Long Rod Penetration," IAT.R 0036, February 1994.

[3] Y. Partom, "Comparison of Penetration Efficiency in Axial and Planar Symmetries," IAT.R 0033, November 1993.

[4] Y. Partom and D. L. Littlefield, "Comparison of AUTODYN/EULER Penetration Runs to CTH," IAT.TN 0029, February 1994.

LONG ROD PENETRATION IN OBLIQUE IMPACT

Gabi Luttwak, Zvi Rosenberg , Yosef Kivity

Rafael Ballistics Center

P.O. Box 2250, Haifa 31021, Israel

The oblique impact and penetration of long rods is investigated both experimentally and numerically. In the experiments a long copper rod impacted obliquely an aluminum plate at a velocity of 850m/sec. We have chosen ductile materials like aluminum and copper to ensure that the results are not sensitive to the failure model used in the calculations. The low density of the aluminum, and the relatively small sizes of the target plates enabled us to resolve the material interfaces in the X-ray shadowgraph taken during the penetration. The numerical simulations were carried out in the multimaterial Euler-with-strength processor of the three dimensional code MSC/DYTRAN. There is a good agreement between the computations and the experimental results, as to the shape of the projectile and target during the penetration process.

INTRODUCTION

When a projectile strikes upon a target a region of high pressure is generated. The emerging shock waves induce a material flow outward from the point of impact. This motion, which is resisted by the material strength leads to the mushrooming of the projectile and the growth of a crater in the target. For the normal impact of a high speed long rod into a thick homogeneous target, the process can be considered to be stationary and assuming incompressible flow, it can be described by the hydrodynamic theory of penetration [1]. At lower impact velocities, this theory has been modified to take into account the effect of material strength [2].

For oblique impact, the reflection of the shock waves from the target free surface, can cause material flow toward the incoming rod. For thin or layered targets, the details of the flow and the wave reflections may be important even in the case of normal impact. In these cases , target penetration can be best studied by performing numerical simulations. While some problems of normal impact can be solved in two dimensional axisymmetric coordinates, for oblique impact, the solution depends on all three spatial coordinates and time.

The three dimensional code MSC/Dytran includes Eulerian , Lagrangian, thin shell and rigid body processors. The different Lagrange, thin shell or rigid body regions may come into contact and interact through a sliding-impacting algorithm. They can be coupled with an Euler region through either general or ALE (Arbitary Lagrangian Eulerian) coupling [9]. The general coupling is an extension of the CEL (Coupled Euler Lagrange) method[10]. The Euler processor can handle moving grid points and general connectivity meshes. It can include several materials, with the interfaces and free surfaces cutting through the zones. The material model may include elastic plastic strength.

During the perforation process, both the target and the projectile are subjected to large deformations. Neighboring points in the target may move far apart from each other. Both the rod and the plate may fracture and separate into several distinct parts. A Lagrangian mesh, which follows the material motion will become highly distorted. Thus, it is preferable to carry out the simulation in stationary, Eulerian coordinates. In an Eulerian calculation, the materials flows through the computational cells, while the boundaries may cut the mesh lines. The multimaterial Euler calculation must adequately resolve these interfaces to prevent any non-physical mixing of the materials. The MMES (multimaterial-Euler-with-stregth) processor of the 3D code MSC/DYTRAN[3], has been already applied [4] to the numerical simulation of oblique impact. Penetration of steel plates by tungsten rods has been considered and the analogy between plate on plate and rod on plate impact has been addressed. The feasibility of using the different modelling choices available in Dytran, like the ALE technique[7−9], or the CEL method[10] has been discussed.

The main purpose of the present work is code calibration and validation of the MMES processor. We considered the 850m/sec , 45°, oblique impact of a long copper rod upon an aluminium plate. We have chosen ductile materials like copper and aluminium, so that results would not be strongly dependant on the material failure model used in the calculations. While the multimaterial Euler technique can easily handle the separation of the target or the projectile following failure, the implementation of an adequate failure model is not trivial and is not considered here. In a related work[5], the high obliquity impact of soft and hard spheres on thin plates has been investigated both experimentally and in numerical simulations carried out using the MMES processor. In another investigation[6], Dytran's Lagrange processor was used to reproduce the deformation of square rods as they hit a rigid wall.

EXPERIMENTAL

The experiments were performed with a 30mm powder gun, using a 450kV flash X-ray tube, which was needed in order to penetrate the aluminum target. The copper projectile was placed in a plastic sabot, using a steel pusher plate. Impact velocity was determined both with laser beams, which are cut by the projectile in flight, and two 150kV flash X-rays. The latter were also used to determine projectile integrity and yaw prior to impact. A short circuiting screen was placed on the target to determine the impact time. A delay time generator was used to control the exact timing of the pulser in order to observe the interaction at the predetermined time. The width of the aluminum (6061-T6) plate was chosen in accord with the penetration capability of the pulser. Static tests

have shown that a width of 75mm can give a good contrast between the copper rod and the aluminum plate during the penetration process.

In the reported experiment, the projectile hit the target at an angle of 45° and the impact velocity was 850m/sec. The 80mm long copper rod had a diameter of 9.5mm. It was composed of a hemi-spherical head, followed by a 75.25mm long cylindrical part. The target was a 300x75x25.4mm aluminium plate. It was hit by the copper rod 125mm up and 27mm to the right from the plate corner. The X-ray shadowgraph shown in Fig.1 was taken $50.7\mu sec$ after the impact. On the picture we marked by dotted lines the leading edge of the back splash region of the aluminum target. This could be resolved on the original film, but cannot be seen in the figure due to lack of contrast.

NUMERICAL SIMULATIONS

In the calculations, the plane defined by the rod and its projection on the target was assumed to be a symmetry plane. Thus, only half of the computational space had to be meshed. This assumption, would have been exact for an infinite target, or provided the rod would have hit at the middle of the plate with its projection parallel to the plate side. We have chosen a simple Cartezian mesh aligned with the impacting rod and with the grid points fixed in space. In this way, most of the advection is normal to the mesh faces and the errors due to diagonal fluxing are minimalised. The general connectivity which is available also in the Euler processor of Dytran is used, to cover with zones only the space occupied by the rod , the target and the expected interaction region. In the interaction region, we had near cubic zones, with a zone size of 1.2mm. The zone sizes were gradually increased going outside this region. This way, a satisfactory resolution is obtained with less than 75000 zones. It took six hours to run this problem until $Time = 75\mu sec$ on a Digital Alpha workstation.

Constant bulk and shear moduli were used in the calculations, as well as constant yield strength and spall strength. The chosen values are given in Table 1.

The material boundaries on the symmetry

plane at a time of $50.7\mu sec$ are shown in Fig. 2. A three dimensional view at $50.7\mu sec$ can be seen in Fig. 3. A small perturbation in the shape of the rod as seen in the figures, is due mainly to the ploting algorithm and not to the calculation. In the pictures, the material boundaries in the multimaterial Euler calculation were obtained as isosurfaces of 0.5 relative volume.

RESULTS AND CONCLUSIONS

The simulated material interfaces at $50.7\mu sec$ are in reasonable agreement with the X-ray picture. All the typical dimensions agree up to 0.5mm. The largest discrepancy is in the size of the mushroom, and this could be fitted better by taking into account work hardening in the material model[6]. Some of the details in the shadowgraph fit more to the 3D perspective view shown in Fig.3. These results, together with the similar agreement found for ball penetration[5] validate the utilisation of the multimaterial-Euler-with-strength processor of the Dytran for terminal ballistics applications.

REFERENCES

[1] Birkhoff G., MacDougall D. P., Pugh E. M., Taylor Sir G.I., "Explosives with Lined Cavities", J. Appl. Phys. **20** , 363 (1949)

[2] Tate A. ,"A Theory of the deceleration of long rods after impact", J. Mech. Phys. Sol. **15**, 387 (1967)

[3] "MSC/DYTRAN, User's Manual", Version 2.3, The MacNeal-Scwendler Corporation, April 1995

[4] Luttwak G., Lenselink H. , "Three Dimensional Numerical Simulation of Oblique Impact and Penetration of Thin Plates by Long Rods", in Proceedings of the $4^{t}h$ Int. High Dynamic Pressures Symposium, Tours, France, June 5-9, 1995, pp.249-257.

[5] Kivity Y. , Mayseless M., Luttwak G. , Stilp A. , Hohler V., Weber K., Florie C., Lenselink H. , Cowler M. , Birnbaum N. , "High Obliquity impact of soft and hard spheres on thin plates", in Proceedings of the $15^{t}h$ Int. Symposium on Ballistics , Jerusalem, May 21-24, 1995, ch. 1, pp 133-142

[6] Luttwak G., Rosenberg Z., Falcovitz J. "Experimental and Computational Study of Taylor Impact with Square Rods ", in Proceedings of the $15^{t}h$ Int. Symposium on Ballistics , Jerusalem, May 21-24, 1995, ch. 1 , pp 291-298

[7] Hirt C. W. , Amsden A. A. , Cook J. L., " An Arbitrary Lagrangian Eulerian calculation method for fluid flows at all speeds", J. Comp. Phys., **14**, 227-253 (1974)

[8] Luttwak G., "Numerical Simulation of Water Jet Penetration", Shock Waves in Condensed Matter-1983, Assay J.R. et al (editors), Elsevier Science Publishers B.V. , pp191-194 , 1984

[9] Luttwak G., Florie C., Venis A., "Numerical simulation of soft body impact", Shock Compression of Condensed Matter-1991, Schmidt S.C. et al (editors) , pp999-1002 , 1992

[10] Noh W. F. , "CEL, a time dependent , two-space dimensional coupled Eulerian Lagrangian code", Meth.in Comp. Phys. **3**, p117, in " The Fundamental Methods in Hydrodynamics " Academic Press, Alder B. et al editors, (1964)

TABLE 1. The material parameters

Material	ρ_0 g/cm^3	c_0 $mm/\mu sec$	G GPA	Y GPA	Spall GPA
Aluminum	2.78	5.33	28.0	0.32	-1.6
Copper	8.93	3.945	48.0	0.20	-3.0

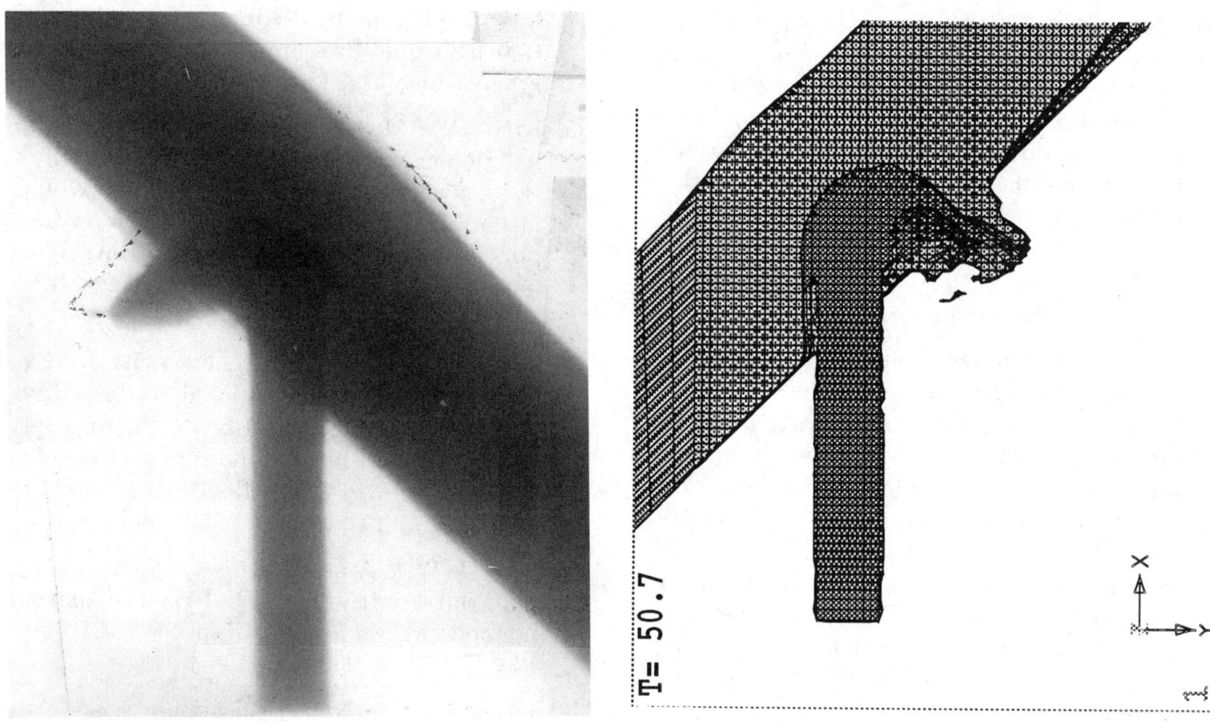

FIGURE 1. The X-ray shadowgraph at $50.7\mu sec$. The leading edge of the aluminum target as seen on the original film is shown by dotted points.

FIGURE 2. The material boundaries on the symmetry plane at $50.7\mu sec$.

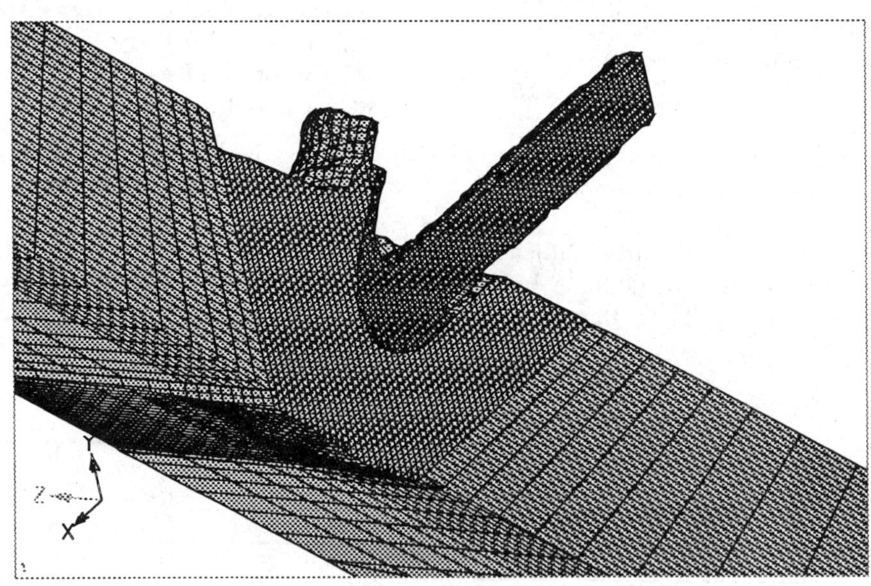

FIGURE 3. The material boundaries at $50.7\mu sec$ in a 3D perspective view.

A STUDY OF ZONING REQUIREMENTS FOR 2-D AND 3-D LONG-ROD PENETRATION

D. L. Littlefield and C. E. Anderson, Jr.

Materials and Structures Division, Southwest Research Institute, San Antonio, Texas 78228

A systematic study is performed to investigate the numerical convergence for the depth of penetration of $L/D = 10$ tungsten alloy projectiles into armor steel. The simulations are carried out using the multi-dimensional Eulerian wavecode CTH. The penetration efficiency P/L is used as the metric for comparison. Two-dimensional axisymmetric and three-dimensional calculations have been performed, and the results are compared and contrasted as the zoning resolution is changed.

INTRODUCTION

A requirement for numerical accuracy is sufficient mesh resolution such that the solution is numerically converged. The discretization process approximates either the differential equations (finite difference) or the geometry (finite element). Specifically, for finite difference algorithms, truncation errors are associated with the discretization process. For example, for some arbitrary variable f that is a function of a coordinate x and time t, the value of the variable can be determined for a nearby location x from the value f at x_o and the partial derivatives of f at x_o using a Taylor series expansion:

$$f(x,t) = f(x_o,t) + f'(x_o,t)(x-x_o) + \frac{f''(x_o,t)(x-x_o)^2}{2!} + \frac{f'''(x_o,t)(x-x_o)^3}{3!} + \dots \tag{1}$$

where the primes denote differentiation with respect to x. With the notation that n represents the nodal point at x_o, and $(n - 1)$ and $(n + 1)$ represent the backward and forward neighboring points of n, respectively, Eqn. (1) can be manipulated to solve for the first derivative at nodal point n:

$$f'(x_n,t) = \frac{f(x_{n+1},t) - f(x_{n-1},t)}{2(\Delta x)} - \frac{f'''(x_n,t)}{6}(\Delta x)^2 + \dots \tag{2}$$

where $\Delta x = (x_{n+1} - x_n) = (x_n - x_n - 1)$. The first term on the right-hand side of Eqn. (2) is the standard central difference formula. By using the Taylor series expansion, it is clearly seen that there is a truncation error associated with the difference approximation, with this truncation error being of the order of $(\Delta x)^2$. The smaller the mesh interval Δx, the more accurate the calculation. On the other hand, the smaller the mesh interval, the more mesh points that have to be processed at each time step. At some point, the second term in Eqn. (2) is sufficiently small that neglecting it does not alter the solution. Therefore, the question of interest is: What mesh resolution provides an "adequately" accurate solution?

The answer to this question, in general, is somewhat problem dependent. For the purposes here, we have selected a canonical problem of interest in penetration mechanics: the depth of penetration of a long-rod projectile into a semi-infinite armor steel target. Both two-dimensional axisymmetric and three dimensional calculations were performed. The three-dimensional calculations are of particular interest for two reasons. First, due to CPU and memory limitations, rarely can 3-D calculations have the same mesh resolution as 2-D calculations; zoning is almost always coarser in 3-D, typically by factors of 3 to 5. In addition, however, there is a fundamental difference in the zoning of 2-D axisymmetric and 3-D problems. In 2-D axisymmetric calculations, assuming that the aspect ratio of the zones is one, the volume of the zones increases with distance from the $r = 0$ axis. The 3-D grid, again assuming constant aspect ratios of one, has constant volume zones. This difference could result in a difference in the mesh dependency of the solution.

THE PROBLEM

Numerical simulations were performed using the Eulerian hydrocode CTH [1]. The projectile used in the simulations was an L/D 10, tungsten alloy cylindrical penetrator with a diameter of 0.787 cm and a hemispherical nose. The target block was semi-infinite armor steel; axial and lateral dimensions of the target were 16.5 and 18.1 cm, respectively. In the 2-D simulations, a cylindrical target was used, so the lateral dimension given here refers to the target diameter. A rectangular target block was used in the 3-D simulations, with square faces in the plane perpendicular to the direction of penetration, and the lateral dimension is the length of this square face. Based on results from previous simulations [2], these target dimensions are sufficiently large to assure semi-infinite confinement in both the lateral and axial directions.

In each of the simulations, the zone size was held constant with an aspect ratio of 1.0 in the strong interaction region where the largest deformations occur in the projectile and target. This region was assumed to be 4 and 12 diameters in the lateral and axial directions into the target, respectively, measured from the initial impact point. Beyond this large deformation region, the zone size was increased at a cumulative rate of 5%.

Equation of state, constitutive and failure models must be specified in the simulations. In this analysis, the Mie-Grüneisen model was the assumed equation of state, and the Johnson-Cook [3] viscoplastic model was used for the constitutive behavior. A simple tensile-void insertion model was used to model failure, where void is inserted into a computational cell whenever any one of the three components of the principal stress exceeds a predefined value σ_{fail}.

Values of the material properties for the projectile and target materials are listed in Table 1. The projectile properties are characteristic values for 90% by weight tungsten alloy with a Ni-Fe matrix, and the target properties are for a "hard" RHA [4], having a Brinell hardness of about 360. Equation of state parameters listed in the table are the reference density and Grüneisen coefficient, ρ and Γ_o, respectively; the bulk sound speed c_o, the slope of the shock-particle velocity curve s, and the specific heat c_v. Elastic deformation is specified with one additional parameter, Poisson's ratio ν, since the bulk modulus is determined from the equation of state. The parameters A, B, C, n, and m are material constants in the Johnson-Cook model. T_{melt} is the melt temperature for the material, and the failure stress σ_{fail} is a dynamic value estimated from the tensile stresses for the two metals.

TABLE 1. Target and Projectile Material Properties

Property	Projectile	Target
ρ (g/cm^3)	17.3	7.85
c_o (m/s)	3850	4500
s	1.44	1.49
Γ_o	1.58	2.17
c_v (J/kg K)	135	450
A (GPa)	1.507	1.225
B (GPa)	0.177	1.575
C	0.016	0.0049
n	0.12	0.768
m	1.0	1.09
T_{melt} (K)	1748	1783
σ_{fail} (GPa)	1.75	1.42
ν	0.30	0.29

RESULTS AND DISCUSSION

Results from the simulations are given in Fig. 1. Shown in the figure are the values for P/L, the final depth of penetration for the projectile normalized by the initial length, for different values of N, the number of zones across the projectile radius. The "converged" normalized depth of penetration is consistent with experimental values for penetration efficiency based on historical data for penetration of tungsten alloy rods into semi-infinite steel targets [5].

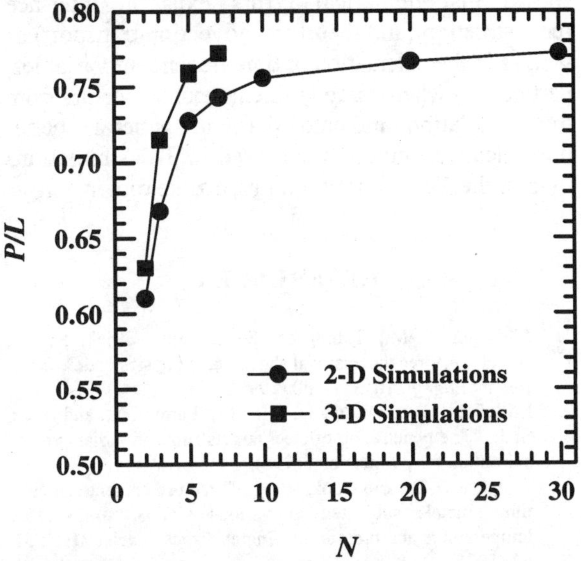

FIGURE 1. Penetration efficiency vs. number of zones across projectile radius.

As can be seen from the results in the figure, the penetration efficiency increases with the number of zones across the projectile radius, and tends to approach a limiting value of about 0.77. For any given value for N, the 3-D results appear to be converging faster than the 2-D values. This seems to indicate that the grid dimension required in 2-D simulations to obtain any given resolution is smaller than the value required in 3-D.

The effect of zone size on the specific details of the numerical solutions can be seen more readily with and examination of the pressure and material interface contours in the problem. Shown in Figs. 2, 3 and 4 are pressure and plastic strain contours in the target at 70 µs after impact for N = 2, 3 and 5, respectively, taken from the 3-D simulations. Figure 2 shows that the shape of the projectile

becomes severely distorted by the interface tracking algorithm when a coarse grid resolution is used. This is particularly evident near the projectile tail, which exhibits considerable change in shape even though yielding and plastic flow has not occurred. Values for the pressure and plastic strain appear to be considerably affected by the zone size in the region near the projectile/target interface, where the gradients in these quantities are the largest. In addition, the mushroom head on the projectile from the coarser zoning calculation in Fig. 2 is significantly larger in size than the mushroom heads from the other two simulations.

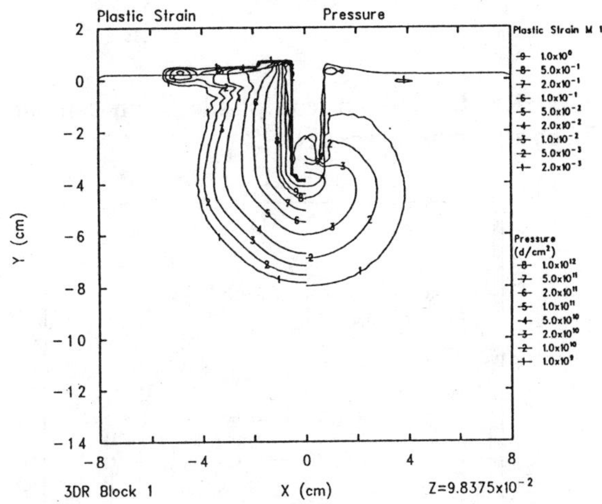

FIGURE 2. Pressure and plastic strain contours 70 µs after impact when N = 2.

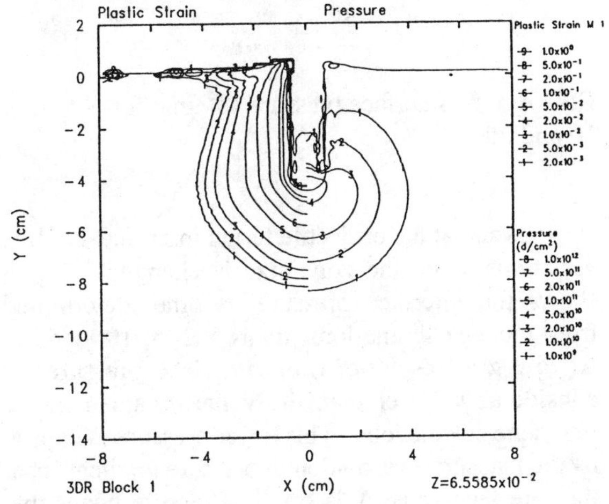

FIGURE 3. Pressure and plastic strain contours 70 µs after impact when N = 3.

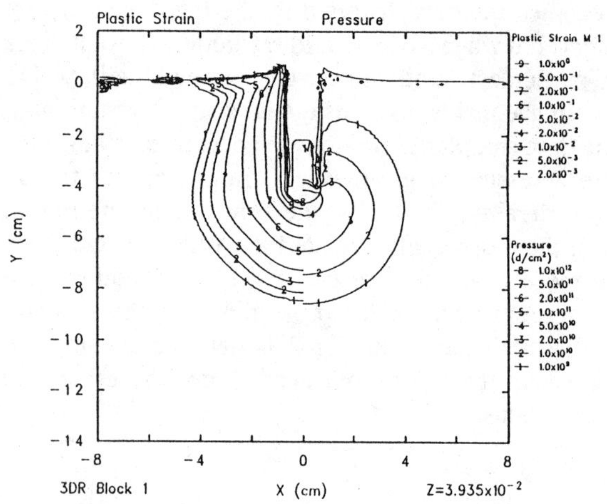

FIGURE 4. Pressure and plastic strain contours 70 μs after impact when N = 5.

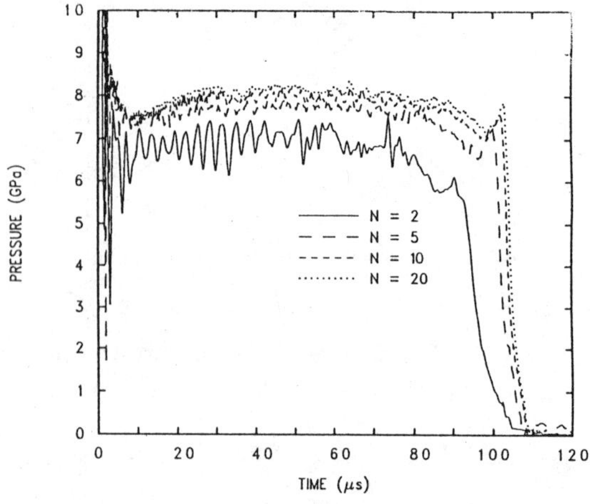

FIGURE 5. Interface pressure vs. time for N = 2, 5, 10 and 20.

Pressure at the projectile/target interface exhibits variations when the zone size is changed. Fig. 5 shows the interface pressure vs time, determined from the 2-D simulations, for N = 2, 5, 10, and 20. At low grid resolutions, the interface pressure is considerably lower than the value obtained in the fine-zone calculations. This lower pressure is caused by the inaccurate resolution of pressure gradients near the interface when N is small. Pressure peaks that normally occur near the interface effectively become

averaged out, thereby lowering the interface pressure. The lower pressure also causes a lower penetration velocity, which ultimately reduces the penetration depth observed in the coarsely zoned simulations.

CONCLUSIONS

The effect of zone size on numerical convergence in 2-D and 3-D Eulerian simulations of long-rod penetration was investigated in the study. It was shown that numerical errors exist in interface reconstruction (and therefore, advection transport) as well as in the calculation of time-dependent variables, particularly where large gradients exist. Results from the simulation indicate that, in general, better numerical resolution is achieved in 3-D simulations than in the 2-D model for a given zone dimension.

REFERENCES

1. McGlaun, J. M., Thompson, S. L., and Elrick, M. G., "CTH—A three dimensional shock wave physics code", *Int. J. Impact Engng.*, **10**, 351-360 (1990).
2. Littlefield, D. L., Anderson, C. E. Jr., Partom, Y., and Bless, S. J., "The penetration of steel targets finite in radial extent," submitted for publication (1995).
3. Johnson, G. R., and Cook, W. H., "Fracture characteristics of three metals subjected to various strains, strain rates, temperatures and pressures", *Engng. Fract. Mech.*, **21**(1), 31-48 (1985).
4. Gray, G. T. III, Chen, S. R., Wright, W., and Lopez, M. F.,, "Constitutive Equations for Annealed Metals under Compression at High Strain Rates and High Temperatures," LANL-12669-MS, Los Alamos National Laboratory, Los Alamos, NM, Jan (1994).
5. Hohler, V., and Stilp, A. J., in *A Penetration Mechanics Database*, C. E. Anderson, Jr., B. L. Morris, and D. L. Littlefield, SwRI Report 3593/001, Southwest Research Institute, San Antonio, TX, Jan (1992).

AN ANALYTIC EXPRESSION FOR *P/L* FOR WA LONG RODS INTO ARMOR STEEL

C. E. Anderson, Jr. and J. D. Walker

Materials and Structures Division, Southwest Research Institute, San Antonio, TX 78228

Analytical expressions for normalized penetration of tungsten alloy long rods into semi-infinite armor steel as a function of impact velocity have been analyzed. Adjustable parameters were determined by least-squares regression to experimental data for impact velocities from ~ 0.4 km/s to ~ 4.5 km/s. Further work has been performed to generalize the procedure to account for target hardness and projectile aspect ratio.

INTRODUCTION

Normalized penetration, the depth of penetration P normalized by the initial projectile length L, is nominally linear with impact velocity V in the ordnance velocity range $(1.0 \leq V \leq 1.8 \text{ km/s})$ e.g., see Ref. [1]:

$$\frac{P}{L} = a + bV + c \ln\left(\frac{L}{D}\right) \quad (1)$$

For long-rod projectiles, generally defined as projectiles with aspect ratios $L/D \geq 10$, normalized penetration tends to saturate to a limiting value at high velocities, e.g., at ~ 3.0 km/s for tungsten alloy rods into armor steel [2,3]. At very low impact velocities, $V \leq 0.5$ km/s, the depth of penetration goes to zero. Silsby proposed an expression to express P/L versus V over the entire velocity range [4]

$$\frac{P}{L} = a_1 + a_2V + a_3[1.0 + a_4V + a_5V^2]^{1/2} \quad (2)$$

where the a_i's are found from least-squares expression to the experimental data. Figure 1 shows a compilation of experimental data from various investigators [5] for L/D 20 projectiles; the curve fits for Eqn. (2) is shown in the figure. (Only tungsten alloy projectiles in armor steels are considered in this paper.)

Equation (2) is better behaved at high impact velocities than a polynomial fit to the data (for approximately the same number of curve fit parameters), but Eqn. (2) does not capture the change of curvature observed for low impact velocities. Neither does Eqn. (2) account for other parameters that have an effect on P/L, specifically target strength and projectile aspect ratio. This paper will examine

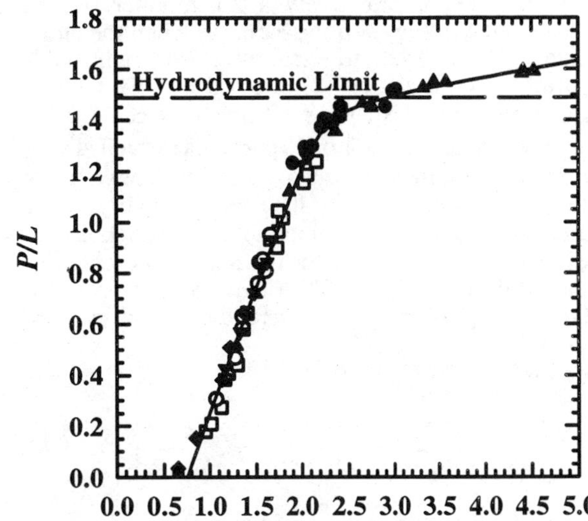

FIGURE 1. *L/D* 20 data with fit to Eqn. (2).

alternative analytical expressions for P/L, and in particular, investigate the dependence of P/L on target strength and projectile aspect ratio, L/D.

GENERALIZED EXPRESSION FOR *P/L*

Analytical expressions have been examined for their suitability to represent P/L as a function of V for long-rod penetration subject to the following constraints: 1) a reasonably good fit to the experimental data; 2) reasonable asymptotic limits for both high and low V; and 3) a minimum number of curve fit constants consistent with the first two requirements. The simplest analytical expressions were selected subject to the constraints above.

An expression that has received considerable interest of late is the modified expression of Lanz and Odermatt [6,7]:

$$\frac{P}{L} = a\sqrt{\rho_p/\rho_t}\ \exp[-2S/\rho_t V^2] \qquad (3)$$

where ρ_p and ρ_t are the densities of the projectile and target, respectively, and S is related to target strength. Rapacki, *et al.* [7], have found that S is proportional to $\mathrm{BHN}^{1/3}$ where BHN is the Brinell hardness. Equation (3) is effectively a 2-parameter fit; the dashed line in Fig. 2 shows Eqn. (2) when the entire data set is used to determine the adjustable parameters. Equation (2) does an excellent job at representing the experimental data between $1.0 \leq V \leq 2.0$ km/s, but it does not do particularly well at very low impact velocities, and overestimates penetration for $V > 2.0$ km/s. If the data set for the regression analysis is limited to points with $V \leq 2.5$ km/s, an excellent fit is obtained for the data, represented by the solid line in Fig. 2. The fitted standard error σ is 0.067 and 0.042, respectively, for the dashed and solid lines.

An alternate expression

$$\frac{P}{L} = \frac{a}{1+(V/b)^{-c}} = \frac{a\,V^c}{b^c+V^c} \qquad (4)$$

is shown in Fig. 3. It is seen that Eqn. (4) represents the experimental data over the entire velocity range extremely well. As $V \to 0$, $P/L \to 0$ smoothly; and as $V \to \infty$, $P/L \to a$. The fitted standard error is 0.0388 for Eqn. (4) against the *L/D* 20 data in Fig. 3. Figures 4 and 5 show data for *L/D* 10 and 30, respectively, along with the fit to Eqn. (4) (σ is 0.0327 and 0.0658, respectively, for the two regression fits). It is observed that the high velocity asymptote, a, increases for decreasing *L/D*, which is consistent with the observations of Ref. [10]. Also, although the exponent c does change, there is a consistent trend with the parameter b, which increases with increasing *L/D*. For the same impact velocity, *P/L* decreases with increasing *L/D*.

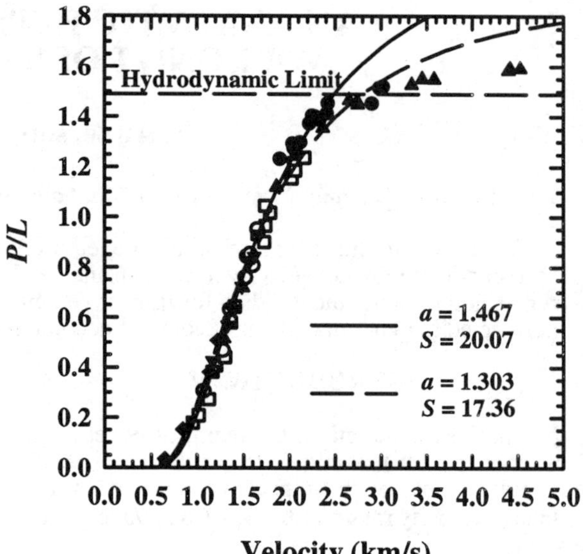

FIGURE 2. *L/D* 20 data with fits to Eqn. (3).

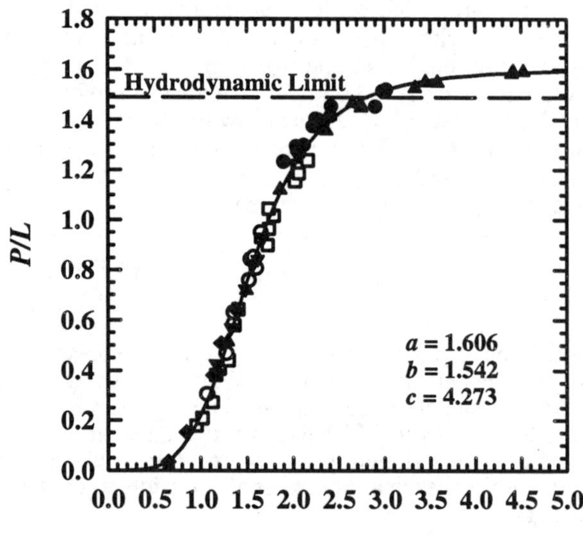

FIGURE 3. *L/D* 20 data with fit to Eqn. (4).

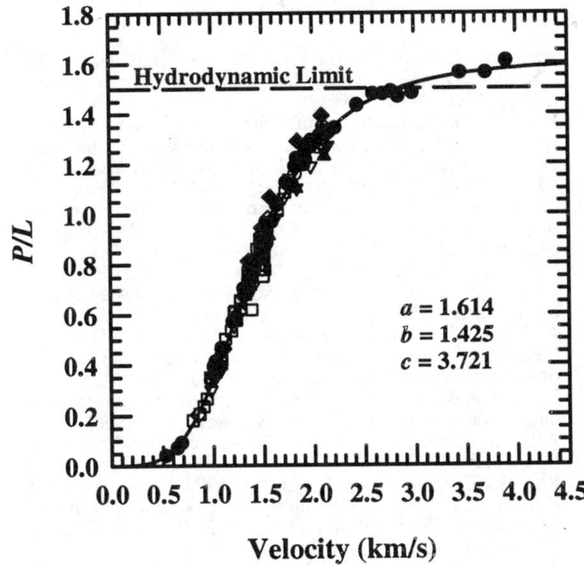

FIGURE 4. *L/D* 10 data with fit to Eqn. (4).

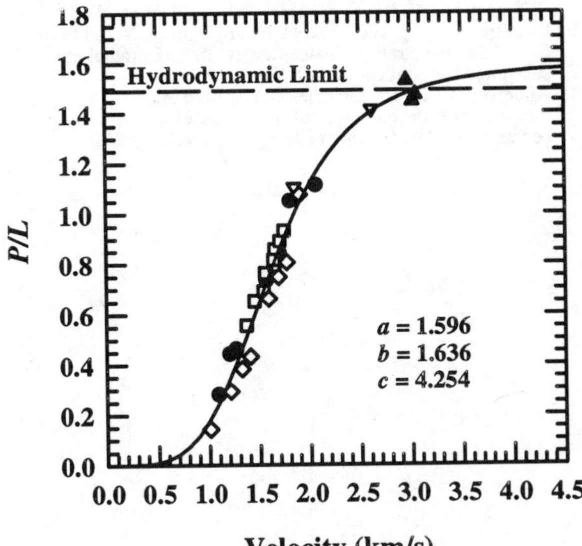

FIGURE 5. *L/D* 30 data with fit to Eqn. (4).

Equation (4) has a relatively linear region during the rapid increase in velocity due to the fact that the maximum slope occurs in that range (thus, the second derivative is zero and a linear expansion about that point is good to third order). The velocity where the maximum slope occurs is

$$V = \left(\frac{c-1}{c+1}\right)^{1/c} b \qquad (5)$$

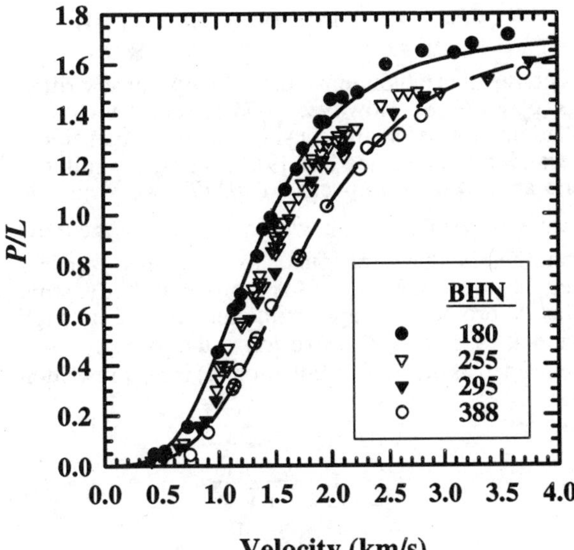

FIGURE 6. *L/D* 10 data with fit to Eqn. (5) for target hardness.

For large c, the maximum slope occurs at roughly $V = b$ where $P/L = a/2$. In this sense, b, a velocity, is located near the center of the region where the velocity increases linearly. The maximum slope is

$$\frac{d(P/L)}{dV} = \frac{a}{4bc}(c+1)^{1+1/c}(c-1)^{1-1/c} \qquad (6)$$

For large c, this approaches the value of the slope at $V = b$, namely $ac/4b$. This allows an estimate of the size of the domain over which the rapid increase in velocity occurs. If the linear approximation is used around the point $V = b$, then the velocity required for P/L to go from 20% to 80% of a is given by $2.4b/ac$. Thus, a larger value for c decreases the size of the velocity region where the depth of penetration undergoes most increase (in other words, a larger c leads to a steeper slope). Thus, the three parameters in Eqn. (4) have been given physical descriptions: a is the high velocity P/L limit; b is roughly the center of the velocity increase region; and c determines the slope of the P/L vs. velocity curve in the region where the velocity increase is greatest.

STRENGTH EFFECTS

P/L depends upon target strength, and it is desirable to rewrite Eqn. (4) to account for strength effects. The target yield strength Y_t is correlated to BHN through the expression:

$$Y_t(GPa) \;=\; 3.483x10^{-3} \cdot (BHN - 11.24).$$

A convenient nondimensional velocity can be written as $(\rho_p V^2/Y_t)$ [8]. However, good correlation with the data cannot be obtained using this nondimensional term, for example, see Ref. [9]. Therefore, similar to the approach by Rapacki, *et al.* [7], we write the strength term as $b_1 Y_t^{b_2}$, where b_1 and b_2 are now parameters that are found from a least squares regression to the data. Also, since hydrodynamic theory provides an approximation for P/L at high velocities, namely $P/L \approx (\rho_p/\rho_t)^{1/2}$, the parameter a is redefined as $a(\rho_p/\rho_t)^{1/2}$. Equation (4) is now rewritten as:

$$\frac{P}{L} \;=\; \frac{a\sqrt{\rho_p/\rho_t}}{1+\overline{V}^{-c}} \tag{7}$$

$$\text{where } \overline{V} \;=\; \frac{\rho_p V^2}{b_1 Y_t^{b_2}}$$

The least-squares regression analysis was performed to determine the fit parameters: $a = 1.168$, $b_1 = 43.47$, $b_2 = 0.677$, and $c = 1.72$, where Y_t is in GPa and V is in km/s. The regression correlation coefficient, r^2, is 0.9955; the standard error is 0.0323. Figure 6 shows data from Hohler and Stilp for L/D 10 projectiles into targets of different strengths. The solid and dashed lines were computed from Eqn. (7) using the above coefficients for BHN hardnesses of 180 and 388, respectively. The curve passes through the data quite well.

SUMMARY

Equation (4) provides a remarkably good fit to the experimental data over the entire velocity range of long-rod projectiles. Physical descriptions were given for the three parameters a, b, and c, in the fit. It was observed that the parameter b depends on both L/D and target hardness. Equation (4) was modified slightly, i.e., Eqn. (7), to account for target hardness, although at this stage, the approach is completely empirical. However, a generalized expression that expresses P/L as a function of both L/D and target strength, subject to the constraints described earlier in the paper, remains to be found.

REFERENCES

1. Anderson, C. E., Jr., Walker, J. D., Bless, S. J., and Partom, Y., "On the L/D effect for long-rod penetration," *Int. J. Impact Engng.*, **18**(1), 1-18, (1996).
2. Hohler, V., and Stilp, A. J., "Influence of the length-to-diameter ratio in the range from 1 to 32 on the penetration performance of rod projectiles," *Proc. 8th Int. Symp. Ballistics*, pp. IB13-IB19, Orlando, FL, Oct. 23-25 (1984).
3. Silsby, G. F., "Penetration of semi-infinite steel targets by tungsten long rods at 1.3 to 4.5 km/s," *Proc. 8th Int. Symp. on Ballistics*, Vol. II, pp. TB31-36, Orlando, FL, 23-25 Oct (1984).
4. Sorensen, B. R., Kimsey, K. D., Silsby, G. F., Scheffler, D. R., Sherrick, T. M., and deRosset, W. S., "High velocity penetration of steel targets," *Int. J. Impact Engng.*, **11**(1), 107-119 (1991).
5. Anderson, C. E., Jr., Morris, B. L., and Littlefield, D. L., "A Penetration Mechanics Database," SwRI Report 3593/001, Southwest Research Institute, San Antonio, TX (1992).
6. Lanz, W. and Odermatt, W., "Penetration limits of conventional large caliber anti tank guns/kinetic energy projectile," *Proc. 13th Int. Symp. Ballistics*, Vol. 3, pp. 225-233, Stockholm, Sweden, June 1-3 (1992).
7. Rapacki, E. J., Jr., Frank, K., Leavy, R. B., Keele, M. J., and Prifti, J. J., "Armor steel hardness influence on kinetic energy penetration," *Proc. 15th Int. Symp. on Ballistics*, Vol. 1, pp. 323-330, Jerusalem, Israel, 21-24 May (1995).
8. Anderson, C. E., Jr., Littlefield, D. L., Blaylock, N. W., Bless, S. J. and Subramanian, R., "The penetration performance of short L/D projectiles," *High-Pressure Science and Technology—1993*, (Edited by S. C. Schmidt, J. W. Shaner, G. A. Smara, and M. Ross), pp. 1809-1812, AIP Press, NY (1994).
9. Littlefield, D. L., Anderson, C. E., Jr., Partom, Y., and Bless, S. J., "The penetration of steel targets finite in radial extent," submitted for publication (1995).
10. Anderson, C. E., Jr., Littlefield, D. L., and Walker, J. D., "On the velocity dependence of the L/D effect for long-rod penetrators," *Int. J. Impact Engng.*, **17**, 13-24 (1995).

MODELING ARMOR THAT USES INTERFACE DEFEAT

James Dehn

ARL, Weapons Technology Directorate, Aberdeen Proving Ground, MD 21005-5066

It is possible to erode completely certain projectiles at the face of a ceramic block, provided the ceramic is confined suitably and is considerably stronger than the projectile. Lower projectile speeds, densities and length-to-diameter ratios favor this phenomenon. Heavy ceramic confinement and the use of shock attenuators also favor projectile interface defeat. In this paper we report on our use of the wave code HULL to study some of the factors which govern this phenomenon.

INTRODUCTION

Because ceramics have high hardnesses and low densities, they have long been considered as candidate armor materials. However, ceramic armors have consistently performed below expectations. In fact, increasing the thickness of a ceramic does not usually produce a corresponding increase in performance. Because ceramics have high wave speeds and are brittle, both direct and reflected waves damage the target and decrease its resistance to penetration well ahead of the projectile.

It is common experience that a snowball will flatten against a solid wall of almost any material. It is also known that soft copper or lead bullets will splatter against hard ceramic blocks at low striking speeds (variations of the Taylor anvil test). Less well-known, however, is the fact that high-density rods striking at ordnance speeds above 1,500 m/s can behave in a similar manner, provided ceramic targets are suitably confined (ref. 1, 2 and 3). Such projectile threats and their defeat are of potential military interest, and we want to study the factors which control the radial spread of rod erosion products and retard axial penetration. In this paper we describe some computational simulations we have carried out using the HULL code (ref. 4) on a CRAY-2 computer. Our eventual goal is to learn more about the geometric and material thresholds which govern the phenomenon of interface defeat.

Figure 1 shows an arrangement used by Hauver and co-workers (refs. 2 and 3) to defeat a 65 gram L/D=20 tungsten alloy rod striking at 1,600 m/s. The titanium diboride ceramic disk at the center of the target is 5 cm thick and 15 cm in diameter. Its front face is covered by a layer of graphite 2 mm thick and a layer of copper 3 mm thick. The graphite is soft enough to allow radial flow of the eroded rod material, yet (in combination with the copper) is strong enough to maintain high pressure on the face of the ceramic to keep it intact. The rolled homogeneous armor (RHA) steel confinement is shrunk-fit to the ceramic. In addition, there is a 2 cm thick RHA layer between the flow layer and a shock-absorbing plug consisting of alternate layers of aluminum and plastic. This plug is also confined by RHA.

Figure 2 at 200 microseconds after impact (after all motion has ceased) shows most of the rod spread over the interface. Figure 3 shows the velocities of the rod nose and tail versus time. The lower curve shows an intial drop in velocity as the nose strikes the attenuator plug, a second drop as it encounters the RHA, a slight increase as the soft flow layer is reached, followed by a decrease to zero at the ceramic front face. The dwell time, during which the rod flows radially, lasts from 70 to 140 microseconds when all motion ceases. The upper curve gives the time history of the rod tail velocity. Figure 4 shows the positions of the rod nose (upper curve) and tail (lower curve). The maximum depth of penetration of the nose is almost 7 cm . Since the ceramic front face was originally located at 6.6 cm, a small excavation 4 mm deep is left in the ceramic. The final position of the nose is about even with the original position of the ceramic front face.

Figure 5 shows the same arrangement as Figure 1 except for the removal of the attenuator plug. At 200 microseconds (in Figure 6) we see a combination of radial flow and penetration into the ceramic. Figure 7 shows the velocities of rod nose and tail versus time. In particular, we note that the nose spreads radially during a zero-velocity dwell time which lasts from 60 to 100 microseconds, after which penetration begins. Figure 8 shows the positions of rod nose and tail versus time.

SUMMARY

It is possible to defeat high-density rods striking at ordnance velocities by using well-confined ceramics. It is also possible to simulate this phenomenon computationally.

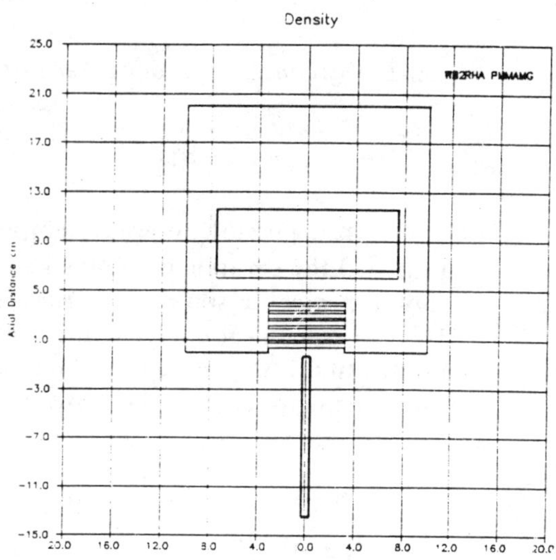

FIGURE 1. Interface defeat target versus 65 gram L/D=20 W alloy rod at 1600 m/s. Time zero.

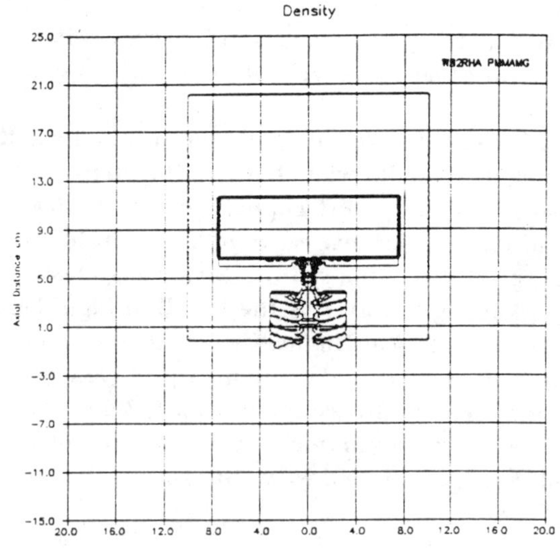

FIGURE 2. Total defeat of rod after 200 microseconds.

REFERENCES

(1) Bless, S. J., M. Benyami, L. S. Apgar, and D. Aylon, "Impenetrable Ceramic Targets Struck by High Velocity Tungsten Long Rods", In *Proceedings of the Second International Conference on Structures under Shock and Impact. (Portsmouth, UK, June 16-18)*, 1992.

(2) Hauver, G. E., P. H. Netherwood, R. F. Benck, and L. J. Kecskes, "Ballistic Performance of Ceramic Targets", In *Proceedings of the Army Symposium on Solid Mechanics. (Plymouth, MA, August 17-19)*, 1993.

(3) Hauver, G. E., P. H. Netherwood, R. F. Benck, and L. J. Kecskes, "Enhanced Ballistic Performance of Ceramics", In *Proceedings of the 19th Army Science Conference. (Orlando, FL, June 20-24)*, 1994.

(4) Matuska, D. A., and R. E. Durrett, AFATL TR 78-125. U. S. Air Force Armament Laboratory, Eglin AFB, FL, 1978.

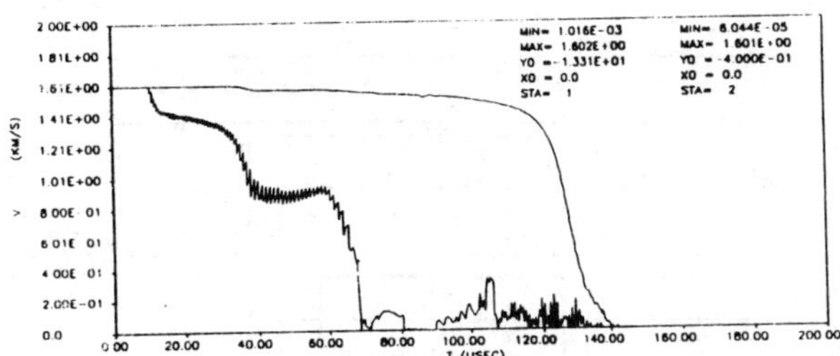

FIGURE 3. Velocities of rod nose and tail versus time.

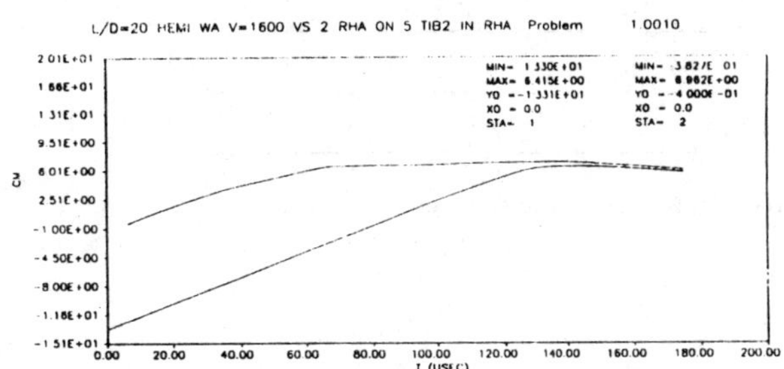

FIGURE 4. Positions of rod nose and tail versus time.

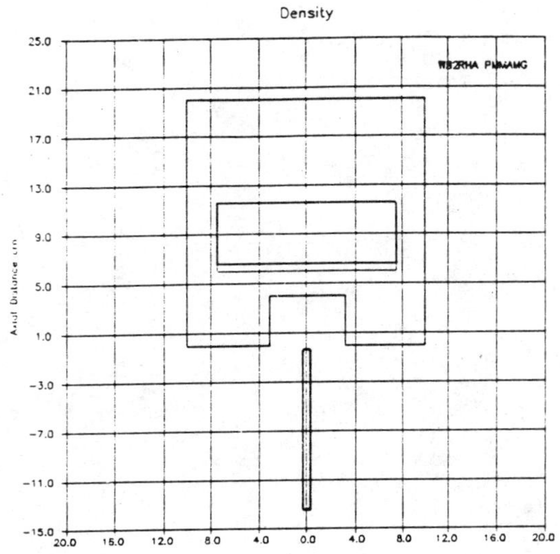

FIGURE 5. Interface defeat target with attenuator plug removed. Time zero.

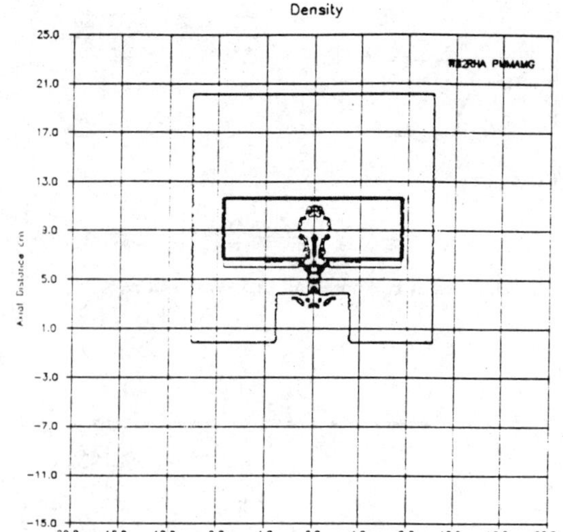

FIGURE 6. Figure 5 200 microseconds later.

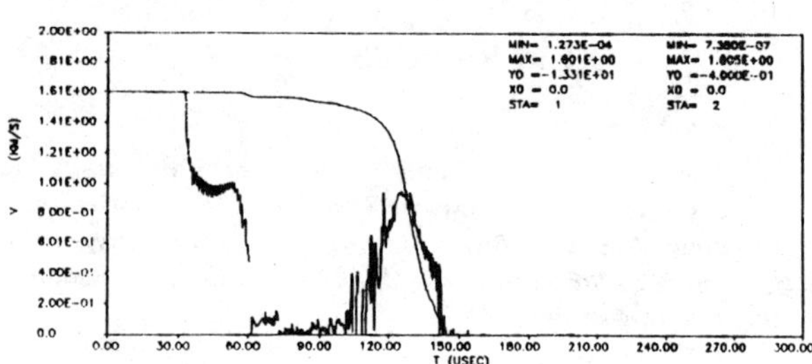

FIGURE 7. Velocities of rod nose and tail versus time for target without attenuator.

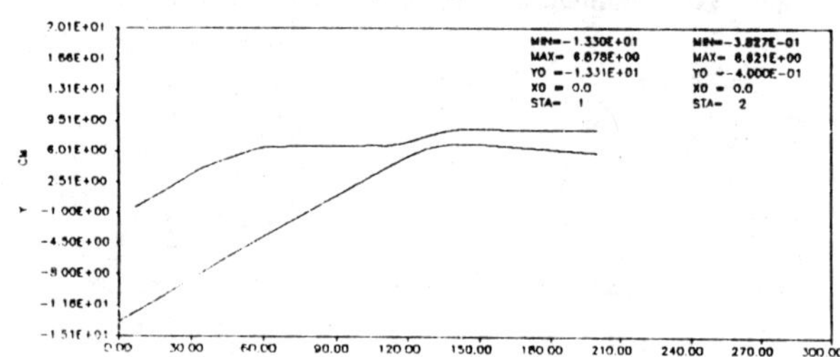

FIGURE 8. Positions of rod nose and tail versus time for target without attenuator.

EFFECTS OF PULVERIZED MATERIAL STRENGTH ON PENETRATION RESISTANCE OF CERAMIC TARGETS

D. J. Grove

University of Dayton Research Institute, Dayton, Ohio, 45469-0182

A. M. Rajendran

Army Research Laboratory, Aberdeen Proving Ground, Maryland 21005-5066

Constitutive/damage model constants are traditionally determined from quasi-static, split Hopkinson bar, and plate impact experimental data. While some constants can be directly determined from these data, others must be calibrated through a trial and error procedure involving numerical simulations of various experimental configurations. In the Rajendran-Grove ceramic failure model, essentially one parameter is used to describe the strength of the pulverized ceramic material. Since the pulverized material strength significantly influences the penetration process in a confined ceramic target, this model parameter can be effectively determined through numerical simulations of ballistic penetration experiments. Using the 1995 version of EPIC, we simulated three penetration experiments in which the thicknesses of the confined ceramic plates were 25.4 mm, 38.1 mm, and 50.8 mm. The Rajendran-Grove model was used to describe the ceramic material's response to the ballistic impact. The pulverized material strength model parameter was adjusted until the numerical simulation results matched reasonably well with the experimental depth-of-penetration measurements. The effect of imposing an upper limit on the pulverized material strength was also investigated.

INTRODUCTION

In the Rajendran-Grove ceramic failure model (1), the parameters used to describe the microcrack evolution process can be determined indirectly using stress and/or particle velocity profiles from ceramic plate impact experiments. Using a shock wave propagation code, each experiment is simulated, and the microcracking constants are adjusted until the numerical results compare well with the experimental data. However, the data from this one dimensional (uniaxial strain) configuration fails to provide information concerning the compressive strength of the pulverized ceramic material. Ballistic penetration experiments, wherein a ceramic target plate is confined laterally and backed by a thick steel plate, provide depth-of-penetration data that can be used to calibrate the model parameter that controls the compressive strength of the pulverized ceramic material. This paper illustrates the pulverized strength model calibration technique and investigates the pulverized material strength of silicon carbide.

CONSTITUTIVE/DAMAGE MODEL

An accompanying paper (2) briefly describes the constitutive relationships in the Rajendran-Grove ceramic failure model, and also discusses how each of the model parameters affects the evolution of microcrack damage.

Pulverized Material Strength

In the Rajendran-Grove model, the comminuted ceramic strength (Y_p) varies with confinement pressure (P). The compressive strength of the pulverized ceramic is described by the following Mohr-Coulomb type relationship:

$$Y_p = \alpha + \beta P , \qquad (1)$$

where α and β are model parameters that control the strength of the pulverized ceramic material. To simplify the calibration scheme, α is generally assumed to be zero; in this case, the post-fracture strength of the ceramic is directly proportional to the applied pressure. Since the one dimensional plate impact experimental configuration is relatively insensitive to β, the calibration of this model parameter requires a series of finite element simulations of ballistic penetration experiments involving laterally confined ceramic target plates. This paper illustrates the calibration of β for silicon carbide.

CALIBRATION OF PULVERIZED MATERIAL STRENGTH PARAMETER β

We implemented the Rajendran-Grove ceramic failure model in the 1995 version of EPIC (3). Using this model to describe the dynamic response of the ceramic material, we simulated a ballistic penetration configuration involving a laterally confined ceramic target plate backed by a thick steel plate; experimental depth-of-penetration measurements were available for three different ceramic plate thicknesses. Assuming $\alpha = 0$, we adjusted the value of β until the simulated depths-of-penetration all matched reasonably well with those measured in the experiments.

Experimental Configuration

The penetration experiment configuration consisted of a square silicon carbide tile confined laterally by a steel frame. The entire target assembly was mechanically clamped to a thick steel backup plate. The ceramic tiles were 152.4 mm wide, and three different tile thicknesses were investigated: 25.4, 38.1, and 50.8 mm. The projectile was a tungsten rod, 79 mm long and 7.9 mm in diameter (L/D = 10), with an impact velocity of 1500 m/s. The experimental data consisted of the rod's depth-of-penetration into the backup plate.

We used the 2D axisymmetric option in EPIC to approximate the above geometry. With this assumption, the target plates and confinement frame were modeled as being circular instead of square. In the simulations, the Rajendran-Grove model described the dynamic response of the ceramic, while standard EPIC material models were used to describe the behavior of the tungsten rod and the steel backup plate. Figure 1 shows the configuration for the 25.4 mm thick ceramic tile.

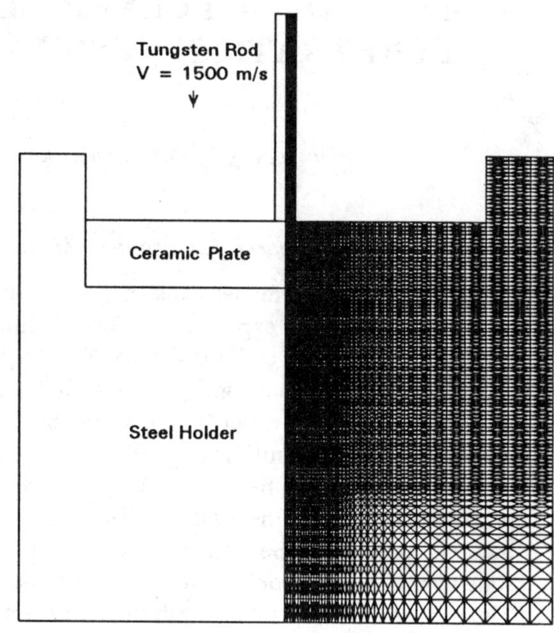

FIGURE 1. EPIC finite element grid for penetration configuration (with 25.4 mm thick ceramic tile), used to calibrate the pulverized material strength parameter β.

Calibration Results

The Rajendran-Grove model employs a generalized Griffith criterion to initiate microcrack growth under both tensile and compressive stress states. In this model, the crack density (a state variable) increases as the microcracks extend. When the crack density reaches a critical value, the damaged ceramic material is assumed to pulverize (under compression). After pulverization, the material has no tensile strength, and its compressive strength is governed by Equation (1).

Based on velocity profiles from high velocity plate impact experiments, the model parameters that characterize the growth of microcracks had previously been determined for silicon carbide (4). Unfortunately, as reference (2) indicates, the pulverized strength parameter β cannot be determined from plate impact experimental data. We calibrated β by adjusting it until the simulated depths-of-penetration matched the experimental results. It is important to note that the microcrack evolution parameters can also significantly affect the ceramic target's simulated penetration resistance. For this study, however, we employed the values of those parameters that had been determined from the plate impact data.

For $\beta = 1$ ($Y_p = P$), the calculated penetration depths were significantly less than the experimental measurements. With no upper bound on the strength of the pulverized material, its penetration resistance was apparently unrealistically high. Reducing β to a value of 0.6 resulted in very good agreement between the simulated and measured depth-of-penetration data, as shown in Figs. 2-4. The dashed lines in these figures indicate the experimentally measured depths-of-penetration.

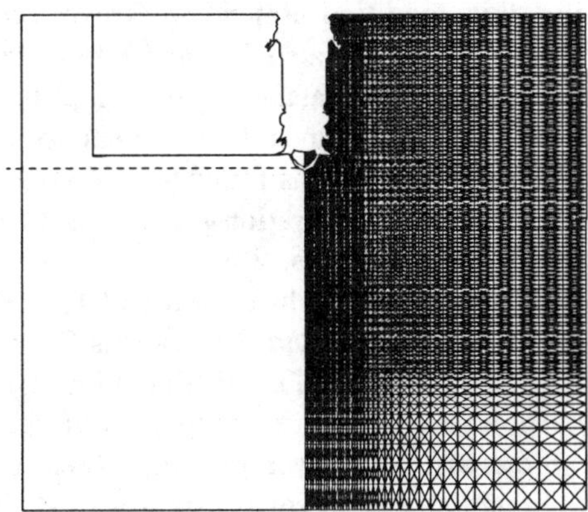

FIGURE 4. Comparison of calculated depth-of-penetration with experimental measurement for 50.8 mm ceramic tile ($\beta = 0.6$).

EFFECT OF IMPOSING UPPER LIMIT ON STRENGTH OF PULVERIZED MATERIAL

When Equation (1) is used to describe the compressive strength of the pulverized ceramic material, the strength is permitted to increase (without bound) as the pressure increases. Such behavior may be reasonable for lower pressures. However, for very high velocity penetration configurations, the pressures at the projectile/target interface can become extremely high. Intuitively, there must be a limiting value for the compressive strength of the pulverized ceramic; certainly it must be lower than that of the intact ceramic material. Accordingly, the Johnson-Holmquist constitutive model for brittle materials (5) defines a strain rate dependent maximum strength for the fractured material.

To investigate the effect of imposing an upper limit on the strength of the pulverized ceramic material, we implemented the following simple expression in the Rajendran-Grove model to describe the compressive strength of the post-fractured material:

$$Y_p = \min\left[Y_{max}, \; (\lambda P) \right], \qquad (2)$$

where Y_{max} and λ are model parameters. Y_{max} is a constant that defines the upper limit on the strength of the pulverized material, and λ is the slope of the linear strength-pressure relationship below Y_{max}.

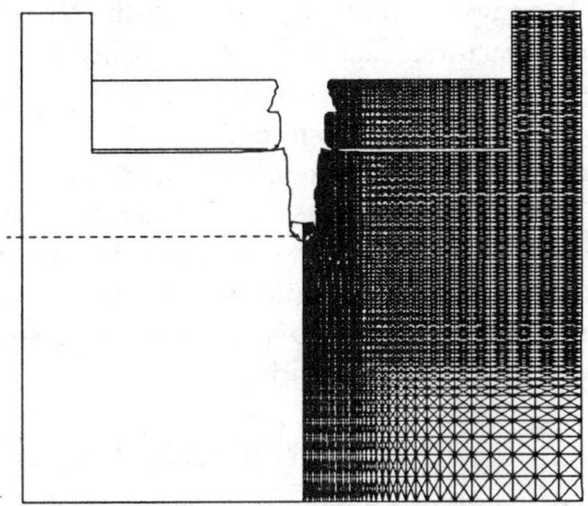

FIGURE 2. Comparison of calculated depth-of-penetration with experimental measurement for 25.4 mm ceramic tile ($\beta = 0.6$).

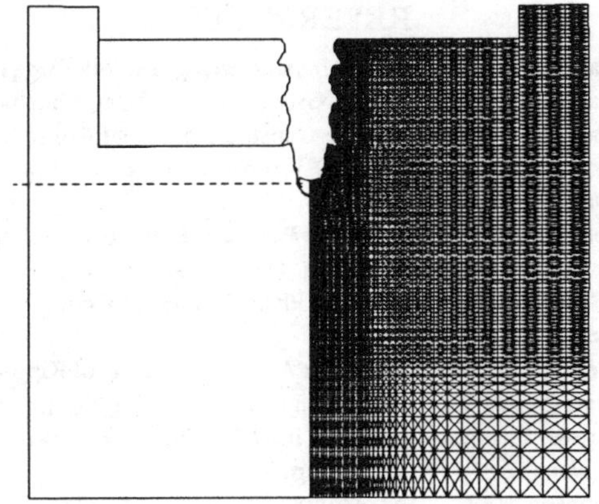

FIGURE 3. Comparison of calculated depth-of-penetration with experimental measurement for 38.1 mm ceramic tile ($\beta = 0.6$).

Using this modified Rajendran-Grove ceramic failure model, we calibrated Y_{max} and λ to match the depth-of-penetration data. To simplify this procedure, λ was assumed to be equal to 1; the optimum value for Y_{max} was found to be 4 GPa. The computed depths-of-penetration were almost identical to those illustrated in Figs. 2-4.

Using the calibrated values for β and Y_{max}, Fig. 5 compares Equations (1) and (2). In this figure, the dashed (Equation (1)) and solid (Equation (2)) lines indicate how strength varies with pressure for the two expressions. At lower pressures (below 6.67 GPa), Equation (1) predicts significantly lower strength for the confined ceramic powder. Beyond 6.67 GPA, the expression predicts continuously increasing strength with increasing pressure; this means that the material's penetration resistance will also increase indefinitely with the pressure. Equation (2), on the other hand, predicts higher strength for the pulverized material at lower pressures, and then imposes a limit on the strength at higher pressures. This strength model, which also limits the penetration resistance of the confined ceramic powder, seems more appropriate for describing the penetration process in a confined ceramic target.

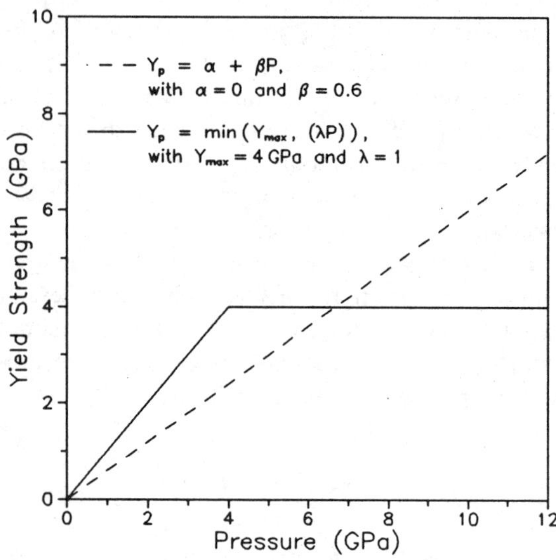

FIGURE 5. Comparison of pulverized material strength as a function of pressure between Equation (1) ($\alpha = 0$ and $\beta = 0.6$) and Equation (2) ($\lambda = 1$ and $Y_{max} = 4$ GPa).

SUMMARY AND RECOMMENDATIONS

The Rajendran-Grove ceramic failure model was implemented in the 1995 version of EPIC. The modified EPIC code was subsequently used to simulate an experimental configuration consisting of a long tungsten rod penetrating a silicon carbide ceramic target plate that was laterally confined and backed by a massive steel holder.

Employing previously determined model parameters to describe the evolution of microcrack damage in the ceramic, the model's pulverized strength parameter was calibrated to match the experimentally measured depth-of-penetration data for three different ceramic target thicknesses.

The pulverized strength equation was revised to include an upper limit, or "cap", on the strength of the post-fractured material. After calibration, this model predicted penetration depths nearly identical to those of the original strength model.

Other penetration configurations should be simulated to investigate the generality of the proposed Rajendran-Grove ceramic failure model constants. Higher penetration velocities would be useful in confirming the upper limit on the strength of the post-fractured material.

ACKNOWLEDGMENTS

The authors gratefully acknowledge the experimental depth-of-penetration data supplied by Pat Woolsey of ARL and the funding support of Dr. James Thompson of TACOM in Warren, MI.

REFERENCES

1. Rajendran, A.M., *Int. J. Impact Engng* **15**, 749-768 (1994).
2. Rajendran, A.M., and Grove, D.J., "Determination of Rajendran-Grove Ceramic Constitutive Model Constants," in *Proceedings of the 1995 APS Shock Compression Conference (this proceedings)*, 1995.
3. Johnson, G.R., Stryk, R.A., Petersen, E.H., Holmquist, T.J., Schonhardt, J.A., and Burns, C.R., "User Instructions for the 1995 Version of the EPIC Research Code," Alliant Techsystems Inc., 1994.
4. Grove, D.J., and Rajendran, A.M., "Modeling of Kipp-Grady Plate Impact Experiments on Ceramics Using the Rajendran-Grove Ceramic Model," in *High-Pressure Science and Technology - 1993*, 1994, pp. 749-752.
5. Johnson, G.R., and Holmquist, T.J., "A Computational Constitutive Model for Brittle Materials Subjected to Large Strains, High Strain Rates, and High Pressures," in *Shock Wave and High Strain-Rate Phenomena in Materials*, 1992.

FRAGMENTATION HISTORY OF GLASS PROJECTILES IMPACTING ALUMINUM TARGETS AT 0.5 TO 1.5 KM/S

Ronald P. Bernhard[1] and Friedrich Hörz[2]

[1]*Lockheed Martin, 2400 NASA Road 1, Houston, Texas 77058*
[2]*NASA - Johnson Space Center, Houston, Texas 77058*

This report expands on a subset of low velocity (0.5 to 1.5 km/s) cratering experiments from a previous study (Bernhard and Hörz, 1994) that employed spherical soda lime glass projectiles of 3.2 mm diameter at impact velocities from 1- 7 km/s. This earlier study presented detailed crater morphologies as a function of impact speed and addressed the spatial distribution and progressive melting of projectile materials residing inside the craters. At velocities between 1 and 1.5 km/s a brittle deformation of projectile remnants were observed, precipitating additional experiments at V < 1 km/s. The purpose of the present report is to describe these additional experiments and to summarize the deformation of glass impactors at 0.5 to 1.5 km/s. Two dramatically different fragmentation regimes seem evident, one a large, relatively unfractured, substantially coherent core of glass, surrounded by the second regime, an intensely fractured annulus of fine-grained material. The annulus material displays substantial structure, as distinct planes of shear and intense comminution alternate with comminution products. Plastic deformation of both the aluminum target and the glass projectile is confined to a very thin liner at the immediate target/projectile contact. Some of these deformations are somewhat consistent with current models describing the shock-isobar distribution and associated collisional fragmentation history of spherical impactors, yet important details are not currently part of these models.

INTRODUCTION

Inspection of spacecraft hardware exposed to the natural meteoriod and man-made orbital debris particulate environment in low Earth orbit (LEO) has led to the documentation of tens of thousands of hypervelocity impacts into aluminum surfaces (refs. 1 and 2). These craters and penetrations range from sub-micron size to several millimeters in diameter and many contain residue from the colliding projectile (ref. 3). We are conducting systematic cratering and penetration studies over a wide range of impact conditions (impactor size, velocity, impact angle, target and projectile materials) for an improved understanding of the dimensional relationships between the projectile and the resulting craters or penetration holes. Also, the detailed investigation of physical distribution and chemical alterations of the projectile residues that are associated with impact feature were conducted.

To this effect we have previously reported on a series of cratering experiments in aluminum 1100 targets using 3.2 mm diameter soda-lime glass spheres at velocities (V) between 1 and 7 km/s (ref. 4). It was observered that under experimental conditions where $V < 2$ km/s, large amounts of deformed and comminuted impactor materials where produced inside the craters. The present report elaborates on these craters and additional experiments conducted at $V < 1.5$ km/s. These impacts are characterized by shock-stresses that fall well below the solid/liquid phase transition of both the projectile and target material. The purpose of this study is to present observations regarding the solid state deformation of the spherical soda lime glass projectiles at relatively modest shock stresses.

RESULTS

An overview of some crater cross-sections is seen in Fig 1. The quantity of projectile remnants detected in the crater correlates with velocity. Note the lack of any projectile material retained in the crater bottoms at $V = 0.60$ km/s. Total projectile mass residing in the craters increases at $V > 0.7$ km/s to form substantial, highly comminuted plugs at 0.83 and 1.23 km/s, the subject of this report. The present experiments are characterized by a distinct maximum in retained impactor mass at $0.8 < V < 1.2$ km/s. The lack of projectile at the very low velocities is attributed to elastic rebound, while increasingly higher, shock-induced particle velocities cause progressively larger fractions of the impactor to be ejected from the growing crater cavity at $V > 1.3$ km/s.

At the lower velocities ($V < 0.78$) the projectile is fractured and the majority of the glass is then ejected from the crater. The ejecta maintains a coherent mass of fractured glass, possessing larger grains of less fractured glass in the central portion. The resulting crater is shallow and retains only small amounts (< 1 % of initial projectile mass) of finely powdered glass. The mass of unmelted, highly fragmented projectile material which occupies the crater at velocities between 0.8 and 1.2 km/s are similar to the glass ejected at lower velocities.

As illustrated in Fig 2 at 1.1 km/s, the grain-size distribution of the fractured silicate projectile is distinctly bimodal. A relatively intact, central core is surrounded by a very fine-grained annulus characterizing the peripheral parts of the spherical impactor. The contact between these fragmentation regimes is sharp, suggesting that dramatically different mechanisms of material failure and / or stress histories are manifested during impact.

The central core is characterized by blocky to platy fragments produced by horizontal fractures and display very little internal deformation. The core width remains somewhat constant, and is approximately 20% of the initial projectile diameter (slightly less at the bottom). The fracture spacing in the glass core decreases with proximity to the projectile/target interface, eventually resulting in increasingly thin slivers of stacked glass plates.

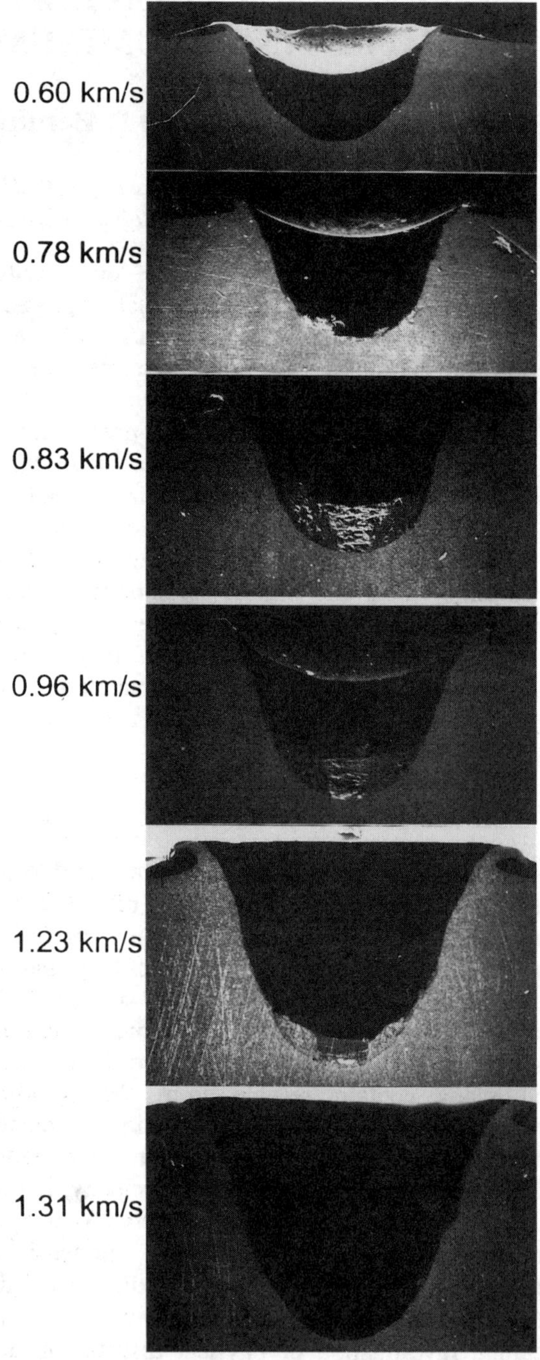

FIGURE 1. Cross-sectional view of representative craters from the low velocity impact series. The 0.6 km/s impact reveals almost no retained projectile material. As the velocity increases, the crater size increases, and the projectile residue retained becomes much more evident until a velocity of about 1.30 km/s.

The central core is surrounded by an annulus of distinctly more fine-grained material. The contact

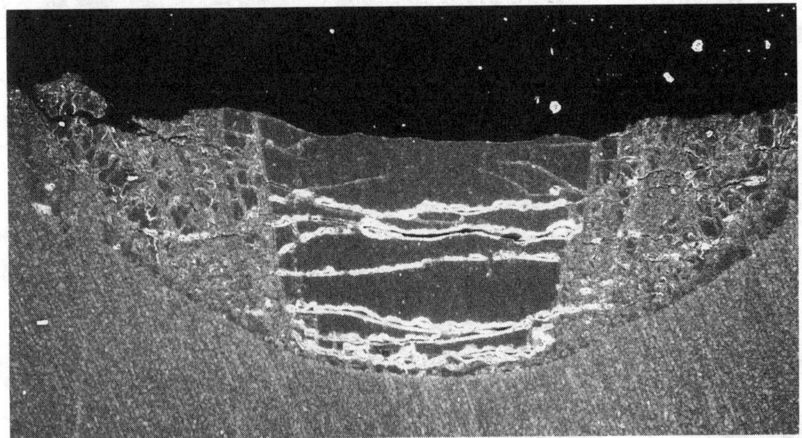

FIGURE 2. This SEM micrograph depicts the projectile residues from the 1.11 km/sec impact crater. The predominately unbroken central plug is 550 microns tall by 680 microns in diameter.

between these two grain size regimes is sharp and lacks any gradational transition. In addition, the fine grained annulus itself displays considerable structure. Highly linear zones of intensely comminuted material alternate with massive comminution products. We interpret these linear zones of intense comminution as shear bands. They parallel the vertical contact with the coarse core initially, and become increasingly curved with radial distance from the core contact to ultimately parallel the curvature of the crater walls. The term shear bands introduced above is largely a descriptive term and does not necessarily imply in situ comminution by shear.

The top surface of the residue is a newly created fracture surface. Total volume represented by the entire projectile plug is < 20% of the initial impactor (M_0) for this 1.1 km/s experiment, and the central core alone is only $\approx$ 0.02 M_0. Most of the projectile was lost by ejection.

As illustrated by Fig 3 at lower velocity (0.78 km/s), these shear bands seem to delineate zones of

discrete relative motion between individual (large) fragments. The abundance of horizontal fracture lines extending from the core into the annulus seem to be equivalents of and initially part of the internally undeformed, platy core fragments. The annulus blocks appear vertically offset from the core plates along fractures that are filled with highly comminuted material. These bands appear to be a product of downward movement of the unfractured central core through the fractured matrix. During collision the forward movement of the unfractured projectile pushes through the shattered glass producing the shear planes and zoning. With increasing velocity (higher shock stress) the comminution of the annulus increases until shear bands are no longer apparent and intense mixing occurs. Conversely, at lower velocities the fracturing of the annulus is subdued as seen in the 0.78 km/s impact crater. Examination of the central core also reveals that the central glass core remains relatively unbroken and the surrounding glass annulus is fractured along the vertical shear bands.

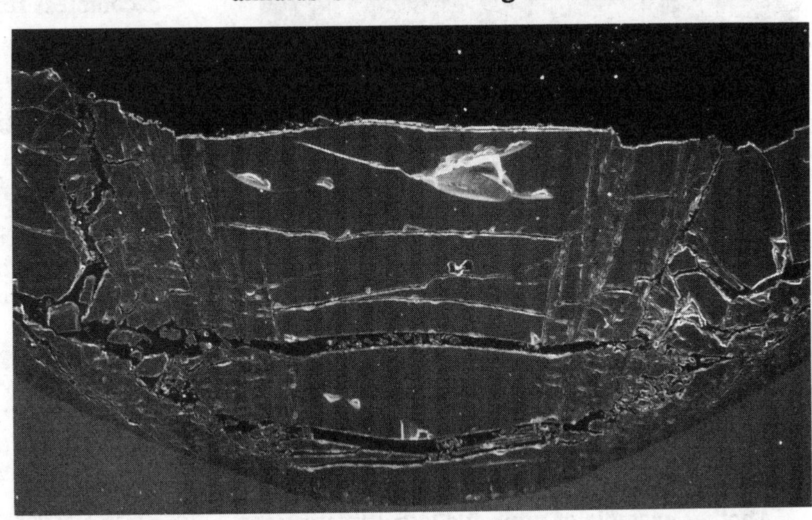

FIGURE 3. Projectile residue seen in the 0.78 km/s impact crater has a less fractured annulus around the central glass core. The shear zones are easily detected, although seperation between the central plug and the surrounding projectile matrix is not as evident as in the figure 2 crater.

Interaction between the glass projectile and the aluminum target was also examined. Higher magnification analysis of the contact zone reveals that in most cases there is a distinct separation among the projectile and target materials. During collision large scale plastic deformation is taking place within the aluminum target, considering the presence of well developed crater rims, combined with our earlier measurements (ref. 4) that show crater volumes to be on the order of 10 times larger at $V < 2$ km/s than the physically ejected and dislodged volume, the latter deduced from target mass-measurements before and after the event.

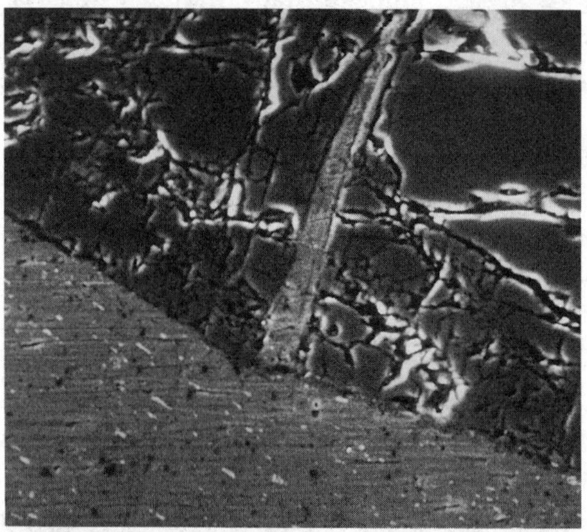

FIGURE 4. The aluminum target is plastically deformed and pushed into a fracture in the glass projectile at 1.07 km/s.

Figures 4 and 5 illustrate evidence for plastic deformation and interactions between impactor and the evolving crater walls. Notice the lighter colored aluminum which has been injected into the darker colored glass projectile at 1.07 km/s. Not all fractures are invaded by aluminum, implying that fractures in the silicate material are produced over some finite time interval, some while the aluminum was still plastic, others at later stages. Also very small (white) grains of iron which are present in the pristine aluminum alloy have become elongated, with their long axis parallel to the crater wall.

Another example of projectile/target interactions is seen in Fig 5. Most fractures display a modest curvature, indicating differential movement by plastic flow of a thin liner of silicate material that is

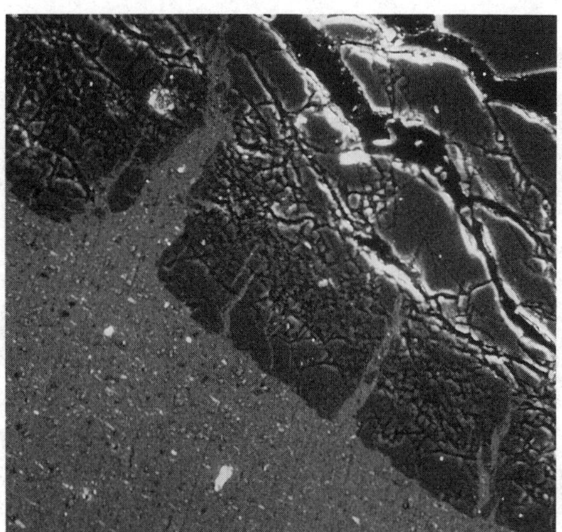

FIGURE 5. Aluminum invading the fractured glass at 0.78 km/s.

close to the target/projectile. The aluminum-filled micro-cracks have widely variable widths, yet similar lengths, none extending into the projectile region where high crack densities were produced by brittle failure.

DISCUSSION

These studies produce contours of shock isobars that are strongly affected by the projectile's free surfaces, which initiate rarefaction waves that decompress the projectile. However, neither shock-isobars per se, nor the tensile stress regimes set up close to the projectile surface reflect how the projectile will ultimately fail (Ref 5). Information on the stress-distribution in the form of shock-isobar contours in spherical impactors as they encounter and penetrate infinite half-space targets. The sharp, *vertical* contact between core and annulus cut across classical stress contours, as do the large, *horizontal* fracture systems typical of the core plates. The distinctly bimodal distribution of comminution products can be understood by the movement of unfractured glass core through the fractured annulus. Nevertheless, the observations presented are difficult to understand in detail from existing models of the stress distribution in shocked spheres.

REFERENCES

1. See, T. et al., *JSC Report 24608*, 1990, p. 561.
2. Levine, A. E. ed. *NASA CP 3194*, 1993, p. 1572.
3. Horz et al.), *NASA CP 3194*, 1993, pp. 415-430
4. Bernhard, R.P. and Horz, F., *Int. J. of Impact Engng.*, in press.
5. O'Keefe, J. et al, *Proc. Lunar Sci. Conf.*, 1975 pp. 2831-2844.

ANALYSIS OF MICROSTRUCTURES AND PROJECTILE RESIDUES IN HYPERVELOCITY IMPACTS ON FUSED SILICA GLASS

Ronald P. Bernhard[1] and Eric Christiansen[2]

[1] *Lockheed Martin, 2400 NASA Road 1, Houston, Texas 77058*
[2] *NASA - Johnson Space Center, Houston, Texas 77058*

This study was conducted to determine the origin of hypervelocity projectiles that have collided with space shuttle window surfaces while exposed to the low Earth orbit (LEO) environment. Findings from these and other investigations (e.g., Solar Max, PALAPA, LDEF) determined that not all on orbit impacts are meteoritic in nature, but some damage is caused by collision with man-made orbital debris. To date, 49 shuttle windows have been replaced due to impact damage features. Many of these impacts were examined by high magnification scanning electron microscopy (SEM) to determine the morphologic structure of the damage and energy dispersive X-ray analysis (EDXA) to determine the chemical composition of the projectile residues retained in the fused silica glass craters. The morphology of the crater and chemistry of the projectile can be used to estimate relative velocity of impact, approximate angle of impact, size of particle, and sometimes density of projectile..

INTRODUCTION

Spacecraft windows have been a source for determining the flux of hypervelocity microparticles in low Earth orbit (LEO) for many decades. In the 1960's fourteen Gemini windows were examined by H. Zook (ref. 1) for impacts. Command module windows of the Apollo/Skylab III and IV missions were examined by Cour-Palais (ref. 2) in the 1970's to determine particle flux and were examined by scanning electron microscopy (SEM) and energy dispersive X-ray analysis (EDXA) by Clanton et al (ref. 3) to determine the origin of the particles. Findings from these and other investigations determined that all impacts were not meteoritic in nature, but some damage was caused by man-made orbital debris. Since then investigators have examined many exposed spacecraft surfaces (e.g. Solar Max, PALAPA, LDEF and Orbiters) to further define the small particulate population in LEO. With the completion of LDEF studies relatively few experiments have been flown to maintain proper monitoring of orbital debris & meteoroid (M&OD) particle flux, therefore the space shuttle surfaces are used to collect this data.

Through STS-66 (Nov. 1994) there have been 193 pits observed by optical inspection on the Space Shuttle windows. This damage is a product of collision with orbital debris and meteoroids. Although low velocity impacts from atmospheric debris during launch and landing, and ground damage could also be a source for window damage, such damage creates bruises, dents, and scrapes which differ fundamentally from high-velocity collisions from M&OD. A significant number of these damages were categorized to be severe enough to warrant removal of the window (ref. 4). Through STS-66 there have been 49 windows replaced due to impact features. The frequency of occurrence and degree of damage is closely related to mission operational parameters such as altitude, mission duration, shuttle attitude, and mission date.

To help correlate the damage found on the Orbiters to actual particulate flux in space, hypervelocity impact tests by the Hypervelocity Impact Test Facility (ref. 5) have been made into Orbiter windshields and other shuttle surfaces with varying projectile velocities, densities, and sizes.

Impacts into glass are more complex then other shuttle surfaces. Because of its brittle nature the glass tends to spall, breaking off large areas of glass around the impact area. This property produces large craters with respect to projectile diameter, in impact tests at 7 km/sec a ratio between 40 and 50 to 1 is observed.

The 15 window samples examined in this report were acquired from 10 different STS missions. Table 1 lists the mission and vehicle from which each of the impacts were derived, as well as the window serial number, location, the impact pit size in microns (1000 microns equals 1.0 millimeter), and EDXA results. Each of these impacts have been examined by high magnification optical microscopy to determine the macrostructure of the damage. The morphology of the crater sometimes reveals information about the relative velocity of impact, approximate angle of impact, approximate size of particle, and sometimes density of projectile. Morphologies of interest are: symmetry of the crater or spall; location of crater pit within the spall zone; amount of pulverization which takes place; and the size ratio between spall and pit diameters, or whether there is any pit retained at all. These characteristics in fused silica glass are the basic criteria used when distinguishing in-flight impacts from low energy collisions with low velocity objects. The damage sites on the replaced windows were cored and analyzed by SEM/EDXA to determine the chemical composition of the projectile. Because of the brittle nature of the fused glass, relatively large quantities of glass are ejected upon hypervelocity impact. The displacement of the spall may result in the loss of the central pit area as well as the projectile residues. In these cases the origin of the damage (meteoroid or debris) is very difficult (or even impossible) to determine by methods used. Likewise if the sample has been handled in a manner which contaminates the impact with foreign materials or destroys the existing projectile residues, SEM/EDX analysis determinations are difficult.

Morphological examination of the craters reveals that a wide range of physical characteristics are produced during hypervelocity collision. This is consistent with the impact conditions one would expect to encounter from LEO particles. Impact velocity and trajectory, as well as particle size and density vary dramatically in this environment. Spacecraft attitude and altitude during the shuttle missions also influence the number of impact events which take place on any given surface of the orbiter. The following SEM micrographs illustrate the typical and non-typical structures detected in this study. All of the impacts are caused by hypervelocity particles less than one millimeter in diameter and none of the impact damages completely penetrated the fused silica glass pane.

TABLE 1. Window impact samples used in this study taken from the space shuttle windshields.

Mission (Flight)	Window Serial #	Window Location	Crater Size (microns)	Fracture Size (Microns)	EDXA Results
STS-7	112	right middle	470 x 650 x 130 deep	2000 x 1750 x 250 deep	paint
STS-41D	106	left side	1900 x 1600 x 200 deep		unknown
STS-41G	104	right forward	2360 x 2000 x 290 deep		meteoritic
STS-41G	117	left side	1400 x 1270 x 180 deep		unknown
STS-61B	120	left side	635 x 380 x 30 deep		meteoritic
STS-35	122	left side	1800 x1100x 350 deep	3800 x 3600 x 460 deep	unknown
STS-48	132	right side	900 x800x 150 deep	2750 x 1575 x 210 deep	aluminum
STS-44	129-26	right side	600 x 500 x 75 deep	675 x 500 100 deep	stainless
STS-44	129-15	right side	1250 x 700 x 155 deep	1400 x 825x 220 deep	unknown
STS-44	129-1	right side	275 x 260 x 50 deep		meteoritic
STS-50	118	right forward	3300 x 2700 x 570 deep	7200 x 6800 x 600 deep	Ti Metal
STS-50	131	right side	900 x 800 x 80 deep		aluminum
STS-50	106	left overhead	1700 x 1300 x 160 deep		meteoritic
STS-51	134	left side	875 x 340 x 210 deep	900 x 660 x 250 deep	unknown
STS-59	105	side hatch	7800 x 6100 x 565 deep	12000 x 8900 x 1150	paint

The glass windshield impact crater from STS-61B had a diameter of 635 microns, with an underlying internal fracture of about 1.3 mm. Three deep fracture zones which extend to the surface are seen in the SEM micrograph labeled figure 1.

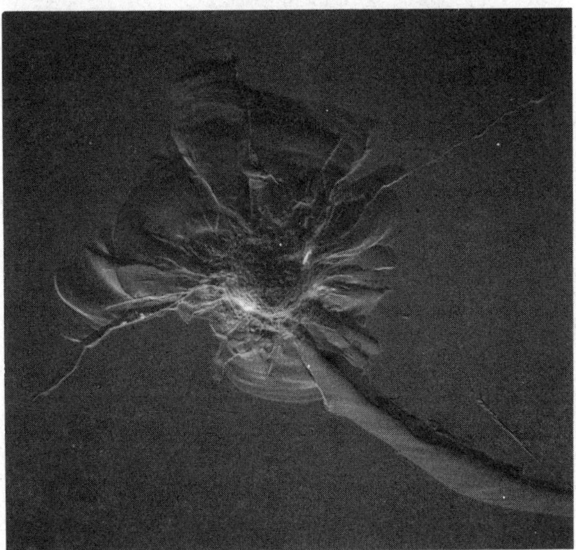

FIGURE 1. Crater from STS-61B, measures about 635 microns in diameter and was caused from meteoritic impact in the side window. The internal fractures underlying the crater measure about 1.3mm.

Morphologically, the high velocity impact crater has a relatively large portion of the central pit area consisting of well pulverized glass. The central pit damage was retained due to the lack of spall by the interior fractures. SEM/EDX analysis of the crater interior identified residues containing Magnesium, Aluminum, Silicon, Iron, and Sulfur which is consistent with the major components in meteoritic materials.

Impact damage on the right side windshield #132, from STS-48 resulted in removal of the pane from the shuttle Discovery. The micrograph in figure 2 shows particles of dirt and contamination as well as residues of aluminum projectile material. The orbital debris residues of aluminum alloy material (with minor Mg and Fe) are melted into the central pit lip, covering a portion of the crater about 100 microns wide and about 2.0 microns thick. The SEM/EDXA analysis detected more aluminum alloy type orbital debris residues in several other areas around the central pit region of the 900 micron diameter and about 150 micron deep impact crater.

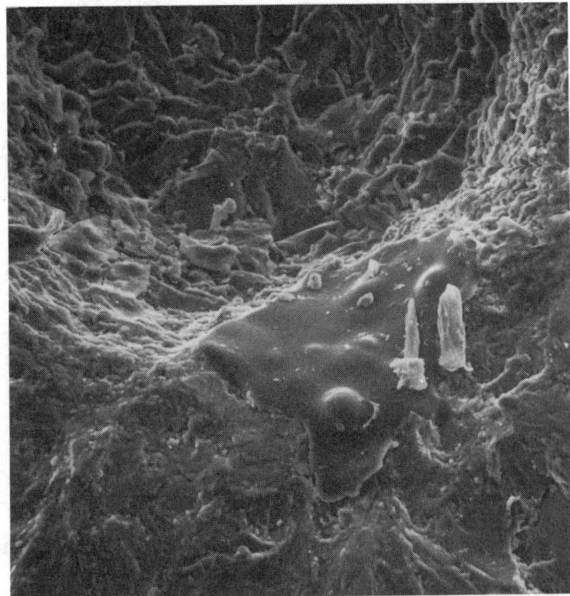

FIGURE 2. Impact from STS-48 is about 3mm in diameter, with residues from the aluminum alloy projectile detected in crater wall.

The impact crater labeled #129-1 from STS-44 measured about 275 microns in diameter, and in the central pit area retained a molten plug of silica glass and meteoritic residue from the colliding projectile (figure 3). SEM/EDXA analysis of these residues reveals Magnesium, Silicon, Aluminum, Sulfur, Calcium, and Iron, typical meteoritic mineralogy.

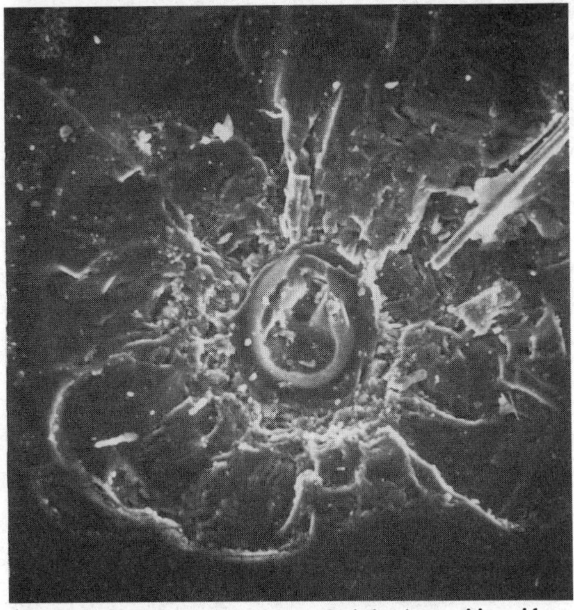

FIGURE 3. Close-up view of the melted glass/meteoritic residues found in the impact crater on the right side window from STS-44.

The impact crater sample from STS-51 has a typical hypervelocity impact morphology into glass. Its asymmetry in terms of central pit location relative to the expelled glass spall, and having a relatively deep central pit area (figure 4) suggests that the colliding particle was about 50 microns in diameter and impacted the glass at an oblique angle. Extensive examination of the sample and energy dispersive X-ray analysis of the particles associated with the impact could not confirm any of the residues detected as being projectile remnants. Although several grains of contamination from handling and processing were detected in and around the region of the impact feature.

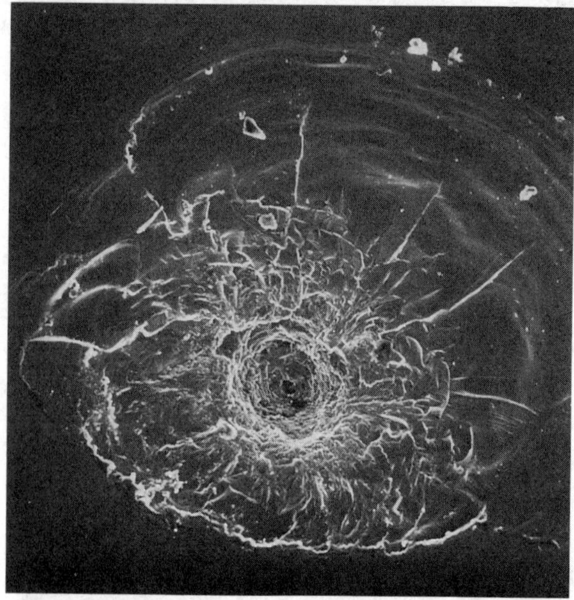

FIGURE 4. Scanning electron microscope image of the almost 1mm diameter crater from STS-51, produced by an oblique impact.

The 1.2cm diameter impact from STS-59 with maximum fracture damage depth being 1.15mm (sub-surface cracks) is the largest glass damage and was caused by impact from spacecraft paint (ref. 6). The slightly elliptical crater formed by expelled glass material is 6.1mm wide by 7.8mm long, and 0.565mm deep. The central pit has diameter dimensions of 1.001mm by 0.988mm and is 0.565mm deep. Around the inside edge of the impact several radial fractures are observed, these small spider like features are not seen in any of the other impacts and are not observed in experimentally produced craters into the silica glass.

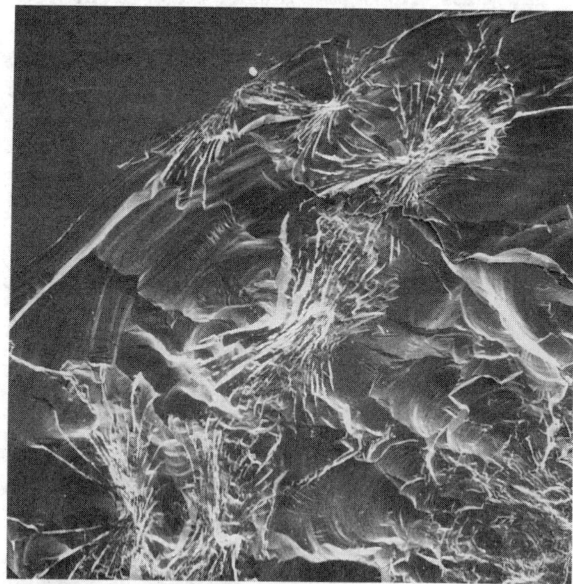

FIGURE 5. Radial fractures detected on the inside edge of 1.2 cm impact crater from side hatch window impact, on mission STS-59.

CONCLUSIONS

Overall, the damage severity was close to the predicted level. Although, there was 3-4 times more orbital debris impacts found than expected. The origin of 5 impact samples could not be determined by methods used ("unknown") and 1 window was not sampled. This 66% identification rate from the shuttle windows is comparable with results from LDEF analysis, and proves to be a viable source for LEO flux data acquisition.

REFERENCES

1. Zook H. A., Flaherty R. E., Kessler D. J. (1970) Meteoroid Impacts on Gemini Windows. Planetary Space Science. 18, pp 953-964.
2. Cour-Palais B. G., (1979) Results of the Examination of the Skylab/Apollo windows for Micrometeoroid Impacts. Proc. of the 10th Lunar Planetary Science Conference. p 1665.
3. Clanton U. S., Zook H. A., Schultz R. A., (1980) Hypervelocity impacts on Skylab/Apollo windows. . Proc. of the 11th Lunar Planetary Science Conference. pp 177-189.
4. Edelstein, K.S., (1992) Hypervelocity Impact Damage Tolerance of Fused Silica Glass, IAF-92-0334, August 28-September 5, 1992.
5. Crews, J.L. and Christiansen, E.L., (1992) The NASA JSC Hypervelocity Impact Test Facility (HIT-F), AIAA Paper No. 92-1640, AIAA Space Programs and Technologies Conference, Huntsville, AL, March 24-27, 1992.
6. Bernhard, R.P., Christiansen E.L., (1994) STS-59 Side Hatch Window Impact Damage Scanning Electron Microscope Analysis. NASA Interagency Report, Dec. 14, 1994.

INTACT HYPERVELOCITY PARTICLE CAPTURE IN AEROGEL IN THE LABORATORY

M.J.Burchell, R.Thomson

University of Kent, Canterbury, Kent CT2 7NR, United Kingdom.

We have investigated in the laboratory the capture in aerogel of soda glass and olivine particles travelling at approximately 5.3 km s^{-1}. Projectile sizes cover the range 75-250 μm. We report upon the penetration depth, hole size, spallation area and the fraction of the incident projectile that is captured. The data shows large scatter, although some trends are evident. We conclude that in such impacts intact capture is possible, although it would be difficult from the impact features to deduce the original projectile size to better than a factor of two.

INTRODUCTION

Since the earliest days of the space era the small particle population in Earth orbit has been of direct interest. The flux has been measured at many times and altitudes with the experiments continuing today. As well as flux however, the composition of the particles is of direct interest. Measurement of the composition is made difficult by the relative speed of a typical spacecraft and a particle. This is dictated by the natural orbital velocities of unpowered objects in Earth or Solar orbits. The impact velocities on the LDEF satellite were between 13 and 21 km s^{-1} depending on the surface impacted (1), with lower velocities for impacts by man-made debris. Such hypervelocity impacts on relatively thick surfaces result in craters with virtually no macroscopic remnants of the impactor. To permit study of the particle composition several schemes are followed. The simplest is to use the Earth's atmosphere as a capture medium and trawl the upper atmosphere using 'sticky' plates deployed from aircraft. However, the desire to study particles in space itself is strong. Two approaches are used. The first is an indirect one, measuring residues found at impact sites on pure metal surfaces exposed in Earth orbit and then retrieved for laboratory study. The second method has been to fly low density materials as capture media. These are either aerogels or under-dense foams. The normal shock of a hypervelocity impact on a surface is thus avoided, and the particle penetrates the medium, dissipating its energy along a track. Depending on the particle's original size, energy and composition it may be captured fairly intact. A general discussion of this topic can be found in (2). Some experimental work in this field has already been carried out by various authors e.g. (2), (3), (4) and (5). A theoretical discussion of capture mechanisms in underdense foams can be found in (6). In this paper we report on a series of experiments capturing olivine and soda glass projectiles in aerogel at just over 5 km s^{-1}.

EXPERIMENTS AND RESULTS

The projectiles used were soda glass spheres (density 2450 ± 50 kg m^{-3}) and powdered olivine (3100 ± 50 kg m^{-3}). The particle size ranges over 75 to 250 μm. The target was aerogel of density 93.1 ± 2.6 kg m^{-3}, and was manufactured by Henning Airglass in Staffanstorp, Sweden. Target sizes were 5 x 5 cm^2 surface area and 3 cm depth. The typical void size in the aerogel was found to be 70 nm. The projectiles were accelerated with a 2-stage light gas gun. In this a shotgun cartridge propels a piston which compresses hydrogen gas. At a preset pressure this was released by a bursting disc and accelerated a sabot down a rifled barrel. After acceleration the sabot was discarded in flight, and its load continued to the target. The velocity was measured by passage through light curtains, and by pzt's mounted on a stop-plate for the discarded sabot and the target mounting plate. For each shot the sabot was loaded with several projectiles in a given size range. Averaged over all shots reported here the velocity was 5.29 ± 0.19 km s^{-1}. Given this small spread we ignore velocity effects in this work.

We report on four shots using soda glass (particle diameters 82.5 ± 7.5 µm twice, 150.9 ± 6.9 µm and 172.7 ± 4.8µm) and four with olivine (particle diameters 82.5 ± 7.5 µm, 107.5 ± 17.5 µm, 137.5 ± 12.5 µm and 231 ± 19 µm). Each shot resulted in several observable impacts, but for the olivine shot with 107.5 ± 17.5 µm only surface features are visible (the tracks in the interior were obscured by other damage to the aerogel sample). We have made a shot with olivine of 450 ± 25 µm, but this shattered our aerogel target so no results are presented here for this.

After shooting, the aerogel was examined under optical microscopes. The surface features were typical of impacts in brittle materials like glass or ice. For each impact there was a central hole, plus a large spallation area of damaged material. The central hole was approximately circular. The spallation area was of irregular shape, divided into 'pizza' like slices by cracks radiating from the centre (typically 6-8 per impact). Each slice may extend a different distance from the centre, can have an angular extent from 20 to 180°, and may be slightly depressed both compared to its neighbours and the original surface level.

Given that each slice of the spallation region can have a different radius, it is hard to characterise the spallation zone by a single quantity. In Fig. 1 we show minimum and maximum spallation radii for each impact. A large spread in these values is seen, but as a general trend both increase with projectile size almost linearly. No difference is seen between soda glass and olivine.

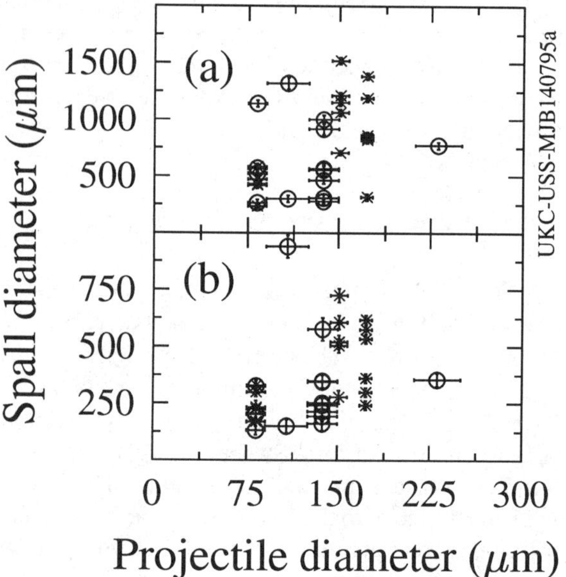

Figure 1. Radius of spallation zone per impact vs. projectile diameter. (a) is maximum and (b) minimum radius. × is soda glass, o is olivine.

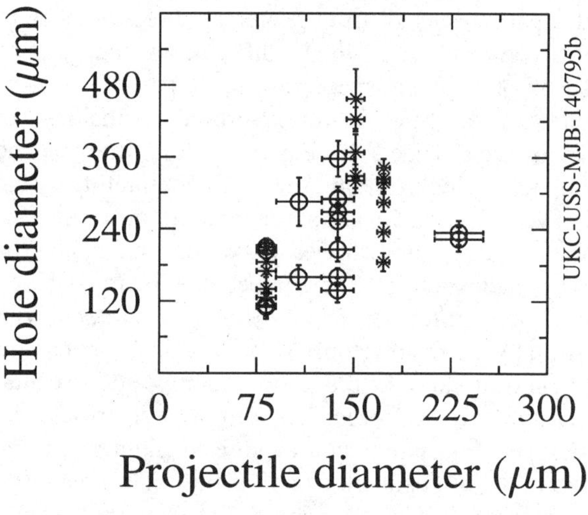

Figure 2. Diameter of impact hole vs. projectile size. × is soda glass, o is olivine.

The central hole in each impact zone is characterised as circular. The mean diameter of the hole for each impact is shown in Fig. 2. We can see that hole size is similar for both types of projectile, and is typically 2 ± 1 times the projectile diameter.

The tracks left by the particles were divided into two main types. The first were carrot shaped (as noted by previous authors). The tracks diminish in diameter as they penetrate into the aerogel, and at the end the captured projectile was seen. A variation on this was also seen, namely carrot shaped tracks with a forked ending (with 2 prongs). One prong contained a large part of the projectile, the other a smaller fragment or nothing. For the bifurcating tracks the second prong separated from the main one before the projectile stopped, and then continued to virtually the same depth as the main projectile (to within 15%). In both cases we measure the track length as that achieved by the larger part of the projectile. These track lengths are shown in Fig 3. For a given shot there is a scatter in individual track lengths, but a clear trend of increasing track length with larger projectile size is seen. Projectiles penetrate to 60 ± 20 times their diameter at the velocities used here. Again olivine and soda glass are indistinguishable.

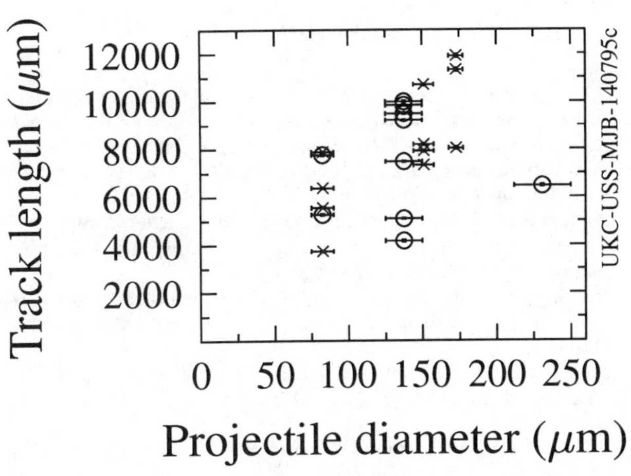

Figure 3. Length of track in aerogel vs. projectile diameter. × is soda glass, o is olivine.

In Figure 4 we show the fraction of the original projectile captured in the carrot shaped tracks. It can be seen that typically the diameter of the captured particle is (70 ± 20)% of the original, i.e. 34% of the original mass is captured as an intact particle. There is no clear difference in this fraction with projectile size or type. However a large scatter is evident on the data.

Another type of track is also observed. In terms of entrance hole and spallation damage they are indistinguishable from other tracks. However in length they are much shorter than the other tracks (perhaps 1/4 the length). These tracks terminate in a 'star-burst' like pattern of shot tracks, the main carrot-track splitting into typically 5 short sub-tracks, with the main track itself ending at that point. The sub-tracks are at some 30° to the direction the original track would have proceeded in. Each sub-track has a very small remnant of the projectile at its end. Approximately 1 in 10 tracks behave like this, and we suggest that the projectiles are breaking up early in the impact process.

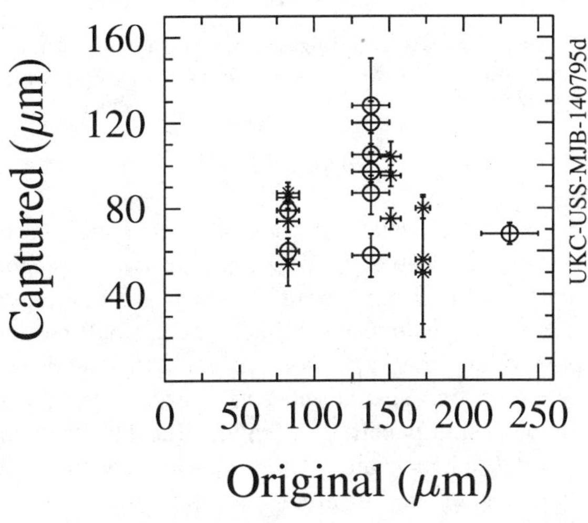

Figure 4. Diameter of captured projectile vs. original diameter. × is soda glass, o is olivine.

We can compare our results with those of (2,5), where Fosterite (a form of olivine) was fired at 6-7 km s^{-1} into aerogels of varying density. The projectiles used were of approximately 100 μm diameter. In (2,5) instead of track length the ratio track length/projectile diameter is used and eight data points are given (without errors). In Fig. 5 we show this data, along with our own data. Despite the difference in velocities it can be seen that the data are totally compatible.

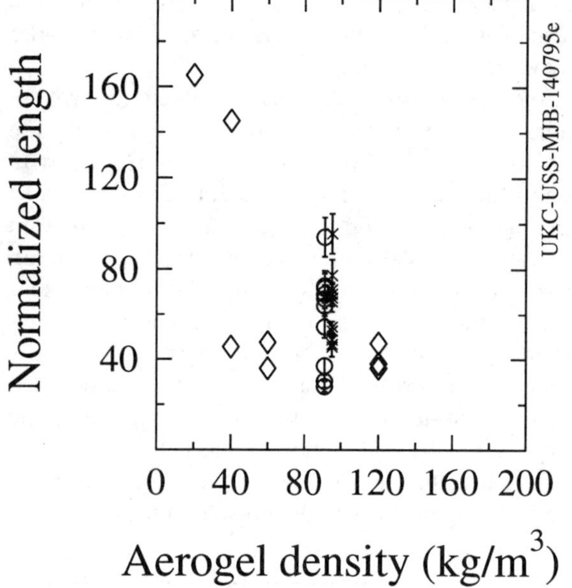

Figure 5. The track length/original projectile diameter is shown vs. aerogel density. × is soda glass, o is olivine, ◊ is data for olivine from (2,5).

DISCUSSION

It can be seen from our results that there is no systematic difference between the two projectile materials used here, both of which are considered reasonable analogues for much of the small particle population in space. We can see that trends are apparent in spallation zone size, central pit diameter and penetration length. But the spread on this data is such that at the velocity used here the diameter of the original projectile can only be deduced to within a factor of two. Looking directly at the captured particle provides a similar uncertainty in the original size. By volume, some 30% of the original projectile is captured intact. Thus although capture is occurring, the projectile is still being subject to extensive processing (i.e. high temperatures and shock, see (2)). Further, before proceeding to analysis of material captured in space it must be remembered that the impact velocity in space will be different and this must be allowed for. Light gas guns do not operate at velocities of some 20 km s^{-1}, so laboratory work is unable to provide a direct calibration.

The capture mechanism in aerogel is not just a function of projectile density and velocity. In (6) drag of molten material formed as a shell around the projectile is suggested as an important stopping mechanism in aerogel. This would naively depend on an effective surface area and not the volume of the projectile. However, we note that our data shows that break-up of projectiles sometimes occurs, indicating that shock processes in the projectile may not be negligible in the capture process. Further, if such drag were the dominant capture process track length might be expected to scale with projectile diameter squared (i.e. cross-sectional area); we find that the dependence is roughly linear.

ACKNOWLEDGEMENTS.

This work has been supported by grants from the Particle Physics and Astronomy Research Council of the United Kingdom.

REFERENCES

1.   Zook H.A., Deriving the velocity distribution of meteoroids from the measured meteoroid impact directionality on the various LDEF surfaces, in *LDEF-69 Months in Space: First Post-Retrieval Symposium*, pp 569-579, ed. A.S.Levine, NASA CP-3134 (1991).

2.   Barrett R.A., Zolensky M.E., Hörz F, Lindstrom D.J. and Gibson E.K., Suitability of Silica Aerogel as a Capture Medium for Interplanetary Dust, *in Proceedings of Lunar and Planetary Science*, 1992, vol. 22, pp 203-212.

3.   Tsou P. et al., Effectiveness of Intact Capture Media, *Abstracts of XX Lunar and Planetary Science Conference*, 1990, pp 1132-1133.

4.   Tsou P. et al., Intact Capture of Cosmic Dust Analogs in Aerogel, *Abstracts of XXI Lunar and Planetary Science*, 1991, pp 1264-1265.

5.   Zolensky M.E., Barrett R.A and Hörz F., The Use of Silica Aerogel to Collect Interplanetary Dust in Space, in *Proceedings of the Workshop on Particle Capture, Recovery, and Velocity/Trajectory Measurement Technologies*, 1994, LPI Tech. Rpt 94-05, Lunar and Planetary Institute, Houston, pp 94-98.

6.   Anderson W. and Ahrens T.J., Physics of Interplanetary Dust Capture via Impact into Organic Foams, *Journal of Geophysical Research* E, vol. 99, pp 2063-2071, 1994.

CONSTITUTIVE MODELING FOR HYPERVELOCITY CRATERING

R. F. Davidson and M. L. Walsh

Los Alamos National Laboratory, Los Alamos, New Mexico 87545

The Hypervelocity Microparticle Impact program conducted cratering experiments on copper and aluminum targets at impact velocities up to 30 km/s (1). These data are presented with minor corrections in a form that will be useful to a variety of investigators. Hydrocode analysis of the data was performed to gain understanding of the "plateau" features in the data. Analyses using conventional strength models were not satisfactory due to the very high strain rates of these experiments (strain rates on the order of 10^{10} /s). Calculations using the PTW strength model rate dependence (2) gave very accurate reproduction of the experimental results, and allowed computational examination of the "data plateaus." These results show that the "data plateaus" may be due to strain-rate effects and experimental variation in the data.

BACKGROUND

The Van de Graaff facility at Los Alamos was used to conduct the Hypervelocity Microparticle Impact (HMI) experiments on a variety of materials(1). Careful examination of these data for aluminum and copper targets led to the discovery that the data values contained slight errors. A reanalysis of the original data was conducted paying careful attention to determining the variation in the data.

There are two interesting data features that were investigated using computations. The strain rates were extremely high (on the order of 10^{10} /s), and the data exhibited an unexplained erratic velocity dependence, sometimes called "data plateaus" (1). Computations on aluminum targets were conducted to test the hypothesis that material strain-rate effects combined with data variation could explain the "data plateaus." The high strain-rate dependence of the PTW model was used to accurately depict the strain-rate effect on the flow stress.

EXPERIMENTAL OVERVIEW AND DATA REANALYSIS

In the Van de Graaff facility, iron microparticles were charged and then accelerated down the beam line. Those particles that met the desired characteristics of velocity and mass were "selected" to be directed onto the target. Tests were conducted for a variety of target materials with impact velocities up to 30 km/s and particle masses of about 0.1 to 1000 fg (1 fg = 10^{-15} g). Because the craters were very small (on the order of a few μm), the dimensions were measured using electron photomicrographs. To date, only crater diameters have been measured, and that is why D^3 (crater diameter cubed) was chosen as the best available estimate of the crater volume.

The data from the reanalysis are given in Figs. 1 and 2 and in Table 1. The primary independent variable D^3/M (crater diameter cubed divided by impactor particle mass) was chosen as an estimate of the crater volume related to the particle mass - thus allowing comparison to data over a wide range of impactor sizes (masses). Additional data is given in the tables so that other independent variables may be examined for these test data.

Each test had a number of particle impactors and a number of craters in the target. It was not possible to determine which particle made which crater. In fact, the number of particles associated with each test was not the same as the number of craters found in the targets (see Table 1). This was due to difficulties in the particle selection apparatus that could have allowed "leakers" onto the target or caused a "selected" particle to be deflected off the target. Because the particles could not be correlated to specific craters it was not possible to determine the standard deviation of D^3/M exactly.

TABLE 1. Aluminum and Copper Data from Reanalysis

Run[a]	Material[b]	Part. Num[c]	Cr. Num[d]	V (m/s)[e]	SD V[f]	M (kg)[g]	SD M[f]	D (m)[h]	SD D[f]	D^3/M (m³/kg)[i]	SD D^3/M (ρ=0.7)[f]
9023310	Aluminum	22	24	6752	138	2.136e-15	166.e-18	2.837e-06	156e-09	0.01069	0.00132
9023309	Aluminum	19	28	8797	125	616.4e-18	55.2e-18	2.185e-06	114e-09	0.01692	0.00193
9023308	Aluminum	10	13	10658	171	2.885e-15	153.e-18	4.182e-06	113e-09	0.02535	0.00147
9023307	Aluminum	17	19	12521	149	1.028e-15	81.3e-18	2.901e-06	95.e-09	0.02374	0.00168
9023306	Aluminum	10	11	14240	287	580.5e-18	90.3e-18	2.527e-06	195e-09	0.02781	0.00459
9023305	Aluminum	18	40	16217	137	276.5e-18	36.7e-18	2.015e-06	182e-09	0.02957	0.00598
9023201	Aluminum	23	29	17787	253	290.4e-18	14.8e-18	2.484e-06	66.e-09	0.05279	0.00301
9023301	Aluminum	22	31	19841	309	100.0e-18	15.5e-18	1.703e-06	118e-09	0.04938	0.00737
9023302	Aluminum	17	21	21429	335	108.3e-18	30.8e-18	1.772e-06	231e-09	0.05137	0.01436
9023304	Aluminum	7	11	23663	233	87.46e-18	26.3e-18	1.820e-06	189e-09	0.06893	0.01634
9023401	Aluminum	19	38	26017	746	86.06e-18	37.4e-18	1.859e-06	318e-09	0.07471	0.02797
9022103	Copper	29	27	5770	214	1.509e-15	130.e-18	2.583e-06	95.0e-09	0.01142	0.00091
9022104	Copper	23	22	8046	194	240.9e-18	31.4e-18	1.507e-06	87.9e-09	0.01421	0.00178
9022105	Copper	21	22	10390	218	375.2e-18	107.e-18	2.020e-06	210.e-09	0.02197	0.00511
9022107	Copper	20	15	12167	357	1.990e-15	506.e-18	3.982e-06	365.e-09	0.03173	0.00653
9022110	Copper	19	8	14241	374	877.0e-18	124.e-18	3.384e-06	171.e-09	0.04418	0.00504
9022203	Copper	30	26	16032	254	535.2e-18	25.5e-18	2.749e-06	101.e-09	0.03881	0.00325
9022201	Copper	24	22	17884	311	364.6e-18	22.8e-18	2.521e-06	125.e-09	0.04396	0.00503
9022504	Copper	16	23	19630	777	245.6e-18	46.5e-18	2.377e-06	130.e-09	0.05465	0.00759
9022209	Copper	20	25	21810	369	117.3e-18	8.22e-18	1.843e-06	52.9e-09	0.05333	0.00332
9022501	Copper	25	24	23043	561	91.66e-18	13.5e-18	1.673e-06	98.9e-09	0.05112	0.00658

[a]Run number designates a specific test and target.
[b]Target plate material.
[c]Number of impactor particles that met the particle selection criteria for this Run.
[d]Number of craters found on the target for this Run.
[e]Average ("selected") particle velocity.
[f]SD = standard deviation.
[g]Average ("selected") particle mass.
[h]Average crater diameter.
[i]D^3/M is an estimate of (two times) the crater volume normalized to the particle mass.

To approximate the error in D^3/M, the correlation coefficient (ρ) between crater diameter (D) and particle mass (M) was estimated. The correlation coefficient varies from 0 (independent or not correlated) to 1 (exactly correlated). A value of ρ = 0.7 was used, which means that the crater diameter was assumed to be strongly, but not exactly correlated with the particle mass. Thus, as the particle mass increases, the crater diameter tends to increase as well. In Figs. 1 and 2, the error estimate for D^3/M is plus or minus one standard deviation with ρ = 0.7. The tabulated statistics for D (crater diameter), V (particle velocity), and M (particle mass) are straightforward. Details of the data reanalysis are documented in an informal report that is available from the authors.

COMPUTATIONAL OVERVIEW

All the computations were made with the 1994 EPIC hydro-code (3). A mesh study was conducted which showed the mesh to be accurate and sufficiently large. The same mesh was scaled for different projectile (particle) sizes. Care was taken so that the computed craters were all measured in a consistent manner. Details of the calculations are documented in an informal report that is available from the authors.

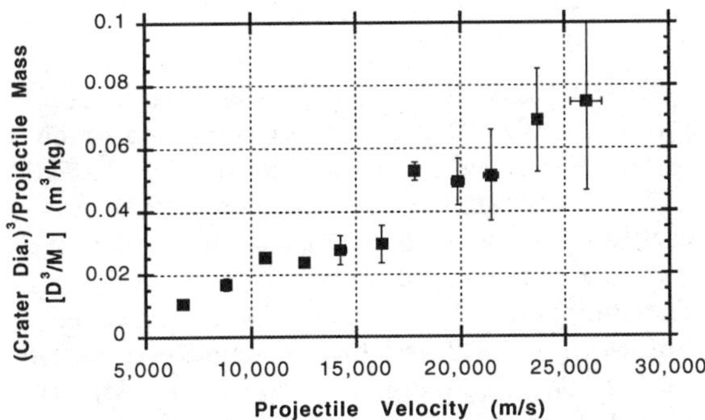

FIGURE 1. Iron Microparticle Impacts onto Aluminum Targets - HMI Data.

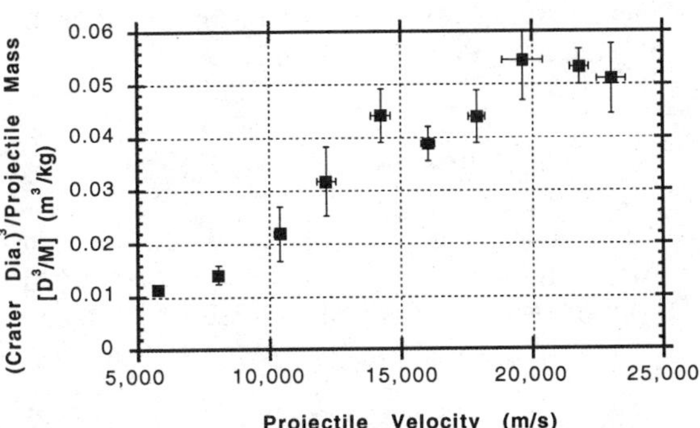

FIGURE 2. Iron Microparticle Impacts onto Copper Targets - HMI Data.

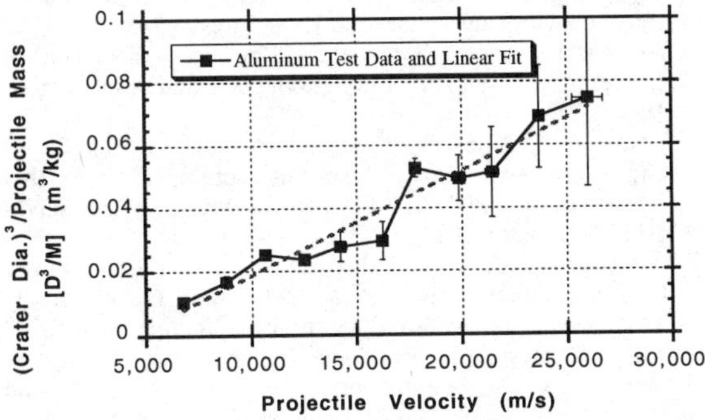

FIGURE 3. Iron Microparticle Impacts onto Aluminum Targets Showing the Data Plateaus and a Linear Fit to the Data.

The computations considered only the aluminum targets. Figure 3 shows the "data plateaus" at 11-16 and 18-22 km/s. Figure 3 also shows that a simple linear fit to the data does not touch all the data error bars.

Figure 4 shows the problem of using a strain-rate hardening law based on data at rates lower than occurred during these tests. The Johnson-Cook (JC) strength model for 1100 aluminum includes rate dependent flow stress behavior up to 10^4/s, and predicts a crater volume that is much too large. The HMI tests had strain-rates that were typically 10^{10} /s. To match the HMI data at 6 km/s, the JC flow stress was multiplied by 5 (5). However, the simple multiplication did not match the crater volume data at higher velocities (See "JC w/ Yo x 5" in Fig. 4).

By examining over-driven shock data, Preston, et. al (2) found a strong increasing rate dependence for copper and aluminum at strain-rates of 10^5 to 10^{10} /s. This increasing rate dependence is included in the PTW (Preston, Tonks, Wallace) strength model. This same increasing rate dependence was added to the JC model. The modified JC model was called the "JC/HSR-PTW" model to indicate that the high strain-rate (HSR) dependence of the JC model was modified to look similar to the rate dependence in the PTW model. This model came reasonably close to matching the HMI data (Fig. 4). The computations tend to bound the data on the low side - computed crater diameters were consistently (slightly) smaller than the test data. The "numerical erosion technique" used in the EPIC calculations is known to produce craters that are too small in diameter (4). Therefore, the JC/HSR-PTW model results give excellent comparison to the test data.

The different particle sizes in the tests introduced some additional strain-rate differences. By making computations using both a small projectile and a large projectile typical of the particle size range in the tests, the difference in crater volume due to particle size could be investigated. These two sets of computations (small projectile - high strain-rate and large projectile - low strain-rate) were adjusted to be centered on the linear fit to the data. The spread between the high-rate and low-rate curves shows the scatter due to particle size differences (Fig. 5). Since the two curves intersect all the data within one standard deviation, this plot shows that rate

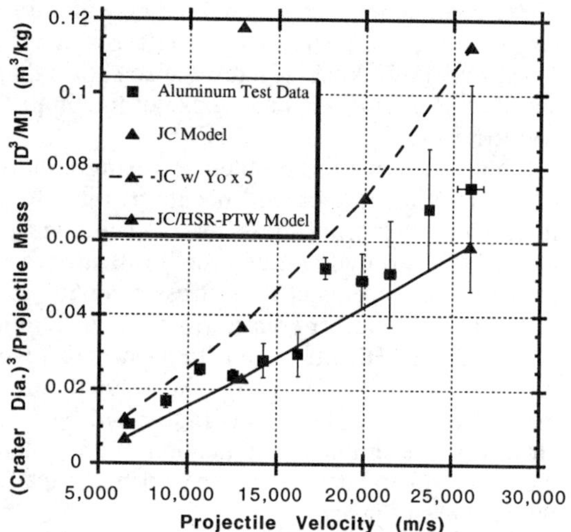

FIGURE 4. Iron Microparticle Impacts onto Aluminum Targets. HMI Data and Computational Results for 3160 fg Particles.

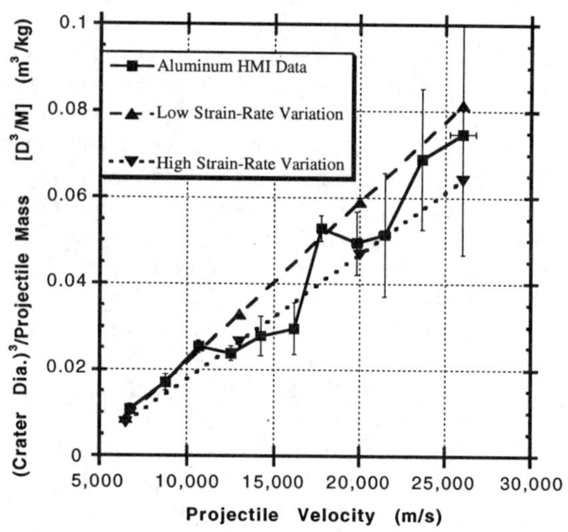

FIGURE 5. Iron Microparticle Impacts onto Aluminum Targets. HMI Data and Computed Variation Due to Strain-rate Effects.

dependence effects alone are sufficient to model the HMI experimental data. In other words, the "data plateaus" can be explained by the expected error in the data (one standard deviation) and the strain-rate dependence of the target material.

CONCLUSIONS

The corrected HMI data are presented in Figs. 1 and 2 and Table 1. The data are presented so that a variety of independent variables can be examined.

The strain-rate dependence of the PTW model was found to accurately describe the behavior of the HMI data at these extremely high strain-rates. The aluminum "data plateaus" can be explained by strain-rate effects and the expected error in the data, as shown in Fig. 5. Careful investigation would probably support similar conclusions regarding the copper data.

The "data plateaus" have commonly been attributed to phase change effects (5). This work does not refute this or other explanations for the "data plateaus," but simply shows that strain-rate effects and the expected error in the data are sufficient to explain the "data plateaus."

ACKNOWLEDGMENTS

This work was performed at Los Alamos National Laboratory and was sponsored by the Departments of Energy and Defense.

REFERENCES

1. Stradling, G. L., Idzorek, G. C., Shafer, B. P., et al., "Ultra-High Velocity Impacts: Cratering Studies of Microscopic Impacts from 3 km/s to 30 km/s," Los Alamos National Laboratory report LA-UR-92-3811 (1992) and Proceedings of the Hypervelocity Impact Symposium, 1992.
2. Preston, D. L., Tonks, D. L., and Wallace, D. C., "The Rate Dependence of the Saturation Flow Stress of Cu and 1100 Al," Proceedings of the American Physical Society Topical Conference on Shock Compression of Condensed Matter, 1991.
3. Johnson, G. R., Stryk, R. A., Petersen, E. H., Pratt, D. E., and Schonhardt, J. A., "User Instructions for the 1994 Version of the EPIC Code," Alliant Techsystems, Inc., report H11229 (January 1994).
4. Raftenberg, M. N., and Kennedy, E. W., "Behind-Armor Debris Characteristics from Steel Plates Perforated by Tungsten Rods of Small Length-to-Diameter Ratios," Proceedings of the Hypervelocity Impact Symposium, 1994.
5. Wingate, C. A., Stellingwerf, R. F., Davidson, R. F., and Burkett, M. W., "Models of High Velocity Impact Phenomena," Los Alamos National Laboratory report LA-UR-92-1982 (1992) and Proceedings of the Hypervelocity Impact Symposium, 1992.

GENERATION OF PSEUDOTACHYLITES IN SHOCK EXPERIMENTS: IMPLICATIONS FOR IMPACT CRATERING PRODUCTS AND PROCESSES

P. S. Fiske[a], W. J. Nellis[a], H. Lorenzana[a], M. Lipp[a], M. Kikuchi[b], Y. Syono[b]

[a] L-299, Lawrence Livermore National Laboratory, Livermore, CA 94550, U.S.A.

[b] Institute for Materials Research, Tohoku University, Katahira 2-1-1, Aoba-ku, Sendai 980, JAPAN

Meteorite impacts produce enormous pressure and strain in rocks. While the role of pressure on the formation of shock metamorphic features has been well studied, the role of strain and strain rate has not been fully appreciated. We shock loaded single-crystal quartz in Al capsules up to 56 GPa using a novel capsule design that allows for significant strain of the sample but 100% recovery of material. We have made features analogous to type A pseudotachylites at pressures of 42-56 GPa. These pseudotachylites contain Al, Si and minor Al_2O_3 in a matrix of SiO_2 glass and cut the sample along radial and concentric fractures. Our results suggest that strain heating is an important energy sink in the formation of large impact craters.

INTRODUCTION

Rocks from impact craters have a variety of distinctive microscopic and macroscopic "shock features" formed by the impact. Some shock features, such as shocked quartz, are found in rocks from both large and small craters have been synthesized in laboratory shock experiments (1). Other features, however, such as shatter cones and pseudotachylites (2), are difficult or impossible to reproduce in laboratory shock experiments and are found in abundance only in larger impact structures, suggesting a significant dependence on scale. While the pressures experienced by rocks in small and large impacts are comparable, the principal difference is in the amount of strain to which the rocks are subjected.

We have conducted a series of experiments (3) using novel shock recovery systems that enable us to vary the strain experienced by the sample during shock-loading. Our experiments have produced glassy veins of black material in quartz that are analogous to pseudotachylites found in the central portion of large impact craters and suggest that the strain produced by large impacts may cause substantial heating of the rocks below the crater floor.

EXPERIMENTAL PROCEDURE

Disks of high-purity, synthetic, single-crystal quartz were tightly enclosed in Al and steel capsules and shock-loaded to 42-56 GPa using the 6.5 m two-stage light-gas gun at Lawrence Livermore National Laboratory. Our experiments employ a sample container similar in geometry to those used in previous studies (4) but made of aluminum instead of steel. The sample and capsule are contained by a large fixture made of a strong titanium alloy. Aluminum, with a density of 2.7 g/cm^3 has a shock impedance much closer to that of silicates (in this case quartz, density of 2.65 g/cm^3) than of steel. The shock impedance match produces a single loading pulse in the sample, rather than a reverberation or "ring up" to high pressure as is the case with silicates held in steel capsules.

This single wave loading produces a greater rise in temperature than the quasi-isentropic loading of a ring-up experiment and more realistically approximates the single-shock pressure rise during meteorite impact. Lateral deformation of the sample takes place on release from high pressure, analogous to the excavation stage of crater development. The critical feature of the aluminum capsule is its lower strength as compared to steel. Despite being held in the strong Ti-alloy fixture, samples held in aluminum capsules undergo more than twice the radial strain of samples held in steel and shocked to the same pressure (Fig. 1).

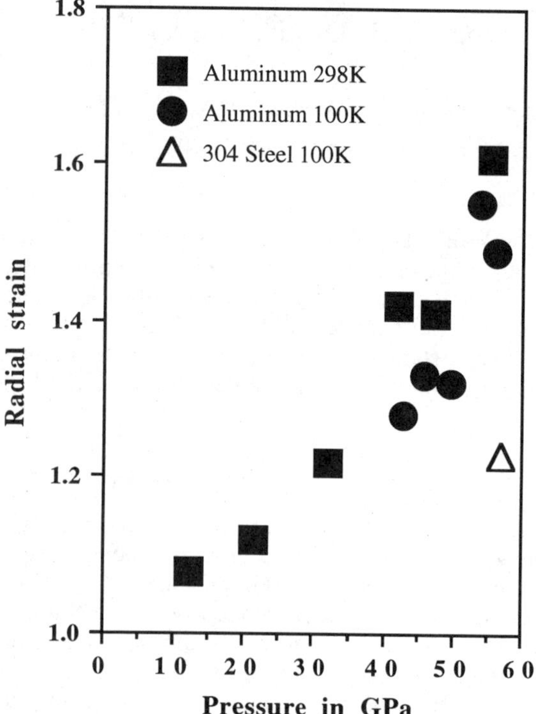

Figure 1. Radial strain of quartz samples held in Al capsules at room T (black squares), held in Al at 100 K (black circles), and held in steel at room T (open triangle). Strain is measured as the ratio of the final to the initial diameter of the sample.

RESULTS

Samples recovered from aluminum capsules consist of transparent amorphous SiO_2 cut by radial and concentric veins that are filled with a black glassy material (Fig. 2). The transparent regions of the sample show variable low birefringence under cross-polarized light indicating significant residual strain in the amorphous material. The regions with the highest birefringence have the densest spacing of veins. Overall, the density of veins increases with pressure. In contrast, a sample shock-loaded to 56 GPa in steel capsules is transparent throughout, shows no birefringence under cross-polarized light, and has only a few small, irregular fractures that are localized in the central portion of the sample (Fig.

2). The total radial strain (final diameter over initial diameter) of the 56 GPa sample held in Al was 1.6 whereas it was only 1.2 for the sample held in steel and shock-loaded to the same pressure.

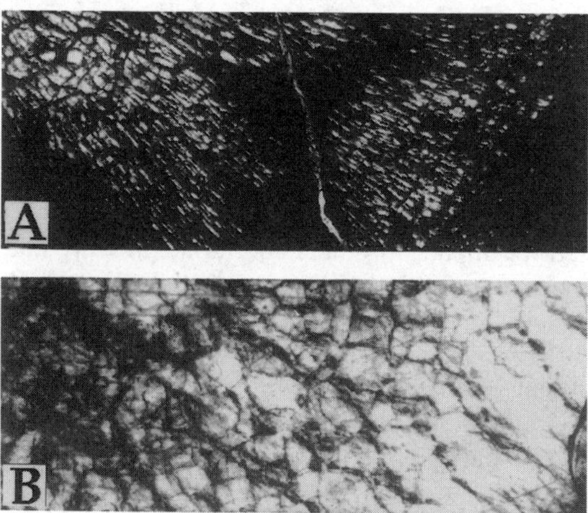

Figure 2. Transmitted light image of portions of three disks of shock-loaded cross-cut quartz viewed in the direction of shock wave propagation. Images are 7 mm by 2 mm. The center of each disk is in the upper left corner, and the edge is in the lower right corner.
(A) Shock loaded to 56 GPa in Al. The sample consists of clear regions of strained amorphous SiO_2 cut by radial and concentric veins filled with black glass. This sample contains 1-2% Si by weight.
(B) Shock-loaded to 56 GPa in steel. Sample is optically clear with a few irregular fractures in the center of the disk. No Si was detected in this sample.

Analyses of the samples by X-ray diffraction and Raman spectroscopy show that the black glass consists of a mixture of nanocrystalline Al and Si and amorphous SiO_2. The amount of Si and Al increases with pressure, with the highest pressure sample containing 1-2% Si by weight. TEM images (Fig. 3) and electron diffraction patterns confirm the presence of Si and Al, along with a small proportion of Al_2O_3, and show that the Si and Al occur as round clasts 10 to 400 nm in size in a matrix of amorphous SiO_2. The melting of SiO_2 provides a lower limit on the temperature of the pseudotachylite in our experiments (2000 K at P=0) and these temperatures are similar to those required to reduce SiO_2 in soils struck by lightning (5).

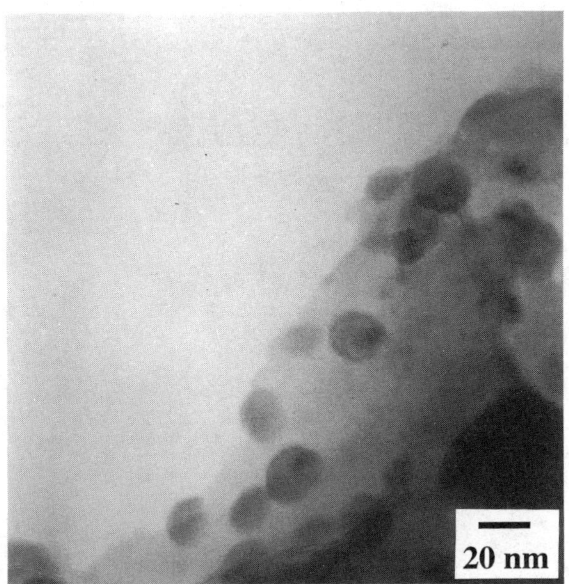

Figure 3. Transmission electron microscope image of black glass formed in quartz shock-loaded to 46 GPa in Al. The black glass consists of spheres of crystalline Si and Al (which appear darker in this image) in a matrix of $SiO2$ glass. Crystallites range in size from 20 to 400 nm.

DISCUSSION

Our results suggest that plastic deformation and the production of pseudotachylites may be an important and little appreciated process in meteorite impacts. Our highest pressure sample contained 1-2% Si by weight. TEM images suggest that the Si represents roughly 10-20% of the volume of pseudotachylite-like material produced. Thus a substantial fraction of the total sample may have been pseudotachylite. The formation of dense networks of type A pseudotachylites in the central proportions of impact structures may similarly leave the rocks significantly hotter than that predicted from Hugoniot measurements or post-shock temperature measurements. This may cause the central portions of large impact structures to have a substantial static thermal metamorphic overprint. In addition, friction melting may be an important contribution to the total volume of impact melt in larger craters and may explain the reported increase in the proportion of impact melt with crater size (6).

Shock wave profile measurements show that quartz shock-loaded above the dynamic elastic limit (12 GPa) is nearly strengthless on release from high pressure (7). The formation and persistence of melt-lubricated fractures may be a principal contributor to the loss of strength on pressure release, and the formation of type A pseudotachylites during compression and crater excavation may substantially weaken the basement rock in the center of the crater and allow for nearly hydrostatic rebound of the crater floor.

Finally, shear heating and the formation of pseudotachylites might have a significant effect on estimations of shock devolatilization of planetary materials based on shock experiments. Experiments employing steel capsules (8) may have prevented significant strain of the sample thus producing lower temperatures and an underestimation of the amount of shock devolatilization with pressure. Furthermore, numerical modeling of cometary impacts (9) may underestimate the role of strain heating in the projectile and thus may overestimate the potential for delivery of organic material by impact.

Work at LLNL was performed under the auspices of the U. S. Department of Energy under Contract No. W-7405-Eng-48.

REFERENCES

1. Stöffler, D. and Langenhorst, F., *Meteoritics* **29**, 155-181 (1994).
2. Thompson, L. M., and Spray, J. G., in *Large Meteorite Impacts and Planetary Evolution*, B. O. Dressler, R. A. F. Grieve, V. L. Sharpton, Eds. (Geological Society of America, Boulder, Co., 1994), vol. 293, pp. 275-288.
3. Fiske, P. S., Nellis, W. J., Lipp, M., Lorenzana, H., Kikuchi, M., and Syono, Y., *Science* (submitted).
4. Cordier, P. et al., *Phys. Chem. Min.* **21**, 133-139 (1994).
5. Essene, E. J., and Fisher,D. C., *Science* **234**, 189-193 (1986)
6. Grieve, R. A. F., and Cintala, M. J., *Meteoritics* **27**, 526-538 (1992).
7. Grady, D. E., *J. Geophys. Res.*, **85**, 913-924 (1980).
8. Chen, G., Tyburczy, J. A., and Ahrens, T. J., *Earth Planet. Sci. Lett.* **128**, 615-628 (1994).
9. Chyba, C. F., Thomas, P. J., Brookshaw, L., and Sagan, C., *Science* **249**, 366-373 (1990).

HYDROCODE MODELING OF
ADVANCED DEBRIS SHIELD DESIGNS

J. H. Kerr, E. L. Christiansen, J. L. Crews

Space Science Branch, NASA Johnson Space Center, Houston, Texas 77058

The NASA Johnson Space Center (JSC) Hypervelocity Impact Test Facility (HIT-F) has developed several low mass, high performance shielding concepts to protect spacecraft from orbital debris and meteoroid impact. Development testing requires shield concept validation in the impact velocity regime from <1km/s to ~14.5km/s. Current two-stage light gas gun testing limits maximum impact velocities to 8km/s; therefore, Sandia National Laboratories and Southwest Research Institute have developed advanced launchers capable of accelerating non-spherical shaped masses to ~15km/s. Since the shape of the impactor influences final rear wall damage, hydrocodes are employed to evaluate the so called shape effect at velocities greater than 8km/s. A series of 14 hypervelocity impact simulations were conducted using the CTH hydrocode. Simulations modeled spherical aluminum (Al) and Al flat plate projectiles of various masses impacting double bumper all Al Whipple shields (DB). Experimental results at ~7km/s are compared with simulation and ballistic limit curves are constructed for the DB Whipple shield in the velocity regime greater than 7km/s. Comments are also made on the shape effect mass ratio for spherical and flat plate projectiles.

INTRODUCTION

For years researchers at the NASA Johnson Space Center (JSC) Hypervelocity Impact Test Facility (HIT-F) (1) have developed low mass shielding designs to protect spacecraft from meteoroid and orbital debris (M/OD) impact. Shielding systems must defeat orbital debris and meteoroids which can impact spacecraft at velocities as great as ~14.5 and ~72km/s respectively. The Multi-shock shield (2), Mesh double-bumper shield (3), and Stuffed Whipple shield (4) each offer greater spacecraft protection than the standard Whipple shield approach. The HIT-F has performed hundreds of tests on these shield designs using spherical impactors. By varying projectile diameter, velocity, and impact angle, shield performance may be assessed. These data were used by Christiansen to develop shield design and performance equations (4, 5). Since two-stage light gas guns were employed for testing, impact velocities were limited to less than 8km/s.

In order to assess shield performance at velocities greater than 8km/s, the NASA JSC HIT-F sponsored development of techniques such as Sandia National Laboratory's (SNL) hypervelocity launcher (HVL) (6) and Southwest Research Institute's (SwRI) inhibited shape charge launcher (ISCL) (7). Although these devices produce high velocity projectiles, the projectile's shape is non-spherical. Impactors of equal mass and dissimilar shape will have different effects on shield performance. This shape effect induces difficulties in comparing shield performance data obtained from non-spherical impactors to data produced by spherical projectiles. Using hydrocodes, such as SNL's CTH, the shape effect may be quantified. Since the HVL accelerates flat plates (FP) to velocities greater than 11.3km/s, we must determine the FP shape effect to compare HVI test results with spherical data.

This paper describes hydrocode work in progress at the NASA JSC HIT-F. A total of 14 HVI simulations were conducted using the CTH

TABLE 1. Experimental Data for All Aluminum Double Whipple Shield

Test Number	Impact Angle (deg)	Impact Vel. (km/s)	Proj. Size (Dia. x thick) (mm)	Proj. Mass (g)	Equivalent Sphere Dia. (mm)	Rear Wall Damage
SWBS-3	0	10.2	18.9x1.003	0.746	8.07	Perforation
B812	0	6.83	6.35	0.362	6.35	No perforation
B813	0	6.90	7.54	0.628	7.54	Perforation

hydrocode. Simulations modeled spherical aluminum (Al) and Al FP projectiles of various masses impacting double bumper (DB) all Al Whipple shields. The spherical impactor series provides a baseline of comparison with the FP projectile simulations. Experimental results at ~7km/s are compared with simulation and ballistic limit curves are constructed for the DB Whipple shield in the velocity regime greater than 7km/s. Comments are also made on the shape effect mass ratio for spherical and FP projectiles.

HYDROCODE RESULTS & DISCUSSION

For this study a single Al double bumper Whipple shield configuration was selected. Figure 1 shows the DB geometry. A 0.127cm thick Al 6061-T6 bumper is followed 4.75cm behind by a 0.2032cm thick Al 2024-T3 intermediate bumper. The rear wall is 2.16cm from the intermediate bumper and is 0.3175cm thick Al 2219-T87. This DB Whipple shield was tested at the HIT-F at 7km/s and using the SNL HVL at 10.2km/s. Results for all experiments are contained in Table 1. CTH was implemented to simulate spherical and FP projectiles impacting the DB Whipple shield. For the FP simulations, the thickness of the plate equals the outer bumper thickness (1.27mm). Table 2 summarizes the final simulation results.

Authors have characterized the performance of spacecraft M/OD shields by developing semi-empirical ballistic limit equations. Using test data in Table 1, we may form a DB Whipple shield ballistic limit equation by adjusting the Stuffed Whipple ballistic limit (BL) equation developed by Christiansen (4). Experimental results have established a critical spherical particle diameter near 7mm at an impact velocity of 7km/s. Momentum

scaling is applied for velocities greater than 7km/s and the Christiansen BL equation coefficient is changed such that the BL curve matches experiment. The new DB Whipple shield equation for a 0° impact angle and an impact velocity greater than 6.5km/s is as follows:

$$d_c = 1.328 V^{-\frac{1}{3}} \qquad (1)$$

where d_c is the critical particle diameter in cm, and V is the impact velocity in km/s.

Figure 2 shows spherical experimental and simulation data plotted with the semi-empirical BL curve from Equation (1). FP simulation results are also plotted in Fig. 2. The figure shows good correlation of experimental and simulation results at 7km/s. The plot also shows that spherical simulation data are non-conservative compared with momentum scaling used in Equation (1).

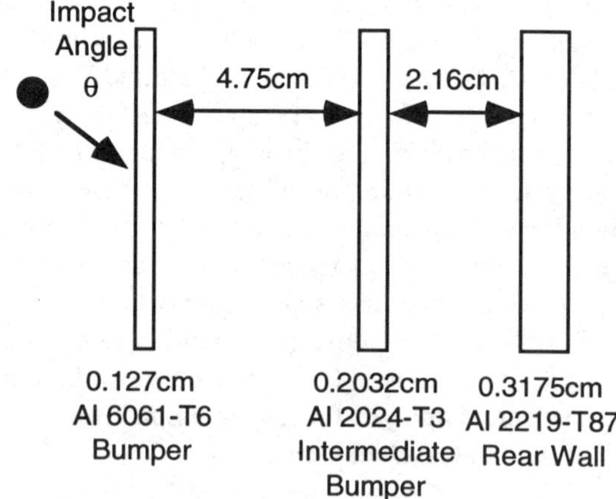

FIGURE 1. The Al double bumper Whipple Shield

TABLE 2. Simulation Data for All Aluminum Double Whipple Shield

Simulation Number	Impact Velocity (km/s)	Proj. Shape (Sphere/FP)	Proj. Size (mm)	Proj. Mass (g)	Equivalent Sphere Dia. (mm)	Rear Wall Damage
S1	7.0	Sphere	10.01	1400	10.01	Perforation
S2	7.0	Sphere	9.51	1200	9.51	Perforation
S3	7.0	Sphere	8.95	1000	8.95	Perforation
S4	7.0	Sphere	7.54	598	7.54	Perforation
S5	7.0	Sphere	6.35	357	6.35	No perforation
S6	14.5	Sphere	10.01	1400	10.01	Perforation
S7	14.5	Sphere	9.51	1200	9.51	Perforation
S8	14.5	Sphere	7.10	500	7.10	No perforation
S9	7.0	FP	9.70	250	5.64	Perforation
S10	7.0	FP	8.12	175	5.01	No perforation
S11	11.18	FP	9.70	250	5.64	Perforation
S12	11.18	FP	8.12	175	5.01	No perforation
S13	14.5	FP	8.12	175	5.01	Perforation
S14	14.5	FP	6.14	100	4.16	No perforation

Clearly, FPs are much more damaging than equal mass spheres. This shape effect can be quantified as the ratio between spherical and FP impactor masses which produce equivalent damage to the rear wall. Typical simulation results for spherical (simulation S7) and FP (simulation S13) impactors are shown in Fig. 3. Notice that the resulting rear wall damage in these simulations is qualitatively similar. The sphere mass is 6.8 times greater than the mass of the FP causing similar damage.

The sphere to FP shape effect ratio equals ~2.2 at 7km/s and ~5-6 at 14.5km/s compared to spherical simulation results. If we compare the FP results to the Equation (1) BL, then the sphere to FP shape effect ratio equals ~2.2 at 7km/s and ~1.6 at 14.5km/s. The latter shape effect ratios justify a conservative design strategy applied to M/OD shielding.

CONCLUSIONS

FP impactors are clearly more damaging than equal mass spheres. Preliminary CTH simulations conducted at the JSC HIT-F indicate that spheres must be more massive than FPs by a factor of ~2.2 at 7km/s and ~1.6 at 14.5km/s to produce similar damage. These shape effect ratios are conservative estimates based on preliminary simulation results. Future testing and impact simulations at the JSC HIT-F will allow for a more complete determination of the FP shape effect.

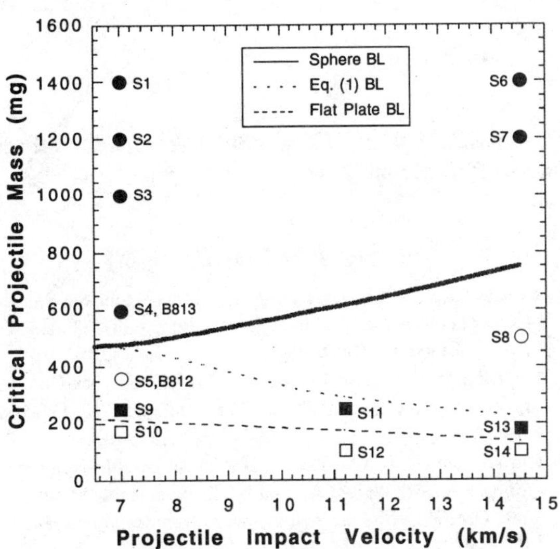

FIGURE 2. Ballistic limit curve for the Al DB Whipple shield. Circles and boxes represent spherical and FP simulation results respectively. A black circle or box represents perforation.

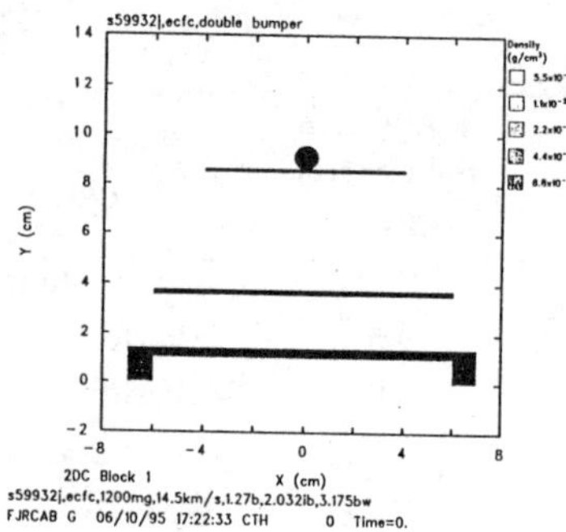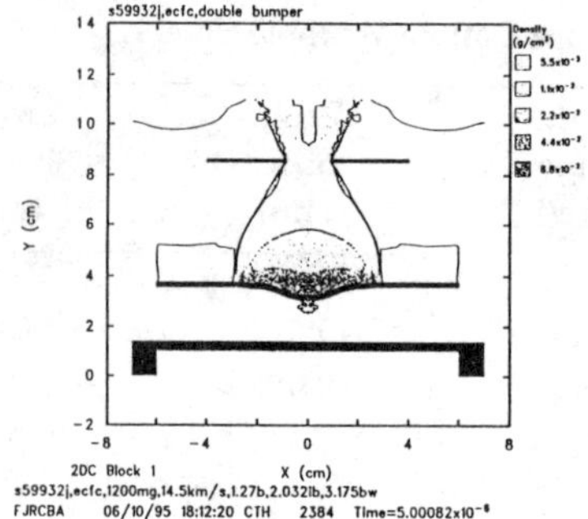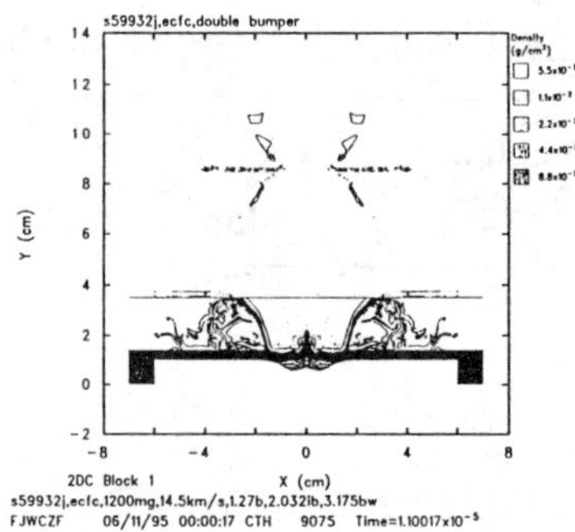

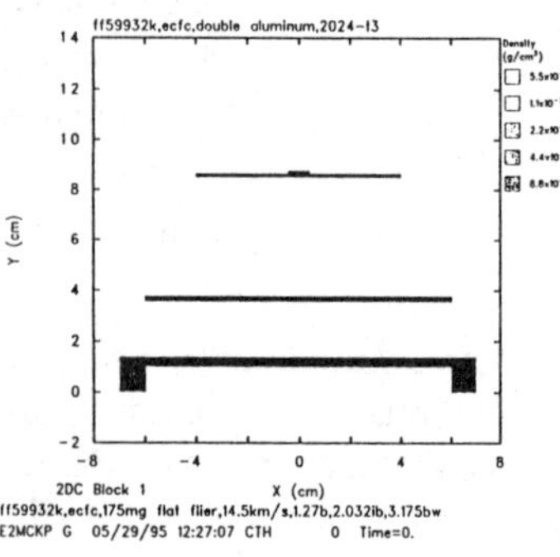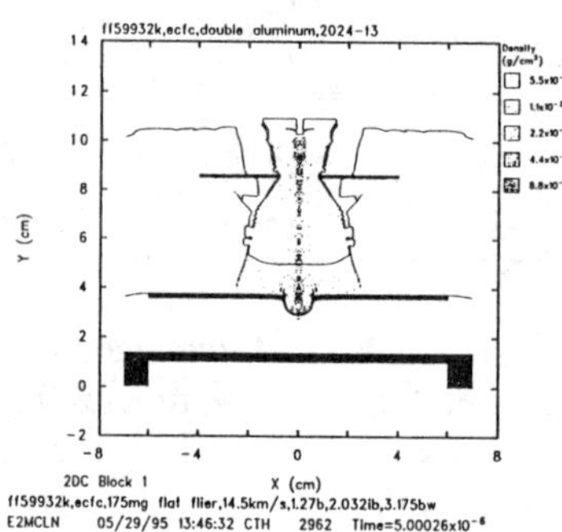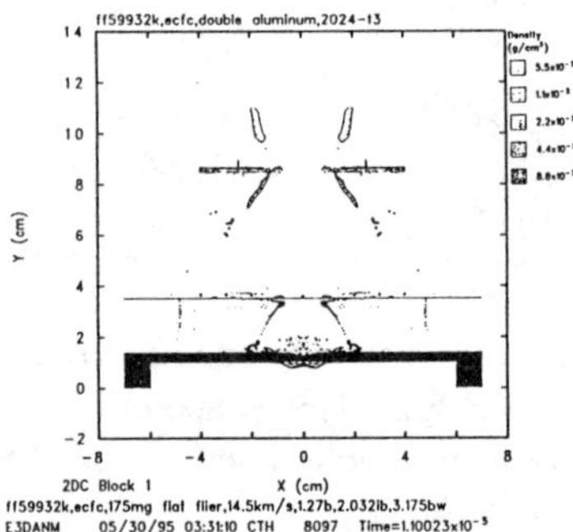

FIGURE 3. Typical CTH simulation results for spherical (simulation S7) and flat plate (simulation S13) impactors at three different time steps.

REFERENCES

1. Crews, J. L., and Christiansen, E. L., "NASA Johnson Space Center Hypervelocity Impact Test Facility (HIT-F)," AIAA Paper Number 92-1640, 1992.
2. Cour-Palais, B. G., and Crews, J. L., "A Multi-Shock Concept for Spacecraft Shielding," *Int. J. Impact Engng,* **10**, 135-146 (1990).
3. Christiansen, E. L., and Kerr, J. H., "Mesh Double-Bumper Shield: A Low-Weight Alternative for Spacecraft Meteoroid and Orbital Debris Protection," *Int. J. Impact Engng,* **14**, 169-180 (1993).
4. Christiansen, E. L., Crews, J. L., Williamsen, J. E., Robinson, J. H., and Nolen, A. M., "Enhanced Meteoroid and Orbital Debris Shielding," *Int. J. Impact Engng,* **17**, To be published (1995).
5. Christiansen, E. L., "Design and Performance Equations for Advanced Meteoroid and Debris Shields," *Int. J. Impact Engng,* **14**, 145-156 (1993).
6. Chhabildas, L. C., Dunn, J. E., Reinhart, W. D., and Miller, J. M., "An Impact Technique to Accelerate Flier Plates to Velocities over 12km/s," *Int. J. Impact Engng,* **14**, 121-132 (1993).
7. Christiansen, E. L., Crews, J. L., Kerr, J. H., Chhabildas, L. C., "Hypervelocity Impact Testing Above 10 km/s of Advanced Orbital Debris Shields," to be published in the proceedings of the 1995 APS Topical Conference on Shock Compression of Condensed Matter, 1995.
8. Reimerdes, H. G., Stecher, K. H., Lambert, M., "Ballistic Limit Equations for the Columbus-Double Bumper Shield Concept," in Proceedings of the First European Conference on Space Debris, 1993, pp. 433-439.

SCALING RELATIONSHIPS FOR EVALUATION OF PERFORMANCE OF ADVANCED SPACE DEBRIS SHIELDS

D. L. Littlefield and B. G. Cour-Palais

Materials and Structures Division, Southwest Research Institute, San Antonio, TX 78228

In a recent paper [1], the authors proposed a scaling law that predicts how a low melting point, high density projectile at a low impact velocity could be used to simulate the damage on a typical spacecraft debris shield by a high velocity fragment. This scaling law differs from the traditional approach to velocity scaling for debris shield impact, since it permits the use of the original shield materials at the lower impact speed. The present work shows how a high sound speed, moderate density, low velocity fragment might also be used to simulate the high velocity impact. Numerical simulations are used to compare and contrast the damage caused by the impacts from the different projectiles.

INTRODUCTION

Several papers (e.g., see refs. 2-5) have appeared in the literature over the past several years concerned with the application of *velocity scaling* to simulate the impact of high-velocity, low density fragments using high density, low melting point, low velocity substitute materials. For example, it has been shown that a zinc projectile impacting a zinc plate at 4.7 km/s can be used to simulate the response of an aluminum/aluminum impact at 10 km/s. This technique allows the simulation of high velocity impacts from aluminum fragments, in ranges that are typically encountered in orbital debris space shields, at much lower velocities that are achievable in the laboratory. The method has been extensively used to simulate the damage incurred on single whipple - backwall arrangements.

However, realistic debris shields will contain more complex components than the simple plate - backwall configuration, including multiple bumper shields and configurations stuffed with momentum-dispersing materials. For such complex shield configurations, it may not be realistic or even possible to alter the shield materials and still obtain a similar response using a low velocity impact. In this paper, an alternate approach to the velocity scaling concept is studied, that permits the retention of the original shield materials in the impact event. Damage characteristics in the shield are retained when impacted by lower velocity, higher density projectiles. The requirements on the projectile material properties are examined in detail.

SCALING ANALYSIS

The foundations of the velocity scaling concept lies in the application of the Buckingham-Pi theorem. The theorem states that the number of dimensionless parameters i needed to completely describe any physical system is related to the number of variables n and the rank of the dimensional matrix r as $i = n - r$. Thus, if the high velocity impact of a projectile on a debris bumper can be described with the nine variables

$$V, \quad l_p, \quad \rho_p, \quad u_{sp}, \quad e_p, \quad l_b, \quad \rho_b, \quad u_{sb}, \quad e_b, \quad (1)$$

where V is the impact velocity; l is a characteristic dimension; ρ, u_s, and e are the density, shock velocity and specific phase change energy; respectively; and the subscripts p and b refer to the projectile and bumper, then the following six dimensionless parameters can be formed:

$$\pi_1 = \frac{l_p}{l_b}; \qquad \pi_2 = \frac{\rho_p}{\rho_b}; \qquad \pi_3 = \frac{e_p}{V^2};$$

$$\pi_4 = \frac{e_b}{V^2}; \qquad \pi_5 = \frac{u_{sp}}{V^2}; \qquad \pi_6 = \frac{u_{sb}}{V^2}; \qquad (2)$$

where π_i denotes a dimensionless parameter. Eq. (2) states that any two physical systems possessing identical values for π_1 - π_6 should also exhibit similar responses. For example, the impact response of a high velocity aluminum projectile on an aluminum shield can be simulated at a lower velocity if the materials involved are substituted with lower melting point, lower sound speed materials. Eq. (2) is the basis for an abundance of numerical and experimental studies [2-5] concerned with the use of velocity scaling to simulate the response of space debris shields.

If the bumper material is held fixed, then Eq. (2) yields no useful scaling information, since it requires the projectile material and impact velocity to remain fixed. Fortunately, however, the π groups given in Eq. (2) are not a unique set of independent dimensionless parameters corresponding to the physical variables given in Eq. (1). An alternate set of π groups can also be derived by replacing certain variables in Eq. (1) by more integrated quantities, such as

$$\rho_p V, l_p, P, u_{sp}, E_p, l_b, \rho_b, u_{sb}, E_b, \qquad (3)$$

where P is the impact pressure and E is a phase change energy per unit volume. Then the six π groups that can be derived from these variables are:

$$\pi_1 = \frac{l_p}{l_b}; \qquad \pi_2 = \frac{\rho_b P}{(\rho_p V)^2}; \qquad \pi_3 = \frac{\rho_b E_p}{(\rho_p V)^2};$$

$$\pi_4 = \frac{\rho_b E_b}{(\rho_p V)^2}; \qquad \pi_5 = \frac{\rho_b^2 u_{sp}^2}{(\rho_p V)^2}; \qquad \pi_6 = \frac{\rho_b^2 u_{sb}^2}{(\rho_p V)^2}. \quad (4)$$

Eq. (4) allows the projectile material to be changed even when the bumper material remains fixed. It states that similar response will be attained if a substitute projectile is used that generates the same impact pressure and has the same momentum per unit volume $\rho_p V$. In addition, Eq. (4) indicates that the shock velocity and phase change energy per unit volume should remain fixed. Thus, Eq. (4) suggests that the impact of a high velocity aluminum fragment could be simulated using higher density, lower melting point projectile at a lower velocity.

However, finding a material that satisfies all six π groups in Eq. (4) when the bumper material is held constant proves to be quite challenging. For instance, if the bulk sound speed for the substitute material is well below the impact velocity, then the impact pressure changes approximately as $\rho_p V^2$. However, Eq. (4) requires both the impact pressure and $\rho_p V$ to remain fixed, so it is not possible to satisfy π_2 using a low- to moderate-sound speed material. On the other hand, if a high sound speed material is used, then π_5 cannot be satisfied, since it requires the shock velocity in the projectile to remain fixed if $\rho_p V$ is held constant. Furthermore, since high sound speed materials typically possess high phase change energies, it also becomes difficult to satisfy π_3.

An alternative to finding a material that satisfies all the parameters in Eq. (4) is to use a material that reasonably satisfies what should be the most important parameters. For example, if polycrystalline Al_2O_3 impacting aluminum at 7.6 km/s is used to simulate an aluminum/aluminum impact at 10km/s, then π_1, π_2, π_4, and π_6 are all closely satisfied. Even better agreement can be achieved if a material with a higher sound speed is used. For example, if an Al_2O_3-*like* material with a bulk sound speed of 13 km/s is used instead, then a 7.1 km/s impact will result in an even better agreement with π_1, π_2, π_4, and π_6 when used to simulate a 10 km/s aluminum/aluminum impact. Although this sound speed is very high, it is not completely an unrealistic value for certain materials, such as a monocrystalline Al_2O_3. Furthermore, if the phase change energy of the material can be reduced to a level somewhat less than the value for Al_2O_3, then π_1 - π_4 and π_6 can all be satisfied.

SIMULATIONS

Numerical simulations were performed using the Eulerian hydrocode CTH [6]. The impact of a 0.969 cm diameter spherical projectile impacting a 0.159 cm thick aluminum bumper was simulated for projectile materials and impact velocities of aluminum at 10 km/s, Al_2O_3 at 7.6 km/s, and an Al_2O_3-*like* material (hereafter referred to as the surrogate Al_2O_3) at 7.1 km/s. An aluminum backwall 0.5 cm thick was placed 30.48 cm behind the bumper in order to record the momentum and damage exerted by the resulting debris clouds.

The mesh used in the simulations consisted of six zones through the thickness of the aluminum bumper, and zones were maintained square radially to a distance of 0.78 cm. Beyond this distance, the radial dimensions of zones were increased at a cumulative rate of 2%. In the 30.48 cm gap region between the two plates, the axial zone dimension was gradually increased at a rate of 0.4% from the bumper and backwall to the midpoint between them. Nineteen zones were used through the thickness of the backwall.

In order to simulate the impact of the surrogate Al_2O_3, it is necessary to build a thermodynamically consistent equation of state. This can be achieved using the ANEOS analytical equation of state [7]. ANEOS uses a series of analytical functions that describe the material states in the solid, liquid, vapor and ionization regimes, including solid-solid, solid-liquid, and liquid-gas phase transitions. Thermodynamic consistency is assured using

ANEOS since all thermodynamic properties are derived from derivatives of the Helmholtz free energy.

ANEOS parameters for the surrogate Al_2O_3 were determined by altering selected parameters from the standard ANEOS Al_2O_3 model to achieve the desired thermodynamic response. This was a nontrivial task that was complicated by the unusual coupling behavior that sometimes occurs when only a single input parameter in the model is changed, causing a "snowball" effect that alters the thermodynamic response over very wide ranges of temperature and pressure. Indeed, using certain combinations for the input parameters in the model can produce very unphysical behavior. The parameters that were changed are listed in Table 1. These changes produced the desired thermodynamic behavior in the surrogate material: a higher sound velocity and lower phase change energy.

Shown in Figs. 1, 2, and 3 are the debris clouds resulting from impacts of the aluminum, Al_2O_3 and surrogate Al_2O_3 projectiles, respectively, at 10 μs after impact. Velocity magnitudes and directions are shown along with the density distribution in the debris cloud. The magnitude of the linear momentum contained each of the clouds is given in the first column of Table 2. As is evident from the figures, there are some distinct differences in the debris generated from the three impacts. The aluminum debris bubble, for instance, is moving faster and is more elongated than either of the Al_2O_3 clouds. However, the linear momentum in each of the debris clouds is remarkably identical, as is shown in the first column of Table 2.

After the debris cloud propagates through the 30 cm gap, it impacts the backwall. The momentum imparted to the backwall is given in the second column of Table 2. Values for the momentum are all within a few percent of each other. Damage to the backwall for the aluminum impact was limited to slight denting, and was confined to a region of about 20 cm^2 in area situated in the center of the backwall. Damage patterns were more localized for the Al_2O_3 projectile, due to impacts from solid particles contained in the debris cloud, but the damage was generally distributed over the same area as the

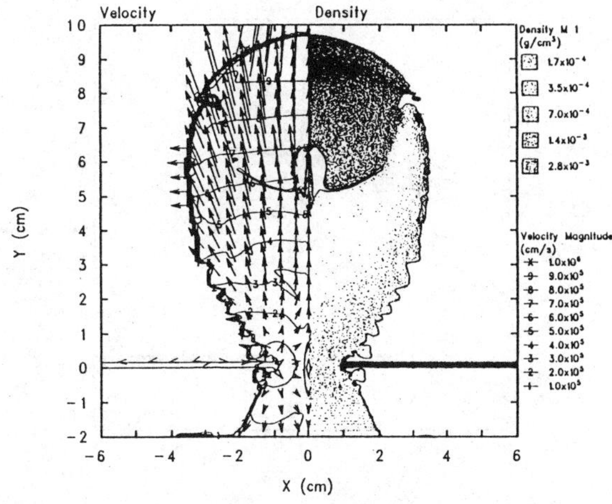

FIGURE 1. Debris cloud generated from impact of an Al sphere at 10 km/s on an Al bumper, 10 μs after impact.

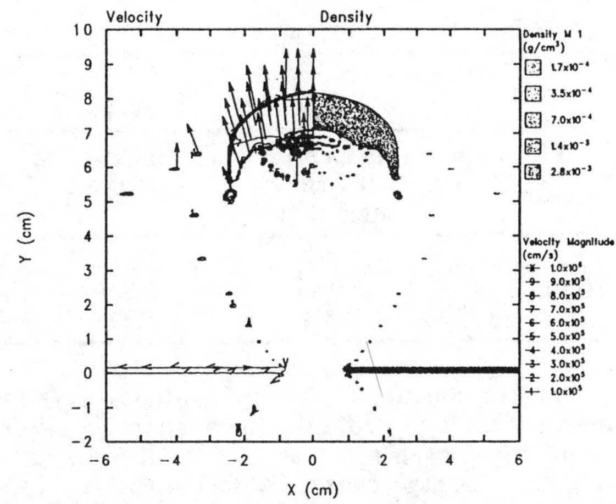

FIGURE 2. Debris cloud generated from impact of an Al_2O_3 sphere at 7.6 km/s on an Al bumper, 10 μs after impact.

TABLE 1. Changes to the standard ANEOS Al_2O_3 model for the surrogate Al_2O_3

Property	Al_2O_3	Surrogate Al_2O_3
Bulk sound speed (km/s)	8.14	13.0
Melt temperature (K)	3482	573
Latent Heat of Fusion (J/kg)	*	5.881 x 10^5
Density ratio at melting	*	0.98

* The solid-liquid phase transition was not included in the standard ANEOS Al_2O_3 model.

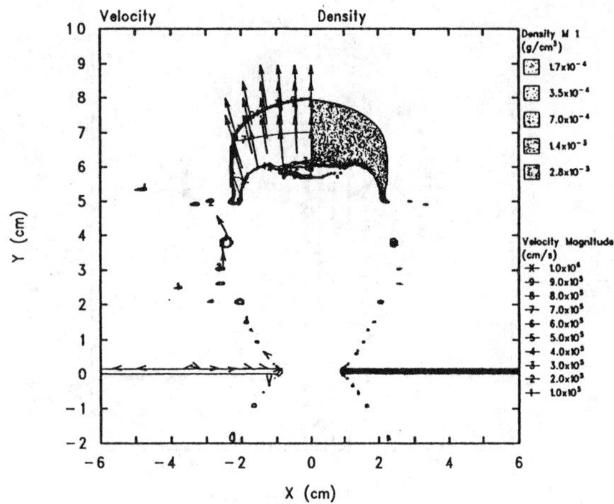

FIGURE 3. Debris cloud generated from impact of a surrogate Al_2O_3 sphere at 7.1 km/s on an Al bumper, 10 μs after impact.

TABLE 2. Momentum values

Projectile	Momentum (Mg cm/s)	
	In Debris Cloud after 10 μs	On Backwall at Late times
Al	1.360	1.367
Al_2O_3	1.485	1.360
Surrogate Al_2O_3	1.390	1.278

aluminum impact. These solid particles were the result of the high melting temperature for the Al_2O_3, which does not change phase as a result of the initial impact. Damage caused by the surrogate Al_2O_3, which had a lower phase change energy, was more uniform than in the Al_2O_3 case, since the impact energy was sufficient to cause partial melting of the projectile. The pattern and extent of the damage was closer to the situation observed in the aluminum impact.

CONCLUSIONS

Results from the simulations have shown how a high sound speed, moderate density projectile at a lower impact velocity could be used to simulate the high velocity impact of a low density fragment. The bumper materials can be held fixed when this methodology is applied. Admittedly, holding the bumper material constant severely limits the range of physical properties that the substitute projectile material can have. However, reasonable agreements

in debris cloud momentum and damage imparted to a backwall were obtained using Al_2O_3 and Al_2O_3-*like* materials at a lower velocity to simulate a high speed aluminum impact. Velocity enhancement factors of about 40% were verified for the Al_2O_3 materials.

REFERENCES

1. Cour-Palais, B. G., Littlefield, D. L., and Piekutowski, A. J., "Using Dense, Low Melting Point Projectiles to Simulate Hypervelocity Impacts on Typical Spacecraft Shields", accepted for publication in *Int. J. Impact Engng.* (1995).
2. Chhabildas, L. C., Hertel, E. S., and Hill, S. A., "Hypervelocity Impact Tests and Simulations of Single Whipple Bumper Shield Concepts at 10 km/sec," *Int. J. of Impact Engng.*, **14** (1-4), pp. 133-144 (1993).
3. Mullin, S. A., Littlefield, D. L., Anderson, Jr., C. E., and Tsai, N. T., "Velocity Scaling of Impacts into Spacecraft Targets at 8 to 15 km/s," *Classified Proceedings of the Hypervelocity Impact Symposium*, (1993).
4. Schmidt, R. M., Housen, K. R., Piekutowski, A. J., and Poormon, K. L., "Cadmium Simulations of Orbital Debris Shield Performance to 18 km/s," *Classified Proceedings of the Hypervelocity Impact Symposium*, (1993).
5. Mullin, S. A., Littlefield, D. L., Anderson, C. E., and Chhabildas, L. C., "An Examination of Velocity Scaling", accepted for publication in *Int. J. Impact Engng.* (1995).
6. McGlaun, J. M., Thompson, S. L., and Elrick, G. A., "CTH-A Three-Dimensional Shock Wave Physics Code," *Int. J. Impact Engng.*, **14** (1990).
7. Thompson, S. L., and Lausen, A. L., "Improvements to the Chart D Radiation-Hydrodynamic Code III: Revised Analytic Equations of State", Sandia National Laboratories Technical Report No. SC-RR-71 0714, (1971).

SPACECRAFT OUTER THERMAL BLANKETS AS HYPERVELOCITY IMPACT BUMPERS

B. G. Cour-Palais

Southwest Research Institute, San Antonio, Texas, U.S.A.

A thermal barrier consisting of a woven fabric outer layer followed by several layers of aluminized mylar insulation has been the primary impact protection against micrometeoroid and orbital impacts for many spacecraft currently in orbit. This paper examines its effectiveness as a hypervelocity "bumper" based on the performance of a NASA space suit. In this case, the thermal barrier consisted of a fabric layer followed by five layers of the aluminized mylar, which shielded either an aluminum rear wall or a rubberized pressure garment. The total areal density of the fabric and mylar layers was 0.052 g/cm^2 and the fabric stand-off was 4 mm from the protected surfaces, with the aluminized mylar filling the space. Test results obtained with hypervelocity aluminum projectile impacts up to 8.5 km/s on the thermal barrier and aluminum wall are described, and a semi-empirical equation for this type of shielding is suggested.

INTRODUCTION

The earliest requirement to demonstrate the hypervelocity impact characteristics of a thermal barrier was during the development of the spacesuits that were worn by the Gemini astronauts during the first American extra-vehicular activities, (EVA). The impact tests conducted at that time by the NASA Johnson Space Center, (NASA-JSC), are described in [1]. Photographic evidence of hypervelocity pyrex glass projectiles being shattered by woven Nomex cloth bumpers, together with the debris impact patterns on aluminum witness sheets, were given in the referenced report. They clearly showed the potential for "soft" bumpers to act like their metallic counterparts, at impacts above 7 km/s. Later tests involved the development of the Space Shuttle spacesuit, [2], and the tests to determine its impact characteristics were done primarily at the NASA-JSC Hypervelocity Imapct Test-Facility, (HIT-F). It is these later tests, described in detail in [3], that led to the subject under discussion in this report. The spacesuit outer thermal barrier also serves as the micrometeoroid and orbital debris impact protection, and for this reason is known as the Thermal Meteoroid Garment or TMG. The cross-section of the TMG considered, shown in Fig. 1, consists of an "orthofabric" outer layer, (0.048 g/cm^2), five layers of

aluminized mylar, (0.0047 g/cm^2), and a liner made from neoprene-coated nylon, (0.031 g/cm^2). In the tests described in this report the TMG was mounted directly in front of aluminum sheets representing "hard" areas of the space suit, or the pressure garment layers representing the "soft" areas. These suit configurations were tested in an evacuated impact chamber attached to either the 1.78 mm and or the 4.32 mm bore two-stage light gas guns at HIT-F, depending on the size of the aluminum or nylon projectile used for the test. Further details of the test procedures are given in [3].

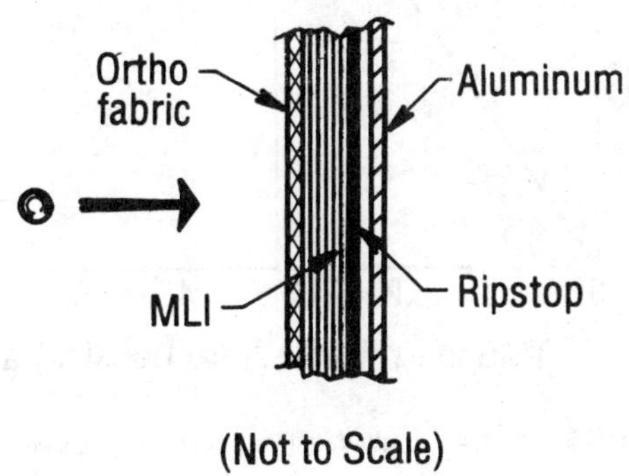

FIGURE 1. Typical cross-section of a test set-up showing the TMG and aluminum rear-wall.

TEST RESULTS

Three tests were conducted in the Gemini series, using a fabric which had an areal density of 0.023 g/cm^2. Pyrex glass spheres 0.037, 0.079 and 0.157 cm diameter, impacted the fabric "bumper" at 7.29, 7.39 and 7.43 km/s, respectively. An 1100-0 aluminum witness sheet was placed 12.7 cm behind the bumper in each test, and the crater distributions on their surfaces are also shown in [1]. The maximum diameters of the area covered by the larger craters were scaled off the photographs in [1], and the total included angle, from an origin at the bumper, was obtained as a function of the outer layer to projectile areal density ratio. A plot of the result is shown in Fig. 2. The reason for discussing the Gemini fabric outer layer bumper effect on projectile breakup in detail is that the Shuttle space suit outer layer will be assumed to act in the same manner.

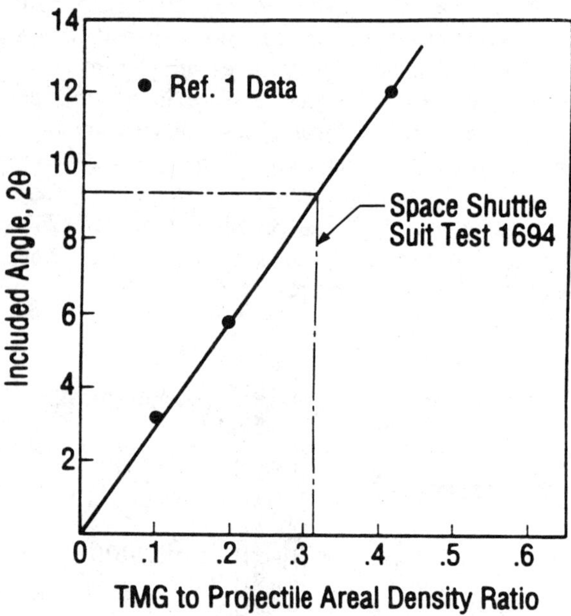

FIGURE 2. Included angle data from the Gemini space suit.

If we consider No. 1694 , one of the Shuttle space suit tests that were conducted at HIT-F with the TMG in front of aluminum sheets, and assume that the 0.079 cm aluminum projectile at 7.46 km/s will breakup on impacting the Shuttle woven fabric bumper, then from Fig. 2 we see that the total angle subtended should be about 9 degrees, for a bumper to projectile areal density ratio of 0.31. Although the debris will not disperse much in the 0.4 cm distance it travels to the aluminum sheet wall, it will be finely divided and heated. In addition, it is contended that the debris from the bumper impacts the intervening MLI layers, which act to repeatedly shock the debris to higher thermal states, like the multi-shock bumper system described in [4] and [5]. As a result, the final effect on the aluminum sheet is an impulsive load distribution which is more Poissonian than Gaussian. The evidence that this happens is shown in the line drawings, Figs. 3, 4 and 5, which depict cross-sections through the impact center-lines of three tests. In the first two figures, the effect of the TMG is

Test No. 1879

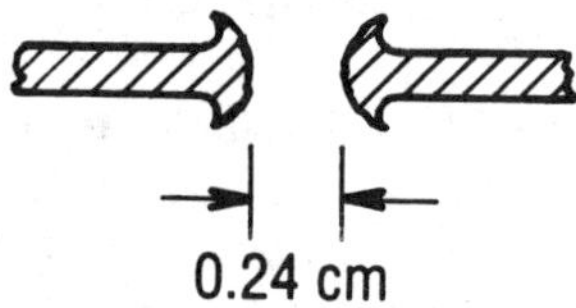

FIGURE 3. Typical hole in a 0.41 mm aluminum 6061-T6 sheet without the TMG caused by a 0.79 mm aluminum sphere at 7.27 km/s.

seen in the very different hole characteristics under almost identical conditions. In Fig. 3, the hole has the typical hypervelocity, double-lipped formation seen in a soft aluminum alloy. On the other hand, the concentrated impulsive load created by the TMG

Test No. 1694

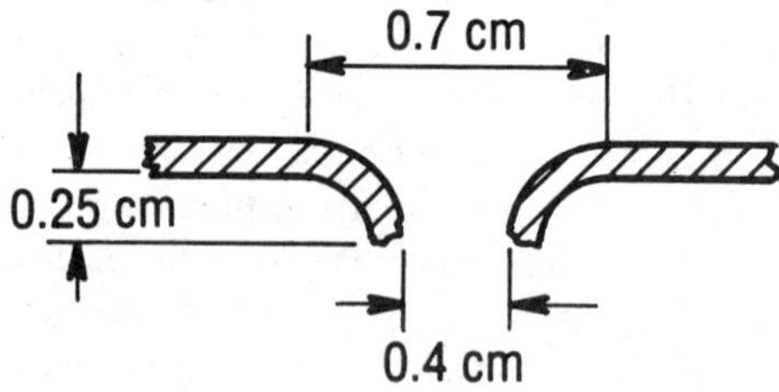

FIGURE 4. TMG effect on the same 0.41 mm sheet caused by a 0.79 mm aluminum sphere at 7.46 km/s. Note the impulsive load and tensile failure

multiple-shock process caused a deeply drawn indentation, followed by tensile failure of the "cap", as seen in Fig. 4. Notice that the presence of the TMG has increased the size of the hole in the 0.041 cm thick aluminum from 0.24 cm to 0.4 cm. Figure 5 shows the effect of the impulsive load on a thicker aluminum sheet placed behind the TMG caused by nearly three times the initial impact momentum as for the others. Now we have a shallow lipped crater, which shows that the debris cluster impacted at a velocity high enough to cause material flow, and an irregularly-shaped rear spall. The combined effect is a hole with an equivalent diameter of 0.22 cm. Even though the projectile diameter is larger than for test No. 1694, the bumper to projectile areal density is 0.28, and the debris included angle is still about 9 degrees from Fig. 2. Hence, it is assumed that the multi-shocking process is still at work for test No. 1750, and the result shown in Fig. 5 is still caused by an intense impulsive load.

Test No. 1750

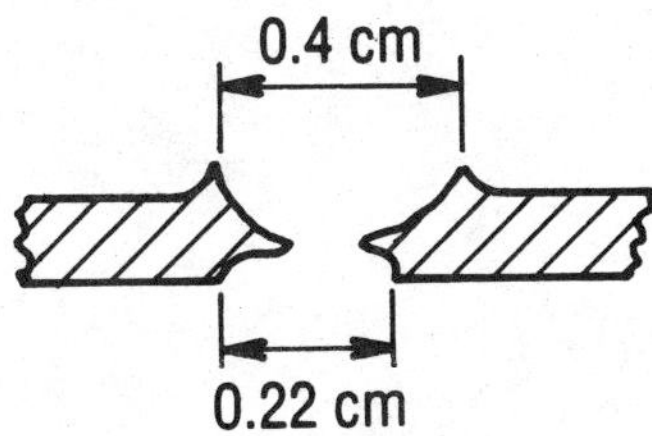

FIGURE 5. TMG effect on a 2.3 mm aluminum 6061-T6 sheet showing more usual crater and spall combination. The 1.15 mm aluminum projectile impacted at 6.75 km/s.

ANALYSIS

The results obtained using the Shuttle TMG with the aluminum sheets just described suggests that the impact process can be considered as multi-shock caused by the presence of the aluminized mylar layers behind the initial fabric bumper. As a result it is valid to use the same expression derived in [4] to determine the areal density or thickness of the aluminum or the combination of the TMG plus aluminum. The areal density of an aluminum wall that is not perforated or spalled is given by the equation:

$$m_w = C \times M_p \times V_p \times S^{-2} \times [2.76 \times 10^8 / Y_t]^{0.5} \; ; \; g/cm^2 \quad (1)$$

where: M_p = projectile mass, g ; V_p = projectile velocity, km/s ; S = bumper to wall distance, cm ; Y_t = yield stress, N/m^2 ; C = coefficient, s/km.

In the equation, the initial impact momentum is transferred as an impulse to the aluminum wall, because it is assumed that very little mass is added to the debris by the bumper or the mylar insulation. The term in the square brackets is a material factor that allows the use of aluminum alloys other than the 6061-T6 used in the tests. The coefficient C was experimentally determined from the shuttle suit tests that resulted in no perforation or detached rear spall. The calculated results using Eq. 1 are given in Fig. 6 with a curve for the same aluminum thicknesses without a TMG, for the same failure conditions. The unprotected aluminum thicknesses were calculated using the single sheet hypervelocity equation given in [3], for the no perforation, attached spall case, and validated by two tests which were done at HIT-F.

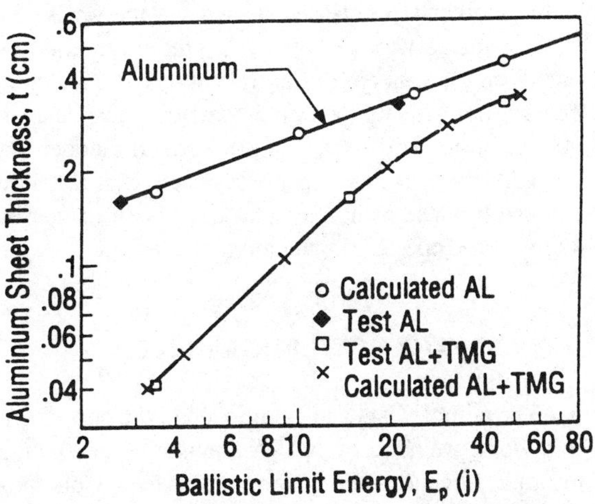

FIGURE 6. Comparison aluminum sheet with and without the TMG. Both curves are seen to fit test data.

It is obvious from Fig. 6 that the TMG is very effective with thin aluminum sheets, but that its effectiveness decreases as the aluminum thickness increases, as was seen in Figs. 4 and 5. However, a residual TMG effect remains as indicated by the two curves becoming parallel but offset.

The discussion so far has been for the results for the Shuttle TMG, placed in front of aluminum sheets of varying thicknesses. An equation to calculate the thickness of aluminum to stop a given hypervelocity threat without spallation has been shown to be a valid predictor. The coefficients used in the equation were derived from a few test points and finally a design curve such as Fig. 6 was needed to complete the picture. This same procedure can be followed for any thermal barrier and aluminum wall combination by using the total areal density of the combination instead of the aluminum wall areal density, m_w, in equation 1.

CONCLUSIONS

It has been shown that the use of a thermal barrier placed in front of an aluminum wall element is very effective in reducing the thickness of the aluminum required to withstand hypervelocity impact. Although the thermal barrier affect is more pronounced for the thinner aluminum sizes, a residual benefit remains even at the thicker sizes. Based on work done on NASA space suits, it was shown that a predictive equation based on the concept of multiple-shock shields could be applied to the thermal barrier. The use of non-metallic walls such as a pressurized bladder was not addressed in this paper. However, it is a natural extension of the principles that have been discussed, and will be reported on at a another time.

ACKNOWLEDGEMENTS

The author wishes to acknowledge the support for this work provided by NASA Johnson Space Center through Lockheed Engineering Services Company, Houston, TX, under Subcontract No. 02N0100754, and the technical support and advice received from Ms. Jeanne Crews, NASA, and Dr. Charles Anderson, Jr, Southwest Research Institute.

REFERENCES

1. W. E. McAllum, "Development of Meteoroid Protection for Extravehicular-Activity Space Suits," AIAA Hypervelocity Impact Conference Proceedings (1969) Paper No. 69-366.
2. B. Remington, D. Cadogan and J. Kosmo, "Enhanced Softgoods Structures for Spacesuit Micrometeoroid/Debris Protective Systems," SAE 22nd International Conference on Environmental Systems (1992) Paper No. 921258.
3. B. G. Cour-Palais, "Review of the EMU Ballistic Limit Tests for Spacesuit Meteoroid and Orbital Debris Protection Capability," SwRI Technical Report (# TBD), 1995.
4. B. G. Cour-Palais and J. L. Crews, "A Multi-Shock Concept for Spacecraft Shielding," *Int. J. Impact Eng.* 10 (1989) 135-146.
5. B. G. Cour-Palais and A. J. Piekutowski, "The Multi-Shock Impact Shield, in: Shock Compression of Condensed Matter," 1991, eds Schmidt, Dick, Forbes and Tasker (Elsevier Science Publishers B.V, (1992).

PROJECTILE SHAPE INFLUENCE ON BALLISTIC LIMIT CURVES AS DETERMINED BY COMPUTATIONAL SIMULATION

E. S. Hertel, Jr. and L. C. Chhabildas

Computational Physics Research and Development Department, 1431
Sandia National Laboratories
Albuquerque, New Mexico, USA 87185-0819

A requirement for an effective debris shield is that it must protect a spacecraft from impacts by irregularly shaped particles. A series of numerical simulations has been performed using the multi-dimensional shock physics code CTH to numerically determine the ballistic limit curve for a Whipple bumper shield. Two different projectile shapes are considered for the numerical simulations, flat plates of varying diameters with a constant thickness and spheres of varying diameters. The critical diameter or ballistic limit was determined over a velocity range from 4 km/s to 15 km/s. We have found both experimentally and numerically that the particle shape has a significant effect on the debris cloud distribution, ballistic limit curve, and penetration capability.

INTRODUCTION

A requirement for an effective debris shield is that it must protect a spacecraft from impacts, both from the meteoroid and orbital space debris environments. The micrometeoroid environment is thought to consist of dust size particles with an average impact velocity of 20 km/s, while the orbital debris environment is believed to be millimeter to centimeter size particles with an average velocity of 10 km/s. The orbital debris environment (1) is thought to be more hazardous because of the relatively large mass. The Whipple bumper (2) is a space shield designed to protect a spacecraft from the more hazardous orbital space debris environment. Design of these bumper shields is currently done by a combination of experiments at accessible velocities and analytical models developed from scaling theories (3).

In this paper, calculated ballistic limit curves and experimental and computational debris cloud distributions will be presented and discussed in detail. The use of simulation tools allows us to perform numerical experiments that are not tractable in the laboratory. Even though we do not consider CTH (4) to be fully validated for the class of problems discussed here, it has been shown to replicate many important features of high velocity impacts.

COMPUTATIONAL SCHEME

Since only normal impacts were considered, the computational geometry was two-dimensional axisymmetric. The computational scheme represents a simple Whipple bumper shield with a sheet of 1.27 mm 6061-T6 aluminum. The rear sheet consists of 3.175 mm 2219-T87 aluminum with a void space of 101.6 mm between the bumper and rear sheet. The projectile was modelled as 6061-T6 aluminum.

Our experience with CTH has shown that four computational zones across the thickness of a material is the minimum number necessary to resolve the shock structure in that material. For these simulations, we chose to increase the computational resolution to ~8 zones across the thickness of the bumper shield. In the region of initial impact, the zones are square (0.015 cm x 0.015 cm). The zones are gradually increased in size radially. Longitudinally, the zones are first increased in size and then decreased, so that in the region of the secondary impact (where the debris cloud impacts the rear sheet), the zones are again square (0.015 cm x 0.015 cm).

The different types of aluminium modelled in this simulation were treated using an elastic-plastic approximation for material strength with a Von Mises yield surface. Failure in tension was modelled by a simple cutoff scheme that relieves tension by introducing void into a computational cell when the mean principal stress exceeds the tensile strength. Identical equations-of-state were used for the three types of aluminum, namely, the SESAME table developed by Kerley (5). This table was scaled to take into account the higher density of the 2219-T87 aluminum (2.851 gm/cm^3 versus 2.7 gm/cm^3 for 6061-T6).

For each velocity, the diameter of the plate or sphere was varied until penetration of the rear sheet occurred, with the minimum change in diameter being 0.025 cm and 0.02 cm for the plate and sphere, respectively. For all simulations, computational penetration was characterized as an intact rear sheet, that is, one with no perforations. This definition allows a rear sheet to have detached spall and still be characterized as not penetrated computationally. This definition may result in some systematic differences between the current ballistic limit curves as developed from experiments at accessible velocities and analytical models (3) that are velocity dependent. Computationally, spall is a major component to rear sheet damage at velocities below ~7 km/s. If the computational no-penetration condition did not allow for detached spall, the computational ballistic limit would be lowered below 7 km/s, but would not change significantly at high velocities.

COMPUTATIONAL RESULTS

Figure 1 shows the ballistic limit as determined by CTH for normally impacting plates and spheres in the velocity range of 4 to 15 km/s. The ballistic limit curve is drawn halfway between the penetration/no-penetration locations as determined by CTH. The differences between the two curves are striking. The ballistic limit for flat plate impacts shows little variation as a function of velocity. An inspection of the debris clouds from flat plate impacts shows amazing consistency over the range of impact velocities. The debris clouds due to a flat plate impacting normally are collimated with minimal lateral dispersion (see Fig. 2 for a typical debris cloud). Even though the total momentum of the debris

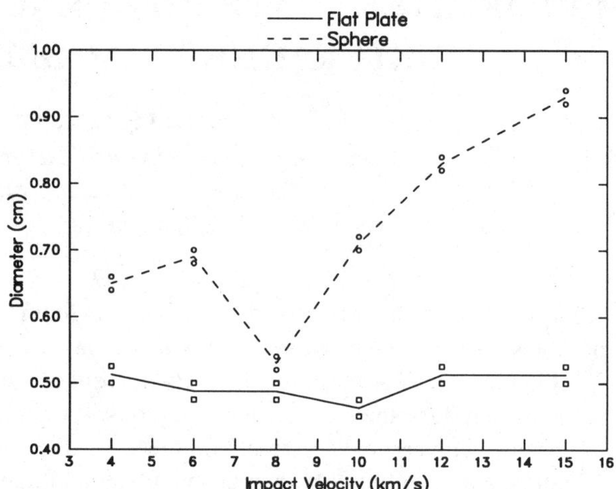

FIGURE 1. Computational ballistic limit for plate and spheres

cloud increases with increasing velocity, Fig. 1 implies that the impulse is relatively constant over the range of impact velocities. The converse of this statement is true for spherical normal impacts (see Fig. 3 for a typical debris cloud). The lateral dispersion for the spherical projectiles depends significantly on the impact velocity. At low velocities (<7 km/s), the projectile and bumper fracture and exhibit little lateral dispersion. At impact velocities near 8 km/s, the projectile and bumper start to melt and lateral dispersion increases. At higher impact velocities (>10 km/s), vaporization of the projectile and bumper occur and lateral dispersion increases.

The ballistic limit for spherical projectiles shows reasonable agreement with experimental data (6) over the 4-7 km/s velocity range accessible by light gas gun technology. CTH tends to over-predict rear sheet damage in this range, most likely due to the relatively simple material strength model. A strain hardening strength model would decrease predicted damage and may improve the correlation with experiment. The current extrapolations (3) beyond the experimental regime rely on analytic scaling assumptions. It is in this velocity regime that the curve determined using CTH and the analytic extrapolation techniques differ significantly. In this velocity range, CTH is predicting that the amount of shock induced vaporization increases as the velocity increases. The increased vaporization results in increased lateral dispersion and a reduction in damage due to the initial impulse.

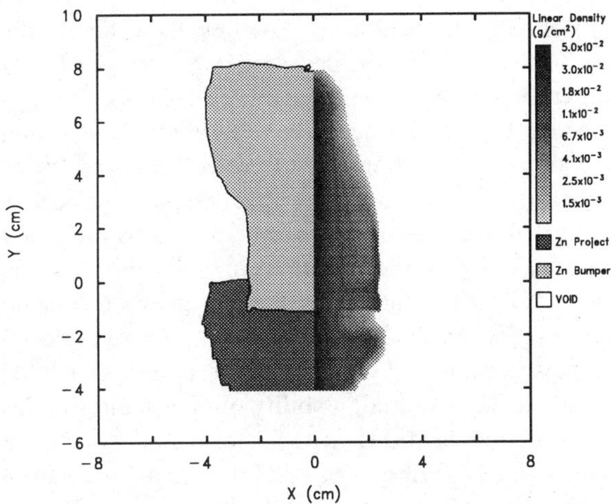

FIGURE 2. CTH debris cloud for a plate impact.

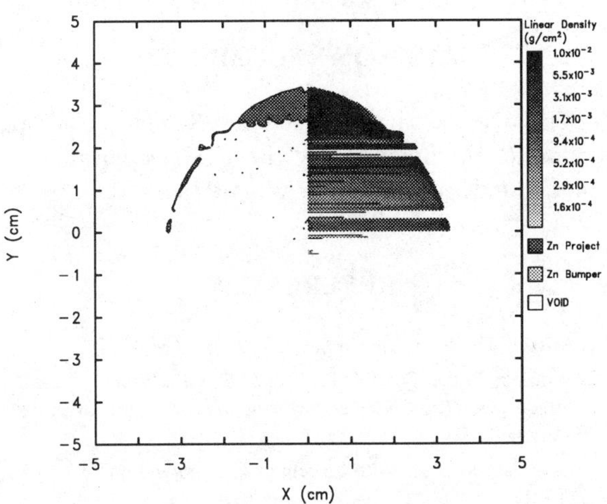

FIGURE 3. CTH debris cloud for a spherical impact.

Figure 4 displays a radiograph taken when a thin plate impacts a sheet similar to that of a Whipple bumper shield. Figure 5 displays a similar radiograph taken when a sphere impacts the same sheet. Experimentally, significant differences between the two debris clouds can be seen. The plate impacts results in a collimated debris structure with a high density remnant at the leading edge of the cloud. The

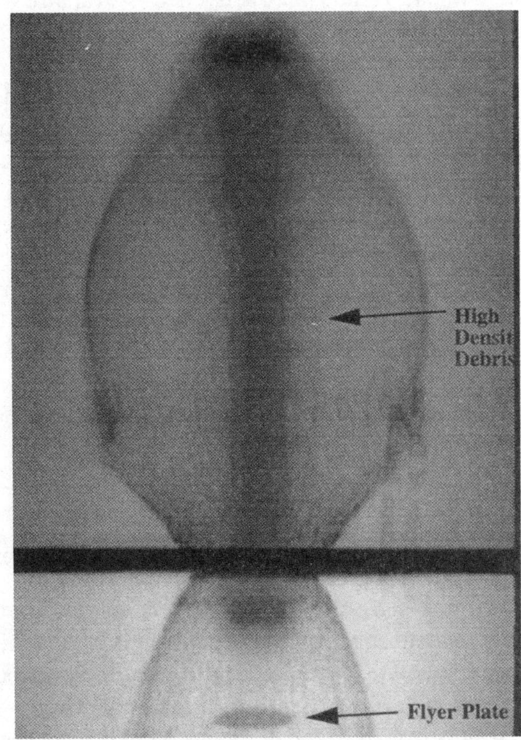

FIGURE 4. Experimental debris cloud for a disk impacting a thin sheet.

FIGURE 5. Experimental debris cloud for a sphere impacting a thin sheet.

1181

sphere impact results in a diffuse debris structure with a higher density leading edge.

The CTH simulations similar to these are seen in Figures 2 and 3. The right and left side of the figures represent different information from CTH. The right side of each figure displays a grey scale that is proportional to the density integrated along the line of sight, similar to the data collected by a radiograph. The left side of each figure displays the material locations by color, where light grey represents the target contribution to the debris cloud and dark grey represents the projectile contribution. Due to a limited computational domain used in these simulations, some portion of the trailing debris structure is lost.

In Fig. 2, the results from right side of the image indicate a collimated debris cloud and a closed tulip shape of the lower density debris that is similar to the data from Fig. 4. Although not visible in the figure a closer examination of the results indicates a high density structure at the leading edge of the predicted debris cloud also similar to the radiograph. In Fig. 3, the results from left side of the image indicate a shell-like leading edge of the cloud and a higher density structure in the front third of the cloud that is also similar to the data from Fig. 5.

SUMMARY

CTH has been utilized extensively for a wide variety of impact and high explosive related applications (7). In recent years, work has also been done with CTH related to debris generation and secondary impacts (8,9) similar to those that are in this paper. This Whipple bumper shield simulation may be the most difficult problem to confront in an Eulerian hydrocode. The wide range of impact velocities entails a wide range of physics that controls the secondary penetration. For all velocities, the wide range of dimensions (from ~1 mm bumper thicknesses to ~300 mm bumper to rear sheet separations) inherent in these simulations can cause significant numerical difficulties. At low velocities, the fracture dynamics and material strength models are important, but at high velocities, the equation-of-state dominates the predictions. Phase change phenomena and phase separation may need to be included in hydrocodes before better agreement with experiment can be expected.

Below ~7 km/s impact velocities, the CTH results for spherical projectiles agree with available experimental data. At higher velocities, CTH can predict the differences in debris structure and secondary penetration due to projectile shape (8) as noted in the previous section. The difference in penetrating ability of plates and spheres as predicted here by CTH is significant. At higher velocities, the lack of experimental data to develop equation-of-state information limits the confidence that can be placed on these simulations. Even with these limitations, the difference in penetrating ability of plates and spheres as predicted by CTH is significant and should be investigated further. In addition, the deviation between the CTH predictions for spherical projectiles and the analytic extrapolations is substantial in the high velocity regime.

ACKNOWLEDGMENTS

This work performed at Sandia National Laboratories supported by the U.S. Department of Energy under contract DE-AC04-94AL85000.

REFERENCES

1. Kessler, D. J., *Adv. Space Res.*, 5 (1985)

2. Whipple, F. L., "Possible Hazards to Satellite Vehicles from Meteorites," *The Collected Contributions of Fred L. Whipple*, Volume 2.

3. Christiansen, E. L., NASA Technical Memorandum SN3-91-42 (1991)

4. McGlaun, J. M., Thompson, S. L., Kmetyk, L. N., and Elrick, M. G., Sandia National Laboratories SAND89-0607 (1990)

5. Kerley, G. I., *Int. J. Impact Engng*, 5 (1987)

6. Christiansen, E. L., Personal Communication (1991)

7. Hertel, E. S., Sandia National Laboratories SAND90-0610 (1990)

8. Hertel, E. S., R. L. McIntosh, and Patterson, B. C., 1994 AAIA Space Programs and Technologies Conference, Paper 105, September (1994)

9. Hertel, E. S., R. L. McIntosh, and Patterson, B. C., *Int. J. Impact Engng*, 5 (1994)

HYPERVELOCITY IMPACT TESTING ABOVE 10 KM/S OF ADVANCED ORBITAL DEBRIS SHIELDS

Eric L. Christiansen[1], Jeanne Lee Crews[1], Justin H. Kerr[1], and
Lalit C. Chhabildas[2]

[1]*NASA Johnson Space Center, SN3, Houston, TX 77058*
[2]*Sandia National Laboratories, 1433, Albuquerque, NM 87185*

NASA has developed enhanced performance shields to improve the protection of spacecraft from orbital debris and meteoroid impacts. One of these enhanced shields includes a blanket of Nextel[TM] ceramic fabric and Kevlar[TM] high strength fabric that is positioned midway between an aluminum bumper and the spacecraft pressure wall. As part of the evaluation of this new shielding technology, impact data above 10 km/sec has been obtained by NASA Johnson Space Center (JSC) from the Sandia National Laboratories HVL ("hypervelocity launcher") and the Southwest Research Institute inhibited shaped charge launcher (ISCL). The HVL launches flyer-plates in the velocity range of 10 to 15 km/s while the ISCL launches hollow cylinders at ~11.5 km/s. The >10 km/s experiments are complemented by hydrocode analysis and light-gas gun testing at the JSC Hypervelocity Impact Test Facility (HIT-F) to assess the effects of projectile shape on shield performance. Results from the testing and analysis indicate that the Nextel[TM]/Kevlar[TM] shield provides superior protection performance compared to an all-aluminum shield alternative.

INTRODUCTION

Man-made orbital debris has grown over the past several decades to become the dominate hypervelocity threat to low Earth orbit (LEO) satellites and spacecraft. The flux of orbital debris now exceeds the natural meteoroid environment for particle sizes above ~1mm diameter at altitudes between ~300km and ~1500km. New hypervelocity impact protection strategies and shielding concepts have been developed by NASA to meet the orbital debris threat.

Several low-weight, enhanced performance shields have been developed by the NASA HIT-F to improve spacecraft protection from meteoroid/orbital debris (M/OD) impact over that offered by conventional 2-sheet Whipple shields (1, 2). For instance, the baseline shielding for those areas of the space station modules that are exposed to the highest concentration of M/OD impacts will incorporate a blanket between the outer aluminum bumper and inner pressure wall that combines two materials: Nextel[TM] ceramic fabric and Kevlar[TM] high strength fabric. These shields are referred to as "stuffed Whipple" shields. Figure 1 shows a typical example although many different stuffed Whipple shield configurations have been tested at the JSC HIT-F (3).

Extensive testing of Nextel[TM]/Kevlar[TM] stuffed Whipple and other multi-wall shields have been performed using light gas gun (LGG) to launch projectiles up to velocities of ~8 km/s. To extend the impact database to higher velocities, NASA JSC sponsored a multi-year effort to obtain test data at >10 km/s to assess the performance of advanced spacecraft shields. This paper will describe results of >10 km/s tests on advanced shielding concepts using two launcher systems: the Sandia National Laboratories (SNL) flyer plate hypervelocity launcher (HVL) (4,5) and the Southwest Research Institute (SwRI) inhibited shaped-charge launcher (ISCL) (6). The projectiles from both of these launchers are non-spherical which influence shielding performance. JSC HIT-F is conducting light-gas gun tests with non-spherical shapes as well as hydrocode simulations to determine the effects of projectile shape in the HVL/ISCL tests and to use this data to extend the ballistic limit curves for spherical projectiles to velocities above 10 km/s.

HIGH-SPEED DATA (>10 KM/S)

The stuffed Whipple and all-aluminum test configurations are shown in Fig.2. The stuffed Whipple test article is a 67%-scale version of the stuffed Whipple shield given in Fig.1. In one test, the target was a 50%-scale version.

Table 1 provides data from five SNL flyer plate tests on stuffed Whipple and all-aluminum shields. Undeformed dimensions of the flyer plate used in these experiments are tabulated in Table 1. The flyer is a thin aluminum or titanium disk that is subject to both deformation and tilting. In some cases, the flyer remains relatively flat in a slightly bowed (i.e., cupped) shape while in others the deformation is more significant and the plate folds into a more compact shape. Orthogonal flash x-rays are used to evaluate shape and integrity of the projectile prior to impact (Fig.3). A measure of the extent of flyer deformation and tilting is given in Table 1 by the ratio of maximum projectile length (along line of flight) to diameter (or width) normal to the line of flight. For instance, the flyer in test SWBS-2 is folded nearly in-half along one axis (~180° bend) and is bent ~120° along an orthogonal axis. The resulting shape has a maximum length to diameter of ~2 which from hydrocode (7,8) and experimental results (9) is more penetrating to multi-wall shields than a flat plate impact.

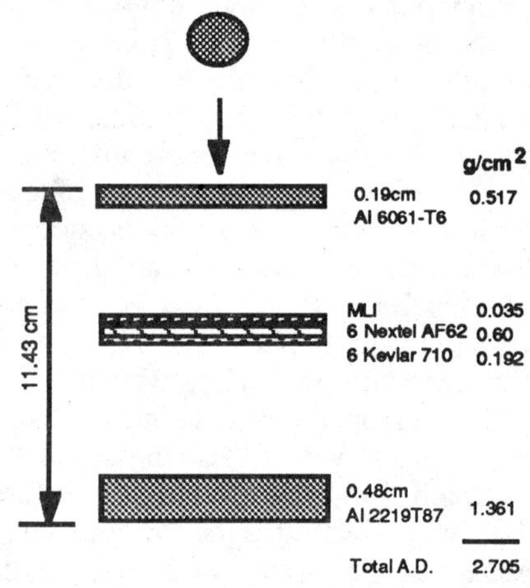

FIGURE 1. Stuffed Whipple Shield (Generic Full-Scale)

The SNL test results indicate the stuffed Whipple shield provides superior protection compared to the all-aluminum shield alternative (compare SWBS-2 and SWBS-3). The all-aluminum shield suffered a large perforation whereas the stuffed Whipple did not (Fig.4). Tthe all-aluminum shield is completely penetrated 20 µs after impact, and the perforation continues to grow with time. The stuffed Whipple rearwall shows a growing bulge through 100 µs after

impact, but no perforation. The bright spot in the center of the growing bulge in Fig.4 appears to be the result of reflection from lighting for the camera; it does not resemble the fast expansion and typical pattern of hot gas exiting a perforation.

The flyer plate tests also demonstrate that the stuffed Whipple shield is more effective at oblique impact angles than for impacts normal to the shield. This characteristic is particularly advantageous for spacecraft shielding since most orbital debris and meteoroid impacts will be oblique (10). However, this property is not shared by some shields such as the Whipple shield, which can be more vulnerable to oblique impact than normal impacts in the 4-11 km/s velocity range (2). Further information on the HVL tests can be found elsewhere (11).

An inhibited shaped charge capable of launching aluminum projectiles in excess of 11 km/s has been developed at SwRI over a number of years (6). NASA sponsored a number of ISCL tests on the stuffed Whipple and all-aluminum shield configurations given in Fig.2. The ISCL launches an aluminum projectile at velocities that are consistently between 11 and 11.5 km/s. The projectile is in the shape of a hollow cylinder (or thick-walled pipe) with length to outside diameter ratio (L/D) of from 1 to 3. Projectile mass typically ranges from 0.8g to 1.5g. Projectile mass is determined by SwRI from digitized X-ray images of the projectile. Two sets of orthogonal flash x-ray images capture the size, shape, and orientation of the projectile prior to impact. General observations from the ISCL tests are as follows:

(1) The stuffed Whipple shield defeated a 0.85g projectile at 0° (normal impact angle) while the all-aluminum shield was perforated.

(2) In 45° impacts, the stuffed Whipple provided successful protection from 0.6g to 1.5g ISCL projectiles (5 tests), while the all-aluminum shield failed in 3 tests from 0.9g to 1.0g projectiles.

(3) The stuffed Whipple provided better protection from 45° oblique impacts than normal impacts.

Details and results of the ISCL tests conducted to date on the stuffed Whipple shields are given elsewhere (3,6).

Impact data (45°) from the HVL and ISCL tests is compared in Fig.5 with predicted ballistic limits for the full-scale stuffed Whipple shield (after scaling diameter by 1.5 to account for the 67% scale test articles). The >10 km/s data shows that non-penetrations have occurred for the stuffed Whipple shield above the predicted ballistic limit (BL)

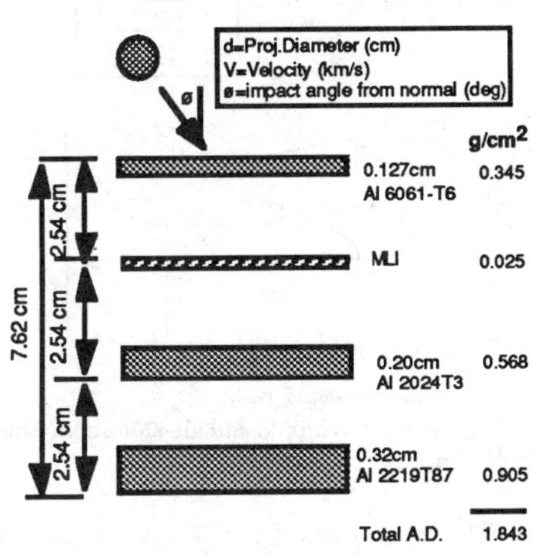

FIGURE 2. Test Configuration, 67%-scale shields: (a) Stuffed Whipple (b) All-Aluminum

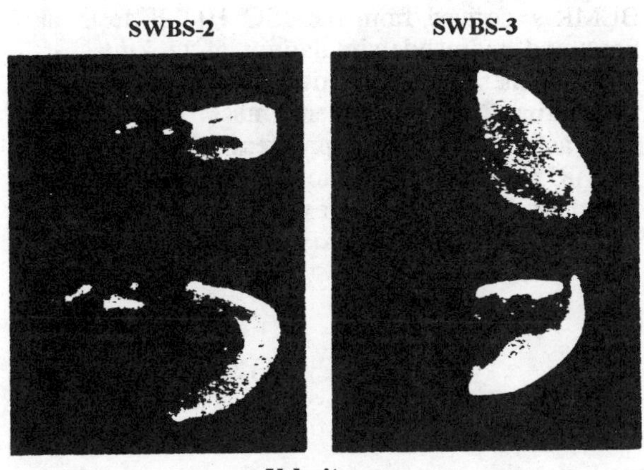

FIGURE 3. Orthogonal x-ray views of the flyer for SNL tests SWBS-2 and SWBS-3. The flyer in SWBS-2 is deformed and fragments follow closely behind projectile which increase the damage to the target. Even so, the target rearwall held (no perforation) in SWBS-2 (Stuffed Whipple), whereas SWBS-3 (all-aluminum shield) failed.

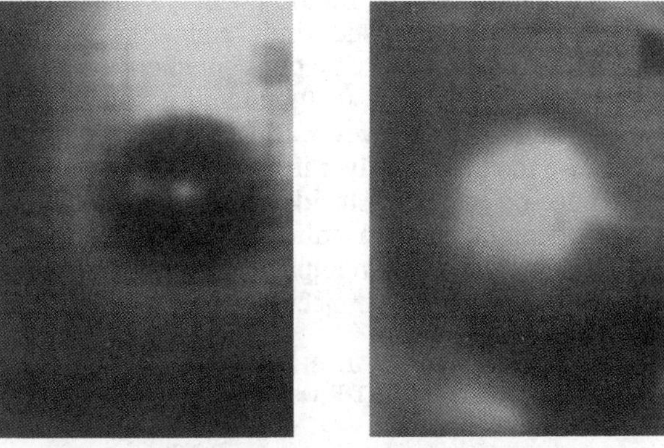

FIGURE 4. High-speed camera images of back of rearwall for all-aluminum shield from SNL tests SWBS-2 & 3 at 60μs. Stuffed Whipple shield (SWBS-2) shows buldge while all-aluminum shield (SWBS-3) has large perforation. (1cm squares on back of rearwall).

TABLE 1. HVL Flyer Plate Tests

SNL No.	Shield Type*	Impact Angle (deg)	Impact Vel. (km/s)	Flyer Dia. x thk. (mm)	Flyer Mass (g)	Flyer Max Length/ Width	Flyer Integrity	Rear Wall Damage
SWBS-2	SW-67%	0	9.9	18.9x1.0	0.75	1.7	V. bowed, trailing fragments	No Perf (large bulge)
SWBS-5	SW-67%	0	10.0	17.4x0.9	0.57	0.4	Slightly bowed	No Perf (large bulge)
SWBS-6	SW-67%	45	10.1	17.3x0.9	0.56	0.5	Slightly bowed	No Perf (small bulge)
SWBS-4	SW-50%	0	14.7	5.92x1.0	0.12	6.1	Flat, tilted on-edge	Perforated (pinhole)
SWBS-3	All-Al	0	10.2	18.9x1.0	0.75	1.0	Bowed, tilted	Perforated (large)

*note: SW=Stuffed Whipple, All-Al=all-aluminum 67%-scale shield

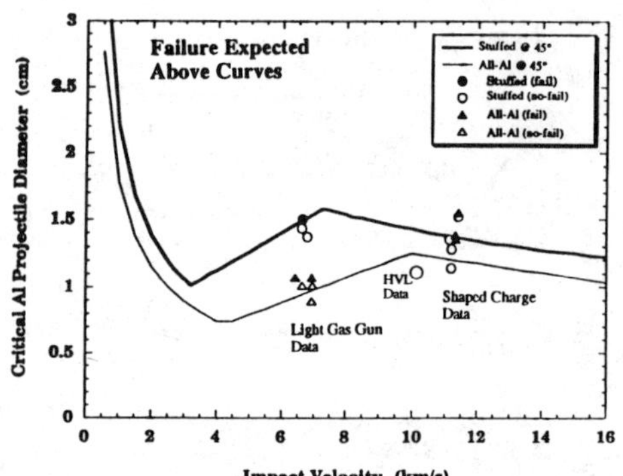

FIGURE 5. Full-Scale stuffed Whipple and all-aluminum shield ballistic limits for 45° impacts (3)

indicating the BL equations in (3) are somewhat conservative for the stuffed Whipple shield at 45°. All HVL and ISCL data on the stuffed Whipple shield was predicted accurately (or conservatively) by the current equations (3), except for test SWBS-6 which resulted in a perforation with a 0.12g titanium flyer while the ballistic limit equations predicted non-penetration (the equations predict the shield should defeat up to a 0.20g titanium sphere at 10 km/s and 0°). However, this result is expected because the titanium flyer in SWBS-6 rotated 90° to hit edge-on into the shield which is known to be more penetrating than solid spheres of equivalent mass (9). The edge-on impact orientation results in a relatively long L/D projectile.

PROJECTILE SHAPE EFFECTS

The effect of projectile shape must be understood to complete the assessment of the ISCL and HVL data. The JSC HIT-F is conducting research on shape effects on shield performance in two areas: LGG testing and hydrocode investigations. The HIT-F has launched hollow aluminum cylinders at 7 km/s similar in size and mass to the ISCL projectiles. This data and LGG data for spherical projectiles is used to determine the ballistic limit mass ratio (BLMR) which is the ratio of the mass of a sphere on the ballistic limit of a shield to the mass of a non-spherical projectile at the ballistic limit. The BLMR is expected to vary with projectile shape and impact velocity. A BLMR above 1 indicates the particular non-spherical shape is more penetrating than a solid spherical projectile. Table 2 lists BLMR's derived from the JSC HIT-F tests at 7 km/s and hydrocode simulations at 11 km/s (3, 7, 8). In general, non-spherical projectiles are more penetrating than equivalent mass solid spheres. This activity is still in progress. Eventually, the BLMR's will be applied to the ballistic limit mass found by the HVL/ISCL tests, which will allow the ballistic limit equations to be raised accordingly in the V>10km/s region.

TABLE 2. BLMR's (Ballistic Limit Mass Ratios) for 3-wall shields from JSC HIT-F tests and hydrocodes.

Shield Type	Proj. Shape	Velocity (km/s)	BLMR	Method
All-aluminum	Hollow Cylinder	7	1.3	Test
All-aluminum	Hollow Cylinder	11	1.9	Hydrocode
All-aluminum	Flat Plate	11	1.7	Hydrocode
Stuffed Whipple	Hollow Cylinder	7	1.2	Test

CONCLUSIONS

Both SNL HVL and the SwRI inhibited shaped charge launcher provide valuable data and are effective techniques to assess the performance of meteoroid/debris shielding at velocities in excess of 10 km/s.

REFERENCES

1. Crews, J.L. and Christiansen, E.L., NASA Johnson Space Center Hypervelocity Impact Test Facility (HIT-F), AIAA Paper No.92-1640, 1992.
2. Christiansen, E.L., *Int. J. Impact Engng*, **14**, 145-156 (1993).
3. Christiansen, E.L., Crews, J.L., Williamsen, J.E., Robinson, J.H., and Nolen, A.M., *Int. J. Impact Engng*, **17**, Proceedings of the 1994 Hypervelocity Impact Symposium (1995).
4. Boslough, M.B., Ang, J.A., Chhabildas, L.C., Reinhart, W.D., Hall, C.A., Cour-Palais, B.G., Christiansen, E.L., and Crews, J.L., *Int. J. Impact Engng*, **14**, 95-106 (1993).
5. Chhabildas, L.C., Dunn, J.E., Reinhart, W.D., and Miller, J.M., *Int. J. Impact Engng*, **14**, 121-132 (1993).
6. Walker, J.D., Grosch, D.J., and Mullin, S.A *Int. J. Impact Engng*, **17**, Proceedings of the 1994 Hypervelocity Impact Symposium (1995).
7. Kerr, J.H. *et al.*, Hydrocode Simulations of Advanced Orbital Debris Shielding, presented at the 1995 APS Shock Compression of Condensed Matter Conference, August 1995.
8. Cykowski, E., Hydrocode Modeling of Projectile Shape Effects on a Double Bumper All-Aluminum Whipple Shield, LESC 31649, JSC HIT-F, April 1995.
9. Cour-Palais, B.G., Piekutowski, A.J., Dahl, K.V., and Poorman, K.L., *Int. J. Impact Engng*, **14**, 193-204 (1993).
10. Christiansen, E.L., Hyde, J., and Snell, G., Spacecraft Survivability in the Meteoroid and Debris Environment, AIAA Paper No. 92-1409, 1992.
11. Chhabildas, L.C., Results of HVL tests at 10 km/s to 15 km/s on NASA Johnson Space Center Advanced Debris Shields, SNL test report (draft), 1995.

IMPACT-GENERATED ATMOSPHERIC PLUMES: OBSERVATIONS ON JUPITER AND IMPLICATIONS FOR EARTH

M. B. Boslough and D. A. Crawford

Sandia National Laboratories, Albuquerque, New Mexico

Computational simulations of the impacts of comet Shoemaker-Levy 9 (SL9) fragments on Jupiter provide a framework for interpreting the observations. A reasonably consistent picture has emerged, along with a more detailed understanding of atmospheric collisional processes. The knowledge gained from the observations and simulations of SL9 has led us to consider the threat of impact-generated plumes to satellites in low-Earth orbit (LEO). Preliminary simulations suggest that impacts of a size that recur about once per century on Earth generate plumes that rise to nearly 1000 km over an area thousands of km in diameter. Detailed modeling of such plumes is needed to quantify this threat to satellites in LEO. Careful observations of high-energy atmospheric entry events using both satellite and ground-based instruments would provide validation for these computational models.

INTRODUCTION

The multiple impacts of comet SL9 fragments with Jupiter in July 1994 provided an historic opportunity to directly observe the phenomena resulting from hypervelocity collisions on a planet. Detailed analysis of this event has advanced our understanding of comets, of Jupiter, and of the collisional processes that shaped the solar system. This improved understanding can now be used to develop better models for the assessment of the impact threat to Earth.

The principal reason we performed computational simulations of the SL9 impacts was to take advantage of a "natural experiment" to validate Sandia's shock physics codes, CTH and PCTH, for an impact involving velocities, masses, and kinetic energies many orders of magnitude higher than had ever before been witnessed. By simulating a natural astronomical event, the validation could be based on observational, rather than on experimental data. Additional reasons for our work were to provide predictions to help guide astronomical observations of the event (see pre-impact prediction papers, 1-3), and to assist astronomers in interpreting the observational data (4-5). More recently, we have added another objective: to apply the lessons learned from this event to issues of national and economic security. In this paper we briefly describe our interpretation of observations of the SL9 impact plumes, and present a pre-liminary analysis of the implications of a much smaller impact event on Earth. Details of all our calculations can be found in the papers listed in the References section.

OBSERVATIONS ON JUPITER

Figure 1 presents a side-by-side comparison of a 3-D simulation of plume evolution following the impact of a 3 km diameter ice fragment with a sequence of images collected by the Hubble Space Telescope during the impact of one of the largest fragments (6). In this simulation, the plume reaches an altitude nearly twice that observed for several plumes observed by Hubble, suggesting that the actual diameter of the fragments (or swarms of particles) was somewhat smaller than 3 km at the time of entry into Jupiter's atmosphere.

The geometry and timing of the series of impacts could hardly have been better for making useful observations from Earth. With the impact location only a few degrees beyond Jupiter's limb (horizon), the hot debris (fireball) ejected by each collision had to rise only a few hundred kilometers to become visible. It could then be seen in profile, making it possible to observe its shape and size. The vantage point from Earth was close to perpendicular to the trajectory of the fragments, so that the effect of impact obliquity could be seen. Because the impact point was beyond the limb, the time of arrival of debris above the line-of-sight altitude could be measured. Combining this information with the time of impact extracted from direct measurements from Galileo (and in some cases from Earth), the fireball trajectory can be determined. The position of Jupiter (near quadrature) put the luminous debris in shadow when it first rose into view, making it possible to determine its brightness. This configuration also means that additional trajectory information can, in principle, come from the time of arrival of the fireball into sun-light, as it is condensing and evolving into a high plume. Morphology information can be extracted from the shadow-line on the plume. Furthermore, each impact site was on the side of Jupiter (near local dawn) that immediately rotated into view from Earth, giving the fireball a velocity component toward the limb, and making it possible to observe the pattern of debris and wave phenomena immediately after impact (see discussion in 4). This best-case impact configuration allows many direct comparisons to be made between simulations and observations.

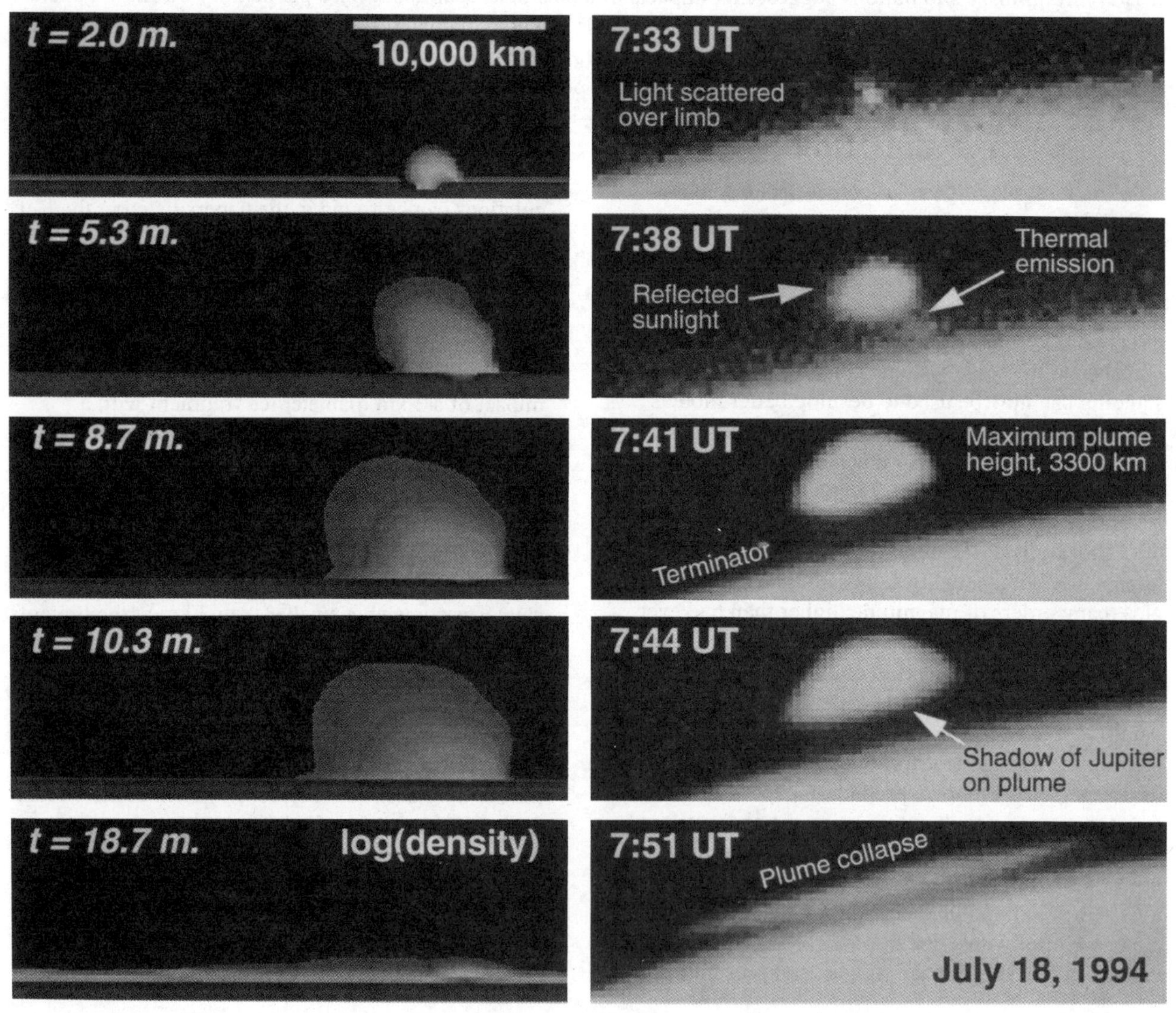

FIGURE 1. Left: Simulation of 3-D fireball/plume evolution after the impact of a 3-km diameter fragment. Shading indicates log(density) with a cutoff at 10^{-12} g/cm^3; times are in minutes after impact. Right: Sequence of G plume images collected by Hubble Space Telescope (6).

IMPLICATIONS FOR EARTH

Figure 2 suggests that the physics of atmospheric entry and plume generation is similar over many orders of magnitude in the scale of impactor kinetic energy and physical size. The figure makes a direct comparison between the 3-D fireball simulation of Crawford et al. (5) and an eyewitness artist's depiction of the February 12, 1947, Sikhote-Alin fireball in Siberia. The Sikhote-Alin impact energy was 10-20 kilotons, whereas the fireball simulated in (5) was for a 6 million megaton impact event, nearly a billion times as energetic.

The series of impacts on Jupiter has shown that ballistic impact fireballs and plumes are ejected to very high altitudes, and that explosive expansion of shocked atmosphere along the entry column is highly directional and poorly modeled by point explosions. These observations suggest that satellites in low-Earth orbit (LEO) may be vulnerable to ejection of material into their environment by an impact into the atmosphere. Because of the high orbital velocities of these satellites (about 7 km/s), even a very low-density plume ejected into their path would be catastrophic.

To test this idea, we have performed preliminary 2-D simulations of the plume generated by a 34-m diameter stone (density=3 g/cm^3) impacting at 20 km/s with vertical incidence. The kinetic energy of the impactor is equivalent to an explosive yield of 3 megatons of TNT, and the expected frequency of such an event is about once per century. The details of this calculation are given by Boslough and Crawford (7). For comparison, the evolution of the buoyant fireball generated by a point-source explosion of the same magnitude is shown.

This preliminary analysis suggest that satellites at low altitude are indeed at risk from plumes due to impacts as small as a few megatons. This newly-rec-

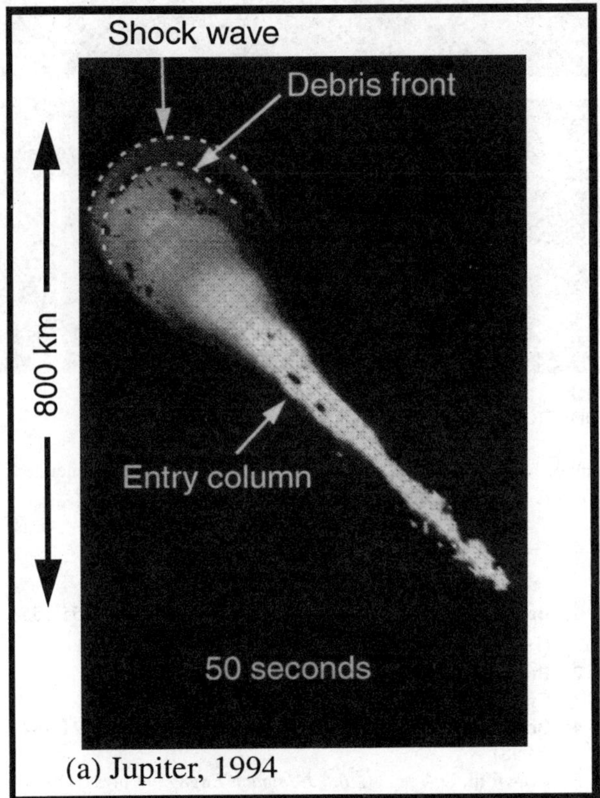

FIGURE 2. Comparison of (a) 3-D simulation of impact of 3-km diameter fragment on Jupiter, 50 seconds after entry (Crawford et al., 5) with (b) artist's depiction of 1947 Sikhote-Alin impact.

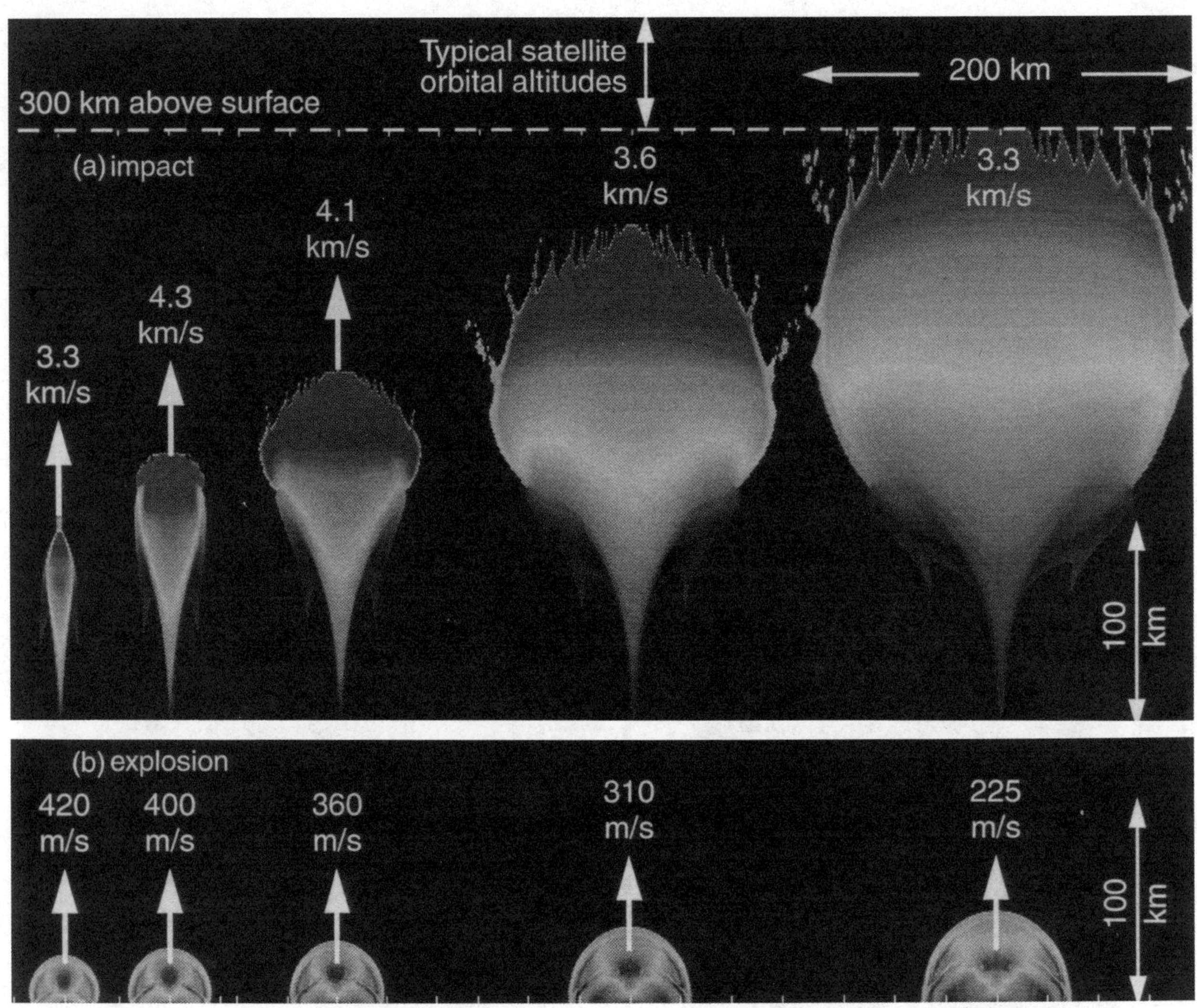

FIGURE 3. Comparison on the same scale of evolution of (a) ballistic fireball generated by 3 megaton impact to (b) buoyant fireball generated by a 3 megaton explosion at same altitude of maximum energy deposition (7 km). Shading indicates velocity magnitude of the air.

ognized threat should be examined by further modeling and by extensive observational validation by gathering data on the continuous impact flux of smaller objects.

ACKNOWLEDGMENTS

This work was supported by the United States Department of Energy under Contract DE-AC04-94AL85000, and funded by the LDRD program.

REFERENCES

1. Boslough, M.B., et al., *Geophys. Res. Lett.* **21**, 1555-1558, (1994).
2. Boslough, M.B., et al., *Eos* **75**, 305-310, (1994).
3. Crawford, D.A., et al., *Shock Waves* **4**, 47-50, (1994).
4. Boslough, M.B., et al. *Geophys. Res. Lett.* **22**, 1821-1824, (1995).
5. Crawford, D.A., et al., *Int. J. Impact. Engng.*, in press, (1995).
6. Hammel, H.B., et al., *Science* **267**, 1288-1296, (1995).
7. Boslough, M.B., and Crawford, D.A., *Procs. Planetary Defense Conf.*, in press (1995).

CHAPTER XVI

HIGH VELOCITY LAUNCHERS AND SHAPED CHARGES

A TRIAL OF THE THREE-STAGE LIGHT-GAS GUN WITH A PREHEATING STAGE

K. Kondo, Y. Hironaka, and H. Ito,
Res. Lab. Engr. Matls., Tokyo Institute of Technology, Nagatsuta, Midori, Yokohama 226, Japan
H. Sugiura,
Faculty of Science, Yokohama City University, Seto, Kanazawaku, Yokohama 236, Japan
S. Ozaki,
TRY Engineering Inc., 1-1-5 Kaizuka, Kawasaki-ku, Kawasaki 210, Japan
and
A. Takeba and M. Katayama
CRC Research Institute Inc., 2-7-5 Minamisuna, Koto, Tokyo 136, Japan

A three-stage light-gas gun having a preheating and filling stage of gas was designed and installed in order to independently regulate the initial energy of the light gas. Although the operation and the performance of gun become so complicated, the computer simulations for the gun indicate many advantages; e.g., a scale of the gun much decreases in comparison with a conventional two-stage light-gas gun at the same value of the maximum obtainable projectile velocity.

INTRODUCTION

Two-stage light-gas guns have been used widely in the world for more than three decades. As they are available for many purposes, e.g., the generation of high shock-pressures and the impact and penetration of projectile, and so on, many types of guns are designed and installed. However, their principle and basic structure are the same and have been unchanged so far.

A representative two-stage light-gas gun operates as follows. A powder charge is burned at one end of a large and long pump tube, accelerating a relatively heavy piston to typically in the range 0.2-1.0 km/s. The piston whose kinetic energy is being transferred from chemical energy of the powder compresses a working gas, usually helium or hydrogen, through volumetric and/or shock-wave compression, producing high pressures of thousands of atmospheres and temperatures estimated at several thousand degrees in Kelvin. A burst diaphragm which is placed just behind the projectile sustains the pressure fairly high. This reservoir of hot, high-pressure gas then accelerates a relatively small projectile in a gun barrel. Many parameters of instrumentation and operation, e.g., burning rate and mass of powder, piston mass, gas fill-pressure, burst diaphragm opening pressure, projectile mass, and so on, must be adjusted for obtaining the optimum performance of the gun.

The maximum velocity routinely being achieved by using hydrogen gas is limited to be less than 7-8 km/s. Although there still exist many unknown parameters for energy losses such as friction, heat conduction, viscosity, and so on, various types of computer simulation codes already are developed and used to optimize the gun performance(1-3).

The physical limits to projectile velocity are the sound speed of the working gas and dissipative losses in the flow. The engineering limits are the stresses in the tapered section as a reservoir and in the barrel. Some new concepts of multi-stage gun have already been proposed for increasing the projectile velocity (4,5).

In order to simply achieve a higher projectile

velocity, temperatures or energy of the gas must be increased with the increase of compression ratio, so that the ratio of diameters between the pump tube and the barrel may be increased. In this case, however, it is difficult to regulate the timing between the increasing rate of pressure in the reservoir and the movement of the projectile running away, because a little piston-displacement induces rapid changes in gas pressures. Contrarily, kinetic energy of the piston quickly loses because of reactant of gas pressure effective to a large area of the piston. This situation makes the engineering problems serious.

Therefore, the conventional gun for obtaining a velocity higher than 8 km/s, such as NASA gun (2), consists of a relatively long and massive pump tube which is 15.18-m long and 64.4 mm in inner diameter for a barrel 3.87-m long and 12.7-mm in inner diameter. However, such a long pump tube induces the difficulty of operation and maintenance. A reduction in pump tube length and mass, while maintaining gun performance, would be very desirable.

The purpose of this study is, as a trial, to construct a three-stage light-gas gun having a pre-

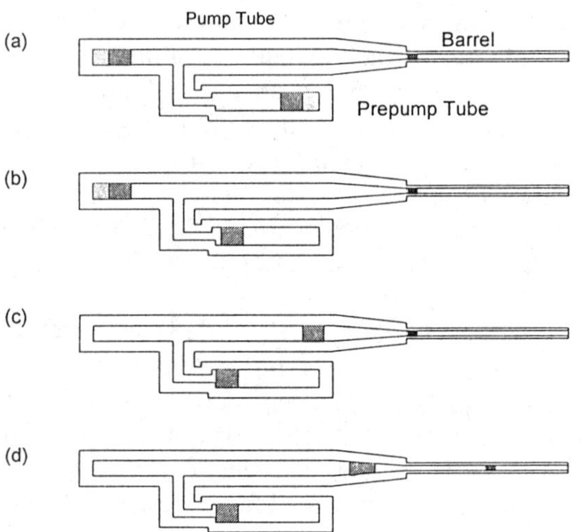

FIGURE 1. Schematic illustrations for the principle and performance of the three-stage light-gas gun.

pump tube for preheating and filling the gas and to examine the effects of independent changes in the initial energy of the working gas.

PRINCIPLE AND DESIGN

Figure 1 shows schematic illustrations for the performance of the three-stage light-gas gun installed. The working gas is filled at a certain pressure into a prepump tube and a pump tube. By firing powder in the prepump breech, a big piston is accelerated and sweeps the working gas into the pump tube through a connection port, while adiabatically compressing the gas. When the big piston approaches the final part of the prepump tube, another breech is moved by air cylinders and the pump tube is sealed. After that, powder in the breech is fired, and the gun is operated as well as a conventional two-stage light-gas gun. The gas pressure and the gas temperature are measured in the connection port, and the signals provide a timing signal for those actions.

Figure 2 shows the outline of the three-stage light-gas gun fabricated in this study. All the unit are mounted on a considerably heavy steel base 10.5-m long and 0.9-m wide. A prepump tube, the first stage, is 2.4-m long and 200 mm in inner diameter. A big piston which is made of plastics and aluminum is 200-mm long and 200 mm in diameter. A pump tube, the second stage, is 4.2-m long and 50 mm in inner diameter. The pump tube is divided at a distance of 0.6 m from a breech. This short pump tube with the breech is moved quickly by two air cylinders and is connected and locked in a connection port. A reservoir has a straight section of 110-mm long and a tapered section 240-mm long with a taper of 4-degrees. A launch barrel, the third stage, is 3.58-m long and 11.7 mm in inner diameter. A target chamber has an approximately 400-l volume, while the total volume of the prepump tube, the connection port, and the pump tube is 78 l. The maximum precompression ratio is calculated to be 9.0 by subtracting the remaining volume, so that the maximum filling temperatures may be 1300 and 720 K for helium and hydrogen, respectively.

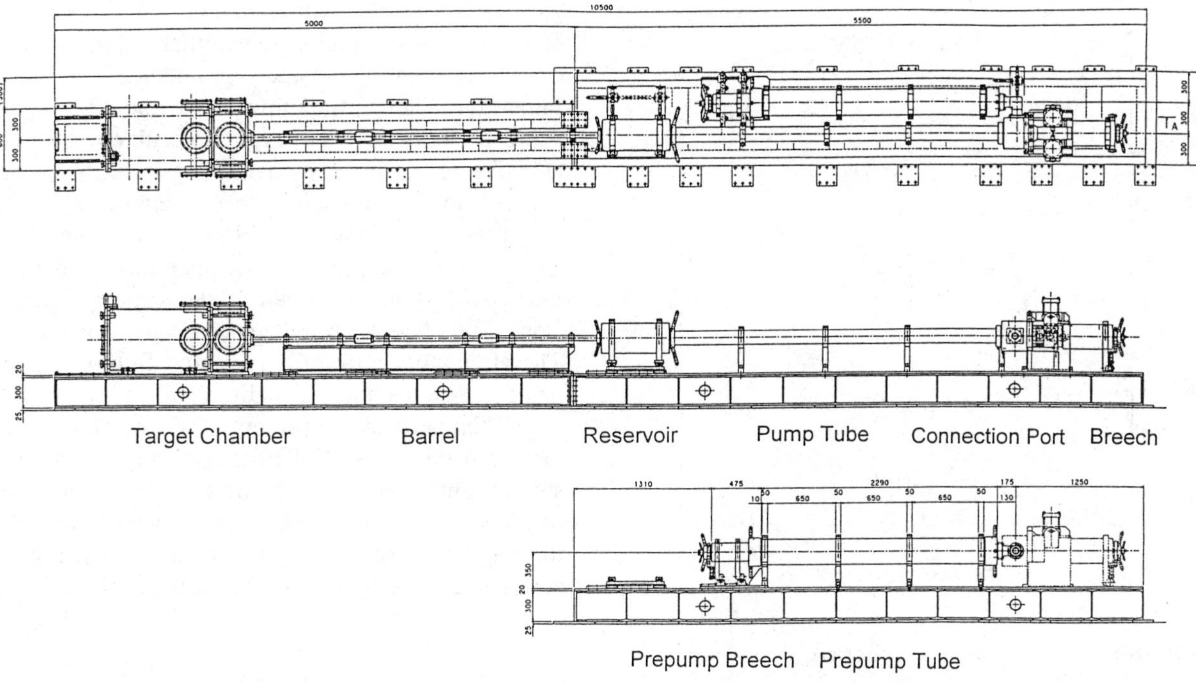

FIGURE 2. Outline of the three-stage light-gas gun fabricated and installed in this study. the total length is 10.5 m.

SIMULATION

A convenient computer code (3) developed by K. Kondo well represents the performance of the conventional two-stage light-gas gun of Tokyo Institute of Technology, HS-3B, which has a size similar to the present new gun. The code is confirmed by AUTODYN[TM]-2D simulations and by the combustion-gas pressure profiles which were measured at the several points in the pump tube of HS-3B under routinely operated conditions (6).

An example of the typical profiles is shown in Fig.3. Each profile from pressure gauge for the combustion gas pressure appears just after piston passing, and the decay profile completely corresponds to one another except for the first one. Since the gauge hole for the first signal is sealed by refuse, the pressure can decay slowly. These data imply that the combustion gas pressure is uniform behind the piston over the pump tube. Since characteristics of the powder used here and values of the loss parameters for the piston movement al-

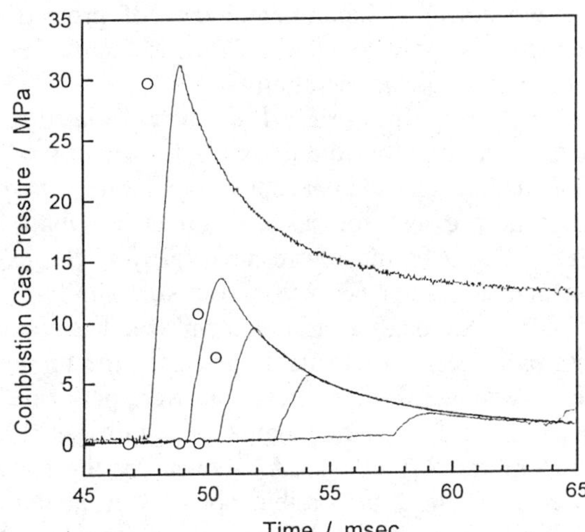

FIGURE 3. Profiles of the combustion gas pressure measured at five points in the pump tube of the conventional two-stage light-gas gun, HS-3B.

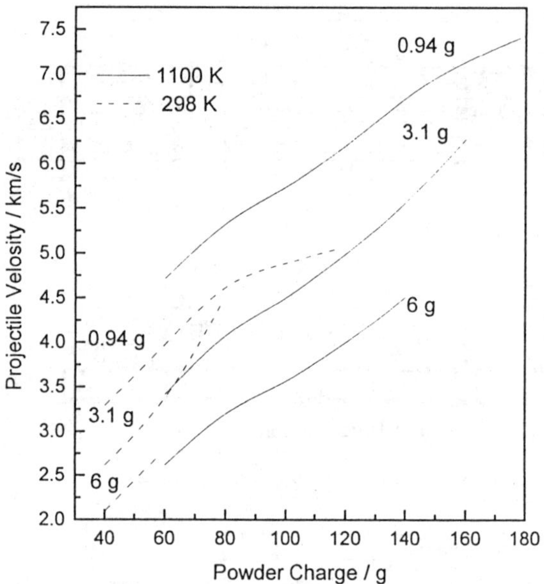

FIGURE 4. Expected projectile velocities for some different mass as a function of powder charge in the case of helium gas whose peak pressures are less than 1 GPa.

ready were obtained from the data and utilized in the calculation, the code well represented the piston movement. Experimental data of projectile velocity also provide values of the loss parameters for the projectile acceleration.

The broken lines in Fig.4 are the calculated results changing projectile mass in the case of 0.87-mole helium gas filled at room temperature, where the peak pressures in the reservoir is less than 1 GPa. The lines may correspond fairly to the experimental results because of the size similar to HS-3B. Although it is unknown whether these loss parameters are available or not in the higher velocity range, similar calculation were performed in the case of the same amount of helium gas at a temperature of 1100 K. As shown by the solid lines in Fig.4, it is possible not only to achieve higher velocities at the same powder charge but also to utilize powder charges much more than those at room temperature even though the peak pressure is less than 1 GPa. As the result, it is possible to attain much higher projectile velocity.

Helium gas has some advantage as a working gas rather than the hydrogen, e.g., safety problem, large molecular weight, and higher value of specific heat ratio. The safety problem is the most important factor in the university without any specialized engineer. The large molecular weight is effective to increase momentum as a pusher but negative to increase sound speed. However, since the specific heat ratio of helium, 1.67, is higher than that of hydrogen, 1.4, temperature increases more easily than hydrogen at the same compression, and hence, sound speed also increases. Therefore, this type of three-stage light-gas gun has a great advantage for helium gas.

All the parts of the present new apparatus have just been fabricated and installed, but the control system and gas pressurization system must be completed before firing tests which determine both the movement of the big piston and the pressure and temperature profiles of the working gas.

ACKNOWLEDGMENT

The authors are grateful to Prof. Y. Syono of Tohoku University and Prof. H. Mizutani of the Institute of Space and Astronautical Science for their helpful advice. This work was supported by a Grant-in-Aide for Scientific Research, #05505005, from the Ministry of Education and Culture.

REFERENCES

1. A. C. Mitchell, W. J. Nellis, and B. Monahan, "Enhanced performance of a two-stage light-gas gun," in *Shock Waves in Condensed Matter-1981*, New York: AIP Press, 1982, pp.184-187.
2. L. A. Glenn, "Performance Analysis of the Two-Stage Light Gas Gun," in *Shock Waves in Condensed Matter-1987*, New York: Elsevier, 1988, pp.653-656.
3. K. Kondo, "Parameters for the Design and the Operation of Two-Stage Light-Gas Gun," in Special Issue of The Review of High Pressure Science and Technology, 3, 158(1994).
4. L. A. Glenn, A. L. Latter, and A. Martinelli, "Multistage Gas dynamic Launchers," in *Shock Waves in Condensed Matter-1989*, New York: Elsevier, 1990, pp.977-984.
5. D. W. Bogdanoff, AIAA Journal, **28**, 483-491(1990).
6. H. Itoh, T. Okino, K. Kondo, A. Takeba, and M. Katayama, "Numerical Analysis of the Two-Stage Light-Gas Gun by AUTODYN^{TM}-2D," in Special Issue of The Review of High Pressure Science and Technology, 3, 159(1994).

LAUNCH CAPABILITIES TO 16 KM/S

L. C. Chhabildas, L. N. Kmetyk, W. D. Reinhart, C. A. Hall

Experimental Impact Physics, Sandia National Laboratories, P.O. Box 5800, Albuquerque, NM 87185-0821

A systematic study is described that has led to the successful launch of thin flier plates to velocities of 16 km/s. In this paper, we describe a novel technique that has been implemented to enhance the performance of the Sandia HyperVelocity Launcher (HVL). This technique of creating an impact-generated acceleration reservoir, has allowed the launch of 0.5 mm to 1.0 mm thick titanium (Ti-6Al-4V) and aluminum (6061-T6) alloy plates to record velocities up to 15.8 km/s. These are the highest metallic projectile plate velocities ever achieved for macroscopic masses in the range of 0.1 g to 1 g.

BACKGROUND

A schematic of the HVL configuration [1,2] used to launch flier plates to hypervelocities is indicated in Figure 1. The structured time-dependent pressure pulse needed to launch plates to hypervelocities is obtained by impacting a graded-density material on the two-stage, light-gas gun onto a plastic-buffered thin flier plate. TPX, a plastic with initial density of 0.81 g/cm^3, is used as a buffer in these studies. Impact velocities over the range of 6 km/s to 7.35 km/s have been used in the past studies. The flier-plate velocity resulting from impact is ~ 1.6 times the impact velocity. Yet, higher flier-plate velocities can be obtained using the graded-density impact technique by impacting the stationary flier plate at an impact velocity of 10 km/s. This would be the simplest way of obtaining higher flier-plate velocities to ~ 16 km/s. In practice, however, such high velocities with the necessary projectile mass cannot be achieved on the two-stage light-gas guns. A technique is therefore needed to enhance the capability of the hypervelocity launcher such that the flier-plate velocity is amplified in excess of 1.6 times the impact velocity. It is the purpose of this paper to report the technical methods that have been used to obtain the additional velocity boost. Implementation of an impact-generated third-stage reservoir is the key to the enhanced HVL velocity capability. This requires that the flier-plate be confined in a smaller diameter high-impedance barrel, as indicated in Figure 2. It has allowed the launch of a 1-mm titanium alloy flier plate to a velocity of 14.4 km/s, and a 0.5 mm titanium flier plate to a velocity of 15.8 km/s.

EXPERIMENTAL TECHNIQUE & RESULTS

The novel experimental arrangement used to enhance the flier plate velocity by confining it in a high-impedance barrel is indicated in Figure 2. A lexan projectile housing a multi-ply graded-density impactor is used in these studies. This impactor is fabricated by bonding a series of thin plates in order of increasing shock-impedance from the impact surface. The series of layered materials used in these studies [2,3] consists of TPX-plastic, magnesium, aluminum, titanium, copper, and tantalum. The thickness of each layer is precisely controlled to tailor the time-dependent stress pulse required to launch the flier plate intact. When these graded-density materials are used to impact a titanium alloy flier plate at a velocity of ~ 6.7 km/s, an initial shock of ~ 60 GPa, followed by a ramp loading to over 100 GPa is introduced into the flier plate.

As shown in Figure 2, the flier plate is made to fit exactly into the expendable tungsten barrel. In all experiments, the flier-plate diameter is equal to that of the tungsten barrel. No guard ring was used because CTH simulations predicted [3] minimal velocity dispersion across the diameter of the flier-plate. Unlike the geometry used in previous HVL studies (see Figure 1), two-dimensional effects due to radial reshock waves (generated upon impact of the tungsten barrel) emanating at the edges of the plate cause the edges to travel faster than the center of the flier plate initially. At later times the velocity dispersion across the flier-plate diameter is reduced resulting in intact though slightly deformed launches. Because the flier-

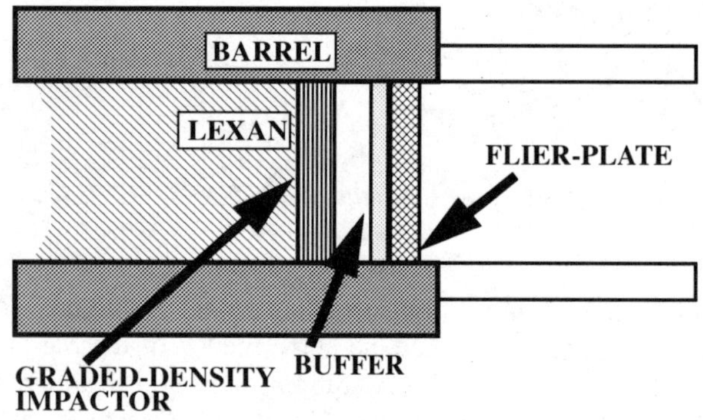

Figure 1: The HyperVelocity Launcher (HVL). Experimental configuration used to launch flier-plates to hypervelocities [1,2].

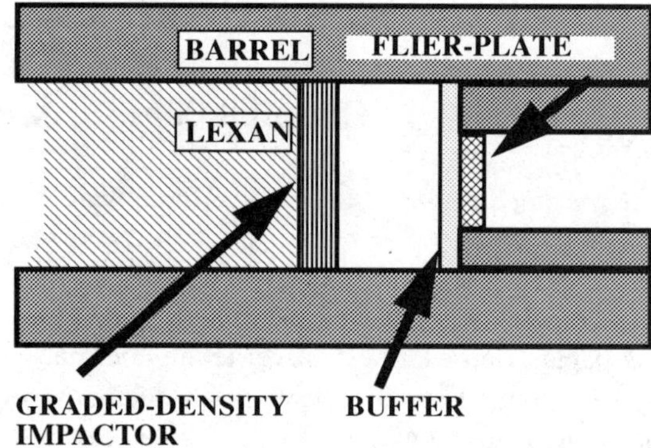

Figure2: Enhanced HyperVelocity Launcher, EHVL. Experimental configuration used to launch confined flier-plates in a tungsten barrel to hypervelocities.

plate material is ductile, and is at one-half the melt temperature [3] large deformations also tend to shape the projectile. The flier-plate materials used in this study were titanium (Ti-6Al-4V), and aluminum (6061-T6) alloys. As shown in Figure 2, a TPX buffer is used in all experiments because it (1) further cushions the input pressure pulse, (2) minimizes the tensile stress, and (3) lowers the flier-plate residual temperatures.

Following impact, up to nine flash x-rays are taken of the flier-plate while it is in motion. They are used to determine the velocity of the flier-plate and to check for its integrity after exit from the tungsten barrel. Three of these flash x-rays are taken within the first few microseconds after the flier-plate exits the muzzle of the barrel. The other x-rays are taken at various positions down range located up to 1.4 m from the impact position. Radiographic pictures of the flier-plate taken in flight over these large distances, and shown in Figure 3, allow an accurate velocity measurement to within 1%. Due to the hypervelocities achieved in this study, the 25 ns pulse duration of the x-ray source can cause a 400 µm blurring of the flier-plate while in flight.

Experiment CP3: Radiographs of the experiment CP3 are shown in Figure 3. The intact titanium flier-plate (1-mm thick by 6-mm diameter) is travelling at a velocity of 14.4 km/s. The flier-plate velocity has been determined using radiographic measurements over a flight distance of 1400 mm. Radiographs 1, 2, 3, 7, and 8 are views from the top, while radiographs 4, 5, 6, and 9 are side views. The flier-plate is tumbling as it travels.

Experiment CP5: This experiment is identical to experiment CP3, except that the titanium flier-plate thickness is reduced to 0.5 mm. The graded-density impact occurs at a velocity of 6.75 km/s resulting in a flier-plate speed of 15.8 km/s. The measurements are made over a distance of ~ 900 mm. The edges of the flier-plate are seen to travel much faster, causing them to separate from the flier-plate (see Figure 4). The center plate, however, appears to be intact. The flier-plate is tumbling as it travels.

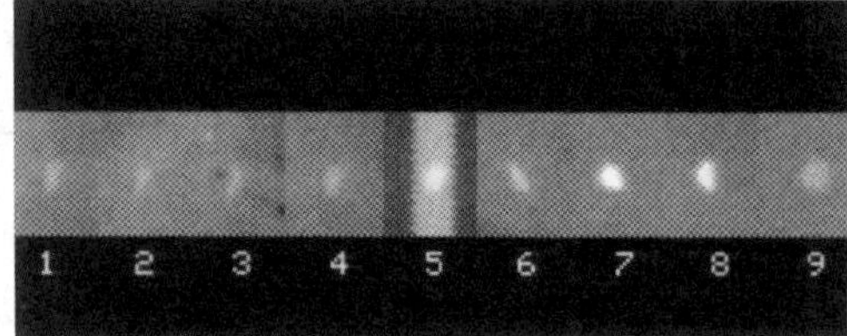

Figure 3: Radiographs of experiment CP3. The flier-plate is travelling from left to right at a speed of 14.4 km/s, over a flight distance of up to 1400 mm.

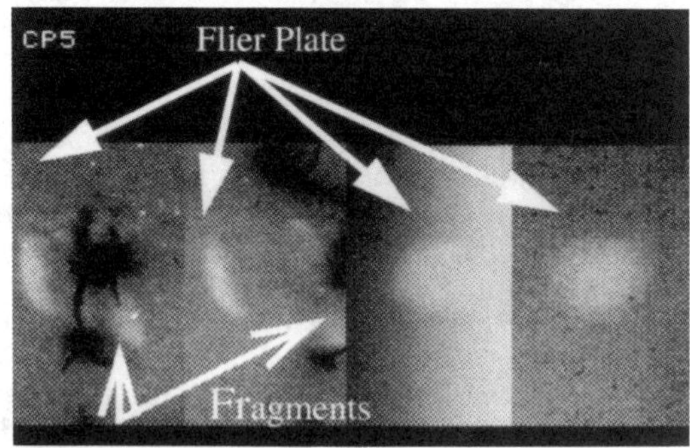

Figure 4: Radiographs of experiment CP5. The plate is travelling from left to right over a flight distance of 900 mm at 15.8 km/s. The fragments—originating at the edges—are moving faster than the intact center plate.

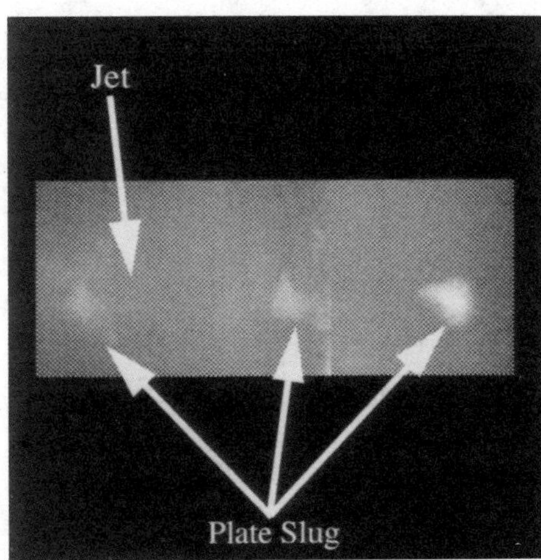

Figure 5: Radiographs of experiment CP6. Aluminum flier-plate traversing at a velocity of 13.8 km/s. Notice the formation of a jet in the first frame. There is evidence of flier-plate tumbling over a flight distance of 500 mm.

Experiment CP6: In this experiment, a 6061-T6 aluminum alloy is used as the flier-plate material. Because the melting point of aluminum is considerably lower than titanium, the impact velocity is reduced to 6 km/s. A low density foam [3] is also used as the first layer of the graded-density impactor. The aluminum plate is deformed during acceleration and is shaped into a "chunky" slug, achieving a speed of 13.8 km/s (see Figure 5). The velocity measurements are made over a flight distance of ~ 500 mm.

DISCUSSION

Very high pressures are needed to launch flier-plates to hypervelocities. In addition, this loading must be nearly shockless, structured, and uniform over the entire surface. To satisfy these criteria, graded-density materials are used to launch flier-plates to high velocities. When this graded-density material is used as an impactor on a two-stage light-gas gun, nearly shockless 100 GPa pressure pulses are introduced into the flier-plate. This time-dependent pressure pulse subsequently accelerates the flier-plates to high velocities. The resultant plate velocity is ~1.53 times the impact velocity for a titanium plate, and 1.63 times the impact velocity for an aluminum plate for the HVL configuration [1,2] depicted in Figure 1.

The current studies yield flier-plate velocities that are now 1.64 times to 2.3 times the impact velocity (Table 1). This improvement is due to the new experimental geometry (see Figure 2) used in this investigation. Specifically, the flier-plate diameter, d_{fp} is small, while the diameter of the graded-density impactor D_{gdi} is significantly larger. In the earlier HVL studies, the flier-plate diameter is the same as the graded-density impactor diameter, *i.e.*, the ratio d_{fp}/D_{gdi} is 1. In Figure 6, the variation of flier-plate velocity is shown as a function of the ratio of the flier-plate diameter to the graded-density impactor diameter. The same results are shown in Figure 7, with the flier-plate velocity normalized with respect to the impact velocity. In each of these figures, notice that the flier-plate velocity increases with decreasing d_{fp}/D_{gdi}. The values shown at $d_{fp}/D_{gdi} = 1$ are

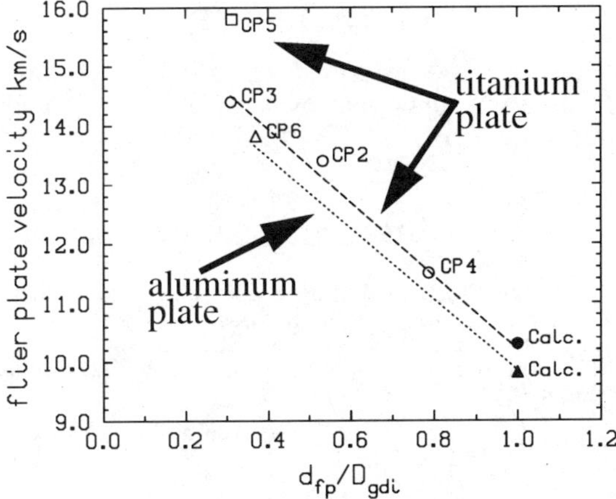

Figure 6: Variation of flier-plate velocity with the ratio of the barrel diameter to the graded-density impactor diameter. The solid points are calculated estimates.

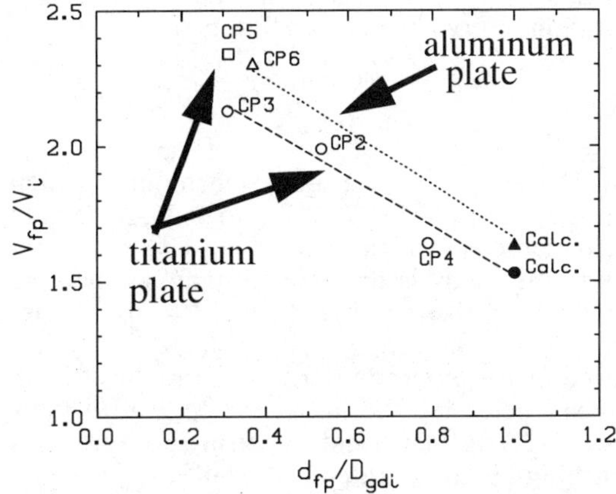

Figure 7: Variation of ratio of flier-plate velocity to impact velocity with the ratio of the barrel diameter to the graded-density impactor diameter. The solid points are calculated estimates.

TABLE 1. Results of Graded-Density Impactor/Confined Flier-Plate Experiments

Expt. No.	Graded-Density Impactor Diameter D_{gdi} (mm)	Flier-Plate Diameter/Material d_{fp} (mm)	Flier-Plate Thickness/Mass (mm)/(g)	Impact Velocity V_i (km/s)	Flier-Plate Velocity V_{fp} (km/s)	d_{fp}/D_{gdi}	V_{fp}/V_i
CP2	18.75	9.994/Ti	1.001/0.337	6.75	13.4	0.533	1.99
CP3	19.08	5.922/Ti	0.994/0.115	6.75	14.4	0.310	2.13
CP4	12.70	9.995/Ti	0.982/0.339	7.00	11.5	0.787	1.64
CP5	19.07	5.960/Ti	0.560/0.068	6.75	15.8	0.313	2.34
CP6	27	9.99/Al	1.003/0.208	6.00	13.8	0.370	2.30

calculated values based on previous HVL studies [1,2]. Thus, to enhance flier-plate velocities, a necessary requirement is to impact a step-down barrel as indicated in Figure 2.

Upon impact, the step down barrel itself is loaded to high pressures. The stress states at the barrel/flier boundary are now higher than those found in the original HVL configuration shown in Figure 1. This prevents the release of the driving pressure behind the flier-plates and sustains higher loading stresses over a longer duration. As the graded-density impactor material enters the step-down barrel, it speeds up, further maintaining an efficient push behind the flier-plate. Both these factors contribute significantly toward accelerating the flier-plate to yet higher velocities. The experimental geometry indicated in Figure 2, therefore, operates as an (impact-generated) acceleration reservoir to launch flier plates.

SUMMARY

This paper presents a novel technique which allows the launching of ~ 0.07 g to 0.35 g metallic plates to record velocities approaching 16 km/s. What is new in this study is that relatively heavy metallic plates are propelled to hypervelocities. To obtain higher velocities, the impact velocity may be increased or the plastic buffer may be replaced with hydrogen. Increasing the impact velocity to 8 km/s will, *in principle,* yield flier-plate velocities approaching 19 km/s. (In practice, however, it is not simple to obtain such high velocities for a "heavy" two-stage light-gas gun projectile required for these studies.) This would, however, require a proper design of the graded-density impactor and the barrel

to ensure that the plate would not melt or fracture. Additional calculational studies were made to determine if hydrogen, with its high sound speed, could be used to further enhance flier velocities. By replacing the plastic buffer in Figure 1 with a 5.0-mm thick frozen hydrogen plate, the velocity of a 0.2-mm titanium flier was predicted to be 13 km/s for a graded-density impact at 6.5 km/s. The expansion velocity of hydrogen compressed to very high pressures is extremely high resulting in an efficient push on the flier plate and can further enhance the flier-plate velocity. Alternately, frozen hydrogen buffers in combination with higher impact velocities could offer a way to achieve flier-plate velocities in excess of 20 km/s.

ACKNOWLEDGMENT

This work performed at Sandia National Laboratories, supported by the U.S. DOE under contract number DE-AC04-94AL85000.

REFERENCES

1. L. C. Chhabildas, L.M. Barker, J.R. Asay, T.G. Trucano, G.I. Kerley and J.E. Dunn, In Shock Waves in Condensed Matter - 1991, (S.C. Schmidt, R.D. Dick, J.W. Forbes. D.G. Tasker, eds.), pp 1025, 1992. Elsevier Science Publishers B.V.

2. L.C. Chhabildas, J.E. Dunn, W.D. Reinhart, and J.M. Miller,. *Intl.. J. Impact Engng.*, 14, pp. 121-132., 1994

3. L. C. Chhabildas, L. N. Kmetyk, W. D. Reinhart, C. A. Hall , *Intl.. J. Impact Engng.*, 17, 1995 (in print).

The NIRIM Two-stage Light-gas Gun: Performance Test Results

T. Sekine, S. Tashiro[1], T. Kobayashi, T. Matsumura[2]

National Institute for Research in Inorganic Materials, Tsukuba 305, Japan

A two-stage light-gun has been installed at the NIRIM in order to investigate the high pressure behavior of materials. For operation and safety test, we used helium and carried out performance test shots. Piston velocity in the pump tube and projectile velocity during free flight are measured by means of gas-pressure profile records at fixed locations and x-ray beam cutting method, respectively.

INTRODUCTION

It is well known that two-stage light gas can accelerate a heavy projectile over 7 km/sec(1). A relatively large two-stage light-gas gun is required to generate greater area plane shock waves than a few cm² and high pressures over 100 GPa.

Recently we designed and installed a 22 m long two-stage light-gas gun at NIRIM for investigations of new, nobel materials and properties of materials under strong shock compressions(2). The gun composes mainly of the following five parts, propellant chamber, pump tube, accelerating reservoir, launcher tube, and impact chamber. The hot burning gas formed in the propellant chamber accelerate a heavy piston to compress isentropically light gas into the accelerating reservoir(AR). At that time when the pressure of hot, compressed driving gas attains to a given value in the AR, a diaphragm separating the AR and launcher tube opens up and the driving gas starts to push a projectile within the launcher tube. Hydrogen is going to be used after test shots in order to obtain higher projectile velocity (3, 4). We compare the measured projectile velocity with the numerical calculations.

---

1. Present address, Ibaraki University, Hitachi 316, Japan
2. Present address, National Institute of Materials and Chemical Research, Tsukuba 305, Japan

EXPERIMENTS

Simulation analysis

The specification of the NIRIM two-stage light gas gun(NIRIM 2ST-1) is given in Table 1. The launcher of 25 mm diameter is used presently here. Operation conditions of two-stage light gas gun contain many parameters such as type and mass of propellant, mass of piston, initial fill pressure of light gas, burst pressures of diaphragms, mass of projectile, inner diameters of launcher and pump tubes and others. The numerical simulation method has been described by Matsumura and Takayama(5). In this calculation, burning of the smokeless powder (Nippon Oil & Fat Co, NY-500) is assumed to obey the following equation,

$$V(t) = \beta P^{\alpha} \quad (\alpha = 0.732, \ \beta = 2.0 \times 10^{-5}) \tag{1}$$

where $V(t)$ is burning-rate at time t and P is pressure. The gas flow is modeled by one-dimensional unsteady compressible Euler equation and numerically solved by the random choice method. According to the numerical simulation, the piston velocity at the sensor locations gives almost the peak and is a function of masses of piston and propellant, but little affected by the type of light gas, its initial fill prressure, and burst pressure setting of diaphragms. The setting of the burst pressure within the AR is sensitive to the

Table 1. Specification of the two-stage light gas gun. (NIRIM 2ST-1)

Pump tube length (m)	8.0
Pump tube diameter (mm)	81.0
locations of pressure sensors (m)	6.60, 6.80, and 7.00
Launch tube length (m)	7.0
Launch tube diameter (mm)	25.0 and 18.0
Piston mass (kg)	2 ~ 8
Projectile mass (g)	10 ~ 20
Initial light gas fill pressure (MPa)	up to 1.0
Propellant mass (g)	up to 1000
Free flight distance (mm)	8 6 5

projectile velocity. The projectile final velocity is a function of the initial fill pressure of light gas, projectile mass and piston velocity. The results of numerical simulations in case of hydrogen and helium driving gas give velocities of 11 g projectiles over 7 km/sec and up to 6 km/sec, respectively.

Performance test

The outer appearance is given in Fig. 1. The impact

chamber and launcher tube, total volume of about 4.5 m³, are evacuated down to 10 Pa before shot and nitrogen gas is filled to 0.1 MPa immediately after the shot.

The piston velocity is read by measuring the gas pressure change at fixed locations in the pump tube. The change is caused by the movement of piston. According to a numerical calculation, the pressure of light gas in front of piston displays multiply abrupt increases but does not show smooth, gradual increase. Experimentally observed pressure profiles are illustrated in Fig. 2. Pressure increases observed at given locations are too complicated to read the arrival time of piston. The pressure drops at the time of passage of piston tail are sharp enough to read out. They are marked by arrows in the figure. The observed profiles at the sensor locations indicate that hot burning gas decompresses adiabatically for 5 to 7 msec and that the reflected waves in the burning gas from the base of the piston interact to give abrupt increase after that time. The range of measured piston velocities are about 400 m/sec.

The projectile velocity is measured by the cw x-

FIGURE 1. Outer appearance of NIRIM 2ST-1

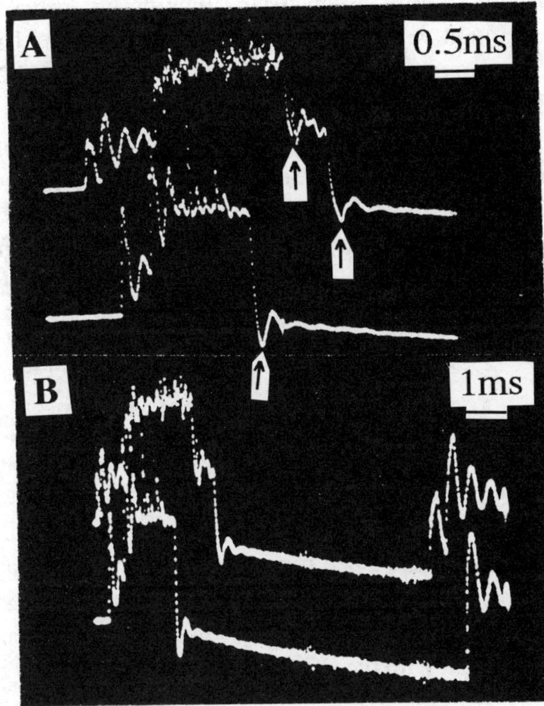

FIGURE 2. Pressure change record at fixed locations in the pump tube. Each arrow indicates the time of passage of piston base at the corresponding gauge.

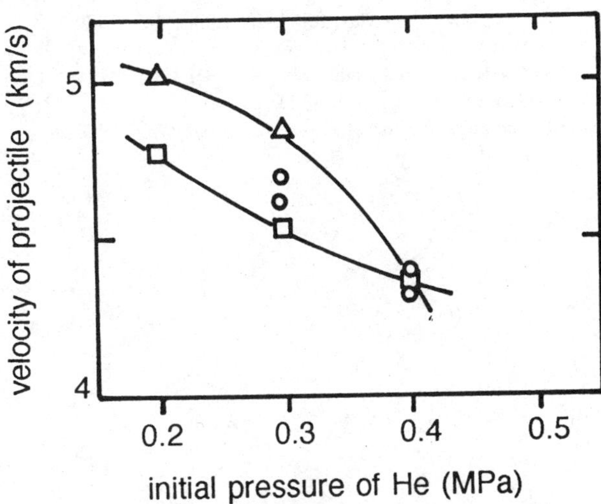

FIGURE 3. Relationship between initial fill He pressure and projectile velocity. Propellant mass used :250 g, piston: 3.92 kg, projectile: 11g, circles are measured experimentally, squares and triangles are calculated for burst pressure setting within the AR of 75 MPa and 100 MPa, respectively.

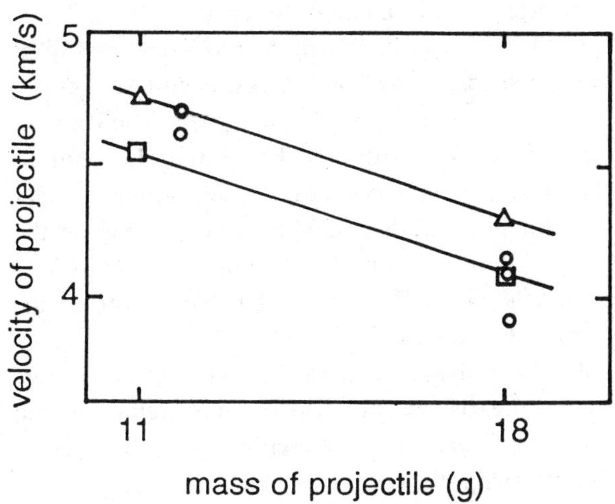

FIGURE 4. Relationship between mass and velocity of projectile. Propellant mass used: 250 g, piston: 3.92 kg, initial fill He pressure 0.3 MPa, circles are measured experimentally, and squares and triangles are calculated burst pressure setting within AR of 75MPa and 100 MPa, respectively.

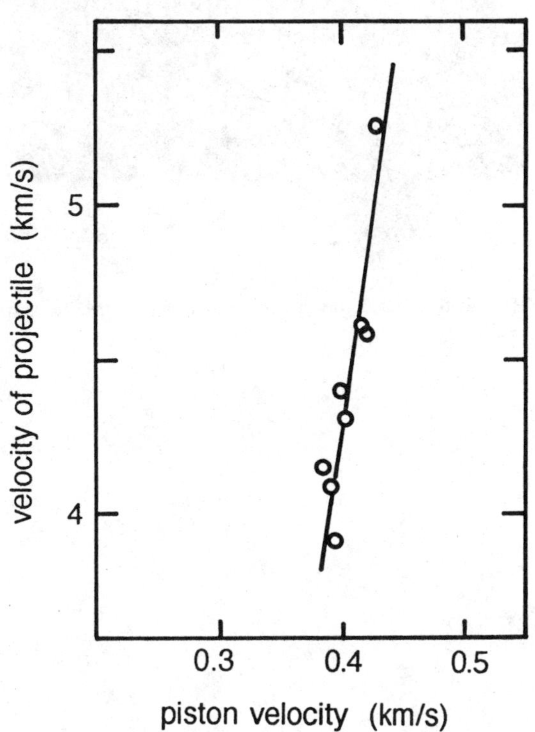

FIGURE 5. Relationship between measured velocities of piston and projectiles.

ray beam cutting method during free flight. Three parallel beams are used to detect acceleration, coupled with NaI scintillators and PMTs. Each PMT output also is used to make triggering signals. The velocity can be measured within ±0.3% relative. In addition, to make a synchronized trigger signals, self-adjustable pulser is available which counts up the time between the two fixed points and then count down the time difference to target depending on the projectile velocity.

Figure 3 illustrates calculated and measured results indicating relationship between the initial fill He pressure and the velocities of projectiles. Relationship between mass and final velocity of projectile is given in Fig. 4, together with calculated

FIGURE 6. Flash X-ray shadowgraph of a projectile during free flight.

results. Figure 5 depicts the relationship between measured velocities of piston and projectiles. The maximum projectile velocity so far observed is 5.25 km/sec.

The shape of projectile during free flight is monitored by a flash x-ray shadowgraphy. Figure 6 is an example of a typical shadowgraph. The tilt of projectile has been checked using four shorting pins on the target surface. The tilting may occur at the time of starting free flight out of the launcher tube and may keep its initial angle.

CONCLUSIONS

Using helium as a driving gas, the NIRIM two-stage light gas gun is subjected to performance test shots. The velocity meters for piston and projectile work well and make trigger signals for flash x-ray, Xe flash lamp, and high speed camera. The final projectile velocities, experimentally measured are consistent with the one numerically calculated.

REFERENCES

1.Jones, A.H., Isabell, W.M.,and Maiden, C.J., Jour. Appl. Phys., **37**, 3493-3499 (1966).
2.Sekine, T., Rev. High Press. Sci. Tech. 3 (special Issue), 156, (1994)
3.Munson, D.E., and May, R.P., AIAA Jour., **14**, 235-242 (1976)
4.Mitchell, A.C., and Neilis, W.J., Rev. Sci, Instrum., **52**, 347-359 (1981)
5.Matsumura, T. and Takayama, K., Shock Waves '93, 407 (1993)

OPTIMAL VELOCITY AMPLIFICATION
IN A SYSTEM OF COLLIDING PLATES

Roger W. Minich

Lawrence Livermore National Laboratory, Livermore, California 94550

The kinetic energy ($1/2\, m_i v_i^2$) of a flyer plate may be transferred to a plate of lower mass (m_{i+1}), accelerating it from rest to velocity, v_{i+1}. Velocity amplification results when $v_{i+1} > v_i$ and depends on the mass ratio ($\Gamma_i = m_{i+1}/m_i$) and shock induced losses in the collision. This process may be continued in a sequence of n-1 binary collisions to achieve a desired velocity, v_n for the n th plate. The method of Lagrange multipliers is used to derive an analytic expression for the discrete mass distribution for the n plates and the corresponding minimum initial kinetic energy necessary to produce the velocity amplification, v_n/v_1. It is found that there exists an optimum number of plates for which the initial kinetic energy of the flyer plate is an absolute minimum.

INTRODUCTION

One method for generating high pressures to create matter at high energy density involves the acceleration of a flyer plate to a high velocity. The pressure generated in a collision with a stationary object scales as the kinetic energy density $1/2\,\rho v_i^2$ of the flyer plate. When the method for accelerating the flyer plate utilizes chemical explosive, the maximum flyer plate velocity is comparable to the detonation velocity of the explosive material and sets a limit on the characteristic pressures that are achievable. This limitation may be overcome if the flyer plate is allowed to collide with another plate of lower mass (m_{i+1}) initially at rest. The ideal kinematic amplification of the flyer plate velocity in a perfectly elastic collision is,

$$\frac{v_{i+1}}{v_i} = \frac{2}{1+\Gamma_i}, \qquad \Gamma_i \equiv \frac{m_{i+1}}{m_i} \tag{1}$$

When $\Gamma_i = \Gamma_0 = $ constant, the velocity amplification after n-1 collisions becomes

$$\frac{v_n}{v_1} = \left(\frac{2}{1+\Gamma_0}\right)^{(n-1)} \tag{2}$$

And when $\Gamma_0 = (m_n/m_1)^{1/(n-1)}$, the velocity amplification is optimized for a a fixed number of plates and m_n/m_1 ratio. Also, the velocity amplification increases asymptotically with the number of plates as,

$$\lim_{n \to \infty} \frac{v_n}{v_1} = \sqrt{\frac{m_1}{m_n}} \tag{3}$$

In contrast, when velocity dependent losses are present, the optimal velocity amplification occurs for $\Gamma_i \neq$ constant and for a finite number of plates. The method of Lagrange multipliers commonly used in statistical mechanics(1) is employed here to derive an analytic expression for the mass ratios (Γ_i) and the corresponding masses m_i for each plate in a system of n plates. The variational technique is applied to the system of equations:

$$\frac{v_n}{v_1} = \prod_{i=1}^{n-1}\left(\frac{2}{1+\Gamma_i}\right) e^{-\sigma\,(1+\Gamma_i)\,f(v_i)}$$

$$\frac{m_n}{m_1} = \prod_{i=1}^{n-1} \Gamma_i \tag{4}$$

The exponential factor in Equations (4) takes into account shock induced losses and includes a scalar constant $\sigma \ll 1$ which characterizes the magnitude of the internal energy loss in a collision and a function, $f(v_i)$, containing the velocity dependence of the internal energy loss. Equations of state of the polytropic or Gruneisen form, are well represented by,

$$f(v_i) = \log \frac{v_i}{v_0} \quad , \tag{5}$$

where v_0 is a reference velocity near or equal to the speed of sound. The factor of $1 + \Gamma_i$ takes into account the phase space constraint on the amount of kinetic energy available to be converted to internal energy due to the finite mass masses of the plates. The precise form of the shock induced losses is not developed here, since the analytical techniques presented here can accommodate a wide range of functional forms for $f(v_i)$. There is only one constraint in Equations (4), and is taken to be the desired velocity amplification, v_n/v_1, and the mass ratios Γ_i will be sought that maximizes the m_n/m_1 ratio.

ANALYTIC SOLUTION
The Maximum m_n/m_1 Ratio

The exact solution to the variational problem may be obtained for the following system of equations:

$$\frac{v_n}{v_1} = \prod_{i=1}^{n-1} Ae^{\alpha(1 - \Gamma_i)} e^{-\sigma(1 + \Gamma_i)\log\frac{v_i}{v_0}} \tag{6}$$

$$\frac{m_n}{m_i} = \prod_{i=1}^{n-1} \Gamma_i$$

Here, A and α are scalar constants describing the amplification in the absence of shock induced losses, and approximates $2/(1+\Gamma_i)$ when $A=2/e$, $\alpha=1.0$ and $\Gamma_i \ll 1$. These values are the values used for the numerical work in the last section of this paper. There is only a single constraint, namely, the velocity

amplification and its associated Lagrange multiplier, λ. Additional constraining relationships would require the introduction of new Lagrange multipliers, one for each new constraint. The Lagrange multiplier in the present case is chosen such that variation of the mass ratios, $\delta\Gamma_i$ results in the following condition,

$$\sum_{i=1}^{n-1} \left\{ \frac{1}{\Gamma_i} - \lambda\left(\alpha + \sigma \log\frac{v_i}{v_0} \right) \right\} \delta\Gamma_i = 0 \tag{7}$$

Since, λ is chosen such that the $\delta\Gamma_i$ are nonzero the expression within the brackets must be zero. That is,

$$\Gamma_i = \frac{1/\lambda}{\alpha + \sigma\log\frac{v_i}{v_0}} \tag{8}$$

When this expression is substituted for Γ_i in Equation (6), a recursion relationship for the velocity is obtained,

$$\frac{v_{i+1}}{v_0} = Ae^{\alpha - \frac{1}{\lambda}}\left(\frac{v_i}{v_0} \right)^{1-\sigma} \quad , \tag{9}$$

which has the solution,

$$\frac{v_k}{v_0} = e^{\left[\alpha + \log A - \frac{1}{\lambda}\right]\left[\frac{1 - (1-\sigma)^{k-1}}{\sigma}\right]}\left(\frac{v_1}{v_0} \right)^{k-1} \tag{10}$$

The Lagrange multiplier may now be expressed in terms of the velocity amplification, when $k=n$,

$$\frac{1}{\lambda} = \alpha + \log A - \sigma\log\frac{v_1}{v_0} - \frac{\sigma\log\frac{v_n}{v_0}}{1 - (1-\sigma)^{n-1}} \tag{11}$$

The maximum m_n/m_1 ratio can now be obtained using Equation (8),

$$\frac{m_n}{m_1} = \prod_{i=1}^{n-1} \Gamma_i = \left(\frac{1}{\lambda}\right)^{n-1} \prod_{i=1}^{n-1} \frac{1}{\alpha + \sigma \log \frac{v_i}{v_0}} \qquad (12)$$

Finally, Equations (11) and (12) are combined to give a complete expression for the maximum m_n/m_1 ratio,

$$\frac{m_n}{m_1} = \frac{\left[\alpha + \log A - \sigma \log \frac{v_1}{v_0} - \frac{\sigma \log \frac{v_n}{v_1}}{1 - (1-\sigma)^{n-1}}\right]^{n-1}}{\prod_{i=1}^{n-1}\left\{\alpha + \sigma \log \frac{v_1}{v_0} + \sigma \log \frac{v_n}{v_1}\left[\frac{1 - (1-\sigma)^{i-1}}{1 - (1-\sigma)^{n-1}}\right]\right\}}$$

Once the velocity amplification is specified along with the loss parameter, σ, there exists sharp peaks in the value of m_n/m_1 with respect to plate number. If in addition, the physical application requires a well defined m_n, the peak value of m_n/m_1 reflects the minimum flyer plate mass, m_1, needed to achieve the specified velocity amplification in the n th plate. Also, since the efficiency (ε) with which the kinetic energy of the flyer plate is transferred to the n th plate is

$$\varepsilon = \left(\frac{m_n}{m_1}\right)\left(\frac{v_n}{v_1}\right)^2 \quad , \qquad (13)$$

the peak in m_n/m_1 may be interpreted as the highest efficiency system for achieving a particular velocity amplification.

Recursion Relation for Γ_i and the Discrete Mass Distribution

The expression for m_n/m_1 following Equation (12) corresponds to the optimized values of Γ_i that are determined using Equation (8) to eliminate the dependence on v_i/v_0. The recursion relation that results is

$$\Gamma_{i+1} = \frac{\Gamma_i}{1 - \sigma + \Gamma_i \sigma \left[(2\alpha + \log A)\lambda - 1\right]} \qquad (14)$$

This recursion relation is known as the Riccati difference equation and has the solution (2),

$$\Gamma_i = \frac{\Gamma_1 \Gamma_{fp}}{(\Gamma_{fp} - \Gamma_1)(1-\sigma)^{i-1} + \Gamma_1} , \qquad (15)$$

where,

$$\Gamma_1 = \frac{\alpha + \log A - \sigma \log \frac{v_1}{v_0} - \frac{\sigma \log \frac{v_n}{v_1}}{1 - (1-\sigma)^{n-1}}}{\alpha + \sigma \log \frac{v_1}{v_0}}$$

$$\Gamma_{fp} = \frac{\Gamma_1\left(\alpha + \sigma \log \frac{v_1}{v_0}\right)}{\alpha + \sigma \log \frac{v_1}{v_0} + \frac{\sigma \log \frac{v_n}{v_0}}{1 - (1-\sigma)^{n-1}}} . \qquad (16)$$

The Γ_1 physically represents the mass ratio between the first two masses in the n-plate system and Γ_{fp} corresponds to the fixed point solution of the recursion relationship, i.e. when $\Gamma_{i+1} = \Gamma_i$. It also represents the mass ratio for which the velocity is at a fixed point. Once the mass ratios are known it is straight forward to determine the mass for the i th plate. In terms of the above definitions,

$$\frac{m_i}{m_1} = \prod_{j=1}^{i-1} \frac{\Gamma_1 \Gamma_{fp}}{\Gamma_1 + (\Gamma_{fp} - \Gamma_1)(1-\sigma)^{j-1}} . \qquad (17)$$

DISCUSSION AND NUMERICAL RESULTS

It has been common practice to approximate the optimal mass ratios as a constant (3), which is valid as long as the internal energy losses are independent of velocity and the mass ratio. However, this assumption is invalid for collisions between finite mass plates when entropy producing shock waves are present. If an amplifying system of plates is designed with a constant mass ratio, the efficiency of the system is not optimum and a higher initial flyer plate mass will be required to achieve a particular velocity amplification. Figure (1) shows the factor, Q, by which the flyer plate mass must be increased for fixed mass ratio system compared to a system designed with the optimized mass ratios given in Equation(15). The dependence on Q increases non-linearly with the loss parameter, σ. The magnitude of this effect increases as the desired velocity amplification increases and decreases as α increases.

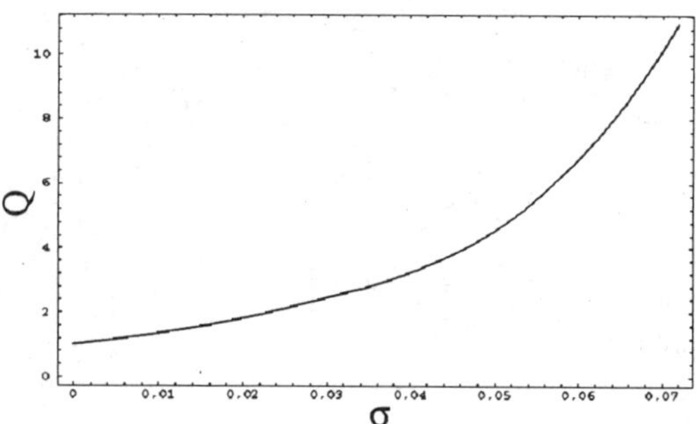

FIGURE 1. A fixed mass ratio solution requires an initial flyer plate mass Q times larger than the minimum flyer plate mass corresponding to the optimized mass ratio distribution of Eq. (15). Q is plotted versus σ for a 10 plate system and velocity amplification, $v_n/v_1 = 100$.

Figure 2 shows that the minimum flyer plate mass, m_1, (assuming m_n fixed and $\sigma=0.03$) increases rapidly with increasing velocity amplification and occurs in systems with a larger number of plates.

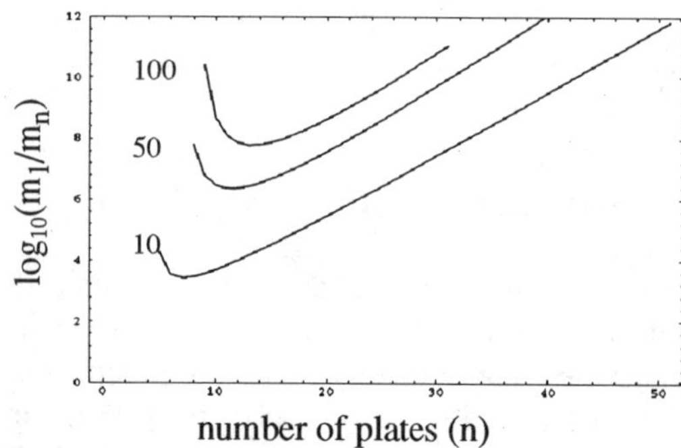

FIGURE 2. The base 10 logarithm of the ratio m_1/m_n versus the number of plates for $\sigma=0.03$ and for velocity amplification, $v_n/v_1=(10, 50, 100)$.

Figure 3 shows how the minimum flyer plate mass increases with the loss parameter σ for a velocity amplification of 100. As expected the minimum flyer

plate mass increases with internal energy losses and occurs in a system with a large number of plates. It may also be observed in Fig. 2 and Fig. 3 that the widths of the curves become narrower as either the velocity amplification or internal energy loss become greater, making it more difficult to design a practical amplifying system near the optimum design parameters. If the plate masses are not optimized, the efficiency of the amplifying system is greatly reduced by as much as an order of magnitude for moderate shock induced losses (Fig.1).

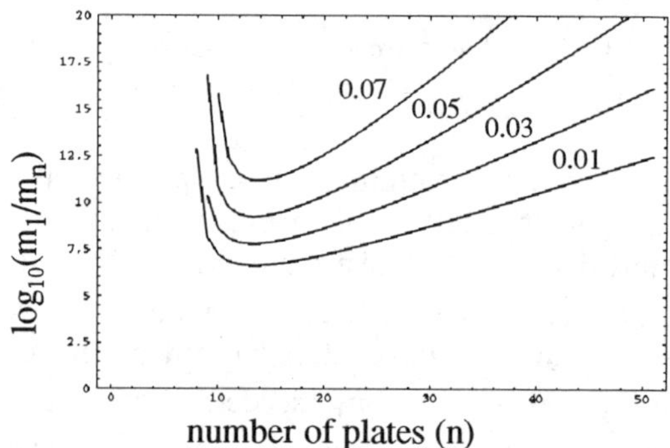

FIGURE 3. The base 10 logarithm of the ratio m_1/m_n versus the number of plates for a velocity amplification, $v_n/v_1 = 100$ and $\sigma = (0.01, 0.03, 0.05, 0.07)$.

ACKNOWLEDGEMENTS

I would like to thank Harry Vantine for bringing this problem to my attention. This work is supported by DOE and the support of many at LLNL. The numerical work made use of the computational tool Mathematica. I would also like to thank my precious wife, Sharon, for her support on this our anniversary.

REFERENCES

1. D. ter Haar, *Elements of Statistical Mechanics*, New York,1954, pp. 444-445.
2. Agarwal, Ravi P., *Difference Equations and Inequalities*, New York,1992, ch. 3, pp. 15-18.
3. Bragowski,J., Derentowicz,H., andLuckner H., *Physica* **139&140B**, 575-581 (1986).

INVESTIGATION OF THIN LASER-DRIVEN FLYER PLATES USING STREAK IMAGING AND STOP MOTION MICROPHOTOGRAPHY[*]

Alan M. Frank

Lawrence Livermore National Laboratory, Livermore, CA 94551-9900

and

Wayne M. Trott

Sandia National Laboratories, Albuquerque, NM 87185-0834

The dynamic behavior of laser-driven flyers has been studied using high-speed streak imaging in combination with stop motion microphotography. With very thin targets, melting and plasma penetration of the flyer material occur in rapid sequence. The time delay from onset of motion to flyer breakup increases with flyer thickness and decreasing incident energy. Materials examined include pure aluminum and composite samples containing an insulating layer of Al_2O_3. While flyer breakup is observed in both cases, the Al_2O_3 barrier significantly delays the deleterious effects of deep thermal diffusion.

INTRODUCTION

Experimental methods for generating laser-driven flyers provide a promising approach to the development of a laboratory-scale capability for quantitative studies of material response to short pulse shock loading. Laser-launched aluminum flyers can easily achieve velocities > 3 km-s^{-1}. Hence, as impactors, flyers can generate shock pressures well in excess of 10 GPa in many materials. Shock durations, on the other hand, are typically only a few ns or less (1). Properly configured, a laser driver and optical transmission system (typically using optical tapers and/or fibers to couple the high-intensity radiation to the flyer target) can produce a reasonably uniform spatial intensity profile at the target plane, promoting nearly planar launch conditions with precisely variable impact velocities and very accurate (~1 ns) timing. Thus, laser-driven flyers are well suited for generation and analysis of shock-induced processes that occur with very short (i.e., nanosecond) charact-

eristic time scales. Miniature laser-driven plates are also potentially useful impactors in the examination of rare, expensive or highly toxic materials.

The dynamic behavior of laser-accelerated flyers is another important factor in determining their utility in shock compression studies. In addition to providing acceptable planarity on impact, it is essential that the flyer arrive at the impact plane either fully intact and at solid density or with dynamic physical properties that can be adequately characterized. Recent high-speed imaging studies (2) have shown that complex, multidimensional effects can occur on a small scale during flyer acceleration, including "tilt" (due to a slightly uneven distribution of optical intensity at the fiber output) and fine-structure nonplanarity (arising from modal noise in the intensity distribution). Additional complications can arise from the high temperatures (~50000 K) generated in the driving plasma (3). Rapid heat transfer may be expected to result in melting/vaporization of some or all of the flyer (depending on the target thickness), quickly leading to a situation where the expanding plasma pushes against fluid material at higher density. These conditions favor the growth of instabilities. In this paper, we discuss tests in which high-

---

[*]This work supported by the U. S. Department of Energy at SNL under contract DE-AC04-94AL85000 and at LLNL under contract W-7405-ENG48.

speed streak imaging and simultaneous stop motion microphotography have been used to examine the limiting case of very thin flyers. The data indicate that rapid melt does occur under these conditions, followed by plasma penetration of the fluid material. Streak imaging allows characteristic time scales for these phenomena to be determined. Results have been obtained as a function of both target thickness and driving energy. In addition, we compare the response of pure aluminum flyers vs. composite materials containing an insulating layer of Al_2O_3.

EXPERIMENTAL

Flyers were generated using techniques described in detail previously (4). The driving laser employed in this work was a Q-switched Cr:Nd:GSGG laser (Allied Signal) that provided an output >100 mJ with a pulse duration near 13 ns. Absorbing neutral density filters were used to vary the energy incident on the flyer targets. These targets were prepared by physical vapor deposition on polished output ends of 400-μm-diameter multimode optical fibers. Tests were performed on pure aluminum ranging from 0.25-2.6 μm in thickness as well as composite samples spanning a more limited range (0.5-2.0 μm).

Principal elements of the optical assembly used for simultaneous streak imaging and stop motion microphotography have also been discussed previously (2). Briefly, the setup utilized a large-format framing camera and a streak camera that viewed the flyer target through a common, fast photographic objective in combination with separate magnifying lens assemblies. Illumination for the framing camera was provided by a short-pulse (~1 ns FWHM) broad band dye laser. The optical properties of this source enabled acquisition of stop-motion images of high-velocity flyers with negligible interference from laser speckle. The dye laser energy was focused to a circular region matching the field of view of the framing camera. The target surface was viewed at an angle that directed the specular reflection of the illumination beam just outside the aperture of the objective. This orientation yielded a high-contrast image in which extremely small angular perturbations on the surface were highly visible. For streak imaging, illumination was supplied by a cw Ar^+ laser. This source was focused to a line matching the entrance

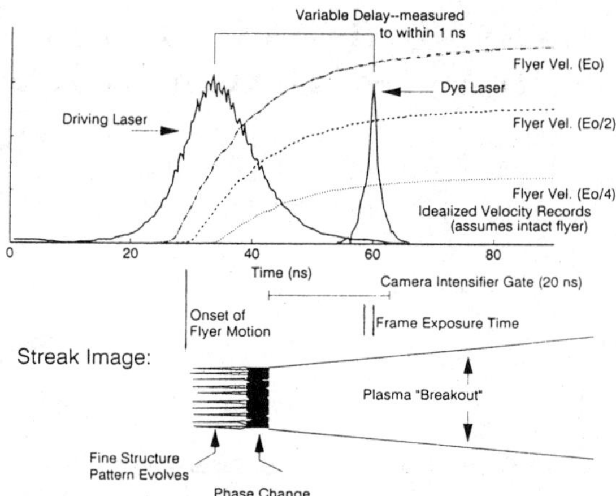

FIGURE 1. Schematic diagram illustrating timing of framing and streak diagnostics relative to onset of flyer motion. At lower driving energies, flyer motion occurs later in the incident pulse.

slit of the streak camera. Both streak and framing images were recorded on TMAX-400 film.

Principal elements of the timing sequence for these tests are illustrated in Fig. 1. Digital delay generators were used to set timing for the driving laser, pulsed dye laser, framing camera intensifier gate, etc. Propagation times in the driving laser and dye laser optical transmission systems were carefully measured in order to derive the actual delay between arrival of the two lasers at the target plane from pulse records acquired by on-line beam monitors. In addition, a dye laser fiducial pulse was recorded on the streak camera trace in order to relate the timing of the two diagnostics. The onset of flyer motion relative to the driving laser pulse was not explicitly measured in these experiments; however, a reasonably accurate estimate of the time for this event could be inferred from velocity interferometer measurements in a similar experimental system (5). In this manner, the temporal relationship of all relevant phenomena was established.

RESULTS AND DISCUSSION

Figure 2 displays the response of 1-μm Al targets to two different driving energies as recorded by streak imaging. Important physical effects in these records can be compared to the schematic representation in Fig. 1. Within a few ns of the onset of motion, a fine pattern of light and dark regions develops in the image of Ar^+ laser light reflected from the target surface. The spatial frequency in this pattern is consistent with stop-motion photographs of the fine-structure nonplanarity driven by modal noise at the fiber output (2). The pattern steadily develops a coarser structure and then evolves into a condition of very low reflectivity. After another short period, the image becomes dominated by plasma light. The latter effects appear to correspond to melting of the flyer followed by plasma penetration of the fluid material. The onset of a transition to the "dark phase" can be seen in the photograph ("face-on" view) shown in Fig. 3. Extended dark regions are evident in the midst of the structure due to dynamic surface roughness (exaggerated by the imaging technique). Other data clearly show fluid flow associated with the "dark phase"(2) and temporal correspondence between localized onset of low reflectivity and plasma appearance (see below).

As illustrated in Fig. 2, the time delay to phase change and plasma "breakout" generally increases (weakly) with decreasing incident energy. The brightness of the emerging plasma is a strong function of energy. Also, higher energy conditions result in obvious venting and radial expansion of the plasma at the perimeter of the accelerating flyer plate. Although the transitions are fairly abrupt, the luminous plasma tends to emanate at discrete points, especially at lower driving energies. Very early appearance of plasma light was observed at one location in the flyer launched at higher energy (Fig. 2a). We attribute this effect to a pinhole in the flyer target. Visual inspection of the films used in these experiments occasionally revealed a small number of these defects, especially with the thinner materials.

In view of the capability of streak imaging to track physical changes arising from rapid heat transfer processes in flyer acceleration, this technique was used to examine the role of a thin, thermally insulating layer of Al_2O_3 (deposited in the target material) in enhancing flyer performance. It has been shown (1,6) that composite materials of this type yield increased efficiency in converting incident optical energy to flyer kinetic energy, especially near the threshold for flyer motion, and can deliver a shock pulse that shows less evidence of flyer ablation or thinning. Figure 4 compares the response of a 2.0-µm-thick composite film to that of a 2.1-µm target of pure Al. Two effects are evident. First, the composite flyer

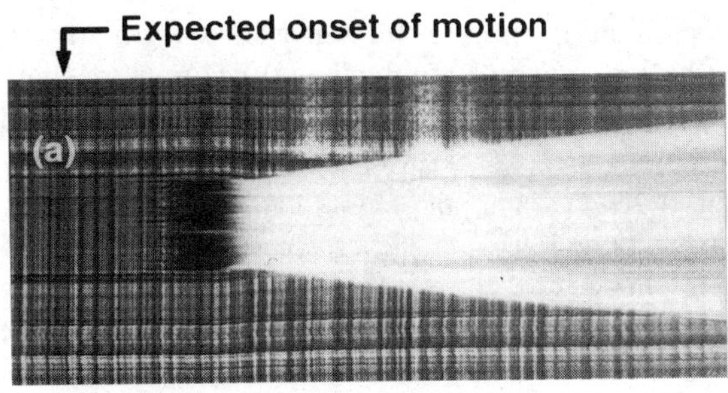

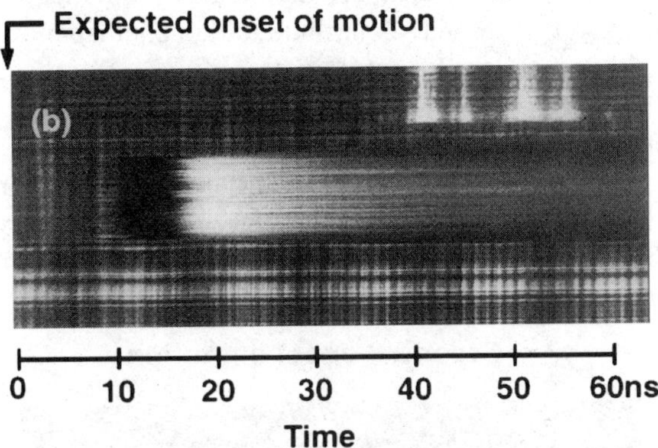

FIGURE 2. Streak records of the response of 1.0-µm-thick Al films to two different driving energies: (a) 29 mJ; (b) 6.8 mJ. The 400-µm diameter of the flyer essentially corresponds to the vertical extent of the dark area.

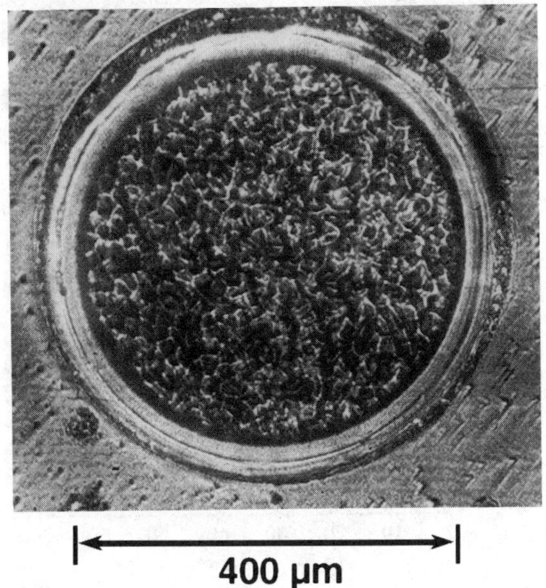

FIGURE 3. Frame image of 2.6-µm-thick Al flyer; 27.7 mJ incident energy. Image acquired ~12 ns after onset of motion.

resists melting and plasma penetration for a signifi-cantly longer period of time than the aluminum flyer. Second, the transitions are less uniform spatially in the composite case with breakout seemingly occur-ring at fewer locations across the face of the flyer. Plasma penetration is clearly linked temporally to prior changes in flyer reflectivity, indicative of an early melt in these regions. Similar effects were observed over a wide range of incident energy as well as over the available range of flyer thickness. Defects in the mechanical or thermal properties of the Al_2O_3 layer may contribute to this phenomenon. Alternatively, the less uniform transitions may reflect differences in the growth of instabilities in portions of the flyer in contact with the hot, driving plasma.

Figure 5 presents a plot of the time required for plasma breakout over the flyer face vs. initial target thickness. Clearly, a thermally insulating Al_2O_3 layer can not prevent melt and plasma penetration of the

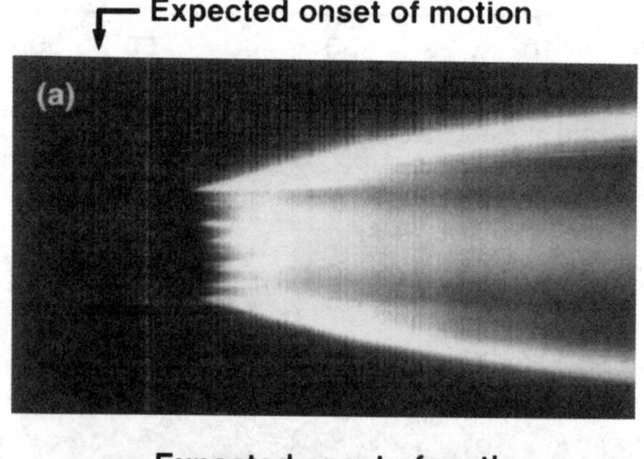

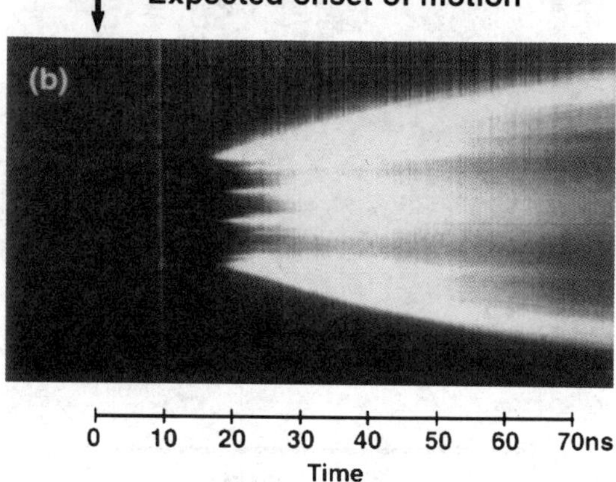

FIGURE 4. Streak records of the response of two different flyer target materials: (a) 2.1-μm-thick aluminum; (b) 2.0-mm-thick composite. Incident energy near 28 mJ in both cases.

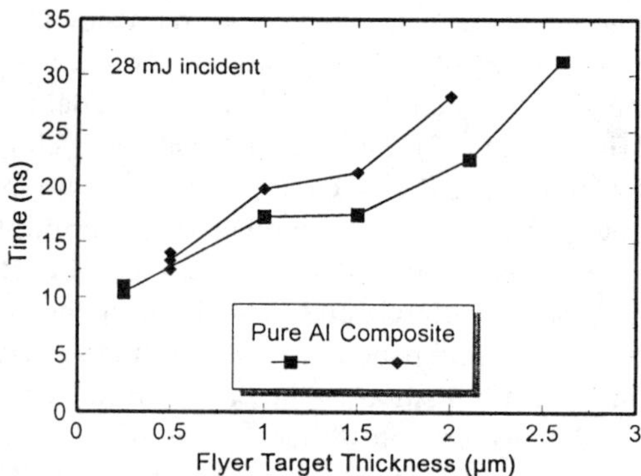

FIGURE 5. Plot of time delay from the onset of flyer motion to the latest plasma breakout across the face of the flyer vs. initial flyer target thickness.

thin materials examined in this work; however, the barrier does significantly delay the harmful effects of deep thermal diffusion. The increased delay becomes more apparent as thickness increases.

The data presented here highlight some important considerations in the use of laser-driven flyers as impactors for shock compression studies. In princi-ple, high impact velocities can be obtained from very thin flyers; however, these materials are subject to extremely rapid melt and plasma penetration. Inclu-sion of a thermally insulating layer in the flyer target can delay but does not prevent flyer breakup. With thicker targets (>2.5 μm), a practical "time window" of acceptable flyer integrity can be obtained.

REFERENCES

1. Trott, W. M., "Investigation of the Dynamic Behavior of Laser-Driven Flyers," in *High-Pressure Science and Technology--1993*, eds. S. C. Schmidt, et al., New York: AIP Press, 1994, pp. 1655-1658.
2. Frank, A. M., and Trott, W. M., "Stop Motion Microphotogra-phy of Laser Driven Plates," in *SPIE Proceedings No. 2273*, San Diego, CA, 1994, pp. 196-206.
3. Farnsworth, Jr., A. V., "Laser Acceleration of Thin Flyers" (this volume).
4. Trott, W. M., and Meeks, K. D., *J. Appl. Phys.* **67**, 3297-3301 (1990).
5. Trott, W. M., "Studies of Laser-Driven Flyer Acceleration Using Optical Fiber Coupling," in *Shock Waves in Condensed Matter--1991*, eds. S. C. Schmidt, et al., New York: Elsevier Science Publishers, 1992, pp. 829-832.
6. D. L. Paisley, "Laser-Driven Miniature Flyer Plates for Shock Initiation of Secondary Explosives," in *Shock Waves in Con-densed Matter--1989*, eds. S. C. Schmidt, et al., New York: Elsevier Science Publishers, 1990, pp. 733-736.

SHOCKS INDUCED BY LASER DRIVEN FLYER PLATES
1 - EXPERIMENTS

J.L. Labaste, M. Doucet, P. Joubert*

C.E.A.. Centre d'Etudes du Ripault B.P. 16 37260 Monts France
**C.E.A.. Centre d'Etudes Scientifiques et Techniques d'Aquitaine B.P. 2 33114 Le Barp France*

To study laser-driven miniature flying plates accelerated by a laser impact, we have performed several types of measurements to determine the reflected energy, acceleration profiles, projectile planarity and impact pressure. A 10 ns Nd : YAG laser pulse at power densities 1-10 GW/cm² accelerates several materials deposited on a UV - fused quartz window. The measurement of the reflectivity, resolved in time by means of an elliptic mirror, is used to make an energy evaluation. A Doppler-laser interferometry system is used to record the flyer acceleration profile and to measure the impact pressure. A good agreement between experiments and numberical results issued from a 1-D laser-matter interaction code is obtained.

INTRODUCTION

To initiate secondary explosives, the best way is to create a shock by means of a flyer plate impact. Two methods are used to launch these plates:

- an exploiding foil, crossed by a high current density,
- a laser beam focused on a window with deposited target material.

The laser driven flyer plates method have been recently investigated[1-4].The laser beam generates a shock wave on the deposited material, and a part of the thickness of the foil is converted into a plasma. This plasma accelerates the remaining part of the material toward a given target.

A 1-D laser-matter interaction numerical code[5] has been developed for these low fluences. To validate the results, we have performed several experiments.

We measure the energy balance of the laser-matter interaction, the acceleration of the projectile and its planarity. The results are compared with the values given by numerical simulations.

EXPERIMENTAL METHOD

The driving laser used is a Q-switched Nd:YAG laser delivering, at 1.06 μm wavelength, a 8 ns FWHM pulse by 850 mJ energy. The experimental set-up is presented in Fig. 1.

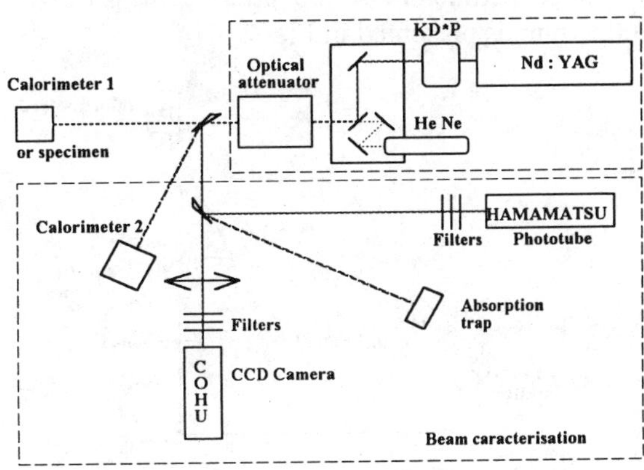

FIGURE 1. Experimental set-up used to generate the laser energy.

We adjust the pulse energy by means of an optical attenuator. For each test, we record:

- the energy by means of calorimeter 2 (it has been calibrated with the effective energy delivered to the specimen),
- the spatial beam profile with a CCD camera and a Beamcode® software,
- the temporal profile with a phototube.

This laser beam is focused on a UV fused silica deposited foil. This transparent window confines the plasma and allows to obtain higher pressures. The power density, in our experiments, is below the dielectric breakdown of the window[6].

We are able to accelerate 2-8 μm thick, 1-2 mm diameter aluminum and copper flyer plates, at given power densities.

ENERGY BALANCE MEASUREMENT

We want to measure the effective part of the energy used in the laser-matter interaction.

For an opaque specimen, the relation between the reflectivity (R_λ) and the absortivity (A_λ) is given by the KIRCHOFF law :

$$R_\lambda + A_\lambda = 1 \qquad \lambda \text{ is the wavelength}$$

The experimental set-up used to measure the reflectivity is presented in Fig. 2 :

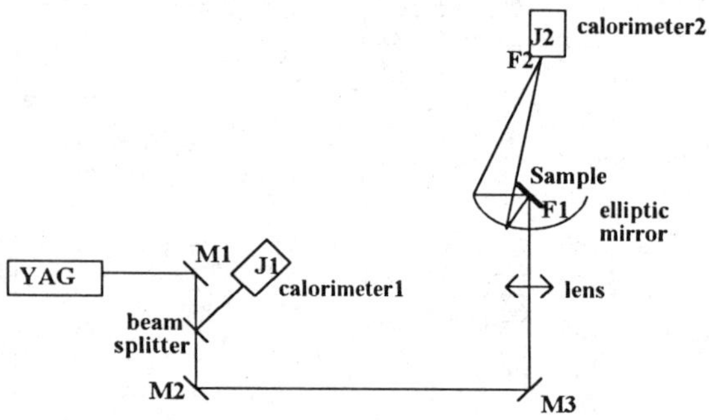

FIGURE 2. Experimental set-up for reflectivity measure

This method is based on the use of the elliptic mirror properties. The specimen, impacted by the laser beam, is set on an elliptic mirror focus. The specular and scattered reflected light is collected on a second mirror focus.

The loss of energy due to the laser injection hole is less than 1%.

A typical reflectivity measurement is presented using Fig. 3.

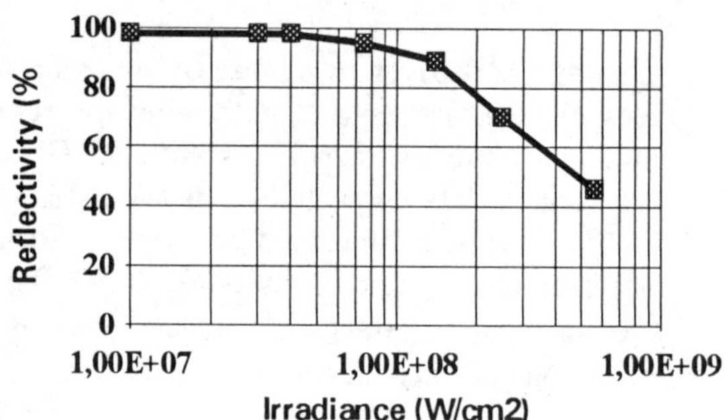

FIGURE 3. Measure of the reflectivity of copper versus irradiance.

In comparison with the integration sphere technique[7], this system leads to some advantages:

- higher level signal,
- the equality of the optical distance focus1-mirror-focus2, for any reflection angle, which allows to measure the reflectivity versus time

FLYER PLATES MEASUREMENT

To determine the velocity history of the flyer plates, we use a Doppler Laser Interferometry technique[8]. The particularity of this technique is the use of a bandwidth dye laser and a Fabry-Perot interferometer acting as a modulator and a demodulator. The fringe intensity, compared with that delivered by the usual CW single mode argon laser, is increased by more than an order of magnitude, and we do not need a cooling system.

The performances in terms of peak flyer velocity of aluminum and cooper targets have been obtained for various incident fluences. Some results are illustrated in Fig. 4.

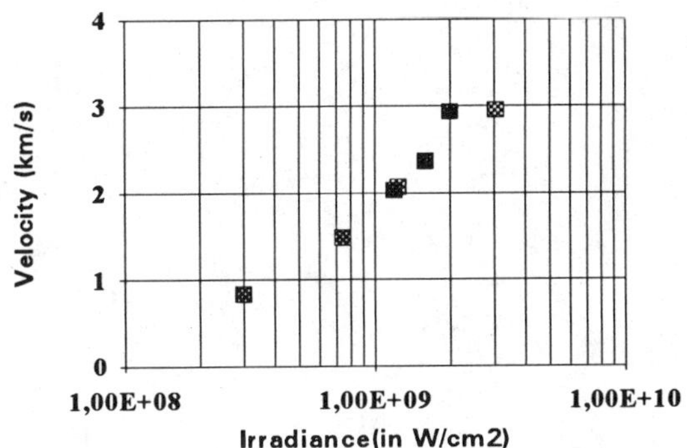

FIGURE 4. Peak flyer velocities as a function of irradiance for aluminum

For low thicknesses of aluminum, we have difficulties in performing DLI measurements, because the flyer does not remain in solid state (the plasma expasion heats the rear surface of the projectile). This is one of the reasons for the use of multi-layer materials [1-3].

We have also performed experiments to improve the planarity of the flyer plate when it is impacting a PMMA target.These experiments have been previously described by numerous authors[1,2].

CONCLUSION

Before optimizing the launch of flyer plates, we try to validate, with a simple layer material, a numerical simulation that is described in an other paper presented at this meeting[5]. We have performed reflectivity and hydrodynamic measurements. The first comparison between experiments and simulation values shows a good agreement in term of reflectivity as well as velocity[5].

REFERENCES

1. Paisley D.L., "Laser driven flyer plates for shocks initiation of secondary explosives",presented at the APS conference on Compression of Condensed Matter, August 14-17, 1989.

2. Trott W.M., Meeks K.D., "Acceleration of thin targets using fiber-coupled optical pulses", presented at the APS conference on Compression of Condensed Matter, August 14-17, 1989.

3. Paisley D.L., Warnes R.H., Kopp R.A., "Laser-driven flat plate impacts to 100 GPa with subnanosecond pulse duration and resolution for material properties studies", presented at the APS conference on Compression of Condensed Matter, June 17-20, 1991.

4. Paisley D.L. "Fiber optic mounted laser-driven flyer plates", United State Patent N°5,029,528, July 9, 1991.

5. Cazalis B., Boissière C., Sibille G., "Shocks induced by laser driven flyer plates: 2. Numerical simulations", presented at the APS conference on Compression of Condensed Matter, August 14-18, 1995.

6. Deveaux D., Fabro R., Tollier L., Bartnicki E., "Generation of shock waves by laser-induced plasma in confined geometry", *Journal Applied Physics,***74(4)**, (1993).

7. David , Goebel , "Generalised integration sphere theory", *Journal Applied Optics,***6**, (1967).

8. Veaux J., Mercier P., Béhar H., Cavailler C., "Multiple-line laser doppler velocimetry", presented in 33rd International Technical Symposium on Optical and Optoelectronic Applied Science and Engineering, August 1989.

SHOCKS INDUCED BY LASER DRIVEN FLYER PLATES : 2. NUMERICAL SIMULATIONS

B. Cazalis, C. Boissière, G. Sibille

C.E.A, Centre d'Etudes de Limeil-Valenton, 94195 Villeneuve Saint-Georges Cedex, France.

We present numerical simulations of shocks and damage induced by laser-driven flyer plates impacting metallic targets. A theoretical description of the different physical processes is included in the 1-D laser-matter numerical code FCI1. Good correlations between numerical and experimental results are obtained. Flyer plate velocities of few km/s and impact pressures greater than 1 Mbar are reached. In relation with experiments, this code can predict spall threshold of materials for high strain rate flow ($>10^6$ s^{-1}).

INTRODUCTION

Laser-driven flyer plates devices seem to be very flexible and well suited to determine high strain rate spall strength of materials [1]. In relation with experiments described in part 1 [2], this paper presents numerical simulations of the physical processes involved during the laser-driven flyer plates experiment : low fluence laser-matter interaction, thin flyer launch, impact and propagation of the shock wave, spalling of metallic targets. These simulations permit to optimize the multilayer deposit configuration, to evaluate the projectile velocity, the impact pressure and tension magnitude near the rear surface of the target, for different laser energy densities and target materials.

PHYSICAL PROCESSES DESCRIPTION

Laser-matter interaction

When a low-fluence laser beam (10^8-10^{12} W/cm^2) interacts a metallic target, the thermal energy deposited at the metal surface depends strongly on the absorptivity of the metal, which is function of temperature and laser wavelength.

To describe the solid-liquid absorption, we introduced the free electrons Drude-Zener model plus the interband contribution in the 1-D, hydrodynamic lagrangian code FCI1. For the Drude contribution, we use a phenomenologic collision frequency, which is in first approximation the electron-phonon collision frequency. A two plane wave pseudo-potential model issued from Ascroft and Sturm works [3], including temperature evolution from Mathewson and Myers results [4], allows to evaluate the interband conductivity and then the interband dielectric constant and reflectivity [5].

In vapor or plasma phases, electrons absorb laser energy mainly by electron-neutral and electron-ion inverse bremsstrahlung. The attenuation of the incident laser intensity is given by an absorption coefficient, which is the sum of these two contributions. Population of each species (electrons, neutrals, ions) and ionization degree are given by the Saha model or, for a wider range of laser irradiance, by the SESAME ionization tables.

Radiative effects of the plasma are taken into account with multi-group diffusion.

Validation of the code with all these low-fluence absorption models were performed in relation with a serie of experiments on the evolution of aluminum reflectivity over a wide range of laser intensity, where all the phases of the metal appear, from solid to plasma. Figure 1 shows that our absorption models, insert in the FCI1 code, permit to restitute with a good accuracy these measurements.

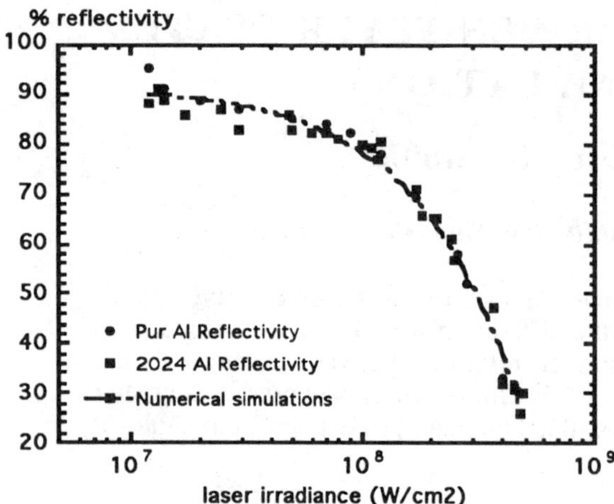

FIGURE 1. Evolution of Aluminum reflectivity in function of laser irradiance. Comparison experiment-numerical simulations.

Flyer plate motion

When the laser fluence becomes suffisant, a vapor and then a dense, cold plasma appears between the metal and the confinement window. An important mechanical coupling is produced, due to the hydrodynamic separation of the metallic deposit: the non-ablated part of the deposit is launched at high velocity by conservation of momentum.

To simulate these phenomena, we resolve the hydrodynamic equations including radiative transfer, laser energy deposition occuring as a heat source term :

- conservation of the mass :

$$\frac{d\rho}{dt} + \rho \, \nabla u = 0$$

- conservation of the momentum :

$$\rho \frac{du}{dt} = - \nabla \, (p + q_a) + f_r$$

- conservation of the energy :

$$\rho \frac{dE_e}{dt} = - P_e \, \nabla.u - \nabla . \, \chi_e^{eff} \nabla . \, T_e + \rho \, c_{ve} \, (T_e - T_i \,) / \tau_{ei}$$

$$+ c \int_0^{\infty} [\, \mu_v^E E_v \, - \mu_v^P B_v \, T_e \,] dv \; + W_{las}$$

$$\rho \frac{dE_i}{dt} = - P_i \, \nabla.u - \nabla . \, \chi_i^{eff} \nabla . \, T_i + \rho \, c_{ve} \, (T_e - T_i \,) / \tau_{ei}$$

where u is the fluid velocity, ρ the volumic mass, p the pressure, q_a the von Neumann-Rychtmyer artificial viscosity, f_r the radiative pressure, χ^{eff} the thermal conductivity, μ_v^E et μ_v^P the emission and absorption opacities, W_{las} the laser energy deposition. The e sign refers to electrons and i to ions.

For laser pulse durations of interest in this study

(about 10 ns) a description with T= T_e =T_i is quite suffisant. Resolution of these equations is realised in relation with the equation of state of the medium.

Figure 2 shows the numerical simulation of the velocity for a 7.5 µm aluminum projectile with an incident laser fluence of 10 J/cm^2 and a laser pulse duration of 10 ns. Good correlation with Doppler laser interferometry measurements is obtained.

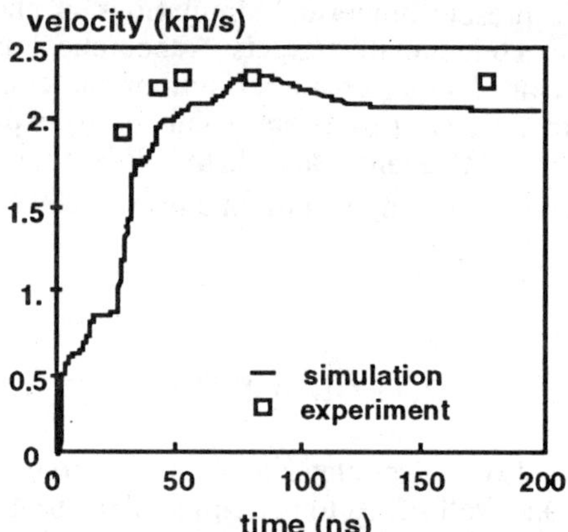

FIGURE 2 Evolution of the velocity for a 7.5 µm Aluminum flyer plate. Comparison experiment-numerical simulations.

Impact, shock propagation and spalling

The amplitude of the shock wave generated by impact depends on flyer velocity and on shock impedances of both projectile and target. Mie-Grüneisen equations of state are well suited to determine the impact pressure.

During its propagation in the target, the peak stress decreases as its duration increases. For these stress durations, the effects produced (plasticity, damage) are generated at very high strain rate (10^6-10^7 s^{-1}). Elastoplastic models describe stress evolution versus strain, strain rate, pressure and eventually temperature [6]. Numerical simulations have shown that, for a large variety of materials, amplitude and duration of the generated stresses are suffisant to display the elastic precursor detachment or a damage for submillimeter target thicknesses.

Within the target, the tension resulting from the interaction of two rarefaction waves leads, after nucleation of defects or voids, their growth and coalescence, to the formation of a rupture zone and spalling. Damage depends on tension magnitude and duration. To describe fracture apparition, we usually employ simple models as limit tension, cumulative model as Tuler-Butcher [7] or more complex model as N.A.G [8].

NUMERICAL RESULTS

Different kinds of metallic layers have been evaluated [9, 10] : metallic monolayers (Cu, Al, Mg) or multilayers. Paisley [10] has shown that multilayer devices leads to a better coupling efficiency by separating each layer function (metallic layer for ablation, dielectric layer to reduce thermal diffusion, metallic layer as projectile). After numerical simulations, we selected a configuration of this type (typically 0.2μm copper/0.2 μm alumina/1 μm copper). Others parameters for computations are:
- laser wavelength : 1.06 μm
- laser pulse width : 9 ns
- silica window thickness : 1 mm
- flight distance : 0.2 mm
- copper target thickness : 100 μm
- laser fluence : from 10 to 70 J/cm^2

Laser-matter interaction

During laser-matter interaction, 5 to 15% of the incident light is reflected, according to the laser energy density. For a laser fluence of 30 J/cm^2, plasma maximum temperature is about 12 eV and ablated thickness is 0.2 μm. A peak pressure of 30 kbars is induced in the plasma, with a half-width duration of 18 ns, twice that of the the laser pulse owing to confinement. Figure 3 illustrates the spatial distribution of the temperature just before impact on the target. The role of alumina layer as thermal barrier is clearly shown. Copper projectile flies then in a quasi-total solid state.

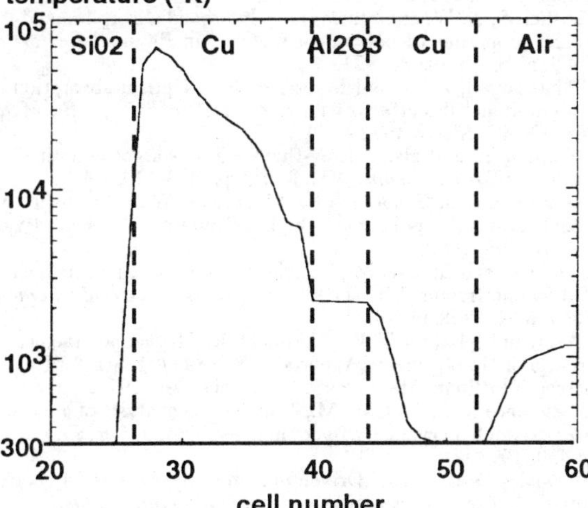

FIGURE 3. Temperature distribution in the multilayer device before impact on target.

Flyer plate motion

As shown in Figure 4, flyer plate velocity increases rapidly in function of the laser fluence.

Velocities greater than 5 km/s are reached when E_L > 50 J/cm^2.

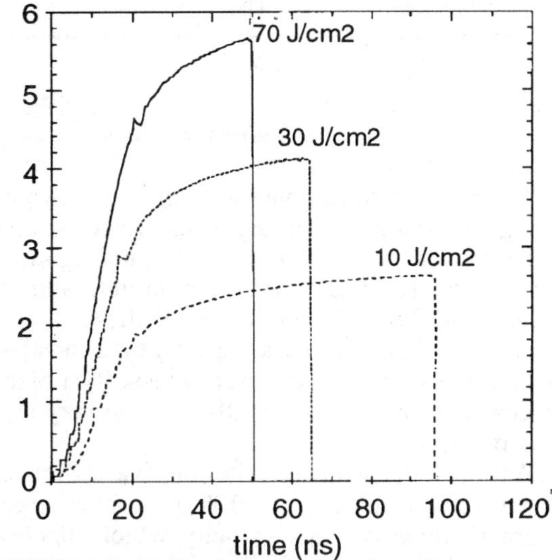

FIGURE 4. Evolution of velocity for a 1 μm Copper projectile at different laser fluences.

Impact

At impact, high pressure is generated in the target. For example, peak pressure of 1.3 Mbar is obtained for E_L = 30 J/cm^2 but its duration is quite small, of the order of 0.4 ns. Figure 5 shows impact pressures in copper versus laser fluence.

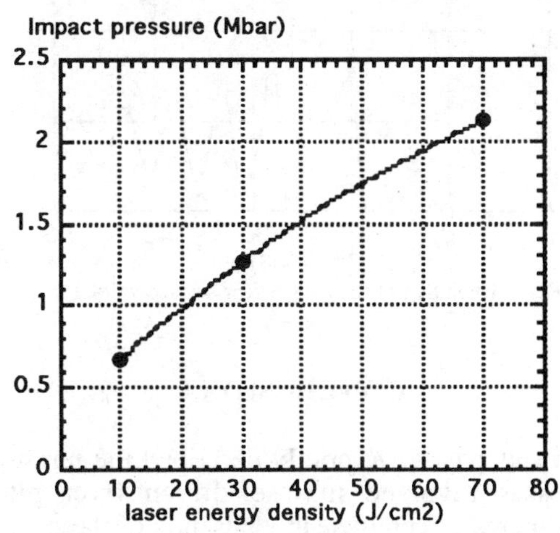

FIGURE 5. Impact pressure in a Copper target for different laser fluences.

TABLE 1 Numerical results of a 1 µm Copper flyer plate impact on different targets for a 30 J/cm2 laser fluence.

target	thickness (µm)	impact pressure (kbar)	max. tensile (kbar)	tensile duration (ns)	distance to rear surface (µm)
Copper	100	1300	125	3.7	13
Tantalum	100	1600	140	3.5	11
Titanium	100	850	90	4	13
Aluminum	100	700	90	4.5	17

Induced effects

We performed computations of shock waves propagation in different target materials, without spall criterion, to estimate the tension magnitude that can be reached within the target, and its localization. Table 1 shows, for all these metals, tensile peak stresses which are greater than standard spall limit tensions. It confirms that laser flyer plates devices are well suited for spall studies at very high strain rate flow.

With a Tuler-Butcher criterion, we calculated the rear surface velocity of a 100 µm copper target. Figure 6 illustrates this velocity which displays spalling of the target. Spall thickness is 9 µm, showing that spalling occurs before maximum of tension (see Table 1).

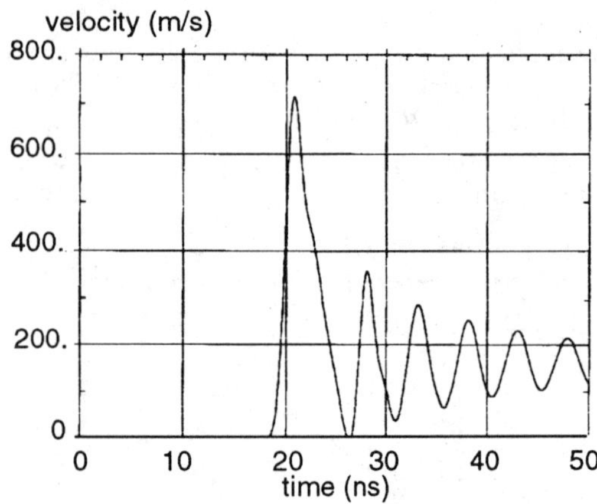

FIGURE 6 Evolution of the copper rear surface velocity.

CONCLUSION

In this paper, we briefly described the physical processes involved in laser-driven flyer plate experiments. A complete modelisation of these phenomena is introduced in the FCI1 code (from laser energy deposition to damage induced in target). Correlations of computations with future experimental results would permit to improve behaviour models in the range of high strain rate flow (10^6 - 10^7 s^{-1}) and provide parameters for these models.

ACKNOWLEDGMENTS

The authors gratefully acknowledge Mss Reisse C. and Wagon F. for their technical assistance in the development of the numerical code FCI1.

REFERENCES

1. Paisley D. L., Warnes R. H., Kopp R. A., "Laser Driven flat plate impacts to 100 GPa with sub-nanosecond pulse duration and resolution for material property studies" in *Proceedings of the Conference on Compression of Condensed Matter*, 1991, pp 825.
2. Labaste J. L, Doucet M. and Joubert P., "Shocks induced by laser-driven flyer plates : 1- Experiments" presented at the APS Conference on Compression of Condensed Matter, Seattle, August 14-18, 1995.
3. Ashcroft, N. W, and Sturm, K., "Interband absorption and the optical properties of polyvalent metals," in *Physical Review B*, Vol 3, n° 6, 15 march 1971.
4. Mathewson, A. G, and Myers, H. P., "Optical absorption in aluminum and the effect of temperature," in *J. Phys. F : Metal Phys.*, Vol 2, March 1972.
5. Combis P. and als., "Low-fluence laser-target coupling" in *Laser and Particle Beams*, Vol. 9, n°2, pp 403-420, 1991.
6. Steinberg D. J, Cochran G. G., Guinan M. W., "A constitutive model for metals applicable at high strain rate" in *J. Appl. Phys.*, 51, pp 1498, 1980.
7. Tuler R. and Butcher M., "A criterium for the time dependence of dynamic fracture" in *The Int. Journal Fracture Mechanics.*, Vol 4, n° 4, 1968, pp 431.
8. Seaman L., Barbee T. W., Curran D. R., "Dynamic Fracture Criteria of Homogeneous Materials", AD 893701, Stanford Research Institute, Menlo Park, California, Dec. 1971.
9. Lawrence R. J., Trott W. M., "Theoretical analysis of a pulsed-laser-driven hypervelocity flyer launcher," in *I. J. Imp. Eng.*, 14, pp 439, 1993.
10. Paisley D. L., "Laser Driven miniature flyer plates for shock initiation of secondary explosives," in *Proceedings of the Conference on Compression of Condensed Matter*, 1989, pp 733.

A STUDY OF LASER-DRIVEN FLYER PLATES

D.J. Hatt and J.A. Waschl

Aeronautical and Maritime Research Laboratory, DSTO, PO Box 4331 Melbourne 3001, Australia

Empirical relationships for the velocity of laser-driven flyer plates versus radiant exposure have been found for a number of single layer and multiple layer thin (< 4 μm) film targets. Single layer Al and multiple layer targets were evaluated with an optical time-of-arrival (TOA) technique for measuring flyer plate velocity. The materials employed in the multiple layer targets included metallic layers of Al, Mg, and Cu and dielectric layers of Al_2O_3, MgF_2, and ZnS. The TOA velocity measurements were compared for various target constructions. Choice of layer thickness for the targets was based on the laser Gurney model. It was found that $Mg/MgF_2/Cu$ and Al/Al_2O_3 multiple layer targets were the most efficient, producing flyer plate velocities of about 4.4 mm/μs and 3.6 mm/μs, respectively, for a radiant exposure of approximately 15 J/cm^2. For some multiple layer targets, dielectric failure was detected as the radiant exposure level was increased. Flyer plate breakup was also detected for some of the multiple layer targets.

INTRODUCTION

Laser-driven flyer plate performance can be successfully examined with the aid of a modified VISAR velocity interferometer (1) known as ORVIS (2,3). The temporal resolution of ORVIS is commensurate with the rapid acceleration ($>10^{10}$ m/s^2) (4) of these flyer plates. As we do not have easy access to ORVIS, we developed a time-of-arrival (TOA) technique to measure average flyer plate velocity for various target constructions and over a range of laser radiant exposures. Comparisons were made using laser Gurney model (LGM) (5) predictions.

EXPERIMENTAL

A pulsed (6 ns FWHM) Nd:YAG laser operating at 1.064 μm provided the incident irradiation. The radiant energy at the target was adjusted by an attenuator consisting of a half-wave plate and Glan laser prism. The energy per pulse was monitored by reflecting a small portion (<1%) of the laser output off a beamsampler and into an energy probe. A spot size of 1 mm at the target was employed.

Targets consisted of either single layer or multiple layer films deposited on UV-grade fused silica substrates. Al was employed for the single layer targets while the multiple layer targets consisted of an initial metal layer, a dielectric layer (Al_2O_3, MgF_2, or ZnS) and usually a third metal layer. An important advantage gained by employing multiple layer targets is a reported increase in the kinetic energy of the flyer plate of ~30% (6).

For the multiple layer targets, either Al or Mg was employed as the initial layer. From the LGM, a Mg initial layer would be expected to produce a higher velocity than Al. To achieve maximum pressure at impact, the third layer of the target should consist of material of high shock impedance. Our choice of materials was limited to Al and Cu.

The construction of the multiple layer targets was based on LGM predictions for a single material target and the fact that the target construction beyond the ablation layer is irrelevant to the LGM, provided that the ablation layer material in the single and multiple layer targets is the same and that the areal density of the two targets is equal. For each initial layer material, the initial layer thickness (ablation layer thickness) was calculated for a velocity of ~ 3 mm/μs using the LGM. For each tri-layer target, an arbitarily chosen dielectric layer thickness of 0.25 μm was used.

TOA measurements were performed by allowing the flyer plate to impact a Perspex (polymethyl methacrylate) witness plate. The witness plate was machined with a recess and held against the metal film target as shown in Fig. 1. The recess provided a known standoff distance for the flyer plate travel and the witness plate enabled clear flyer impact impressions to be recorded. Nominal standoff distances of 0.2 mm and 0.5 mm were employed.

This TOA technique utilizes the pulse of light emitted as the flyer plate impacts the witness plate. The light pulse is generated by the shock wave travelling ahead of the flyer plate compressing the air (7). As the flyer plate approaches the witness plate, shock reverberations between the flyer plate and the witness plate increasingly compress the air

to a maximum just prior to impact. The light output was collected by a lens in a fibre-optic probe located about 30 mm behind the witness plate. Reflected light from the front surface of the target was collected by a fibre-optic cable. After detection by a photomultiplier tube (PMT), both light signals were recorded on a digital storage oscilloscope.

Initial motion of the flyer plate was taken as the time at which maximum intensity occurred for the reflected light from the front surface. The moment of impact was assumed to be concurrent with the maximum light intensity as seen by the fibre-optic probe. Average velocity was calculated from the time difference between these two signals (after correcting for cable lengths) and the known standoff distance, d.

RESULTS AND DISCUSSION

Figure 2 shows typical timing marks used for determining the average velocity of a single layer flyer plate.

To facilitate intercomparisons between the different targets, multiple layer targets were

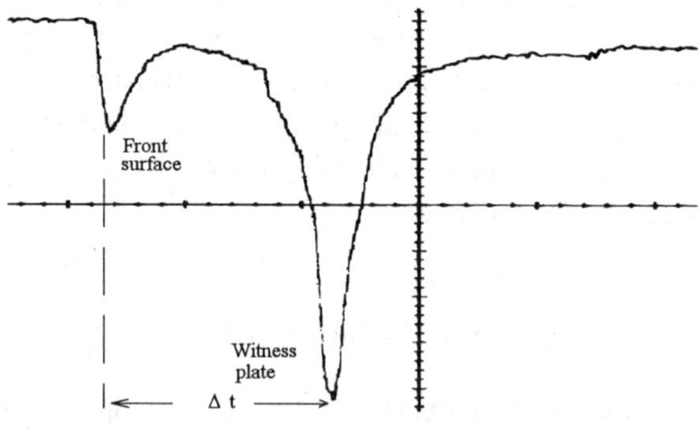

FIGURE 2. Plot of the typical timing marks produced by a single layer target. The peaks represent the reflected light from the front surface of the target and the light produced by shock compression at the witness plate. Intensity increases to the bottom of the page and Δt is the time difference.

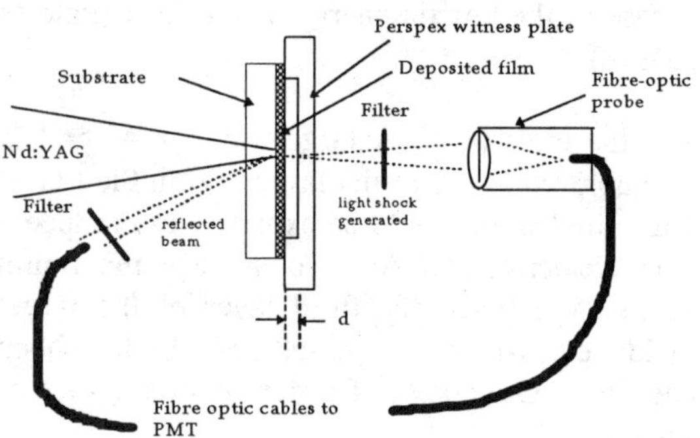

FIGURE 1. Target holder for average velocity measurements. Standoff distance, d, was either 0.2 mm or 0.5 mm.

described in terms of an equivalent thickness and a single material, i.e., a single layer target. The equivalent thickness is the thickness of a target composed of only the material of the initial layer of the multiple layer target being described, calculated to give the same areal density as that of the multiple layer target.

It has been reported (6) that 90% of the maximum velocity of laser-driven flyer plates is achieved in approximately two pulse widths (12 ns in this study), implying that for a peak velocity of 5 mm/μs the distance travelled is ~ 0.05 mm. Thus, although the TOA technique provides an average velocity, for the velocity range investigated, the high acceleration of the flyer plates ensures that the measurement is within 10% of the true peak velocity. This was supported by good agreement between TOA data collected at 0.2 mm and 0.5 mm standoff distances for a 2 μm Al target. The data also indicated that the flyer plate was stable and intact for a travel distance of at least 0.5 mm.

Figure 3 shows a comparison between a single layer 4 μm thick Al target and two Al multiple layer targets with 4 μm and 3.6 μm equivalent thicknesses. One of these targets has an Al_2O_3 dielectric layer followed by an Al layer while the other has the Al_2O_3 dielectric layer only.

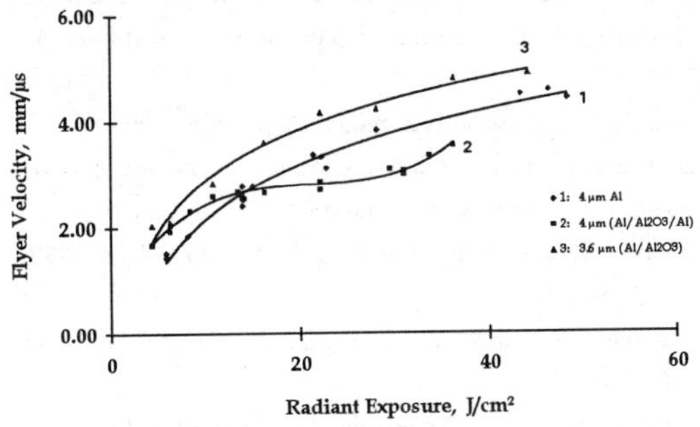

FIGURE 3. Comparison between various Al targets, showing the effect of a dielectric layer failure near 15 J/cm². For the multiple layer targets, the thickness value refers to the equivalent target thickness while the layer materials are shown in parentheses.

The design velocity (i.e., ~3 mm/μs) for these multiple layer targets occurred at a maximum radiant exposure of 16 J/cm². The crossover of the $Al/Al_2O_3/Al$ and single layer Al curves, at approximately 15 J/cm², is consistent with having achieved a radiant exposure high enough to ablate the initial metal layer and subsequently cause failure of the dielectric layer with a consequent loss in the pressure advantages that such a layer provides. Thereafter, the $Al/Al_2O_3/Al$ curve continues to lie beneath the single layer Al curve as the dielectric effectively thwarts efficient coupling between the laser and the outer layer of Al. No similar trend occurs for the Al/Al_2O_3 target as the dielectric layer is much thicker and can presumably withstand much higher radiant exposures without failure.

Allowing for the difference in the target thicknesses, the 3.6 μm thick multiple layer target achieves approximately 17% higher velocity than does the single layer 4 μm thick Al target. This is consistent with the data reported by Paisley (6). The 4 μm thick $Al/Al_2O_3/Al$ target initially achieves equivalent performance to that of the 3.6 μm thick target, then the performance gradually declines as the radiant exposure is increased.

In Fig. 4, the performance of two Mg multiple layer targets employing different dielectric layers are compared with the performance of a 2 μm thick Al single layer target (an appropriate single

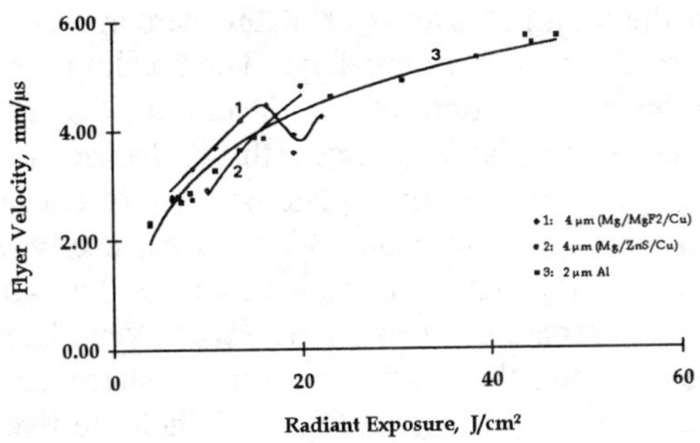

FIGURE 4. Comparison between two Mg targets with either MgF_2 or ZnS dielectric layers and a 2 μm thick single layer Al target.

layer Mg target could not be manufactured). The multiple layer targets design velocity occurred at a radiant exposure of 6 J/cm^2, however, maximum velocity is achieved near 15 J/cm^2 for the MgF$_2$ based target. The maximum for the ZnS based target was not determined.

Based on the equivalent thickness criterion, the 4 μm multiple layer targets should be compared to a 2.6 μm thick Al single layer target; a 2 μm thick Al target was the closest available. The LGM predicts that the velocity for the 2 μm thick Al target would be approximately 15% greater than that for a 2.6 μm thick Al target. Thus, the performance of the Mg targets is better than is apparent in Fig. 4. In particular, the MgF$_2$ based target is expected to show a 27% improvement in velocity compared with that of the equivalently thick 2.6 μm Al target for a radiant exposure of 15 J/cm^2.

When a ZnS dielectric layer is coupled with Mg as the initial layer, the performance is not as good. The reason for the different performance found when using the MgF$_2$ and ZnS dielectric layers is not known, however, it may be related to porosity differences in the dielectric layers. As the radiant exposure level increases, the failure level of the dielectric materials becomes important. It would appear that the radiant exposure failure level of the ZnS dielectric is higher than either the Al$_2$O$_3$ or MgF$_2$.

For the multiple layer flyer plates there appeared to be evidence of flyer breakup. The breakup may have been separation of the layers due to the impedance mismatch between the dielectric and the remainder of the flyer plate or fracture due to the thinness of the plate. This evidence was available in the number of registrations in the light output at impact. When single layer flyer plates are employed, the output is usually sharp and single peaked as shown in Fig. 2. When the flyer plate is a multiple layer one, often two or more distinct registrations are apparent in the light signal.

CONCLUSION

Multiple layer targets can provide enhanced performance over single layer targets. Near our design velocity (~ 3 mm/μs), the most efficient performance was found for Mg/MgF$_2$/Cu and Al/Al$_2$O$_3$ targets. Despite their good performance, there is concern about the integrity of the flyer plates produced from the multiple layer targets. The TOA technique indicated that flyer breakup may be more common in multiple layer targets.

ACKNOWLEDGMENTS

Brian Jones assisted with the experiments and manufactured several components.

REFERENCES

1. Barker, L.M. and Hollenbach, R.E., *J. Appl. Phys.*, **43**, 4669-4675 (1972).

2. Sheffield, S.A. and Fisk, G.A., Particle velocity measurements in laser irradiated foils using ORVIS, *APS Shock Waves in Condensed Matter eds. Asay, J.R., Graham, R.A., and Straub, G.K.* North-Holland 1983, pp. 243-246.

3. Paisley, D.L., Montoya, N.I., Stahl, D.B., Garcia, I.A. and Hemsing, W.F., Velocity interferometry of miniature flyer plates with sub-nanosecond time resolution, presented at the SPIE High Speed Conference, San Diego, USA, July 8-13, 1990.

4. Paisely, D.L. Laser-driven miniature plates for one-dimensional impacts at 0.5 -≥ 6 km/s. *Shock-wave and high-strain-rate phenomena in materials eds.* Meyers, M.A., Murr, L.E., and Staudhammer, K.P., Marcel Dekker 1992, pp.1131-1141.

5. Lawrence, R.J. and Trott, W.M., *Int. J. Impact Eng.*, **14** 439-449 (1993).

6. Paisley, D.L., Laser-driven miniature flyer plates for shock initiation of secondary explosives, *APS Shock Compression of Condensed Matter eds. Schmidt, S.C., Johnson, J.N., and Davison, L.W.*, North-Holland 1989, pp. 733-736.

7. Maccoll, J.W., *Proc. Roy. Soc. Lon.*, **A159**, 459-472 (1937).

Laser Acceleration of Thin Flyers

Archie V. Farnsworth, Jr.

Sandia National Laboratories, Albuquerque, N.M. 87185-0820

Laser energy delivered through an optical fiber has been used to accelerate a thin metallic foil to high velocity. Subsequent impact of the foil onto an explosive charge can initiate an explosion. The present computational study addresses the physical processes of laser absorption, energy transport, flyer acceleration, and foil impact on HE or on diagnostic materials in associated experiments. The objective has been to gain understanding that will allow optimizing the system for practical HE initiation. The structure of the foil, especially the presence of a thermally insulating layer near the ablation surface, significantly influences foil effectiveness as an initiator. These simulations show a marked effect on the acceleration process by that layer, which influences both the onset of full laser power absorption, and the physical competence of the foil on arrival at the HE. The effect on laser absorption is especially noteworthy at low laser intensity, where foil launch is marginal. A role of the glass fiber in absorbing laser energy and contributing ablated material is seen in the calculations and confirmed by experiments.

Introduction

This paper addresses the processes involved when laser energy is delivered through a small diameter optical fiber to a thin metallic foil in contact with the fiber end face, causing ablation and acceleration of that foil across a small gap to an explosive charge, for the purpose of initiating an explosion[1,2]. In this computational study, we have sought to understand these processes in sufficient detail to allow a practical initiator of explosives (HE) to be designed and optimized. Another goal has been to develop the tools to provide some predictive capability for present and future studies of this subject.

The Computational Tools

In a previous paper[1] we presented some numerical computations using the LASNEX hydrocode. Despite much excellent laser modeling capability in this code, it lacked capabilities important to this work: the capability to transmit laser energy through a transparent material, and to absorb and reflect the appropriate quantities of laser energy at a cold metal interface with such a material. We briefly describe here the CTH hydrocode[3] and special additions to it that have been made to create a straight-forward laser energy transport and absorption package having the features required for this work.

The CTH hydrocode is an Eulerian code with a fixed grid which is distorted during a Lagrangian step, and then remapped back to the original grid for each hydrodynamic cycle. It contains numerous models[3] for handling material properties and strength, and it accesses a substantial material property database. It also allows energy transport by thermal diffusion.

The additions to the code include a grid-following laser transport package that allows materials to be either transparent or absorbing. The absorption routines are appropriate to metallic absorption in cool materials, or plasma absorption when the materials become sufficiently hot. As laser energy is transported along a row of cells, each cell and material is treated as transparent, absorbing, or reflecting. If absorbing, energy is deposited in the cell and the balance conveyed to the next cell. If the material has a free electron number density that exceeds the critical value, energy is deposited in the vicinity as appropriate to a reflecting beam, and the direction of transport is reversed for the reflected component. The unsolved physics problem of the ionization of dense vapors is avoided by using a simple criterion for transition between metallic and plasma-like behavior. We found sufficient accuracy for present purposes by using the plasma formulation when the density-temperature product $(\rho_0/\rho)\sqrt{T}$ exceeded the value 2, where ρ_0 and ρ are the solid and current densities, and T is temperature measured in electron volts (eV). The ionization properties of metals are treated using routines from Lee and More[4] and for non-metal plasmas by the simple perfect gas formula $z = (PA)/(\rho kT) - 1$ where z is the ionization per atom, P is pressure, A is the average atomic mass, ρ is the density, k is the Boltzmann constant and T is the temperature. The success of these simple relations derives in part from the fact that for the present problem the transition to the plasma state occurs very rapidly, once it begins.

Physics Insights

The role of the glass fiber material

When the code modifications described above became available, one of the noteworthy results that was seen immediately was the substantial role played by the glass substrate in the ablation process driving the metallic foil. Very soon after the formation of a plasma in the aluminum, the heat transported from the metal plasma to the glass produced a small amount of glass plasma. The laser light then began to deposit energy in glass, with only the residue reaching the foil. The process evolved with additional glass being heated so that the beginning point of absorption of laser power receded into the glass fiber during the laser pulse. Indeed, it was seen that for the typical laser conditions most of the absorption occurred in glass, and correspondingly, the glass provided most of the driving plasma. This unanticipated result has been confirmed by comparing the glass removed from the fiber after a shot with that predicted by the code to have been vaporized, with excellent agreement. Typical glass removal is about 0.4 μm. Additional confirmation is found in the spectroscopic evidence obtained by Trott[5]. The participation of the glass as a

pusher of the foil suggested that some improvement in performance might be obtained by choosing a different substrate material for these experiments. Calculations were made using fused silica (the usual material), polycrystalline quartz, and a plastic (see Fig. 1). The performance of the foil using the

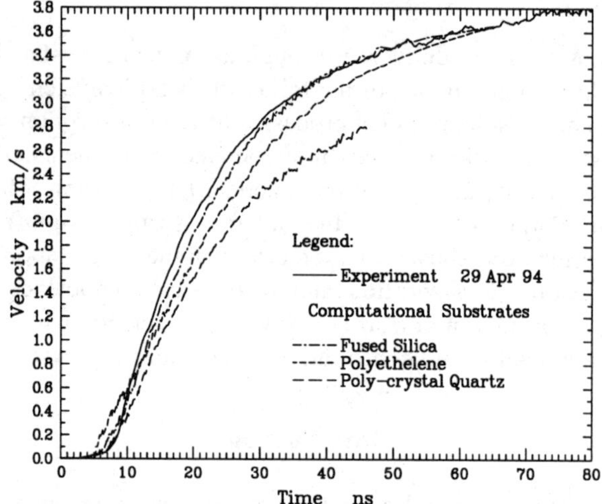

FIGURE 1. The velocity time history of a foil placed on a fused silica substrate (used in experiments) is compared with calculated results made assuming substrates of fused silica, polyethylene, and polycrystalline quartz.

polycrystalline quartz was much inferior to that obtained with the usual fused silica. The velocity obtained using polyethylene began early but lagged at intermediate times. It appears from the figure that it might exceed the velocity using fused silica at late times. For reasons that will be discussed below, it was thought undesirable to allow longer flight distances and longer times for these foils, so no present benefit was seen for changing the substrate material.

The composite foil

It has been observed in the experimental program that composite foils (consisting of a thin layer of aluminum followed by a thin layer of Al_2O_3, and the remainder of aluminum, with the first aluminum layer adjacent to the glass fiber) performed substantially better as impactors for igniting HE than do simple pure aluminum foils. Indeed, for very thin foils, experiments by Harris[6] show failure to ignite the test HE, even at quite large laser energies, while the composite foils consistently ignite the HE at moderate laser energy in comparable tests. Furthermore, at low incident laser energy, Trott[2] has shown that the composite flyer is accelerated to much greater velocities than that achieved by the pure aluminum foil but that the velocity difference becomes negligibly small at larger energies. We believe that these observed phenomena can both be explained by results obtained from our computational simulations.

We first address the difference seen at larger incident laser energy. In Fig. 2, we show the temperature distributions in the aluminum impactor computed for each foil, at a time just before impact on HE would normally occur. The heating of the foil, of

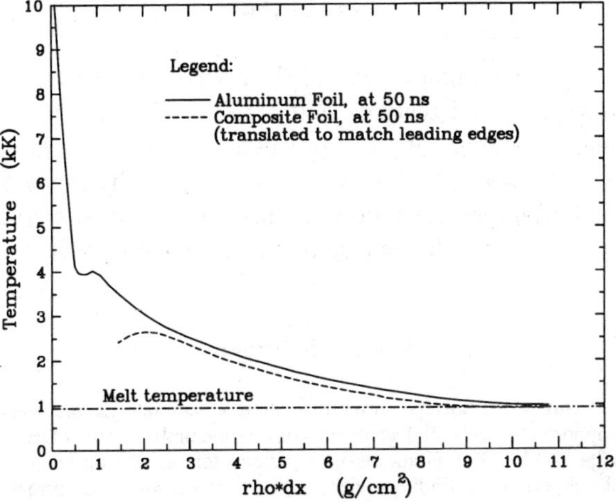

FIGURE 2. Temperature distributions of impactors for aluminum and composite foils near the time for impact on HE.

course, continues throughout the entire flight time which is typically more than twice the laser pulse length (here the pulse width (fwhm) is 15 ns).

The temperature of the pure aluminum foil is seen to be above the melt temperature throughout, while part of the composite foil remains at or below the melt temperature, at the impact side of the foil. A low density plasma accelerating a molten metal at normal densities is clearly Rayleigh-Taylor unstable. One must expect the interpenetration of the materials at the gas-liquid interface. The presence of solid material in the composite case would clearly dampen the instability growth, resulting in a more competent impactor striking the HE. This interpretation is consistent with the findings of Harris, and also with impact velocity data of Trott[2], who found a significant

difference in the pulse width of the impact velocity records for the impacts of composite and aluminum foils onto a LiF window. The insulation afforded the impact layer by the thin layer of alumina between the metal layers delays the heating of the impactor, for the composite foil design.

The near melt conditions of the impactor computed for virtually all foils suggested trying a foil design that would minimize that condition. A short analytical survey of potential replacements for the impactor metal, emphasizing high melt temperatures was performed. Titanium was found to deliver results comparable to aluminum, from that study, and was tried as an impactor. The computations produced comparable results as an effective impactor, assuming a composite design in which the impactor aluminum was replaced with an equal mass of titanium. Using the ignition criterion of Setchell[7] for fine grained HNS explosive, we produced the results shown in Figs. 3 and 4. Clearly, these results are comparable to

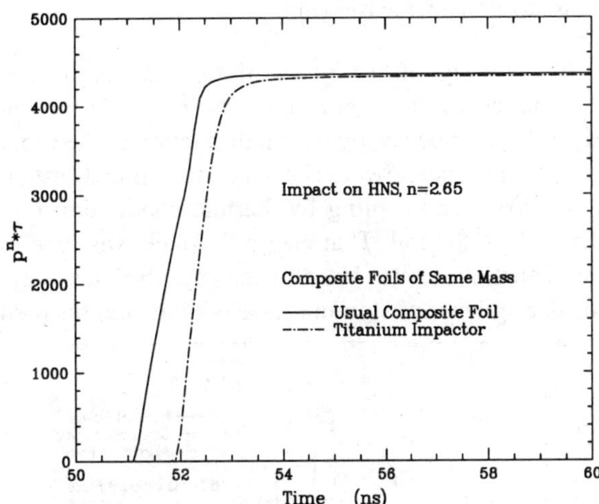

FIGURE 3. The integral of pressure in the explosive to the nth power times the time, as a function of time.

aluminum for impactor effectiveness, but the temperature of the titanium was not more favorable with respect to its melt temperature than was the aluminum; indeed, it is poorer in the computation, as seen in Fig. 4 despite the higher melt temperature.

We have noted previously[1] that the onset of high level absorption begins with the formation of plasma at the metal-glass interface. Before that time, most of the energy is reflected back into the glass fiber. This fact is critical in understanding the higher velocity obtained by the composite foils in comparison with aluminum foils, for lower energy pulses.

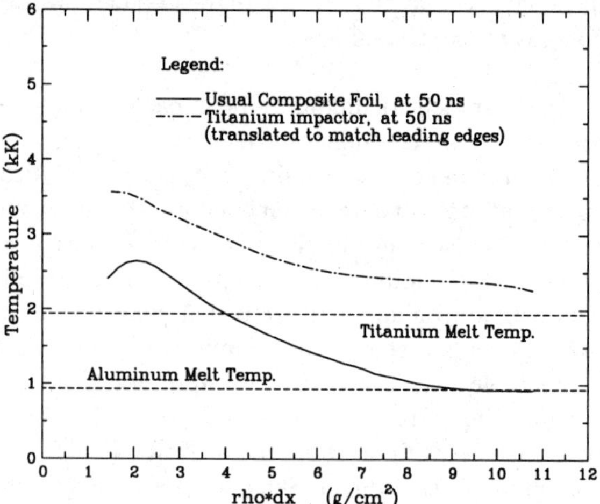

FIGURE 4. Temperatures of the impactors of composite foils having aluminum or titanium impactors, shown in relation to the melt temperatures for each material.

The timing of the onset of plasma creation is determined by the energy balance in the first metal layer; heating occurs by the small fraction of the input power that is not reflected back into the optical system (~4% here) and cooling by thermal conduction into the bulk of the foil. That energy balance was seen in the calculations to be affected by the thermally insulating layer of alumina at early time, and resulted

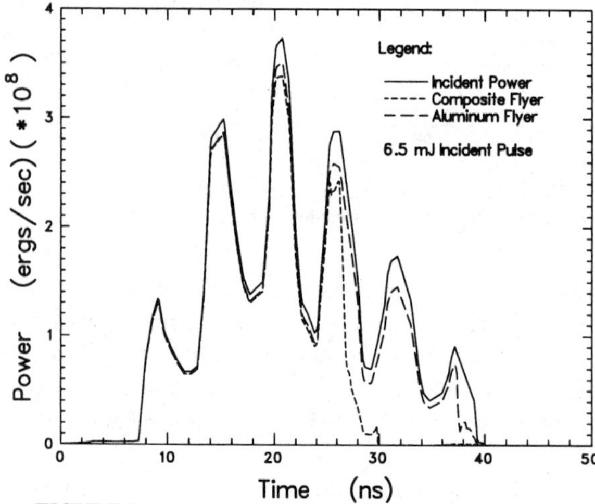

FIGURE 5. Incident laser power pulse, and reflected power lost (wasted) for the cases of the aluminum foil and the composite foil.

in a highly absorbing plasma being formed at an earlier time during the input power pulse. This reduced the energy wasted from the first part of the pulse, and, for low energy pulses, resulted in a greatly enhanced velocity. In Fig. 5 the incident laser power and the computed reflected power are shown for both the aluminum and composite foils. The resultant peak flyer velocities are 45 m/s and 600 m/s respectively, a somewhat extreme example showing very marginal launch of the pure aluminum foil. This trend is consistent with Trott's velocity[2] data, and with his measurements of reflected light for such cases.

Conclusions

These computations and their agreement with data have afforded insights that build confidence in our understanding of the processes underlying laser acceleration of thin foils. The CTH hydrocode, including present modifications for laser deposition, offer a tool with some actual predictive capability.

Acknowledgments

This work was performed at Sandia National Laboratories, supported by the U.S. Department of Energy, under contract DE-A-C04-94AL85000. Gratitude is expressed for permission to quote the experimental findings of Steve Harris and Wayne Trott prior to their publication.

References

1. Farnsworth, A. V., Jr., and Lawrence, R. J.,"Numerical and Analytical Analysis of Thin Laser-Driven Flyers," in *Shock Compression of Condensed Matter*, 1991, pp. 821-824.

2. Trott, W. M. "Investigation of the Dynamic Behavior of Laser-Driven Flyers," in *Shock Compression of Condensed Matter*, 1993, pp.1655-1658.

3. Hertel, E. S., Jr., Bell, R. L., Elrick, M. G., Farnsworth, A. V., Kerley, G. I., McGlaun, J. M., Petne, S. V., Taylor, P. A., and Yarrington, P., "CTH: A Software Family for Multi-Dimensional Shock Physics Analysis", in *Shock Waves @Marseille I*, 1993, pp. 377-382.

4. Lee, Y. T., and More, R. M., *Phys. Fluids* **27**, 1273-1286 (1984).

5. Trott, W. M. "High-Speed Optical Studies of the Driving Plasma in Laser Acceleration of Flyer Plate Studies".this vol.

6. Harris, S. M., Sandia National Lab., private communication.

7. Setchell, R. E. Sandia National Lab., private communication.

IMPACT EFFECTS OF A TWO-PARTS HIGH VELOCITY JET ON HOMOGENEOUS TARGET

T. S. Vong - J. P. Leyrat - H. C. Pujols*

C.E.A., Centre d'Etudes de Vaujours-Moronvilliers, B.P. n° 7, 77181 Courtry, France
**C.E.A., Centre d'Etudes Scientifiques et Techniques d'Aquitaine, B.P. N° 2, 33114 Le Barp, France*

Very fast jets were produced by launching a flyer plate upon a disk with a cylindrical cavity (ϕ20mm) machined in the opposite face. Experiments performed on aluminum showed that tip jet velocities of 9.7km/s and 21.5km/s were attained respectively for hemispherical and flat bottom cavities. For hemispherical bottom, unique jet is formed (shaped charge principle). For flat bottom one's, in addition to the latter jet, a faster but thinner jet propagates ahead. The influence of this very fast dart is pointed out by studying the jet's impact on a tungsten target. The calculations show that the dart penetrates deeply in the tungsten, and creates a cylindrical crater large enough to permit a penetration without resistance of the thicker part of the jet. Consequently the impact cratering and perforation effects of this thick jet take place in the very heart of the target, and the penetration ability of this type of jet is therefore more important than for a classical jet created in hemispherical bottom cavity. This impact effect can be multiplied by machining several cavities in the same disk. The 3D calculations show the coupling effects between jets, for hexagonal lay-out of seven identical cavities.

INTRODUCTION

In a classical shaped charge with a metallic liner, the velocity of the jet cannot exceed a limit (1,2) which is about 10km/s for copper and 13km/s for aluminum. For higher velocity jets, sometimes exceeding 20km/s, we have proposed an experimental device (3) using another concept previously mentioned by Asay *et al.* (4). By launching a flyer plate upon a disk with a cavity machined in the opposite face, very fast jets are produced under the combined effects of the increased projection at the cavity bottom and the implosion of this cavity.

Experiments have been performed on different materials and geometries of cavity, showing the jet shapes and their impact effects. This impact effect can be multiplied by machining several cavities in the same disk. The influence between jets is then directy dependent to the cavity spacing and setting. Some experimental data of single and mutiple jets are shown here.

In this paper we present new numerical simulation with CEL-2D in studying the impact effects of a two-parts very fast jet compared with a "classical" one. The 3D simulation with HESIONE shows the coupling between "classical" jets, for an hexagonal lay-out of seven identical cylindrical cavities.

GENERATION OF JETS

Using an explosive cylinder, where a plane detonation wave is initiated, an aluminum disk (8mm thick) is then projected at 4.9km/s velocity upon another aluminum disk target (25mm thick) with cavity on the opposite face.

For a cylindrical cavity with hemispherical bottom (20mm diameter, 20mm depth) a single jet is created with a protuberance at the tip (9,7km/s velocity), as predicted by "classical" shaped charge principle. Figure 1 shows the jet vizualised by high speed camera with argon flashes put close to.

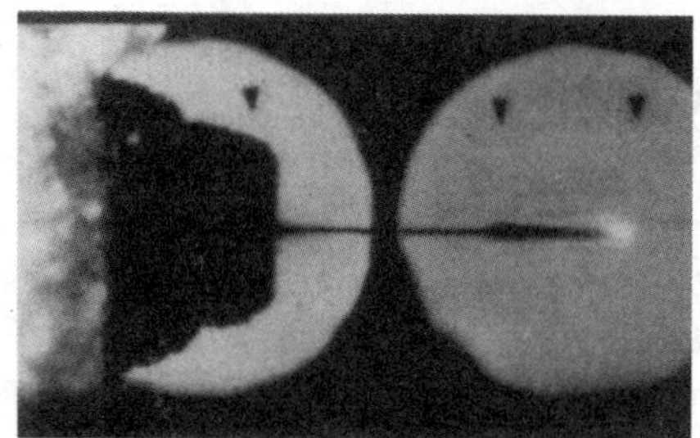

FIGURE 1. "Classical" jet (hemispherical bottom cavity).

For a flat bottom cylindrical cavity (20mm diameter, 20mm depth), in addition to the above jet, a faster but thinner jet propagates ahead with 21.5km/s velocity (Fig. 2).

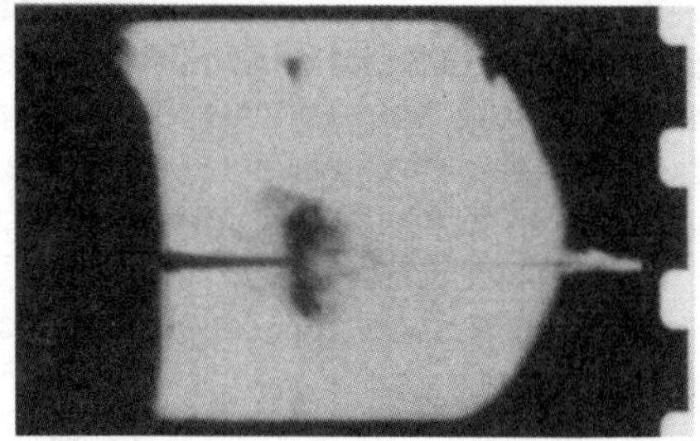

FIGURE 2. Two-parts high velocity jet (flat bottom cavity).

Calculations performed with CEL-2D code in eulerian scheme show that in the early stage, the jet formation mechanisms are very different between the two geometries of cavity. However at later time the difference appears principally in the presence of the high velocity dart in the jet front (Fig. 3).

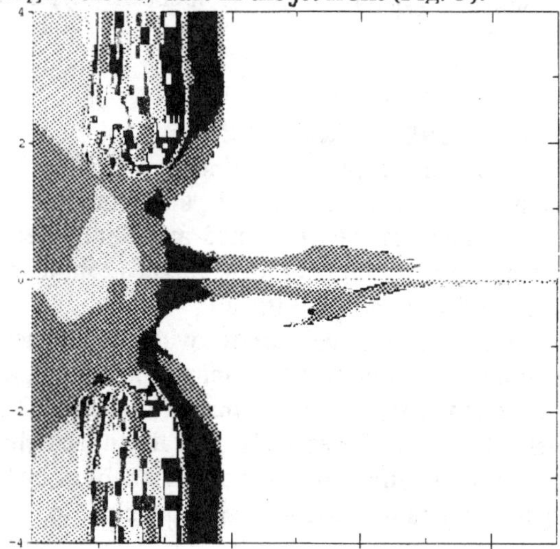

FIGURE 3. Shape difference between the two jets (CEL-2D code).

IMPACT EFFECT OF CLASSICAL JET

In order to point out the influence of this very fast dart we simulate a tungsten target (60mm thick) placed at 200mm in front of the cavity.

At 20.75µs, the jet created from the impacted hemispherical cavity reaches the target. At the impact point a crater is formed and spreads spherically (Fig. 4).

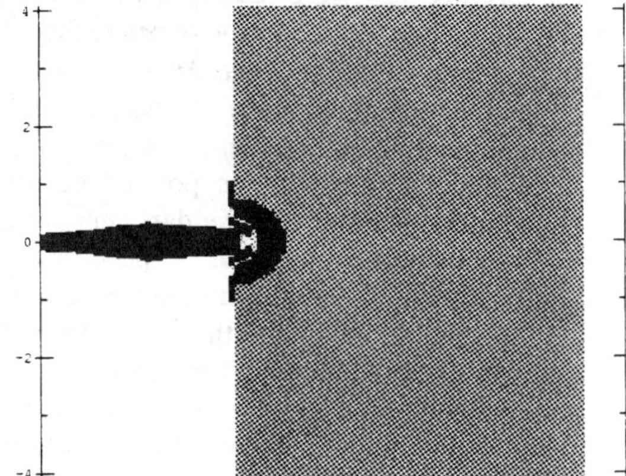

FIGURE 4. Unique jet impact : cratering phase.

Follows then a penetration phase characterized by a cylindrical crater with hemispherical cap. Once the protuberance totally "consumed", the constant section part of the jet penetrates the crater bottom creating then a conical form ahead (Fig. 5).

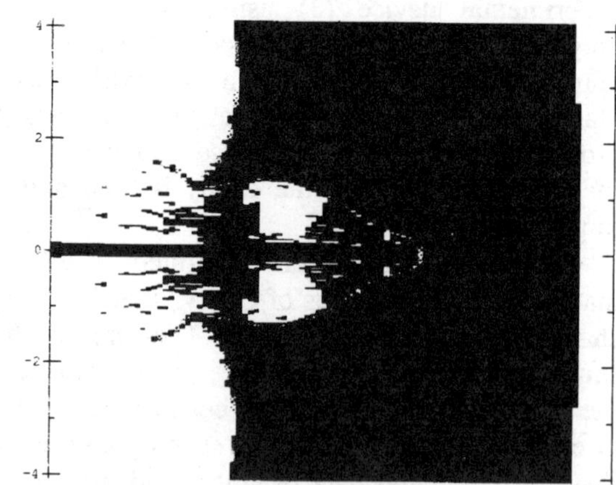

FIGURE 5. Unique jet impact : penetration phase.

After perforation of the tungsten target at 52µs, the tip jet has a residual velocity of 7km/s (Fig. 6).

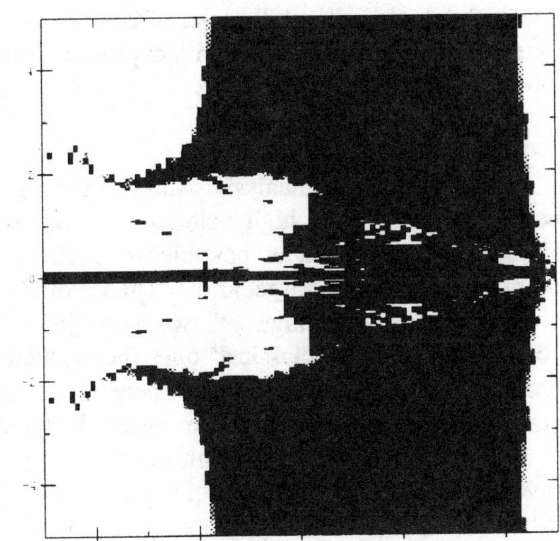

FIGURE 6. Unique jet impact : target perforation.

IMPACT EFFECT OF 2-PARTS JET

The very fast dart (0.6mm diameter) impacts the target at 13.75µs. Because of its very high velocity (19km/s), the dart penetrates without cratering 30mm of tungsten (Fig. 7). At this time of 22.5µs the target aperture (8mm diameter) is greater than the section of the lower velocity part of the jet.

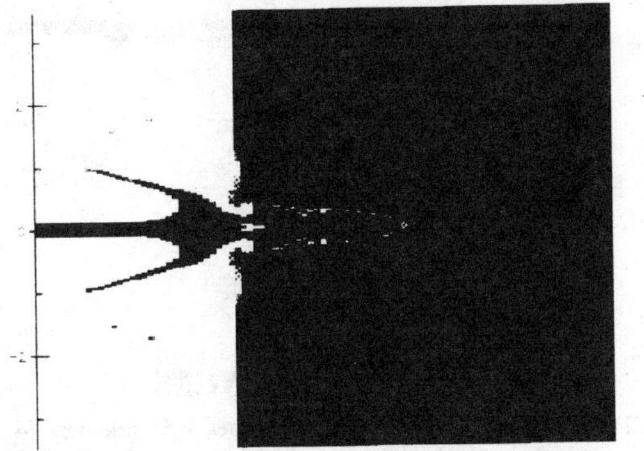

FIGURE 7. Two-parts jet impact : dart penetration phase.

At 26µs, the thick constant section part of the jet penetrates without resistance and reaches the bottom of the crater (Fig. 8) and impacts the "undamaged" tungsten (initial density).

Afterwards the cratering mechanisms follow the same phases as described for the equivalent "classical" jet: spherical expansion of the crater at impact, and conical terminal penetration of the jet.

The target is perforated at 38µs (Fig. 9).

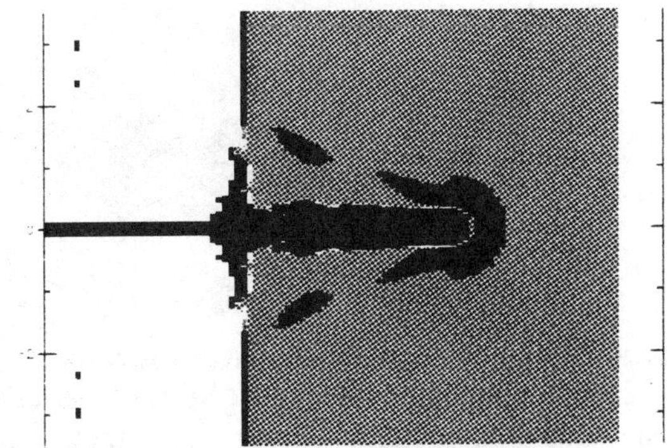

FIGURE 8. Two-parts jet impact : thick part jet impact.

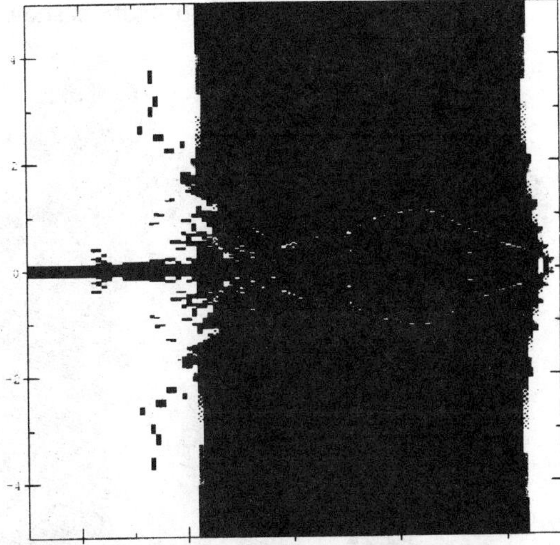

FIGURE 9. Two-parts jet impact : inside expansion effect.

We obtain the same residual tip jet velocity of 7km/s as for "classical" jet : this confirms the great similarity between these two jets.

The principal effect of the high velocity dart is to increase damage deeper inside the target. In fact, the rear part of the jet behaves as if it penetrates a thinner target, with initial thickness shortened by the depth penetration realised by the dart.

GENERATION OF MULTIPLE JETS

By machining in the same plane disk seven identical cavities (with hemispherical bottom), in an hexagonal lay-out with a spacing between cavity axes of 30mm, we generate simultaneously seven "classical" jets (Fig. 10).

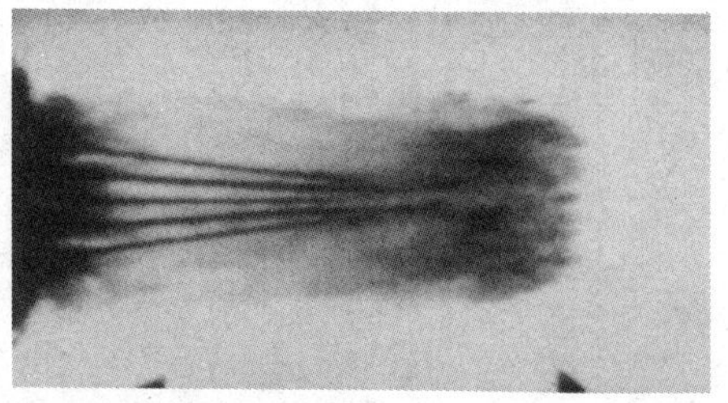

FIGURE 10. Seven simultaneous "classical" jets (plane disk).

A 3D calculation with HESIONE code performed on this experiment gives us then more understanding in the mechanisms of these coupled jets. Figure 11 shows the generated jets at 5μs and 17.5μs, where only half of the structure is shown.

FIGURE 11. Seven simultaneous "classical" jets (HESIONE code).

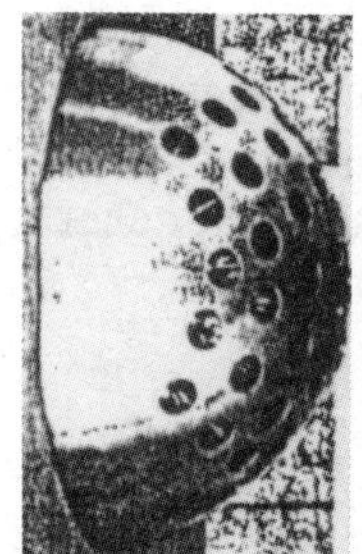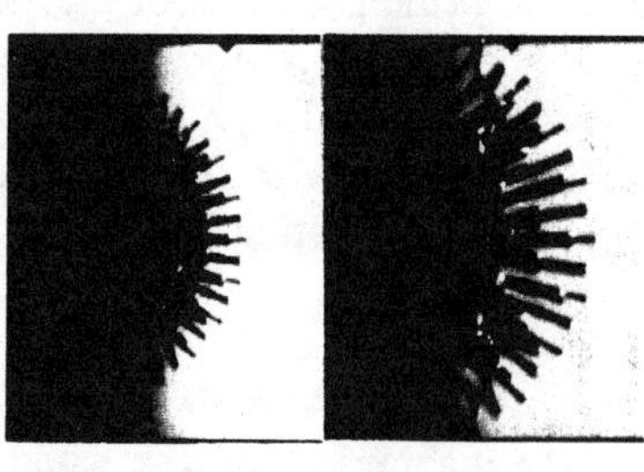

FIGURE 12. Multiple "classical" jets (hemispherical disk).

Figure 12 presents another experiment of multiple jets where 61 hemispherical bottom cavities are machined in an hemispherical disk.

The disk has an internal diameter of 208mm and an external diameter of 290mm, with explosive inside.

CONCLUSION

The numerical analysis shows the great penetration ability of high velocity jets. So, with aluminum jets it seems possible to perforate a 60mm thick tungsten target at the distance of 20cm.

The major advantage of two-parts jets (flat bottom cavity), over "classical" ones (hemispherical bottom cavity), is that impact penetration and cratering effects take place deeper inside the target.

These effects can be multiplied by setting several cavities in compact lay-out.

REFERENCES

1. Defourneau, M., Sciences et Techniques de l'Armement **44**, tome 2 (1970).
2. Walters, W. P., and Zukas, J. A., Fundamentals of shaped charges, John Wiley and Sons Ed., (1989).
3. Leyrat, J. P., Charvet, E., and Pujols, H. C., Int. J. Impact Engng **14**, 467-477 (1993).
4. Asay, J. P., Mix, L. P., and Perry, C., Applied Physics Letters **29**, 284-287 (1976).

EFFECT OF STRAIN RATE ON THE TENSILE STRENGTH OF COPPER SHAPED CHARGE JET

V. V. Silvestrov and N. N. Gorshkov

Lavrentyev Institute of Hydrodynamics, Novosibirsk, 630090 Russia

The jet is produced by a cylindrical 45-mm shaped charge inside a conical copper liner with 120° apex angle. The data on rotating shaped charge penetration are used to estimate the strength of a copper jet under radial tension due to the action of centrifugal force. The value of 0.07 to 0.15 GPa is obtained, which is close to the static yield strength for deformed copper. The strength of the jet estimated when the jet breaks into separate fragments under tensile load along the axis with the strain rate $\sim 2 \cdot 10^4$ s^{-1} achieves substantially higher value from 1.5 GPa.

There are two ways of obtaining information on the strength σ_j of shaped charge jet material (1). First applies the jet breakup into separate fragments due to the axial velocity gradient. The strength of the jet material is estimated from the velocity gradient and linear size of fragments. Simple procedure for analyzing the data on jet breakup was proposed in (2), and for copper jet formed from a hemispherical cavity in metal block it was obtained the value $\sigma_j \simeq 1.1$ GPa with strain rate $\dot{\epsilon} \sim (1\text{--}4) \cdot 10^4$ s^{-1}. In numerical simulation of necking instability in stretching jet the following values are used $\sigma_j = 0.2\text{--}0.27$ GPa at $\dot{\epsilon} \sim 10^5$ s^{-1} (1). Comparatively high strength from 1.4 to 2.4 GPa was required in calculations of shaped-charge jet formation (3).

Second estimation approach applies the data obtained when studying rotating shaped charges, which allow for the estimation of jet material strength with respect to the action of centrifugal force. For example, the values of about 0.1 GPa are cited in (1, 4). These values are much lower, probably due to the lower strain rate which is close to zero in this case.

All results are obtained with different designs of shaped charges. The purpose of this paper is to estimate the strength σ_j of a copper shaped-charge jet which is formed under explosive collapse of the same "low" conical liner using the two approaches of phenomenological estimate: from the spinning jet breakup and from the stretching jet fragmentation due to the axial velocity gradient.

SHAPED CHARGE AND RESULTS

We used a cylindrical shaped charge confined in 1.5–mm thick steel casing with outer diameter $d = 45$ mm. Explosive charge of cast TNT/RDX 50/50 was of diameter equal to that of liner $d_l = 42$ and 100 mm long. Conical liner of thickness 2 mm was made of annealed copper, with 120° apex angle. Using flash radiography we determined the velocity and shape of jet elements, as well as the character of its breakup under stretching at later stages and during rotation with angular velocity N up to 1000 rps. The penetration depth L of shaped-charge jet into a mild steel target was determined.

Under explosive deformation of "low" liner a shaped charge jet is formed in the regime, in which the most part of liner material migrates into the jet (following the estimates based on radiographs, up to 80% of liner mass). A low-velocity thick jet is formed in this case, which is analogous to the jets formed under collapse of hemispherical liners.

Unrotating Shaped Charge

Tip velocity of shaped charge jet equals 4.2 ± 0.2 km/s, tail velocity is 1.8 ± 0.2 km/s, slug ve-

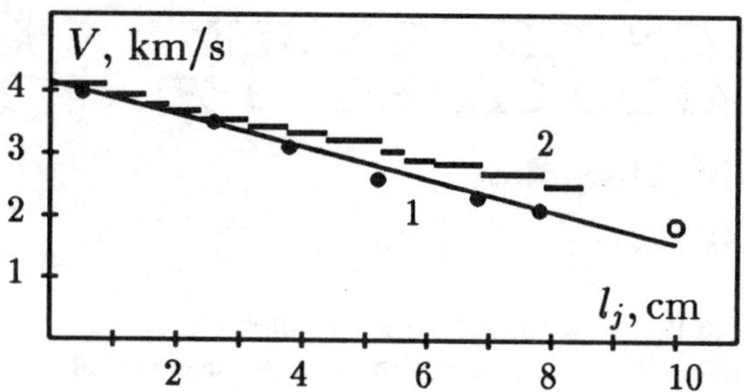

FIGURE 1. Velocity distribution along the jet.
1, 80 μs ($\bullet$, lead tracers method; o, jet tail);
2, 150 μs (jet "is composed" of separate fragments)

locity is 1.5 ± 0.1 km/s. Using 0.1–0.2 mm thick lead tracers applied to inner surface of the cone we determined velocity distribution $V(l_j)$ for this jet at the time 80 μs after blasting (Fig. 1), where l_j is the distance from the jet head. At the same moment the jet length achieves 100 mm, velocity distribution is close to linear, and average velocity gradient $dV/dl_j \simeq 0.24$ km/(s·cm). The jet form is close to a prolate cone of diameter 1.5 mm at the head and 8 mm at the tail. Jet diameter at half-length is $d_j \simeq 3.5$–4 mm.

To 110–120 μs the jet is stretched up to 140 mm. Jet breakup into separate fragments of length δ_i begins from the jet head. At this moment the velocities of the jet head and tail do not differ from the above values, and the estimate of the velocity gradient yields $dV/dl_j \simeq 0.16$ km/(s·cm). The velocity V_i and length of separate fragments for the first half of the jet (from its head) were determined in three experiments using radiographs obtained at 140–150 μs (Table 1). Thus, the velocity distribution was partially reconstructed (bars 2 in Fig. 1 correspond to the data of one experiment). The mean velocity gradient equals 0.18 ± 0.02 km/(cm·s).

The boundaries of separate jet fragments do not bear evidence of tapering due to neckings and are directed at an angle to the jet axis, which points to brittle fracture of this jet under tension. Linear size of 8-12 fragments, into which the jet desintegrates, varies from 3 to 10 mm, with mean size $\bar{\delta}_i = 5$–8 mm. The breakup pattern is similar to that realized under the breakup of jets formed from hemispherical cavity in a metal block (2).

TABLE 1. Velocity gradient and size of fragments for broken jet for three tests

Shot	1	2	3
dV/dl_j, $10^4 s^{-1}$	1.8	1.7	2.0
$\delta_i \pm 1$, mm	5.4	7.8	8.3
$\bar{\sigma}_z$, GPa	0.9	1.2	1.5

Optimum standoff for the considered charge is $3d$. Maximum penetration L_0 equals 145 ± 5 mm ($L_0/d = 3.2d$). As the standoff increases up to $6d$ the relative penetration decreases to $2.8d$. When more powerful cast TNT/HMX 34/66 is used optimum standoff increases to $4d$, and the penetration achieves $3.8d$.

Effect of Rotation

Figure 2 shows the dependence of relative penetration depth L_N/L_0 on spin velocity. Over the range 500 rps the depth L_N/L_0 decreases by 30% much slower than for "high" cones (5). With the increase of rotation velocity from 600 to 900 rps the penetration depth decreases to about d and less. The character of target damage also changes: if with $N \leq 500$ rps one deep crater is formed, with $N \geq 580$ rps several relatively shallow craters surrounding the main one are observed.

One can see in radiographs of spinning jet that under the effect of centrifugal force the process of jet formation is considerably decelerated. At the same time moments its length is less than for unrotating jet, its diameter increases, head and tail

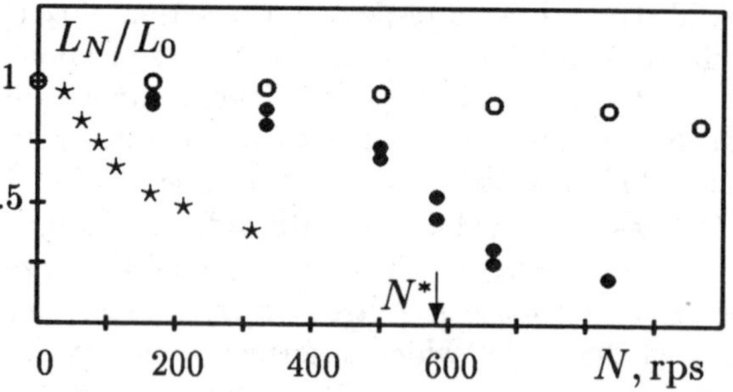

FIGURE 2. Dependence of the relative penetration on spin velocity of shaped charge. "Low" cone: $\bullet$ — copper, o — nickel; $\star$ — "high" steel cone (5).

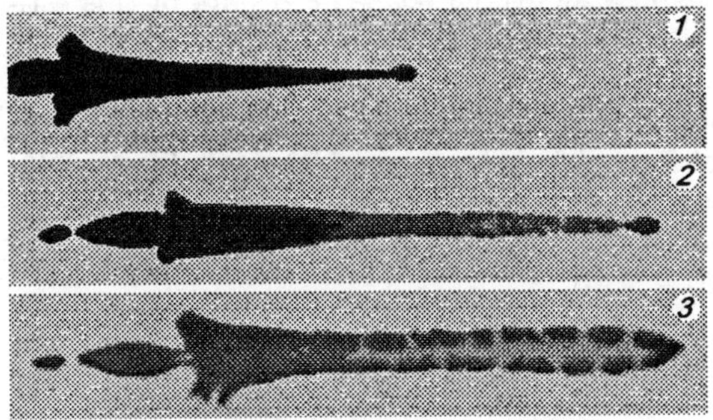

FIGURE 3. X-rays of spin copper jet, 670 rps. 1, 55 μs, 2, 70 μs, 3, 84 μs from the HE ignition

velocities decrease. But with $N \leq 500$ rps the jet remains intact until it interacts with the target. At higher rotation velocities the jet decomposes into separate fragments with characteristic size from 2 to 5 mm and the jet elements are intensely ejected in radial direction with velocity ~ 150 m/s (Fig. 3). Critical rotation velocity, with which the jet breaks before the interaction with the target, is equal to $N^* \simeq 580$ rps. The jet breakup, which can be seen in radiographs, occurs at 70–65 μs with $N \simeq 580$–670 rps. At 60–55 μs, before the beginning of its breakup, the jet is a cone of length 80–70 mm and base diameter $d_0 = 12 \pm 0.5$ mm. Jet diameter measured at its half-length is $d_j = (3.8$–$5.4) \pm 0.8$ mm.

ESTIMATE OF THE JET STRENGTH
From the Critical Spin Velocity

We assume that for a rotating shaped charge under the liner collapse the entire liner mass migrates to the jet, the jet rotates as a whole with spin velocity N_j and angular momentum is retained. These assumptions are rather strong, since due to internal friction a part of angular momentum is lost, and it is probable that jet parts rotate with respect with one another. However, it is impossible to take qualitative account of these effects. Therefore, the estimates of jet spin velocity N_j presented below are overestimated, and the jet material strength are probably underestimated. Under these assump-

tions the rotation velocity of conical jet equals

$$N_j = \frac{5}{3}\left(\frac{d_l}{d_0}\right)^2 N \simeq 20.4N. \qquad (1)$$

To estimate stresses we neglect the effect of the axial force and will consider the jet to be *a long solid continuous rod* with diameter equal to jet diameter at its half-length d_j. In this case maximum stress (circumferential and radial) is

$$\sigma_{max} = \frac{3 - 2\mu}{32(1 - \mu)}\rho N_j^{\,2} d_j^{\,2}, \qquad (2)$$

where $\mu = 0.34$ is Poisson's coefficient for copper. Taking $d_l = 42$ mm, $d_j = 3.8$–5.4 mm, we obtain the estimate of the stress σ_r in spinning "cylindrical" jet (Table 2).

At critical rotation velocity the centrifugal stresses are 0.07–0.15 GPa. It is reasonable to assume that acting stress is close to the ultimate strength for the jet material. The value obtained coincides with analogous estimate of strength $\sigma_r \sim 0.1$ GPa for spinning copper shaped-charge jets from "high" cone cited in (1, 4).

Mean value of strength $\sigma_r \simeq 0.11$ GPa is close to static yield strength $\sigma_y = 0.07$–0.1 GPa for deformed copper. Hence follows the qualitative conclusion that critical rotation velocity is

$$N^* \sim \sqrt{\sigma_y}\,, \qquad (3)$$

and will increase when stronger materials will be used for liner. This conclusion is supported by the data for the charges with nickel liners with $\sigma_y \simeq 0.2$ GPa, for which the critical spin velocity increases substantially (Fig. 2): even when rotation velocity is up to 970 rps nickel shaped charge jet does not fail, and relative decrease of the cavern depth in a steel target is 17%, there is one cavern in the target. For nickel estimated critical spin velocity is at least

$$N_{Ni}^* \sim \sqrt{20/7}N_{Cu}^* \simeq 1015 \text{ rps}$$

TABLE 2. Stresses in spinning jet (in 10 MPa)

d_j, mm	N, rps		
	500	580*	670
3.8	5.6	7.5	10
4.6	8.2	11	14.7
5.4	11.3	15.2	20.3

(no account has been taken of the fact that in the case of nickel shorter and thicker jets are formed, $L_0 = 2.3d$). These data verify qualitative dependence (3) and indirectly the correctness of the estimate of copper jet strength with respect to the effect of centrifugal force.

From the Fragment Size and Velocity Gradient

We used the method of estimating the shaped charge jet strength σ_z proposed in (2). Physical basements of the method are thoroughly analyzed in (2), therefore we note only that the method is based on three principal assumptions:

- under axial tension the jet breaks into pieces of maximum length δ, with which kinetic energy of the piece T in the system of center of inertia does not exceed the work of its tension A to the ultimate strength, i. e., $T \leq A$;

- it is assumed that jet fragmentation is brittle, and the work of deformation of jet material to the ultimate strength is determined by the expression $A = \sigma_z^2/(2E\rho)$, where ρ and E are the density and Young's modulus of the jet material;

- it is assumed additionally that there is a linear velocity distribution along the jet, and the jet is broken into the pieces of length δ with approximately equal difference of velocities at the piece boundaries Δv, obviously connected with the velocity gradient $\Delta v \simeq (dv/dl)\delta$.

Under these assumptions specific kinetic energy of longitudinal relative motion of parts of the whole piece is determined as $T = (\Delta v)^2/24$. From the above expressions it follows that the estimate for the jet material strength is as follows

$$\sigma_z \geq \sqrt{\frac{E\rho}{12}}(dv/dl)\bar{\delta}. \qquad (4)$$

The shaped charge jet strength σ_z under tension in axial direction at strain rate $\dot{\epsilon} \sim 2 \cdot 10^4\, \text{s}^{-1}$ was estimated from the mean size of the pieces $\bar{\delta}$ into which the jet breaks (in experiments without rotation) and characteristic value of velocity gradient along the jet. The results of this estimate are presented in the last line of Table 1. Considerably higher value $\bar{\sigma}_z \simeq 0.9$–$1.5$ GPa is obtained, which is close to analogous data for the jet formed from hemispherical cavity in metal (2) and the data on spall fracture strength for copper at this strain rate, which were systemized in (6).

CONCLUSIONS

In view of a series of simplifying assumptions, the estimates of strength of a copper shaped charge jet are correct to the coefficient of the order of unity. However, the difference in the values of σ_j at different strain rates achieves 10–20 times. This allows one to state that although the liner material is subject to intense plastic deformation in the cumulative node and dislocations are intensely accumulated in the material, a strong dependence of copper shaped charge jet material on the strain rate is observed under tension. Thus, the jet material displays the properties characteristic of hard cooper.

REFERENCES

1. Walters W. P., Fundamentals of Shaped Charges. *High Velocity Impact Dynamics*, Edited by J. F. Zukas, USA, John Wiley & Sons Inc., 1990, ch. 11.

2. Mikhailov A. N., and Trofimov V. S., *Combustion, Explosion and Shock Waves* **15**(5), 670–674 (1979).

3. Van Thiel M., and Levatin JoAnue, *J. Appl. Phys.* **12**, 6107–6114, (1980).

4. Von Holle W. G., and Trimble J. J., "Shaped Charge Temperature Measurement", in *Proceedings Sixth Symposium on Detonation*, 1976, pp. 691–699.

5. Singh S., *J. Appl. Phys.* **31**(3), 578–581 (1960).

6. Paisley D. L., Warnes R. H., and Kopp R. A., "Laser-Driven Flat Plates Impacts to 100 GPa with Sub-Nanosecond Pulse Duration and Resolution for Material Property Studies", in *Shock Compression of Condensed Matter–1991*, pp. 825–828.

MICROTEXTURAL CHARACTERIZATION OF COPPER SHAPED CHARGE JET FRAGMENTS

S. I. Wright[*], J. F. Bingert[*], L. Zernow[†]

Los Alamos National Laboratory, Los Alamos, New Mexico 87545
Zernow Technical Services Inc., San Dimas, CA 91773

The microstructures of two soft-caught copper shaped charge jet particles were investigated. In particular, the spatial distributions of crystallographic texture within the particles were characterized using point specific measurements of crystallographic orientation. Significant variations in preferred orientation were observed. These results are discussed in light of previous computer simulations of the jetting process which showed significant radial gradients in both strain and strain rate.

INTRODUCTION

The formation of a shaped charge jet from an explosively collapsed metal liner produces very large strains and very high strain–rates. It is, therefore, difficult to experimentally characterize the time and space dependent deformation processes which the material undergoes. One approach to gaining more understanding of the jetting process has been to examine the microstructures of jet particles recovered from low density recovery media such as polystyrene (1). There are several caveats to this approach. Since substantial heating occurs as a result of the plastic work done on the jet material during jet formation, elongation and particulation, the individual jet particles reach elevated (but equilibrated) temperatures before entering the recovery media. It has been shown experimentally, for the specific particles studied here, that the jet particles enter the recovery medium at approximately 500°C and then cool to about 250° C (the melting point of polystyrene) in 15 to 20 seconds (2). Further cooling from 250°C to near ambient temperature is estimated to require another 20 to 30 seconds. With this temperature–time cooling history, the recovered particles may have undergone significant static recrystallization. Thus, the question arises whether the microstructures observed in the retrieved fragments will be the same as those which existed during and/or after particulation. This question is presently unanswered. Another uncertainty in the interpretation of the microstructures of the soft–caught fragments arises from the fact that, at the time these particles were recovered, the precise locations of the particles in the jet stream were not precisely known. It was, however, known that these particles came from the rear half of the jet. It is worth noting that techniques are now available for more precisely determining the location of individual recovered jet particles along the jet stream. Computer simulations of jet formation using Lagrangian tracer particles to evaluate linear strains and strain–rates (3, 4) indicate that the strain and strain–rate histories of material from jet particles located at different positions along the jet may be quite different.

Despite these potential difficulties in fully understanding data obtained from softly recovered jet particles, several microstructural examinations (5, 6, 7, 8) of such particles have been performed with the aim of extracting information regarding the jetting process. Each of these studies have reported observations of significant microstructural changes (mostly involving grain size reduction) from the original liners to the recovered jet particles. In addition systematic annular radial inhomogeneities were observed in many of the jet particles. An early study (9) of similar particles with annular rings visible in their transverse sections, using selected area x-ray diffraction with a small spot size, indicated the presence of strongly oriented [100] structures in the central region and in the second annular ring. In a more recent study(10), providing much better resolution, the local distribution of preferred orientation (micro-texture) in a softly recovered jet particle was investigated using an automated electron diffraction technique (11, 12). A strong correlation was found between the micro-structural inhomogeneities and the microtexture measurements. In this work, microtexture measurements obtained from still another softly recovered particle (sample B). compared to the results previously reported in (9) (sample A).

EXPERIMENTAL RESULTS

Soft-caught fragments of ETP (electrolytic tough pitch) copper shaped charge jets were characterized using optical microscopy and electron diffraction in the scanning electron microscope. The section planes examined were approximately normal to the long axes of the fragments. Individual measurements of crystallographic orientation were made using an automatic technique based on computer indexing of electron backscatter diffraction patterns (11, 12). Point-by-point measurements of crystallographic orientation were made on sample A on a 2000 μm $\times$ 1200 μm hexagonal grid with 10 μm spacing between measurement points (~28,000 measurements). Measurements on sample B were made on a 1000 μm $\times$ 1200 μm hexagonal grid with 10 μm spacing (~14,000 measurements). At each measurement point, the orientation, the grid coordinates and a parameter (IQ) describing the quality of the diffraction pattern were recorded.

Optical Microscopy

In both fragments, a fine-grained region (~10μm) was observed near the center surrounded by a ring of coarser grains. Another ring of finer grains (although not as fine as the center region) encloses this coarser–grained region. However, the shape and size of these regions are quite different in the two samples.

The central fine-grained region in sample A is approximately 400μm in diameter; whereas, in sample B it is less than 100μm. The average grain diameters in the coarse-grained regions are ~25μm in sample A and ~100μm in sample B. In addition, a void exists at the center of the fine-grained region in sample A.

Crystallographic Texture

Sample A was partitioned into three zones, labeled zones 1, 2 and 3. Zone 1 corresponds to the fine-grained central region, zone 2 to the coarse-grained ring surrounding zone 1 and zone 3 refers to all measurements outside zone 2. Orientation distribution functions (ODFs) were calculated for each zone using the harmonic series expansion method (13) to an order of $l = 16$. Each orientation was represented as a Gaussian with a 5° half-width (14, 15).

Sample A exhibited a fiber texture approximately aligned with the long axis of the particle. Inverse pole figures determined from the ODF calculations are shown in FIGURE 1. Inverse pole figures give the distribution of crystal directions aligned with a particular sample direction (in this case a direction approximately normal to the transverse plane). The texture in zone 1 is a strong (100) fiber texture (the other weaker features are an artifact of using harmonics to resolve this sharp texture). Zone 2 shows a mix of (100) and (211) fibers and zone 3 shows primarily a weak (100) fiber texture with some (111).

Sample B did not lend itself to such convenient partitioning as sample A. However, the measurements on sample B were interrogated according to "grain" size. A "grain" was defined by grouping together neighboring orientation measurements which did not differ in orientation by more than 15°. The measurement set was then partitioned into a set of large "grains" (containing more than four orientation measurements per "grain") and small "grains" (containing at most four measurements). Inverse pole figures for the large and small "grains" are shown in FIGURE 2. These pole figures show that both the large and small "grains" have a (211) type texture. However, the small–"grain" texture is considerably weaker. The central fine-grained region contained only 125 measurement points. An inverse pole figure for these measurements is

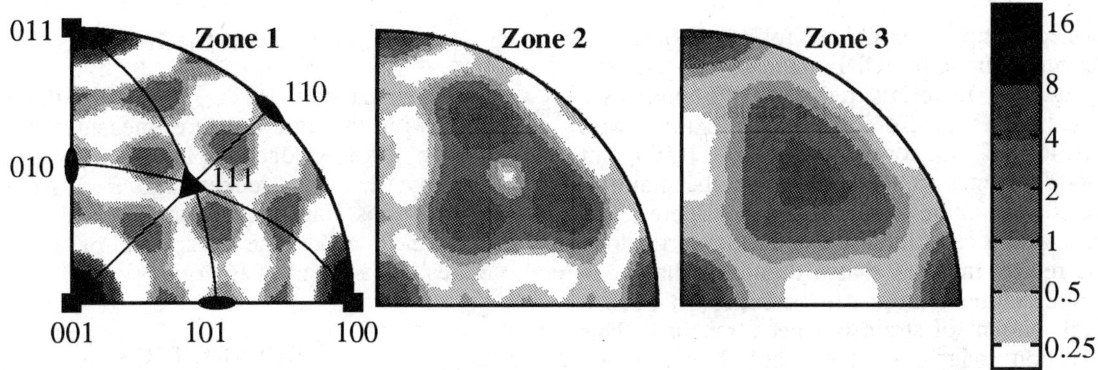

FIGURE 1. Inverse pole figures for sample A.

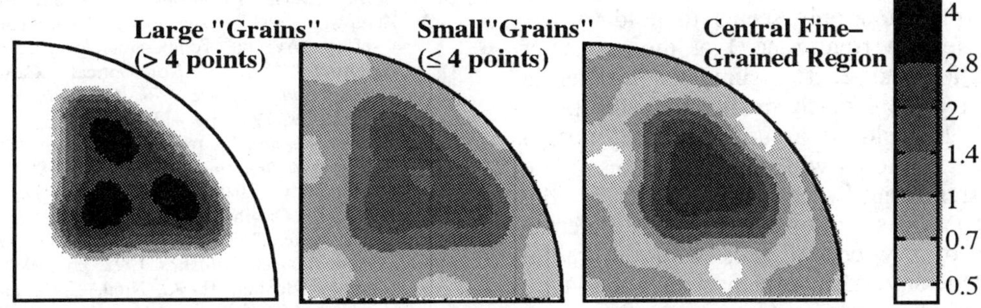

FIGURE 2. Inverse pole figures for sample B.

also given in FIGURE 2 and exhibits, in contrast to sample A, a weak (111) type texture.

DISCUSSION

The textures observed in the fragments are consistent with recrystallization textures observed in drawn copper wires (16) where a mix of <100>, <111> and <112> texture components generally develops during annealing. The <100> fiber is greatly enhanced after annealing. The <112> texture has been observed to be a coarsening texture which is in direct agreement with the microtexture results for the coarse-grained regions in both samples.

Computer simulations of the jetting process (3, 4) predict strong radial gradients in both strain and strain rate. In general, the simulations predicted the strain at the center of the jet to be extremely large and to decrease rapidly with increasing distance from the jetting axis. The strains varied in magnitude and direction from nearly axial to nearly radial depending on location (in both the jetting and radial directions). A direct one-to-one correlation of the jet fragments to these simulations can not be made as the relative positions of the fragments along the jet axis are unknown. However, an indirect correlation to the simulations can be made; namely, the center of a jet particle undergoes more strain than the outer regions. In addition, the material at the center of a particle undergoes axial stretching; whereas, the predicted strains in the material away from the jet axis varied to near radial directions.

No indications of the sharp microstructural boundaries observed in the fragments exist in the simulated radial strain gradients. However, there is evidence in the literature that suggests that relatively slight increases in deformation can produce sharp changes in recrystallization texture. Necker et al. (17) rolled OFE copper samples to height reductions of 58%, 73% and 90% (Von Mises strains of 1.0, 1.5 and 2.7) and then fully recrystallized the samples. The deformed samples exhibited rolling textures of increasing strength. ODFs measured on the recrystallized materials reduced 58% and 73% deviated somewhat from the rolling textures with intensities at cube orientations, {001}(100), of 1.2 and 2.1 times random; however, the material reduced 90% exhibited a nearly complete cube texture with a cube intensity of 20 times random. These results indicate that a dramatic change in recrystallization texture can be achieved once a critical amount of strain has been reached. This observation suggests that the large strain gradients predicted by the simulations could produce the distinct changes in recrystallization texture observed in the fragment.

This explanation only seems to hold for the change in texture from zone 1 at the center of sample A to zone 2. The simulated strains in zones 2 and 3 were much smaller than the those in zone 1. It could be argued that the inertial confinement of the jet will produce a stress state with a significant shear component in the peripheral zones. However, no preferred alignment of any crystal axes with the radial direction nor any other evidence of shear–type textures was observed.

The absence of the strong (100) type texture in the fine-grained region in sample B suggests that this fragment comes from a position along the axis of the jet where the critical strains required to generate the (100) type recrystallization texture were not reached.

CONCLUSIONS

The inhomogeneities observed in the microstructures of soft–caught copper shaped charge jet fragments can be correlated to the spatial distribution of crystallographic orientation. The strong (001) fiber texture observed at the center in one of the fragments seems to confirm previous numerical predictions of extreme strain in the central region of the jet. In addition, the distinct changes in texture and microstructure in the radial direction observed in the fragment may be due to large radial strain gradients generated during jet formation.

ACKNOWLEDGMENTS

K. Kunze and B. L. Adams are thanked for help with the microtexture measurements. The authors acknowledge helpful discussions with G. T. Gray III, C. T. Necker, and P. J. Maudlin of Los Alamos National Laboratory. This work was performed under the auspices of the United States Department of Energy.

REFERENCES

1. Zernow, L., "Recovery of Shaped Charge Jet Samples for X-Ray Diffraction Studies," in *Proceedings of the 10th International Symposium on Ballistics*, 1987.
2. Zernow, L., and Zernow, R. H., "Experimental Estimation of the Average Temperature of Shaped Charge Jet Particles by a Non–Optical Calorimetric", in *Proceedings of the 15th International Symposium of Ballistics*, 1995, pp. 413-412.
3. Zernow, L., and Chapyak, E. J., *International Journal of Impact Engineering*. **14**, 863–875 (1993).
4. L. Zernow, L., Chapyak, E. J., Meyer, K., and Zernow, R. H., "The Origins of Liner Material in a Shaped Charge Jet Particle," in *Proceedings of the 13th International Symposium on Ballistics*, 1992, pp. 309–317.
5. Murr, L. E., Shih, H. K., Niou, C.-S., and Zernow, L., *Scripta Metallurgica et Material ia*, **29**, 567-572 (1993).
6. Zernow, L., "Metallurgical, X-Ray Diffraction and S.E.M. Studies of Individual Shaped Charge Jet Particles Captured by Soft Recovery," in *Proceedings of the International Conference on Ballistics*, 1988, pp. 22–27.
7. Zernow, L., and Lowry, L., "High Strain Rate Deformation of Copper in Shaped Charge Jets," in *International Conference on Shock-Wave and High-Strain Rate Phenomena in Materials*, 1992, pp. 508-519.
8. Thiagarajan, S., Gurevitch, A., Murr, L. E., Varma, S. K., Fisher, W. W., and Advani, A., "Development of Microstructures in Pure Copper at Various Strain Rates," in *Modeling the Deformation of Crystalline Solids*, 1991, pp. 159–171.
9. Lowry, L., Selected area x-ray diffraction study of a recovered copper jet particle, private communication, 1993.
10. Wright, S. I., Bingert J. F., and Zernow, L., *Materials Science and Engineering*. (submitted June 1995).
11. Wright, S. I., *Journal of Computer Assisted Microscopy*, **5**, 207–221 (1993).
12. Adams, B. L., Wright, S. I., and Kunze, K., *Metallurgical Transactions*, **24A**, 819–831 (1993).
13. Bunge, H.-J., *Texture Analysis in Materials Science. Mathematical Methods* London: Butterworths, 1982.
14. Wagner, F., Wenk, H. R., Esling C., and Bunge, H. J., *Physica Status Solidi (b)*, **67**, 269–285 (1981).
15. Wright, S. I., and Adams, B. L., *Textures and Microstructures*, **12**, 65–76 (1990).
16. Dillamore, L. L., and Roberts, W. T., *Metallurgical Reviews*, **10**, 271–380 (1965).
17. Necker, C. T., Doherty, R. D., and Rollett, A. D., *Textures and Microstructures*, **14-18**, 635–640 (1991).

CHAPTER XVII

LASER AND PARTICLE BEAM-MATTER INTERACTION

EXPERIMENTAL OBSERVATIONS OF STATE OF MATTER HEATED BY THE ION BEAM

A. V. Utkin and G.I. Kanel

Russian Academy of Sciences, Institute of Chemical Physics,Chernogolovka 142432, Russia

K. Baumung

Forschungszentrum Karlsruhe, D-76021 Karlsruhe, Germany

The interaction of a high-power proton beam with double layered aluminum targets has been investigated. The experiments were carried out at the Karlsruhe Light Ion Facility (KALIF) using laser Doppler velocimetry with subnanosecond temporal resolution. The results allow us to determine with high accuracy the range of 1.53 MeV protons in the hot aluminum ablation plasma. Experiments with impact of accelerated foils on a transparent window confirmed the influence of the thermo-condactivity on thickness of the ablation zone. Analysis of wave dynamic in the target was made to find a correlation between the profile of free surface velocity and the parameters of the ion beam.

INTRODUCTION

Pulsed high-power particle beams generate intense shock-wave phenomena in condensed targets. Measurement and analysis of wave dynamics in both the energy deposition zone and the residual cold part of the target permits to investigate the response of the target matter as well as properties of the beam itself. In a previous paper [1] the authors estimated the power density in the front of the proton beam. The objective of this work is to investigate the evolution in time of the ion beam parameters during the whole rise time of the pulse. For this purpose we have performed a series of experiments with composite targets consisting of two aluminum foils separated by a large gap of ~2 mm. Figure 1 shows the scheme of the experiments with composite aluminum targets. The first foil acts as an absorbing filter that reduces the energy of the protons and absorbs a part of them. The interaction of the penetrating protons with the second foil - the actual target - induces shock-wave phenomena that are the subject of the present investigations.

The experiments were performed at the Karlsruhe Light Ion Facility "KALIF" [2]. KALIF is a 1.7 MV, 500 kA, 50 ns full width at half maximum pulsed power accelerator. For our investigations we used the "B_Θ-diode" [3] as an ion source. A focal spot was from 6 to 8 mm diameter.

The compression wave, generated in the target foil by the transmitted part of the ion pulse, is recorded by an ORVIS-type laser Doppler velocimeter [4]. The maximum time resolution of the velocity measurements was ~200 ps. The accuracy of the velocity measurements is estimated to be from 10 to 50 m/s.

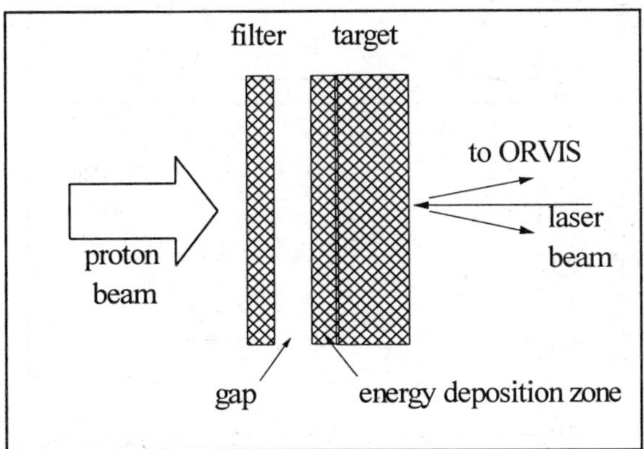

FIGURE 1. Scheme of the target arrangement

EXPERIMENTAL RESULTS

Figure 2 shows the free surface velocity profiles for different composite aluminum targets. Numbers in brackets indicate filter and target thickness in micrometers. In shot 3439 after the first compression pulse several steps can be discerned which are related to the wave reflections between the rear free surface and the boundary to the ablation plasma. The thickness of the evaporated layer in the target foil can be inferred from the period of the wave reverberation. But the accuracy of this method is not sufficient to reveal the expected difference between the range in cold aluminum and in the hot dense plasma where free electrons enhance the stopping power [6].

The range of the protons with the maximum energy can be determined more accurately by increasing the filter thickness. Only a weak velocity pulse was registered in shot 3442 using 22.5-mm-thick filter. The maximum ion energy in this shot was 1.54 MeV. No response of the target was detected in an analogous shot with the maximum ion energy of 1.52 Mev. Therefore, the range of 1.53 MeV protons in the hot plasma is 22.5 µm, that is 4 µm less than for the cold material [5].

The temperature in the ablation plasma is in the order of 10 eV. Hence, a thermal wave propagates into the adjacent cold material. To reveal the contribution of the thermal conductivity to the thickness of the ablation zone, the experiments with impact of aluminum foils, accelerated by the ion beam, on a thick LiF window were made (Fig.3).

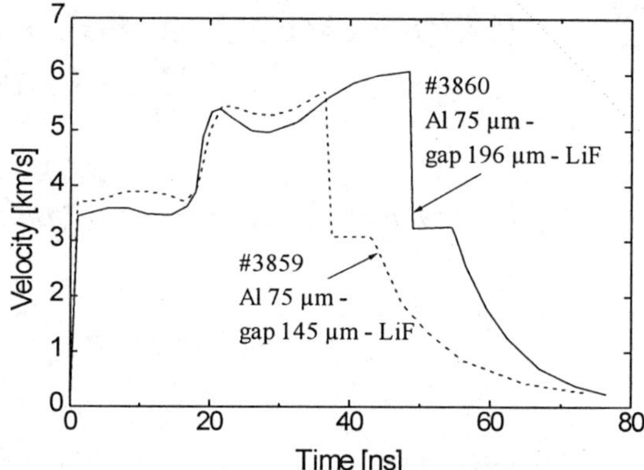

FIGURE 3. Velocity histories of ablatively accelerated aluminum flyers impacting thick LiF windows.

The duration of the velocity plateau after impact on a window is defined by the residual thickness of the foil. Within the limits of experimental error these values are the same in the shots 3859 and 3860. This result is unexpected for a number of reasons. First, in the shot 3860 the acceleration time is 12 ns greater than in the shot 3859 and a thermal wave has to decrease the residual thickness. Second, the energy deposition zone for the shot 3860 exceeds

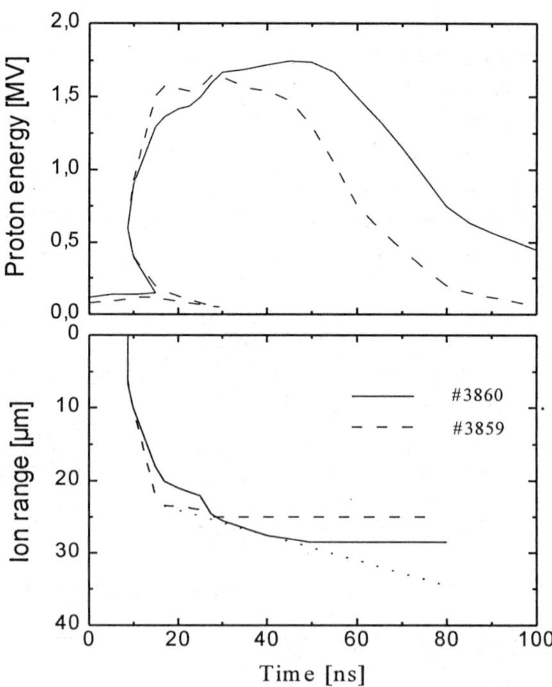

FIGURE 4. Time-of-flight compressed proton energy histories on target and associated dependence of ion range on the time.

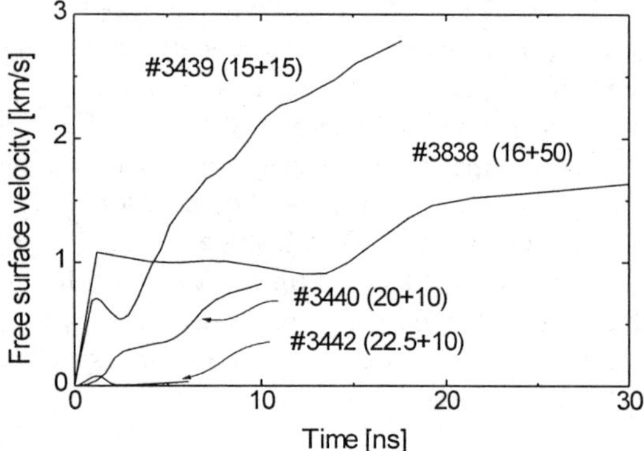

FIGURE 2. Free surface velocity profiles for different filter + target compositions.

one in the shot 3859. It follows from the determination of the accelerator voltage as a function of time, calculated for the position of the target. Figure 4 shows "bunched" voltage curves and the dependence of the ion range on the time. The proton ranges were determined for cold material and than decreased by 17% according to experimental results described above. Thus to explain the experimental results we have to take into account a thermal wave.

There is a critical value of a thermal wave velocity which is equal to 0.2 µm/ns (dotted line in Fig.4). If the thermal wave propagates with a smaller velocity the difference in the thicknesses of the ablation zones remains about 5 µm. The situation changes abruptly when it is equal to 0.2 µm/ns: in this case the ablation zones for the both shots have the same size for $t>27$ ns and the difference of the velocity plateau duration after impact on a window is due to the difference of the acceleration time only. It is smaller than 0.3 ns and can't be defined from the experiments. From this fact it transpires that the critical velocity is a lower estimation of the rate of evaporation produced by the thermal wave.

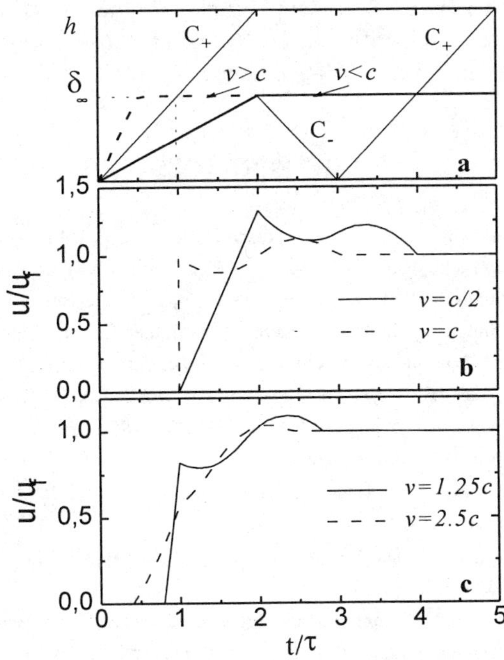

FIGURE 5. Space-time diagram in Lagrangian coordinates of the wave phenomena considered in the analytical model (**a**) and particle velocity at $h=\delta_\infty$ for sub- (**b**) and hypersonic (**c**) regimes of the increase rate of the deposition zone thickness.

ACOUSTIC ANALYSIS

To understand the details in the free surface velocity profile an analysis of the wave phenomena in the ablation plasma associated with the fast bulk energy deposition was performed using an acoustic approximation. The initial hydrodynamic response in shots with filters is caused by low energy ions since they lost nearly all of their energy in the filter. Therefore, we consider the model in which the thickness of deposition zone δ increases linearly in time t with rate v from zero to maximum δ_∞ and then remains constant for $t \geq \tau_\infty = \delta_\infty/v$. From the beginning of the process unloading waves C+ and C- start from the free surface of the energy deposition zone and from its boundary to the undisturbed material (see Fig.5a). The mass flow in the target is described by conservation laws of mass and momentum. In acoustic approximation the equation of state has the next form:

$$P = \rho_0^2 c^2 (1/\rho_0 - V) + \rho_0 \Gamma E(t,h))$$

where h is the mass coordinate, P and V are the pressure and the specific volume, ρ_0 and Γ are the initial density and Grüneisen parameter. It is assumed that the specific energy E is an assigned function of coordinate and time so the energy conservation law should not be taken into account.

The system of acoustic equations was solved like it was made in paper [1]. If the specific power density is uniformly distributed over coordinate h, and the ion beam power density at the target surface $I(t)$ is a linear function of time $I(t)=I_1 t$, the next results follow from this solution.

1. $v \leq c$: Let us consider the time dependence of the particle velocity at $h=\delta_\infty$. The information of increasing the pressure in the deposition zone arrives at this position at time $\tau=\delta_\infty/c$, and the particle velocity begins to increase linearly in time until the maximum proton range is reached (curve in Fig.5b, that corresponds to the case $v = c/2$). At that instant, the velocity is maximum $u_m=\Gamma I_1\tau/(\rho_0 c(c+v))$ and then, a characteristic oscillation is observed. For $t \geq \tau_\infty+2\tau$ the velocity is constant and equal to $u_f=\Gamma I_1\tau/(2\rho_0 c^2)$. As v approaches the sound velocity, the rise time of the compression decreases, the amplitude of the oscillations increases, and, finally, a shock wave is formed when $v= c$ (Fig.5b).

2. *v > c :* In this case, the particle velocity at $h=\delta_\infty$ begins to increase at time τ_∞. If $v<c(\sqrt{5}+1)/2$ the local maximum is reached at $t=\tau$, and the further variation of the velocity qualitatively corresponds to the subsonic regime (curve in Fig. 5b for $v = 1.25c$). However, if $v>c(\sqrt{5}+1)/2$, a qualitative change occurs: the sharp bend at $t=\tau$ smoothes and gives way to a continuous increase (curve in Fig. 5b for $v = 2.5\,c$). A further increase of the growth rate v (corresponding to $\tau_\infty\to\infty$) shifts the maximum toward 2τ, its amplitude tending to u_f: the velocity profile approaches the monotonously increasing function of time that was found in Ref. 1 for a constant thickness of the deposition zone.

Let us compare the results of acoustic analysis with experimental data shown in Fig.2.The particle velocity at $h=\delta_\infty$ discussed above corresponds to half the measured free surface velocity. Notice that the initial part of the velocity history of shot 3439 in Fig. 2 is similar to solid curve in Fig.5b: at first it increases linearly in time, passes the maximum, and begins to decrease. Taking the shock front rise time ≈1 ns and the resulting thickness of the deposition zone of ≈6 μm, we obtain a growth rate of the proton range of $v = 3$ km/s $\approx c/2$.

The minimum predicted by the acoustic model can not be observed, if the rise time is comparable to, or longer than the period of one wave reverberation in the cold part of the target. To settle this question, in the shot 3888 the thickness of the target was increased up to 50 μm. As is seen from Fig.2 the velocity profile has clearly defined oscillations after the shock front. The parameters of ion beam in shots 3888 and 3439 were different and the amplitudes of the velocity were different too.

The value of the free surface velocity after the first increase can be used to estimate the growth rate of beam power density at the irradiated surface of the target: $I_1 \approx 2\cdot10^{-3}$ TW/ (cm^2·ns) in shot 3439. The increase rate I_1 obtained in Ref. 1 from the analysis of the acceleration of a 33-μm-thick aluminum foil in a shot without filter is one order of magnitude greater than the value obtained here. Assuming that in both shots the beam power was approximately the same, we can conclude that a 15-μm-thick filter cuts off the main part of the radiation and only ~10% of the initial ion power is absorbed in the target.

CONCLUSION

The experimental results presented and the analysis of wave dynamics demonstrate that composite targets provide improved diagnostics of ion beam parameters. Measurements allowed us to determine the range of 1.53 MeV protons in hot aluminum ablation plasma with high accuracy, and to reveal the influence of the thermal conductivity on the ablation zone thickness. An acoustic model was established to interpret the foil acceleration as a function of the power history.

ACKNOWLEDGMENT

The authors appreciate the continuous interest and support of their joint work by Prof. G. Kessler, director of the INR, Forschungszentrum Karlsruhe, and V.E. Fortov, director of the HEDRC of Russian Academy of Sciences, Moscow. We would also like to acknowledge the assistance by J. Singer, O. Stoltz, and the KALIF operators.

This work was supported by the Russian-German Scientific and Technical Co-Operation Program and the NATO Science Program.

REFERENCES

1. Baumung K., Karow H.U., Rusch D., Kanel G.I., Utkin A.V., and Licht V., *J. Appl. Phys.* **75**, 7633 7638, (1994).

2. Bluhm H., Böhnel K., Buth L., Hoppé P., Karow H.U., Klumpp A., Rusch D., Scherer T. , Schülken U. , and Singer J., " Experiments on KfK's light ion accelerator KALIF," in *Digest of Technical Papers , 5th IEEE Pulsed Power Conference,* Arlington, VA, June 10-12, 1985 (IEEE, New York, 1985), pp. 114-117.

3. Schimassek W., Bauer W., and Stoltz O., Rev. Sci. Instrum. **62**, 168 (1991).

4. Bloomquist D. D. and Sheffield S. A., J. Appl. Phys. **54**, 1717 1722, (1983).

5. *Handbook of Range Distributions for Energetic Ions in all Elements,* edited by J. F. Ziegler, Pergamon, New York, 1980 ,Vol. 6 of *The Stopping and Ranges of Ions in Matter.*

6. Goel B. and Bluhm H., Journ. de Physique - Coll. **49**, C7-169 (1988).

NUMERICAL ANALYSIS OF FOIL ACCELERATION EXPERIMENTS AT KALIF

B.Goel, K.Baumann, W.Höbel
Forschungszentrum Karlsruhe,
Post Box 3640, D-76021 Karlsruhe, Germany

O.Yu. Vorobiev, A.V. Shutov, V.E. Fortov
High Energy Research Center,
142432 Moscow, Russia

The present report contains an investigation of the dynamics of thin-foils accelerated by light-ion beams. The experiments were performed at KALIF. Equations of State used in the present calculations is in a simple analytical form. The empirical parameters of this EOS have been adjusted to adequately describe experimental data in one dimensional calculations as well as shock compression data. In this paper we redo these calculations with a moving grid 2D code. 2D computations are much more time consuming then 1D especially for problems containing different time and space scales. Efficient numerical algorithms are required to perform these computations. We present in this report 2D applications of Godunov Moving Grid Code to beam-target interaction problems.

Introduction

Intense ion beams have been used to generate ultrahigh pressures in condensed matter [1], [2]. Light-ion beams generated at the *Karlsruhe Light Ion Facility* (KALIF) typically have a focus of less than 1 cm diameter. The maximum power density at KALIF achieved to date is of the order of 1 TW/cm^2 [3]. In the energy deposition zone of the targets, the 1.5 MeV protons of the KALIF beam deposit energy of the order of 100 TW/g or 5 MJ/g. A strongly coupled, hot, dense plasma is thus created by the action of a light ion beam on matter. The description of such plasma requires sophisticated theoretical tools, and experimental information is scarce in this region of plasma. A successful interpretation of beam-target interaction experiments, in the long run, can serve as a check of theoretical models and of the equation of state data based on these models.

Mathematical Model

The Euler equations of motion are used to describe the dynamics of matter under the impact of a beam. They express conservation laws for mass, momentum and energy. The caloric equation of state is used to close the equations of motion. The radiative energy transfer is treated in the radiation heat conduction approximation. In this case, the energy flow is proportional to the temperature gradient, the proportionality coefficient being expressed by the Rosseland mean free path. Thus the conductive energy flow consists of two terms: electron conductivity and radiation conductivity. The electron conductivity coefficient [4] and the Rosseland mean free path [5] are functions of density and temperature.

The maximum target temperature reached in the KALIF experiments analysed in this report does not exceed a few 10's of eV. Therefore, the energy of the radiation field is not very important in the calculations reported here and the simple approximation of radiation heat conduction is deemed to be adequate to calculate the radiative heat transfer. The absorption of the beam energy by the target is calculated in a single-particle approximation. Stopping power is calculated by semi-empirical formulas developed in [4]. These formulas describe the enhancement of the stopping power as a function of heating and the expansion of matter and have been checked against the results of more complex models.

Numerical method

The numerical method used in the present calculations is the Godunov method in moving grids and is described in detail in reference [6]. Our physical domain is divided basically in 3 different zones. There is a zone in which particle energy is deposited. A fine mesh is required in this zone to accurately simulate the details of energy deposition in the target. To calculate efficiently this energy source an additional cartesian mesh is introduced in this zone. This region gets heated and its density drops to less than one hundredth of the solid density. The high pressure generated in this zone drives a shock wave into the cold and condensed part of the target. This part of the target is again divided into two subregions, one behind the shock and one ahead of it. There is another zone with no hydromotion. The boundary of this zone also changes with time. The interfaces move with the local fluid velocity. This procedure corresponds to a Lagrangean treatment. Within each subregion Godunov scheme is used to compute the evolution of the conservative variables. After each movement cycle new boundaries of subregions are calculated. The gridline orthogonality at the boundaries is provided by the redistribution of boundary points.

Semiempirical Equation of State

For the numerical simulation of foil acceleration by KALIF beam it is necessary to use an equation of state (EOS) describing, with reasonable accuracy, the shock compression of matter and the plasma thermodynamics in the energy deposition region. In the following calculations, a semi-empirical equation of state in the form of generalized Mie-Grüneisen equation of state is developed which describes the experimental data on shock compression of the solid samples in the pressure range up to several Mbar is used [7]. Following the usual procedure, the internal energy of matter is divided into two parts: The cold energy E_c is only a function of matter density, while the energy of thermal motion E_t is a function both of temperature and density.

$$E(p, \rho) = E_c(\rho) + E_t(p, \rho) \qquad (1)$$

To describe Hugoniot data for solid matter we choose the cold pressure in traditional form. The thermal pressure is related to the thermal energy through the Grüneisen parameter γ, which is represented as a function of energy (temperature) and density

$$P_t = \gamma(E, \rho)E_t\rho \qquad (2)$$

We use the following dependence of Grüneisen parameter on the internal energy

$$\gamma = \gamma(\rho)_\infty + \frac{\gamma(\rho)_0 - \gamma(\rho)_\infty}{1 + (\frac{E}{E_0})^\alpha}, \qquad (3)$$

where E_0 and α are fitting parameters. $\gamma(\rho)_0$ tends to $\gamma(\rho)_\infty$ for hot matter. To describe asymptotic behaviour of γ in the limit of high compressions (Fermi gas) and expansions (ideal gas) we choose both functions γ_0 and γ_∞ having asymptotic value 2/3 for $\rho \to \infty$ and $\rho \to 0$.

$$\gamma(\rho)_0 = \frac{2}{3} + (\gamma_s - \frac{2}{3})\frac{2}{(\frac{\rho_0}{\rho} + \frac{\rho}{\rho_0})} \qquad (4)$$

$$\gamma(\rho)_\infty = \frac{2}{3} + (\gamma_{min} - \frac{2}{3})(\frac{2}{x + 1/x})^\beta , \qquad (5)$$

where $x = \frac{\rho_0}{\rho}\xi$ and γ_s is Grüneisen coefficient at normal conditions, ρ_0 is the normal density, γ_{min} is some minimum value of γ in dense plasma and β and ξ are fitting parameters. E_0 can be matched by fitting shock-wave experiments with porous samples. To describe compression of porous samples the thermal component of the pressure in EOS must be corrected for densities close to normal density.

Experiments

Experiments analysed in this paper have been described extensively in [8]. Thin foils of aluminum of different thickness were irradiated with KALIF beam. The velocity of rear surface was measured by the ORVIS system. Two types of experiments are analysed here:

(1)*Foil acceleration experiments* : In these experiments, thin superrange (i.e. the thickness of the target was larger than the range of protons) aluminum foils of various thickness were irradiated by the KALIF beam. The back surface velocity was recorded with ORVIS. The thickness of the targets varied between 22 μm and 75 μm.

(2)*Pressure measurements* : In these experiments the aluminum foils, whose thickness only slightly exceeded the range of protons, were covered with LiF windows to avoid shock reflections. The impedance of LiF is approximately the same as that of aluminum. The velocity at the Al-LiF interface was recorded as a pressure signal.

The maximum power density in all these experiments was 0.15 $\pm$0.05 TW/cm^2. These experiments were conducted with the so called B_θ-diode.

Simulations

In our previous analysis we have found that available equation of states (SESAME [9] or Bushman EOS [10]) do not describe the foil acceleration experiments satisfactorily [7]. Results of these calculations show that both EOS give inadequate results for the time interval between 10 and 20 ns, when the density drops to an order of magnitude less than its solid value and temperature increases to about 10 eV. These states of matter correspond to the region of strongly coupled plasma.

To rectify this situation we used our semiempirical EOS model described earlier. With this EOS we can reproduce all acceleration experiments performed at KALIF. Results of numerical simulations are shown in Fig.1 together with experimental data for foils of different thickness.

In Fig.2 we show the simulation using our Moving Grid 2D code. The beam parameters in these calculations are a peak power density of 1 TW/cm^2 and a focus of 1 mm radius. These parameters were chosen to highlighten the 2D effects. For the sake of better understanding we plot in this pictures only the evolution of 2D

disturbance. The subregion division was done, as depicted, during the calculation. In the beginning of beam interaction this 2D disturbance is located at the beam boundary and is represented in a 2x12 mesh. As the time increases the number of grid cells is increased. The region interfaces are chosen as described before. In the final run when the 2D disturbance reaches the center after 90 ns the number of grid cells in all regions except the region of no hydromotion has increased. Before dividing a subregion the user can modify the subregion interfaces, introduce new boundaries or combine subregions.

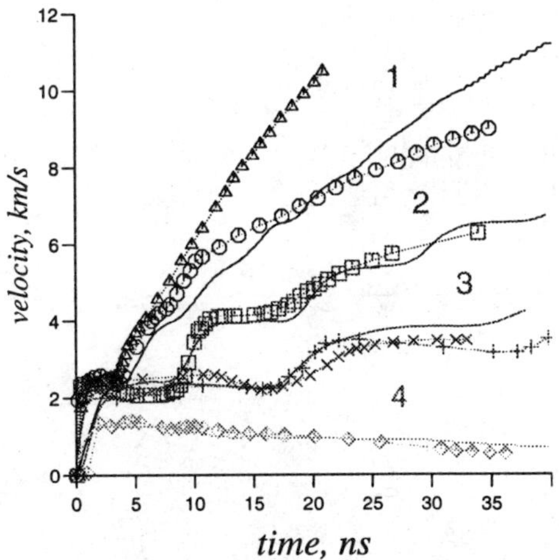

Fig.1: Results of 1D simulations(lines) and experimental data on aluminum foil acceleration.Foil thickness: 1- 33 μm, 2-50 μm, 3-75 μm and 4- 33 μm with LiF window.

The KALIF pulse is less than 80 ns long. Therefore, during the pulse duration the target axis do not observe any 2D effects. It is pointed out that in these simulations beam focus was assumed to be of 1 mm radius. In reality this is of the order of 4 mm. Thus if the diagnostic sits at the center and the focus do not move with time 1D calculations will give reasonable results.

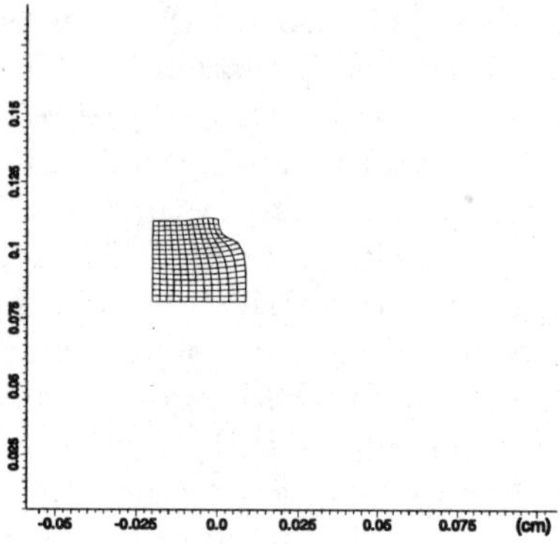

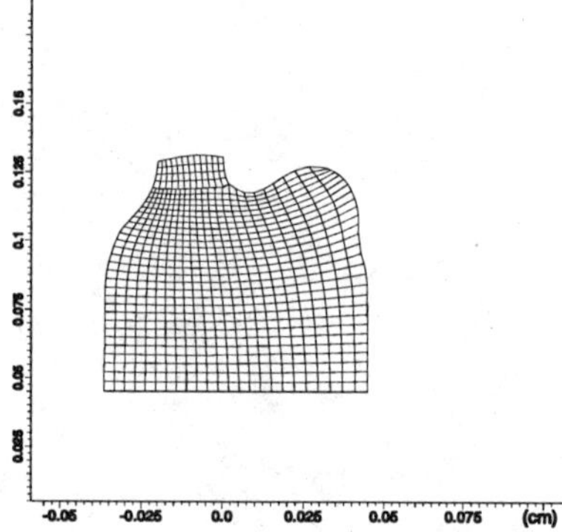

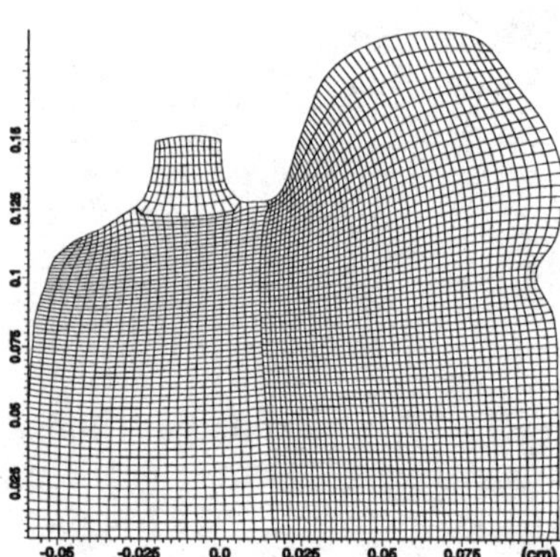

Fig.2: Results of 2D simulations: grid at times 20.267 ns, 54.333 ns and 90.223 ns.

Conclusions

We have presented 2D calculation of foil acceleration experiments using Godunov method in a moving grid algorithm. This method has been successfully applied to simulate beam target in-

teraction with realistic equation of state. 2D disturbance originating at the edges of the KALIF beam reaches the target axis only after the beam is switched off. Thus in the present experiments 1D analysis is sufficient. As a next step we will use this code to study Rayleigh-Taylor instability in the target.

Acknowlegments

This work was partly supported by the HCM programm of the European community and the German-Russian-Agreement on scientific and tecnical cooperation.

References

[1] Baumung, K. et al., Hydrodynamic Target Experiments with Proton Beams at KALIF, Proc. Ninth International Conference on High-Power Particle Beams, (ed. D.Mosher, G.Cooperstein), May 25-30,1992, Washington DC, USA , Vol. I. p.68.

[2] Goel,B., Vorobjev,O.Yu.,Ni, A.L., Generation of Ultra High Pressure with Light Ion Beams, Proc. Ninth International conference on High-Power Particle Beams, (ed. D.Mosher, G.Cooperstein), May 25-30,1992, Washington DC, USA , Vol. II. p.957.

[3] Bluhm, H., et al., Focussing Properties of a Strongly Insulated of a Applied B Proton Diode with a Preformed Anode Plasma, Proc. Ninth International Conference on High-Power Particle Beams, (ed. D.Mosher, G.Cooperstein), May 25-30,1992, Washington DC, USA , Vol. I. p.51.

[4] Polyshchuk,A.Ya., Fortov,V.E., Khloponin,V.s., Sov. J. Plasma Phys. 17 (1991) 523.

[5] Tsakiris,G.D., Eidman, K., JQSRT 38 (1987) 353.

[6] Fortov, V.E., Goel, B., Munz, C.-D., Ni, A.L., Shutov, A.V., Vorobiev, O.Yu., Numerical Simulation of Nonstationary Fronts and Interfaces by the Godunov Method in Moving Grid, To appear in Nucl. Sci. & Eng.

[7] Vorobiev, O.Yu., Fortov,V.E.,Goel, B., Numerical simulation of shock wave generation with KALIF beam.

[8] Baumung, K., et al., Light-ion Beam-Target Interaction Experiments on KALIF, Nuovo Cimento 106 A (1993) 1771

[9] Lawrence Livermore National Laboratory, Report UCID-118574-82-2 (1982).

[10] Bushman,A.V., Fortov,V.E., Lomonosov,I.V., Proc. Enrico Fermi School 1989, Elsevier Publ. 1989 p. 249

PARTICLE DENSITY AND TEMPERATURE DISTRIBUTION IN THE EARLY STAGE OF LASER-PLASMA INTERACTION

V.G. Molinari, D. Mostacci, M. Sumini

Laboratorio di Ingegneria Nucleare di Montecuccolino
Università degli Studi di Bologna
via dei Colli 16 - I-40136 Bologna - Italy

Abstract

Starting from the Boltzmann-Vlasov equation, particle density and temperature profiles and their time evolution are calculated, valid for the initial stage of laser-plasma interaction. The main focus of the work is on the analysis of the classical problem of two semi-infinite media kept at different temperatures and on the ensuing system evolution. Preliminary analytical results are presented.

INTRODUCTION

A well known class of problems in the physics of laser produced plasmas is related to the density and temperature distributions that follow the energy deposition taking place in a thin surface shell. The extremely sharp gradients generated in the interaction of ultra-short laser pulses with matter cannot be investigated with the classical method of Spitzer and Härm (which corresponds to first order spherical harmonics expansion) (1). Several methods have been proposed instead in recent years (2). A different method is proposed in this paper, based on the integral form of the Boltzmann-Vlasov equation (3, 4), which, to the authors' knowledge, has not been used yet in this context: the reason for resorting to an integral equation is that initial and boundary conditions are intrinsically taken into account and so is also any discontinuity, so that no assumptions or approximations are made on the expected shape of the distribution function. In this paper the method is presented and applied to the following simple but representative case. Consider two semi-infinite media composed of field particles kept at different temperatures T_0 and $T_{01} > T_0$, with the interface between the two media at x=0. Let the electrons in the two zones (zones I and II, see Fig. 1) be separately at equilibrium with the respective field particles, with local Maxwellian distribution functions at the respective temperatures T_{01} and T_0 (they will be denoted here as $M(T_{01}, \mathbf{v})$ and $M(T_0, \mathbf{v})$).

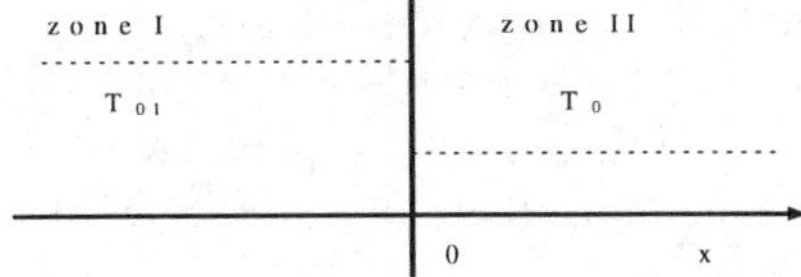

FIGURE 1 The simplified physical problem.

At time t=0 the interface is "open" and the subsequent evolution of the system is studied. Figure 1 above shows the geometry of the system (it must be noted that the system possesses planar symmetry, and its only spatial dependence is on x: however, velocity remains three-dimensional).

This problem, albeit largely simplified, is essentially that of energy deposition by an ultrashort laser pulse: a shell of some depth at the surface of the target is heated suddenly, and then hot electrons carry the energy over into the cold gas (thermal diffusion effect).

THE INTEGRAL FORMULATION

The integral form of the Boltzmann-Vlasov equation giving the evolution in time of a known initial distribution function can be found by the following argument (method of characteristics): consider a particle that at time t is in $\mathbf{r}$ with velocity $\mathbf{v}$ with the self-consistent electric field $\mathbf{E}$ being the only force acting on it. From the equations of motion

$$\frac{\partial \mathbf{R}(u)}{\partial u} = \mathbf{V}(u) \qquad (1\text{-}a)$$

$$\frac{\partial \mathbf{V}(u)}{\partial u} = -\frac{e}{m}\mathbf{E}(\mathbf{R}(u), u) \qquad (1\text{-}b)$$

with the conditions $\mathbf{R}(u=t) = \mathbf{r}$ and $\mathbf{V}(u=t) = \mathbf{v}$, the position $\mathbf{R}(u)$ and velocity $\mathbf{V}(u)$ at all times $u \in [0,t]$ can be calculated, and the trajectory of the particle in the phase space can be drawn. Knowing $\mathbf{R}(u)$ and $\mathbf{V}(u)$, the collision frequency $v[\mathbf{R}(0), \mathbf{V}(0)]$ can also be found for all points of the trajectory: this can be written as $v(u)$. Consider first the collisionless case. The initial point of the trajectory being $[\mathbf{R}(0), \mathbf{V}(0)]$ (that is, at time u=0), all those particles possessing that initial position and velocity, and only those particles, will be in $[\mathbf{r}, \mathbf{v}]$ at time u=t, therefore, for a collisionless plasma,

$$f(\mathbf{r}, \mathbf{v}, t) = f(\mathbf{R}(u=0), \mathbf{V}(u=0), 0) \qquad (2)$$

When collisions are considered, not all those particles in $[\mathbf{R}(0), \mathbf{V}(0)]$ at time u=0 will reach $[\mathbf{r}, \mathbf{v}]$ at time u=t, as some will be removed from the trajectory by collisions. The surviving "uncollided" distribution function $f^{(0)}$ is

$$f^{(0)}(\mathbf{r}, \mathbf{v}, t) = f(\mathbf{R}(0), \mathbf{V}(0), 0) e^{-\int_0^t v(u)du} \qquad (3)$$

Furthermore, some types of collision events (which might or might not be present in the system considered) lead indeed to a terminal removal of the particle from the system, but scattering events merely knock particles from one trajectory to another, and the contribution to $f(\mathbf{r}, \mathbf{v}, t)$ of those particles that are knocked to the $(\mathbf{r}, \mathbf{v}, t)$ trajectory must be added. At time u, such a "source" term is

$$\int_{\Re^3} K[\mathbf{R}(u), \mathbf{v}' \to \mathbf{V}(u)] f(\mathbf{R}(u), \mathbf{v}', u) d\mathbf{v}' \qquad (4)$$

but, again, only a fraction $\exp\left[-\int_u^t v du'\right]$ survives on the trajectory until time u=t. Collecting these contributions for all points on the trajectory (i.e., all values of u), the "collided" distribution function is obtained:

$$f^{(coll)}(\mathbf{r}, \mathbf{v}, t) = \int_0^t du \left\{ e^{-\int_u^t v du'} \times \right. \qquad (5)$$

$$\times \int_{\Re^3} K[\mathbf{R}(u), \mathbf{v}' \to \mathbf{V}(u)] f(\mathbf{R}(u), \mathbf{v}', u) d\mathbf{v}'$$

The seeked distribution function is the sum of the "uncollided" and "collided" terms (2, 3):

$$f(\mathbf{r}, \mathbf{v}, t) = f(\mathbf{R}(0), \mathbf{V}(0), 0) e^{-\int_0^t v du'} +$$

$$+ \int_0^t \left\{ e^{-\int_u^t v du'} \int_{\Re^3} K[\mathbf{R}(u), \mathbf{v}' \to \mathbf{V}(u)] \times \right.$$

$$\times f(\mathbf{R}(u), \mathbf{v}', u) \right\} du \qquad (6)$$

where the symbols have their usual meaning.

From the distributions function so obtained, particle current $\mathbf{J}$, temperature T and heat flow $\mathbf{Q}$ can ben computed:

$$\mathbf{J}(\mathbf{r}, t) = \int_{\Re^3} \mathbf{v} f(\mathbf{r}, \mathbf{v}, t) d\mathbf{v} \qquad (7\text{-}a)$$

$$T(\mathbf{r}, t) = \int_{\Re^3} \frac{mv^2}{2} f(\mathbf{r}, \mathbf{v}, t) d\mathbf{v} \qquad (7\text{-}b)$$

$$\mathbf{Q}(\mathbf{r}, t) = \int_{\Re^3} \frac{mv^2}{2} \mathbf{v} f(\mathbf{r}, \mathbf{v}, t) d\mathbf{v} \qquad (7\text{-}c)$$

With reference to the one-dimensional problem outlined (see Fig. 1), consider a location x>0 at a time t: a particle that was at X(u=0)=0 with velocity $V_x(0)=0$ can reach x at time u=t if and only if $V_x(0)$ equals a specific velocity, which here will be referred to as V_{xl}. Particles with $V_x(0) > v_{xl}$ will reach x too early, and conversely. This particular velocity V_{xl} , calculated from the equations of motion , is found to be

$$V_{xl} = \frac{x}{t} + \frac{1}{t}\frac{e}{m}\int_0^t du \int_0^u E du' \qquad (8)$$

The same particle, will have one specific velocity v, and faster particles will have started with a larger $V_x(0)$ from $X(0) < 0$, slower particles with a smaller $V_x(0)$ from $X(0) > 0$; meaning that faster particles originate from zone I, where they were part of a Maxwellian distribution at $T = T_{01}$ $(M(T_{01},\mathbf{v}))$, slower ones from zone II, from a Maxwellian distribution at $T=T_0$ $(M(T_0,\mathbf{v}))$. Therefore, the uncollided distribution function becomes:

$$f^{(0)}(x,\mathbf{v},t) = \qquad (9)$$

$$= \begin{cases} e^{-vt}M(T_0,\mathbf{V}(0)) & V_x(0) < V_{xl} \\[2ex] e^{-vt}M(T_{01},\mathbf{V}(0)) & V_x(0) > V_{xl} \end{cases}$$

and the uncollided electron density results:

$$n^{(0)}(x,t) = \int_{\Re^3} f^{(0)}(x,\mathbf{v},t)d\mathbf{v} = \qquad (11)$$

$$= \frac{n_0}{2}e^{-vt}\left[2 - erf\left(\sqrt{\frac{m}{2T_{01}}}V_{xl}\right) + erf\left(\sqrt{\frac{m}{2T_0}}V_{xl}\right)\right]$$

It can be observed that for $v \to 0$ this is the solution of the Vlasov equation. In fact, however, the self-consistent electric field is as yet unknown. The contribution to the distribution functions due to the uncollided electrons given by the first term in Eq. (6), is particularly significant as we are interested in the early times of the evolution of the system, i.e., when t is small compared to the

inverse of the collision frequency. Under these conditions, the contribution of the collided part of the distribution function can be neglected: the self-consistent electric field, in particular, can be calculated from Gauss equation with the density $n^{(0)}$ as given by Eq. (11). The problem, though, is non-linear, insofar as $n^{(0)}$ contains E through the solutions of the motion equations, under the form of time integrals from 0 to t of the field. Again, as the focus here is on small values of t, and physical considerations dictate that the electric field initially be zero, an iterative procedure can be envisioned on the system formed by Eq. (11) and Gauss equation.   The first iteration can be done analytically, and can be considered sufficient for the small values of t considered here. The electric field so calculated is shown in Fig. 2, and the density $n^{(0)}$ (x,t) calculated inserting this field into Eq. (11) is plotted in Fig. 3 as a function of x for several different values of t.

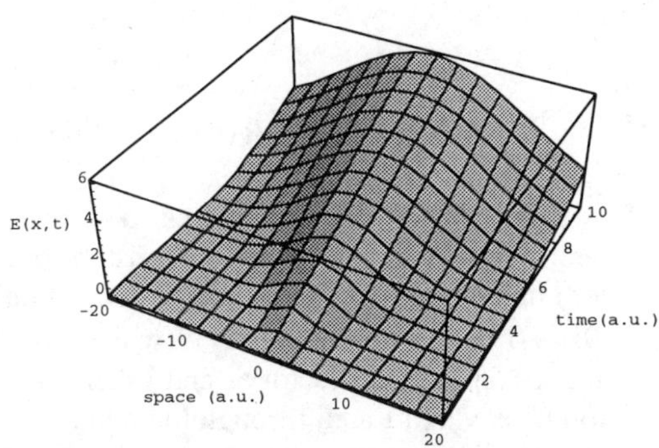

FIGURE 2 x-t dependence of the self-consistent electric field.

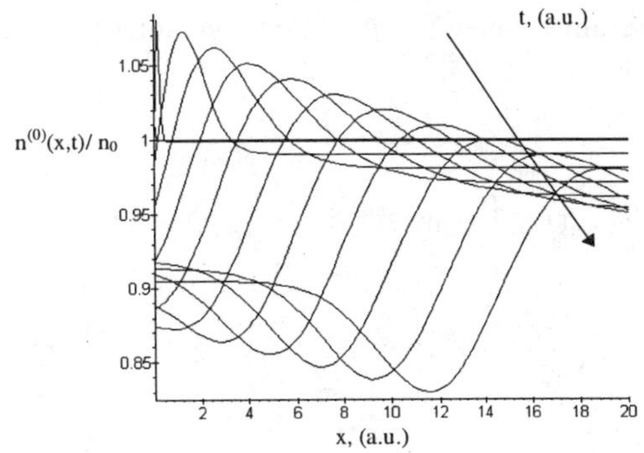

FIGURE 3 Electron density vs x, at different times.

THE COLLIDED TERM

To obtain the first order distribution function, the integral in the r.h.s of Eq. (6) must be calculated giving the distribution of the collided particles. To this end the scattering kernel must be specified: the BGK approximation will be adopted here, i.e.:

$$\int_{\Re^3} K[\mathbf{r}, \mathbf{v}' \to \mathbf{v}] f(\mathbf{r}, \mathbf{v}', u) d\mathbf{v}' = \nu M(T(\mathbf{r}, t), \mathbf{v}) =$$

$$= \nu n(\mathbf{r}, t) g(T(\mathbf{r}, t), \mathbf{v}) \tag{12}$$

Inserting the BGK kernel in Eq. (6), the density of the collided particles is found to be:

$$n(\mathbf{r}, t) = \int_{\Re^3} d\mathbf{v} \times \tag{13}$$

$$\times \int_0^t e^{-\nu(t-u)} \nu n(\mathbf{R}(u), u) g(T[\mathbf{R}(u)], \mathbf{V}(u)) du$$

As we are considering small times, $u \approx t$ can be assumed. However, an exception must be made for the temperature, due to the sharp discontinuity in x=0. To deal with it, a time τ (which will be dependent on the trajectory and therefore a function of x, $\mathbf{v}$ and t and, through the equations of motion, of E) can be defined, such that the trajectory of the particle crosses the x=0 plane at u=t-τ. Then the integral over u can be split in two parts as follows (again, only the x-dependence will be retained, due to the planar symmetry of the problem):

$$n^{(coll)}(x, t) = \int_{\Re^3} d\mathbf{v} n(x, t) \times \left[g(T_I(x, t), \mathbf{v})(1 - e^{-\nu\tau}) + \right.$$

$$+ g(T_{II}(x, t), \mathbf{v})(e^{-\nu\tau} - e^{-\nu t}) \right] \tag{13}$$

where T_I and T_{II} are the (unknown) temperatures in zones I and II respectively.

Again, an iterative procedure must be set up to calculate the electric field, together with $\mathbf{J}$, $\mathbf{Q}$ and T_I and T_{II} (see Eqs. 7). Approximating the electric field with the one calculated from the uncollided density (see previous section), we can obtain a formal expression for the collided density:

$$n^{(coll)}(x, t) = \frac{A}{1 - A} n^{(0)}(x, t) \tag{15}$$

where the quantity A, function of x,t and, through the electric field, of $n^{(0)}$, is given by

$$A\left(x, t; E\left(x, t; n^{(0)}\right)\right) = \int_{\Re^3} d\mathbf{v} \left[g(T_I(x, t), \mathbf{v})(1 - e^{-\nu\tau}) \right.$$

$$+ g(T_{II}(x, t), \mathbf{v})(e^{-\nu\tau} - e^{-\nu t}) \right] \tag{16}$$

The procedure outlined is very well suited to be translated into a numerical algorithm: a computer code is at present under development.

ACKNOWLEDGEMENTS

This work has been partly supported by the Italian National Institute for the Physics of Matter (INFM).

REFERENCES

1. Spitzer, L. and Härm, R.,
   Phis. Rev., **89,** 997 (1953).
2. "Handbook of Plasma Physics, vol. 3, Physics of Laser Plasma", Edited by A. Rubenchik and S. Witkowski (1991), articles by:
   Galeev, A.A. and Natanzon, A.M., p. 549
   Key, M.H., p. 575.
3. Molinari, V.G. and Peerani, P.,
   Nuovo Cimento, **5,** 527-540 (1985).
4. Molinari, V.G., Sumini , M. and Ganapol, B.D.,
   Progress in Astronauthics and Aeronauthics, **116,** 102-114, (1988).

GENERATION OF SHOCK WAVES BY SOFT X-RADIATION FROM Z-PINCH PLASMA

V. Fortov, M. Lebedev, K. Dyabilin, O. Vorob'ev,
V. Smirnov, E. Grabovskij

High Energy Density Research Center, Izhorskaya 13/19, Moscow, 127412, Russia

Behavior of matter under intensive soft x-radiation is consided. Impulse of x-radiation (about $2*10^{12}$ W/cm^2) produced in the sample strong shock wave, the velocity of this shock wave was measured. The shock compression about 300 GPa in the lead target was achieved. Experiments allowed to verify the simulation of the interaction of x-radiation with matter.

INTRODUCTION

A fundamental problem in the use of concentrated fluxes of charged particles and laser light in the dynamic physics of high energy densities [1] is the substantial spatial nonuniformity of the power which is released. This nonuniformity disrupts the symmetry of the spherical compression of the fusion fuel [2] and hinders the excitation of plane shock waves in the experiments on the behavior of matter under extreme conditions. One of the most effective ways to solve this problem is to use the x-ray emission from a plasma with an approximately thermal spectrum which arises when directed energy fluxes are applied to a target [3] or during the electrodynamics compression of cylindrical shells in a Z-pinch geometry [4]. The plane shock waves excited by this radiation, which is an extremely simple type of self-similar hydrodynamic flow [5], might be a more nature and highly rich source of experimental information on both the intensity of the incident x-radiation and of the thermophysical properties of matter under action of this radiation with condensed targets.

Z-pinch plasma, produced in the big installations by the cylindrical liners electrodynamics compression seems to be one of the most favorable candidate for the source of the such x-radiation.

The paper presents the measurements of the shock waves intensities, generated by soft x-radiation in Al, Sn, Fe and Pb targets. The scheme with the conversion of the laser to soft x-radiation, described in [6,7], is different from that used here: the soft x-radiation was induced by the dynamic compression and heating of the plasma in the cylindrical Z-pinch geometry in the ANGARA-5-1 installation [8,9]. As a result, the radiation pulse duration was about an order of magnitude more, with the power level being nearly the same as in [6,7].

EXPERIMENT

Z-pinch plasma radiated soft X-R with the typical temperatures about 60-120 eV. The x-ray pulse duration was 30 ns FWHM. This radiation was coming on the planar target which was positioned under the internal liner 1 mm apart [10]. The experimental set-up is shown in Fig. 1. The targets were made as step 16... 30 μm Al and 80... 200 μm Pb, or pure 180 μm Pb, or stepped 16 μm Sn and 180 μm Pb plates being connected together. Such large thickness allowed to eliminate the thermal preheating of the target. The diameter of target was about 5 mm. The velocity of the shock

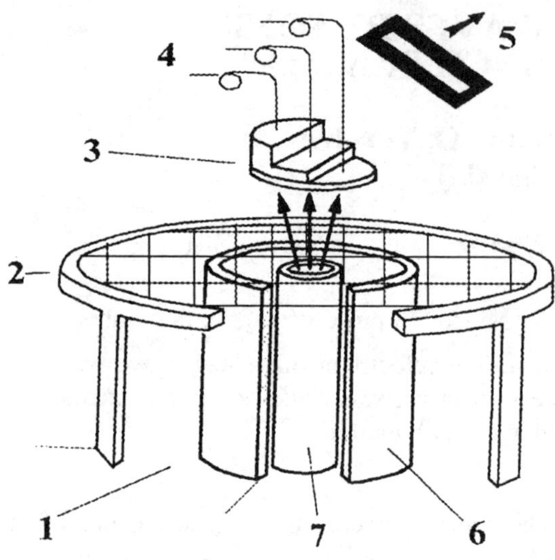

FIGURE 1. Experimental set-up: 1 - cathode; 2 - anode; 3 - step target; 4 - optical fibers; 5 - orientation of streak camera's slit; 6 - xenon annulal (outer liner); 7 - inner liner (three arrow - soft x-radiation)

wave was defined by the optical base method as the difference between the moments in the flashes of light emitted upon the shock waves break out at the rear surface of a sample. The sample was imaged with the help of a f/10 (f=1.5 m) objective and a f=3 m objective with 2 - fold magnification onto photocathode of a SNFT-2 streak camera. Spatial resolution corresponds to 20 lp/mm in sample plane, temporal resolution less than 0.3 ns was provided. In another base method the optical fibers (quartz-polymer, length 80 m, 0.4 Db dumping, bandwidth 2 GGz) were used, providing the high enough noise defense of the experiment apparatus. Optical fibers (diameter 400 mm) were closely connected with the target. To eliminate the influence of the hard x-ray, which can induce the light inside the fibers, they were positioned inside the steel tube. For the detecting of the optical radiation from the fibers, the silicon photodiodes with the time resolution less than 1 ns were used. Under the experimental data processing, the time difference between the signals fronts was used, so

the measurements temporal resolution less than 1 ns in this fiber method was provided.

Typical streak camera records are shown in Fig. 2.

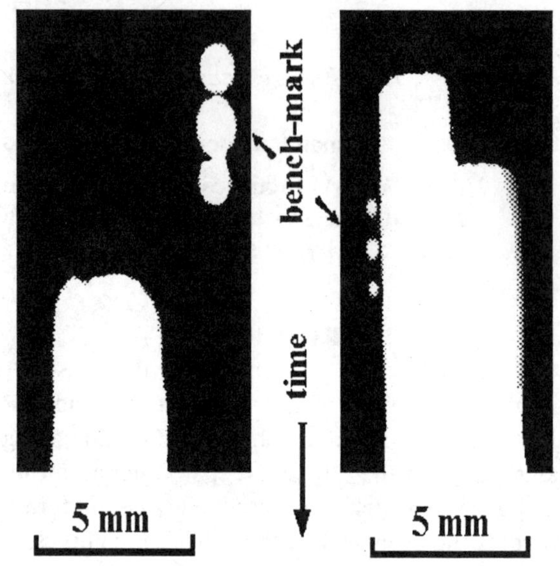

FIGURE 2. Streak camera records (positive) of shock break out from sample obtained in ANGARA-5-1 experiments: left - plane target: lead-180 µm; right - stepped target: tin-16 µm, lead-180 µm. Between bench-marks - 13 ns

SIMULATION

To describe the dynamics of the target under x-ray action Euler equation closed by wide range semiempirical EOS [11,12] were used, taking into account melting, vaporization and ionization of matter. Energy transfer by x-ray radiation was treated in multigroup diffusive approximation, which allow one to replace quasi-stationary transfer equations by equations of radiation diffusion. Spectral opacities were used calculated in the frame of modified Hartri-Fock-Sletter model wide range of temperatures and densities. To validate opacity model Rosseland mean opacities coefficients were compared with other semiempirical formulas. Euler equation were integrated by Godunov method. For calculation of energy transfer an implicit numerical scheme was employed.

The result of calculations are shown in Fig. 3 and Fig 4.

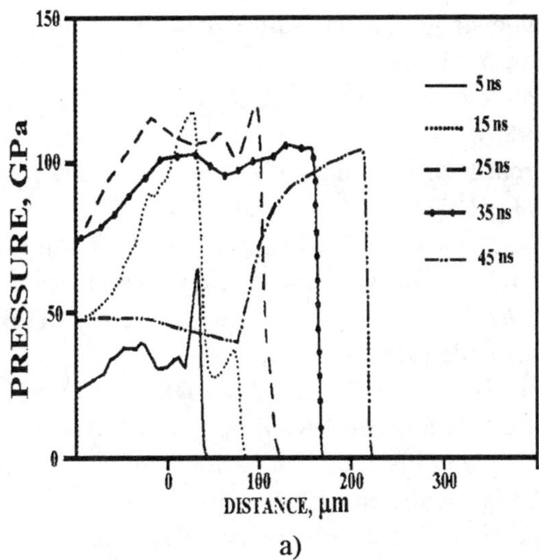

a)

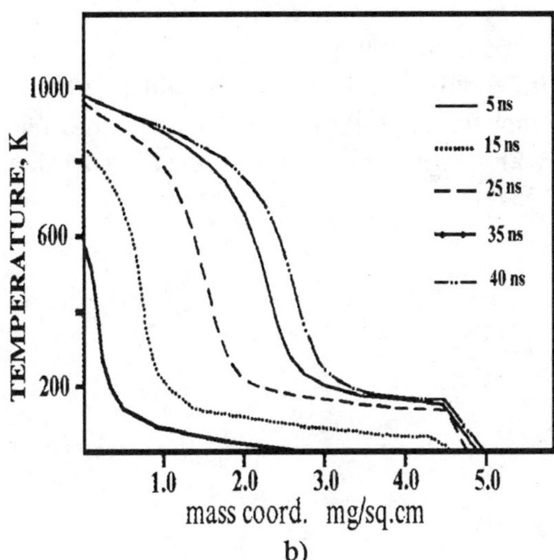

b)

FIGURE 3. . Numerical simulation of soft x-ray interaction with condensed Al target: a) Pressure evaluation; b) temperature evaluation. Power level $2*10^{12}$ W/cm^2

RESULTS

The averaged (over the target volume) shock wave velocity for Al plus Pb stepped target is $(7.3\pm0.6)*10^3$ m/s for 80 μm Pb thickness, and $(4.6\pm0.3)*10^3$ m/s for 200 μm. In accordance with the Huguenot lead adiabate this means that the

shock compression pressures are 300 GPa and 90 GPa correspondingly [12]. In stepped tin plus lead target shock compression of lead (thickness 180 mm) about 120 Gpa was measured.

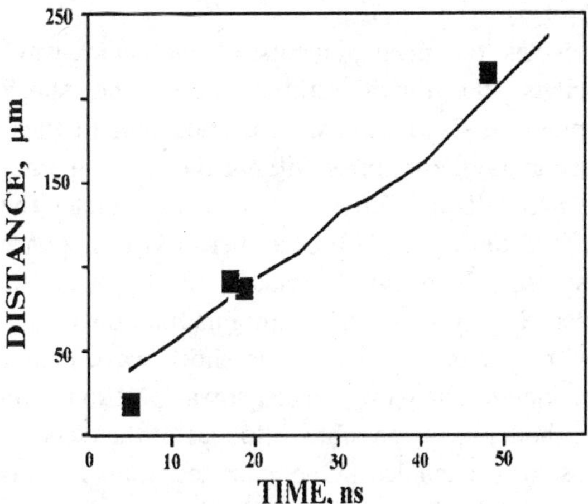

FIGURE 4. a) Shock front vs. time position for stepped "sandwich" (Al 30 μm + Pb 200 μm) target, power level $1.2*10^{12}$ W/cm^2, squares - experiment, solid line - calculation

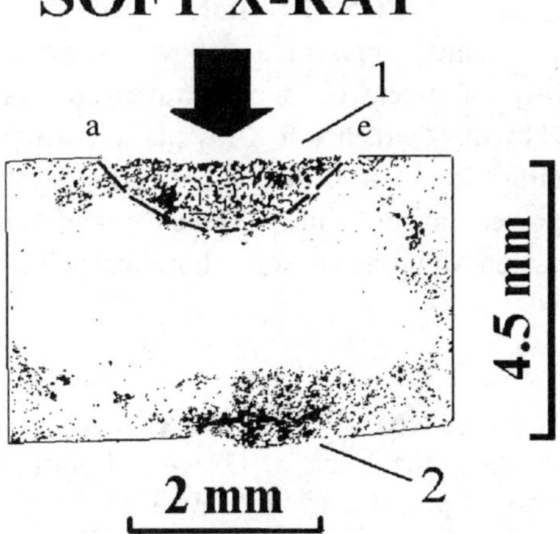

FIGURE 5. Microphotography of polish section of pure iron target after intensive soft x-radiation interaction (power level $0.5*10^{12}$ W/cm^2) : 1 - phase (a-e) transition region, shoick wave pressure more than 13 GPa; 2 - split

The important question under the study was the space uniformity of irradiated power. As shown in Fig. 2, the spatial uniformity of the shock wave front were derived from streak photographs with plane and step samples. Across the 4 mm diameter of shock sample we estimated a uniformity better than ±2%.

Besides the measurements of the shock-wave velocities, the direct estimations of the shock compression size region were carried out. In these experiments, the regimes without the internal liner were used. As it is known [11], in iron under the pressures more than 13 GPa there exists a phase transition, which can be visible on the metallic surface. Fig. 5 shows such an imagination of the phase transition induced by the shock wave in our experiments. The pure iron target was positioned on the cathode, so the possible influence of the electron beams was excluded. One can see that in this experimental scheme, when only the external liner was compressed on itself, the typical radial radiative region size is about 2 mm. One should note that in a standard double-liner scheme, this size is to be somewhat more. But the direct measurements in a standard scheme are impossible because of the target damage due to the strong radiative flux.

CONCLUSION

The results, presented above shows the possibility of intensive shock waves generation induced by the Z-pinch soft X-ray plasma radiation. The uniformity of shock wave is very high. At the flux power about some TW/cm^2 the shock compression value about some hundreds GPa was achieved.

ACKNOWLEDGMENTS

We wish to thank the ANGARA-5-1 staff. This work was supported by the Russian Fund for Fundamental Research (project 94-02-03430-a).

REFERENCES

1. Anisimov S., Fortov V. and Prohorov A., *Sov. Phys. Usp.* **27**, 181 (1984).
2. Duderstadt J. J., *Inertial Confinement Fusion*, New York: 1982
3. Matthews D. L. et al., *J. Appl. Phys.* **54** 4260 (1983).
4. Turchi P. J. and Baker W. L., *J. Appl. Phys.*, **44**, 4936 (1973).
5. Zel'dovich Ya. B. and Raizer Yu. P., *Physics of Shock Waves and High Temperature Hydrodynamic Phenomena*, New York: Academic Press, 1966.
6. Endo T. et al., *Phys. Rev. Lett.*, **60**, 1022 (1988).
7. Lower Th. and Sigel R., *Proc. 7th Int. Workshop of the Physics on Nonideal Plasma,*. Markgrafenheide, (1993).
8. Smirnov V. et al., *Proc. 8th Int. Conf. BEAMS-90,* Novosibirsk, (1990).
9. Gasilov V., Zakharov S. and Smirnov V., *JETP Lett.*, **53**, 83 (1991).
10. Grabovskij E. et al., *JETP Lett.*, **60**, 3 (1994)
11. Rinehart I. and Pearson J., *Behavior of Metals under Impalsiv Loads*, ASM, Clevelend, (1954).`

SHOCK WAVE GENERATION BY ULTRA-SHORT LASER PULSES

N.E. Andreev, V.E. Fortov, V.V. Kostin

High Energy Density Research Center of RAS, 127412, Moscow, Izhorskaya 13/19, Russia

The interaction of an intense ultra-short normally incident laser pulses with the solid targets is investigated for pulse duration 0.1-0.5 ps, beam intensity more then 1 TW/cm^2 and wavelength less then microm. Pressured area is created by the following mechanisms: ablation pressure, ponderomotive force, electron and ion components of the target material heating.

INTRODUCTION

Recent developments in the generation of high power femtosecond laser pulses offer new possibilities in the studies of ultra fast phenomena. In contrast to longer pulses, a femtosecond pulse deposits its energy in the target surface before significant hydrodynamic motion orthogonal to the surface occurs [1, 2]. The photoexcited region is not only strongly heated, but also pressured into the Mbar regime because it is heated at constant volume [2, 3].

RESEARCH METHODS

Light photons are absorbed by the electron component of the target matter. Energy exchange between electron and ion components of the matter takes place due to electron-ion relaxation.

The interaction of ultra-short intense normally incident laser pulses with the solid target is investigated by us of 1-D Lagrange computer code [3]. The laser pulse parameters described above allow consideration of the laser energy deposition by the normal high frequency skin-effect theory [3].

Computer modeling allows accounting for: laser irradiation absorption and ponderomotive force action, electron-ion energy relaxation, ionization and heating of material by thermal conduction, elastoplastic properties and destruction of the target material [3]. The radiative relaxation was omitted because of the small pulse du ration [3]. Estimation of the heating time to the temperatures when the produced plasma can be treated an ideal is less 10 fs for these pulse intensities which is much less than the pulse duration [3]. Due to this fact the target

material was described in limits of ideal gas approximation (thermophysical properties and equation of state). The laser field absorption was calculated on the real electron density profile determined by the hydrodynamic motion of the laser plasma [3]. Electrodynamic wave, thermoconductivity and hydrodynamic motion equations were solved simultaneously to obtain the electron and ion temperature profiles in the target and the distribution of plasma density and pressure.

RESULTS DISCUSSION

The intense, lineary polarized light wave strikes the overdence plasma and penetrates a thin layer somewhat larger than c/w_p due to the relativistic mass change [5]. Even for normal incidence, there is a very strong, high frequency electrostatic field operating on the plasma in this interface. This field is simply the oscillating component of the ponderomotive force [5]. Although generally nonresonant, this field can accelerate electrons which interact nonadiabatically with it in the steep interface. The frequency of the oscillating ponderomotive force is $2w_0$. This means that twice every cycle, a group of electrons will be accelerated into the plasma by this ponderomotive force.

The pressurized area is formed by the following mechanisms: due to plasma motion (ablation pressure), electron and ion components heating and ponderomotive force. At first, the pressurized area corresponds to the thermal wave in the target material. The main mechanism of pressurization in this case is the heating of the electron component. Electron temperature relaxation occurs due to the thermal wave propagation, plasma expansion, radiative heat transfer and energy exchange with ion component. As it was mentioned above the radiative heat transfer may be omitted at the dearly stages of the discussed problems. Growth of the ion component temperature of the target material increases the lattice pressure. In the early stages, the front of high pressure area and its profile corresponds to the thermal wave region. Then the front of pressurized area begins to outstrip the thermal wave due to difference in the propagation velocities. Its form begins to change configuration.

A shock wave was observed at a time of about 1 ps after the start of the laser pulse (For pulse duration 100 fs, Intensity 10^{17} W/ cm^2, wavelength 0.35 microm). It has a "classical" strength front surface and "tail." Increasing the pulse duration leads to growth of time for shock wave formation, but it is not a proportional connection. Pulse intensity change does not influence directly on this time.

SUMMARY

Obtained results allow us to draw a conclusion that the shock wave generation for these ultra-short pulse durations have a different character than do long pulses. The main mechanism at the early stage of pressurized area formation is not ablation pressure, but material heating and ponderomotive force action. From the other side, the existence of the very hot plasma with normal density gives an interesting instrument for ultra-fast phenomena investigation.

REFERENCES

[1] Fortov V.E., Kostin V.V., Vorob'ev O.Yu.//BEAM's-94, v.1, 1883-1886 (1994)
[2] Wange X.y., Downer M.C. // OPTIC LETTERS, v.17, 20, 1450-1452 (1992)
[3] Andreev N.E., Fortov V.E., Kostin V.V. et all // Rus. J. Physica Plasmy, 21, 7, 1-8 (1995)
[4] Wilks S.C., Kruer W. L., Tabak M., Langdon A.B. // Physical Rev. Lett., 69, 9, 1383-1386 (1992)
[5] Sprangle P., Esary E., Ting A. // Phys. Rev. Lett., 64, 2011-2013 (1990).

UNIFORM SHOCK WAVES DRIVEN BY THERMAL RADIATION FROM LASER-HEATED CAVITIES

Th. Löwer[1] and R. Sigel[2]

[1] *Max-Planck-Institut für Quantenoptik, D-85748 Garching, Germany*
[2] *Technische Hochschule Darmstadt, Institut für Angewandte Physik, D-64289 Darmstadt, Germany; on leave from Max-Planck-Institut für Quantenoptik, D-85748 Garching, Germany*

The generation of uniform and preheat-free shock waves in solids by laser-generated thermal x-rays requires careful design of the configuration of the cavity, laser beams and sample. With a new cavity design high-quality shock waves with pressures up to 14 Mbar could be generated with a single laser beam of 280 J energy.

INTRODUCTION

It has been demonstrated during the past few years that powerful pulsed lasers can generate very intense blackbody radiation when focused inside a millimetre size cavity made of a high-Z material like gold. Temperatures in the range 100-300 eV have been achieved, limited only by the power available from existing lasers (1,2).

If solid matter is exposed to such intense thermal soft x-ray radiation, surface material is strongly heated and ablated and a pressure wave is launched into the material. In dynamic high-pressure physics the possibility to generate spatially homogeneous pressure waves by the uniform deposition of energy in the form of incoherent radiation is of great interest for studies of the properties of matter at extremely high pressures. Pressures up to about 100 Mbar have now been achieved (3,4), exceeding the pressures accessible by classical methods by at least one order of magnitude. Pressures approaching the Gigabar range can be reached in the impact of flyers accelerated by laser-generated thermal radiation (5).

In the laser approach to the dynamic generation of high pressure the available driver energy is orders of magnitude lower than in the classical approach based on chemical explosives. This means that the experiments have to be miniaturised and that cavity and sample should form a compact, efficiently coupled device. This may result in problems with the irradiation uniformity of the sample and with preheat. In this paper we report our experience with cavities of different shape and present a new type of cavity which is capable of uniform irradiation of a sample when heated by a single laser beam only.

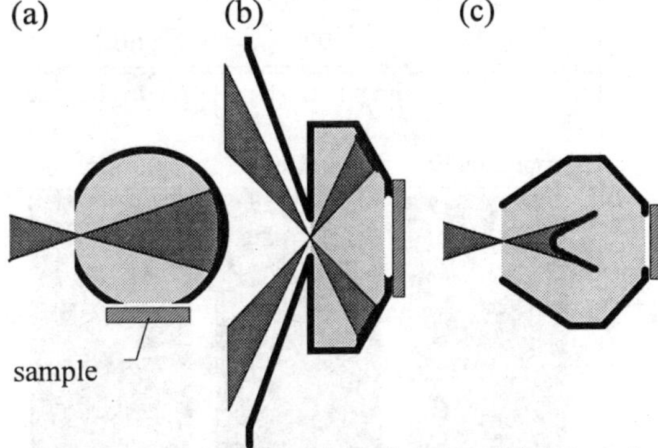

FIGURE 1. Design of laser-heated cavities. (a) Spherical cavity used at the single beam ASTERIX laser. (b) D-shaped, rotationally symmetric cavity heated with four beams of the GEKKO laser. (c) New Labyrinth cavity used at the ASTERIX laser.

LASER-HEATED CAVITIES

We have performed shock wave experiments with three types of cavities as shown in Fig.1(a)-(c). Fig.1(a) shows a spherical gold cavity, originally developed for studies of radiation confinement (1). It was heated with the single beam of the ASTERIX laser at MPQ with a 200J/ 0.45 ns/ 0.44 μm laser pulse. The cavity shown in Fig.1(b) was irradiated by

4 beams of the GEKKO XII laser at Osaka university with a 2.2 kJ/ 0.8 ns/ 0.35 µm pulse (4). Fig.1(c) shows the new design (called Labyrinth cavity) where a single beam heats a conical converter located inside the cavity. Upon incidence onto the converter the laser light is partially converted into x-rays which heat the cavity.

The cavities (a) and (b) had both an inner surface area equal to that of a 1 mm diameter sphere. Cavity (b) was used in three sizes equivalent to 1, 2, and 3 mm spheres.

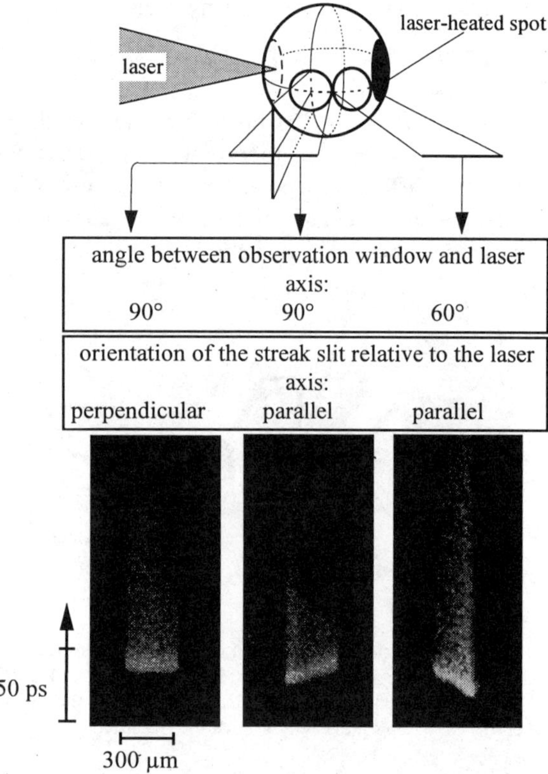

FIGURE 2. Streak camera records of the shock breakout taken with the spherical cavity heated by ASTERIX. The emission of the shock sample has been observed at different positions on the cavity and under different orientations of the streak camera slit. The variation of the shock arrival time along the slit indicates gradients of the heating x-ray radiation across the sample.

SHOCK UNIFORMITY AND SAMPLE PREHEAT

Using the spherical cavity shown in Fig. 1(a) experiments were performed (6) with the sample loca-

ted at two different angles of 60° and 90° with respect to the laser axis (see Fig.2). Observations of shock breakout with a streak camera for visible light, its slit being oriented parallel to the laser axis, revealed a variation of the shock arrival time across the 300 µm diameter gold sample (3 or 5 µm thick). At 60° the shock arrives earlier on that side of the sample which is close to the laser-irradiated area (which constitutes the radiation source for the heating of the cavity). At 90° the situation is reversed: the shock arrives earlier on the side facing the laser entrance hole. Obviously the driving x-ray intensity varies inside the cavity and goes through a minimum as one moves along the wall of the cavity from the laser-heated area to the entrance hole. Uniform irradiation of the sample is obviously problematic in this configuration.

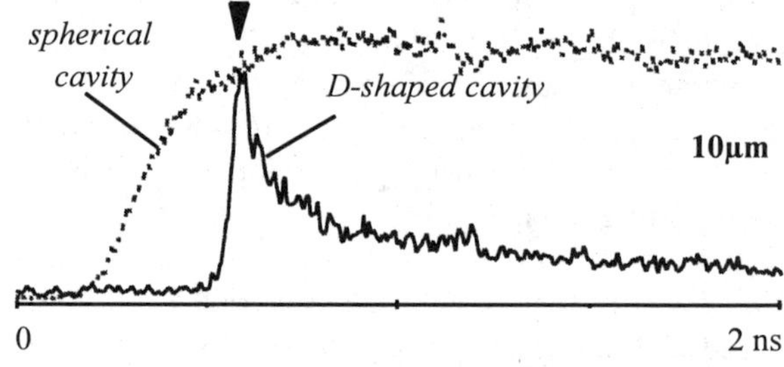

FIGURE 3. Streak record traces of 10 µm thick aluminium samples obtained with the spherical and D-shaped cavity under similar driver conditions. The strong precursor observed before shock arrival (marked by the arrow) in the case of the spherical cavity is generated by radiation preheat.

It may be interesting to note that, for reasons of symmetry, the x-ray intensity should be constant on parallels of latitude (with respect to the laser axis). Thus, if one orients the streak slit perpendicular to the laser axis, the shock wave should be found to arrive simultaneously across the sample. This is indeed observed as shown in Fig.2. Nevertheless, for reasons of simplicity, one would like to use in experiments an arrangement where the shock speed is the same everywhere on the sample.

With the spherical cavity the streaks obtained with aluminium foils showed strong luminosity on the rear side of the sample before shock arrival (see Fig.3), indicating x-ray preheating of the sample. The preheat was attributed to the primary x-rays from the laser-irradiated area. The relatively steep angle of incidence (around 45°) favours the penetration of the x-rays to the rear of the sample. The results obtained with the simple spherical cavity clearly show the necessity for more elaborate designs.

A major improvement was achieved with the cavity shown in Fig. 1(b). With this cavity the shock speed was found to be uniform across the sample (examples of streaks have been given in Ref.(4)) and preheat become undetectable. The improvement in uniformity is attributed to the location of the sample on the axis of symmetry of the cavity where an extremum for the driving x-ray intensity is expected. The reduction in preheat is mainly due to the flat angle of incidence of the primary x-rays. But it should be kept in mind that these improvements came along with the complication of using several laser beams.

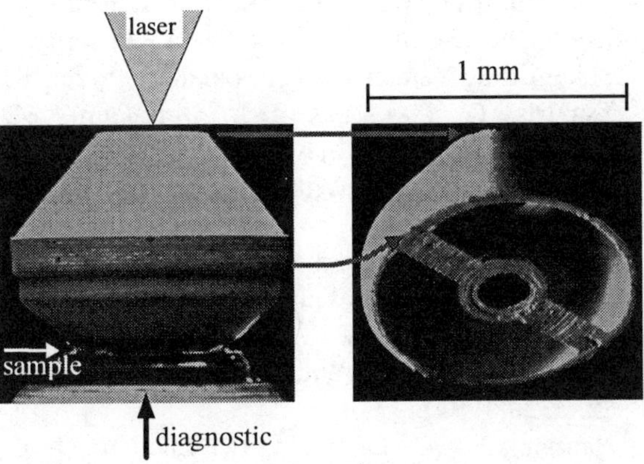

FIGURE 4. Scanning electron microscope photographs of the Labyrinth cavity. Left: target arrangement with cavity, sample and part of the shielding cone of the diagnostic. Right: top part of the cavity with the converter cone (the cavity is built from two parts fastened together).

CAVITY FOR UNIFORM DRIVE WITH A SINGLE LASER BEAM

For single beam facilities (like ASTERIX) we developed the Labyrinth cavity shown in Fig. 1(c) which allows uniform, preheat-free drive by a single laser beam. In this configuration the sample is located on the symmetry axis for uniformity and the sam-

ple is shielded from preheat due to primary x-rays by the gold wall of the converter. A photograph of the closed cavity is shown in the left part of Fig. 4. The right part shows a view into the upper part of the cavity which contains the converter cone. The sample and the cavity are glued onto the hole in the flat top of a cone. The cone is carried by a metallic tube, which leads to the streak camera as shielding against stray light and can be precisely adjusted. In the experiments the unit comprising cavity, sample, and cone can be exchanged and adjusted within a few minutes without breaking the vacuum.

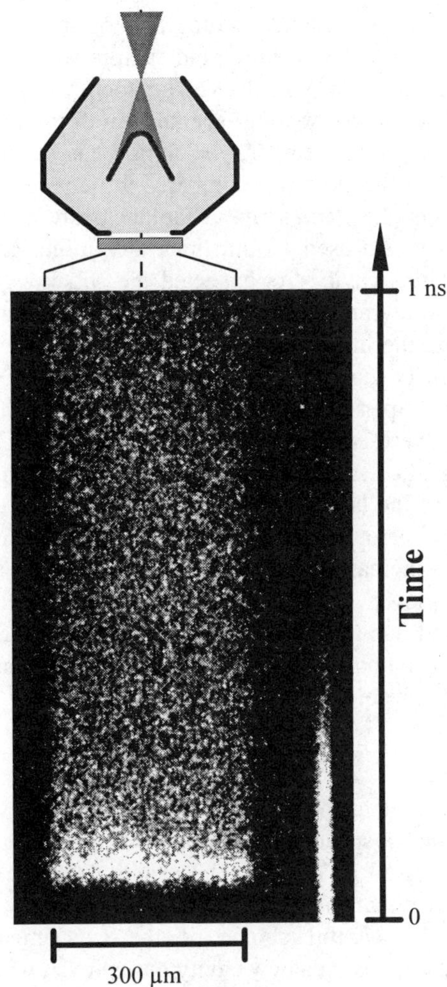

FIGURE 5. Streak camera record of the shock induced light emission in an experiment with the Labyrinth geometry. The signal at the right is the time fiducial generated by laser light.

TABLE 1. Summary of Results.

Type of cavity	Number of beams	Laser energy	Uniformity	Preheat	Shock pressure
(a)	1	200 J	poor	yes	7 Mbar
(b)	4	2.2 kJ	good	no	20 Mbar
(c)	1	280 J	good	no	14 Mbar

Fig.5 shows a streak photograph of the shock breakout from a 3 µm thick gold sample which has been obtained in an ASTERIX experiment using this cavity. The shock wave is spatially uniform (note that the spatial and temporal resolution has been increased by a factor of two compared to our previous experiments (4,6)) and arrives simultaneously within ± 8 ps. We also used 10 µm thick aluminium samples with this cavity. As expected the shock signal does not show any indication of preheat (the signal from this experiment is equal to the signal of the D-shaped cavity shown in Fig.3).

It was experimentally observed that the highest pressures were achieved when the focal spot of the laser was adjusted slightly in front of the cavity such that part of the laser light would hit the rear wall of the cavity surrounding the sample. This effect may be due to increased laser conversion efficiency because of a lower laser intensity in the laser-produced plasma and due to improved heating of the rear part of the cavity containing the sample. The defocusing had no observable effect on uniformity or preheat.

SUMMARY

The laser data and achieved pressures (as derived from the shock speed using published Hugoniot data) are summarised in Table I. The results demonstrate in particular that high quality shock waves may be achieved by careful choice of the experimental configuration. Using a new cavity design even with a single beam of modest energy uniform and preheat-free shock waves with the pressure in excess of 10 Mbar could be generated.

ACKNOWLEDGEMENTS

This work was supported in part by the Commission of the European Communities in the framework of the Euratom-IPP Association.

REFERENCES

1. Nishimura, H., Kato, Y., Takabe, H., Endo, T., Kondo, K., Shiraga, H., Sakabe, S., Jitsuno, T., Takagi, M., Yamanaka, C., Nakai, S., Sigel, R., Tsakiris, G. D., Massen, J., Murakami, M., Lavarenne, F., Fedosejevs, R., Meyer-ter-Vehn, J., Eidmann, K., and Witkowski, S., *Phys. Rev. A* **44**, 8323(1991)

2. Kaufmann, R. L., Suter, L. J., Darrow, C. B., Kilkenny, J. D., Kornblum, H. N., Montgomery, D. S., Phillion, D. W., Rosen, M. D., Theissen, A. R., Wallace, R. J., and Ze, F., *Phys. Rev. Lett.* **72**, 2320(1994).

3. Hammel, B. A., Griswold, D., Landen, O. L., Perry, T. S., Remington, B. A., Miller, P. L., Peyser, T. A., and Kilkenny, J. D., *Phys. Fluids B* **5**, 2259(1993).

4. Löwer, Th., Sigel, R., Eidmann, K., Földes, I. B., Hüller, S., Massen, J., Tsakiris, D., Witkowski, S., Preuß, W., Nishimura, H., Shiraga, H., Kato, Y., Nakai, S., and Endo, T., *Phys. Rev. Lett.* **72**, 3186(1994).

5. Cauble, R., Phillion, D. W., Hoover, T. J., Holmes, N. C., Kilkenny, J. D., and Lee, R. W., *Phys. Rev.Lett.* **70**, 2102(1993).

6. Th. Löwer, "Production of Multimegabar shock waves in solids by laser-generated thermal radiation," PhD dissertation, Universität Frankfurt, Germany, 1994 (in German).

EXPERIMENTAL AND NUMERICAL STUDY OF LASER-DRIVEN SPALLATION WITH VISAR DIAGNOSTIC

L. Tollier, E. Bartnicki, R. Fabbro

LALP (ETCA-CNRS) - 16 bis Avenue Prieur de la Côte d'Or - 94114 ARCUEIL - FRANCE*
** : Laboratoire d'Applications des Lasers de Puissance*

Laser-driven shock experiments in combination with VISAR technique have been performed to study the ablation pressure, the dynamic damage and spallation for Aluminum and Copper targets with thicknesses in the 25-500 µm range. Shock-waves up to 100 kbar have been generated by laser irradiation intensities from 10^{10} - 10^{12} W/cm^2 with a wavelength of 1,06 µm and pulses duration of 20 - 30 ns. The pressure profile has been determined using the laser-matter interaction code FILM, its amplitude has been compared with the one inferred by the VISAR velocity measurements. This temporal profile has been also used as a boundary condition applied at the front face of the target in the hydrodynamic code EFHYD-2D. Recorded free surface velocities from VISAR measurements exibiting spallation features have been compared with numerical EFHYD simulations to assess a continuous kinetic model of ductile spallation implemented in the code. Good agreement has been found between measured and predicted rear surface velocities for irradiation conditions leading to damage from void nucleation to complete spallation. Recovered samples have been examined by means of metallographic methods to compare the simulated damage with the experimental one in terms of spall thickness and damage zone size.

INTRODUCTION

The purpose of the work presented here is to investigate the spallation behavior of aluminum and copper using the laser-driven shock-wave technique. The spallation is caused by tension induced within the target by the crossing of two rarefaction waves : the first-one coming from the front surface at the end of the pressure loading and the second one coming from the rear surface due to the reflection of the incident shock-wave.

The VISAR diagnosis enables the study of the loading profile induced by the laser-matter interaction and the detection of damage growth in the sample. Comparisons between numerical and experimental velocity profiles allow us to validate a continuous kinetic model of ductile spallation inplemented in a 2D simulation code.

EXPERIMENTAL CONFIGURATION

These experiments were performed in the direct irradiation configuration with the Nd-glass pulsed laser of the LALP laboratory (ETCA - France). This laser delivers a maximum output energy of 80 Joules with gaussian pulses of 25 ns FWMH (full width at half maximum) at 1,06 µm wavelength. The sketch of Figure 1 representes the experimental set-up used during this study.

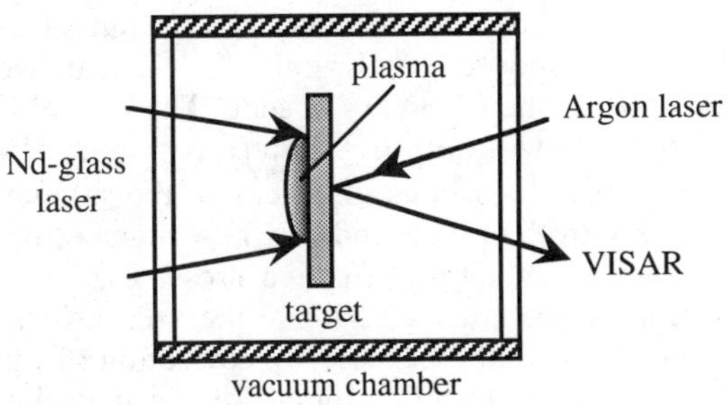

Figure 1. sketch of the experimental set-up

Different focal spot diameters in the range of 1 - 3 mm allowed us to obtain laser irradiation intensities from 10^{10} to 5×10^{11} W/cm^2. Aluminum and copper targets with thicknesses in the range of 25 - 250 µm were used, more than four times smaller than the spot size in order to ensure one-dimensional configuration. Samples were placed in a vacuum chamber to avoid air ionization which takes place at those incident laser intensities.

The velocity of the rear surface of the sample was recorded using a laser-Doppler interferometer VISAR (1). The time resolution of this velocimeter is about 2 - 3 ns, the precision is 1 %.

After they have been irradiated, the targets were sectioned and polished to allow the damage evaluation made using an optical microscope.

NUMERICAL CODES

Two simulation codes were used during this study. The first-one is the laser-matter interaction code FILM (LULI - Ecole Polytechnique - France). This one-dimensionnal hydrodynamic Lagrangian code simulates the laser energy deposition and the pressure profile induced at the front face of the sample.

The second numerical code is the hydrodynamic code EFHYD-2D that we used to calculate the waves propagation and the damage created by tension in the material. We simulate an elastic-plastic behavior of aluminum with an elastic limit determined by the Von-Mises criterion and a Johnson-Cook behavior of copper. The code uses the Mie-Grüneisen equation of state.

Dynamical failure effects in ductile materials were described by a continuous kinetic model of spallation proposed by Kanel (2, 3) that we inplemented in the EFHYD-2D code. This model is based on the evolution of a failure parameter V_t equal to specific volume of cracks in the sample. The crack growth rate depends on the volume of the cracks formed and on the effective stress. The stress relaxation on the cracks begins at the void growth start and is taken into account by correction of the shear modulus and yield strength of the material. Correction of the volume is also made by using the compact volume (without voids volume) in the equation of state.

RESULTS AND DISCUSSION

Determination of the ablation pressure profile

The pressure profile has been determined using the code FILM. Figure 2 shows a simulated temporal pressure profile resulting from a Gaussian pulse irradiation of 25 ns duration. We can see on the graph of Fig. 2 that the laser pulse duration is a

few nanoseconds smaller than the pressure pulse and the plasma release is much longer than the laser decrease.

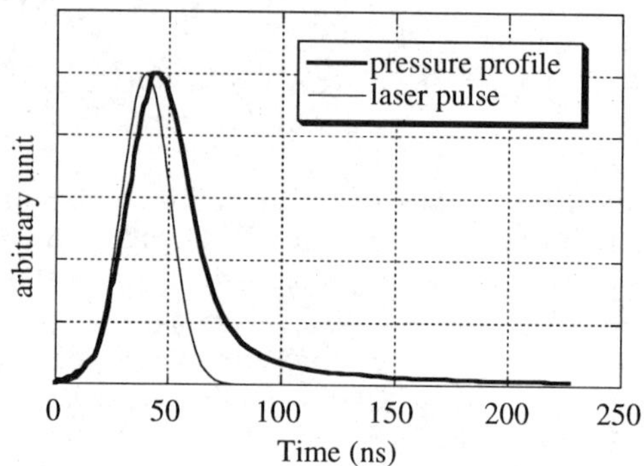

Figure 2. Pressure profile at the front face of an aluminum target induced by a 25 ns laser pulse

This pressure profile has been used as a boundary condition applied at the front face of the target in the code EFHYD-2D. Figure 3 presents a velocity profile measured with the VISAR on an aluminum target compared with a simulation realized with EFHYD-2D. The successive peaks are due to reloadings formed by the reflection of the incident shock-wave between the front and the rear surface of the 250 μm thick foil.

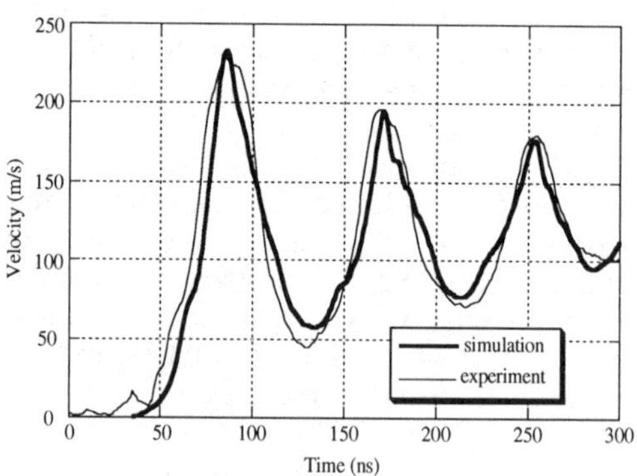

Figure 3. Comparison of experimental and numerical velocity profiles at the back face of a 250 μm thick Al foil P = 20 kbar - ϕ = 66 GW/cm^2

Good agreement is observed between experimental and numerical results, so the predicted pressure profile applied at the front face of the sample seems to be validated.

By using thin foil targets (< 250 μm), there is no significant attenuation of the peak pressure between the front and back faces of the foil with laser pulse duration of 25 ns. So the peak pressure P is directly related to the amplitude of the first peak U_s observed on the VISAR velocity measurements by the Hugoniot equation of the material :

$$P = \rho \cdot D \cdot \frac{U_s}{2} \quad \text{with} \quad D = C_0 + S \cdot \frac{U_s}{2}$$

ρ : material density
C_0 : bulk sound speed at initial state
S : material parameter

The experimental results of induced peak pressure in Al and Cu samples as a function of laser intensity obtained from VISAR records are reported and compared with FILM simulations on Figure 4.

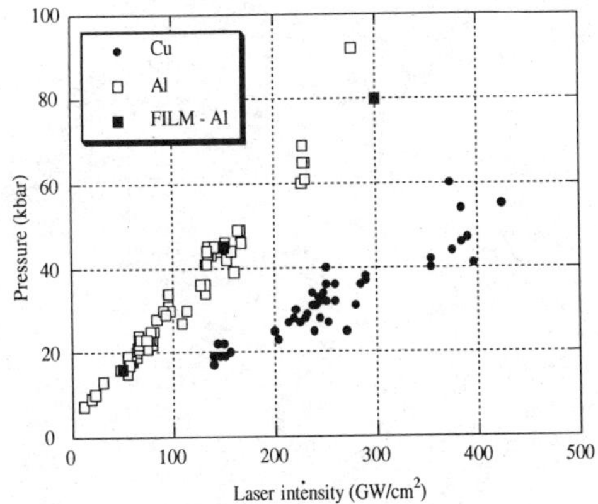

Figure 4. Ablation peak pressure in 250 μm Al and Cu targets versus laser incident intensity compared with FILM results on Al.

The comparison shows a good agreement between FILM calculations and experimental results. The important difference between Al and Cu results from the higher values of atomic coefficients Z and A of copper and from its higher reflectivity.

Study of spallation with VISAR diagnostic

The VISAR is a very sensitive instrument for detection of void growth in the sample. This is also a precise means for determination of low level of damage like voids nucleation and coalescence.

Figures 5 and 6 show an example of an Al foil irradiated at a laser intensity near the spallation threshold.

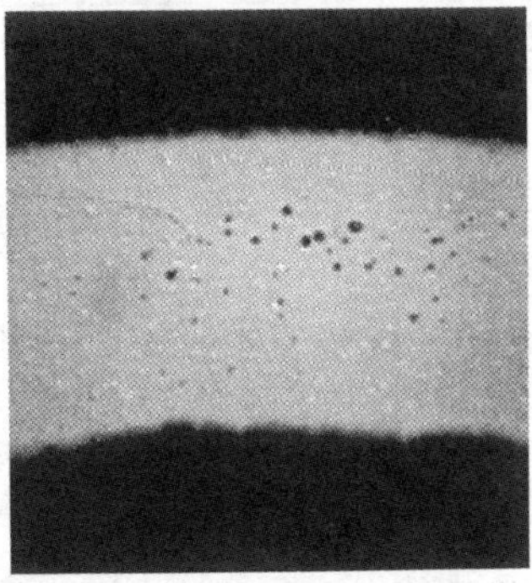

Figure 5. Photography of a section of a 250 μm thick Al sample irradiated with the following conditions
$\phi = 95$ GW/cm^2 - $\tau = 25$ ns

We can observe on Figure 5 a section of the target that shows the voids formed by traction in the material. The release wave coming from the rear surface is reflected by those voids in a compression wave. This wave reaches the rear surface at a time corresponding to the velocity peak encircled on Figure 6.

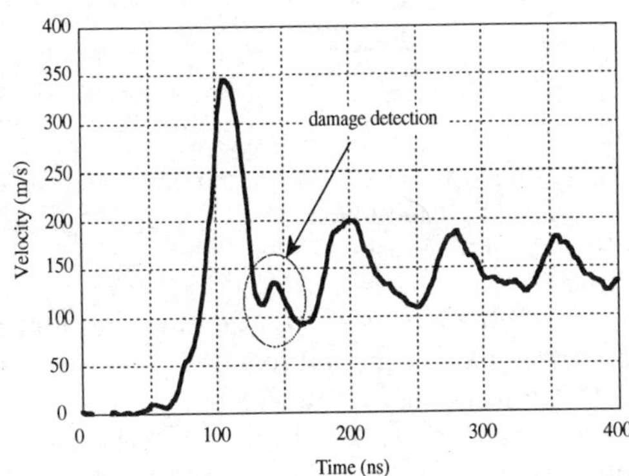

Figure 6. VISAR velocity record corresponding to the sample presented on Fig. 5

This level of damage produces a higher attenuation during the reverberations of the waves within the sample than the one we can observe on Figure 3. Table 1 reportes the values of incident laser intensity ϕ and peak pressure P at the front face of the target corresponding to the spallation threshold of the material.

TABLE 1.

Material	Al 250 µm	Cu 250 µm
φ (GW/cm^2)	80	250
P (kbar)	23	32

Kanel's damage model has been used to simulate the ductile spallation of Al and Cu. We made comparisons between numerical EFHYD-2D simulations and recorded free surface velocities exibiting damage features obtained with the VISAR. The graph of Figure 7 shows a low stage of material failure while Figure 8 illustrates a VISAR record at a higher stage of damage.

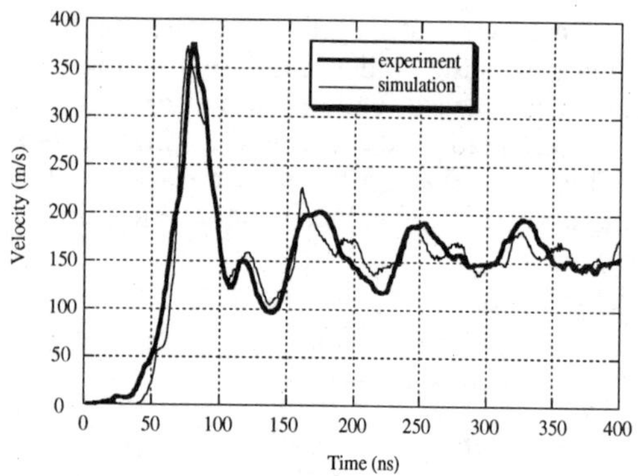

Figure 7. Comparison of EFHYD-2D simulation and VISAR record on a 250 µm Al foil
φ = 95 GW/cm^2 - P = 34 kbar - τ = 25 ns

The agreement between the damage model predictions and experimental results is good in both cases. The chronology of the velocity signals and the maximum velocity level are well reproduced. We can observe that the reflection coming from the voids reaches the rear surface earlier when the pressure is increasing, so it means that the spall width is decreasing.

CONCLUSION

We have been interested in two fields of research in this study. The first-one was the calibration of the laser-matter interaction in terms of ablation pressure profile induce in the material versus incident laser intensity. We have validated the amplitude and time history of ablation pressure by comparing experiments and simulations.

The understanding of this laser-driven loading has been used in the second research field which was the numerical and experimental study of spallation in ductile materials under laser irradiation. For both Al and Cu materials the spallation threshold conditions were determined. By means of computer simulation the history of damage formation recorded on VISAR velocity measurements has been quite well reproduced.

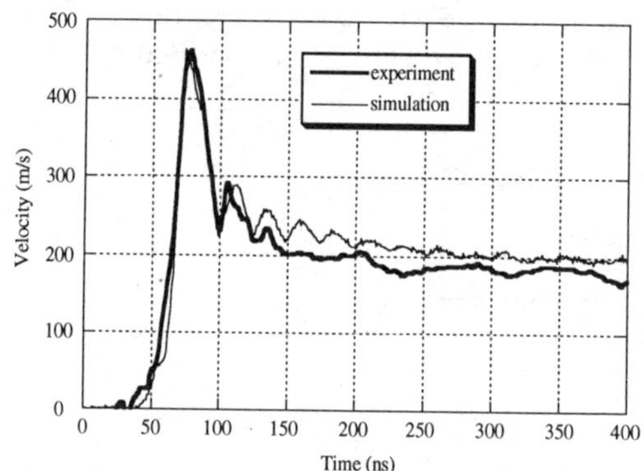

Figure 8. Comparison of EFHYD-2D simulation and VISAR record on a 250 µm Al foil
φ = 135 GW/cm^2 - P = 42 kbar - τ = 25 ns

REFERENCES

1. Barker L.M., Hollenbach R.E.,
"Laser interferometer for measuring high velocities of any reflecting surface"
Journal of Applied Physics 43(11), 1972, pp. 4669

2. Kanel G.I.,
Sowiet J. Physiva gorenija i vsriva 3, 77, 1983

3. Fortov V.E., Kostin V.V., Eliezer S.,
"Spallation of metals under laser irradition"
Journal of Applied Physics 70(8), 1991, pp. 4524

COMPUTED AND OBSERVED DOUBLE PULSE PHENOMENA IN ALUMINUM PRS EXPERIMENTS

Robert E. Tokheim

SRI International, Menlo Park, CA 94025-3493

George C. Williams

APTEK, Inc., Colorado Springs, CO 80906-3578

A complicated wave structure develops in an irradiated target during energy deposition that arises from the relief depth of the deposition profile. For low doses that produce vaporization of aluminum in plasma-radiation-source (PRS) experiments, this complicated wave structure reduces to a double pulse stress wave. At high doses, only a single pulse is observed. In this paper, we show correlations of computations with quartz gage data obtained from the Saturn Argon experiments. We then describe how the double pulse phenomenon originates in the stress-internal-energy space-time continuum.

INTRODUCTION

In solid materials exposed to irradiative energy deposition, only a single transmitted stress pulse is usually seen. However, we have observed a double pulse when the peak deposited energy is only somewhat above the incipient vaporization energy, under conditions where the energy deposition profile is shallow and the deposition time is short. This double pulse phenomenon is further complicated by "fine structure" of the stress relief of the deposition profile during deposition. But in this paper we concern ourselves only with the time-integrated stress generation.

EXPERIMENTS

Experiments were performed on the Saturn plasma-radiation source using argon gas at two fluences. The configuration consisted of an Al6061-T6 target bonded to a quartz gage. One experiment (Shot #1907-25,26 with 0.81 mm aluminum thickness) at a fluence of 3 cal/cm^2 showed a relatively small vapor stress pulse in addition to the usual main pulse, as shown in Figure 1. The duplicate experiments show the second pulse occurring after the main pulse that is not explainable as the front-surface reflected pulse from the gage interface, which occurs at a much later

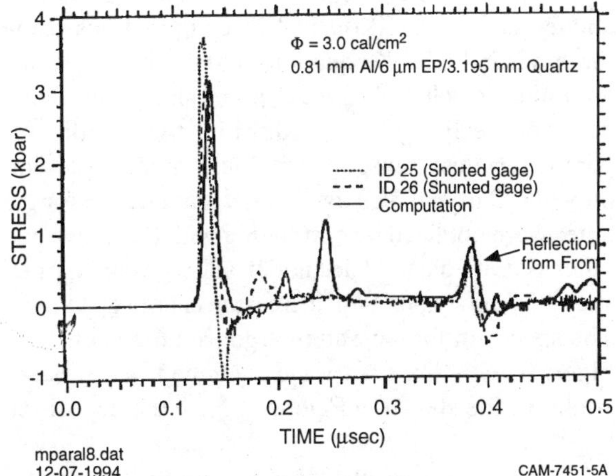

FIGURE 1. Correlation for 3 cal/cm^2 Saturn argon Shot 1907-25, 26.

time (0.38 µs). The other experiment (Shot #1729-65 with 1.57 mm aluminum thickness) had a fluence of 25 cal/cm^2 with much vapor generated stress, as shown in Figure 2. In this case, no double pulse distinguishable from unloading behavior was observed (it would be expected before the debris pulse), although there is slight evidence of the front-surface reflected pulse from the gage interface occurring at 0.76 µs.

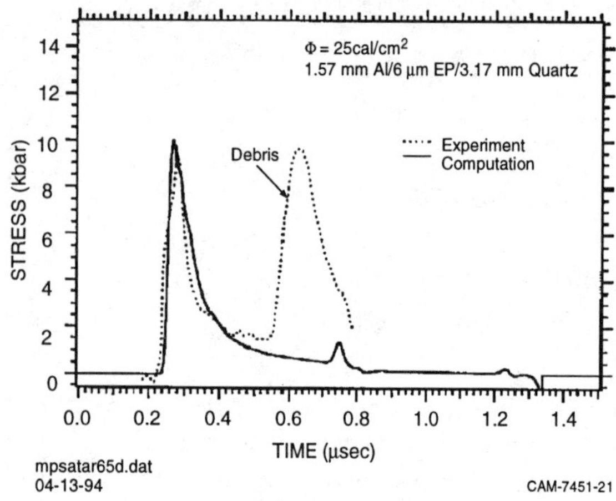

FIGURE 2. Correlation for 25 cal/cm² Saturn argon Shot 1729-65.

CORRELATIONS

Correlations are also shown in Figures 1 and 2 above and are based on PUFF[1] computations using the modified GRAY equation-of-state model for the aluminum (including critical point and steam dome data for tie lines).[2,3] In addition, the familiar 2-parameter elastic-viscoplastic yield model was employed (20 ns and 5E9 dyne/cm^2). Parameters for the latter were obtained by correlating with the above and other Saturn data. The model allowed no thermal strength degradation with increasing energy up to melt. The spectrum for the Saturn argon source is given in Figure 3, with flux history and computed energy deposition profile shown in Figure 4. Note the deposition

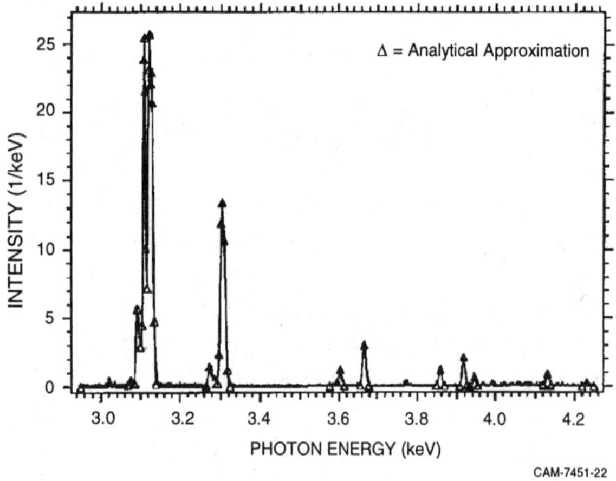

FIGURE 3. Saturn argon x-ray spectrum.

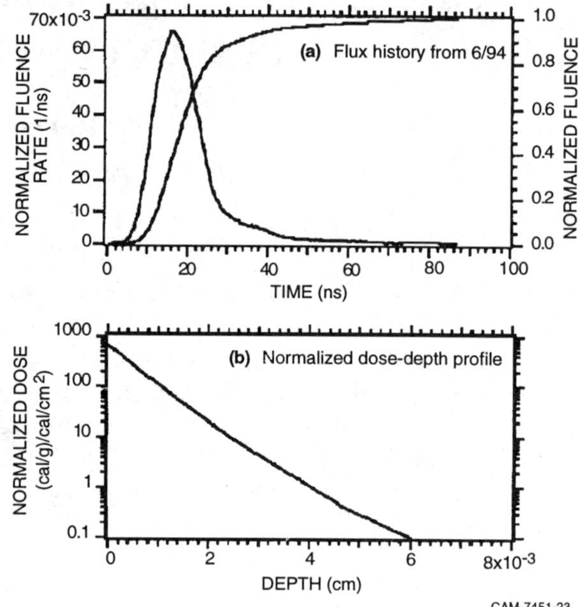

FIGURE 4. Flux history and aluminum target energy profile for Saturn argon.

profile is very shallow, with most of the energy deposited within 10 µm, and in a FWHM deposition time of 13 ns.

The PUFF correlations are good for the main pulses and adequate to show the presence of the vapor pulse of interest for the low fluence case. No attempts have yet been made to change the model parameters to accommodate the different arrival time and amplitude of the small vapor pulse in the low fluence case.

We found that the above detailed equation-of-state model was required to predict the double pulse phenomenon.

EVOLUTION OF STRESS PULSES

We can conveniently examine the evolution of stress generation in stress-Lagrangian coordinate-internal energy space with time. This approach allows the observation of stresses propagating in "hot" as well as "cold" material that cannot be seen in the usual stress-specific volume-internal energy space. The "hot" stresses in the energy deposition region show up mostly in the energy direction because of the shallow deposition with respect to the coordinate scale that is shown. Computations of pulse evolution for both experiments are shown in Figures 5 and 6. The first sub-figure in each case indicates the location of the gage interface and the melt and incipient vaporization energies.

Figure 5 shows pulse evolution for the low fluence shot (#1907). Note that there are two distinct regions of stress pulse development during energy deposition time—below and above vaporization. The first pulse develops below the vaporization energy, where unloading from the front surface during deposition produces a pulse that moves into cold material. The second pulse develops above the vaporization energy as the material vaporizes and rapidly expands, until a pressure builds up in equilibirum with the melt interface; then the second pulse is propagated into colder material after a time delay. At low pressures (just above vaporization energy) the second wave does not overtake the first wave. Although there is additional fine structure arising from stress relief, the basic two regimes develop separate consolidated pulses. We can also see the reflection of the main pulse from the gage interface and subsequent propagation and return from the front surface.

Figure 6 shows pulse evolution for the high fluence shot (#1729). In this case, the energy deposition is so high that the vapor pulse catches up with and dominates the low energy pulse. At high pressures, the second wave overtakes the first wave because it propagates at a higher speed in the cold material. Again, we can see the reflection of the main pulse from the interface and subsequent propagation and return from the front surface.

CONCLUSIONS

A distinct double stress pulse develops in a solid material when the peak deposited energy is only somewhat above the incipient vaporization energy, but not at much higher energies. Our computational model using the GRAY equation of state confirms this phenomenon that has been observed in Saturn argon x-ray experiments.

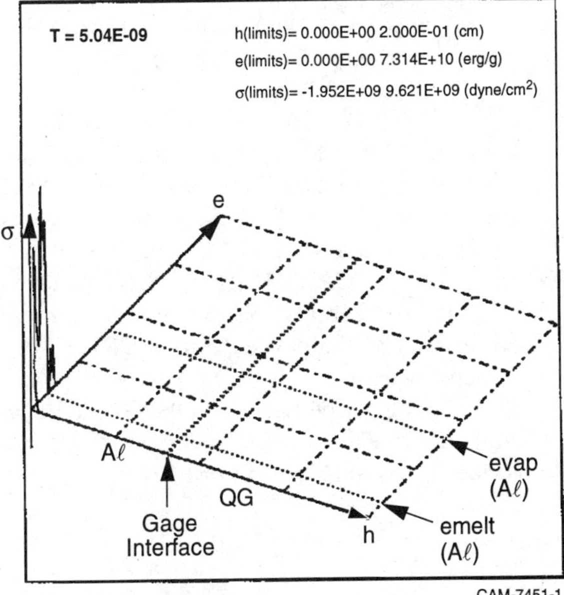

FIGURE 5a. First plot of pulse evolution for 3 cal/cm² Saturn argon Shot 1907-25, 26.

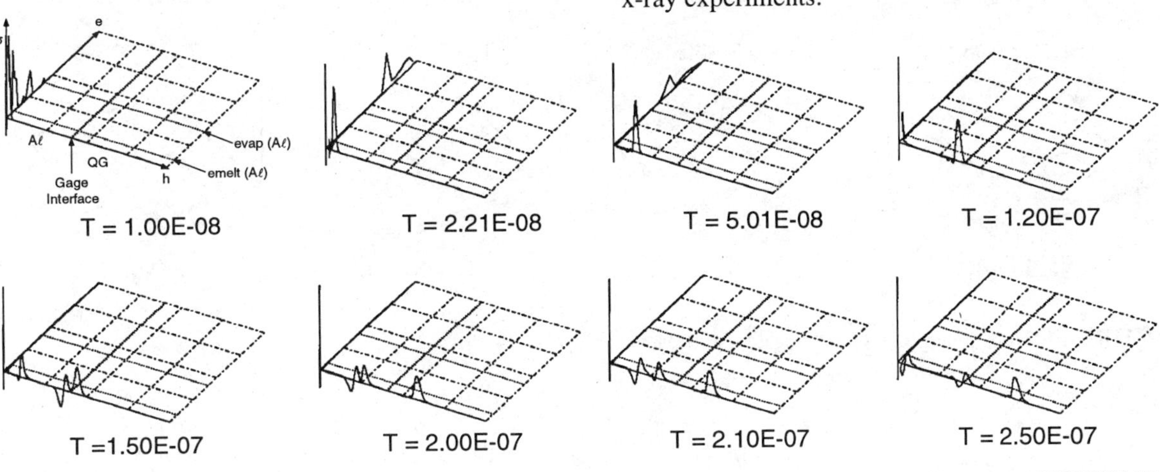

FIGURE 5b. Continuing plots of pulse evolution shown in Figure 5a.

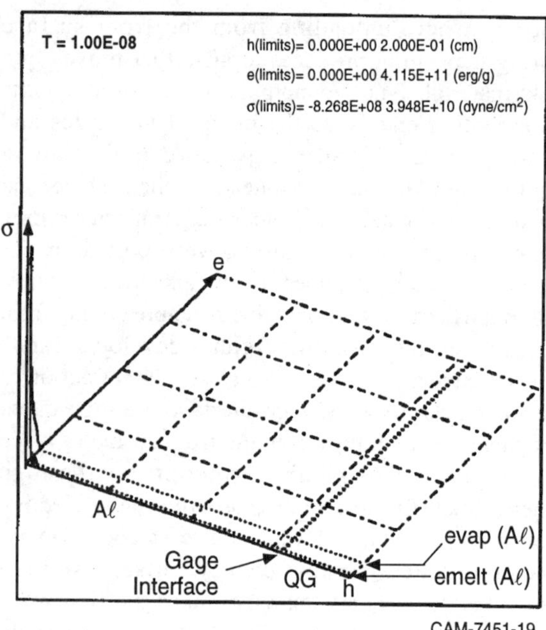

FIGURE 6a. First plot of pulse evolution for 25 cal/cm² Saturn argon Shot 1729-65.

ACKNOWLEDGMENTS

We acknowledge Defense Nuclear Agency's sponsorship of this work and are indebted to T. Kennedy for programmatic guidance. The Department of Defense review of this paper does not imply factual accuracy or opinion. We appreciate the close cooperation of the experimental staff at the Sandia National Laboratory Saturn facility in Albuquerque, NM. Thanks go to R. L. Donovan and A. W. Raskob of APTEK for their assistance in conducting the Saturn experiments, and also to B. Lew at SRI International for computational support.

REFERENCES

1. L. Seaman and D. R. Curran, "SRI PUFF 8 Computer Program for One-Dimensional Stress Wave Propagation," Vol. II, U.S. Army Ballistics Research Laboratory, MD (August 1978).

2. E. B. Royce, "GRAY, A Three-Phase Equation of State for Metals," Lawrence Livermore National Laboratory, Report UCRL-51121 (3 September 1971).

3. D. A. Young, "Modification of the GRAY Equation of State in the Liquid-Vapor Region," Report UCRL-51575 (15 April 1974).

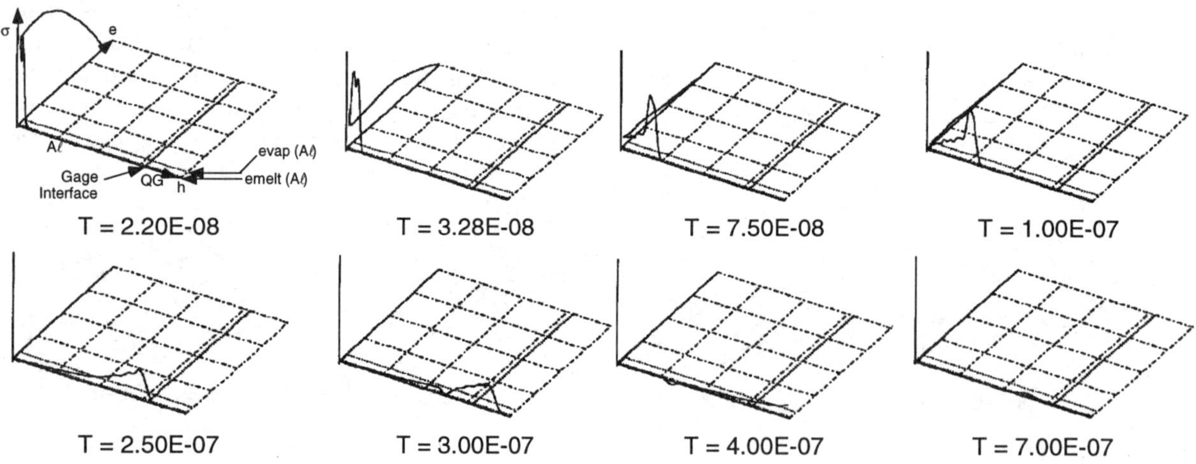

FIGURE 6b. Continuing plots of pulse evolution shown in Figure 6a.

THE DERIVATION OF MATERIAL PROPERTIES FROM MEASUREMENTS OF RADIATION INDUCED STRESS-TIME HISTORIES

F.W. Davies, D.L. Reeder, D.E. Johnson, L.M. Lee

Ktech Corporation, 901 Pennsylvania NE, Albuquerque, NM, 87110

Material response experiments using the SATURN soft x-ray source in conjunction with hydrocode modelling have been used to determine equation-of-state parameters for aluminum. In these experiments, stresswaves in irradiated targets have been measured using both quartz and PVDF piezoelectric stress gauges for a wide range of incident fluences. The induced stress profiles are sensitive functions of: (1) the deposition profiles in the target material, (2) the temporal history of the radiation source, and (3) the target material properties. Analysis of the experimental data has shown that observed changes in the shape of the induced stresswave as fluence is varied can be directly correlated to phase transitions. The unique combination of x-ray energy and irradiation time allow phenomena important for determining equation-of-state parameters to be studied separately. By temporally separating the development of thermomechanically induced stress from stress induced by material vaporization a very sensitive measurement of vaporization energy is possible.

INTRODUCTION

The measurement of stress time histories in materials irradiated with short duration intense pulses of x-rays and other penetrating radiation has been used as a method to determine material behavior and thermal properties for some time[1,2]. In current plasma radiating source (PRS) facilities the irradiation times and deposition profiles allow material thermal response phenomena to be studied separately in a single experiment. The effects of material vaporization are separated in time from stress due to thermomechanical response (Gruneisen effects). Studying these effects separately provides an opportunity to assess both parts of the equation of state for a material. In this case a very sensitive measure of the vaporization energy for the material is available.

EXPERIMENT DESCRIPTION

Results from experiments in the Saturn PRS facility were used with a modified version of the WONDY V computer program to study the vaporization energy for aluminum. Figure 1 shows the experimental configuration in which an aluminum 6061-T6 sample was bonded to a quartz piezoelectric stress gauge. The experiment was sized to generate uniaxial strain configuration for the duration of the measurements. Stress data from the quartz gauge[3] mounted on a 950-μm thick aluminum 6061-T6 sample which was exposed to ~5.6 cal/cm^2 of the output from an argon source are shown in Fig. 2. The initial peak results from a purely thermomechanical response (Gruneisen stress) prior to melt or vaporization of the material. The smaller second peak results from material vaporization.

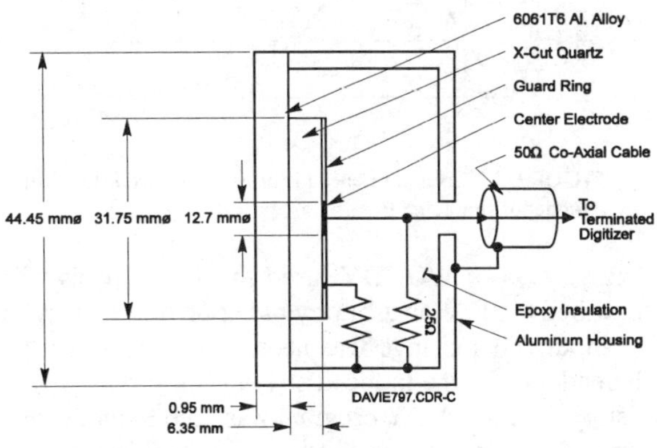

FIGURE 1. Shunted Quartz gauge configuration used to examine response of 6061-T6 aluminum.

RESULTS

In Fig. 3, the flux and fluence time histories for the Saturn x-rays are shown with the energy deposition profile in aluminum. The deposition depth has been divided by the speed of a relief wave in aluminum. The time for the heated depth to be stress relieved by rarefaction waves from the front surface is quite small relative to the irradiation time. The short time required to relieve the heated material makes the peak stress markedly dependent on the details of the flux history.

The early time stress development and stress wave propagation can be seen in Fig. 4. The legend of this figure gives the time in nanoseconds of each stress profile. The peak stress is developed by about 7.5 ns at which time only about 12 percent of the fluence has been delivered. The front surface begins to melt at ~ 5 ns and at 7.5 ns the melt front extends to ~ 7 µm. The melt front does not catch up with the stress wave moving into the material. At ~21 ns the front surface begins to vaporize and significant blow off induced stress is developed by 24 ns (Fig. 5).

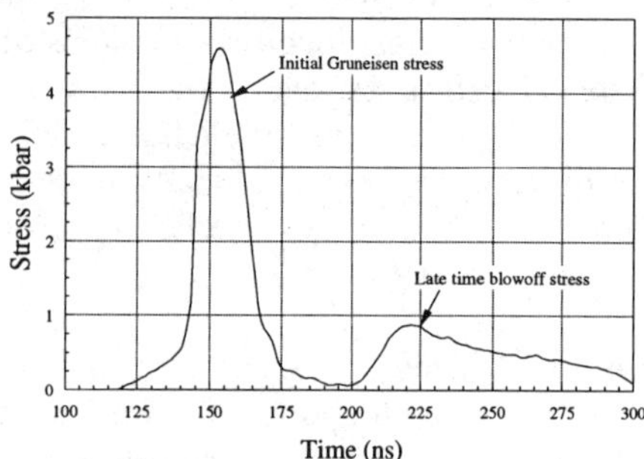

FIGURE 2. Stress history from Al6061 irradiated in Saturn.

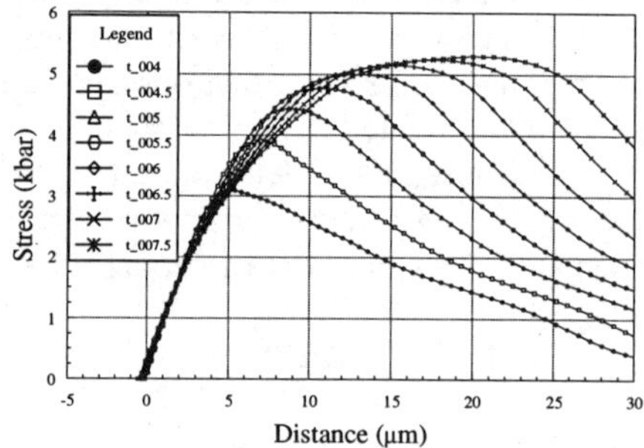

FIGURE 4. Early time development of stress near front surface.

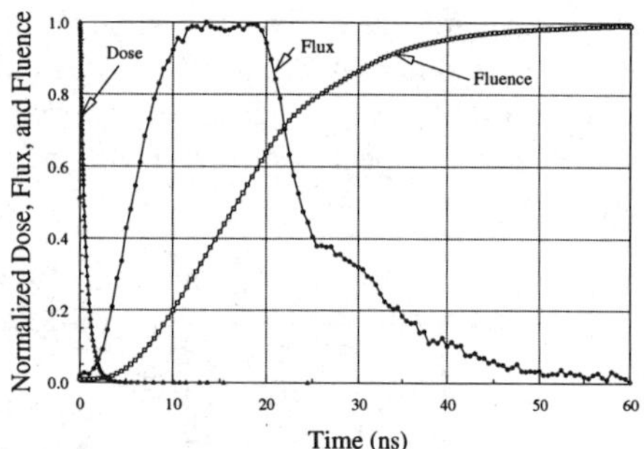

FIGURE 3. Flux and fluence histories for Saturn PRS source with deposition relief time.

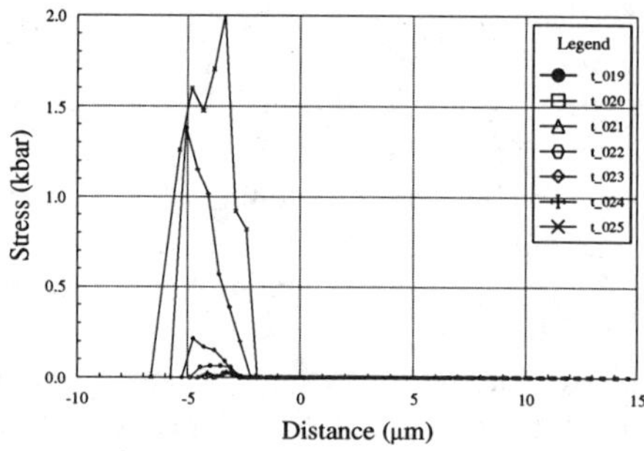

FIGURE 5. Development of stress at front surface due to vaporization.

The version of WONDY[4] used for this analysis was modified to allow an arbitrary radiation flux history to be input. This change was necessary because of the dependence of the induced stress profile on the flux history. The WONDY program was used to model this experiment to gain a better understanding of the phenomena interactions occurring and to derive effective materials property data. Measured flux histories were used as input to these calculations.

The vaporization and blowoff continue for the duration of the irradiation. The melt front is ~16 µm deep in the aluminum at the end of irradiation so the compressive blowoff stress must propagate across the molten material into the solid. In Fig. 6, the vaporization induced stress pulse is shown shortly after the irradiation ended. The timing and magnitude of this stress pulse can be modified in the WONDY analysis by varying the material vaporization energy.

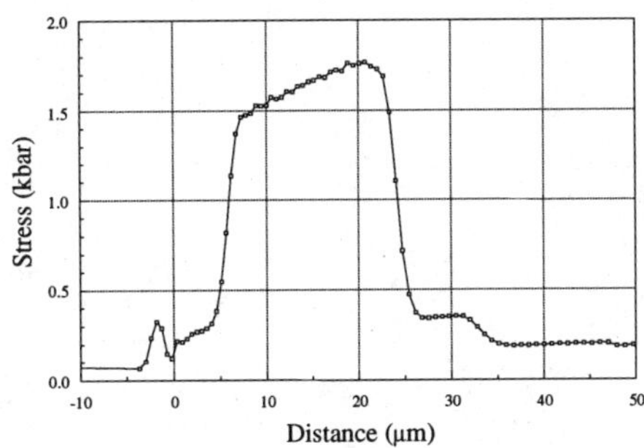

FIGURE 6. Stress wave from blowoff after irradiation terminates.

This sensitivity is shown in Fig. 7 where the legend defines the vaporization energy (cal/gm) used in the Mie Grüneisen equation of state description of the aluminum alloy. As expected the initial peak (Grüneisen stress) is unaffected by changes in vaporization energy. The second peak is very sensitive, particularly since the difference in vaporization energy for the curves shown in Fig. 7 is only 2.4 cal/g. Comparison of the calculations with the measured results allows the vaporization energy to be defined by matching the predicted and measured second stress pulse. A vaporization energy of 3067 cal/g for 6061-T6 aluminum is deduced by matching the amplitude of the predicted and measured stress. As shown in Fig. 8, the shape and timing of the second peak do not agree well with the data.

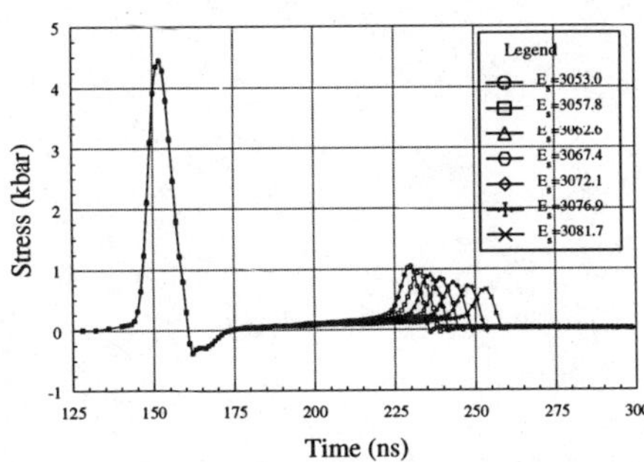

FIGURE 7. Sensitivity of calculated stress to material vaporization energy.

More detailed analysis of the calculated results indicate that deficiencies in the representation of liquid behavior in the WONDY V equation of state affected the results. The analyses need to be performed with a more realistic equation-of-state for liquids to accurately capture the stress pulse resulting from late time blowoff. We are currently working to make the ANEOS[5] multiphase equation of state package in WONDY V to support these analyses.

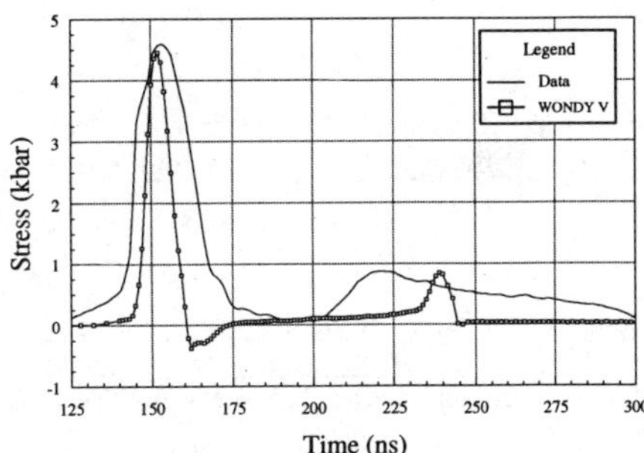

FIGURE 8. Data with WONDY V calculation showing both stress peaks.

CONCLUSIONS

In a limited fluence range the irradiation of aluminum by a PRS Argon source (3.1 keV) results in the propagation of two stress waves. The initial peak results from purely thermomechanical response prior to melt or vaporization of the material. The smaller second peak results from material vaporization. This second wave propagates through melted material.

The present analyses demonstrate that experiments using short duration soft x-rays can provide a sensitive measure of the vaporization energy, a critical equation of state parameter. Furthermore, the potential for examining the behavior of melted materials exists.

REFERENCES

1. Schmidt, R.M., Davies, F.W., Lempriere, B.M., Holsapple, K.A.. *J. Physics & Chem. of Solids* **39**,375-385.

2. Lempriere, B.M., and F.W. Davies. <u>Extendible Nozzle Exit Cone (ENEC) Hardening Technology</u>. Defense Nuclear Agency Report No. DNA 5103F (AD-B055640L): 1979.

3. Graham, R.A., Nielson, F.W. and Benedick, W.B., *J. Appl. Phys.* **36(5)**, 1775-1783 (1965).

4. Kipp, M.E., and R.J. Lawrence. <u>WONDY V - A One-Dimensional Finite Difference Wave Propagation Code.</u> Sandia National Laboratories Report No. SAND81-0930. 1982.

5. Thompson, S.L. and H.S. Lauson. <u>Improvements in the Chart D Radiation - Hydrodynamic Code III: Revised Equations of State.</u> Sandia National Laboratories Report No. SC-RR-71-0714. 1972.

DAMAGE AND FAILURE MECHANISMS ASSOCIATED WITH PHOTOABLATION OF BIOLOGICAL TISSUES

Tarabay Antoun, Lynn Seaman, Donald Curran

Poulter Laboratory, SRI International, 333 Ravenswood Avenue, Menlo Park, CA 94025

Michael Glinsky

Lawrence Livermore National Laboratory, 7000 East Avenue, P.O. Box 808, Livermore, CA 94551

This paper aims to examine the processes associated with failure of the cornea and other collagenous tissues during photoablation. Two different constitutive models are applied to simulate a series of laser deposition experiments into porcine reticular dermis (1), a biological tissue similar to the cornea in composition and photoablation characteristics. The first of our constitutive models, DFRACT, is a physically motivated, micromechanical model based on the nucleation and growth of spherical voids (2). The second is a relatively simple model that allows the material to vaporize and thermally soften. The simulation results reproduce the prominent features observed experimentally thereby shedding a new light on the operative mechanisms during photoablation. The good qualitative agreement between the simulated stress histories and the stress histories measured during the experiments also demonstrates the effectiveness of micromechanical damage and failure modeling as a viable tool for optimizing existing laser surgery procedures and designing new ones.

INTRODUCTION

Despite the widespread use of lasers in medical applications, the physical phenomena associated with laser-tissue interactions are still not very well understood. This is partially due to the complex structure of biological tissue and to the influence this structure has on the absorption characteristics of the tissue. In this paper, we attempt to shed new light on the damage and failure mechanisms associated with photoablation of biological tissues. This is accomplished by simulating the response of porcine reticular dermis to the deposition of laser energy and comparing the simulation results with the experimental data of Venugopalan (1).

EXPERIMENTAL STUDY

The experimental configuration of the laser deposition experiments performed by Venugopalan on biological tissue is shown schematically in Fig. 1. This configuration consisted of a porcine dermis layer approximately 400-μm-thick, and a thick, impedance-matched layer of saline solution. A thin

PVDF film sandwiched between the two layers was used to record the stress history at the back surface of the tissue sample. The laser beam diameter was 1 mm or larger, thus ensuring 1-D strain conditions throughout the experiment.

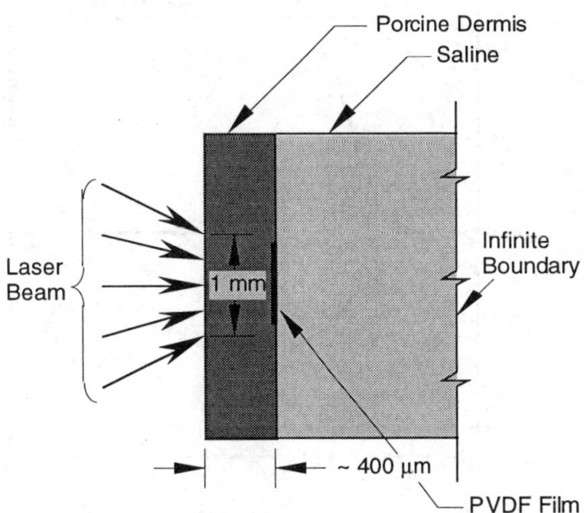

FIGURE 1. Experimental configuration of the laser deposition experiments.

Porcine dermis, as well as other collagenous tissues, is composed of two primary energy-absorbing components: water, which accounts for about 70% of the total mass, and collagen fibers, which form the reticular extracellular matrix (ECM) and account for the majority of the remaining 30% of the total mass. Two lasers, carefully chosen to selectively target either the water or the ECM of the tissue, were used in the experiments simulated in the present study. The first is a 248-nm Kr-F excimer laser which selectively targets the extracellular matrix of the tissue. This laser has an absorption depth of 30 μm and a pulse duration of 24 ns. The second is a 10.6-μm TEA CO_2 laser which targets the tissue water. It has an absorption depth of 20 μm and a pulse duration of 30 ns.

Stress histories at several different levels of radiant exposure are shown in Fig. 2 and Fig. 3 for tissue samples irradiated with the Kr-F excimer laser and the TEA CO_2 laser, respectively. With increasing fluence, both figures appear to indicate a gradual loss of strength as evident by the progressive deterioration of the ability of the material to propagate the tensile component of the bipolar stress pulse. This is an indication that failure has occurred in the interior of the sample. The transitions from fully elastic behavior to inelastic behavior then failure in Fig. 2 and Fig. 3 proceed in two different manners as evidenced by the different characteristics of the intermediate stress pulses in the two figures. In the remainder of this paper, we describe two constitutive models and associated simulation results that will help explain the trends in the experimental data in terms of underlying damage and failure mechanisms.

CONSTITUTIVE MODELS

Two constitutive models were used to represent porcine dermis in our hydrocode simulations. For the case where the extracellular matrix (ECM) is the primary chromophore of the laser radiation, we modeled the material using DFRACT - a physically motivated, micromechanical model based on the nucleation and growth of spherical voids and the manner in which they affect the strength and subsequent deformation of the material (2). The damage evolution , in terms of the number of voids

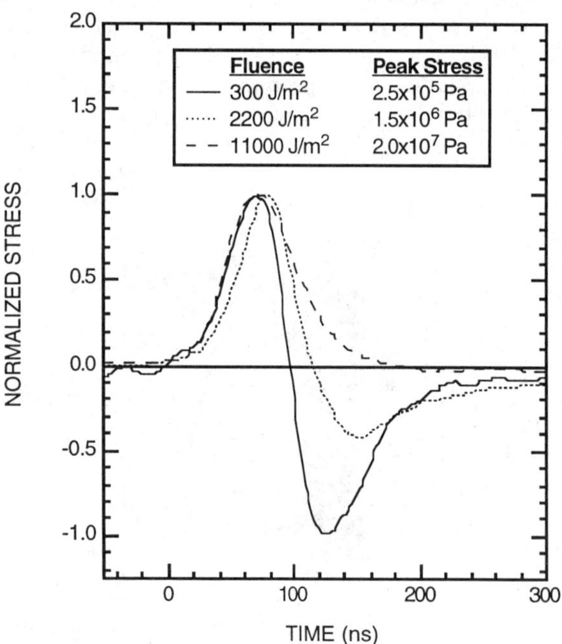

FIGURE 2. Stress histories recorded with a PVDF transducer 400 μm away from the surface of porcine dermis samples irradiated using a 248-nm Kr-F excimer laser (1).

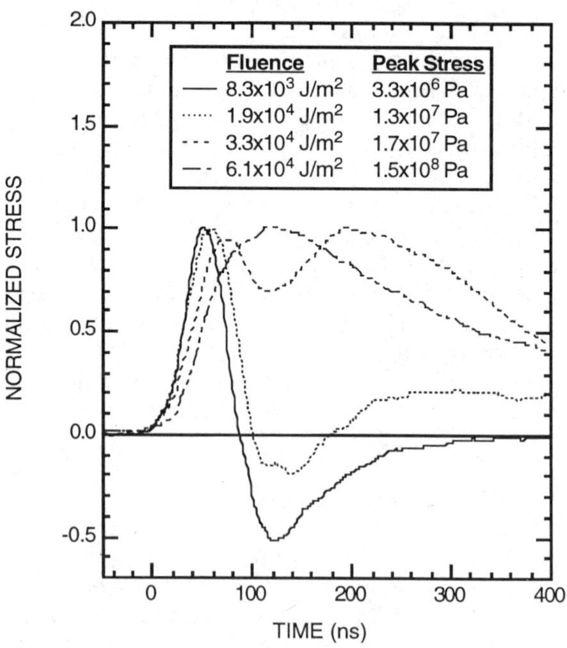

FIGURE 3. Stress histories recorded with a PVDF transducer 400 μm away from the surface of porcine dermis samples irradiated using a 10.6-μm TEA CO_2 laser (1).

per unit volume and the void size, is monitored in the model with two rate equations that depend on the yield strength, the number of nucleation sites, the initial flaw size, and the viscosity of the material. These model parameters were chosen to be representative of the properties of collagen fibers.

For the case where tissue water is the primary chromophore of the laser energy, the material was represented by a simple model that allows the biological tissue to vaporize and thermally soften. In the model, vaporization occurs when the internal energy exceeds the sublimation energy, the value of which was chosen to reproduce the trends in the data. The thermal softening model simply reduces the yield strength of the tissue from its initial value at room temperature to zero when the internal energy becomes equal to the melt energy (assumed to be equal to the sublimation energy).

The pressure-volume response of the tissue was modeled using the equation of state of water, and the energy deposition profile was described using the Beer-Lambert equation. The simulations were performed using the one-dimensional finite difference hydrocode SRI PUFF (3).

HYDROCODE SIMULATION RESULTS

The stress histories simulated using the DFRACT model and the thermal softening and vaporization model are shown in Fig. 4 and Fig. 5, respectively. The radiant exposures (i.e., fluences) used in the simulations are the same as those used in the experimental studies the results of which are shown in Fig. 2 and Fig. 3. A comparison of the simulation results in Fig. 4 with the experimental data in Fig. 2 shows a good qualitative agreement, as does a similar comparison of Fig. 5 and Fig. 3.

In Fig. 4, the transition from elastic behavior at a fluence of 300 J/m^2, to complete fracture at a fluence of 11000 J/m^2 is due to nucleation and growth of spherical voids. This is further illustrated in Fig. 6 where the void volume fraction at the end of the simulation for each of the three fluence levels is plotted (on a logarithmic scale) as a function of Lagrangian distance. As can be seen in this figure, the void volume fraction associated with the lowest fluence is on the order of 10^{-8} indicating an essentially elastic behavior. On the other hand, the void volume fraction associated with the highest fluence level is approximately 0.25

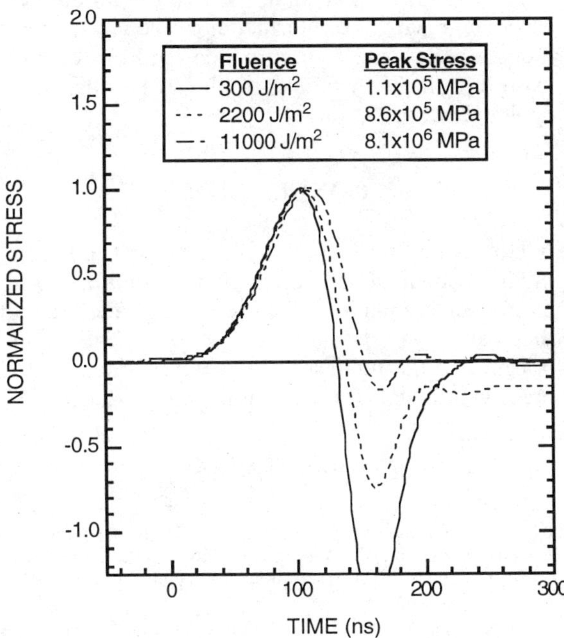

FIGURE 4. Simulated stress histories 400 μm away from the surface of porcine dermis samples irradiated using a 248-nm Kr-F excimer laser . Results obtained by allowing the material to fail by void nucleation and growth under tension.

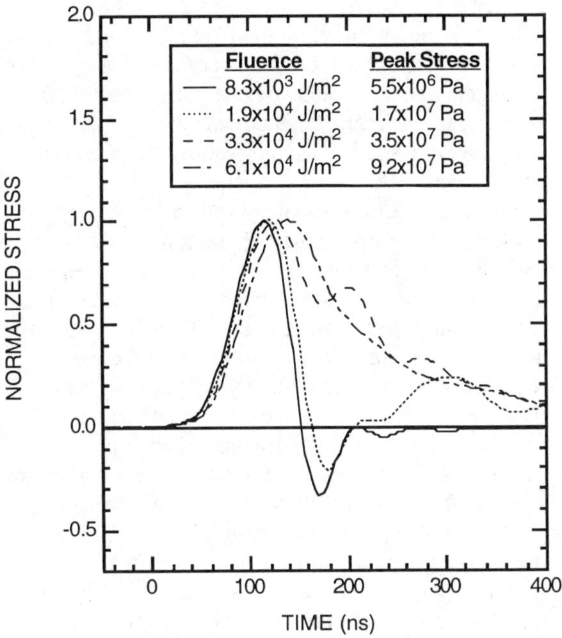

FIGURE 5. Simulated stress histories 400 μm away from the surface of porcine dermis samples irradiated using a 10.6-μm TEA CO_2 laser. Results obtained by allowing the material to vaporize and/or thermally soften.

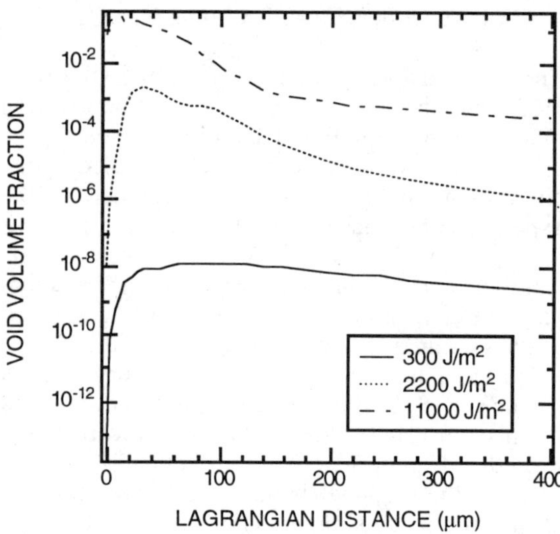

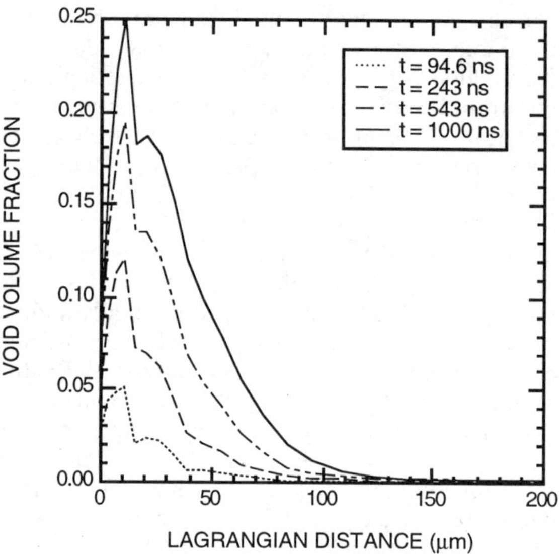

FIGURE 6. Simulated void volume fraction as a function of distance for the porcine dermis samples irradiated using a 248-nm Kr-F excimer laser .

FIGURE 7. Simulated damage (void volume fraction) evolution in the porcine dermis sample irradiated using a 248-nm Kr-F excimer laser at a fluence of 1.1×10^4 J/m^2.

indicating that voids occupy about 25% of the volume. At this fluence level, Fig. 4 shows that the ability of the material to support tensile loads is greatly diminished.

Fig. 7 shows snapshots of damage distribution across the first half of the specimen. As shown, damage is concentrated in a region about 100 μm wide near the front surface. The material at and near the gage section remains elastic throughout the simulation.

The wave structures observed in Fig. 5 can be explained in terms of thermal softening and front surface vaporization. At the lowest fluence, the tissue does not vaporize, but the energy level is close enough to the melt energy that the material near the front surface thermally softens causing a reduction in the peak tensile stress magnitude. As the fluence increases, a layer of tissue vaporizes near the front surface of the specimen. The usually-bipolar thermoelastic stress pulse is perturbed by two additional stress disturbances: one originating at the vapor-liquid boundary and the other at the boundary between the layer of material within the laser deposition depth and the rest of the specimen. The latter of these disturbances dominates the response at a fluence of 1.9×10^4 J/m^2 and the former dominates the response at a fluence of 3.3×10^4 J/m^2 leading to the "double-pulse" structure in the wave profile.

As the fluence is increased to 6.1×10^4 J/m^2, the whole layer of material within the deposition depth vaporizes. As such, the two disturbances associated with the intermediate fluence levels merge into a single disturbance leading to the "single-pulse" structure in the wave profile depicted in Fig. 5.

CONCLUSION

The study described here demonstrates that micromechanical damage modeling combined with hydrocode simulations is a useful and effective tool that can be used to interpret experimental data, optimize existing laser surgery procedures, and provide guidance for designing new procedures.

REFERENCES

1. Venugopalan, V., "The Thermodynamic Response of Polymers and Biological Tissues to Pulsed Laser Irradiation," PhD Dissertation, Massachusetts Institute of Technology, May 1994.

2. Seaman, L., Curran, D. R., Aidun, J. B., and T. Cooper, "A Microstatistical Model for Ductile Fracture with Rate Effects," *Nuclear Eng. and Design* **105**, 35-42 (1987).

3. Seaman, L. and Curran, D. R, "SRI PUFF 8 Computer Program for One-Dimensional Stress Wave Propagation," SRI International Final Report (vol. II) prepared for the Army Ballistics Research Laboratory, August, 1978.

SHOCK PHYSICS WITH THE NOVA LASER FOR ICF APPLICATIONS

B.A. Hammel, R. Cauble, P. Celliers, L.B. DaSilva, T.R. Dittrich, D. Griswold, S.W. Haan, N. Holmes, O.L. Landen, T. Orzechowski, and T.S. Perry

Lawrence Livermore National Laboratory , Livermore, CA 94550

Accurate predictions of NIF target performance depend critically on shock timing and material compressibility. Current Nova experiments are designed to test our understanding of shock physics in actual target materials. Shock breakout measurements test shock timing through samples of ablator material. Experiments to measure the compressibility of D_2 ice are planned to further increase the margin for ignition in NIF target designs.

The physics of high pressure shocks plays a central role in Inertial Confinement Fusion (ICF). In indirect drive ICF, x-rays from a gold cavity (hohlraum) are used to ablatively drive a series of high pressure shocks into a spherical target (capsule). The shell converges toward the center, compressing the fuel and forming a hot dense core. The target performance, such as peak fuel density and temperature and neutron yield, depends critically on shock timing and material compressibility.

A typical target for the proposed National Ignition Facility (NIF) is composed of an outer ablator and an inner fuel region (Fig. 1) (1).

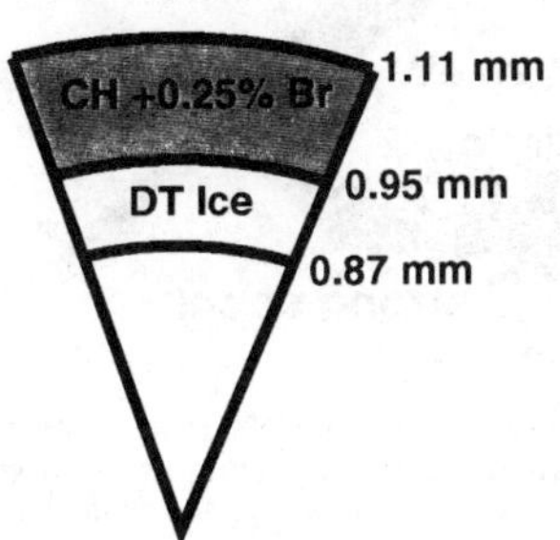

FIGURE 1. Pie section of typical NIF capsule. In this case the outer ablator section is made of CH doped with Br. Other designs have an ablator made of Cu-doped Be or B_4C. The inner fuel region is frozen DT.

The ablator is typically CH, Be or B_4C. It is doped with a high-Z material (Ge, Br, Cu) in order to minimize x-ray preheat of the DT fuel. The fuel region is a frozen layer of DT on the inner wall of the ablator. The NIF x-ray drive, which consists of a two step "foot" (~80 eV and ~120 eV) followed by a ramp up to a peak (~300 eV), drives multiple shocks into the ablator with pressures in the range from ~1-300 Mbar (Fig. 2).

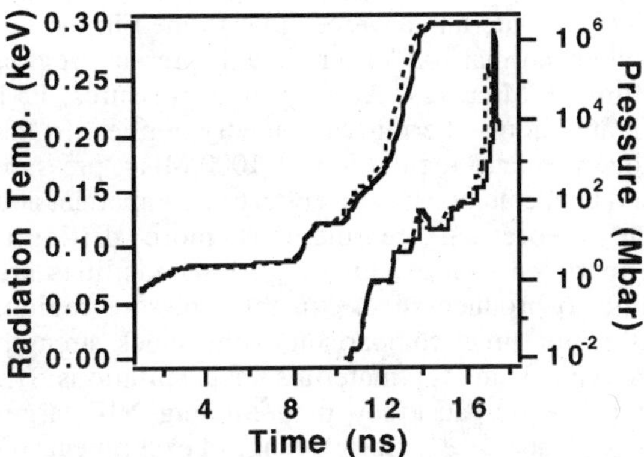

FIGURE 2. Optimal radiation drive and fuel pressure vs. time for a particular NIF capsule design (solid). Also shown is the same drive shifted earlier by 0.5 ns during the rise following the "foot" (dashed).

Proper timing of the shocks, to better than 10%, is crucial to minimize the fuel entropy and maximize the capsule neutron yield. The two simulations shown in Fig. 2 illustrate this sensitivity: advancing the timing of the "ramp" section of the drive 0.5 ns from the optimal design advances the third shock 0.5 ns and increases the fuel isentrope by roughly a factor of 2 during the compression phase of the implosion (Fig. 3) which ultimately results in a factor of ~50 reduction in neutron output. Clearly, an accurate knowledge of the EOS and opacity of these materials is an essential part of designing and modeling NIF experiments.

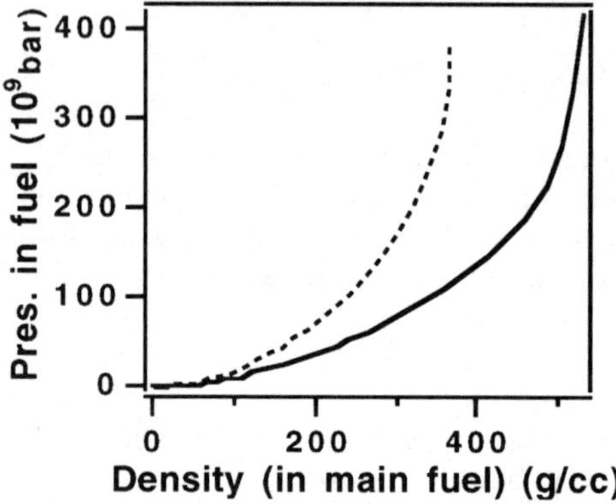

FIGURE 3. Resulting fuel isentrope for the optimal and shifted drive during compression. The target with optimally timed drive reaches higher density at a given pressure.

Extensive experimental measurements of the EOS of materials have been performed in the so-called "normal" or "experimental" pressure region below 1 Mbar (2). At very high pressures, P>1 Gbar, Thomas-Fermi-Dirac theory applies. In the "intermediate" region, from 1-1000 Mbar, pressure and temperature ionization effects are important and the theoretical treatment is more difficult. Experiments on existing large laser facilities are able to produce shocks in this pressure region, allowing direct comparison of the shock strength in typical ablator materials to calculations. To improve our accuracy in predicting NIF target performance, we perform a range of experiments on the Nova laser to test our understanding of the important physical processes. Specifically, we study; 1) shock strength in typical ablator materials, and 2) hydrodynamic instabilities in ablatively accelerated targets (3,4). In this proceedings article we summarize two techniques for determining properties of x-ray driven shock in typical target materials.

Indirect drive experiments on Nova are performed by heating a cylindrical gold hohlraum (typically ~3 mm long and 1.6 mm in diam.) with a total of ~ 30 kJ of 351 nm light (Fig. 4). The ten Nova beams are directed into the hohlraum though holes in the ends, typically 1.2 mm diameter. The laser energy is absorbed in the high-Z wall and reradiated as x-rays, producing a

uniform and nearly Plankian source with radiation temperature, T_R, ~225 eV. Radiation temperatures up to ~300 eV have been achieved using smaller scale hohlraums (5). A target exposed to this x-ray drive inside the hohlraum is strongly shocked by the pressure generated from the ablating target material. The shock pressure can be roughly estimated as (6):

$$p \approx \frac{1}{2} \frac{\sigma T_R^{\,4}}{\sqrt{RT_R/\mu}}, \qquad (1)$$

where T_R is the radiation temperature, σ and R are the Stefan-Boltzmann and gas constants respectively, and μ is the mean molecular weight per free particle,

$$\mu = A/(Z+1). \qquad (2)$$

For fully ionized CH ($\mu = 13/9$),

$$p(Mbar) \approx 2 \times 10^4 T_R(keV)^{3.5}. \qquad (3)$$

Typical radiation temperatures on Nova hohlraums therefore produce shocks with peak pressures of ~100 Mbar in plastic targets.

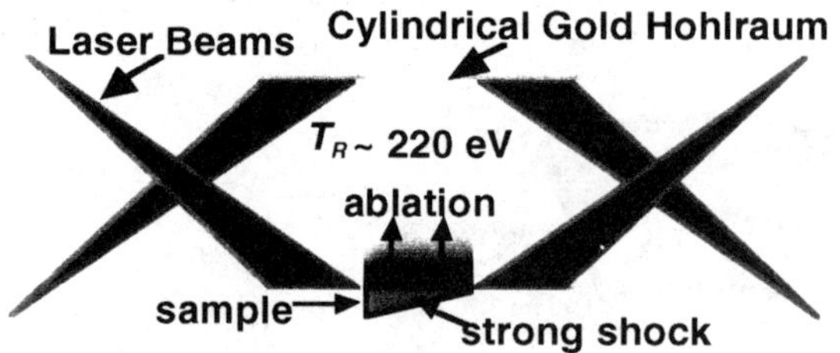

FIGURE 4. Schematic of hohlraum target. X-rays from the laser heated cavity ablate material from a sample or target exposed to the drive, driving a strong shock into the material.

In our experiments we directly measure the x-ray driven shock strength in typical ablator materials by observing the shock transit time through a slab of the material that is mounted over a hole in the side of the hohlraum (Fig. 4). The slab is made with "steps" of known thickness or as a wedge of increasing thickness, providing a range of the shock transit times for different times during the duration of the x-ray drive. The breakout of a shock on the rear side of the ablator material results in a flash of ultraviolet light from the shock heating of the material on the rear surface to ~10 eV. The UV light is detected by a Cassegrain telescope that images the emission onto the slit of the streak camera. Typical data from shock breakout through a wedge shaped sample is shown in Fig. 5. The current accuracy of the shock velocity measurement by this technique is ~ 2-3%.

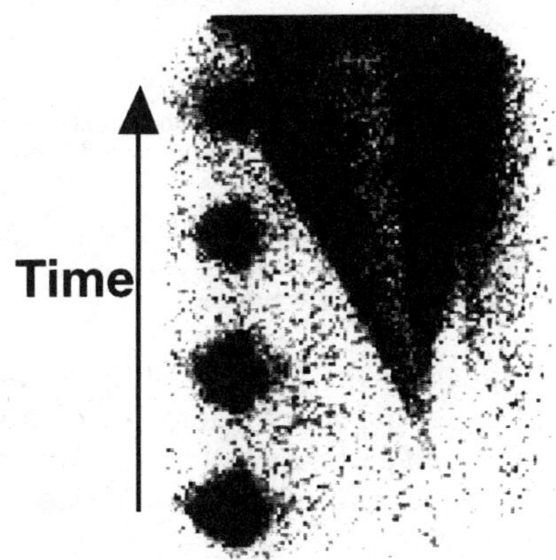

FIGURE 5. Typical shock breakout data from wedge shaped target showing emission from progressively thicker regions of the wedge shaped target at later times.

The measured and calculated shock breakout through a wedged sample of CH doped with 3% Br is shown in Fig. 6. This material is currently used on Nova targets as an ablator for NIF-like implosion experiments. In this measurement, the hohlraum was driven with a time varying laser pulse, resulting in the radiation temperature shown in Fig. 7.

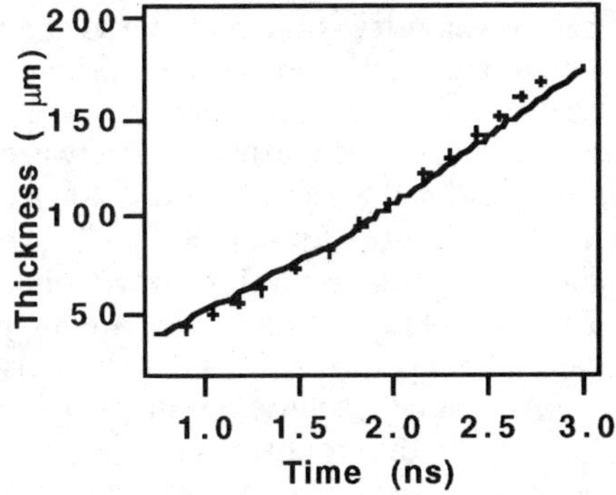

FIGURE 6. Measured (+) and calculated (line) shock breakout time for a Br doped CH wedged sample driven with the time dependent radiation temperature shown in Fig. 6.

Although this technique provides a direct measurement of shock timing for a given laser

drive, which is ultimately what is required for tuning a NIF capsule to reach ignition, it does not directly measure the EOS of the material because of uncertainties in the ablator opacity and consequently the ablation pressure. In some cases, such as predicting the compression of the DT ice during the "foot" of the NIF pulse, it is desirable to measure just the EOS. Recent calculations for D_2 ice, that include molecular dissociation, indicate a softer EOS than earlier calculations for pressures in the 1-10 Mbar regime (7). This softer EOS predicts a higher fuel density during the "foot" of the NIF drive and results in a significant increase in ignition margin in current designs (8). The EOS of ablator materials is also important in determining the evolution of hydrodynamic instability growth (Richtmyer-Meshkov and Rayleigh-Taylor), of initial ablator and DT ice surface imperfections (9).

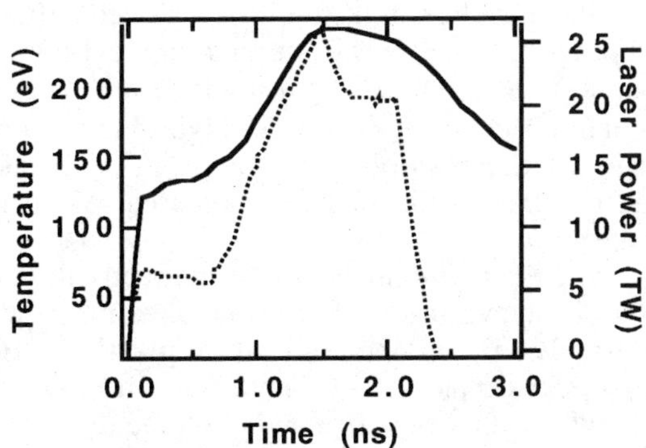

FIGURE 7. Laser power (dotted) and radiation temperature (solid) for shock break out measurements.

The EOS can be isolated by conducting experiments where two materials are driven side-by-side, a "test sample" of interest and a "reference standard" such as Al. These experiments would use a common ablator material to generate an equivalent shock into both materials. This so-called "impedance matching" technique relies on the fact that the shock pressure P and the particle velocity, u_p, are conserved when the shock crosses from one material into another. By applying an equivalent steady pressure to both materials, a point on the Hugoniot of the "test sample" can be inferred from the shock velocity, u_s, through the "sample" and the "standard", if the initial conditions are known. Accurate measurements, however, place very strict requirements on shock planarity and steadiness, and x-ray preheat. In addition to standard ablator materials, these experiments could also be performed with frozen D_2 as the "sample".

An alternate technique that may be useful for determining the EOS of target materials is x-ray radiography. In this approach, a shocked sample is backlit with x-rays, produced by focusing one or two beams of the laser, delayed relative to the drive beams, onto a backlighter disk. The backlighter x-rays transmitted through the sample are viewed normal to the direction of shock propagation. Shock compression of the sample increases the areal density, and consequently the optical depth, through the package in the viewing direction, resulting in a decrease in the transmitted backlighter x-ray intensity. A series of radiographic images yield the position of the shock front in the material as a function of time and hence the shock velocity. For probing solid density ablator materials, with typical sample dimensions of ~1 mm, the backlighter is typically a metal disk, resulting in K-shell x-ray emission in the 4-7 keV spectral region. For a ~1 mm solid D_2 sample, the optimum backlighter energy for high transmission through the sample and a high contrast measurement of the shock compressed region is ~1 keV.

A radiographic image of an indirectly driven target is shown in Fig. 8. In this experiment, we ablatively drove one end of a plastic (CH) cylindrical sample (0.5 - 0.7-mm diam and 1.5-mm in length) with x-ray drive and observed the shock propagate into the material. These samples were mounted from the wall of the hohlraum as shown in Fig. 2 and were viewed from a direction perpendicular to their length, and therefore transverse to the axial direction of shock propagation. The radiographic images were recorded with an 8× magnification pinhole camera coupled to a gated microchannel plate detector, providing 16 time-resolved "snapshot" x-ray radiographs of the sample (10). The position of the shock front is clearly visible. In this experiment a thin (150 μm diam.) Al wire was located on the axis of the cylindrical plastic sample for preliminary experiments on shear flow (11).

A quantitative measurement of the transmission through the shock compressed region, along with knowledge of the opacity of the material, yields the compressed density (11). The particle velocity can also be measured by doping a region of the material with a small amount of high-Z material so that it is radiographically

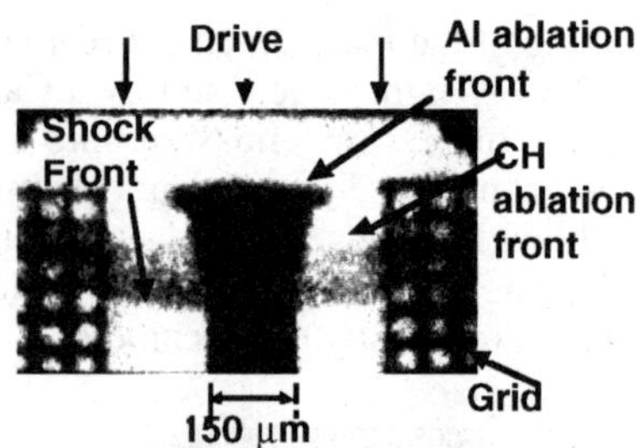

FIGURE 8. An x-ray radiographic "snap-shot" image of a shock propagating into a cylindrical plastic sample. The shock is driven by x-ray ablation from above. This sample has an aluminum wire embedded along its axis for a series of experiments aimed at studying shear flow. The compression of the plastic is seen as a region of decreased x-ray transmission (darker) behind the shock front. The shock pressure was ~12 Mbar at the time that this image was recorded.

opaque, or by imaging the position of an opaque pusher. For the case of a steady shock, knowledge of both the shock and particle velocity provides an independent measurement of the compression of the shocked material.

To test existing EOS models for D_2 we are planning a series of experiments to measure the shock speed and compressed density of D_2 ice in the 1-10 Mbar regime. Since the predicted difference in shock speed for current models differs by at most 1-2% we are investigating the possibility of extending optical interferometric techniques used extensively in the gas gun community to laser driven shocks. We have considered both standard Michelson interferometers (12) and VISAR (velocity interferometer system for any reflector) (13) techniques. Along with being able to measure shock speed to the necessary accuracy to test our models, these techniques allow us to accurately quantify shock planarity and steadiness which have historically been a primary concern in laser shock wave experiments.

Although the shock speed differences between current EOS models for D_2 are small, the predicted compression differs by as much as 20%. To directly measure the compression were must optimize our radiographic techniques. The accuracy of radiography measurements is primarily determined by the spatial resolution of the imaging diagnostic. In the radiographic measurements described above, hard x-ray imaging was necessary to penetrate the target and consequently pinhole optics with spatial resolutions of ~10 μm were used. The optimal backlighting energy for D_2 is ~1 keV, which allows us to use multi-element imaging systems that have the potential for 1-

3 μm resolution. An aberration corrected Kirkpatrick-Baez microscope with 3 μm resolution has already been demonstrated by a group at OSAKA (14). Another option is to design a multilayer coated aspherical mirror system that can operate at higher angles. Since we can tolerate some astigmatism it should be possible to achieve the necessary spatial resolution with a single mirror system. We estimate that density measurements with accuracies of better than 5% should be possible with these radiographic techniques.

SUMMARY

Accurate predictions of NIF target performance depends critically on shock timing and material compressibility. Current measurement techniques enable us to accurately determine shock timing in planar samples of ablator material as a function of laser drive. Although this technique does not separately address uncertainties in material EOS and opacity, it does allow us to tune the laser drive until the desired shock timing is achieved. Experiments to directly address the EOS of D_2 ice are planned to further increase the margin for ignition in current target designs.

REFERENCES

1. Steven W. Haan, Stephen M. Pollaine, John D. Lindl, Laurance J. Suter, Richard L. Berger, Linda V. Powers, W. Edward Alley, Peter A. Amendt, John A. Futterman, W. Kirk Levedahl, Mordecai D. Rosen, Dana P. Rowley, Richard A. Sacks, Aleksei I. Shestakov, George L. Strobel, Max Tabak, Steven V. Weber, George B. Zimmerman, William J. Krauser, Douglas C. Wilson, Stephen V. Coggeshall, David B. Harris, Nelson M. Hoffman and Bernhard H. Wilde, Phys. Plasmas **2**, 2480 (1995).

2. B. K. Godwal, S. K. Sikka and R. Chidambarah, Phys. Reports **102**, 121 (1983) and references therein.

3. J.D. Kilkenny, Phys. Fluids B 2, 1400 (1990).

4. B. A. Remington, S. W. Haan, S. G. Glendinning, J. D. Kilkenny, D. H. Munro, and R. J. Wallace, Phys. Rev. Lett. **67**, 3259 (1991).

5. R. L. Kauffman, L. J. Suter, C. B. Darrow, J. D. Kilkenny, H. N. Kornblum, D. S. Montgomery, D. W. Phillion, M. D. Rosen, R. Thiessen, R. J. Wallace, and F. Ze, Phys. Rev. Lett. **73**, 2320 (1994).

6. See National Technical Information Service document no DE 92006882, *Ablation Gas Dynamics of Low-Z Materials Illuminated by Soft X-rays*, S. P. Hatchett (1991). Copies may be ordered from the National Technical Information Service, Springfield, Virginia 22161.

7. N. Holmes, these proceedings

8. S.W. Haan, private communication

9. S.V. Weber, B.A. Remington, S.W. Haan, B.G. Wilson, and J.K. Nash, Phys. Plasmas **1**, 3652 (1994).

10. See for example, D. K. Bradley, P. M. Bell, J. D. Kilkenny, R. Hanks, O. L. Landen, P. A. Jaanimagi, P. W. McKenty and C. P. Verdon, Rev. Sci. Instrum. **63**, 4813 (1992).

11. B. A. Hammel, D. Griswold, O. L. Landen, T. S. Perry, B. A. Remington, P. L. Miller, T. A. Peyser, J. D. Kilkenny, Physics of Fluids B (Plasma Physics), **5**, 2259-64, (1993). B. A. Hammel, J. D. Kilkenny, D. Munro, B. A. Remington, H. N. Kornblum, T. S. Perry, D. W. Phillion, R. J. Wallace, Physics of Plasmas, **1**, 1662, (1994)

12. L. M. Barker and R. E. Hollenbach, *Rev. Sci. Instrum.* , **36**, 1617-1620 (1965).

13. L. M. Barker and R. E. Hollenbach, *J. Appl. Phys.* , **43**, 4669-4675 (1972).

14. R. Kodama, N. Ikeda, Y. Kato, Y, Katori, T. Iwai, K. Takeshi , SPIE Proc., Vol. 2525, 165-173, (1995).

AUTHOR INDEX

interface
- friction, sliding 307
- thermal relaxation 279

interferometry
- astigmatism 1007
- Doppler laser 955, 1011, 1015, 1213,
    1243, 1247
- Fabry Perot 61, 133, 1213
- line imaging 1015
- Michelson 593
- ORVIS 1015, 1221, 1243, 1247
- two channel 623
- variable sensitivity displacement 1011
- VISAR 9, 93, 97, 201, 235, 311, 401,
    495, 503, 507, 523, 527, 535, 555,
    619, 627, 631, 647, 815, 863, 879,
    947, 1003, 1007, 1015, 1019, 1101,
    1109, 1221, 1265
- white light Doppler 1003

ion beam
- driven flyer 1015, 1243, 1247
- driven shock 1033

iron
- asteroid 929
- cratering 1159
- defects 175
- deformation 303
- dislocation 619
- equation of state 51, 963
- fracture 175
- granular 971
- hardness 697
- Hugoniot 117, 303, 721, 929, 963
- hypervelocity impact 1159
- magnetization 929
- microstructure 619, 697
- molecular dynamics 175
- optical depth 917
- optical pyrometry 279, 917
- oxide 3, 721
- oxide/aluminum powder 681
- oxide/zinc oxide powder 721
- phase transition 51, 929
- plasma 1255
- porous 117, 697, 963, 971
- powder 697
- projectile 1159
- shear stress 515
- shock 971
- strength 619
- temperature 279, 917
- twining 619
- x-ray driven shock 1255

isentropic compression, ruby 897

jet
- fragment 1233, 1237
- initiation 867
- microstructure 1237
- shaped charge 1229, 1233, 1237
- tensile strength 1233

jetting, hypervelocity impact 1229

kimberlite
- constitutive model 335
- triaxial test 335

- yield behavior 335

LAX112 explosive
- detonation distance 803
- Hugoniot 803
- wedge test 803

LX-17 explosive
- cylinder test 449
- detonability 377
- sensitivity 445
- thermicity 377

LX-19 explosive
- initiation 891
- reaction 891
- wedge test 891

Lagrange
- analysis 357, 1057, 1065
- code 133, 295, 299, 331, 409, 563,
    749, 775, 799, 993, 1037, 1127,
    1259

laser driven flyers
- alumina 1221, 1225
- aluminum 21, 921, 1209, 1213, 1217,
    1221, 1225
- aluminum /alumina 921, 1209, 1225
- aluminum/polyethylene 1225
- aluminum/quartz 1225
- copper 495, 1213, 1217, 1221
- magnesium 1221
- magnesium fluoride 1221
- multiple layer 1221
- quartz 1213, 1217, 1225
- titanium 1225
- zinc sulfide 1221

laser driven shocks
- aluminum 943, 1097, 1259, 1265
- copper 77, 943, 1097, 1265
- deuterium/ 459, 1281
- focal spot smoothing 943
- gold 943
- instabilities 459
- model 459, 1097, 1259
- Parylene-C 77
- polymethylmethacrylate 905
- silicon 997
- TATB explosive 905

laser
- generated plasma 921, 997, 1251
- matter interaction 299, 459, 921, 1209,
    1213, 1217, 1221, 1225, 1251,
    1259, 1261, 1265, 1277, 1281

lattice disorder, quartz 97

lead
- /bismuth 985
- constitutive modeling 295
- critical point 81
- equation of state 51, 81, 207
- Hugoniot 207
- interface instability 985
- phase transition 51, 81, 207, 503
- plasma 1255
- shock 81, 295, 503, 985
- spall 295, 503

- temperature 81
- /tin 985
- x-ray driven shock 1255

Lexan, initiation, projectile, jet 867

Lindemann melting law 109, 207, 267,
    295

liquids (see fluids)

lithium
- equation of state 51
- phase transition 51

lithium fluoride
- failure 9
- window 495, 1243

luggage container 463

luminescence, ruby 897

magnesium
- additive, pyrotechnic powder 745, 749
- equation of state 207
- flyer, laser driven 1221
- flyer, multiple layer 1221
- Hugoniot 207
- optical pyrometry 279
- phase transition 207
- temperature 279

magnesium fluoride
- flyer, laser driven 1221
- flyer, multiple layer 1221

magnetic gauge 771, 851, 855, 879,
    1057

magnetic properties
- $Bi_2Sr_2CaCu_2O_{8+\delta}$ powder 741
- Curie temperature 729, 929
- iron 929
- mercuric oxide powder 733
- rhodium oxide 901
- $Sm_2Fe_{17}N_\gamma$ 729
- ultrahigh field 897

manganin gauge 9, 97, 101, 219, 323,
    409, 413, 491, 515, 567, 577, 677,
    693, 763, 767, 795, 875, 883, 887,
    891, 967, 985, 1045, 1049, 1053,
    1061

many-body effects
- chemical reaction 183, 187, 191, 195
- defects 175
- dense fluid 167
- iron 175
- TATB explosive 179

mass spectroscopy, TATB explosive
    871

mechanical properties
- Bauschinger effect 311, 639
- bulk modulus 109, 475, 479
- crack growth 171, 511, 531, 551, 647,
    651, 745, 925